Hazardous Substances Resource Guide

NOTICE TO THE READER

GALE
ENVIRONMENTAL
LIBRARY

Hazardous Substances Resource Guide

Richard P. Pohanish and Stanley A. Greene, Editors

REFERENCE MATERIALS *Gale Research Inc.* · DETROIT · LONDON

83962

Richard P. Pohanish and Stanley A. Greene, *Editors*

Gale Research Inc. Staff

Lawrence W. Baker, *Senior Developmental Editor*
Carol DeKane Nagel, *Developmental Editor*

Mary Beth Trimper, *Production Director*
Evi Seoud, *Assistant Production Manager*
Mary Winterhalter, *Production Assistant*

Cynthia Baldwin, *Art Director*
Bernadette M. Gornie, *Graphic Designer*
Yolanda Latham, C. J. Jonik, Nicholas Jakubiak, *Keyliners*
Arthur Chartow, *Technical Design Services Manager*
Robert J. Huffman, *Photographer*

Theresa A. Rocklin, *Supervisor of Systems and Programming*
Charles Beaumont, *Programmer*

™
∞ This book is printed on acid-free paper that meets the minimum requirements of American National Standard for Information Sciences—Permanence Paper for Printed Library Materials, ANSI Z39.48-1984.

This book is printed on recycled paper that meets Environmental Protection Agency standards.

ISBN 0-8103-8494-6
Printed in the United States of America

Published simultaneously in the United Kingdom
by Gale Research International Limited
(An affiliated company of Gale Research Inc.)

Contents

Part 3: Indexes

Part 4: Glossary and Appendix

Tables

A Word About Gale and the Environment

We at Gale would like to take this opportunity to publicly affirm our commitment to preserving the environment. Our commitment encompasses not only a zeal to publish information helpful to a variety of people pursuing environmental goals, but also a rededication to creating a safe and healthy workplace for our employees.

In our effort to make responsible use of natural resources, we are publishing all books in the Gale Environmental Library on recycled paper. Our Production Department is continually researching ways to use new environmentally safe inks and manufacturing technologies for all Gale books.

In our quest to become better environmental citizens, we've organized a task force representing all operating functions within Gale. With the complete backing of Gale senior management, the task force reviews our current practices and, using the Valdez Principles* as a starting point, makes recommendations that will help us to: reduce waste, make wise use of energy and sustainable use of natural resources, reduce health and safety risks to our employees, and finally, should we cause any damage or injury, take full responsibility.

We look forward to becoming the best environmental citizens we can be and hope that you, too, have joined in the cause of caring for our fragile planet.

The Employees of Gale Research Inc.

---

*The Valdez Principles were set forth in 1989 by the Coalition for Environmentally Responsible Economics (CERES). The Principles serve as guidelines for companies concerned with improving their environmental behavior. For a copy of the Valdez Principles, write to CERES at 711 Atlantic Avenue, 5th Floor, Boston, MA 02111.

Preface

Publishing has been described as an "accidental profession." Shortly after starting our working careers in chemistry, we accidentally found ourselves at different publishing firms. Among the professional, scientific, and law books, newsletters, and electronic information that we worked with were some leading hazardous materials references.

We became aware that most of the available reference works on hazardous materials were written for technicians who must work with hazardous substances, respond to chemical accidents, and evaluate threats to humans and the environment. However, a growing public concern about the environmental and health effects of chemical materials and the expansion of government "right-to-know" programs prompted us to consider the need for an easy-to-use and comprehensive book about hazardous materials. The book began to grow, and we realized that anyone concerned about chemicals found in and around the home, workplace, and community might benefit from our research.

In addition to introductory chapters on chemical hazards and extensive lists of resources to help the reader find additional information, this book contains profiles outlining the potential dangers and storage of more than 1,000 hazardous substances. These chemical profiles are provided so that individuals who do not have collections of material safety data sheets (MSDS), professional reference books, and computer resources might also have easy access to basic information on the chemical components of commercial products. Arranged in alphabetical order, the chemical profiles present each chemical substance with a minimum of technical and medical jargon. Care has been made to refrain from oversimplifying the vital information presented. Readers using the profiles should realize that the available literature may not provide complete data on human effects from exposure to hazardous substances. For example, nearly all reference sources assume that except for contaminated food, ingestion of many solid and liquid materials in the workplace is improbable.

This book does not include information on general safety and is not meant to be a substitute for workplace hazard communication programs required by the Occupational Safety and Health Administration (OSHA) and/or other government agencies. The user of craft materials, automotive products, paint strippers for household furniture, or any other chemical substance should take time to learn about the dangers of these materials, seek proper training, and use this book to find less hazardous substitute products.

R.P. and S.G.

Acknowledgments

The editors wish to thank the many individuals and organizations that contributed in one way or another in making this book possible. Although it would be impossible to acknowledge everyone, special thanks are due the following, who contributed directly or indirectly to this project:

Kathy Deck, Technical Information Specialist, National Center for Environmental Health and Injury Control, Centers for Disease Control, for pointing us to information sources in this field and for releasing to us her bibliographic references.

Deborah Fanning, Executive Vice President, and Laurie Doyle, Associate Director, Art and Craft Materials Institute, for preparing the chapter on Art and Craft Materials.

David V. Galvin, Supervisor, Hazardous Waste Management Program, Seattle Metro, for permission to incorporate portions of the report *Toxicants in Consumer Products,* published by the Household Hazardous Waste Disposal Project, Metro Toxicant Program, Municipality of Metropolitan Seattle. Also to Selma Matte for her assistance in obtaining material for our work.

Anne Gravereaux, Director, Inquiry Services, Canadian Centre for Occupational Health and Safety, for permission to use their report *How to Work Safely with Dangerously Reactive Liquids and Solids* in this work.

Jane Kochersperger and the National Coalition Against the Misuse of Pesticides for permission to use material from their publication *Safety at Home.*

Richard J. Lewis, former editor of the *NIOSH Registry of Toxic Effects of Chemical Substances,* author of *Carcinogenically Active Chemicals,* and author with Irving Sax of several major reference works including *Dangerous Properties of Industrial Materials, Food Additives Handbook, Rapid Guide to Hazardous Chemicals in the Workplace,* and *Hazardous Chemicals Desk Reference* for his expert advice on portions of this manuscript.

Geoffrey Lomax and the National Environmental Law Center for permission to use tables and charts from their publication *Toxic Truth and Consequences.*

David Steinman for permission to use portions of his article "Glossary of Common Cosmetic Terms and Ingredients" that appeared in *Greenkeeping* magazine.

George W. Ware for permission to use charts from his book *The Pesticide Book,* published by Thomson Publications.

James L. Wood, Ph.D., developer of the CHEMTOX database of regulated chemicals and co-author of the widely used textbook *Chemistry, Man and Society,* for his advice, suggestions, and encouragement to produce this manuscript.

We also wish to thank these other individuals and organizations for guiding us through the maze of information that is available from government and private sources, and in many instances providing us with reports and unpublished material: Eva M. Brownawell, Information Specialist, Center for Air Environmental Studies, Pennsylvania State University; Ronadin Carey, Pesticide Information Specialist, National Pesticide Telecommunication Network, Texas Tech University Health Sciences Center; Barbara Goodman, Librarian of the Philadelphia Office of the U.S. Occupational Safety and Health Administration (OSHA); and Diana McCreary, Librarian, Philadelphia Office of the U.S. Environmental Protection Agency (EPA).

Also, the American Chemical Society, Office of Legislative and Regulatory Programs; Center for Science in the Public Interest; Hope L. Barrett, Technical Information Specialist, National Injury Information Clearinghouse; Annie Berthold-Bond, editor and publisher, *Greenkeeping* magazine; Robert P. Benedetti, Senior Flammable Liquids Engineer, National Fire Protection Association; California Department of Health Services, Occupational Health Program; Center for Hazardous Materials Research, University of Pittsburgh Applied Science Center; Center for Environmental

Research Information; Health Effects Institute/Asbestos Research; Chemical Manufacturers Association; Dennis P. Curtin, author and computer expert; Environmental Hazards Management Institute; Environmental Research Foundation; Honor Publications; Pennsylvania Environmental Council, Inc.; Dr. Theodore G. Tong, Executive Secretary, American Association of Poison Control Centers; and the U.S. Public Interest Research Group (USPIRG).

Thank you...

ALA Task Force on the Environment (TFOE)

In developing this and other environment-related publications, Gale Research Inc. seeks to work closely with members of the Task Force on the Environment of the American Library Association. Of primary concern to us as publishers is designing publications to best meet the user's needs for useful and timely environmental reference information. We appreciate the availability of the members of TFOE and their willingness to answer questions and provide advice on our environmental publications. At the same time, we recognize that the ultimate responsibility for these publications is ours as publisher, and that the TFOE and ALA are in no way officially involved in their preparation or endorse their purchase.

Highlights

Citizens concerned with hazardous substances encountered in the home, community, and workplace can turn to the *Hazardous Substances Resource Guide* to help identify these toxic chemicals as well as organizations and references that can further provide information about the substance. The *Hazardous Substances Resource Guide* presents this crucial data on 1047 hazardous substances:

- Name
- Chemical Abstract Service (CAS) number
- Other names
- Danger profile
- Uses
- Appearance
- Odor
- Effects of exposure
- Long-term exposure
- Storage
- First aid guide, where available

Resource Listings Point User in the Right Direction

The *Hazardous Substances Resource Guide* includes up-to-date information on more than 1500 organizations, agencies, publications, and services concerned with toxic chemicals and their effects on human health and the environment. These resources include:

- Hotlines and clearinghouses
- Poison control centers
- Federal and state government organizations
- National and international organizations
- Books
- Magazines and newsletters
- Publishers
- Databases
- Database Producers

Three Indexes Speed Access to Information

The Chemical Name Index contains the name of each chemical profiled in the guide, as well as its synonyms. The Chemical Abstract Service (CAS) Number Index provides easy access to a chemical when only its CAS number is available. The Resource Index combines all listed organizations, agencies, publications, and information services into one comprehensive index.

Glossary and Appendix Add to Ready Reference Convenience

For quick reference, the *Hazardous Substances Resource Guide* includes a glossary of more than 435 terms applying to safety, occupational health, and the environment. An appendix contains a list of oxidizers for quick reference to these common and often-hazardous substances.

Introduction

The age of innocent faith in science and technology may be over ... every major advance in the technological competence of man has enforced revolutionary change in the economic and political structure of society.

—Barry Commoner

We have met the enemy and he is us.

—Pogo, Earth Day 1971

We live in a society of chemical expedience and excess. Each year in the United States over 350 billion pounds of toxic chemicals are produced, 8 billion pounds of hazardous substances are shipped, 2,200 new pesticide and 500 new cosmetic products are introduced, and more than 20,000 chemically based products are produced by the cosmetic industry. The average American household generates 15 pounds of hazardous waste per year. Nearly 5 million poisonings occur in the United States annually, resulting in more than 5,000 deaths.

As this book is being written, 5,000 residents have been evacuated from their homes because of a railroad tank car spill in Wisconsin. A federal appeals court cast a legal cloud over worker safety by throwing out Occupational Safety and Health Administration (OSHA) worker exposure limits for 428 toxic chemicals. Researchers at the National Cancer Institute report in *The American Journal of Public Health* that women who dye their hair increase the risk of lymphoma by 50 percent. And *The Washington Post* reports that long-term consumption of chlorinated water is being linked to bladder and rectum cancer. The chlorination of drinking water is the single most important public health measure in history, according to many public health experts.

In less than one decade, chemical safety and protection of the environment have become major public issues, especially since the chemical tragedy in Bhopal, India, where more than 12,000 people were killed or injured, and the horror of nuclear disaster in Chernobyl, Ukraine. Worldwide, the concerns are similar: industrial chemical accidents and spills, water and air quality, holes in the ozone layer, global warming, acid rain, toxic waste disposal, pollution of waterways from agricultural pesticide run-off, and the threat to the rain forests. Given public concern over environmental and workplace safety, chemical hazards have become political hot buttons. The public is looking for tough lawmakers, and from every indication regulatory impact will increase.

Before Congress passed the Emergency Planning and Community Right-to-Know Act (EPCRA), interested citizens wanting to know more about chemicals manufactured or used in their communities had to rely on industry cooperation. Although some firms shared information willingly, many others flatly refused to cooperate with the public. One of the important messages of EPCRA is that hazardous chemical information has become too important to be left solely to the scientist or the industrial manager. The public now has a right to know about chemicals used and made in their living environment, and the public is expected to work as a partner with industry, public-interest organizations, and government to make EPCRA work. In the words of Lee M. Thomas, former administrator of the U.S. Environmental Protection Agency, "The more each citizen learns about, understands and participates in managing chemical hazards, the safer the environment will be for everyone."

A nominal level of chemical understanding is vital in a world seemingly overwhelmed by chemical products. Over the next decades many of us will have to make personal decisions about dangerous substances, and the chemical "literates" could well be the survivors. Hopefully, this book will be a valuable reference to those individuals interested in a better world and a safer place to live.

R.P.

User's Guide

Hazardous Substances Resource Guide is divided into four parts:

Part 1: Chemical Profiles
Part 2: Resource Listings
Part 3: Indexes
Part 4: Glossary and Appendix

Part 1: Chemical Profiles

Entries are arranged alphabetically by most common chemical name. Eleven data items, when available, are provided for each material:

Name: A single chemical material may be known by many names. The name used here is the most common name of the material being described.

CAS: The Chemical Abstract Service (CAS) number is a unique identifier assigned to chemical materials by the Chemical Abstract Service of the American Chemical Society. This number can assist the reader in the identification of an ingredient or material regardless of the name used.

Other names: Chemical "synonyms" for the same material include other common names, commercial product names, generic names, foreign names, and code names. In some cases, common names, such as "paint thinner," may be designated to more than one material.

Danger profile: This is a brief summary of the hazards associated with each hazardous material. If the hazards are not self-explanatory, definitions may be found in the glossary. In many instances chemicals have not been studied for hazardous or toxicological effects; in cases of conflicting information, the editors have chosen to use the most conservative estimates of a material's dangers. The terms "flammable" and "combustible" have specific technical definitions, which are currently undergoing change; after conferring with National Fire Protection Association staff, the editors have chosen to use the word "combustible" to describe a material that will burn.

Uses: This section lists commercial uses of the material.

Appearance: Physical description of the material in terms of form and color is provided.

Odor: Describes how the material smells to aid in identification. It should be noted that smell is not a safe means of identification; if you can smell the chemical, you may be past the point of safety.

Effects of exposure: This section gives a summary of the symptoms caused by acute, or short-term, exposure to a particular material when it is breathed and swallowed and comes in contact with eyes and skin.

Long-term exposure: A summary of the symptoms caused by chronic, or long-term, exposure to a particular material is described here.

Storage: Provided is a brief summary of precautions for safe storage, including chemical incompatibilities and fire and explosion hazards.

First aid guide: This section gives standardized first aid procedures based primarily on the U.S. Department of Transportation *1990 Emergency Response Guidebook.* It is not meant to replace medical or hazardous materials books published for qualified medical personnel.

Part 2: Resource Listings

Arrangement of entries within individual sections varies as appropriate to the material being presented. Entries may be arranged alphabetically or geographically.

1. Hotlines and Clearinghouses

- **Content and Selection Criteria:** Includes the name, address, telephone number, contact name, and description, where available, of more than 60 hotlines and clearinghouses located in the United States and Canada.
- **Arrangement:** Alphabetical, first by topic, then by organization name.
- **Indexed by:** Organization name and keywords.

2. Poison Control Centers

- **Content and Selection Criteria:** Includes the name, address, telephone number, contact name, and certification information, where available, of nearly 85 poison control centers located in the United States.
- **Arrangement:** Alphabetical, first by state, then by organization name.
- **Indexed by:** Organization name and keywords.

3. Federal Government Organizations

- **Content and Selection Criteria:** Includes the name, address, telephone number, contact name, and description, where available, of more than 120 U.S. and Canadian federal government agencies and commissions concerned with hazardous substances.
- **Arrangement:** Alphabetical by organization name.
- **Indexed by:** Organization name and keywords.

4. State Government Organizations

- **Content and Selection Criteria:** Includes the name, address, telephone number, contact name, and description, where available, of nearly 200 state government agencies and offices concerned with hazardous substances.
- **Arrangement:** Alphabetical, first by state, then by organization name.
- **Indexed by:** Organization name and keywords.

5. Organizations

- **Content and Selection Criteria:** Includes the name, address, telephone number, contact name, and description, where available, of nearly 380 U.S., Canadian, and multinational organizations focussing on hazardous substances and the environment.
- **Arrangement:** Alphabetical by organization name.
- **Indexed by:** Organization name and keywords.

6. Books

- **Content and Selection Criteria:** Includes book title, author, publisher with location, and description, where available, of more than 225 general interest and reference books, including directories, dictionaries, handbooks, sourcebooks, manuals, bibliographies, and government publications, concerning hazardous substances and the environment published in the United States, Canada, and England.
- **Arrangement:** Alphabetical by publication title.
- **Indexed by:** Publication title and keywords.

7. Magazines

- **Content and Selection Criteria:** Includes publication title, editor, publisher name, address, and telephone number, and description, where available, of more than 50 general interest and professional magazines focusing on hazardous substances and the environment published in the United States and Canada.
- **Arrangement:** Alphabetical by publication title.
- **Indexed by:** Publication title and keywords.

8. Newsletters

• **Content and Selection Criteria:** Includes newsletter title, editor, publisher name, address, and telephone number, and description, where available, of 180 newsletters devoted to hazardous substances published in the United States and Canada.
• **Arrangement:** Alphabetical by publication title.
• **Indexed by:** Publication title and keywords.

9. Publishers

• **Content and Selection Criteria:** Includes the name, address, telephone number, and description, where available, of more than 100 U.S., Canadian, and British publishers who specialize in environmental books, directories, reports, newsletters, and other materials.
• **Arrangement:** Alphabetical by publisher name.
• **Indexed by:** Publisher name.

10. Online Databases

• **Content and Selection Criteria:** Includes the database name, producer name, address, and telephone number, and description of 85 U.S., Canadian, and British online databases that can provide electronic access to information on hazardous substances.
• **Arrangement:** Alphabetical by database name.
• **Indexed by:** Database name.

11. Online Database Producers

• **Content and Selection Criteria:** Includes the name, address, telephone number, and products of nearly 75 U.S., Canadian, and British online database producers who provide electronic access to information on hazardous substances.
• **Arrangement:** Alphabetical by producer name.
• **Indexed by:** Producer name.

Part 3: Indexes

The **Chemical Name Index** includes the name of each chemical profiled, as well as the synonyms appearing in each **Other names** rubric. Index citations contain entry numbers rather than page numbers to provide the most specific access. Entry numbers of profiled chemicals will appear in boldface. This index also contains chemical references from the introductory material, with these index citations containing the roman numeral page number on which the reference can be found.

The **Chemical Abstract Service (CAS) Number Index** is arranged numerically by the CAS number of each profiled chemical for easy access to a chemical name when only its CAS number is available.

The **Resource Index** combines all listed organizations, agencies, publications, and information services into one comprehensive index. Index citations contain entry numbers rather than page numbers to provide specific access. All *HSRG* entries are indexed by name of organization or entity listed. If parent organizations are included in entry headings, then both the parent group and the unit name are included in the index citations for that entry. In addition to this alphabetical arrangement of entries, citations may also appear under keywords for quick access by environmental topic.

Part 4: Glossary and Appendix

The **Glossary** defines more than 435 terms and phrases as they apply in the restricted sense to the fields of safety, occupational health, and the environment. Terms are arranged alphabetically and are cross-referenced.

The appendix comprises a list of **Common Oxidizing Materials,** presented first in alphabetical order and second by Chemical Abstract Service (CAS) number, for easy access to the names of these hazardous substances, which are often cited in the chemical profiles.

Editors' Note

In order to present hazardous materials information in the simplest terms, technical data has been, for the most part, omitted from this book. The information presented here was compiled from U.S. Government agencies and has been checked against various private sources. To the best of our knowledge and belief, this information is correct. Nonetheless, individuals using these materials should refer to product labels and material safety data sheets, if obtainable and accessible, for applicable warnings and instructions for safe use.

R.P. and S.G.

Abbreviations

ACGIH	American Conference of Governmental Industrial Hygienists
ACS	American Chemical Society
AHERA	Asbestos Hazard Emergency Response Act
AIHA	American Industrial Hygiene Association
ANSI	American National Standards Institute
CAS	Chemical Abstract Service
CCOHS	Canadian Centre for Occupational Health and Safety
CDC	Centers for Disease Control
CERCLA	Comprehensive Environmental Response, Compensation, and Liability Act (Superfund)
CFCs	Chlorofluorocarbons
CFR	Code of Federal Regulations
CPSC	U.S. Consumer Product Safety Commission
CSA	Canadian Standards Association
DOT	U.S. Department of Transportation
EPA	U.S. Environmental Protection Agency
EPCRA	Emergency Planning and Community Right-to-Know Act
FDA	U.S. Food and Drug Administration
FHSA	Federal Hazardous Substance Act
FIFRA	Federal Insecticide, Fungicide, and Rodenticide Act
HCS	Hazard Communication Standard
HMIS	Hazardous Materials Identification System
IPM	Integrated Pest Management
LEPC	Local Emergency Planning Committees
LHAMA	Labeling of Hazardous Art Materials Act
MSDS	Material Safety Data Sheets
NCAMP	National Coalition Against the Misuse of Pesticides
NFPA	National Fire Protection Association
NIOSH	National Institute for Occupational Safety and Health
NPTN	National Pesticide Telecommunications Network
OSHA	Occupational Safety and Health Administration
PCBs	Polychlorinated biphenyls
PCP	Pentachlorophenol
PPE	Personal Protective Equipment
RCRA	Resource Conservation and Recovery Act
RTECS	Registry of Toxic Effects of Chemical Substances
SARA	Superfund Amendment and Reauthorization Act
SERC	State Emergency Response Commissions
TRI	Toxic Release Inventories
TSCA	Toxic Substance Control Act
VOCs	Volatile Organic Compounds
WHMIS	Workplace Hazardous Materials Information System (Canada)

Understanding Hazardous Substances

Everything in the world, including your own body, is composed of chemicals. Our food contains small amounts of copper, chromium, fluorine, iodine, and selenium. Some of these chemicals, in tiny amounts, are essential to proper health; in large doses, however, they may be toxic or carcinogenic, even dangerous if they come in contact with other substances.

Every week the news media disturbs our peace with reports about a harmful substance found or used in our homes, communities, or workplaces. These news items frequently create more questions than they answer.

- How do toxic substances get into our bodies?

- How will a chemical release or accident threaten my personal health and the environment?

- What actions can I take to reduce my own exposure?

- What government agencies have jurisdiction over chemical accidents or releases? What is being done to regulate chemical use and to protect citizens?

- Should I question "educational" information distributed by industry associations and lobbying groups promoting the use of chemical products?

Our access to information about accidents, pollutants, and illnesses caused by hazardous substances has heightened our concern about the health effects of the chemicals in our water, air, food, and household products. Generally, there is a low risk to health from exposure; many potentially dangerous and toxic chemicals are present in the environment in very low concentrations. Sometimes, though, the health risk is high, as when some non-toxic substances are not handled properly, or when you are exposed to low doses of a toxic substance over a period of time.

Hazardous substances are naturally occurring (radon, lead, some plants) or synthetic (pesticides, cleaners, solvents), and may be in the form of solid, liquid, or gas. Through better education, labeling, media publicity, information provided by community groups, and particularly the development of poison control centers throughout the country, accidents and deaths from hazardous substances have decreased over the past ten years. The American Association of Poison Control Centers maintains accurate records of reported inci-

dents, as reported in Table 1. Cleaning substances lead the list of substances most frequently involved in human poison exposures, while many incidents involve a variety of pharmaceuticals, which are not covered in this book.

Table 1

SUBSTANCES MOST FREQUENTLY INVOLVED IN
HUMAN POISON EXPOSURES (1990)

SUBSTANCE	NUMBER	PERCENTAGE
Cleaning substances	180,096	10.5%*
Analgesics	172,278	10.1
Cosmetics	146,274	8.5
Plants	102,254	6.0
Cough/cold preparations	97,277	5.7
Pesticides (inc. rodenticides)	66,677	3.9
Hydrocarbons	63,131	3.7
Topicals	63,030	3.7
Bites/envenomations	62,509	3.6
Foreign bodies	59,205	3.5
Antimicrobials	56,347	3.3
Chemicals	55,084	3.2
Sedatives/hypnotics/ antipsychotics	54,578	3.2
Alcohols	49,097	2.9
Food poisoning	48,383	2.8
Vitamins	40,407	2.4

Source: "1990 Annual Report of the American Association of Poison Control Centers National Data Collection System." *American Journal of Emergency Medicine.* Used with permission.

These substances are not necessarily the most toxic, but are more readily available, especially to children.

*Percentages are based on the total number of human exposures (1,541,744) rather than the total number of substances.

What Makes a Chemical Toxic?

Scientists, legislators, and government regulators use a variety of terms to describe different kinds of substances that have harmful health effects. For example, a local factory may be reported as having released a "toxic chemical" in to the air, while a "hazardous material" may have been released by an overturned tanker truck.

The toxicity of a substance is measured by its ability to cause a harmful effect on humans or animals when swallowed, inhaled, or absorbed through the skin, depending on its chemical nature, the extent to which the substance is absorbed by the body, and the body's ability to eliminate the substance or change it into a less toxic substance. It may cause visible damage, or decrease the performance of a body function measurable only by a test.

A material is said to be hazardous when there is a practical likelihood that it will cause harm, depending on several factors:

Toxicity: How much of the substance is required to cause harm.

Route of exposure: How the substance entered the body.

Dose: How much entered the body.

Duration: The length of time the body was exposed to the harmful substance.

Reaction and interaction: Other substances that the body was exposed to.

Sensitivity: How the exposed body reacts to the substance compared to other people.

The chemicals profiled in this book are hazardous because of their potential toxicity as well as the risk of fire or explosion—safety hazards—associated with them.

Hazardous substances come from either natural or man-made sources. Man-made sources are created by human activity and include commercial facilities that make, treat, store, use, or dispose of hazardous chemicals; sewage and water treatment plants; and consumer products such as pesticides, paints, solvents, household cleaners, and gasoline. Some natural sources of hazardous substances are metals, minerals, gases such as coal gas, petroleum fumes, and radon, and naturally occurring pesticides found in plants.

Why Are Some Chemicals More Harmful Than Others?

The most important factor in the measurement of toxicity of a substance is its chemical structure—what it is made of, that is, what atoms and molecules it contains and how they are arranged. Substances with similar structures often cause similar health problems. However, slight differences in chemical structure can lead to great differences in the health effects produced. For example, silica in one form (amorphous) has little effect on health and is acceptable in the workplace at relatively high levels. After silica is heated, however, it turns into another form (crystalline) that can cause serious lung damage and is allowed to be present only at very low levels.

Routes of Exposure

Hazardous substances enter your body though your lungs from the air and vapors you breathe, your digestive tract from foods or liquids you swallow, or through your skin from contact with a chemical.

Inhalation. Breathing a substance into the lungs is the most common type of exposure. Some chemicals are irritating to the nose or throat, or may cause discomfort, coughing, or chest pain when they are inhaled into the air passages and the lungs. Other chemicals can be inhaled without warning symptoms but are still dangerous. Some substances can be present in the air as small particles such as dust or mist, and when they enter the air passages and lungs may be coughed out. Some particles may remain in the lungs and cause long-term damage, or will be dissolved and absorbed into the blood stream and carried elsewhere in the body to have delayed health effects on the bones and vital organs.

Skin contact. Even though the skin acts as a protective barrier against chemicals, some substances can easily pass through the skin and enter the bloodstream or attack the nerves. Some strong acids and alkalies will burn the skin upon contact. Strong solvents are capable of dissolving the natural oils in the skin, leaving it dry and cracked and susceptible to infection and absorption of chemicals. Many families of those who contacted chemicals in the workplace have become affected because of substances that had been brought home on clothing.

Eye contact. Any eye contact with chemicals should be considered serious. Some chemicals may irritate or burn the eye, and/or may be absorbed through the eye into the bloodstream.

Swallowing. A common source of exposure to chemicals in the home is by swallowing, usually by children who are unsupervised, or by adults who misread labels. Chemicals can be ingested if they are left on hands, clothing, or beards or accidentally contaminate food. In the workplace, metal dusts such as lead or cadmium are easily ingested.

Doses and How They Are Measured

The amount of a hazardous substance taken in by a person in one day is called the daily dose. The measurement of acceptable daily dosage is complicated and subjected to many variables involving the substance itself and the individual exposed to the substance. When you inhale a toxic chemical, the dose you receive depends on: (1) the concentration of the chemical in the air; (2) how hard you are breathing; (3) how much of the inhaled chemical stays in your lungs and is absorbed by the bloodstream; and (4) how long you were exposed.

To find out how much of a hazardous substance you take into your body through a particular exposure, the Environmental Protection Agency (EPA) has established conversion factors for different types of exposures. The concentration of the hazardous substance is multiplied by the appropriate conversion factor to arrive at the dose to which the individual was exposed. The following are examples of conversion factors that have been used in some EPA studies:

Water drunk per day

By adult	2 liters (approximately eight 8-ounce glasses, including the water in coffee, soda, etc.)
By child	1 liter (approximately four 8-ounce glasses)
Air breathed per day	20 cubic meters (approximate volume of air in a small bedroom or den)
Fish eaten per week (1 portion)	224 grams (approximately 7 ounces)
Soil consumed per day by a child	100 milligrams (0.004 ounces)

These are reasonable estimates made by the EPA based on observations of normal human behavior. Over a lifetime the daily dose may vary for many reasons, such as where you live and for how long, your lifestyle, what you eat, and your long-term exposure to hazardous substances in the environment. People may be more susceptible to health effects of doses at different ages. Adults with greater body weights may be less affected than children with smaller bodies and lungs.

Duration of Exposure

The longer you are exposed to a chemical, the more likely you are to be affected by it. At very low levels you may not experience any effects no matter how long you are exposed. At higher concentrations you may not be affected following a short-term exposure, but repeated exposure over a long period of time is often hazardous because some chemicals can accumulate in the body or the damage does not have a chance to be repaired. The combination of the dose and the duration of the exposure is called the rate of exposure.

The liver, kidneys, and lungs are able to change chemicals to a less-toxic form and eliminate them from the body. A chemical will accumulate in the body if the rate of exposure exceeds the rate at which it can be eliminated. For example, if you work with a chemical for

eight hours each day, you will have the sixteen hours remaining in the day to eliminate it from your body before returning to work the next day. If the chemical cannot be eliminated entirely in sixteen hours and you continue to be exposed, the chemical will accumulate each day.

While some chemicals, such as ammonia and formaldehyde, do not accumulate at all, others, unfortunately, are stored in the body for a long time before their effects become noticeable. Lead is stored in the bones, calcium in the liver and kidneys, and polychlorinated biphenyls (PCBs) in fat. Asbestos fibers, once deposited in the lungs, remain forever.

An Example of How Exposure Rates Are Used

What is your exposure to methylene chloride, a cancer-causing chemical, released from a factory near your home?

If you live and work within a mile downwind of a chemical factory that routinely releases 183,000 pounds of methylene chloride into the air each year, the concentration in your air may be as high as 0.12 milligrams per cubic meter if you live and work less than 250 yards from the factory, according to mathematical modeling estimates. Assuming a breathing rate of 20 cubic meters per day, at these concentrations your daily dose would be 2.4 milligrams per day. On the other hand, if you live and work between one and two miles from the factory, modeling results indicate that your daily dose would be no more than 0.011 milligrams per day. Your exposure falls rapidly as your distance from the source increases, and if you spend less time near the source.

Acute and Chronic Effects

An acute toxic effect occurs following a single or few exposures over a short period of time and becomes apparent immediately, in seconds or days. Examples of acute toxic effects are acid burns, rashes, and breathing difficulties. A chronic exposure is one that is repeated over a long period of time, causing long-term effects which may not show up until months or years that have passed. Chronic toxic effects include cancers, degenerative illnesses to bones, liver, kidneys, and lungs, and genetic damage.

A chemical may cause both acute and chronic effects. Continuously inhaling solvents in the workplace or in a craft studio can cause an acute effect of headaches and dizziness that will pass at the end of the day. The chronic effect may be liver and kidney damage that may not appear for months, and then linger for a long time.

Differences between Acute and Chronic Effects of Exposure

ACUTE

Occurs immediately or soon after exposure.

Often involves a high exposure over a short period.

Often reversible after the exposure stops.

Can be minor or severe. E.g., a small amount of ammonia can cause throat or eye irritation; larger amounts can be serious or even fatal.

Relationship between chemical exposure and symptoms is generally, although not always, obvious.

Knowledge often based on human exposure.

CHRONIC

Occurs over time.

Often involves a low exposure over a long period.

Many effects are not reversible.

Chronic effects are still unknown for many chemicals. Most chemicals have not been tested for cancer or reproductive effects.

May be difficult to establish the relationship between chemical exposure and illness because of the long time delay.

Knowledge often based on animal studies.

Different Forms of Toxic Substances

Toxic substances can be solids, liquids, gases, vapors, dusts, fumes, fibers, and mists. The same material may take different forms at different temperatures, each form presenting a different kind of hazard.

Solid. A material that retains its form and is generally not likely to be absorbed by the body.

Liquid. A material that flows freely and is the form that many hazardous substances take at normal temperatures. Liquids can damage the skin and some have the property of passing through the skin and entering the body.

Gas. When individual molecules disperse in the air at normal temperature and pressure it is a gas, and can be toxic, flammable, or explosive. Many gases are colorless and odorless and do not cause any immediate irritation; consequently, they may be difficult to detect, but extremely hazardous.

Vapor. When a liquid evaporates it produces vapors, which can be readily absorbed into the lungs and irritate the eyes, skin, or respiratory tract. Vapors, especially organic solvents, can also be flammable or explosive and/or highly toxic.

Dust. Small solid particles in the air are called dust. Smaller dust particles can reach the respiratory tract beyond the nose and damage the lungs or be absorbed in the bloodstream. Lead dust, for example, may be absorbed in the bloodstream and deposited eventually in the bones, not to be discovered for months or even years. Grain dust can be extremely explosive when it reaches high concentrations.

Fume. A fume consists of very small, fine, solid particles in the air that form when solid chemicals are heated to very high temperatures, evaporate to vapor, and finally become solid again. They are hazardous because they are easily inhaled.

Fiber. A solid particle whose length is a least three times its width. Asbestos is a small fiber that can lodge in the lungs and cause serious harm. Larger fibers are usually trapped in the respiratory tract and coughed or sneezed out.

Mist. A mist consists of liquid particles that are produced when a liquid is sprayed or agitated. Pesticide sprays and metal-working fluids are typical mists.

Household Cleaning Products

Cleaning products are perhaps the most complex— and common—hazardous substances found in the home. Because they are so widely used and come in a variety of physical forms and chemical formulations, we forget that they contain ingredients that can be extremely dangerous to humans and pets. Cleaning products lead the list of hazardous products involved in home accidents reported to poison control centers.

Included in this section are detergents, disinfectants, deodorizers, oven and drain cleaners, dry cleaning fluids and spot removers, floor and furniture polishes, shoe polishes, bathroom cleaners, mothballs, and general household cleaning products.

Detergents. Detergents are synthetic soaps that clean by lowering the surface tension and provide wetting, emulsifying, and dispersing action. Active ingredients in detergents are surfactants, builders, brighteners, and compounds that make suds, foam, and scents and provide bulk and alkalinity.

Surfactants loosen the soil by lowering the surface tension of the cleaning solution. There are three kinds of surfactants, "anionic," "cationic," and "nonionic," each of which has a different chemical function in the detergent.

Builders or **sequestrants** remove the calcium and magnesium in hard water, allowing the surfactants to work and preventing the soil from redepositing on the material being cleaned. Builders also have ingredients that prevent the minerals in hard water from corroding the enamel on the inside of washing machines.

Brighteners in detergents make fabrics appear whiter by converting invisible ultraviolet light to more visible blue light (thus the term "bluing").

Disinfectants. Disinfectants in cleaners *temporarily* remove germs and bacteria but in no way can disinfectants make anything "germ-free," which can only be done by sterilization or treating with special cleansers or preparations. The very properties of disinfectants that enable them to kill microbes also make them toxic and harmful to life.

At one time cresols and phenols were widely used in their raw state as antiseptics in household preparations and in surgical and hospital use. However, these two chemicals are now recognized as toxic substances, and less-toxic compounds of phenols that are safer than simple phenol are used.

Undefined Ingredients. Labels of cleaning products do not always identify the ingredients, or they simply lump the contents in terms describing their functions, such as "coloring agent" or "polishing agent." Sometimes only a broad category, such as "petroleum distillates," is shown. The words "inert" and "active" are frequently used on labels to describe the contents, leaving the reader to determine the nature of the substances being used and without the guidance whether or not they may be harmful in normal use.

Hazardous Substances

Household cleaners are almost always acutely toxic to humans and will cause skin, throat, and eye irritation, depression of the central nervous system if swallowed, injury to the respiratory tract, and perhaps death. With the exception of organic solvents found in many cleaners, it is generally accepted that chronic exposure or long-term toxic effects present a low hazard to humans. But some of the common organic solvents are able to produce serious chronic effects, such as cancer and birth defects, and are harmful to the environment. Drain cleaners and septic tank cleaners, because they reach the ground water or waste system directly and may degrade very slowly, are especially harmful to groundwater and aquatic life. The organic solvents most prevalent in cleaners are described in Table 2.

Label Warnings

The Federal Hazardous Substance Act (FHSA) sets the labeling requirements for household cleaners that contain "hazardous substances" as determined by the EPA. A product that contains a designated "hazardous substance" must have a label with specified "signal words," a list of active ingredients, first aid instructions, and other information designated for that product. If cleaners are flammable, corrosive, or toxic, the word "danger" must be on the label; if it contains a highly toxic ingredient, "poison" must also be on the label. The signal words "caution" and "warning" in decreasing order of hazard must be on products that contain hazardous ingredients. Labeling requirements are confusing, in the least, and not all household cleaning agents are required to list their ingredients on the labels.

Table 2

HOUSEHOLD CLEANERS INGREDIENTS: USES, HEALTH HAZARDS, AND CHARACTERISTICS

CHEMICAL	USES	HUMAN HEALTH HAZARDS	PHYSICAL CHARACTERISTICS/ ENVIRONMENTAL EFFECTS
Aliphatic and aromatic hydrocarbons			
Benzene	*No longer used in consumer products.* May be found in old dry cleaning fluids, spot removers, paint and varnish removers, airplane glue, nail polish removers, and pesticides. Used in gasoline additives, chemical manufacturing, photography chemicals.	Highly toxic. Carcinogen. Inhaled vapors accumulate in fat, liver, bone marrow. Chronic exposure causes blood abnormalities. May be teratogenic; may cause reduced fertility.	EPA hazardous substance. EPA priority pollutant. EPA hazardous waste. A light yellow liquid, pleasant odor. Acutely toxic to aquatic organisms.
Chlorinated hydrocarbons			
Carbon tetrachloride	*No longer used in consumer products.* May be found in old dry cleaning fluids and spot removers. Used in industrial degreasing, fire extinguishing agents, and fumigants.	Carcinogenic. Highly toxic if inhaled, swallowed, or absorbed through the skin. Hazard of systemic effects increased when used in conjunction with alcohol.	EPA hazardous substance. EPA priority toxic pollutant. EPA hazardous waste. Nonflammable. Incompatible with chemically active metals.
Paradichlorobenzene	Used in solid toilet bowl deodorizers and moth balls.	Vapor irritates eyes, skin, and throat.	
Glycols			
Diethylene glycol	Used in pre-wash spot and stain removers, window cleaners, disinfectants, floor cleaners, etc.	Slightly less toxic than ethylene glycol. Causes CNS depression and liver and kidney damage.	Combustible. Dangerous to aquatic life in high concentrations.
Triethylene glycol	Used in disinfectants.	Less toxic than diethylene glycol. Very low acute and chronic toxicity. Symptoms similar to ethylene glycol.	

Table 2 (contd)

HOUSEHOLD CLEANERS INGREDIENTS: USES, HEALTH HAZARDS, AND CHARACTERISTICS

CHEMICAL	USES	HUMAN HEALTH HAZARDS	PHYSICAL CHARACTERISTICS/ ENVIRONMENTAL EFFECTS
Acids and bases			
Sodium hydroxide	Used in drain and oven cleaners and bleaches.	Alkaline and very corrosive. Repeated exposure to diluted solutions may result in dermatitis. Contact with eyes, skin and throat may cause extensive damage, burns, blindness. Systemic effects may result from tissue injury.	Corrosive. EPA hazardous substance. Incompatible with water, acids, flammable liquids, organic halogens, metals, nitromethane, and nitro compounds.
Potassium hydroxide	Used in drain and oven cleaners and bleaches.	A corrosive poison. See Sodium hydroxide.	See Sodium hydroxide.
Hydrochloric acid	Used in toilet bowl cleaners and drain cleaners.	Highly corrosive to eyes, skin, and mucous membranes. May produce burns, scarring, blindness. Swallowing may be fatal. Inhaled vapors may result in respiratory damage and death.	Acid in water or a nonflammable gas. Incompatible with most metals, alkalies, or active metals, and may produce flammable gas. High concentrations dangerous to aquatic life.
Hydrofluoric acid	Used in aluminum cleaners and rust removers.	Extremely dangerous. Can cause burns to the flesh down to the bone and extensive damage. Irritates the eyes and can cause immediate blindness.	Colorless and highly volatile.
Sodium hypochlorite	Used in disinfectants, bleaches, deodorizers, and drain cleaners.	Corrosive to skin, mucous membranes, and eyes. Harmful if swallowed.	Corrosive. Decomposes in fire to chlorine gas. Very low concentrations harmful to aquatic life. Will decompose on exposure to air, becoming less toxic.
Sulfuric acid	Used in wet cell batteries, toilet bowl and drain cleaners.	Highly dangerous to eyes. Contact can cause damage to skin and mucous membranes.	Highly corrosive. Colorless, odorless, oily liquid.

Table 2 (contd)

HOUSEHOLD CLEANERS INGREDIENTS: USES, HEALTH HAZARDS, AND CHARACTERISTICS

CHEMICAL	USES	HUMAN HEALTH HAZARDS	PHYSICAL CHARACTERISTICS/ ENVIRONMENTAL EFFECTS
Calcium hypochlorite	Used in toilet bowl cleaners, bathroom cleaners, and chlorine bleaches.	See Sodium Hypochlorite.	See Sodium hypochlorite. Never mix with anything other than soap and water.
Oxalic acid	Used in toilet bowl cleaners, dry cleaning fluids and spot removers.	Corrosive of mouth, esophagus, stomach; kidney and liver damage; irritates upper respiratory tract, skin, and eyes.	Incompatible with strong oxidizers and silver. Fire may release toxic gases.

Surfactants

CHEMICAL	USES	HUMAN HEALTH HAZARDS	PHYSICAL CHARACTERISTICS/ ENVIRONMENTAL EFFECTS
Linear alkyl benzene sulfonate	Used in laundry and other detergents, bathroom and rug cleaners, and all-purpose cleaners.	Swallowing will cause vomiting.	Biodegradable.
Sodium dodecylbenzene sulfonate	Used in laundry and other household detergents, bathroom, rug and all-purpose cleaners.	Prolonged skin contact may cause irritation. Swallowed high doses may result in death.	
Sodium lauryl glyceryl ether sulfonate	Used in laundry and other household detergents, bathroom, rug and all-purpose cleaners.		
Alkyl dimethyl benzyl ammonium chloride	Used in deodorizers, bacteriacides, sanitizers.	Irritates skin and mucous membranes. Swallowing causes burning pain, weakness and confusion, and CNS depression.	
Alkyl ethoxylate	Used in air fresheners, dish and laundry detergent, bathroom cleaners.		
Alkyl phenoxy poly-ethoxy ethanol	Used as a wetting agent and emulsifier in air fresheners, laundry and dishwashing detergents, and bathroom cleaners.	Rarely irritates skin even at full concentrations.	

Table 2 (contd)

HOUSEHOLD CLEANERS INGREDIENTS: USES, HEALTH HAZARDS, AND CHARACTERISTICS

CHEMICAL	USES	HUMAN HEALTH HAZARDS	PHYSICAL CHARACTERISTICS/ ENVIRONMENTAL EFFECTS
Builders (detergents)			
Trisodium phosphate	Used in abrasive powders, laundry and dishwashing detergents.	Water solutions highly alkaline and may produce caustic burns.	Phosphates encourage algae growth, which results in depleted oxygen, which can lead to fish kills.
Sodium tripolyphosphate	Used in laundry and dishwashing detergents and coffee cleaners.	Strong concentrations attack skin, eyes, and lungs. If swallowed, probable nausea, vomiting, diarrhea.	See Trisodium phosphate.
Sodium bisulfate	Used as a water softener and in detergents.	Corrosive in most bodily tissues.	Aqueous solutions are strongly acid.
Sodium carbonate	Used as washing soda.	Highly concentrated solutions can damage mucous membranes.	
Sodium metasilicate	Used in low-phosphate and phosphate-free laundry and dishwashing detergent.	The most alkaline and corrosive substance in detergents; dangerous to skin, mucous membranes, eyes.	
Fungicides			
o-Phenyl phenol	Used in disinfectants	Irritates eye, may cause corneal injury; not absorbed in acutely toxic amounts through skin.	Disinfectants send some heavy metals into wastewater.
o-Benzyl parachlorophenol	A germicide used in disinfectants and furniture polish.	Mild skin irritation after prolonged contact.	Low solubility in water; photodegradation in water.
4-Chloro-2-cyclopentylphenol	Used in disinfectants.	Vapors may irritate lungs. Concentrated phenols penetrates skin, cause CNS depression, narcosis, cardiac or respiratory death.	Low solubility in water. Photodegradation in water.

Table 2 (contd)

HOUSEHOLD CLEANERS INGREDIENTS: USES, HEALTH HAZARDS, AND CHARACTERISTICS

CHEMICAL	USES	HUMAN HEALTH HAZARDS	PHYSICAL CHARACTERISTICS/ ENVIRONMENTAL EFFECTS
Propellants			
Dichlorodifluoromethane	Aerosol propellant in spray spot removers and disinfectants.	Short term exposure: dizziness, trembling, unconsciousness and death. May cause irregular heartbeat if liquid gets in eyes or on skin.	Incompatible with chemical active metals; toxic gases may be released when decomposes; depletes ozone in upper atmosphere.
Isobutane	Aerosol propellant in rug cleaners, air fresheners, and furniture polish.	Vapor irritates eyes.	Flammable. Very volatile.
Trichlorotrifluoroethane	Propellant, especially in rug and upholstery cleaners.	Irritates throat, skin; causes drowsiness; narcotic; high concentrations may cause irregular heartbeat and death.	Incompatible with chemically active metals; may cause depletion of ozone layer; toxic gases and vapors released when decomposes.
Propane	Propellant, especially in air fresheners and furniture polish.	Causes asphyxiation, dizziness, disorientation, frostbite that affects CNS.	Flammable. Incompatible with strong oxidizers.
Other cleaners			
Ammonia	Used in window cleaners, oven and all-purpose cleaners, floor cleaners and waxes.	Vapors extremely irritating to eyes, respiratory tract, skin; high gas concentration may cause severe eye damage and temporary blindness. Long term exposure: chronic irritation to eyes and respiratory tract.	Incompatible with strong alkalies, bleach and any product containing chlorine. Contact with strong oxidizers may cause explosion and fire.
Ethylene diamine tetracitic acid	Used in disinfectants, toilet bowl cleaners, rug and upholstery cleaners.	Irritates eyes and skin.	Nonflammable. Fouls shoreline; may be dangerous if enters water intakes.
Pine oil	Used in disinfectants.	Irritates eyes and mucous membranes; causes hemorrhagic gastritis; systemic effect: CNS depression, respiratory failure.	Combustible liquid.

Source: Susan M. Ridgley, *Toxicants in Consumer Products* (Seattle: Municipality of Seattle, 1982). Used with permission.
CNS=central nervous system

Special label requirements have been set for household cleaners that contain petroleum distillates, acids, alkalies, and ammonia, in which their percent of concentration must be shown. The FHSA has also banned from cleaning products chemicals that have been shown to be extremely toxic. Those include carbon tetrachloride, benzene, flammable water repellents, and liquid drain cleaners with more than 10 percent potassium hydroxide. Although aerosols are not technically a product but a vehicle for delivering a liquid mist, the use of vinyl chloride monomers in aerosols has been banned.

Storage and Disposal Precautions

Household cleaners used for dishes, clothes, windows, drains, bathrooms, and general cleaning that do not contain organic solvents are easily soluble in water and can be poured down the drain with plenty of water. However, there are some chemicals that can be harmful to septic tanks. Always read the labels to determine if the cleaners contain chemicals that can harm the septic system.

When using cleaners containing chlorine, such as bleaches, care should be taken to *never* mix with ammonia or ammonia-based cleaners. To be perfectly safe, never mix different cleaning solutions and read the labels for specific precautions.

Carbon tetrachloride, the original ingredient in the spot remover Carbona, had been the most-used dry cleaning fluid because of its nonflammability. In 1970, however, it was banned from household use because of its high toxicity. When absorbed through the skin or inhaled, it can damage the liver, kidneys, and the central nervous system. Users should be cautioned against using old containers of products containing carbon tetrachloride.

Environmental Effects

Many people use household cleaners without giving thought as to their effects on the environment. Used for washing and cleaning, they normally are flushed into the sewer system and are forgotten. Yet the pervasive use of cleaners containing toxic ingredients has resulted in long-term harm to the environment. Petroleum-based ingredients break down very slowly in septic and sewer systems and contain impurities that contaminate air, water, and soil. The overuse of detergents containing complex phosphates causes algae to excessively bloom and thus compete for the oxygen in water with fish, depleting that population in certain areas. Ingredients containing chlorine combine with petroleum-based products to form compounds that break down slowly and are eventually stored in the fatty tissue of wildlife, eventually killing them, affecting their reproductive capabilities, or making them valueless as a commercial product for human or animal consumption.

Non-toxic Alternatives

Everyone has favorite methods for general cleaning or removing messy spills and resultant stains and odors. Baking soda can be used as a non-toxic oven cleaner, drain opener, rug cleaner, laundry bleach, toiler bowl cleaner, and refrigerator deodorizer. Other common household materials can be mixed into a paste with water, vinegar, or lemon juice to make non-toxic cleaners and stain removers. Following is a sampling of non-toxic cleaning methods from a variety of sources:

Blood stains. Apply cold water or club soda immediately, absorb with a paper towel and repeat. Hydrogen peroxide or a paste of cornstarch are other options.

Floor and furniture polish. Shake well one part lemon juice and two parts olive or vegetable oil.

Fruit stains. Rinse cotton fabric with boiling water. Apply lemon juice to stubborn spots and expose to sunlight for several hours. Also try soda water, vinegar, or milk.

Grease and oil. For grease on cotton fabric, pour on boiling water then rub with a solution of washing soda and water. For suede, sponge with vinegar, let dry, then brush with a suede brush. For other materials, blot with a towel, dampen the stain, then rub with soap and baking soda. Wash in hot water and extra soap. Also try rubbing with bread, cornstarch, a dry bar of soap, or a paste of salt and vinegar or lemon juice. To remove grease and oil from concrete garage floors, sprinkle with dry cement or cat-box litter, sweep around and allow to absorb, then sweep up.

Ink stains. For ballpoint stains, sponge with rubbing alcohol, rub with soap, then rinse and wash. For felt-tip stains, rub with soap then rinse and wash. Also try lemon juice or soaking in milk or a solution of milk, vinegar, and cornstarch.

Metal stains. Make a paste containing a mild abrasive such as table salt or baking soda together with a mild acidic liquid such as lemon juice or vinegar (but do not use baking soda on aluminum products). Soak the metal in the paste for some time to let the acid work. Rinse well with hot water. A lemon and salt or lemon and baking soda paste makes a fine brass and copper polish. Use apple-cider vinegar for polishing chrome. Try cooking rhubarb in a stained pot. Clean tarnished pans or utensils by stewing over a low heat with slices of lemon or grapefruit, both acidic. Try using toothpaste to clean copper and brass.

Mildew. Sponge on vinegar and let stand in sunlight. Use paste wax on leather.

Mothballs. Use cedar chips, newspapers, or lavender flowers.

Non-oily stains. Add one teaspoon white vinegar and one teaspoon liquid detergent into one pint of lukewarm water. Rub into the stain gently, rinse, and blot with a paper towel. Vinegar can bleach, so test first.

Pet stains. Quickly absorb as much of the moisture as possible with paper towels. Rinse with warm water and sponge with a solution of vinegar and liquid soap; rinse and blot dry. Remove odors by sprinkling with baking soda or a solution of baking soda and water; allow the baking soda to sit for several hours, then vacuum.

Rust stains. Wash with lemon juice or a paste of lemon juice and salt or baking soda. Slightly acidic vegetables such as rhubarb or tomatoes can be applied.

Scratches. To remove scratches from wood, rub on mineral oil or oily nuts such as walnuts, pecans, or peanuts.

Water marks. To remove white water marks from wood surfaces, apply mayonnaise; wipe in and let stand for one hour.

Paint and Wood Preservatives

Paint

"Paint" refers to a wide variety of complex chemical mixtures that can cause extensive personal injury, pollute landfills, rivers, and aquifers, and contaminate the air. With annual production estimated at five hundred million gallons, paint ranks as one of the most potentially hazardous substances in common use. Paint formulas may contain several hazardous ingredients such as solvents, toxic heavy metals, polychlorinated biphenyls (PCBs), and pesticides that may not be identified on the label. These components vary with the type of paint, the manufacturer, the specific application, and the environment in which it will be used.

Paint includes a broad category of coatings that include alkyd and latex paints, varnishes, urethanes, polyurethanes, epoxies, lacquers, primers, stains, seals and sealants, adhesives, and specialty coatings. Paint is also described as either "oil-based" or "water-based," depending on the primary solvent used. Alkyd, or "oil-based" paints contain hydrocarbon solvents (turpentine, mineral spirits, "paint thinner," methylene chloride, methyl ethyl ketone [MEK], toluene, etc.), while latex paint is soluble in water.

Paint consists of four primary components: resins, solvents, pigments, and additives.

Resins. This is the binding material, once only natural but now being replaced by synthetic alkyds and oils derived from petroleum and water-soluble compounds that hold the pigments and other ingredients together and form the coating properties of the paint.

Acrylic latex resins account for 60-65 percent of exterior and 25 percent of interior latex, or water-based, house paint. They are replacing solvent-based paints because they are easy to apply, have a low odor, dry fast, and are easy to clean up with water.

Alkyd resins, once called oil-based paint or oil paint, are formulated in mineral spirits; because they are solvent-based, they are losing their popularity to latex, water-based paint. Their most important applications around the home are in high gloss interior and exterior enamels, for they form a shiny and very hard coating that protects against water and moisture. Many professional house painters prefer alkyd exterior or flat paints for their high resistance to water.

Vinyl resins are also water soluble and are used for their resistance to chemicals and water, fire-retardant properties, durability, lack of odor and taste, low cost, and versatility. Other types of resins, or coatings, include urethanes, nitrocellulose, styrene-butadiene, polyester, phenolic, amino, linseed oil, and natural resins.

Solvents. The function of solvents is to dissolve the other constituents in the paint so they can be applied in liquid form. Some solvents are used exclusively in the production of paint, others are used separately as thinners or furniture strippers. Oil-based paint solvents are called volatile organic compounds (VOCs) and are responsible for most of the human health hazards and environmental problems associated with paint. As VOCs evaporate, they help form low-atmosphere ozone, smog, and acid rain, not to mention vapors that can damage eyes and the respiratory tract. Increasing federal and state regulations to control the VOCs in the atmosphere have hastened the development of more versatile water-based acrylic latex paint.

Pigments. Pigments give paint its color and opacity. Colors include metallic pigments, such as chromate salts, iron oxide, carbon black, cadmium, and titanium dioxide. Pigment fillers such as talc, clay, silica, and calcium carbonate provide bulk, durability, and mildew control.

Additives. Different additives are used to enhance certain applications, extend the shelf life of the paint, prevent mildew, hasten drying, and improve the viscosity or brushability of the paint, prevent freezing, and aid the manufacturing process. Wood preservatives, which are actually pesticides, are important ingredients added to repel or kill pests that destroy wood.

Lead in paint

While the pigments and solvents present in paint pose the greatest health hazards, the most prevalent of which are described in Table 5, lead-based paint is the major source of lead poisoning for children and can also affect adults. Lead poisoning, which is irreversible, causes mental retardation, reading and learning problems, lowered IQ, and kidney and liver damage. It also leads to coma, convulsions, and death. Children who have some amount of lead poisoning may show initial symptoms of unusual irritability, poor appetite, stomach pains or vomiting, persistent constipation, sluggishness, or drowsiness. More blatant symptoms include anemia, cramps, or convulsions.

In adults, lead poisoning can cause irritability, poor muscle coordination, nerve damage to sense organs and nerves controlling the body, and problems with reproduction. It may also increase blood pressure. Thus young children, fetuses, infants, and adults with high blood pressure are the most vulnerable to the effects of lead-based paint.

A study by the Oak Ridge National Laboratory reports that lead in the environment from all sources is having an adverse health effect on three to four million children each year, and an additional four hundred thousand fetuses in the U.S. are probably harmed before they are born. One of nine children under the age of six has enough lead in his blood to place him at risk, placing lead poisoning as the primary environmental threat to children.

Children are exposed to lead by eating paint chips, but the more common routes of exposure are eating or inhaling lead dust that is created as lead-based paint chalks, chips, or is improperly removed during home renovations. Lead poisoning has even been traced to a sandbox that was contaminated by sanding paint from a house's exterior.

Lead paint has not been used commonly in homes since 1950 and has been completely banned in housing paint since 1978. Yet, it is seldom removed from homes; it remains in an estimated fifty-seven million dwellings, about 75 percent of all private housing built before 1980. Lead dust pervades the environment, but there has not been a genuine technology breakthrough for an expedient way to clean up the dangerous sources of lead poisoning. According to Dr. Jane S. Lin-Fu, Director of the Childhood Lead Poisoning Prevention Program, Maternal and Child Health Bureau of the Department of Health and Human Welfare, "each year tens of thousands of children with increased levels of lead in their blood continue to be discovered by screening programs. An unknown number are not identified due to lack of screening in their communities."

Federal law requires the government to pay for much of the lead cleanup in homes it owns or insures, or where subsidized tenants live. Even though the government acquired thousands of houses through foreclosure during the 1991-92 recession, it has been slow to act upon the lead-paint problem, and it is possible to take legal action against federal departments responsible for the units.

The administrative responsibility for the prevention of lead poisoning in children was given to the U.S. Public Health Service in 1981, with state health agencies taking on the major effort. A concise legislative history of the battle to identify and control the sources of lead poisoning, including the Lead-Based Paint Poisoning Prevention Act of 1971 (Public Law 91-695) and the Maternal and Child Health (MCH) Services Block Grant Act of 1981 (Public Law 97-36), is available from the National Center for Education in Maternal and Child Health, which publishes *A Resource Directory of Childhood Lead Poisoning Prevention*. The directory also contains extensive lists of state and local agencies to contact for information concerning medical treatment, lead poisoning screening, paint chip testing, home and environmental inspections. To order, contact: National Maternal and Child Health Clearinghouse (NMCHC), 38th and R Streets, N.W., Washington, DC 20067, telephone: (202) 625-8410 or (703) 821-8955.

The U.S. Consumer Product Safety Commission has issued a safety alert, "What You Should Know about Lead-Based Paint in Your Home," which takes you through the steps on how to have paint tested for lead, how to reduce exposure to paint dust, and how to locate and use a professional to remove lead-based paint. To order, write: U.S. Government Printing Office #1990-726-058, Consumer Product Safety Alert September 1990, U.S. Consumer Product Safety Commission, Washington, DC 20207. The National Lead Watch offers lead testing kits and information. To order, contact: National Lead Watch, P.O. Box 2236, Fairfield, IA 52556, telephone: 800-531-6886.

Mercury in paint

In recent years mercury has been added to water-based latex paint as a pesticide to prevent mildew and as a preservative. In August, 1990, following a publicized case of mercury poisoning of a child in Detroit, Michigan, the Environmental Protection Agency (EPA) banned the use of mercury in all interior paint products; in May, 1991, mercury was banned from all outdoor paint products. Nevertheless, as much as one third of all latex paint made before August, 1990, contained mercury, which was allowed to remains on dealers' shelves. In most cases, mercury is an unlisted ingredient on the paint can, so consumers should investigate carefully whether interior latex paint was produced prior to the EPA ban. To ensure that the paint your using does not contain mercury, call the EPA National Pesticide Telecommunications Network (NPTN) hotline: 800-858-7378. Have the manufacturer's name and any serial numbers that are on the label handy.

Storage and Handling Precautions

Follow these general practices when storing and handling paint and other wood preservatives:

- Always use paint and strippers in a well-ventilated area, preferably outdoors; avoid inhaling the

Table 3

PAINT INGREDIENTS: USES, HEALTH HAZARDS, CHARACTERISTICS

CHEMICAL	USES	HUMAN HEALTH HAZARDS	PHYSICAL CHARACTERIS-TICS/ ENVIRONMENTAL EFFECTS
Acrylic resins			
Ethyl acrylate	Used in water-based resins for paints and coatings	Irritates eyes, skin, respiratory system; vapors may cause drowsiness, headache, extreme irritation of respiratory tract, nausea; short exposure to liquid causes first degree burns	Flammable; incompatible with oxidizers, peroxides, strong alkalies, moisture, polymerizers
Methyl methacrylate	Same as Lucite	EPA hazardous substance; irritates eyes, nose, throat; may cause dermatitis and narcosis	See Ethyl acrylate
n-Butyl acrylate	Used in acrylic resins of water-based acrylic paints	Moderately irritating to skin; possible tissue degeneration in eye if unwashed	Flammable; combustible; floats on water; fouls shorelines
Vinyl resins			
Vinyl acetate	Used to produce vinyl resins	EPA hazardous substance; vapors in low concentration a primary irritant to upper respiratory tract and eyes	Flammable liquid
2-Ethylhexyl acrylate	Used to produce vinyl acetate for resins in vinyl acrylic paints	May cause smarting and skin rash if in contact with skin or clothes; vapors may cause drowsiness and convulsions; liquid irritates eyes; swallowing causes drowsiness and convulsions	Colorless liquid; sharp odor; combustible; containers may explode in fire; fouls shorelines
Ethylene	Used to produce vinyl resins for paint	Vapors cause headache, dizziness, loss of consciousness; liquid will cause frostbite (do not rub)	Flammable; floats and boils on water, producing a flammable, visible vapor

Table 3 (contd)

PAINT INGREDIENTS: USES, HEALTH HAZARDS, CHARACTERISTICS

CHEMICAL	USES	HUMAN HEALTH HAZARDS	PHYSICAL CHARACTERISTICS/ ENVIRONMENTAL EFFECTS
Urethane resins			
Urethane alkyds	Primarily used in resins in paints and varnishes		Combustion can produce hydrogen cyanide and carbon monoxide
Cellulosic resins			
Nitrocellulose	Used to produce paint resins		Flammable
Styrene-butadiene resins			
Styrene	Used in interior latex flat paint	Short-term exposure: irritates eyes, nose, throat, skin; drowsiness and death caused by high concentrations; long-term exposure may cause skin rash	Corrodes copper and dissolves rubber; may explode if not stabilized by an inhibitor; incompatible with oxidizing agents and catalysts such as peroxides, acids, or aluminum chloride; fire may release toxic gas and vapor
Butadiene	Used in interior latex flat paint	Overexposure may irritate eyes, nose and throat; cause drowsiness, light-headedness; high concentrations may cause unconsciousness and death; skin irritation and frostbite from prolonged contact	Flammable gas; incompatible with strong oxidizers, copper, or copper alloys; fire may release toxic gas and vapor
Pigment fillers			
Silicates	Synonym for talc, kaolin, bentonite, fuller's earth, ball clay; used as filler in pigments	Fine inert powder	

Table 3 (contd)

PAINT INGREDIENTS: USES, HEALTH HAZARDS, CHARACTERISTICS

CHEMICAL	USES	HUMAN HEALTH HAZARDS	PHYSICAL CHARACTERISTICS/ ENVIRONMENTAL EFFECTS
Silica	Used as filler in paint pigments	Chemically and biologically inert when swallowed	Incompatible with fluorine, oxygen difluoride, chlorine trifluoride
Calcium carbonate	Synonym for limestone, chalk, dolomite, marble; used mostly for filler in house paints	Used as anti-acid	
Pigment colors			
Carbon black	Used in all pigmented paints	No harmful effects	Incompatible with strong oxidizers such as chromates, bromates, nitrates
Chrome	Used in outdoor and special paints and primers, and in various forms for yellows, oranges, and greens	Some forms of chrome metals may cause higher risk of respiratory cancer	Flammable; incompatible with strong oxidizers, may cause fires and explosions
Cuprous oxide	Used mostly in marine antifouling paints	Copper oxide irritates eyes; short-term exposure to mist and dust causes chills; long-term exposure irritates skin, discolors skin and hair	Incompatible with acetylene gas and magnesium, producing explosive and flammable gases; copper does not accumulate in the environment
Iron oxide	Exterior house paints, wood fillers, stains	Exposure over period of years may cause changes in lungs	Contact of fumes with calcium hypochlorite may cause explosion
Titanium dioxide	Used as whitener	Mildly irritating to lungs	
Zinc oxide	Widely used in all kinds of paints	Low toxicity to humans; high level of exposure may cause dry throat, coughing, metallic taste in mouth	May react violently with chlorinated rubber

Source: Susan M. Ridgley, *Toxicants in Consumer Products* (Seattle: Municipality of Seattle, 1982). Used with permission.

vapors; take frequent breaks; never smoke while painting or paint near an open flame or intense heat.

- Store paint in a tightly sealed container, away from heat and children and pets.

- Paint keeps for many years if tightly sealed; if it can be mixed, it can be used. Keep it from freezing. Latex (water-based) paint will keep up for to ten years, but it is not usable if permitted to become lumpy. Oil paint keeps even longer and can be rejuvenated by adding a solvent.

- The best rule is to use up the paint, by mixing it with similar paint for undercoating other projects or by using it in less-obvious places. Allow latex or water-based paint to evaporate to a hard substance and dispose of in a lightly sealed can in the trash. Unusable oil paint should be held for the hazardous waste collection. If local standards permit, evaporate unusable oil paint outdoors and discard the solidified contents in a tightly sealed can in the trash.

- After using thinner to clean brushes, let the solids settle, then strain or decant the liquid and reuse. Concentrated residue should be disposed of through household hazardous waste collection.

Environmental Effects

The solvents and heavy metal pigments in paint pose the greatest dangers to the environment. Lead and cadmium pigments, once used extensively in architectural paint, have been replaced with non-toxic alternatives but still present latent dangers from old paint-covered surfaces. Heavy metals such as cobalt, mercury, lead, and vanadium are used less extensively in household paint, yet are used in art supplies, which should be handled carefully and disposed of with caution.

Paint solvents are all toxic to some degree and many are highly flammable, combustible, and vaporize easily. Many solvents, especially chlorinated hydrocarbons, are persistent in the environment: they do not break down into less-toxic substances and remain trapped in groundwater. Fortunately, most used paint remains in its original cans, tightly sealed, with little evaporation into the environment.

Non-toxic alternatives

Non-toxic paints do exist but they have drawbacks. For large scale use, they are more expensive and take longer to dry because they lack synthetic drying agents. They also lack artificial binders, resulting in reduced covering qualities and requiring a more thorough preparation of the surface to be covered. Their colors tend to be less dramatic than alkyd and latex paint, and they

have a shorter shelf life. Notwithstanding, there are a number of companies that make "organic" or natural paints, strippers, thinners, and varnishes that are non-toxic to a major degree.

Other than paint made from natural substances, there exist few non-toxic alternatives. Latex or water-based paint can be substituted for oil-based paint, eliminating the toxic solvents in the paint and thinners for cleaning brushes. Elbow grease and sandpaper can replace furniture strippers if one has the time and patience.

Wood Preservatives

Wood preservatives are actually pesticides, meant to kill pests in wood and prevent rot. They contain "inert" ingredients such as solvents and other substances that enable the preservative to be applied. Wood preservatives are generally applied alone or as ingredients in exterior paint.

The wood preservatives in paint are more damaging to the environment and humans than are the solvents and pigments. In fact, wood preservatives are some of the most hazardous chemicals likely to be found in household products. The EPA estimates that one billion pounds of active ingredients of wood preservatives were used in the United States in 1988, almost equal to the combined amount of all other pesticides. The products that contain chlorinated phenols (hydrocarbons with chlorine attached) are particularly toxic to humans, mammals, and aquatic life. Pentachlorophenol (PCP) is very persistent. Its vapors can remain airborne for more than a month, and when waste liquid is washed into the soil, it does not dissolve easily, remaining hazardous for more than a year.

Hazardous Substances

The hazardous compounds that are generally contained in wood preservatives are summarized in Table 4. PCP, the most toxic wood preservative, is contaminated with dioxins that accumulate in organisms as they move up the food chain. Creosote and inorganic compounds of arsenic are used extensively in wood preservatives and are also extremely toxic. Since November, 1986, use of preservatives containing these ingredients has been restricted to applicators who have been certified and licensed by their states.

Storage and disposal precautions

Because all wood preservatives have the potential for acute and long-term human poisoning, even more caution should be exercised in handling them than ordinary pesticides, especially for products containing inorganic compounds of arsenic, creosote, and PCP. Seal decks and furniture treated with creosote or PCP with several coats of shellac, urethane, or other sealants to prevent evaporation, and check all products that have already

Table 4

WOOD PRESERVATIVES: USES, HEALTH HAZARDS, CHARACTERISTICS

CHEMICAL	USES	HUMAN HEALTH HAZARDS	PHYSICAL CHARACTERISTICS/ ENVIRONMENTAL EFFECTS
Chlorinated phenols	Used in wood preservatives.	Highly toxic if swallowed or touched, causing dermatitis, CNS depression, damage to kidneys, liver, spleen, pancreas, lungs.	Fish tainted at low concentrations. Some forms of chlorinated phenols emit highly toxic fumes when heated. Persists in the atmosphere and soil.
Copper naphthenate	Used in wood preservatives and paint driers.	Possible carcinogen. Irritates lungs, eyes, skin, stomach; causes CNS excitement.	Combustible. Low concentrations harmful to aquatic life and shorelines.
Creosote	Used in wood preservatives.	Skin carcinogen. Irritates eyes and nose. If swallowed causes serious systemic effects. May cause cancer and linked to genetic damage.	Can migrate from treated wood into the environment. Low evaporation rate. Incompatible with strong oxidizers.
Magnesium fluorosilicate	Used in wood preservatives.		
Naphthenic acid	Used in wood preservatives.	Irritates eyes, nose, throat, and can cause dermatitis. Moderately toxic if swallowed.	Combustible. Low concentrations harmful to aquatic life.
Pentachlorophenol (PCP)	Used in wood preservatives. Commercial applications may be contaminated with more toxic dioxin compounds. Also used as pesticide.	May cause headache, appetite loss, cramps, muscle weakness, severe burns. Concentrates in liver, kidneys, and intestines. Chronic exposure may cause liver damage and skin rash. Associated with birth defects and fetal toxicity.	EPA Hazardous Substance. EPA priority pollutant. Fire may release toxic gases and vapors. Incompatible with strong oxidizers. Persists in warm moist soil for twelve months.
Zinc naphthenate	Used in wood preservatives.	Low skin irritant.	

Source: Susan M. Ridgley, *Toxicants in Consumer Products* (Seattle: Municipality of Seattle, 1982). Used with permission.
CNS=central nervous system

been treated with wood preservatives to ascertain that they have been sealed.

Never burn wood that has been treated with preservatives, for the fumes may be highly toxic. Leave the disposal of old decking, outdoor furniture, old telephone poles, and the sawdust from your projects to professionals. It is better to use up wood preservatives on your project, but if the label indicates that it contains creosote, inorganic arsenic compounds, or PCP, take it to a hazardous substance disposal facility.

Unfortunately there are no alternatives to wood preservatives, so if you have to use them, do so in a well-ventilated place and observe all precautions for handing and disposal.

Solvents

A solvent is a liquid that dissolves other substances. There are two kinds of solvents: aqueous solvents, which include water and water-base substances such as acids, alkalies, and detergents; and organic solvents, which are derived from petroleum, and which dissolve substances that water solvents cannot. Consequently, solvents are used extensively as critical ingredients in commonly used household, commercial, and industrial products and as intermediates in the manufacture of chemicals. Following is a list of common products that contain organic solvents:

Common Household Products
Acetone
Cresol
p-Dichlorobenzene
Ethylene glycol
Isopropyl alcohol
Methanol
Methyl ethyl ketone (MEK)
Methyl isobutyl ketone (MIBK)
Methylene chloride
o-Phenyl phenol
Toluene
1,1,1-Trichloroethane
Trichloroethylene
Xylenes

Dry Cleaning Fluids
Acetone
Ethylene dichloride
Mineral spirits
Perchloroethylene
1,1,1-Trichloroethane
Trichloroethylene

Fumigants
p-Dichlorobenzene
1,3-Dichloropropane
1,2-Dibromo-3-dichloropropane (DBCP)
Ethylene dibromide (EDB)
Ethylene dichloride (EDC)
Methylene chloride (pesticide carrier)
o-Dichlorobenzene
Ethylene dichloride (EDC)
n-Hexane
Methyl ethyl ketone (MEK)
Methylene chloride
Perchloroethylene
Styrene
1,1,1-Trichloroethane
Trichloroethylene
Xylenes

Metal Degreasers
Ethylene glycol monobutyl ether
Methylene chloride
Perchloroethylene
Trichloroethylene
1,1,1-Trichloroethane

Paint and Coating
Acetone
Ethylene glycol
Methyl ethyl ketone (MEK)
Methyl isobutyl ketone (MIBK)
Mineral spirits
o-Phenyl phenol
Toluene
Xylenes

Paint and Varnish Removers
Acetone
Kerosene
Methanol
Methyl ethyl ketone (MEK)

Gasoline
Benzene
Ethylene dichloride (EDC)
Ethylene dibromide (EDB)
Toluene
Xylenes

Manufacturing
Benzene
Carbon tetrachloride
Chloroform
Methylene chloride
Mineral spirits
Petroleum distillates
Toluene
Trichloroethane
Xylene

Photographic Chemicals
Ethylene dichloride (EDC)
Methylene chloride

Hazardous Substances

Most organic solvents are highly flammable, volatile, and toxic. Commonsense precautions should be taken when using them, such as reading the labels to know what you are using, having adequate ventilation, using proper enclosures in manufacturing processes, and taking protective measures recommended by the manufacturers, employers, and union organizations.

All organic solvents have the potential to affect the

brain and the central nervous system, causing narcosis, a "high" feeling similar to feeling drunk and dizzy. Low stages of symptoms, called acute systems, include fatigue, mild depression, anxiety, and the inability to concentrate. Advancing levels of exposure can cause headache, nausea, even unconsciousness and death.

Solvents also are irritants to the skin, eyes, and respiratory tract. Most solvents are easily absorbed through the skin, and may lead to liver, kidney, and heart damage; they have the ability to cause long-term (chronic) effects such as cancer and reproductive and birth defects. (See Table 5.)

Some solvents are considered less hazardous than others. Alcohols, ketones, and esters, for example, are moderately toxic. It would require exposure to a large quantity of them over a period of time to feel the effects equivalent to smaller doses of more toxic solvents. The most toxic solvents include benzene, toluene, xylene, methylene chloride, trichlorobenzene, and 1,1,1-trichloroethane. These are hydrocarbons, and tend to cause chronic health hazards in addition to immediate (acute) injury because they are absorbed through the skin, the way a user is most likely to be exposed. Methylene chloride has been cited by the Consumer Products Safety Commission as "one of the highest cancer risks ever calculated for a chemical in a consumer product."

Warning Labels

Consumers come in contact with solvents when they are used as diluents in products such as paints, varnishes, and rug cleaners, and also when they are used as stand-alone products such as paint brush cleaners. When used in paints and other products as the vehicle to carry the pigments and resins, solvents are listed on the product label. Stand-alone solvents are usually mixtures of several solvents.

The Occupational Safety and Health Administration (OSHA) sets the standards for many solvents and other chemical substances used in industry. They are called "Permissible Exposure Limits," or PELs, and primarily indicate the acute or immediate exposure dangers, although many PELs also alert users to chronic, or long-term, effects. Users should be alert to the dangers of combining two or more chemicals to achieve a more powerful solvent because labels do not cover such applications. In the workplace, the National Institute for Occupational Safety and Health (NIOSH), the American Conference of Governmental Industrial Hygienists (ACGIH), and the American Industrial Hygiene Association (AIHA) also have recommended exposure limits that many believe to be more protective and up-to-date than OSHA limits.

Cautionary labeling is a highly technical and complex matter because a product warning must be tailored to that individual product, taking into consideration not only the components but other factors such as physical properties, flammability, interaction of the ingredients, type and size of the container, and method and place of use. Yet not all household cleaning agents are required to have their ingredients listed on their labels. However, consumer and industrial solvents, degreasers, and similar fluids designated as hazardous by the U.S. Environmental Protection Agency (EPA) must have "signal words" on labels developed for each product. Non-pesticide products must say "danger" if they are highly flammable, corrosive, or toxic; highly toxic substances must include the word "poison" on the label. On the label must appear the principal acute hazards involved in using the product, for example, irritation to the skin, eyes, or gastrointestinal or respiratory systems; however, long-term effects of exposure need not be listed.

Storage and Handling Precautions

When storing solvents, as with other hazardous products, keep them in a cool, dry place in original containers with the lids tight to prevent spilling and vaporization. Check containers frequently for leaks and deterioration. Separate incompatible products.

When working with solvents, have fumes blown away from the work area to avoid inhalation. Keep children and pets away from the work area as they may be effected by even low concentrations of fumes. Wear the proper gloves for the chemicals being handled. Absorb spills with an absorbent such as kitty litter. Dispose of leftover or waste material in a sealed container outdoors.

When disposing of products that contain solvents, follow label directions carefully. Some solvents will not be absorbed in septic systems, for example, and will travel through the system and damage groundwater downstream. It is always best to buy only the amount of a solvent you need for a job and use it up entirely. Paint brush cleaners and paint thinners used to clean brushes can be decanted and reused. Products such as paints, varnishes, stains, and nail polish should be evaporated out-of-doors and away from children and animals. The sludge should be allowed to harden, and then be disposed of in the trash.

Take protective measures when working with solvents at home and in the workplace:

- Never weld or apply heat in areas where chlorinated solvents are used. They can produce extremely deadly phosgene gas.

- Never weld or use tools that could cause sparks where flammable solvent vapors may be present.

- Never eat where solvents are being used.

Table 5

COMMON SOLVENTS: USES, HEALTH HAZARDS, CHARACTERISTICS

CHEMICAL	USES	HUMAN HEALTH HAZARDS	PHYSICAL CHARACTERISTICS/ ENVIRONMENTAL EFFECTS
Alcohols			
n-Butyl alcohol	Used in paints, organic solvents, paint and varnish removers, brush cleaners.	Irritant to eyes, nose, throat; causes mild headaches, dizziness. Long-term exposure: dermatitis. No chronic systemic effects on humans.	EPA hazardous waste. Flammable liquid; vapors and gases may result in fire; may react with aluminum at high temperature; may attack rubber, plastics, and coatings.
Isopropyl alcohol	Used as starting material for acetone; in industrial processes; diluents, cleaners, rubbing alcohol, disinfectants, glass cleaners.	Short term: mild irritant to mucous membrane, skin, and eyes. Swallowing causes nausea, vomiting, drowsiness, and narcosis. Long-term: drying of skin.	Flammable and colorless. Fire releases toxic gases and vapors; may attack plastics, rubber, and coatings; incompatible with strong oxidizers.
Ethyl alcohol	Used in industrial processes; solvent in shellac, thinners, and nitrocellulose lacquers. The alcohol in beer and liquor.	Denatured form almost always toxic; very high concentrations irritate nose and eyes; heart disease and liver damage from chronic ingestion; prolonged inhaling can cause headache, drowsiness, fatigue, and tremors.	Flammable and colorless. Dangerous to aquatic life in high concentrations.
Methyl alcohol	Used in paints, antifreeze, inks, cement, strippers, windshield washing fluids, dyes, Sterno fluid, gasoline additives.	Readily absorbed to cause headaches, dizziness, confusion, cramps, weakness, and liver damage. Swallowing generates more toxic formaldehyde, causing blindness and death.	Flammable and volatile.
Aliphatic and aromatic hydrocarbons			
Mineral spirits	Used in general solvents, medium oils, alkyl resin paints, paint thinners.	Irritates skin, eyes, nose, throat, lungs; may damage liver and blood; narcotic; very high concentrations may cause unconsciousness and death. Swallowing may result in CNS depression.	Flammable. Incompatible with strong oxidizing agents; fire may release toxic gases and vapors; attacks some forms of plastics, rubber, and coatings.

Table 5 (contd)

COMMON SOLVENTS: USES, HEALTH HAZARDS, CHARACTERISTICS

CHEMICAL	USES	HUMAN HEALTH HAZARDS	PHYSICAL CHARACTERISTICS/ ENVIRONMENTAL EFFECTS
Petroleum ether	Consists mainly of pentane, hexane, heptane; usually mixed with other aliphatic solvents.	Irritates skin, eyes, nose, throat, lungs; damages nervous system; narcotic; can cause skin burns; dangerous if breathed or swallowed. May cause heart disease.	Highly volatile and combustible in liquid form. Incompatible with strong oxidizers.
Toluene	Used as gasoline additive; to produce benzene and other chemicals; paint, paint and varnish removers,cleaners, nail polish, glues, pesticides, spot removers.	Acute exposure: CNS depression, irritant to nose, throat; narcosis. Long-term exposure: skin drying and cracking. Possible chronic liver, kidney, blood damage and genetic changes. More narcotic than benzene.	EPA hazardous substance. EPA priority pollutant. EPA hazardous waste. Acutely toxic to freshwater fish. Extremely flammable, volatile and subject to photochemical degradation in the atmosphere.
Petroleum spirits	Used mainly in medium oils, oil alkyd paints, solvents, waxes, plastics, dry cleaning fluids, varnish, pesticides. May contain benzene.	Short-term exposure: CNS depression, irritates, throat, eyes, and skin. Long-term exposure: skin drying and cracking.	Extremely flammable, combustible, and volatile.
Xylene	Used as gasoline additive, solvent; in paints, paint and varnish removers, cleaners, glues, pesticides, nail polish, spot removers; to make other chemicals.	Irritant to eyes, nose, throat. High concentration: narcosis, drowsiness, nausea, vomiting, abdominal pain. Damages liver and kidneys. May cause blood changes, menstrual problems, risk of heart attack, birth defects. Less toxic than benzene.	Extremely volatile, flammable, and unstable at high temperatures. With strong oxidizers can result in fire, explosion; fire releases toxic gases and vapors. Will attack some forms of rubber, plastic, and coatings. Harmful to aquatic life in very low concentrations.

Ketones and esters

CHEMICAL	USES	HUMAN HEALTH HAZARDS	PHYSICAL CHARACTERISTICS/ ENVIRONMENTAL EFFECTS
Acetone	Used in wash thinners, strippers, acrylic and nitro-cellulose lacquers,varnish, paints, cleaning fluids, nail polish removers. One of the most common and least-toxic solvents.	Repeated exposure can cause chronic conjunctivitis, bronchitis, gastritis,gastroduodenitis . Headache from prolonged inhalation. No long-term effects known. Contact can cause rash and nail splitting.	EPA hazardous waste. Flammable. Sweet odor. Incompatible with oxidizing materials and acids; dangerous to aquatic life in high concentrations.

Table 5 (contd)

COMMON SOLVENTS: USES, HEALTH HAZARDS, CHARACTERISTICS

CHEMICAL	USES	HUMAN HEALTH HAZARDS	PHYSICAL CHARACTERISTICS/ ENVIRONMENTAL EFFECTS
Butyl acetate	Used in nitrocellulose and other lacquers.	Mild irritant and allergen; irritant to eyes, nose, throat. Long-term effect: skin irritation. Severe over-exposure: weakness, drowsiness, unconsciousness. Mild narcotic.	EPA hazardous substance. Flammable. Toxic gases and vapors may be released in fire. Dissolves many plastics, resins, and coatings. Incompatible with nitrates, strong oxidizers, strong alkalies, and strong acids.
Methyl isobutyl ketone	Used as thinner for cellulose and other synthetic resins that require strong solvents.	Mild irritant. High concentrations can result in narcosis, coma. Long-term exposure: dermatitis, heart, liver, and kidney damage.	EPA hazardous waste. Flammable. Incompatible with strong oxidizers. May be dangerous if enters water intakes.
Ethyl acetate	Sold in solution with ethyl alcohol. Used as solvent in cellulose resins, shellac, and vinyl resins.	Irritant to mucous membranes, eyes, gums, respiratory tract; mildly narcotic.	EPA hazardous waste. Volatile and flammable. Incompatible with nitrates and strong oxidizers, alkalies, and acids. May be dangerous if enters water intakes.
Methyl ethyl ketone	Used as thinner in paints, glues, strippers, and coatings, degreasers.	Strong irritant of mucous membranes; affects peripheral nervous system and CNS. Easily absorbed by the skin and inhalation and may cause narcosis. Long-term exposure: dermatitis.	Highly flammable. Very soluble in water and may be dangerous if enters water intakes; high concentrations dangerous to aquatic life.

Other solvents

CHEMICAL	USES	HUMAN HEALTH HAZARDS	PHYSICAL CHARACTERISTICS/ ENVIRONMENTAL EFFECTS
Carbon disulfide	*Never use as a solvent.* Deadly. Used in laboratories.	High exposure causes unconsciousness and death. Chronic exposure causes brain damage, mental illness, heart disease.	Extremely flammable and combustible.

Table 5 (contd)

COMMON SOLVENTS: USES, HEALTH HAZARDS, CHARACTERISTICS

CHEMICAL	USES	HUMAN HEALTH HAZARDS	PHYSICAL CHARACTERISTICS/ ENVIRONMENTAL EFFECTS
Cresol	Used in solvents and disinfectants.	Corrosive to tissues; dangerous to eyes. Repeated or prolonged exposure in small concentrations may cause skin rash, vomiting, headache, diarrhea, fainting, appetite loss, mental disturbance. Extensive exposure may cause death.	EPA hazardous substance. EPA hazardous waste. Corrosive and combustible. Sweet odor; solid or liquid. Toxic gases may be produced in fire; incompatible with strong oxidizers. Harmful to aquatic life in very low concentrations.
Ethylene dichloride	Used in resins, paint and varnish removers, dry cleaning fluid; fumigant; gasoline additive.	Inhaling or swallowing high concentrations may cause vomiting, dizziness, confusion, lung, skin, and eye damage. Prolonged exposure may cause liver and kidney damage, possible cancer.	Flammable. Pleasant odor.
Ethylene glycol	Used in water-based (latex) paints; major component of antifreeze. Some forms used in brake fluid.	Vapor is non-irritating, harmless to skin unless heated. Inhalation can result in CNS depression and dysfunction of blood system. Fatal kidney and brain damage if swallowed; 3 oz. can be fatal. Can cause cardiac and respiratory failure.	Combustible. Sweet tasting and a dangerous attraction to pets. Can be dangerous if enters water intakes.
Ethylene glycol monoethyl ether	Used in acrylic paints.	Easily absorbed through the skin, causing lung, eye irritation, drowsiness, fatigue, blood disorders, kidney and liver damage, birth defects, damage to male reproductive system.	Low volatility. Combustible.

Table 5 (contd)

COMMON SOLVENTS: USES, HEALTH HAZARDS, CHARACTERISTICS

CHEMICAL	USES	HUMAN HEALTH HAZARDS	PHYSICAL CHARACTERISTICS/ ENVIRONMENTAL EFFECTS
Methylene chloride	Used in paint and varnish removers, aerosol products, degreasers, septic tank cleaners, pesticides, shoe polish, oven cleaners, Christmas bulb lights, brush cleaners, graffiti removers, disinfectants.	"One of the highest cancer risks ... in a consumer product" (CPSC). Repeat contact can result in dermatitis. Liquid and vapor irritates eyes and lungs in high concentrations. May cause burns if liquid held in contact with skin. Mild narcotic; chronic inhalation may cause mental confusion, nausea, headache, liver and kidney injury. If continues, causes unconsciousness and death. May cause fatal heart attack if used by people with heart conditions.	EPA priority pollutant. EPA hazardous waste. Colorless, volatile, ether-like odor; nonflammable. May release toxic phosgene gas in fire. Incompatible with strong oxidizers, strong caustics, and chemically active metals. Will attack some plastics, rubber, and coatings.
Perchloroethylene	Used in dip and spray coatings, paint removers, cleaners, dry cleaning fluids, degreasers, septic tank cleaners, shoe polish. Used instead of carbon tetrachloride in household spotremovers.	Suspected carcinogen. Toxic by inhalation, swallowing, and prolonged contact with skin. Chronic exposure at high levels causes mild intoxication, liver dysfunction, heart failure. Systemic effects: CNS depression, kidney damage, and anesthetic death.	EPA priority pollutant. EPA hazardous waste. Volatile and nonflammable. Incompatible with strong oxidizers and chemically active metals. Reacts violently with concentrated nitric acid. Fat soluble and accumulates in animals and the environment.
Propylene glycol	Used in water-based (latex) paints.	Vapors non-irritating, harmless to skin.	Combustible.
1,1,1-Trichloroethane	Used as dip and spray coatings, paint removers, spot removers, oven cleaners, drain openers, furniture polish, cleaning fluids.	Liquid and vapor irritating to eyes on contact. Inhalation can cause psychophysiologic effects. High concentrations damage liver and kidneys. Narcotic. Acute exposure can result in CNS depression; may cause cardiac arrest.	EPA priority pollutant. EPA hazardous waste. Volatile. Incompatible with strong oxidizers and chemically active metals. Upon contact with hot metal or exposure to UV radiation, decomposes to produce irritating gases (hydrochloric acid, phosgene, and dichloroacetylene). Moderately toxic to aquatic life. Extremely persistent in groundwater.

Table 5 (contd)

COMMON SOLVENTS: USES, HEALTH HAZARDS, CHARACTERISTICS

CHEMICAL	USES	HUMAN HEALTH HAZARDS	PHYSICAL CHARACTERISTICS/ ENVIRONMENTAL EFFECTS
Turpentine	Used in paint and varnish thinners, brush cleaners.	Irritates eyes, nose, throat, skin. Vapors cause headache, confusion, CNS depression, respiratory distress. Repeated exposure in high concentration can cause nephritis, predisposition to pneumonia. Poisonous if swallowed.	Flammable and combustible. Incompatible with strong oxidizers and chlorine; may release toxic gases in fire. Will attack some rubber, plastics, and coatings. Dangerous to aquatic life in high concentrations.

Source: Susan M. Ridgley, *Toxicants in Consumer Products* (Seattle: Municipality of Seattle, 1982). Used with permission.
CNS=central nervous system

- When cleaning up, use hand cleaners, not solvents.

- Never bring clothing soiled on the job home.

- Large quantities of solvents should not be used for cleaning machinery or floors unless proper ventilation, proper respirators, and protective clothing are used.

Environmental Effects

Many organic solvents, particularly those containing halogens such as chlorine, are resistant to breaking down into less-harmful components. This makes them more liable to long-term exposure to humans, animals, and plants.

Many solvents used in industrial processes find their way untreated into the ground and waterways where they damage aquatic life and pose dangers to drinking water and well water. These solvents pose one of the major problems in Superfund sites and professional help should be sought to ascertain their levels of contamination and proper course of containment and cleanup.

Art and Craft Materials

Both children and adults enjoy creating artworks and crafts using a wide variety of media. In recent years, however, increased attention has focused on the possible hazards of painting, drawing, silk screening, sculpting, and ceramics crafting, and many have become increasingly concerned about how to practice their arts without harming themselves or their children.

The majority of art and craft materials are non-toxic, primarily because many of the toxic ingredients have been replaced with safer ingredients. Yet, there are still some potentially hazardous products on the market because some key ingredients cannot be substituted with non-toxic alternatives. These potentially hazardous products, though, can still be used safely, as long as label instructions are followed and safety precautions heeded.

Hazardous Substances

Though lead has been banned from use in household paint since 1978, artists' paints, certain glazes, and certain enamels are permitted to contain it. (Lead poisoning, which is irreversible, can cause in children mental retardation, kidney and liver damage, even coma and death; in adults it can cause nerve damage and reproduction problems.) Lead fumes are apt to be given off during processes requiring soldering, ceramic firing, and lead casting and melting, and when working with stained glass. Pottery glazes, besides containing lead and copper, may sometimes contain arsenic, a poison. The most common hazardous ingredients and the products in which they are commonly found are described in Table 6.

Table 6

ART AND CRAFT MATERIALS: USES, HEALTH HAZARDS, PRECAUTIONS

INGREDIENT	USES	EXPOSURE ROUTES	HAZARDS	PRECAUTIONS
Lead	Ceramic glazes; flake white oil pigment; oil and water lead chromate colors; jewelry making; stained glass; fumes from soldering and casting.	Breathing, swallowing.	Highly toxic. May cause damage to CNS, kidneys, bone marrow; harm to developing fetus; effects mental development in children.	Don't eat, drink, smoke when using; wear apron; wash hands immediately after use; do not apply by spray; do not use if pregnant or contemplating pregnancy; if swallowed, get prompt medical attention.
16 soluble heavy metals (barium, copper, nickel, cadmium, cobalt, etc.)	Pigments in ceramic glazes; oil and acrylic colors; dry ground pigments; etching inks; mordants; stained glass.	Breathing, swallowing, absorption.	May damage organs, CNS, bone marrow; may cause anemia, harm to the developing fetus, allergic sensitization.	See Lead.
40 various dyes	Glazes; markers; printing inks; dyes.	Breathing, swallowing, absorption.	Allergic reactions; irritates eyes; damage to organs; may lead to cancer; anemia; may cause harm to the developing fetus.	See Lead.

Table 6 (contd)

ART AND CRAFT MATERIALS: USES, HEALTH HAZARDS, PRECAUTIONS

INGREDIENT	USES	EXPOSURE ROUTES	HAZARDS	PRECAUTIONS
Solvents (toluene, xylene, hexane, acetone, 1,1,1-trichloroethane, turpentine, etc.)	Sprays and fixatives; markers; mediums and varnishes; silkscreen ink; etching grounds; rubber cement and other adhesives; enamels; lacquers; markers; correction fluids.	Breathing, swallowing, absorption.	Chronic damage to kidneys, testes, CNS; possible birth defects.	Avoid breathing vapors; do not use near flames, heat, potential sparks; wash immediately after use; use mineral spirits in place of turpentine, especially for brush cleaning.
Methyl alcohol	Paint strippers; shellac thinner; dyes.	Breathing, swallowing, absorption.	Highly toxic. Can cause acute poisoning and blindness; may cause harm to the developing fetus.	Substitute ethyl or isopropyl alcohol. Don't breath fumes; don't eat, drink, smoke in working area.
Petroleum distillates	Solvents, including mineral spirits.	Swallowing, breathing.	May be harmful or fatal if swallowed; nausea, headache, confusion or instability; irritates eyes, respiratory tract.	See Solvents.
Methylene chloride	Sprays.	Breathing, swallowing.	May damage fetus; experimental carcinogen; induces angina in susceptible people.	Don't eat, drink, smoke when using; remove vapors; wash hands immediately after use; avoid using when pregnant or contemplating pregnancy.
Crystalline silica	Dry clays; working with stone; jewelry making.	Breathing.	May cause lung damage (silicosis) or lung cancer.	Handle dry material in hood or sealed box; use NIOSH-certified mask for dust.
Acrylate monomers	Acrylics.	Absorption.	May cause allergic reactions.	Avoid prolonged contact with skin. Wash hands immediately after use.
Ethylene glycol	A few acrylic colors, water colors, and gouaches.	Swallowing.	May damage kidneys.	Don't eat, drink, smoke when using. Wash hands immediately after use.

Table 6 (contd)

ART AND CRAFT MATERIALS: USES, HEALTH HAZARDS, PRECAUTIONS

INGREDIENT	USES	EXPOSURE ROUTES	HAZARDS	PRECAUTIONS
Dibutyl phthalate	Enamels and lacquers.	Swallowing.	May damage kidneys or testes; may harm developing fetus; may cause birth defects.	Don't eat, drink, smoke when using; wash hands immediately after use; do not use if pregnant or contemplating pregnancy.
Powdered gum arabic	Powdered paints.	Breathing of powdered form only.	May cause respiratory allergies.	Use NIOSH-certified mask for dust.
Dusts (wood, plastic, stone, metal, leather)	Working with wood, plastics, metal, stone, leather.	Breathing, swallowing.	May cause eye, lung, and respiratory irritation, allergic reaction, asthma, and chronic lung disease, nasal cancer.	Always use proper masks in well-ventilated area; wet mop all areas.

Source: Compiled from various publications of The Art and Craft Materials Institutes, Inc.
CNS=central nervous system

Label Warnings

Knowledge of materials and their proper use ensures safe practices in the art field. It is important to read the labels of all products you use so you will know they have been evaluated by a qualified toxicologist and are non-toxic, or require special handling to avoid health hazards from misuse. In November, 1990, the Federal Labeling of Hazardous Art Materials Act (LHAMA) was enacted. This law requires that all art and craft materials be evaluated by a qualified toxicologist and uniformly labeled for chronic health hazards, if necessary, according to the chronic hazard labeling standard of the American Society for Testing and Materials, ASTM D-4236. First approved in 1983, ASTM D-4236 is believed to be the first standard to provide for chronic hazard labeling of any consumer product.

Products that are potentially hazardous require the following information on their labels:

- A statement of conformance to ASTM D-4236, which must appear on non-toxic products as well.

- A signal word such as "Warning" or "Caution."

- A listing of ingredients in the product that are at a toxic level.

- A listing of how the product may hurt you if not used properly, such as "May cause lung cancer" or "May cause harm to the developing fetus."

- Instructions on how to use the product properly and safely, such as "Do not eat, drink, or smoke," "Use a respirator," or "Wear gloves."

- An appropriate telephone number—usually of the manufacturer or importer—to call for advice or information on the product.

- A statement that the product is inappropriate for use by children.

Material Safety Data Sheets

Most workplaces, including schools in some states, require material safety data sheets (MSDS) under worker right-to-know laws. MSDSs provide helpful information on products, such as spill procedures, waste disposal, and methods for extinguishing fires if the product is flammable. But MSDSs can be misleading. They are generally designed for industry use; consequently, information on an MSDS is written in highly technical language and is often geared toward industrial-size products. Therefore, a great deal of the information that appears on an MSDS may not be useful or appropriate to a consumer. Fortunately, accurate information on art and craft products—ingredients, proper use, potential hazards—generally appears on their labels.

Certification by the Art and Craft Materials Institute

One way to tell if a product is properly labeled is to look for one of the seals of the Art and Craft Materials Institute, Inc. (ACMI), a nonprofit association of manufacturers of art and craft materials that was formerly known as the Crayon, Water Color and Craft Institute, Inc. The ACMI certifies the majority of art and craft materials sold in the United States in accordance with the chronic hazard labeling standard, ASTM D-4236, and the Federal labeling law, LHAMA. There are four levels of certification, each denoted with a ACMI seal. The Approved Product Nontoxic seal (Figure 1) marks approved products as non-toxic. The Certified Product Nontoxic seal (Figure 2) marks non-toxic products that meet specific requirements of material, workmanship, working qualities, and color developed by ACMI and other standards organizations. The Health Label Cautions Required seal (Figure 3), used on both toxic and non-toxic products, notes that a product's label features specific information on any known health risks and safe and proper use.

Figure 2

Figure 1   Figure 3

Following are some general precautions to take when handling and storing art and craft materials:

- Always read the label.

- Always use products that are appropriate for the individual user. Children in grades six and under and adults who cannot read and understand safety labels should use only non-toxic materials.

- Do not use products that have passed their expiration date.

- Do not eat, drink, or smoke while using art and craft materials.

- Wash yourself and your supplies after use.

- Never use products for skin painting or food preparation unless it is indicated that they are intended to be used for these purposes.

- Do not transfer art materials to other containers, or you will lose the valuable safety information that is on the original package.

Following are precautions to take when using products that have cautionary labeling:

- Keep materials out of reach of children.

- Keep your work area clean. Vacuum or wet mop; do not sweep or dry mop. Follow suggested disposal methods.

- Point your pen, brush, other tools, and materials away from your mouth and eyes. Avoid skin contact and swallowing materials.

- Use all protective equipment such as gloves, safety glasses, and masks specified on the label. The wrong type of equipment can do as much harm as using no equipment at all.

- Protect cuts or open sores with gloves or other coverings.

- Work in a well-ventilated area. Handle certain dry materials in a locally exhausting hood or a sealed box; spray certain materials in a spray booth with filters.

- Do not store or use flammable products near heat, sparks, or flame. Do not heat above the temperature specified on the label. If material is flammable, combustible, or highly volatile, use explosion-proof electric switches and an exhaust fan with an explosion-proof motor.

- Do not mix different dinnerware-safe glazes together; the balance of ingredients in the mixed glaze will be changed, resulting in a glaze that may be unsafe for dinnerware.

For more information on the safe use of art materials or particular products that have been certified through the ACMI program, contact: The Art and Craft Materials Institute, Inc., 100 Boylston St., Suite 1050, Boston, MA 02116, telephone: (617)426-6400, fax: (617)426-6639. Another source for information on art and craft materials and safe practices in the performing arts is: The Center for Safety in the Arts, 5 Beekman St., Suite 1030, New York, NY 10038, telephone: (212)227-6220. For information on the Federal Labeling of Hazardous Art Material Act, contact: Chuck Jacobson, Compliance Officer, Division of Regulatory Management, U.S. Consumer Product Safety Commission, 111 18th St., N.W., Washington, DC 20207, telephone: (301)504-0400.

Non-toxic Alternatives

There are a number of non-toxic alternatives to hazardous art and craft materials. Instead of powdered clay, use talc-free, pre-mixed clay; instead of leaded ceramic glazes, use glazes that are mostly lead-free. For synthetic dyes, substitute plant and vegetable dyes; for petroleum-based solvents, substitute water-based paints, inks, or soap or non-toxic cleaners. Instead of epoxy glue, try white glue or wheat paste.

—*By Deborah Fanning, Executive Vice President, and Laurie A. Doyle, Associate Director, The Art and Craft Materials Institute, Inc., Boston, MA*

Automotive Fluids

The products used to drive and lubricate automobile engines are so commonly used and potentially toxic they present a great threat to health and the environment. In addition to the fluids an automobile owner comes in contact with daily—gasoline and lubricating fluids—other products used in vehicles, such as gasoline additives, antifreeze, brake and transmission fluids, and batteries, are also classified as hazardous. Brake fluid is composed mainly of glycol ether; automobile batteries, or wet batteries, are composed of sulfuric acid and lead, all of which are highly toxic.

Hazardous Substances

Gasoline and motor oils are derived from crude petroleum, which is distilled and refined with additives to give it properties for combustion, in the case of gasoline, or to lubricate and clean the moving parts of engines. Automotive fluids are organic compounds and have similar properties as solvents that come from petroleum: they are volatile, extremely flammable and combustible, and are highly hazardous to human health and the environment. Added to gasoline are numerous chemicals to make it burn more efficiently, raise the octane levels, reduce "knock," keep it clean, and keep the water contents from freezing. There are even additives to remove other, unburned additives.

Used motor lubricants contain heavy metals such as lead, magnesium, copper, and zinc that have been flushed from the burned gasoline in the engine, which pose human health problems and danger to the environment. Heavy metals cause human health problems that are cumulative over a period of time, causing chronic damage to the nervous system, kidneys, and liver, sterility, miscarriages, and retardation of mental development in children.

The 1970 Clean Air Act and the new act of 1990 targeted automobiles as the major source of air pollution and directed that engine emissions be dramatically reduced. The shift from leaded to unleaded gasoline, cleaner burning engines, and catalytic converters to clean exhausts have resulted in lower emissions. Yet lead as an additive to make gasoline more efficient has been replaced with additives such as benzene, which may be more dangerous to human health.

Antifreeze solutions used as engine coolants contain ethylene glycol, which is also used as an industrial solvent. When antifreeze is used in older automobile radiators that have lead solder seams, the used coolant will contain waste lead that is highly toxic and difficult to dispose. It should be handled with care and taken to a hazardous disposal center. Table 7 lists common automotive products and their health hazards and characteristics.

Storage and Handling Precautions

Petroleum products are flammable, combustible, and highly toxic and should be used with caution. They should not be flushed into the sewer system, where they will eventually form surface films that kill waterfowl and fish or become sludge that destroys aquatic plants. Take used motor oil to service stations or recycling centers where it can be recycled into heating oil, road surfacing materials, wood preservatives, and compressed artificial "logs" for fireplaces. Used motor oil may also be "re-refined" into usable lubricating products, but the economics of the process and the environmental regulations make this a limited alternative for the consumer who needs to dispose of waste lubricants.

Following are precautions to take when disposing of waste motor oil, transmission and brake fluids, and diesel fuel:

- Do not put used oil in the trash, where it will eventually pollute the ground and waterways.

- Do not add to it volatile wastes such as gasoline, paint thinner, or naphtha. This will increase the flammability of the used oil, causing it to fall under more stringent hazardous waste regulations and increase the cost of disposal.

- Do not pour used oil into the sewer system or on the ground.

- Do not burn used motor oil where heavy metal contaminants will leach into the soil or be released into the atmosphere.

- If you cannot reach a hazardous waste collection center, purchase a special kit for disposal, or pour the waste oil into a metal container of cat-box litter or sawdust, seal, and discard in the trash.

Following are special precautions to take when handling gasoline:

- Burn up the gasoline in the engine or take it to a hazardous waste collection center.

- Gasoline is combustible and dangerous when stored;

Table 7

AUTOMOTIVE PRODUCTS: USES, HEALTH HAZARDS, CHARACTERISTICS

CHEMICAL	USES	HUMAN HEALTH HAZARDS	PHYSICAL CHARACTERIS-TICS/ ENVIRONMENTAL EFFECTS
Gasoline: Petroleum hydrocarbons			
Benzene	An antiknock fuel additive and a natural component of gasoline.	Carcinogen. Main exposure route is respiratory and accumulates in fat, bone marrow, liver. Long-term exposure leads to blood abnormalities. May be teratogenic and cause reduced fertility.	EPA Priority Pollutant; EPA Hazardous Substance; EPA Hazardous Waste. Acutely toxic to aquatic life. Requirements for unleaded fuel caused increased use of benzene.
Gasoline: Additives			
Amine and amine derivatives	Anti-rust agent; metal deactivator; antioxidant to prevent gasoline from forming gum resins.	Liquid and vapor irritates skin and mucous membranes; forms strong caustic solutions.	
Ethylene dibromide (EDB)	Removes lead oxides.	Positive carcinogen. Prolonged contact may cause blistering and skin ulcers; vapor irritates eyes, mucous membranes. Breathing causes severe respiratory injury, CNS depression, vomiting, liver and kidney damage.	EPA Hazardous Substance. EPA Hazardous Waste. Incompatible with chemically active metals and strong oxidizers. Use of unleaded gasoline has resulted in reduced use of this chemical.
Ethylene dichloride	Antiknock agent; removes lead from fuel.	Positive carcinogen. Breathing high concentrations may cause nausea, confusion, dizziness, pulmonary edema and death; long-term exposure leads to neurologic change, appetite loss, mucous membrane irritation, liver and kidney damage, death.	EPA Priority Pollutant. EPA Hazardous Substance. EPA Hazardous Waste. Flammable.
Organic phosphorous compounds	Increases gasoline performance.	Different compounds irritate eyes, skin; causes dizziness, anorexia, vomiting, chest pains, pulmonary edema, respiratory illnesses.	Corrosive.

Table 7 (contd)

AUTOMOTIVE PRODUCTS: USES, HEALTH HAZARDS, CHARACTERISTICS

CHEMICAL	USES	HUMAN HEALTH HAZARDS	PHYSICAL CHARACTERISTICS/ ENVIRONMENTAL EFFECTS
Tetraethyl lead (TEL)	Antiknock agent.	Swallowing or touching small amounts can be fatal.	EPA Hazardous Substance. EPA Hazardous Waste. Incompatible with strong oxidizers; decomposes in sunlight. Lead is a major contributor to air pollution and is no longer being used in gasoline.
Toluene	Used as gasoline additive and to produce benzene.	Acute exposure: CNS depression, irritant to nose, throat; narcosis. Long-term exposure: skin drying and cracking. Possible chronic liver, kidney, blood damage, and genetic changes. More narcotic than benzene.	EPA Hazardous Substance. EPA Priority Pollutant. EPA Hazardous Waste. Acutely toxic to freshwater fish. Extremely flammable, volatile, and subject to photochemical degradation in the atmosphere.
Xylene	Used as gasoline additive and solvent and to make other chemicals.	Irritant to eyes, nose, throat. High concentration causes narcosis, drowsiness, nausea, vomiting, abdominal pain. Damages liver and kidneys. May cause blood changes, menstrual problems, risk of heart attack, birth defects. Less toxic than benzene.	Extremely volatile, flammable, and unstable at high temperatures. With strong oxidizers can result in fire, explosion; fire releases toxic gases and vapors. Will attack some forms of rubber, plastic, and coatings. Harmful to aquatic life in very low concentrations.

Lubricating oil: Additives

Tricresyl phosphate	Additive in gasoline and lubricating oil.	High concentrations can cause paralysis, gastrointestinal distress, muscle soreness, numbness.	Low concentrations dangerous to aquatic life.

Lubricating oil: Contaminants

Barium compounds	Contaminant.	Acute exposure to barium salts may cause uncontrolled contractions, paralysis, heart problems.	EPA Hazardous Waste.

Table 7 (contd)

AUTOMOTIVE PRODUCTS: USES, HEALTH HAZARDS, CHARACTERISTICS

CHEMICAL	USES	HUMAN HEALTH HAZARDS	PHYSICAL CHARACTERISTICS/ ENVIRONMENTAL EFFECTS
Lead	See Tetraethyl lead. Also used in wet cell batteries.	Causes chronic CNS illness.	EPA Hazardous Waste. See Tetraethyl lead.
Antifreeze solution			
Ethylene glycol	The major component of antifreeze. Some forms used in brake fluid.	Vapor is non-irritating, harmless to skin unless heated. Inhalation can result in CNS depression and dysfunction of blood system. Fatal kidney and brain damage if swallowed; 3 oz. can be fatal. Can cause cardiac and respiratory failure.	Combustible. Sweet tasting and a dangerous attraction to pets. Can be dangerous if enters water intakes.
Batteries			
Sulfuric acid	Used in wet cell batteries.	Highly dangerous to eyes. Contact can cause damage to skin and mucous membranes.	Highly corrosive. Colorless, odorless, oily liquid.
Lead	Used in wet cell batteries.	See Tetraethyl lead.	See Tetraethyl lead.
Windshield washer solution			
Methyl alcohol	Used in antifreeze, windshield washing fluids, gasoline additives.	Readily absorbed to cause headaches, dizziness, confusion, cramps, weakness, and liver damage. Swallowing generates more toxic formaldehyde, causing blindness and death.	Flammable and volatile.

Source: Susan M. Ridgley, *Toxicants in Consumer Products* (Seattle: Municipality of Seattle, 1982). Used with permission.
CNS=central nervous system

however, if you must store, use a metal container with a built-in flame arrestor and keep it tightly closed. Do not fill container completely; allow space for the gasoline to expand. Store in a cool place away from open flame and heat.

Following are special precautions to take when handling antifreeze and coolants:

- The main ingredient in antifreeze is ethylene glycol, a sweet-tasting liquid that attracts pets and can kill them. If connected to a sewage system, dilute with lots of water and dispose of in the drain. If you drain your radiator yourself, flush the old coolant with lots of water and keep pets away.

- If connected to a septic system, dispose sparingly into the drain over a period of time, with plenty of water.

- If antifreeze has been used in an older car with lead solder in the radiator, dispose of it in a hazardous waste center.

Following is a special precaution to take when handling windshield washing fluid:

- Contains methyl alcohol (methanol), which is highly toxic. Use up entirely or flush into drain with lots of water.

Pesticides

Because pests are present everywhere, and because they create a nuisance and sometimes severe financial loss and a threat to health, we are increasingly turning to pesticides to control them. The U.S. Environmental Protection Agency (EPA) estimates that every year for the past decade about twenty-one million people applied pesticides to residential lawns and gardens, using twenty-five million pounds of herbicides and thirty million pounds of insecticides.

Pests can be anything from ants and cockroaches to algae, mildew, rodents, and plant insects. Thus, "pesticides" is a broad term applying to a class of chemicals that kills or otherwise controls pests of any kind, and includes insecticides, herbicides, fungicides, and rodenticides. Around the home, insect sprays and weed killers are thought of as pesticides, yet disinfectants and wood preservatives are also pesticides because they kill or control organisms that cause germs and decay wood. Moreover, pesticides are frequently incorporated in products not generally associated with insect control: wood preservatives and fungicides are mixed with paints to prevent rot and mildew, while disinfectants are a vital component of many detergents.

Consumer use of pesticides has been receiving more scrutiny from federal and state lawmakers. The National Academy of Science has found that homeowners use four to five times as many chemical pesticides per acre as farmers, a statistic inspiring the EPA to examine more closely the vulnerability of water systems and groundwater to contamination by pesticides and nitrates, while a Senate subcommittee is exploring legislation that would require residents to notify neighbors before using toxic sprays on lawns or trees. Some communities in Florida, Illinois, and Pennsylvania already have such public notification laws.

Most state and federal laws class pesticides as "economic poisons" and define them as "any substance used for controlling, preventing, destroying, repelling, or mitigating any pest." The active chemical ingredients of pesticides are usually mixed with inert substances such as liquid solvents, powders, and suspending agents or with aerosol propellants and applied as a water dilution spray, dust, aerosol, granules, gaseous fumigant, repellant, or baits.

Types of Pesticides

Pesticides are identified by specific names according the kinds of pests they control (see Table 8).

Table 8

PESTICIDE CLASSES AND THEIR USES

PESTICIDE CLASS	USES
Acaricide	Kills mites
Algicide	Kills algae
Attractant	Attracts insects
Avicide	Kills or repels birds
Bactericide	Kills bacteria
Chemosterilant	Sterilizes insects or pest vertebrates (birds, rodents)
Defoliant	Removes leaves
Desiccant	Speeds drying of plants
Disinfectant	Destroys or inactivates harmful microorganisms
Fungicide	Kills fungi
Growth regulator	Stimulates or retards growth of insects or plants
Herbicide	Kills weeds
Insecticide	Kills insects
Larvicide	Kills larvae (e.g., mosquito)
Miticide	Kills mites
Molluscicide	Kills snails and slugs
Nematicide	Kills nematodes
Ovicide	Destroys eggs
Pediculicide	Kills lice (head, body, crab)
Pheromone	Attracts insects or vertebrates
Piscicide	Kills fish
Predicide	Kills predators (e.g., coyotes)
Repellent	Repels insects, mites, ticks, and pest vertebrates (e.g., dogs, deer)
Rodenticide	Kills rodents
Silvicide	Kills trees and brush
Slimicide	Kills slime
Termiticide	Kills termites

Source: George W. Ware, *The Pesticide Book*, 3rd ed. (Fresno: Thomson, 1989). Used with permission.

Pesticides are also typed according to their chemical structures and sub-structures. These include synthetic organics, inorganics, botanicals derived from plants, and biological agents derived from living organisms. Herbicides and fungicides also are typed according to their chemical composition and the pests they target.

Trade Names of the Most Common Pesticides

Many chemicals are made by several manufacturers and, combined with inert ingredients, fillers, and propellants, are sold under a variety of trade names, the most common of which are summarized in Table 9. The health effects of the most commonly used insecticides, herbicides, and fungicides are listed in Tables 11 and 12.

Table 9

TRADE NAMES OF THE MOST COMMONLY USED PESTICIDES

TRADE NAME	CHEMICAL NAME
Aatrex	Atrazine
Accord	Glyphosate
Amaze	Isofenphos
Amine	MCPA
Amiral	Triadimefon
Anthon	Trichlorfon
Aquathol	Endothall
Arrhenal	DSMA
Arsinyl	DSMA
Balan	Benefin/Benfluralin
Banex	Dicamba
Banlene Plus	MCPA
Banvel	Dicamba
Banzine	Benefin/Benfluralin
Baygon	Propoxor
Bayleton	Triadimefon
Banlate	Benomyl
Betesan	Bensulide
Bordermaster	MCPA
Bravo	Chlorothalonil
Brimestone	Sulfur
Brush Buster	Dicamba
Cardona	Stirofos
Casoron	Dichlobenil
Chlorothal	Decthal
Cythion	Malathion
Daconil	Chlorothalonil
D-CON	Warfarin
DDVP	Dichlorvos
Dibron	Naled
Dipterex	Trichlorfon
DMDT	Methoxychlor
Dursban	Chlorpyrifos
Dylox	Trichlorfon
Dymid	Diphenamid
Dyrene	Anilazine
Earthcide	PCNB
Ficam	Bendiocarb
Flexidor	Isoxaben
Gallery	Isoxaben
Halizan	Metaldehyde
Herb-All	MSMA
Hydrothal	Endothall
Kerb	Pronamide
Kleen-up	Glyphosate
Knox-out	Diazinon
Korlan	Ronnel
Kwell	Lindane
Lorsban	Chlorpyrifos
Marlate	Methoxychlor
Mecopex	MCPP/Mecoprop
Metason	Metaldehyde
Methoxone	MCPP/Mecoprop
Neguvon	Trichlorofon
Phaltan	Folpet
Oftenol	Isofenphos
Orotran	Ovex
Orthocide	Captan
Orthone	Acephate
Plantgard	2,4-D
Prefar	Bensulide
Prentox	Methoxychlor
Primatol A	Atrazine
Primaze	Atrazine
Prowl	Pendimethalin
Proxol	Trichlorfon
Pryfon	Isofenphos
Rabon	Stirofos
Rodeo	Glyphosate
Round-up	Glyphosate
Sevin	Carbaryl
Spectracide	Diazinon
Superman	Meneb
Target MSMA	MSMA
Terraclor	PCNB
Thiopal	Folpet
Treflan	Trifluralin
Trolene	Ronnel
Tubothane	Meneb
Turcam	Bendiocarb
Turfcide	PCNB
Turflon	Triclopyr
Vapona	Dichlorvos
Warf	Warfarin
Weed-B-Gon	2,4-D
Weed-Feed	2,4-D
Weed-Pro	2,4-D
Weed-Rhap	MCPA
Zirberk	Ziram
Ziride	Ziram

Sources: National Coalition Against the Misuse of Pesticides, *Safety at Home: A Guide to the Hazards of Lawn and Garden Pesticides and Safer Ways to Manage Pests* (Washington, DC: NCAMP, 1991); Susan M. Ridgley, *Toxicants in Consumer Products* (Seattle: Municipality of Seattle, 1982); National Pesticides Telecommunications Network.

Federal Regulations of Pesticides

The EPA regulates thousands of pesticide products for agricultural, forestry, household, industrial, and other uses. The Federal Insecticide, Fungicide, and Rodenticide Act (FIFRA; 40 CFR 152-180) requires registration of all pesticides, classifies them for general (household) or restricted use based on their degree of hazard, and specifies requirements for labeling. When FIFRA was amended in 1972, it set requirements for the review and reregistration of all existing pesticides—some eleven hundred chemical ingredients contained in nearly forty-five thousand registered products. The 1988 FIFRA amendments established an accelerated program to reregister all pesticides by 1997 and authorized the collection of fees to support the program.

The EPA states that intensive scientific reviews have been done on the approximately two-hundred pesticides that account for most pesticide use in the United States, including 90 percent of agriculture pesticide use. Nevertheless, the review process for pesticides applying for reregistration has been a slow one, and coupled with the increasingly high cost of developing new products, has resulted in reducing the total number on the market. Many manufacturers, unwilling to change their formulations or to invest in new products with questionable payback, have voluntarily withdrawn their products from the market.

The EPA determines whether or not, and under what conditions, a pesticide can be used without adverse effects. Each pesticide product is approved for use to control specific pests, on specific plants or crops in designated geographical areas, and at specific concentrations and frequency of application. The EPA also may restrict the use of certain pesticides to certified pesticide applicators and require special precautions and labels. State governments also have regulatory authority, and often place restrictions beyond those of the EPA.

When questions arise about the proper application for a pesticide or for any restrictions that have been placed upon its general use, the EPA or your state pesticide office should be contacted directly. The EPA Office of Pesticides and Toxic Substances maintains a list of pesticides that have been canceled, suspended, voluntarily withdrawn from the market, or are waiting further review. Many pesticides have been restricted to use for specific applications and under labeling restrictions, which are included in the EPA publication, along with citations to the rules that appear in the Federal Register. Table 10 is a partial list of these pesticides.

Hazardous Substances

Pesticides are chemicals that are made to kill insects, weeds, microorganisms, vertebrate pests such as rodents and birds, and predatory animals. Not only do they accomplish their mission outright, but in humans they can cause chemical sensitivity or chronic poisoning, in which exposure to low levels of toxic substances over a long period of time causes various

---

Table 10

CANCELED, SUSPENDED, OR RESTRICTED PESTICIDES: A PARTIAL LIST

Aldicarb	Creosote	Metaldehyde
Aldrin	Cyanazine	Methyl bromide
Aluminum	Cycloheximide	Mirex
Amitraz	Cyhexatin	Oxyflurofen
Arsenic trioxide	2,4-D	PCNB
Azinphos methyl	Daminozide	Parathion (ethyl
Benomyl	DBCP	and methyl)
BHC	DDD (TDE)	2,4-D Paraquat
Bithionol	DDT	Pentachlorophenol
Bromoxynil	Diallate	Phosphide
Bromoxynil	Dieldrin	Quaternary ammo-
butyrate	Dimethoate	nium com-
Cadmium	Dinocap	pounds
Calcium arsenate	Dinoseb	Safrole
Calcium cyanide	Dioxathion	Silvex/2,4,5-T
Captafol	Diisulfoton	Sodium arsenate
Captan	EDB	Sodium arsenite
Carbofuran	Endrin	Sodium cyanide
Carbon tetrachlo-	EPN	Sodium mono
ride	Ethoprop	fluroacetate
Chloranil	Fluroacetamide	Strobane
Chlordane	Heptachlor	Strychnine
Chlordimeform	Kepone	Thallium sulfate
Chlorfenvinphos	Lead arsenate	TOK
Chlorbenzilate	Lindane	Toxaphene
Chloropicrin	Magnesium phos-	Tributyltin (TBT)
Copper arsenate	phide	Vinyl chloride
(basic)	Mercury	Zinc phosphide

Source: U.S. Environmental Protection Agency, Office of Pesticides and Toxic Substances, "Suspended, Canceled, and Restricted Pesticides" (Washington, DC: EPA, February, 1990).

nonspecific medical problems such as chronic fatigue, nausea, headaches, rashes, disorientation, and general malaise. Popular magazines recently have carried accounts of mysterious health symptoms being traced to the prolonged use of pesticides in household sprays for insects, termites, and other pests. Children and pets, particularly, are more prone to chemical sensitivity from chronic exposure because their lesser weight makes them more susceptible to lower doses of pesticides than adults.

Table 11

COMMONLY USED LAWN AND GARDEN INSECTICIDES AND THEIR HEALTH EFFECTS

INSECTICIDE	CANCER	BIRTH DEFECTS	REPRODUCTIVE EFFECTS	NEUROTOXICITY	KIDNEY/LIVER DAMAGE	SENSITIZER/ IRRITANT
Acephate	C			X		X
Bendiocarb				X		X
Carbary		X	X	X	X	X
Chlorpyrifos			X	X	X	X
DDVP	C			X		
Diazinon				X	X	X
Isazophos				X		X
Isofenphos				X		X
Malathion		X		X		X
Methoxychlor						X
Trichlorfon		X		X		X

Source: National Coalition Against the Misuse of Pesticides, *Safety at Home: A Guide to the Hazards of Lawn and Garden Pesticides and Safer Ways to Manage Pests* (Washington, DC: NCAMP, 1991). Used with permission.
C=possible human carcinogen; X=adverse health effect.

Label Warnings

Pesticide label information is regulated by FIFRA and must be written to exacting specifications and appear on every container. It is illegal to use a pesticide in ways that are inconsistent with its labeling. Information on the label must include:

- Name, brand, and trademark.

- EPA registration number, which assures that the product is legally registered.

- Directions for use, which will protect the health of the user and the environment.

- Precautions. Always look for one of the following signal words on the label describing how hazardous a pesticide is if swallowed, inhaled, or absorbed through the skin:

   "DANGER": the substance is highly poisonous.

   "WARNING": the substance is moderately toxic.

   "CAUTION": the substance is less toxic, though still dangerous.

Other warnings may not always ensure the safety of the product but reflect what the EPA considers adequate warning information and proper use. For exam-

Table 12

COMMONLY USED HERBICIDES AND FUNGICIDES AND THEIR HEALTH EFFECTS

HERBICIDE	CANCER	BIRTH DEFECTS	REPRODUCTIVE EFFECTS	NEUROTOXICITY	KIDNEY/LIVER DAMAGE	SENSITIZER/ IRRITANT
Atrazine	C	X		X	X	X
Benefin						X
Bensulide				X	X	X
2,4-D	X*	X	X	X	X	X
DSMA				X		X
Dacthal	X**				X	X
Dicamba			X		X	X
Diphenamid						
Endothall		X				
Glyphosate						X
Isoxaben	C				X	
MCPA		X	X	X		X
MCPP		X	X	X	X	X
MSMA				X		X
Pronamide	C				X	X
Siduron						X

B2=probably human carcinogen; C=possible human carcinogen; X=adverse effect demonstrated; X*=based on National Cancer Institute epidemiological evidence; X**=based on contamination by TCDD (dioxin) and hexachlorobenzene (HCB).

Table 12 (contd)

COMMONLY USED HERBICIDES AND FUNGICIDES AND THEIR HEALTH EFFECTS

FUNGICIDE	CANCER	BIRTH DEFECTS	REPRODUCTIVE EFFECTS	NEUROTOXICITY	KIDNEY/LIVER DAMAGE	SENSITIZER/ IRRITANT
Benomyl	C	X	X	X		X
Chlorothalonil	B2				X	X
Meneb	B2	X		X	X	X
PCNB		X				X
Sulfur						X
Triadimefon		X				
Ziram				X		X

Source: National Coalition Against the Misuse of Pesticides, *Safety at Home: A Guide to the Hazards of Lawn and Garden Pesticides and Safer Ways to Manage Pests* (Washington, DC: NCAMP, 1991). Used with permission.
B2=probably human carcinogen; C=possible human carcinogen; X=adverse effect demonstrated; X*=based on National Cancer Institute epidemiological evidence; X**=based on contamination by TCDD (dioxin) and hexachlorobenzene (HCB).

ple, "Keep out of reach of children" must be on every pesticide label.

One can obtain information about a pesticide, including its ingredients and health and environmental effects and if it has been recalled or suspended from particular applications through a variety of agencies. Contact a pesticide product manager in the EPA's Office of Pesticide Programs (telephone: 703-557-5447) and request a copy of the EPA registration and fact sheet on the product. File a Freedom of Information Act (FOIA) request by writing to the Public Information Officer, FOIA Office (H7506C), 401 M St., S.W., Washington, DC 20460. Contact the pesticide office in your state or the local farm extension service in your county. (Frequently, state requirements are more stringent than the EPA's and are more readily available.) Or request a material safety data sheet (MSDS) from the maker of the pesticide; it will contain information on its hazardous ingredients, health effects, and safe storage and disposal.

Storage and Handling Precautions

Pesticides are produced specifically because they are toxic to some thing, but they are inherently unsafe to the user, neighbors, pets, and wildlife. Therefore, be sure to read all instructions on the label carefully and be sure there is an EPA registration number. Following are storage and handling precautions to take when working with pesticides:

- Do not use a pesticide for any purpose other than for what it was intended and keep herbicides away from non-target plants. Use inconsistent with label directions is subject to civil and/or criminal penalties. This is an important item for inquiry should injury or damage results from a commercial application of a pesticide.

- Do not use a "restricted use" pesticide unless you are a formally trained, certified pesticide applicator, licensed by your state.

- Wear long sleeves and pants, gloves, socks and shoes, and protective masks; use on a windless day; cover bird baths, pools, and pet food dishes, and avoid using near wells; close house windows if spraying outdoors. Clean any spills on yourself immediately and shower and shampoo and wash your clothes after using pesticides.

- Mix only the amount you will use outdoors or in a well-ventilated area. If applying indoors, allow adequate ventilation and leave the area for at least the length of time prescribed on the label. Keep children and pets off lawns until liquid pesticides have dried, granules have washed into the soil, or the prescribed time on the label has passed.

- The best way to dispose of pesticides is to use them up entirely. Keep them in their original containers, and never mix pesticides for multiple applications without professional direction.

OSHA Hazard Communication Standard

Use of pesticides, as with all hazardous substances, by businesses, farmers, ranchers, and similar agricultural occupations is regulated by the Occupational Safety and Health Administration (OSHA) Hazard Communication Standard, which in 1988 was extended to all businesses with one or more employees. Each employer must have a written program that includes material safety data sheets, a written plan to inform employees, workers, and outside contractors about the hazardous chemicals to which they may be exposed, and a training program to implement the plan. Workers who believe they run the risk of exposure to pesticides have the right to all company data about the chemicals they are using; if the company is unwilling to release the information, the local federal or state OSHA office should be contacted. Several states have pesticide laws that are more strict than federal OSHA or Department of Transportation laws, so always check your state pesticide agency in addition to OSHA.

Obtaining Medical Help for Pesticide Exposure Victims

Persons who run the greatest risk of poisoning are those who mix, load, or apply pesticides. The most common route of poisoning is absorption through the skin, and the most toxic pesticides have resulted in death through this route of exposure.

Speed is essential in treating a pesticide poisoning victim, whether human or animal. First remove the victim from the poison and poisoned area and administer first aid, following the directions on the label. Then get medical attention immediately. Carefully take the pesticide container with you so the doctor can know what the poison is before prescribing treatment.

A good resource for obtaining instructions for human or animal treatment is the National Pesticide Telecommunication Network (NPTN), which maintains a toll-free 24-hour telephone service (800-858-7378). If necessary, they can transfer inquiries directly to affiliated poison control centers. See Part 2, Section 2 of this guide for a list of poison control centers.

The EPA is interested in receiving information on adverse effects from pesticide exposure. If you have such information, contact the EPA: Pesticide Incident Response Officer, Field Operations Division (H-7506C), Office of Pesticide Programs, 401 M St., S.W., Washington, DC 20460. The National Coalition Against the Misuse of Pesticides (NCAMP) also monitors pesticide incidents and compiles annual data based upon reports they receive from victims. They ask that information, including the victim's name and address, the date and location and description of the incident, the pesticide(s) to which the victim was exposed and the targeted pests, the health and/or property damage that resulted and whether legal action was or will be taken regarding the incident, the names of the medical professionals as well as the local, state, and/or federal agencies contacted, and medical history, be sent to: National Coalition Against the Misuse of Pesticides, 701 E St., S.E., Washington, DC 20003.

Environmental Effects

Because pesticides are so widely used and many of them break down very slowly, low levels of pesticide residues are found throughout the environment. Their most serious effect on the environment occurs through contaminated groundwater. The EPA's National Pesticide Survey provides evidence that pesticides are leaching into groundwater and effecting wells and drinking water, crops, and aquatic life far more commonly than was believed just a decade ago. Out of thirteen hundred wells sampled nationwide, the EPA found pesticides in 10 percent of community drinking water wells and in 4 percent of rural wells.

Pesticides enter the soil and groundwater from such non-agricultural incidents as accidental spills, improper disposal of containers, drain water from washing fruits and vegetables from home gardens, air mists from drifting sprays, and runoffs from improper applications. The higher use of pesticides on suburban lawns result in higher runoffs into storm sewers and consequently streams, rivers, and lakes. High incidents of bird and fish kills have been traced to the use of specific pesticides and in some cases have resulted in the mandated or voluntary withdrawal of that pesticide from the market. Diazinon was heavily used on golf courses and home lawns, but the EPA canceled its use on golf turf in 1986 because it had been traced to a high rate of bird deaths. (It has not been withdrawn from home lawn use.)

The NCAMP reports that in 1990, an aerial application of malathion for mosquito control in Massachusetts killed more than five-hundred thousand fish. The environmental effect of pesticides is most keenly felt at the local level, where spraying is proposed to control a situation that presents an immediate danger to health (mosquitos, rodents, giant snails, and predators), beauty and quality of life (gypsy moths, highway brush, pine rust), or the economy (Mediterranean fruit fly).

Non-toxic Alternatives

There are scores of methods for ridding the home garden and farm of insects, weeds, rodents, mildew, and other pests. Integrated pest management (IPM) methods are new approaches that apply commonsense solutions calling for careful identification of the pests involved, ascertaining the seriousness of the problem, exploring appropriate remedies (which may be toxic or non-toxic), applying treatment, and monitoring the results. IPM techniques have resulted in marked decreases in the use of insecticides on cotton crops in Texas, on sweet corn in Connecticut, and on pears in California.

Following is a sampling of non-toxic alternatives to various forms of pesticides using materials readily available around the home:

Fungicide alternatives. Do not overwater the area; keep it clean and dry.

Synthetic pesticides alternatives. Botanical (naturally derived) pesticides such as pyrethrum, rotenone, sabadilla, and nicotine.

Botanical (naturally derived) pesticides alternatives. Insecticidal soap, which breaks down the insects protective coatings.

House plant insecticide alternatives. Mixture of bar soap and water or old dishwater. Spray on leaves, then rinse.

Flea collar and spray alternatives. Herbal collar or ointment (eucalyptus or rosemary) or brewer's yeast in pet's food.

Roach and ant killer alternatives. For roaches: traps or baking soda and powdered sugar mix; for ants: chili powder on trails, boiling water on mounds.

Rat and mouse poison alternatives. Live traps; remove food supply.

Other non-toxic home remedies include spreading borax in infected areas to kill roaches, setting beer traps and milky spore disease (a bacterium) on grubs, and simply maintaining a healthy lawn to discourage the growth of weeds.

Volatile Organic Compounds

The Clean Air Act of 1990 mandated that the U.S. Environmental Protection Agency (EPA) and the states adopt regulations that would reduce the emission of pollutants into the atmosphere, the states being permitted to implement standards that exceed those set by the federal agency. Chemicals that are the largest potential air pollutants are called volatile organic compounds, or VOCs. VOCs contain carbon and easily vaporize to form emissions that pollute the air. While the common perception of air pollution is smog, haze, acid rain, and foul air, indoor air pollution is also receiving attention. We now recognize that VOC emissions, or the harmful gasses given off by such common materials as building products, carpeting, cleaning compounds, paints, pressed wood products, adhesives, chalks, and deodorants, cause "sick building syndrome"—or irritation of the eyes, nose, and throat, headaches, mental fatigue, and respiratory distress to affected buildings' residents.

Many states are enacting regulations to reduce air pollution by limiting the amount of VOCs that come from vehicle emissions, manufacturing plants, dry cleaning establishments, and even from common household substances such as aerosol-driven products.

Table 13

VOC EMISSIONS RANKED BY PRODUCT CATEGORY

PRODUCT CATEGORY		VOC EMISSIONS LBS./DAY
1.	Hair sprays	92,000
2.	Windshield washer fluids	48,000
3.	Air fresheners (inc. dual-use disinfectants)	17,000
4.	General purpose cleaners	9,800
5.	Engine degreasers	9,000
6.	Furniture maintenance products	5,600
7.	Floor polishes	5,200
8.	Glass cleaners	4,600
9.	Laundry pre-wash products	4,000
10.	Nail polish removers	2,200
11.	Oven cleaners	2,000
12.	Hair mousse	1,160
13.	Bathroom and tile cleaners	900
14.	Insect repellents (aerosol)	880
15.	Hair styling gels	820
16.	Shaving creams	520
	Total	204,000

Source: Consumer Product Survey (State of California Air Resources Board, 1990).

The state of California has been in the forefront in enacting clean air regulations. The California Clean Air Act of 1988 requires new regulations to reduce VOC emissions from consumer products. As part of the study to set emission objectives, the California Air Resources Board conducted an industry survey to determine the products that had the most VOC emissions, which were expressed in "pounds per day" of emissions in California. The environmental impact of VOCs in consumer products in California, as shown in Table 13, also indicates the relative importance of products containing VOCs to the national clean air campaign.

Chlorofluorocarbons

Chlorofluorocarbons (CFCs) are a class of chemicals that have found extensive use in household products, manufacturing processes, and construction. They are highly volatile and, when inhaled, can cause considerable damage to the respiratory and central nervous systems. Their notoriety arises from their widespread use as refrigerants, aerosol propellants, cooling "charges" in home and auto air conditioners, styrofoam packaging (which does not decompose in landfills), and insulation, where its vaporization has been blamed as a major contributor to air pollution.

CFC vapors do not decompose in the atmosphere; their slow rise into the stratosphere, which can take as long as eight years, has been cited as the major cause of the depletion of the ozone layer, a layer of gasses that shields the earth from the harmful effects of the sun's ultraviolet radiation. The depletion of the ozone layer is thought to cause global warming, or "the greenhouse effect," worldwide changes in temperature and sea levels that can have adverse effects on the environment.

According to findings based on data from a National Aeronautics and Space Administration (NASA) satellite and reported by EPA head William K. Reilly, a 4-5 percent depletion of the ozone layer over the United States has taken place since 1978. The depletions are as much as five times greater than predicted by computer models and considerably greater than reported by NASA in its 1989 assessment. The reported data point to sizable depletions over an area that roughly includes the United States and Europe.

CFCs have been banned since 1978 from aerosol products, and under the Montreal Protocol of 1987 and the U.S. Clean Air Act of 1990, all CFC production must cease by the year 2000. Yet CFCs still rank as a major outdoor pollutant. According to the DuPont Company, CFCs are used in about thirty-five hundred different applications.

Formaldehyde

Formaldehyde, though one of the most offensive of all chemicals found in household and building products, is an important ingredient in adhesives, preservatives, and wood care formulations. Formaldehyde emissions, which have been traced to foam insulation, particle board and pressed wood used in furniture, wallboard in dry walls, rugs and carpets, permanent-pressed clothing, waxes, dyes, polishes, plaster, and paper products, are a cause of "sick building syndrome," and contact with them brings about headaches, mental disorientation, respiratory disorders, and nausea. Formaldehyde is also a by-product of gasoline combustion.

Asbestos

Asbestos is the name for a group of natural minerals found in rocks that separate into strong, microscopic fibers. Asbestos is fireproof, heat and chemical resistant, and extremely durable, qualities that make it desirable for construction, shipbuilding, and the fabrication industry. But asbestos is associated with several serious and often debilitating health problems, usually with children exposed to the substance over a long period of time in schools and workers exposed in construction, mining, and shipbuilding industries.

Between 1900 and 1980, about thirty million tons of asbestos were put in place in buildings, shipping vessels, and products. However, since the 1970s, asbestos use has declined significantly. Still, each year about two hundred thousand tons of asbestos are mined and processed into an estimated three thousand products in the United States. The U.S. Environmental Protection Agency (EPA) estimates that asbestos-containing materials can be found in approximately 31,000 schools and 733,000 other public and commercial buildings in the United States.

Federal and state courts are overloaded with cases alleging that manufacturers of asbestos products knew of the potential dangers from the mineral and failed to alert workers of harmful effects from working with the substance. Nearly sixty-thousand cases are pending in various state courts; another twenty-thousand cases have been consolidated in the federal court in Philadelphia.

Commonly Used Asbestos Names

Asbestos is a common name for any combination of the following substances: actinolite, amosite, anthrophyllite, anthrophylite, chrysotile, crocidolite, tremolite, and asbestos not otherwise specified. Nearly 95 percent of all asbestos used today in commercial products is chrysotile. It is "friable," meaning it can easily be crumbled with the hand into a fine dust of microscopic fibers that can readily be inhaled or swallowed. When tightly bonded or incorporated into a finished product such as vinyl floor tile, the material is considered "nonfriable" and will not be released unless it is sawed, cut, scraped, drilled, or otherwise disturbed.

Health Hazards

The physical properties that give asbestos its resistance to heat and decay are linked with several adverse human health effects. Unbonded asbestos can crumble into a dust of microscopic fibers that can remain sus-

pended in the air for a long time. They can easily be inhaled or swallowed and penetrate body tissues without detection and, because of their durability, remain unchanged in the body for many years.

Most people exposed to small amounts of asbestos do not develop any related health problems, but studies of workers in factories and shipyards have verified that exposure to asbestos fibers over a period of time can lead to an increased risk of several chronic illnesses:

> *Asbestosis,* in which the lungs become scarred with fibrous tissue, makes breathing progressively more difficult and can lead to death. Early signs of asbestosis usually appear ten years after initial exposure. From 1979 to 1987, it ranked first among the most common occupational diseases according to the *New England Journal of Medicine*. Sixty-five thousand suffer from this disease.

> *Mesothelioma,* a cancer of the lining of the chest and abdominal membranes, almost never occurs without exposure to asbestos, and it is almost always fatal. It may not appear until forty years after exposure.

> *Lung cancer* takes fifteen to thirty-five years to develop.

> *Other cancers,* primarily of the digestive tract, have been associated with exposure to asbestos and can remain undetected for forty years.

Exposure to asbestos over long periods of time can be more dangerous than once believed possible, and the risk to smokers is as much as ninety-two times greater. Even though most people exposed to small amounts of asbestos at home do not develop health problems, it has been shown that family members have developed mesothelioma from asbestos dust carried home on work clothes.

Although asbestos was first linked to disease among workers in the 1920s (though ancient Romans noted asbestos-related diseases among the slaves who wove asbestos fiber into cloth for cremation wrappings), it was not until the early 1970s that legislation was enacted to regulate its use. The first Clean Air Act, of 1970, gave the EPA the authority to regulate hazardous air pollutants, with asbestos being one of the first three substances selected. In 1971 the Occupational Safety and Health Administration (OSHA) first set exposure limits

(12 fibers, greater than 5 microns, per cubic centimeter of air averaged over an eight-hour shift), which had been lowered to 0.2 fibers per cubic centimeter by 1986.

In 1973 the EPA banned the spraying of asbestos material for insulation, fire protection, and decorative soundproofing, and by 1989 it enacted regulations to curtail all but 6 percent of asbestos uses by 1997.

Where Asbestos Can be Found

Until the 1970s, many types of building products and insulation materials contained asbestos. Most products made today do not contain asbestos, and the few that still contain fibers that can be inhaled are required to be labeled as such. Table 14 lists products in factories and households that were previously manufactured with asbestos and the actions that could release asbestos fibers.

Maintenance and custodial workers are at high risk from asbestos-containing material. These workers often unknowingly clean areas where damaged asbestos is present, or repair older ceilings, floor tiles, wallboard, and pipe insulation without taking proper safety precautions. Others at risk are public employees and teachers because asbestos was used in most public buildings built between the 1940s and the early 1970s, boiler room workers, automotive repair workers who work on brakes, sanitation workers who work around material disposed of improperly, road workers, and families of exposed workers.

If damaged asbestos must be removed, the American Lung Association recommends that "asbestos should not be handled, sampled, removed or repaired by anyone other than a qualified professional."

For information on asbestos analysis and removal and how to contact certified asbestos removal specialists, contact: U.S. EPA Toxic Substance Control Act (TSCA) Hotline, (202)554-1404; U.S. EPA Office of Asbestos and Small Business Ombudsman, to help with asbestos-in-schools issues, (800)368-5888; Regional Asbestos Coordinators in the Regional Offices of the EPA or the state environmental agency; and Asbestos Information Association/North America, 1745 Jefferson Davis Highway, Suite 509, Arlington, VA 22202, telephone: (703)979-1150.

The National Institute of Standards and Technology maintains a list of laboratories receiving accreditation to perform bulk asbestos analysis. The laboratories are recognized by the National Voluntary Laboratory Accreditation Program (NVLAP) to perform asbestos tests. For more information, contact: National Voluntary Laboratory Accreditation Program, Laboratory Accreditation, ADMIN A527, National Institute of Standards and Technology, Gaithersburg, MD 20899, telephone: (301)975-4016; NVLAP electronic bulletin board: (301)948-2058.

Federal Regulatory Programs

Several federal agencies regulate the use and disposal of asbestos and asbestos-containing products. In 1973 the EPA launched the National Emission Standards for Hazardous Air Pollutants (NESHAPs) program, which regulates emissions from asbestos mills and manufacturing operations, the use of asbestos in road surfacing and insulation material, and asbestos sprays. It also requires the removal of asbestos that would be disturbed or damaged during renovation or demolition. In 1986 the EPA's Asbestos Hazard Emergency Response Act (AHERA) was enacted, protecting children and employees from exposure to asbestos in public and nonprofit private elementary and secondary schools. It requires building inspections, asbestos management programs, and notification programs for parents, teachers, and other school employees.

The U.S. Occupational Safety and Health Administration (OSHA) sets limits for worker exposure to asbestos on the job; it also prohibits smoking where there is occupational levels of exposure to asbestos. The U.S. Food and Drug Administration (FDA) is responsible for preventing asbestos contamination in food, drugs, and cosmetics. The U.S. Consumer Product Safety Commission (CPSC) regulates asbestos in consumer products except those under the FDA. The Department of Transportation (DOT) regulates the transport and labeling of asbestos-containing material and wastes. The Mine Safety and Health Administration (MSHA) regulates the mining and milling of asbestos.

Additionally, state, county, and municipal regulations have been enacted governing such subjects as certification of asbestos management and control professionals and disclosure requirements for real estate sales. There have been legal actions in some cities for tax abatements where the cost of forced removal of asbestos from old buildings under renovation cannot be recovered from the sale or rental of the building.

Table 14

PRODUCTS PREVIOUSLY MANUFACTURED WITH ASBESTOS AND ACTIONS THAT COULD RELEASE FIBERS

PRODUCTS	DANGEROUS CONDITIONS
Steam pipes, boilers, furnace ducts insulated with asbestos blankets or asbestos tape; furnace cement.	Material is damaged, repaired, or removed improperly. Recent studies indicate that asbestos insulation that is in good condition is best left alone, that removal may release more fibers than if kept in place. This type of insulation was manufactured from 1920 to 1972.
Vinyl asbestos, vinyl linoleum, asphalt, and rubber floor tile; backing on vinyl sheet flooring; floor tile adhesive.	Sanding, cutting, or scraping can release fibers; sanding and scraping the backing of vinyl flooring.
Cement sheet and millboard in wall and ceiling material; insulation around furnaces and wood-burning stoves; asbestos insulation sandwiched between plaster walls.	Repairing or removing appliances; sawing, cutting, drilling, sanding, or tearing insulation; home improvements. Used in homes built between 1930 and 1950.
Door gaskets in furnaces; wood and coal stoves.	Worn seals. Used between 1940 and the early 1980s.
Soundproofing and decorative material sprayed on walls and ceilings.	Sanding, drilling, scraping; loose, crumbly, or water-damaged material.
Textured paints; patching and joint compounds (spackle) made before 1978.	Sanding, scraping, drilling. If in good condition, leave alone. CPSC banned asbestos-containing patching compounds and textured paint in 1978.
Asbestos cement roofing, shingles, siding.	Not likely to release fibers unless sawed, drilled, or cut. Avoid disturbing these products.
Artificial ashes and embers for gas-fired fireplaces.	Damaged or crumbled products.
Older products, such as fireproof gloves, stove-top pads, ironing board covers, certain hair dryers, toasters, ovens, ranges, clothes dryers, electric blankets, and dishwashers.	Damaged, cut, worn-out products. Normal use unlikely to release fibers from asbestos components. Hair dryers with asbestos heat shields were recalled in 1979. All consumer products with asbestos require labeling since 1986. Most asbestos products banned by 1996.
Automobile brake pads and linings, clutch facings, and gaskets.	Cutting, grinding, sanding.

Source: *Asbestos in the Home* (Washington D.C.: U.S. Consumer Product Safety Commission and U.S. Environmental Protection Agency, August 1989); *Asbestos Fact Book* (Washington D.C.: U.S. Environmental Protection Agency, May 1986); *A Guide to Asbestos in the Home for New Jersey Residents* (Trenton, NJ; New Jersey Department of Health, 1989).

Radon

Radon is a colorless, odorless, radioactive gas produced in nature by the radioactive decay of radium, a decay product of uranium. Radon can be found in high concentrations in soils and rocks containing uranium and in granite, shale, phosphate rock, and pitchblende. It can also be found in well water contaminated from the surrounding soil and rocks.

In outdoor air, radon is diluted to such low concentrations that it is usually not harmful. However, once inside an enclosed space, radon can accumulate to a concentration that can be hazardous to your health. The U.S. Environmental Protection Agency (EPA) estimates that between eight million and ten million homes in the United States have radon levels greater than four picocuries per liter of air (pCi/L), the highest level that is considered safe. The EPA also estimates that twenty-two thousand die each year from radon.

Health Hazards

Radon gas can seep into buildings through dirt floors, cracks or joint openings in concrete floors and walls, floor drains, sump pumps, and tiny cracks or pores in hollow-block walls. Radon can also enter water through private wells and be released into a home when the water is used. As building materials rarely cause radon problems by themselves, radon in large community water supplies would likely be released into the outside air before the water reaches the home.

When you breathe air containing even slightly elevated levels of radon over a long period of time, there is an increased risk of developing lung cancer, and the time between exposure and the onset of the disease may be many years. This type of exposure may present a greater danger than exposure to a significantly elevated level for a short time. But not everyone exposed to radon will develop lung cancer, and, like with other environmental pollutants, there is some uncertainty of radon's health risks. But more is known about radon risks than risks from most other cancer-causing substances because radon risks are based upon studies of cancer in underground miners.

Your chances of getting lung cancer from radon depends mostly on how much radon is in your home, the amount of time you spend there, and whether you are a smoker or have ever smoked. Studies have shown that smokers exposed to radon run a greater risk of developing lung cancer than those who do not smoke. At the radon exposure level of 20 picocuries per liter of air (pCi/L), approxi-mately 135 of every 1000 smokers will develop cancer; approximately 8 non-smokers will. At 4 pCi/L, about 29 of every 1000 smokers will develop cancer; approximately 2 non-smokers will. At 2 pCi/L, approximately 15 of every 1000 smokers will develop cancer; about 1 non-smoker will. And at 1.3 pCi/L (the average indoor radon level), about 9 smokers will develop cancer; less than 1 non-smoker will. If you are a former smoker, your risk will be higher than that of a non-smoker.

There are so many different variables that influence the risk from radon that the scientific community does not agree with the EPA estimates of the number of homes with radon or of the degree of risk from radon-causing cancer deaths. For example, the EPA maintains that if your home has as little as four picocuries of radon per liter of air, your lifetime risk of lung cancer is 1 percent. But that risk is based upon living in the same polluted house for seventy years; in our mobile society, the average person moves every seven years. Also, differences among individuals, such as age, genetic inheritance, diet, and personal habits, make it impossible to determine individual risks accurately.

Detecting Radon

Since radon cannot be seen or smelled, special equipment is needed to detect it. The two most popular commercially available radon detectors are the charcoal canister and the alpha track detector. These devices are exposed to the air in a home or building for a specified period of time and sent to a laboratory for analysis. There are other, more sophisticated techniques for measuring radon that require operation by trained personnel.

If the results of testing show that the average radon concentration is more than 4 pCi/L, you should determine which products are emitting radon and remove or fix them. Readings between 2 and 4 pCi/L deserve consideration of removal or fixing; readings of 2 or less pCi/L are normal and probably cannot be reduced much more (reducing levels below 2 is difficult).

The EPA conducts a Radon Measurement Proficiency Program, a voluntary program that allows laboratories and businesses to demonstrate their capabilities in measuring indoor radon. The names of firms participating in this program can be obtained from your EPA regional office or your state radiation protection office. Your state office also can provide you with guides to radon testing companies in your area and information on how to evaluate testing results, reduce your radon health risks, and select devices to prevent or remove radon gas from new or older homes. Another source for information about radon is the American Association of Radon Scientists and Technologists, P.O. Box 70, Park Ridge, NJ 07656, telephone: (201)391-6445. Following is a list of state radon contacts:

Alabama	(800)582-1866
Alaska	(800)478-4845
Arizona	(602)255-4845
Arkansas	(501)661-2301
California	(800)745-7236
Colorado	(800)846-3986
Connecticut	(203)566-3122
Delaware	(800)544-4636
District of Columbia	(202)727-5728
Florida	(800)543-8279
Georgia	(800)745-0037
Hawaii	(808)586-4700
Idaho	(800)445-8647
Illinois	(800)325-1245
Indiana	(800)272-9723
Iowa	(800)383-5992
Kansas	(913)296-1560
Kentucky	(502)564-3700
Louisiana	(800)256-2494
Maine	(800)232-0842
Maryland	(800)872-3666
Massachusetts	(413)586-7525
Michigan	(517)335-8190
Minnesota	(800)798-9050
Mississippi	(800)626-7739
Missouri	(800)669-7236
Montana	(406)444-3671
Nebraska	(800)334-9491
Nevada	(702)687-5394
New Hampshire	(800)852-3345
New Jersey	(800)648-0394
New Mexico	(505)827-4300
New York	(800)458-1158
North Carolina	(919)571-4141
North Dakota	(701)221-5188
Ohio	(800)523-4439
Oklahoma	(405)271-5221
Oregon	(503)731-4014
Pennsylvania	(800)237-2366
Puerto Rico	(809)767-3563
Rhode Island	(401)277-2438
South Carolina	(800)768-0362
South Dakota	(605)773-3351
Tennessee	(800)232-1139
Texas	(512)834-6688
Utah	(801)538-6734
Vermont	(800)640-0601
Virginia	(800)468-0138
Washington	(800)323-9727
West Virginia	(800)922-1255
Wisconsin	(608)267-4795
Wyoming	(800)458-5847

Cosmetics and Personal Care Products

American consumers spend $20 billion annually on cosmetics and personal care products, purchasing twenty thousand different products with more than fifty thousand brand names. Until recently, consumer concerns about the health and safety of cosmetics and personal care items were centered on allergic reactions to the hundreds of ingredients that enhance our eyes, lips, skin, hair, nails, and teeth. But now consumer awareness of hazardous ingredients in the products we use has turned to cosmetics and the mounting evidence that the creams and lotions we spread upon ourselves can be toxic and cause long-term health effects. We now know that cosmetic ingredients are capable of causing cancer, chromosomal damage, and other chronic diseases far more serious than the normal allergic reactions that effect 1 percent of cosmetic users.

Cosmetics and personal care products are regulated by a division of the Food and Drug Administration (FDA) under the Federal Food, Drug and Cosmetic Act of 1938. Yet only one page of the 156-page Act is devoted to cosmetics. Essentially, the Act exempts cosmetic and personal care products from most of the regulations, relying instead on the industry for self-regulation.

The greatest toxic effect on humans from personal care products arises from absorption through the skin of seemingly benign solvents, coloring agents, and fragrances. Posing lesser danger is talc, which may be inhaled and cause injury to the lungs, while mists of fragrances, deodorizers, antiperspirants, and hair sprays may also irritate the eyes and respiratory tract. Personal care products present slight harm to the groundwater environment, for they are formulated to be washed away through normal sewage and septic systems. But the great dependency of consumers on aerosols, especially those containing chlorofluorocarbons (CFCs) that pollute the upper atmosphere and reduce the protective ozone layer of the earth, create the greatest environmental threat from these products. The elimination of CFCs from aerosols by the mid-1990s has been mandated by international agreement, and chemical companies are racing to develop alternative propellants less damaging to the environment.

Hazardous Substances

Of the thousands of ingredients that comprise cosmetics and personal care products, most are meant to make them smell nice, look pretty, and prolong their shelf lives. The most common reaction to a specific chemical is an allergy, causing skin rash and eye irritation, something most consumers have experienced and which they learn to avoid by trying other products with different ingredients. The allergens formaldehyde and Quaternium 15 are frequently used in fragrances and perfumes to enhance shampoos, conditioners, lotions, and creams. Quaternium 15 particularly, a preservative also used in mascara, is responsible for more cases of contact dermatitis than any other cosmetic ingredient.

The product cited as most likely to cause health problems is mascara, not so much from the ingredients, but from contaminated products and the misuse of the applicator. But mascara is used so close to the eye that even a small speck on the eyeball can lead to conjunctivitis, an inflammation of the mucous membrane that lines the inner surface of the eyelid. Women who wear contact lenses should be particularly careful when using mascara to avoid contamination of the lens.

Coloring dyes have proven to be severely toxic to humans as well and some, such as FD&C Red No. 4, are banned for use in foods. Yet, they are permitted in personal care products. Caution should be exercised when using them, especially around the eyes.

Review of Cosmetic Ingredients

Although controlled by the Food and Drug Administration, the cosmetic industry is self-regulating. To ensure the safety of personal care products, the Cosmetic, Toiletry, and Fragrance Association (CTFA) established an organization to review information on ingredients used in the formulations of products. The Cosmetic Ingredients Review (CIR) is an independent panel of professionals, none of whom are connected with companies in the industry, whose findings and reports are not subject to prior review by the industry.

By 1990, CIR had issued reviews covering 321 ingredients out of 1,005 personal care products that were of concern to a congressional subcommittee. Twenty-five percent of the total ingredients reviewed by the subcommittee were found to be already approved by the FDA for use in food or drugs, while nearly 40 percent have little or no use in cosmetics.

CIR relies primarily on unpublished safety test data by the cosmetic industry for its evaluations. It also acquires information and evaluations from the National Institute for Occupational Safety and Health (NIOSH) Registry of Toxic Effects of Chemical Substances

Table 15

COSMETICS AND PERSONAL CARE PRODUCTS INGREDIENTS: USES AND HARMFUL EFFECTS

INGREDIENT	PRODUCTS/USES	HARMFUL EFFECTS
Ammonium carbonate	Permanent wave products	Sensitizes the face, scalp, and hands; causes contact dermatitis
Butyl stearate	Nail care products	Possible allergin; acne
Coal tar	Dandruff shampoos, hair dyes	Carcinogenic
Diethanolamine (DEA)	Used in many products, including shampoos, hair conditioners, dyes, permanent wave products	Can become contaminated with nitrosamines, which have caused cancer in animals; is absorbed through skin
Dimethicone	A silicone oil used in many products	Allergenic
Ethyl alcohol	Used in many products as a solvent, particularly mouthwashes, shaving and hair care products, and lotions	Drying to skin and hair ("SD" denatured alcohol has chemicals added making it poisonous to drink)
FD&C Blue No. 1	Coloring agent in toothpaste, hair coloring	Allergenic; causes cancer in animals
FD&C Red No. 4	Coloring for shampoo, hair conditioners; no longer permitted for food coloring	Can be extremely toxic; don't use around eyes; may damage adrenal glands and bladder
FD&C Yellow No. 5	Coloring agent	May cause severe asthmatic symptoms and allergic reaction to aspirin-sensitive persons
FD&C Yellow No. 10	Coloring agent	Potential allergenic; can be contaminated with betanaphthylamine, a carcinogen
FD&C No. 17	Nail care products	Causes cancer in animals
Formaldehyde	Solvent and preservative in nail care products, shampoos, fragrances	Eye irritant; causes cancer in animals; long term exposure may depress the CNS, cause drowsiness
Fragrances	Combination of synthetic and/or natural chemicals; used in many products	Skin irritant and other allergic reactions
Nitrosamine	Contaminates ingredients in preservatives; formed by reacting with DEA and TEA	Causes cancer in animals

Table 15 (contd)

COSMETICS AND PERSONAL CARE PRODUCTS INGREDIENTS: USES AND HARMFUL EFFECTS

INGREDIENT	PRODUCTS/USES	HARMFUL EFFECTS
Parabans (methyl—; propyl—; butyl—)	Synthetic preservatives in hair conditioners	Allergenic
PEG components	Used as binders, emollients	May be contaminated with 1,4-dioxane, an animal carcinogen; is absorbed through skin
Quaternium compounds	Emulsifiers, preservatives, softeners; shampoo, hair conditioners; fragrances, perfumes; mascara	A leading cause of allergic reactions; can release formaldehyde; dangerous to eyes in concentrations as low as .5%
Sodium lauryl sulfate	Emulsifier and detergent; used in may products, including shampoos, toothpaste, lotions, creams	Cause skin drying; eye irritant
Solvents (toluene, acetone, butyl acetate, ethyl acetate, amyl acetate, etc.)	Used in many products including nail care products, permanent wave products	Skin irritant; breathing high concentrations can depress CNS
Talc	Cosmetic base; baby and talcum powder	Prolonged inhaling can lead to lung disease
Triethanolamine (TEA)	Used heavily as emulsifier, pH adjuster, and preservative in shampoos, hair conditioners, dyes, permanent wave products	Can become contaminated with nitrosamines, which have caused cancer in animals; is absorbed through skin

Source: David Steinman, "Glossary of Common Cosmetic Terms and Ingredients." *Greenkeeping*, March/April, 1992. Used with permission. CNS=central nervous system

(RTECS), the CTFA *Cosmetic Ingredient Dictionary*, the Research Institute for Fragrance Material (RIFM), and published reports. All final reports of CIR are published in the *Journal of the American College of Toxicology* or the *Journal of Environmental Pathology and Toxicology* and are available to the public from CIR. For an index of all ingredients for which final reports have been published or those in preparation, and for information about ordering collections of reports, literature reviews, and tentative reports, contact the Cosmetic Ingredient Review (CIR), 1110 Vermont Ave., N.W., Suite 810, Washington, DC 20005, telephone: (202) 331-0651. For information about personal care products and cosmetics, contact the Cosmetic, Toiletry, and Fragrance Association, Office of Public Affairs, 1101 17th St.,
N.W., Suite 300, Washington, DC 20036, telephone: (202) 331-1770, fax: (202) 331-1969.

Non-toxic Alternatives

Many people are allergic to specific ingredients in cosmetics and personal care products, but labeling can be so non-specific that only by trial and error can an uncomfortable product be eliminated. "Hypoallergenic" is a term used to indicate a product that is least likely to cause an allergic reaction, and it often simply means that the product is fragrance-free. Nevertheless, such products can frequently contain other allergens, irritants, and carcinogens whose effects may only be apparent over a long period of time.

Hazardous Substance Regulations

In the past two decades, many laws have been passed to protect the environment and citizens from exposure to hazardous substances. These federal and state rules and regulations are geared toward reducing specific areas of health risks such as polluted air, water, or land; environmental dangers from products such as pesticides and chemicals; and harmful exposures to toxic chemicals in the workplace.

With many federal and state agencies involved in various aspects of hazardous substances, it is logical to question which agency has jurisdiction over a specific problem. Environmental laws are enforced by several different federal, state, and local agencies, and in many instances where reports must be filed, it may be necessary to file the same report in two or more agencies.

The U.S. Environmental Protection Agency (EPA) administers most laws concerning pollution in the environment and provides information on pollutants in the indoor air of buildings. It also maintains a registry of hazardous substances and conducts studies to determine how many pollutants are released into the environment. Among the acts under the EPA are the Comprehensive Environmental Response, Compensation, and Liability Act (CERCLA, or Superfund), the Toxic Substances Control Act (TSCA), the Clean Air and Clean Water Acts, and the Community Right-to-Know Act (Title III of the Superfund Amendment and Reauthorization Act). The Office of Pesticide Programs regulates pesticide products used in the home and the information presented on their labels.

The U.S. Occupational Safety and Health Administration (OSHA) regulates the use of hazardous substances in the workplace and ensures safe working conditions. It mandates the use of material safety data sheets (MSDS) for communicating the hazards of chemicals found in the workplace. OSHA sets the guidelines for determining whether a product used in the workplace is considered toxic.

The U.S. Department of Transportation (DOT) regulates hazardous materials while they are in transit and develops procedures for dealing with accidents and spills involving hazardous and toxic materials.

The U.S. Consumer Products Safety Commission (CPSC) promulgates safe products and proper labeling. Under the Hazardous Substance Act, CPSC regulates the labeling of consumer products other than food, drugs, pesticides, cosmetics, alcohol, and tobacco.

The U.S. Food and Drug Administration (FDA) regulates drug and cosmetic products and food products except for meat and poultry. Cosmetics such as hair sprays, creams, shaving lotions, nail polishes, and perfumes are required to list all ingredients on their labels but, except for a few exceptions, labels do not have to include warnings on health hazards.

The U.S. Department of Agriculture (USDA) inspects and provides safe handling and labeling guidelines for meat and poultry products.

The major federal laws pertaining to hazardous substances and the areas they address are summarized in Table 16.

The U.S. Environmental Protection Agency (EPA) administers most laws concerning pollution in the outdoor environment and provides information on indoor air pollutants. EPA's responsibilities include:

- Setting and enforcing standards under environmental laws;

- Developing and testing new methods to reduce the sources of environmental risks;

- Requiring the cleanup of sites where damage from hazardous substances has already occurred (under Superfund);

- Authorizing state and local environmental agencies to conduct their own environmental programs, overseeing and insuring effectiveness of state and local programs;

- Providing information to the public and businesses about regulatory requirements, environmental programs, procedures to reduce exposures to hazardous substances, and the health effects of hazardous substances;

- Assisting state and local governments in planning for emergencies; and

- Coordinating the efforts of local government groups.

Emergency Planning and Community Right-to-Know Act

The gravity of the 1984 Bhopal, India, disaster—more

Table 16

MAJOR LEGISLATION CONTROLLING HAZARDOUS SUBSTANCES

AREA	LAW	DESCRIPTION OF LAW	RESPONSIBLE AGENCY
Air	Clean Air Act, amended in 1990	Sets and monitors air quality standards; limits emissions (including radioactive emissions into outdoor air) from factories, power plants, vehicles, and other major sources of air pollution. The basic authority for the nation's air pollution control programs.	EPA, state agencies
Chemicals	Toxic Substances Control Act, 1976 (TSCA)	Provides authority for EPA to regulate the manufacture, distribution and use of old and new chemical substances.	EPA
Drinking water	Safe Drinking Water Act, amended in 1977	Sets national standards for contaminant levels in public drinking water systems, including lead. Regulates discharges into groundwater.	EPA, state agencies
Food	Food, Drug, and Cosmetic Act	Regulates contamination of food.	USDA, FDA
	Federal Insecticide, Fungicide, and Rodenticide Act (FIFRA)	Regulates pesticide residue on food marketed in the U.S.	EPA
Pesticides	Federal Insecticide, Fungicide, and Rodenticide Act (FIFRA)	Regulates the manufacture, distribution, and use of pesticides and mandates that EPA conduct research into their health and environmental effects; can ban pesticides and limit manufacture of unregistered pesticides; regulates labels of pesticides used in the home.	EPA

Table 16 (contd)

MAJOR LEGISLATION CONTROLLING HAZARDOUS SUBSTANCES

AREA	LAW	DESCRIPTION OF LAW	RESPONSIBLE AGENCY
Products	Consumer Product Safety Act (CPSC)	Sets safety standards and regulates labeling for consumer products other than food, drugs, pesticides, cosmetics, alcohol, and tobacco; can ban unsafe products.	CPSC
	Hazardous Substances Labeling Act	Sets labeling requirements for consumer products.	CPSC
Right-to-Know	Emergency Planning and Community Right-to-Know Act, 1986 (EPCRA; Title III of the Superfund Amendment and Reauthorization Act [SARA]).	Requires businesses and factories to report on chemicals, inventories, environmental releases of certain chemicals, and emergency planning.	EPA, state and local agencies
Transportation	Hazardous Materials Transportation Act	Sets standards for shipping hazardous materials.	DOT
Waste treatment, storage, and disposal	Resource Conservation and Recovery Act, 1976 (RCRA)	Regulates disposal of solid and hazardous wastes and groundwater contamination around treatment, storage, and disposal facilities.	EPA
	Comprehensive Environmental Response, Compensation, and Liability Act (CERCLA, or Superfund), 1980	Requires cleanup of hazardous substances released into the air, soil, groundwater, and surface water; responds to hazardous substance spills. Provides funds for cleaning up hazardous cites and imposes liability.	EPA
Water	Clean Water Act	Enforces quality standards for interstate and intercostal waterways to make them fishable and swimmable; sets effluent standards for each industry. Requires discharge permits for factories, sewage treatment plants, and storm runoff.	EPA, state agencies

Table 16 (contd)

MAJOR LEGISLATION CONTROLLING HAZARDOUS SUBSTANCES

AREA	LAW	DESCRIPTION OF LAW	RESPONSIBLE AGENCY
	Marine Protection, Research, and Sanctuaries Act, 1972	Limits dumping of all types of waste material in U.S. ocean waters.	EPA, Coast Guard, Army
Workplace	Occupational Safety and Health Act (OSHA), 1970	Sets health and safety standards for workplace environments; requires Material Safety Data Sheets for hazardous chemicals used in products or the workplace.	OSHA, state agencies
	Federal Insecticide, Fungicide, and Rodenticide Act (FIFRA)	Limits worker exposure to pesticides.	EPA
	Toxic Substances Control Act (TSCA)	Requires makers of toxic substances to provide notification of potential hazardous substances in the workplace.	EPA

than twenty-five hundred people dead and tens of thousands more injured from a release of toxic methyl isocyanate from a Union Carbide plant—opened the eyes of the world to the dangers of chemical accidents. Bhopal, and a subsequent accident involving a similar chemical in West Virginia, demonstrated that the general public needed to know more about the hazardous chemicals used in the manufacturing plants in and around their communities and what preparations were being made to cope with accidents. In the past, citizens who wanted to know more about toxic chemicals in their communities had to rely on the cooperation of industry for information. If a company was not willing to share information about its operations, nothing much could have been done about it.

That situation changed in 1986 when the Emergency Planning and Community Right-to-Know Act established requirements for federal, state, and local governments and industry regarding emergency planning and reporting on hazardous and toxic chemicals. The new law, also known as Title III of the Superfund Amendment and Reauthorization Act (SARA), made many vol-

untary programs already in existence mandatory and required that detailed information about the nature of hazardous substances in or near communities be made available to the public. It also provided stiff penalties for companies that did not comply, and it allowed lawsuits against companies and government agencies to force them to obey the law.

Information available under the Title III law includes:

Material Safety Data Sheets (MSDSs). Available at State Emergency Response Commissions (SERC), Local Emergency Planning Committees (LEPC), and local fire departments for any facility required to complete MSDSs under OSHA.

Chemical Inventories. Available from SERC, LEPC, and local fire departments. Annual inventories of hazardous chemicals for which MSDSs have been prepared reported in the aggregate or detailed, containing quantities and locations of chemicals, average daily amount of use, and maximum amounts of chemicals on hand during the past year.

Toxic Release Inventories (TRI). Annual reports about routine releases of chemicals for facilities that use toxic chemicals above specified thresholds. Provides how chemicals are used, maximum amount present during the year, disposal methods and their efficiencies, and amounts routinely released in the air, water, and soil.

Emergency Planning Information. Available on request from the LEPC from any facility using or storing extremely hazardous substances above a specified threshold.

Emergency Planning Notification. Reports of emergency releases of any extremely hazardous substance.

The Title III law also involves four complementary activities that encompass this information:

Emergency Planning (Sections 301-303). These sections require the governor of each state to appoint a SERC which, in turn, designates LEPCs to collect and organize chemical information and develop emergency response plans for their communities. Facilities notify SERCs and LEPCs if they have any of 366 Extremely Hazardous Substances (EHS) present above Threshold Planning Quantities (TPQ) as listed by the EPA (Code of Federal Regulations; 40 CFR 300 and 355).

Emergency Notification (Section 304). Facilities must report accidental releases of hazardous substances above certain amounts to SERCs and LEPCs, which must make that information available to the public. Chemicals subject to this requirement are not only those on the list of 366 Extremely Hazardous Substances (40 CFR 300 and 355), but also more than 725 hazardous substances subject to the emergency notification requirements under CERCLA Section 103(a) (40 CFR 302.4). Some chemicals appear on both lists.

Community Right-to-Know Reporting (Sections 311-312). You have the right to information about the amounts, locations, and potential effects of hazardous chemicals being used or stored in certain quantities in your community. Facilities that are required to prepare MSDSs under OSHA regulations must also submit them to their LEPCs, SERCs, and local fire departments. This information must be made available to the public. There is no specific list of chemicals requiring MSDSs; as many as fifty thousand may fit the definition of "hazardous chemical": essentially, any chemical that poses a physical or health hazard. Section 311 calls for submitting MSDS data; Section 312 requires annual reporting of chemical inventories (40 CFR 370).

Toxic Chemical Release Reporting (Section 313). Along with all of the information on hazardous chemicals use, storage, and accidental releases, the public has the right to know if certain manufacturing plants are routinely releasing any of more than 320 toxic chemicals into the air, water, or soil. Facilities that release these must report the emissions to the EPA and to state officials. Reports must be filed by July 1 of each year to the EPA, which compiles them into the Toxic Release Inventory (TRI) database, which must be made available to the public through computer telecommunications or other means (40 CFR 350; 40 CFR 372).

To obtain more information on the Emergency Planning and Community Right-to-Know Act, contact the Local Emergency Planning Committee (LEPC) in your community or the State Emergency Response Commission (SERC) in your state. If they are unable to answer any technical or regulatory questions, contact your nearest EPA regional office or call the Emergency Planning and Community Right-to-Know (Title III) Information Hotline, (800)535-0202; in Washington, D.C. (202)479-2449.

Toxic Release Inventory (TRI) data can be obtained from the U.S. Environmental Protection Agency, P.O. Box 70266, Washington, DC 20024-0266. For an introductory videotape on Title III titled "Emergency Planning and Community Right-to-Know: What It Means to You," contact your nearest EPA regional office, the above Title III Hotline number, or write: Emergency Planning and Community Right-to-Know Information (OS-120), U.S. Environmental Protection Agency, Washington, DC 20246. The Emergency Planning and Community Right-to-Know Information office also makes available four publications that provide excellent overviews of Title III and the chemicals incorporated under Section 313: *Chemicals in Your Community: A Guide to the Emergency Planning and Community Right-to-Know Act* (EPA, September 1988); *Community Right-to-Know and Small Business: Understanding Sections 311 and 312 of the Emergency Planning and Community Right-to-Know Act of 1986* (EPA, Office of Solid Waste and Emergency Response, September 1988); *Title III List of Lists: Consolidated List of Chemicals Subject to Reporting under the Emergency Planning and Community Right-to-Know Act* (EPA, Office of Pesticides and Toxic Substances, 560/4-91-011, January 1991); and *Common Synonyms for Chemicals Listed Under Section 313 of the Emergency Planning and Community Right-to-Know Act* (EPA, Office of Pesticides and Toxic Substances, 560/4-91-005, January 1991). Also worthwhile is *Using Community Right-to-Know: A Guide to a New Federal Law* from OMB Watch, 2001 O St., N.W., Washington, DC 20036; telephone (202)659-1711.

U.S. Occupational Safety and Health Administration

The Occupational Safety and Health Administration (OSHA) was created within the Department of Labor in

1970 to protect employees against workplace safety and health hazards, and with respect to the use of potentially hazardous substances in the work environment, it developed mandatory job safety and health standards and procedures for reporting and enforcing their use. OSHA guarantees employers and employees the right to be fully informed about health hazards in their workplace and to participate actively in the formulations of standards and appeal actions that may occur.

OSHA encourages states to develop their own job safety and health plans, but they must be at least as effective as those of the federal program. A quick overview of OSHA, including how it develops its standards, its requirements for record keeping and reporting, and how the appeals process works, is provided in the booklet *All about OSHA* (OSHA #2056, 1991, revised), available from regional OSHA offices. OSHA also issues a guide for small businesses, *OSHA Handbook for Small Businesses* (OSHA #2209, 1990, revised), that contains checklists for all manners of operations, including environmental controls, flammable and combustible materials, hazardous chemical exposure, and hazardous substance communication.

Hazard Communication Standard

Of all of the activities of OSHA, the Hazard Communication Standard (HCS) sets forth regulations that most affect the information available to workers about hazardous chemicals in the workplace (29 CFR 1910.1200). Initially, the HCS applied only to manufacturers in Standard Industrial Classification (SIC) codes 20 to 39; however, in 1987 OSHA amended the regulation to include all businesses regardless of classification or size. As a result, even "small businesses" may now be subject to OSHA reporting requirements.

Public employees in most states are covered by state Right-to-Know laws that generally contain provisions similar to the non-public sector. In states without Right-to-Know laws, public employers follow the standards set by OSHA's Hazard Communication Standard.

Material Safety Data Sheets

Under the HCS, chemical manufacturers, importers, and distributors must research the chemicals or products they produce or distribute to determine if they present any of the physical and health hazards specified in the HCS. If the substance does present a hazard, it must be communicated to all employees as well to all "downstream" users of the product in the form of the material safety data sheet (MSDS). MSDSs provide information about the chemical, including physical characteristics, safety hazards and risks, safe handling procedures, first aid measures, and what to do in the case of spills, fire, overexposure, or environmental releases.

Employers must maintain an MSDS for every hazardous

chemical they use in their processing operations or incorporate into final products. There is no specific format to follow, but certain information must be included. The format recommended by OSHA is reproduced here and is widely used (see Figure 4). Regardless of the presentation, the following information must be included in every MSDS:

Section I	**Identity of the Chemical and Manufacturer.**
Section II	**Hazardous Ingredients and Identity Information.** Also safe exposure limits.
Section III	**Physical and Chemical Characteristics.** Included here are such data as boiling and melting points, vapor pressure and density, specific gravity, water solubility, and its appearance and odor.
Section IV	**Fire and Explosion Hazard Data.** Its flash point and flammable limits and special fire fighting procedures.
Section V	**Reactivity Data.** What happens when the chemical is mixed with air, water, or other chemicals and what conditions to avoid.
Section VI	**Health Hazard Data.** How the chemical can enter your body and affect your health and emergency and first aid procedures.
Section VII	**Precautions for Safe Handling and Use.** The proper way to handle, store, and dispose of the chemical and what to do if it spills or is released into the air, ground, or water.
Section VII	**Control Measures.** How to protect yourself, and information on equipment, ventilation, and special work and hygiene practices.

Though the HCS contains no formal list of chemicals, any of roughly five hundred thousand products may trigger the reporting requirement. While the responsibility for issuing current MSDSs rests with chemical manufacturers, distributors, and importers, the chemical user must maintain complete MSDS files.

For information about OSHA Hazard Communication Standard and material safety data sheets, contact your state or local OSHA office or the OSHA Office of Information Consumer Affairs in Washington, D.C., at (202)523-8151. Another excellent source for compliance information about employee Right-to-Know rules are unions that represent employees in non-public companies and public institutions.

Material Safety Data Sheet
May be used to comply with
OSHA's Hazard Communication Standard,
29 CFR 1910.1200. Standard must be
consulted for specific requirements.

U.S. Department of Labor
Occupational Safety and Health Administration
(Non-Mandatory Form)
 Form Approved
 OMB No. 1218-0072

IDENTITY *(As Used on Label and List)*

Note: Blank spaces are not permitted. If any item is not applicable, or no information is available, the space must be marked to indicate that.

Section I

Manufacturer's Name	Emergency Telephone Number
Address *(Number, Street, City, State, and ZIP Code)*	Telephone Number for Information
	Date Prepared
	Signature of Preparer *(optional)*

Section II — Hazardous Ingredients/Identity Information

Hazardous Components (Specific Chemical Identity; Common Name(s))	OSHA PEL	ACGIH TLV	Other Limits Recommended	% *(optional)*

Section III — Physical/Chemical Characteristics

Boiling Point		Specific Gravity (H$_2$O = 1)	
Vapor Pressure (mm Hg.)		Melting Point	
Vapor Density (AIR = 1)		Evaporation Rate (Butyl Acetate = 1)	

Solubility in Water

Appearance and Odor

Section IV — Fire and Explosion Hazard Data

Flash Point (Method Used)	Flammable Limits	LEL	UEL

Extinguishing Media

Special Fire Fighting Procedures

Unusual Fire and Explosion Hazards

Section V — Reactivity Data

Stability	Unstable		Conditions to Avoid
	Stable		

Incompatibility (*Materials to Avoid*)

Hazardous Decomposition or Byproducts

Hazardous Polymerization	May Occur		Conditions to Avoid
	Will Not Occur		

Section VI — Health Hazard Data

Route(s) of Entry: Inhalation?   Skin?   Ingestion?

Health Hazards (*Acute and Chronic*)

Carcinogenicity: NTP?   IARC Monographs?   OSHA Regulated?

Signs and Symptoms of Exposure

Medical Conditions
Generally Aggravated by Exposure

Emergency and First Aid Procedures

Section VII — Precautions for Safe Handling and Use

Steps to Be Taken in Case Material Is Released or Spilled

Waste Disposal Method

Precautions to Be Taken in Handling and Storing

Other Precautions

Section VIII — Control Measures

Respiratory Protection (*Specify Type*)

Ventilation	Local Exhaust		Special
	Mechanical (*General*)		Other

Protective Gloves	Eye Protection

Other Protective Clothing or Equipment

Work/Hygienic Practices

Hazardous Substance Labeling

It is the goal of government to protect the health and well being of its citizens and to protect the environment. Over the past twenty years these objectives have led to a series of important statutes including the Consumer Product Safety Act and Occupational Safety and Health Administration Hazard Communication Standard. In order to increase consumer and worker awareness of the dangers associated with hazardous materials and products, these and other equally important laws require cautionary instructions and warning labels on consumer and industrial products.

Labels are used for at least eight distinctly different kinds of information. They reveal the name of the product, its manufacturer, important marketing information, what the product is supposed to do, how to use the product, actions to be taken (or avoided), hazard/risk warnings, and information such as nutritional value. Labels influence both the behavior and the attitude of those who read them.

Many labels must conform to a series of regulatory requirements that might include anything from a product's ingredients to a "sell by" date, a bar code used for pricing and inventory control, an illustration that must not be misleading to prospective purchasers, statements of safety and health, and/or transportation markings and hazard class.

Regulatory Control of Labels

Choosing the right label for a hazardous material, chemical, or product is confusing. The United Nations Committee of Experts on the Transport of Dangerous Goods, various government agencies and departments, and the chemical industry itself have established guidelines on labeling. These guidelines provide for the different needs of a label user. Some of these guidelines carry information on the display of properties about a product that may be harmful or even fatal, some deal with the transportation of materials and substances, and others tell how to describe health hazards, first aid, and storage and disposal information.

While the U.S. Environmental Protection Agency (EPA), the U.S. Department of Transportation (DOT), the Consumer Product Safety Commission (CPSC), and the Department of Labor's Occupational Safety and Health Administration (OSHA) all create regulations for labeling hazardous substances, there exists an entire other world of non-mandatory advisory (and strongly recommended) industry schemes. For example, we have the following:

CMA	Chemical Manufacturing Association
ANSI	American National Standards Institute
NFPA	National Fire Protection Association
NPCA	National Paint & Coatings Association
ISO	International Standards Organization
ASTM	American Society of Testing Materials
EIA	Electronic Industries Association
ATA	Air Transport Association
CGA	Compressed Gas Association
NIST	National Institute of Standards Technology
SCUSA	Standards Council of the United States of America
API	American Petroleum Institute.

These organizations—and they are only a few of a long list—are responsible for industry advisory standards that often precede government standards. Normally, in the interest of safety, these groups suggest more stringent and different guidelines than required by the government; these differences can offer very special comprehension and training challenges.

Each new set of rules can be its own special communication puzzle. Each of these groups, private or government, usually has very strong opinions both for and against the use of graphic communications symbols.

A cluttered label can have a significantly negative effect because it will be harder to perceive the critical individual elements. The more hazards that appear on a label, the more confusion and anxiety that label will produce. To overcome this problem, hazard statements must be strong and clear.

No labeling system meets every need. Millions of dollars have been spent developing and testing labels. The regulated use of labels continues to mushroom, and chemical labeling has become a task informed by what makes a warning label effective and ineffective and what is correct and adequate from a regulatory perspective.

U.S. Department of Transportation

In the United States, the U.S. Department of Transportation (DOT) is responsible for regulating the transportation of hazardous materials by air, water, highway, rail, and pipeline. Hazardous materials regulations are contained in the Code of Federal Regulations (CFR), the annually revised official compilation of general and permanent rules published by agencies of the federal government, including the DOT. The CFR is divided into fifty titles, or sections, that represent broad areas subject to federal regulation. Title 49 covers transportation, including every phase of shipment and whether a material may or may not be transported.

Because of a movement towards worldwide cooperation in the use of regulations, the DOT's rules are based on the *Recommendations on the Transport of Dangerous Goods,* a publication of the United Nations (UN). The source for hazardous materials regulations at the UN is found within a small policy group, the Economic and Social Committee (ECOSOC), which reports to the General Assembly. The primary groups that comprise ECOSOC are the Economic Commission for Europe (ECE) and the Committee of Experts on the Transport of Dangerous Goods, comprising ten nations, including the United States and Canada, which meet at least twice a year to discuss the rules for hazardous materials transportation.

The DOT defines a hazardous material as "a substance or material, including a hazardous substance, which has been determined by the Secretary of Transportation to be capable of posing an unreasonable risk to health, safety and property when transported in commerce" (49 CFR 172.8). In dealing with the variety of hazardous materials and the uncertainty surrounding their behavior under varying conditions, the DOT has adopted the UN system of classification, under which hazardous materials are grouped into nine hazard classes based on the material's chemical and physical properties. Definitions of the hazard classes are found at 49 CFR 173. Where materials meet the definition of more than one hazard class, the primary and secondary (subsidiary risk) hazards are assigned by using the precedence of hazard table 49 CFR 173.2.

The hazard class assigned to a chemical determines the label required by the DOT. An alphabetical table of proper DOT shipping names, a list of DOT regulated substances along with their UN and North American (NA) four-digit identification numbers, hazard classes, label-placard assignment, and packaging requirements, is located in 49 CFR 172.101.

The DOT has evolved a system of labels, placards, and package markings that must be placed on the packaging or on the ends and sides of vehicles that transport the packaging. The DOT strictly regulates the design, dimension, type style and size, color tolerances, durability, and use of its labels. They are diamond shaped (square-on-point), 100 mm per side. Text (when used), symbols, numbers, and inner border are black, but the DOT allows white to be used on labels with a one-color background of green, red, or blue. Primary hazard class labels *always* show their hazard class number in the lower half. This number is *always* removed for subsidiary risk labels. Illustrations of DOT labels are shown in Figure 5. Full details of DOT labels can be found in 49 CFR 172.400 et seq.

Figure 5

U.S. Environmental Protection Agency

The U.S. Environmental Protection Agency (EPA) also has specific guidelines for labeling. Pesticides must be labeled in accordance with the Federal Insecticide, Fungicide, and Rodenticide Act (FIFRA). Under this rule, an EPA certification must be obtained for a pesticide prior to its sale, based on a registration that defines the chemical in terms of any preliminary tests and studies available. The EPA must also approve label copy for registered substances.

Label copy is largely based on the toxicity category of the product as assigned by the EPA. There are four routes of entry: ingestion (oral), absorption through the skin (dermal), inhalation, and eye/skin irritation. There are also four categories of severity, Category I being the most toxic, Category IV the least. Where a Toxicity Category I is assigned to a material, the front panel of the label for the material must show the signal word "DANGER." Where the category is assigned because of the material's oral, dermal, or inhalation toxicity, the front panel must also show the signal word "POISON" in red letters. The universally recognized skull-and-crossbones symbol must also appear. Category II must carry the signal word "WARNING," with Category III and IV displaying the signal word "CAUTION." All labels prepared under FIFRA must carry the child hazard warning with which we have become so familiar: "KEEP OUT OF REACH OF CHILDREN." The EPA also regulates the type size of the signal words, dependent on the overall size of the front label panel.

Other elements on an EPA-regulated label include the product identity, name of the manufacturer, product registration number, manufacturer's registration number, net weight or volume, ingredients, and first aid.

Once a material is determined to be a hazardous waste by the EPA, each package of it requires specific labeling before transport, in accordance with DOT regulations on hazardous materials (49 CFR 172). Other marking requirements in 40 CFR 262.32 state that before transporting a hazardous waste off-site in containers of 110 gallons or less, each container must be marked: "HAZARDOUS WASTE—Federal Law Prohibits Improper Disposal. If found, contact the nearest police or public safety authority or the U.S. Environmental Protection Agency." (See Figure 6.)

Marking Requirements for Polychlorinated Biphenyls

Polychlorinated biphenyls (PCBs) were first commercially produced in 1929. For the next fifty years, millions of pounds of PCBs were used as fire-resistant fluid in transformers, capacitors and heat transfer and hydraulic equipment. In 1971, after the Council on Environmental Quality (CEQ) reported that PCB contamination should be restricted to essential uses with minimal human exposure, manufacturers voluntarily stopped using PCBs in most new applications and products. New uses of PCBs were banned in 1978, and existing uses were regulated by the EPA, in 40 CFR 761.40 et seq.

In 1982, the EPA announced the "Electric Rule," which established three categories for transformers containing PCBs. Non-PCB transformers contain PCBs in concentrations of less than 50 parts per million (ppm); these are not regulated. PCB contaminated transformers contain concentrations between 50 and 499 ppm; these units have no restrictions on in-service use, but disposal of the PCB-based coolant is regulated. PCBs with over 500 ppm of PCBs are heavily regulated with additional restrictions. The EPA requires specific exterior labeling of PCB contaminated transformers and capacitors. (See Figure 7.)

U.S. Occupational Safety and Health Administration

The U.S. Occupational Safety and Health Administration (OSHA) has also provided guidelines for labeling in the workplace. These guidelines are "performance oriented" and are as specific as the previously shown labels from

Figure 6

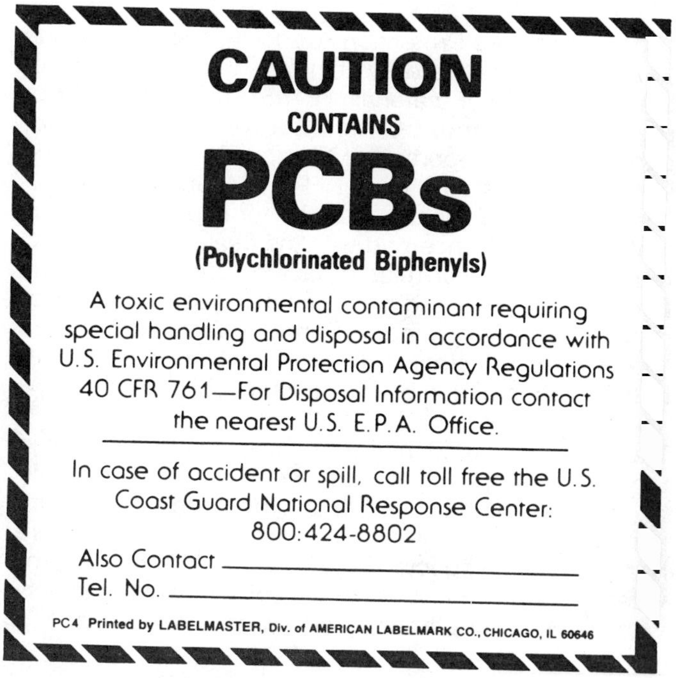

Figure 7

DOT or EPA. OSHA's guidelines are set out in the *Hazard Communication Standard* (29 CFR 1910.1200), which is commonly called the Worker's Right-to-Know Law.

In the Hazard Communication Standard (HCS), OSHA requires that a label list the chemical identity (which must be the same as name appearing on the material safety data sheet [MSDS]); list appropriate hazard warnings; and list the name and address of the manufacturer.

This labeling sounds easy, but often individuals must rely on their own judgment and knowledge to distinguish what is and what is not appropriate hazard warnings. In some instances, OSHA does provide specific wording for labels, but only for very specific substances, such as asbestos, carcinogens, arsenic, vinyl chloride, polyvinyl chloride, acrylonitrile, and dibromochloropropane. A complete list of these substances can be found in 29 CFR 1910.1000 et seq.

Because of the performance nature of OSHA's rule, industry has established systems used for information transfer through labels. Chief among these are the National Fire Protection Association's NFPA 704, the National Paint and Coating Association's Hazardous Materials Identification System (HMIS), and the American National Standards Institute's (ANSI) Z535 standards.

National Fire Protection Association

The National Fire Protection Association (NFPA) began developing a standard system for the identification of the fire hazards of materials in 1957, using the work of the NFPA Sectional Committee on Classification, Labeling, and Properties of Flammable Liquids begun in 1952. The system, titled NFPA 704, applies to facilities for the manufacture, storage, or use of hazardous materials. It is concerned with the health, flammability, reactivity, and other related hazards created by short-term exposure as might be encountered under fire or related emergency conditions. The program was intended for industrial and institutional facilities, but not transportation or use by the general public.

The NFPA system identifies the hazards of a material in terms of three principal categories: health, flammability, and reactivity. It indicates the order of severity numerically by five divisions, ranging from 4 (for severe hazard) to 0 (no special hazard). A fourth space is provided on the NFPA label to indicate unusual reactivity with water, or for other specific hazard information.

The National Paint and Coatings Association

The National Paint and Coatings Association also recognized the need for effective hazard communication. The association anticipated that a uniform, federal regulatory approach would come about, and began the

development of the Hazardous Materials Identification System (HMIS) in 1976. There are several elements to this total system: hazard assessment, hazard communication, and job safety training based on workplace hazards. The label is designed to communicate information on chemical identity, the degree of acute health, flammability and reactivity hazards, proper personal protective equipment, and chronic health hazards.

Health, flammability, and reactivity hazards are rated as in the NFPA system, by numerical ratings of 0 to 4. An alphabetical designation is used to denote proper combinations of personal protective equipment. The primary differences between HMIS and NFPA labels are that the NFPA is intended for use in short-term exposure of the type which a fire fighter might meet; HMIS is intended for normal or long-term exposure in a general workplace situation.

American National Standards Institute

The American National Standards Institute (ANSI) is a consensus organization made up of those who are substantially concerned with the scope and provisions of industry standards. Standards are intended only as a guide to aid manufacturers, employers, consumers and the general public. The scope of the committee that worked on the Z535 Standards was "to develop standards for the design, application, and use of signs, colors, and symbols intended to identify and warn against specific hazards and for other accident prevention purposes."

The ANSI Z535 (1991) series contains:

ANSI Z535.1   Safety Color Code

ANSI Z535.2   Environmental and Facility Safety Signs

ANSI Z535.3   Criteria for Safety Symbols

ANSI Z535.4   Product Safety Signs and Labels

ANSI Z535.5   Accident Prevention Tags

Together, these five standards contain information needed to specify formats, colors, and symbols for safety signs used in environmental and facility applications, product applications, and accident prevention tags. Z535.4 sets out a hazard communication system developed specifically for product safety signs and labels. It uses a number of graphic approaches to present product hazard information in an orderly and visually consistent manner for effective communication.

No single label is able to carry all of the information thought to be necessary, and ANSI style labels are no exception. What ANSI labels attempt to do is to provide a reasonably simple, non-technical format that can communicate pertinent information as a situation warrants. Where

NFPA 704 and HMIS are very visual in the method they use to convey information, the ANSI labels are textual and are meant to be read and followed in the workplace.

Consumer Product Safety Commission

The Consumer Product Safety Commission (CPSC) is responsible for the labeling of household products. These rules can be found in 16 CFR 1000-1799. CPSC has established a set of criteria to determine whether or not a material is hazardous. You will find definitions for "extremely flammable," "flammable," "combustible," "toxic," "highly toxic," and other terms in these regulations. The definitions refer to the physical and chemical properties of a material or substance, and must be checked in addition to any qualifications by DOT, OSHA, or EPA.

CPSC requires the following information on their labels:

Identification: Name of the manufacturer, packer, seller, distributor; identification of the hazardous materials.

Signal words: Signal words for "extremely flammable," "highly toxic," or "corrosive" must show the signal word "DANGER"; "highly toxic" must also show the signal word "POISON"; all other hazard classes must bear the signal "WARNING" or "CAUTION."

Hazard description: The hazard description tells the user about the health hazards of the product.

Precautionary: Precautionary information includes both the user instructions and first-aid treatments (if any). The warning, "KEEP OUT OF REACH OF CHILDREN" must also appear on each label.

There are many questions that could possibly arise in regard to labeling. To find more information you will need to refer to the various industry guides or to the CFR. Remember, the regulated use of labels continues to mushroom and chemical labeling has truly become an undertaking that demands professional attention.

There are many sources that offer consultation to labelers in the United States. Some include: *Recommendations on the Transport of Dangerous Goods* (United Nations Orange Source Book); Code of Federal Regulations: Title 49—Transportation (DOT), Title 40—Environmental Protection Agency (EPA), Title 29—Department of Labor (OSHA), Title 16—Consumer Product Safety Commission (CPSC), and Title 16—Food and Drug Administration (FDA); *Technical Instructions for the Safe Transport of Dangerous Goods by Air* (International Civil Aviation Organization [ICAO]); *Dangerous Goods Regulations* (International Air Transport Association [IATA]); International Maritime Dangerous Goods Code (International Maritime Organization [IMO]); and *Transport of Dangerous Goods Regulations* (Transport Canada). These resources can be obtained from: American Labelmark Company, 5724 N. Pulaski Rd., Chicago, Illinois 60646-6797, telephone: (800)621-5808. American National Standards Institute (ANSI) Standards can be obtained from: National Electrical Manufacturers Association, 2101 L Street, N.W., Suite 300, Washington, DC 20037-1580, telephone: (202)457-8400.

—By Patrick McConnell, Labelmaster division of American Labelmark Company

Working with Dangerously Reactive Liquids and Solids

Dangerously reactive liquids and solids are present in many workplaces, and everyone who works with these materials must be aware of their hazards and know how to work safely with them. This section provides general information on how to work safely with dangerously reactive liquids and solids. It is intended for workplaces where these chemicals are present in storage containers holding fifty-five gallons (250 liters) or less. It does not cover bulk storage tanks, piping systems, dangerously reactive compressed gases, peroxides formed during the storage of chemicals such as ethers, or specific processes using dangerously reactive liquids and solids.

In most jurisdictions there are legal requirements, such as occupational health and safety regulations and fire and building codes, that apply to the storage, handling, and use of dangerously reactive materials. For details, contact occupational health and safety officials in the jurisdiction where you work and your local fire and building departments.

Read the container label and material safety data sheet (MSDS) before starting to work with a hazardous material. In most cases, the Occupational Safety and Health Administration (OSHA) Hazard Communication Standard (HCS) (29 CFR 1910.1200) and Canada's Workplace Hazardous Materials Information System (WHMIS) regulations require the material's supplier or manufacturer to supply a current MSDS if the material is imported or sold for use in a workplace. Canadian suppliers also provide MSDSs for their products to the Canadian Centre for Occupational Health and Safety (CCOHS). CCOHS makes these MSDSs readily available through its computerized online and compact disc information systems (CCINFOline and CCINFOdisc).

Dangerously Reactive Liquids and Solids

Dangerously reactive liquids and solids are defined as those that can undergo vigorous polymerization, condensation, or decomposition; become self-reactive under conditions of shock or increase in pressure or temperature; or react vigorously with water to release a lethal gas.

Vigorous Polymerization

Polymerization is a chemical reaction in which many small molecules (monomers) join together to form a large molecule (polymer). Often the reaction produces heat and pressure. Industry carries out these process-es under closely monitored conditions. Other chemicals (catalysts and initiators) and controlled amounts of heat, light, and pressure are often involved. Under carefully controlled conditions, monomers such as styrene can be safely changed into useful polymers such as polystyrene.

Vigorous polymerization is potentially hazardous because the reaction may get out of control. Once started, the reaction is accelerated by the heat that it produces. The uncontrolled build-up of heat and pressure can cause a fire or an explosion, or can rupture closed containers. Depending on the material, temperature increases, sunlight, ultraviolet (UV) radiation, X-rays, or contact with incompatible chemicals can trigger such reactions.

Many pure substances can undergo vigorous polymerization quite easily by themselves when they are heated slightly or exposed to light. These include acrylic acid, acrylonitrile, cyclopentadiene, diketene, ethyl acrylate, hydrocyanic acid, methacrylic acid, methyl acrylate, styrene, and vinyl acetate.

An inhibitor is a chemical that is added to a material to slow down or prevent an unwanted reaction, such as polymerization. Inhibitors are added to many materials that can polymerize easily when they are pure.

Inhibitor levels in materials may gradually decrease during storage, even at recommended temperatures. At storage temperatures higher than recommended, inhibitor levels can decrease at a much faster rate. At temperatures lower than recommended, the inhibitors may separate out. This can result in some part of the material having little or no inhibitor.

Some inhibitors need oxygen to work effectively. Chemical suppliers may recommend checking oxygen and inhibitor levels regularly in stored materials and adding more if levels are too low.

Vapors from inhibited materials do not contain inhibitors. If these vapors condense and form polymers, they can block vents or flame arresters in process equipment or containers.

Vigorous Condensation

Condensation is a chemical reaction in which two or more molecules join together to form a new substance. Water or some other simple substance may be given off as a by-product. Some polymers, such as nylon, can be formed by condensation reactions.

Vigorous condensation can produce more energy than the surroundings can safely carry away. This could cause a fire or explosion or rupture closed containers.

Few common pure chemicals undergo vigorous condensation by themselves. Some members of the aldehyde chemical family, including butyraldehyde and acetaldehyde, condense vigorously, but bases or sometimes strong acids must also be present.

Some commercial products sold to be mixed for specialized applications may undergo vigorous condensation if they are not stored, handled, and used as directed by the chemical supplier.

Vigorous Decomposition

Decomposition is a chemical change in which a molecule breaks down into simpler molecules. Vigorous decomposition is potentially hazardous because large amounts of energy can be released very quickly. This could result in a fire or explosion, or rupture a closed container causing the release of dangerous decomposition products.

Some pure materials are so chemically unstable that they vigorously decompose at room temperature by themselves. For example, some organic peroxide products are relatively safe only when refrigerated or diluted.

Self-Reactivity

Materials can react vigorously and, in some cases, explosively, under conditions of mechanical shock such as a hammer blow or even slightly elevated temperature or pressure. Materials in this category include ammonium perchlorate, acetylides, fulminates, many organic peroxides, nitro and nitroso compounds, perchloric acid solutions (over 72.5% by weight), triazines, azo and diazo compounds, azides, hydrogen peroxide solutions (91% by weight), nitrate esters, picric acid, picrate salts, and some epoxy compounds.

Vigorous Reactivity

Some materials can react vigorously with water to rapidly produce gases that are lethal at low airborne concentrations. For example, sodium or potassium phosphide release phosphine gas when they contact water. Alkali metal cyanide salts, such as sodium or potassium cyanide, slowly release hydrogen cyanide gas on contact with water. This can result in a serious problem in confined spaces or poorly ventilated areas. The cyanide salts of alkaline earth metals, such as calcium or barium cyanide, react at a faster rate with water to produce hydrogen cyanide gas.

Large amounts of hydrogen chloride gas are rapidly released when water reacts with aluminum chloride, phosphorous trichloride, tin chloride, and chlorosilane compounds. When water contacts thionyl chloride or sulphuryl chloride, they decompose rapidly, giving off sulphur dioxide gas and hydrogen chloride gas.

Hazards

Dangerously reactive liquids and solids can be extremely hazardous. Accidental or uncontrolled chemical reactions are important causes of severe personal injury and property damage. Highly reactive chemicals may undergo vigorous, uncontrolled reactions that can cause an explosion or a fire, or rupture sealed reaction vessels or storage containers. Even slow reactions can be hazardous if they involve large amounts of material or if the heat and gases are confined, such as in a sealed storage drum. Drums that are swollen and distorted from over-pressurization are potentially very dangerous. They may rupture at any time without warning.

Some dangerously reactive liquids such as styrene, methyl acrylate, and acrylonitrile are also flammable liquids. They give off enough vapor at normal workplace temperatures to form flammable mixtures with air. They can be serious fire hazards at temperatures lower than those at which they would begin to polymerize or decompose. Fires involving dangerously reactive materials can be more hazardous than normal. The heat from the fire can lead to violent, uncontrolled chemical reactions and potentially explosive ruptures of sealed containers.

Rapid release of very toxic or corrosive gases occurs when water contacts some dangerously reactive materials. In addition, many dangerously reactive materials are themselves toxic or very toxic. Depending on the material, route of exposure (inhalation, eye or skin contact, swallowing), and dose, they could harm the body.

Many dangerously reactive materials can also undergo dangerous reactions from direct contact with other, incompatible materials. Incompatibility hazards can be complicated. The chance of a dangerous reaction depends not just on the different combinations of chemicals involved, but on the amounts of each, the surrounding conditions such as temperature, and whether the substances are enclosed in a sealed container.

The MSDSs and the container labels should explain all of the hazards of the dangerously reactive liquids and solids that you work with.

Substitution

Substitution can be the best way to avoid or reduce a hazard. But it is not always easy or even possible to find a less-hazardous substitute for a particular dangerously reactive material used for a certain job. Speak to the chemical supplier to find out if safer substitutes are available. For materials that polymerize easily, use a product that contains a polymerization inhibitor instead of a pure product whenever possible; check for any limitations associated with the inhibitor. Obtain MSDSs for all possible substitutes. Find out about all of the hazards (health, fire, corrosivity, chemical reactivity) of these materials before making any changes.

Sometimes process changes or modification can improve control of the hazards from working with a dangerously reactive material. These could include the installation of alarms or automatic shut-off switches on equipment to warn of equipment failure, high temperatures, or high pressures.

Ventilation

Well-designed and -maintained ventilation systems remove airborne, dangerously reactive materials from the workplace and reduce their hazards. The amount and type of ventilation needed depends on such things as the type of job, the kind and amount of materials used, and the size and layout of the work area.

An assessment of the specific ways a workplace stores, handles, uses, and disposes of its dangerously reactive materials is the best way to find out if existing ventilation controls (and other hazard control methods) are adequate. Some workplaces may need a complete system of hoods, ducts, and fans to provide acceptable ventilation. Others may require a single, well-placed exhaust fan. No special ventilation system may be needed to work with small amounts of dangerously reactive materials which do not give off airborne contaminants.

Make sure ventilation systems for dangerously reactive materials are designed and built so that they do not result in an unintended hazard. Ensure that hoods, ducts, air cleaners and fans are made from materials compatible with the dangerously reactive substance. Systems may require explosion-proof electrical equipment. Ensure that the system is designed to avoid build-ups of dusts or condensation of vapors. The vapors of inhibited liquids are not inhibited. When they condense, the liquid could polymerize or decompose easily.

Keep systems for dangerously reactive materials separate from other systems exhausting incompatible substances.

Locate exhaust openings near floor level, if necessary, to remove heavier-than air vapors from storage areas.

Periodic inspection of ventilation systems will help keep them in good operating condition.

Storage

Store dangerously reactive liquids and solids according to the occupational health and safety regulations and fire and building codes that apply to your workplace. These laws may specify the kinds of storage areas, such as storage rooms and buildings, allowed for different materials. They may also specify how to construct these storage areas and the amounts of dangerously reactive materials that can be stored in each storage area.

Inspect all incoming containers before storing to ensure that they are undamaged and properly labeled. Do not accept delivery of defective containers.

Store dangerously reactive materials in containers that the chemical supplier recommends. Normally, these are the same containers in which the material was shipped. Repackaging can be dangerous, especially if contaminated or incompatible containers are used. For example, strong hydrogen peroxide solutions can decompose explosively if placed in a container with rusty surfaces. Bottles for light-sensitive materials are often made of dark blue or brown glass to protect the contents from light. Containers for water-sensitive compounds should be waterproof and tightly sealed to prevent moisture in the air from reacting with the material.

Make sure containers are suitably labeled. For materials requiring temperature control, the recommended storage temperature range should be plainly marked on the container. It is also a good practice to mark the date that the container was received and the date it was first opened.

Protect containers against impact or other physical damage that might cause shock. Do not use combustible pallets, such as wood, for storing oxidizing materials or organic peroxides.

Normally keep stored containers tightly closed. This helps to avoid contamination of the material or evaporation of solvents used to dilute substances, such as some organic peroxides, to safer concentrations.

Some dangerously reactive liquids, such as strong hydrogen peroxide solutions or certain organic peroxide products, gradually decompose at room temperature and give off gas. These liquids are shipped in containers with specially vented caps. These vent caps relieve the normal build-up of gas pressure that could rupture an unvented container. Check vent caps regularly to ensure that they are working properly. Keep vented containers in the upright position. Do not stack vented containers on top of each other.

Storage Area

Store dangerously reactive liquids and solids separately, away from processing and handling areas and from incompatible materials. Some dangerously reactive materials are incompatible with each other; do not store these beside each other. (Check the reactivity data and storage requirements sections of the MSDS for details about what substances are incompatible with a specific dangerously reactive material.) Separate storage can minimize personal injury and damage caused by fires, spills, or leaks.

Construct walls, floors, shelving, and fittings in storage areas from suitable materials. Use non-combustible building materials in storage areas for dangerously reactive oxidizers or organic peroxides. Use corrosion-resistant materials for dangerously reactive corrosives.

Ensure that floors in storage areas are watertight and without cracks in which spilled materials can lodge. Contain spills or leaks by storing smaller containers in trays made of compatible materials. For larger containers, such as drums or barrels, provide dikes around storage areas and sills or ramps at door openings.

Store smaller containers at a convenient height for handling, below eye level if possible, to reduce the risk of dropping them. Avoid overcrowding in storage areas. Do not store containers in out-of-the-way locations where they could be forgotten.

Store containers away from doors. Although it is convenient to place frequently used materials next to the door, they could cut off the escape route if an emergency occurs.

Store dangerously reactive materials in areas that are:

- well ventilated;

- supplied with adequate fire fighting equipment including sprinklers (sprinklers may not be allowed in areas where materials that react dangerously with water are present);

- supplied with suitable spill clean-up equipment and materials;

- free of ignition sources such as sparks, flames, burning tobacco or hot surfaces;

- accessible at all times; and

- labelled with suitable warning signs.

Storage Temperature

Store dangerously reactive materials in dry, cool areas, out of direct sunlight and away from steam pipes, boilers, or other heat sources. Follow the chemical supplier's recommendations for maximum and minimum temperatures for storage and handling. Higher temperatures can be hazardous since they can start and speed up hazardous chemical reactions. In many cases, inhibitors can be rapidly depleted at higher-

than-recommended storage temperatures. Loss of inhibitor can result in dangerous reactions.

Some dangerously reactive materials must be kept at low temperatures in refrigerators or freezers. Use only approved or specially modified units. These are generally known as "laboratory safe." Standard domestic refrigerators and freezers contain many ignition sources inside the cabinet.

It can also be hazardous to store dangerously reactive materials at less than the recommended temperature. For example, acrylic acid is normally supplied with an inhibitor to prevent polymerization. Acrylic acid freezes at 55°F (13°C). At temperatures less than this, it will partly solidify. The solid part contains little or no inhibitor; the inhibitor remains in the liquid portion. The uninhibited acrylic acid can be safely stored below the freezing point, but it may polymerize violently if it is heated to warmer temperatures.

Some organic peroxides are sold dissolved or dispersed in solvents, including water, to make them less shock sensitive. If these are cooled to below their freezing points, crystals of the pure, very sensitive organic peroxide may be formed.

Alarms that indicate when storage temperatures are higher or lower than required may be needed.

Follow the chemical supplier's directions about inhibitors used in a particular product. Where appropriate, check inhibitor and oxygen levels and add more as needed according to the supplier's instructions. Do not keep a material for longer than the chemical supplier recommends.

Following are general precautions to take when storing dangerously reactive liquids and solids:

- Allow only trained, authorized people into storage areas.

- Keep the amount of dangerously reactive materials in storage as small as possible.

- Inspect storage areas regularly for any deficiencies, including damaged or leaking containers and poor housekeeping.

- Correct all deficiencies as soon as possible.

Dispensing

Open and dispense containers of dangerously reactive materials in a special room or area outside the storage area. Do not allow any ignition sources in the vicinity. Take care that the dangerously reactive materials do not contact incompatible substances.

When transferring materials from one container to another, avoid spilling material and contaminating your skin or clothing. Spills from open, unstable, or breakable containers during material transfer have caused serious accidents. Use containers and dispensing equipment, such as drum pumps, scoops, or spatulas, that the chemical supplier recommends. These items must be made from materials compatible with the chemicals with which they are used. Keep them clean to avoid contamination.

Never transfer liquids by pressurizing their usual shipping containers with air or inert gas. The pressure may damage ordinary drums and barrels. If air is used, it may also create a flammable atmosphere inside containers of flammable or combustible liquids.

Glass containers with screw-cap lids or glass stoppers may not be acceptable for friction-sensitive materials. Avoid using ordinary screw-cap bottles with a cardboard liner in the cap for moisture-sensitive chemicals. Airborne moisture can diffuse slowly but steadily through the liner. Never transfer materials stored in a vented container into a tightly sealed, non-vented container. The build-up of gas pressure could rupture it.

Dispense from only one container at a time. Finish dispensing and labeling one material before starting to dispense another. Keep containers closed after dispensing to reduce the risk of contaminating their contents.

Never return any unused material, even if it does not seem to be contaminated, to the original container.

If a dangerously reactive material freezes, do not chip or grind it to break up lumps, or heat it to thaw it out. Follow the chemical supplier's advice.

Avoid dropping, sliding, or skidding heavy metal containers such as drums or barrels of friction- or shock-sensitive material.

Make sure that all areas where dangerously reactive liquids and solids are used are clean and free of incompatible materials and ignition sources. Do not allow temperatures in these areas to become hot enough to cause a hazardous reaction.

Processing Equipment

Ensure that processing equipment is clean, properly designed, and made from materials compatible with the dangerously reactive material used. For example, some steels and aluminum alloys, zinc, and galvanized metal can cause rapid decomposition of certain organic peroxides.

Accidents have happened when reactive materials came in contact with incompatible heat exchange fluids or fluids used in instruments to monitor processes.

Reactive substances have, on occasion, leaked and soaked into equipment insulating materials. Insulators have good heat-keeping ability. Once a reaction begins within the insulating material, the heat given off from the reaction can rapidly build up to hazardous levels and may result in fire.

Diluting Dangerously Reactive Materials

Some jobs require that dangerously reactive materials be diluted prior to use. Always follow the chemical supplier's advice. Using the wrong solvent or a contaminated solvent could cause an explosion. Using reclaimed solvents of unknown purity can be hazardous; they might contain dangerous concentrations of contaminants that are incompatible with the dangerously reactive material.

Hazardous Operations

Some operations involving dangerously reactive materials can be especially hazardous. Many accidents have occurred during distillation, extraction, or crystallization because these processes concentrated reactive substances. Sieving dry, unstable materials might result in static electricity sparks which could cause ignition. Filtering friction- or shock-sensitive chemicals with materials and devices that produce frictional heat, such as sintered glass filters, can be hazardous.

Before using a new material in an operation, find out as much as possible about the potential hazards of the particular chemical and operation.

Follow these general practices when handling dangerously reactive materials:

- Inspect containers for damage or leaks before handling them.

- Handle containers carefully to avoid damaging them.

- Keep containers tightly closed except when actually using the material.

- Avoid returning used chemicals to containers of unused materials.

- Keep only the smallest amounts possible (never more than one day's supply) of dangerously reactive materials in the work area.

- Return unopened containers to the proper storage area and opened containers to a dispensing area at the end of the day.

- Check that all containers are properly labeled, and handle containers so that the label remains undamaged and easy to read.

Regular workplace inspections can help to spot situations in which dangerously reactive materials are stored, handled or used in potentially hazardous ways.

Disposal

Dangerously reactive wastes are hazardous. Dispose of unwanted or contaminated reactive chemicals promptly using a method the chemical supplier recommends. Consider any reactive materials accidentally mixed with an unknown or foreign material as contaminated, and dispose of them. Never attempt to salvage spilled or contaminated dangerously reactive materials.

"Empty" drums, bottles, bags, and other containers usually contain hazardous residues. Never use these "empty" containers for anything else, no matter how clean they seem to be. Treat them as dangerously reactive wastes. Follow the chemical supplier's advice for safely handling or decontaminating "empty" containers.

Store reactive waste in the same way as unused dangerously reactive materials. Never dispose of these wastes in ordinary garbage or down sinks or drains. Dispose of them according to the supplier's advice, or through hazardous waste collection and disposal companies. In all cases, dispose of dangerously reactive wastes according to the environmental laws that apply to your jurisdiction. Contact the appropriate environmental officials for details.

Good Housekeeping

Maintain good housekeeping at all times in the workplace. Follow these general guidelines:

- Clean-up any spills promptly and safely according to directions in the MSDS.

- Use suitable clean-up materials (e.g., some commercial sorbent materials used for spill clean-up may initiate polymerization in some monomers, including styrene; do not use sawdust or other combustible sweeping compounds to clean up spills of oxidizers or organic peroxides).

- Properly dispose of unlabeled or contaminated materials.

- Promptly remove combustible wastes, including wood, paper, or rags from work area.

- Avoid any build-up of chemical dusts on ledges or other surfaces.

- Ensure that all waste containers are compatible with the reactive materials used, properly marked, and located close to the job.

Personal Cleanliness

Personal cleanliness helps protect people working with dangerously reactive materials. Follow these general guidelines:

- Wash hands before eating, drinking, smoking, or going to the toilet.

- Remove contaminated clothing and footwear since they can be a severe fire or health hazard.

- Wash contaminated clothing and footwear thoroughly before re-wearing or discarding. Check the MSDS or contact the chemical supplier for details.

- Do not wear or carry contaminated items into areas having ignition sources or where smoking is allowed.

- Store food and tobacco products in uncontaminated areas.

- Clean yourself thoroughly at the end of the workday.

Equipment Maintenance

Regular equipment maintenance can prevent hazardous conditions in the workplace. Follow these general guidelines:

- Ensure that maintenance personnel know the possible hazards of the materials they may encounter and any special procedures and precautions needed before they begin work.

- Prevent leaks of grease or other lubricants from equipment where dangerously reactive materials are used.

- Do not allow materials, such as cleaning liquids, paints, or thinners, to come into contact with dangerously reactive materials.

- Check to ensure that the job can be done safely before performing shock-, heat-, or friction-generating operations, such as hammering, grinding, cutting, or welding, on containers or equipment used for dangerously reactive materials.

- Comply with applicable regulations and contact the chemical supplier for advice.

Personal Protective Equipment

If other methods, such as engineering controls, are not available or effective in controlling exposure to dangerously reactive materials, wear suitable personal protective equipment (PPE). Choosing the right PPE for a particular job is essential. OSHA's requirements for PPE are outlined in 29 CFR 1910.132, and MSDSs should provide general guidance. Also obtain help from someone who knows how to evaluate the hazards of a specific job and how to select the proper PPE.

Avoid Skin Contact

When using materials that are harmful by skin contact, wear protective gloves, aprons, boots, hoods, or other clothing depending on the risk of skin contact. Choose clothing made of materials that resist penetration or damage by the chemical. The MSDS should recommend appropriate materials; if it does not, contact the chemical supplier for specific information.

Protect Your Eyes and Face

Always wear eye protection when working with dangerously reactive chemicals. Avoid ordinary safety glasses; use chemical safety goggles instead. In some cases, you should also wear a face shield to protect your face from splashes. The current Canadian Standards Association (CSA) Standard Z94.3, "Industrial Eye and Face Protectors," provides advice on selection and use of eye and face protectors. In the United States, eye and face protective devices follow standards set forth by the American National Standards Institute (ANSI) Manual Z87.1-1968, "Occupational and Educational Eye and Face Protection."

Avoid Breathing Dust, Vapor or Mist

If respirators must be used for breathing protection, there should be a written respiratory protection program to follow. The current CSA Standard Z94.4, "Selection, Care, and Use of Respirators," gives guidance for developing a program. Follow all legal requirements for respirator use and approvals. These may vary between jurisdictions in Canada. The selection of respirators in the United States should be made in accordance with standards established by the ANSI in Z88.2-1969.

Sorbents in respirator cartridges or canisters must be compatible with the chemical they are supposed to protect against. For example, oxidizable sorbents such as activated charcoal may not be acceptable if high airborne concentrations of oxidizers or organic peroxides are present. A hazardous reaction might occur.

Emergency Procedures

Act fast in emergencies like chemical leaks, spills, and fires. Follow these general guidelines:

- Evacuate the area at once if you are not trained to handle the problem or if it is clearly beyond your control.

- Alert other people in the area to the emergency.

- Call the fire department immediately.

- Report the problem to the people responsible for handling emergencies where you work.

- Obtain first aid if you have been exposed to harmful chemicals and remove all contaminated clothes.

Locate emergency eyewash stations and safety showers wherever accidental exposure to chemicals that can damage skin or eyes might occur.

Only specially trained and properly equipped people should handle the emergency. Nobody else should go near the area until it is declared safe.

Planning, training, and practicing for emergencies help people to know what they must do. Prepare a written emergency plan. Update it whenever conditions in the workplace change. Use the MSDSs for the materials used as a starting point for drawing up an emergency plan. MSDSs have specific sections on spill clean-up procedures, first aid instructions, and fire and explosion hazards, including suitable fire extinguishing equipment and methods. If the directions in each MSDS section are not clear or seem incomplete, contact the material's supplier for help.

It is very important to know the best ways to fight fires involving dangerously reactive materials. For example, using water on water-reactive chemicals can cause the rapid release of lethal gas or, in some cases, violent explosions. The "built-in" supply of oxidizer in oxidizing materials makes fire fighting methods based on smothering, such as foam or carbon dioxide, ineffective.

Many other sources can also help develop emergency plans. Local fire departments can assist with fire emergency plans and training. Occupational health and safety and environmental enforcement agencies, state or provincial safety associations, St. John Ambulance in Canada, insurance carriers, professional societies in occupational health and safety, labor unions, trade associations, some colleges and universities, and OSHA and CCOHS can supply useful information at little or no cost. Specialized private consultants are also available.

Basic Safe Practices

Following these basic safe practices will help protect you from the hazards of dangerously reactive liquids and solids:

- Read the material safety data sheets (MSDSs) and labels for all of the materials you work with.

- Know all of the hazards (fire, explosion, health, corrosivity, chemical reactivity) of the materials you work with.

- Know which of the materials you work with are dangerously reactive.

- Store dangerously reactive materials in suitable, labeled containers (usually their shipping containers) in a cool, dry area.

- Store, handle, and use dangerously reactive materials in well-ventilated areas and away from compatible materials.

- Follow the chemical supplier's advice on maximum and minimum temperatures for storage and use.

- Follow the chemical supplier's advice on checking and maintaining inhibitor and dissolved oxygen levels where appropriate.

- Eliminate ignition sources (sparks, smoking, flames, hot surfaces) when working with dangerously reactive materials.

- Handle containers carefully to avoid damaging them or shocking their contents.

- Keep containers closed when not in use.

- Dispense dangerously reactive materials carefully into acceptable containers, using compatible equipment.

- Do not subject dangerously reactive materials to any type of friction or impact.

- Be careful when performing operations such as separations or distillations that concentrate dangerously reactive materials.

- Never return unused or contaminated dangerously reactive materials to their original containers.

- Handle and dispose of dangerously reactive wastes safely.

- Wear the proper personal protective equipment for each of the jobs you do.

- Know how to handle emergencies (fires, spills, personal injury) involving the dangerously reactive materials with which you work.

- Follow the health and safety rules that apply to your job.

If you need more information on how to work safely with dangerously reactive materials, in Canada call the CCOHS Inquiries Service at (800)263-8466; in the U.S. call the OSHA Hotline at (800)-321-OSHA.

Further Reading

Bretherick, L. *Handbook of Reactive Chemical Hazards,* 3rd ed. London: Butterworths, 1985.

Controlled Products Regulations (1987) 122 Canada Gazette: Part II, 551, (1987) 12 31, am. SOR/88-66.

Cote, A. E., et al., eds. *Fire Protection Handbook,* 16th ed. Quincy, MA: National Fire Protection Association, 1986.

Fawcett, H. H. et al. *Safety and Accident Prevention in Chemical Operations,* 2nd ed. New York: Wiley, 1982.

Fire Protection Guide on Hazardous Materials, 9th ed. Quincy, MA: National Fire Protection Association, 1986.

Meidl, J. H. *Explosive and Toxic Hazardous Materials.* Beverly Hills, CA: Glencoe Press, 1970.

National Fire Code of Canada, 5th ed. Ottawa: National Research Council of Canada, 1985.

Yoshida, T. *Safety of Reactive Chemicals* (Industrial Safety Series, 1). Amsterdam: Elsevier Science Publishers, 1987.

—By Michael Lowther CIH, ROH, Canadian Centre for Occupational Health and Safety

The editors are grateful to the Canadian Centre for Occupational Health and Safety for permission to reprint this article. It has been edited to include U.S. references. For reprints of the original article and for a list of CCOHS publications and magnetic media, contact the Canadian Centre for Occupational Health and Safety, 250 Main Street East, Hamilton, Ontario, Canada L8N 1H6, telephone: (416)572-2981; fax: (416)572-2206.

Hazardous Substances at a Glance

Classification of Chemicals as Human Health and Environmental Threats

In 1990 the U.S. Environmental Protection Agency (EPA) updated the toxicity table of health and environmental threats for the Superfund Amendments and Reauthorization Act (SARA; 1986) Title III Section 313 Toxic Release Inventory (TRI) Chemicals. The 1990 list contains 319 specific compounds plus twenty-four broad categories. The chemicals each were evaluated for the following threats:

Carcinogenicity (C): the ability to cause cancer.

Heritable genetic and chromosomal mutation (H): the ability to produce in human germ cells mutations that can be passed from generation to generation.

Developmental toxicity (D): the ability to damage children's development in the womb or after birth, causing such problems as structural defects, prenatal death, learning disorders, growth retardation, etc.

Reproductive toxicity (R): the ability to damage men's or women's capability to reproduce by sterility, an inability to have sex, an inability to produce milk, etc.

Acute toxicity (A): the ability to cause death after short-term exposure.

Chronic toxicity (Ch): the ability to cause adverse effects (other than cancer) after long-term exposure, such as damage to the kidneys, lungs, liver, or bones.

Neurotoxicity (N): the ability to damage the central or peripheral nervous system after long-term exposure.

Envirotoxicity (E): the ability to seriously and significantly damage the environment and harm wildlife and plants. Two related factors are persistence, which is a chemical's longevity in the environment, and bioaccumulation, which is the tendency of a chemical to be retained or accumulate in an organism and then be eaten and retained by other organisms that are higher up on the food chain.

Table 17

CHEMICALS AND THEIR HEALTH EFFECTS

CAS Number	Chemical Name	C	H	D	R	A	Ch	N	E
75-07-0	acetaldehyde	X					X		X
67-64-1	acetone						X		X
75-05-8	acetonitrile					X	X	X	
79-10-7	acrylic acid					X	X		X
107-13-1	acrylonitrile	X		X	X	X	X		X
134-32-7	alpha-naphthalamine	X							

Table 17 (contd)

CHEMICALS AND THEIR HEALTH EFFECTS

CAS Number	Chemical Name	C	H	D	R	A	Ch	N	E
7429-90-5	aluminum								
7664-41-7	ammonia					X	X		X
62-53-3	aniline	X				X	X		X
120-12-7	anthracene						X		X
7440-36-0	antimony and compounds (pigments)				X		X		
1332-21-4	asbestos	X					X		
7440-39-3	barium			X			X		
98-87-3	benzal chloride	X		X		X	X	X	X
71-43-2	benzene	X		X	X		X		X
94-36-0	benzoyl peroxide								
100-44-7	benzyl chloride	X		X		X	X	X	X
7440-41-7	beryllium (alloys)	X				X	X		
92-52-4	biphenyl			X			X		X
103-23-1	bis(2-ethylhexyl) adipate	X	X						
141-32-2	butyl acrylate								X
85-68-7	butyl benzyl phthalate						X		X
7440-43-9	cadmium	X		X	X	X	X		X

C=Cancer; H=Heritable; D=Developmental; R=Reproductive; A=Acute; Ch=Chronic; N= Neurotoxicity; E=Envirotoxicity.

Table 17 (contd)

CHEMICALS AND THEIR HEALTH EFFECTS

CAS Number	Chemical Name	C	H	D	R	A	Ch	N	E
133-06-2	captan	X	X	X	X	X	X		X
75-15-0	carbon disulfide			X	X		X	X	X
7782-50-5	chlorine					X	X		X
108-90-7	chlorobenzene				X		X		X
67-66-3	chloroform	X		X	X	X			X
126-99-8	chloroprene		X	X	X	X	X	X	
7440-47-3	chromium (pigments)	X				X	X		X
7440-48-4	cobalt (compounds in metal alloy)						X		
1319-77-3	cresol (mixed isomers)					X	X		X
98-82-8	cumene						X	X	X
1163-19-5	decabromodiphemyl oxide	X		X			X		X
117817	di-(2-ethylhexyl) phthalate (DEHP)	X	X	X	X	X	X		X
101-80-4	4,4'-diaminodiphenyl ether	X					X		
84-74-2	dibutyl phthalate			X	X		X		X
95-50-1	1,2-dichlorobenzene					X	X		X
107-06-2	1,2-dichloroethane	X	X	X	X				X
75-09-2	dichloromethane	X					X		

Table 17 (contd)

CHEMICALS AND THEIR HEALTH EFFECTS

CAS Number	Chemical Name	C	H	D	R	A	Ch	N	E
84-66-2	diethyl phthalate								X
131-11-3	dimethyl phthalate					X			X
77-78-1	dimethyl sulfate	X	X			X	X	X	X
121-14-2	2,4-dinitrotoluene	X			X		X	X	X
106-89-8	epichlorohydrin	X	X		X	X	X	X	X
110-80-5	2-ethoxyethanol								
100-41-4	ethylbenzene			X	X		X		X
74-85-1	ethylene						X		
107-21-1	ethylene glycol						X		
75-21-8	ethylene oxide	X	X	X	X	X	X	X	X
50-00-0	formaldehyde	X	X		X	X	X		X
76-13-1	freon 113					X	X		X
302-01-2	hydrazine	X				X	X		X
7647-01-0	hydrochloric acid					X	X		
7664-39-3	hydrogen fluoride		X	X	X	X	X		
123-31-9	hydroquinone					X	X		X

C=Cancer; H=Heritable; D=Developmental; R=Reproductive; A=Acute; Ch=Chronic; N= Neurotoxicity; E=Envirotoxicity.

Table 17 (contd)

CHEMICALS AND THEIR HEALTH EFFECTS

CAS Number	Chemical Name	C	H	D	R	A	Ch	N	E
67-63-0	isopropyl alcohol	X					X	X	
7439-92-1	lead	X		X	X		X	X	X
7439-96-5	manganese			X	X	X	X		
7439-97-6	mercury			X	X	X	X	X	X
67-56-1	methanol							X	
109-86-4	2-methoxyethanol								
78-93-3	methyl ethyl ketone			X	X		X	X	
108-10-1	methyl isobutyl ketone						X	X	
80-62-6	methyl methacrylate			X	X		X		
1634-04-4	methyl tert-butyl ether								
71-36-3	n-butyl alcohol						X		
117-84-0	n-dioctyl phthalate					X	X		
91-20-3	naphthalene			X			X		X
7440-02-0	nickel (pigments)	X		X	X	X	X		X
7697-37-2	nitric acid					X			
95-53-4	ortho toluidine	X					X		X
106-44-5	p-cresol						X		X

Table 17 (contd)

CHEMICALS AND THEIR HEALTH EFFECTS

CAS Number	Chemical Name	C	H	D	R	A	Ch	N	E
106-42-3	p-xylene			X	X		X		X
108-95-2	phenol			X		X	X		X
75-44-5	phosgene					X			
7664-38-2	phosphoric acid								
85-44-9	phthalic anhydride						X	X	X
1336-36-3	polychlorinated biphenyls	X		X	X	X	X		
75-56-9	propylene oxide	X	X	X	X	X	X	X	X
78-92-2	sec-butyl alcohol								
100-42-5	styrene	X	X	X			X		X
7664-93-9	sulfuric acid					X	X		X
127-18-4	tetrachloroethylene	X		X	X		X	X	X
961-11-5	tetrachlorvinphos	X					X		X
78-00-2	tetraethyl lead								
108-88-3	toluene			X	X				X
584-84-9	toluene-2,4-diisocyanate	X				X	X		X
91-08-7	toluene-2,6-diisocyanate					X	X		
79-01-6	trichloroethylene	X		X	X		X	X	

C=Cancer; H=Heritable; D=Developmental; R=Reproductive; A=Acute; Ch=Chronic; N= Neurotoxicity; E=Envirotoxicity.

Table 17 (contd)

CHEMICALS AND THEIR HEALTH EFFECTS

CAS Number	Chemical Name	C	H	D	R	A	Ch	N	E
71-55-6	1,1,1-trichloroethane			X	X				X
79-00-5	1,1,2-trichloroethane	X					X		X
95-63-6	1,2,4-trimethylbenzene								X
51-79-6	urethane	X	X						
108-05-4	vinyl acetate						X		X
75-01-4	vinyl chloride	X	X	X	X	X	X		
1330-20-7	xylene (mixed isomers)			X	X		X		X
7440-66-6	zinc			X	X		X		X

C=Cancer; H=Heritable; D=Developmental; R=Reproductive; A=Acute; Ch=Chronic; N= Neurotoxicity; E=Envirotoxicity.

Table 18

Toxic Products: Ingredients and Health Hazards

Product	Ingredients	Health Effects under Normal Use
Antifreeze	Ethylene glycol	Irritates eyes and skin; headaches, dizziness, and nausea if inhaled; absorbed through skin and may cause abdominal pain.
Art and craft supplies	Acetone, cresol, phenol, toluene, xylene, metal pigments, dyes	Lead, metal pigments, and solvents may damage CNS, kidneys, bone marrow, and retard mental development; allergic reactions; harmful or fatal if swallowed.
Automotive fluids	Hydrocarbons (benzene, toluene, etc.), heavy metals (lead, chromium, etc.)	Irritates skin, nose, respiratory tract; headache, dizziness, nausea; prolonged breathing may cause liver, kidney, respiratory damage, CNS disorder, possible cancer and birth defects.
Batteries, automotive	Sulfuric acid, lead	Sulfuric acid severely irritates eyes and may cause blindness upon contact; irritates skin, mucous membranes, and lungs; long-term exposure to lead may cause depression of CNS, retard mental development in children.
Batteries, household	Cadmium, lithium, mercury, nickel, zinc	Toxic if swallowed; fumes from burning are highly toxic, causing kidney, CNS, and thyroid damage, loss of appetite, weakness, nausea, irritability.
Building products	Asbestos, fiberglass, formaldehyde, isocyanates, acetone, glues	Causes severe lung damage; lung, stomach, and colon cancer; tightness in chest; long-term exposure causes CNS depression, dizziness, headache.
Cleaning products	Ammonia, carbon tetrachloride, paradichlorobenzene, pine oil, paradichloroethylene, 1,1,1-trichloroethane, acids and alkalies, propellants	Irritates respiratory tract and eyes; coughing, shortness of breath, bronchitis, skin rash, conjunctivitis. Some products combustible and flammable.
Cosmetics and personal care products	Acetone, toluene, xylene, phenols, n-butyl alcohol, formaldehyde, propellants, ammonia	Nail polish and polish removers have solvents which may cause dizziness and nausea if inhaled; deodorants and similar products may irritate skin; formaldehyde is possible cause of cancer.

Table 18 (contd)

TOXIC PRODUCTS: INGREDIENTS AND HEALTH HAZARDS

PRODUCT	INGREDIENTS	HEALTH EFFECTS UNDER NORMAL USE
Disinfectants and air fresheners	Diethylene or methylene glycol, chlorine, cresol, paradichloroben zene, phenols, sodium hypochlorite, pine oil, propellants	Prolonged breathing may cause liver and kidney damage or possible cancer.
Drain openers	Potassium or sodium hydroxide (lye), petroleum distillates, hypochloric acid	May cause severe damage to skin and eyes.
Floor and furniture polish	Petroleum distillates, diethylene glycol, methyl ethyl ketone, nitrobenzene, ammonia	Irritates respiratory tract, eyes, skin; may cause vomiting and shortness of breath if breathed or absorbed through skin; possible liver, kidney, and blood damage from prolonged exposure.
Glues and adhesives	Solvents (acetone, acetaldehyde, ethylene chloride, methyl alcohol, phenol, naphthalene, 1,1,1-tichloroethane, toluene, methyl ethyl ketone), epoxy	Irritates skin, nose, respiratory tract; headache, dizziness, nausea; prolonged breathing may cause liver, kidney, respiratory damage, CNS disorder, possible cancer and birth defects.
Metal primers	Toluene, methylene chloride, petroleum distillates, chromic acid, nitric acid, sulfuric acid	Irritates skin, nose, mucous membrane, respiratory tract; headache, dizziness, nausea; prolonged breathing may cause liver, kidney, respiratory damage, CNS disorder, possible cancer and birth defects.
Metalworking	Cadmium, lead, beryllium, zinc, ammonia, epoxy, fluorides, hydrochloric acid	Irritation of respiratory tract, eyes, skin; may cause emphysema and chemical pneumonia; bone and kidney damage from long-term exposure.
Mothballs	Paradichlorobenzene, naphthalene	Irritates respiratory tract. Prolonged breathing may cause depression and liver and kidney damage.
Oven cleaners	Ammonia, sodium or potassium hydroxide (lye)	Breathing ammonia irritates lungs; may severely burn skin and eyes.
Paint strippers	Acetone, benzene, toluene, xylene, methyl ethyl ketone, methylene chloride, methyl alcohol	Irritates skin, nose, respiratory tract; headache, dizziness, nausea; prolonged breathing may cause liver, kidney, respiratory damage, CNS disorder, possible cancer and birth defects; methyl alcohol may cause blindness.

Table 18 (contd)

TOXIC PRODUCTS: INGREDIENTS AND HEALTH HAZARDS

PRODUCT	INGREDIENTS	HEALTH EFFECTS UNDER NORMAL USE
Paint thinners	Acetone, methyl ethyl ketone, methyl isobutyl ketone, n-butyl alcohol, methyl alcohol, toluene, distillates	Irritates skin, nose, respiratory tract; headache, dizziness, nausea; prolonged breathing may cause liver, kidney, respiratory damage, CNS disorder, possible cancer and birth defects.
Paints, oil-based	Pigments (lead, chromates, etc.), hydrocarbon solvents (ethylene, toluene, methyl ethyl ketone, etc.), talc, fungicides	Irritates skin, nose, respiratory tract; headache, dizziness, nausea; prolonged breathing may cause liver, kidney, respiratory damage, CNS disorder, possible lung cancer and birth defects.
Paints, water-based	Pigments, fungicides (mercury, etc.), acrylic and vinyl resins, styrene, talc	Irritates eyes, skin, respiratory tract; vapors may cause drowsiness, headache, nausea; acute exposure may cause first degree burns.
Pesticides	Organophosphates, chlorinated hydrocarbons, carbamates, naphthalene, xylene, and other toxins, propellants.	Irritates eyes, skin, nose, throat; cause dizziness, nausea, headache; prolonged breathing may injure liver, kidneys, respiratory tract, CNS, possible cancer and birth defects.
Stains and finishes	Glycol ethers, petroleum distillates, halogenated hydrocarbons, lead, naphtha, acetone, ketones	Irritates skin, nose, respiratory tract; headache, dizziness, nausea; prolonged breathing may cause liver, kidney, respiratory damage, CNS disorder, possible cancer and birth defects.
Wood preservatives	Creosote, distillates, arsenic compounds, chromates, phenols, naphthenic acid	Irritates skin, nose, respiratory tract; headache, dizziness, nausea; prolonged breathing may cause liver, kidney, respiratory damage, CNS disorder, possible cancer and birth defects. Arsenic compounds and creosote are known causes of cancer.

Source: Beth Impson, ed., *Guide to Hazardous Products around the Home* (Springfield, MO; Household Hazardous Waste Project, 1989); various other sources.
CNS=central nervous system.

References

Understanding Hazardous Substances

Chemicals in Your Community: A Guide to the Emergency Planning and Community Right-to-Know Act. Washington, DC: U.S. Environmental Protection Agency (EPA), Public Information Center, 1988.

Cohrsson, J. and V. T. Covello, *Risk Analysis: A Guide to Principles and Methods for Analyzing Health and Environment Risks.* National Technical Information Service, 1989.

Lomax, Geoffrey, Marc Osteen, and William Ryan, *Toxic Truth and Consequences: The Magnitude of and the Problems Arising from America's Use of Toxic Chemicals.* Boston: National Environmental Law Center and the U.S. Public Interest Research Group, 1991.

Marczewski, A. E., and M. Kamrin, *Toxicology for the Citizen.* 2nd ed. East Lansing: Michigan State University, Center for Environmental Toxicology, 1991.

Ream, Kathleen, *Chemical Risks: A Primer.*: Washington D.C.: American Chemical Society, Department of Government Relations and Science Policy, 1984.

Routes of Exposure to Environmental Chemicals. New York: EPA, Region II, 1989.

Technical Guidance for Hazards Analysis: Emergency Planning for Extremely Haxardous Substances. Washington, DC: EPA, 1987.

Toxic Chemicals: What They Are, How They Affect You. New York: EPA, Region II, 1989.

Turner, Susan, *Chemical Risks: Personal Decisions.* Washington, D.C.: American Chemical Society, Department of Government Relations and Science Policy, 1989.

Household Cleaning Products

Aslett, Don, *Don Aslett's Stainbuster's Bible: A Complete Guide to Spot Removal.* New York: Plume, 1990.

Berthold-Bond, Annie, *Clean and Green.* Woodstock, NY: Ceres Press, 1990.

Dadd, Debra Lynn, *Nontoxic and Natural.* Los Angeles: Jeremy P. Tarcher, 1984.

Dadd, *The Nontoxic Home.* Los Angeles: Jeremy P. Tarcher, 1986.

A Database for Safer Substitutes of Hazardous Household Products, Phase One Report. Seattle: Washington Toxics Coalition, July, 1990.

A Database for Safer Substitutes of Hazardous Household Products, Phase Two Report. Seattle: Washington Toxics Coalition, March, 1991.

Florman, Monte, and Marjorie Florman, et al., *How to Clean Practically Anything,* revised ed. Mt. Vernon, NY: Consumers Union, 1984.

"A Guide to Household Toxics," *Garbage,* March/April, 1990.

Household Hazardous Waste Project, *Guide to Hazardous Products Around the House,* 2nd ed. Springfield, MO: Southwest Missouri State University, 1989.

"Household Hazardous Waste: What You Should Know and Shouldn't Do." Alexandria, VA: Water Pollution Control Federation, 1987.

"The Laundry Quandry," *Green-keeping.* March/April, 1991, May/June, 1991.

Lord, John, "Hazardous Waste from Homes." Santa Monica: Enterprise for Education, 1988.

Ridgley, Susan M., *Toxicants in Consumer Products.* Household Hazardous Waste Disposal Project, Metro Toxicant Program, Report 1B. Seattle: Municipality of Metropolitan Seattle, 1982.

Paint and Wood Preservatives

Alliance to End Childhood Lead Poisoning, *Childhood Lead Poisoning: A Resource Directory,* 2nd ed. Washington, DC: National Center for Education in Maternal and Child Health, Maternal and Child Health Bureau, Department of Health and Human Services, 1991.

Ridgley, Susan M., *Toxicants in Consumer Products.* Household Hazardous Waste Disposal Project, Metro Toxicant Program, Report 1B. Seattle: Municipality of Metropolitan Seattle, 1982.

U.S. Consumer Product Safety Commission, "Consumer Product Safety Alert: What You Should Know About Lead-Based Paint in Your Home." Washington, DC: U.S. Government Printing Office (GPO), 1990. No. 1990-726-058.

Waldman, Steven, "Lead and Your Kids." *Newsweek,* July 15, 1991, p. 42.

"Wood Preservatives." *Journal of Pesticide Reform,* Spring, 1991.

Solvents

"AFSMCE Fact Sheet: Solvents." Washington, DC: American Federation of State, County, and Municipal Employees, 1988.

Berthold-Bond, Annie, *Clean and Green.* Woodstock, NY: Ceres Press, 1990.

Dadd, Debra Lynn, *The Nontoxic Home.* Los Angeles: Jeremy P. Tarcher, 1986.

Dadd, *Nontoxic and Natural.* Los Angeles: Jeremy P. Tarcher, 1984.

Household Hazardous Waste Project, *Guide to Hazardous Products Around the House,* 2nd ed. Springfield, MO: Southwest Missouri State University, 1989.

"Human Toxic Chemical Exposure, The Bulletin of Pacific Toxicology Laboratories: Organic Solvents." Pacific Toxicology Laboratories. Vol. 1, No. 3. (1988).

Lord, John, "Hazardous Waste from Homes." Santa Monica: Enterprise for Education, 1988.

Ridgley, Susan M., *Toxicants in Consumer Products.* Household Hazardous Waste Disposal Project, Metro Toxicant Program, Report 1B. Seattle: Municipality of Metropolitan Seattle, 1982.

Art and Craft Materials

Arena, Jay M., *Child Safety Is No Accident,* revised ed. New York: Berkeley Press, 1987.

Arena, *Poisoning: Toxicology, Symptoms, Treatments,* 5th ed. Springfield, IL: Charles C. Thomas, 1986.

Burros, Marian, "Lead in Ceramics, More Questions, Some Answers." *New York Times,* Feb. 26, 1992.

"Guidelines for Safe Use of Hobby Ceramic Art Materials." Appendix to ASTM C-1023. Philadelphia: American Society for Testing and Materials, 1987.

Health Hazards in Arts and Crafts. New York: American Lung Association, 1985.

McCann, Michael, *Artist Beware: The Hazards and Precautions in Working with Art and Craft Materials.* New York: Watson-Guptill, 1979.

McCann, et al., *Health Hazards Manual for Artists,* 3rd ed. New York: Watson-Guptill, 1985.

Oltman, Debra L., *Pennsylvania Classroom Guide to Safety in the Visual Arts.* Harrisburg: Pennsylvania Department of Education, 1990.

Qualley, Charles, *Safety in the Artroom.* Worcester, MA: Davis, 1986.

"The Safe and Successful Use of Art Materials" (videotape). Clifton, NJ: National Art Materials Trade Association, 1986.

Seeger, N., *A Ceramist's Guide to the Safe Use of Materials.* Chicago: The Art Institute of Chicago, 1982.

Seeger, *A Painter's Guide to the Safe Use of Materials.* Chicago: The Art Institute of Chicago, 1982.

Stopford, Woodhall, "Safety of Lead-Containing Hobby Glazes." *North Carolina Medical Journal,* January 1988.

What You Need to Know about Art and Craft Materials, Safety, and Labeling. Boston: Art and Craft Materials Institute, 1992.

Automotive Fluids

Cancer at the Pump: The Toxic Chemicals in Gasoline. Citizens Fund, November, 1988.

Lord, John, *Hazardous Waste from Homes.* Santa Monica: Enterprise for Education, 1988.

Ridgley, Susan M., *Toxicants in Consumer Products.* Household Hazardous Waste Disposal Project, Metro Toxicant Program, Report 1B. Seattle: Municipality of Metropolitan Seattle, 1982.

Pesticides

"Child-Resistant Packages for Pesticides." Washington, DC: U.S. Environmental Protection Agency (EPA), February, 1985.

"Citizen's Guide to Pesticides," 4th ed. Washington, DC: EPA, April, 1990. EPA 20T-1003.

"Highlights of the 1988 Pesticide Law." Washington, DC: EPA, Office of Pesticides and Toxic Substances, December, 1988.

Kourik, Robert, "Combatting Household Pests without Chemical Warfare." *Garbage,* March/April, 1990.

Lawn Care Pesticides: Risks Remain Uncertain While Prohibited Safety Claims Continue. Washington, DC: General Accounting Office, March 1990. GAO/RCED-90-134.

Mott, Lawrie, and Karen Snyder, *Pesticide Alert, A Guide to Pesticides in Fruits and Vegetables.* San Francisco: Sierra Club Books, 1987.

"Pesticides In Drinking-Water Wells." Washington, DC: EPA, Office of Pesticides and Toxic Substances, September, 1989.

"Pesticides Information Pamphlet." Washington, DC: American Chemical Society, Department of Government Relations, 1987.

"Pesticide Safety for Farmworkers." Washington, DC: EPA, Office of Pesticide Programs (in English and Spanish), April, 1985.

"Pesticide Safety for Non-Certified Mixers, Loaders, and Applicators." Washington, DC: EPA, Office of Pesticide Programs (in English and Spanish), September, 1986.

Recognition and Management of Pesticide Poisoning, 4th ed. Washington, DC: EPA, March 1989. EPA-540/9-88-001.

Ridgley, Susan M., *Toxicants in Consumer Products.* Household Hazardous Waste Disposal Project, Metro Toxicant Program, Report 1B. Seattle: Municipality of Metropolitan Seattle, 1982.

Safety at Home: A Guide to the Hazards of Lawn and Garden Pesticides and Safer Ways to Manage Pests. Washington, DC: National Coalition Against the Misuse of Pesticides, 1991.

Stein, Dan, *Dan's Practical Guide to Least Toxic Home Pest Control.* Eugene, OR: Hulogosi Communications, Inc., 1991.

"Suspended, Canceled, and Restricted Pesticides." Washington, DC: EPA, Office of Pesticides and Toxic Substances, February, 1990.

Thomson, W. T., *Agricultural Chemicals Book.* Volume I: *Insecticides.* Volume II: *Herbicides.* Volume III: *Fumigants, Growth Regulators, Repellents, Rodenticides.* Volume IV: *Fungicides.* Fresno: Thomson Publications, 1988–90.

Ware, George W., *Complete Guide to Pest Control with and without Chemicals,* 2nd ed. Fresno, CA: Thomson Publications, 1988.

Ware, *The Pesticide Book,* 3rd ed. Fresno, CA: Thomson Publications, 1989.

Volatile Organic Compounds

Otto, David, Kenneth H. Hudnell, Dennis E. House, Lars Molhave, and David Counts, "Exposure of Humans to a Volatile Organic Mixture." *Archives of Environmental Health,* January/February, 1992.

Solvents Control Section, *Proposed Regulation to Reduce Volatile Organic Compound Emissions from Consumer Products Staff Report,* Sacramento: State of California Air Resources Board, Stationary Source Division, August 1990.

Asbestos

The ABCs of Asbestos in Schools. Wasington, DC: EPA, Office of Pesticides and Toxic Substances, June 1989.

"Asbestos: A Natural Substance for Modern Needs," Asbestos Information Association/North America, 1988.

Asbestos Fact Book, 3rd ed. Washington, DC: EPA, May, 1986.

"Asbestos Handbook, A Closer Look at Asbestos in Buildings." Atlanta: National Asbestos Council, 1989.

Asbestos in Buildings: What Owners and Managers Should Know, 1st ed. Washington, DC: Safe Building Alliance, 1988.

Asbestos in Public and Commercial Buildings: A Literature Review and Synthesis of Current Knowledge. Cambridge, MA: Health Effects Institute—Asbestos Research, 1991.

"Asbestos in Your Home," Baltimore: American Lung Association, Consumer Product Safety Commission, EPA, 1990.

Consumer Product Safety Commission and Environmental Protection Agency, *Asbestos in the Home,* revised ed. Washington, DC: GPO, August, 1989.

Environmental and Occupational Health Information Program, University of Medicine and Dentistry of New Jersey, *A Guide to Asbestos in the Home for New Jersey Residents.* Trenton: New Jersey Department of Health, 1989.

"Lung Hazards on the Job: Asbestos," Baltimore: American Lung Association, 1986.

New England Journal of Medicine, March 1, 1990.

"Recommended Practices for Handling Asbestos Fiber." Asbestos Information Association/North America, 1986.

Sawyer, Michael F., "Asbestos in the Home: The Homeowner's Dilemma." *NAC Journal,* Fall, 1989.

"What You Should Know About Asbestos Safety," South Deerfield, MA: Channing L. Bete Co., 1989.

Radon

A Citizen's Guide to Radon, Second Edition: The Guide to Protecting Yourself and Your Family from Radon. Washington, DC: EPA, May 1992. No. 402-K92-001.

Facts and Recommendations on Exposure to Radon. Trenton: New Jersey Department of Health, Division of Occupational and Environmental Health, 1987.

"Radon: Are You at Risk?" *Parade Magazine,* November 4, 1990.

Radon Reduction Methods, A Homeowner's Guide, 3rd ed. Washington, DC: EPA, July 1989. No. RD-681.

Reporting on Radon, A Journalist's Guide to Covering the Nation's Second-Leading Cause of Lung Cancer. Washington, DC: EPA, October, 1989. No. 520/1-89-026.

Cosmetics and Personal Care Products

Begoun, Paula, *Don't Go to the Cosmetic Counter Without Me.* Seattle: Beginning Press, 1991.

Chase, Deborah, *The New Medically Based No-nonsense Beauty Book.* New York: Avon Books, 1990.

Cosmetic Ingredient Handbook. Washington, DC: Cosmetic, Toiletry, and Fragrance Association, 1988.

Conry, Tom, *Consumer's Guide to Cosmetics.* Garden City, New York: Anchor Press/Doubleday, 1980.

CTFA Labeling Manual. Washington, DC: Cosmetic, Toiletry, and Fragrance Association, 1990.

Davis, William J., *The Columbia University College of Physicians and Surgeons Complete Home Medical Guide,* 2nd ed. New York: Crown Publishers, 1989.

International Cosmetic Ingredient Dictionary, 4th ed. Washington, DC: Cosmetic, Toiletry, and Fragrance Association, 1991.

Stabile, Toni, *Everything You Want to Know About Cosmetics.* New York: Dodd, Mead, 1984.

Steinman, David, "Glossary of Common Cosmetic Terms and Ingredients." *Greenkeeping,* March/April, 1992.

Winter, Ruth, *Consumer's Dictionary of Cosmetic Ingredients,* 3rd, revised ed. New York: Crown Publishing, 1989.

Winter, *A Consumer's Dictionary of Food Additives,* 3rd ed. New York: Crown Publishing, 1989.

Hazardous Substances at a Glance

Bellafante, Ginia, "Minimizing Household Hazardous Waste." *Garbage,* March/April, 1990.

Bergin, Edward J., and Grandon, Donald E. *How to Survive in Your Toxic Environment.* New York: Avon Books, 1984.

Dadd, Debra Lynn, *The Nontoxic Home.* Los Angeles: Jeremy P. Tarcher, 1986.

Dadd, *Nontoxic, Natural, and Earthwise.* Los Angeles: Jeremy P. Tarcher, 1990.

Fritsch, Albert J., general ed., *The Household Pollutants Guide,* Garden City, NY: Anchor Books, 1978.

"Household Hazardous Waste Wheel." Durham, NH: Environmental Hazardous Management Institute, 1987.

Hunter, Linda Mason, *The Healthy Home.* Emmaus, PA: Rodale Press, 1989.

Lord, John, *Hazardous Waste from Homes.* Santa Monica: Enterprise for Education, 1988.

A Survey of Household Hazardous Wastes and Related Collection Programs. Washington, DC: EPA, 1986.

Chemical Profiles

Baldwin, David G., *Chemical Safety Handbook For the Semiconductor/Electronics Industry.* Boston: QEM Health Information, 1991.

Brookes, Vincent J., and Morris B. Jacobs, *POISONS: Properties, Chemical Identification, Symptoms and Emergency Treatment,* 2nd ed. New York: Van Nostrand Reinhold, 1958.

Chemical Fact Sheets. Albany: New York State Department of Health, Bureau of Toxic Substance Assessment, various issues.

CHRIS Hazardous Chemical Data. Washington, DC: U.S. Department of Transportation, U.S. Coast Guard, 1985.

Clansky, Kenneth B., ed., *Suspect Chemical Sourcebook: A Guide to Industrial Chemicals Covered under Major Federal Regulatory and Advisory Programs.* Bethesda, MD: Roytech Publications, 1990.

Department of Health and Human Services (NIOSH) Publication N0. 89-104, Supplement II-OHG. Cincinnati: U.S. Department of Health and Human Services, 1988.

Glanze, Walter D, ed., *The Mosby Medical Encyclopedia.* New York: C.V. Mosby, 1985.

Grant, Roger, and Claire Grant, *Grant and Hackh's Chemical Dictionary,* 5th ed. New York: McGraw-Hill, 1987.

Hampel, Clifford A., and Gessner G. Hawley, *Glossary of Chemical Terms,* 2nd ed. New York: Van Nostrand Reinhold, 1982.

Hazardous Substance Fact Sheets. Trenton, NJ: New Jersey Department of Health, Right to Know Project, 1985–91.

Lefevre, Marc J., and Shirley A. Conibear, *First Aid Manual for Chemical Accidents,* 2nd ed. New York: Van Nostrand Reinhold, 1989.

Lewis, Richard J., *Hazardous Chemical Desk Reference,* 2nd ed. New York: Van Nostrand Reinhold, 1991.

Lewis, *Hazardous Chemicals In the Workplace,* 2nd ed. New York: Van Nostrand Reinhold, 1990.

National Institute of Occupational Safety and Health, *NIOSH Pocket Guide to Chemical Hazards.* Washington, DC: GPO, 1990.

1990 Emergency Response Guidebook. Washington, DC: Research and Special Programs Administration, U.S. Department of Transportation, 1990.

Occupational Safety and Health Guidelines for Chemical Hazards. Cincinnati: U.S. Department of Health and Human Services, 1981.

Proctor, Nick H., James P. Hughes, and Michael L. Fischman, *Chemical Hazards in the Workplace.* Philadelphia: Lippincott, 1987.

Public Health Service, Center for Disease Control, National Institute for Occupational Safety and Health. Registry of Toxic Effects of Chemical Substances (NIOSH Pub. No. 80-102). Cincinnati: U.S. Department of Health and Human Services, 1980.

Sax, N. Irving, *Hazardous Chemical Information Annual, No. 2.* New York: Van Nostrand Reinhold, 1987.

Sax and Lewis, *Dangerous Properties of Industrial Materials,* 7th ed. New York: Van Nostrand Reinhold, 1988.

Sax and Lewis, *Hawley's Condensed Chemical Dictionary,* 11th ed. New York: Van Nostrand Reinhold, 1987.

Spellman, Jeanne M., and Susan M. Daum, *Work is Dangerous to Your Health.* New York: Vintage Books, 1973.

Verschueren, Karel, *Handbook of Environmental Data on Organic Chemicals,* 2nd ed. New York: Van Nostrand Reinhold, 1983.

Weiss, G., *Hazardous Chemical Data Book,* 2nd ed. Park Ridge, NJ: Noyes Data Corporation, 1986.

Winter, Ruth, *Consumer's Dictionary of Cosmetic Ingredients,* 3rd ed. New York: Crown Publishers, 1989.

Wood, James L., CHEMTOX Database. Brentwood, TN: CHEMTOX Systems, Division of Resource Consultants, Inc., 1992.

Part 1: Chemical Profiles

Chemical Profiles

A

★1★
ACETAL
CAS: 105-57-7
Other names: ACETAAL (DUTCH); ACETAL DIETHYLIQUE (FRENCH); ACETALE (ITALIAN); ACETEHYDE; 1,1-DIAETHOXY-AETHAN (GERMAN); DIAETHYLACETAL (GERMAN); 1,1-DIETHOXY-ETHAAN (DUTCH); 1,1- DIETHOXYE-THANE; DIETHYL ACETAL; 1,1-DIETOSSIETANO (ITALIAN); ETHYLIDENE DIETHYL ETHER; USAF DO-45

Danger profile: Fire hazard, narcotic, acrid fumes in fire.
Uses: Solvent in synthetic perfumes; cosmetics; fruit flavoring; hypnotic in medicine; making other chemicals.
Appearance: Colorless liquid.
Odor: Agreeable, nutty. The odor is a warning of exposure; however, no smell does not mean you are not being exposed.
Effects of exposure:
  Breathed: May be poisonous if inhaled. Vapors may cause dizziness, respiratory depression, cardiovascular collapse, high blood pressure, suffocation, death.
  Eyes: May cause irritation. See *Breathed* and *Skin.*
  Skin: May cause irritation. Passes through the unbroken skin; can increase exposure and the severity of the symptoms listed.
  Swallowed: Moderately toxic. See *Breathed.*
Storage: Store in a tightly closed, airtight container in a dry, cool, well-ventilated place. Keep away from heat, sources of ignition, oxidizing materials and water. Contact with air produces a heat-sensitive explosive. Containers may explode in fire. Vapor may explode if ignited in an enclosed area.
First aid guide: Move victim to fresh air and call emergency medical care; if not breathing, give artificial respiration; if breathing is difficult, give oxygen. In case of contact with material, immediately flush eyes with running water for at least 15 minutes. Wash skin with soap and water. Remove and isolate contaminated clothing and shoes at the site.

★2★
ACETALDEHYDE
CAS: 75-07-0
Other names: ACETIC ALDEHYDE; ACETEHYD (GERMAN); ACETEHYDE; ACETIC EHYDE; EHYDE ACETIQUE (FRENCH); EIDE ACETICA (ITALIAN); ETHANAL; ETHYL ALDEHYDE; ETHYL EHYDE; FEMA NO. 2003; NCI-C56326; OCTOWY EHYD (POLISH)

Danger profile: Suspected carcinogen, mutagen, suspected reproductive hazard, fire and explosion hazard, on Community Right-to-Know List, toxic fumes produced in fire, harmful to aquatic life in very low concentrations, exposure limit established.
Uses: In photography; as a preservative in food products and leather. Making synthetic flavorings, perfumes, dyes, plastics, synthetic resins, gelatin, glue, casein products, synthetic rubber and many other chemicals. Silvering mirrors.
Appearance: Colorless watery liquid or gas.
Odor: Penetrating, fruity, sharp, pungent. The odor is a warning of exposure; however, no smell does not mean you are not being exposed.
Effects of exposure:
  Breathed: Vapor may cause intense irritation to nose, throat, mouth and lungs, mucous membrane. Symptoms of exposure include watering of the eyes, coughing, difficult breathing, headache, nausea, vomiting, muscular weakness, unconsciousness. High exposure can lead to drowsiness, coma and death; buildup of fluid in the lungs (pulmonary edema) may cause death. Pulmonary edema is a medical emergency that may be delayed by one to two days following exposure.
  Eyes: Severe irritant. Contact may cause burning and stinging, tearing, irritation and redness and swelling of the eyelids, intense pain; can burn the eyes and lead to blurred vision, chronic irritation and other permanent damage.
  Skin: Severe irritant. Contact may cause burning or stinging feeling, redness, swelling, intense pain, blisters, irritation, rash. Repeated exposure may lead to skin allergy and chronic irritation.
  Swallowed: May cause burning feeling of the mouth and throat, intense thirst, throat swelling, stomach cramps, nausea, vomiting, diarrhea and unconsciousness, kidney damage, and severe breathing difficulties. Breathing may stop. Watch for delayed symptoms.
Long-term exposure: May result in chronic irritation of the eyes and skin, and an allergic skin rash. May cause cancer and reproductive damage.
Storage: Store in a tightly closed, airtight container in a cool, dark, well-ventilated place. Keep away from all other materials, as contact may cause violent reactions, and from heat, sources of ignition, dust, strong oxidizing or reducing substances, strong acids and bases. Vapor may explode if ignited in an enclosed area.
First aid guide: Move victim to fresh air and call emergency medical care; if not breathing, give artificial respiration; if breathing is difficult, give oxygen. In case of contact with material, immediately flush eyes with running water for at least 15 minutes. Wash skin with soap and water. Remove and isolate contaminated clothing and shoes at the site.

★3★
ACETALDEHYDE OXIME
CAS: 107-29-9
Other names: ACETALDOXIME; ALDOXIME; ALDOXIME, ETHYLDENEHYDROXYLAMINE; ETHANAL OXIME; ETHYLIDENEHYDROXYLAMINE; USAF AM-5

Danger profile: Poisonous, dangerous fire hazard, vapor-air mixtures are explosive, toxic fumes when burned.
Uses: Making other chemicals.
Appearance: Colorless liquid or crystalline solid.
Effects of exposure:
  Breathed: May be poisonous. Contact may irritate the nose, mouth, throat and lungs.
  Skin: May be poisonous if absorbed. May cause irritation and redness.
Long-term exposure: May effect lungs.
Storage: Store in a tightly closed container in a cool, well-ventilated place. Avoid contact with heat, oxidizers and strong acids. The flash point of this chemical is at room temperature, making it a dangerous fire hazard. Vapor may explode if ignited in an enclosed area.
First aid guide: Move victim to fresh air and call emergency medical care; if not breathing, give artificial respiration; if breathing is difficult, give oxygen. In case of contact with material, immediately flush eyes with running water for at least 15 minutes. Wash skin with soap and water. Remove and isolate contaminated clothing and shoes at the site.

★4★
ACETAMIDE
CAS: 60-35-5
Other names: ACETIC ACID AMIDE; ACETIMIDIC ACID; AMID KYSELINY OCTOVE; ETHANAMIDE; METHANECARBOXAMIDE; NCI-C02108

Danger profile: Suspected carcinogen, combustible, suspected mutagen, on Community Right-to-Know List, toxic and irritating fumes in fire.
Uses: Lacquers; explosives; soldering flux; a stabilizer, plasticizer and solvent.
Appearance: Colorless crystals.
Odor: Odorless when pure; ''mousey'' odor reported. The odor is a warning of exposure; however, no smell does not mean you are not being exposed.
Effects of exposure:
  Breathed: May irritate the eyes, nose and throat.
  Eyes: May cause irritation and burns.
  Skin: Passes through the unbroken skin; can increase exposure and the severity of the symptoms listed. Low toxicity. May cause irritation and burns.
Long-term exposure: Repeated exposure may cause liver damage. May cause cancer of the liver (has caused it in animals), lungs and stomach; reproductive damage.
Storage: Store in tightly closed containers in a dry, cool, well-ventilated place. Keep away from heat, sources of ignition, strong acids, strong oxidizers, strong bases and strong reducing agents.

★5★
ACETIC ACID
CAS: 64-19-7
Other names: ACETIC ACID (AQUEOUS SOLUTION); ACETIC ACID, GLACIAL; ACIDE ACETIQUE (FRENCH); ACIDO ACETICO (ITALIAN); AZIJNZUUR (DUTCH); ESSIGSAEURE (GERMAN); ETHANOIC ACID; ETHYLIC ACID; FEMA NO. 2006; GLACIAL ACETIC ACID; METHANE CARBOXYLIC ACID; OCTOWY KWAS (POLISH); VINEGAR ACID

Danger profile: Poisonous, suspected mutagen and reproductive hazard, highly corrosive, combustible, explosion hazard, toxic and corrosive fumes when heated, harmful to aquatic life in low concentrations, exposure limit established.
Uses: A solvent for making gums, acetates, nylon, rayon, acrylic fibers, plastics, volatile oils, pharmaceuticals (aspirin, vitamins, antibiotics, hormones), dyes, cosmetics (hand lotions, hair dyes, skin-bleaching agent), rubber, latex, insecticides, photographic chemicals, other chemicals; textile printing; in the dry cleaning industry to remove rust from garments; a laboratory chemical. Used as a food preservative and in the canning industry as a food additive or flavorant for pickles, fish, meat, candy and glazes; etching compound for engraving; deliming agent during leather tanning.
Appearance: Clear, colorless watery liquid or solid.
Odor: Strong, pungent, vinegar-like. The odor is a warning of exposure; however, no smell does not mean you are not being exposed.
Effects of exposure:
  Breathed: May be poisonous. Vapor may cause extreme irritation of nose mouth and throat, lungs. May cause coughing, shortness of breath, swelling of the throat, and buildup of fluid in the lungs.
  Skin: Severe irritant. Liquid may cause pain and reddening of broken skin. High concentrations may cause severe burns.
  Eyes: Severe irritant. Liquid or vapor may cause immediate pain, swelling and irritation. Glacial acetic acid (100%) can cause severe burns, permanent damage and loss of eyesight.
  Swallowed: May cause pain and irritation in weak concentrations. Strong solutions may cause corrosion of the throat, mouth and intestines. Vomiting, diarrhea with blood, coma and death may occur.
Long-term exposure: Repeated or prolonged exposure may cause darkening, cracking, thickening and irritation of the skin, erosion of exposed front teeth, and chronic inflammation of the nose, throat, and bronchial tubes. May cause reproductive damage. Animal studies show the potential for fertility problems in males. Has caused cancer in mice and rats.
Storage: Store in a well-ventilated area, away from heat, in a tightly closed container. Keep away from open flame, high temperatures, strong oxidizers and alkaline substances. Reacts with most metals including stainless steel. Protect from freezing; glacial acetic acid freezes at 62°F and expands when frozen. May rupture glass container. Container may explode in fire.

★6★
ACETIC ANHYDRIDE
CAS: 108-24-7
Other names: ACETIC ACID, ANHYDRIDE (9CI); ACETIC OXIDE; ACETYL ANHYDRIDE; ACETYL ETHER; ACETYL OXIDE; ANHYDRIDE ACETIQUE (FRENCH); ANIDRIDE ACETICA (ITALIAN); AZIJNZUURANHYDRIDE (DUTCH); ESSIGSAEUREANHYDRID (GERMAN); ETHANOIC ANHYDRATE; OCTOWY BEZWODNIK (POLISH); ETHANOIC ANHYDRIDE

Danger profile: Violent reaction with water or steam vapor, corrosive, combustible, toxic fumes when heated, harmful to aquatic life in very low concentrations, exposure limit established.
Uses: Making cellulose acetate fibers, food starch, plastics, lacquers, protective coating solutions, photographic film, cigarette filters, magnetic tape, vinyl acetate, pharmaceutical, dyes, perfume chemicals, explosives, weed killers; chemical treatment of papers, and aspirin.
Appearance: Clear, colorless watery liquid.
Odor: Strong, vinegar-like. The odor is a warning of exposure; however, no smell does not mean you are not being exposed.
Effects of exposure:

Breathed: Vapor may cause intense irritation to nose, throat, mouth and lungs, mucous membrane. Symptoms of exposure include watering of the eyes, coughing, difficult breathing, headache, nausea, vomiting, muscular weakness, unconsciousness. High exposure can lead to drowsiness, coma and death; buildup of fluid in the lungs (pulmonary edema) may cause death. Pulmonary edema may be delayed by one to two days following exposure.

Eyes: Severe irritant. Contact may cause burning and stinging, tearing, irritation and redness and swelling of the eyelids, intense pain; can burn the eyes and lead to blurred vision, chronic irritation and other permanent damage. Allergy could develop over time.

Skin: Severe irritant. Contact may cause burning or stinging feeling, redness, swelling, intense pain, blisters, irritation, rash. Skin may turn white and wrinkled. A burning sensation is not felt; however, severe burns may follow contact. Repeated exposure may lead to skin allergy and chronic irritation. Allergy could develop over time.

Swallowed: Poisonous. May cause burning feeling of the mouth and throat, intense thirst, throat swelling, stomach cramps, nausea, vomiting, diarrhea, severe breathing difficulties and unconsciousness. Tissue damage may result. Breathing may stop. Nausea and vomiting may develop after exposure. Watch for delayed symptoms.

Long-term exposure: May cause skin allergy and chronic lung irritation; kidney damage.

Storage: Store in a cool, well ventilated place with non-wood floors. Keep away from sources of ignition, heat, water, oxidizers, acids, alcohols, amines and strong caustics. Remote storage is recommended.

First aid guide: Move victim to fresh air and call emergency medical care; if not breathing, give artificial respiration; if breathing is difficult, give oxygen. In case of contact with material, immediately flush skin or eyes with running water for at least 15 minutes. Speed in removing material from skin is of extreme importance. Remove and isolate contaminated clothing and shoes at the site. Keep victim quiet and maintain normal body temperature.

★7★
ACETONE

CAS: 67-64-1

Other names: ACETON (DUTCH, GERMAN, POLISH); DIMETHYLFORMEHYDE; DIMETHYLKETAL; DIMETHYL KETONE; KETONE, DIMETHYL; KETONE PROPANE; BETA-KETOPROPANE; METHYL KETONE; PROPANONE; 2-PROPANONE; PYROACETIC ACID; PYROACETIC ETHER; DIMETHYLFORMALDEHYDE

Danger profile: Fire hazard, narcotic, exposure limit established, on Community Right-to-Know List.

Uses: Paint, paint removers, paint thinners; varnish; markers; plastics; rubber, oil, wax, fats and lacquer solvent; airplane glue solvent; nail polish removers and nail finishes; fabric cement; cleaning fluids. Used in industry to clean and dry parts of precision equipment and to manufacture other chemicals.

Appearance: Clear, colorless, watery liquid.

Odor: Strong, sweet, resembling mint or fruit. The odor is a warning of exposure; however, no smell does not mean you are not being exposed.

Effects of exposure:

Breathed: Vapor may cause headache, dizziness, general weakness, slight irritation of eyes, nose and throat. Exposure to a large dose depresses the central nervous system; can cause victim to appear drunk, excitable with loss of coordination, mental confusion; fatigue; nausea and vomiting; unconsciousness, coma and death.

Eyes: Severe irritant. Contact with vapor or liquid may cause irritation, painful stinging and burning sensation, tearing, inflammation of the eyelids; eyes may become sensitive to light; permanent eye damage and blindness.

Skin: Severe irritant. Contact with the skin may cause irritation, dryness, and possibly dermatitis.

Swallowed: Moderately toxic. May cause gastro-intestinal irritation, vomiting, stomach pain and possible death. See symptoms under *Breathed.*

Long-term exposure: May result in liver, kidney, brain and nerve damage. Repeated skin exposure can cause chronic dryness and irritation.

Storage: Store in tightly closed containers in a cool, well-ventilated place. Keep away from heat, sources of ignition, oxidizing materials and acids. Container may explode in fire. Vapor may explode if ignited in an enclosed area.

First aid guide: Move victim to fresh air and call emergency medical care; if not breathing, give artificial respiration; if breathing is difficult, give oxygen. In case of contact with material, immediately flush eyes with running water for at least 15 minutes. Wash skin with soap and water. Remove and isolate contaminated clothing and shoes at the site.

★8★
ACETONE CYANOHYDRIN

CAS: 75-86-5

Other names: ACETONCIANIDRINA (ITALIAN); ACETON-CIANHIDRINEI (RUMANIAN); ACETONCYAANHYDRINE (DUTCH); ACETONCYANHYDRIN (GERMAN); ACETONECY-ANHYDRINE (FRENCH); ACETONKYANHYDRIN (CZECH); CYANHYDRINE D'ACETONE (FRENCH); 2-CYANO-2-PROPANOL; ALPHA-HYDROXYISOBUTYRONITRILE; 2-HYDROXYISOBUTYRONITRILE; HYDROXY ISOBUTYRO NITRITE; 2-HYDROXY-2-METHYLPRO PIONITRILE; 2-METHYLLACTONITRILE; PROPANENITRILE,2-HYDROXY-2-METHYL-; USAF RH-8

Danger profile: Poisonous, combustible, on Community Right-to-Know List and EPA extremely hazardous substance list, toxic fumes in fire, harmful to aquatic life in very low concentrations.

Uses: Making pesticides and other chemicals.

Appearance: Colorless, watery liquid.

Odor: Distinct, mild almond. The odor is a warning of exposure; however, no smell does not mean you are not being exposed.

Effects of exposure:

Breathed: Poison. High exposure can cause sudden fatal poisoning. Symptoms of lower exposure may include headache, vomiting weakness, confusion, nausea, irritation of eyes, nose, throat and skin; pounding heart beat. Breathing rate and depth will usually be increased at the beginning; and, at later stages, become slow and gasping.

Eyes: Contact may cause burns and fatal poisoning. See *Breathed.*

Skin: Poison if absorbed. Contact may cause burns. See *Breathed* and *Eyes.*

Swallowed: Liquid may cause fatal poisoning. See all above.

Long-term exposure: May cause fatal cyanide poisoning. May interfere with normal thyroid function and cause enlargement; may cause nervous system damage; changes to the blood cell count.

Storage: Store in a tightly closed container in a cool, well-ventilated place. Do not store for long periods; decomposes to acetone and hydrogen cyanide. Keep away form sulfuric acid and other strong acids, strong oxidizers, strong reducers and strong bases.

First aid guide: Move victim to fresh air and call emergency medical care; if not breathing, give artificial respiration; if breathing is difficult, give oxygen. In case of contact with material, immediately flush skin or eyes with running water for at least 15

minutes. Speed in removing material from skin is of extreme importance. Remove and isolate contaminated clothing and shoes at the site. Keep victim quiet and maintain normal body temperature. Effects may be delayed; keep victim under observation.

★9★
ACETONITRILE
CAS: 75-05-8
Other names: ACETONITRIL (DUTCH, GERMAN); CYANOMETHANE; CYANURE DE METHYL (FRENCH); ETHANENITRILE; ETHYL NITRILE; METHANECARBONITRILE; METHANE, CYANO-; METHYL CYANIDE; NCI-C60822; USAF EK-488; RCRA U003

Danger profile: Poisonous, toxic, suspected mutagen, fire hazard, on Community Right-to-Know List, reacts with water, poisonous gases (including hydrogen cyanide) are produced in fire, exposure limit established.
Uses: Making other chemicals, pesticides; solvent; furniture polish; floor polish; pharmaceuticals; water proofing compounds; antistatic agents; detergents; polymers; chemical research; preparation of vitamins, perfumes, water softeners, brighteners for metals.
Appearance: Colorless, watery liquid.
Odor: Sweet, ether-like. The odor is a warning of exposure; however, no smell does not mean you are not being exposed.
Effects of exposure:
   Breathed: Poison. May cause headache, eye, nose and throat irritation, wheezing, foaming at the mouth, general irritability, nausea, chest tightness, shortness of breath, vomiting blood, respiratory depression, weakness, chest and abdominal pains, dizziness, diarrhea, tremors, convulsions, shock, unconsciousness, weak and irregular pulse and death. Reactions may appear hours following overexposure.
   Eyes: May cause severe irritation, tearing, chemical burns, and permanent damage.
   Skin: Poison. Passes through the unbroken skin; can increase exposure and the severity of the symptoms listed. May cause severe irritation, redness, stinging and blisters.
   Swallowed: Poison. May cause severe irritation and chemical burns to the tongue, mouth, throat and stomach as well as many of the symptoms described under *Breathed.*
Long-term exposure: Exposure can cause fatal cyanide poisoning, damage to the central nervous system, liver, kidneys, lungs; and may cause enlargement of the thyroid gland.
Storage: Store away from heat and sources of ignition in a dry, cool, well-ventilated place. Keep away from water, strong oxidizers such as chlorine, bromine, and fluorine. Vapors may travel to sources of ignition and flash back.
First aid guide: Move victim to fresh air and call emergency medical care; if not breathing, give artificial respiration; if breathing is difficult, give oxygen. In case of contact with material, immediately flush skin or eyes with running water for at least 15 minutes. Remove and isolate contaminated clothing and shoes at the site. Keep victim quiet and maintain normal body temperature. Effects may be delayed; keep victim under observation.

★10★
2-ACETYLAMINOFLUORINE
CAS: 53-96-3
Other names: AAF; 2-AAF; 2-ACETAMINOFLUORENE; ACETAMIDE, N-9H-FLUOREN-2-YL- (9CI); ACETAMIDE, N-FLUOREN-2-YL-; 2-ACETAMIDOFLUORENE; ACETOAMINO-FLUORENE; 2-ACETYLAMINO-FLUOREN (GERMAN); N-ACETYL-2-AMINOFLUORENE; 2-(ACETYLAMINO)FLUORENE; AZETYLAMINOFLUOREN (GERMAN); FAA; 2-FAA; 2-FLUORENYLACETAMIDE; N-2-FLUORENYLACETAMIDE; N-FLUOREN-2-YLACETAMIDE

Danger profile: Carcinogen, mutagen, suspected reproductive hazard, toxic fumes when burned, on Community Right-to-Know List, exposure limit established.
Uses: Laboratory research.
Appearance: Tan powder or solid.
Effects of exposure: Any exposure is considered extremely dangerous. Substance can effects victim by breathing and by passing through the unbroken skin. Chemical has limited use; little is known about short-term effects at this time.
Long-term exposure: May cause reduced function of liver kidneys, bladder and pancreas; cancer of the bladder, kidney or liver; birth defects.
Storage: Store in a marked and regulated place; in tightly closed containers in a cool, well-ventilated place. Keep away from cyanides.

★11★
ACETYL BENZOYL PEROXIDE
CAS: 644-31-5
Other names: ACETOZONE; BENZO-BENZONE; BENZOZONE; PEROXIDE, ACETYL BENZOYL

Danger profile: Combustible, poisonous gas when burned.
Uses: Drugs; germicide and disinfectant; for bleaching flour.
Appearance: White crystalline solid.
Effects of exposure:
   Breathed: May cause irritation of the nose, throat, and lungs, coughing, difficult breathing, chest tightness and buildup of fluid in the lungs (pulmonary edema), which can cause death. Pulmonary edema can occur one to two days following exposure.
   Skin: May cause severe irritation and burning.
   Eyes: May cause severe irritation, burning and permanent damage.
   Swallowed: Toxic. See *Breathed.*
Long-term exposure: May cause chronic skin irritation and rash; lung damage.
Storage: Store away from heat, sources of ignition, moisture, reducing agents and combustibles. Containers may explode in fire.
First aid guide: Move victim to fresh air; call emergency medical care. In case of contact with material, immediately flush eyes with running water for at least 15 minutes. Wash skin with soap and water. Remove and isolate contaminated clothing and shoes at the site. Keep victim quiet and maintain normal body temperature.

★12★
ACETYL BROMIDE
CAS: 506-96-7
Other names: ETHANOYL BROMIDE

Danger profile: Corrosive, poisonous gases produced in fire, reacts violently with water, forming corrosive and toxic fumes.
Uses: For making dyes and organic chemicals.
Appearance: Colorless fuming liquid that turns yellow in air.
Odor: Sharp, acrid. The odor is a warning of exposure; however, no smell does not mean you are not being exposed.
Effects of exposure:
   Breathed: May cause irritation of the nose, throat, air passages and lungs. Higher levels may cause coughing, difficult breathing, chest tightness and buildup of fluid in the lungs (pulmonary edema), which can cause death. Pulmonary edema is a medical emergency that can be delayed one to two days following exposure.
   Eyes: Passes through the mucous membrane of the eye; can increase exposure and the severity of the symptoms listed.

May cause irritation and burns that result in permanent damage.

Skin: Passes through the unbroken skin; can increase exposure and the severity of the symptoms listed. May cause severe irritation and burns. Repeated exposure may cause chronic skin irritation; blistering is not uncommon.

Swallowed: Toxic. Decomposes violently in water forming bromic and acetic acid.

Long-term exposure: May cause lung problems. Repeated skin exposure may cause chronic skin irritation.

Storage: Store in tight container in a dry, cool, well-ventilated place with non-wood floors. Keep away from heat, sources of ignition, combustible materials, metals and moisture (water and steam). Combustible hydrogen gas may collect in enclosed space. Containers may explode in fire.

First aid guide: Move victim to fresh air; call emergency medical care. In case of contact with material, immediately flush skin or eyes with running water for at least 15 minutes. Remove and isolate contaminated clothing and shoes at the site. Keep victim quiet and maintain normal body temperature.

★ 13 ★
ACETYL CHLORIDE
CAS: 75-36-5
Other names: ACETIC ACID CHLORIDE; ACETIC CHLORIDE; ETHANOYL CHLORIDE

Danger profile: Fire hazard, explosion hazard, reactive, corrosive, reacts violently with water, poisonous gases produced in fire.
Uses: Making dyestuffs, pharmaceuticals and pesticides.
Appearance: Colorless, fuming liquid.
Odor: Pungent. The odor is a warning of exposure; however, no smell does not mean you are not being exposed.
Effects of exposure:
  Breathed: May be poisonous. May cause irritation of the nose, throat and lungs. May cause coughing. Higher levels may cause difficult breathing, chest tightness and buildup of fluid in the lungs (pulmonary edema), which can cause death. Pulmonary edema is a medical emergency that can be delayed one to two days following exposure.
  Eyes: Contact may cause irritation and burns can cause permanent damage.
  Skin: May cause severe irritation and burns. Repeated exposure can cause chronic skin irritation.
Long-term exposure: May injure lungs.
Storage: store in a tightly closed container in a cool, dry, well-ventilated place. Keep away from water, alcohols, dimethylsulfoxide and phosphorus trichloride. Highly corrosive to metals.
First aid guide: Move victim to fresh air and call emergency medical care; if not breathing, give artificial respiration; if breathing is difficult, give oxygen. In case of contact with material, immediately flush skin or eyes with running water for at least 15 minutes. Remove and isolate contaminated clothing and shoes at the site. Keep victim quiet and maintain normal body temperature.

★ 14 ★
ACETYL CYCLOHEXANE SULFONYL PEROXIDE
CAS: 3179-56-4
Other names: PEROXIDE, ACETYL CYCLOHEXYLSULFONYL

Danger profile: Fire hazard, reactive, poisonous gas when burned.
Uses: Making rubber, paints and plastics.
Appearance: White solid, often used in a liquid solution.
Effects of exposure:

Breathed: May cause irritation of the nose, throat and lungs. Can cause coughing, difficult breathing, chest tightness. High exposures can cause fluid to rapidly build up in the lungs (pulmonary edema) and possibly cause death. Pulmonary edema is a medical emergency that can be delayed one to two days following exposure.

Eyes: May cause severe irritation and burns can cause permanent damage.

Skin: Passes through the unbroken skin; can increase exposure and the severity of the symptoms listed. May cause severe irritation and burns.

Swallowed: See **Breathed** and **First aid guide.**

Long-term exposure: May cause lung problems.

Storage: Store in tightly closed containers in a cool, well-ventilated, detached area with non-wood floors. Keep away from heat, sources of ignition and combustible material such as paper wood, and oil. Containers may explode in fire.

First aid guide: Move victim to fresh air; call emergency medical care. In case of contact with material, immediately flush eyes with running water for at least 15 minutes. Wash skin with soap and water. Remove and isolate contaminated clothing and shoes at the site. Keep victim quiet and maintain normal body temperature.

★ 15 ★
ACETYLENE
CAS: 74-86-2
Other names: ACETYLEN; ACETYLENE, DISSOLVED; ETHINE; ETHYNE; NARCYLEN; WELDING GAS

Danger profile: Fire hazard, toxic, narcotic, explosion hazard, reactive, exposure limit established.
Uses: In welding, cutting and brazing of metals; as an illuminant; making many other chemicals.
Appearance: Colorless gas. Usually sold in pressurized containers partially filled with acetone.
Odor: Distinctive, mild, garlic-like, due to impurities. The odor is a warning of exposure; however, no smell does not mean you are not being exposed.
Effects of exposure:
  Breathed: Mildly toxic. May be poisonous. Depending on concentration, may cause a need for air; rapid, occasionally irregular breathing, headache, intoxication, staggered walk, general lack of coordination, fatigue, exhaustion and loss of consciousness. In extremely high doses results in complete anesthesia with increased blood pressure and rapid breathing. Acts as an asphyxiant (causes suffocation) by lack of oxygen, and may cause death. Technical grade contains highly toxic impurities that should be taken into account. These include arsine, phosphine and hydrogen sulfide.
  Skin: Feeling of intense cold; inability to feel pain. May cause frostbite.
  Eyes: May cause pain, watering, irritation of eyelids and frostbite.
Long-term exposure: Contaminated acetylene, produced by mixing water and calcium carbide, may contain arsine, phosphine, or hydrogen sulfide. Exposure may result in long-term effects from contaminants. If contamination is possible, refer to entries for contaminants.
Storage: Explosion hazard. Store outdoors in a cool, well ventilated, detached and isolated structure. Keep away from sources of ignition, fluorine, chlorine, bromine, iodine and potassium.
First aid guide: Move victim to fresh air and call emergency medical care; if not breathing, give artificial respiration; if breathing is difficult, give oxygen. In case of frostbite, thaw frosted parts with water. Keep victim quiet and maintain normal body temperature.

★ 16 ★
TRANS-ACETYLENE DICHLORIDE
CAS: 156-60-5
Other names: 1,2-TRANS-DICHLOROETHYLENE; DIO-
FORM; ETHENE, TRANS- 1,2-DICHLORO-; TRANS-
DICHLOROETHYLENE, TRANS-1,2-DICHLOROETHYLENE

Danger profile: Fire hazard, explosion hazard, possible muta-
gen, toxic fumes when burned.
Appearance: Colorless liquid.
Odor: Pleasant, ether-like. The odor is a warning of exposure;
however, no smell does not mean you are not being exposed.
Effects of exposure:
  Breathed: Mildly toxic. May be poisonous. May cause hallu-
cinations, sleep, nausea, vomiting, weakness, cramps, trem-
ors. High concentrations may cause central nervous system
depression and narcotic effect.
  Eyes: May cause irritation.
  Skin: May cause irritation and dermatitis.
  Swallowed: See *Breathed.*
Storage: Store in tightly closed containers in a dry, cool, well-
ventilated place. Keep away from heat, sources of ignition, oxi-
dizers.

★ 17 ★
ACETYLENE TETRABROMIDE
CAS: 79-27-6
Other names: MUTHMANN'S LIQUID; TBE; TETRABRO-
MOETHANE; 1,1,2,2-TETRABROMAETHAN (GERMAN);
TETRABROMOACETYLENE; 1,1,2,2-TETRABROMOETANO
(ITALIAN); S-TETRABROMOETHANE; 1,1,2,2-
TETRABROOMETHAAN (DUTCH); SYMMETRICAL TETRA-
BROMOETHANE; SYM-TETRABROMOETHANE; ETH-
ANE,1,1,2,2-TETRABROMO-

Danger profile: Possible carcinogen, possible mutagen, poisonous, narcotic, highly toxic fumes when heated, exposure limit
established.
Uses: Solvent; synthetic fibers, additive in flame-proof, flame retardant polystyrenes, polyurethanes and polyolefins; separating
minerals by specific gravity; processing and separation of fats,
oils, and waxes; as a mercury substitute in the manufacture of
gauges; in microscopy as a refractive index liquid.
Appearance: Pale yellow liquid.
Odor: Pungent, sweet, similar to camphor or chloroform. The
odor is a warning of exposure; however, no smell does not mean
you are not being exposed.
Effects of exposure:
  Breathed: Poison. May cause severe headache, loss of appetite, nausea, dizziness, light-headedness, abdominal pain,
severe irritation of the eyes, nose, throat, fatigue, kidney and
liver damage with such symptoms as dark urine and yellow
jaundice. High concentrations may cause lung damage. May
cause victim to pass out and death.
  Eyes: May cause irritation. See *Breathed.*
  Skin: Passes through the unbroken skin; can increase exposure and the severity of the symptoms listed. Prolonged or
repeated contact might cause drying and cracking, irritation
and burning. See *Breathed.*
  Swallowed: Poison. See *Breathed.*
Long-term exposure: Can cause liver, kidney and lung damage.
Storage: Store in a cool, well-ventilated place, away from heat
and moisture. Keep away from chemically active metals and
strong caustics. In the presence of steam, contact with hot iron,
aluminum and zinc may cause formation of toxic vapors. May attack some forms of plastics, rubber and coatings.
First aid guide: Move victim to fresh air and call emergency
medical care; if not breathing, give artificial respiration; if breath-

ing is difficult, give oxygen. In case of contact with material, immediately flush skin or eyes with running water for at least 15
minutes. Remove and isolate contaminated clothing and shoes
at the site. Effects should disappear after individual has been exposed to fresh air for approximately 10 minutes.

★ 18 ★
ACETYL IODIDE
CAS: 507-02-8

Danger profile: Corrosive, toxic gas when burned, reacts with
water or steam to produce highly toxic and corrosive fumes.
Uses: Making other chemicals.
Appearance: Colorless liquid. Turns brown and fumes on contact with air.
Odor: Suffocating. The odor is a warning of exposure; however,
no smell does not mean you are not being exposed.
Effects of exposure:
  Breathed: May cause lung irritation and quick, shallow
breathing and coughing. High exposures may cause fluid to
rapidly build up in the lungs (pulmonary edema) and possibly
cause death. Pulmonary edema is a medical emergency that
can be delayed one to two days following exposure.
  Eyes: May cause severe burns and possibly permanent damage.
  Skin: May cause severe burns.
Long-term exposure: May effect the lungs.
Storage: Store in a tightly closed container in a cool, well-
ventilated place. Keep away from water and steam.
First aid guide: Move victim to fresh air; call emergency medical
care. In case of contact with material, immediately flush skin or
eyes with running water for at least 15 minutes. Remove and isolate contaminated clothing and shoes at the site. Keep victim
quiet and maintain normal body temperature.

★ 19 ★
ACETYL PEROXIDE
CAS: 110-22-5
Other names: DIACETYL PEROXIDE

Danger profile: Possible tumorigen, spontaneously explosion
hazard, reacts with water or steam, toxic fumes when exposed
to acid.
Uses: Manufacturing resins.
Appearance: Colorless solid, crystal or liquid.
Odor: Sharp, pungent. The odor is a warning of exposure; however, no smell does not mean you are not being exposed.
Effects of exposure:
  Eyes: Contact with vapor or liquid may cause severe irritation.
  Skin: Contact with vapor or liquid may cause severe irritation.
  Swallowed: May cause severe irritation to mouth, throat and
stomach.
Storage: Store in an airtight container in a cool, dry, remote
place with non-wood floors. Keep away from heat, sources of ignition and combustible materials. Can explode spontaneously,
on contact with combustibles such as wood, and will react with
water or steam. Containers may explode in fire. Toxic fumes are
produced on contact with acid.
First aid guide: Move victim to fresh air; call emergency medical
care. In case of contact with material, immediately flush eyes
with running water for at least 15 minutes. Wash skin with soap
and water. Remove and isolate contaminated clothing and
shoes at the site. Keep victim quiet and maintain normal body
temperature.

★ 20 ★
ACETYLSALICYLIC ACID
CAS: 50-78-2
Other names: ACETATE SALICYCLIC ACID; ACETICYL; AC-ETILUM ACIDILATUM; ACETOL; ACETOPHEN; ACETOSAL; ACETOSALIC ACID; ACETOSALIN; O-ACETOXYBENZOIC ACID; 2-ACETOXYBENZOID ACID; ACETYLIN; ACETYSAL; ACIDUM ACETYLSALICYLICUM; ACYLPRIN; ASA; A.S.A. EMPRIN; ASPIRIN; ASPIRINE; ASPRO; ASTERIC; AC-ENTERIN; ACETICYL; ACETYL SAL; ACIMETTER; ASA-TARD; BENZOIC ACID,2-(ACETYLOXY)-; CARPIN; COL-FARIT; CONTR'HEUMA RETARD; DELGESIC; DURAMAX; ECOTRIN; ECM; ENDYDOL; ENTROPHEN; ENTEROSA-RINE; NEURONILCA; EMPIRIN; HELICON; MEASURIN; RHODINE; SALACETIN; SALCETOGEN; SALETIN; SALICYL-IC ACID, ACETATE; SOLPYRON; XAXA

Danger profile: Possible reproductive hazard, exposure limit established.
Uses: Medicine (the active ingredient in aspirin), analgesic, anti-inflammatory, antipyret.
Appearance: White crystalline powder.
Odor: Usually odorless, but in moist air acquires the odor of acetic acid or vinegar. Slightly bitter taste. The odor is a warning of exposure; however, no smell does not mean you are not being exposed.
Effects of exposure:
　Breathed: May cause irritation of the throat and lungs.
　Eyes: May cause irritation, and burns that may result in scarring.
　Skin: Contact may cause irritation.
Long-term exposure: Can result in the development of allergies, hives, difficult breathing and/or rapid drop in blood pressure. Decreases the ability of the blood to clot.
Storage: Store in a tightly closed container in a cool, well-ventilated place.

★ 21 ★
ACROLEIN
CAS: 107-02-8
Other names: ACREHYDE; ACROLEINA (ITALIAN); ACROL-EINE (DUTCH, FRENCH); ACRYLEHYD (GERMAN); AC-RYLEHYDE; ACRYLIC ALDEHYDE; ACRYLALDEHYDE; ACRYLIC EHYDE; AKROLEIN (CZECH); AKROLEINA (POL-ISH); ALLYLALDEHYDE; EHYDE ACRYLIQUE (FRENCH); EIDE ACRILICA (ITALIAN); ALLYL EHYDE; AQUALIN; AQUA-LINE; BIOCIDE; ETHYLENE EHYDE; NSC 8819; PROPENAL; PROPENAL (CZECH); 2-PROPENAL; PROP-2-EN-1-AL; 2-PROPEN-1-ONE; SLIMICIDE; ALLYLEHYDE

Danger profile: Poison, possible carcinogen, mutagen, combustible, explosion hazard, toxic gases in fire, exposure limit established.
Uses: Making plastics, artificial resins, synthetic fibers, and polyurethane foams; drugs and pharmaceuticals; herbicide to control aquatic weeds in drainage and irrigation channels; tear gas; fungicide to prevent slime; denaturant in alcohol.
Appearance: Clear colorless or yellow liquid.
Odor: Very sharp, piercing, disagreeable. Causes tears. The odor is a warning of exposure; however, no smell does not mean you are not being exposed.
Effects of exposure:
　Breathed: Poison. May cause irritation of the eyes, nose and throat, lungs, and skin; a feeling of pressure in chest, headache, dizziness and upset stomach, shortness of breath, nausea. Death can rapidly take place if high concentrations are breathed. High exposures can cause fluid to rapidly build up in the lungs (pulmonary edema) and possibly cause death.

Pulmonary edema is a medical emergency that is often delayed by one to two following exposure.
　Eyes: Severe irritant. Vapors may cause burning at low concentrations. May cause violent irritation, tearing or burns at high concentrations. Contact may cause burning and stinging, tearing, irritation and redness and swelling of the eyelids, intense pain; can burn the eyes and lead to blurred vision, chronic irritation and other permanent damage.
　Skin: Poison if absorbed. Severe irritant. Contact may cause reddening, burning sensation, skin burns and blistering; possible shock, coma and death.
　Swallowed: May cause burning feeling of the mouth and throat, intense thirst, throat swelling, stomach cramps, nausea, vomiting, diarrhea and unconsciousness, kidney damage, and severe breathing difficulties. Breathing may stop. Watch for delayed symptoms. See *Breathed.*
Long-term exposure: Skin irritation and occasionally skin allergy appearing as hives or rash. May cause lung damage, mutations (genetic changes) and may have a cancer risk.
Storage: Store in tightly closed containers in a cool, well-ventilated place. Keep away from heat, sources of ignition, oxidizing agents, acids, alkalies and ammonia. Containers may explode in heat of fire. Vapors may travel to sources of ignition and flash back.
First aid guide: Move victim to fresh air and call emergency medical care; if not breathing, give artificial respiration; if breathing is difficult, give oxygen. In case of contact with material, immediately flush skin or eyes with running water for at least 15 minutes. Remove and isolate contaminated clothing and shoes at the site. Keep victim quiet and maintain normal body temperature. Effects may be delayed; keep victim under observation.

★ 22 ★
ACRYLAMIDE
CAS: 79-06-1
Other names: ACRYLAMIDE MONOMER; ACRYLIC AMIDE; AKRYLAMID (CZECH); ETHYLENECARBOXAMIDE; ETHYL-ENE MONOCLINIC TABLETS CARBOXAMIDE; PROPENA-MIDE; 2-PROPENAMIDE

Danger profile: Carcinogen, suspected mutagen, poisonous, combustible, on EPA extremely hazardous substance list and Community Right-to-Know List, toxic and acrid fumes when burned, exposure limit established.
Uses: Making dyes, plastics, adhesives, sizing agents, paper, textiles and permanent-press fabrics, printing, nail enamels, cosmetic face masks, molded parts; soil conditioners; water, sewage and waste treatment; photography; ore processing; chemical grouting in drilling of oil well drill holes, basements, tunnels, mine shafts, caissons and dams; additive in concrete; soil stabilizer in wet situations.
Appearance: Colorless or white powder or crystals. Can also be found in solution.
Effects of exposure:
　Breathed: May be poisonous. May cause irritation of the eyes, nose, throat and lungs, tears and runny nose, coughing, sore throat, difficult breathing; lethargy, fatigue, and drowsiness; slurred speech, shaking, and dizziness; burning, itching of fingers, cold and tender skin, numbness and sweating of legs and arms, muscle weakness of hands and feet, loss of sensations, weak reflexes, urinary retention, constipation, hallucinations, tremors, muscle atrophy and paralysis. Has caused central nervous system injury which is partially reversible. May cause shortness of breath, headache, nausea, vomiting, dizziness, diarrhea and a dangerous build-up of fluids in the lungs (pulmonary edema), which can cause death. Pulmonary edema is a medical emergency that can be delayed one to two days following exposure.

Eyes: May be poisonous. May cause irritation, redness, tearing, pain, inflamed eyelids, and eye injury; burns in corneas and possible loss of eyesight.

Skin: Toxic. Even weak solutions (1%) in water may cause painful irritation, redness, and skin peeling. Repeated or prolonged skin contact may cause painful irritation, redness, and blisters; may also cause the legs to feel numb.

Swallowed: May cause irritation of the mouth and throat; stomach and abdominal pain; loss of muscular coordination and mental confusion; convulsions. Animal studies suggest that death may occur from swallowing less than an ounce. Also see symptoms for *Breathed.*

Long-term exposure: May cause lung and skin cancer. May cause chronic poisoning which can lead to midbrain disturbances. Can damage the nervous system causing numbness, "pins and needles," and weakness in the hands and feet. There is limited evidence that chemical may damage male testes. Workers exposed to substance have developed some symptoms after one month to one year. Repeated or prolonged skin contact may cause irritation of the skin and may also cause the legs to feel numb.

Storage: Store in tightly closed containers in a dry, cool, well-ventilated place. Keep away from heat, sources of ignition, oxidizers and nitrates.

First aid guide: Move victim to fresh air and call emergency medical care; if not breathing, give artificial respiration; if breathing is difficult, give oxygen. In case of contact with material, immediately flush skin or eyes with running water for at least 15 minutes. Speed in removing material from skin is of extreme importance. Remove and isolate contaminated clothing and shoes at the site. Keep victim quiet and maintain normal body temperature. Effects may be delayed; keep victim under observation.

★ 23 ★
ACRYLIC ACID
CAS: 79-10-7
Other names: ACROLEIC ACID; ACRYLIC ACID, INHIBITED; ACRYLIC ACID, GLACIAL; ETHYLENECARBOXYLIC ACID; GLACIAL ACRYLIC ACID; KYSELINA AKRYLOVA; PROPENE ACID; PROPENOIC ACID; 2-PROPENOIC ACID; VINYLFORMIC ACID

Danger profile: Mutagen, possible carcinogen, poisonous, corrosive, suspected reproductive hazard, fire hazard, reactive, poisonous gas produced in fire, exposure limit established.

Uses: Making plastic products, acrylic polymers and resins; leather treatment; paper coatings; polyacrylic and polymethacrylic acids.

Appearance: Colorless watery liquid.

Odor: Acrid, irritating, rancid, sweet. The odor is a warning of exposure; however, no smell does not mean you are not being exposed.

Effects of exposure:
Breathed: May be poisonous. May cause irritation of the eyes, nose and throat; tearing.

Eyes: Contact may cause severe eye burns and permanent damage.

Skin: Contact may cause irritation and severe skin burns.

Swallowed: May cause severe damage to the gastrointestinal tract, nausea, convulsions and coughing.

Long-term exposure: May cause skin allergy, rash and itching; kidney and lung damage; birth defects by damaging the fetus.

Storage: Store in tightly closed containers in a cool, well-ventilated place. Keep away from heat, sources of ignition. Container may explode in fire. Vapors may travel to sources of ignition and flash back.

First aid guide: Move victim to fresh air and call emergency medical care; if not breathing, give artificial respiration; if breathing is difficult, give oxygen. In case of contact with material, immediately flush skin or eyes with running water for at least 15 minutes. Remove and isolate contaminated clothing and shoes at the site. Keep victim quiet and maintain normal body temperature.

★ 24 ★
ACRYLONITRILE
CAS: 107-13-1
Other names: ACRYLNITRIL (GERMAN, DUTCH); ACRYLON; ACRYLONITRILE MONOMER; AKRYLONITRYL (POLISH); CARBACRYL; CIANURO DI VINILE (ITALIAN); CYANOETHYLENE; CYANURE DE VINYLE (FRENCH); ENT 54; FUMIGRAIN; MILLER'S FUMIGRAIN; NITRILE ACRILICO (ITALIAN); NITRILE ACRYLIQUE (FRENCH); PROPENENITRILE; 2-PROPENENITRILE; TL 314; VCN; VENTOX; VINYL CYANIDE; VINYL CYANIDE, PROPENENITRILE

Danger profile: Carcinogen, mutagen, poisonous, fire hazard, permit may be required to use as a pesticide, toxic (cyanide) gas produced when heated, on Community Right-to-Know List and EPA extremely hazardous substance list, exposure limit established.

Uses: Making synthetic fibers, plastics, surface coatings, adhesives, nitrile rubber, cotton; pesticide; grain fumigant.

Appearance: Clear, watery, colorless to pale yellow liquid.

Odor: Unpleasant, irritating, onion or garlic-like. NOTE: Odor can only be detected above the legal permissible exposure limit (PEL). The odor is a warning of exposure; however, no smell does not mean you are not being exposed.

Effects of exposure:
Breathed: Poison. May cause headache, eye, nose and throat irritation, wheezing, foaming at the mouth, general irritability, nausea; vomiting, diarrhea, tremors, dizziness, convulsions, unconsciousness, weak and irregular pulse, death.

Eyes: May cause severe irritation, tearing and chemical burns.

Skin: Poison. May cause severe irritation, redness, stinging and blisters. Absorption through the unbroken skin may contribute to other symptoms listed.

Swallowed: Poison. May cause severe irritation and chemical burns to the tongue, mouth, throat and stomach as well as many of the symptoms described under *Breathed.*

Long-term exposure: May cause cancer; damage to the developing fetus; kidney and liver irritation; liver damage, anemia and blood abnormalities.

Storage: Store in a cool, dark, well-ventilated place. Keep away from strong acids, strong alkalies, bromine, tetrahydrocarbazole, copper, ammonia, oxygen. Strong light and concentrated alkalies may cause violent spontaneous polymerization.

First aid guide: Move victim to fresh air and call emergency medical care; if not breathing, give artificial respiration; if breathing is difficult, give oxygen. In case of contact with material, immediately flush skin or eyes with running water for at least 15 minutes. Remove and isolate contaminated clothing and shoes at the site. Keep victim quiet and maintain normal body temperature. Effects may be delayed; keep victim under observation.

★ 25 ★
ACTIVATED CARBON
CAS: 64365-11-3
Other names: ACTIVATED CHARCOAL; ACTIVE CARBON; CARBORAFFIN; CARBORAFINE; GAC; KARBORAFIN; NUCHAR 722; NUCHAR C190N; CHARCOAL, ACTIVATED; VEGETABLE CARBON; ANIMAL CARBON; WOOD CARBON; SHELL CARBON; MINERAL CARBON

Danger profile: Combustible, poisonous gas produced in fire.

Uses: Water and air purification, decolororizing sugar, water treatment, removal of sulfur dioxide from stack gases and clean rooms, deodorant, catalyst, brewing, air conditioning, solvent recovery.
Appearance: Black powder, lumps or granules.
Effects of exposure:
Breathed: May cause irritation of the nose, throat, lungs and mucous membrane.
Skin: May cause irritation.
Eyes: May cause irritation.
Swallowed: May cause irritation of the mouth, throat and stomach. Nontoxic; actually used in therapy of poisoning cases.
Storage: Store in tightly closed containers in a cool, well-ventilated place. Keep away from heat and sources of ignition.

★26★
ADIPIC ACID
CAS: 124-04-9
Other names: ACIFLOCTIN; ACINETTEN; ADILAC-TETTEN; ADIPINIC ACID; 1,4-BUTANEDICARBOXYLIC ACID; FEMA NO. 2011; HEXANEDIOIC ACID; 1,6-HEXANEDIOIC ACID; KYSELINA ADIPOVA (CZECH), MOLTEN ADIPIC ACID

Danger profile: Poisonous, combustible, poisonous and acrid fumes and smoke produced in fire.
Uses: Used in food additives; resins, plastics, urethanes; hair color rinses; baking powder. Making lubricating oil additives, nylon, adhesives and pharmaceuticals.
Appearance: Fine, white crystals or powder.
Effects of exposure:
Breathed: May cause irritation of the eyes, nose, mucous membrane, throat and lungs; coughing sneezing and difficult breathing.
Eyes: Liquid may cause severe eye irritation and burns.
Skin: Contact may cause irritation and burns; a pronounced drying effect on the skin; may cause dermatitis.
Swallowed: Based on animal studies, a lethal dose is approximately one half pound.
Storage: Store in tightly closed containers in a well-ventilated place. Keep away from heat, flame, sources of ignition, strong acids alkalies and oxidizers.

★27★
ADIPONITRILE
CAS: 111-69-3
Other names: ADIPIC ACID DINITRILE; ADIPIC ACID NITRILE; ADIPODINITRILE; 1,4-DICYANOBUTANE; HEXANEDINITRILE; HEXANEDIOIC ACID, DINITRILE; NITRILE ADIPICO (ITALIAN); TETRAMETHYLENE CYANIDE

Danger profile: Poisonous, combustible, on EPA extremely hazardous substance list and Community Right-to-Know List, toxic fumes in fire.
Uses: Making nylon.
Appearance: Colorless to light yellow liquid.
Odor: Almost odorless. The odor is a warning of exposure; however, no smell does not mean you are not being exposed.
Effects of exposure:
Breathed: Poison. Vapor may cause wheezing, foaming at the mouth, headache, convulsions and death.
Eyes: May cause irritation, redness, inflammation of the eyelids and burns. Passes through the mucous membrane of the eye, increasing severity and symptoms listed in *Breathed* and *Swallowed.*
Skin: Poisonous. Passes through the unbroken skin; can increase exposure and the severity of the symptoms listed. Be-

haves as a cyanide when absorbed through the skin. May cause irritation, inflammation and burns.
Swallowed: Poison. May cause metallic taste, disturbed and rapid respiration and circulation, mental confusion, increased heart beat, irritation of the stomach and intestines, stomach pain, nausea and vomiting, diarrhea, weakness, convulsions, coma and death.
Storage: Store in a well-ventilated place. Keep away from oxidizers, heat and source of ignition.
First aid guide: Move victim to fresh air and call emergency medical care; if not breathing, give artificial respiration; if breathing is difficult, give oxygen. In case of contact with material, immediately flush skin or eyes with running water for at least 15 minutes. Speed in removing material from skin is of extreme importance. Remove and isolate contaminated clothing and shoes at the site. Keep victim quiet and maintain normal body temperature. Effects may be delayed; keep victim under observation.

★28★
ALDICARB
CAS: 116-06-3
Other names: ALDECARB; ALDECARBE (FRENCH); AMBUSH; CARBANOLATE; ENT-27093; 2-METHYL-2-METHYLTHIO-PROPIONALDEHYD-O-(N- METHYL-CARBAMOYL)-OXIM (GERMAN); 2-METIL-2-TIOMETIL-PROPIONALDEID- O-(N-METIL-CARBAMOIL)-OSSIMA (ITALIAN); NCI-CO8640; OMS-771; PERMETHRIN; PROPANAL,2-METHYL-2-(METHYTHIO)-,O- ((METH-YLAMINO)CARBONYL)OXIME; TEMIC; TEMIK; TEMIK G10; UC21149 (UNION CARBIDE)

Danger profile: Poisonous; suspected mutagen; EPA extremely hazardous substance list; very toxic fumes in fire; use is restricted.
Uses: Restricted use pesticide. Insecticide and nemacide for soil; on flowers and fruit crops; cabbage, cauliflower and sprouts.
Appearance: White crystalline solid.
Effects of exposure:
Breathed: Extremely poisonous. May cause headache, dizziness, profuse sweating, nausea, vomiting, reduced heart beat, stomach cramps, diarrhea, loss of coordination, slow and weak breathing, fever, loss of consciousness and death. May cause shortness of breath and a dangerous build-up of fluids in the lungs (pulmonary edema), which can cause death. Pulmonary edema is a medical emergency that can be delayed one to two days following exposure.
Eyes: Rapidly absorbed; poisoning can happen very rapidly. See *Breathed.*
Skin: Poison. May cause irritation. Absorption through the unbroken skin is significant and may cause or increase severity of symptoms listed in *Breathed* and *Swallowed.*
Swallowed: Powerful, deadly poison. See *Breathed.*
Long-term exposure: May cause cancer; nerve damage including loss of coordination.
Storage: Store in a regulated, specially marked place in tightly closed containers in a dry, well-ventilated place. Keep away from water, strong alkalis. Containers may explode in fire.

★ 29 ★
ALDRIN
CAS: 309-00-2
Other names: ALDREX; ALDRINE (FRENCH); ALDRITE; AL-DROSOL; ALTOX; COMPOUND 118; DRINOX; ENT 15949; HEXACHLOROHEXAHYDRO-ENDO-EXO-DIMETH AN-ONAPHTHALENE; 1,2,3,4,10,10-HEXACHLORO-1,4,4A,5,8,8A-HEXAHYDRO-1,4,5,8-DIMETHA NONAPHTHA-LENE; 1,2,3,4,10,10-HEXACHLORO-1,4,4A,5,8,8A-HEXAHYDRO-EXO-1,4-ENDO-5,8-D IMETHANONAPHTHA-LENE; 1,2,3,4,10,10-HEXACHLORO-1,4,4A,5,8,8A-HEXAHYDRO-1,4-ENDO-EXO-5, 8-DIMETHANONAPHTHALENE; HHDN; NCI-C00044; OCTALENE; SEEDRIN; 1,2,3,4,10-10-HEXACHLORO-1,4,4A,5,8,8A-HEXAHYDRO-1,4,5,8-ENDO,EX O-DIMETHANONAPHTHALENE

Danger profile: Suspected carcinogen, suspected mutagen, suspected reproductive hazard, poisonous, toxic fumes in fire, on EPA extremely hazardous substance list and Community Right-to-Know List, exposure limit established.
Uses: Insecticide. Use not permitted in many areas.
Appearance: Colorless to dark brown crystalline solid.
Odor: Mild chemical. The odor is a warning of exposure; however, no smell does not mean you are not being exposed.
Effects of exposure:
   Breathed: May cause headache, dizziness; excessive irrita-bility, confusion and excitability; weakness, numb arms and legs; shallow breathing, convulsions, blue skin and lips, coma and death.
   Eyes: May cause irritation and redness. See symptoms in *Breathed* and *Swallowed.*
   Skin: Poison. May cause irritation. Absorption through the unbroken skin is significant and may cause or increase sever-ity of symptoms listed in *Breathed* and *Swallowed.*
   Swallowed: Poison. May cause nausea, vomiting, excessive irritability, convulsions, rapid and irregular heartbeat, and death. In an adult, some of these symptoms may occur after swallowing of as little as 1/30 ounce (one gram), and may be delayed from 2 to 12 hours. Short exposure may result in tem-porary and reversible liver and kidney injury.
Long-term exposure: Continued exposure can result in perma-nent liver damage.
Storage: Store outdoors in tightly sealed containers. Keep away from strong acids, strong oxidizers and phenol. Avoid heat, may give off toxic fumes.
First aid guide: Move victim to fresh air and call emergency medical care; if not breathing, give artificial respiration; if breath-ing is difficult, give oxygen. In case of contact with material, im-mediately flush skin or eyes with running water for at least 15 minutes. Speed in removing material from skin is of extreme im-portance. Remove and isolate contaminated clothing and shoes at the site. Keep victim quiet and maintain normal body tempera-ture. Effects may be delayed; keep victim under observation.

★ 30 ★
ALLYL ACETATE
CAS: 591-87-7
Other names: ACETIC ACID ALLYL ESTER; ACETIC ACID, 2-PROPENYL ESTER; ALKYL ACETIC ACID; 2-PROPENYL METHANOATE; 3-ACETOXYPROPENE

Danger profile: Fire hazard, explosion hazard, poison, poison-ous gases and acrid fumes produced in fire.
Uses: Flavoring additive in beverages, ice cream, candy, baked goods and dairy products.
Appearance: Colorless liquid.
Effects of exposure:

   Breathed: Moderately toxic. Vapors may cause irritation of the eyes, nose, throat and bronchial tubes, nosebleeds, hoarseness, cough and phlegm, and tightness of the chest. Higher exposures could cause shortness of breath and a dangerous build-up of fluids in the lungs (pulmonary edema), which can cause death. Pulmonary edema is a medical emer-gency that can be delayed one to two days following expo-sure.
   Eyes: Moderately toxic. Contact may cause irritation.
   Skin: Moderately toxic. May cause irritation. Repeated or prolonged contact may cause redness, rash and itching.
   Swallowed: Poison. May be fatal.
Long-term exposure: Repeated or long term skin contact may cause rash. May cause lung damage.
Storage: Store in tightly closed containers in a cool, well-ventilated place. Keep away from heat, sources of ignition, oxi-dizers and peroxides. Containers may explode in fire. Flash point is at room temperature. Vapors may travel to sources of ignition and flash back.
First aid guide: Move victim to fresh air and call emergency medical care; if not breathing, give artificial respiration; if breath-ing is difficult, give oxygen. In case of contact with material, im-mediately flush skin or eyes with running water for at least 15 minutes. Remove and isolate contaminated clothing and shoes at the site. Keep victim quiet and maintain normal body tempera-ture. Effects may be delayed; keep victim under observation.

★ 31 ★
ALLYL ALCOHOL
CAS: 107-18-6
Other names: ALCOOL ALLILCO (ITALIAN); ALCOOL AL-LYLIQUE (FRENCH); ALLILOWY ALKOHOL (POLISH); AA; ALLYL AL; ALLYLALKOHOL (GERMAN); ALLYLIC ALCO-HOL; 3-HYDROXYPROPENE; ORVINYLCARBINOL; PR-OPENOL; 2-PROPENOL; 2-PROPEN-1-OL; PROPEN-1-OL-3; 1-PROPEN-3-OL; PROPENYL ALCOHOL; 2-PROPENYL AL-COHOL; SHELL UNKRAUTTED A; VINYL CARBINOL; VINYL CARBINOL,2-PROPENOL; WEED DRENCH

Danger profile: Poisonous, fire hazard, explosion hazard, poi-son, poisonous gas is produced in fire, on EPA extremely haz-ardous list, exposure limit established.
Uses: Fungicide, herbicide, and nematicide; military poison gas; manufacture of resins, plasticizers, drugs, glycerol, acrolein, and other chemicals; refining and dewaxing of mineral oil. Deriva-tives used in perfumes, flavorings, and pharmaceuticals.
Appearance: Mobile, colorless liquid.
Odor: Pungent, mustard-like. The odor is a warning of exposure; however, no smell does not mean you are not being exposed.
Effects of exposure:
   Breathed: Poison. Vapors may cause irritation of the eyes, nose, throat and lungs, causing coughing, vomiting, short-ness of breath. Exposure can cause dizziness, nausea, weakness and may cause victim to pass out.
   Eyes: Poison. Severe irritant. May pass through the unbro-ken skin or eyes and cause severe burns, blurred vision, pain in the eyeballs and permanent damage. Pain may not begin until after six hours following exposure.
   Skin: Poison. Passes through the unbroken skin; can in-crease exposure and the severity of the symptoms listed. May cause deep-seated muscle pain in region where absorp-tion occurred, swelling, and muscle spasms. Liquid may cause first and second degree burns with blister formation; underlying part will become swollen and painful, and local muscle spasms may occur. Person with pre-existing skin dis-orders may be more susceptible to the effects of this sub-stance. See *Eyes.*
   Swallowed: Poison. See *First aid guide.*

Long-term exposure: Long term or high exposure may cause kidney and liver damage; may effect the lungs.

Storage: Store in a tightly closed container in a cool, well-ventilated place. Keep away from heat, sources of ignition, strong oxidizers, acids, oleum, chlorosulfonic acid, and carbon tetrachloride. Do not smoke or eat in areas where liquid is handled, processed or stored. Containers may explode in heat of fire.

First aid guide: Move victim to fresh air and call emergency medical care; if not breathing, give artificial respiration; if breathing is difficult, give oxygen. In case of contact with material, immediately flush skin or eyes with running water for at least 15 minutes. Speed in removing material from skin is of extreme importance. Remove and isolate contaminated clothing and shoes at the site. Keep victim quiet and maintain normal body temperature. Effects may be delayed; keep victim under observation.

★ 32 ★
ALLYL BROMIDE
CAS: 106-95-6
Other names: BROMALLYLENE; 3-BROMOPROPENE; 3-BROMOPROPYLENE

Danger profile: Possible mutagen, poisonous, dangerous fire hazard, explosion hazard, exposure limit established, toxic fumes when burned.
Appearance: Colorless to light yellow liquid.
Odor: Irritating. The odor is a warning of exposure; however, no smell does not mean you are not being exposed.
Effects of exposure:
Breathed: Mildly toxic. Vapors may cause irritation of the nose, throat, mucous membranes and lungs; headache, dizziness, coughing, shortness of breath, difficult breathing; lightheadedness, drowsiness and victim may pass out.
Eyes: Contact with liquid may cause irritation, severe burns and possible permanent eye damage.
Skin: Passes through the unbroken skin; can increase exposure and the severity of the symptoms listed. Contact may cause irritation and severe burns.
Swallowed: Poison. May cause irritation of the mouth and stomach.
Long-term exposure: May damage the developing fetus.
Storage: Store in a tightly closed container in a cool, well-ventilated place. Keep away from heat, sources of ignition and strong oxidizers. Combustible liquid; containers may explode in fire.
First aid guide: Move victim to fresh air and call emergency medical care; if not breathing, give artificial respiration; if breathing is difficult, give oxygen. In case of contact with material, immediately flush skin or eyes with running water for at least 15 minutes. Speed in removing material from skin is of extreme importance. Remove and isolate contaminated clothing and shoes at the site. Keep victim quiet and maintain normal body temperature. Effects may be delayed; keep victim under observation.

★ 33 ★
ALLYL CHLORIDE
CAS: 107-05-1
Other names: ALLILE (CLORURO DI) (ITALIAN); ALLYL-CHLORID (GERMAN); ALLYLE (CHLORURE D') (FRENCH); CHLORALLYLENE; 3-CHLOROPRENE; 1-CHLORO PROP-ENE-2; 3-CHLOROPROPENE-1; 3-CHLOROPROPENE; 1-CHLORO-2-PROPENE; 3-CHLOROPROPYLENE; 3-CHLORPROPEN 10 (GERMAN); NCI-C04615; 3-CHLOROPROPENE,1-CHLORO-2-PROPENE; 3-CHLORO-1-PROPENE

Danger profile: Suspected carcinogen, suspected mutagen, combustible, explosion hazard, poison, poisonous gas when burned, on Community Right-to-Know List, exposure limit established.
Uses: Preparation of allyl alcohol and other allyl compounds; resins for varnishes, plastics, adhesives; manufacture of pharmaceuticals, insecticides and other chemicals.
Appearance: Colorless, yellow or purple liquid.
Odor: Pungent, unpleasant, penetrating, irritating.
Effects of exposure:
Breathed: Moderately toxic. May cause irritation of the lungs, mucous membrane, coughing, shortness of breath. Symptoms may appear several hours after exposure and may cause permanent damage. Higher exposures may cause shortness of breath and a dangerous build-up of fluids in the lungs (pulmonary edema), which can cause death. Pulmonary edema is a medical emergency that can be delayed one to two days following exposure.
Eyes: Moderately toxic. Vapor can cause severe irritation of eyes (with eye pain), nose and throat. Liquid can cause severe eye burns, pain, internal injury and possible permanent damage. Contact lenses should not be worn when working with this chemical.
Skin: Moderately toxic. Passes through the unbroken skin; can increase exposure and the severity of the symptoms listed. May cause irritation, skin burns, numbness, muscular aches and pain in the bones.
Swallowed: Poison. Immediately seek medical attention. See *First aid guide.*
Long-term exposure: May damage the developing fetus; chronic exposure may result in liver and kidney damage; may effect the lungs; long-term exposure may cause drying and cracking of the skin.
Storage: Store in a tightly closed containers in a cool, well-ventilated place. Keep away from heat, sources of ignition and open flame, strong oxidizers, strong acids, aluminum, zinc, iron, aluminum, peroxides, chlorides, magnesium. Do not smoke or eat in areas where liquid is handled, processed or stored. High concentrations of the vapor may cause explosion. Containers may explode in heat of fire.
First aid guide: Move victim to fresh air and call emergency medical care; if not breathing, give artificial respiration; if breathing is difficult, give oxygen. In case of contact with material, immediately flush skin or eyes with running water for at least 15 minutes. Speed in removing material from skin is of extreme importance. Remove and isolate contaminated clothing and shoes at the site. Keep victim quiet and maintain normal body temperature. Effects may be delayed; keep victim under observation.

★ 34 ★
ALLYL GLYCIDYL ETHER
CAS: 106-92-3
Other names: ALLIL-GLICIDIL-ETERE (ITALIAN); AGE; 1-ALLILOSSI-2,3 EPOSSIPROPANO (ITALIAN); ALLYLGLYCI-DAETHER (GERMAN); ALLYL 2,3-EPOXYPROPYL ETHER; 1-ALLYLOXY-2,3-EPOXY-PROPAAN (DUTCH); 1-ALLYLOXY-2,3-EPOXYPROPAN (GERMAN); 1-(ALLYLOXY)-2,3-EPOXYPROPANE; ETHER, ALLYL 2,3-EPOXYPROPYL; NCI-C56666; OXYDE D'ALLYLE ET DE GLYCIDYLE (FRENCH); PROPANE, 1-(ALLYLOXY)-2,3-EPOXY-; OXIRANE,((2-PROPENYLOXY)METHYL-); 1-ALLYLOXY-2,3-EPOXY-PROPANE

Danger profile: Poisonous, reproductive hazard, combustible, irritating and poisonous fumes when burned, exposure limit established.
Uses: Making resins, other chemicals, rubber, and vinyl products.
Appearance: Colorless liquid.

Odor: Strong, sweet, pleasant. The odor is a warning of exposure; however, no smell does not mean you are not being exposed.

Effects of exposure:

Breathed: Moderately toxic. May irritate the eyes, nose and lungs; coughing, shortness of breath, dizziness, lightheadedness, and victim may pass out. Exposure can result in depression of the central nervous system, shortness of breath and build-up of fluid in the lungs (pulmonary edema), and possibly death. Pulmonary edema is a medical emergency that may be delayed for one to two days following exposure.

Eyes: Severe irritant. May cause burns and permanent eye damage. See *Breathed.*

Skin: Toxic. Passes through the unbroken skin; can increase exposure and the severity of the symptoms listed. May cause painful irritation, redness, and blisters. Repeated or prolonged skin contact may cause irritation of the skin and may also cause the legs to feel numb.

Swallowed: May cause irritation of the mouth and throat; stomach and abdominal pain; loss of muscular coordination and mental confusion; convulsions. Also see symptoms for *Breathed.*

Long-term exposure: May result in damage to the liver and kidneys; skin allergy with itching and skin rash; skin drying and cracking; high concentrations of vapor may cause eye damage.

Storage: Store in tightly closed containers in a cool, well-ventilated place. Keep away from heat, sources of ignition, and strong oxidizers. Vapors may travel to sources of ignition and flash back. Containers may explode in fire.

First aid guide: Move victim to fresh air and call emergency medical care; if not breathing, give artificial respiration; if breathing is difficult, give oxygen. In case of contact with material, immediately flush skin or eyes with running water for at least 15 minutes. Remove and isolate contaminated clothing and shoes at the site. Keep victim quiet and maintain normal body temperature.

★ 35 ★
ALLYL IODIDE
CAS: 556-56-9
Other names: 3-IODO-1-PROPENE; 1-IODO-2-PROPENE; PROPENE,3-IODO; 3- IODO-1-PROPENE; 3-IODOPROPYLENE; 3-IODOPROPENE

Danger profile: Poisonous, possible mutagen, possible reproductive hazard, combustible, corrosive, poisonous fumes when burned.

Uses: Making other chemicals and polymers.

Appearance: Yellowish liquid. Darkens on contact with air.

Odor: Unpleasant, irritating, pungent. The odor is a warning of exposure; however, no smell does not mean you are not being exposed.

Effects of exposure:

Breathed: May cause irritation of the eyes, nose, throat and respiratory system; tearing. Higher exposures may cause shortness of breath and a dangerous build-up of fluids in the lungs (pulmonary edema), which can cause death. Pulmonary edema is a medical emergency that can be delayed one to two days following exposure.

Eyes: Both liquid and vapors can cause severe irritation and tearing. Contact may cause serious eye surface burns.

Skin: Severe irritant. Passes through the unbroken skin; can increase exposure and the severity of the symptoms listed. Contact may cause severe burns.

Swallowed: Poisonous. See *First aid guide.*

Long-term exposure: May cause genetic changes (mutations); damage of the liver and kidneys. May effect the lungs.

Storage: Store in tightly closed containers in a dark, dry, well-ventilated place. Keep away from light, strong oxidizers and

source of ignition. Containers may explode in fire. Vapors may cause explosion.

First aid guide: Move victim to fresh air and call emergency medical care; if not breathing, give artificial respiration; if breathing is difficult, give oxygen. In case of contact with material, immediately flush skin or eyes with running water for at least 15 minutes. Remove and isolate contaminated clothing and shoes at the site. Keep victim quiet and maintain normal body temperature.

★ 36 ★
ALLYL ISOTHIOCYANATE
CAS: 57-06-7
Other names: ALLYL ISORHODANIDE; AITC; ALLYL ISO-SULFOCYANATE; ALLYL MUSTARD OIL; ALLYLSENEVOL; ALLYLSENFOEL (GERMAN); CARBOSPOL; FEMA NO. 2034; ISOTHIOCYANATE D'ALLYLE (FRENCH); 3-ISOTHIOCYANATO-1-PROPENE; MUSTARD OIL; NCI-C50464; OIL OF MUSTARD, ARTIFICIAL; OLEUM SINAPIS VOLATILE; PROPENE,3- ISOTHIOCYANATE; REDSKIN; SENF OEL (GERMAN); SYNTHETIC MUSTARD OIL; VOLATILE OIL OF MUSTARD

Danger profile: Suspected carcinogen, poisonous, mutagen, allergen, contact dermatitis, combustible, reactive, toxic (cyanide) fumes when burned and on contact with acid.

Uses: Ointment, mustard plasters, fumigants.

Appearance: Colorless to pale yellow, oily liquid.

Odor: Pungent, irritating. The odor is a warning of exposure; however, no smell does not mean you are not being exposed.

Effects of exposure:

Breathed: May cause cough, watery eyes, sneezing, irritation of the eyes, nose and throat.

Eyes: Irritation, burns. See *Breathed.*

Skin: Poison. Passes through the unbroken skin; can increase exposure and the severity of the symptoms listed. Prolonged exposure can cause chemical burns and blisters; may cause dermatitis.

Swallowed: Poisonous. See *First aid guide.*

Long-term exposure: May cause cancer of the bladder; may damage the developing fetus; asthma-like allergy, watery eyes, cough, wheezing and tightness of the chest. Once allergy develops even small exposures may cause symptoms listed.

Storage: Keep away from acids. Store in an airtight container away from heat, acids, and oxidizing materials. Containers may explode in fire. Poisonous gas may be produced in fire. Vapor may explode if ignited in an enclosed area.

First aid guide: Move victim to fresh air and call emergency medical care; if not breathing, give artificial respiration; if breathing is difficult, give oxygen. In case of contact with material, immediately flush skin or eyes with running water for at least 15 minutes. Speed in removing material from skin is of extreme importance. Remove and isolate contaminated clothing and shoes at the site. Keep victim quiet and maintain normal body temperature. Effects may be delayed; keep victim under observation.

★ 37 ★
ALLYL PROPYL DISULFIDE
CAS: 2179-59-1
Other names: 4,5-DITHIA-1-OCTENE; DISULFIDE, ALLYL PROPYL; DISULFIDE, 2-PROPENYL PROPYL; ONION OIL; PROPYL ALLYL DISULFIDE

Danger profile: Combustible, exposure limit established, poisonous gases are produced in fire.

Uses: The main ingredient in onion oil.

Appearance: Pale yellow liquid.

Odor: Pungent, irritating. The odor is a warning of exposure; however, no smell does not mean you are not being exposed.
Effects of exposure:
Breathed: May cause irritation of the eyes, nose, throat and lungs, tears and runny nose, coughing, sore throat, difficult breathing; lethargy, fatigue, and drowsiness; slurred speech, shaking, and dizziness. Higher exposures could cause shortness of breath, headache, nausea, vomiting, dizziness, diarrhea and a dangerous build-up of fluids in the lungs (pulmonary edema), which can cause death. Pulmonary edema is a medical emergency that can be delayed one to two days following exposure.
Eyes: May cause irritation, redness, tearing, pain, inflamed eyelids, and eye injury; burns in corneas and possible loss of sight.
Skin: Toxic. Passes through the unbroken skin; can increase exposure and the severity of the symptoms listed. May cause painful irritation, redness, and blisters. Repeated or prolonged skin contact may cause irritation of the skin and may also cause the legs to feel numb.
Swallowed: May cause irritation of the mouth and throat; stomach and abdominal pain; loss of muscular coordination and mental confusion; convulsions. Also see symptoms for *Breathing.*
Storage: Store in tightly closed containers in a dry, cool, well-ventilated place. Keep away from heat, sources of ignition and oxidizers.

★ 38 ★
ALLYL TRICHLOROSILANE
CAS: 107-37-9
Other names: ALLYL TRICHOROSILANE, STABILIZED; ALLYLTRICHLOROSILANE; SILANE, TRICHLOROALLYL-; SILANE,TRICHLORO-2-PROPENYL-; ALLYLSILICONE TRICHLORIDE; TRICHLOROALLYLSILANE

Danger profile: Poisonous, corrosive, combustible, toxic fumes when burned, reactive with water and steam (produces poisonous gas).
Uses: Making silicones; glass fiber finishes.
Appearance: Colorless liquid.
Odor: Pungent; irritating. The odor is a warning of exposure; however, no smell does not mean you are not being exposed.
Effects of exposure:
Breathed: May be poisonous. Irritation of the eyes, nose and lungs. Can cause coughing and shortness of breath. Exposure may result in a build-up of fluids in the lungs (pulmonary edema), and result in death. Pulmonary edema is a medical emergency that may not appear for one to two days following exposure.
Eyes: Irritation, and may result in severe burns and permanent damage. See *Breathed.*
Skin: Can cause irritation and severe chemical burns.
Swallowed: Poisonous.
Long-term exposure: May cause lung damage.
Storage: Store in a tightly closed container in a cool, well-ventilated place. Keep away from water, metals, sources of ignition and flame. Container may explode in fire. Poisonous gases are produced in fire. Vapor may explode if ignited in an enclosed area.
First aid guide: Move victim to fresh air and call emergency medical care; if not breathing, give artificial respiration; if breathing is difficult, give oxygen. In case of contact with material, immediately flush skin or eyes with running water for at least 15 minutes. Remove and isolate contaminated clothing and shoes at the site. Keep victim quiet and maintain normal body temperature.

★ 39 ★
ALUMINUM
CAS: 7429-90-5
Other names: A 00; A 95; A 99; A 995; A 999; AA 1099; AA 1199; AD 1; AD1M; ADO; AE; ALAUN (GERMAN); ALLBRI ALUMINUM PASTE AND POWDER; ALUMINA FIBRE; ALUMINIUM; ALUMINIUM BRONZE; ALUMINIUM FLAKE; ALUMINUM 27; ALUMINUM DEHYDRATED; ALUMINUM, METALLIC, POWDER; ALUMINUM POWDER; AO A1; AR2; AV00; AV000; C.I. 77000; EMANAY ATOMIZED ALUMINUM POWDER; JISC 3108; JISC 3110; L16; METANA; METANA ALUMINUM PASTE; NORAL ALUMINUM; NORAL EXTRA FINE LINING GRADE; NORAL INK GRADE ALUMINUM; NORAL NON-LEAFING GRADE; PAP-1

Danger profile: Powder can be combustible and explosive, on Community Right-to-Know List (dust or fume.)
Uses: Building and construction; corrosion-resistant chemical equipment for desalination plants; metal alloys (die-cast auto parts); power transmission lines; photoengraving plates; permanent magnets; cryogenic technology; miscellaneous food-processing equipment; tubes for ointments, toothpaste, shaving cream, etc. As powder, in paints, inks and protective coatings; as rocket fuel, ingredient of incendiary mixtures such as thermite; foamed concrete. As foil, in packaging, cooking, decorative stamping.
Appearance: Silvery white metallic solid or powder.
Effects of exposure:
Breathed: Fine dust may cause irritation of the nose, sneezing, coughing and shortness of breath. Also see *Swallowed.*
Eyes: Extremely irritating. Contact may cause scratches.
Swallowed: No symptoms.
Long-term exposure: May cause cancer, reproductive damage; scarring of the lungs (pulmonary fibrosis) cough and shortness of breath; brain damage.
Storage: Keep powder in tightly closed containers away from source of ignition. Aluminum powder is combustible and can form an explosive mixture in air. Moisture can make the hazard even greater. Do not use water on fire.

★ 40 ★
ALUMINUM CHLORIDE
CAS: 7446-70-0
Other names: ALLUMINIO(CLORURO DI) (ITALIAN); ALUMINIUMCHLORID (GERMAN); ALUMINUM CHLORIDE (1:3); ALUMINUM CHLORIDE ANHYDROUS; ALUMINUM TRICHLORIDE; CHLORURE D'ALUMINUM (FRENCH); PEARSALL; TRICHLOROALUMINUM

Danger profile: Suspected mutagen, suspected reproductive hazard, corrosive, a violent reaction and poisonous and corrosive fumes result from contact with water or steam, poisonous and corrosive hydrochloric acid may be produced in fire.
Uses: Making other chemicals, dyes; antiperspirants; petroleum and rubber industries.
Appearance: White, grey, yellow or green powder or liquid.
Odor: Strong, acidic, irritating. The odor is a warning of exposure; however, no smell does not mean you are not being exposed.
Effects of exposure:
Breathed: Dust may cause severe irritation to the nose, throat and lungs, causing coughing, difficult breathing, shortness of breath, bluish skin and lips. May cause a build-up of fluids in the lungs (pulmonary edema) and possible death. Pulmonary edema is a medical emergency that may be delayed from one to two days following exposure.
Skin: May cause severe irritation with thermal and acid burns, especially if skin is wet.

Eyes: May cause severe irritation, burns, pain and possible blindness.
Swallowed: May cause severe burns of the mouth, throat and stomach, salivating, vomiting, general weakness, watery or bloody diarrhea, coma and death. Damage to kidneys and liver, collapse and convulsions can result.
Long-term exposure: May cause damage to the developing fetus, and the effects of long-term exposure can last for months or years.
Storage: Store in tightly closed containers in a cool, dry place. Keep away from water, steam, sunlight, combustible materials, acids, bases, metals, allyl chloride, benzene, ethylene oxide, nitrobenzene, phenol, nitro-methane, and perchloryl fluoride. Container may explode in fire. Old containers may explode when opened.

★ 41 ★
ALUMINUM HYDROXIDE
CAS: 21645-51-2
Other names: ALCOA 331; ALCOA C 30BF; AF 260; ALUMI-GEL; ALUMINA HYDRATE; ALUMINA HYDRATED; ALUMINA TRIHYDRATE; ALPHA-ALUMINA TRIHYDRATE; ALUMINIC ACID; ALUMINIUM HYDROXIDE; ALUMINUM HYDRATE; ALUMINUM (III) HYDROXIDE; ALUMINUM HYDROXIDE GEL; ALUMINUM OXIDE-3H2O; ALUMINUM OXIDE HYDRATE; ALUMINUM OXIDE TRIHYDRATE; ALUMINUM TRIHYDRAT; ALUMINUM TRIHYDROXIDE; ALUSAL; AMBEROL ST 140F; AMPHOJEL; BACO AF 260; BRITISH ALUMINUM AF 260; C 31; C 33; C 31C; C 4D; C 31F; C-31-F; C.I. 77002; GHA 331; GHA 332; H 46; HIGILITE; HIGILITE H 32; HIGILITE H 42; HIGILITE H 31S; HYCHOL 705; HYDRAL 705; HYDRAL 710; HYDRATED ALUMINA; MARTINAL; P 30BF; TRIHYDRATED ALUMINA; TRIHYDROXYALUMINUM

Danger profile: Exposure limit established, poisonous gas produced in fire.
Uses: In ceramics, glass; flame retardants; mattress batting; paper coating; cosmetics.
Appearance: White crystalline powder, balls or granules.
Effects of exposure:
    Breathed: May cause irritation of the nose, coughing and sneezing.
    Eyes: May cause irritation, redness, tearing, pain and inflammation of the eyelids.
    Swallowed: No symptoms.
Storage: Store in tightly closed containers in a dry, cool, well-ventilated place. Containers may explode in fire.

★ 42 ★
ALUMINUM NITRATE
CAS: 13473-90-0
Other names: ALUMINUM (III) NITRATE (1:3); ALUMINUM NITRATE NONHYDRATE; ALUMINUM SALT OF NITRIC ACID; ALUMINUM TRINITRATE; NITRAM; NITRIC ACID, ALUMINUM SALT; NITRIC ACID, ALUMINUM (3); NORWAY SALTPETER; VARIOFORM I

Danger profile: Poisonous, oxidizer, toxic fumes when burned, exposure limit established.
Uses: Making fertilizers, explosives, insecticides, and other chemicals; textiles; tanning leather; incandescent filaments for light bulbs; antiperspirants; anticorrosion agent; rocket propellants. Used in veterinary medicine.
Appearance: White, light grey or brown pellets, flakes, or crystalline material often used in liquid solution.
Effects of exposure:
    Breathed: Dust may cause irritation to the nose, throat and lungs. May cause coughing and difficult breathing.

Skin: May cause irritation and burns.
Eyes: May cause severe irritation, burns and possible damage.
Swallowed: May cause gastric irritation, nausea, vomiting, and frequent urination.
Long-term exposure: Repeated exposure may cause skin irritation. May cause lung damage.
Storage: Store in a well-ventilated non-combustible place (with automatic sprinkler system) away from source of ignition. Keep away from combustable material, acetic acid, powdered metals, ammonium chloride, phosphorus, sodium hypochlorite, sodium chlorate, heat, sulfur and strong alkalies (releases toxic ammonia gas). Contact with heat—especially in confined areas—can cause explosion and the production of poisonous gases. Protect containers from physical damage or shock. Containers may explode in fire.
First aid guide: Move victim to fresh air; call emergency medical care. In case of contact with material, immediately flush skin or eyes with running water for at least 15 minutes. Remove and isolate contaminated clothing and shoes at the site.

★ 43 ★
ALUMINUM OXIDE
CAS: 1344-28-1
Other names: ACTIVATED ALUMINUM OXIDE; A 1 (SORBENT); ALCOA F 1; ALMITE; ALON; ALON C; ALOXITE; ALUMINA; ALPHA-ALUMINA; GAMMA-ALUMINA; BETA-ALUMINA; ALUMINITE 37; ALUMINUM OXIDE (2:3); ALPHA-ALUMINUM OXIDE; BETA-ALUMINUM OXIDE; GAMMA-ALUMINUM OXIDE; ALUMINUM SESQUIOXIDE; ALUMINUM TRIOXIDE; ALUMITE; ALUMITE (OXIDE); ALUNDUM; ALUNDUM 600; BAUXITE; BAYERITE; BOEHMITE; BROCKMANN, ALUMINUM OXIDE; CAB-O-GRIP; CATAPAL S; COMPALOX; CONOPAL; CORUNDUM; DIALUMINUM TRIOXIDE; DIASPORE; DISPAL; DISPAL ALUMINA; DISPAL M; DOTMENT 324; DOTMENT 358; EMERY; EXOLON XW 60; F 360 (ALUMINA); FASERTON; FASERTONERDE; G 0 (OXIDE); G2 (OXIDE); GIBBSITE; GK (OXIDE); HYPALOX II; JUBENON R; KA 101; KETJEN B; KHP 2; LA 6; LUCALOX; LUDOX CL; MARTOXIN; MICROGRIT WCA; PORAMINAR; PS 1; PS1 (ALUMINA); Q-LOID A 30; RC 172DBM

Danger profile: Possible carcinogen; on Community Right-to-Know List.
Uses: Making aluminum, abrasive, refractories, ceramics, cosmetic colors where opacity is desired, electrical insulators, paper; spark plugs; laboratory equipment; adsorbing gases and water vapors; chromatographic analysis; fluxes; light bulbs; artificial gems; heat-resistant fibers; food additive; antacid.
Appearance: White to red powders, balls or lumps.
Effects of exposure:
    Breathed: Dust particles may cause irritation, cough, sneezing, nasal symptoms, chest pain, shortness of breath, and possible lung damage. Inhalation of high dust concentration can damage the respiratory tract.
    Eyes: May cause irritation, tearing, pain, and inflammation of the eyelids.
    Skin: May cause irritation, redness and itching.
    Swallowed: One form—aluminum hydroxide gel—is used as an antacid and may cause constipation.
Long-term exposure: Exposure is usually accompanied by exposure to silicates and iron compounds and these are often linked to respiratory diseases including emphysema. Repeated exposure to high levels may cause scarring of the lungs (pulmonary fibrosis) with shortness of breath. This condition may be fatal.
Storage: Store in tightly closed containers in a cool, well-ventilated place. Keep away from chlorine trifluoride and ethylene oxide.

★ 44 ★

ALUMINUM PHOSPHATE

CAS: 7784-30-7
Other names: ALUMINOPHOSPHORIC ACID; ALUPHOS; PHOSPHAGEL; PHOSPHORIC ACID, ALUMINUM SALT (1:1)

Danger profile: Corrosive, exposure limit established, poisonous gases produced in fire.
Uses: Making paper, ceramics, cosmetics as a gelling agent, dental cements, drugs, paints, papers, pharmaceuticals and varnishes.
Appearance: White crystals.
Effects of exposure:
Breathed: Exposure may cause irritation of the eyes, nose, throat and lungs.
Skin: Irritant. The liquid may cause severe irritation and burns on contact.
Eyes: Dust may cause irritation. The liquid may cause severe burns on contact.
Long-term exposure: May cause lung damage.
Storage: Store in tightly closed containers in a dry, cool, well-ventilated place. Keep away from water, steam, moisture, combustible materials and source of ignition.

★ 45 ★

ALUMINUM PHOSPHIDE

CAS: 20859-73-8
Other names: AL-PHOS; AIP; ALUMINUM FOSFIDE (DUTCH); ALUMINUM PHOSPHITE; ALUMINUM MONOPHOSPHIDE; CELPHOS (INDIAN); DELICIA; DELICIA GASTOXIN; DETIA; DETIA GAS EX-B; DETIA-EX-B; FOSFURI DI ALLUMINIO (ITALIAN); PHOSPHURES D'ALUMIUM (FRENCH); PHOSTOXIN; QUICKPHOS

Danger profile: Poisonous, toxic gas when burned, contact with water or acid produces highly toxic gas, EPA extremely hazardous substances list, exposure limit established.
Uses: Insecticide; fumigant for grain; manufacture of semiconductors.
Appearance: Dark grey or dark yellow crystals.
Effects of exposure:
Breathed: May cause restlessness, headache, dizziness, fatigue, nausea, vomiting, coma, convulsions. Higher exposures may cause lowered blood pressure, shortness of breath and a dangerous build-up of fluids in the lungs (pulmonary edema), which can cause death. Pulmonary edema is a medical emergency that can be delayed one to two days following exposure.
Eyes: May cause irritation. See *Breathed.*
Swallowed: Poisonous. See *Breathed.*
Long-term exposure: Repeated exposure may lead to bone damage and teeth loss. Phosphine gas released by aluminum phosphide may cause damage to the liver, kidneys or lungs.
Storage: Store in tightly closed containers in a dry, cool, well-ventilated place. Keep away from water, steam, alkalis. On contact with moisture or acid highly toxic phosphine gas is released. See entry for phosphine, which is spontaneously combustible in air.

★ 46 ★

ALUMINUM SODIUM FLUORIDE

CAS: 15096-52-3
Other names: CRYOLITE; ENT 24,984; KOYOSIDE; KRYOCIDE; KRYOLITH (GERMAN); NATRIUMALUMINIUMFLUORID (GERMAN); NATRIUMHEXAFLUOROALUMINATE (GERMAN); SODIUM ALUMINOFLUORIDE; SODIUM ALUMINUM FLUORIDE; SODIUM FLUOALUMINATE; SODIUM HEXAFLUOROALUMINATE; VILLIAUMITE

Danger profile: Exposure limit established, poisonous gas produced in fire.
Uses: Making pesticides, ceramics, glass, polishes, aluminum.
Appearance: Colorless or white powder or solid.
Effects of exposure:
Breathed: May cause irritation of the nose, throat and breathing passages. High exposures may cause shortness of breath and a dangerous build-up of fluids in the lungs (pulmonary edema), which can cause death. Pulmonary edema is a medical emergency that can be delayed one to two days following exposure.
Eyes: May cause severe irritation, burns and permanent damage.
Skin: May cause irritation and burns.
Long-term exposure: May cause lung damage; joint and ligament damage with stiffness and possible disability.
Storage: Store in tightly closed containers in a dry, cool, well-ventilated place. Keep away from strong acids.

★ 47 ★

ALUMINUM SULFATE

CAS: 10043-01-3
Other names: ALUM; ALUMINUM ALUM; ALUMINUM SULFATE (2:3); ALUMINUM TRISULFATE; CAKE ALUM; DIALUMINUM SULFATE; DIALUMINUM TRISULFATE; FILTER ALUM; PATENT ALUM; PERL ALUM; PAPER MAKER'S ALUM; PICKEL ALUM; SULFURIC ACID, ALUMINUM SALT (3:2)

Danger profile: Toxic, suspected reproductive hazard, exposure limit established, poisonous gases when burned.
Uses: Tanning leather, paper sizing, antiseptic, astringent lotions, skin fresheners, dyeing process (for fixing colors), purifying water, fire proofing and waterproofing cloth, clarifying oils and fats, treating sewage, in antiperspirants to prevent skin irritation from aluminum chloride, after shave lotions, styptic to stop bleeding, pesticides, manufacturing other chemicals.
Appearance: Gray-white, or green powder, granules or crystals. Often used in a water solution; in liquid form sulfuric acid is present.
Effects of exposure:
Breathed: Exposure may cause irritation of the eyes, nose, mouth, throat, bronchial tubes and lungs.
Skin: May cause irritation. If chemical is wet consult entry for SULFURIC ACID.
Eyes: May cause irritation, and possibly cause eye damage. If chemical is wet consult entry for SULFURIC ACID.
Swallowed: May cause stomach irritation, nausea, vomiting and diarrhea. Concentrated dose may cause gum damage. See entry for SULFURIC ACID.
Long-term exposure: May cause kidney and lung damage.
Storage: Store in tightly closed containers in a cool, dry, well-ventilated place. Keep away from water, steam, combustible materials and sources of fire or ignition. Forms sulfuric acid when wet.

★48★
2-AMINO-5-AZOTOULENE
CAS: 97-56-3
Other names: AMINOAZOTOLUENE; O-AMINOAZOTOLUENE; O-AMINOAZOTOLUOL (GERMAN); 3,2-AMINO-5-AZOTOLUENE; AAT; O-AAT; OAAT; BUTTER YELLOW; C.I. SOLVENT YELLOW; C.I. 11160B; FAST OIL YELLOW; FAST YELLOW AT; OIL YELLOW; OIL YELLOW 21; OIL YELLOW 2681; OIL YELLOW AT; OIL YELLOW A; OIL YELLOW C; OIL YELLOW 1; OIL YELLOW 2R; OIL YEL-LOW T; ORGANO YELLOW 25; SOMALIA YELLOW R; SUDAN YELLOW RRA; TOLUAZOTOLUIDINE; O-TOLUENE-AZO-O-TOULUIDENE; TULA-BASE FAST GARNET GB; TULABASE FAST GARNET GBC; WAZAKOL YELLOW NL

Danger profile: Carcinogen, suspected mutagen, poisonous, suspected reproductive hazard, toxic fumes when burned, on Community Right-to-Know List.
Uses: Dyes, drugs.
Appearance: Reddish-brown to yellow crystals.
Effects of exposure:
   Swallowed: Poison.
Long-term exposure: May cause cancer. Mutation data reported.
Storage: Store in tightly closed containers in a cool, dry, well-ventilated place.

★49★
2-AMINO-4-CHLOROPHENOL
CAS: 95-85-2
Other names: P-CHLORO-O-AMINOPHENOL; PHENOL, 2-AMINO-4-CHLORO-

Danger profile: Poisonous, poisonous fumes produced in fire.
Uses: Making dyes and other chemicals.
Appearance: Light brown or grayish crystals.
Effects of exposure:
   Breathed: Poisonous. Exposure may lower the ability of the blood to carry oxygen (methemoglobinemia). This may cause headache, dizziness, fatigue, shortness of breath, failure of muscular control, difficult breathing, fast heart rate, and a bluish color to the skin and lips. High exposure could lead to collapse and death.
   Eyes: Contact may cause irritation of eyes and eyelids. See *Breathed.*
   Skin: Contact may cause irritation. See *Breathed.*
   Swallowed: Poisonous. May cause nausea and vomiting. See *Breathed.*
Storage: Store in tightly closed containers in a dry, cool, well-ventilated place. Keep away from water and moisture, iron and heat (keep temperature below 110°F).

★50★
2-AMINO-5-DIETHYLAMINOPENTANE
CAS: 140-80-7
Other names: 1,4-PENTANEDIAMINE, N1,N1-DIETHYL-; TETRAMETHYLENEDIAMINE,N,N-DIETHYL-4-METHYL-; N,N-DIETHYL-4- METHYLTETRAMETHYLENEDIAMINE

Danger profile: Combustible, can cause skin allergy, poisonous gas when burned.
Uses: Making drugs, especially those used for malaria.
Appearance: Liquid.
Odor: Fishy. The odor is a warning of exposure; however, no smell does not mean you are not being exposed.
Effects of exposure:
   Breathed: May cause irritation of the nose and throat.
   Eyes: Contact may cause severe irritation and burns.
   Skin: Passes through the skin; exposure may be increased. Contact may cause strong irritation of the eyes and skin; burns.
Long-term exposure: Allergy can develop and future exposures, even if low, can cause rash and itching. Very irritating substance; may injure the lungs.
Storage: Store in a tightly closed container in a cool, well-ventilated place. Keep away from heat and sources of ignition. Containers may explode in fire; poisonous gas produced in fire.
First aid guide: In case of contact with material, immediately flush eyes with running water for at least 15 minutes. Wash skin with soap and water. Remove and isolate contaminated clothing and shoes at the site.

★51★
4-AMINODIPHENYL
CAS: 92-67-1
Other names: P-AMINOBIPHENYL; 4-AMINOBIPHENYL; 4-AMINODIFENIL (SPANISH); ANILINOBENZENE; BIPHENYL-AMINE; 4-BIPHENYLAMINE; 1-1'-BIPHENYL-4-AMINE; P-BIPHENYLAMINE; PARAMINODIPHENYL; P-PHENYLANILINE; 4-PHENYLANILINE; XENYLAMIN (CZECH); XENYLAMINE

Danger profile: Carcinogen, mutagen, poisonous, on Community Right-to-Know List, toxic fumes when burned.
Uses: Laboratory chemical.
Appearance: Colorless crystals that turn purple when exposed to air.
Odor: Floral. The odor is a warning of exposure; however, no smell does not mean you are not being exposed.
Effects of exposure:
   Breathed: Exposure may lower the ability of the blood to carry oxygen (methemoglobinemia). This may cause headache, dizziness, fatigue, shortness of breath, failure of muscular control, difficult breathing, fast heart rate, and a bluish color to the skin and lips; urinary bladder irritation, urinary burning, inflammation of the bladder and bloody urine. High exposure could lead to collapse and death.
   Skin: Passes through the skin; exposure may be increased. May cause irritation as well as symptoms listed in *Breathed.*
   Eyes: May cause irritation. Also see *Breathed.*
   Swallowed: Poison. Also see *Breathed.*
Long-term exposure: Chemical may cause bladder cancer in humans; may damage the developing fetus. Similar solvents and petroleum-based chemicals have caused brain and other nerve damage. Effects may include reduced memory and concentration, personality changes including irritability and withdrawal; fatigue, sleep disturbances, reduced coordination and "pins and needles" in the arms and legs.
Storage: Store in accordance with OSHA standard 1910.1011. Storage area should be marked, regulated and maintained under negative pressure. Keep away from heat and sources of ignition, oxidizers.

★52★
1-AMINO-2-METHYL-ANTHRAQUINONE
CAS: 82-28-0
Other names: ACETATE FAST ORANGE R; ACETOQUI-NONE LIGHT ORANGE JL; 1-AMINO-2-METHYL-9,10-ANTHRACENEDIONE; 9,10-ANTHRACENEDIONE, 1-AMINO-2-METHYL-; ARTISIL ORANGE 3RP; CELLITON ORANGE R; C.I. 60700; C.I. DISPERSE ORANGE 11; CILLA ORANGE R; DISPERSE ORANGE; DURANOL ORANGE G; 2-METHYL-1-ANTHRAQUINONYLAMINE; MICROSETILE ORANGE RA; NCI-C01901; NYLOQUINONE ORANGE JR; PERLITON OR-ANGE 3R; SERISOL ORANGE YL; SUPRACET ORANGE R

Danger profile: Carcinogen, on Community Right-to-Know List, poisonous gases produced in fire.
Uses: Dyes; making textiles and thermoplastics.
Appearance: Orange crystals.
Effects of exposure:
Breathed: May cause irritation of the eyes, nose and throat.
Eyes: Contact may cause irritation.
Skin: Contact may cause irritation.
Long-term exposure: May cause cancer and reproductive damage.
Storage: Store in tightly closed containers in a dry, cool, well-ventilated, marked and regulated place.

★53★
4-AMINO-2-NITROANILINE
CAS: 5307-14-2
Other names: 1,4-DIAMINO-2-NITROBENZENE; C.I. 76070; C.I. OXIDATION BASE 22; DURAFUR BROWN; DURAFUR BROWN 2R; DYE GS; FOURAMINE 2R; FOURRINE 36; FOURRINE BROWN 2R; NCI-C02222; 2NDB; 2-NITRO-1,4-BENZENEDIAMINE; 2-NITRO-1,4-DIAMINOBENZENE; NITRO-P-PHENYLENEDIAMINE; O-NITRO-P-PHENYLENEDIAMINE; 2-NITRO-1,4-PHENYLENEDIAMINE; 2-NITROL-P-PHENYLENEDIAMINE; 2-NP; 2-NPPD; 2-N-P-PDA; OXIDATION BASE 22; P-PHENYLENEDIAMINE,2-NITRO-; URSOL BROWN RR; ZOBA BROWN RR

Danger profile: Carcinogen, suspected birth defects, toxic fumes when burned.
Effects of exposure:
Swallowed: Moderately toxic.
Skin: Passes through the unbroken skin; can increase exposure and the severity of the symptoms listed.
Long-term exposure: May cause cancer.
Storage: Store in tightly closed containers in a dry, cool, well-ventilated place.

★54★
AMINOPHENOL
CAS: 27598-85-2
Other names: M-AMINOPHENOL (591-27-5); 3-AMINOPHENOL (591-27-5); O-AMINOPHENOL (95-55-6); 2-AMINOPHENOL (95-55-6); P-AMINOPHENOL (123-30-8); 4-AMINOPHENOL (123-30-8)

Danger profile: Suspected mutagen, possible reproductive hazard, poisonous gases produced in fire.
Uses: Dyes for hair, fur and feathers.
Appearance: White to reddish-yellow crystals that turn pink or violet when exposed to air.
Effects of exposure:
Breathed: Aminophenols may lower the ability of the blood to carry oxygen (methemoglobinemia); results in a bluish color of the skin and lips, headaches, dizziness; with higher exposures, collapse and death.
Eyes: Contact may cause irritation. Chemical passes through the skin.
Skin: Contact may cause irritation, burning sensation and rash.
Long-term exposure: May cause genetic changes (mutations); asthma-like allergy. There is evidence that chemical causes birth defects in animals by damaging the fetus.
Storage: Store in tightly closed containers in a dry, cool, well-ventilated place. Keep away from strong oxidizers, heat, and sources of ignition.
First aid guide: Move victim to fresh air and call emergency medical care; if not breathing, give artificial respiration; if breathing is difficult, give oxygen. In case of contact with material, im-

mediately flush skin or eyes with running water for at least 15 minutes. Speed in removing material from skin is of extreme importance. Remove and isolate contaminated clothing and shoes at the site. Keep victim quiet and maintain normal body temperature. Effects may be delayed; keep victim under observation.

★55★
2-AMINOPYRIDINE
CAS: 504-29-0
Other names: O-AMINOPYRIDINE; ALPHA-AMINOPYRIDINE; AMINO-2 PYRIDINE; 1,2-DIHYDRO-2-IMINO-PYRIDINE; ALPHA-PYRIDINAMINE; ALPHA-PYRIDYLAMINE; 2-PYRIDYLAMINE; PYRIDINE,2-AMINO; 2-PYRIDINAMINE

Danger profile: Poisonous, highly poisonous gases produced in fire.
Uses: Making drugs, especially antihistamines, dyes, lubricants and herbicides.
Appearance: Colorless or white powder, leaflets or crystals.
Odor: Unique, characteristic of pyridine. The odor is a warning of exposure; however, no smell does not mean you are not being exposed.
Effects of exposure:
Breathed: Poisonous. Chemical has poor warning properties and exposure could take place before odor is detected. Dust or vapors may cause intoxication, headache, dizziness, nausea and weakness. It may also cause flushing of the arms and legs, convulsions, coma and death.
Skin: Chemical may enter the body through the unbroken skin. Contact may cause irritation, weakness, intoxication, dizziness, severe headache, respiratory distress, elevated blood pressure, convulsions and a stuporous state that can last for several days. Caused death of a worker who continued to work in contaminated clothing.
Storage: Store in tightly closed containers in a cool, well-ventilated area. Keep away from oxidizers, heat and sources of ignition. Do not eat or smoke in areas where substance is handled, processed or stored.

★56★
AMITROL
CAS: 61-82-5
Other names: AMEROL; AMINOTRIAZOLE; 2-AMINOTRIAZOLE; 3-AMINOTRIAZOLE; 3-AMINO-S-TRIAZOLE; 3-AMINO-1,2,4-TRIAZOLE; 2-AMINO-1,3,4-TRIAZOLE; 3-AMINO-1H-1,2,4-TRIAZOLE; AMINO TRIAZOLE WEEDKILLER 90; AMITOL; AMITRIL; AMITROLE; AMITROL 90; AMITROL-T; AMIZOL; AT; ATA; AT LIQUID; AZAPLANT; AZOLAN; AZOLE; CYTROL; DIUROL; DOMATOL; ELMASIL; EMISOL; ENT 25445; FENAMINE; FENAVAR; HERBICIDE TOTAL; HERBIZOLE; KLEER-LOT; ORGA-414; RADOXONE TL; RAMIZOLE; TRIAZOLAMINE; USAF XR-22; VOROX; WEEDAR ADS; WEEDEX GRANULAT; WEEDAZIN; WEEDAZOL; WEEDOCLOR; X-ALL; X-ALL LIQUID

Danger profile: Carcinogen, mutagen, poisonous, suspected reproductive hazard, exposure limit established, toxic fumes when burned.
Uses: Herbicide; defoliant; photographic chemicals; plant growth regulator.
Appearance: Crystalline solid.
Effects of exposure: No short term symptoms or health effects can be found at this time.
Long-term exposure: May cause cancer, enlarged thyroid gland (goiter) and/or underactive thyroid; liver damage.
Storage: Store in tightly closed containers in a dry, cool, well-ventilated place.

★ 57 ★

AMMONIA

CAS: 7664-41-7
Other names: ANHYDROUS AMMONIA; AM-FOL; AMMONIA ANHYDROUS; AMMONIAC (FRENCH); AMMONIACA (ITALIAN); AMMONIA GAS; AMMONIALE (GERMAN); AMONIAK (POLISH); AMMONIUM HYDROXIDE; AQUA AMMONIA; NITRO-SIL; R717; SPIRIT OF HARTSHORN

Danger profile: Poisonous, suspected mutagen, corrosive; toxic fumes when exposed to heat, explosion hazard when exposed to flame, EPA extremely hazardous list; on Community Right-to-Know List.
Uses: Among the five highest-volume chemicals produced in the U.S. Fertilizer; making fertilizers and many other chemicals. In synthetic fibers, dyestuffs, explosives, rocket fuel, nitrocellouse, cosmetics, permanent wave and hair bleaches, metallic hair dyes, hair straighteners, melamine and other plastics; as a solvent in the manufacture of textiles, leather, pulp and paper processing; blueprinting and film developing; electroplating; laboratory chemical; refrigerant in food installations, making of ice, cold storage, deicing; dyeing; making rubber; flame retardants. Found in some cleaning solutions and household cleaners, feminine douches, furniture and floor polishes, mouth washes, oven cleaners.
Appearance: At normal temperatures and pressures, a colorless gas. Chemical may be liquified by refrigeration or by high pressure (anhydrous ammonia). Also a solution in water.
Odor: Sharp, intensely irritating, pungent, suffocating. The odor is a warning of exposure; however, no smell does not mean you are not being exposed.
Effects of exposure:
  Breathed: May cause lung and eye irritation; coughing, shortness of breath; headache and loss of the sense of smell; permanent injury may result if prompt remedial measures are not taken. Large doses may result in long term, chronic, eye, nose, mouth, throat, or respiratory problems, and can cause immediate death from spasm, inflammation, or build up of fluids in the lungs (pulmonary edema) or larynx. This can cause death. Pulmonary edema is a medical emergency that can be delayed one to two days following exposure.
  Eyes: Severe irritant; vapors may cause severe burns. Contact may cause temporary blindness or result in partial or complete blindness.
  Skin: Severe irritant. Contact of liquid solution may cause burns and blisters that may be slow to heal. Anhydrous ammonia can cause frostbite, skin freezing and produce a caustic burn. Also see *Breathed.*
  Swallowed: Pain, burning of the throat and stomach, vomiting. A small dose of a strong solution may cause death.
Long-term exposure: Repeated exposure to gas may cause chronic irritation of the eyes and upper respiratory tract.
Storage: Store in a cool, dry, well-ventilated place. Keep away from heat, direct sunlight, chlorine, chlorates including hypochlorates, chlorine bleach, scouring powders, bromine, iodine, pool chemicals, mercury, gold and silver. Corrosive to copper, tin, zinc, galvanized surfaces and many other metals and alloys. Containers may explode in fire.
First aid guide: Move victim to fresh air and call emergency medical care; if not breathing, give artificial respiration; if breathing is difficult, give oxygen. In case of contact with material, immediately flush skin or eyes with running water for at least 15 minutes. Remove and isolate contaminated clothing and shoes at the site. Keep victim quiet and maintain normal body temperature. Effects may delayed; keep victim under observation.

★ 58 ★

AMMONIUM ACETATE

CAS: 631-61-8
Other names: ACETIC ACID, AMMONIUM SALT

Danger profile: Poisonous, poisonous gases when burned.
Uses: As an antidote in formaldehyde poisoning; laboratory chemical; textile dyeing; food preservative; making drugs, foam rubbers, vinyl plastics and explosives.
Appearance: Colorless to white crystalline solid.
Odor: Weak ammonia. The odor is a warning of exposure; however, no smell does not mean you are not being exposed.
Effects of exposure:
  Breathed: May cause irritation of the nose, mouth and throat. Higher levels of exposure can irritate the lungs causing cough, difficult breathing, tightness of the chest, shortness of breath and a build-up of fluid in the lungs (pulmonary edema) and death. Pulmonary edema is a medical emergency that may be delayed one to two days following exposure.
  Eyes: Contact with dust may cause irritation and eye damage.
  Skin: Contact may cause skin irritation.
  Swallowed: May cause irritation of mouth and stomach.
Long-term exposure: May effect the lungs.
Storage: Store in a tightly closed container in a cool, well-ventilated place. Keep away from sodium hypochlorite, potassium chlorate and sodium nitrate.

★ 59 ★

AMMONIUM BICARBONATE

CAS: 1066-33-7
Other names: AMMONIUM BICARBONATE (1:1); ACID AMMONIUM CARBONATE; AMMONIUM CARBONATE; AMMONIUM HYDROGEN CARBONATE; CARBONIC ACID, MONOAMMONIUM SALT; MONOAMMONIUM CARBONATE

Danger profile: Poisonous, toxic fumes when burned.
Uses: Making ammonium salts; dyes; baking powders, leavening agent for cookies, crackers, cream puff dough; cold permanent wave lotions; fire-extinguisher; pharmaceuticals; expectorant and to break up intestinal gas; degreasing textiles; blowing agent for foam rubber; boiler scale removal; compost treatment to accelerate decomposition.
Appearance: White powder or crystals.
Odor: Weak ammonia. The odor is a warning of exposure; however, no smell does not mean you are not being exposed.
Effects of exposure:
  Breathed: May cause irritation of the eyes, nose, throat, breathing passages and lungs causing difficult breathing and cough.
  Eyes: Contact may cause irritation.
  Skin: Contact may cause irritation.
  Swallowed: May cause irritation of the mouth, chest and stomach, with cramps, nausea and vomiting.
Long-term exposure: May cause lung damage.
Storage: Store in tightly closed containers in a cool, well-ventilated place. Keep away from caustics, heat and sources of ignition. May attack copper, nickel, zinc.

★ 60 ★

AMMONIUM BIFLUORIDE

CAS: 1341-49-7
Other names: AMMONIUM HYDROGEN FLUORIDE; ACID AMMONIUM FLUORIDE; AMMONIUM HYDROGEN BIFLUORIDE; AMMONIUM HYDROGEN DIFLUORIDE; AMMONIUM HYDROFLUORIDE

Danger profile: Poisonous gases produced in fire.

Uses: Ceramics; laboratory chemical; etching glass; sterilizer for brewery, dairy and other equipment; in electroplating and glass industries; making and processing other chemicals.
Appearance: White crystals, commonly found in liquid solution.
Effects of exposure:
   Breathed: May cause severe irritation of nose, throat and respiratory system; sore and nosebleeds. High levels may cause coughing, wheezing, chest tightness; serious irritation of the lungs with burns; shortness of breath and a dangerous build-up of fluids in the lungs (pulmonary edema), which can cause death. Pulmonary edema is a medical emergency that can be delayed one to two days following exposure.
   Eyes: Contact may cause severe irritation, burning pain, redness, tearing, blurred vision, burns; permanent damage and blindness.
   Skin: Contact may cause severe irritation, burns and skin rash. Burns may be deep and slow to heal. High concentrations of fluorine have been reported in urine following skin irritation.
   Swallowed: May cause irritation of the mouth and stomach, vomiting, abdominal pain, convulsions, collapse and acute toxic nephritis.
Long-term exposure: May cause liver and kidney damage; irritation of the lungs, bronchitis with shortness of breath, coughing and phlegm. Prolonged or repeated exposure may cause chronic irritation of the eyes, redness, tearing; nose and throat discomfort; chronic irritation of the skin with redness and rash; weight loss, fatigue, brittle bones and stiff joints.
Storage: Store in tightly closed, plastic, rubber or paraffined containers in a dry, cool, well-ventilated place. Keep away from strong oxidizers. Corrodes glass, cement, most metals
First aid guide: Move victim to fresh air; call emergency medical care. In case of contact with material, immediately flush skin or eyes with running water for at least 15 minutes. Remove and isolate contaminated clothing and shoes at the site. Keep victim quiet and maintain normal body temperature.

★ 61 ★
AMMONIUM CARBAMATE
CAS: 1111-78-0
Other names: AMMONIUM AMINOFORMATE; CARBAMIC ACID, MONOAMMONIUM SALT; CARBAMIC ACID, AMMONIUM SALT

Danger profile: Possible carcinogen, poisonous gases produced in fire, harmful to aquatic life in very low concentrations.
Uses: Fertilizers; in fire extinguishes, permanent wave solutions and creams; in baking powders; processing woolens; medicine (an expectorant).
Appearance: White crystals or powder.
Odor: Ammonia-like. The odor is a warning of exposure; however, no smell does not mean you are not being exposed.
Effects of exposure:
   Breathed: May cause irritation of the nose, throat, breathing passages and lungs; cough.
   Eyes: May cause irritation, redness and tearing.
   Skin: May cause irritation.
Long-term exposure: Carbamates are suspected carcinogens of the lungs and bone marrow. May effect lung function.
Storage: Store in tightly closed containers in a cool, well-ventilated place. Keep away from heat and sources of ignition.
First aid guide: In case of contact with material, immediately flush eyes with running water for at least 15 minutes. Wash skin with soap and water. Remove and isolate contaminated clothing and shoes at the site.

★ 62 ★
AMMONIUM CARBONATE
CAS: 506-87-6
Other names: AMMONIUMCARBONAT (GERMAN); CARBONIC ACID, AMMONIUM SALT; CARBONIC ACID, DIAMMONIUM SALT; DIAMMONIUM CARBONATE; CRYSTAL AMMONIA; HARTSHORN

Danger profile: Poisonous, toxic fumes when burned.
Uses: Medicine (an expectorant); baking powders; smelling salts; fire extinguishings; pharmaceuticals; to fix dyes in textiles; making wine and other chemicals; ceramics; processing wool.
Appearance: Colorless plates or crystals.
Odor: Pungent, suffocating. The odor is a warning of exposure; however, no smell does not mean you are not being exposed.
Effects of exposure:
   Breathed: May cause irritation of nose, eyes and throat; coughing; difficult breathing.
   Skin: Contact with wet skin may cause irritation, redness and swelling; burning feeling.
   Eyes: Contact may cause irritation, pain, tearing; serious injury may result.
   Swallowed: May cause burning feeling in the mouth and stomach; mouth and gastric irritation; stomach cramps.
Storage: Decomposes on exposure to air with loss of ammonia and carbon dioxide. Store in tightly closed containers in a dry, cool, well-ventilated place.
First aid guide: In case of contact with material, immediately flush eyes with running water for at least 15 minutes. Wash skin with soap and water. Remove and isolate contaminated clothing and shoes at the site.

★ 63 ★
AMMONIUM CHLORIDE
CAS: 12125-02-9
Other names: AMCHLOR; AMMONERIC; AMMONIUMCHLORID (GERMAN); AMMONIUM MURIATE; CHLORID AMONNY (CZECH); DARAMMON; SAL AMMONIA; SAL AMMONIAC; SALAMMONITE; SALMIAC

Danger profile: Poisonous, extremely reactive, toxic fumes when burned, exposure limit established.
Uses: To keep snow from melting on ski slopes. Dry batteries; in permanent wave solutions; fixing dyes and inks; eye lotions; soldering flux; making other chemicals; explosives; fertilizer; pickling agent in zinc coating and tinning; electroplating; as a skin wash; washing powders; resins; bakery products; in medicine (urinary acidifier and diuretic).
Appearance: White crystals.
Effects of exposure:
   Breathed: My cause irritation of nose and throat.
   Skin: Contact may cause irritation.
   Eyes: Contact may cause severe irritation.
   Swallowed: May cause gastric irritation.
Storage: Store in tightly closed containers in a dry, cool, well-ventilated place. Protect containers from physical damage. Keep away from acids, alkalies, silver salts and all other chemicals.

★64★
AMMONIUM CHLOROPLATINATE
CAS: 16919-58-7
Other names: AMMONIUM HEXACHLOROPLATINATE (IV); AMMONIUM PLATINIC CHLORIDE; DIAMMONIUM HEXA-CHLOROPLATINATE (2-); 1-HEXADECANAMINIUM N-ETHYL-N,N-DIMETHYL-, BROMIDE (9CI); PLATINATE (2-), HEXACHLORO-, DIAMMONIUM; PLATINATE (2-), HEXA-CHLORO-, DIAMMONIUM, (OC-6-11)-(9CI); PLATINIC AM-MONIUM CHLORIDE; QUATERNIUM-17; PLATINATE (2-1), HEXACHLORO-, DIAMMONIUM

Danger profile: Possible mutagen, poisonous, explosive, exposure limit established, very toxic fumes when burned.
Uses: Platinum plating; making platinum sponge.
Appearance: Cubic yellow crystals, red-orange crystals or yellow powder.
Effects of exposure:
Breathed: Poisonous. May cause irritation of the nose, throat, eyes, breathing passages and lungs; ulcers of the nose.
Eyes: May cause irritation.
Skin: May cause irritation.
Swallowed: Poisonous.
Long-term exposure: May cause genetic changes (mutations) that can lead to birth defects, miscarriages or cancer; asthma-like allergy. Repeated or prolonged exposure may cause sores in the nose lining; scar tissue to form on the lungs; allergic skin rash, sometimes with rash.
Storage: Explosively unstable. Store in tightly closed containers in a cool, well-ventilated place. Keep away from heat and sources of ignition, potassium hydroxide.

★65★
AMMONIUM CHROMATE
CAS: 7788-98-9
Other names: AMMONIUM CHROMIUM OXIDE; CHROMIC ACID, DIAMMONIUM SALT; AMMONIUM CHROMATE(IV); DIAMMONIUM CHROMATE; NEUTRAL AMMONIUM CHRO-MATE

Danger profile: Carcinogen, poisonous, explosion hazard, on Community Right-to-Know List, toxic fumes when heated.
Uses: Textile dyeing; photography; laboratory chemical; corrosion inhibitor.
Appearance: Yellow needles or crystals.
Effects of exposure:
Breathed: Burning sensation, coughing, wheezing, laryngitis, shortness of breath, headache, nausea, vomiting.
Long-term exposure: May cause cancer. May cause skin allergy; bronchitis with cough and phlegm; lung allergy with coughing and shortness of breath; kidney damage; a hole in the "bone" dividing the inner nose.
Storage: Store in tightly closed containers in a cool, well-ventilated place with non-wood floors. Keep away from heat, sources of ignition, combustible materials, reducing agents. Keep away from any condition that may cause shock. Containers may explode when shocked or heated.

★66★
AMMONIUM CITRATE
CAS: 3012-65-5
Other names: AMMONIUM CITRATE, DIBASIC; CITRIC ACID, AMMONIUM SALT; CITRIC ACID, DIAMMONIUM SALT; DIAMMONIUM CITRATE; 1,2,3-PROPANE TRICAR-BOXYLIC ACID, 2-HYDROXY-, AMMONIUM SALT

Danger profile: Acrid smoke and poisonous gases when burned.
Uses: Drugs; cosmetic preservative and astringent, freckle and nail bleaches, cleansing creams, eye lotions, hair colorings and rinses and waving lotions, bath products; rust proofing; textile printing; laboratory chemical; foam inhibitor; making plastics, paints, varnishes.
Appearance: White powder or crystals.
Odor: Weak ammonia. The odor is a warning of exposure; however, no smell does not mean you are not being exposed.
Effects of exposure:
Breathed: May cause nose, throat and respiratory irritation.
Eyes: May cause irritation.
Skin: May cause irritation.
Swallowed: May cause diarrhea.
Storage: Store in tightly closed containers in a cool, well-ventilated place. Keep away from sodium hydroxide, potassium hydroxide and other caustics.

★67★
AMMONIUM DICHROMATE
CAS: 7789-09-5
Other names: AMMONIO (BICROMATO DI) (ITALIAN); AMMONIO (DICROMATO DI) (ITALIAN); AMMONIUMBICHRO-MAAT (DUTCH); AMMONIUMDICHROMAT (GERMAN); AMMONIUM BICHROMATE; AMMONIUMDICHROMAAT (DUTCH); AMMONIUM (DICHROMATE D') (FRENCH); AMMONIUM DICHROMATE(VI); BICHROMATE D'AMMONIUM (FRENCH); DICHROMIC ACID, DIAMMONIUM SALT; CHRO-MIC ACID, DIAMMONIUM SALT

Danger profile: Carcinogen, poisonous, oxidizer, poisonous gas produced in fire, on Community Right-to-Know List, exposure limit established.
Uses: Fixing dyes on textiles; pigments; manufacturing of other chemicals, perfumes; oil purification; leather tanning; photography; process engraving and lithography; explosives, fireworks.
Appearance: Bright red-orange needles or crystals.
Effects of exposure:
Breathed: Inhalation may cause irritation or ulceration of mucous membranes of nose, throat, respiratory tract. Prolonged inhalation of dust may cause respiratory irritation, produce symptoms resembling asthma and may cause a hole in the bone dividing the inner nose (perforation of nasal septum). Early signs of perforation of the nasal septum include bleeding and formation of a crust in the inner nose.
Eyes: Contact can cause irritation, conjunctivitis, severe damage including loss of vision.
Skin: Contact may cause irritation, skin burns, rash or external ulcers. If contact persists, the chemical may enter the body through the skin wounds.
Swallowed: Irritates mucous membrane and causes vomiting.
Long-term exposure: May include skin allergy, bronchitis with cough and phlegm, lung allergy with coughing and shortness of breath, and kidney damage.
Storage: Store in a tightly closed container in a cool, well-ventilated place with non-wood floors. Keep away from heat, sources of ignition, combustible materials such as wood, paper, oil, fuels, etc. Containers may explode in fire.
First aid guide: Move victim to fresh air; call emergency medical care. In case of contact with material, immediately flush skin or eyes with running water for at least 15 minutes. Remove and isolate contaminated clothing and shoes at the site.

★68★
AMMONIUM FLUORIDE
CAS: 12125-01-8
Other names: AMMONIUM FLUORURE (FRENCH); NEUTRAL AMMONIUM FLUORIDE

Danger profile: Poisonous, exposure limit established, poisonous gases produced in fire.
Uses: Laboratory chemical; in brewing; glass etching; printing and dyeing textiles; wood preserving; moth proofing.
Appearance: Colorless or white crystals.
Effects of exposure:
Breathed: Dust may cause irritation of nose, throat and respiratory system. High exposures could cause a build-up of fluids in the lungs (pulmonary edema), which can cause death. Pulmonary edema is a medical emergency; the symptoms may be delayed for one to two days following exposure.
Eyes: Contact may cause irritation of mucous membrane, and chemical burns.
Skin: Contact may cause chemical burns. Concentrations of fluorine in urine have been reported following skin contact.
Swallowed: May be fatal if relatively small quantities are swallowed. Ingestion produces nausea, salivation, vomiting, abdominal pain, diarrhea, hemorrhage, vascular collapse. Increased respiration is followed by depression, and possibly death.
Long-term exposure: The following effects may not occur with prescribed levels in drinking water or cavity prevention dental use. Repeated exposure may cause loss of appetite, nausea, constipation or diarrhea; brittle bones, stiff joints and muscles, and eventual crippling; scarring of the lungs, shortness of breath and reduced lung functions.
Storage: Store in a tightly closed container in a cool, well-ventilated place. May corrode glass, cement, and most metals. Keep away from acids and alkalis.
First aid guide: Move victim to fresh air; call emergency medical care. In case of contact with material, immediately flush eyes with running water for at least 15 minutes. Wash skin with soap and water. Remove and isolate contaminated clothing and shoes at the site.

★69★
AMMONIUM HEXAFLUOROSILICATE
CAS: 1309-32-6
Other names: AMMONIA FLUOSILICATE; AMMONIUM FLUOSILICATE; AMMONIUM SILICOFLUORIDE; CRYPTOHALITE; DIAMMONIUM HEXAFLUOROSILICATE; FLUOSILICATE DE AMMONIUM (FRENCH)

Danger profile: Poisonous, very poisonous fumes produced in fire or on contact with acid.
Uses: Pesticide; soldering flux; etching of glass.
Appearance: White powder or granules.
Effects of exposure:
Breathed: May cause difficult breathing, burning of the mouth, nose and throat. May cause nose bleeds. At high levels may cause nausea, vomiting, heavy sweating and extreme thirst.
Eyes: May cause severe irritation.
Skin: May cause itching, chemical burns and rash or sores.
Swallowed: Poisonous. May cause nausea, diarrhea and vomiting. Higher doses of one gram or more may cause stomachache, burning sensation, sores of the mouth, throat and digestive tract, tremors, convulsions, shock and death.
Long-term exposure: May result in an increase in bone density, stimulation of new bone growth and calcium deposits in the ligaments. Teeth damage may occur.
Storage: Store in a cool, dry well-ventilated place. Keep away from acids, source of heat and ignition.

★70★
AMMONIUM HYDROXIDE
CAS: 1336-21-6
Other names: AMMONIA AQUEOUS; AMMONIA SOLUTION; AQUA AMMONIA, AQUEOUS AMMONIA; AMMONIA WATER

Danger profile: Suspected mutagen, corrosive, poisonous, toxic gases when burned.
Uses: Textiles; making rayon, rubber, metallic hair dyes and straighteners, fertilizers, inks; refrigeration; photography; pharmaceuticals; soaps; protective skin creams, lubricants; fire proofing wood; explosives; ceramics; other chemicals; detergents food additive; household cleanser.
Appearance: Clear, colorless liquid.
Odor: Penetrating, pungent, suffocating. The odor is a warning of exposure; however, no smell does not mean you are not being exposed.
Effects of exposure:
Breathed: May cause irritation of the eyes, nose, throat and lungs; burning sensation; coughing, wheezing and shortness of breath. Swelling around the voice box may cause suffocation and death. Exposure could cause shortness of breath and a dangerous build-up of fluids in the lungs (pulmonary edema), which can cause death. Pulmonary edema is a medical emergency that can be delayed one to two days following exposure.
Eyes: Contact may cause severe irritation, burns and permanent injury.
Skin: Contact of liquid or vapor may cause irritation or burns.
Swallowed: May cause burning pain in the mouth, throat, stomach and thorax; pain when swallowing, constriction of throat, salivation, coughing. This is soon followed by vomiting of blood or coffee-ground looking material; stomach cramps; passage of loose stools, possibly containing blood. Breathing difficulty, convulsions, shock may result. Ingestion of 3-4 ml may be fatal.
Long-term exposure: May cause chronic bronchitis with cough, phlegm and/or shortness of breath. Repeated or prolonged skin contact may cause dryness, itching and redness.
Storage: Store in a tightly sealed container in a cool (below 77°F), well-ventilated place. Keep away from acrolein, acrylic acid, chlorosulfonic acid, dimethyl sulfate, fluorine, hydrofluoric acid, hydrochloric acid, nitric acid, sulfuric acid, beta-propiolactone, silver nitrate, silver oxide, ethyl alcohol, silver permanganate, nitromethane; halogens, copper, aluminum alloys, galvanized surfaces.
First aid guide: Move victim to fresh air; call emergency medical care. In case of contact with material, immediately flush skin or eyes with running water for at least 15 minutes. Remove and isolate contaminated clothing and shoes at the site. Keep victim quiet and maintain normal body temperature.

★71★
AMMONIUM METAVANADATE
CAS: 7803-55-6
Other names: AMMONIUM VANADATE; VANADIC ACID, AMMONIUM SALT; VANADATE (V031-), AMMONIUM

Danger profile: Poisonous, possible reproductive hazard, possible mutagen, toxic gases when burned.
Uses: Making dyes; varnishes; printing; in inks and paints; photography; laboratory chemical.
Appearance: White or slightly yellow crystals or powder.
Effects of exposure:
Breathed: Dust may be poisonous. May cause irritation of the eyes, nose and throat; coughing; wheezing and shortness of breath. High levels may result in lung irritation (pneumonitis) or accumulation of fluid in the lungs (pulmonary edema). If severe this could cause suffocation and death.

Pulmonary edema is a medical emergency that may be delayed for one to two days following exposure.

Eyes: Contact may cause irritation.

Skin: Contact may cause irritation or rash.

Swallowed: Poison.

Long-term exposure: Repeated or prolonged contact with the skin may cause irritation, itching, rash; skin allergy may develop. Exposure may cause asthma-like lung allergy, with wheezing and cough. Repeated overexposure may cause kidney or lung damage.

Storage: Store in tightly closed containers in a cool, well-ventilated place. Keep away from heat.

First aid guide: Move victim to fresh air; call emergency medical care. In case of contact with material, immediately flush skin or eyes with running water for at least 15 minutes. Remove and isolate contaminated clothing and shoes at the site.

★72★

AMMONIUM MOLYBDATE

CAS: 13106-76-8

Other names: AMMONIUM PARAMOLYBDATE; DIAMMONIUM MOLYBDATE; MOLYBDIC ACID, DIAMMONIUM SALT

Danger profile: Poisonous, exposure limit established, toxic fumes when burned.

Uses: Laboratory chemical; making ceramics and other chemicals; in pigments.

Appearance: White or greenish-yellow powder or crystals.

Effects of exposure:

Breathed: May cause irritation of eyes, nose and throat.

Eyes: Contact may cause irritation.

Skin: Passes through the unbroken skin; can increase exposure and the severity of the symptoms listed.

Swallowed: Poisonous.

Long-term exposure: May cause kidney and liver damage.

Storage: Store in a dry, cool, well-ventilated place. Keep away from water, potassium, magnesium, sodium, zinc and other chemically active metals.

★73★

AMMONIUM NITRATE

CAS: 6484-52-2

Other names: AMMONIUM (1) NITRATE; (1:1) AMMONIUM NITRATE; NITRAM; NITRIC ACID, AMMONIUM SALT; AMMONIUM SALT OF NITRIC ACID; NORWAY SALTPETER; VARIOFORM; VARIOFORM I

Danger profile: Powerful oxidizer, allergen, ignites on contact with many common materials, on Community Right-to-Know List, poisonous gases when heated and in fire.

Uses: Fertilizers; explosives, fireworks; matches; herbicides and insecticides; making other chemicals; in freezing mixtures; in solid rocket propellants; nutrient for antibiotics and yeast; in veterinary practice.

Appearance: Colorless, white, light gray or brown pellets or flakes.

Effects of exposure:

Breathed: May cause irritation of eyes, nose and throat and mucous membranes. Coughing and difficult breathing may result. Overexposure may cause nausea and vomiting; flushing of head and neck, headache, weakness, faintness, and collapse. With severe overexposure the ability of the blood to carry oxygen (methemoglobinemia) is effected. This results in a bluish color to the skin and lips (cyanosis), dizziness, possible collapse and death. See **Swallowed.**

Eyes: May cause irritation and burns. See **Breathed.**

Skin: May cause irritation and burns.

Swallowed: May cause frequent urination; acid in the urine and blood disorders. Large exposure may cause systemic acidosis and metheglobinemia (abnormal hemoglobin).

Long-term exposure: May cause allergic reaction.

Storage: Store in a tightly sealed container in a dry, cool, well-ventilated place with non-wood floors. Protect containers from physical damage. Keep away from all other materials, especially combustable materials, fuels, powdered metals, sodium, acetic acid, sawdust, oil, charcoal, combustible liquids, acetic acid, alkali metals, powdered aluminum, copper, iron, lead, nickel, zinc, brass, stainless steel. Container may explode in confinement; in heat or fire. When heated breaks down into highly poisonous nitrogen oxide gasses.

★74★

AMMONIUM OXALATE

CAS: 1113-38-8

Other names: ETHANEDIOIC ACID DIAMMONIUM SALT; OXALIC ACID, DIAMMONIUM SALT; DIAMMONIUM OXALATE; AMMONIUM OXALATE HYDRATE

Danger profile: Poisonous, poisonous fumes in fire.

Uses: Laboratory chemical; safety explosives; manufacture of other chemicals, including those for rust removal.

Appearance: Colorless crystals or prisms.

Effects of exposure:

Breathed: Poison. May cause irritation of the eyes, nose, throat, breathing passages and lungs; coughing and shortness of breath. Excessive inhalation of dust causes systemic poisoning; possible symptoms include pain in throat, esophagus, stomach; mucous membranes turn white; vomiting, severe purging, weak pulse, cardiovascular collapse, neuromuscular symptoms.

Eyes: Poison. Contact with eyes causes irritation and severe burns.

Skin: Poison. Contact with skin causes irritation, deeper sores (ulcer) or severe burns.

Swallowed: Poison. Dust may cause systemic poisoning; possible symptoms include pain in throat, esophagus, stomach; mucous membranes turn white; vomiting, severe purging, weak pulse, cardiovascular collapse, neuromuscular symptoms.

Long-term exposure: May cause bronchitis, with cough, phlegm and/or shortness of breath. Repeated contact with skin may cause chronic inflammation, cracking and slowly-healing sores. May lead to kidney stones and kidney damage.

Storage: Store in tightly closed containers in a cool, well-ventilated place.

First aid guide: Move victim to fresh air; call emergency medical care. In case of contact with material, immediately flush eyes with running water for at least 15 minutes. Wash skin with soap and water. Remove and isolate contaminated clothing and shoes at the site.

★75★

AMMONIUM PERCHLORATE

CAS: 7790-98-9

Other names: PERCHLORIC ACID, AMMONIUM SALT

Danger profile: Combustible, explosion hazard, poisonous gas produced in fire.

Uses: Explosives; pyrotechnics; analytical chemistry; etching and engraving agent; smokeless rocket and jet propellant.

Appearance: White, odorless solid.

Effects of exposure:

Breathed: May cause irritation of the nose and eyes; sneezing and nose ulcerations.

Eyes: May cause irritation, tearing, inflammation of the eyelids, and burns.
Skin: May cause slight irritation, inflammation and burns.
Swallowed: May cause stomach pains, nausea, vomiting, diarrhea, blue color the skin, lips and fingertips, rapid breathing, dizziness and mental confusion.
Storage: Powerful oxidizer. Store in tightly closed containers in a dry, cool, well-ventilated place. Keep away from heat, flame, sources of ignition, sugar, charcoal, carbon, aluminum and copper compounds, copper chromate, copper oxide, iron oxide, ethylene dinitrate, and combustible materials. Containers may explode in fire.

★ 76 ★
AMMONIUM PERMANGANATE
CAS: 13446-10-1
Other names: PERMANGANIC ACID, AMMONIUM SALT

Danger profile: Dangerous explosion hazard, highly reactive, on Community Right-to-Know List, poisonous gases produced in fire.
Appearance: Violet-brown or dark purple rhombs or crystalline solid.
Effects of exposure:
Breathed: Toxic. May cause irritation of eyes, nose, throat, lungs and mucous membranes. Coughing and difficult breathing may result.
Eyes: May cause irritation and burns. See *Breathed.*
Skin: May cause irritation and burns. See *Breathed.*
Swallowed: Toxic.
Long-term exposure: May effect the lungs.
Storage: Store in an isolated, cool, well-ventilated place. Keep away from heat and sources of ignition, friction, combustible materials, oxidizing materials, fuels. May explode if heated above 140°F. Containers may explode in heat of fire, contamination or from shock.
First aid guide: Move victim to fresh air; call emergency medical care. In case of contact with material, immediately flush skin or eyes with running water for at least 15 minutes. Remove and isolate contaminated clothing and shoes at the site.

★ 77 ★
AMMONIUM PERSULFATE
CAS: 7727-54-0
Other names: AMMONIUM PEROXYDISULFATE; PERSULFATE D'AMMONIUM (FRENCH); PEROXYDISULFANIC ACID, DIAMMONIUM SALT

Danger profile: Powerful oxidizer, poisonous, toxic fumes when burned.
Uses: Bleaching and cleaning agents; photography; making printed circuit boards and other chemicals and dyes; etching copper; electroplating; food preservative.
Appearance: Colorless to light yellow crystalline solid.
Odor: Unpleasant, slightly acrid. The odor is a warning of exposure; however, no smell does not mean you are not being exposed.
Effects of exposure:
Breathed: May cause irritation of the eyes, nose, throat and lungs; cough and difficult breathing. Exposure may trigger an allergy-like reaction; there may be hives on the skin with itching, hay fever-like eye tearing, sneezing and nasal congestion with runny nose, asthma-like wheezing and difficulty in breathing and possibly life threatening shock.
Eyes: Contact may cause irritation and burns.
Skin: Contact may cause irritation, burns and skin rash.
Swallowed: Moderately toxic. See *Breathed.*

Long-term exposure: May cause allergy-like reaction; dermatitis, urticaria, rhinitis, "baker's" asthma.
Storage: Store in a tighly sealed container in a cool, dry, well-ventilated place with non-wood floors. Keep away from combustible materials, fuels, water, sodium peroxide, powdered aluminum, zinc, iron, solutions of ammonia, sulfuric acid.
First aid guide: Move victim to fresh air; call emergency medical care. In case of contact with material, immediately flush skin or eyes with running water for at least 15 minutes. Remove and isolate contaminated clothing and shoes at the site.

★ 78 ★
AMMONIUM PHOSPHATE
CAS: 7783-28-0
Other names: AMMONIUM DIHYDROGEN PHOSPHATE; AMMONIUM BIPHOSPHATE; MONOAMMONIUM PHOSPHATE

Danger profile: Toxic fumes when burned, irritant.
Uses: Textile, wood and paper fire proofing; soldering flux; fertilizer; baking powder; food additives.
Appearance: Colorless or white crystals or powder.
Odor: Slight ammonia-like; salty taste. The odor is a warning of exposure; however, no smell does not mean you are not being exposed.
Effects of exposure:
Breathed: Fumes may cause nose, throat and lung irritation. High levels may result in accumulation of fluid in the lungs (pulmonary edema), suffocation and death. Pulmonary edema is a medical emergency that may be delayed for one to two days following exposure.
Eyes: Contact may cause irritation.
Skin: Contact with may cause irritation.
Swallowed: May cause sagging of facial muscles, tremors, anxiety, difficult in controlling muscles, stupor and coma. Large doses may cause calcium imbalance and increased urination.
Storage: Store in tightly sealed containers in a cool, well-ventilated place. Keep away from heat and alkalies. Heat may cause toxic fumes.

★ 79 ★
AMMONIUM PHOSPHATE DIBASIC
CAS: 7722-76-1
Other names: AMMONIUM HYDROGEN PHOSPHATE

Danger profile: Toxic fumes when burned.
Uses: Textile, wood and paper fire proofing; soldering flux; fertilizer; baking powder; food additives.
Appearance: Colorless or white crystals or powder.
Odor: Slight ammonia-like; salty taste. The odor is a warning of exposure; however, no smell does not mean you are not being exposed.
Effects of exposure:
Breathed: Fumes may cause nose, throat and lung irritation. High levels may result in accumulation of fluid in the lungs (pulmonary edema), suffocation and death. Pulmonary edema is a medical emergency that may be delayed for one to two days following exposure.
Eyes: Contact may cause irritation.
Skin: Contact may cause irritation.
Swallowed: May cause sagging of facial muscles, tremors, anxiety, difficult in controlling muscles, stupor and coma. Large doses may cause calcium imbalance and increased urination.
Storage: Store in tightly sealed containers in a cool, well-ventilated place. Keep away from heat and alkalies. Heat may cause toxic fumes.

★80★
AMMONIUM PICRATE
CAS: 131-74-8
Other names: AMMONIUM PICRATE, WET; AMMONIUM PICRATE, DRY; AMMONIUM PICRATE WETTED WITH LESS THAN 10% WATER; AMMONIUM PICRATE WETTED WITH MORE THAN 10% WATER; OBELINE PICRATE; PHENOL,2,4,6-TRINITRO-, AMMONIUM SALT (9CI); AMMONIUM PICRONITRATE; EXPLOSIVE D; PICTAROL; PICRIC ACID, AMMONIUM SALT

Danger profile: Explosive, powerful oxidizer, fire hazard, allergen, poisonous gases are produced in fire.
Uses: In fireworks, explosives; rocket propellent.
Appearance: Bright yellow crystalline solid.
Effects of exposure:
  Breathed: May cause irritation of the nose, throat, lungs and mucous membrane. High levels of exposure can cause kidney, liver and red cell damage; urine may become red, fatigue, drowsiness, coma and possible death.
  Eyes: May cause irritation and burns. See *Breathed.*
  Skin: May cause irritation and burns.
Long-term exposure: Repeated exposure may cause the eyes and skin to turn yellow; high exposure may cause liver, kidney and blood cell damage. May cause skin allergy with rash and itching.
Storage: Store in sealed containers in a dry, cool, remote and detached well-ventilated place. Keep away from all other materials. Keep away from heat, flame and sources of ignition, metals, oxidizers and reducing agents. Combustible and a dangerous explosive when heated or shocked. Containers may explode in fire.

★81★
AMMONIUM SULFAMATE
CAS: 7773-06-0
Other names: AMCIDE; AMICIDE; AMMAT; AMMATE; AMMATE HERBICIDE; AMMONIUM AMINOSULFONATE; AMMONIUM SULPHAMATE; AMMONIUM AMIDOSULPHATE; AMMONIUM SALZ DER AMIDOSULFONSAURE (GERMAN); AMS; IKURIN; MONOAMMONIUM SULFAMATE; MONOAMMONIUM SALT OF SULFAMIC ACID; SULFAMATE; SULFAMINSAURE (GERMAN); SULFAMIC ACID, MONOAMMONIUM SALT

Danger profile: Possible explosive, oxidizer, exposure limit established, toxic fumes when burned.
Uses: Weed and brush killer; making fire-retardant for textiles and paper products, and electroplating solutions.
Appearance: Colorless, white or bright yellow-orange crystalline solid.
Effects of exposure:
  Breathed: May cause irritation of the eyes, nose, throat and lungs. Exposure to very high levels may cause nausea and vomiting.
  Eyes: Contact may cause irritation and burns.
  Skin: Contact may cause irritation and burns.
  Swallowed: Moderately toxic. May cause cough, difficult breathing and gastrointestinal disturbances.
Storage: Store in tightly sealed containers in a dry, cool, well-ventilated place with non-wood floors. Keep away from strong oxidizers, acids, heat, hot water, potassium, potassium chlorate, sodium nitrite, metal chlorates and hot acid solutions.

★82★
AMMONIUM SULFIDE, SOLID
CAS: 12124-99-1
Other names: AMMONIUM HYDROSULFIDE; AMMONIUM SULFIDE

Danger profile: Corrosive, fire hazard, poisonous and combustible gases produced.
Uses: Synthetic flavorings; textile manufacturing; developer in photography; coloring metals.
Appearance: Yellow powder or crystals commonly found in solution.
Odor: Strong; rotten eggs. The odor is a warning of exposure; however, no smell does not mean you are not being exposed.
Effects of exposure:
  Breathed: May cause irritation of the nose, throat, lungs; causing sneezing, cough and difficult breathing; headache, dizziness, nausea and vomiting, cold sweat, diarrhea, fatigue, drowsiness. Very high levels may cause a build up in the lungs of fluid (pulmonary edema). Pulmonary edema is a medical emergency that can result in death. Symptoms of pulmonary edema may be delayed one to two days following exposure. Inhalation of 500 ppm for 30 minutes produces headache, dizziness and bronchial pneumonia; 600 ppm for 30 minutes can cause death.
  Eyes: May cause severe irritation, tearing, redness and eyelid swelling, burns and permanent damage.
  Skin: May cause severe irritation, inflammation, pain and color change of skin. May be absorbed through the skin and cause hydrogen sulfide poisoning. See symptoms under *Breathed.*
  Swallowed: May cause severe irritation of the stomach and mucous membranes.
Long-term exposure: May effect or damage the lungs.
Storage: Store in tightly closed containers in a dry, cool, well-ventilated place with non-wood floors. Keep away from acids and acid fumes, water and moisture, heat, and sources of ignition. Containers may explode in heat of fire. Vapors may travel to source of ignition and flash back.

★83★
AMMONIUM SULFIDE, SOLUTION
CAS: 12135-76-1
Other names: AMMONIUM MONOSULFIDE; AMMONIUM SULFIDE; TRUE AMMONIUM SULFIDE

Danger profile: Corrosive, fire hazard, poisonous and combustible gases produced.
Uses: Synthetic flavorings; textile manufacturing; developer in photography; coloring metals.
Appearance: Yellow powder or crystals commonly found in solution.
Odor: Strong; rotten eggs. The odor is a warning of exposure; however, no smell does not mean you are not being exposed.
Effects of exposure:
  Breathed: May cause irritation of the nose, throat, lungs; causing sneezing, cough and difficult breathing; headache, dizziness, nausea and vomiting, cold sweat, diarrhea, fatigue, drowsiness. Very high levels may cause a build up in the lungs of fluid (pulmonary edema). Pulmonary edema is a medical emergency that can result in death. Symptoms of pulmonary edema may be delayed one to two days following exposure. Inhalation of 500 ppm for 30 minutes produces headache, dizziness and bronchial pneumonia; 600 ppm for 30 minutes can cause death.
  Eyes: May cause severe irritation, tearing, redness and eyelid swelling, burns and permanent damage.
  Skin: May cause severe irritation, inflammation, pain and color change of skin. May be absorbed through the skin and

cause hydrogen sulfide poisoning. See symptoms under *Breathed.*
Swallowed: May cause severe irritation of the stomach and mucous membranes.
Long-term exposure: May effect or damage the lungs.
Storage: Store in tightly closed containers in a dry, cool, well-ventilated place with non-wood floors. Keep away from acids and acid fumes; water and moisture; heat and sources of ignition. Containers may explode in heat of fire. Vapors may travel to source of ignition and flash back.
First aid guide: Move victim to fresh air and call emergency medical care; if not breathing, give artificial respiration; if breathing is difficult, give oxygen. In case of contact with material, immediately flush skin or eyes with running water for at least 15 minutes. Remove and isolate contaminated clothing and shoes at the site. Keep victim quiet and maintain normal body temperature. Effects may be delayed; keep victim under observation.

★ 84 ★
AMMONIUM SULFITE
CAS: 10196-04-0
Other names: SULFUROUS ACID, DIAMMONIUM SALT; DIAMMONIUM SULFITE

Danger profile: Poisonous gas produced in fire.
Uses: Making drugs and other chemicals; permanent wave solutions; photography; metal lubricants.
Appearance: Colorless crystals.
Odor: Acrid, with a sulfurous taste. The odor is a warning of exposure; however, no smell does not mean you are not being exposed.
Effects of exposure:
Breathed: Dust may cause irritation of nose, throat, sinuses, bronchial tubes and lungs; sneezing, cough and/or difficult breathing.
Eyes: Dust may cause irritation and burns.
Skin: Contact may cause irritation and burns.
Swallowed: May cause irritation of mouth, stomach; nausea.
Long-term exposure: May cause an asthma-like allergy. Once allergy develops, future exposures—even small ones—could result in asthma attacks, coughing, difficult breathing with shortness of breath and tightness in the chest. Severe general reaction may occur and result in death.
Storage: Store in tightly closed containers in a cool, dry well-ventilated place. Keep away from acids.

★ 85 ★
AMMONIUM TARTRATE
CAS: 3164-29-2
Other names: AMMONIUM D-TARTRATE; BUTANEDIOIC ACID, 2,3-DIHYDROXY-, DIAMMONIUM SALT; DIAMMONIUM TARTRATE; 2,3-DIHYDROXY-BUTANEDIOIC ACID, DIAMMONIUM SALT (9CI); L-TARTARIC ACID, AMMONIUM SALT; TARTARIC ACID, DIAMMONIUM SALT

Danger profile: Poisonous, very toxic gases produced in fire.
Uses: Making textiles; drugs.
Appearance: Colorless white crystals or white granules.
Effects of exposure:
Breathed: Exposure to high levels may irritate the nose throat and lungs.
Eyes: Contact may cause irritation and burns.
Skin: Contact may cause irritation and burns.
Storage: Store in tightly closed containers in a cool, well-ventilated place. Keep away from potassium chlorate and sodium nitrite. Chemical can release ammonia when exposed to air.

★ 86 ★
AMMONIUM TETRACHLOROPLATINATE
CAS: 13820-41-2
Other names: TETRAMINE PLATINUM(II) CHLORIDE; PLATINATE(2-),TETRACHLORO-, DIAMMONIUM

Danger profile: Carcinogen, exposure limit established.
Uses: In photography.
Appearance: Ruby-red crystals.
Effects of exposure:
Breathed: May irritate the nose, throat and air passages; asthma with cough and shortness of breath; runny nose.
Eyes: Contact causes burning and irritation. Skin rash and hives may develop.
Skin: Contact may cause irritation.
Long-term exposure: May cause asthma-like allergy, lung scarring, irritability and seizures. Repeated exposure may cause sores or ulcers of the nose. Allergic skin rash and hives may develop.
Storage: Store in tightly closed containers in a dry, cool, well-ventilated place.

★ 87 ★
AMMONIUM THIOCYANATE
CAS: 1762-95-4
Other names: AMMONIUM RHODANATE; AMMONIUM RHODANIDE; AMMONIUM SULFOCYANATE; AMMONIUM SULFOCYANIDE; AMMONIUM RHODANTATE; AMTHIO; AMMONIUMTHIOCYANATE; RHODANID; RHODANIDE, AMMONIUM SALT; THIOCYANIC ACID, AMMONIUM SALT; TRANS-AID; USAF EK-P-433; WEEDAZOL TL

Danger profile: Poisonous, poisonous gases produced in fire.
Uses: Many uses in chemical manufacturing and photography. Fertilizers; ingredient of freezing solutions, liquid rocket propellants; fabric dyeing; weed killer and defoliant; adhesives; curing resins; electroplating; zinc coating; soil sterilizer.
Appearance: Colorless solid or crystals which can absorb water from the air and become liquid.
Effects of exposure:
Breathed: Dust may cause irritation of nose and throat.
Skin: Passes through the unbroken skin; can increase exposure and the severity of the symptoms listed. Contact may cause irritation. Prolonged contact may produce various skin eruptions, dizziness, cramps, nausea, mild to severe disturbance of nervous system.
Swallowed: Causes dizziness, cramps, nervous disturbances. Dust irritates eyes. Chemical has caused fatal poisoning with symptoms of vomiting, confusion, convulsions.
Long-term exposure: May cause abdominal problems, weight loss, weakness, skin rashes, fatigue, and runny nose. Prolonged exposure may effect the thyroid gland and the blood cells. May cause personality and mood change.
Storage: Store in tightly closed containers in a dry, cool, well-ventilated place. Keep away from water, moisture, acid, acid fumes, chlorine, potassium chlorate and lead nitrate. Chemical in water may be corrosive to metals.

★ 88 ★
N-AMYL ACETATE
CAS: 628-63-7
Other names: ACETATE D'AMYLE (FRENCH); ACETIC ACID, AMYL ESTER; AMYL ACETATE; AMYL ACETIC ESTER; AMYL ACETIC ETHER; AMYLAZETAT (GERMAN); BANANA OIL; BIRNENOEL; OCTAN AMYLU (POLISH); PEAR OIL; PENT-ACETATE; PENT-ACETATE 28; 1-PENTANOL ACETATE; PENTYL ACETATE; N-PENTYL ACETATE; 1-PENTYL ACETATE; PENTYL ESTER OF ACETIC ACID; PRIMARY AMYL ACETATE; ACETIC ACID, PENTYL ESTER

Danger profile: Combustible, exposure limit established, poisonous gases produced in fire.
Uses: Solvent for paints, lacquers, lacquer thinners, cements, nail enamel, nail polish removers; in making plastics, toys, eyeglass frames, combs and novelties; photographic film; furniture polish, perfuming leather and shoe polishes; warning odor; artificial fruit-flavoring agents for food and beverages; manufacture of textile finishing compounds; printing and finishing fabrics; manufacture of fluorescent lamps, quick-drying inks, metallic inks and transfer inks; production of antibiotics. Pre-spotting in dry cleaning industry. Stiffening agent in the manufacture of straw hats.
Appearance: Colorless liquid.
Odor: Persistent, banana-like, fruity. The odor is a warning of exposure; however, no smell does not mean you are not being exposed.
Effects of exposure:
  Breathed: Mildly toxic. May be poisonous. May cause irritation of eyes, nose and throat; headache, drowsiness, excitability, drunken behavior, mental confusion, fatigue. Very high levels of exposure may cause dizziness, lightheadedness and cause victim to pass out, go into coma and die.
  Eyes: Contact may cause irritation, pain, burning, stinging, tearing, inflammation of the eyelids. Light may cause pain.
  Skin: May remove oil from the skin and cause dryness. Persons with pre-existing skin disorders may be more susceptible to the effects of this substance.
  Swallowed: May be poisonous. May cause gastro-intestinal irritation. See Breathed.
Long-term exposure: Prolonged or repeated contact may cause irritation of the skin and dermatitis; nervous system changes. High exposure may injure the liver, cause reduced memory and concentration, fatigue, sleep disturbance, reduced coordination and effects on nerves supplying vital organs and/or nerves to the arms or legs, causing a feeling of "pins and needles."
Storage: Highly reactive. Store in a cool, well-ventilated place, preferably a detached shed or building. Keep away from direct sunlight, heat, any source of ignition, nitrates, strong oxidizers, alkalis and acids. Vapors are explosive in heat or flame. Containers may explode in fire.
First aid guide: Move victim to fresh air and call emergency medical care; if not breathing, give artificial respiration; if breathing is difficult, give oxygen. In case of contact with material, immediately flush eyes with running water for at least 15 minutes. Wash skin with soap and water. Remove and isolate contaminated clothing and shoes at the site.

★ 89 ★
SEC-AMYL ACETATE
CAS: 626-38-0
Other names: ACETIC ACID, 2-PENTYL ESTER; 2-ACETOXYPENTANE; 1-METHYLBUTYL ACETATE; 2-PENTANOL, ACETATE; 2-PENTYL ACETATE

Danger profile: Fire hazard, reactant, toxic, exposure limit established, acrid smoke and irritating fumes when burned.
Uses: Solvent; in cements, coated paper, lacquers, leather finishes, nail enamels, wood filler, textile sizing.
Appearance: Clear colorless liquid.
Odor: Mild, fruity. The odor is a warning of exposure; however, no smell does not mean you are not being exposed.
Effects of exposure:
  Breathed: Mildly toxic. May cause irritation of eyes, nose and throat; headache, drowsiness, excitability, drunken behavior, mental confusion, fatigue. Very high levels of exposure may cause dizziness, lightheadedness and cause victim to pass out, go into coma and die.
  Eyes: Contact may cause irritation, pain, burning, stinging, tearing, inflammation of the eyelids. Light may cause pain.
  Skin: May remove oil from the skin and cause dryness. Persons with pre-existing skin disorders may be more susceptible to the effects of this substance.
  Swallowed: May cause gastro-intestinal irritation. See Breathed.
Long-term exposure: Prolonged or repeated contact may cause irritation of the skin and dermatitis, and nervous system changes.
Storage: Store in a tightly closed container in a cool, well-ventilated place. Keep away from heat and sources of ignition, nitrates, strong oxidizers, strong alkalis and strong acids. Vapors are explosive in heat or flame. Vapors may travel to sources of ignition and flash back. Containers may explode in fire.

★ 90 ★
AMYL ALCOHOL
CAS: 71-41-0
Other names: ALCOOL AMYLIQUE (FRENCH); N-AMYL ALCOHOL; AMYL ALCOHOL, NORMAL; AMYLOL; N-BUTYLCARBINOL; N-AMYLALKOHOL (CZECH); PENTANOL; PENTANOL-1; N-PENTANOL; PENTAN-1-OL; PENTASOL; PRIMARY AMYL ALCOHOL; 1-PENTANOL

Danger profile: Mutagen, narcotic, fire hazard.
Uses: Solvent; in nail lacquers; making other chemicals.
Appearance: Clear, colorless liquid.
Odor: Mild, sweet, camphor-like. The odor is a warning of exposure; however, no smell does not mean you are not being exposed.
Effects of exposure:
  Breathed: Mildly toxic. Vapor may cause irritation of the eyes nose and throat, dizziness, vertigo, cough, diarrhea, double vision, deafness, delirium, confusion, headache, nausea, vomiting. Very high levels of exposure may cause fatal poisoning preceded by severe nervous symptoms; dizziness, lightheadedness and cause victim to pass out, go into coma and die.
  Eyes: Contact may cause irritation, pain, burning, stinging, tearing, inflammation of the eyelids. Light may cause pain.
  Skin: May remove oil from the skin and cause dryness. Persons with pre-existing skin disorders may be more susceptible to the effects of this substance.
  Swallowed: May cause gastro-intestinal irritation; double vision, deafness, delirium, and occasionally fatal poisoning, preceded by severe nervous symptoms, have been reported.
Long-term exposure: Prolonged or repeated contact may cause irritation of the skin and dermatitis, and nervous system changes.
Storage: Store in tightly closed containers in a dry, cool, well-ventilated place. Keep away from heat, sources of ignition and hydrogen trisulfide. Vapors may travel to sources of ignition and flash back. Containers may explode in fire.
First aid guide: Move victim to fresh air and call emergency medical care; if not breathing, give artificial respiration; if breath-

ing is difficult, give oxygen. In case of contact with material, immediately flush eyes with running water for at least 15 minutes. Wash skin with soap and water. Remove and isolate contaminated clothing and shoes at the site.

★91★
AMYL NITRATE
CAS: 1002-16-0
Other names: N-AMYL NITRATE; NITRATE D'AMYLE (FRENCH); NITRIC ACID, PENTYL ESTER

Danger profile: Combustible, oxidizer, toxic fumes when burned.
Uses: Additive in diesel fuels.
Appearance: Clear, colorless to light straw liquid.
Odor: Ether-like. The odor is a warning of exposure; however, no smell does not mean you are not being exposed.
Effects of exposure:
  Breathed: Moderately toxic. Exposure may effect the ability of the blood to carry oxygen. May cause dizziness, headache, rapid and difficult breathing, mental confusion, blue coloration (cyanosis) of the lips, nose, and ear lobes, drop in blood pressure, fast pulse rate. If the lack of oxygen becomes severe, a person may have drowsiness, nausea and vomiting. If lack of oxygen is very severe it may cause convulsions, unconsciousness and even death.
  Eyes: May cause irritation, redness and swelling of the eyelids, tearing and burning feeling; looking into light may cause pain; severe eye damage. Passes through the skin; exposure may be increased. See *Breathed.*
  Skin: Toxic when absorbed through the skin. May cause irritation, redness, rash, small blisters and burning feeling. See *Breathed.*
  Swallowed: May cause irritation of the mouth, and stomach; cramps and diarrhea. See *Breathed.*
Long-term exposure: Repeated exposure may cause a low blood count. After repeated exposure tolerance to chemical may develop. If exposure suddenly stops, chest pains and heart attack may follow.
Storage: Store in a dry, cool, well-ventilated place with a non-wood floor. Keep away from heat, flame and sources of ignition and combustable materials. May attack some plastics. Vapors may travel to sources of ignition and flash back. Containers may explode in fire.
First aid guide: Move victim to fresh air and call emergency medical care; if not breathing, give artificial respiration; if breathing is difficult, give oxygen. In case of contact with material, immediately flush eyes with running water for at least 15 minutes. Wash skin with soap and water. Remove and isolate contaminated clothing and shoes at the site.

★92★
AMYL NITRITE
CAS: 110-46-3
Other names: 1-NITROPENTANE; NITROUS ACID, PENTYL ESTER; PENTYL NITRITE; NITROUS ACID, 3-METHYL BUTYL ESTER; ISOPENTYL ALCOHOL NITRATE

Danger profile: Possible mutagen. Combustible, reactive, explosion hazard, oxidizer, poisonous gases produced in fire.
Uses: Making pharmaceuticals, perfumes and other chemicals.
Appearance: Yellowish liquid.
Effects of exposure:
  Breathed: Moderately toxic. Exposure may effect the ability of the blood to carry oxygen. May cause dizziness, headache, rapid and difficult breathing, mental confusion, blue coloration (cyanosis) of the lips, nose, and ear lobes, drop in blood pressure, fast pulse rate. If the lack of oxygen becomes

severe, a person may have drowsiness, nausea and vomiting. If lack of oxygen is very severe it may cause convulsions, unconsciousness and even death.
  Eyes: May cause irritation, redness, blurred vision and swelling of the eyelids, tearing and burning feeling; looking into light may cause pain; severe eye damage. Passes through the skin; exposure may be increased. See *Breathed.*
  Skin: May cause irritation, redness, rash, small blisters and burning feeling. See *Breathed.*
  Swallowed: Moderately toxic. May cause irritation of the mouth, and stomach; cramps and diarrhea. See *Breathed.*
Long-term exposure: Repeated exposure may cause a low blood count (anemia). May cause skin allergy, itching and rash. After repeated exposure a tolerance to the chemical may develop. If exposure suddenly stops, chest pains and heart attack may follow.
Storage: Store in a dry, cool, well-ventilated place with a non-wood floor. Keep away from air, light, heat, flame, and sources of ignition; oxidizers, reducing agents, alcohols, antipyrine, caustic alkaline materials, alkaline carbonates, potassium iodide, bromide and ferrous salts. Containers may explode in fire. Vapors explode when heated. Vapors may travel to sources of ignition and flash back. May attack some plastics.
First aid guide: Move victim to fresh air and call emergency medical care; if not breathing, give artificial respiration; if breathing is difficult, give oxygen. In case of contact with material, immediately flush eyes with running water for at least 15 minutes. Wash skin with soap and water. Remove and isolate contaminated clothing and shoes at the site.

★93★
N-AMYL NITRITE
CAS: 463-04-7
Other names: AMYL NITRITE; 1-NITROPENTANE, NITROUS ACID, PENTYL ESTER; NITRAMYL; PENTYL ALCOHOL, NITRITE; PENTYL NITRITE

Danger profile: Mutagen, combustible liquid, oxydizer, toxic fumes when burned.
Uses: Making other chemicals; perfumes.
Appearance: Colorless to light yellow liquid.
Odor: Pleasant, fragrant, fruity, ether-like. The odor is a warning of exposure; however, no smell does not mean you are not being exposed.
Effects of exposure:
  Breathed: May be poisonous. May cause flushing of face, pulsating headache, disturbing tachycardia, cyanosis (methemoglobinemia), weakness, confusion, restlessness, faintness, collapse.
  Eyes: Contact may cause irritation, flushing of skin, rapid pulse, headache, and fall in blood pressure.
  Skin: May be poisonous if absorbed. Contact may cause irritation, flushing of skin, rapid pulse, headache, and fall in blood pressure.
  Swallowed: May cause flushing of face, pulsating headache, disturbing tachycardia, cyanosis (methemoglobinemia), weakness, confusion, restlessness, faintness, collapse.
Storage: Store in tightly closed containers in a dry, cool, well-ventilated place. Keep away from metals when wet, strong oxidizers, water, light, heat (vapors explode when heated). Decomposes on exposure to air, light and water, giving off toxic oxides of nitrogen. Vapors may travel to sources of ignition and flash back. Containers may explode in fire.

★94★
AMYLTRICHLOROSILANE
CAS: 107-72-2
Other names: AMYL TRICHLOROSILANE; PENTYLTRI-CHLOROSILANE; SILANE,TRICHLOROPENTYL-; TRI-CHLOROPENTYLSILANE; SILANE, PENTYLTRICHLORO-

Danger profile: Corrosive, explosion hazard, reactive, toxic fumes when burned.
Uses: Making silicones.
Appearance: Colorless to yellowish liquid.
Odor: Sharp, irritating. The odor is a warning of exposure; however, no smell does not mean you are not being exposed.
Effects of exposure:
  Breathed: May cause irritation of mucous membranes and lungs, causing coughing and difficult breathing. Higher exposures could cause shortness of breath and a dangerous build-up of fluids in the lungs (pulmonary edema), which can cause death. Pulmonary edema is a medical emergency that can be delayed one to two days following exposure.
  Eyes: Corrosive irritant; contact may cause severe burns.
  Skin: Moderately toxic. Corrosive irritant; contact may cause severe burns.
  Swallowed: Moderately toxic. May cause severe burns of the mouth and stomach, painful cramps. See *Breathed.*
Long-term exposure: May effect the lungs.
Storage: Store in tightly closed containers in a dry, cool, well-ventilated place. Keep away from water, heat, sources of ignition. Corrodes metals; reacts vigorously to produce toxic hydrogen chloride. Vapors may travel to sources of ignition and flash back. Containers may explode in water and in fire.
First aid guide: Move victim to fresh air and call emergency medical care; if not breathing, give artificial respiration; if breathing is difficult, give oxygen. In case of contact with material, immediately flush skin or eyes with running water for at least 15 minutes. Remove and isolate contaminated clothing and shoes at the site. Keep victim quiet and maintain normal body temperature.

★95★
ANILINE
CAS: 62-53-3
Other names: AMINOBENZENE; AMINOPHEN; ANILIN (CZECH); ANILINA (ITALIAN, POLISH); ANILINE OIL; ANILINE OIL, LIQUID; ANYVIM; BENZENAMINE; BENZENE, AMINO; BENZIDAM; BLUE OIL; C.I. 76000; C.I. OXIDATION BASE 1; CYANOL; HUILE D'ANILINE (FRENCH); KRYSTALLIN; KYANOL; NCI-C03736; PHENYLAMINE; AMINOBENZENE

Danger profile: Carcinogen, poisonous, mutagen, toxic fumes when burned, EPA extremely hazardous substance list, exposure limit established, on Community Right-to-Know List.
Uses: Making dyes, hair dyes, inks, colored pencils, crayons, lithographic and other printing inks, perfumes, pharmaceuticals and medicinals, nylon fibers, resins, varnishes, explosives, shoe black; rubber processing; industrial solvent; photographic chemicals; for rigid polyurethanes and urethane foams; petroleum refining; herbicides, fungicides; artificial sweetening agents; corrosion inhibitors; dyeing furs.
Appearance: Colorless to brown oily liquid. Darkens on exposure to air and light.
Odor: Weak amine-like. The odor is a warning of exposure; however, no smell does not mean you are not being exposed.
Effects of exposure:
  Breathed: May cause headache, weakness, dizziness, confusion, disorientation, drowsiness and coma. Convulsions may occur. Blood effects result from a decreased ability to carry oxygen and may include moderate to severe headache,

nausea, vomiting, blue coloring of skin (cyanosis), decreased blood pressure and irregular heart beat. Poisonous; if treatment is not given promptly, death can occur.
  Eyes: May cause severe irritation and damage.
  Skin: Poison. Passes through the unbroken skin; can increase exposure and the severity of the symptoms listed. Even small amounts absorbed from clothes or shoes may cause toxic symptoms, and severe irritation. May cause weakness, headache, blue coloration of skin, fingertips, lips, nose and cheeks; vomiting and collapse. See *Breathed.*
  Swallowed: May cause many of the symptoms listed. Serious poisoning can result from a small fraction of a teaspoon, and a full teaspoon could lead to death.
Long-term exposure: Loss of appetite, dizziness, insomnia, tremors, malignant bladder growths, liver damage and jaundice. Anemia has been reported. Has been linked to bladder cancer according to a NIOSH study.
Storage: Store in a tightly closed container in a cool, dark, dry place. Keep away from heat, sources of ignition, strong oxidizers, ozone, acids and alkalis and all other chemicals. Containers may explode in fire.
First aid guide: Move victim to fresh air and call emergency medical care; if not breathing, give artificial respiration; if breathing is difficult, give oxygen. In case of contact with material, immediately flush skin or eyes with running water for at least 15 minutes. Speed in removing material from skin is of extreme importance. Remove and isolate contaminated clothing and shoes at the site. Keep victim quiet and maintain normal body temperature. Effects may be delayed; keep victim under observation.

★96★
O-ANISIDINE
CAS: 90-04-0
Other names: O-AMINOANISOLE; ORTHO-ORTHO-AMINOANISOLE; 2-ANSIDINE; O-METHOXYANILINE; AMINE (9CI), O-METHOXYPHENYLAMINE

Danger profile: Carcinogen, mutagen, poisonous, on Community Right-to-Know list, exposure limit established, toxic fumes in fire.
Uses: Making dyes and other chemicals; hair dyes; corrosion inhibitors for steel.
Appearance: Red or yellow oily liquid. Becomes brown on exposure to air.
Odor: Amine-like. The odor is a warning of exposure; however, no smell does not mean you are not being exposed.
Effects of exposure:
  Breathed: Exposure may effect the ability of the blood to carry oxygen. May cause dizziness, headache, rapid and difficult breathing, mental confusion, blue coloration (cyanosis) of the lips, nose, and ear lobes. If the lack of oxygen becomes severe, a person may have drowsiness, nausea and vomiting. If lack of oxygen is very severe it may cause convulsions, unconsciousness and even death.
  Eyes: May cause irritation, redness and swelling of the eyelids, tearing and burning feeling; looking into light may cause pain; severe eye damage. See *Breathed.*
  Skin: Passes through the unbroken skin; can increase exposure and the severity of the symptoms listed. Toxic when absorbed through the skin. May cause irritation, redness, rash, small blisters and burning feeling. See *Breathed.*
  Swallowed: Moderately toxic. May cause irritation of the mouth, and stomach; cramps and diarrhea. See *Breathed.*
Long-term exposure: May cause cancer, skin allergy, nerve damage and low blood count.
Storage: Store in a dry, cool, well-ventilated place. Keep away from strong oxidizers, heat. Liquid will attack some forms of plastics, rubber and coatings. Containers may violently explode in heat of fire.

★ 97 ★
P-ANISIDINE
CAS: 104-94-9
Other names: P-AMINOANISOLE; 4-AMINOANISOLE; 1-AMINO-4-METHOXYBENZENE; PARA-AMINOANISOLE; 4-ANISIDINE; P-METHOXYANILINE; 4-METHOXYANILINE; 4-METHOXYBENZENAMINE; 4-METHOXYBENZENEAMINE

Danger profile: Suspected carcinogen, on Community Right-to-Know List, exposure limit established, toxic fumes in fire.
Uses: Making dyes, including hair dyes; preparation of organic compounds; corrosion inhibitors.
Appearance: Yellow to brown crystalline solid.
Odor: Amine-like. The odor is a warning of exposure; however, no smell does not mean you are not being exposed.
Effects of exposure:
Breathed: Exposure may effect the ability of the blood to carry oxygen. May cause dizziness, headache, rapid and difficult breathing, mental confusion, blue coloration (cyanosis) of the lips, nose, and ear lobes. If the lack of oxygen becomes severe, a person may have drowsiness, nausea and vomiting. If lack of oxygen is very severe it may cause convulsions, unconsciousness and even death.
Eyes: May cause irritation, redness and swelling of the eyelids, tearing and burning feeling; looking into light may cause pain; severe eye damage. Passes through the skin; exposure may be increased. See *Breathed.*
Skin: Toxic when absorbed through the skin. May cause irritation, redness, rash, small blisters and burning feeling. See *Breathed.*
Swallowed: Moderately toxic. May cause irritation of the mouth, and stomach; cramps and diarrhea. See *Breathed.*
Long-term exposure: May cause cancer, skin allergy, nerve damage and low blood count.
Storage: Store in a dry, cool, well-ventilated place. Keep away from strong oxidizers, heat. Liquid will attack some forms of plastics, rubber and coatings. Containers may explode in heat of fire.

★ 98 ★
ANTHRACENE
CAS: 120-12-7
Other names: ANTHRACEN (GERMAN); ANTHRACIN; GREEN OIL; PARANAPHTHALENE; TETRA OLIVE N2G

Danger profile: Suspected carcinogen, mutagen, allergen, combustible, explosive, on Community Right-to-Know List, exposure limit established.
Uses: Dyes; calico printing; wood preservative; smoke screens; semiconductor research.
Appearance: Colorless crystals or white crystalline flakes with bluish or violet fluorescence.
Odor: Weak, aromatic. The odor is a warning of exposure; however, no smell does not mean you are not being exposed.
Effects of exposure:
Breathed: Contact or heated fumes may cause irritation of the nose, throat and breathing passages.
Eyes: Contact may cause irritation and burns.
Skin: Passes through the skin; exposure may be increased. May cause severe irritation, redness and swelling; skin allergy which is aggravated by sunlight may develop.
Swallowed: May cause gastrointestinal irritation.
Long-term exposure: May cause cell mutations and cancer. If allergy of skin or eyes develop the condition can be greatly aggravated by sunlight exposure. Repeated skin exposure may cause skin thinning, yellow-brown pigment changes, loss of skin pigment, skin warts, pimples and skin cancer. Breathing fumes may cause chronic bronchitis, cough and phlegm.

Storage: Store in tightly closed containers in a cool, well-ventilated place. Keep away from heat, flame and sources of ignition, oxidizing materials, fluorine, calcium hypochlorite, chromic acid.

★ 99 ★
ANTIMONY
CAS: 7440-36-0
Other names: ANTIMONY BLACK; ANTIMONY POWDER; ANTIMONY, REGULUS; ANTYMON (POLISH); C.I. 77050; STIBIUM

Danger profile: Possible carcinogen, poisonous, on Community Right-to-Know List, exposure limit established, poisonous gases produced in fire.
Uses: Hardening lead, especially for storage batteries; bearing metal; bullets; type metal; solder; semiconductors; fireworks. Used medicinally as an emetic and to combat worms.
Appearance: Silver-white metal; lustrous, scale-like crystals.
Effects of exposure:
Breathed: Dust may be poisonous. May cause irritation of the nose, throat and mouth; metallic taste, coughing, dizziness, headache, nausea, vomiting, diarrhea, stomach cramps, insomnia, anorexia, unable to smell properly. Higher levels of exposure may cause lung congestion, and the heart to beat irregularly or stop.
Eyes: May cause eye irritation, tearing, burning sensation. There is a risk of serious injury.
Skin: Passes through the unbroken skin; can increase exposure and the severity of the symptoms listed. Contact may cause irritation, itching and rash.
Swallowed: Poisonous. May cause irritation of the mouth and throat; burning pains in the stomach, nausea, vomiting and diarrhea.
Long-term exposure: Repeated exposure may cause headache, loss of appetite, dry throat and insomnia, heart and liver damage; prolonged or repeated contact may cause sores and ulcers.
Storage: Store in tightly closed containers in a cool, well-ventilated place. Keep away from strong oxidizers, acids, heat, halogens, bromine trifluoride, chloric acid, chlorine monoxide, bromoazide. Moderate explosion hazard in the form of dust or vapor when exposed to flame.

★ 100 ★
ANTIMONY LACTATE
CAS: 58164-88-8
Other names: ANTIMONY LACTATE SOLID; LACTIC ACID, ANTIMONY SALT; 2-HYDROXY-,TRIANHYDRIDE WITH ANTIMONIC ACID; PROPANIC ACID, 2-HYDROXY-,ANTIMONY (3) SALT (3:1)

Danger profile: Poisonous, on Community Right-to-Know List, exposure limit established, poisonous gases produced in fire.
Uses: Fabric dyeing.
Appearance: Tan solid.
Effects of exposure:
Breathed: May cause irritation of the nose, throat and mouth; metallic taste, coughing, dizziness, headache, nausea, vomiting, diarrhea, stomach cramps, insomnia, anorexia, unable to smell properly. Higher levels of exposure may cause lung congestion, and the heart to beat irregularly or stop.
Eyes: May cause eye irritation, tearing, burning sensation. There is a risk of serious injury.
Skin: Passes through the unbroken skin; can increase exposure and the severity of the symptoms listed. Contact may cause irritation, itching and rash.

Swallowed: May cause irritation of the mouth and throat; burning pains in the stomach, nausea, vomiting and diarrhea.
Long-term exposure: Repeated exposure may cause headache, loss of appetite, dry throat and insomnia, heart and liver damage; prolonged or repeated contact may cause sores and ulcers. Chemical may contain contaminated arsenic or other toxic substances.
Storage: Store in tightly closed containers in a cool, well-ventilated place. Keep away from strong oxidizers; toxic fumes when exposed to acids and heat; halogens, bromine trifluoride, chloric acid, chlorine monoxide, bromoazide. Noncombustible in bulk form, but a moderate explosion hazard in the form of dust or vapor when exposed to flame. Contact with acids will produce stabine, a deadly gas.
First aid guide: Move victim to fresh air; call emergency medical care. In case of contact with material, immediately flush skin or eyes with running water for at least 15 minutes. Remove and isolate contaminated clothing and shoes at the site.

★ 101 ★
ANTIMONY PENTACHLORIDE
CAS: 7647-18-9
Other names: ANTIMONIC CHLORIDE; ANTIMONIO (PENTACLORURO DI) (ITALIAN); ANTIMONPENTACHLORID (GERMAN); ANTIMONY (V) CHLORIDE, ANTIMONY PERCHLORIDE; ANTIMOONPENTACHLORIDE (DUTCH); BUTTER OF ANTIMONY; PENTACHLOROANTIMONY; TENTACHLORURE D'ANTIMOINE (FRENCH); PERCHLORURE D'ANTIMOINE (FRENCH)

Danger profile: Mutagen, poisonous, corrosive, on Community Right-to-Know List, exposure limit established, reacts with water, poisonous gases produced in fire.
Uses: Dyeing, organic chemical reactions, coloring metals.
Appearance: Colorless to reddish-yellow, oily liquid.
Odor: Offensive. The odor is a warning of exposure; however, no smell does not mean you are not being exposed.
Effects of exposure:
Breathed: May cause irritation of the nose, throat and mouth; metallic taste, coughing, dizziness, headache, nausea, vomiting, diarrhea, stomach cramps, insomnia, anorexia, unable to smell properly. Higher levels of exposure may cause lung congestion, and the heart to beat irregularly or stop.
Eyes: May cause eye irritation, tearing, burning sensation. There is a risk of burns and serious injury.
Skin: Passes through the unbroken skin; can increase exposure and the severity of the symptoms listed. Contact may cause irritation, burns, itching and rash.
Swallowed: May cause irritation of the mouth and throat; burning pains in the stomach, nausea, vomiting and diarrhea.
Long-term exposure: Repeated exposure may cause headache, loss of appetite, dry throat and insomnia, heart and liver damage; prolonged or repeated contact may cause sores and ulcers.
Storage: Store in tightly closed containers in a dry, cool, well-ventilated place with non-wood floors. Keep away from organic or combustible materials, water, moisture and heat. Reacts with water.

★ 102 ★
ANTIMONY PENTAFLUORIDE
CAS: 7783-70-2
Other names: ANTIMONY FLUORIDE; ANTIMONY (V) FLUORIDE; ANTIMONY (V) PENTAFLUORIDE; PENTAFLUOROANTIMONY

Danger profile: Poisonous, corrosive, on Community Right-to-Know List, exposure limit established, EPA extremely hazardous list, reacts with water, poisonous gases produced in fire.
Uses: Making other chemicals.
Appearance: Colorless, oily liquid.
Effects of exposure:
Breathed: May be poisonous. May cause irritation of the nose, throat and mouth; metallic taste, coughing, dizziness, headache, nausea, vomiting, diarrhea, stomach cramps, insomnia, anorexia, unable to smell properly. Higher levels of exposure may cause lung congestion, and the heart to beat irregularly or stop.
Eyes: May cause eye irritation, tearing, burns and serious injury.
Skin: May be poisonous. Passes through the unbroken skin; can increase exposure and the severity of the symptoms listed. Contact may cause irritation, itching, rash, and burns.
Swallowed: May be poisonous. May cause irritation of the mouth and throat; burning pains in the stomach, nausea, vomiting and diarrhea.
Long-term exposure: Repeated exposure may cause headache, loss of appetite, dry throat and insomnia, heart and liver damage; prolonged or repeated skin contact may cause sores and ulcers.
Storage: Store in tightly closed containers in a dry, cool, well-ventilated place with a non-wood floor. Keep away from phosphorus, phosphates, combustible or organic materials; toxic fumes when exposed to acids and heat. Reacts with water. Contact with acids will produce stabine, a deadly gas.
First aid guide: Move victim to fresh air and call emergency medical care; if not breathing, give artificial respiration; if breathing is difficult, give oxygen. In case of contact with material, immediately flush skin or eyes with running water for at least 15 minutes. Remove and isolate contaminated clothing and shoes at the site. Keep victim quiet and maintain normal body temperature. Effects may be delayed, keep victim under observation.

★ 103 ★
ANTIMONY POTASSIUM TARTRATE
CAS: 28300-74-5
Other names: ANTIMONATE (2-), BIS .MU.-2,3-DIHYDROXYBUTANEDIOATA (4-)- 01,02:03,04DI-, DIPOTASSIUM, TRIHYDRATE, STEREOISOMER; ANTIMONYL POTASSIUM TARTRATE; POTASSIUM ANTIMONYL-D-TARTRATE; POTASSIUM ANTIMONT TARTRATE; EMETIQUE (FRENCH); ENT 50,434; POTASSIUM ANTIMONYL TARTRATE; TARTARIC ACID, ANTIMONY POTASSIUM SALT; TARTAR EMETIC; TARTARIZED ANTIMONY; TARTRATED ANTIMONY; TASTOX

Danger profile: Poisonous, corrosive, on Community Right-to-Know List, exposure limit established, poisonous gases produced in fire.
Uses: Dyeing; insecticide; in medicine.
Appearance: Colorless crystals, or white powder.
Effects of exposure:
Breathed: Poison. May cause irritation of the nose, throat and mouth; metallic taste, coughing, dizziness, headache, nausea, vomiting, diarrhea, stomach cramps, insomnia, anorexia, unable to smell properly. Higher levels of exposure may cause lung congestion, and the heart to beat irregularly or stop.
Eyes: May cause eye irritation, tearing, burns; serious injury.
Skin: Passes through the unbroken skin; can increase exposure and the severity of the symptoms listed. Contact may cause irritation, itching, rash, and burns.
Swallowed: Poison. May cause irritation of the mouth and throat; burning pains in the stomach, nausea, vomiting and diarrhea.

Long-term exposure: Repeated exposure may cause headache, loss of appetite, dry throat and insomnia, heart and liver damage; skin sores and ulcers. Large doses may cause severe liver damage.

Storage: Store in tightly closed containers in a dry, cool, well-ventilated place.

First aid guide: Move victim to fresh air; call emergency medical care. In case of contact with material, immediately flush skin or eyes with running water for at least 15 minutes. Remove and isolate contaminated clothing and shoes at the site.

★ 104 ★
ANTIMONY TRIBROMIDE
CAS: 7789-61-9
Other names: STIBINE, TRIBROMO-

Danger profile: Poisonous, corrosive, on Community Right-to-Know List, exposure limit established, poisonous gases produced in fire.

Uses: Making other chemicals; dyeing; laboratory chemical.

Appearance: Yellow crystals that absorb water.

Effects of exposure:
Breathed: May be poisonous. May cause irritation to the nose, throat and mouth; metallic taste, coughing, dizziness, headache, nausea, vomiting, diarrhea, stomach cramps, insomnia, anorexia, unable to smell properly. Higher levels of exposure may cause lung congestion; fluid in the lungs and the heart to beat irregularly or stop.

Eyes: May cause eye irritation, tearing, burns; serious injury.

Skin: Passes through the unbroken skin; can increase exposure and the severity of the symptoms listed. Contact may cause irritation, itching, rash, and chemical burns.

Swallowed: May be poisonous. May cause irritation of the mouth and throat; burning pains in the stomach, nausea, vomiting and diarrhea.

Long-term exposure: Repeated exposure may cause headache, loss of appetite, dry throat and insomnia, heart and liver damage; prolonged or repeated contact may cause sores and ulcers. Large doses may cause severe liver damage.

Storage: Store in tightly closed containers in a dry, cool, well-ventilated place. Keep away from water, moisture, heat, potassium, sodium, bases and combustable materials.

First aid guide: Move victim to fresh air; call emergency medical care. In case of contact with material, immediately flush skin or eyes with running water for at least 15 minutes. Remove and isolate contaminated clothing and shoes at the site. Keep victim quiet and maintain normal body temperature.

★ 105 ★
ANTIMONY TRICHLORIDE
CAS: 10025-91-9
Other names: STIBINE, TRICHLORO-; ANTIMONIUS CHLORIDE; ANTIMONY (III) CHLORIDE; TRICHLORO STIBINE; ANTIMOINE (TRICHLORURE D') (FRENCH); ANTIMONIO (TRICHLORURO DI) (ITALIAN); ANTIMONY BUTTER; ANTIMOONTRICHLRIDE (DUTCH); BUTTER OF ANTIMONY; CHLORID ANTIMONITY; C.I. 77056; STIBINE, TRICHLORO-; TRICHLOROSTIBINE; TRICHLORURE D' ANTIMOINE (FRENCH)

Danger profile: Poisonous, corrosive, on Community Right-to-Know List, exposure limit established, poisonous gases produced in fire.

Uses: Drugs; dyeing; chlorinating agent; fire proofing for textiles; organic chemistry.

Appearance: Clear, colorless crystals that absorb water.

Effects of exposure:

Breathed: May be poisonous. May cause irritation to the nose, throat and mouth; metallic taste, coughing, dizziness, headache, nausea, vomiting, diarrhea, stomach cramps, insomnia, anorexia, unable to smell properly. Higher levels of exposure may cause lung congestion; fluid in the lungs and the heart to beat irregularly or stop.

Eyes: May cause severe eye irritation, tearing, burns; serious injury.

Skin: Passes through the unbroken skin; can increase exposure and the severity of the symptoms listed. Contact may cause severe irritation, itching, rash, and deep chemical burns.

Swallowed: May be poisonous. May cause irritation of the mouth and throat; burning pains in the stomach, nausea, vomiting and diarrhea with bloody stools, slow pulse, shallow breathing; coma and convulsions sometimes followed by death.

Long-term exposure: Repeated exposure may cause headache, loss of appetite, dry throat and insomnia, heart and liver damage; prolonged or repeated contact may cause sores and ulcers. Large doses may cause severe liver damage.

Storage: Store in tightly closed containers in a dry, cool, well-ventilated place. Keep away from water, moisture, heat, aluminum, potassium, sodium, and bases.

First aid guide: Move victim to fresh air; call emergency medical care. In case of contact with material, immediately flush skin or eyes with running water for at least 15 minutes. Remove and isolate contaminated clothing and shoes at the site. Keep victim quiet and maintain normal body temperature.

★ 106 ★
ANTIMONY TRIFLUORIDE
CAS: 7783-56-4
Other names: ANTIMONY (III) FLUORIDE (1:3); ANTIMOINE FLUORURE (FRENCH); ANTIMONOUS FLUORIDE; TRIFLUOROANTIMONY, STIBINE,TRIFLUORO-

Danger profile: Poisonous, corrosive, on Community Right-to-Know List, exposure limit established, poisonous gases produced in fire.

Uses: Making pottery and porcelain; dyeing; as a fluorinating agent.

Appearance: White to gray crystals.

Effects of exposure:
Breathed: May be poisonous. May cause irritation to the nose, throat and mouth; metallic taste, coughing, dizziness, headache, nausea, vomiting, diarrhea, stomach cramps, insomnia, anorexia, unable to smell properly. Higher levels of exposure may cause lung congestion; fluid in the lungs and the heart to beat irregularly or stop.

Eyes: May cause eye irritation, tearing, burns; serious injury.

Skin: Passes through the unbroken skin; can increase exposure and the severity of the symptoms listed. Contact may cause irritation, itching, rash, and burns.

Swallowed: May be poisonous. May cause irritation of the mouth and throat; burning pains in the stomach, nausea, vomiting and diarrhea.

Long-term exposure: Repeated exposure may cause headache, loss of appetite, dry throat and insomnia, heart and liver damage; prolonged or repeated contact may cause sores and ulcers. Large doses may cause severe liver damage.

Storage: Store in tightly closed containers in a dry, cool, well-ventilated place. Keep away from water, moisture, heat, hot perchloric acid.

First aid guide: Move victim to fresh air; call emergency medical care. In case of contact with material, immediately flush skin or eyes with running water for at least 15 minutes. Remove and isolate contaminated clothing and shoes at the site. Keep victim quiet and maintain normal body temperature.

★ 107 ★
ANTIMONY TRIOXIDE
CAS: 1327-33-9
Other names: ANTIMONIOUS OXIDE; ANTIMONY(3) OXIDE; ANTIMONY PEROXIDE; ANTIMONY SESQUIOXIDE; ANTIMONY WHITE; ANTOX; A 1530; A 1582; A 1588LP; AP 50; CHEMETRON FIRE SHIELD; C.I. 77052; C.I. PIGMENT WHITE 11; DECHLORANE A-O; DIANTIMONY TRIOXIDE; EXITELITE; FLOWERS OF ANTIMONY; NCI-C55152; NYACOL A 1530; THERMOGUARD B; THERMOGUARD S; TIMONOX; SENARMONTITE; VALENTINITE; WEISSPIESSGLANZ; ANTIMONY (III) OXIDE

Danger profile: Carcinogen, poisonous gas produced in fire, exposure limit established, on Community Right-to-Know List.
Uses: Flame proofing of textiles, paper, and plastics; paint pigments; ceramic; staining iron and copper; glass decolorizer.
Appearance: White, colorless, crystalline powder.
Effects of exposure:
  Breathed: Inhalation may cause sore throat, irritation of breathing passages, cough, inflammation of upper and lower respiratory tract, including pneumonitis. Higher exposures may cause shortness of breath and a dangerous build-up of fluids in the lungs (pulmonary edema), which can cause death. Pulmonary edema is a medical emergency that can be delayed one to two days following exposure.
  Eyes: Contact may cause irritation, burns and conjunctivitis.
  Skin: Contact may cause irritation, dermatitis and rhinitis.
  Swallowed: Ingestion causes irritation of mouth, nose, stomach and intestines; vomiting, purging with bloody stools; slow pulse and low blood pressure; slow, shallow breathing; coma and convulsions sometimes followed by death.
Long-term exposure: May cause lung and liver cancer, damage of the developing fetus and miscarriage. Repeated exposure can cause headache, loss of appetite, dry throat, insomnia. Frequent and high exposures may cause liver and heart damage; possible lung damage.
Storage: Store in tightly closed containers in a cool, well-ventilated place. Keep away from heat, sources of ignition, bromine trifluoride, acids.

★ 108 ★
ANTU
CAS: 86-88-4
Other names: ALPHANAPHTHYL THIOUREA; ALPHANAPHTYL THIOUREE (FRENCH); ALRATO; ANTURAT; BANTU; CHEMICAL 109; DIRAX; KILL KANTZ; KRYSID; KRYSID PI; 1-NAFTIL-TIOUREA (ITALIAN); 1-NAFTYLTHIOUREUM (DUTCH); ALPHA-NAPHTHOTHIOUREA; ALPHA-NAPHTHYLTHIOCARBAMIDE; 1-NAPHTHYL-THIOHARNSTOFF (GERMAN); 1-NAPHTHYL THIOUREA; N-(1-NAPHTHYL)-2-THIOUREA; ALPHA-NAPHTHYLTHIOUREA; 1-(1-NAPHTHYL)-2-THIOUREA; 1-NAPHTHYL-THIOUREE (FRENCH); NAPHTOX; RATTRACK; RAT-TU; U-5227; USAF EK-P-5976; ALPHA-NAPHTHYLTHIOUREA; THIOUREA, 1-NAPHTHALENYL-; UREA,1-(1-NAPHTHYL)-2-THIO-

Danger profile: Possible carcinogen, possible mutagen, poisonous, EPA extremely hazardous substances list, toxic fumes when burned.
Uses: Rat poison.
Appearance: White crystalline or gray powder.
Effects of exposure:
  Breathed: Poisonous. May cause cough, shortness of breath or difficult breathing, pulmonary congestion, blue coloration of skin, fingertips, lips, nose and cheeks (cyanosis), malaise, nausea and vomiting. Chronic toxicity known to cause drug rashes and a decrease in white blood cells.

Swallowed: Poisonous. Large doses may cause vomiting, shortness of breath or difficult breathing, blue coloration of skin, fingertips, lips, nose and cheeks (cyanosis), mild liver damage.
Storage: Keep away from strong oxidizers, silver nitrate.

★ 109 ★
ARSENIC
CAS: 7440-38-2
Other names: ARSEN (GERMAN, POLISH); ARSENICALS; ARSENIC-75; ARSENIC BLACK; ARSENIC, SOLID; COLLOIDAL ARSENIC; GREY ARSENIC; METALLIC ARSENIC; RUBY ARSENIC; REALGAR; BUTTER OF ARSENIC

Danger profile: Carcinogen, mutagen, suspected reproductive hazard, poisonous, combustible in form of dust, on Community Right-to-Know List, exposure limit established, poisonous gases produced in fire.
Uses: Making alloys, especially of lead and copper, semiconductors, germanium and silicon solid state products, certain types of glass; to harden alloys; in the medical (toxicology) field as a radioactive tracer and to treat spirochetal infections, blood disorders, skin diseases; in hair tonics, hair dyes; pesticides, insecticides and rodenticides; in taxidermy; drugs; solders.
Appearance: Silver, gray or black, shiny metal.
Effects of exposure:
  Breathed: Poison. May cause irritation and ulceration of nasal septum, coughing, chest pains, difficult breathing, giddiness, headache, nausea, vomiting, diarrhea; breakdown of nasal tissue.
  Eyes: May cause irritation.
  Skin: May cause irritation, redness, blisters, and dermatitis. Some compounds are readily absorbed causing symptoms listed under Breathed.
  Swallowed: Poison. May cause stomach pains; vomiting and diarrhea (possibly with blood); inflammation of the throat, leg cramps, restlessness, paralysis, shock and death. As little as a teaspoon can be fatal.
Long-term exposure: May cause weakness, loss of appetite, irritation and inflammation of the eyes, nose and throat; breakdown of nasal tissue, excess itching; development of thickened skin and warts on hands and feet, skin discoloration; paralysis of the hands and feet; skin, lung and liver (tumors) cancer. There has been an increase of chromosome abnormalities in workers exposed to arsenic.
Storage: Store in a cool, well-ventilated place away from combustible materials and food products. Keep away from hydrogen gas, strong oxidizers; forms highly toxic fumes on contact with acids and active metals such as iron, aluminum and zinc. Dust is combustible. Keep away from food products.

★ 110 ★
ARSENIC ACID
CAS: 7778-39-4
Other names: METAARSENIC ACID; ORTHOARSENIC ACID; ZOTOX

Danger profile: Possible carcinogen, possible mutagen, suspected reproductive hazard, poisonous, on Community Right-to-Know List, exposure limit established, poisonous gases produced in fire.
Uses: Wood treatment, drying agent, soil sterlant; to manufacture other chemicals (arsenates).
Appearance: White semi-transparent crystal.
Effects of exposure:
  Breathed: Poison. May cause ulceration of nasal septum, coughing, chest pains, difficult breathing, giddiness, headache, loss of appetite, nausea, vomiting, diarrhea and break-

down of nasal tissue. Higher exposures could cause shortness of breath and a dangerous build-up of fluids in the lungs (pulmonary edema), which can cause death. Pulmonary edema is a medical emergency that can be delayed one to two days following exposure.

Eyes: May cause redness, tearing and irritation.

Skin: May cause irritation, redness, blisters, and dermatitis. Some compounds are readily absorbed causing symptoms listed under *Breathed.*

Swallowed: Poison. May cause irritation of the mouth and throat, stomach pains; vomiting and diarrhea (possibly with blood) that may cause burning pain when passed; inflammation of the throat, leg cramps, restlessness, paralysis, shock and death.

Long-term exposure: May cause cancer, weakness, loss of appetite, irritation and inflammation of the eyes, nose and throat, breakdown of nasal tissue, excess itching; development of thickened skin and warts on hands and feet, skin discoloration; paralysis of the hands and feet. There has been an increase of chromosome abnormalities in workers exposed to arsenic.

Storage: Store in tightly closed containers in a cool, well-ventilated place. Keep away from combustible materials, heat, food products, hydrogen gas, strong oxidizers; forms highly toxic fumes on contact with acids and chemically active metals such as iron, aluminum and zinc. Keep away from food products.

★ 111 ★
ARSENIC OXIDE
CAS: 1327-53-3

Other names: ACIDE ARSENIEUX (FRENCH); ANHYDRIDE ARSENIEUX (FRENCH); ARSENIC BLANC (FRENCH); ARSENIC (III) OXIDE; ARSENIC SESQUIOXIDE; ARSENIC TRIOXIDE, SOLID; ARSENICUM ALBUM; ARSENIGEN SAURE (GERMAN); ARSENIOUS ACID; ARSENIOUS OXIDE; ARSENIOUS TRIOXIDE; ARSENITE; ARSENOLITE; ARSENOUS ACID; ARSENOUS ACID ANHYDRIDE; ARSENOUS ANHYDRIDE; ARSENOUS OXIDE; ARSENOUS OXIDE ANHYDRIDE; ARSENTRIOXIDE; ARSODENT; CLAUDELITE; CLAUDETITE; CRUDE ARSENIC; DIARSENIC TRIOXIDE; WHITE ARSENIC; ARSENIC (III) OXIDE

Danger profile: Carcinogen, mutagen, poisonous, EPA extremely hazardous list, on Community Right-to-Know List, exposure limit established, poisonous gases produced in fire.

Uses: Pigments, ceramic enamels, aniline colors; decolorizing agent in glass; insecticide; rodenticide; herbicide; sheep and cattle dip; hide preservative; wood preservative; making other chemicals.

Appearance: Colorless or white crystals or powder.

Effects of exposure:

Breathed: Poison. May cause irritation and ulceration of nasal septum, coughing, chest pains, difficult breathing, giddiness, headache, nausea, vomiting, diarrhea; breakdown of nasal tissue.

Eyes: May cause irritation.

Skin: May cause irritation, redness, blisters, and dermatitis. Some compounds are readily absorbed causing symptoms listed under *Breathed.*

Swallowed: Poison. May cause irritation of mucous membrane, weakness, loss of appetite, gastrointestinal disturbances. Overdose can cause arsenic poisoning but symptoms may be delayed.

Storage: Store in tightly closed containers in a cool, well-ventilated place. Keep away from food products.

★ 112 ★
ARSENIC PENTOXIDE
CAS: 1303-28-2

Other names: ANHYDRIDE ARSENIQUE (FRENCH); ARSENIC (V) OXIDE; ARSENIC ACID ANHYDRIDE; DIARSENIC PENTOXIDE; ARSENIC ANHYDRIDE; FOTOX

Danger profile: Carcinogen, mutagen, poisonous, EPA extremely hazardous list, on Community Right-to-Know List, exposure limit established, poisonous gases produced in fire.

Uses: Making fungicides, insecticides, wood preservatives; printing; dyeing.

Appearance: White lumpy solid or powder.

Effects of exposure:

Breathed: Poison. May cause ulceration of nasal septum, coughing, chest pains, difficult breathing, giddiness, headache, loss of appetite, nausea, vomiting, diarrhea and breakdown of nasal tissue. Higher exposures could cause shortness of breath and a dangerous build-up of fluids in the lungs (pulmonary edema), which can cause death. Pulmonary edema is a medical emergency that can be delayed one to two days following exposure.

Eyes: May cause redness, tearing and irritation.

Skin: May cause irritation, burning, itching, redness, rash and pigment change. Some compounds are readily absorbed causing symptoms listed under *Breathed.*

Swallowed: Poison. May cause irritation of the mouth and throat, stomach pains; vomiting and diarrhea (possibly with blood) that may cause burning pain when passed; inflammation of the throat, leg cramps, restlessness, paralysis, shock and death.

Long-term exposure: May cause nerve damage, with "pins and needles," burning, numbness followed by weakness in the arms and legs, breakdown of nasal tissue; can cause an ulcer or hole in the nasal septum; can cause thickened skin, skin discoloration; skin allergy can develop. Has been shown to cause skin and lung cancer. There has been an increase of chromosome abnormalities in workers exposed to arsenic.

Storage: Store in tightly closed containers in a cool, well-ventilated place. Keep away from metals, acid or acid mists. Keep away from food products.

First aid guide: Move victim to fresh air; call emergency medical care. In case of contact with material, immediately flush skin or eyes with running water for at least 15 minutes. Remove and isolate contaminated clothing and shoes at the site.

★ 113 ★
ARSENIC TRICHLORIDE
CAS: 7784-34-1

Other names: ARSENOUS CHLORIDE; ARSENOUS TRICHLORIDE; TRICHLOROARSINE; BUTTER OF ARSENIC

Danger profile: Possible carcinogen, possible mutagen, poisonous, EPA extremely hazardous list, on Community Right-to-Know List, exposure limit established, poisonous gases are produced in fire.

Uses: Making pharmaceuticals, pesticides, ceramics and other chemicals.

Appearance: Colorless or pale yellow, oily liquid.

Effects of exposure:

Breathed: Poison. May cause ulceration of nasal septum, coughing, chest pains, difficult breathing, giddiness, headache, loss of appetite, nausea, vomiting, diarrhea and breakdown of nasal tissue. Higher exposures could cause shortness of breath and a dangerous build-up of fluids in the lungs (pulmonary edema), which can cause death. Pulmonary edema is a medical emergency that can be delayed one to two days following exposure.

Eyes: May cause redness, tearing and irritation.

Skin: May cause irritation, itching, redness, rash and severe skin burns; pigment change. Some compounds are readily absorbed causing symptoms listed under *Breathed.*
Swallowed: Poison. May cause irritation of the mouth and throat, stomach pains; vomiting and diarrhea (possibly with blood) that may cause burning pain when passed; inflammation of the throat, leg cramps, restlessness, paralysis, shock and death.
Long-term exposure: May cause nerve damage, with "pins and needles," burning, numbness followed by weakness in the arms and legs, breakdown of nasal tissue; can cause an ulcer or hole in the "bone" dividing the inner nose; can cause thickened skin, skin discoloration; skin allergy can develop. Arsenic compounds have been determined to cause cancer. There has been an increase of chromosome abnormalities in workers exposed to arsenic.
Storage: Store in tightly closed containers in a cool, dry, well-ventilated place. Keep away from sodium, potassium, powdered aluminum, acid or acid mists, heat and water. Containers may explode in fire. Keep away from food products.
First aid guide: Move victim to fresh air and call emergency medical care; if not breathing, give artificial respiration; if breathing is difficult, give oxygen. In case of contact with material, immediately flush skin or eyes with running water for at least 15 minutes. Speed in removing material from skin is of extreme importance. Remove and isolate contaminated clothing and shoes at the site. Keep victim quiet and maintain normal body temperature. Effects may be delayed; keep victim under observation.

★ 114 ★
ARSENIC TRIOXIDE
CAS: 1327-53-3
Other names: ARSENOUS ACID ANHYDRIDE; CRUDE ARSENIC; DIARSENIC TRIOXIDE; WHITE ARSENIC

Danger profile: Carcinogen, mutagen, poisonous, EPA extremely hazardous list, on Community Right-to-Know List, exposure limit established, poisonous gases produced in fire.
Uses: Pesticide, wood and tanning preservatives; manufacture of other chemicals and glass.
Appearance: White powder or colorless crystals.
Effects of exposure:
    Breathed: Poison. May cause coughing, chest pains, difficult breathing, giddiness, headache, loss of appetite, nausea, vomiting, diarrhea ulceration of nasal septum, and breakdown of nasal tissue. Higher exposures could cause shortness of breath and a dangerous build-up of fluids in the lungs (pulmonary edema), which can cause death. Pulmonary edema is a medical emergency that can be delayed one to two days following exposure.
    Eyes: May cause redness, tearing, irritation and eye damage.
    Skin: May cause irritation, itching, redness, rash and burns; pigment change. Some compounds are readily absorbed causing symptoms listed under *Breathed.*
    Swallowed: Poison. May cause irritation of the mouth and throat, stomach pains; vomiting and diarrhea (possibly with blood) that may cause burning pain when passed; inflammation of the throat, leg cramps, restlessness, paralysis, shock and death.
Long-term exposure: May cause nerve damage, with "pins and needles," burning, numbness followed by weakness in the arms and legs, breakdown of nasal tissue; can cause an ulcer or hole in the "bone" dividing the inner nose; can cause thickened skin, skin discoloration; skin allergy can develop. Arsenic compounds have been determined to cause cancer. Arsenic trioxide may cause skin and/or liver cancer. There has been an increase of chromosome abnormalities in workers exposed to arsenic.

Storage: Store in tightly closed containers in a cool, well-ventilated place. Keep away from chlorine trifluoride, fluorine, hydrogen fluoride, oxygen difluoride, sodium chlorate, mercury, acid or acid mists. Containers may explode in fire. Keep away from food products.
First aid guide: Move victim to fresh air; call emergency medical care. In case of contact with material, immediately flush skin or eyes with running water for at least 15 minutes. Remove and isolate contaminated clothing and shoes at the site.

★ 115 ★
ARSENIC TRISULFIDE
CAS: 1303-33-9
Other names: ARSENIC SULFIDE; ARSENIC SESQUISULFIDE; ARSENIC YELLOW; DIARSENIC TRISULFIDE; KING'S YELLOW

Danger profile: Carcinogen, mutagen, suspected reproductive hazard, poisonous, fire hazard, on Community Right-to-Know List, exposure limit established, poisonous gases produced in fire.
Uses: Used in the manufacture of glass, oil cloth, linoleum, electrical semi-conductors, fireworks, and as a pigment.
Appearance: Yellow powder.
Effects of exposure:
    Breathed: Poison. May cause coughing, chest pains, difficult breathing, giddiness, headache, loss of appetite, nausea, vomiting, diarrhea ulceration of nasal septum, and breakdown of nasal tissue. Higher exposures could cause shortness of breath and a dangerous build-up of fluids in the lungs (pulmonary edema), which can cause death. Pulmonary edema is a medical emergency that can be delayed one to two days following exposure.
    Eyes: May cause redness, tearing, irritation and eye damage.
    Skin: May cause irritation, itching, redness, rash and burns; pigment change. Some compounds are readily absorbed causing symptoms listed under *Breathed.*
    Swallowed: Poison. May cause irritation of the mouth and throat, stomach pains; vomiting and diarrhea (possibly with blood) that may cause burning pain when passed; inflammation of the throat, leg cramps, restlessness, paralysis, shock and death.
Long-term exposure: May cause nerve damage, with "pins and needles," burning, numbness followed by weakness in the arms and legs, breakdown of nasal tissue; can cause an ulcer or hole in the "bone" dividing the inner nose; can cause thickened skin, skin discoloration; skin allergy can develop. Arsenic compounds have been determined to cause cancer. Arsenic trioxide may cause skin and/or liver cancer. There has been an increase of chromosome abnormalities in workers exposed to arsenic.
Storage: Store in tightly closed containers in a dry, cool, well-ventilated place. Keep away from water, steam, acid or acid mists, oxidizers. Containers may explode in fire. Keep away from food products.
First aid guide: Move victim to fresh air; call emergency medical care. In case of contact with material, immediately flush skin or eyes with running water for at least 15 minutes. Remove and isolate contaminated clothing and shoes at the site.

★ 116 ★
ARSINE
CAS: 7784-42-1
Other names: ARSENIC TRIHYDRIDE; ARSENIC HYDRIDE; ARSENIURETTED HYDROGEN; ARSENOUS HYDRIDE (POLISH); ARSENWASSERSTOFF (GERMAN); HYDROGEN ARSENIDE

Danger profile: Carcinogen, mutagen, suspected reproductive hazard, poisonous, fire hazard, on Community Right-to-Know List, exposure limit established, poisonous gases produced in fire.

Uses: Military poison gas; manufacture of semiconductors and solid state electronic components.

Appearance: Colorless gas.

Odor: Garlic-like. The odor is a warning of exposure; however, no smell does not mean you are not being exposed.

Effects of exposure:

Breathed: Poison. Immediately seek medical attention. May cause giddiness, headache, and nausea. More severe exposure may cause malaise, weakness, shivering, dizziness; dyspnea; abdominal and back pain, nausea, vomiting; bronze or yellow coloration of skin, jaundice; garlic breath; dark red urine; accumulation of fluid in the lungs, delirium, coma and death. Symptoms may be delayed for one to two days following exposure.

Eyes: May cause irritation, watering, burning sensation and burns. High levels of exposure may cause permanent damage.

Skin: May cause painful irritation, inflammation, blisters, burns and deep wounds (ulcer). Liquid may cause frostbite.

Swallowed: Poison. Immediately seek medical attention. Unlikely route of exposure as most cases of poisoning occur from accidental generation of the gas.

Long-term exposure: There has been an increase of chromosome abnormalities in workers exposed to arsenic. May cause cancer in humans.

Storage: Store in tightly closed containers in a dry, cool, well-ventilated place. Keep away from heat, acids, strong oxidizers, open flame, shock, chlorine, nitric acid, mixtures of potassium and ammonia. Arsine can be formed when inorganic arsenic reacts with freshly formed (nascent) hydrogen gas. Vapors may travel to sources of ignition and flash back. Store away from food products. Containers may explode in fire.

First aid guide: Move victim to fresh air and call emergency medical care; if not breathing, give artificial respiration; if breathing is difficult, give oxygen. In case of contact with material, immediately flush skin or eyes with running water for at least 15 minutes. Remove and isolate contaminated clothing and shoes at the site. Keep victim quiet and maintain normal body temperature. Effects may be delayed; keep victim under observation.

★117★
ASBESTOS

CAS: 1332-21-4

Other names: ACTINOLITE (CAS 77536-67-5); AMOSITE (CAS 12172-73-5); AMIANTHUS; AMOSITE (OBS); AMPHIBOLE; AMBEST (GERMAN); ANTHOPHYLITE (CAS 77536-67-5); ANTHOPHYLLITE (17068-78-9); ASBESTOSE (GERMAN); 7-45 ASBESTOS (CAS 12001-29-5); ASBESTOS FIBER; ASBESTOS FIBRE; ASBESTOS, WHITE (CAS 12001-29-5); ASCARITE; AVIBEST C (CAS 12001-29-5); AZBILLEN ASBESTOS (17068-78-9); AZBOLEN ASBESTOS (CAS 77536-67-5); BLUE ASBESTOS (CAS 12001-28-4); CHRYSOTILE (CAS 12001-29-5); CHRYSOTILE A (CAS 12001-29-5); CROCIDOLITE (CAS 12001-28-4); FERROANTHOPHYLLITE (CAS 77536-67-5); FIBROUS GRUNERITE; MYSORITE (CAS 12172-73-5); KROKYDOLITH (CAS 12001-28-4); NCI-C08991 (CAS 14567-73-8); NCI-C60253A (CAS 12172-73-5); NCI C09007 (CAS 12001-28-4); SERPENTINE; TREMOLITE (CAS 14567-73-8); TREMOLITE ASBESTOS (CAS 14567-73-8)

Danger profile: Carcinogen, on Community Right-to-Know List.

Uses: Fire proofing; brake linings; gaskets; roofing compositions; electrical and heat resistant insulation; paint filler; chemical filters; reinforcing agent in rubber and plastics.

Appearance: Asbestos is the common name for a group of mineral fibers. Fine, slender, flaxy fiber which can range in color: white or greenish (chrysotile), blue (crocidolite), or gray-green (amosite).

Effects of exposure: Although there are no known short-term (acute) health effects, short-term exposure to asbestos fibers has been shown to cause an increased risk of developing several forms of cancer and other chronic lung disease. Those exposed and who develop serious or fatal disease later in life may feel fine at the time of exposure. Long-term (chronic) health effects can occur months or years (from 7-30) after exposure to asbestos. There is a very large increase in the risk of developing lung cancer among smokers who are exposed to asbestos. This risk of lung cancer may be more than 90 times higher for smokers who are exposed than for those without both smoking and asbestos exposure.

Long-term exposure: No level or duration of exposure can be assumed to be free of risk and any exposure may contribute to development of cancer of the lungs, stomach, colon, rectum, vocal chords, and kidneys. Mutation data reported.

Storage: Anyone working with asbestos may have to be trained—by law—in its proper storage and handling. Store an isolated, enclosed, marked, regulated area in closed, heavy-gauge, impervious plastic bags in sealed rigid containers protected from physical damage. The material should be kept wet with special materials and water. Do not smoke, eat, or drink in the area where asbestos is being stored or used.

First aid guide: In case of contact with material, immediately flush eyes with running water for at least 15 minutes. Wash skin with soap and water. Remove and isolate contaminated clothing and shoes at the site.

★118★
ASPHALT

CAS: 8052-42-4

Other names: ASPHALTUM; BITUMEN; JUDEAN PITCH; MINERAL PITCH; PETROLEUM PITCH; ROAD ASPHALT; ROAD TAR

Danger profile: Suspected carcinogen or may carcinogenic components, possible reproductive hazard, combustible, acrid and irritating fumes when burned.

Uses: Road paving; roofing; special paints; adhesive in electrical laminates and hot-melt compositions; in certain rubber products; for radioactive waste disposal; underground pipeline and cable coating; rust-preventive hot-dip coatings; base for synthetic turf; in converting petroleum components to protein.

Appearance: Dark brown to black thick solid or semi-solid.

Odor: Faint, pitch-like. The odor is a warning of exposure; however, no smell does not mean you are not being exposed.

Effects of exposure:

Breathed: Fumes can cause irritation of the nose, throat, air passages and lungs; increase in coughing and spitting; burning sensation in throat and chest, hoarseness, headache and runny nose.

Eyes: May cause irritation.

Skin: May cause the skin to darken, irritation, rash and burns. May cause dermatitis. Symptoms similar to those under *Breathed* may be seen. Hot asphalt can burn the skin.

Long-term exposure: Long term exposure can cause dermatitis, acne-like sores, thickening, and yellow discoloration of the skin. Fumes contain hydrogen sulfide which is combustible and highly toxic, refer to that entry. Fumes contain substances known to cause cancer in humans.

Storage: Store in tightly closed containers in a cool, well-ventilated place. Combustible, may be explosive in closed spaces. When heated gives off gasses which are combustible. Keep away from heat, sources of ignition, naphtha and other volatile solvents. Containers may explode in fire.

First aid guide: Move victim to fresh air and call emergency medical care; if not breathing, give artificial respiration; if breathing is difficult, give oxygen. In case of contact with material, immediately flush eyes with running water for at least 15 minutes. Wash skin with soap and water. Remove and isolate contaminated clothing and shoes at the site.

★ 119 ★
ATRAZINE
CAS: 1912-24-9
Other names: ARSENOUS ACID ANHYDRIDE; CRUDE ARSENIC; DIARSENIC TRIOXIDE; WHITE ARSENIC

Danger profile: Possible carcinogen, mutagen, poisonous, possible reproductive hazard, on Community Right-to-Know List, exposure limit established, poisonous gases produced in fire.
Uses: Pesticide, wood and tanning preservatives; manufacture of other chemicals and glass.
Appearance: White powder or colorless crystals.
Effects of exposure:
Breathed: Poison. May cause coughing, chest pains, difficult breathing, giddiness, headache, loss of appetite, nausea, vomiting, diarrhea ulceration of nasal septum, and breakdown of nasal tissue. Higher exposures could cause shortness of breath and a dangerous build-up of fluids in the lungs (pulmonary edema), which can cause death. Pulmonary edema is a medical emergency that can be delayed one to two days following exposure.
Eyes: May cause redness, tearing, irritation and eye damage.
Skin: May cause irritation, itching, redness, rash and burns; pigment change. Some compounds are readily absorbed causing symptoms listed under *Breathed.*
Swallowed: Poison. May cause irritation of the mouth and throat, stomach pains; vomiting and diarrhea (possibly with blood) that may cause burning pain when passed; inflammation of the throat, leg cramps, restlessness, paralysis, shock and death.
Long-term exposure: May cause nerve damage, with "pins and needles," burning, numbness followed by weakness in the arms and legs, breakdown of nasal tissue; can cause an ulcer or hole in the "bone" dividing the inner nose; can cause thickened skin, skin discoloration; skin allergy can develop. Arsenic compounds have been determined to cause cancer. Arsenic trioxide may cause skin and/or liver cancer. There has been an increase of chromosome abnormalities in workers exposed to arsenic.
Storage: Store in tightly closed containers in a cool, well-ventilated place. Keep away from chlorine trifluoride, fluorine, hydrogen fluoride, oxygen difluoride, sodium chlorate, mercury, acid or acid mists. Containers may explode in fire. Keep away from food products.

B

★ 120 ★
BACILLUS SUBTILIS BPN
CAS: 1395-21-7
Other names: ALCALASE; ALK; BACO; BACILLOMYCIN (8CI, 9CI); BACILLOMYCIN R; FUNGOCIN; MAXATASE; PROTEASE 150; SUBTILISINS; SUBTILISIN BPN

Danger profile: Exposure limit established, toxic gases when burned.
Uses: In laundry detergent. Enzyme from bacteria.
Effects of exposure:

Breathed: May cause irritation of the respiratory tract, constriction of the bronchia, runny nose, breathlessness, wheezing, sore throat, congested noses, headache, persistent cough. Respiratory allergies may develop.
Eyes: May cause severe irritation.
Skin: Contact may cause irritation, especially in sweaty areas.
Long-term exposure: May cause chronic lung problems; asthma-like allergy may develop; shortness of breath, cough, chest tightness.
Storage: Store in tightly closed containers in a dry, cool, well-ventilated place.

★ 121 ★
BARIUM
CAS: 7440-39-3
Other names: BARIUM METAL

Danger profile: Community Right-to-Know List, irritant, fire hazard, explosion hazard, poisonous, violent reaction when exposed to water (dangerous when wet), exposure limit established, poisonous gas produced in fire.
Uses: Making photographic papers, lubricating oils, dyes and other chemicals; pesticides, rodenticide; explosives; refining vegetable and animal oils; tanning; embalming; spark plugs, engine rod bearings.
Appearance: Silver-white or yellowish metal powder.
Effects of exposure:
Breathed: May cause eye, nose, throat, bronchial tube and lung irritation; coughing. High exposure can cause poisoning, cold sweat, vomiting, diarrhea, rapid respiration, tremor, lumbar pain, hypertension, slow and irregular heart beat, muscular weakness, paralysis of limbs, convulsions, shock and death. See *Swallowed.*
Eyes: Exposure may cause irritation, tearing, burning sensation and may result in permanent damage.
Skin: Contact may cause irritation and dermatitis.
Swallowed: Stomach acids may react with chemical and cause severe stomach pains, diarrhea, colic, slow irregular pulse rate, irregular heart beat, dizziness, convulsions, muscle spasms, hypertension, tremors and paralysis, and death within hours or days.
Storage: Store in sealed containers under an inert gas or petroleum liquid. Keep away from heat, sources of ignition, and other chemicals, especially water, oxidizers, strong acids, halogenated hydrocarbons and ammonia. Container may explode in fire. Powder may ignite spontaneously in air or other gases, and possibly at room temperature.
First aid guide: Move victim to fresh air; call emergency medical care. Wipe material from skin immediately; flush skin or eyes with running water for at least 15 minutes. Remove and isolate contaminated clothing and shoes at the site.

★ 122 ★
BARIUM ACETATE
CAS: 543-80-6
Other names: ACETIC ACID, BARIUM SALT; BARIUM DIACETATE; OCTAN BARNATY (CZECH)

Danger profile: Exposure limit established, Community Right-to-Know List, poisonous gas produced in fire.
Uses: Laboratory chemical; dyeing textiles; in paints and varnishes.
Appearance: White crystals.
Effects of exposure:
Breathed: May cause eye, nose, throat, bronchial tube and lung irritation; coughing. High exposure can cause poisoning, cold sweat, vomiting, diarrhea, rapid respiration, tremor, lum-

bar pain, hypertension, slow and irregular heart beat, muscular weakness, paralysis of limbs, convulsions, shock and death. See **Swallowed.**
Eyes: Exposure may cause irritation, tearing, burning sensation and may result in permanent damage.
Skin: Contact may cause irritation and dermatitis.
Swallowed: May cause nausea and vomiting, excessive salivation, severe abdominal pain, violent purging with bloody stools, slow and often irregular pulse, increased blood pressure, ringing in the ears, giddiness and vertigo, confusion, increasing somnolence, without coma, collapse and death from respiratory failure and cardiac arrest. Has a digitalis-like effect on the heart.
Storage: Store in tightly closed containers in a dry, cool, well ventilated place.

★ 123 ★
BARIUM AZIDE
CAS: 18810-58-7

Danger profile: Poisonous, exposure limit established, Community Right-to-Know List, explosive, spontaneously combustible in air, poisonous gas produced in fire.
Uses: Making high explosives, pesticides, rodenticides.
Appearance: White crystals or powder.
Effects of exposure:
Breathed: May cause irritation of the eyes, nose, throat and air passages, drop in blood pressure, weakness, headache, faintness and unconsciousness.
Eyes: May cause irritation and permanent damage. See **Breathed.**
Swallowed: See **Breathed.**
Long-term exposure: May cause nerve damage in the arms and legs and vision loss. Some barium compounds may be contaminated with silica. Barium can build up in the body. See SILICA and BARIUM entries.
Storage: Easily explodes and may result in serious bodily injury. Store in tightly closed containers in a cool, well-ventilated place. Explodes when heated or shocked. Keep away from heat, sources of ignition, carbon disulfide, and anything that could shock chemical. Keeping chemical wet greatly reduces its fire and explosion hazard. Container may explode in fire.

★ 124 ★
BARIUM BROMATE
CAS: 13967-90-3
Other names: BROMIC ACID, BARIUM SALT

Danger profile: Poisonous, oxydizer, Community Right-to-Know List, fire hazard, very reactive, exposure limit established, poisonous gas produced in fire.
Uses: Corrosion inhibitor, analytical laboratory chemical, oxidizer, water treatment.
Appearance: White crystals, or crystalline powder.
Effects of exposure:
Breathed: Dust or mist may cause of irritation of the nose, throat, and bronchial tubes, upper respiratory system, causing coughing and phlegm, headache, weakness, abdominal pain; dark urine, jaundice and possible kidney damage. High exposure can change the blood's hemoglobin, reduce the oxygen supply to body organs and cause the skin and lips to turn blue (cyanosis). Higher levels can cause breathing troubles, collapse and even death.
Eyes: Contact may cause severe irritation.
Skin: Contact may cause severe irritation, dryness and cracking.
Swallowed: Poisonous. May cause excessive salivation, vomiting, colic, diarrhea, convulsive tremors; slow, hard

pulse; and elevated blood pressure. Hemorrhages may occur in stomach, intestines and kidneys. Muscular paralysis may follow. See **Breathed.**
Long-term exposure: Repeated skin contact may cause chronic dryness and cracking. Some barium compounds may be contaminated with silica. Silica can cause gradual lungs scarring. Barium can build up in the body. See SILICA and BARIUM entries.
Storage: Store in tightly closed containers in a cool, well-ventilated place with non-wood floors. Keep away from heat, combustible materials, water, sources of ignition, aluminum, arsenic, carbon, copper, phosphorus, and other oxidizable materials. Containers may explode in fire.
First aid guide: Move victim to fresh air; call emergency medical care. In case of contact with material, immediately flush skin or eyes with running water for at least 15 minutes. Remove and isolate contaminated clothing and shoes at the site.

★ 125 ★
BARIUM CARBONATE
CAS: 513-77-9

Danger profile: Poisonous, Community Right-to-Know List, exposure limit established, poisonous gas produced in fire.
Uses: Rodenticide; making optical glass, radiation-resistant glass for color television tubes and ceramics.
Appearance: Colorless prisms and white powder.
Effects of exposure:
Breathed: May cause eye, nose, throat, bronchial tube and lung irritation; coughing. High exposure can cause poisoning, cold sweat, vomiting, diarrhea, rapid respiration, tremor, lumbar pain, hypertension, slow and irregular heart beat, muscular weakness, paralysis of limbs, convulsions, shock and death. See **Swallowed.**
Eyes: Exposure may cause irritation, tearing, burning sensation and may result in permanent damage.
Skin: Contact may cause irritation and dermatitis.
Swallowed: May cause nausea and vomiting, excessive salivation, severe abdominal pain, violent purging with bloody stools, slow and often irregular pulse, increased blood pressure, ringing in the ears, giddiness and vertigo, confusion, increasing somnolence without coma, collapse and death from respiratory failure and cardiac arrest.
Long-term exposure: Some barium compounds can be contaminated with silica. Silica may cause gradual lungs scarring. Barium can build up in the body. See SILICA and BARIUM entries.
Storage: Store in tightly closed containers in a cool, well-ventilated place. Keep away from foods.

★ 126 ★
BARIUM CHLORATE
CAS: 13477-00-4
Other names: BARIUM CHLORATE MONOHYDRATE; CHLORIC ACID, BARIUM SALTCHLORIC ACID, BARIUM SALT

Danger profile: Poisonous, oxydizer, fire hazard, explosion hazard, Community Right-to-Know List, exposure limit established, poisonous gas produced in fire.
Uses: Fireworks, explosives, textile dying, rodenticide, pesticides, and to make other chemicals.
Appearance: White powder, crystals, or crystalline powder.
Effects of exposure:
Breathed: Dust or mist may cause irritation of the nose, throat, and bronchial tubes, upper respiratory system; may cause coughing and phlegm, headache, weakness, abdominal pain; dark urine, jaundice and possible kidney damage.

High exposure can change the blood's hemoglobin, reduce the oxygen supply to body organs and cause the skin and lips to turn blue (cyanosis). Higher levels can cause breathing troubles, collapse and even death.

Eyes: Contact may cause severe irritation.

Skin: Contact may cause severe irritation, dryness and cracking.

Swallowed: Poisonous. May cause excessive salivation, vomiting, colic, diarrhea, convulsive tremors; slow, hard pulse; and elevated blood pressure. Hemorrhages may occur in stomach, intestines and kidneys. Muscular paralysis may follow.

Long-term exposure: Repeated skin contact may cause chronic dryness and cracking. Repeated or high exposure may result in kidney damage. Some barium compounds may be contaminated with silica. Barium can build up in the body. See SILICA and BARIUM entries.

Storage: Highly reactive. Store in tightly closed containers in a cool, well-ventilated place with non-wood floors. Can form explosive mixture with combustibles. Can be ignited by friction. Keep away from combustible materials, sulfur, aluminum, copper, ammonium salts and other oxidizable materials. Containers may explode in fire.

First aid guide: Move victim to fresh air; call emergency medical care. In case of contact with material, immediately flush skin or eyes with running water for at least 15 minutes. Remove and isolate contaminated clothing and shoes at the site.

★ 127 ★
BARIUM CHLORIDE
CAS: 10361-37-2
Other names: BARIUM DICHLORIDE; NCI-C61074; SBA 0108E

Danger profile: Exposure limit established, Community Right-to-Know List, poisonous gas produced in fire.
Uses: Making other chemicals and leather; laboratory chemical; lubricating oil additives; textile dyeing; pigments.
Appearance: Colorless crystals, granules or powder.
Effects of exposure:
Breathed: May cause eye, nose, throat, bronchial tube and lung irritation; coughing. High exposure can cause poisoning, cold sweat, vomiting, diarrhea, rapid respiration, tremor, lumbar pain, hypertension, slow and irregular heart beat, muscular weakness, paralysis of limbs, convulsions, shock and death. See *Swallowed.*
Eyes: Exposure may cause irritation, tearing, burning sensation and may result in permanent damage.
Skin: Contact may cause irritation and dermatitis.
Swallowed: Bitter salty taste. Water and stomach acids react with chemical and may cause nausea and vomiting, excessive salivation, severe abdominal pain, violent purging with bloody stools, slow and often irregular pulse, increased blood pressure, ringing in the ears, giddiness and vertigo, confusion, increasing somnolence, without coma, collapse and death from respiratory failure and cardiac arrest.
Storage: Store in tightly closed containers in a dry, cool, well ventilated place. Containers may explode in fire.

★ 128 ★
BARIUM CHROMATE(VI)
CAS: 10294-40-3
Other names: BARIUM CHROMATE; BARIUM CHROMATE (1:1); BARIUM CHROMATE OXIDE; BARYTA YELLOW; CHROMIC ACID, BARIUM SALT (1:1); C.I. 77103; C.I. PIGMENT YELLOW 31; LEMON CHROME; LEMON YELLOW; PERMANENT YELLOW; STEINBUHL YELLOW; ULTRAMARINE YELLOW

Danger profile: Carcinogen, poisonous, possible mutagen, Community Right-to-Know List, poisonous gas produced in fire.
Uses: Making explosives; safety matches, corrosion inhibitor, enamels, paint pigments and metal primers, printing inks, ceramics.
Appearance: Yellow crystalline powder.
Effects of exposure:
Breathed: May cause irritation of eyes, nose and throat.
Eyes: May cause irritation.
Skin: May cause irritation.
Swallowed: Poisonous. May cause excessive salivation, vomiting, colic, diarrhea, convulsive tremors; slow, hard pulse; and elevated blood pressure. Hemorrhages may occur in stomach, intestines and kidneys. Muscular paralysis may follow.
Long-term exposure: Some barium compounds may be contaminated with silica. Silica can cause gradual lungs scarring. Barium can build up in the body. See SILICA and BARIUM entries.
Storage: Store in tightly closed containers in a dry, cool, well-ventilated place. Containers may explode in fire.

★ 129 ★
BARIUM CYANIDE
CAS: 542-62-1
Other names: BARIUM CYANIDE, SOLID; BARIUM DICYANIDE

Danger profile: Poisonous, Community Right-to-Know List, poisonous fumes in fire, contact with acids and acid salts forms deadly combustible hydrogen cyanide gas, exposure limit established, harmful to aquatic life in very low concentrations.
Uses: Electroplating, case hardening of steel in metallurgy, welding aluminum, aluminum and sodium refining.
Appearance: Solid white crystals or powder.
Effects of exposure:
Breathed: Deadly poison. Exposure can cause headaches, weakness, nausea, vomiting, confusion, dizziness, vertigo, difficult breathing, palpitation, paralysis, respiratory arrest, cyanosis, collapse and death.
Eyes: May cause irritation.
Skin: Passes through the unbroken skin; can increase exposure and the severity of the symptoms listed. May cause itching, rash, top layer of skin to come off in tiny pieces (desquamation).
Swallowed: Deadly poison. See *Breathed.*
Long-term exposure: Some barium compounds may be contaminated with silica. Barium can build up in the body. See SILICA and BARIUM entries.
Storage: Store in tightly closed containers in a cool, dry, well-ventilated place. Keep away from acid, acid mists, acid salts, carbon dioxide, and strong oxidizers. Cyanides in water are corrosive to metals.
First aid guide: Move victim to fresh air; call emergency medical care. In case of contact with material, immediately flush skin or eyes with running water for at least 15 minutes. Remove and isolate contaminated clothing and shoes at the site.

★ 130 ★
BARIUM FLUORIDE
CAS: 7787-32-8
Other names: BARYUM FLUORURE (FRENCH)

Danger profile: Exposure limit established, Community Right-to-Know List, poisonous gas produced in fire.
Uses: Making carbon brushes for electrical motors, glass and other chemicals; dry lubricants; electronic products.

Appearance: White powder.
Effects of exposure:
Breathed: May cause eye, nose, throat, bronchial tube and lung irritation; coughing. High exposure can cause poisoning, cold sweat, vomiting, diarrhea, rapid respiration, tremor, lumbar pain, hypertension, slow and irregular heart beat, muscular weakness, paralysis of limbs, convulsions, shock and death. See *Swallowed.*
Eyes: Exposure may cause irritation, tearing, burning sensation and may result in permanent damage.
Skin: Contact may cause irritation and dermatitis.
Swallowed: Water and stomach acids react with chemical and may cause severe stomach pains, diarrhea, colic, slow irregular pulse rate, irregular heart beat, dizziness, convulsions, muscle spasms, hypertension, tremors and paralysis, and death within hours or days. May cause nausea and vomiting, excessive salivation, severe abdominal pain, violent purging with bloody stools, slow and often irregular pulse, increased blood pressure, ringing in the ears, giddiness and vertigo, confusion, increasing somnolence, without coma, collapse and death from respiratory failure and cardiac arrest.
Storage: Store in tightly closed containers in a dry, cool, well ventilated place. Keep away from oxidizers. Containers may explode in heat of fire.

★ 131 ★
BARIUM HYPOCHLORITE
CAS: 13477-10-6
Other names: HYPOCHLORIC ACID, BARIUM SALT

Danger profile: Poisonous gases in fire, reactive, Community Right-to-Know List, exposure limit established.
Uses: Bleaching agent and antiseptic.
Appearance: Colorless crystals.
Effects of exposure:
Breathed: Dust or mist may irritate the nose, throat and breathing passages, causing coughing and phlegm.
Eyes: Contact may cause severe irritation and burns.
Skin: Contact may cause severe irritation, burns, dryness and cracking.
Long-term exposure: Repeated skin contact may cause chronic dryness and cracking. Some barium compounds may be contaminated with silica. Barium can build up in the body. See SILICA and BARIUM entries.
Storage: Store in tightly closed containers in a cool, dry, well-ventilated place with non-wooden floors. Keep away from water, steam, heat, fuels, combustible materials, sources of ignition, acids and urea. Containers may explode in fire.
First aid guide: Move victim to fresh air and call emergency medical care; if not breathing, give artificial respiration; if breathing is difficult, give oxygen. In case of contact with material, immediately flush skin or eyes with running water for at least 15 minutes. Remove and isolate contaminated clothing and shoes at the site. Keep victim quiet and maintain normal body temperature. Effects may be delayed; keep victim under observation.

★ 132 ★
BARIUM NITRATE
CAS: 10022-31-8
Other names: NITRIC ACID, BARIUM SALT; BARIUM DINITRATE; DUSICNAN BARNATY (CZECH); NITRATE DE BARYUM (FRENCH); NITRIC ACID, BARIUM SALT; NITRO-BARITE

Danger profile: Poisonous, poisonous gases in fire, reactive, Community Right-to-Know List, exposure limit established.

Uses: Making other chemicals, fireworks, electronics industry, ceramic glazes, rodenticide.
Appearance: Lustrous white crystals.
Effects of exposure:
Breathed: Dust or mist may irritate the nose, throat and breathing passages, causing coughing and phlegm. High concentrations can cause lung damage, may increase muscle excitability and cause paralysis. May cause change to the blood and hemorrhages.
Eyes: Contact may cause severe irritation and burns.
Skin: Contact may cause severe irritation, burns dryness and cracking.
Swallowed: May cause excessive salivation, vomiting, colic, diarrhea, convulsive tremors; slow, hard pulse; and elevated blood pressure. Hemorrhages may occur in stomach, intestines and kidneys. Muscular paralysis may follow. See *Breathed.*
Long-term exposure: Repeated skin contact may cause chronic dryness and cracking. May cause lung damage. Some barium compounds may be contaminated with silica. Barium can build up in the body. See SILICA and BARIUM entries.
Storage: Store in tightly closed containers in a cool, dry, well-ventilated place with non-wooden floors. Keep away from heat, oxidizing agents, combustible materials, sources of ignition, aluminum and magnesium. Containers may explode in fire.
First aid guide: Move victim to fresh air; call emergency medical care. In case of contact with material, immediately flush skin or eyes with running water for at least 15 minutes. Remove and isolate contaminated clothing and shoes at the site.

★ 133 ★
BARIUM OXIDE
CAS: 1304-28-5
Other names: BARIUM MONOXIDE; BARIUM TOXIDE; BARSITO; CALCINED BARSITO; BARIUM PROTOXIDE; BARYTA; CALCINED BARYTA; OXYDE DE BARYUM (FRENCH)

Danger profile: Poisonous, reacts violently with water, poisonous gases in fire, reactive, Community Right-to-Know List, exposure limit established.
Uses: Detergent for lubricating oils; to dry gases and solvents.
Appearance: White to yellowish-white powder.
Effects of exposure:
Breathed: Poison. Dust or mist may irritate the nose, throat, and breathing passages, causing cough and phlegm. High exposures may cause a build-up of fluid in the lungs (pulmonary edema). Pulmonary edema is a medical emergency that can be delayed one to two days following exposure. It can come on rapidly and cause death.
Eyes: Contact may cause severe irritation and burns with possible loss of vision.
Skin: Contact may cause severe irritation.
Swallowed: Poison. High exposure can change the blood's hemoglobin, reduce the oxygen supply to body organs and cause the skin and lips to turn blue (cyanosis). Higher levels can cause severe abdominal pains, vomiting, rapid pulse and breathing, paralysis of the arms and legs, collapse and even death.
Long-term exposure: Repeated skin contact may cause chronic dryness and cracking. May cause lung damage. Some barium compounds may be contaminated with silica. Barium can build up in the body. See SILICA and BARIUM entries.
Storage: Store in tightly closed containers in a cool, dry, well-ventilated place. Keep away from water, hydrogen sulfide, carbon dioxide, hydroxlamine, nitrogen tetroxide, sulfur trioxide, and triuranium.
First aid guide: Move victim to fresh air; call emergency medical care. In case of contact with material, immediately flush skin or

eyes with running water for at least 15 minutes. Remove and isolate contaminated clothing and shoes at the site.

★ 134 ★
BARIUM PERCHLORATE
CAS: 13465-95-7
Other names: PERCHLORIC ACID, BARIUM SALT; BARIUM PERCHLORATE TRIHYDRATE; PERCHLORIC ACID, BARIUM SALT 3H2O

Danger profile: Poisonous, poisonous gases in fire, reactive, Community Right-to-Know List, exposure limit established.
Uses: Making explosives; rodenticide; pesticides; rocket fuel.
Appearance: White crystals.
Effects of exposure:
Breathed: Poison. Dust or mist may irritate the nose, throat, and breathing passages, causing cough and phlegm. High exposure can change the blood's hemoglobin, reduce the oxygen supply to body organs and cause the skin and lips to turn blue (cyanosis). Higher levels can cause breathing troubles, collapse and even death.
Eyes: Contact may cause severe irritation and burns.
Skin: Contact may cause severe irritation and burns.
Swallowed: Poison. May cause excessive salivation, vomiting, colic, diarrhea, convulsive tremors; slow, hard pulse; and elevated blood pressure. Hemorrhages may occur in stomach, intestines and kidneys. Muscular paralysis may follow.
Long-term exposure: Repeated skin contact may cause chronic dryness and cracking. Repeated or overexposure to perchlorates may interfere with normal functioning of the thyroid gland and may cause goiter and severe damage to the bone marrow. Some barium compounds may be contaminated with silica. May cause lung damage. Barium can build up in the body. See SILICA and BARIUM entries.
Storage: Store in tightly closed containers in a cool, well-ventilated place. Explodes when heated or shocked. Keep away from heat, sources of ignition, combustible materials, metal powders, especially magnesium and aluminum, sulfur, calcium hydride and strontium hydride.
First aid guide: Move victim to fresh air; call emergency medical care. In case of contact with material, immediately flush skin or eyes with running water for at least 15 minutes. Remove and isolate contaminated clothing and shoes at the site.

★ 135 ★
BARIUM PERMANGANATE
CAS: 7787-36-2
Other names: PERMANGANIC ACID, BARIUM SALT; BARIUM MANGANATE (VIII)

Danger profile: Poisonous, fire hazard, Community Right-to-Know List, exposure limit established, poisonous gas in fire.
Uses: Disinfectant, making other chemicals, depolarizing dry cells.
Appearance: Brownish-violet to black crystals.
Effects of exposure:
Breathed: Poison. Dust or mist may irritate the nose, throat, and breathing passages, causing cough and phlegm. High exposure can change the blood's hemoglobin, reduce the oxygen supply to body organs and cause the skin and lips to turn blue (cyanosis). Higher levels can cause breathing troubles, collapse and even death.
Eyes: Contact may cause severe irritation and burns.
Skin: Contact may cause severe irritation and burns.
Swallowed: Poison. May cause abdominal pain, nausea, vomiting, pallor, shortness of breath.
Long-term exposure: Repeated skin contact may cause chronic dryness and cracking. Some barium compounds may be

contaminated with silica. May cause lung damage. Barium can build up in the body. See SILICA and BARIUM entries.
Storage: Store in tightly closed containers in a cool, well-ventilated place with non-wooden floors. Keep away from heat, sources of ignition, combustible materials, acetic acid, acetic anhydride. May spontaneously ignite.
First aid guide: Move victim to fresh air; call emergency medical care. In case of contact with material, immediately flush skin or eyes with running water for at least 15 minutes. Remove and isolate contaminated clothing and shoes at the site.

★ 136 ★
BARIUM PEROXIDE
CAS: 1304-29-6
Other names: BARIO (PEROSSIDO DI) (ITALIAN); BARIUM BINOXIDE, BARIUM DIOXIDE, BARIUMPEROXID (GERMAN); BARIUM PEROXYDE (DUTCH); BARIUM SUPEROXIDE; DIOXYDE DE BARYUM (FRENCH); PEROXYDE DE BARYUM (FRENCH)

Danger profile: Poisonous, oxidizer, reactant, Community Right-to-Know List, exposure limit established, poisonous gas in fire.
Uses: Aluminum welding; textile dyeing, bleaching agent, decolorizing glass; making other chemicals.
Appearance: Grayish-white or light tan powder.
Effects of exposure:
Breathed: Poison. Dust or mist may irritate the nose, throat, and breathing passages, causing cough and phlegm. High exposure can change the blood's hemoglobin, reduce the oxygen supply to body organs and cause the skin and lips to turn blue (cyanosis). Higher levels can cause breathing troubles, collapse and even death.
Eyes: Contact may cause severe irritation and burns.
Skin: Contact may cause severe irritation and burns.
Swallowed: Poison. May cause excessive salivation, vomiting, colic, diarrhea, convulsive tremors; slow, hard pulse; and elevated blood pressure. Hemorrhages may occur in stomach, intestines and kidneys. Muscular paralysis may follow. See *Breathed.*
Long-term exposure: Repeated skin contact may cause chronic dryness and cracking. Some barium compounds may be contaminated with silica. May cause lung damage. Barium can build up in the body. See SILICA and BARIUM entries.
Storage: Store in tightly closed containers in a cool, dry well-ventilated place with non-wooden floors. Keep away from water, steam or other forms of moisture, heat, sources of ignition, combustible, organic or other easily oxidized materials, acetic anhydride, powdered aluminum, peroxyformic acid, hydrogen sulfide, and solutions of hydroxylamine.
First aid guide: Move victim to fresh air; call emergency medical care. In case of contact with material, immediately flush skin or eyes with running water for at least 15 minutes. Remove and isolate contaminated clothing and shoes at the site.

★ 137 ★
BARIUM SULFATE
CAS: 7727-43-7
Other names: ACTYBARYTE; ARTIFICIAL BARITE; ARTIFICIAL HEAVY SPAR; BAKONTAL; BARIDOL; BARITE; BARITOP; BARIUM SULFATE (1:1); BARIUM SULPHATE; BAROSPERSE; BAROTRAST; BARYTA WHITE; BARYTES; BARYTES 22; BAYRITES; BLANC FIXE; C.I. 77120; C.I. PIGMENT WHITE 21; CITOBARYUM; COLONATRAST; ENAMEL WHITE; ESOPHOTRAST; EWEISS; E-Z-PAQUE; FINEMEAL; LACTOBARYT; LIQUIBARINE; MACROPAQUE; NEOBAR; ORATRAST; PERMANENT WHITE; PRECIPITATED BARIUM SULPHATE; RAYBAR; REDI-FLOW; SOLBAR; SULFURIC ACID, BARIUM SALT (1:1); SUPRAMIKE; TRAVAD; UNIBARYT

Danger profile: Possible carcinogen, Community Right-to-Know List, exposure limit established, poisonous fumes in fire.
Uses: Paper coatings, paints, making textiles, rubber, plastics, printing inks, as an opaque medium in medical radiography, x-ray photography.
Appearance: White or yellowish powder.
Effects of exposure:
  Breathed: Dust or mist may irritate the nose, throat, and breathing passages, causing cough and phlegm.
  Eyes: Contact may cause irritation.
  Skin: Contact may cause irritation.
  Swallowed: The insoluble chemical used in radiography is not believed to be acutely (short-term) toxic.
Long-term exposure: Repeated skin contact may cause chronic dryness and cracking. Some barium compounds may be contaminated with silica. May cause lung damage. Barium can build up in the body. See SILICA and BARIUM entries.
Storage: Store in tightly closed containers in a cool, dry well-ventilated place. Keep away from heat, sources of ignition, aluminum and phosphorus.

★ 138 ★
BENDIOCARB
CAS: 22781-23-3
Other names: BENCARBATE; BENDIOCARBE; 1,3-BENZODIOXOLE,2,2-DIMETHYL-4-(N-METHYLAMINOCARBOXYLATO)-; 1,3-BENZODIOXOLE,2,2-DIMETHYL-4-(N-METHYLCARBAMATO)-; 1,3-BENZODIOXOL-4-OL,2,2-DIMETHYL-,METHYLCRBAMATE; BICAM ULV; CARBAMIC ACID, METHYL-,2,3-(ISOPROPYLIDENEDIOXY)PHENYL ESTER; CARBAMIC ACID, METHYL-,2,3-(DIMETHYLMETHYLENEDIOXY)PHENYL ESTER 2,2-DIMETHYL-1,3-BENZODIOXOL-4-OL METHYL-CARBAMATE; 2,2-DIMETHYL-1,3-BENZODIOXOL-4-YL-N-METHYLCARBAMATE; 2,2-DIMETHYLBENZO-1,3-BENZODIOXOL-4-YL-N-METHYLCARBAMATE; 2,2-DIMETHYL-4-(N-METHYLAMINOCARBOXYLATO)-1,3-BENXODIOXOLE; 2,2-DIMETHYLBENZO-1,3-DIOXOL-4-YL METHYLCARBAMATE; DYCARB; FICAM; FICAM D; FICAM ULV; FICAM W; FICAM 80W; FUAM; GARVOX; MC 6897; 2,3-ISOPROPYLIDENE-DIOXYPHENYL METHYLCARBAMATE; METHYLCARBAMIC ACID 2,3-(ISOPROPYLIDENEDIOXY)PHENYL ESTER; MULTAMAT; MULTIMET; NC6897; NIOMIL; OMS-1394; ROTATE; SEEDOX; TATTOO; TURCAM

Danger profile: Poisonous, poisonous gases (methyl isocyanate) in fire.
Uses: Insecticide.
Appearance: White crystals.
Effects of exposure:
  Breathed: Poison. See *Skin* and *Swallowed.*
  Eyes: Contact may cause blurred vision.

Skin: Moderately toxic and passes through the skin. Contact may cause weakness, nausea, vomiting sweating and stomach pain.
Swallowed: Poison. May cause weakness, nausea, vomiting sweating and stomach pain. Higher levels can cause muscle twitching, loss of concentration, and breathing to stop.
Storage: Store in tightly closed containers in a cool, well-ventilated place. Slightly combustible. Keep containers away from flames and sources of ignition.

★ 139 ★
BENOMYL
CAS: 17804-35-2
Other names: AGROCIT; ARILATE; BBC; BENLAT; BENLATE; BENLATE 50; BENLATE 50 W; BNM; 1-(BUTYLAMINO)CARBONYL-1H-BENZIMIDAZOL-2-YL-, METHYL ESTER; 1-(BUTYLCARBAMOYL)-2-BENZIMIDAZOLEC ARBAMIC ACID, METHYL ESTER; 1-(N-BUTYLCARBAMOYL)-2-(METHOXY-CAR BOXAMIDO)-BENZIMIDAZOL (GERMAN); CARBAMIC ACID, 1- (BUTYLAMINO)CARBONYL- 1H-BENZIMIDAZOL-2YL, METHYL 20ESTER; D 1991; DUPONT 1991; F 1991; FUNGICIDE 1991; MBC; METHYL 1-(BUTYLCARBAMOYL)-2-BENZIMIDAZOLYLCARBAMATE; TERSAN 1991; BENLATE 40 W; 2-BENZIMIDAZOLECARBAMIC ACID, 1-(BUTYLCARBAMOYL)-, METHYL ESTER; 1-(N-BUTYLCARBAMOYL)-2-(METHOXY-CAR BOXAMIDO)-BENZAMIDAZOL (GERMAN); METHYL 1-(BUTYLCARBAMOYL)-2-BENZIMIDAZOYLYLCARBAMATE

Danger profile: Poisonous, mutagen, possible reproductive hazard, irritant, toxic fumes produced in fire.
Uses: Fungicide for a wide range of molds and fungi. Used on fruits and vegetable.
Appearance: White crystals.
Effects of exposure:
  Breathed: Mildly toxic.
  Eyes: May cause irritation.
  Skin: Contact may cause skin irritation, itching and rash; allergy may develop.
  Swallowed: Poison. May cause redness or swelling of the mucous membrane, salivation, sweating, reproductive effects, loss of muscular coordination, nausea, vomiting, abdominal cramps, a cramping pain in the chest (angina pectoris), central nervous system depression, inflammation of the cornea of the eye (keratitis) and a state of exhaustion.
Long-term exposure: May cause damage to the male reproductive system; fertility problems (decreased sperm count), genetic changes, mutations. If skin allergy develops, ever low exposures to the chemical in the future may cause irritation, itching and skin rash.
Storage: Store in tightly closed containers in a cool, well-ventilated place. Slightly combustible. Keep containers away from food.

★ 140 ★
BENTONITE
CAS: 1302-78-9
Other names: ALBAGEL PREMIUM USP 4444; ALUMINUM SILICATE, MONTMORILLONITE; BENTONITE BC; BENTONITE 2073; BENTONITE MAGMA; HI-JEL; IMVITE I.G.B.A.; MAGBOND; MONTMORILLONITE; OTAYLITE; PANTHER CREEK BENTONITE; SOUTHERN BENTONITE; TIXOTON; VOLCAY; VOLCLAY; WILKNITE; WILKONITE

Danger profile: Poisonous, possible carcinogen, possible tumorogen.

Uses: Base for plasters; cosmetics; polishes and abrasives; food additive.

Appearance: Light yellow, yellow-green, cream, pink, gray or black powder.

Effects of exposure:

Breathed: Dust may cause irritation of the nose, throat and lungs.

Eyes: Contact may cause irritation.

Skin: Poison by intervenenous route and may cause blood to clot.

Long-term exposure: Dust may cause irritation and bronchial asthma.

Storage: Store in tightly closed containers in a cool, well-ventilated place.

★ 141 ★
BENZALCHLORIDE
CAS: 98-87-3

Other names: BENZAL CHLORIDE; BENZYL DICHLORIDE; BENZYLENE CHLORIDE; BENZYLIDENE CHLORIDE; CHLOROBENZAL; CHLORURE DEBENZYLIDENE (FRENCH); (DICHLOROMETHYL)BENZENE; ALPHA,ALPHA-DICHLOROTOLUENE; TOLUENE, ALPHA,ALPHA-DICHLORO-; BENZENE, DICHLORO METHYL-

Danger profile: Suspected carcinogen, mutagen, combustible, poisonous, strong irritant, extremely poisonous gases produced in fire.

Uses: Dyes; making other chemicals.

Appearance: Colorless, oily liquid. Fumes in air.

Odor: Faint, pungent. The odor is a warning of exposure; however, no smell does not mean you are not being exposed.

Effects of exposure:

Breathed: Poison. May irritate the nose, throat and lungs, cause coughing, shortness of breath and difficult breathing. Higher levels can cause dizziness, unconsciousness; a build-up of fluids in the lungs (pulmonary edema), which can cause death. Pulmonary edema is a medical emergency that may be delayed one to two days following exposure.

Eyes: Contact may cause irritation, excessive tearing and burns.

Skin: Poison. May cause irritation and burns.

Swallowed: Poison. May be fatal.

Long-term exposure: May cause lung damage.

Storage: Store in tightly closed containers in a cool, well-ventilated place. Keep away from heat, sources of ignition, acids, and acid fumes. Containers may explode in fire.

★ 142 ★
BENZALDEHYDE
CAS: 100-52-7

Other names: ALMOND ARTIFICIAL ESSENTIAL OIL; ARTIFICIAL ALMOND OIL; BENZENE CARBALDEHYDE; BENZENECARBONAL; BENZOIC ALDEHYDE; NCI-C56133; OIL OF BITTER ALMOND

Danger profile: Mutagen, combustible, allergen, poisonous gas produced in fire.

Uses: Solvent; making dyes, flavors, synthetic perfumes, other chemicals, drugs, photographic chemicals, dyes.

Appearance: Colorless to yellow liquid.

Odor: Bitter almond. The odor is a warning of exposure; however, no smell does not mean you are not being exposed.

Effects of exposure:

Breathed: Vapor may cause irritation of the eyes, nose and throat, dizziness and lightheadedness. Higher levels may

cause seizures, central nervous system depression, lung injury, unconsciousness and death.

Eyes: Contact may cause severe irritation.

Skin: Moderately toxic. Passes through the skin. May cause irritation. Repeated contact may cause skin rash.

Swallowed: Poison. See *Breathed* for symptoms.

Long-term exposure: May cause dermatitis. Vapors may cause eye and lung injury. Narcotic in high concentration.

Storage: Store in tightly closed containers in a dark, cool, well-ventilated place with a non-wooden floor. Keep away from light, strong oxidizers, strong acids, combustible materials. Containers may explode in fire.

First aid guide: Move victim to fresh air and call emergency medical care; if not breathing, give artificial respiration; if breathing is difficult, give oxygen. In case of contact with material, immediately flush eyes with running water for at least 15 minutes. Wash skin with soap and water. Remove and isolate contaminated clothing and shoes at the site.

★ 143 ★
BENZENE
CAS: 71-43-2

Other names: BENZOL; BENZOLE; BENZELENE; CARBON OIL; CARBON NAPTHA; COAL TAR NAPHTHA; CYCLOHEXATRIENE; PHENYL HYDRIDE; PHENE; COAL NAPHTHA; COAL NAPTHA, PHENYL HYDRIDE; BENZELENE; MINERAL NAPTHA; MOTOR BENZOL; PYROBENZOL

Danger profile: Carcinogen, mutagen, possible reproductive hazard, dangerous fire hazard, exposure limit established, Community Right-to-Know List, poisonous gases produced in fire.

Uses: Perfumes, paints and coatings. Processing nylon and photographic chemicals; making gasoline, styrene, pesticides, plastics, resins, synthetic rubber, aviation fuel, dyes, explosives, and other chemicals including phenol.

Appearance: Colorless to pale yellow watery liquid.

Odor: Strong, pleasant. Similar to gasoline. The odor is a warning of exposure; however, no smell does not mean you are not being exposed.

Effects of exposure:

Breathed: Poison. May cause irritation of the eyes, nose and throat, excitation, giddiness, headache, staggered gait, irritability, stomach pains, irritation of the eyes, nose and reparatory tract, dizziness, nausea, slurred speech, tremors. Higher exposure may cause convulsions, coma, possible sudden death from irregular heartbeat.

Eyes: May cause severe irritation and burning feeling.

Skin: Poison. Passes through the unbroken skin; can increase exposure and the severity of the symptoms listed. May cause irritation, drying and scaling.

Swallowed: Toxic. Immediately seek medical attention. Symptoms similar to those listed under *Breathed.* May cause irritation of the mouth, throat and stomach. A dose as small as one tablespoon may cause pneumonia, bronchitis and death.

Long-term exposure: A cancer causing agent in humans. Repeated or prolonged exposure has been linked to Hodgkin's disease, blood damage including aplastic anemia and leukemia; liver and heart damage. Causes birth defects by damaging the fetus. Skin exposure may cause chronic drying and scaling.

Storage: Store away from heat in a cool, well-ventilated place. Keep away from heat, sources of ignition, strong oxidizers, many fluorides and perchlorates, and nitric acid. Vapors may travel to sources of ignition and flash back. Containers may explode in fire.

First aid guide: Move victim to fresh air and call emergency medical care; if not breathing, give artificial respiration; if breathing is difficult, give oxygen. In case of contact with material, immediately flush eyes with running water for at least 15 minutes.

Wash skin with soap and water. Remove and isolate contaminated clothing and shoes at the site.

★ 144 ★
BENZENE HEXACHLORIDE
CAS: 608-73-1
Other names: BHC; COMPOUND-666; DBH; ENT 8.601; GAMMEXANE; HCCH; HEXA; HEXACHLOR; HEXACHLO-RAN; HEXACHLOROCYCLOHEXANE; 1,2,3,4,5,6- HEXA-CHLOROCYCLOHEXANE; HEXYLAN; LINDANE, TRI-6

Danger profile: Carcinogen, poisonous, possible mutagen, possible reproductive hazard; very toxic gases (phosgene) in fire.
Appearance: White, or light brown crystals.
Odor: Musty. The odor is a warning of exposure; however, no smell does not mean you are not being exposed.
Effects of exposure:
Breathed: May cause irritation of the eyes, nose and throat, headache, nausea, vomiting and fever.
Eyes: Contact may cause irritation.
Skin: Moderately toxic. Contact may cause irritation, smarting of the skin, dermatitis, hives, skin eruption marked by wheals of varying shapes and sizes.
Swallowed: Poison. May cause hyperirritability, central nervous system excitation: vomiting, restlessness, loss of equilibrium, muscle spasms, ataxia, and depression. Subsequent central nervous system depression, leading to respiratory failure.
Long-term exposure: May cause cancer, tumors, aplastic anemia; birth defects by damaging the fetus; dermatitis.
Storage: Store in a cool, well-ventilated place. Keep away from heat and sources of combustion. Vapors may travel to sources of ignition and flash back. Containers may explode in fire.

★ 145 ★
BENZENE HEXACHLORIDE-ALPHA ISOMER
CAS: 319-84-6
Other names: ALPHA-BENZENEHEXACHLORIDE; ALPHA-BHC; CYCLOHEXANE 1,2,3,4,5,6-HEXACHLORO-; ENT 9,232; ALPHA-HCH; ALPHA- HEXACHLORANE; ALPHA-HEXACHLORAN; HEXACHLOROCYCLOHEXANE; HEXA-CHLOROCYCLOHEXAN (GERMAN); ALPHA-LINDANE

Danger profile: Carcinogen, mutagen, poisonous, toxic fumes in fire.
Uses: Pesticide.
Appearance: Light to dark brown crystals or powder.
Odor: Musty. The odor is a warning of exposure; however, no smell does not mean you are not being exposed.
Effects of exposure:
Breathed: Poison. May cause irritation of the eyes, nose and throat, headache, nausea, vomiting and fever.
Eyes: Contact may cause irritation.
Skin: Moderately toxic. Contact may cause irritation, smarting of the skin, dermatitis, hives, skin eruption marked by wheals of varying shapes and sizes.
Swallowed: Poison. May cause hyperirritability, vomiting, restlessness, loss of equilibrium, muscle spasms, ataxia, and depression. Subsequent central nervous system depression, and possible respiratory failure.
Long-term exposure: May cause cancer, tumors, aplastic anemia; birth defects by damaging the fetus; dermatitis.
Storage: Store in a cool, well-ventilated place. Keep away from heat and sources of combustion.

★ 146 ★
BENZENE HEXACHLORIDE-GAMMA ISOMER
CAS: 58-89-9
Other names: AALINDAN; AFICIDE; AGRISOL G-20; AGRO-CIDE; AGROCIDE 2; AGROCIDE 7; AGROCIDE 6G; AGRO-CIDE III; AGROCIDE WP; AGRONEXIT; AMEISENATOD; AM-EISENMITTEL MERCK; APARASIN; APHTIRIA; APLIDAL; ARBITEX; BBH; BEN-HEX; BENTOX 10; GAMMA-BENZENE HEXACHLORIDE; BEXOL; BHC; GAMMA-BHC; CELANEX; CHLORESENE; CODECHINE; DBH; DETMOL-EXTRAKT; DETOX 25; DEVORAN; DOL GRANULE; DRILL TOX-SPEZIAL AGLUKON; ENT 7,796; ENTOMOXAN; EXAGAMA; FORLIN; GALLOGAMA; GAMACID; GAMAPHEX; GAMENE; GAMMAHEXA; GAMMAHEXANE; GAMMALIN; GAMMALIN 20; GAMMATERR; GAMMEX; GAMMEXANE; GAMMOPAZ; GEXANE; HCCH; HCH; GAMMA-HCH; HECLOTOX; HEXA; HEXACHLORAN; GAMMA-HEXACHLORAN; HEXACHLO-RANE; GAMMA-HEXACHLORANE; GAMMA-HEXACHLOROBENZENE; 1-ALPHA,2-ALPHA,3-BETA,4-ALPHA,5-ALPHA,6-BETA-HEXACHLOROCYCLOHEX ANE; GAMMA-HEXACHLOROCYCLOHEXANE; GAMMA-1,2,3,4,5,6-HEXACHLOROCYCLOHEXANE; HEXACHLORO-CYCLOHEXANE, GAMMA-ISOMER; 1,2,3,4,5,6-HEXACHLOROCYCLOHEXANE, GAMMA-ISOMER; HEXA-TOX; HEXAVERM; HEXICIDE; HEXYCLAN; HGI; HORTEX; INEXIT; ISOTOX; JACUTIN; KOKOTINE; KWELL; LENDINE; LENTOX; LIDENAL; LINDAFOR; LINDAGAM; LINDAGRAIN; LINDAGRANOX; LINDANE; GAMMA-LINDANE; LINDANE; LINDAPOUDRE; LINDATOX; LINDOSEP; LINTOX; LOREX-ANE; MILBOL 49; MSZYCOL; NCI-C00204; NEO-SCABICIDOL; NEXEN FB; NEXIT; NEXIT-STARK; NEXOL-E; NICOCHLORAN; NOVIGAM; OMNITOX; OVADZIAK; OWADZIAK; PEDRACZAK; PFLANZOL; QUELLADA; SANG GAMMA; SILVANOL; SPRITZ-RAPIDIN; SPRUEHPFLANZOL; STREUNEX; TAP 85; TRI-6; VITON; HEXACHLOROCYCLO-HEXANE, GAMMA ISOMER; 1,2,3,4,5,6-HEXACHLOR-CYCLOHEXANE; GAMMA-HEXACHLORO-CYCLOHEXANE

Danger profile: Carcinogen, mutagen, EPA extremely hazardous list, Community Right-to-Know List, extremely toxic gases in fire.
Uses: To pretreat seeds; in veterinary and human medicine; as a component in insecticides. May require a permit to use as insecticide or pesticide.
Appearance: Pure substance is a colorless solid. Technical grade is light to dark brown crystals, powder or solution.
Odor: Musty. Pure material is odorless. The odor is a warning of exposure; however, no smell does not mean you are not being exposed.
Effects of exposure:
Breathed: May cause irritation of the eyes, nose and throat, headache, nausea, vomiting and fever. Exposure to high levels can cause chills, weakness, headache, profuse sweating, pain in the bones or joints, loss of equilibrium and depth perception.
Eyes: Contact may cause irritation.
Skin: Poison. May be absorbed through the skin to cause or contribute to the severity of symptoms listed. Contact may cause irritation, smarting of the skin, dermatitis, hives, skin eruption marked by welts or wheals of varying shapes and sizes, fatigue, dizziness, mental slowness, stomach and muscle pain, diarrhea, inflammation of the mouth lining, mental confusion, blindness, speech impediment from muscle paralysis, convulsions, liver and kidney damage, and death.
Swallowed: Poison. May cause convulsions, dyspnea, cyanosis, hyperirritability, central nervous system excitation: vomiting, restlessness, loss of equilibrium, muscle spasms, ataxia, and depression. Subsequent central nervous system depression, leading to respiratory failure.

Long-term exposure: May cause cancer, tumors, aplastic anemia; birth defects by damaging the fetus; dermatitis; altered menstrual cycles; irregular heartbeat.

Storage: Store in a cool, dry, well-ventilated place. Keep away from light, heat and sources of combustion, strong alkalis such as sodium hydroxide. Containers may explode in fire.

★ 147 ★
TRANS-ALPHA-BENZENEHEXACHLORIDE
CAS: 319-85-7
Other names: BETA-BENZENEHEXACHLORIDE; BETA-BHC; BETA-HEXACHLOROBENZENE; 1-ALPHA, 2-BETA, 3-ALPHA, 4-BETA, 5-ALPHA, 6-BETA-HEXACHLOROCYCLOHEXANE; BETA-HEXACHLOROCYCLOHEXANE; BETA-1,2,3,4,5,6-HEXACHLOROCYCLOHEXANE; BETA-ISOMER; BETA-LINDANE; BETA-LINDANE 6

Danger profile: Carcinogen, very toxic fumes in fire.
Uses: Pesticide.
Appearance: Light to dark brown crystals or powder.
Odor: Musty. The odor is a warning of exposure; however, no smell does not mean you are not being exposed.
Effects of exposure:
   Breathed: May cause irritation of the eyes, nose and throat, headache, nausea, vomiting and fever.
   Eyes: Contact may cause irritation and burns.
   Skin: Moderately toxic. Contact may cause irritation, smarting of the skin; possible burns.
   Swallowed: Moderately toxic. See BENZENE and HEXACHLORIDE entries.
Long-term exposure: May cause dermatitis and hives.
Storage: Store in a cool, well-ventilated place. Keep away from heat and sources of combustion.

★ 148 ★
BENZENE SULFONYL CHLORIDE
CAS: 98-09-9
Other names: BENZENESULFONYL CHLORIDE; BENZENE SULFONECHLORIDE; BENZENE SULFONE-CHLORIDE; BENZENESULFONIC (ACID) CHLORIDE; BENZENOSULFO-CHLOREK (POLISH); BENZENOSULPHOCHLORIDE; BSC-REFINED D; BENZENESULFONIC ACID CHLORIDE

Danger profile: Corrosive, poisonous, dangerous storage hazard, explosive hazard, toxic fumes in fire.
Uses: Manufacturing olefins.
Appearance: Colorless to brown oily liquid.
Effects of exposure:
   Breathed: Poison. Breathing this chemical may cause liver damage. May irritate the nose, throat, lungs and air passages. May cause cough, phlegm, difficult breathing, chest tightness, shortness of breath. Fluid may build up in the lungs (pulmonary edema) and death may follow. Pulmonary edema is a medical emergency that may be delayed one to two days following exposure.
   Eyes: Contact may cause severe irritation, burns and permanent damage.
   Skin: Poison. Passes through the skin, and may cause severe irritation and burns.
   Swallowed: Poison.
Long-term exposure: May cause chronic irritation of the air passages and lungs with cough and phlegm. Repeated, prolonged exposure may cause liver damage. Repeated contact can cause dry skin, chronic irritation, redness, rash and sores.
Storage: Store in tightly closed containers in a dry, cool, well-ventilated place. Keep away from water, dimethyl sulfoxide, methyl formamide, strong acids and strong bases.

First aid guide: Move victim to fresh air and call emergency medical care; if not breathing, give artificial respiration; if breathing is difficult, give oxygen. In case of contact with material, immediately flush skin or eyes with running water for at least 15 minutes. Remove and isolate contaminated clothing and shoes at the site. Keep victim quiet and maintain normal body temperature. Effects may be delayed, keep victim under observation.

★ 149 ★
BENZIDINE
CAS: 92-87-5
Other names: BENZIDIN (CZECH); BENZIDINA (ITALIAN); BENZYDYNA (POLISH); P,P-BIANILINE; 4,4'-BIANILINE; (1,1'-BIPHENYL)-4,4'-DIAMINE (9CI); 4,4'-BIPHENYLDIAMINE; BIPHENYL, 4,4'-DIAMINO-; 4,4'-BIPHENYLENEDIAMINE; C.I. 37225; C.I. AZOIC DIAZO COMPONENT 112; P,P'-DIAMINOBIPHENYL; 4,4'-DIAMINOBIPHENYL; 4,4'-DIAMINO-1,1'-BIPHENYL; P-DIAMINODIPHENYL; 4,4'-DIAMINODIPHENYL; P,P'-DIANILINE; 4,4'-DIPHENYLENEDIAMINE; NCI-C03361; (1,1'-BIPHENYL)-4,4'DIAMINE; FAST CORINTH BASE B; C.I. AZOIC DIAZO COMPONENT 112

Danger profile: Carcinogen, mutagen, poisonous, Community Right-to-Know List, exposure limit established, poisonous gas produced in fire.
Uses: Making dyes, rubbers; detection of blood stains; as a stain in microscopy; laboratory chemical.
Appearance: Grayish-yellow, crystals; white or slightly reddish powder. Darkens on exposure to air and light and slowly changes from a solid to a gas.
Effects of exposure:
   Breathed: Poison. Any exposure is considered extremely dangerous. Symptoms may be the same as listed under swallowed.
   Eyes: Any exposure is considered extremely dangerous. May cause irritation.
   Skin: Passes through the unbroken skin; can increase exposure and the severity of the symptoms listed. Any exposure is considered extremely dangerous. Contact may cause irritation, rash or a burning feeling. Skin allergy may develop; future exposure can cause itching and rash.
   Swallowed: Poison. Immediately seek medical attention. May cause nausea, vomiting, damage to the blood, liver and kidneys; decrease in liver, kidney and body weight; and increase in spleen weight; swelling of the liver; blood in the urine.
Long-term exposure: May cause cancer. It has been shown to cause bladder tumors; the average latency period between first exposure and tumor appearance is 17 years. May cause skin allergy. May cause damage to blood, including bone marrow depression.
Storage: Store in tightly closed containers in a dry, cool, well-ventilated place. Keep away from heat; emits highly toxic fumes when heated.
First aid guide: Move victim to fresh air; call emergency medical care. In case of contact with material, immediately flush skin or eyes with running water for at least 15 minutes. Remove and isolate contaminated clothing and shoes at the site.

★ 150 ★
BENZO(A)PYRENE
CAS: 50-32-8
Other names: BENZO(D,E,F)CHRYSENE; BENZOPYRENE; BAP; BP; 3,4-BENZOPYRENE; 3,4-BENZOPIRENE (ITALIAN); 6,7-BENZOPYRENE; 3,4-BENPYREN (GERMAN); 3,4-BENZ(A)PYRENE

Danger profile: Possible carcinogen, possible reproductive hazard, exposure limit established, acrid fumes in fire.
Uses: Laboratory chemical; found in water food and smoke; common air contaminant and by-product of burning various carbon products (tar soot, pitch, carbon black, coal, coke, etc.).
Appearance: Pale yellow crystals.
Effects of exposure:
  Eyes: Contact may cause irritation, redness, tearing and burning sensation, and burns. Sunlight can make these conditions worse.
  Skin: Contact may cause irritation, redness, rash and burns. Sunlight can make these conditions worse.
Long-term exposure: May cause cancer of the skin, lung and bladder; may damage the developing fetus. Chemical may be transferred to infant through exposed mother's milk. May effect male reproductive organs. Repeated exposure may cause bronchitis; acne, skin darkening and thickening; loss of color, warts.
Storage: Store in tightly closed containers in a dry, cool, well-ventilated, marked and regulated place. Keep away from oxidizers.

★ 151 ★
BENZOIC ACID
CAS: 65-85-0
Other names: ACIDE BENZOIQUE (FRENCH); BENZENE-CARBOXYLIC ACID; BENZENEFORMIC ACID; BENZENEME-THANOIC ACID; BENZOATE; CARBOXYBENZENE; DRAC-YLIC ACID; KYSELINA BENZOOVA (CZECH); PHENYLCARBOXYLIC ACID; PHENYLFORMIC ACID; RE-TARDER BA; RETARDEX; SALVO LIQUID; SALVO POWDER TENN-PLAS

Danger profile: Mutagen, poisonous, combustible, acrid smoke and irritating fumes produced in fire.
Uses: Making plastic products tougher and more flexible; making other chemicals; alkyd resins; food preservative; processing tobacco; flavors; perfumes; tooth paste and cleaners; laboratory chemical; antifungal agent.
Appearance: White crystals, powder or flakes.
Odor: Faint, pleasant; slight aromatic. The odor is a warning of exposure; however, no smell does not mean you are not being exposed.
Effects of exposure:
  Breathed: At elevated temperatures, fumes may cause irritation of eyes, nose, throat and breathing passages.
  Eyes: Contact may cause severe irritation and burns.
  Skin: Contact may cause irritation, burning feeling, rash and burns.
  Swallowed: Moderately toxic.
Long-term exposure: May damage the developing fetus.
Storage: Store in tightly closed containers in a dry, cool, well-ventilated place. Keep away from heat, sources of ignition, strong oxidizers, strong bases. Containers may explode in fire; dust may explode.

★ 152 ★
BENZONITRILE
CAS: 100-47-0
Other names: BENZENE, CYANO-; BENZOIC ACID NITRILE; CYANOBENZENE; PHENYL CYANIDE; BENZENENITRILE

Danger profile: Combustible, Community Right-to-Know List, poisonous gases produced in fire.
Uses: Making other chemicals; rubber chemicals; solvent for nitrile rubber, lacquers, many resins, polymers and metal salts.
Appearance: Colorless liquid.
Odor: Almond-like. The odor is a warning of exposure; however, no smell does not mean you are not being exposed.

Effects of exposure:
  Breathed: May be poisonous. May cause irritation of the nose, throat and lungs. Earliest symptoms of cyano intoxication may be weakness, headaches, confusion, nausea and vomiting, sleepiness. Respiratory rate and depth will usually be increased at beginning and at later stages become slow and gasping. Blood pressure is usually normal, especially in mild or moderately severe cases, although the pulse rate is usually more rapid than normal. Higher levels may cause dizziness, difficult breathing, convulsions, coma and death.
  Eyes: Contact may cause irritation. See *Breathed.*
  Skin: May be poisonous. Passes through the unbroken skin; can increase exposure and the severity of the symptoms listed in *Breathed.* Contact may cause irritation. Prolonged contact may cause blistering.
  Swallowed: May be poisonous.
Long-term exposure: May cause damage to the liver and central nervous system.
Storage: Store in tightly closed containers in a dry, cool, well-ventilated place. Keep away from strong acids, strong bases, strong reducing agents and strong oxidizers, heat and sources of ignition. Containers may explode in fire. May attack some plastics.
First aid guide: Move victim to fresh air and call emergency medical care; if not breathing, give artificial respiration; if breathing is difficult, give oxygen. In case of contact with material, immediately flush skin or eyes with running water for at least 15 minutes. Speed in removing material from skin is of extreme importance. Remove and isolate contaminated clothing and shoes at the site. Keep victim quiet and maintain normal body temperature. Effects may be delayed; keep victim under observation.

★ 153 ★
BENZOYL CHLORIDE
CAS: 98-88-4
Other names: BENZALDEHYDE, ALPHA-CHLORO-; BENZENECARBONYL CHLORIDE; BENZOIC ACID, CHLORIDE; ALPHA-CHLOROBENZALDEHYDE

Danger profile: Mutagen, corrosive, combustible, reacts violently with water producing corrosive and poisonous fumes.
Uses: Making dyes; other chemicals; laboratory chemical.
Appearance: Colorless to slightly brown watery liquid.
Odor: Strong, pungent, penetrating. The odor is a warning of exposure; however, no smell does not mean you are not being exposed.
Effects of exposure:
  Breathed: Poison. Vapor may cause irritation to nose, throat, mouth and lungs; mucous membrane. Symptoms of exposure include watering of the eyes, coughing, difficult breathing, headache, nausea, vomiting, muscular weakness, unconsciousness. High exposure can lead to drowsiness, coma, buildup of fluid in the lungs (pulmonary edema), and may cause death. Pulmonary edema may be delayed by one to two days following exposure.
  Eyes: Severe irritant. Contact may cause burning and stinging, tearing, irritation and redness and swelling of the eyelids, intense pain; can burn the eyes and lead to blurred vision, chronic irritation and other permanent damage.
  Skin: Severe irritant. Contact may cause burning or stinging feeling, redness, swelling, pain, blisters, rash and burns.
  Swallowed: Poison. May cause burning feeling of the mouth and throat, intense thirst, throat swelling, stomach cramps, nausea, vomiting, diarrhea, severe breathing difficulties and unconsciousness. Tissue damage may result. Breathing may stop. Watch for delayed symptoms.
Long-term exposure: Possible risk of cancer; lung damage. May lead to chronic sore throat; chronic sinus problems; reduced sense of smell; chronic skin rash.

Storage: Store in tightly closed containers in a dry, cool, well-ventilated place with non-wood floors. Keep away from water, steam, strong oxidizers, strong bases, alcohols, dimethyl sulfoxide and combustible materials. Causes slow corrosion of metals.
First aid guide: Move victim to fresh air and call emergency medical care; if not breathing, give artificial respiration; if breathing is difficult, give oxygen. In case of contact with material, immediately flush skin or eyes with running water for at least 15 minutes. Speed in removing material from skin is of extreme importance. Remove and isolate contaminated clothing and shoes at the site. Keep victim quiet and maintain normal body temperature.

★ 154 ★
BENZOYL PEROXIDE
CAS: 94-36-0
Other names: ACETOXYL; ACETOWYL; ACNEGEL; AZTEC BPO; BENZOIC ACID PEROXIDE; BENZOPEROXIDE; BENZOYLPEROXID (GERMAN); BENZOYLPEROXYDE (DUTCH); DIBENZOYL PEROXIDE; BENZOYL SUPEROXIDE; BENOXYL; BENZAKNEW; BZF-60; CADET; CADOX; CLEARASIL BENZOYL PEROXIDE LOTION; CLEARASIL BP ACNE TREATMENT; CUTICURA ACNE CREAM; DEBROXIDE; DIBENZOYLPEROXID (GERMAN); EPICLEAR; FOSTEX; GAROX; INCIDOL; LOROXIDE; LUCIDOL; LUPERCOL; LUPEROX FL; NOROX; BPZ-250; NOVADELOX; OXY-5; OXY-10; OXYLITE; OXY WASH; PAN OXYL; PANOXYL; PEROXYDE DE BENZOYLE (FRENCH); PERSADOX; PEROXIDE, DIBENZOYL; SULFOXYL; THERADERM; XERAC; QUINOLOR COMPOUND; TOPEX; VANOXIDE

Danger profile: Possible carcinogen, poisonous, allergen, irritant, highly combustible, highly reactive, fire hazard, explosion hazard, exposure limit established, Community Right-to-Know List, poisonous gases produced in fire.
Uses: Making plastics, rubber, inks, pharmaceuticals, cosmetics and skin creams, acetate yarns, cheese, vinyl flooring; bleaching agent for flour, fats, oils, and waxes; rubber vulcanization.
Appearance: White solid powder or granules.
Effects of exposure:
Breathed: May cause irritation of the eyes, nose, throat and breathing passages.
Eyes: Contact may cause irritation, burns and permanent damage.
Skin: Contact may cause irritation, redness, rash and burns.
Long-term exposure: May cause skin allergy, itching and rash; dermatitis. May effect the lungs.
Storage: Store in tightly closed containers in a cool, well ventilated place. Keep away from heat, sources of ignition, combustible materials, alcohols, lithium aluminum hydride, dimethyl aniline, amines, and metallic naphthenates. Containers may explode in fire. May attack some plastics, rubber and coatings, producing fire or explosions. Protect containers from shock and friction.
First aid guide: Move victim to fresh air; call emergency medical care. In case of contact with material, immediately flush eyes with running water for at least 15 minutes. Wash skin with soap and water. Remove and isolate contaminated clothing and shoes at the site. Keep victim quiet and maintain normal body temperature.

★ 155 ★
BENZYL BROMIDE
CAS: 100-39-0
Other names: BENZENE, (BROMOMETHYL)-; (BROMOMETHYL)BENZENE; BROMOPHENYLMETHANE; ALPHA-BROMOTOLUENE; OMEGA-BROMOTOLUENE; P-(BROMOMETHYL)NITROBENZENE

Danger profile: Mutagen, corrosive, poisonous gases produced in fire.
Uses: Making foaming and frothing agents, and other chemicals.
Appearance: Clear liquid.
Odor: Pleasant. The odor is a warning of exposure; however, no smell does not mean you are not being exposed.
Effects of exposure:
Breathed: Poison. May cause irritation of the eyes, nose and throat. Large doses may cause central nervous system depression; shortness of breath and a dangerous build-up of fluids in the lungs (pulmonary edema), which can cause death. Pulmonary edema is a medical emergency that can be delayed one to two days following exposure.
Eyes: Contact may cause severe irritation, tearing, burns and permanent damage.
Skin: May be poisonous. Contact may cause severe irritation, redness, swelling, rash or sores and burns.
Swallowed: Poison. May cause irritation of the mouth and stomach.
Long-term exposure: May cause genetic damage in living cells; birth defects, miscarriage and cancer. May effect the lungs. Repeated skin contact may cause chronic irritation, dry skin, redness and rash.
Storage: Store in tightly closed containers in a dry, cool, well-ventilated place. Keep away from water and nearly all common metals.
First aid guide: Move victim to fresh air and call emergency medical care; if not breathing, give artificial respiration; if breathing is difficult, give oxygen. In case of contact with material, immediately flush skin or eyes with running water for at least 15 minutes. Remove and isolate contaminated clothing and shoes at the site. Keep victim quiet and maintain normal body temperature. Effects may be delayed, keep victim under observation.

★ 156 ★
BENZYL CHLORIDE
CAS: 100-44-7
Other names: BENZENE, (CHLOROMETHYL)-; BENZILE (CLORURO DI) (ITALIAN); BENZYLE (CHLORURE DE) (FRENCH); BENZYLCHLORID (GERMAN); CHLOROMETHYLBENZENE; CHLOROPHENYLMETHANE; ALPHA-CHLOROTOLUENE; OMEGA-CHLOROTOLUENE; ALPHA-CHLORTOLUOL (GERMAN); CHLORURE DE BENZYLE (FRENCH); NCI-C06360; TOLYL CHLORIDE; ALPHA-CHLOROTOLUENE; BENZENE, CHLOROMETHYL-

Danger profile: Carcinogen, mutagen, poisonous, combustible, explosive, Community Right-to-Know List, EPA extremely hazardous list, exposure limit established, poisonous gas produced in fire.
Uses: Making plastics, dyes, other chemicals, lubricants, gasoline, perfumes, pharmaceuticals, photographic developer.
Appearance: Colorless to yellow watery liquid.
Odor: Pungent, irritating. The odor is a warning of exposure; however, no smell does not mean you are not being exposed.
Effects of exposure:
Breathed: May be poisonous. Vapor may cause intense irritation to nose, throat, mouth and lungs; mucous membrane. Symptoms of exposure include watering of the eyes, coughing, difficult breathing, headache, nausea, vomiting, muscular weakness, unconsciousness. High exposure can lead to drowsiness, coma and death; buildup of fluid in the lungs (pulmonary edema), and may cause death. Pulmonary edema may be delayed by one to two days following exposure.
Eyes: Severe irritant. Contact may cause; burning and stinging, tearing, irritation and redness and swelling of the eyelids,

intense pain; can burn the eyes and lead to blurred vision, chronic irritation and other permanent damage.

Skin: Severe irritant. May be poisonous. Contact may cause burning or stinging feeling, redness, swelling, intense pain, blisters, irritation, rash. Repeated exposure may lead to skin allergy and chronic irritation.

Swallowed: May be poisonous. May cause burning feeling of the mouth and throat, intense thirst, throat swelling, stomach cramps, nausea, vomiting, diarrhea and unconsciousness, kidney damage, and severe breathing difficulties. Breathing may stop. Watch for delayed symptoms.

Long-term exposure: May cause cancer and genetic changes in living cells; damage to the developing fetus and miscarriage. May cause skin allergy, itching and skin rash; liver damage; possible lung damage.

Storage: Store in tightly closed containers in a dry, cool, well-ventilated place. Keep away from water, active metals, oxidizing materials, dimethyl sulfoxide. Containers may explode in fire.

First aid guide: Move victim to fresh air and call emergency medical care; if not breathing, give artificial respiration; if breathing is difficult, give oxygen. In case of contact with material, immediately flush skin or eyes with running water for at least 15 minutes. Remove and isolate contaminated clothing and shoes at the site. Keep victim quiet and maintain normal body temperature. Effects may be delayed, keep victim under observation.

★ 157 ★
BENZYL TRICHLORIDE
CAS: 98-07-7
Other names: BENZENE, TRICHLOROMETHYL-; BENZENYL CHLORIDE; BENZENYLTRICHLORIDE; BENZOIC TRICHLORIDE; BENZYLIDYNECHLORIDE; BENZOTRICHLORIDE; CHLORURE DE BENZENYLE (FRENCH); PHENYL CHLOROFORM; PHENYLCHLOROFORM; PHENYLTRICHLOROMETHANE; TOLUENE TRICHLORIDE; TRICHLOORMETHYLBENZEEN (DUTCH); TRICHLORMETHYLBENZOL (GERMAN); TRICHLOROMETHYLBENZENE; 1-(TRICHLOROMETHYL)BENZENE; TRICHLOROPHENYLMETHANE; ALPHA,ALPHA,ALPHA-TRICHLOROTOLUENE; OMEGA,OMEGA,OMEGA-TRICHLOROTOLUENE; TRICLOROMETILBENZENE (ITALIAN); TRICLOROTOLUENE (ITALIAN)

Danger profile: Carcinogen, possible mutagen, corrosive, EPA extremely hazardous substance list.
Uses: To make synthetic dyes.
Appearance: Colorless to yellow liquid. Fumes when exposed to air.
Odor: Penetrating. The odor is a warning of exposure; however, no smell does not mean you are not being exposed.
Effects of exposure:
Breathed: Vapors may irritate the nose, throat and lungs; headache, weakness, fatigue, irritability; poor appetite. Higher exposures may cause shortness of breath and a dangerous build-up of fluids in the lungs (pulmonary edema), which can cause death. Pulmonary edema is a medical emergency that can be delayed one to two days following exposure.
Eyes: Contact may cause severe irritation, burns and permanent damage.
Skin: Passes through the unbroken skin; can increase exposure and the severity of the symptoms listed. Contact may cause severe irritation and burns.
Swallowed: May be harmful.
Long-term exposure: May cause cancer of the lungs, stomach and skin; possible reproductive damage; birth defects and miscarriage.
Storage: Store in tightly closed containers in a dry, cool, well-ventilated, regulated and marked place with non-wood floors.

Keep away from heat, sources of ignition and combustible materials.

★ 158 ★
BERYLLIUM
CAS: 7440-41-7
Other names: BERYLLIUM-9; BERYLLIUM METAL POWDER; GLUCINIUM; GLUCINUM; BERYLLIUM DUST

Danger profile: Carcinogen, Community Right-to-Know List, poisonous gas produced in fire.
Uses: Extensively used in the making electrical components, ceramics, other chemicals; in computer parts; solid propellant and rocket fuel.
Appearance: Hard, brittle, gray-white, silver solid.
Effects of exposure:
Breathed: Dust is extremely toxic. May cause irritation to the breathing passages and lungs with nasal discharge, chest tightness, shortness of breath, difficult breathing, coughing; possible fever. Symptoms are often delayed for days following exposure. Death may occur.
Eyes: Contact may cause irritation, itching and burning feeling. Allergy may develop.
Skin: Contact may cause irritation, redness and rash. Lumps and ulcers may develop if substance gets under the skin.
Long-term exposure: May cause cancer; scars on the lungs and possibly other organs; development of kidney stones. Skin and eye allergy with redness and rash may develop. Symptoms may include fatigue, shortness of breath, poor appetite and weight loss.
Storage: Store in tightly closed containers in a dry, cool, well-ventilated place. Keep away from heat, sources of ignition, oxidizers, strong acids and water. Forms an explosive mixture in air.

★ 159 ★
BERYLLIUM CHLORIDE
CAS: 7787-47-5
Other names: BERYLLIUM DICHLORIDE

Danger profile: Carcinogen, mutagen, poisonous, Community Right-to-Know List, exposure limit established, poisonous fumes produced in fire.
Uses: Making other chemicals.
Appearance: White to faint green or yellow crystals or needles.
Odor: Sharp, acrid. The odor is a warning of exposure; however, no smell does not mean you are not being exposed.
Effects of exposure:
Breathed: Dust may be poisonous. May cause irritation of the nasal passages and throat; pneumonitis; cough.
Eyes: Contact may cause redness, inflammation of the eyelids.
Skin: Contact may cause irritation.
Swallowed: May be poisonous. May cause irritation of mouth, stomach.
Long-term exposure: May cause beryllium disease. Any dramatic, unexplained weight loss should be considered as possible first indication of this disease. Pulmonary hypertension, right heart failure, pulmonary insufficiency, cachexia followed by death, renal failure and cardiac arrest. Symptoms may be delayed as much as 10 years. Joint pains, sarcoid-like skin lesions and renal calculi; nasopharyngitis, tracheobronchitis, dyspnea, chronic cough.
Storage: Store in tightly closed containers in a dry, cool, well-ventilated place. Keep away from water. Corrodes most metals in presence of moisture. Combustible and explosive hydrogen gas may collect in enclosed spaces.
First aid guide: Move victim to fresh air; call emergency medical care. In case of contact with material, immediately flush skin or

eyes with running water for at least 15 minutes. Remove and isolate contaminated clothing and shoes at the site.

★ 160 ★
BERYLLIUM FLUORIDE
CAS: 7787-49-7
Other names: BERYLLIUM DIFLUORIDE

Danger profile: Carcinogen, poisonous, Community Right-to-Know List, exposure limit established, poisonous gases produced in fire.
Uses: Making glass, other chemicals; in nuclear reactors.
Appearance: Colorless solid.
Effects of exposure:
Breathed: Inhalation causes irritation of nose, throat, lungs, severe pneumonitis, and/or a dangerous build-up of fluids in the lungs (pulmonary edema), which can cause death. Pulmonary edema is a medical emergency that can be delayed one to two days following exposure.
Eyes: Contact with eyes may cause severe irritation and burns.
Skin: Contact with skin may cause irritation, redness, dermatitis with non-healing ulcers.
Swallowed: May cause fatigue, weakness, loss of appetite.
Long-term exposure: May cause beryllium disease. Any dramatic weight loss should be considered possible first indication of this disease. Pulmonary hypertension, right heart failure, pulmonary insufficiency, cachexia followed by death, renal failure and cardiac arrest. Symptoms may be delayed as much as 10 years. Joint pains, sarcoid-like skin lesions and renal calculi.
Storage: Store in tightly closed containers in a dry, cool, well-ventilated place.
First aid guide: Move victim to fresh air; call emergency medical care. In case of contact with material, immediately flush skin or eyes with running water for at least 15 minutes. Remove and isolate contaminated clothing and shoes at the site.

★ 161 ★
BERYLLIUM NITRATE
CAS: 13597-99-4
Other names: BERYLLIUM DINITRATE; NITRIC ACID, BERYLLIUM SALT

Danger profile: Carcinogen, Community Right-to-Know List, poisonous, oxidizer, exposure limit established, poisonous gases produced in fire.
Uses: Making other chemicals.
Appearance: White-yellowish crystals.
Effects of exposure:
Breathed: May be poisonous. May cause irritation of the nasal passages and throat.
Eyes: Contact may cause irritation, redness and inflammation of the eyelids.
Skin: Contact may cause irritation, dermatitis and non-healing ulcers.
Swallowed: May be poisonous. May cause loss of appetite, fatigue, weakness, malaise.
Long-term exposure: May cause beryllium disease. Any dramatic weight loss should be considered possible first indication of this disease. Pulmonary hypertension, right heart failure, pulmonary insufficiency, cachexia followed by death, renal failure and cardiac arrest. Symptoms may be delayed as much as ten years. Joint pains, sarcoid-like skin lesions and renal calculi; pneumonotis, nasopharyngitis, tracheobronchitis, dyspnea, chronic cough.
Storage: Store in tightly closed containers in a dry, cool, well-ventilated place. Keep away from combustable materials. In

presence of moisture can damage wood and corrode most metals.
First aid guide: Move victim to fresh air; call emergency medical care. In case of contact with material, immediately flush skin or eyes with running water for at least 15 minutes. Remove and isolate contaminated clothing and shoes at the site.

★ 162 ★
BERYLLIUM OXIDE
CAS: 1304-56-9
Other names: BERYLLIA; BROMELLITE; BERYLLIUM MONOXIDE; THERMALOX

Danger profile: Carcinogen, Community Right-to-Know List, exposure limit established, poisonous gases produced in fire.
Uses: Making glass, ceramics and plastics; electron tubes, other chemicals; resistor cores; transistor mountings; high-temperature reactor systems.
Appearance: White solid; light amorphous powder.
Effects of exposure:
Breathed: May cause irritation of the nasal passages and throat.
Eyes: Contact may cause irritation , redness and inflammation of the eyelids.
Skin: Contact may cause irritation, dermatitis and non-healing ulcers.
Swallowed: May cause loss of appetite, fatigue, weakness, malaise.
Long-term exposure: May cause cancer. May cause beryllium disease. Any dramatic weight loss should be considered possible first indication of this disease. Pulmonary hypertension, right heart failure, pulmonary insufficiency, cachexia followed by death, renal failure and cardiac arrest. Symptoms may be delayed as much as 10 years. Joint pains, sarcoid-like skin lesions and renal calculi; pneumonotis, nasopharyngitis, tracheobronchitis, dyspnea, chronic cough.
Storage: Store in tightly closed containers in a dry, cool, well-ventilated place. Keep away from heat, magnesium and acids.

★ 163 ★
BHT
CAS: 128-37-0
Other names: ADVASTAB 401; AGIDOL; ANTIOXIDANT DBPC; ANTIOXIDANT 29; ANTRACINE 8; AO 29; AO4K; CAO 1; BUTYLATED HYDROXYTOLUENE; 2,6-DI-T-BUTYL-4-METHYLPHENOL; CAO 3; CATALIN CAO-3; CHEMANOX 11; DBMC; 2,6-DI-TERT-BUTYL-P-CRESOL; 4-HYDROXY-3,5-DI-T-BUTYL-TOLUENE; HYDROXYTOLUENE; IMPRUVOL; IONOL; IONOL CP; KERABIT; 4-METHYL-2,6-DI-T-BUTYLPHENOL; NCI-C03598; NONOX TBC; PARABAR 441; B-SANTALOL; SUSTANCE; SUSTANE; TENOX BHT; TOPANOL A-SANTALOL; VANLUBE PCX

Danger profile: Mutagen, combustible, possible reproductive hazard, exposure limit established, poisonous gases produced in fire.
Uses: Antitoxidant; stabilizer in rubber, plastic and petroleum; food preservative and additive.
Appearance: Colorless or white crystals.
Odor: Faint, characteristic. The odor is a warning of exposure; however, no smell does not mean you are not being exposed.
Effects of exposure:
Breathed: May cause irritation of the nose and throat.
Eyes: May cause irritation.
Skin: May cause irritation.
Swallowed: Moderately toxic.

Long-term exposure: May damage the developing fetus; liver effects; reduce blood's ability to clot; behavior and learning effects.
Storage: Store in tightly closed containers in a dry, cool, well-ventilated place. Keep away from heat, sources of ignition and oxidizers.

★ 164 ★
BIPHENYL
CAS: 92-52-4
Other names: BIBENZENE; 1,1'-BIPHENYL; DIPHENYL; 1,1'-DIPHENYL; LEMONENE; PHENADOR-X; PHENYLBENZENE; PHPH; XENENE

Danger profile: Mutagen, combustible, exposure limit established, Community Right-to-Know List, poisonous gas produced in fire.
Uses: Making other chemicals; heat-transfer agent; fungistat in packaging of citrus fruit; plant disease control; dyeing polyesters.
Appearance: Colorless white to pale yellow crystals.
Odor: Characteristic odor is peculiar, but pleasant. The odor is a warning of exposure; however, no smell does not mean you are not being exposed.
Effects of exposure:
  Breathed: May cause irritation to the eyes, nose throat and lungs; headache, stomach pain, nausea, indigestion, fatigue and aching arms and legs. Very high levels may cause severe poisoning; brain and nerve damage; bronchitis; severe liver damage; possible death.
  Eyes: May cause irritation.
  Skin: Passes through the unbroken skin; can increase exposure and the severity of the symptoms listed. May cause irritation and skin allergy to develop.
  Swallowed: Moderately toxic. See *Breathed.*
Long-term exposure: May cause birth defects, miscarriage or cancer. Repeated exposure may cause brain damage; nerve damage with trembling, weakness, fatigue, numbness, headache, sleeplessness and mood changes; loss of appetite; skin allergy; lung damage and bronchitis with cough, phlegm and /or shortness of breath.
Storage: Store in tightly closed containers in a dry, cool, well-ventilated place. Keep away from heat, sources of ignition and oxidizers.

★ 165 ★
BIS (AMINOPROPYL) PIPERAZINE
CAS: 7209-38-3
Other names: AMINOPROPYLPIPERAZINE; PIPERAZINE, 1,4-BIS (3- AMINOPROPYL)-; 1,4-BIS(AMINOPROTYL)PIPERAZINE; BIS(AMINOPROPYL)PIPERAZINE

Danger profile: Corrosive, extreme irritant, poisonous gases produced in fire.
Uses: Corrosion inhibitor; insecticide.
Appearance: Colorless, needle-like crystals.
Odor: Weak, fishy. The odor is a warning of exposure; however, no smell does not mean you are not being exposed.
Effects of exposure:
  Breathed: May cause irritation of the nose, throat, breathing passages and lungs; shortness of breath and coughing. Higher exposures may cause shortness of breath and a dangerous build-up of fluids in the lungs (pulmonary edema), which can cause death. Pulmonary edema is a medical emergency that can be delayed one to two days following exposure.
  Eyes: May cause severe irritation.

  Skin: May cause severe irritation.
Long-term exposure: May cause skin rash and allergy. May effect the lungs, and possibly cause allergy.
Storage: Store in tightly closed containers in a dry, cool, well-ventilated place with non-wood floors. Keep away from combustible materials.

★ 166 ★
BIS (BETA-CHLOROETHYL) METHYLAMINE
CAS: 51-75-2
Other names: BIS(2-CHLOROETHYL)METHYLAMINE; N,N-BIS(2-CHLOROETHYL)METHYLAMINE; CARYOLYSIN; CHLORAMINE; CHLORMETHINE; CLORAMIN; DICHLOR-AMINE; DICHLOREN (GERMAN); BETA,BETA'-DICHLORODIETHYL-N-METHYLAMINE; DI(2-CHLOROETHYL)METHYLAMINE; 2,2'-DICHLORO-N-METHYLDIETHYLAMINE; EMBICHIN; ENT-25294; HN2; MBA; MECHLORETHAMINE; METHYLBIS(BETA-CHLOROETHYL)AMINE; N-METHYL-BIS-CHLORAETHYLAMIN (GERMAN); N-METHYL-BIS(BETA-CHLOROETHYL)AMINE; N-METHYL-BIS(2-CHLOROETHYL)AMINE; MECHLORETHAMINE; N-METHYL-2,2'-DICHLORODIETHYLAMINE; METHYLDI(2-CHLOROETHYL)AMINE; N-METHYL-LOST; MUSTARGEN; MUSTINE; NITROGEN MUSTARD; N-LOST (GERMAN); NSC 762; TL 146; CHLORETHAZINE; DIETHYLAMINE,2,2'-DICHLORO-N-METHYL-(8CI); ETHANAMINE,2-CHLORO-N-(2-CHLOROETHYL)-N-METHYL-(9CI); METHYLBIS(2-CHLOROETHYL)AMINE; N,N-BIS(2-CHLOROETHYL)METHYLAMINE; N,N-DI(CHLOROETHYL)METHYLAMINE

Danger profile: Carcinogen, mutagen, possible reproductive hazard, poisonous, powerful irritant, EPA extremely hazardous substance list, Community Right-to-Know List, very poisonous gas produced in fire.
Uses: Blistering agent in warfare.
Appearance: Dark, mobile liquid.
Odor: Faint, fishy. The odor is a warning of exposure; however, no smell does not mean you are not being exposed.
Effects of exposure:
  Breathed: Deadly poison. Any exposure is considered extremely dangerous. May cause death.
  Eyes: Deadly poison. Any exposure is considered extremely dangerous. Powerful irritant.
  Skin: Deadly poison. Any exposure is considered extremely dangerous. Powerful irritant.
  Swallowed: Deadly poison. Any exposure is considered extremely dangerous.
Long-term exposure: Skin contact may cause tumors; cancer and mutations.
Storage: Store in tightly closed containers in a dry, cool, well-ventilated place.

★ 167 ★
BIS (2-CHLOROETHYL) SULFIDE
CAS: 505-60-2
Other names: BIS(BETA-CHLOROETHYL)SULFIDE; BIS(2-CHLOROETHYL)SULPHIDE; 1-CHLORO-2-(BETA-CHLOROETHYLTHIO)ETHANE; BETA,BETA'-DICHLOROETHYL SULFIDE; 2,2'-DICHLORODIETHYL SULFIDE; DI-2-CHLOROETHYL SULFIDE; BETA,BETA-DICHLOR-ETHYL-SULPHIDE; 2,2'-DICHLOROETHYL SULPHIDE; DISTILLED MUSTARD; KAMPSTOFF "LOST"; MUSTARD GAS; MUSTARD HD; MUSTARD VAPOR; SCHWEFEL-LOST; S-LOST; S MUSTARD; SULFUR MUSTARD GAS; SULFUR MUSTARD; SULPHUR MUSTARD; SULPHUR MUSTARD GAS; 1,1'-THIOBIS(2-CHLOROETHANE); YELLOW CROSS LIQUID; YPERITE

Danger profile: Combustible, carcinogen, mutagen, poisonous, poisonous gas produced in fire and on contact with water or steam and acid or acid fumes.
Uses: Solvent; making lubricating oils and other chemicals; processing textiles and dyeing; paints, varnishes, lacquers; paint and finish strippers; dry cleaning; soil fumigant; military blistering gas.
Appearance: Colorless to light yellow, oily liquid.
Odor: Weak, sweet odor or garlic-like odor. The odor is a warning of exposure; however, no smell does not mean you are not being exposed.
Effects of exposure:
   Breathed: May cause severe irritation to the nose, throat, breathing passages and lungs. Higher exposures may cause shortness of breath and a dangerous build-up of fluids in the lungs (pulmonary edema), which can cause death. Pulmonary edema is a medical emergency that can be delayed one to two days following exposure. Pulmonary lesions may develop, and are often fatal.
   Eyes: May cause severe irritation and blindness.
   Skin: Passes through the unbroken skin; can increase exposure and the severity of the symptoms listed. Penetrates deeply and may cause injury to blood vessels. May cause intense inflammation with pain. Secondary infections are possible.
   Swallowed: May cause severe gastric disturbances; nausea and vomiting.
Long-term exposure: May cause cancer, mutations. In Japan worker exposure was studied in a dyestuff factory. Lung cancer cases in these workers were found after 23 months of employment (*Hazchem Alert*, 5/20/91.)
Storage: Store in tightly closed containers in a dry, cool, well-ventilated place. Keep away from heat, flame and sources of ignition, water and steam, acid, acid fumes, bleach and oxidizing materials. Containers may explode in fire.

★ 168 ★
BIS(CHLOROMETHYL) ETHER
CAS: 542-88-1
Other names: CHLORO(CHLOROMETHOXY)METHANE; CHLOROMETHYL ETHER; DICHLORODIMETHYL ETHER; DICHLORDIMETHYLAETHER (GERMAN); SYM-DICHLORODIMETHYL ETHER; SYM-DICHLOROMETHYL ETHER; DIMETHYL-1,1'-DICHLOROETHER; OXYBIS(CHLOROMETHANE); BCME; BIS-CME

Danger profile: Fire hazard, carcinogen, mutagen, poisonous, fire hazard, Community Right-to-Know List, EPA extremely hazardous substance list, poisonous gases produced in fire.
Uses: Laboratory chemical; making ion exchange resins.
Appearance: Volatile liquid.
Effects of exposure:

   Breathed: May cause irritation of the eyes, nose and nasal passages.
   Eyes: May cause irritation, redness and inflammation of the eyelids.
   Skin: Contact may cause irritation.
   Swallowed: Toxic.
Long-term exposure: May cause cancer and mutations.
Storage: Store in tightly closed containers in a dry, cool, well-ventilated place. Keep away from heat, flame and sources of ignition.

★ 169 ★
BIS (2-CHLOROMETHYL) ETHER
CAS: 542-88-1
Other names: BIS (CHLOROMETHYL) ETHER; METHANE OXYBIS (CHLORO-); BCME; BIS-CME; CHLORO(CHLOROMETHOXY)METHANE; DICHLORODIMETHYL ETHER; DICHLORDIMETHYLAETHER (GERMAN); SYM-DICHLORODIMETHYL ETHER; DIMETHYL-1,1'-DICHLOROETHER; OXYBIS (CHLOROMETHANE)

Danger profile: Carcinogen, mutagen, poisonous, fire hazard, Community Right-to-Know List, EPA extremely hazardous substances list, poisonous gases produced in fire.
Uses: Making plastics and ion exchange resins; laboratory chemical.
Appearance: Volatile liquid.
Effects of exposure:
   Breathed: Poison. Any exposure is considered extremely dangerous. May cause irritation of the nose, throat, breathing passages and lungs; shortness of breath and coughing. Exposure may cause loss of appetite, nausea and fatigue. Higher exposures may cause shortness of breath and a dangerous build-up of fluids in the lungs (pulmonary edema), which can cause death. Pulmonary edema is a medical emergency that can be delayed one to two days following exposure.
   Eyes: Poison. Any exposure is considered extremely dangerous. Contact may cause irritation, tearing, burning feeling, burns and permanent damage.
   Skin: Poison. Any exposure is considered extremely dangerous. Contact may cause irritation, redness and burns.
   Swallowed: Poison. Any exposure is considered dangerous.
Long-term exposure: May cause cancer; may cause reproductive damage; birth defects and miscarriage.
Storage: Store in tightly closed containers in a dry, cool, well-ventilated, marked and regulated place. Keep away form heat and sources of ignition.

★ 170 ★
BIS (2-ETHYLHEXYL) ADIPATE
CAS: 103-23-1
Other names: HEXANEDIOIC ACID, BIS (2-ETHYLHEXYL) ESTER; BEHA; DEHA; DOA; DIOCTYL ADIPATE; OCTYL ADIPATE; DI-2-ETHYLHEXYL ADIPATE

Danger profile: Possible carcinogen, possible mutagen, possible reproductive hazard, combustible, poisonous gases produced in fire.
Uses: Making plastics including polyvinyl tougher and more flexible; solvent; aircraft lubricants.
Appearance: Clear or slightly yellow oily liquid.
Effects of exposure:
   Breathed: Any exposure is considered dangerous. May cause irritation to the eyes, nose and throat.
   Eyes: Any exposure is considered dangerous. Contact may cause irritation, tearing and burns.

Skin: Any exposure is considered dangerous. Contact may cause irritation, redness and burns.
Swallowed: Any exposure is considered dangerous.
Long-term exposure: May cause cancer; damage to the developing fetus; a decrease in fertility of both males and females.
Storage: Store in tightly closed containers in a dry, cool, well-ventilated place. Keep away from heat, sources of ignition and nitrates.

★ 171 ★
BISPHENOL A DIGLYCIDYL ETHER
CAS: 1675-54-3
Other names: 2,2-BIS(4-(2,3-EPOXYPROPYLOXY)PHENYL)PROPANE; BIS(4-GLYCIDYLOXYPHENYL)DIMETHYAMETHANE; 2,2-BIS(P-GLYCIDYLOXYPHENYL)PROPANE; BIS(4-HYDROXYPHENYL)DIMETHYLMETHANE DIGLYCIDYL ETHER; 2,2-BIS(4-HYDROXYPHENYL)PROPANE,DIGLYCIDYL ETHER; D.E.R. 332; DIGLYCIDYL ETHER OF BISPHENOL A; EPI-REZ 508; EPI-REZ 510; EPON 828; EPOXIDE A; ERL-27774; 4,4'ISO-PROPYLIDENEDIPHENOL DIGLYCIDYL ETHER; 2,2'-(1-METHYLETHYLIDENE)BIS(4,1- PHENYLELEOXYMETHYL-ENE)BISOXIRANE

Danger profile: Possible carcinoge; possible mutagen, combustible, poisonous gases produced in fire.
Uses: Basic active ingredient of epoxy resins.
Appearance: Amber liquid.
Effects of exposure:
Breathed: May be poisonous. Vapors may cause irritation of the eyes, nose, throat and bronchial tubes; nosebleeds, coughing and tightness in the chest.
Eyes: May cause irritation and burns.
Skin: Passes through the unbroken skin; can increase exposure and the severity of the symptoms listed. May cause irritation burns and rash; possible skin allergy.
Swallowed: May be poisonous.
Long-term exposure: May cause skin cancer, birth defects, miscarriage, skin allergy.
Storage: Store in tightly closed containers in a dry, cool, well-ventilated place.

★ 172 ★
BLADEX
CAS: 21725-46-2
Other names: 2-CHLORO-4-(1-CYANO-1-METHYLETHYLAMINO)-6-ETHYLAMINO- 1,3,5-TRIAZINE; 2-CHLORO-4-ETHYLAMINO-6-(1-CYANO-1- METH-YL)ETHYLAMINO-S-TRIAZINE; 2-((4-CHLORO-6-(ETHYLAMINO)-S- TRIAZIN-2-YL)AMINO)-2-METHYLPROPIONITRILE 2-(4-CHLORO-6-(ETHYLAMINO)-1,3,5-TRIAZIN-2-YL)AM-2-METHYL-; BLADEX 80WP; CLEAV-AL; COUPLER; CYANAZINE; DW3418; EF737; ENVOY; FORTROL; FF6135; HOLTOX; METHYL PROPIONITRILE; PAYZE; PROPIONITRILE, BLADEX R; SD 45418; SFO 6266; STAY-KLEEN; TOPSHOT; WL 19805

Danger profile: Poisonous, possible mutagen, possible reproductive hazard, Community Right-to-Know List, poisonous fumes produced in fire.
Uses: Pesticide; herbicide, to control weeds and grass; weedkiller for wheat, barley, maize, broad beans and peas, cotton, sugar cane, potatoes, bulbs, onions.
Appearance: White to tan crystals.
Effects of exposure:

Breathed: May cause irritation of the nose, throat and air passages; difficult breathing, weakness, nausea. High levels may cause convulsions and coma.
Eyes: Contact may cause irritation.
Skin: Contact may cause irritation, redness and rash.
Swallowed: Poison. Very toxic; symptoms may include lethargy, labored breathing, blood stained saliva.
Long-term exposure: May effect the liver, thyroid gland. May cause goiter; skin allergy and rash. Mutation data reported. Harmful to fish, wildlife and domestic cattle.
Storage: Store in tightly closed containers in a dry, cool, well-ventilated place. Keep away from heat.

★ 173 ★
BORIC ACID
CAS: 10043-35-3
Other names: BORACIC ACID; BOROFAX; BORSAURE (GERMAN); NCI-C56417; ORTHOBORIC ACID; THREE ELE-PHANT

Danger profile: Poisonous, poisonous gas produced in fire.
Uses: Electroplating; medicinal ointments and eye washes; making glass, porcelain enamels and other chemicals; flame retardant.
Appearance: White solid.
Effects of exposure:
Breathed: Absorbed through the mucous membrane.
Eyes: Dust or aqueous solutions may cause severe irritation.
Skin: Dust and solutions are absorbed through burns and open wounds but not through unbroken skin. Dust or aqueous solutions may cause irritation especially if skin is wet.
Swallowed: May cause diarrhea, stomach cramps, nausea, vomiting, irregular heart beat, blue skin and fingertips, delirium, convulsions and coma. Ingestion of five grams or more may irritate gastrointestinal tract and affect central nervous system.
Long-term exposure: No chronic effects have been recognized, but continued contact should be avoided. Mutation data reported.
Storage: Store in tightly closed containers in a dry, cool, well ventilated place. Keep away from potassium.

★ 174 ★
BORON OXIDE
CAS: 1303-86-2
Other names: ANHYDROUS BORIC ACID; BORIC ANHY-DRIDE; BORIC OXIDE; BORON SEQUIOXIDE; BORON TRI-OXIDE; FUSED BORIC ACID

Danger profile: Exposure limit established.
Uses: Pesticide and herbicide; making other chemicals; heat-resistant glassware; fire-resistant materials and paints; electronics.
Appearance: Colorless, semi-transparent glassy granules, powder, flakes, lumps or hard white crystals.
Effects of exposure:
Breathed: May irritate the eyes, nose and throat.
Eyes: Contact may cause irritation.
Skin: Contact may cause irritation.
Swallowed: Moderately toxic.
Storage: Store in tightly closed containers in a dry, cool, well-ventilated place.

★ 175 ★
BORON TRIBROMIDE
CAS: 10294-33-4
Other names: BORANE, TRIBROMO-; BORON BROMIDE; BORON TRIBROMIDE 6; TRONA

Danger profile: Corrosive, poisonous, exposure limit established, corrosive, irritant, poisonous gases produced in fire, toxic, corrosive and explosive fumes produced on contact with water and steam.
Uses: Making other chemicals.
Appearance: Colorless fuming liquid.
Odor: Sharp, irritating. The odor is a warning of exposure; however, no smell does not mean you are not being exposed. usually found as a solution in n-hexane or methylene chloride.
Effects of exposure:
   Breathed: Poison. May cause lung irritation, shortness of breath, coughing; burns of the eyes, nose throat and lungs. Higher exposures may cause shortness of breath and a dangerous build-up of fluids in the lungs (pulmonary edema), which can cause death. Pulmonary edema is a medical emergency that can be delayed one to two days following exposure.
   Eyes: May cause irritation, inflammation of the mucous membrane, tearing and burns.
   Skin: May cause irritation, redness, burns and permanent damage.
   Swallowed: Poison. May cause burns of the mouth and stomach. Also see symptoms under *Breathed.*
Long-term exposure: May cause irritation of the lungs, bronchitis, shortness of breath, cough and phlegm. Repeated exposure may cause upset stomach, runny nose, bad breath and cause the tongue to turn brown.
Storage: May explode when heated. Store in tightly closed containers in a dry, cool, well-ventilated place. Keep away from heat, water and steam, potassium, sodium, and alcohol. Containers may explode in heat or fire.
First aid guide: Move victim to fresh air and call emergency medical care; if not breathing, give artificial respiration; if breathing is difficult, give oxygen. In case of contact with material, immediately flush skin or eyes with running water for at least 15 minutes. Remove and isolate contaminated clothing and shoes at the site. Keep victim quiet and maintain normal body temperature. Effects may be delayed, keep victim under observation.

★ 176 ★
BORON TRIFLUORIDE
CAS: 7637-07-2
Other names: BORON FLUORIDE; FLUORURE DE BORE (FRENCH); BORANE, TRIFLUORO-; TRIFLUORO BORANE; TRIFLUORO BORON

Danger profile: Poisonous, corrosive, strong irritant, explosion hazard, exposure limit established, EPA extremely hazardous substances list, poisonous fumes produced in fire.
Uses: Making other chemicals; soldering fluxes; brazing.
Appearance: Colorless gas; smoky in moist air.
Odor: Pungent, irritating, suffocating. The odor is a warning of exposure; however, no smell does not mean you are not being exposed.
Effects of exposure:
   Breathed: Poison. Vapor may cause intense irritation to nose, throat, mouth and lungs; mucous membrane; watering of the eyes, coughing, difficult breathing, headache, nausea, vomiting, muscular weakness, unconsciousness. High exposure can lead to drowsiness, coma and death; buildup of fluid in the lungs (pulmonary edema), and may cause death. Pul-

monary edema is a medical emergency that may be delayed by one to two days following exposure.
   Eyes: Severe irritant. Contact may cause, burning and stinging, tearing, irritation and redness and swelling of the eyelids, intense pain; can burn the eyes and lead to blurred vision, chronic irritation and permanent damage.
   Skin: May be poisonous if absorbed through the skin. Severe irritant. Contact may cause burning or stinging feeling, redness, swelling, intense pain, blisters, irritation, rash. Repeated exposure may lead to skin allergy and chronic irritation.
   Swallowed: May cause burning feeling of the mouth and throat, intense thirst, throat swelling, stomach cramps, nausea, vomiting, diarrhea and unconsciousness, kidney damage, and severe breathing difficulties. Breathing may stop. Watch for delayed symptoms.
Long-term exposure: May cause damage to the kidneys, lungs, teeth and bones; nose lining. Repeated exposure may cause nose dryness and nose bleeding.
Storage: Store in tightly closed cylinders in a dry, cool, well-ventilated place. Keep away from water, heat, sources of combustion, explosives, combustible materials, alkalies including metals. Protect containers from shock and temperature extremes. Maintain temperature above -20°F and below 130°F. Cylinders may explode in heat of fire.
First aid guide: Move victim to fresh air and call emergency medical care; if not breathing, give artificial respiration; if breathing is difficult, give oxygen. In case of contact with material, immediately flush skin or eyes with running water for at least 15 minutes. Remove and isolate contaminated clothing and shoes at the site. Keep victim quiet and maintain normal body temperature. Effects may delayed; keep victim under observation.

★ 177 ★
BROMACIL
CAS: 314-40-9
Other names: 5-BROMO-3-SEC-BUTYL-6-METHYLURACIL; ALPHA BROMACIL 80 WP; BOROCIL EXTRA; BOREA; BROMACIL 1.5; BROMAZIL; 5-BROMO-6-METHYL-3-(1-METHYLPROPYL)-2,4(1H,3H)-PYRIMIDINEDIONE; 3-SEK-BUTYL-5-BROM-6-METHYLURACIL (GERMAN); BROMAX; CROPTEX ONYX; CYNOGAN; DUPONT HERBICIDE 976; EEREX; EEREX GRANULAR WEED KILLER; EEREX WATER SOLUBLE GRANULAR WEED KILLER; FENOCIL; HERBICIDE 976; HYDON; HYVAR; HYVAREX; HYVAR X; HYVAR X BROMACIL; HYVAR X WEED KILLER; KROVAR II; NALKIL; URACIL, 5-BROMO-3-SEC-BUTYL-6-METHYL; URAGAN; URAGON; UROX; UROX B WATER SOLUBLE CONCENTRATE WEED KILLER; UROX B; UROX-HX; UROX HX GRANULAR WEED KILLER

Danger profile: Mutagen, possible reproductive hazard, exposure limit established, poisonous fumes produced in fire.
Uses: Herbicide, weed killer; on citrus groves and pineapple fields.
Effects of exposure:
   Breathed: May cause irritation of the eyes, nose, nasal passages and respiratory system.
   Eyes: May cause irritation.
   Skin: Passes through the unbroken skin; can increase exposure and the severity of the symptoms listed. May cause irritation.
   Swallowed: Moderately toxic.
Long-term exposure: May cause mutations, lead to birth effects, miscarriage and cancer; evidence of thyroid abnormalities in animal tests. Dangerous to wildlife.

★ 178 ★
BROMINE
CAS: 7726-95-6
Other names: BROM (GERMAN); BROME (FRENCH); BROMO (ITALIAN); BROOM (DUTCH)

Danger profile: Corrosive, exposure limit established, poisonous gases produced in fire.
Uses: Making other chemicals, anti-knock gasoline; bleach; water purification; solvent; fumigants; fire-retardant for plastics; dyes; pharmaceuticals; photography; shrink-proofing wool.
Appearance: Heavy, red-brown fuming liquid; rhombic crystals.
Odor: Sharp, harsh, penetrating with a choking, irritating odor, causes tears. The odor is a warning of exposure; however, no smell does not mean you are not being exposed.
Effects of exposure:
Breathed: Vapor may cause intense irritation to nose, throat, mouth and lungs; mucous membrane. Symptoms of exposure include watering of the eyes, coughing, difficult breathing, headache, nausea, vomiting, muscular weakness, unconsciousness. Higher exposures may cause windpipe swelling and suffocation; shortness of breath and a dangerous build-up of fluids in the lungs (pulmonary edema), which can cause death. Pulmonary edema is a medical emergency that can be delayed one to two days following exposure. Brief exposure to 100 ppm may be fatal
Eyes: Severe irritant. Contact may cause, burning and stinging, tearing, irritation and redness and swelling of the eyelids, intense pain; can burn the eyes and lead to blurred vision, chronic irritation and other permanent damage.
Skin: Severe irritant. Contact may cause burning or stinging feeling, redness, swelling, intense pain, blisters, irritation, rash, acne and slow-healing ulcers. Repeated exposure may lead to skin allergy and chronic irritation.
Swallowed: Poison. May cause burning feeling of the mouth and throat, intense thirst, throat swelling, stomach cramps, nausea, vomiting, diarrhea and unconsciousness, kidney damage, and severe breathing difficulties. Breathing may stop. Probable lethal oral dose for adult is 1 ml. Watch for delayed symptoms.
Long-term exposure: Exposure may cause acne-like rash and lung damage. Repeated high exposures may cause headache, pains in the chest and joints; indigestion.
Storage: Store in tightly closed containers in a dry, cool, well-ventilated place with non-wood floors. Keep away from heat, combustible and organic materials, oxidizers, aluminum, titanium, mercury, potassium, diethylzinc, hydrogen, germane, silane, trimethylamine, ammonia and all other chemicals. Chemical attacks rubber and some plastics. Containers may explode in fire.
First aid guide: Move victim to fresh air and call emergency medical care; if not breathing, give artificial respiration; if breathing is difficult, give oxygen. In case of contact with material, immediately flush skin or eyes with running water for at least 15 minutes. Remove and isolate contaminated clothing and shoes at the site. Keep victim quiet and maintain normal body temperature. Effects may be delayed, keep victim under observation.

★ 179 ★
BROMINE PENTAFLUORIDE
CAS: 7789-30-2
Other names: BROMINE FLUORIDE

Danger profile: Poisonous, corrosive, reactive with water, oxidizer, poisonous gases produced in fire.
Uses: Fluorinating agent and oxidizer.
Appearance: Colorless fuming liquid.
Effects of exposure:
Breathed: Poison. May cause irritation of the nose, throat and lungs; coughing, gasping, choking, shortness of breath;

bloody phlegm. Higher exposures may cause shortness of breath and a dangerous build-up of fluids in the lungs (pulmonary edema), which can cause death. Pulmonary edema is a medical emergency that can be delayed one to two days following exposure.
Eyes: May cause severe irritation, burns, permanent damage and blindness.
Skin: Poison. May cause severe irritation, rash and burns.
Swallowed: Poison. Water and stomach acids react with chemical and may cause severe stomach pains, diarrhea, colic, slow irregular pulse rate, irregular heart beat, dizziness, convulsions, muscle spasms, hypertension, tremors and paralysis, and death.
Long-term exposure: May cause lung, kidney and liver damage. Repeated exposure may cause a build up of bromine or fluorine in the body, both can cause serious health problems.
Storage: Violent reaction with water. Store in tightly closed containers in a dry, cool, well-ventilated place with non-wood floors. Extremely dangerous. Keep away from water, steam, combustible and organic materials, and all other chemicals.
First aid guide: Move victim to fresh air and call emergency medical care; if not breathing, give artificial respiration; if breathing is difficult, give oxygen. In case of contact with material, immediately flush skin or eyes with running water for at least 15 minutes. Remove and isolate contaminated clothing and shoes at the site. Keep victim quiet and maintain normal body temperature. Effects may be delayed; keep victim under observation.

★ 180 ★
BROMINE TRIFLUORIDE
CAS: 7787-71-5
Other names: BROMINE FLUORIDE

Danger profile: Poisonous, corrosive, highly reactive, exposure limit established, poisonous gases produced in fire.
Uses: Fluorinating agent and electrolytic solvent.
Appearance: Colorless fuming liquid.
Effects of exposure:
Breathed: Poison. May cause irritation of the nose, throat and lungs; coughing, gasping, choking, shortness of breath; bloody phlegm. Higher exposures may cause shortness of breath and a dangerous build-up of fluids in the lungs (pulmonary edema), which can cause death. Pulmonary edema is a medical emergency that can be delayed one to two days following exposure.
Eyes: May cause severe irritation, burns, permanent damage and blindness.
Skin: Poison if absorbed. May cause severe irritation and burns.
Swallowed: Poison. Water and stomach acids react with chemical and may cause severe stomach pains, diarrhea, colic, slow irregular pulse rate, irregular heart beat, dizziness, convulsions, muscle spasms, hypertension, tremors and paralysis, and death.
Long-term exposure: May cause lung damage. Repeated exposure may cause a build up of bromine or fluorine in the body, both can cause serious health problems. Repeated or prolonged contact may cause skin rash.
Storage: Store in tightly closed containers in a dry, cool, well-ventilated place with non-wood floors. Very dangerous. Keep away from water, combustible materials, chloride and bromine salts, metals and all other chemicals.
First aid guide: Move victim to fresh air and call emergency medical care; if not breathing, give artificial respiration; if breathing is difficult, give oxygen. In case of contact with material, immediately flush skin or eyes with running water for at least 15 minutes. Remove and isolate contaminated clothing and shoes at the site. Keep victim quiet and maintain normal body temperature. Effects may be delayed; keep victim under observation.

★ 181 ★
BROMOFORM
CAS: 75-25-2
Other names: BROMOFORME (FRENCH); BROMOFORMIO (ITALIAN); METHENYL TRIBROMIDE; NCI-C55130; METHANE, TRIBROMO-; METHYL TRIBROMIDE; TRIBROMMETHAAN (DUTCH); TRIBROMMETHAN (GERMAN); TRIBROMOMETAN (ITALIAN); TRIBROMOMETHANE

Danger profile: Poisonous, possible mutagen, narcotic, Community Right-to-Know List, exposure limit established, very poisonous gases produced in fire.
Uses: Solvent for fats, waxes and oils; making other chemicals and drugs.
Appearance: Heavy colorless liquid; becomes yellow on decomposition.
Odor: Sweet, chloroform-like. The odor is a warning of exposure; however, no smell does not mean you are not being exposed.
Effects of exposure:
Breathed: May cause irritation of the eyes, nose, throat and air passages; coughing, headache; increased saliva; reddening of the face; dizziness, lightheadedness, fatigue, depression, loss of reflexes; loss of consciousness, shock, coma, irregular heartbeat, death. If the victim recovers watch for the possibility of a buildup of fluid in the lungs (pulmonary edema). Pulmonary edema is a medical emergency that can be delayed one to two days following exposure.
Eyes: Contact may cause irritation, tearing, inflammation of the eyelids; burning sensation.
Skin: Passes through the unbroken skin; can increase exposure and the severity of the symptoms listed. May cause irritation, redness, burning feeling, swelling; dry skin; rash or blisters.
Swallowed: Poison. Sweet taste. Breath may smell of chloroform. May cause irritation of the mouth, gastrointestinal tract, nausea, vomiting, diarrhea with possible blood; drowsiness; loss of consciousness; irregular heartbeat and death.
Long-term exposure: May cause liver damage and birth defects. May effect the lungs. May cause dryness of the skin and rash.
Storage: Store in tightly closed containers in a dry, cool, well-ventilated place. Keep away from heat, bases, sodium, potassium, calcium, powdered aluminum, zinc, magnesium and other chemically active metals, potassium hydroxide and acetone.
First aid guide: Move victim to fresh air and call emergency medical care; if not breathing, give artificial respiration; if breathing is difficult, give oxygen. In case of contact with material, immediately flush skin or eyes with running water for at least 15 minutes. Remove and isolate contaminated clothing and shoes at the site. Effects should disappear after individual has been exposed to fresh air for approximately 10 minutes.

★ 182 ★
1,3-BUTADIENE
CAS: 106-99-0
Other names: BIETHYLENE; BIVINYL; BUTADIEEN (DUTCH); BUTA-1,3-DIEEN (DUTCH); BUTADIEN (POLISH); BUTA-1,3-DIEN (GERMAN); BUTADIENE; BUTA-1,3-DIENE; ALPHA-GAMMA-BUTADIENE; DIVINYL; ERYTHRENE; NCI-C50602; PYRROLYLENE; VINYLETHYLENE

Danger profile: Carcinogen, possible reproductive hazard, corrosive, fire hazard, exposure limit established, Community Right-to-Know List.
Uses: Making other chemicals, rubber and resins; chemical intermediate.
Appearance: Colorless liquefied compressed gas.

Odor: Mildly aromatic gasoline-like odor. The odor is a warning of exposure; however, no smell does not mean you are not being exposed.
Effects of exposure:
Breathed: May be poisonous. Exposure to the gas may irritate the eyes, nose, throat and lungs; coughing, shortness of breath; sleepiness, lightheaded feeling. High exposures may cause victim to pass out and die.
Eyes: Liquid may cause irritation and frostbite on contact.
Skin: Liquid may cause irritation and frostbite on contact.
Long-term exposure: May cause cancer, birth defects or damage to the fetus. May cause damage to the ovaries and testes.
Storage: Store in tightly closed containers in a dry, cool, well-ventilated place. Keep away from heat, sources of ignition, strong oxidizers, aluminum tetrahydroborate, sodium nitrite, copper and copper alloys. Vapors may travel to sources of ignition and flash back. Containers may explode in heat of fire.

★ 183 ★
BUTANE
CAS: 106-97-8
Other names: N-BUTANE; BUTANEN (DUTCH); BUTANI (ITALIAN); DIETHYL; DIETHYL, LIQUIFIED PETROLEUM GAS; METHYLETHYLMETHANE; METHYL ETHYL METHANE

Danger profile: Asphyxiant, extremely dangerous fire hazard, explosive, exposure limit established.
Uses: Fuel for household and for many industrial purposes; food additive; propellant in aerosols; raw material for synthetic rubber and high octane liquid fuels; making other chemicals; solvent; refrigerant.
Appearance: Colorless liquified compressed gas.
Odor: Faint, disagreeable; sometimes described as gasoline-like. The odor is a warning of exposure; however, no smell does not mean you are not being exposed.
Effects of exposure:
Breathed: Mildly toxic. Gas may cause coughing, dizziness, drowsiness, difficult breathing or suffocation.
Eyes: Contact may cause severe frostbite.
Skin: Contact may cause severe frostbite.
Storage: Store in tightly closed containers in a dry, cool, well-ventilated place. Keep away from heat, flame, sources of ignition and oxidizers. Vapors may travel to sources of ignition and flash back. Containers may explode in fire. See OSHA regulation 1910.111 concerning storage and handling of liquified petroleum gases.
First aid guide: Move victim to fresh air and call emergency medical care; if not breathing, give artificial respiration; if breathing is difficult, give oxygen. In case of frostbite, thaw frosted parts with water. Keep victim quiet and maintain normal body temperature.

★ 184 ★
2,3-BUTANEDIONE
CAS: 30031-64-2
Other names: BIACETYL; BUTANEDIONE; 2,2'-BIOXIRANE; 1,2:3,4-DIEPOXYBUTANE; BUTADIENE DIEPOXIDE; BUTADIENE DIOXIDE; BIOXIRAN; DIACETYL (FCC); 2,3-DIKETOBUTANE; DIMETHYL DIKETONE; DIMETHYL GLYOXAL; FEMA NO. 2370

Danger profile: Possible carcinogen, poisonous, mutagen, fire hazard, poisonous gases produced in fire.
Uses: Food additive; aroma producer for food products.
Appearance: Yellow-green liquid.
Odor: Strong, chlorine-like. The odor is a warning of exposure; however, no smell does not mean you are not being exposed.

Effects of exposure:
Breathed: May be poisonous. May cause irritation of the eyes, nose and throat, dizziness and suffocation.
Eyes: Contact may cause irritation and burns.
Skin: May be poisonous if absorbed. Contact may cause irritation and burns.
Swallowed: Moderately toxic.
Long-term exposure: May cause lung cancer; birth defects, miscarriage. May effect the central nervous system and blood count.
Storage: Store in tightly closed containers in a cool, well-ventilated place. Keep away form heat and sources if ignition. Vapors may travel to sources of ignition and flash back. Containers may explode in fire.

★ 185 ★
1-BUTENE
CAS: 25167-67-3
Other names: BUTYLENE; ALPHA-BUTYLENE; N-BUTENE; ETHYLETHYLENE

Danger profile: Fire hazard, explosion hazard, poisonous gases produced in fire, asphyxiant.
Uses: Used to make other chemicals and gasoline.
Appearance: Colorless gas.
Effects of exposure:
Breathed: Chemical is an asphyxiant; may cause suffocation. May cause dizziness, lightheadedness, rapid and irregular breathing, headache, fatigue, mental confusion, nausea and vomiting, loss of consciousness, convulsions and death.
Eyes: Contact may cause stinging pain, tearing, inflammation of eyelids, frostbite cloudiness in eyes.
Skin: Contact may cause feeling of intense cold, frostbite.
Swallowed: May cause breath to have a sweet smell, confusion, fatigue, headache, loss of consciousness, coma and death.
Storage: Keep away form heat, sources of ignition, oxidizers and aluminum tris-tetrahydroborate. Containers may explode in fire.

★ 186 ★
2-BUTOXYETHANOL
CAS: 111-76-2
Other names: BUTOKSYETYLOWY ALKOHOL (POLISH); 2-BUTOSSI-ETANOLO (ITALIAN); 2-BUTOXY-AETHANOL (GERMAN); BUTOXYETHANOL; BUTOXY ETHANOL; 2-BUTOXY-1-ETHANOL; BUTYL CELLOSOLVE; O-BUTYL ETHYLENE GLYCOL; BUTYL GLYCOL; BUTYLGLYCOL; BUTYL OXITOL; BUCS; DOW ANOL EB; EKTASOLVE EB; ETHANOL, 2-BUTOXY-; ETHYLENE GLYCOL-N-BUTYL ETHER; GAFCOL EB; GLYCOL BUTYL ETHER; GLYCOL ETHER EB; GLYCOL ETHER EB ACETATE; GLYCOL MONOBUTYL ETHER; JEFFERSON EB; MONOBUTYL GLYCOL ETHER; 3-OXA-1-HEPTANOL; POLY-SOLV EB

Danger profile: Possible reproductive hazard; irritant, combustible, exposure limit established, Community Right-to-Know List, poisonous gases produced in fire.
Uses: Solvent; making plastics.
Appearance: Colorless liquid.
Odor: Pleasant; mild. The odor is a warning of exposure; however, no smell does not mean you are not being exposed.
Effects of exposure:
Breathed: May cause irritation of the eyes, nose, throat and lungs, tears and runny nose, coughing, sore throat, difficult breathing. Higher exposures could cause shortness of breath, headache, nausea, vomiting, dizziness, diarrhea and a dangerous build-up of fluids in the lungs (pulmonary

edema), which can cause death. Pulmonary edema is a medical emergency that can be delayed one to two days following exposure.
Eyes: May cause irritation, redness, tearing, pain, inflamed eyelids; blurred vision.
Skin: Toxic. Passes through the unbroken skin; can increase exposure and the severity of the symptoms listed. May cause painful irritation, redness, and rash.
Swallowed: Poison. May cause irritation of the mouth and throat; stomach and abdominal pain. Also see symptoms for *Breathed.*
Long-term exposure: May cause anemia; liver and kidney damage. May effect the lungs.
Storage: Store in tightly closed containers in a dry, cool, well-ventilated place. Keep away from heat, sources of ignition, strong oxidizers, strong caustics. Containers may explode in fire.

★ 187 ★
2-BUTOXYETHYL ACETATE
CAS: 112-07-2
Other names: 2-BUTOXYETHANOL ACETATE; 2-BUTOXYETHYL ESTER ACETIC ACID; BUTYL CELLO-SOLVE ACETATE; EKTASOLVE EB ACETATE; ETHANOL, 2-BUTOXY-, ACETATE; ETHYLENE GLYCOL MONOBUTYL ETHER ACETATE; GLYCOL MONOBUTYL ETHER ACE-TATE

Danger profile: Combustible, Community Right-to-Know List, poisonous gases produced in fire.
Uses: Solvent.
Appearance: Colorless liquid.
Odor: Weak, fruity. The odor is a warning of exposure; however, no smell does not mean you are not being exposed.
Effects of exposure:
Breathed: Concentrated vapor may cause headache, nausea, dizziness.
Eyes: Liquid may cause irritation.
Skin: Moderately toxic. May cause mild irritation.
Swallowed: Moderately toxic. May cause headache, nausea and dizziness.
Storage: Store in tightly closed containers in a dry, cool, well-ventilated place. Keep away from heat, flame, sources of ignition and oxidizers.

★ 188 ★
BUTOXYL
CAS: 4435-53-4
Other names: ACETIC ACID, 3-METHOXYBUTYL ESTER; 3-METHOXYBUTYL ACETATE; METHYL-1,3-BUTYLENE GLY-COL ACETATE; 1-BUTANOL, 3-METHOXY-, ACETATE; 3-METHOXY-1-BUTANOL ACETATE

Danger profile: Combustible, poisonous gas produced in fire.
Uses: Cleaning solvent; in varnishes.
Appearance: Colorless liquid.
Odor: Acrid, bitter taste. The odor is a warning of exposure; however, no smell does not mean you are not being exposed.
Effects of exposure:
Breathed: May be poisonous. Vapor may irritate the eyes, nose and throat. Very high levels may cause dizziness.
Eyes: Contact may cause irritation.
Skin: May be poisonous if absorbed. Passes through the unbroken skin; can increase exposure and the severity of the symptoms listed. Contact may cause irritation.
Swallowed: Mildly toxic.
Storage: Store in tightly closed containers in a dry, cool, well-ventilated place. Keep away from heat, sources of ignition and oxidizing materials. Containers may explode in fire.

First aid guide: Move victim to fresh air and call emergency medical care; if not breathing, give artificial respiration; if breathing is difficult, give oxygen. In case of contact with material, immediately flush eyes with running water for at least 15 minutes. Wash skin with soap and water. Remove and isolate contaminated clothing and shoes at the site.

★ 189 ★
BUTYL ACETATE
CAS: 123-86-4
Other names: ACETATE DE BUTYLE (FRENCH); ACETIC ACID N-BUTYL ESTER; BUTILE (ACETATI DI) (ITALIAN); BUTYLACETAT (GERMAN); N-BUTYL ACETATE; 1-BUTYL ACETATE; BUTYLACETATEN (DUTCH); BUTYLE (ACETATE DE) (FRENCH); BUTYL ETHANOATE; FEMA NO. 2174; OCTAN N-BUTYLU (POLISH); M-BUTYL ACETATE; ACETIC ACID, BUTYL ESTER; N-BUTYL ESTER OF ACETIC ACID

Danger profile: Experimental teratogen; combustible, exposure limit established; poisonous gas produced in fire.
Uses: Leather dressings, perfumes, flavoring extracts; solvent gums, resins and lacquers; airplane glue.
Appearance: Colorless watery liquid.
Odor: Pleasant fruity in low concentrations; strong, disagreeable in high concentrations. The odor is a warning of exposure; however, no smell does not mean you are not being exposed.
Effects of exposure:
   Breathed: Mildly toxic. May cause irritation of eyes, nose and throat; headache, drowsiness, excitability, drunken behavior, mental confusion, fatigue. Very high levels of exposure may cause dizziness, lightheadedness; unconsciousness; coma and death.
   Eyes: Contact may cause irritation, pain, burning, stinging, tearing, inflammation of the cornea and eyelids. Light may cause pain.
   Skin: Passes through the unbroken skin; can increase exposure and the severity of the symptoms listed. May cause burning feeling and rash. May remove oil from the skin and cause dryness. Persons with pre-existing skin disorders may be more susceptible to the effects of this substance.
   Swallowed: May cause gastrointestinal irritation.
Long-term exposure: Prolonged or repeated exposure to skin may cause drying and cracking; skin allergy may develop. May cause birth defects, miscarriage or cancer. Many solvents and other petroleum-based chemicals may cause nerve and brain damage; memory loss, personality changes, fatigue, sleep disturbances, reduced coordination, a feeling of "pins and needles" in the arms and legs.
Storage: Store in tightly closed containers in a dry, cool, well-ventilated place. Keep away from heat, sources of ignition, oxidizers, strong acids, and potassium tert-butoxide. Containers may explode in fire.
First aid guide: Move victim to fresh air and call emergency medical care; if not breathing, give artificial respiration; if breathing is difficult, give oxygen. In case of contact with material, immediately flush eyes with running water for at least 15 minutes. Wash skin with soap and water. Remove and isolate contaminated clothing and shoes at the site.

★ 190 ★
SEC-BUTYL ACETATE
CAS: 105-46-4
Other names: ACETATE DE BUTYLE SECONDAIRE (FRENCH); ACETIC ACID, 2-BUTOXY ESTER; ACETIC ACID, 1-METHYLPROPYL ESTER (9CI); S-BUTYL ACETATE; 2-BUTYL ACETATE; SEC-BUTYL ALCOHOL ACETATE; 1-METHYL PROPYL ACETATE

Danger profile: Exposure limit established, allergen, fire hazard, poisonous gases produced in fire.
Uses: A solvent with many uses.
Appearance: Colorless, watery liquid.
Odor: Pleasant, fruity. The odor is a warning of exposure; however, no smell does not mean you are not being exposed.
Effects of exposure:
   Breathed: May be poisonous. Mildly toxic. May cause irritation of eyes, nose and throat; headache, drowsiness, excitability, drunken behavior, mental confusion, fatigue. Very high levels of exposure may cause dizziness, lightheadedness; unconsciousness; coma and death.
   Eyes: Contact may cause irritation, pain, burning, stinging, tearing, inflammation of the cornea and eyelids. Light may cause pain.
   Skin: May be poisonous if absorbed. Passes through the unbroken skin; can increase exposure and the severity of the symptoms listed. May cause burning feeling and rash. May remove oil from the skin and cause dryness. Persons with pre-existing skin disorders may be more susceptible to the effects of this substance.
   Swallowed: May cause gastrointestinal irritation.
Long-term exposure: Prolonged or repeated exposure to skin may cause drying and cracking; skin allergy may develop. Many solvents and other petroleum-based chemicals may cause nerve and brain damage; memory loss, personality changes, fatigue, sleep disturbances, reduced coordination, a feeling of "pins and needles" in the arms and legs.
Storage: Store in tightly closed containers in a dry, cool, well-ventilated place. Keep away from heat, sources of ignition, strong oxidizers, strong alkalis and strong acids. Vapors may travel to sources of ignition and flash back. Containers may explode in fire.

★ 191 ★
TERT-BUTYL ACETATE
CAS: 540-88-5
Other names: ACETIC ACID-TERT-BUTYL ESTER; ACETIC ACID, 1,1-DIMETHYLETHYL ESTER (9CI); T-BUTYL ACETATE; TEXACO LEAD APPRECIATOR; TLA; ACETIC ACID, TERTIARY BUTYL ESTER-

Danger profile: Exposure limit established, mutagen, possible reproductive hazard, fire hazard, poisonous gases produced in fire.
Uses: A solvent with many uses and as a gasoline additive.
Appearance: Colorless-watery liquid.
Odor: Fruity. The odor is a warning of exposure; however, no smell does not mean you are not being exposed.
Effects of exposure:
   Breathed: May be poisonous. May cause irritation of eyes, nose, throat and breathing passages; nosebleeds; hoarseness, cough, phlegm; headache, drowsiness, tightness in the chest; excitability, drunken behavior, mental confusion, fatigue. Very high levels of exposure may cause dizziness, lightheadedness; unconsciousness; coma and death.
   Eyes: Contact may cause irritation, pain, burning, stinging, tearing, inflammation of the cornea and eyelids. Light may cause pain.
   Skin: May be poisonous if absorbed. Passes through the unbroken skin; can increase exposure and the severity of the symptoms listed. May cause burning feeling and rash. May remove oil from the skin and cause dryness. Persons with pre-existing skin disorders may be more susceptible to the effects of this substance.
   Swallowed: May cause gastrointestinal irritation. See *Breathed.*
Long-term exposure: May effect the liver; cause dry skin and rash. Many solvents and other petroleum-based chemicals may

cause nerve and brain damage; memory loss, personality changes, fatigue, sleep disturbances, reduced coordination, a feeling of "pins and needles" in the arms and legs.
Storage: Store in tightly closed containers in a dry, cool, well ventilated place. Keep away from heat, sources of ignition, potassium and sodium alloys, oxidizers, strong acids, and alkalis. Vapors may travel to sources of ignition and flash back. Containers may explode in fire.

★ 192 ★
N-BUTYL ACRYLATE
CAS: 141-32-2
Other names: ACRYLIC ACID, BUTYL ESTER; ACRYLIC ACID N-BUTYL ESTER; BUTYLACRYLATE, INHIBITED; BUTYL ACRYLATE; BUTYL 2-PROPENOATE; 2-PROPENOIC ACID, BUTYL ESTER

Danger profile: Possible reproductive hazard, irritant, combustible, exposure limit established, Community Right-to-Know List, poisonous gases produced in fire.
Uses: Making paints, polymers and resins.
Appearance: Colorless watery liquid.
Odor: Sharp, fragrant. The odor is a warning of exposure; however, no smell does not mean you are not being exposed.
Effects of exposure:
   Breathed: Vapor may cause irritation of the eyes, nose and throat; tears and runny nose, coughing, sore throat.
   Eyes: May cause irritation, redness, tearing, pain, inflamed eyelids, and eye injury; burns in corneas and possible loss of sight.
   Skin: May be poisonous if absorbed. Passes through the unbroken skin; can increase exposure and the severity of the symptoms listed. May cause irritation, redness and rash.
   Swallowed: May cause irritation of the mouth and throat; stomach and abdominal pain; loss of muscular coordination and mental confusion; convulsions. Also see symptoms for *Breathed.*
Long-term exposure: May cause skin allergy with itching and rash. May effect the lungs and liver. Many solvents and other petroleum-based chemicals may cause nerve and brain damage; memory loss, personality changes, fatigue, sleep disturbances, reduced coordination, a feeling of "pins and needles" in the arms and legs.
Storage: Store in tightly closed containers in a dry, cool, well-ventilated place. Keep away from heat, sources of ignition and oxidizing materials. Vapors may travel to sources of ignition and flash back. Containers may explode in fire.

★ 193 ★
N-BUTYL ALCOHOL
CAS: 71-36-3
Other names: ALCOOL BUTYLIQUE (FRENCH); BUTANOL; 1-BUTANOL; N-BUTANOL; BUTYL ALCOHOL; PROPYL CARBINOL; BUTAN-1-OL; BUTANOLEN (DUTCH); BUTANOLO (ITALIAN); BUTYL HYDROXIDE; BUTYLOWY ALKOHOL (POLISH); CCS 203; FEMA NO. 2178; 1-HYDROXYBUTANE; METHYLOLPROPANE; PROPYLCARBINOL; PROPYLMETHANOL

Danger profile: Possible mutagen, poisonous, combustible, Community Right-to-Know List, exposure limit established, poisonous gases produced in fire.
Uses: Preparation of other chemicals; solvent for waxes, shellac, resins, gums and varnishes; dyeing; hydraulic fluids; detergents; plastics.
Appearance: Colorless, watery liquid.

Odor: Alcohol-like; pungent; strong. The odor is a warning of exposure; however, no smell does not mean you are not being exposed.
Effects of exposure:
   Breathed: May be poisonous. May cause irritation of eyes, nose and throat; headache, drowsiness, excitability, drunken behavior, mental confusion, fatigue. Very high levels of exposure may cause dizziness, lightheadedness; unconsciousness; coma and death.
   Eyes: Contact may cause irritation, pain, burning, stinging, tearing, inflammation of the cornea and eyelids. Light may cause pain.
   Skin: May be poisonous if absorbed. Passes through the unbroken skin; can increase exposure and the severity of the symptoms listed. May cause burning feeling and rash. May remove oil from the skin and cause dryness. Persons with pre-existing skin disorders may be more susceptible to the effects of this substance.
   Swallowed: May cause gastrointestinal irritation.
Long-term exposure: Prolonged or repeated contact may cause skin drying and cracking; irritation of the skin and dermatitis, and nervous system changes. May cause liver and kidney damage; hearing problems and may effect victim's sense of balance. May cause birth defects, miscarriages and cancer. Many solvents and other petroleum-based chemicals may cause nerve and brain damage; memory loss, personality changes, fatigue, sleep disturbances, reduced coordination, a feeling of "pins and needles" in the arms and legs.
Storage: Store in a tightly closed container in a cool, well-ventilated place. Keep away from heat and sources of ignition, acid anhydrides, reducing agents, aluminum, chromium trioxide, copper and copper alloys. Vapors may travel to sources of ignition and flash back. Containers may explode in fire.

★ 194 ★
SEC-BUTYL ALCOHOL
CAS: 78-92-2
Other names: ALCOOL BUTYLIQUE SECONDAIRE (FRENCH): SEC-BUTANOL; BUTAN-2-OL; BUTANOL-2; 2-BUTANOL; TANOL SECONDAIRE (FRENCH); 2-BUTYL ALCOHOL; BUTYLENE HYDRATE; CCS 301; ETHYLMETHYL CARBINOL; 2-HYDROXYBUTANE; METHYLETHYLCARBINOL; S.B.A.; 1-METHYPROPYL ALCOHOL; METHYL ETHYL CARBINOL; ETHYL METHYL CARBINOL

Danger profile: Poisonous, fire hazard, Community Right-to-Know List, exposure limit established, poisonous gases produced in fire.
Uses: Making flavors, perfumes and dyestuffs; in paint removers and cleaners.
Appearance: Colorless, watery liquid.
Odor: Strong, pleasant, alcohol-like. The odor is a warning of exposure; however, no smell does not mean you are not being exposed.
Effects of exposure:
   Breathed: May be poisonous. May cause irritation of eyes, nose and throat; headache, drowsiness, excitability, drunken behavior, mental confusion, fatigue. Very high levels of exposure may cause dizziness, lightheadedness; unconsciousness; coma and death.
   Eyes: Contact may cause severe irritation, pain, burning feeling, stinging, tearing, inflammation of the cornea and eyelids; burns. Light may cause pain.
   Skin: May be poisonous if absorbed. Passes through the unbroken skin; can increase exposure and the severity of the symptoms listed. May cause burning feeling and rash. May remove oil from the skin and cause dryness. Persons with

pre-existing skin disorders may be more susceptible to the effects of this substance.

Swallowed: May cause gastrointestinal irritation. See *Breathed.*

Long-term exposure: May cause drying and cracking of the skin. Many solvents and other petroleum-based chemicals may cause nerve and brain damage; memory loss, personality changes, fatigue, sleep disturbances, reduced coordination, a feeling of "pins and needles" in the arms and legs.

Storage: Store in tightly closed containers in a dry, cool, well-ventilated place. Keep away from heat, sources of ignition, chromium trioxide, strong oxidizers. Vapors may travel to sources of ignition and flash back. Containers may explode in fire.

★ 195 ★
TERT-BUTYL ALCOHOL

CAS: 75-65-0

Other names: ALCOOL BUTYLIQUE TERTIAIRE (FRENCH); TERT-BUTANOL; BUTANOL TERTIAIRE (FRENCH); TERT-BUTYL HYDROXIDE; 1,1-DIMETHYLETHANOL; METHANOL, TRIMETHYL-; 2-METHYL-, 2-PROPANOL; NCI-C55367; 2-PROPANOL, 2-METHYL-; TRIMETHYLCARBINOL; 2-METHYL-2-PROPANOL; TBA

Danger profile: Possible mutagen, fire hazard, Community Right-to-Know List, exposure limit established, poisonous gases produced in fire and on contact with acid.

Uses: Making perfumes, flavors; paint remover; solvent for drugs and pharmaceuticals; in leaded gasoline.

Appearance: Colorless, oily liquid.

Odor: Camphor; mothball-like; pungent. The odor is a warning of exposure; however, no smell does not mean you are not being exposed.

Effects of exposure:

Breathed: May be poisonous. Vapor is narcotic in action. May cause irritation of eyes, nose and throat; headache, drowsiness, excitability, drunken behavior, mental confusion, fatigue. Very high levels of exposure may cause dizziness, lightheadedness; unconsciousness; coma and death.

Eyes: Contact may cause irritation, pain, burning, stinging, tearing, inflammation of the cornea and eyelids. Light may cause pain.

Skin: May be poisonous if absorbed. Passes through the unbroken skin; can increase exposure and the severity of the symptoms listed. May cause burning feeling and rash. May remove oil from the skin and cause dryness. Persons with pre-existing skin disorders may be more susceptible to the effects of this substance.

Swallowed: May cause gastrointestinal irritation. See *Breathed.*

Long-term exposure: May cause drying and cracking of the skin. May cause birth defects, miscarriages and cancer. Many solvents and other petroleum-based chemicals may cause nerve and brain damage; memory loss, personality changes, fatigue, sleep disturbances, reduced coordination, a feeling of "pins and needles" in the arms and legs.

Storage: Store in tightly closed containers in a dry, cool, well-ventilated place. Keep away from heat, sources of ignition, hydrochloric acid, potassium and sodium alloys, oxidizing materials and hydrogen peroxide. Combustible gas produced on contact with acid. Vapors may travel to sources of ignition and flash back. Containers may explode in fire.

★ 196 ★
BUTYL ALDEHYDE

CAS: 123-72-8

Other names: ALDEHYDE BUYYRIQUE (FRENCH); EHYDE BUTYRIQUE (FRENCH); EIDE BUTIRRICA (ITALIAN); BUTAL; BUTALYDE; BUTANAL; N-BUTANAL (CZECH); BUTANEHYDE; BUTYL EHYDE; N-BUTYL EHYDE; BUTYRAL; BUTYREHYD (GERMAN); BUTYRALDEHYDE (CZECH); N-BUTYLALDEHYDE; FEMA NO. 2219; NCI-C56291

Danger profile: Combustible, community right-to-know list, poisonous gas produced in fire.

Uses: Plasticizers, rubber accelerators, solvents, high polymers.

Appearance: Clear colorless liquid.

Odor: Intense. The odor is a warning of exposure; however, no smell does not mean you are not being exposed.

Effects of exposure:

Breathed: May be poisonous. Vapor may cause intense irritation to nose, throat, mouth and lungs; mucous membrane. Symptoms of exposure include watering of the eyes, coughing, difficult breathing, headache, nausea, vomiting, muscular weakness, unconsciousness. High exposure can lead to drowsiness, coma and death; buildup of fluid in the lungs (pulmonary edema), and may cause death. Pulmonary edema may be delayed by one to two days following exposure.

Eyes: Severe irritant. Contact may cause, burning and stinging, tearing, irritation and redness and swelling of the eyelids, intense pain; can burn the eyes and lead to blurred vision, chronic irritation and other permanent damage.

Skin: May be poisonous if absorbed. Severe irritant. Contact may cause burning or stinging feeling, redness, swelling, intense pain, blisters, irritation, rash. Repeated exposure may lead to skin allergy and chronic irritation.

Swallowed: May cause burning feeling of the mouth and throat, intense thirst, throat swelling, stomach cramps, nausea, vomiting, diarrhea, severe breathing difficulties and unconsciousness. Tissue damage may result. Breathing may stop.

Storage: Store in tightly closed containers in a dry, cool, well ventilated place. Keep away from heat, flame and sources of ignition, chlorosulfonic acid and oxidizing materials.

First aid guide: Move victim to fresh air and call emergency medical care; if not breathing, give artificial respiration; if breathing is difficult, give oxygen. In case of contact with material, immediately flush eyes with running water for at least 15 minutes. Wash skin with soap and water. Remove and isolate contaminated clothing and shoes at the site.

★ 197 ★
BUTYLAMINE

CAS: 109-73-9

Other names: 1-AMINO-BUTAAN (DUTCH); 1-AMINOBUTAN (GERMAN); 1-AMINOBUTANE; 1-BUTANAMINE; N-BUTILAMINA (ITALIAN); N-BUTYLAMIN (GERMAN); N-BUTYLAMINE; MONOBUTYLAMINE; MONO-N-BUTYLAMINE; NORVALAMINE

Danger profile: Possible carcinogen, mutagen, fire hazard, severe irritant, poisonous gases produced in fire.

Uses: Making rubber and drugs.

Appearance: Clear colorless liquid.

Odor: Ammonia or fish-like. The odor is a warning of exposure; however, no smell does not mean you are not being exposed.

Effects of exposure:

Breathed: May be poisonous. May cause irritation of the nose, throat, and eyes; burning sensation, flushed feeling, difficult breathing; nausea, vomiting, headache, faintness, severe coughing, chest pains. Higher exposures may cause severe shortness of breath and a dangerous build-up of fluids in the lungs (pulmonary edema), which can cause death. Pulmonary edema is a medical emergency that can be delayed one to two days following exposure.

Eyes: May cause painful irritation of the eyes and eyelids; tearing, severe burns; rapid permanent damage.
Skin: Poison if absorbed. May cause irritation, itching, painful burning, blisters and sores, shock. The skin may feel slippery.
Swallowed: Poison. May cause burning feeling in the mouth, throat, and stomach; painful swallowing, swelling of the throat area, nausea, vomiting, stomach cramps, rapid breathing diarrhea and bloody stool, risk of stomach perforation. Also see *Breathed.*
Long-term exposure: Repeated exposure may cause skin rash and itching; may effect the lungs.
Storage: Store in tightly closed containers in a dry, cool, well-ventilated place. Keep away from heat, sources of ignition, strong acids, oxidizers and perchloryl fluoride. Containers may explode in fire.
First aid guide: Move victim to fresh air and call emergency medical care; if not breathing, give artificial respiration; if breathing is difficult, give oxygen. In case of contact with material, immediately flush skin or eyes with running water for at least 15 minutes. Remove and isolate contaminated clothing and shoes at the site. Keep victim quiet and maintain normal body temperature.

★198★
TERT-BUTYLAMINE
CAS: 75-64-9
Other names: 2-AMINOISOBUTANE; 2-AMINO-2-METHYLPROPANE; BUTYLAMINE, TERTIARY; 1,1-DIMETHYLETHYLAMINE; 2-METHYL-2-PROPANAMINE; TRI-METHYLAMINOMETHANE

Danger profile: Fire hazard, poisonous, poisonous gases produced in fire.
Uses: Intermediate for rubber accelerators, insecticides, fungicides, dyestuffs, pharmaceuticals.
Appearance: Colorless liquid.
Odor: Strong ammonia-like. The odor is a warning of exposure; however, no smell does not mean you are not being exposed.
Effects of exposure:
Breathed: May be poisonous. May cause irritation of the nose, mouth and lungs, dizziness and suffocation. High exposures may cause shortness of breath and a dangerous build-up of fluids in the lungs (pulmonary edema), which can cause death. Pulmonary edema is a medical emergency that can be delayed one to two days following exposure.
Eyes: May cause irritation and burns.
Skin: May be poisonous. May cause irritation and burns.
Swallowed: May cause irritation of the stomach.
Storage: Store in tightly closed containers in a dry, cool, well-ventilated place. Keep away from heat, flame, sources of ignition, 2,2-dibromo-1,3-dimethylcyclopropanoic acid. Vapors may travel to sources of ignition and flash back. Containers may explode in fire.

★199★
BUTYL BROMIDE
CAS: 109-65-9
Other names: BUTANE, 1-BROMO-; 1-BROMOBUTANE; BUTYL BROMIDE, NORMAL; N-BUTYL BROMIDE

Danger profile: Fire hazard, poisonous gases produced in fire.
Uses: Making pharmaceuticals and other chemicals.
Appearance: Colorless to pale yellow liquid.
Effects of exposure:

Breathed: May be poisonous. Exposure to high levels may cause dizziness, lightheadedness, and unconsciousness. Very high levels may cause death.
Eyes: Contact may cause irritation and burns.
Skin: Contact may cause irritation and burns.
Long-term exposure: May cause kidney and liver damage. Possible teratogen; may cause birth defects by damaging the fetus.
Storage: Dangerous fire hazard. Store in tightly closed containers in a dry, cool, well-ventilated place. Keep away from heat, sources of ignition, bromo benzene, sodium and oxidizers. Vapors may travel to sources of ignition and flash back. Containers may explode in fire.
First aid guide: Move victim to fresh air and call emergency medical care; if not breathing, give artificial respiration; if breathing is difficult, give oxygen. In case of contact with material, immediately flush skin or eyes with running water for at least 15 minutes. Remove and isolate contaminated clothing and shoes at the site. Keep victim quiet and maintain normal body temperature.

★200★
BUTYL CHLORIDE
CAS: 109-69-3
Other names: BUTANE, 1-CHLORO-; N-BUTYL CHLORIDE; 1-CHLOROBUTANE; CHLORURE DE BUTYLE (FRENCH); NCI-C06155; N-PROPYLCARBINYL CHLORIDE

Danger profile: Mutagen, combustible, highly toxic fumes produced in fire.
Uses: Solvent; medicine to control worms; making other chemicals.
Appearance: Clear colorless liquid.
Odor: Pungent. The odor is a warning of exposure; however, no smell does not mean you are not being exposed.
Effects of exposure:
Breathed: May be poisonous. May cause irritation of the eyes, nose and throat. High levels may cause dizziness, cause victim to pass out, soffocate and die.
Eyes: Contact may cause irritation of the cornea.
Skin: May be poisonous if absorbed. Contact may cause irritation and burns.
Swallowed: Moderately toxic.
Long-term exposure: May cause mutations; central nervous system depression.
Storage: Store in tightly closed containers in a dry, cool, well-ventilated place. Keep away from heat, flame and sources of ignition and oxidizers. Vapors may travel to sources of ignition and flash back. Containers may explode in fire.
First aid guide: Move victim to fresh air and call emergency medical care; if not breathing, give artificial respiration; if breathing is difficult, give oxygen. In case of contact with material, immediately flush eyes with running water for at least 15 minutes. Wash skin with soap and water. Remove and isolate contaminated clothing and shoes at the site.

★201★
1,2-BUTYLENE OXIDE
CAS: 106-88-7
Other names: OXIRANE, ETHYL-; 1,2-EPOXYBUTANE; 2-ETHYLOXIRANE; PROPYL OXIRANE

Danger profile: Mutagen, combustible, poisonous gases produced in fire.
Uses: Making other chemicals such as gasoline additives; stabilizer.
Appearance: Watery-white liquid.
Effects of exposure:

Breathed: May be poisonous. May cause irritation of the eye, nose, throat and lungs; coughing and shortness of breath. High exposure may cause lightheadedness and cause victim to pass out.

Eyes: Contact may cause irritation and burns.

Skin: May be poisonous if absorbed. Contact may cause irritation, redness, burns and blisters.

Long-term exposure: May cause a decrease of fertility in females. Prolonged or repeated contact may cause skin blisters.

Storage: Store in tightly closed containers in a dry, cool, well-ventilated place. Keep away from water, heat, sources of ignition, oxidizers, bases. Vapors may travel to sources of ignition and flash back. Containers may explode in fire.

★ 202 ★
BUTYL ETHER
CAS: 142-96-1
Other names: 1-BUTOXYBUTANE; N-BUTYL ETHER; DI-N-BUTYL ETHER; DIBUTYL ETHER; DIBUTYL OXIDE; ETHER BUTYLIQUE (FRENCH); 1,1'-OXYBIS(BUTANE); BUTYL OXIDE; BUTANE, 1,1'-OXYBIS-; 1-BUTOXYBUTONE

Danger profile: Fire hazard, poisonous gases produced in fire.
Uses: Solvent; for metals separation.
Appearance: Colorless liquid.
Odor: Fruity; sweet. The odor is a warning of exposure; however, no smell does not mean you are not being exposed.
Effects of exposure:

Breathed: May be poisonous. Vapor may cause irritation of the nose, throat, breathing passages; nose bleeding, hoarseness, cough, phlegm and tightness in chest. High exposure may cause dizziness, headache, lightheadedness, slurred speech; shuddering, staggered gait; blurred vision; unconsciousness and death.

Eyes: Contact may cause irritation.

Skin: May be poisonous if absorbed. Contact may cause irritation, redness and rash.

Swallowed: Mildly toxic.

Long-term exposure: Repeated or prolonged skin contact may cause rash.

Storage: Store in tightly closed containers in a dry, cool, detached, well-ventilated place. Keep away from heat, sources of ignition and oxidizers. Vapors may travel to sources of ignition and flash back. Containers may explode in fire.

First aid guide: Move victim to fresh air and call emergency medical care; if not breathing, give artificial respiration; if breathing is difficult, give oxygen. In case of contact with material, immediately flush eyes with running water for at least 15 minutes. Wash skin with soap and water. Remove and isolate contaminated clothing and shoes at the site.

★ 203 ★
BUTYL GLYCIDYL ETHER
CAS: 2426-08-6
Other names: N-BUTYL GLYCIDYL ETHER; AGEFLEX BGE; BGE; 2,3-EPOXYPROPYL BUTYL ETHER; ETHER, BUTYL 2,3-EPOXYPROPYL; ETHER, BUTYL GLYCIDYL; GLYCIDYL BUTYL ETHER; GLYCIDYLBUTYLETHER; PROPANE,1-BUTOXY-2,3-EPOXY-

Danger profile: Carcinogen, mutagen, poisonous gases produced in fire.
Appearance: Clear to pale yellow liquid.
Odor: Irritating. The odor is a warning of exposure; however, no smell does not mean you are not being exposed.
Effects of exposure:

Breathed: May be poisonous. May cause irritation of the eyes, nose, throat and lungs, tears and runny nose, cough-

ing, sore throat, difficult breathing; lethargy, fatigue, and drowsiness; slurred speech, shaking, and dizziness. Higher exposures could cause shortness of breath, headache, nausea, vomiting, dizziness, diarrhea and a dangerous build-up of fluids in the lungs (pulmonary edema), which can cause death. Pulmonary edema is a medical emergency that can be delayed one to two days following exposure.

Eyes: May cause irritation, redness, tearing, pain, inflamed eyelids, and eye injury; burns in corneas and possible loss of sight.

Skin: May be poisonous if absorbed. Passes through the unbroken skin; can increase exposure and the severity of the symptoms listed. May cause painful irritation, redness, and blisters Repeated or prolonged skin contact may cause irritation of the skin and may also cause the legs to feel numb.

Swallowed: Toxic. May cause irritation of the mouth and throat; stomach and abdominal pain; loss of muscular coordination and mental confusion; convulsions.

Long-term exposure: May cause cancer; birth defects, miscarriage.

Storage: Store in tightly closed containers in a dry, cool, well-ventilated place.

★ 204 ★
N-BUTYL LACTATE
CAS: 138-22-7
Other names: BUTYL LACTATE; BUTYL ALPHA-HYDROXYPROPIONATE; PROPANOIC ACID, 2-HYDROXY-, BUTYL ESTER (9CI); LACTIC ACID, BUTYL ESTER; 2-PROPANOIC ACID

Danger profile: Irritant, combustible, poisonous gas produced in fire, exposure limit established.
Uses: Making paints, inks, stencil pastes, perfumes, adhesives; solvent for oils, dyes, natural gums and many synthetic polymers; lacquers; varnishes; dry cleaning fluids; adhesives; making other chemicals.
Appearance: Colorless liquid.
Odor: Mild. The odor is a warning of exposure; however, no smell does not mean you are not being exposed.
Effects of exposure:

Breathed: May cause irritation of the eyes, nose, throat and breathing passage. Toxic for humans in concentration of about 4 ppm; regulated exposure limit (time weighted average) is 5 ppm.

Eyes: May cause irritation.

Skin: May cause severe irritation.

Long-term exposure: May cause headache, coughing, sleepiness, nausea, vomiting following exposure.

Storage: Store in tightly closed containers in a dry, cool, well-ventilated place. Keep away from heat, sources of ignition, strong acids, bases and oxidizers.

★ 205 ★
BUTYL MERCAPTAN
CAS: 109-79-5
Other names: BUTANETHIOL; BUTANE-THIOL; NCI-C60866; N-BUTANETHIOL; N-BUTYL THIOALCOHOL; N-BUTYL MERCAPTAN; 1-MERCAPTOBUTANE; 1-BUTANETHIOL; NCI-C60866

Danger profile: Fire hazard, poisonous gas produced in fire, exposure limit established.
Uses: Solvent.
Appearance: Colorless to yellow liquid.
Odor: Strong skunk-like. The odor is a warning of exposure; however, no smell does not mean you are not being exposed.
Effects of exposure:

Breathed: May be poisonous. May irritate the nose and throat; cause a loss of sense of smell. High concentrations may cause muscular weakness, nausea, dizziness, headache, confusion and suffocation. Very high levels may cause convulsions, respiratory paralysis, unconsciousness and death.
Eyes: Contact may cause irritation.
Skin: May be poisonous if absorbed. Contact may cause irritation.
Swallowed: May cause nausea.
Long-term exposure: May cause skin rash.
Storage: Store in tightly closed containers in a dry, cool, well-ventilated place. Keep away from heat, sources of ignition, strong oxidizers and acids. Vapors may travel to sources of ignition and flash back. Containers may explode in fire.
First aid guide: Move victim to fresh air and call emergency medical care; if not breathing, give artificial respiration; if breathing is difficult, give oxygen. In case of contact with material, immediately flush eyes with running water for at least 15 minutes. Wash skin with soap and water. Remove and isolate contaminated clothing and shoes at the site.

★ 206 ★
BUTYL METHACRYLATE
CAS: 97-88-1
Other names: 2-PROPENIC ACID, 2-METHYL-, BUTYL ESTER; 2-METHYL-BUTYLACRILATE; BUTYL-2-METHYL-2-PROPENOATE; BUTILMETACRILATO (ITALIAN); BUTYL METHACRYLAAT (DUTCH); METHACRYLATE DE BUTYLE (FRENCH); METHACRYLSAEURE BUTYL ESTER (GERMAN); N-BUTYL METHACRYLATE

Danger profile: Possible reproductive hazard, irritant, combustible, explosion hazard.
Uses: Dental materials; resins; solvent coatings; adhesives; textiles
Appearance: Colorless liquid.
Odor: Ester. The odor is a warning of exposure; however, no smell does not mean you are not being exposed.
Effects of exposure:
Breathed: May be poisonous. May cause dizziness. High exposure may be suffocating.
Eyes: May cause irritation.
Skin: May be poisonous if absorbed. May cause irritation, redness, itching and rash. Allergy may develop.
Swallowed: Mildly toxic.
Long-term exposure: There is experimental evidence that chemical may cause reproductive or birth defects by damaging the fetus. May cause skin allergy with rash and itching.
Storage: Store in tightly closed containers in a dark, dry, cool, well-ventilated place. Keep away from heat, moisture and water, sources of ignition, light, and strong oxidizers. Vapors may travel to sources of ignition and flash back. Containers may explode in fire.
First aid guide: Move victim to fresh air and call emergency medical care; if not breathing, give artificial respiration; if breathing is difficult, give oxygen. In case of contact with material, immediately flush eyes with running water for at least 15 minutes. Wash skin with soap and water. Remove and isolate contaminated clothing and shoes at the site.

★ 207 ★
BUTYL PHENOL
CAS: 28805-86-9

Danger profile: Possible carcinogen, poisonous gas produced in fire.

Uses: In synthetic lubricants, insecticides, and specialized starches; plasticizer.
Appearance: White crystals.
Effects of exposure:
Breathed: May be poisonous and fatal. May cause irritation of the eyes, nose throat and lungs.
Eyes: Contact may cause irritation, pain and burns.
Skin: May be poisonous and fatal if absorbed. Passes through the unbroken skin; can increase exposure and the severity of the symptoms listed. Contact may cause irritation, burning feeling, rash or burns.
Swallowed: May be poisonous and fatal. May cause gastrointestinal irritation.
Long-term exposure: May cause cancer; possible reproductive hazard. May effect the lungs; cause liver damage. Repeated or prolonged contact with the skin may cause chronic rash or ulcers and permanent loss of skin pigment in affected areas.
Storage: Store in tightly closed containers in a dry, cool, well-ventilated place. Keep away from heat and sources of ignition and oxidizers. Containers may explode in fire.

★ 208 ★
O-SEC-BUTYLPHENOL
CAS: 89-72-5
Other names: 2-SEC-BUTYLFENOL (CZECH); 2-SEC-BUTYLPHENOL; PHENOL, 2- (1-METHYLPROPYL)

Danger profile: Possible carcinogen, poisonous, severe irritant, combustible, poisonous fumes produced in fire, exposure limit established.
Uses: Making other chemicals, resins and plasticizers.
Appearance: Colorless liquid.
Effects of exposure:
Breathed: May be poisonous. May cause irritation of the eyes, nose throat and lungs.
Eyes: Contact may cause irritation, pain.
Skin: May be poisonous if absorbed. Passes through the unbroken skin; can increase exposure and the severity of the symptoms listed. Contact may cause irritation, burning feeling, rash or burns.
Swallowed: May be poisonous. May cause gastrointestinal irritation.
Long-term exposure: May cause cancer; possible reproductive hazard. May effect the lungs; cause liver damage. Repeated or prolonged contact with the skin may cause chronic rash or ulcers and permanent loss of skin pigment in affected areas.
Storage: Store in tightly closed containers in a dry, cool, well-ventilated place. Keep away from heat, sources of ignition, and oxidizers. Containers may explode in fire.

★ 209 ★
P-TERT-BUTYLPHENOL
CAS: 98-54-4
Other names: P-TERT-BUTYLFENOL (CZECH); 4-TERT-BUTYLPHENOL; 4-T-BUTYLPHENOL; 4-(1,1-DEMETHYLETHYL)PHENOL; BUTYLPHEN; 1-HYDROXY-4-TERT-BUTYLBENZENE; UCAR BUTYLPHENOL 4-T

Danger profile: Questionable carcinogen, severe irritant, combustible, poisonous gases produced in fire.
Uses: Plasticizer for cellulose acetate; making other chemicals, petroleum oils and plastics; synthetic lubricants; insecticides; industrial odorants; motor oil additives.
Appearance: White crystals or flakes.
Odor: Disinfectant-like. The odor is a warning of exposure; however, no smell does not mean you are not being exposed.
Effects of exposure:

Breathed: May cause irritation of the eyes, nose throat and respiratory tract.
Eyes: May cause irritation, pain.
Skin: Contact may cause irritation and burns, especially with wet skin.
Swallowed: May cause irritation of the mouth, stomach and gastrointestinal tract.
Long-term exposure: Questionable carcinogen with limited evidence; however, any exposure may be considered dangerous.
Storage: Store in tightly closed containers in a dry, cool, well-ventilated place. Keep away from heat, sources of ignition and oxidizers.

★ 210 ★
P-TERT-BUTYLTOLUENE
CAS: 98-51-1
Other names: P-METHYL-TERT-BUTYLBENZENE; 1-METHYL-4-TERT-BUTYLBENZENE; TBT

Danger profile: Exposure limit established, poisonous gas produced in fire.
Uses: Solvent; making other chemicals.
Appearance: Colorless liquid.
Odor: Aromatic gasoline-like. The odor is a warning of exposure; however, no smell does not mean you are not being exposed.
Breathed: May cause irritation of the eyes, nose, throat and air passages; coughing, headache; increased saliva; reddening of the face; dizziness, lightheadedness, fatigue, nausea and vomiting, loss of consciousness, shock, coma, irregular heartbeat, possible death. If the victim recovers watch for the possibility of a buildup of fluid in the lungs (pulmonary edema). Pulmonary edema is a medical emergency that can be delayed one to two days following exposure.
Eyes: Contact may cause imitation, tearing, inflammation of the eyelids; burning sensation.
Skin: May cause irritation, redness, burning feeling, swelling; dry skin; rash or blisters.
Swallowed: Poison. Breath may smell of chloroform. May cause irritation of the mouth, gastro-intestinal tract, nausea, vomiting, diarrhea with possible blood; drowsiness; loss of consciousness; irregular heartbeat and death. See *Breathed.*
Long-term exposure: May cause liver and kidney damage.
Storage: Store in tightly closed containers in a dry, cool, well ventilated place. Keep away from heat, flame, sources of ignition and oxidizing materials.

★ 211 ★
BUTYL TRICHLOROSILANE
CAS: 7521-80-4
Other names: SILANE, BUTYLTRICHLORO-; N-BUTYLTRICHLOROSILANE; TRICHLOROBUTYLSILANE; TRICHLOROALLYLSILANE

Danger profile: Corrosive, poisonous, combustible, reacts with water, poisonous and irritating gases produced in fire.
Uses: Making silicones.
Appearance: Colorless liquid.
Odor: Sharp, hydrochloric acid-like; pungent and irritating. The odor is a warning of exposure; however, no smell does not mean you are not being exposed.
Effects of exposure:
Breathed: May be poisonous. Vapor may irritate the eyes, nose throat and upper respiratory system; coughing and shortness of breath. Higher exposures may cause shortness of breath and a dangerous build-up of fluids in the lungs (pulmonary edema), which can cause death. Pulmonary edema

is a medical emergency that can be delayed one to two days following exposure.
Eyes: Contact may cause, irritation, tearing, inflammation of the eyelids and severe burns; permanent damage.
Skin: Contact may cause irritation and severe burns.
Swallowed: May cause severe irritation of the mouth and stomach.
Long-term exposure: May effect the lungs.
Storage: Store in tightly closed containers in a dry, cool, well-ventilated place. Keep away from heat, sources of ignition, water, steam, common metals and oxidizers. Will react with rubber over clothing. Vapors may travel to sources of ignition and flash back. Containers may explode in fire.
First aid guide: Move victim to fresh air and call emergency medical care; if not breathing, give artificial respiration; if breathing is difficult, give oxygen. In case of contact with material, immediately flush skin or eyes with running water for at least 15 minutes. Remove and isolate contaminated clothing and shoes at the site. Keep victim quiet and maintain normal body temperature.

★ 212 ★
BUTYRALDEHYDE
CAS: 123-72-8
Other names: ALDEHYDE BUTYRIQUE (FRENCH); ALDEIDE BUTIRRICA (ITALIAN); BUTAL; BUTALDEHYDE; BUTA-LYDE; BUTANAL; N-BUTANAL (CZECH); BUTYRAL; BUTYRALDEHYD (GERMAN); BUTYL ALDEHYDE; N-BUTYL ALDEHYDE; BUTYRIC ACID; BUTYRIC ALDEHYDE; FEMA NO. 2219; NCI-C56291

Danger profile: Severe irritant, combustible, Community Right-to-Know List, poisonous gases produced in fire.
Uses: Manufacturing synthetic resins, solvents and plasticizers.
Appearance: Colorless liquid.
Effects of exposure:
Breathed: May be poisonous. Vapor may cause intense irritation to nose, throat, mouth and lungs; mucous membrane. Symptoms of exposure include watering of the eyes, coughing, difficult breathing, headache, nausea, vomiting, muscular weakness, dizziness, unconsciousness. May cause suffocation. High exposure can lead to drowsiness, coma and death; buildup of fluid in the lungs (pulmonary edema), and may cause death. Pulmonary edema may be delayed by one to two days following exposure.
Eyes: Severe irritant. Contact may cause, burning and stinging, tearing, irritation and redness and swelling of the eyelids, intense pain; can burn the eyes and lead to blurred vision and permanent damage.
Skin: May be poisonous if absorbed. Severe irritant. Contact may cause burning or stinging feeling, redness, swelling, intense pain, blisters, irritation, rash.
Swallowed: May cause burning feeling of the mouth and throat, intense thirst, throat swelling, stomach cramps, nausea, vomiting, diarrhea, severe breathing difficulties and unconsciousness. Tissue damage may result. Breathing may stop. Watch for delayed symptoms.
Storage: Store in tightly closed containers in a dry, cool, well-ventilated place. Keep away heat, sources of ignition, from oxidizers, strong bases, strong reducing agents, strong acids. Vapors may travel to sources of ignition and flash back. Containers may explode in fire.
First aid guide: Move victim to fresh air and call emergency medical care; if not breathing, give artificial respiration; if breathing is difficult, give oxygen. In case of contact with material, immediately flush eyes with running water for at least 15 minutes. Wash skin with soap and water. Remove and isolate contaminated clothing and shoes at the site.

★ 213 ★
BUTYRIC ACID
CAS: 107-92-6
Other names: BUTANIC ACID; BUTANOIC ACID; BUTTER-SAEURE (GERMAN); N-BUTYRIC ACID; ETHYLACETIC ACID; FEMA NO. 2221; 1-PROPANECARBOXYIC ACID; PROPYLFORMIC ACID

Danger profile: Mutagen, corrosive, severe irritant, fire hazard, poisonous gases produced in fire.
Uses: Food additive; artificial flavoring; in perfumes; varnishes; pharmaceuticals; disinfectants.
Appearance: Oily liquid.
Odor: Strong, irritating; like spoiled butter. The odor is a warning of exposure; however, no smell does not mean you are not being exposed.
Effects of exposure:
   Breathed: May cause irritation of the nose and throat. High exposure may lead to drowsiness, coma and death; buildup of fluid in the lungs (pulmonary edema), and may cause death. Pulmonary edema may be delayed by one to two days following exposure.
   Eyes: Severe irritant. Contact may cause, burning and stinging, tearing, irritation and redness and swelling of the eyelids, intense pain; can cause severe burns and lead to blurred vision, and permanent damage.
   Skin: Severe irritant. Contact may cause burning or stinging feeling, redness, swelling, intense pain, blisters, irritation, rash, severe burns.
   Swallowed: Moderately toxic. May cause burning feeling of the mouth and throat, intense thirst, throat swelling, stomach cramps, nausea, vomiting, diarrhea, severe breathing difficulties and unconsciousness. Tissue damage may result. See *Breathed.*
Long-term exposure: May effect the lungs.
Storage: Store in tightly closed containers in a dry, cool, well-ventilated place. Keep away from sources of ignition, chromium trioxide, and from oxidizers. Protect containers from physical damage. Containers may explode in fire.
First aid guide: Move victim to fresh air; call emergency medical care. In case of contact with material, immediately flush skin or eyes with running water for at least 15 minutes. Remove and isolate contaminated clothing and shoes at the site. Keep victim quiet and maintain normal body temperature.

C

★ 214 ★
CADMIUM
CAS: 7440-43-9
Other names: C.I. 77180; COLLOIDAL CADMIUM; KADMIUM (GERMAN)

Danger profile: Carcinogen, mutagen, reproductive hazard, poisonous, exposure limit established, Community Right-to-Know List, toxic fumes produced in fire.
Uses: Electroplating; making plastics; storage batteries; television picture tubes; in ceramic glazes and enamels; nuclear reactor rods; fungicide; lithography and photography; electronic devices; electrodes for street lamps; photo-electric cells.
Appearance: Bluish metal or gray powder.
Effects of exposure:
   Breathed: Poison. Dust may cause severe irritation to the nose, throat and lungs, causing coughing, shortness of breath. May cause flu-like symptoms including chills, muscle ache, headache, fever. High exposure may cause nausea,

salivation, vomiting, cramps, diarrhea, chest pains, cough, a build-up of fluids in the lungs (pulmonary edema) and possible death. Pulmonary edema is a medical emergency that may be delayed from one to two days following exposure.
   Skin: May cause severe irritation and thermal and acid burns, especially if skin is wet.
   Eyes: May cause severe irritation, burns, pain and possible blindness.
   Swallowed: Poisonous. May cause severe burns of the mouth, throat and stomach, vomiting, watery or bloody diarrhea. Damage to kidneys and liver, collapse and convulsions can result.
Long-term exposure: May cause cancer, damage to the developing fetus, may effect the female reproductive cycle, kidney damage, kidney stones, emphysema, anemia, loss of smell, fatigue. May cause the teeth to turn yellow.
Storage: Store in tightly closed containers in a cool dry place. Keep away from water, oxidizers, metals, sulfur, selenium, tellurium, nitryl fluoride, ammonium nitrate and hydrazoic acid.

★ 215 ★
CADMIUM ACETATE
CAS: 543-90-8
Other names: ACETIC ACID, CADMIUM SALT; BIS(ACE-TOXY)CADMIUM; CADMIUM (II) ACETATE; CADMIUM DIACETATE, C.I. 77185

Danger profile: Carcinogen, mutagen, reproductive hazard, exposure limit established, Community Right-to-Know List, toxic fumes produced in fire.
Uses: Ceramics; dyeing and printing; electroplating metals; laboratory chemical; making other acetate compounds.
Appearance: Colorless, odorless crystal material.
Effects of exposure:
   Breathed: Poison. Dust may cause severe irritation to the nose, throat and lungs, causing coughing, shortness of breath. May cause flu-like symptoms including chills, muscle ache, headache, fever. High exposure may cause nausea, salivation, vomiting, cramps, diarrhea, chest pains, cough, a build-up of fluids in the lungs (pulmonary edema) and possible death. Pulmonary edema is a medical emergency that may be delayed from one to two days following exposure.
   Skin: May cause severe irritation and thermal and acid burns, especially if skin is wet.
   Eyes: May cause severe irritation, burns, pain and possible blindness.
   Swallowed: Poison. May cause severe burns of the mouth, throat and stomach, vomiting, watery or bloody diarrhea. Damage to kidneys and liver, collapse and convulsions may result.
Long-term exposure: May cause cancer of the kidneys or prostate; damage to the developing fetus, kidney damage, emphysema, loss of smell, lowered blood sugar, low blood count. May cause yellowing of teeth.
Storage: Store in tightly closed containers in a dry, cool, well-ventilated place. Keep away from heat.

★ 216 ★
CADMIUM BROMIDE
CAS: 7789-42-6

Danger profile: Carcinogen, exposure limit established, Community Right-to-Know List, poisonous gases produced in fire.
Uses: Photography; engraving; lithography.
Appearance: White, to yellowish crystalline powder.
Effects of exposure:
   Breathed: Poison. Dust may cause severe irritation to the nose, throat and lungs, causing coughing, shortness of

breath. May cause flu-like symptoms including chills, muscle ache, headache, fever. High exposure may cause nausea, salivation, vomiting, cramps, diarrhea, chest pains, cough, a build-up of fluids in the lungs (pulmonary edema) and possible death. Pulmonary edema is a medical emergency that may be delayed from one to two days following exposure.
Skin: May cause severe irritation.
Eyes: May cause severe irritation, burns, pain and possible blindness.
Swallowed: Poison. May cause severe burns of the mouth, throat and stomach, vomiting, watery or bloody diarrhea. Damage to kidneys and liver; collapse and convulsions may result.
Long-term exposure: May cause cancer, damage to the developing fetus, decrease fertility in men and women; kidney damage, kidney stones, emphysema, lung scarring, anemia, loss of smell. Yellowing of teeth may occur.
Storage: Store in tightly closed containers in a dry, cool, well-ventilated place. Keep away from potassium.

★ 217 ★
CALCIUM
CAS: 7440-70-2
Other names: CALCICAT; CALCIUM METAL, CRYSTALINE; CALCIUM, ALLOYS; CALCIUM METAL

Danger profile: Combustible, explosion hazard, violent reaction with water.
Uses: Making aluminum, lead alloys, copper.
Appearance: Shiny, silver-white metal; turns to grayish-white on exposure to air.
Effects of exposure:
Breathed: May be poisonous. Dust may cause irritation of the lungs and air passages, cough and difficult breathing. High exposure may cause nausea, salivation, vomiting, cramps, diarrhea, chest pains, and a build-up of fluids in the lungs (pulmonary edema) and possible death. Pulmonary edema is a medical emergency that may be delayed from one to two days following exposure.
Eyes: Contact may cause irritation, tearing, redness, burns and permanent damage.
Skin: Contact may cause irritation, redness and chemical burns.
Swallowed: May cause irritation of the mouth and throat, stomach cramps and burning feeling, nausea, dizziness, diarrhea, convulsions, coma and death.
Long-term exposure: May effect the lungs.
Storage: Store in neutral oil such as kerosene in tightly closed containers in a dry, cool, well-ventilated place. Keep containers tightly closed. Keep away from heat, water, steam and other forms of moisture, acid, oxidizers, sulfur, carbonates, dinitrgen tetraoxide, lead chloride, halogens, alkaline hydroxides, alkaline hydroxides. Containers may explode in fire.

★ 218 ★
CALCIUM CARBIDE
CAS: 75-20-7
Other names: ACTYLENOGEN; CALCIUM ACETYLIDE

Danger profile: Fire hazard, explosion hazard, dangerous when wet, poisonous gases produced in fire.
Uses: Making acetylene gas used for welding and other chemicals.
Appearance: Grey or bluish-black crystals or granules.
Odor: Garlic-like. The odor is a warning of exposure; however, no smell does not mean you are not being exposed.
Effects of exposure:

Breathed: May be poisonous. Dust may cause severe irritation of the nose, throat, mouth, tongue, lungs and air passages, cough and difficult breathing; swollen tongue, inflamed lips. High exposure may cause nausea, salivation, vomiting, cramps, diarrhea, chest pains, and a build-up of fluids in the lungs (pulmonary edema) and possible death. Pulmonary edema is a medical emergency that may be delayed from one to two days following exposure.
Eyes: Contact may cause severe irritation, tearing, redness, burns and permanent damage.
Skin: Contact may cause severe irritation, redness, burns and ulcers.
Swallowed: May cause severe irritation of the mouth and throat, stomach cramps and burning feeling, nausea, dizziness, diarrhea, convulsions, coma and death.
Long-term exposure: May cause bronchitis, cough, phlegm, shortness of breath, lung irritation and damage.
Storage: Store in tightly closed containers in a dry, cool, well-ventilated place. Keep away from water, steam, water sprinklers, and other sources of moisture, oxidizers, acids, acid fumes, methanol, silver nitrate and copper salt solutions.
First aid guide: Move victim to fresh air; call emergency medical care. Wipe material from skin immediately; flush skin or eyes with running water for at least 15 minutes. Remove and isolate contaminated clothing and shoes at the site.

★ 219 ★
CALCIUM CARBONATE
CAS: 1317-65-3
Other names: AGRICULTURAL LIMESTONE; AGSTONE; ARAGONITE; ATOMIT; BELL MINE PULVERIZED LIMESTONE; CALCITE; CARBONIC ACID, CALCIUM SALT (1:1); CHALK; DOMOLITE; LIMESTONE; LITHOGRAPHIC STONE; MARBLE; NATURAL CALCIUM CARBONATE; PORTLAND STONE; SOHNHOFEN STONE; CALCIUM(II)CARBONATE (1:1); FRANKLIN; VATERITE

Danger profile: Irritant, exposure limit established.
Uses: Building stone; metallurgy; making lime; source of carbon dioxide; agriculture; cement; cleaning stack gases, and removing sulfur from coal.
Appearance: White powder or crystals.
Effects of exposure:
Breathed: May cause irritation of the nose, coughing and sneezing.
Eyes: May cause irritation, redness, tearing, pain and inflammation of the eyelids.
Storage: Store in tightly closed containers in a dry, cool, well-ventilated place. Keep away from acid, aluminum sulfate, and ammonium salt.

★ 220 ★
CALCIUM CHLORATE
CAS: 10137-74-3
Other names: CALCIUM CHLORATE AQUEOUS SOLUTION; CHLORATE DE CALCIUM (FRENCH); CHLORIC ACID, CALCIUM SALT

Danger profile: Powerful oxidizer, explosion hazard, poisonous gas produced in fire.
Uses: Photography; making fireworks; herbicide.
Appearance: White-yellowish crystals.
Effects of exposure:
Breathed: May cause irritation of the eyes, nose, throat and upper respiratory tract. High exposure may interfere with the ability of the blood to carry oxygen, causing bluish color to

skin and lips; headache, weakness, shortness of breath, dizziness and death.
Eyes: May cause irritation and burns.
Skin: May cause irritation and burns.
Swallowed: May cause abdominal pain, nausea, vomiting, diarrhea, pallor, shortness of breath, unconsciousness.
Long-term exposure: May cause kidney and liver damage.
Storage: Store in tightly closed containers in a dry, cool, well-ventilated place with non-wood floors. Keep away from combustible materials, fuels, acids, chemically active metals, charcoal, ammonium compounds, aluminum, copper, cyanides.
First aid guide: Move victim to fresh air; call emergency medical care. In case of contact with material, immediately flush skin or eyes with running water for at least 15 minutes. Remove and isolate contaminated clothing and shoes at the site.

★ 221 ★
CALCIUM CHROMATE
CAS: 13765-19-0
Other names: CALCIUM CHROMATE (VI); CALCIUM CHROME YELLOW; CALCIUM CHROMIUM OXIDE; CALCIUM MONOCHROMATE; C.I. 77223; C.I. PIGMENT YELLOW 33; GELBIN; YELLOW ULTRAMARINE; CHROMIC ACID, CALCIUM SALT (1:1); C.I. 77223

Danger profile: Carcinogen, mutagen, powerful oxidizer, exposure limit established, Community Right-to-Know List.
Uses: Making pigments; corrosion inhibitor; to depolarize batteries; coating for metal alloys.
Appearance: Yellow crystals.
Effects of exposure:
Breathed: May cause irritation of the nose and throat; coughing, dyspnea, wheezing, chest pain, difficult breathing and fever.
Eyes: Contact may cause severe irritation, redness, tearing inflammation of the inner lining of the eyelids (conjunctiva); severe chemical burns and possible permanent damage and loss of vision.
Skin: Passes through the unbroken skin; can increase exposure and the severity of the symptoms listed. Contact may cause burns, ulcers, deep sores known as "chrome holes" and sensitization dermatitis.
Swallowed: May cause severe sore throat; irritation of the throat, stomach and intestines; circulatory collapse.
Long-term exposure: May cause cancer, damage to the developing fetus; kidney disease; skin allergy; chronic irritation of the bronchial tubes, cough and phlegm; chronic lung allergy with painful and difficult breathing; breakdown of nasal septum tissue; jaundice.
Storage: Store in tightly closed containers in a dry, cool, well-ventilated place with non-wood floors. Keep away from combustible materials and boron. May cause spontaneous fires.

★ 222 ★
CALCIUM CYANAMIDE
CAS: 156-62-7
Other names: AERO-CYANAMID; AERO CYANAMID GRANULAR; AERO CYANAMID SPECIAL GRADE; ALZODEF; CALCIUM CARBIMIDE; CALCIUM CYANAMID; CCC; CY-L 500; CYANAMIDE, CALCIUM SALT (1:1); CYANAMID; CYANAMIDE; CYANAMIDE CALCIQUE (FRENCH); CYANAMID GRANULAR; CYANAMID SPECIAL GRADE; CY-L 500; LIME NITROGEN; NCI-CO2937; NITROGEN LIME; NITROLIME; USAF CY-2

Danger profile: Possible carcinogen, mutagen, poisonous, fire hazard on contact with water, exposure limit established, Community Right-to-Know List, poisonous gases produced in fire.

Uses: Fertilizer; pesticide; making other chemicals; hardening and removing sulfur from steel.
Appearance: Fine gray crystals or powder.
Effects of exposure:
Breathed: May be poisonous. May cause ulceration of the nose and throat; inflammation of the mucous membranes of the nose; nasal discharge; nausea, dizziness, and flushing of the skin. Over exposure may cause vomiting, seizures and possibly death. Drinking alcoholic beverages before or after exposure may increase these effects.
Eyes: May cause irritation, inflammation of the inner surface of the eyelids and burns.
Skin: Passes through the unbroken skin; can increase exposure and the severity of the symptoms listed. May cause irritation of skin varying from redness and swelling (similar to sunburn) to blisters or crusting. In severe cases, ulceration may develop.
Swallowed: May cause gums to be red, swollen and bleeding.
Long-term exposure: May cause nervous system damage, skin allergy, and skin ulcers.
Storage: Store in tightly closed containers in a refrigerator or other dry, cool place. Keep away from water, solvents and calcium carbide.

★ 223 ★
CALCIUM FLUORIDE
CAS: 7789-75-5
Other names: CALCIUM DIFLUORIDE; FLUOSPAR; FLUORSPAR; MET-SPAR

Danger profile: Mutagen, exposure limit established, poisonous gases produced in fire.
Uses: Steel making; in smelting; arc welding; glass and ceramics; fluoridation in drinking water.
Appearance: Colorless crystals or white solid.
Effects of exposure:
Breathed: May cause irritation of the nose, throat and lungs. High exposure can lead to buildup of fluid in the lungs (pulmonary edema), and may cause death. Pulmonary edema may be delayed by one to two days following exposure.
Eyes: Contact may cause irritation, redness, tearing, burns and permanent damage.
Skin: Contact may cause irritation, redness, swelling and burns.
Swallowed: These effects do not occur at the levels used in drinking water treated for dental cavity prevention.
Long-term exposure: May cause nausea, constipation, poor appetite; stiff muscles and joints; scarring of the lungs and shortness of breath.
Storage: Store in tightly closed containers in a dry, cool, well-ventilated place. Keep away from acids and chemically active metals.

★ 224 ★
CALCIUM HYDROXIDE
CAS: 1305-62-0
Other names: BELL MINE; CALCIUM HYDRATE; CARBOXIDE; HYDRATED KEMIKAL; LIME WATER; SLAKED LIME

Danger profile: Mutagen, severe irritant, exposure limit established.
Uses: Mortar, plasters and cements; hair remover; hide processing; whitewash; soil conditioner; disinfectant; water softener; making rubber compounds; petrochemicals; food additive, poultry feed.
Appearance: Colorless crystals or white powder or granules.
Effects of exposure:

Breathed: May cause irritation of nose, eyes and throat; coughing; difficult breathing.

Eyes: Contact may cause irritation, pain, tearing; serious burns, permanent damage and loss of sight may result.

Skin: Contact may cause irritation, redness and swelling; burning feeling. Burns may occur if left on the skin.

Swallowed: May cause burning feeling in the mouth and stomach; mouth and gastric irritation; stomach cramps.

Long-term exposure: May effect the lungs.

Storage: Store in tightly closed containers in a dry, cool, well-ventilated place. Keep away from maleic anhydride, nitroethane, nitroether, nitromethane, nitroparaffins, nitropropane, phosphorus.

★ 225 ★
CALCIUM HYPOCHLORITE

CAS: 7778-54-3

Other names: B-K POWDER; BLEACHING POWDER; CALCIUM CHLOROHYDROCHLORITE; CALCIUM HYPOCHLORIDE; CALCIUM OXYCHLORIDE; CAPORIT; CCH; CHLORIDE OF LIME; CHLORINATED LIME; HTH; HY-CHLOR; HYPOCHLOROUS ACID, CALCIUM; LIME CHLORIDE; LOBAX; LOSANTIN; PERCHLORON; PITTCIDE; PITTCHLOR; SENTRY

Danger profile: Mutagen, oxidizer, very reactive, explosion hazard, reacts with water to produce corrosive and toxic fumes, poisonous gases produced in fire.

Uses: Drinking water purification; swimming pool chemical; bleaching agent in the manufacture of textiles and paper; algicide, bactericide, disinfectant, fungicide, commercial deodorant.

Appearance: White powder, pellets, granules or sticks.

Odor: Like household bleach. The odor is a warning of exposure; however, no smell does not mean you are not being exposed.

Effects of exposure:

Breathed: Dust may cause severe irritation to the nose, throat, voice box, breathing passages and lungs, causing coughing, hoarseness, shortness of breath. May cause flu-like symptoms including chills, muscle ache, headache, fever. High exposure may cause nausea, salivation, vomiting, cramps, diarrhea, chest pains, cough, a build-up of fluids in the lungs (pulmonary edema) and possible death. Pulmonary edema may be delayed from one to two days following exposure. Hypochlorous acid fumes (given off only if compound comes in contact with acid) cause severe respiratory tract irritation and pulmonary edema.

Eyes: May cause severe irritation, burns, pain and possible blindness.

Skin: May cause severe irritation and thermal and acid burns, especially if skin is wet.

Swallowed: May cause severe burns and pain of the mouth, throat and stomach, vomiting (hemorrhaging may cause vomitus to resemble coffee grounds), watery or bloody diarrhea, circulatory collapse, with cold, clammy skin, skin and lips turn blue (cyanosis), shallow respirations; confusion, delirium, coma, swelling of the throat may cause obstruction; perforation of esophagus or stomach, shock, convulsions and collapse can result.

Long-term exposure: May cause genetic changes that may lead to birth defects, miscarriage or cancer; irritation of the lungs, bronchitis, chronic cough and shortness of breath. Prolonged or repeated skin contact may cause blisters (vesicular eruptions) and eczematoid dermatitis.

Storage: Store in tightly closed containers in a dry, cool, well-ventilated place with non-wood floors. Keep away from heat, acids, combustible materials, all other chemicals and especially aniline and other amines, carbon tetrachloride, acetylene, sulfur or organic sulfur compounds. Becomes explosive when heated above 212°F. Protect containers against physical damage. Water or steam produces corrosive and toxic fumes.

★ 226 ★
CALCIUM NITRATE

CAS: 10124-37-5

Other names: CALCIUM (II) NITRATE (1:2); NITRIC ACID, CALCIUM SALT; LIME SALTPETER; NORWEGIAN SALTPETER; NITROCALCITE

Danger profile: Explosion hazard, oxidizer, poisonous gases produced in fire.

Uses: Fireworks; explosives; matches; fertilizers; making other chemicals and industrial products.

Appearance: Colorless crystals. Absorbs water.

Effects of exposure:

Breathed: May cause irritation of the nose, throat, breathing passages and lungs; redness, tearing and pain. Overexposure may cause headache, nausea, vomiting, weakness, bluish skin and lips, shortness of breath, collapse, coma and death.

Eyes: May cause irritation, redness, tearing, inflammation of the mucous membrane of the eyelids.

Skin: May cause irritation, redness, inflammation and swelling and rash.

Swallowed: May cause burning feeling of the mouth and throat, intense thirst, throat swelling, stomach cramps, nausea, vomiting, diarrhea, severe breathing difficulties and unconsciousness. Tissue damage may result. Breathing may stop. Watch for delayed symptoms.

Long-term exposure: May effect the breathing passages and lungs with shortness of breath.

Storage: Store in tightly closed containers in a dry, cool, well-ventilated place with non-wood floors. Keep away from combustible materials, fuels, organics and other readily oxidizable materials, chemically active metals, aluminum and ammonium nitrate. May explode if shocked or heated.

First aid guide: Move victim to fresh air; call emergency medical care. In case of contact with material, immediately flush skin or eyes with running water for at least 15 minutes. Remove and isolate contaminated clothing and shoes at the site.

★ 227 ★
CALCIUM OXIDE

CAS: 1305-78-8

Other names: BURNT LIME; CALCIA; CALX; LIME; LIME, BURNED; LIME, UNSLAKED; OXYDE DE CALCIUM (FRENCH); QUICKLIME; WAPNIOWY TLENEK (POLISH); PEBBLE LIME

Danger profile: Exposure limit established, powerful caustic, poisonous gases produced in fire.

Uses: Making other chemicals, building products and glass; metal processing; sewage treatment; refractory; pulp and paper; poultry feeds; neutralization of acid waste effluents; insecticides and fungicides; hair removal from hides; sugar refining; food additive.

Appearance: White or gray solid.

Effects of exposure:

Breathed: May cause irritation of the eyes, nose and throat with burning feeling; difficult breathing, coughing and sneezing. High exposure can lead to a buildup of fluid in the lungs (pulmonary edema), and may cause death. Pulmonary edema may be delayed by one to two days following exposure.

Eyes: May cause severe irritation with pain in eyes and eyelids; intense watering; burns and damage to the mucous

membrane, perforation of the eyes, permanent damage and blindness.
Skin: May cause irritation, itching, pain, ulceration with slippery feeling; serious, deep chemical burns.
Swallowed: May cause immediate burning feeling in the mouth, throat and stomach; intense pain in swallowing; throat swelling, vomiting (hemorrhaging may cause vomitus that looks like coffee grounds), watery or bloody diarrhea, shallow respirations; confusion, delirium, coma; swelling of the throat may cause obstruction; perforation of esophagus or stomach, shock, convulsions and collapse and death can result.
Long-term exposure: May effect the lungs and cause breakdown of nasal tissue dividing the inner nose; cause skin thickening and cracking and brittle nails.
Storage: Store in tightly closed containers in a dry, cool, well-ventilated place with non-wood floors. Keep away from water, steam and other forms of moisture, phosphorus oxide, chlorine trifluoride, oxidizers, and hydrofluoric acid. May react explosively with water; contact with water may generate enough heat to ignite combustible materials stored nearby.
First aid guide: Move victim to fresh air; call emergency medical care. In case of contact with material, immediately flush skin or eyes with running water for at least 15 minutes. Remove and isolate contaminated clothing and shoes at the site. Keep victim quiet and maintain normal body temperature.

★ 228 ★
CALCIUM PEROXIDE
CAS: 1305-79-9
Other names: CALCIUM DIOXIDE; CALCIUM SUPEROXIDE

Danger profile: Oxidizer.
Uses: Antiseptic, seed disinfectant; dentifrices; conditioners for dough; rubber stabilizer; for bleaching oils.
Appearance: Yellow crystals or white powder.
Effects of exposure:
   Breathed: May cause irritation of the eyes, nose and throat with burning feeling; difficult breathing, coughing and sneezing. High exposure can lead to a buildup of fluid in the lungs (pulmonary edema), and may cause death. Pulmonary edema may be delayed by one to two days following exposure.
   Eyes: May cause severe irritation with pain in eyes and eyelids; intense watering; burns.
   Skin: May cause irritation, itching, pain and burns.
   Swallowed: Irritates mouth and stomach.
Long-term exposure: May effect the lungs; cause skin damage.
Storage: Store in tightly closed containers in a dry, cool, well-ventilated place with non-wood floors. Keep away from combustible materials, fuels and polysulfide polymers.
First aid guide: Move victim to fresh air; call emergency medical care. In case of contact with material, immediately flush skin or eyes with running water for at least 15 minutes. Remove and isolate contaminated clothing and shoes at the site.

★ 229 ★
CALCIUM PHOSPHIDE
CAS: 1305-99-3
Other names: PHOTOPHOR; TRICALCIUM DIPHOSPHIDE

Danger profile: Highly toxic, dangerous fire hazard, explosion hazard, poisonous gases produced when wet and in fire.
Uses: To kill rodents; fireworks; torpedoes; explosives.
Appearance: Reddish-brown crystals.
Odor: Musty. The odor is a warning of exposure; however, no smell does not mean you are not being exposed.
Effects of exposure:

Skin: Contact may be poisonous.
Swallowed: May cause faintness, weakness, nausea, vomiting.
Storage: Store in tightly closed containers in a dry, cool, well-ventilated place. Keep away from moisture and water.
First aid guide: Move victim to fresh air and call emergency medical care; if not breathing, give artificial respiration; if breathing is difficult, give oxygen. In case of contact with material, immediately flush skin or eyes with running water for at least 15 minutes. Remove and isolate contaminated clothing and shoes at the site. Keep victim quiet and maintain normal body temperature.

★ 230 ★
CAMPHENE
CAS: 79-92-5
Other names: 2,2-DIMETHYL-3-METHYLENE-; BICYCLO-(2.2.1)HEPTANE; 2-2-DIMETHYL-3-METHYLENE NORBORANE; 3,3-DIMETHYLENENORCAMPHENE; 3,3-DIMETHYL-2-METHYLENE NORCAMPHONE; FEMA NO. 2229

Danger profile: Mutagen, combustible, combustible vapors and irritating fumes produced when heated.
Uses: Making synthetic camphor; moth proofing; food flavoring; cosmetics and perfume.
Appearance: Colorless to white crystals.
Odor: Camphor-like. The odor is a warning of exposure; however, no smell does not mean you are not being exposed.
Effects of exposure:
   Breathed: Extremely irritating. May cause irritation of the eyes, nose and throat, coughing, difficult breathing and nausea. Higher levels may cause headache, nausea and profuse sweating; confusion, sleepiness and coma.
   Eyes: Contact may cause severe irritation, burning feeling and tears.
   Skin: Contact may cause severe irritation.
Long-term exposure: May cause birth defects, miscarriage or cancer; skin allergy.
Storage: Store in tightly closed containers in a dry, cool, well-ventilated place. Keep away from heat, sources of ignition, and oxidizers. Combustible vapors are released even at room temperature.
First aid guide: Move victim to fresh air and call emergency medical care; if not breathing, give artificial respiration; if breathing is difficult, give oxygen. In case of contact with material, immediately flush skin or eyes with running water for at least 15 minutes. Remove and isolate contaminated clothing and shoes at the site. Effects should disappear after individual has been exposed to fresh air for approximately 10 minutes.

★ 231 ★
CAMPHOR
CAS: 76-22-2
Other names: BICYCLO 2.2.1 HEPTAN-2-ONE,1,7,7-TRIMETHYL-; BORNANE, 2-OXO-; 2-BORNANONE; 2-CAMPHANONE; 2-CAMPHORONE; CAMPHOR, NATURAL; FORMOSA CAMPHOR; GUM CAMPHOR; HUILE DE CAMPHRE (FRENCH); JAPAN CAMPHOR; KAMPFER (GERMAN); 2-KETO-1,7,7-TRIMETHYLNORCAMPHANE; LAUREL CAMPHOR; MATRICARIA CAMPHOR; NORCAMPHOR, SYNTHETIC CAMPHOR; 1,7,7-TRIMETHYL-; 1,7,7-TRIMETHYLBICYCLO(2.2.1)-2-HEPTANONE; 1,7,7-TRIMETHYLNORCAMPHOR

Danger profile: Mutagen, combustible, exposure limit established, poisonous gases produced in fire.

Uses: Making medicines, plastics, lacquers and paints; explosives; insecticides, moth proofing; preventing mildew; tooth powders; flavoring, embalming; fireworks.
Appearance: Colorless, shiny, glassy solid.
Odor: Penetrating aromatic odor. The odor is a warning of exposure; however, no smell does not mean you are not being exposed.
Effects of exposure:
Breathed: may cause irritation of the nose, throat mucous membrane. Higher levels may cause headache, dizziness, nausea, vomiting, excitement, confusion, irrational behavior, seizures (epileptiform convulsions or "fits"), coma and death.
Eyes: Contact may cause irritation, redness, tearing and inflammation of the eyelids.
Skin: Contact may cause irritation and redness.
Swallowed: Poison.
Long-term exposure: May cause birth defects, miscarriages or cancer; kidney damage. Long-term exposure may effect the sense of smell.
Storage: Store in tightly closed containers in a dry, cool, well-ventilated place. Keep away from heat, flame and sources of combustion, oxidizers, and chromic anhydride. Vapors are explosive.
First aid guide: Move victim to fresh air; call emergency medical care. In case of contact with material, immediately flush skin or eyes with running water for at least 15 minutes. Removal of solidified molten material from skin requires medical assistance. Remove and isolate contaminated clothing and shoes at the site.

★ 232 ★
CAPTAFOL
CAS: 2425-06-1
Other names: CAPTATOL; CAPTOFOL; DIFOLATAN; DIFOSAN; FOLCID; 1H-ISOINDOLE-1,3(2H)-DIONE,3A,4,7,7A-TETRAHYDRO-2-(1,1,2,2- TETRACHL OROETHYL)THIO-; ORTHO 5865; SANSPOR; SULFONIMIDE; SULPHEIMIDE; N-(1,1,2,2-TETRACHLORAETHYLTHIO)-CYCLOHEX-4-EN-1,4-DIACARBOXIMID (GERMAN); N-1,1,2,2-TETRACHLOROETHYLMERCAPTO- 4-CYCLOHEXENE-1,2-CARBOXIMIDE; N-((1,1,2,2-TETRACHLOROETHYL)SULFENYL)-CIS-4-CYCLOHEXENE-1,2-DICARBOXIMIDE; N-(1,1,2,2-TETRACHLOROETHYLTHIO)-4- CYCLOHEXENE-1,2-DICARBOXIMIDE; 4-CYCLOHEXENE-1,2-DICARBOXIMIDE, N-(1,1,2,2-TETRACHLOROETHYL)THIOL-

Danger profile: Mutagen, exposure limit established, poisonous fumes produced in fire.
Uses: Fungicide. To kill molds, mildew, etc.
Appearance: White crystals.
Odor: Strong. The odor is a warning of exposure; however, no smell does not mean you are not being exposed.
Effects of exposure:
Breathed: May be poisonous. May cause irritation, cough, wheezing, shortness of breath, difficult breathing and an asthma-like allergy may develop.
Eyes: Contact may cause irritation, redness, tearing and inflammation of eyelids.
Skin: Contact may cause irritation, redness and rash; allergy may develop.
Swallowed: May be poisonous.
Long-term exposure: May cause birth defects, miscarriages or cancer. May cause skin allergy, kidney and liver damage, asthma-like allergy with cough and shortness of breath.
Storage: Store in tightly closed containers in a dry, cool, well-ventilated place. Keep away from acids, oxidizers and heat.

★ 233 ★
CAPTAN
CAS: 133-06-2
Other names: CAPTANCAPTENEET 26,538; CAPTANE; AMERCIDE; CAPTAF; CAPTAF 85W; CAPTEX; ENT 26,538; ESSOFUNGICIDE 406; FLIT 406; FUNGUS BAN TYPE II; GL-YODEX 3722; LE CAPTANE (FRENCH); MERPAN; NCI-0077; ORTHOCIDE; ORTHOCIDE 7.5; ORTHOCIDE 50; ORTHO-CIDE 406; SR406; STAUFFER CAPTAN; N-TRICHLOROMETHYLMERCAPTO-4-CYCLOHEXENE-1,2-DICARBOXIMIDE; N-(TRICHLOROMETHYLMERCAPTO)-DELTA(SUP 4)-TETRAHYDROPHTHALIMIDE; N-TRICHLOROMETHYLTHIOCYCLOHEX-4-ENE-1,2-DICARBOXIMIDE; N-TRICHLOROMETHYLTHIO-CIS-DELTA(SUP4)-CYCLOHEXENE-1,2-DICARBOXIMI DE; N-((TRICHLOROMETHYL)THIO)-4-CYCLOHEXENE-1,2-DICARBOXIMIDE; N-((TRICHLOROMETHYL)THIO)TETRAHYDROPHTHALIMIDE; N-TRICHLOROMETHYLTHIO-3A,4,7,7A-TETRAHYDROPHTHALIMIDE; VANCIDE 89; VANGARD K; VONDCAPTAN

Danger profile: Mutagen, combustible, exposure limit established, reacts with heat and water to produce poisonous gas, poisonous gases produced in fire.
Uses: Fungicide to kill molds and mildew on food crops; fruit preservative and seed treatment; fungicide in paints, plastics, leather, fabrics.
Appearance: White to brown crystalline solid or yellow powder.
Odor: Slight, pungent. The odor is a warning of exposure; however, no smell does not mean you are not being exposed.
Effects of exposure:
Breathed: May cause irritation and tearing. See *Swallowed.*
Eyes: Contact may cause irritation, redness and burns.
Skin: Passes through the unbroken skin; can increase exposure and the severity of the symptoms. Contact may cause irritation and burns.
Swallowed: May cause depression, difficult breathing, tightness in the chest, vomiting and diarrhea.
Long-term exposure: Mutagen, handle it with extreme caution, as a possible cancer-causing substance. May cause birth defects, miscarriage and cancer; skin allergy.
Storage: Store in tightly closed containers in a dry, cool, well-ventilated place. Keep away form water, heat, oil, alkalis and hydrogen sulfide. Heat and water may cause the production of poisonous gas.

★ 234 ★
CARBARYL
CAS: 63-25-2
Other names: CARBAMINE; CARBARIL (ITALIAN); CARBATOX; CARBATOX-60; CARBATOX 75; CARPOLIN; CRAG SEVIN; DENAPON; DICARBAM; ENT 23,969; EXPERIMENTAL INSECTICIDE 7744; GAMONIL; GERMAIN'S; HEXAVIN; KARBARYL (POLISH); KARBASPRAY; KARBATOX; KARBOSEP; N-METHYLCARBAMATE DE 1-NAPHTYLE (FRENCH); METHYLCARBAMATE 1-NAPHTHALENOL; METHYLCARBAMATE 1-NAPHTHALENOL, METHYCARBAMATE; 1-NAPHTHOL; METHYLCARBAMIC ACID, 1-NAPHTHYL ESTER; N-METHYL-1-NAFTYL-CARBAMAAT (DUTCH); CAPROLIN; N-METHYL-1-NAPHTHYL-CARBAMAT (GERMAN); N-METHYL-ALPHA-NAPHTHYLCARBAMATE; N-METHYL-1-NAPHTHYL CARBAMATE; N-METHYL-ALPHA-NAPHTHYLURETHAN; N-METIL-1-NAFTIL-CARBAMMATO (ITALIAN); CARBARYL, NAC; ALPHA-NAFTYL-N-METHYLKARBAMAT (CZECH); 1-NAPHTHOL N-METHYLCARBAMATE; ALPHA-NAPHTHYL N-METHYLCARBAMATE; 1-NAPHTHYLMETHYLCARBAMATE; 1-NAPHTHYL N-METHYLCARBAMATE; 1-NAPHTHYL N-METHYL-CARBAMATE; OMS-29; PANAM; RAVYON; SEPTENE; SEVIMOL; SEVIN; SOK; TERCYL; TRICARNAM; COMPOUND 7744; UC 7744 (UNION CARBIDE)

Danger profile: Mutagen, poisonous, possible reproductive hazard, exposure limit established, community right-to-know list, poisonous gases produced in fire.
Uses: Pesticide.
Appearance: White to gray crystal, powder, liquid or paste.
Odor: Weak. The odor is a warning of exposure; however, no smell does not mean you are not being exposed.
Effects of exposure:
  Breathed: May be poisonous and fatal. May cause sweating, excessive nasal discharge or salivation, sweating, nausea, vomiting, stomach cramps and pain, diarrhea, tremor, blue skin, lips and fingernails, blurred vision, convulsions. Higher exposures may cause shortness of breath and a dangerous build-up of fluids in the lungs (pulmonary edema). This can cause death. Pulmonary edema is a medical emergency that can be delayed one to two days following exposure.
  Eyes: May cause severe irritation, redness, tearing, inflammation of the eyelids and burns.
  Skin: May be poisonous if absorbed. Passes through the unbroken skin; can increase exposure and the severity of the symptoms listed. May cause severe irritation, redness, burning feeling, rash and burns.
  Swallowed: May be poisonous.
Long-term exposure: Mutagen, handle it with extreme caution, as a possible cancer-causing substance. May reduce fertility in men and women and cause birth defects by damaging the fetus. May cause kidney and nervous system damage.
Storage: Store in tightly closed containers in a dry, cool, well-ventilated place. Keep away from oxidizers, heat, flame and sources of ignition. Containers may explode in fire
First aid guide: Move victim to fresh air and call emergency medical care; if not breathing, give artificial respiration; if breathing is difficult, give oxygen. In case of contact with material, immediately flush skin or eyes with running water for at least 15 minutes. Speed in removing material from skin is of extreme importance. Remove and isolate contaminated clothing and shoes at the site. Keep victim quiet and maintain normal body temperature. Effects may be delayed; keep victim under observation.

★ 235 ★
CARBOFURAN
CAS: 1563-66-2
Other names: BAY 70143; CARBAMIC ACID, METHYL-, 2,2-DIMETHYL-2,3-DIHYDROBENZOFURAN-7-YL ESTER; CARBOSIP 5G; CURATERR; D 1221; NEX; CRISFURAN; 2,3-DIHYDRO-2,2-DIMETHYLBENZOFURANYL-7-N-METHYLCARBAMATE; 2,3-DIHYDRO-2,2-DIMETHYL-7-BENZOFURANYL METHYLCARBAMATE; 2,2-DIMETHYL-2,2-DIHYDROBENZOFURANYL-7 N-METHYLCARBAMATE; PILLARFURAN; ENT 27,164; FMC 10242; FURADAN; FURODAN; NIA 10242; NIAGARA 10242; YALTOX

Danger profile: Mutagen, poisonous, exposure limit established, EPA extremely hazardous list, poisonous gases produced in fire.
Uses: Pesticide (carbamate) sprayed on corn, rice, broccoli, sprouts, cabbage, cauliflower, carrot, parsnip, etc.
Appearance: White sugar/sand-like crystalline solid.
Effects of exposure:
  Breathed: Poison. Exposure may cause excessive sweating, nausea, vomiting, abdominal pain, sweating, diarrhea, weakness and blurred vision, loss of coordination, slow heart beat and may cause breathing to stop.
  Eyes: Absorbed by eye contact; can increase exposure and the severity of the symptoms listed. May cause irritation, tearing and blurred vision.
  Skin: Passes through the unbroken skin; can increase exposure and the severity of the symptoms listed.
  Swallowed: Poison. See *Breathed.*
Long-term exposure: Mutagen, handle with extreme caution, as a possible cancer-causing substance. May lead to birth defects, miscarriage or cancer.
Storage: Store in tightly closed containers in a dry, cool, well-ventilated place. Keep away from heat, sources of ignition, acids, and oxidizers. Containers may explode in fire.
First aid guide: Move victim to fresh air and call emergency medical care; if not breathing, give artificial respiration; if breathing is difficult, give oxygen. In case of contact with material, immediately flush skin or eyes with running water for at least 15 minutes. Speed in removing material from skin is of extreme importance. Remove and isolate contaminated clothing and shoes at the site. Keep victim quiet and maintain normal body temperature. Effects may be delayed; keep victim under observation.

★ 236 ★
CARBON BLACK
CAS: 1333-86-4
Other names: ACETYLENE BLACK; C.I. PIGMENT BLACK 7; CHANNEL BLACK; LAMP BLACK; FURNACE BLACK; THERMAL BLACK

Danger profile: Possible carcinogen, mutagen, combustible, exposure limit established.
Uses: Rubber products; plastics; in printing inks; carbon paper; typewriter ribbons; paint pigment; in battery plates; solar energy absorber: A suspension of finely divided carbon particles in compressed air has been researched as a solar energy absorber. The heat is absorbed until the particles vaporize, yielding energy that can be used directly or for power production. The suspension is placed in a transparent container located in a solar concentrator.
Appearance: Black powder, pellets or paste.
Effects of exposure:
  Breathed: May cause irritation of the nose, coughing with phlegm and sneezing.
  Eyes: May cause irritation, redness, tearing, pain and inflammation of the eyelids.
  Swallowed: No symptoms.

Long-term exposure: May contain known carcinogens; effect the lungs; bronchitis. Certain contaminants may cause skin rashes or other changes such as rash.
Storage: Store in tightly closed containers in a dry, cool, well-ventilated place. Keep away from sources of ignition, chlorates, bromates and nitrates.

★ 237 ★
CARBON DIOXIDE
CAS: 124-38-9
Other names: ACID GAS; ANHYDRIDE CARBONIQUE (FRENCH); CARBONIC ACID GAS; CARBONIC ANHYDRIDE; DRY ICE (SOLID); KOHLENDIOXYD (GERMAN); KOHLENS-AURE (GERMAN)

Danger profile: Mutagen, possible reproductive hazard, exposure limit established.
Uses: Refrigeration; beverage carbonation; aerosol propellant; fire extinguishing; water treatment; medicine; in greenhouses; foundry industry; shielding gas for welding; cloud seeding; making other chemicals.
Appearance: Colorless gas. White solid used as dry ice.
Effects of exposure:
  Breathed: May cause increased respiration rate, headache, dizziness, restlessness, sweating, shortness of breath, nausea, vomiting, increased heart rate and pulse pressure, elevated blood pressure, coma. High concentrations can cause convulsions, unconsciousness, suffocation and death.
  Eyes: Liquid may cause frostbite; freezing injury similar to a burn.
  Skin: Solid may cause frostbite. Liquid may cause freezing injury similar to a burn.
Long-term exposure: May cause birth defects, miscarriage; acidosis and changes to the body metabolism.
Storage: Store in tightly closed containers in a dry, well-ventilated place. Keep away from chemically active metals, acrolein, ethyleneimine, sodium dioxide. Containers may explode in fire. Protect containers from physical damage.
First aid guide: Move victim to fresh air and call emergency medical care; if not breathing, give artificial respiration; if breathing is difficult, give oxygen. In case of frostbite, thaw frosted parts with water. Keep victim quiet and maintain normal body temperature.

★ 238 ★
CARBON DISULFIDE
CAS: 75-15-0
Other names: CARBON BISULFIDE; CARBON BISULPHIDE; CARBON DISULPHIDE; CARBONE (SUFURE DE) (FRENCH); CARBONIO (SOLFURO DI) (ITALIAN); CARBON SULFIDE; DITHIOCARBONIC ANHYDRIDE; KOHLENDISUL-FID (SCHWEFELKOHLENSTOFF) (GERMAN); KOOLSTOF-DISULFIDE (ZWAVELKOOLSTOF) (DUTCH); NCI-4591; SCH-WEFELKOHLENSTOFF (GERMAN); SULPHOCARBONIC ANHYDRIDE; WEEVILTOX; SOLFURO DI CARBONIO (ITALIAN); WEEVILTOX; WEGLA DWUSIARCZEK (POLISH)

Danger profile: Mutagen, poisonous, fire hazard, explosion hazard, narcotic, exposure limit established, EPA extremely hazardous list, Community Right-to-Know List, poisonous gases produced in fire.
Uses: Making cellophane, viscose rayon and other chemicals.
Appearance: Clear, colorless to faintly yellow, watery liquid.
Odor: Strong disagreeable; like decaying cabbage. The odor is a warning of exposure; however, no smell does not mean you are not being exposed. The pure chemical is nearly odorless.
Effects of exposure:

Breathed: Poison. Vapor may cause intense irritation to nose, throat, mouth and lungs; mucous membrane. Symptoms of exposure include watering of the eyes, coughing, difficult breathing, garlic-smelling breath, headache, nausea, vomiting, muscular weakness, vertigo, mania, hallucinations, unconsciousness. High exposure can lead to drowsiness, coma and death may occur during coma or after a convulsion; buildup of fluid in the lungs (pulmonary edema). Pulmonary edema is a medical emergency that may cause death an can be delayed by one to two days following exposure.
Eyes: Severe irritant. Contact may cause, burning and stinging, tearing, irritation and redness and swelling of the eyelids, intense pain; can burn the eyes and lead to blurred vision, chronic irritation and other permanent damage.
Skin: Severe irritant. Contact may cause burning or stinging feeling, redness, swelling, intense pain, blisters, irritation, rash. Repeated exposure may lead to skin allergy and chronic irritation.
Swallowed: May cause burning feeling of the mouth and throat, intense thirst, throat swelling, stomach cramps, nausea, vomiting, diarrhea, severe breathing difficulties and unconsciousness. Tissue damage may result. Breathing may stop. Watch for delayed symptoms.
Long-term exposure: May cause fertility problems in both men and women; damage to the developing fetus; skin allergy with itching and rash; an increase in cholesterol and hardening of the arteries (atherosclerosis); nervous system and brain damage with various symptoms of coordination; mood and personality changes including insanity.
Storage: Store in tightly closed containers in a dark, dry, cool, well-ventilated place. Keep away from heat, sunlight, flame and sources of ignition, all other chemicals especially oxidizers, chemically active metals, permanganic acid, fluorine, azides and amines. Vapors may travel to sources of ignition and flash back. Containers may explode in fire.
First aid guide: Move victim to fresh air and call emergency medical care; if not breathing, give artificial respiration; if breathing is difficult, give oxygen. In case of contact with material, immediately flush skin or eyes with running water for at least 15 minutes. Remove and isolate contaminated clothing and shoes at the site. Keep victim quiet and maintain normal body temperature. Effects may be delayed; keep victim under observation.

★ 239 ★
CARBON MONOXIDE
CAS: 630-08-0
Other names: CARBONE (OXYDE DE) (FRENCH); CARBONIC OXIDE; CARBONIO (OSSIDO DI) (ITALIAN); CARBON OXIDE (CO); EXHAUST GAS; FLUE GAS; KOHLEN-MONOXID (GERMAN); KOOLMONOXYDE (DUTCH); OXYDE DE CARBONE (FRENCH); WEGLA TLENEK (POLISH)

Danger profile: Mutagen, exposure limit established, dangerous fire hazard, explosion hazard.
Uses: Making other chemicals; in fuels and gasoline; in metallurgy
Appearance: Colorless, odorless compressed gas or liquefied compressed gas.
Effects of exposure:
  Breathed: May cause headache, dizziness, fatigue; rapid irregular breathing; irregular heartbeat; nausea, vomiting, confusion, loss of consciousness convulsions and death.
  Eyes: Contact with liquid may cause frostbite, pain, stinging, tearing, inflammation of the eyelids and permanent damage.
  Skin: Contact with liquid may cause frostbite, intense cold feeling, skin turns white and feels cold and hard.
  Swallowed: Liquid may cause headache, confusion, drowsiness, unconsciousness and death.

Long-term exposure: May cause heart disease, hardening of the arteries (atherosclerosis); fertility problems in both men and women.

Storage: Store in tightly closed containers in a safe, dry, cool, well-ventilated place. Keep away from heat, sunlight, flame and sources of ignition, and oxidizers, sodium, potassium, iron. Containers may explode in fire.

First aid guide: Move victim to fresh air and call emergency medical care; if not breathing, give artificial respiration; if breathing is difficult, give oxygen. In case of contact with material, immediately flush skin or eyes with running water for at least 15 minutes. Remove and isolate contaminated clothing and shoes at the site. Keep victim quiet and maintain normal body temperature. Effects may be delayed; keep victim under observation.

★240★
CARBON TETRABROMIDE
CAS: 558-13-4
Other names: CARBON BROMIDE; METHANE, TETRABROMIDE; METHANE, TETRABROMO-; TETRABROMOMETHANE; CARBON BROMIDE; TETRABROMIDE METHANE

Danger profile: Poisonous, narcotic, exposure limit established, poisonous gas produced in fire.
Uses: Making other chemical.
Appearance: Colorless powder.
Effects of exposure:
Breathed: May be poisonous. May irritate the eyes, nose throat and breathing passages. Higher exposures may cause shortness of breath and a dangerous build-up of fluids in the lungs (pulmonary edema), which can cause death. Pulmonary edema is a medical emergency that can be delayed one to two days following exposure.
Eyes: Contact may cause irritation, redness, tearing and permanent damage.
Skin: May cause irritation.
Swallowed: May be poisonous.
Long-term exposure: May cause liver and kidney damage, and may effect the lungs.
Storage: Store in tightly closed containers in a dry, cool, well-ventilated place. Keep away from heat, lithium and hexacylcohexyldilead.
First aid guide: Move victim to fresh air; call emergency medical care. In case of contact with material, immediately flush skin or eyes with running water for at least 15 minutes. Remove and isolate contaminated clothing and shoes at the site.

★241★
CARBON TETRACHLORIDE
CAS: 56-23-5
Other names: BENZINOFORM; CARBONA; CARBON CHLORIDE; CARBON TET; CZTEROCHLOREK WEGLA (POLISH); ENT 4705; FASCIOLIN; FLUKOIDS; FREON 10; HALON 104; METHANE TETRACHLORIDE; METHANE, TETRACHLORO-; NECATORINA; NECATORINE; PERCHLOROMETHANE; R 10; STCC; TETRACHLOORKOOLSTOF (DUTCH); TETRACHLOORMETAN; TETRACHLORKOHLENSTOFF, TETRA (GERMAN); TETRACHLORMETHAN (GERMAN); TETRACHLOROCARBON; TETRACHLOROMETHANE; TETRACHLORURE DE CARBONE (FRENCH); TETRACLOROMETANO (ITALIAN); TETRACLORURO DI CARBONIO (ITALIAN); TETRAFINOL; TETRAFORM; TETRASOL; UNIVERM; VERMOESTRICID

Danger profile: Carcinogen, mutagen, exposure limit established, poisonous gases produced in fire.
Uses: Spot remover, dry cleaning agent, refrigerant; making fire extinguisher; metal degreasing; aerosols; pesticide, agricultural

fumigant; production of semiconductors; solvent. The U.S. Food and Drug Administration (FDA) has banned chemical from household use.
Appearance: Clear colorless watery liquid.
Odor: Strong, sweet, similar to chloroform or ether.
Effects of exposure:
Breathed: Poisonous. May cause dizziness, headache, vomiting, nausea, extreme fatigue and drowsiness, nose and throat irritation, confusion, loss of balance, tremors, irregular heart beat. Severe exposure can lead to liver and kidney damage (often resulting in a decrease or stopping of urinary output), coma and death.
Skin: Poisonous if absorbed. Passes through the unbroken skin; can increase exposure and the severity of the symptoms listed. May cause irritation and redness.
Eyes: May cause irritation and redness.
Swallowed: Poisonous. May cause severe stomach pains, vomiting (including blood), diarrhea followed by symptoms listed under breathed.
Long-term exposure: May cause liver cancer; liver and kidney damage. May have the potential for damaging the unborn human or animal. In persons who regularly consume alcoholic beverage, liver damage may be increased. Repeated exposure may cause skin problems.
Storage: Store in tightly closed containers in a dry, cool, well-ventilated place. Keep away from chemically active metals and all other chemicals. Containers may explode in fire.
First aid guide: Move victim to fresh air and call emergency medical care; if not breathing, give artificial respiration; if breathing is difficult, give oxygen. In case of contact with material, immediately flush skin or eyes with running water for at least 15 minutes. Speed in removing material from skin is of extreme importance. Remove and isolate contaminated clothing and shoes at the site. Keep victim quiet and maintain normal body temperature. Effects may be delayed; keep victim under observation.

★242★
CARBONYL SULFIDE
CAS: 463-58-1
Other names: CARBON MONOXIDE MONOSULFIDE; CARBON OXIDE SULFIDE; CARBON OXYGEN SULFIDE; CARBON OXYSULFIDE; OXYCARBON SULFIDE; CARBONYLSULFID-(32)S; SCO

Danger profile: Poisonous, severe irritant, narcotic, fire hazard, Community Right-to-Know List, poisonous gases produced in fire.
Uses: Making other chemicals.
Appearance: Colorless gas or a cold liquid.
Odor: Typical sulfide odor except when pure. The odor is a warning of exposure; however, no smell does not mean you are not being exposed.
Effects of exposure:
Breathed: May cause irritation of the nose, throat, breathing passages and lungs; coughing, sneezing. High exposure may cause headache, dizziness, heavy sweating, salivation, nausea, vomiting, diarrhea, weakness, muscle cramps, irregular and fast heartbeat and death.
Eyes: May cause irritation, redness, tearing and burns. Contact with the liquified gas may cause frostbite.
Skin: Passes through the unbroken skin; can increase exposure and the severity of the symptoms listed. May cause irritation, redness and burns. Contact with the liquified gas may cause frostbite.
Swallowed: May cause irritation of the mouth and stomach. See *Breathed.*
Long-term exposure: May cause liver damage and increased cholesterol level; personality changes, memory loss, personality change and brain damage.

Storage: Store in tightly closed containers in a dry, cool, well-ventilated place. Keep away from all oxidizers. Containers may explode in fire.

First aid guide: Move victim to fresh air and call emergency medical care; if not breathing, give artificial respiration; if breathing is difficult, give oxygen. In case of contact with material, immediately flush skin or eyes with running water for at least 15 minutes. Remove and isolate contaminated clothing and shoes at the site. Keep victim quiet and maintain normal body temperature. Effects may be delayed; keep victim under observation.

★ 243 ★
CATECHOL
CAS: 120-80-9
Other names: BENZENE, O-DIHYDROXY-; O-BENZENEDIOL; 1,2-BENZENEDIOL; CATECHIN; CATE-CHOL; C.I. 76500; C.I. OXIDATION BASE 26; O-DIHYDROXYBENZENE; 1,2-DIHYDROXYBENZENE; O-DIOXYBENZENE; O-DIPHENOL; DURAFUR DEVELOPER C; FOURAMINE PCH; FOURRINE 68; O-HYDROQUINONE; O-HYDROXYPHENOL; 2-HYDROXYPHENOL; NCI-C55856; OXYPHENIC ACID; PELAGOL GREY C; O-PHENYLENEDIOL; PYROCATECHIN; PYROCATECHINE; PYROCATECHINIC ACID; PYROCATECHOL; PYROCAT-ECHUIC ACID

Danger profile: Mutagen, reproductive hazard, exposure limit established, EPA extremely hazardous substances list, Community Right-to-Know List, toxic gases produced in fire.
Uses: Making rubber, photography chemicals, dyes, fats oils, cosmetics, pharmaceuticals and other chemicals; antiseptic; electroplating; ink.
Appearance: Colorless crystals.
Effects of exposure:
    Breathed: Extremely irritating. May cause irritation of the mucous membranes, nose and eyes; headaches, coughing, difficult breathing, nausea, vomiting, tremors, muscle twitching and convulsions; increased blood pressure; convulsions and respiratory failure.
    Eyes: Contact may cause severe irritation, tears, burns and permanent damage.
    Skin: Passes through the unbroken skin; can increase exposure and the severity of the symptoms listed. May cause irritation and burns; allergy may occur. Extensive contact may cause death.
    Swallowed: Poison.
Long-term exposure: May cause cancer or reproductive problems; skin allergy; kidney and liver damage; interference with the ability of the blood to carry oxygen causing blue color to the skin, dizziness and rapid breathing.
Storage: Store in tightly closed containers in a dry, cool, well-ventilated place. Keep away from heat and sources of ignition.

★ 244 ★
CESIUM HYDROXIDE
CAS: 21351-79-1
Other names: CAESIUM HYDROXIDE; CESIUM HYDRATE; CESIUM HYDROXIDE DIMER

Danger profile: Powerful corrosive, exposure limit established.
Uses: Making other chemicals; as an electrolyte in storage batteries used at sub-zero temperatures.
Appearance: Colorless to yellowish sugar/sand-like crystals.
Effects of exposure:
    Breathed: Vapor may cause intense irritation to nose, throat, mouth and lungs; mucous membrane. Symptoms of exposure may include watering of the eyes, coughing, difficult

breathing, headache, nausea, vomiting, muscular weakness, unconsciousness.
    Eyes: Severe irritant. Contact may cause, burning and stinging, tearing, irritation and redness and swelling of the eyelids, intense pain; can burn the eyes and lead to blurred vision, and permanent damage.
    Skin: Severe irritant. Contact may cause burning or stinging feeling, redness, swelling, intense pain, blisters, irritation, rash.
    Swallowed: May cause burning feeling of the mouth and throat, intense thirst, throat swelling, stomach cramps, nausea, vomiting, diarrhea, severe breathing difficulties and unconsciousness. Tissue damage may result. Breathing may stop. Watch for delayed symptoms.
Long-term exposure: May cause lung damage.
Storage: Store in very tightly closed silver or platinum containers in a dry, cool, well-ventilated place. Keep away from air; violent reaction with oxygen.
First aid guide: Move victim to fresh air; call emergency medical care. In case of contact with material, immediately flush skin or eyes with running water for at least 15 minutes. Remove and isolate contaminated clothing and shoes at the site. Keep victim quiet and maintain normal body temperature.

★ 245 ★
CHLORAL
CAS: 75-87-6
Other names: ACETALDEHYDE, TRICHLORO-; TRI-CHLOROACETALDEHYDE; CHLORAL, ANHYDROUS, INHIBITED.

Danger profile: Mutagen, poisonous, combustible, regulated by DEA, poisonous gas produced in fire.
Uses: To make other chemicals and various pesticides; medical sedative.
Appearance: Colorless, oily liquid.
Odor: Sharp. The odor is a warning of exposure; however, no smell does not mean you are not being exposed.
Effects of exposure:
    Breathed: May cause irritation of the eyes, nose, breathing passages, throat and lungs; coughing, difficult breathing and shortness of breath, dizziness, unconsciousness. High levels may cause death.
    Eyes: Contact may cause severe irritation and burns.
    Skin: Contact may cause severe irritation, redness and burns.
    Swallowed: Poisonous. See *Breathed* for symptoms.
Long-term exposure: May cause birth defects, miscarriage; acne-like rash; may effect the lungs and cause respiratory damage. Lab animals show damage to liver, central nervous system,
Storage: Store in tightly closed containers in a dark, dry, cool, well-ventilated locked and secure place. Keep away from moisture, air, acids, and sources of ignition. Containers may explode in fire.

★ 246 ★
CHLORAMBEN
CAS: 133-90-4
Other names: AMBIBEN; AMIBEN; AMIBIN; ACP-M-728; 3-AMINO-2,5-DICHLOROBENZOIC ACID; AMOBEN; CHLORAMBED; CHLORAMBENE; BENZOIC ACID, 3-AMINO-2,5-DICHLORO-; 3-AMINO-2,6- DICHLOROBENZOIC ACID; 2,5-DICHLORO-3-AMINOBENZOIC ACID; NCI-C00055; ORNAMENTAL WEEDER; PST 29084; VEGABEN; VEGIBEN

Danger profile: Possible carcinogen, mutagen, combustible; Community Right-to-Know List; poisonous gases produced in fire and on contact with acid.

Uses: Pesticide; herbicide; weed killer.
Appearance: Colorless sugar/sand-like crystals. May turn purple over time as it is exposed to light.
Effects of exposure:
Breathed: May be poisonous.
Eyes: Contact may cause irritation, redness and tearing.
Skin: Passes through the unbroken skin; can increase exposure and the severity of the symptoms listed. Contact may cause irritation and rash may develop with redness and itching.
Swallowed: Poison. May cause labored breathing, lack of coordination, excessive urination.
Long-term exposure: May cause cancer; birth defects and miscarriage. Prolonged contact with the eyes may produce inflammation of the eyelids. prolonged contact with the skin may cause rash and itching.
Storage: Store in tightly closed containers in a dark, dry, cool, well-ventilated place. Keep away from light, acids and acid fumes.

★ 247 ★
CHLORAMPHENICOL
CAS: 56-75-7
Other names: ACETAMIDE, 2,2-DICHLORO-N-2-HYDROXY-1-(HYDROXYMETHYL)-2-(4- NITROPHENYL)ETHYL-,R-(R*,R*)-; ACETAMIDE, 2,2-DICHLORO-N-(2-HYDROXY-1-(HYDROXYMETHYL)-2-(4-NITROPHENYL)ETHY L)-,; ACETAMIDE, 2,2-DICHLORO-N-(BETA-HYDROXY-ALPHA-(HYDROXYMETHYL)-P-NITROPHENETH YL)-,D-(-)-THREO-; ALFICETYN; AMBOFEN; AMPHENICOL; AMPHICOL; AMSECLOR; AQUAMYCETIN; AUSTRACIL; AUSTRACOL; BIOCETIN; BIOPHENICOL; CAF; CAM; CAP; CATILAN; CHEMICETIN; CHEMICETINA; CHLOMIN; CHLOMYCOL; CHLORAMEX; D-CHLORAMPHENICOL; D-THREO-CHLORAMPHENICOL; D-(-)-THREO-CHLORAMPHENICOL; CHLORAMSAAR; CHLORASOL; CHLORA-TABS; CHLORICOL; CHLOROCAPS; CHLOROCID; CHLOROCIDE; CHLOROCIDIN C; CHLOROCIDIN C TETRAN; CHLOROCOL; CHLOROMYCETIN; CHLOROMYCETIN R; CHLORONITRIN; CHLOROPTIC; CHLORO-25 VTAG; CIDOCETINE; CIPLAMYCETIN; CLORAMFICIN; CLORAMICOL; CLORAMIDINA; CLOROAMFENICOLO (ITALIAN); CLOROCYN; CLOROMISAN; CLOROSINTEX; COMYCETIN; CPH; CYLPHENICOL; DESPHEN; DETREOMYCINE; DEXTROMYCETIN; D-(-)-THREO-2-DICHLOROACETAMIDO-1-P-NITROPHENYL-1,3-PROPANEDIOL; D-THREO-N-DICHLOROACETYL-1-P-NITROPHENYL-2-AMINO-1,3-PROPANEDIOL; D-(-)-THREO-2,2-DICHLORO-N-(BETA-HYDROXY-ALPHA-(HYDROXYMETHYL))-P-NITROPHENETHYLACETAMIDE; D-THREO-N-(1,1'-DIHYDROXY-1-P-NITRO PHENYLISOPROPYL)DICHLOROACETAMIDE; DOCTAMICINA; ECONOCHLOR; EMBACETIN; EMETREN; ENICOL; ENTEROMYCETIN; ERBAPLAST; ERTILEN; FARMICETINA; FENICOL; GLOBENICOL; GLOROUS; HALOMYCETIN; HORTFENICOL; I 337A; INTRAMYCETIN; ISICETIN; ISMICETINA; ISOPHENICOL; ISOPTO FENICOL; JUVAMYCETIN; KAMAVER; KEMICETINA; KEMICETINE; KLORITA; KLOROCID S; LEUKOMYAN; LEUKOMYCIN; LEVOMICETINA; LEVOMYCETIN; LOROMISAN; LOROMISIN; MASTIPHEN; MEDIAMYCETINE; MICLORETIN; MICOCHLORINE; MICOCLORINA; MICROCETINA; MYCHEL; MYCINOL; NCI-C55709; D-(-)-THREO-1-P-NITROPHENYL-2-DICHLORACETAMIDO-1,3-PROPANEDIOL; D-THREO-1-(P-NITROPHENYL)-2-(DICHLOROACETYLAMINO)-1,3-PROPANEDIOL; NORMIMYCIN V NOVOCHLOROCAP; NOVOMYCETIN; NOVOPHENICOL; NSC 3069; OFTALENT; OLEOMYCETIN; OPCLOR; OPELOR; OPHTHOCHLOR; OPHTOCHLOR; OTACHRON; OTOPHEN; PANTOVERNIL; PARAXIN; PENTAMYCETIN; QUEMICETINA; RIVOMYCIN; ROMPHENIL; SEPTICOL; SIFICETINA; SINTOMICETINA; SINTOMICETINE R; STANOMYCETIN; SYNTHOMYCETIN; SYNTHOMYCETINE; SYNTHOMYCINE; TEVCOCIN; TEVCOSIN; TIFOMYCIN; TIFOMYCINE; TREOMICETINA; U-6062; UNIMYCETIN; VETICOL

Danger profile: Carcinogen, mutagen, reproductive hazard, poisonous gases produced in fire.
Uses: Antibiotic; antifungal agent.
Appearance: Grayish-white or yellowish-white crystalline material.
Effects of exposure:
Breathed: Taken as a medication may cause nausea, vomiting and diarrhea.
Skin: Passes through the unbroken skin; can increase exposure and the severity of the symptoms listed.
Swallowed: Moderately toxic.
Long-term exposure: May cause cancer; reproductive damage, damage to the developing fetus; liver damage; leukemia; damage to the bone marrow that may interfere with the production of blood cells and the body's ability to produce platelets for

blood clotting; may lead to an increase of yeast infections and other types of infections.
Storage: Store in tightly closed containers in a dry, cool, well-ventilated and regulated area.

★ 248 ★
CHLORDANE
CAS: 57-74-9
Other names: ASPON-CHLORDANE; BELT; CD 68; CHLOORDAAN (DUTCH); CHLORDAN; GAMMA-CHLORDAN; CHLORINDAN; CHLOR KIL; CHLORODANE; CLORDAN (ITALIAN); CORODANE; CORTILAN-NEU; DICHLORO-CHLORDENE; DOWCHLOR; ENT 9,932; ENT 25,552-X; HCS 3260; KYPCHLOR; M 140; M 410; 4,7-METHANO-1H-INDENE,1,2,4,5,6,7,8,8-OCTACHLORO-2,3,3A,4,7,7A-HEXAHYDRO-; NCI-C00099; NIRAN; OCTACHLOR; OCTACHLORODIHYDRODICYCLOPENTADIENE; 1,2,4,5,6,7,8,8-OCTACHLORO-2,3,3A,4,7,7A-HEXAHYDRO-4,7-METHANOINDENE; 1,2,4,5,6,7,8,8-OCTACHLORO-2,3,3A,4,7,7A-HEXAHYDRO-4,7-METHANO-1H-INDENE; 1,2,4,5,6,7,8,8-OCTACHLORO-3A,4,7,7A-HEXAHYDRO-4,7-METHYLENE INDANE; OCTACHLORO-4,7-METHANOHYDROINDANE; OCTACHLORO-4,7-METHANOTETRAHYDROINDANE; 1,2,4,5,6,7,8,8-OCTACHLORO-4,7-METHANO-3A,4,7,7A-TETRAHYDROINDANE; 1,2,4,5,6,7,8,8-OCTACHLOOR-3A,4,7,7A-TETRAHYDRO-4,7-ENDO-METHANO-INDAAN (DUTCH); 1,2,4,5,6,7,8,8-OCTACHLORO-3A,4,7,7A-TETRAHYDRO-4,7-METHANOINDAN; 1,2,4,5,6,7,8,8-OCTACHLORO-3A,4,7,7A-TETRAHYDRO-4,7-METHANOINDANE; 1,2,4,5,6,7,10,10-OCTACHLORO-4,7,8,9-TETRAHYDRO-4,7-METHYLENEINDANE; 1,2,4,5,6,7,8,8-OCTACHLOR-3A,4,7,7A-TETRAHYDRO-4,7-ENDO-METHANO-INDAN (GERMAN); OCTA-KLOR; OKTATERR; 1,2,4,5,6,7,8,8-OTTOCHLORO-3A,4,7,7A-TETRAIDRO-4,7-ENDO-METANO-INDANO (ITALIAN); ORTHO-KLOR; SD 5532; SHELL SD-5532; SYNKLOR; TAT CHLOR 4; TOPICHLOR 20; TOPICLOR; TOPICLOR 20; TOXICHLOR; VELSICOL 1068

Danger profile: Carcinogen, mutagen, combustible, exposure limit established, Community Right-to-Know List, EPA extremely hazardous list, poisonous gas produced in fire.
Uses: Insecticide; fumigant.
Appearance: Colorless to amber-colored viscous liquid.
Odor: Penetrating; slightly pungent, chlorine-like. The odor is a warning of exposure; however, no smell does not mean you are not being exposed.
Effects of exposure:
　Breathed: Poison. May cause irritation of the eyes, nose mouth and throat; nose bleeding; coughing and sneezing; painful swelling with obstruction to the throat; excitability; convulsions, unconsciousness and death.
　Eyes: May cause painful burning feeling; tearing; redness and inflammation of the eyelids; sensitivity to light.
　Skin: Poison. Passes through the unbroken skin; can increase exposure and the severity of the symptoms listed. May cause irritation, redness and inflammation; painful rash and blisters.
　Swallowed: Poison. Burning feeling in the mouth, throat and stomach; difficult swallowing; nausea, vomiting (possibly with blood); stomach cramps; stomach pain; diarrhea
Long-term exposure: May cause cancer; damage to the developing fetus; decreased fertility in men and women; liver and kidney damage. Prolonged skin contact may produce dermatitis.
Storage: Store in tightly closed containers in a dry, cool, well-ventilated place. Keep away from heat, sources of ignition, oxidizers and alkalis. Containers may explode in heat of fire.

★ 249 ★
CHLORFENVINFOS
CAS: 470-90-6
Other names: BENZYL ALCOHOL,2,4-DICHLORO-ALPHA-(CHLOROMETHYLENE)-, DIETHYL PHOSPHATE; BIRLANE; BIRLANE LIQUID; C8949; C-10015; CFV; CGA 26351; CHLOFENVINPHOS; O-2-CHLOOR-1-(2,4-DICHLOOR-FENYL)-VINYL-O,O-DIETHYLFOSFAAT (DUTCH); O-2-CHLOR-1-(2,4-DICHLOR-PHENYL)-VINYL-O,O-DIAETHYLPHOSPHAT (GERMAN); CHLORFENVINPHOS; 2-CHLORO-1-(2,4-DICHLOROPHENYL)VINYL DIETHYL PHOSPHATE; BETA-2-CHLORO-1-(2',4'-DICHLOROPHENYL) VINYL DIETHYLPHOSPHATE; CHLOROFENVINPHOS; CHLORPHENVINPHOS; CHLORPHENVINPHOS; CLOFENVINFOS; O-2-CLORO-1-(2,4-DICLORO-FENIL)-VIN IL-O,O-DI ETILFOSFATO (ITALIAN); COMPOUND 4072; CVP; O,O-DIETHYLO-(2-CHLORO-1-(2',4'-DICHLOROPHENYL)VINYL) PHOSPHATE; DIETHYL1-(2,4-DICHLOROPHENYL)-2-CHLOROVINYL PHOSPHATE; ENT 24969; GC 4072; OMS 1328; PHOSPHATE DE O,O-DIETHYLE ET DEO-2-CHLORO-1-(2,4-DICHLOROPHENYL) VINYLE (FRENCH); SAPECRON; SAPRECON C; SAPECRON 240; SAPECRON 10FGEC; SD 4072; SD 7859; SHELL 4072; SUPONA; SUPONE; UNITOX; VINYLPHARE; VINYLPHATE

Danger profile: Mutagen, poisonous, EPA extremely hazardous list, poisonous gases produced in fire.
Uses: Insecticide (organophosphate). Used on vegetables, cereals, sugar cane, citrus fruit, corn and rice. Used against cut worms, root flies, root worms and mosquito larvae.
Appearance: Liquid.
Effects of exposure:
　Breathed: Poison. May cause headache, profuse sweating, nausea, vomiting, diarrhea, dizziness, loss of coordination, stomach cramps, convulsions, slow heart beat, coma and death. Higher exposures may cause shortness of breath and a dangerous build-up of fluids in the lungs (pulmonary edema), which can cause death. Pulmonary edema is a medical emergency that can be delayed one to two days following exposure.
　Eyes: Poison. May cause irritation, tearing and blurred vision. followed by symptoms shown above.
　Skin: Poison. Passes through the unbroken skin; can increase exposure and the severity of the symptoms listed, or may cause fatal poisoning.
　Swallowed: Poison. See symptoms under *Breathed.*
Long-term exposure: May cause birth defects, miscarriage; nerve damage; personality change and depression.
Storage: Store in tightly closed original containers in a dry, cool, well-ventilated place. Containers may explode in fire.

★ 250 ★
CHLORINATED DIPHENYL OXIDE
CAS: 55720-99-5
Other names: BENZENE,1,1'-OXY BIS CHLORO; HEXACHLORODIPHENYL OXIDE; ETHER, HEXACHLOROPHENYL; PHENYLETHER HEXACHLORO; BIS(TRICHLOROPHENYL)ETHER

Danger profile: Poisonous, exposure limit established, combustible, poisonous gas produced in fire.
Uses: Making other chemicals; adds flexibility to cellulose products, vinyl resins, and chlorinated rubbers; dry cleaning detergents.
Appearance: White or yellow waxy (viscous) solid or liquid.
Effects of exposure:
　Breathed: Poison. May cause irritation of the nose and mouth.
　Eyes: May cause irritation, tearing and redness.

Skin: May be poisonous. Contact may cause irritation, burning feeling, redness and rash.
Swallowed: Poison.
Long-term exposure: May cause acne-like dermatitis with rash; liver damage.
Storage: Store in tightly closed containers in a dry, cool, well-ventilated place. Keep away from heat, flame and sources of combustion and oxidizing material.

★251★
CHLORINE
CAS: 7782-50-5
Other names: BERTHOLITE; CHLOOR (DUTCH); CHLOR (GERMAN); CHLORE (FRENCH); CHLORINE MOLECULAR; CLORO (ITALIAN); DICHLORINE; MOLECULAR CHLORINE

Danger profile: Mutagen, poisonous, exposure limit established, Community Right-to-Know List, EPA extremely hazardous substance list, poisonous gases produced in fire.
Uses: Chlorine bleach cleaners; making solvents, and other chemicals; water purification; disinfectants; shrink proofing wool; processing of meat, fish, vegetables and fruit; flame-retardant compounds; in lithium and zinc batteries.
Appearance: Greenish-yellow gas.
Odor: Irritating, bleach-like suffocating, choking. The odor is a warning of exposure; however, no smell does not mean you are not being exposed.
Effects of exposure:
Breathed: May be poisonous. May be extremely irritating to the nose, throat, mouth and lungs; mucous membranes. High concentrations can cause watering of the eyes, respiratory distress, difficult breathing, violent coughing, vomiting and retching, sneezing, salivation, headache, nausea, vomiting, muscular weakness, unconsciousness. Death can result from suffocation or buildup of fluid in the lungs (pulmonary edema). Pulmonary edema is a medical emergency that may be delayed by one to two days following exposure.
Eyes: Extremely irritating. Irritation may last for several days. Contact may cause, burning and stinging, tearing, irritation and redness and swelling of the eyelids, intense pain; can burn the eyes and lead to blurred vision, chronic irritation and other permanent damage.
Skin: Severe irritant. Contact with liquid may cause frostbite. May cause irritation and first degree burns on short exposure. Contact may cause burning or stinging feeling, redness, swelling, intense pain, blisters, irritation, rash.
Swallowed: May cause burning feeling of the mouth and throat, intense thirst, throat swelling, stomach cramps, nausea, vomiting, diarrhea, severe breathing difficulties and unconsciousness. Tissue damage may result. Breathing may stop. Watch for delayed symptoms. See *Breathed.*
Long-term exposure: Repeated exposure may lead to birth defects, miscarriage; skin allergy and chronic irritation; lung irritation and possible bronchitis with cough and difficult breathing. Long-term exposure may cause tooth damage.
Storage: Store in tightly closed containers in a dry, cool, well-ventilated place. Keep away from combustible materials, heat, and all other chemicals including petroleum products like gasoline and other fuels, rubber, sulfamic acid, polypropylene, turpentine, alcohols, ammonia, sulfur, acetylene, hydrogen and metal powders. Containers may explode in heat and fire.
First aid guide: Move victim to fresh air and call emergency medical care; if not breathing, give artificial respiration; if breathing is difficult, give oxygen. In case of contact with material, immediately flush skin or eyes with running water for at least 15 minutes. Remove and isolate contaminated clothing and shoes at the site. Keep victim quiet and maintain normal body temperature. Effects maybe delayed; keep victim under observation.

CHLORINE DIOXIDE
CAS: 10049-04-4
Other names: ALCIDE; ANTHIUM DIOXCIDE; CHLORINE PEROXIDE; CHLORINE OXIDE; CHLORINE(IV) OXIDE; CHLOROPEROXYL; CHLORYL RADICAL; DOXCIDE 50

Danger profile: Mutagen, exposure limit established, combustible, violent explosive, powerful oxidizer, Community Right-to-Know List, transportation of undiluted chlorine dioxide is forbidden.
Uses: Bleaching textiles, wood pulp, fats and oils; disinfectant and biocide; treating flour; water purification and taste and odor control; swimming pool chemical.
Appearance: Yellow-green to red-orange gas. Sold in frozen form and looks like orange ice.
Odor: Pungent, sharp. Similar to chlorine and nitric acid. The odor is a warning of exposure; however, no smell does not mean you are not being exposed.
Effects of exposure:
Breathed: May be poisonous. May cause irritation to the eyes, nose, throat and breathing passages; tearing, coughing and wheezing; chest pains. Higher exposures may cause shortness of breath and a dangerous build-up of fluids in the lungs (pulmonary edema), which can cause death. Pulmonary edema is a medical emergency that can be delayed one to two days following exposure.
Eyes: Severe irritant. Contact may cause, burning and stinging, tearing, irritation and redness and swelling of the eyelids, intense pain; can burn the eyes and lead to blurred vision, chronic irritation and other permanent damage.
Skin: Severe irritant. Contact may cause burning or stinging feeling, redness, swelling, intense pain, blisters, irritation, rash. Repeated exposure may lead to skin allergy and chronic irritation.
Swallowed: May cause burning feeling of the mouth and throat, intense thirst, throat swelling, stomach cramps, nausea, vomiting, diarrhea, severe breathing difficulties and unconsciousness. Tissue damage may result. Breathing may stop. Watch for delayed symptoms.
Long-term exposure: May cause birth defects, miscarriage; lung damage; chronic bronchitis with shortness of breath and phlegm.
Storage: Keep frozen and store in airtight containers in a dry, dark, well-ventilated place with non-wood floors. Keep away from sunlight, heat, water and steam, combustible material, fuels, sources of ignition, metal powders, and all other chemicals and materials including ammonia, platinum, mercury, potassium peroxide. Containers may explode in fire.

CHLORINE TRIFLUORIDE
CAS: 7790-91-2
Other names: CHLORINE FLUORIDE; CHLOROTRIFLUORIDE; TRIFLUORURE DE CHLORE (FRENCH)

Danger profile: Poisonous, spontaneously combustible, highly reactive, dangerous explosive, violent reaction with water, exposure limit established, poisonous fumes produced in fire.
Uses: Fluorination; rocket fuels; cutting oil well tubes; processing reactor fuel rods; oxidizer in propellants.
Appearance: Greenish-yellow almost colorless liquid or gas.
Odor: Strong, pungent, sweetish; suffocating and irritating. The odor is a warning of exposure; however, no smell does not mean you are not being exposed.
Effects of exposure:
Breathed: May be poisonous. Vapor may cause intense irritation to nose, throat, mouth and lungs; mucous membrane. Symptoms of exposure include watering of the eyes, cough-

ing, difficult breathing, headache, nausea, vomiting, muscular weakness, unconsciousness. High exposure can lead to drowsiness, coma and death; buildup of fluid in the lungs (pulmonary edema), and may cause death. Pulmonary edema is a medical emergency that may be delayed by one to two days following exposure.

Eyes: Severe irritant. Contact may cause, burning and stinging, tearing, irritation and redness and swelling of the eyelids, intense pain; can burn the eyes and lead to blurred vision, cornea damage and other permanent damage.

Skin: May be poisonous if absorbed. Severe irritant. Contact may cause burning or stinging feeling, redness, swelling, intense pain, blisters, irritation, deep tissue damage.

Swallowed: May be poisonous. May cause burning feeling of the mouth and throat, intense thirst, throat swelling, coughing, stomach cramps, nausea, vomiting, diarrhea, severe breathing difficulties, bluish skin, lips and nails; unconsciousness, coma and death. Tissue damage may result. See symptoms under **Breathed** and watch for delayed symptoms.

Long-term exposure: May effect the lungs.

Storage: Store in tightly closed containers in a dry, cool, well-ventilated place. Very dangerous and reactive. Keep away from water, heat, combustable materials, fuels and all other chemicals and materials.

First aid guide: Move victim to fresh air and call emergency medical care; if not breathing, give artificial respiration; if breathing is difficult, give oxygen. In case of contact with material, immediately flush skin or eyes with running water for at least 15 minutes. Remove and isolate contaminated clothing and shoes at the site. Keep victim quiet and maintain normal body temperature. Effects may be delayed; keep victim under observation.

★ 254 ★
CHLOROACETALDEHYDE
CAS: 107-20-0

Other names: 2-CHLOROACETALDEHYDE; CHLOROACETALDEHYDE MONOMER; 2-CHLORO-1-ETHANAL; MONOCHLOROACETALDEHYDE; ACETALDEHYDE, CHLORO-; 2-CHLOROETHANAL; CHLOROACETALDEHYDE (40% AQUEOUS)

Danger profile: Mutagen, combustible, exposure limit established, reacts with water, poisonous gases produced in fire.

Uses: Making other chemicals; fungicide for controlling bacteria and algae.

Appearance: Clear, colorless liquid.

Odor: Pungent, acrid, penetrating. The odor is a warning of exposure; however, no smell does not mean you are not being exposed.

Effects of exposure:

Breathed: May cause irritation of the eyes, nose throat, breathing passages and lungs. Higher exposures may cause shortness of breath and a dangerous build-up of fluids in the lungs (pulmonary edema), which can cause death. Pulmonary edema is a medical emergency that can be delayed one to two days following exposure.

Eyes: Contact may cause severe irritation, burns and permanent damage.

Skin: Poison. Contact may cause severe irritation, redness, inflammation and burns. Allergy may develop.

Swallowed: Poison. May cause irritation of the mouth, throat and stomach. See **Breathed** and **First aid guide.**

Long-term exposure: May cause birth defects, miscarriage; skin allergy; asthma-like allergy with difficult breathing and cough.

Storage: Store in tightly closed containers in a dry, cool, well-ventilated place. Keep away from water, heat, sources of ignition, oxidizers and acids. Containers may explode in fire.

First aid guide: Move victim to fresh air and call emergency medical care; if not breathing, give artificial respiration; if breathing is difficult, give oxygen. In case of contact with material, immediately flush skin or eyes with running water for at least 15 minutes. Speed in removing material from skin is of extreme importance. Remove and isolate contaminated clothing and shoes at the site. Keep victim quiet and maintain normal body temperature. Effects may be delayed; keep victim under observation.

★ 255 ★
CHLOROACETIC ACID
CAS: 79-11-8

Other names: ACIDE CHLORACETIQUE (FRENCH); ACIDE MONOCHLORACETIQUE (FRENCH); ACIDOMONOCLOROACETICO (ITALIAN); ACETIC ACID, CHLORO-; CHLORACETIC ACID; CHLOROETHANOIC ACID; MCA; MONOCHLOROACETIC ACID; MONOCHLOORAZIJNZUUR (DUTCH); MONOCHLORESSIGSAEURE (GERMAN); NCI-C60231

Danger profile: Mutagen, combustible, corrosive, poisonous gases produced in fire.

Uses: Making other chemicals and dyes; herbicide; preservative; disinfectant.

Appearance: Colorless to light brown crystals.

Effects of exposure:

Breathed: May cause irritation of the nose, throat, breathing passages and lungs; sore throat, nosebleeds, coughing; blurred vision, hallucinations, muscle twitching, convulsions, coma, irregular heartbeat and death. These effects may delayed for several hours following exposure, and come on suddenly without warning.

Eyes: May cause severe irritation, redness, tearing, burns and permanent damage.

Skin: May cause severe irritation, redness, inflammation, rash and burns.

Swallowed: Poison. See symptoms under **Breathed.**

Long-term exposure: May cause genetic changes, birth defects, miscarriage; kidney damage; skin rash. May effect lungs, liver and thyroid.

Storage: Store in tightly closed containers in a dry, cool, well-ventilated place. Keep away from oxidizers, metals and combustible materials.

★ 256 ★
CHLOROACETYL CHLORIDE
CAS: 79-04-9

Other names: ACETYL CHLORIDE, CHLORO-; CHLOROACETIC ACID CHLORIDE; CHLOROACETIC CHLORIDE; CHLORURE DE CHLORACETYLE (FRENCH); MONOCHLOROACETYL CHLORIDE

Danger profile: Poisonous, corrosive, exposure limit established, poisonous gases produced in fire.

Uses: Making other chemicals; tear gas.

Appearance: Colorless to light yellow liquid.

Odor: Sharp, extremely irritating. The odor is a warning of exposure; however, no smell does not mean you are not being exposed.

Effects of exposure:

Breathed: May be poisonous. May cause irritation of the breathing passages and lungs; tearing. Higher exposures may cause shortness of breath and a dangerous build-up of fluids in the lungs (pulmonary edema), which can cause death. Pulmonary edema is a medical emergency that can be delayed one to two days following exposure.

Eyes: May cause irritation, severe burns and permanent damage.

Skin: May be poisonous. Passes through the unbroken skin; can increase exposure and the severity of the symptoms listed. May cause irritation, inflammation, redness and severe burns.

Swallowed: Poison. May cause severe irritation of mouth and stomach. See symptoms under *Breathed* and *First aid guide.*

Long-term exposure: May effect the lungs.

Storage: Store in tightly closed containers in a dry, cool, well-ventilated place with non-wood floors. Keep away from heat and combustible materials.

First aid guide: Move victim to fresh air and call emergency medical care; if not breathing, give artificial respiration; if breathing is difficult, give oxygen. In case of contact with material, immediately flush skin or eyes with running water for at least 15 minutes. Remove and isolate contaminated clothing and shoes at the site. Keep victim quiet and maintain normal body temperature. Effects may be delayed, keep victim under observation.

★257★
CHLOROANILINE
CAS: 27134-26-5
Other names: BENZENENAMINE, CHLORO-

Danger profile: Severe irritant, fire hazard, poisonous gases produced in fire.

Uses: Making insecticides, dyes, and many other chemicals and industrial products.

Appearance: Colorless to yellow liquid; white to yellow solid.

Effects of exposure:

Breathed: May be poisonous. May cause headache, dizziness, fatigue, difficult breathing and bluish color to the skin, lips and fingernails; collapse and death.

Eyes: Contact may cause severe irritation, redness, tearing, burns and permanent damage.

Skin: May be poisonous. Passes through the unbroken skin; can increase exposure and the severity of the symptoms listed. Contact may cause severe irritation, redness and swelling.

Swallowed: May be poisonous. See *Breathed.*

Long-term exposure: The symptoms described in *Breathed* may occur very slowly over a period of weeks.

Storage: Store in tightly closed containers in a dry, cool, well-ventilated place. Keep away from heat, sources of ignition and oxidizers. Containers may explode in fire.

★258★
CHLOROBENZENE
CAS: 108-90-7
Other names: MONOCHLOROBENZENE; CHLOROBENZOL; PHENYL CHLORIDE; MCB; BENZENE, CHLORO-; BENZENE CHLORIDE; CHLOORBENZEEN (DUTCH); CHLORBENZEN; CHLOROBENZOL; CHLOROBENZEN (POLISH); MONOCHLOORBENZEEN (DUTCH); MONOCHLORBENZENE; MONO-CHLORBENZOL (GERMAN); NCI-C54886; PHENYLCHLORIDE; PHENYL CHLORIDE

Danger profile: Mutagen, poisonous, narcotic, fire hazard, explosion hazard, Community Right-to-Know List, exposure limit established, poisonous gases produced in fire.

Uses: Making other chemicals, dyes, pesticides, rubber, dyes; solvent; heat transfer.

Appearance: Colorless, watery liquid.

Odor: Mild, sweet, almond-like; aromatic. The odor is a warning of exposure; however, no smell does not mean you are not being exposed.

Effects of exposure:

Breathed: May be poisonous. May cause irritation of the eyes, nose, throat and air passages; coughing, headache; increased saliva; reddening of the face; dizziness, lightheadedness, fatigue, sleepiness, loss of consciousness, shock, coma, rapid breathing; weak, irregular heartbeat, death. If the victim recovers watch for the possibility of a buildup of fluid in the lungs (pulmonary edema). Pulmonary edema is a medical emergency that can be delayed one to two days following exposure.

Eyes: Contact may cause irritation, tearing, inflammation of the mucous membrane of the eyelids; burning feeling.

Skin: May be poisonous if absorbed. Passes through the unbroken skin; can increase exposure and the severity of the symptoms listed. May cause irritation, redness, burning feeling, swelling; dry skin; rash or blisters.

Swallowed: Poison. Breath may smell of chloroform. May cause irritation of the mouth and lips, gastrointestinal tract, nausea, vomiting, diarrhea with possible blood; drowsiness; loss of consciousness; irregular heartbeat and death.

Long-term exposure: Repeated exposure of skin may cause skin burns. Chronic inhalation of vapors or mist may cause lung, and kidney damage. Similar solvents may cause brain damage, fatigue, irritability and personality change, loss of memory and concentration.

Storage: Store in tightly closed containers in a dry, cool, well-ventilated place. Keep away from heat, sources of ignition powdered sodium, dimethyl sulfoxide and oxidizers. Vapors may travel to sources of ignition and flash back. Containers may explode in fire.

First aid guide: Move victim to fresh air and call emergency medical care; if not breathing, give artificial respiration; if breathing is difficult, give oxygen. In case of contact with material, immediately flush eyes with running water for at least 15 minutes. Wash skin with soap and water. Remove and isolate contaminated clothing and shoes at the site.

★259★
CHLOROBROMOMETHANE
CAS: 74-97-5
Other names: BROMOCHLOROMETHANE; CB; CBM; HALON 1011; METANE, BROMOCHLORO-; METHYLENE CHLOROBROMIDE; MIL-B-4394-B; MONO-CHLORO-MONO-BROMO-METHANE

Danger profile: Mutagen, poisonous, narcotic, exposure limit established, poisonous fumes released in fire.

Uses: Grain fumigant; in fire extinguisher.

Appearance: Colorless to pale yellow liquid.

Odor: Characteristic, sweet-smelling.

Effects of exposure:

Breathed: May cause irritation of the eyes, nose, throat and air passages; coughing, headache; increased saliva; reddening of the face; dizziness, lightheadedness, fatigue, loss of consciousness, shock, coma, irregular heartbeat, death. If the victim recovers watch for the possibility of a buildup of fluid in the lungs (pulmonary edema). Pulmonary edema is a medical emergency that can be delayed one to two days following exposure.

Eyes: Contact may cause irritation, tearing, inflammation of the eyelids; burning sensation.

Skin: Passes through the unbroken skin; can increase exposure and the severity of the symptoms listed. May cause irritation, redness, burning feeling, swelling; dry skin; rash or blisters.

Swallowed: Poison. Breath may smell of chloroform. May cause irritation of the mouth, gastrointestinal tract, nausea, vomiting, diarrhea with possible blood; drowsiness; loss of consciousness; irregular heartbeat and death. See *Breathed.*

Long-term exposure: May cause liver and kidney damage, which may be progressive; skin irritation and cracking. May effect the lungs.
Storage: Store in tightly closed containers in a dry, cool, well-ventilated place. Keep away from chemically active and finely divided metals.

★ 260 ★
CHLORODIFLUOROETHANE
CAS: 75-68-3
Other names: 1-CHLORO-1,1-DIFLUOROETHANE; ALPHA-CHLOROETHYLIDENE FLUORIDE; ETHANE, 1-CHLORO-1-DIFLUORO-; FREON 142; FREON 142B; GENETRON 1016; GENETRON 101; GENETRON 142B; DIFLUORO-1-CHLOROETHANE

Danger profile: Mutagen, fire hazard, toxic gases produced in heat, poisonous gases produced in fire.
Uses: Making other chemicals, refrigerant; solvent.
Appearance: Colorless gas.
Odor: Slight odor. The odor is a warning of exposure; however, no smell does not mean you are not being exposed.
Effects of exposure:
  Breathed: May cause headache, dizziness, fatigue; rapid irregular breathing; nausea, vomiting, confusion, loss of consciousness convulsions and death. High exposure may cause irregular heartbeat, death.
  Eyes: Contact with liquid may cause frostbite. May cause pain, stinging, tearing, inflammation of the eyelids and permanent damage.
  Skin: Contact with liquid may cause frostbite, intense cold feeling, skin turns white and feels cold and hard.
  Swallowed: Liquid may cause headache, confusion, drowsiness, unconsciousness and death.
Long-term exposure: May cause skin allergy to develop.
Storage: Store in tightly closed containers in a dry, cool, well-ventilated place. Keep away from heat and oxidizers. Vapors may travel to sources of ignition and flash back. Vapors may travel to sources of ignition and flash back. Containers may explode in fire.
First aid guide: Move victim to fresh air and call emergency medical care; if not breathing, give artificial respiration; if breathing is difficult, give oxygen. In case of frostbite, thaw frosted parts with water. Keep victim quiet and maintain normal body temperature.

★ 261 ★
CHLORODIFLUOROMETHANE
CAS: 75-45-6
Other names: DIFLUOROCHLOROMETHANE; ALGOFRENE TYPE 6; ARCTON 4; DIFLUOROMONOCHLOROMETHANE; ELECTRO-CF 22; ESKIMON 22; F 22; FLUOROCARBON-22; FREON; FREON 22; FRIGEN; GENETRON 22; ISCEON 22; ISOTRON 22; METHANE, CHLORODIFLUORO-; MONOCHLORODIFLUOROMETHANE; PROPELLANT 22; R-22; REFRIGERANT 22; UCON 22; UCON 22/HALOCARBON 22

Danger profile: Mutagen, exposure limit established, poisonous gases produced in fire.
Uses: Making other chemicals, refrigerant; aerosol propellent; low-temperature solvent.
Appearance: Colorless liquefied compressed gas.
Odor: Faint, carbon tetrachloride-like. The odor is a warning of exposure; however, no smell does not mean you are not being exposed.
Effects of exposure:

Breathed: May cause headache, dizziness, fatigue; rapid irregular breathing; nausea, vomiting, confusion, loss of consciousness convulsions and death. Inhalation at greater than 10% concentration in air may cause narcosis.
Eyes: Contact with liquid may cause frostbite, pain, stinging, tearing, inflammation of the eyelids and permanent damage.
Skin: Contact with liquid may cause frostbite, intense cold feeling, skin turns white and feels cold and hard.
Swallowed: Liquid may cause headache, confusion, drowsiness, unconsciousness and death.
Long-term exposure: May cause birth defects and miscarriage. Repeated contact may cause skin rash; irregular heartbeat; liver and kidney function.
Storage: Store in tightly closed containers in a dry, cool, well-ventilated place. Keep away from flame and sources of ignition. Containers may explode in fire.
First aid guide: Move victim to fresh air and call emergency medical care; if not breathing, give artificial respiration; if breathing is difficult, give oxygen.

★ 262 ★
CHLOROFORM
CAS: 67-66-3
Other names: CHLOROFORME (FRENCH); CLOROFORMIO (ITALIAN); FORMYL TRICHLORIDE; FREON 20; METHANE TRICHLORIDE; METHANE, TRICHLORO-; METHENYL TRICHLORIDE; METHYL TRICHLORIDE; NCI-C02686; R 20; R 20 (REFRIGERANT); TCM; TRICHLOORMETHAAN (DUTCH); TRICHLORMETHAN (CZECH); TRICHLOROFORM; TRICHLOROMETHANE; TRICLOROMETANO (ITALIAN)

Danger profile: Carcinogen, mutagen, reproductive hazard, poisonous, exposure limit established, Community Right-to-Know List, EPA extremely hazardous substances list, poisonous gases produced in fire.
Uses: Making other chemicals including pesticides, drugs and dyes.
Appearance: A clear, colorless liquid.
Odor: Pleasant, sweet. The odor is a warning of exposure; however, no smell does not mean you are not being exposed.
Effects of exposure:
  Breathed: May cause irritation of the eyes, nose, throat and air passages; coughing, headache; increased saliva; reddening of the face; dizziness, lightheadedness, fatigue, loss of consciousness, shock, coma, irregular heartbeat, death. If the victim recovers watch for the possibility of a buildup of fluid in the lungs (pulmonary edema). Pulmonary edema is a medical emergency that can be delayed one to two days following exposure.
  Eyes: Contact may cause irritation, tearing, inflammation of the eyelids; burning sensation.
  Skin: Passes through the unbroken skin; can increase exposure and the severity of the symptoms listed. May cause irritation, redness, burning feeling, swelling; dry skin; rash or blisters.
  Swallowed: Poison. Breath may smell of chloroform. May cause irritation of the mouth, gastrointestinal tract, nausea, vomiting, diarrhea with possible blood; drowsiness; loss of consciousness; irregular heartbeat and death. See *Breathed.*
Long-term exposure: May cause cancer of the liver, kidney and thyroid; birth defects, miscarriage; liver damage; skin drying and cracking; kidney and nervous system damage.
Storage: Store in tightly closed containers in a dry, cool, well-ventilated place. Keep away from caustics, chemically active metals, and all other chemicals including dinitrogen tetraoxide and fluorine. Containers may explode in fire.
First aid guide: Move victim to fresh air and call emergency medical care; if not breathing, give artificial respiration; if breath-

ing is difficult, give oxygen. In case of contact with material, immediately flush skin or eyes with running water for at least 15 minutes. Speed in removing material from skin is of extreme importance. Remove and isolate contaminated clothing and shoes at the site. Keep victim quiet and maintain normal body temperature. Effects may be delayed; keep victim under observation.

★ 263 ★
2-CHLORO-4-METHYLANILINE
CAS: 615-65-6
Other names: BENZENAMINE, 2-CHLORO-4-METHYL-; 2-CHLORO-4-METHYL- BENZENAMINE; 2-CHLORO-P-TOLUIDINE; CHLOROTOLUIDINE; 2-CHLOR-4-TOLUIDIN (CZECH); 4-METHYL-2-CHLOROANILINE; P-TOLUIDINE, 2-CHLORO-

Danger profile: Mutagen, poisonous, poisonous gases produced in fire.
Uses: Making dyes and dyestuffs.
Effects of exposure:
Breathed: May be poisonous. May cause irritation of the kidney, bladder and urethra; frequent urination, possibly painful and bloody; dizziness, nausea, headaches and cause the skin (face, lips, earlobes, fingernails and hands) to turn blue, rapid and difficult breathing, convulsions, coma and death.
Eyes: Contact may cause irritation, redness, tearing, inflammation and swelling of the eyelids, and permanent damage.
Skin: May be poisonous. Passes through the unbroken skin; can increase exposure and the severity of the symptoms listed under *Breathed.* Contact may cause irritation, redness, swelling and burning feeling with rash.
Swallowed: Poison. May cause irritation of the mouth and stomach, stomach cramps, diarrhea. See symptoms under *Breathed.*
Long-term exposure: May effect kidneys, bladder and urethra; possible birth defects, miscarriage. Symptoms under *Breathed* may occur from a single exposure or gradually, over time from repeated exposures.
Storage: Store in tightly closed containers in a dry, cool, well-ventilated place. Containers may explode in fire.

★ 264 ★
3-CHLORO-2-METHYLANILINE
CAS: 87-60-5
Other names: 1-AMINO-2-CHLORO-6-METHYLBENZENE; 2-AMONO-6- CHLOROTOLUENE; 1-AMINO-3-CHLORO-2-METHYLBENZENE; 2-AMINO-6-CHLOROTOLUENE; AZOIC-DIAZO COMPONENT 46; 3-CHLORO-2- METHYL BENZENE-AMINE; 3-CHLOR-2-TOLUIDIN (CZECH); 3-CHLORO-O-TOLUIDINE; FAST SCARLET TR BASE; SCARLET TR BASE; O-TOLUIDINE, 3-CHLORO-

Danger profile: Mutagen, poisonous, poisonous gases produced in fire.
Uses: Making dyes.
Appearance: Colorless or white crystals or liquid. Rapidly turns brown in air.
Effects of exposure:
Breathed: May be poisonous. May cause irritation of the kidney, bladder and urethra; frequent urination, possibly painful and bloody; dizziness, nausea, headaches and cause the skin (face, lips, earlobes, fingernails and hands) to turn blue, rapid and difficult breathing, convulsions, coma and death.
Eyes: Contact may cause irritation, redness, tearing, inflammation and swelling of the eyelids, and permanent damage.
Skin: May be poisonous. Passes through the unbroken skin; can increase exposure and the severity of the symptoms list-

ed under *Breathed.* Contact may cause irritation, redness, swelling and burning feeling with rash.
Swallowed: Poison. May cause irritation of the mouth and stomach, stomach cramps, diarrhea. See symptoms under *Breathed.*
Long-term exposure: May effect kidneys, bladder and urethra; possible birth defects, miscarriage. Symptoms under *Breathed* may occur from a single exposure or gradually, over time from repeated exposures.
Storage: Store in tightly closed containers in a dry, cool, well-ventilated place.

★ 265 ★
3-CHLORO-4-METHYLANILINE
CAS: 95-74-9
Other names: 1-AMINO-3-CHLORO-4-METHYLBENZENE; 4-AMINO-2-CHLOROTOLUENE; 2-CHLORO-4-AMINOTOLUENE; 3-CHLORO-4- METHYL-BENZENAMINE; CHLOROTOLUIDINE; DKC 1347; DRC 1339; NCI-C02040; P-TOLUIDINE, 3-CHLORO-

Danger profile: Mutagen, poisonous, poisonous gases produced in fire.
Uses: As a bait to kill bird populations.
Appearance: Pellets.
Effects of exposure:
Breathed: May be poisonous. May cause irritation of the kidney, bladder and urethra; frequent urination, possibly painful and bloody; dizziness, nausea, headaches and cause the skin (face, lips, earlobes, fingernails and hands) to turn blue, rapid and difficult breathing, convulsions, coma and death.
Eyes: Contact may cause irritation, redness, tearing, inflammation and swelling of the eyelids, and permanent damage.
Skin: May be poisonous. Passes through the unbroken skin; can increase exposure and the severity of the symptoms listed under *Breathed.* Contact may cause irritation, redness, swelling and burning feeling with rash.
Swallowed: Poison. May cause irritation of the mouth and stomach, stomach cramps, diarrhea. See symptoms under *Breathed.*
Long-term exposure: May effect kidneys, bladder and urethra; possible birth defects, miscarriage. Symptoms under *Breathed* may occur from a single exposure or gradually, over time from repeated exposures.
Storage: Store in tightly closed containers in a dry, cool, well-ventilated place. Keep away from flame and sources of ignition. Containers may explode in fire.

★ 266 ★
5-CHLORO-2-METHYL ANILINE
CAS: 95-79-4
Other names: 2-AMINO-4-CHLOROTOLUENE; 1-AMINO-3-CHLORO-6-METHLBENZENE; BENZENAMINE, 5-CHLORO-2-METHYL-; CHLOROTOLUIDINE; 4-CHLORO-2-AMINOTOLUENE; 3-CHLORO-6-METHYLANILINE; 5-CHLORO-O-TOLUIDINE; NCI-CO2051

Danger profile: Carcinogen, mutagen, poisonous gases produced in fire.
Uses: Making dyes
Appearance: Off-white solid or oily liquid.
Effects of exposure:
Breathed: May be poisonous. May cause irritation of the kidney, bladder and urethra; frequent urination, possibly painful and bloody; dizziness, nausea, headaches and cause the skin (face, lips, earlobes, fingernails and hands) to turn blue, rapid and difficult breathing, convulsions, coma and death.

Eyes: Contact may cause irritation, redness, tearing, inflammation and swelling of the eyelids, and permanent damage.
Skin: May be poisonous. Passes through the unbroken skin; can increase exposure and the severity of the symptoms listed under *Breathed*. Contact may cause irritation, redness, swelling and burning feeling with rash.
Swallowed: Poison. May cause irritation of the mouth and stomach, stomach cramps, diarrhea. See symptoms under *Breathed*.
Long-term exposure: May cause cancer; possible birth defects, miscarriage; kidneys, bladder and urethra damage. Symptoms under *Breathed* may occur from a single exposure or gradually, over time from repeated exposures.
Storage: Store in tightly closed containers in a dry, cool, well-ventilated place. Containers may explode in fire.

★ 267 ★
CHLOROMETHYL METHYL ETHER
CAS: 107-30-2
Other names: CHLORDIMETHYLETHER (CZECH); CMME; DIMETHYLCHLOROETHER; ETHER, DIMETHYL CHLOR O; ETHER METHYLIQUE MONOCHLORE (FRENCH); METHYL-CHLOROMETHYL ETHER; METHYL CHLOROMETHYL ETHER, ANHYDROUS; METHANE, CHLOROMETHOXY-

Danger profile: Carcinogen, mutagen, poisonous, fire hazard, corrosive, Community Right-to-Know List, EPA extremely hazardous list, poisonous gas produced in fire.
Uses: Making other chemicals and plastics.
Appearance: Colorless liquid.
Odor: Irritating. The odor is a warning of exposure; however, no smell does not mean you are not being exposed.
Effects of exposure:
Breathed: Poison. May cause sore throat, fever, chills, shortness of breath and difficult breathing; coughing, irritation of the lungs. Higher exposures may cause shortness of breath and a dangerous build-up of fluids in the lungs (pulmonary edema), which can cause death. Pulmonary edema is a medical emergency that can be delayed one to two days following exposure.
Eyes: Contact of liquid may cause severe burns and permanent damage; vapor is powerful tear gas.
Skin: May be poisonous if absorbed. Passes through the unbroken skin; can increase exposure and the severity of the symptoms listed. Contact causes severe burns and tissue damage.
Swallowed: May be poisonous. May cause severe burns of mouth and stomach. See symptoms under *Breathed*.
Long-term exposure: May cause lung and skin cancer; reproductive damage, birth defects and miscarriage; lung irritation, bronchitis, cough and difficult breathing.
Storage: Keep away from open flame and sources of ignition; reactive metals. Containers may explode in fire.

★ 268 ★
1-CHLORO-1-NITROPROPANE
CAS: 600-25-9
Other names: 1,1-CHLORONITROPROPANE; PROPANE, 1-CHLORO-1-NITRO-; KORAX; KORAX 6; LANSTAN

Danger profile: Very reactive, combustible, explosion hazard, exposure limit established, poisonous gas produced in fire.
Uses: Making rubber cement; solvent.
Appearance: Colorless liquid.
Odor: Unpleasant odor that causes tears. The odor is a warning of exposure; however, no smell does not mean you are not being exposed.
Effects of exposure:

Breathed: May cause irritation of the eyes, nose, breathing passages and throat. Higher exposures may cause shortness of breath and a dangerous build-up of fluids in the lungs (pulmonary edema), which can cause death. Pulmonary edema is a medical emergency that can be delayed one to two days following exposure.
Eyes: Contact may cause irritation, redness, tearing and burns.
Skin: Contact may cause irritation, redness, inflammation and burns.
Swallowed: See *Breathed*.
Long-term exposure: May effect the lungs; liver, kidney, blood vessel and heart damage. In animals: Irritation of the lungs, eyes, liver, kidney, heart.
Storage: Store in tightly closed containers in a dry, cool, well-ventilated place. Keep away from heat, sources of ignition and oxidizers. Containers may explode in fire.

★ 269 ★
CHLOROPENTAFLUOROETHANE
CAS: 76-15-3
Other names: ETHANE, CHLOROPENTAFLUORO-; F-115; FLUROCARBON 115; FREON 115; GENETRON 115; MONO-CHLOROPENTAFLUOROETHANE; PENTAFLUOROMONO-CHLOROETHANE; CFC-115

Danger profile: Exposure limit established, poisonous gases produced in fire.
Uses: Refrigerant; propellent in spray cans.
Appearance: Colorless gas or compressed liquid.
Odor: Chloroform-like. The odor is a warning of exposure; however, no smell does not mean you are not being exposed.
Effects of exposure:
Breathed: May cause irritation of eyes, nose, throat; over excitement, intoxication, headache, loss of consciousness, state of insensibility, irregular heart beat, coma, death.
Eyes: May cause irritation, redness and inflammation of the eyelids.
Skin: May cause irritation, frostbite or freezing; inflammation, dryness.
Swallowed: Breath may smell of chloroform. May cause intestinal irritation. Higher exposures may cause a dangerous build-up of fluids in the lungs (pulmonary edema), which can cause death. Pulmonary edema is a medical emergency that can be delayed one to two days following exposure.
Long-term exposure: May effect the lungs.
Storage: Store in tightly closed containers in a dry, cool (less than 125°F), well-ventilated place. Keep away from heat, open flames, sources of ignition, metals, especially those that are chemically active (aluminum, beryllium, zinc, etc.). Containers may explode in fire.
First aid guide: Move victim to fresh air and call emergency medical care; if not breathing, give artificial respiration; if breathing is difficult, give oxygen.

★ 270 ★
2-CHLOROPHENOL
CAS: 95-57-8
Other names: PHENOL, O-CHLORO; O-CHLOROPHENOL; O-CHLORPHENOL (GERMAN); PHENOL, 2-CHLORO-

Danger profile: Carcinogen, mutagen, poisonous, Community Right-to-Know List, poisonous gases produced in fire.
Uses: Making dyes.
Appearance: Colorless to amber liquid or crystals.
Effects of exposure:
Breathed: May be poisonous. May cause restlessness, rapid breathing, tremors, seizures, coma and death.

Eyes: Contact may cause irritation, redness, tearing, inflammation and burns.

Skin: May be poisonous if absorbed. Passes through the unbroken skin; can increase exposure and the severity of the symptoms listed. Contact may cause irritation, redness and burns.

Swallowed: May be poisonous.

Long-term exposure: May cause soft tissue cancer and leukemia; reproduction damage; birth defects, miscarriage; kidney damage; acne-like skin rash that is aggravated by sun light.

Storage: Store in tightly closed containers in a dry, cool, well-ventilated place. Keep away from heat, sources of ignition and oxidizers. Containers may explode in fire.

★ 271 ★
3-CHLOROPHENOL
CAS: 108-43-0
Other names: M-CHLOROPHENOL; CHLOROPHENATE

Danger profile: Carcinogen, mutagen, poisonous, Community Right-to-Know List, poisonous gases produced in fire.
Uses: Making other chemicals.
Appearance: White crystals or clear liquid.
Effects of exposure:
Breathed: May be poisonous. May cause restlessness, rapid breathing, shortness of breath, tremors, seizures, coma and death.
Eyes: Contact may cause irritation, redness, tearing, inflammation and burns.
Skin: May be poisonous if absorbed. Passes through the unbroken skin; can increase exposure and the severity of the symptoms listed. Contact may cause irritation, redness and burns.
Swallowed: May be poisonous. See *Breathed.*
Long-term exposure: May cause skin cancer; birth defects, miscarriage; liver and kidney damage.
Storage: Store in tightly closed containers in a dry, cool, well-ventilated place. Containers may explode in fire.

★ 272 ★
4-CHLOROPHENOL
CAS: 106-48-9
Other names: P-CHLORFENOL (CZECH); CHLOROPHENATE; P-CHLOROPHENOL; PARACHLOROPHENOL

Danger profile: Carcinogen, mutagen, corrosive, poisonous, combustible, Community Right-to-Know List, poisonous gases produced in fire.
Uses: Making other chemicals, topical antiseptic.
Appearance: White to straw colored, needle-like crystals.
Odor: Unpleasant. The odor is a warning of exposure; however, no smell does not mean you are not being exposed.
Breathed: May be poisonous. May cause restlessness, rapid breathing, shortness of breath, tremors, seizures, coma and death.
Eyes: Contact may cause irritation, redness, tearing, inflammation and burns.
Skin: May be poisonous if absorbed. Passes through the unbroken skin; can increase exposure and the severity of the symptoms listed. Contact may cause irritation, redness and burns.
Swallowed: May be poisonous.
Long-term exposure: May cause skin cancer; birth defects, miscarriage; liver and kidney damage.
Storage: Store in tightly closed containers in a dry, cool, well-ventilated place. Keep away from open flame and sources of ignition. Containers may explode in fire.

★ 273 ★
CHLOROPHENYL TRICHLOROSILANE
CAS: 26571-79-9
Other names: CHLOROPHENYLTRICHLOROSILANE; TRICHLORO (CHLOROPHENYL)SILANE

Danger profile: Poisonous, corrosive, combustible, water reactive, poisonous gases produced in fire.
Uses: Making pharmaceuticals and silicones; rodenticide; insecticide.
Appearance: Colorless to pale yellow liquid.
Effects of exposure:
Breathed: Poison. May cause irritation of the eyes, nose, throat and lungs; coughing, difficult breathing, shortness of breath. Higher exposures may cause shortness of breath and a dangerous build-up of fluids in the lungs (pulmonary edema), which can cause death. Pulmonary edema is a medical emergency that can be delayed one to two days following exposure.
Eyes: Contact may cause irritation, redness, tearing, severe eye burns and permanent damage.
Skin: Contact may cause irritation, redness and severe skin damage.
Swallowed: Poison. May cause severe irritation of the mouth, throat and stomach.
Long-term exposure: May cause lung irritation.
Storage: Store in tightly closed containers in a dry, cool, well-ventilated place with non-wood floors. Keep away from combustible materials, water, steam and other forms of moisture.
First aid guide: Move victim to fresh air; call emergency medical care. In case of contact with material, immediately flush skin or eyes with running water for at least 15 minutes. Remove and isolate contaminated clothing and shoes at the site. Keep victim quiet and maintain normal body temperature.

★ 274 ★
CHLOROPICRIN
CAS: 76-06-2
Other names: ACQUINITE; CHLOORPIKRINE (DUTCH); CHLOR-O-PIC; CHLOROPICRIN, ABSORBED; CHLOROPICRINE (FRENCH); CHLORPIKRIN (GERMAN); CHLORPICRINA (ITALIAN); LARVACIDE; METHANE, TRICHLORONITRO-; MYCROLYSIN; NITROTRICHLOROMETHANE; NITROCHLOROFORM; PICFUME; PIC-CHLOR; PICRIDE; PROFUME A; PS; TRICHLORONITROMETHANE; TRICHLOR

Danger profile: Mutagen, possible carcinogen, exposure limit established, very reactive, explosion hazard, poisonous gases produced in fire.
Uses: Making other chemicals; dyes; fumigants; fungicides; insecticide and rodent control; tear gas. Was used as a poison gas in World War I.
Appearance: Colorless, oily liquid.
Odor: Intensely irritating. A powerful tear gas. The odor is a warning of exposure; however, no smell does not mean you are not being exposed.
Effects of exposure:
Breathed: May cause irritation of the eyes, nose, throat and lungs, tears, and runny nose, coughing, sore throat, difficult breathing; lethargy, fatigue, and drowsiness; slurred speech, shaking, and dizziness. Higher exposures could cause shortness of breath, headache, nausea, vomiting, dizziness, diarrhea and a dangerous build-up of fluids in the lungs (pulmonary edema), which can cause death. Pulmonary edema is a medical emergency that can be delayed one to two days following exposure.
Eyes: May cause irritation, redness, tearing, pain, inflamed eyelids, and severe burns.

Skin: May cause painful irritation, redness, and severe burns.
Swallowed: Poison. May cause severe irritation of the mouth, throat and stomach; abdominal pain; loss of muscular coordination and mental confusion; convulsions. Also see symptoms for *Breathed.*

Long-term exposure: Repeated exposure can cause bronchitis, cough and difficult breathing; lung, liver and kidney damage. May cause birth defects, miscarriage.

Storage: Store in tightly closed containers in a dry, cool, well-ventilated place. Keep away from heat, shock, all other chemicals, especially oxidizers. Containers may explode in fire. Can be detonated by shock or friction.

First aid guide: Move victim to fresh air and call emergency medical care; if not breathing, give artificial respiration; if breathing is difficult, give oxygen. In case of contact with material, immediately flush skin or eyes with running water for at least 15 minutes. Speed in removing material from skin is of extreme importance. Remove and isolate contaminated clothing and shoes at the site. Keep victim quiet and maintain normal body temperature. Effects may be delayed, keep victim under observation.

★ 275 ★
CHLOROPLATINIC ACID
CAS: 16941-12-1
Other names: DIHYDROGENHEXACHLOROPLATINATE; DIHYDROGENHEXACHLOROPLATINATE (2-); HEXACHLOROPLATINIC ACID; HEXACHLOROPLATINIC(IV) ACID; HEXACHLOROPLATINIC(4) ACID, HYDROGEN-; HYDROGEN HEXACHLOROPLATINATE(4); PLATINATE, HEXACHLORO-; PLATINIC CHLORIDE

Danger profile: Mutagen, allergen, poisonous, corrosive, exposure limit established, poisonous gases produced in fire.
Uses: Electroplating; photography; etching zinc for printing; indelible ink; producing color on high-grade porcelain; ceramics and may other uses.
Appearance: Red-brown or yellowish crystalline mass.
Effects of exposure:
Breathed: May cause irritation of the nose, throat and breathing passages; coughing and sneezing, runny nose. Ulcers in the nose and lung allergy may develop.
Eyes: May cause irritation, redness, tearing, itching.
Skin: May cause irritation, redness; skin allergy with itching.
Long-term exposure: May cause birth defects or miscarriage. Lung, nasal and skin allergy may develop. Nasal allergy may include runny nose, ulcers and burning feeling from irritation. Repeated or prolonged exposure may lead to pulmonary fibrosis (permanent lung damage), shortness of breath, chest tightness and cough. Skin allergy may include persistent rash and itching. Once an allergy develops a very low future exposure may cause symptoms to appear.
Storage: Store in tightly closed containers in a dry, cool, well-ventilated place.
First aid guide: Move victim to fresh air; call emergency medical care. In case of contact with material, immediately flush skin or eyes with running water for at least 15 minutes. Remove and isolate contaminated clothing and shoes at the site. Keep victim quiet and maintain normal body temperature.

★ 276 ★
O-CHLOROSTYRENE
CAS: 2039-87-4
Other names: BENZENE, 1-CHLORO-2-ETHENYL-; CHLOROSTYRENE; 2-CHLOROSTYRENE

Danger profile: Exposure limit established, poisonous gas produced in fire.
Uses: Making other chemicals, plastics and polymers.

Appearance: Clear liquid.
Effects of exposure:
Breathed: May cause irritation of the eyes, nose and throat.
Eyes: May cause irritation, redness and tearing.
Skin: Mildly toxic. Passes through the unbroken skin; can increase exposure and the severity of the symptoms listed. May cause irritation.
Swallowed: Mildly toxic.
Long-term exposure: Repeated exposures may cause kidney and liver damage.
Storage: Store in tightly closed containers in a dry, cool, well-ventilated place.

★ 277 ★
CHLOROSULFURIC ACID
CAS: 7790-94-5
Other names: CHLOROSULFONIC ACID; MONOCHLOROSULFURIC ACID; SULFONIC ACID, MONOCHLORIDE; SULFURIC CHLOROHYDRIN

Danger profile: Poisonous, extremely corrosive, poisonous gases produced in fire, explosive on contact with water.
Uses: Making pharmaceuticals, pesticides, dyes, resins and detergents.
Appearance: Colorless to light yellow, cloudy, fuming liquid.
Odor: Sharp, acrid, choking. The odor is a warning of exposure; however, no smell does not mean you are not being exposed.
Effects of exposure:
Breathed: May be poisonous. Vapor may irritate the eyes, nose, throat, breathing passages and lungs; coughing, shortness of breath, difficult breathing, unconsciousness. Higher exposures may cause shortness of breath and a dangerous build-up of fluids in the lungs (pulmonary edema), which can cause death. Pulmonary edema is a medical emergency that can be delayed one to two days following exposure.
Eyes: May cause irritation, redness, inflammation of the eyelids, severe burns and permanent damage.
Skin: May cause irritation, redness, severe burns and permanent damage.
Swallowed: May be poisonous. May cause severe irritation of the mouth, esophagus and stomach.
Long-term exposure: May cause damage to lung tissue. Prolonged or repeated contact may cause dermatitis.
Storage: Store in tightly closed containers in a dry, cool, well-ventilated place with non-wood floors. If stored in metal containers, they should be vented to release produced hydrogen. Keep away from water and other forms of moisture, combustible materials, acids, bases, alcohols, metal powder, and all other chemicals especially, phosphorus, silver nitrate. Containers may explode in fire.

★ 278 ★
CHLOROTHALONIL
CAS: 1897-45-6
Other names: BRAVO; BRAVO-W-75; 1,3-BENZENEDICARBONITRILE,2,4,5.6- TETRACHLORO-; CHLOROALONIL; DACONIL; CHLORTHALONIL (GERMAN); DACONIL 2787; EXOTHERM; FORTURF; ISOPHTHALONITRILE, TETRACHLORO-; ISOPHTHALONITRILE,2,4,5,6-TETRACHLORO-; NCI-C00102; SWEEP; TCIN; TERMIL; TETRACHLOROISOPHTHALONITRILE; M-TETRACHLOROPHTHALONITRILE; M-TCPN; TPN

Danger profile: Possible carcinogen, mutagen, poisonous, Community Right-to-Know List, extremely poisonous fumes produced in fire.
Uses: Pesticide, fungicide.

Appearance: White crystalline solid.
Effects of exposure:
Breathed: May be poisonous. May cause irritation of the nose, throat, air passages, and lungs.
Eyes: Contact may cause irritation.
Skin: Passes through the unbroken skin; can increase exposure and the severity of the symptoms listed. Contact may cause irritation.
Swallowed: May be poisonous.
Long-term exposure: May cause cancer; birth defects, miscarriage. Repeated exposure may cause skin rash, nose bleeds; bloody urine, vaginal bleeding.
Storage: Store in tightly closed containers in a dry, cool, well-ventilated place. Keep away from heat.

★ 279 ★
O-CHLOROTOLUENE
CAS: 95-49-8
Other names: BENZENE, 1-CHLORO-2-METHYL- (9CI); 2-CHLORO-1-METHYLBENZENE; 2-CHLOROTOLUENE; 1-METHYL-2-CHLOROBENZENE; 2-METHYLCHLOROBENZENE; TOLUENE, O-CHLORO-; O-TOLYLCHLORIDE

Danger profile: Combustible, exposure limit established, poisonous gas produced in fire.
Uses: Making other chemicals, pharmaceuticals, dyes and synthetic rubber; a solvent.
Appearance: Colorless liquid.
Effects of exposure:
Breathed: May be poisonous. May cause irritation of the eyes, nose and throat. High concentrations may cause dizziness, loss of coordination, convulsions, coma and death.
Eyes: Contact may cause irritation.
Skin: Passes through the unbroken skin; can increase exposure and the severity of the symptoms listed. Contact may cause irritation.
Swallowed: May be poisonous. See symptoms under *Breathed.*
Long-term exposure: May effect the liver.
Storage: Store in tightly closed containers in a dry, cool, well-ventilated place. Keep away from heat and sources of ignition. Containers may explode in fire.

★ 280 ★
6-CHLORO-2-TOLUIDINE
CAS: 87-63-8
Other names: 2-AMINO-3-CHLOROTOLUENE; BENZENEAMINE,2-CHLORO,6-METHYL-; 3-CHLORO-2-AMINOTOLUENE; 6-CHLORO-2-METHYL ANILINE; 6-CHLORO-O-TOLUIDINE; CHLOROTOLUIDINE; O-TOLUIDINE, 6-CHLORO-

Danger profile: Mutagen, poisonous, poisonous gases produced in fire.
Uses: Making make dyes and dyestuffs.
Appearance: Colorless or white crystals.
Effects of exposure:
Breathed: May be poisonous. May cause irritation of the kidney, bladder and urethra; frequent urination, possibly painful and bloody; dizziness, nausea, headaches and cause the skin (face, lips, earlobes and hands) to turn blue, rapid and difficult breathing, convulsions, coma and death.
Eyes: Contact may cause irritation, redness, tearing, inflammation and swelling of the eyelids, and permanent damage.
Skin: May be poisonous. Passes through the unbroken skin; can increase exposure and the severity of the symptoms listed under *Breathed.* Contact may cause irritation, redness, swelling and burning feeling with rash.
Swallowed: Poison. May cause irritation of the mouth and stomach, stomach cramps, diarrhea. See symptoms under *Breathed.*
Long-term exposure: May effect kidneys, bladder and urethra; possible birth defects, miscarriage. Symptoms under *Breathed* may occur from a single exposure or gradually, over time, from repeated exposures.
Storage: Store in tightly closed containers in a dry, cool, well-ventilated place. Containers may explode in fire.

★ 281 ★
4-CHLORO-2-TOLUIDINE HYDROCHLORIDE
CAS: 3165-93-3
Other names: 2-AMINO-5-CHLOROTOLUENE HYDROCHLORIDE; AMARTHOL FAST RED TR BASE; AMARTHOL FAST RED TR SALT; AZANIL RED SALT TRD; AZOENE FAST RED TR SALT; AZOIC DIAZO COMPONENT 11 BASE; AZOGENE FAST RED TR; BENZENEAMINE, 4-CHLORO-2-METHYL-,HYDROCHLORIDE; BRENTAMINE FAST RED TR SALT; CHLORHYDRATE DE 4-CHLOROORTHOTOLUIDINE (FRENCH); 5-CHLORO-2-AMINOTOLUENE HYDROCHLORIDE; 4-CHLORO-2-METHYLBENZENAMINE HYDROCHLORIDE; 4-CHLORO-2-METHYLANILINE HYDROCHLORIDE; 4-CHLORO-6-METHYLANILINE YDROCHLORIDE; 4-CHLORO-O-TOLUIDINEHYDROCHLORIDE; 4-CHLORO-O-TOLUIDINE HYDROCHLORIDE; C.I. 37085; C.I. AZOIC DIAZO COMPONENT 11; DAITO RED SALT TR; DEVOL RED K; DEVOL RED TA SALT; DEVOL RED TR; DIAZO FAST RED TR; DIAZO FAST RED TRA; FAST RED 5CT SALT; FAST RED SALT TR; FAST RED SALT TRA; FAST RED SALT TRN; FAST RED TR SALT; HINDASOL RED TR SALT; KROMON GREEN B; 2-METHYL-4-CHLOROANILINE HYDROCHLORIDE; NATASOL FAST RED TR SALT; NCI-C02368; NEUTROSEL RED TRVA; OFNA-PERL SALT RRA; RED BASE CIBA IX; RED BASE IRGA IX; RED SALT CIBA IX; RED SALT IRGA IX; RED TRS SALT; SANYO FAST RED SALT TR

Danger profile: Suspected carcinogen, mutagen, poisonous, poisonous gases produced in fire.
Uses: Making dyes for various natural and synthetic fabrics.
Appearance: Colorless or white crystals, or beige powder.
Effects of exposure:
Breathed: May be poisonous. May cause irritation of the kidney, bladder and urethra; frequent urination, possibly painful and bloody; dizziness, nausea, headaches and cause the skin (face, lips, earlobes, fingernails and hands) to turn blue, rapid and difficult breathing, convulsions, coma and death.
Eyes: Contact may cause irritation, redness, tearing, inflammation and swelling of the eyelids, and permanent damage.
Skin: Passes through the unbroken skin; can increase exposure and the severity of the symptoms listed under *Breathed.* Contact may cause irritation, redness, swelling and burning feeling with rash.
Swallowed: Poison. May cause irritation of the mouth and stomach, stomach cramps, diarrhea. See symptoms under *Breathed.*
Long-term exposure: Causes cancer in animals. May cause cancer; possible birth defects, miscarriage; kidneys, bladder and urethra damage. Symptoms under *Breathed* may occur from a single exposure or gradually, over time from repeated exposures.
Storage: Store in tightly closed containers in a dry, cool, well-ventilated place.

★ 282 ★
CHLORPYRIFOS
CAS: 2921-88-2
Other names: ALPHA CHLORPYRIFOS 48EC (ALPHA); BRODAN; DETMOL U.A.; O,O-DIETHYL; DOWCO 179; DURSBAN 4; DURSBAN 5G; DURSBAN; DURSBAN F; EF 121; ENT 27311; ERADEX; GLOBAL CRAWLING INSECT BAIT; LORSBAN; O-3,5,6-TRICHLORO-2-PYRIDYLPHOSPHOROTHIOATE; O,O-DIAETHYL-O-3,5,6-TRICHLOR-2-PYRIDYLMONOTHIOPHOSPHAT (GERMAN); MURPHY SUPER ROOT GUARD; 2-PYRIDINOL, 3,5,6-TRICHLORO-,O-ESTER WITH O,O-DIETHYL PHOSPHOROTHIOATE; PYRINEX; SPANNIT; TALON; TWINSPAN

Danger profile: Mutagen, poisonous, exposure limit established, poisonous gases produced in fire.
Uses: Pesticide; control of chinch bugs; tick control on cattle and sheep; control of mosquitos, animal flies; crop pests on pears, strawberries, currants and cereals.
Appearance: White sugar/sand-like crystals.
Odor: Like natural gas. The odor is a warning of exposure; however, no smell does not mean you are not being exposed.
Effects of exposure:
   Breathed: Poisonous. May cause headache, dizziness, profuse sweating, nausea, vomiting, reduced heart beat, stomach cramps, diarrhea, loss of coordination, slow and weak breathing, fever, loss of consciousness and death. May cause shortness of breath and a dangerous build-up of fluids in the lungs (pulmonary edema), which can cause death. Pulmonary edema is a medical emergency that can be delayed one to two days following exposure.
   Eyes: May be poisonous if absorbed. Rapidly absorbed. See *Breathed.*
   Skin: Poisonous. Passes through the unbroken skin; can increase exposure and the severity of the symptoms listed.
   Swallowed: Poison. See symptoms under *Breathed.*
Long-term exposure: May damage the developing fetus or cause birth defects; nerve damage including loss of coordination; liver damage.
Storage: Store in tightly-closed containers in a dry, well-ventilated place. Keep away from heat, acids or acid fumes. Containers may explode in fire.
First aid guide: Move victim to fresh air and call emergency medical care; if not breathing, give artificial respiration; if breathing is difficult, give oxygen. In case of contact with material, immediately flush skin or eyes with running water for at least 15 minutes. Speed in removing material from skin is of extreme importance. Remove and isolate contaminated clothing and shoes at the site. Keep victim quiet and maintain normal body temperature. Effects may be delayed; keep victim under observation.

★ 283 ★
CHROMIC ACETATE
CAS: 1066-30-4
Other names: ACETIC ACID, CHROMIUM (3) SALT; CHROMIC ACETATE (III); CHROMIUM ACETATE; CHROMIUM(III) ACETATE; CHROMIUM TRIACETATE

Danger profile: Carcinogen, mutagen, exposure limit established, Community Right-to-Know-List, poisonous gases produced in fire.
Uses: Textile mordant; tanning; photographic emulsion hardener.
Appearance: Gray-green powder, or blue-green pasty mess.
Effects of exposure:
   Breathed: May cause irritation of the eyes, nose and throat.
   Eyes: Contact may cause irritation, redness and tearing.
   Skin: Contact may cause irritation, redness and tearing.

Long-term exposure: May cause lung or throat cancer; birth defects, miscarriage, skin allergy with redness, itching and rash.
Storage: Store in tightly closed containers in a dry, cool, well-ventilated place. Keep away from heat.

★ 284 ★
CHROMIC ACID
CAS: 7738-94-5
Other names: ACIDE CHROMIQUE (FRENCH); CHROMIC-(VI) ACID; CHROMIUM TRIOXIDE; CHROMIUM ANHYDRIDE

Danger profile: Carcinogen, mutagen, oxidizer, exposure limit established, Community Right-to-Know List, poisonous gas produced in fire.
Uses: Making other chemicals; chromium plating; medicine; engraving; anodizing; ceramic glazes, colored glass; metal cleaning; inks; tanning; paints; textiles; etching plastics.
Appearance: Dark, purplish red, crystalline solid.
Effects of exposure:
   Breathed: May be poisonous. Dust may cause severe irritation to the nose, throat and lungs, causing coughing, shortness of breath. May cause flu-like symptoms including chills, muscle ache, headache, fever. High exposure may cause nausea, salivation, vomiting, cramps, diarrhea, chest pains, cough, a build-up of fluids in the lungs (pulmonary edema) and possible death. Pulmonary edema may be delayed from one to two days following exposure.
   Skin: May cause severe irritation and thermal and acid burns, especially if skin is wet.
   Eyes: May cause severe irritation, burns, pain and possible blindness.
   Swallowed: May be poisonous. May cause severe burns of the mouth, throat and stomach, vomiting, watery or bloody diarrhea. Damage to kidneys and liver, collapse and convulsions can result.
Long-term exposure: May cause lung cancer, birth defects, miscarriage; kidney and liver damage; skin allergy and ulcers; injury to the nasal septum; discoloration of teeth; bronchitis; lung allergy.
Storage: A storage hazard; sealed containers may burst from carbon dioxide release. Store in tightly closed containers in a dry, cool, well-ventilated place with non-wood floors. Keep away from combustible materials, alcohols and acetone. Containers may explode in fire.

★ 285 ★
CHROMIC SULFATE
CAS: 10101-53-8
Other names: CHROMIC SULPHATE; CHROMIUM III SULFATE; CHROMIUM SULFATE (2:3); CHROMIUM SULPHATE; CHROMIUM SULPHATE (2:3); C.I.77305; DICHROMIUM SULFATE; DICHROMIUM SULPHATE; DICHROMIUM TRISULFATE; DICHROMIUM TRISULPHATE; SULFURIC ACID, CHROMIUM (3) SALT (3:2)

Danger profile: Carcinogen, mutagen, exposure limit established, Community Right-to-Know List, poisonous gas produced in fire.
Uses: Making paints (green); inks; ceramics.
Appearance: Powder.
Effects of exposure:
   Breathed: May cause irritation.
   Eyes: Contact may cause irritation, redness and burns.
   Skin: Contact may cause irritation, redness and rash.
Long-term exposure: May cause cancer, birth defects, miscarriage.
Storage: Store in tightly closed containers in a dry, cool, well-ventilated place.

★ 286 ★
CHROMIUM
CAS: 7440-47-3

Danger profile: Carcinogen, mutagen, poisonous, powder is an explosion hazard, exposure limit established, Community Right-to-Know List, poisonous gas produced in fire.
Uses: Plating other metals and plastics; increasing corrosion resistance; nuclear and high-temperature research; inorganic pigments.
Appearance: Steel gray or silvery metal powder.
Effects of exposure:
Breathed: Fumes may cause a flu-like illness called "metal fume fever" with muscle aches, chills, coughing and fever. Will usually last 24 hours.
Eyes: Particles may cause irritation.
Skin: May cause irritation.
Swallowed: Poison. May cause irritation of the throat, mouth and gastrointestinal tract.
Long-term exposure: May cause throat and lung cancer; birth defects, miscarriage; lung allergy with coughing, wheezing and difficult breathing.
Storage: Keep away from flame and other sources of ignition and oxidizers, especially bromine pentafluoride, sulfur dioxide and nitrogen oxide.

★ 287 ★
CHROMIUM NITRATE
CAS: 13548-38-4
Other names: CHROMIC NITRATE; CHROMIUM (III) NITRATE; CHROMIUM (3) NITRATE; CHROMIUM TRINITRATE; NITRIC ACID, CHROMIUM (3) SALT

Danger profile: Carcinogen, mutagen, poisonous, exposure limit established, Community Right-to-Know List, poisonous gas produced in fire.
Uses: Corrosion prevention; textile printing.
Appearance: Pale green powder.
Effects of exposure:
Breathed: Carcinogen. Any exposure is considered extremely dangerous.
Eyes: Contact may cause irritation, redness and itching.
Skin: Contact may cause irritation, redness and itching.
Long-term exposure: May cause cancer; birth defects; skin allergy with rash and itching.
Storage: Store in tightly closed containers in a dry, cool, well-ventilated place. Keep away from reducing agents, fuels and ether. Containers may explode in fire.
First aid guide: Move victim to fresh air; call emergency medical care. In case of contact with material, immediately flush skin or eyes with running water for at least 15 minutes. Remove and isolate contaminated clothing and shoes at the site.

★ 288 ★
CHROMIUM OXIDE
CAS: 1308-38-9
Other names: ANADOMIS GREEN; ANIDRIDE CROMIQUE (FRENCH); CASALIS GREEN; CHROME GREEN; CHROME OXIDE; CHROMIA; CHROMIC ACID; CHROMIC OXIDE; CHROMIUM (III) OXIDE; CHROMIUM (3) OXIDE; CHROMIUM SESQUIOXIDE; CHROMIUM (3) TRIOXIDE; C.I. 77288; C.I. PIGMENT GREEN; DICHROMIUM TRIOXIDE; GREEN CHROMIC OXIDE; GREEN CINNABAR; GREEN ROUGE; LEAF GREEN; LEVANOX; GREEN GA; OIL GREEN; ULTRAMARINE GREEN

Danger profile: Carcinogen, mutagen, poisonous, oxidizer, exposure limit established, Community Right-to-Know List, poisonous gas produced in fire.
Uses: Textile dyeing; paint pigments; ceramic glazes; making other chemicals.
Appearance: Bright green powder.
Effects of exposure:
Breathed: Carcinogen. Any exposure is considered extremely dangerous. May cause irritation of the eyes, nose and throat.
Eyes: Contact may cause severe irritation, redness, and itching.
Skin: Contact may cause severe irritation, redness, itching and rash.
Swallowed: Poison. May cause irritation of the lips, mouth and throat; pain; intense thirst; throat swelling; painful cramps in the stomach; difficult breathing; severe convulsions, coma and death.
Long-term exposure: May cause cancer; birth defects, mutations; skin allergy with itching and rash.
Storage: Store in tightly closed containers in a dry, cool, well-ventilated place. Keep away from oxidizers, oxygen difluoride and glycerol.

★ 289 ★
CHROMIUM OXYCHLORIDE
CAS: 14977-61-8
Other names: CHLORURE DE CHROMYLE (FRENCH); CHROMIC OXYCHLORIDE; CHROMYL CHLORIDE; CHROMIUM CHLORIDE OXIDE; CHROMIUM, DICHLORODIOXO-; CHROMIUM DIOXYCHLORIDE; CHROMIUM DIOXYCHLORIDE DIOXIDE; CHROMIUM(IV) DIOXYCHLORIDE; CHROMYLCHLORID (GERMAN); CROOMOXYLCHLORIDE (DUTCH); CROMILE, CLORURO DI (ITALIAN); CHROMYL CHLORIDE; DIOXODICHLOROCHROMIUM; OXYCHLORURE CHROMIQUE (FRENCH)

Danger profile: Carcinogen, mutagen, poisonous, corrosive, exposure limit established, Community Right-to-Know List, poisonous gas produced in fire, violent reaction with water.
Uses: Making dyes and other chemicals; chlorinating agent.
Appearance: Dark red, fuming liquid.
Odor: Musty, burning. The odor is a warning of exposure; however, no smell does not mean you are not being exposed.
Effects of exposure:
Breathed: Poison. May cause irritation of the nose, throat, breathing passages and lungs; coughing and wheezing; nose sores, nose bleeds and nasal septum damage; nose and lung tumors may develop.
Eyes: Contact may cause severe irritation, damage and vision loss.
Skin: Passes through the unbroken skin; can increase exposure and the severity of the symptoms listed. May cause severe irritation, rash, chemical burns with deep ulcers.
Swallowed: Poison. May cause irritation of the lips, mouth and throat; pain; intense thirst; throat swelling; painful cramps in the stomach; difficult breathing; severe convulsions, coma and death.
Long-term exposure: May cause cancer of the lungs and throat; birth defects, fetus damage, possible miscarriage; skin allergy, with itching and rash; lung allergy with cough, wheezing and difficult breathing; kidney damage; damage to the nasal septum.
Storage: Store in tightly closed containers in a dry, cool, well-ventilated place with non-wood floors. Keep away from water, steam and other forms of moisture, all combustible and combustible materials; all other chemicals including acetone, ammonia, alcohol, ether, turpentine. Containers may explode in fire.

First aid guide: Move victim to fresh air and call emergency medical care; if not breathing, give artificial respiration; if breathing is difficult, give oxygen. In case of contact with material, immediately flush skin or eyes with running water for at least 15 minutes. Speed in removing material from skin is of extreme importance. Remove and isolate contaminated clothing and shoes at the site. Keep victim quiet and maintain normal body temperature.

★ 290 ★
CHROMIUM TRIOXIDE
CAS: 1333-82-0
Other names: ANHYDRIDE CHROMIQUE (FRENCH); ANIDRIDE CHROMICA (ITALIAN); CHROMIC ANHYDRIDE; CHROMIC ACID; CHROMIC(IV) ACID; CHROMIC TRIOXIDE; CHROMIUM OXIDE; CHROMIUM(IV) OXIDE; CHROMIUM(IV) OXIDE (1:3); CHROMIUM TRIOXIDE, ANHYDROUS; CHROMO (TRIOSSIDO DI) (ITALIAN); CHROMSAUREANHYDRID (GERMAN); CHROOMTRIOXYDE (DUTCH); MONOCHROMIUM OXIDE; MONOCHROMIUM TRIOXIDE; PURATRONIC CHROMIUM TRIOXIDE

Danger profile: Carcinogen, mutagen, poisonous, oxidizer, exposure limit established, Community Right-to-Know List, poisonous gas produced in fire.
Uses: Making other chemicals; chromium plating; medicine; process engraving; anodizing; ceramic glazes; colored glass; metal cleaning; inks; tanning; paints; textile mordant; etching plastics.
Appearance: Red powder.
Effects of exposure:
 Breathed: May cause irritation of the nose, throat, breathing passages and lungs; coughing and wheezing; nose sores, nose bleeds and septum damage; nose and lung tumors may develop.
 Eyes: Contact may cause severe irritation, damage and vision loss.
 Skin: Passes through the unbroken skin; can increase exposure and the severity of the symptoms listed. May cause severe irritation, rash, chemical burns with deep ulcers.
 Swallowed: Poison. May cause irritation of the lips, mouth and throat; pain; intense thirst; throat swelling; painful cramps in the stomach; difficult breathing; severe convulsions, coma and death.
Long-term exposure: May cause cancer of the lungs and throat; birth defects, fetus damage, possible miscarriage; skin allergy, with itching and rash; stomach ulcers; bronchial irritation; lung allergy with cough, wheezing and difficult breathing; liver and kidney damage; damage to the nasal septum.
Storage: Store in tightly closed containers in a dry, cool, well-ventilated place with non-wood floors. Highly reactive. Keep away from all combustible material and all other chemicals. Containers may explode in fire.
First aid guide: Move victim to fresh air; call emergency medical care. In case of contact with material, immediately flush skin or eyes with running water for at least 15 minutes. Remove and isolate contaminated clothing and shoes at the site.

★ 291 ★
CHROMOSULFURIC ACID
CAS: 64093-79-4
Other names: BASIC CHROMIC SULFATE; BASIC CHROMIC SULPHATE; BASIC CHROMIUM SULFATE; BASIC CHROMIUM SULPHATE; CHROMIUM HYDROXIDE SULFATE; CHROMIUM SULFATE; CHROMIUM SULFATE, BASIC; CHROMIUM SULPHATE; KOREON; MONOBASIC CHROMIUM SULFATE; MONOBASIC CHROMIUM SULPHATE; NEOCHROMIUM; SULFURIC ACID, CHROMIUM SALT, BASIC; SULFURIC ACID, CHROMIUM SALT

Danger profile: Carcinogen, mutagen, corrosive, poisonous, reacts with water; exposure limit established, Community Right-to-Know List, poisonous gases produced in fire.
Uses: Papermaking; printing; photography; leather tanning; making paint, ink and ceramic glazes; chrome plating; textile dyeing.
Appearance: Violet or dark green powder.
Effects of exposure:
 Breathed: May be poisonous. May cause irritation to the eyes, nose, throat and breathing passages. Higher exposures may cause shortness of breath and a dangerous build-up of fluids in the lungs (pulmonary edema), which can cause death. Pulmonary edema is a medical emergency that can be delayed one to two days following exposure.
 Eyes: Contact may cause irritation, severe burns and permanent damage.
 Skin: Contact may cause irritation and severe burns. Skin allergy with itching, redness and rash may develop.
 Swallowed: May be poisonous. May cause irritation of the lips, mouth and throat; pain; intense thirst; throat swelling; painful cramps in the stomach; difficult breathing; severe convulsions, coma and death.
Long-term exposure: May cause cancer, birth defects, miscarriage; skin allergy with itching and rash; possible lung damage.
Storage: Store in tightly closed containers in a dry, cool, well-ventilated place with non-wood floors. Keep away from water and combustible materials.
First aid guide: Move victim to fresh air and call emergency medical care; if not breathing, give artificial respiration; if breathing is difficult, give oxygen. In case of contact with material, immediately flush skin or eyes with running water for at least 15 minutes. Speed in removing material from skin is of extreme importance. Remove and isolate contaminated clothing and shoes at the site. Keep victim quiet and maintain normal body temperature.

★ 292 ★
CHRYSENE
CAS: 218-01-9
Other names: 1,2-BENZOPHENANTHRENE; BENZO-(A)PHENANTHRENE; 1,2-BENZPHENANTHRENE; BENZ-(A)PHENANTHRENE; 1,2,5,6-DIBENZONAPHTHALENE

Danger profile: Carcinogen, mutagen, exposure limit established, acrid smoke and fumes produced in fire.
Uses: Almost never found by itself. Found in asphalt, tars, coal tar pitch, creosote; organic synthesis.
Appearance: Colorless crystals.
Effects of exposure:
 Breathed: Any exposure is considered extremely dangerous. See CREOSOTE entry.
 Skin: Passes through the unbroken skin; can increase exposure and the severity of the symptoms listed. Contact may cause irritation and a rash. Skin pigment change (similar to sunburn) may occur if skin is exposed to sunlight.
Long-term exposure: May cause cancer; birth defects, miscarriage; changes in skin pigment.
Storage: Store in tightly closed containers in a dry, cool, well-ventilated place.

★ 293 ★
CISPLATIN
CAS: 15663-27-1
Other names: PLATINUM(II),DIAMMINEDICHLORO-,CIS-; CACP; CDDP; CPDC; DDP; CIS-DDP; CIS-DIAMINEDICHLOROPLATINUM; CIS-DIAMMINEDICHLOROPLATINUM; NEOPLATIN; NSC-119875; PEYRONE'S HLORIDE; PLATINOL; CIS-PLATINOUS DIAMMINE DICHLORIDE; CIS-PLATINUM; CIS-PLATINUM(II); CIS-PLATINUM(II) DIAMINE DICHLORIDE; CIS-PT(II)

Danger profile: Carcinogen, mutagen, exposure limit established, poisonous gases produced in fire.
Uses: A drug to treat cancer.
Appearance: White powder.
Effects of exposure:
Breathed: High levels of exposure may cause nausea and vomiting.
Long-term exposure: May cause cancer; damage to the developing fetus; damage to the male reproductive glands (testes); kidney damage; ringing in the ears and hearing impairment; effect on blood cells.
Storage: Store in tightly closed containers in a dry, cool, well-ventilated place. Keep away from aluminum.

★ 294 ★
CLOPIDOL
CAS: 2971-90-6
Other names: 3,5-DICHLORO-2,6-DIMETHYL-4-PYRIDINOL; 3,5-DICHLORO-4- PYRIDNOL; COCCIDIOSTAT C; COYDEN; FARMCOCCID; LERBEK; METHYLCHLOROPINDOL; METHYLCHLORPINDOL; METILCHLORPINDOL

Danger profile: Exposure limit established, dust may explode, very poisonous gases produced in fire.
Uses: Animal antibiotic.
Appearance: White to light brown powder.
Effects of exposure:
Breathed: May cause irritation of the eyes, nose and throat.
Eyes: May cause irritation.
Skin: May cause irritation.
Long-term exposure: Unknown at this time.
Storage: Store in tightly closed containers in a dry, cool, well-ventilated place.

★ 295 ★
COBALT
CAS: 7440-48-4
Other names: AQUACAT; C.I. 77320; COBALT-59; KOBALT (GERMAN, POLISH); SUPER COBALT; NCI-C60311

Danger profile: Carcinogen, fire hazard, combustible, exposure limit established, Community Right-to-Know List, may be radioactive.
Uses: In steel alloys for magnets and high-speed tool steels; jet engines; carbide abrasive; making other chemicals; electroplating; ceramics; lamp filaments; in fertilizers; glass; printing inks, paints and varnishes; colors.
Appearance: Metallic silver-gray, hard magnetic metal.
Effects of exposure:
Breathed: May cause irritation of the eyes, nose, throat, breathing passages and lungs. Higher exposures may cause shortness of breath and a dangerous build-up of fluids in the lungs (pulmonary edema), which can cause death. Pulmonary edema is a medical emergency that can be delayed one to two days following exposure.
Eyes: May cause irritation, redness and itching.
Skin: May cause irritation, redness and dermatitis.

Swallowed: Poison. May cause irritation, nausea and vomiting.
Long-term exposure: May cause cancer; heart damage; a severe allergic reaction of the lungs with coughing, wheezing and difficult breathing; repeated exposure may cause lung scarring which may show no symptoms and may be fatal. Skin allergy may develop.
Storage: Store in tightly closed containers in a dry, cool, well-ventilated place. Keep away from heat, flame and other sources of ignition, oxidizers and acids. Powder spontaneously ignites in air.

★ 296 ★
COBALT NAPHTHENATE
CAS: 61789-51-3
Other names: COBALT NAPHTHENATE POWDER; COBALTOUS NAPHTHENATE; NAPHTENATE DE COBALT (FRENCH); NAPHTHENIC ACID, COBALT SALT; NAPHTHENATE DE COBALT (FRENCH)

Danger profile: Exposure limit established, combustible, poisonous gases produced in fire.
Uses: Paint varnish; bonding rubber to various materials including steel; dryer in ink.
Appearance: Brown powder or bluish-red solid.
Effects of exposure:
Breathed: May cause irritation of the eyes, nose, throat, breathing passages and lungs. Higher exposures may cause shortness of breath and a dangerous build-up of fluids in the lungs (pulmonary edema), which can cause death. Pulmonary edema is a medical emergency that can be delayed one to two days following exposure.
Eyes: May cause irritation, redness and itching.
Skin: May cause irritation, redness and dermatitis.
Swallowed: Poison. May cause irritation, nausea and vomiting.
Long-term exposure: May cause heart damage; enlarged tyroid (goiter); a severe allergic reaction of the lungs with coughing, wheezing and difficult breathing; repeated exposure may cause loss of the sense of smell; lung scarring which may show no symptoms and may be fatal. Skin allergy may develop. Similar solvents can cause nerve and brain damage with loss of memory and personality change, sleep disturbances.
Storage: Store in tightly closed containers in a dry, cool, well-ventilated place. Keep away from heat, flame and other sources of ignition and oxidizers.
First aid guide: Move victim to fresh air; call emergency medical care. In case of contact with material, immediately flush skin or eyes with running water for at least 15 minutes. Removal of solidified molten material from skin requires medical assistance. Remove and isolate contaminated clothing and shoes at the site.

★ 297 ★
COPPER
CAS: 7440-50-8
Other names: ARWOOD COPPER; BRONZE POWDER; ANAC 110; CDA 101; CDA 102; CDA 110; CDA 122; C.I. 77400; COPPER AIRBORNE; COPPER BRONZE; COPPERMILLED; 1721 GOLD; GOLD BRONZE; KAFAR COPPER; MI (COPPER); M2 COPPER; RANEY COPPER

Danger profile: Possible reproductive hazard, exposure limit established, toxic fumes produced in fire.
Uses: Making brass; widely used in may industries and applications including construction, plumbing, heating, electrical; welding.
Appearance: Reddish-brown metal.

Effects of exposure:
Breathed: Dust and fumes may cause irritation of the eyes, nose, throat; coughing wheezing and nose bleeding. May cause an influenza-like illness called "copper fume fever"; metallic taste in the mouth, fever, cough, aching muscles, dry throat and difficult breathing. Symptoms of fever may be delayed and last from 24-48 hours.
Eyes: Dust or other particles may cause severe irritation and permanent damage including blindness.
Long-term exposure: May cause liver damage; chronic nose irritation with sores and ulcers; discoloration of the skin and hair; skin allergy may develop.
Storage: Store in tightly closed containers in a dry, cool, well-ventilated place. Keep away from oxidizers, chemically active metals and acetylene gas. Liquid explodes on contact with water.

★ 298 ★
COPPER ACETOARSENITE
CAS: 12002-03-8
Other names: (ACETATO)TRIMETAARSENITADICOPPER; (ACETATO)(TRIMETAARSENITO)COPPER; ACETOARSENITE DE CUIVRE (FRENCH); BASLE GREEN; COPPER(II) ACETOARSENITE; COPPER ACETATE ARSENITE; CI 77410; C.I. PIGMENT GREEN 21; COPPER ACETOARSENITE; CUPRIC ACETOARSENITE; EMERALD GREEN; ENT 884; FRENCH GREEN; GENUINE RARIS GREEN; IMPERIAL GREEN; KINGS GREEN; MINERAL GREEN; MITIS GREEN; MOSS GREEN; MOUNTAIN GREEN; NEW GREEN; PARIS GREEN; PARROT GREEN; SCHWEINFURTGRUN; SCHWEINFURT GREEN; SWEDISH GREEN; VIENNA GREEN

Danger profile: Carcinogen, poisonous, exposure limit established, Community Right-to-Know List, poisonous gases produced in fire.
Uses: Pigment; wood preservative; insecticide and larvicide; marine anti-fouling paints.
Appearance: Emerald-green powder.
Effects of exposure:
Breathed: May cause irritation of the nose, throat and air passages.
Eyes: Dust may cause irritation, redness and inflammation of the eye lids.
Skin: May cause itching, burning feeling, redness and rash.
Swallowed: Poison. May cause gastric disturbance, tremors, muscular cramps, and nervous collapse which may lead to death.
Long-term exposure: May cause lung cancer; sores and bleeding of the nose, and a hole in the nasal septum (hole in the nose) may develop; skin thickening and skin pigment changes; nerve damage.
Storage: Store in tightly closed containers in a dry, cool, well-ventilated place. Keep away from acid, chemically active metals and acetylene gas.
First aid guide: Move victim to fresh air; call emergency medical care. In case of contact with material, immediately flush skin or eyes with running water for at least 15 minutes. Remove and isolate contaminated clothing and shoes at the site.

★ 299 ★
COPPER ARSENITE
CAS: 10290-12-7
Other names: ACID COPPER ARSENITE; AIR-FLO GREEN; ARSONIC ACID, COPPER(2)SALT (1:1) (9CI); COPPER ORTHOARSENITE; CUPRIC ARSENITE; CUPRIC GREEN; SCHEELES GREEN; SCHEELE'S MINERAL; SWEDISH GREEN

Danger profile: Carcinogen, combustible, exposure limit established, reproductive hazard, Community Right-to-Know List, poisonous gas produced in fire.
Uses: Insecticide; fungicide; pigment; wood preservative; rodenticide.
Appearance: Yellow-green powder.
Breathed: Poison. Vapor may cause wheezing, foaming at the mouth, headache, convulsions and death.
Eyes: May cause irritation, redness and inflammation of the eyelids. Passes through the mucous membrane of the eye increasing severity and symptoms listed in *Breathed* and *Swallowed.*
Skin: Poisonous. Passes through the unbroken skin; can increase exposure and the severity of the symptoms listed. Behaves as a cyanide when absorbed through the skin. May cause irritation and inflammation.
Swallowed: Poison. May cause metallic taste, disturbed and rapid respiration and circulation, mental confusion, increased heart beat; gastric disturbance, tremors, irritation of the stomach and intestines, stomach pain, nausea and vomiting, diarrhea, weakness, convulsions, coma and death.
Long-term exposure: May cause liver damage, loss of appetite; hearing loss.
Storage: Store in tightly closed containers in a dry, cool, well ventilated place.
First aid guide: Move victim to fresh air; call emergency medical care. In case of contact with material, immediately flush skin or eyes with running water for at least 15 minutes. Remove and isolate contaminated clothing and shoes at the site.

★ 300 ★
COPPER CYANIDE
CAS: 14763-77-0
Other names: COPPER(II) CYANIDE; COPPER CYNANAMIDE; CUPRIC CYANIDE; CYANURE DE CUIVRE (FRENCH)

Danger profile: Poisonous, exposure limit established, Community Right-to-Know List, poisonous gases produced in fire.
Uses: Making other chemicals; electroplating; insecticide.
Appearance: Yellowish-green powder.
Effects of exposure:
Breathed: May be poisonous. May cause irritation of the nose, nose bleeds with crusting and sores.
Eyes: Contact may cause severe burns and permanent damage or loss of vision.
Skin: Contact may cause irritation, redness and burns.
Swallowed: Poison. May cause gastric disturbance, tremors, muscular cramps, and nervous collapse which may lead to death.
Long-term exposure: May cause a metallic taste in the mouth; hair to turn green; liver and lung damage; skin allergy; damage to the inner lining of the nose.
Storage: Store in tightly closed containers in a dry, cool, well-ventilated place. Keep away from chemically active metals (magnesium, potassium, sodium, zinc) and acetylene gas.
First aid guide: Move victim to fresh air; call emergency medical care. In case of contact with material, immediately flush skin or eyes with running water for at least 15 minutes. Remove and isolate contaminated clothing and shoes at the site.

★301★
COUMAPHOS
CAS: 56-72-4
Other names: 3-CHLORO-7-HYDROXY-4-METHYL-COUMARIN O,O-DIETHYL PHOSPHOROTHIOATE; 3-CHLORO-7-HYDROXY-4-METHYL-COUMARIN O-ESTER WITH O,O-DIETHYL PHOSPHOROTHIOATE; 3-CHLORO-4-METHYL-7-COUMARINYLDIETHYL PHOSPHOROTHIOATE; O-3-CHLORO-4-METHYL-7-COUMARINYLO,O-DIETHYL PHOSPHOROTHIOATE; 3-CHLORO-4-METHYL-7-HYDROXYCOUMARINDIETHYL THIOPHOSPHORIC ACID ESTER; 3-CHLORO-4-METHYLUMBELLIFERONEO-ESTER WITH O,O-DIETHYL PHOSPHOROTHIOATE; AGRIDIP; AS-UNTHOL; ASUNTOL; AZUNTHOL; BAYER 21/199; BAYMIX; BAYMIX 50; CO-RAL; COUMAFOS; CUMAFOS (DUTCH); O,O-DIAETHYL-O-(3-CHLOR-4-METHYL-CU MARIN-7-YL)-MONOTHIOPHOSPHAT (GERMAN); O,O-DIETHYL-O-(3-CHLOOR-4-METHYL-CUMARIN-7-YL)MONOTHIOFOSFAAT (DUTCH); O,O-DIETHYLO-(3-CHLORO-4-METHYL-7-COUMARINYL) PHOSPHOROTHIOATE; O,O-DIETHYLO-(3-CHLORO-4-METHYLCOUMARINYL-7)THIOPHOSPHATE; O,O-DIETHYLO-(3-CHLORO-4-METHYL-2-OXO-2H-BENZOPYRAN-7-YL)PHOSPHOR OTHIOATE; O,O-DIETHYL3-CHLORO-4-METHYL-7-UMBELLIFERONE THIO-PHOSPHATE; O,O-DIETHYLO-(3-CHLORO-4-METHYLUMBELLIFERYL)PHOSPHOROTHIOATE; DIETH-YL3-CHLORO-4-METHYLUMBELLIFERYL THIONOPHOS-PHATE; DIETHYLTHIOPHOSPHORIC ACID ESTER OF 3-CHLORO-4-METHYL-7-HYDROXYCOUMARIN; O,O-DIETIL-O-(3-CLORO-4-METIL-CUMARIN-7-IL-MONOTIOFOSFATO) (ITALIAN); ENT 17,957; MELDANE; MELDONE; MUSCATOX; NCI-C08662; PHOSPHOROTHIOIC ACID, O,O-DIETHYL ESTER, O-ESTER WITH 3-CHLORO-7-HYDROXY-4-METHYLCOUMARIN; RESITOX; SUNTOL; THIOPHOSPHATE DE O,O-DIETHYLE ET DE O-(3-CHLORO-4-METHYL-7-COUMARINYLE) (FRENCH); UMBETHION; 3-CHLORO-7-HYDROXY-4-METHYL-COUMARIN-O,O-DIETHYLPHOSPHOROTHIONAT E

Danger profile: Mutagen, deadly poison, EPA extremely hazardous substance list; poisonous gases produced in fire.
Uses: Insecticide; animal medicines; control of livestock insects; anthelmintic.
Appearance: White to slightly brownish sugar/sand-like crystals.
Odor: Sulfur-like. The odor is a warning of exposure; however, no smell does not mean you are not being exposed.
Effects of exposure:
Breathed: Extremely poisonous. May cause headache, dizziness, profuse sweating, nausea, vomiting, reduced heart beat, stomach cramps, diarrhea, loss of coordination, slow and weak breathing, fever, loss of consciousness and death. May cause shortness of breath and a dangerous build-up of fluids in the lungs (pulmonary edema), which can cause death. Pulmonary edema is a medical emergency that can be delayed one to two days following exposure.
Eyes: Poisoning can happen very rapidly from absorption. May cause irritation.
Skin: Poisonous; contact may cause fatal poisoning. Passes through the unbroken skin; can increase exposure and the severity of the symptoms listed.
Swallowed: Powerful, deadly poison. See *Breathed.*
Long-term exposure: May damage the developing fetus or cause birth defects; nerve damage including loss of coordination; liver damage.
Storage: Store in a regulated, specially marked place in tightly-closed containers in a dry, well-ventilated place. Keep away from piperonyl butoxide. Containers may explode in fire.
First aid guide: Move victim to fresh air and call emergency medical care; if not breathing, give artificial respiration; if breathing is difficult, give oxygen. In case of contact with material, immediately flush skin or eyes with running water for at least 15 minutes. Speed in removing material from skin is of extreme importance. Remove and isolate contaminated clothing and shoes at the site. Keep victim quiet and maintain normal body temperature. Effects may be delayed; keep victim under observation.

★302★
CREOSOTE
CAS: 8001-58-9
Other names: BRICK OIL; AWPA NO. 1; COAL TAR OIL; COAL TAR CREOSOTE; CREOSOTE, COAL TAR; CREO-SOTE, FROM COAL TAR; CREOSOTE OIL; CREOSOTE P1; CREOSOTUM; CRESYLIC CREOSOTE; DEAD OIL; HEAVY OIL; LIQUID PITCH OIL; NAPHTHALENE OIL; PRESERV-O-SOTE; TAR OIL; WASH OIL

Danger profile: Carcinogen, mutagen, poisonous gas produced in fire.
Uses: Wood preservative; lubricant for dye molds; roofing; animal dips.
Appearance: Heavy, oily yellow, brownish or black liquid.
Odor: Smoky, tarry odor. The odor is a warning of exposure; however, no smell does not mean you are not being exposed.
Effects of exposure:
Breathed: Vapor may cause moderate irritation to nose, throat, mouth and lungs; mucous membrane. Symptoms of high exposure include watering of the eyes, coughing, difficult breathing, headache, nausea, vomiting, muscular weakness, unconsciousness. Very high exposure can lead to drowsiness, coma and death; buildup of fluid in the lungs (pulmonary edema), and may cause death. Pulmonary edema may be delayed by one to two days following exposure.
Eyes: Contact may cause, burning and stinging, tearing, irritation and redness and swelling of the eyelids, intense pain; can cause severe burns and lead to blurred vision, chronic irritation and other permanent damage.
Skin: Contact may cause burning or stinging feeling, redness, swelling, intense pain, blisters, irritation, rash and severe burns; rapid and severe ill feeling, vomiting, convulsions, body organ damage and death.
Swallowed: May cause burning feeling of the mouth and throat, intense thirst, throat swelling, salivation, respiratory difficulties, vertigo, headache, loss of pupillary reflexes, stomach cramps, nausea, vomiting, diarrhea, severe breathing difficulties, cyanosis, mild convulsions and unconsciousness. Breathing may stop. Watch for delayed symptoms.
Long-term exposure: May cause cancer. Repeated exposure may lead to skin allergy, itching, rash and chronic irritation; may cause skin pigment color change to yellow or bronze.
Storage: Store in tightly closed containers in a regulated, dry, cool, well-ventilated place. Keep away from oxidizers. Containers may explode in fire.
First aid guide: Move victim to fresh air and call emergency medical care; if not breathing, give artificial respiration; if breathing is difficult, give oxygen. In case of contact with material, immediately flush eyes with running water for at least 15 minutes. Wash skin with soap and water. Remove and isolate contaminated clothing and shoes at the site.

★ 303 ★
CRESOL
CAS: 1319-77-3
Other names: ACEDE CRESYLIQUE (FRENCH); BACILLOL; CRESOLI (ITALIAN); CRESYLIC ACID; HYDROXYTOLUOLE (GERMAN); KRESOLE (GERMAN); KRESOLEN (DUTCH); KREZOL (POLISH); PHENOL, METHYL- (9CI); METHYLP-HENOL; 2-METHYL PHENOL; 3-METHYL PHENOL; 4-METHYL PHENOL; TEKRESOL; AR-TOLUENOL; TRICRE-SOL

Danger profile: Corrosive, poisonous, combustible, exposure limit established, Community Right-to-Know List, poisonous gas produced in fire.
Uses: Disinfectant and fumigant; making other chemicals, explosives, photographic developers, synthetic resins and herbicides; textile scouring agent; synthetic food flavors.
Appearance: Colorless solid or yellowish liquid.
Odor: Sweet, tarry. The odor is a warning of exposure; however, no smell does not mean you are not being exposed.
Effects of exposure:
Breathed: May be poisonous. Vapor may cause intense irritation to nose, throat, mouth and lungs; mucous membrane. Symptoms of exposure include watering of the eyes, coughing, difficult breathing, headache, nausea, vomiting, muscular weakness, unconsciousness; coma and death.
Eyes: Severe irritant. Contact may cause, burning and stinging, tearing, irritation and redness and swelling of the eyelids, intense pain; can burn the eyes and lead to blurred vision, chronic irritation and other permanent damage.
Skin: May be poisonous if absorbed. Severe irritant. Rapidly passes through the unbroken skin; can increase exposure and the severity of the symptoms listed. Contact may cause burning or stinging feeling, redness, swelling, intense pain, blisters, irritation, burns; nausea, collapse, coma and death.
Swallowed: May be poisonous. May cause burning feeling of the mouth and throat, intense thirst, throat swelling, stomach cramps, nausea, vomiting, diarrhea, severe breathing difficulties and unconsciousness. Tissue damage may result. Breathing may stop. Watch for delayed symptoms.
Long-term exposure: Repeated exposure may lead to skin allergy and chronic irritation; dermatitis; liver, kidney and nervous system damage; poor appetite; personality change.
Storage: Store in tightly closed containers in a dry, cool, well-ventilated place. Keep away from oxidizers, oleum, nitric acid and chlorosulfonic acid. Containers may explode in fire.
First aid guide: Move victim to fresh air and call emergency medical care; if not breathing, give artificial respiration; if breathing is difficult, give oxygen. In case of contact with material, immediately flush skin or eyes with running water for at least 15 minutes. Speed in removing material from skin is of extreme importance. Remove and isolate contaminated clothing and shoes at the site. Keep victim quiet and maintain normal body temperature. Effects may be delayed; keep victim under observation.

★ 304 ★
CRIMIDINE
CAS: 535-89-7
Other names: 2-CHLOOR-4-DIMETHYLAMINO-6-METHYL-PYRIMIDINE (DUTCH); 2-CHLORO-4-METHYL-6-DIMETHYLAMINOPYRIMIDINE; 2-CLORO-4-DIMETILAMINO-6-METIL-PIRIMIDINA (ITALIAN); CRIMIDIN (GERMAN); CRIMIDINA (ITALIAN); CASTRIX; PYRIMIDINE,2-CHLORO-4-(DIMETHYLAMINO)-6-METHYL-; W 491

Danger profile: Deadly poison, EPA extremely hazardous substances list, poisonous gases produced in fire.
Uses: Rodent poison; pesticide.
Appearance: Brown waxy solid; colorless crystals.

Effects of exposure:
Breathed: Poison. Even extremely small exposures may cause restlessness, violent seizures and fits. The seizures may occur without warning and may be triggered by noise or sudden movement of the victim.
Swallowed: Deadly poison. See symptoms under *Breathed.*
Long-term exposure: May cause damage to the central nervous system.
Storage: Store in tightly closed containers in a cool, well-ventilated place. Keep away from acids and acid fumes.

★ 305 ★
CROTONALDEHYDE
CAS: 4170-30-3
Other names: 2-BUTENAL; BETA-METHYLACROLEIN; CROTONIC ALDEHYDE; KROTONALDEHYD (CZECH)

Danger profile: Suspected carcinogen, mutagen, poisonous, combustible, explosion hazard, extremely hazardous substance list, poisonous gas produced in fire.
Uses: Making other chemicals; solvent; insecticides; tear gas; in gas fuels; tanning leather.
Appearance: Colorless to straw-colored liquid.
Odor: Strong, suffocating. The odor is a warning of exposure; however, no smell does not mean you are not being exposed.
Effects of exposure:
Breathed: Poison. May cause irritation of the eyes, nose, throat, breathing passages and lungs; cough, difficult breathing, shortness of breath. Higher exposures may cause a dangerous build-up of fluids in the lungs (pulmonary edema), which can cause death. Pulmonary edema is a medical emergency that can be delayed one to two days following exposure.
Eyes: Contact may cause irritation, extreme watering, inflammation of the eyelids and burns.
Skin: Contact may cause irritation, and may cause allergy to develop.
Swallowed: Poison.
Long-term exposure: May cause cancer or reproductive damage; skin allergy; asthma-like allergy. May reduce fertility in males.
Storage: Store in tightly closed containers in a dry, cool, well-ventilated place. Keep away from heat and sources of ignition, caustics, amines and ammonia, oxidizing materials, nitric acid, mineral acids and 1,3-butadiene. Vapors may travel to sources of ignition and flash back. Containers may explode in fire.
First aid guide: Move victim to fresh air and call emergency medical care; if not breathing, give artificial respiration; if breathing is difficult, give oxygen. In case of contact with material, immediately flush skin or eyes with running water for at least 15 minutes. Remove and isolate contaminated clothing and shoes at the site. Keep victim quiet and maintain normal body temperature. Effects may be delayed; keep victim under observation.

★ 306 ★
(E) CROTONALDEHYDE
CAS: 123-73-9
Other names: ALDEHYDE CROTONIQUE (FRENCH); TRANS-2-BUTENAL; (E)-2-BUTENAL; CROTONAL; CRO-TONALDEHYDE; CROTONALDEHYDE (E)-; CROTONIC AL-DEHYDE; ETHYLENE DIPROPIONATE (8CI); NCI-C56279; PROPYLENE ALDEHYDE; TOPENEL

Danger profile: Mutagen, poisonous, combustible, explosion hazard, extremely hazardous substance list, exposure limit established, poisonous gas produced in fire.
Uses: Making other chemicals; insecticides, solvents, fuels.
Appearance: Colorless to straw-colored liquid.

Odor: Pungent, suffocating. The odor is a warning of exposure; however, no smell does not mean you are not being exposed.

Effects of exposure:

Breathed: May cause irritation of the eyes, nose, throat, breathing passages and lungs; cough, difficult breathing, shortness of breath. Higher exposures may cause a dangerous build-up of fluids in the lungs (pulmonary edema), which can cause death. Pulmonary edema is a medical emergency that can be delayed one to two days following exposure.

Eyes: Contact may cause irritation, extreme watering and burns.

Skin: Contact may cause irritation; allergy may develop.

Swallowed: Poison. See *Breathed.*

Long-term exposure: May cause cancer or reproductive damage; skin allergy; asthma-like allergy. May reduce fertility in males.

Storage: Store in tightly closed containers in a dry, cool, well-ventilated place. Keep away from heat and sources of ignition, caustics, amines and ammonia, oxidizing materials, mineral acids and 1,3-butadiene. Vapors may travel to sources of ignition and flash back. Containers may explode in fire.

★307★
CRUFOMATE

CAS: 299-86-5

Other names: 4-T-BUTYL-2-CHLOROPHENYL METHYL METHYLPHOSPHORAMIDATE; DOWCO 132; ENT 25,602-X; O-METHYL O-2-CHLORO-4-TERT-BUTYLPHENYL N-METHYLAMIDOPHOSPHATE; METHYLPHOSPHORAMIDIC ACID,4-T-BUTYL-2-CHLOROPHENYL METHYL ESTER; MONTREL; PHENOL,4-T-BUTYL-2-CHLORO-, ESTER WITH METHYL METHYLPHOSPHORAMIDATE; PHOS-PHORAMIDIC ACID, 4-TERT-BUTYL-2-CHLOROPHENYLPHOSPHOR AMIDATE; PHOS-PHORAMIDIC ACID, METHYL-,2-CHLORO-4-(1,1-DIMETHYLETHYL)PHENYL METHYL ESTER; RUELENE; RUELENE DRENCH; RULENE; O-(4-TERTBUTYL-2-CHLOOR-FENYL)-O-METHYL-FOSF ORZUUR-N-METHYL-AMIDE (DUTCH); 4-TERT BUTYL 2-CHLOROPHENYL METHYLPHOSPHORAMIDATE DE METHYLE (FRENCH); O-(4-TERT-BUTYL-2-CHLOR-PHENYL)-O-METHYL-PHOSPHORSAEURE-N-METHYLA MID(GERMAN); O-(4-TERZ.-BUTIL-2-CLORO-FENIL)-O-METIL-FOSFORAMMIDE (ITALIAN); PHOSPHORAMIDIC ACID, METHYL-,4-TERT-BUTYL-2-CHLOROPHENYL

Danger profile: Possible reproductive hazard, poisonous, exposure limit established, poisonous gases produced in fire.

Uses: Insecticides, pesticides.

Appearance: Colorless powder. Commercial product is a yellow oil.

Effects of exposure:

Breathed: Extremely poisonous. May cause headache, dizziness, profuse sweating, nausea, vomiting, reduced heart beat, stomach cramps, diarrhea, loss of coordination, slow and weak breathing, fever, loss of consciousness and death. May cause shortness of breath and a dangerous build-up of fluids in the lungs (pulmonary edema), which can cause death. Pulmonary edema is a medical emergency that can be delayed one to two days following exposure.

Eyes: Rapidly absorbed. Poisoning can happen quickly. May cause irritation. See *Breathed.*

Skin: Poisonous. Contact may cause fatal poisoning. Passes through the unbroken skin; can increase exposure and the severity of the symptoms listed.

Swallowed: Powerful, deadly poison. See *Breathed.*

Long-term exposure: May damage the developing fetus or cause birth defects; nerve damage including loss of coordination; liver damage; personality change.

Storage: Store in a regulated, specially marked place in tightly closed containers in a dry, well-ventilated place. Keep away from oxidizers and heat. Containers may explode in fire.

★308★
CUMENE

CAS: 98-82-8

Other names: BENZENE, (1-METHYLETHYL-)-; CUMEEN (DUTCH); CUMOL; 2-FENILPROPANO (ITALIAN); 2-FENYL-PROPAN (DUTCH); ISOPROPYLBENZEEN (DUTCH); ISO-PROPILBENZENE (ITALIAN); ISOPROPYLBENZENE; ISO-PROPYL BENZENE; ISOPROPYLBENZOL; ISOPROPYL-BENZOL (GERMAN); 2-PHENYLPROPANE; (1-METHYLETHYL) BENZENE

Danger profile: Combustible, narcotic, exposure limit established, Community Right-to-Know List, poisonous gas produced in fire.

Uses: Making other chemicals; industrial solvent. Found in petroleum.

Appearance: Colorless liquid.

Odor: Sharp, penetrating; slightly irritating; fragrant; aromatic. The odor is a warning of exposure; however, no smell does not mean you are not being exposed.

Effects of exposure:

Breathed: May cause irritation of the eyes, nose and throat; dizziness, lightheadedness, rapid breathing, mental confusion, unsteady walk, headache, fatigue, nausea, vomiting, unconsciousness, coma and death.

Eyes: May cause burning feeling, tearing, irritation, redness and inflammation of the eyelids.

Skin: Passes through the unbroken skin; can increase exposure and the severity of the symptoms listed. May cause irritation of the skin, dryness, redness and rash.

Swallowed: May cause indigestion, dizziness, fatigue, unconsciousness.

Long-term exposure: May cause drying and cracking of the skin; lung, kidney and liver damage; nerve damage; personality change, sleep disturbance, memory loss. Repeated or high exposure may cause damage to the bone marrow and cause a reduction of red and white blood cells, and platelets. This could result in anemia, reduced resistance to infection and damage to the blood clotting mechanism.

Storage: Store in tightly closed containers in a dry, cool, well-ventilated place. Keep away from oxidizers nitric acid, oleum, chlorosulfonic acid and heat. Vapors may travel to sources of ignition and flash back. Containers may explode in fire.

First aid guide: Move victim to fresh air and call emergency medical care; if not breathing, give artificial respiration; if breathing is difficult, give oxygen. In case of contact with material, immediately flush skin or eyes with running water for at least 15 minutes. Remove and isolate contaminated clothing and shoes at the site. Keep victim quiet and maintain normal body temperature. Effects may be delayed; keep victim under observation.

★ 309 ★
CUMENE HYDROPEROXIDE
CAS: 80-15-9
Other names: CUMEENHYDROPEROXYDE (DUTCH); CUMENT HYDROPEROXIDE; CUMENYL HYDROPEROXIDE; CUMOLHYDROPEROXID (GERMAN); CUMYL HYDROPEROXIDE; ALPHA-CUMYLHYDROPEROXIDE; ALPHA,ALPHA-DIMETHYLBENZYL HYDROPEROXIDE; HYDROPEROXYDE DE CUMENE (FRENCH); HYDROPEROXYDE DE CUMYLE (FRENCH); HYPERIZ; IDROPEROSSIDO DI CUMENE (ITALIAN); IDROPEROSSIDO DI CUMOLO (ITALIAN); ISOPROPYLBENZENE HYDROPEROXIDE; ALPHA, ALPHA-DIMETHYLBENZYLHYDROPEROXIDE; HYDROPEROXIDE, 1-METHYL-1-PHENYLETHYL-; TRIGOROX K 80

Danger profile: Mutagen, poisonous, combustible, explosion hazard, highly reactive, Community Right-to-Know List, poisonous gases produced in fire.
Uses: Making other chemicals.
Appearance: Colorless to light yellow liquid.
Odor: Sharp, irritant, aromatic. The odor is a warning of exposure; however, no smell does not mean you are not being exposed.
Effects of exposure:
Breathed: Extremely irritating. Vapor may cause irritation of the nose, mouth, throat, breathing passages and lungs; nosebleed, headache, sore and burning throat; cough, phlegm; dizziness, unconsciousness. Higher exposures may cause shortness of breath and a dangerous build-up of fluids in the lungs (pulmonary edema), which can cause death. Pulmonary edema is a medical emergency that can be delayed one to two days following exposure.
Eyes: May cause liquid causes severe irritation, burns and permanent damage.
Skin: Passes through the unbroken skin; can increase exposure and the severity of the symptoms listed. May cause burning, throbbing sensation, irritation, blisters.
Swallowed: Poison. May cause irritation of mouth and stomach.
Long-term exposure: May cause mutations; damage to the lungs, kidneys and liver; skin allergy and rash.
Storage: Store in tightly closed containers in a dark, dry, cool, well-ventilated place. Keep away from heat, sources of ignition, oxidizers, acids, reducing agents, organic material mineral acids, copper and copper alloys, lead and lead alloys, cobalt, charcoal. Containers may explode in fire.
First aid guide: Move victim to fresh air; call emergency medical care. In case of contact with material, immediately flush eyes with running water for at least 15 minutes. Wash skin with soap and water. Remove and isolate contaminated clothing and shoes at the site. Keep victim quiet and maintain normal body temperature.

★ 310 ★
CUPRIC ACETATE
CAS: 142-71-2
Other names: ACETATE DE CUIVRE (FRENCH); ACETIC ACID, COPPER (2); ACETIC ACID, CUPRIC SALT; COPPER(2) ACETATE; COPPER(II) ACETATE; COPPER ACETATE; COPPER DIACETATE; COPPER(2) DIACETATE; CRYSTALLIZED VERDIGRIS; CRYSTALS OF VENUS; CUPRIC DIACETATE; NEUTRAL VERDIGRIS; OCTAN MEDNATY (CZECH)

Danger profile: Poisonous, exposure limit established, Community Right-to-Know List, poisonous gas and smoke produced in fire.

Uses: Making other chemicals; fungicide; in ceramics; textile dyeing.
Appearance: Greenish-blue powder or small crystals.
Effects of exposure:
Breathed: May cause irritation of the throat, air passages and lungs. Higher exposures may cause shortness of breath and a dangerous build-up of fluids in the lungs (pulmonary edema), which can cause death. Pulmonary edema is a medical emergency that can be delayed one to two days following exposure.
Eyes: Contact with solutions may cause irritation; contact with solid may cause severe eye surface burns.
Skin: Contact may cause irritation and burns.
Swallowed: Large amounts may cause violent vomiting and purging, intense pain, collapse, coma, convulsions and paralysis.
Long-term exposure: May cause hair and skin to turn a greenish color; thickening skin; nose sores, bleeding and shrinking of the inner lining of the nose.
Storage: Store in tightly closed containers in a dry, cool, well-ventilated place. Keep away from chemically active metals and acetylene gas.

★ 311 ★
CUPRIC NITRATE
CAS: 3251-23-8
Other names: COPPER DINITRATE; COPPER (2) NITRATE; COPPER(II) NITRATE; NITRIC ACID, COPPER (2) SALT; CUPRIC DINITRATE

Danger profile: Exposure limit established, Community Right-to-Know List, poisonous gas and smoke produced in fire.
Uses: Insecticide; wood preservative; in paint and varnishes.
Appearance: Blue or blue-green crystals.
Effects of exposure:
Breathed: May cause irritation of the throat, air passages and lungs. Fumes may cause metal fume fever with flu-like symptoms lasting 24 to 48 hours.
Eyes: Contact with solutions may cause irritation; contact with solid may cause severe eye surface burns.
Skin: Contact may cause irritation and burns.
Swallowed: Large amounts may cause violent vomiting and purging, intense pain, collapse, coma, convulsions and paralysis.
Long-term exposure: May cause hair and skin to turn a greenish color; liver and lung damage; thickening skin; nose sores, bleeding and shrinking of the inner lining of the nose.
Storage: Store in tightly closed containers in a dry, cool, well-ventilated place with non-wood floors. Keep away from combustible materials, chemically active metals and acetylene gas.
First aid guide: Move victim to fresh air; call emergency medical care. In case of contact with material, immediately flush skin or eyes with running water for at least 15 minutes. Remove and isolate contaminated clothing and shoes at the site.

★ 312 ★
CUPRIC OXALATE
CAS: 814-91-5
Other names: COPPER(II) OXALATE; COPPER OXALATE; OXALIC ACID, COPPER(2) SALT

Danger profile: Exposure limit established, Community Right-to-Know List, poisonous gas and smoke produced in fire.
Uses: Making other chemicals.
Appearance: White powder.
Effects of exposure:
Breathed: May cause irritation of the throat, air passages and lungs.

Eyes: Contact may cause irritation and severe eye surface burns.
Skin: Contact may cause irritation and burns.
Swallowed: Large amounts may cause violent vomiting and purging, intense pain, collapse, coma, convulsions and paralysis.
Long-term exposure: May cause metallic taste in mouth; hair and skin to turn a greenish color; lung damage; thickening skin; nose sores, bleeding and shrinking of the inner lining of the nose.
Storage: Store in tightly closed containers in a dry, cool, well-ventilated place. Keep away from chemically active metals. Containers may explode in fire.

★313★
CUPRIC SULFATE
CAS: 7758-98-7
Other names: BLUE COPPER; BLUE STONE; BLUE VITRIOL; BCS COPPER FUNGICIDE; COPPER MONOSULFATE; COPPER SULFATE; COPPER SULFATE (1:1); COPPER(II) SULFATE; COPPER(2) SULFATE; COPPER(2) SULFATE (1:1); CP BASIC SULFATE; CUPRIC SULFATE ANHYDROUS; CUPRIC SULPHATE; INCRACIDE E 51; KUPPERSULFAT (GERMAN); ROMAN VITRIOL; SULFATE DE CUIVRE (FRENCH); SULFURIC ACID, COPPER(2) SALT (1:1); TNCS 53; TRIANGLE

Danger profile: Mutagen, poisonous, exposure limit established, Community Right-to-Know List, poisonous gas and smoke produced in fire.
Uses: Food additives; fungicides and germicides; leather tanning; paint and ceramic pigments.
Appearance: Blue solid.
Effects of exposure:
Breathed: May cause irritation of the throat, air passages and lungs. Fumes may cause metal fume fever with flu-like symptoms that may last 24 to 48 hours.
Eyes: Contact may cause irritation and severe eye surface burns.
Skin: Passes through the unbroken skin; can increase exposure and the severity of the symptoms listed. Contact may cause irritation and burns.
Swallowed: May induce severe gastroenteric distress (vomiting, gastroenteric pain, and local corrosion and hemorrhages). Prostration, anuria, hematuria, anemia, increase in white blood cells, icterus, coma, respiratory difficulties and circulatory failure.
Long-term exposure: May cause mutations, damage the developing fetus and decrease female fertility; skin allergy and rash; kidney, liver and lung damage; metallic taste in mouth; hair and skin to turn a greenish color; thickening skin; nose sores, bleeding and shrinking of the inner lining of the nose.
Storage: Store in tightly closed containers in a dry, cool, well-ventilated place. Keep away from hydroxylamine, magnesium and other chemically active metals and acetylene gas.

★314★
CUPRIETHYLENEDIAMINE
CAS: 13426-91-0
Other names: COPPER, BIS (1,2-ETHENE DIAMINE-N,N')-; 1,2-DIAMINOETHANE COPPER COMPLEX; CUPRIETHYLENE DIAMINE

Danger profile: Corrosive, poisonous, Community Right-to-Know List, poisonous gases produced in fire.
Uses: Solvent; making cellulose products.
Appearance: Purple liquid.

Odor: Ammonia-like. The odor is a warning of exposure; however, no smell does not mean you are not being exposed.
Effects of exposure:
Breathed: May be poisonous. May cause irritation of the throat, air passages and lungs.
Eyes: Contact may cause irritation and severe eye surface burns.
Skin: May be poisonous if absorbed. Contact may cause irritation and burns.
Swallowed: May be poisonous. Large amounts may cause violent vomiting and purging, intense pain, collapse, coma, convulsions and paralysis.
Long-term exposure: May cause metallic taste in mouth; hair and skin to turn a greenish color; lung damage; thickening skin; nose sores, bleeding and shrinking of the inner lining of the nose.
Storage: Store in tightly closed containers in a dry, cool, well-ventilated place.
First aid guide: Move victim to fresh air and call emergency medical care; if not breathing, give artificial respiration; if breathing is difficult, give oxygen. In case of contact with material, immediately flush skin or eyes with running water for at least 15 minutes. Remove and isolate contaminated clothing and shoes at the site. Keep victim quiet and maintain normal body temperature. Effects may be delayed, keep victim under observation.

★315★
CYANAMIDE
CAS: 420-04-2
Other names: AMIDOCYANOGEN; CARBAMONITRILE; CARBIMIDE; CORBODIAMIDE; CYANOAMINE; N-CYANOAMINE; CYANOGENAMIDE; CYANOGEN NITRIDE; HYDROGEN CYANAMIDE; USAF EK-1995; AMIDOCYANAGEN; USAF ED-1995

Danger profile: Poisonous, combustible, highly reactive, explosion hazard, exposure limit established, Community Right-to-Know List, water reactive, poisonous gas produced in fire.
Uses: Making other chemicals.
Appearance: Crystals. Usually found as watery solution.
Effects of exposure:
Breathed: May cause severe irritation of the nose, throat and mucous membrane; headache; nausea, dizziness, flushing of the face, gasping for air, throbbing heart, increased respiration and heart rate, lowered blood pressure and unconsciousness.
Eyes: Contact may cause severe irritation, inflammation of the eyelids and burns.
Skin: Contact may cause severe irritation and burns. Skin allergy may develop.
Swallowed: Symptoms are more likely to occur if alcoholic beverages are consumed before or after exposure. See **Breathed.**
Storage: Store in tightly closed containers in a dry, cool, well-ventilated place. Keep away from heat and sources of ignition. Liquid should be stabilized with phosphoric acid, sulfuric acid or boric acid containers may explode in fire.

★316★
CYANOGEN
CAS: 460-19-5
Other names: CARBON NITRIDE; CYANOGENE (FRENCH); CYANOGEN GAS; DICYAN; DICYANOGEN; ETHANEDINITRILE; MONOCYANOGEN; NITRILOACETONITRILE; OXALIC ACID DINITRILE; OXALONITRILE; OXALYL CYANIDE; PRUSSITE

Danger profile: Deadly poison, fire hazard, explosion hazard, reactive, exposure limit established, Community Right-to-Know List, poisonous gases produced in fire, and on contact with water and acid.

Uses: Making other chemicals; special gas for welding and cutting metals; fumigant; rocket propellant.

Appearance: Colorless liquefied gas.

Odor: Strong almond-like; pungent, penetrating. The odor is a warning of exposure; however, no smell does not mean you are not being exposed. May not be strong enough to provide an adequate warning. The odor is at the deadly level.

Effects of exposure:

Breathed: Poison. May cause headache, eye, nose and throat irritation, wheezing, foaming at the mouth, general irritability, nausea; chest tightness, shortness of breath, vomiting blood, respiratory depression, weakness, chest and abdominal pains, dizziness, diarrhea, tremors, convulsions, shock, unconsciousness weak and irregular pulse and death. Reactions may appear hours following overexposure.

Eyes: May cause severe irritation, tearing and chemical burns.

Skin: Poison. Passes through the unbroken skin; can increase exposure and the severity of the symptoms listed. May cause severe irritation, redness, stinging and blisters.

Swallowed: May cause severe irritation and chemical burns to the tongue, mouth, throat and stomach as well as many of the symptoms described under *Breathed.*

Long-term exposure: May cause enlarged thyroid glands, nervous system damage. Vision loss may result from a single large exposure.

Storage: Store in tightly closed containers in a dry, cool, well-ventilated place. Keep away from heat, flame and sources of ignition, water, steam or other forms of moisture, fluorine, oxygen, acid. Containers may explode in fire.

First aid guide: Move victim to fresh air and call emergency medical care; if not breathing, give artificial respiration; if breathing is difficult, give oxygen. In case of contact with material, immediately flush skin or eyes with running water for at least 15 minutes. Remove and isolate contaminated clothing and shoes at the site. Keep victim quiet and maintain normal body temperature. Effects may be delayed; keep victim under observation.

★317★
CYCLOHEPTENE
CAS: 628-92-2
Other names: SUBERANE; SUBERYLENE

Danger profile: Fire hazard, narcotic, poisonous gas produced in fire.

Uses: Making other chemicals.

Appearance: Colorless, oily liquid.

Effects of exposure:

Breathed: May be poisonous. May cause dizziness, lighthededness and unconsciousness.

Eyes: May cause irritation, redness and burns.

Skin: May be poisonous if absorbed. Passes through the unbroken skin; can increase exposure and the severity of the symptoms listed. Contact may cause irritation, burns drying and cracking.

Long-term exposure: May cause drying and cracking of the skin.

Storage: Store in tightly closed containers in a dry, cool, well-ventilated place. Keep away from heat, flame, sources of ignition, oxidizers. Vapors may travel to sources of ignition and flash back. Containers may explode in fire.

First aid guide: Move victim to fresh air and call emergency medical care; if not breathing, give artificial respiration; if breathing is difficult, give oxygen. In case of contact with material, immediately flush eyes with running water for at least 15 minutes.

Wash skin with soap and water. Remove and isolate contaminated clothing and shoes at the site.

★318★
CYCLOHEXANE
CAS: 110-82-7
Other names: BENZENE, HEXAHYDRO-; CICLOESANO (ITALIAN); CYCLOHEXAAN (DUTCH); CYCLOHEXAN (GERMAN); CYKLOHEKSAN (POLISH); HEXAHYDROBENZENE, HEXAMETHYLENE; HEKSAN (POLISH); HEXAHYDROBENZENE; HEXAMETHYLENE; HEXANAPHTHENE

Danger profile: Mutagen, fire hazard, narcotic, exposure limit established, Community Right-to-Know List, poisonous gases produced in fire.

Uses: Making nylon and other chemicals; industrial solvent for various materials including lacquers and resins; extracting essential oils; paint and varnish remover; solid fuels; fungicides; laboratory chemical.

Appearance: Colorless, watery, mobile liquid.

Odor: Gasoline-like; mild, sweet, pungent. The odor is a warning of exposure; however, no smell does not mean you are not being exposed.

Effects of exposure:

Breathed: May cause irritation of the eyes, nose and throat; dizziness, lightheadedness, rapid breathing, mental confusion, unsteady walk, headache, fatigue, nausea, vomiting, unconsciousness, coma and death by respiratory failure. High concentrations are narcotic.

Eyes: May cause burning feeling, tearing, irritation, redness and inflammation of the eyelids.

Skin: Passes through the unbroken skin; can increase exposure and the severity of the symptoms listed. May cause irritation of the skin, dryness, redness and rash.

Swallowed: Moderately toxic. May cause indigestion, dizziness, fatigue, unconsciousness.

Long-term exposure: May effect the liver and kidneys. Repeated or prolonged exposure to the skin may cause drying and cracking.

Storage: Store in tightly closed containers in a dry, cool, well-ventilated place. Keep away from heat, sources of ignition dinitrogen tetroxide, and oxidizers. Vapors may travel to sources of ignition and flash back. Containers may explode in fire.

First aid guide: Move victim to fresh air and call emergency medical care; if not breathing, give artificial respiration; if breathing is difficult, give oxygen. In case of contact with material, immediately flush eyes with running water for at least 15 minutes. Wash skin with soap and water. Remove and isolate contaminated clothing and shoes at the site.

★319★
CYCLOHEXANOL
CAS: 108-93-0
Other names: ADRONAL; ANOL; CICLOESANOLO (ITALIAN); 1-CYCLOHEXANOL; CYCLOHEXYL ALCOHOL; CYKLOHEKSANOL (POLISH); HEXAHYDROPHENOL; HEXALIN; HYDRALIN; HYDROPHENOL; HYDROXYCYCLOHEXANE; NAXOL; PHENOL, HEXAHYDRO-

Danger profile: Mutagen, combustible, exposure limit established, poisonous gas produced in fire.

Uses: Making soap and other chemicals; plastics; solvent; insecticides; textile finishing; in lacquers, paints and vanishers; finish removers; leather tanning; polishes; germicides.

Appearance: Colorless needles, viscous liquid or sticky solid.

Odor: Mothball-like, camphor-like. The odor is a warning of exposure; however, no smell does not mean you are not being exposed.

Effects of exposure:
Breathed: May cause irritation of the eyes, nose, throat and lungs, tears and runny nose, coughing, sore throat, difficult breathing; lethargy, fatigue, and drowsiness; slurred speech, shaking, and dizziness. Higher exposures could cause shortness of breath, headache, nausea, vomiting, dizziness, diarrhea and a dangerous build-up of fluids in the lungs (pulmonary edema), which can cause death. Pulmonary edema is a medical emergency that can be delayed one to two days following exposure.
Eyes: May cause irritation, redness, tearing, pain, inflamed eyelids, and eye injury; burns in corneas and possible loss of sight.
Skin: Toxic. Passes through the unbroken skin; can increase exposure and the severity of the symptoms listed. May cause painful irritation, redness, and blisters. Repeated or prolonged skin contact may cause irritation of the skin and may also cause the legs to feel numb.
Swallowed: May cause irritation of the mouth and throat; stomach and abdominal pain; loss of muscular coordination and mental confusion; convulsions. Also see symptoms for *Breathed.*
Long-term exposure: Has caused liver, kidney, vascular injury in experimental animals. May cause liver or kidney damage in humans. May cause skin rash.
Storage: Store in tightly closed containers in a dry, cool, well-ventilated place. Keep away from oxidizers. Vapors may travel to sources of ignition and flash back. Containers may explode in fire.

★320★
CYCLOHEXANONE
CAS: 108-94-1
Other names: ANONE; CICLOESANONE (ITALIAN); CYCLOHEXANON (DUTCH); CYKLOHEKSANON (POLISH); CYCLOHEXYL KETONE; HEXANON; KETOHEXAMETHYLENE; NADONE; PIMELIC KETONE; SEXTONE

Danger profile: Mutagen, combustible, exposure limit established, poisonous gases produced in fire.
Uses: Making other chemicals; wood stains; paint and varnish removers; spot removers; degreasing metals; polishes; finishing silk; lube oil additive; industrial solvent.
Appearance: Colorless to light yellow watery liquid.
Odor: Peppermint-like. The odor is a warning of exposure; however, no smell does not mean you are not being exposed.
Effects of exposure:
Breathed: May cause irritation of the eyes, nose, throat and lungs, tears and runny nose, coughing, sore throat, difficult breathing; lethargy, fatigue, and drowsiness; slurred speech, shaking, and dizziness. Higher exposures could cause shortness of breath, headache, nausea, vomiting, dizziness, diarrhea and a dangerous build-up of fluids in the lungs (pulmonary edema), which can cause death. Pulmonary edema is a medical emergency that can be delayed one to two days following exposure.
Eyes: May cause irritation, redness, tearing, pain, inflamed eyelids, and eye injury; burns in corneas and possible loss of sight.
Skin: Toxic. Passes through the unbroken skin; can increase exposure and the severity of the symptoms listed. May cause painful irritation, redness, and blisters. Repeated or prolonged skin contact may cause irritation of the skin and may also cause the legs to feel numb.
Swallowed: May cause irritation of the mouth and throat; stomach and abdominal pain; loss of muscular coordination and mental confusion; convulsions. Also see symptoms for *Breathed.*

Long-term exposure: May cause liver and kidney damage; drying and cracking of the skin; cataracts.
Storage: Store in tightly closed containers in a dry, cool, well-ventilated place. Keep away from flame, sources of ignition, nitric acid, and oxidizers. Vapors may travel to sources of ignition and flash back. Containers may explode in fire.
First aid guide: Move victim to fresh air and call emergency medical care; if not breathing, give artificial respiration; if breathing is difficult, give oxygen. In case of contact with material, immediately flush eyes with running water for at least 15 minutes. Wash skin with soap and water. Remove and isolate contaminated clothing and shoes at the site.

★321★
CYCLOHEXENE
CAS: 110-83-8
Other names: BENZENETETRAHYDRIDE; BENZENE TETRAHYDRIDE; CYKLOHEKSEN (POLISH); HEXANAPTHYLENE; TETRAHYDROBENZENE; 1,2,3,4- TETRAHYDROBENZENE

Danger profile: Combustible, exposure limit established, poisonous gas produced in fire.
Uses: Making other chemicals; solvent; oil extraction.
Appearance: Colorless liquid.
Odor: Sweet. The odor is a warning of exposure; however, no smell does not mean you are not being exposed.
Effects of exposure:
Breathed: May cause irritation of the nose, throat and air passages; dizziness, lightheadedness, loss of coordination; tremors, unconsciousness and death.
Eyes: May cause irritation, redness and inflammation of the eyelids.
Skin: Passes through the unbroken skin; can increase exposure and the severity of the symptoms listed. May cause irritation, redness, dry skin and rash.
Swallowed: Moderately toxic. See symptoms under *Breathed* and *First aid guide.*
Long-term exposure: May cause liver damage by repeated exposure to relatively low levels of exposure, and brain damage with behavioral changes. Similar chemicals may cause irregular heart beat.
Storage: Store in tightly closed containers in a dry, cool, well-ventilated place. Keep away from flame, sources of ignition and oxidizers. Vapors may travel to sources of ignition and flash back. Containers may explode in fire.
First aid guide: Move victim to fresh air and call emergency medical care; if not breathing, give artificial respiration; if breathing is difficult, give oxygen. In case of contact with material, immediately flush skin or eyes with running water for at least 15 minutes. Remove and isolate contaminated clothing and shoes at the site. Keep victim quiet and maintain normal body temperature.

★322★
CYCLOHEXENYL TRICHLOROSILANE
CAS: 10137-69-6
Other names: CYCLOHEXENE, 4-(TRICHLOROSILYL)-; CYCLOHENE, 4-(TRICHLOROSILYL)-; SILANE, (3-CYCLOHEXENYL)TRICHLORO-

Danger profile: Corrosive, combustible, poisonous gas produced in fire and in moist air.
Uses: Making silicones.
Appearance: Colorless, fuming liquid.
Odor: Hydrochloric acid-like. The odor is a warning of exposure; however, no smell does not mean you are not being exposed.
Effects of exposure:

Breathed: May cause irritation of the eyes, nose, throat, breathing passages, lungs; coughing and difficult breathing. Higher exposures may cause shortness of breath and a dangerous build-up of fluids in the lungs (pulmonary edema), which can cause death. Pulmonary edema is a medical emergency that can be delayed one to two days following exposure.
Eyes: Contact may cause irritation, inflammation of the eyelids and severe burns.
Skin: Contact may cause irritation, redness and severe burns.
Swallowed: May cause severe burns of the mouth and stomach. See symptoms under *Breathed.*
Long-term exposure: May effect the lungs.
Storage: Store in tightly closed containers in a dry, cool, well-ventilated place. Vapors may travel to sources of ignition and flash back. Containers may explode in fire.
First aid guide: Move victim to fresh air and call emergency medical care; if not breathing, give artificial respiration; if breathing is difficult, give oxygen. In case of contact with material, immediately flush skin or eyes with running water for at least 15 minutes. Remove and isolate contaminated clothing and shoes at the site. Keep victim quiet and maintain normal body temperature.

★ 323 ★
CYCLOHEXYLAMINE
CAS: 108-91-8
Other names: AMINOCYCLOHEXANE; AMINOHEXAHYDRO-BENZENE; ANILINE, HEXAHYDRO-; CHA; CYCLOHEXANAMINE; HEXAHYDROANILINE; HEXAHYDROBENZENAMINE

Danger profile: Mutagen, fire hazard, exposure limit established, EPA extremely hazardous substance list, poisonous gases produced in fire.
Uses: Making other chemicals, dyes, rubber; dry cleaning chemicals; insecticides.
Appearance: Colorless to yellow liquid.
Odor: Strong, fishy. The odor is a warning of exposure; however, no smell does not mean you are not being exposed.
Effects of exposure:
Breathed: May cause irritation of the eyes, nose, throat, breathing passages and lungs; dizziness, nausea; anxiety, slurred speech, pupillary dilation, slurred speech and drowsiness. Nausea may last for eight to twelve hours following exposure.
Eyes: Poison. May cause irritation, severe burns and permanent damage.
Skin: Poison. Passes through the unbroken skin; can increase exposure and the severity of the symptoms listed. May cause irritation, redness and burns.
Swallowed: Poison. May cause burning feeling of the mouth and throat, intense thirst, throat swelling, stomach cramps, dizziness, nausea, vomiting, diarrhea, severe breathing difficulties and unconsciousness. Tissue damage may result. Breathing may stop. Watch for delayed symptoms.
Long-term exposure: May cause a change in the genetic material of the body (mutations), birth defects, miscarriage, damage to the male testes, reduce fertility in females, cancer. May cause skin allergy; dermatitis; kidney and liver damage.
Storage: Store in tightly closed containers in a flame proof cabinet or an explosion proof refrigerator. Keep away from heat, air, light and oxidizers. Vapors may travel to sources of ignition and flash back. Containers may explode in fire.
First aid guide: Move victim to fresh air and call emergency medical care; if not breathing, give artificial respiration; if breathing is difficult, give oxygen. In case of contact with material, immediately flush skin or eyes with running water for at least 15 minutes. Remove and isolate contaminated clothing and shoes

at the site. Keep victim quiet and maintain normal body temperature.

★ 324 ★
CYCLOHEXYL ISOCYANATE
CAS: 3173-53-3
Other names: CYCLOHEXANE, ISOCYANATO-; CYCLOHEXL ESTER; ISOCYANIC ACID; ISOCYANIC ACID, CYCLOHEXYL ESTER

Danger profile: Mutagen, poisonous, combustible, poisonous gases produced in fire.
Uses: Making agricultural chemicals
Appearance: Clear liquid.
Odor: Sharp. The odor is a warning of exposure; however, no smell does not mean you are not being exposed.
Effects of exposure:
Breathed: May be poisonous. May cause irritation of the eye, nose, throat and lungs. Higher exposures may cause shortness of breath and a dangerous build-up of fluids in the lungs (pulmonary edema), which can cause death. Pulmonary edema is a medical emergency that can be delayed one to two days following exposure.
Eyes: May cause irritation, redness and tearing.
Skin: May be poisonous if absorbed. May cause irritation.
Swallowed: May be poisonous. See symptoms under *Breathed.*
Long-term exposure: May cause a change in the genetic material of the body (mutations), birth defects, miscarriage, cancer. May effect the lungs. Possible skin allergies.
Storage: Store in tightly closed containers in a dry, cool, well-ventilated place. Keep away from water, steam and other sources of ignition, forms of moisture; temperatures above 200°F; bases, alcohol, metal compounds. Containers may explode in fire.
First aid guide: Move victim to fresh air and call emergency medical care; if not breathing, give artificial respiration; if breathing is difficult, give oxygen. In case of contact with material, immediately flush skin or eyes with running water for at least 15 minutes. Speed in removing material from skin is of extreme importance. Remove and isolate contaminated clothing and shoes at the site. Keep victim quiet and maintain normal body temperature. Effects may be delayed; keep victim under observation.

★ 325 ★
CYCLOHEXYL TRICHLOROSILANE
CAS: 98-12-4
Other names: CYCLOHEXANE, 1-(TRICHLOROSILYL)-; CYCLOHEXYLTRICHLOROSILANE; SILANE, TRICHLOROCYCLOHEXYL-

Danger profile: Corrosive, poisonous gases produced in fire and on contact with water, steam and moisture.
Uses: Making silicones.
Appearance: Colorless to pale yellow liquid.
Effects of exposure:
Breathed: May cause irritation of the eye, nose, throat breathing passages and lungs; coughing, difficult breathing, shortness of breath. Higher exposures may cause shortness of breath and a dangerous build-up of fluids in the lungs (pulmonary edema), which can cause death. Pulmonary edema is a medical emergency that can be delayed one to two days following exposure.
Eyes: May cause irritation and severe burns; permanent damage.
Skin: May cause irritation, redness, severe burns.
Swallowed: May cause burning feeling of the mouth and throat, intense thirst, throat swelling, stomach cramps, nau-

sea, vomiting, diarrhea, severe breathing difficulties and unconsciousness. Tissue damage may result. Breathing may stop. Watch for delayed symptoms.
Long-term exposure: May cause lung damage.
Storage: Store in tightly closed containers in a dry, cool, well-ventilated place with non-wood floors. Keep away from water, steam and other forms of moisture, combustible material.

★ 326 ★
CYCLONITE
CAS: 121-82-4
Other names: CYCLOTRIMETHYLENENITRAMINE; CYCLOTRIMETHYLENETRINITRAMINE; ESAIDRO-1,3,5-TRINITRO-1,3,5-TRIAZINA (ITALIAN); HEKSOGEN (POLISH); HEXAHYDRO-1,3,5-TRINITRO-S-TRIAZINE; HEXAHYDRO-1,3,5-TRINITRO-1,3,5-TRIAZIN (GERMAN); HEXAHYDRO-1,3,5-TRINITRO-1,3,5-TRIAZINE; HEXOGEEN (DUTCH); HEXOGEN; HEXOGEN 5W; HEXOLITE; PBX(AF) 108; RDX; T4; TRIMETHYLEENTRINITRAMINE (DUTCH); TRIMETHYLENETRINITRAMINE; SYM-TRIMETHYLENETRINITRAMINE; TRINITROCYCLOTRIMETHYLENE TRIAMINE; 1,3,5-TRINITRO-1,3,5-TRIAZACYCLOHEXANE; HEXAHYDRO-1,3,5-TRINITRO-S-TRIAZINE; S-TRIAZINE, EXAHYDRO-1,3,5-TRINITRO-; SYM-TRIMETHYLENETRINITRINITRAMINE; 1,3,5-TRINITRO-1,3,5-TRIZAZCYCLOHEXANE

Danger profile: Possible reproductive hazard, dangerous explosion hazard (1.5 times more powerful than TNT), poisonous, corrosive, exposure limit established, toxic fumes produced in fire.
Uses: Rat poison; powerful explosive; aerial bombs, torpedoes, mines.
Appearance: White crystalline powder.
Effects of exposure:
   Breathed: May cause irritation of breathing passages and lungs; ulceration of the mucous membrane; headache, irritability, dizziness, nausea, vomiting; salivation, anorexia, asthma, insomnia, unconsciousness, convulsions, seizures and nervous system damage.
   Eyes: May cause irritation, redness and inflammation of the eyelids, burns and permanent damage.
   Skin: Passes through the unbroken skin; can increase exposure and the severity of the symptoms listed. May cause irritation, redness, chemical burns, rash and allergy.
   Swallowed: Poison. May cause burning feeling of the mouth and throat, intense thirst, throat swelling, stomach cramps, dizziness, nausea, vomiting, diarrhea, severe breathing difficulties and unconsciousness. Tissue damage may result. Breathing may stop.
Long-term exposure: Repeated exposure may cause insomnia, fits and seizures; skin allergy, dermatitis.
Storage: Store in tightly closed containers in a detached and remote, cool and well-ventilated place. Keep away from heat, shock, combustible materials, other explosives and oxidizers.

★ 327 ★
CYCLOPENTADIENE
CAS: 542-92-7
Other names: 1,3-CYCLOPENTADIENE; PENTOLE; PYROPENTYLENE; R-PENTINE

Danger profile: Fire hazard, explosion hazard above 32°F, exposure limit established, poisonous gas produced in fire.
Uses: Making other chemicals, insecticides, resins.
Appearance: Colorless liquid.

Odor: Liquid smells like turpentine; solid is camphor-like. The odor is a warning of exposure; however, no smell does not mean you are not being exposed.
Effects of exposure:
   Breathed: May be poisonous. May cause irritation of the eyes, nose throat and breathing passages; dizziness.
   Eyes: May cause irritation and burns.
   Skin: May be poisonous if absorbed. Passes through the unbroken skin; can increase exposure and the severity of the symptoms listed. May cause irritation, redness, rash, burns and allergy.
Long-term exposure: May cause liver and kidney damage; skin allergy.
Storage: Store in tightly closed containers in a refrigerated unit below 32°F. Keep away from heat, flame, oxidizers, acids, potassium hydroxide. Vapors may travel to sources of ignition and flash back. Containers may explode in fire.

★ 328 ★
CYCLOPENTANE
CAS: 287-92-3
Other names: PENTAMETHYLENE

Danger profile: Fire hazard, narcotic, exposure limit established, poisonous gases produced in fire.
Uses: Solvent; motor fuel; laboratory chemical.
Appearance: Colorless watery liquid.
Odor: Mild, sweet; gasoline-like. The odor is a warning of exposure; however, no smell does not mean you are not being exposed.
Effects of exposure:
   Breathed: May be poisonous. May cause dizziness, lightheadedness, unconsciousness. Coma may result from high level of exposure.
   Eyes: Contact may cause irritation.
   Skin: May be poisonous if absorbed. Passes through the unbroken skin; can increase exposure and the severity of the symptoms listed. Contact may cause irritation.
Long-term exposure: May cause skin drying and cracking; possible brain and nerve damage.
Storage: Store in tightly closed containers in a dry, cool, well-ventilated place. Keep away from heat, sources of ignition and oxidizers. Vapors may travel to sources of ignition and flash back. Containers may explode in fire.
First aid guide: Move victim to fresh air and call emergency medical care; if not breathing, give artificial respiration; if breathing is difficult, give oxygen. In case of contact with material, immediately flush eyes with running water for at least 15 minutes. Wash skin with soap and water. Remove and isolate contaminated clothing and shoes at the site.

★ 329 ★
CYCLOPROPANE
CAS: 75-19-4
Other names: CYCLOPROPANE, LIQUIFIED; TRIMETHYLENE

Danger profile: Mutagen, reproductive hazard, narcotic, dangerous fire hazard, explosion hazard, poisonous gas produced in fire.
Uses: Making other chemicals; surgical anesthetic.
Appearance: Colorless gas or liquid under pressure.
Effects of exposure:
   Breathed: May cause dizziness, lightheadedness, unconsciousness; pupil dilation, shallow depth of respirations, decreasing muscle tone, irregular heart beat. Possible suffocation. High levels may cause coma and death.
   Eyes: Contact with liquid may cause irritation and frostbite.

Skin: Contact with liquid may cause irritation and frostbite.

Long-term exposure: May cause a change in the genetic material of the body (mutations), birth defects, miscarriage, damage to the male testes, reduce fertility in females, cancer.

Storage: Store in tightly closed containers in a dry, cool, well-ventilated place. Keep away from flame, sparks or other sources of ignition, oxygen mixtures and oxidizers. Vapors may travel to sources of ignition and flash back. Containers may explode in fire.

First aid guide: Move victim to fresh air and call emergency medical care; if not breathing, give artificial respiration; if breathing is difficult, give oxygen. In case of frostbite, thaw frosted parts with water. Keep victim quiet and maintain normal body temperature.

D

★ 330 ★
2,4-D ACID
CAS: 94-75-7

Other names: ACIDE 2,4-DICHLORO PHENOXYACETIQUE (FRENCH); ACIDO(2,4-DICLORO-FENOSSI)-ACETICO (ITALIAN); AGROTECT; AMIDOX; AMOXONE; AQUA-KLEEN; BH 2,4-D; BRUSH-RHAP; BUSH KILLER; B-SELEKTONON; CHIPCO TURF HERBICIDE ''D''; CHLOROXONE; CROP RIDER; CROTILIN; 2,4-D; D 50; 2,4-D, SALTS AND ESTERS; DACAMINE; DEBROUSSAILLANT 600; DECAMINE; DEDWEED; DED-WEED LV-69; DESORMONE; (2,4-DICHLOOR-FENOXY)-AZIJNZUUR (DUTCH); DICHLOROPHENOXYACETIC ACID; 2,4-DICHLOROPHENOXYACETIC ACID; 2,4-DICHLORPHENOXYACETIC ACID; 2,4-DICHLORPHENOXYACETIC ACID, SALTS AND ESTERS; (2,4-DICHLOR-PHENOXY)-ESSIGSAEURE (GERMAN); DICOPUR; DICOTOX; DINOXOL; DMA-4; DORMONE; 2,4-DWUCHLOROFENOKSYOCTOWY KWAS (POLISH); EMULSAMINE BK; EMULSAMINE E-3; ENVERT 171; ENVERT DT; ESTERON; ESTERON 99; ESTERON 76 BE; ESTERON 99 CONCENTRATE; ESTERONE FOUR; ESTERON 44 WEED KILLER; ESTONE; FERNESTA; FERNIMINE; FERNOXONE; FERXONE; FOREDEX 75; FORMULA 40; HEDONAL; HEDONAL (THE HERBICIDE); HERBIDAL; IPANER; KROTILINE; LAWN-KEEP; MACRONDRAY; MIRACLE; MONOSAN; MOXONE; NETAGRONE; NETAGRONE 600; NSC 423; PENNAMINE; PENNAMINE D; PHENOX; PIELIK; PLANOTOX; PLANTGARD; RHODIA; SALVO; SPRITZ-HORMIN/2,4-D; SPRITZ-HORMIT/2,4-D; SUPER D WEEDONE; SUPERORMONE CONCENTRE; TRANSAMINE; TRIBUTON; U 46; U-5043; U 46DP; VERGEMASTER; VERTON; VERTON D; VERTON 2D; VERTRON 2D; VIDON 638; VISKO; VISKO-RHAP; VISKO-RHAP LOW DRIFT HERBICIDES; VISKO-RHAP LOW VOLATILE 41; WEED-AG-BAR; WEEDAR; WEEDAR-64; WEED-B-GON; WEEDEZ WONDER BAR; WEEDONE; WEEDONE LV4; WEED-RHAP; WEED TOX; WEEDTROL

Danger profile: Suspected carcinogen, mutagen, reproductive hazard, poison, combustible, exposure limit established, Community Right-to-Know List, poisonous gases produced in fire.

Uses: Fungicide, herbicide, weed killer, defoliant; plant growth regulator; to control fruit dropping from trees.

Appearance: White or tan powder or crystals.

Effects of exposure:

Breathed: Poisonous. May cause irritation to the mouth, nose, throat and air passages. Higher exposure may cause headache, nausea, loss of appetite, perspiration, vomiting, fever, diarrhea; muscle weakness, twitching, poor coordina-

tion, swelling of the feet and legs; possible nerve, liver and kidney damage. Severe exposure may result in death.

Eyes: Dust may cause severe irritation. Also see *Skin.*

Skin: Poisonous. Contact may cause severe irritation. Liquid or dust left in contact with the skin for several hours passes through the unbroken skin; can increase exposure and the severity of the symptoms listed. Symptoms may be delayed. These symptoms may last for months or years.

Swallowed: Poisonous. May cause gastroenteric distress, nausea, vomiting, diarrhea, mild central nervous system depression, difficult swallowing and possible liver and kidney injury. Death has resulted from as little as 1/5 ounce. Survival for more than 48 hours is usually followed by complete recovery although symptoms may last for several months.

Long-term exposure: May cause mutations, cancer, birth defects, fetus damage; decreased fertility in males; skin allergy; weakness, rapid fatigue, vertigo, liver damage, low blood pressure and slowed heartbeat. Some evidence of breast cancer in animals, and non-Hodgkin's lymphoma in farmers exposed to the chemical.

Storage: Store in tightly closed containers in a dark, dry, cool, well-ventilated place. Keep away from heat, light and oxidizers. Containers may explode in fire.

★ 331 ★
DAUNOMYCIN
CAS: 20830-81-3

Other names: ACETYLADRIAMYCIN; CERUBIDIN; DAUNAMYCIN; DAUNORUBICIN; DAUNORUBICINE; DM; FI6339; LEUKAEMO-MYCIN C; 5,12-NAPHTHACENEDIONE,8-ACETYL-10-(3-AMINO-2,3,6-TRIDEOXY-ALPHA-L-LYXO-HEXOPYRANOSYL)OXY-7,8,9,10-TETRAHYDRO-6,8,11-TRIHYDROXY-1-METHOXY-, (8S-CIS)-; NCI-C04693; NSC-82151; RP 13057; RUBIDOMYCIN; RUBIDOMYCINE; RUBOMYCIN C; RUBOMYCIN C 1; STREPTOMYCES PEUCETIUS

Danger profile: Carcinogen, mutagen, possible reproductive hazard, poison, poisonous gas produced in fire.

Uses: For treating cancer.

Appearance: Thin, red needles.

Effects of exposure:

Breathed: Repeated or prolonged exposure to dust may cause irregular heart beat, heart muscle damage and may effect bone marrow, red blood cells and blood platelets. Bone marrow damage may cause a reduction in white blood cells and reduce resistance to infection. Blood platelet damage may effect bleeding. Red cell damage may cause anemia.

Long-term exposure: May cause cancer, birth defects, damage to the fetus, miscarriage; slow and permanent damage to the heart muscle and congestive heart failure.

Storage: Store in tightly closed containers in a dry, cool, well-ventilated place.

★332★
DDT
CAS: 50-29-3
Other names: BENZENE,1,1'-(2,2,2-TRICHLOROETHYLIDENE)BIS (4-CHLORO); ALPHA,ALPHA-BIS(P-CHLOROPHENYL)-BETA,BETA,BETA-TRICHLORETHANE; 1,1-BIS-(P-CHLOROPHENYL)-2,2,2-TRICHLOROETHANE; 2,2-BIS(P-CHLOROPHENYL)-1,1-TRICHLOROETHANE; DICHLORODIPHENYL TRICHLOROETHANE 2,2-BIS(P-CHLOROPHENYL)-1,1,1-TRICHLOROETHANE; AGRITAN; ANOFEX; ARKOTINE; AZOTOX; BOSAN SUPRA; BOVIDERMOL; CHLOROPHENOTHAN; CHLOROPHENOTHANE; CHLOROPHENOTOXUM; CITOX; CLOFENOTANE; P,P'-DDT; 4,4' DDT; DEDELO; DEOVAL; DETOX; DETOXAN; DIBOVAN; DICHLORODIPHENYLTRICHLOROETHANE; P,P'-DICHLORODIPHENYLTRICHLOROETHANE; 4,4'-DICHLORODIPHENYLTRICHLOROETHANE; DICOPHANE; DIDIGAM; DIDIMAC; DIPHENYLTRICHLOROETHANE; DODAT; DYKOL; ENT 1,506; ESTONATE; GENITOX; GESAFID; GESAPON; GESAREX; GESAROL; GUESAROL; GYRON; HAVERO-EXTRA; IVORAN; IXODEX; KOPSOL; MUTOXIN; NCI-C00464; NEOCID; PARACHLOROCIDUM; PEB1; PENTACHLORIN; PENTECH; PPZEIDAN; RUKSEAM; SANTOBANE; 1,1,1-TRICHLOOR-2,2-BIS(4-CHLOORFENYL)-ETHAAN (DUTCH); 1,1,1-TRICHLOR-2,2-BIS(4-CHLORPHENYL)-AETHAN (GERMAN); TRICHLOROBIS(4-CHLOROPHENYL)ETHANE; 1,1,1-TRICHLORO-2,2-BIS(P-CHLOROPHENYL)ETHANE; 1,1,1-TRICHLORO-2,2-DI(4-CHLOROPHENYL)-ETHANE; 1,1,1-TRICLORO-2,2-BIS(4-CLORO-FENIL)-ETANO (ITALIAN); ZEIDANE; ZERDANE

Danger profile: Carcinogen, mutagen, reproductive hazard, exposure limit established.
Uses: Insecticide. Use has been restricted by the U.S. Environmental Protection Agency.
Appearance: Colorless, white or slightly off-white powder. available as granules, powders, smoke candles and in other forms.
Odor: Weak or slightly aromatic chemical. The odor is a warning of exposure; however, no smell does not mean you are not being exposed.
Effects of exposure:
Breathed: May be poisonous. With smaller doses, symptoms usually appear within 30 minutes to 6 hours following exposure. Exposure may cause a tingling feeling in the lower face, mouth and tongue, headache, nausea, vomiting, dizziness, confusion, weakness, loss of muscular control and muscular tremors. Higher exposures can cause convulsions and may cause death.
Eyes: See *Skin.* May cause irritation.
Skin: May be poisonous if absorbed. If dissolved in solvents or vegetable oils passes through the unbroken skin; can increase exposure and the severity of the symptoms listed. Contact with dust may cause irritation.
Swallowed: May be poisonous. Very large doses are followed promptly by vomiting; delayed vomiting or diarrhea may occur. With smaller doses, symptoms usually appear 2-3 hours following ingestion. These include tingling of lips, tongue, face; feeling of weakness, headache, sore throat, fatigue, coarse tremors of neck, head and eyelids; apprehension, staggering walk, poor balance and confusion. Convulsions may alternate with periods of coma and partial paralysis. Vital signs are essentially normal, but in severe poisoning the pulse may be irregular and abnormally slow; ventricular fibrillation and sudden death may occur at any time during acute phase.
Long-term exposure: DDT builds up in the body and remains for a long period of time. Exposure to both DDT and aldrin may increase retention of DDT in the body. May cause liver cancer; damage to the developing fetus; decrease fertility in males; liver and kidney damage.
Storage: Store in a cool, dry well-ventilated place in a non-iron container. Keep away from high temperatures and oxidizers. Containers may explode in fire.
First aid guide: Move victim to fresh air and call emergency medical care; if not breathing, give artificial respiration; if breathing is difficult, give oxygen. In case of contact with material, immediately flush skin or eyes with running water for at least 15 minutes. Speed in removing material from skin is of extreme importance. Remove and isolate contaminated clothing and shoes at the site. Keep victim quiet and maintain normal body temperature. Effects may be delayed; keep victim under observation.

★333★
DECABORANE
CAS: 17702-41-9
Other names: BORON HYDRIDE; DECABORANE(14); DECARBORON TETRADECAHYDRIDE

Danger profile: Combustible, explosion hazard, exposure limit established, poisonous gases produced in fire.
Uses: Making other chemicals, corrosion inhibitor; fuel additive; stabilizer; processing rayon; moth proofing agent; dye stripping agent; rocket propellant.
Appearance: White crystalline solid.
Odor: Sharp, bitter. The odor is a warning of exposure; however, no smell does not mean you are not being exposed.
Effects of exposure:
Breathed: Vapor may cause irritation of the eyes, nose and throat. May cause headache, nausea, light-headedness, drowsiness, nausea, dizziness, restlessness, nervousness, lack of coordination, tremor; muscle spasms and generalized convulsions may occur. The onset of symptoms frequently delayed until one or two days after exposure.
Eyes: Contact with dust may cause irritation, severe burns; clouding of the eyes with loss of vision.
Skin: Passes through the unbroken skin; can increase exposure and the severity of the symptoms listed. Dust may cause irritation, severe burns and rash.
Swallowed: May cause headache, nausea, light-headedness, drowsiness, nervousness, lack of coordination, tremor; muscle spasms and generalized convulsions may occur. The onset of symptoms frequently delayed until one or two days after exposure.
Long-term exposure: High or repeated exposure may damage the liver and kidneys; damage to the nervous system with spasms; permanent eye damage.
Storage: Store in tightly closed containers in a dry, cool, well-ventilated and detached place. Containers may explode in fire. Keep away from sources of ignition, heat, water and oxidizers. Corrosive to natural rubber, as well as to some synthetic rubbers, greases and lubricants.
First aid guide: Move victim to fresh air; call emergency medical care. In case of contact with material, immediately flush skin or eyes with running water for at least 15 minutes. Remove and isolate contaminated clothing and shoes at the site.

★ 334 ★
DEMETON
CAS: 8065-48-3
Other names: DEMETON ODEMETON S; DEMOX; DENOX; BAY 10756; BAYER 8169; O,O-DIETHYL-2-ETHYLMERCAPTOETHYL THIOPHOSPHATE, DIETHOX-YTHIOPHOSPHORIC ACID; E-1059; ENT 17295; MERCAP-TOPHOS; PHOSPHOROTHIOIC ACID,O,O-DIETHYL O-2-(ETHYLTHIO)ETHYL ESTER, MIXED WITH O,O-DIETHYL S-2-(ETHYLTHIO)ETHYL PHOSPHOROTHIOATE; SYSTOX; SYSTEMOX

Danger profile: Mutagen, deadly poison, EPA extremely hazardous substance list, poisonous gases produced in fire.
Uses: Insecticide (a permit may be required for use), pesticide.
Appearance: Dense, brown to yellow oily liquid.
Odor: Faint sulfur-like. The odor is a warning of exposure; however, no smell does not mean you are not being exposed.
Effects of exposure:
  Breathed: Extremely poisonous. May cause headache, dizziness, profuse sweating, nausea, vomiting, reduced heart beat, stomach cramps, diarrhea, loss of coordination, slow and weak breathing, fever, loss of consciousness and death. May cause shortness of breath and a dangerous build-up of fluids in the lungs (pulmonary edema), which can cause death. Pulmonary edema is a medical emergency that can be delayed one to two days following exposure.
  Eyes: Rapidly absorbed. Poisoning can happen very rapidly. See *Breathed.*
  Skin: Poisonous. Passes through the unbroken skin; can increase exposure and the severity of the symptoms listed. Fatal poisoning can occur from skin contact even if there is no feeling of irritation following contact.
  Swallowed: Powerful, deadly poison. May produce nausea, vomiting, abdominal cramps, diarrhea and excessive salivation. Depending on dose other symptoms may include blurred vision, tearing, loss of muscle coordination, slurred speech, mental confusion, weakness, drowsiness, convulsions, coma and death.
Long-term exposure: May damage the developing fetus or cause birth defects; nerve damage including loss of coordination; liver damage. Repeated exposure may cause depression, irritability and personality changes.
Storage: Store in a regulated, specially marked place in tightly closed containers in a dry, well-ventilated place. Keep away from strong oxidizers.

★ 335 ★
DIACETONE ALCOHOL
CAS: 123-42-2
Other names: ACETONYLDIMETHYLCARBINOL; DIACE-TONALCOHOL (DUTCH); DIACETONALCOOL (ITALIAN); DIACETONALKOHOL (GERMAN); DIACETONE-ALCOOL (FRENCH); DIKETONE ALCOHOL; 4-HYDROXY-2-KETO-4-METHYLPENTANE; 4-HYDROXY-4-METHYL-2-PENTANONE; 4-HYDROXY-4-METHYL-PENTAN-2-ON (GERMAN, DUTCH); 4-HYDROXY-4-METHYLPENTAN-2-ONE; 4-IDROSSI-4-METIL-PENTAN-2-ONE (ITALIAN); 4-METHYL-4-HYDROXY-2- PENTANONE; 2-METHYL-2-PENTANOL-4-ONE; 2-PENTANONE, 4-HYDROXY-4-METHYL-; TYRANTON

Danger profile: Combustible, exposure limit established, poisonous gases produced in fire.
Uses: Solvent for nitrocellulose, cellulose acetate, resins, waxes, fats, oils, dyes, tars. In lacquers, dopes; wood preservatives; stains; rayon and artificial leather; imitation gold leaf; dyeing mixtures; antifreeze mixtures; metal-cleaning compounds; hydraulic fluids; stripping agent (textiles); laboratory chemical.
Appearance: Colorless to light yellow watery liquid.

Odor: Mild, pleasant.
Effects of exposure:
  Breathed: May be poisonous. May cause irritation of the eyes, nose, throat and may cause tightness of the chest; nausea, dizziness, lightheadedness and unconsciousness. Very high concentrations may have a narcotic effect.
  Eyes: Vapor is irritating to mucous membrane of eye.
  Skin: May be poisonous if absorbed. Liquid not highly irritating, but can cause dermatitis.
  Swallowed: See *Breathed* and *First aid guide.*
Long-term exposure: May cause liver and kidney damage; anemia; dermatitis.
Storage: Store in tightly closed containers in a dry, cool, well-ventilated place. Keep away from heat, sources of ignition and oxidizers. Vapors may travel to sources of ignition and flash back. Containers may explode in fire.
First aid guide: Move victim to fresh air and call emergency medical care; if not breathing, give artificial respiration; if breathing is difficult, give oxygen. In case of contact with material, immediately flush eyes with running water for at least 15 minutes. Wash skin with soap and water. Remove and isolate contaminated clothing and shoes at the site.

★ 336 ★
DIALLATE
CAS: 2303-16-4
Other names: AVADEX; BIS(1-METHYLETHYL) CARBA-MOTHIOIC ACID, S-(2,3-DICHLORO-2-PROPENYL)ESTER; CARBAMOTHIOIC ACID, BIS(1-METHYLETHYL)-,S-(2,3-DICHLORO-2-PROPENYL) ESTER; CP 15,336; DATC; 2,3-DCDT; DIALLAAT (DUTCH); DIALLAT (GERMAN); S-(2,3-DICHLOR-ALLYL)-N,N-DIISOPROPYL-MONOTHIOCARBAMAAT (DUTCH); S-(2,3-DICHLORO-ALLIL)-N,N-DIISOPROPIL-MONOTIOCARBAMMATO (ITALIAN); DICHLOROALLYLDIISOPROPYLTHIOCARBAMATE; S-2,3-DICHLOROALLYLDIISOPROPYLTHIOCARBAMATE; 2,3-DICHLOROALLYL N,N-DIISOPROPYLTHIOLCARBAMATE; 2,3-DICHLORO-2-PROPENE-1-THIOL, IISOPROPYLCARBA-MATE; DIISOPROPYLTHIOCARBAMIC ACID, -(2,3-DICHLOROALLYL) ESTER; DI-ISOPROPYLTHIOLOCARBAMATE DES-(2,3-DICHLORO ALLYLE) (FRENCH); 2-PROPENE-1-THIOL, 2,3-DICHLORO-,DIISOPROPYLCARBAMATE; S-2,3-DICHLOROALLYL IISO-PROPYLTHIOCARBAMATE

Danger profile: Possible carcinogen, possible mutagen, poison, mutagen, Community Right-to-Know List, poisonous gases produced in fire.
Uses: Pesticide, herbicide.
Appearance: Brown liquid.
Effects of exposure:
  Breathed: may be poisonous. May cause irritation of the nose, throat and lungs; difficult breathing and chest tightness. Higher exposures may cause nausea, vomiting, diarrhea, abdominal pain, reduced muscle coordination, twitching of the muscles, excess saliva, blurred vision, convulsions, seizures, and even death.
  Eyes: May cause irritation.
  Skin: May be poisonous if absorbed. Passes through the unbroken skin; can increase exposure and the severity of the symptoms listed. May cause irritation and burns.
  Swallowed: Poisonous. See *Breathed.*
Long-term exposure: High or repeated exposure may cause liver and kidney damage; liver cancer; skin allergy. Carbamates are suspected carcinogens of the lungs and bone marrow.
Storage: Store in tightly closed containers in a dry, cool, well-ventilated place. Containers may explode in fire.

★337★
2,4-DIAMINOANISOLE SULFATE
CAS: 39156-41-7
Other names: 2,4-DIAMINOANISOLE SULPHATE; 2,4-DIAMINO-ANISOL SULPHATE; 2,4-DIAMINO-1-METHOXYBENZENE; 4-METHOXY-1,3-BENZENEDIAMINE SULFATE; 4-METHOXY-1,3-BENZENEDIAMINE SULPHATE; 4-METHOXY-M-PHENYLENEDIAMINE SULFATE; 4-METHOXY-M-PHENYLENEDIAMINE SULPHATE; NCI-C01989

Danger profile: Carcinogen, mutagen, poison, Community Right-to-Know List, very poisonous gases produced in fire.
Uses: In hair and fur dyes.
Appearance: Off-white to violet powder.
Effects of exposure:
Breathed: High exposures may cause poisoning with symptoms of trembling, nausea, diarrhea, difficult breathing, convulsions and death.
Eyes: Contact may cause irritation and possible damage.
Skin: May be poisonous if absorbed. Passes through the unbroken skin; can increase exposure and the severity of the symptoms listed.
Swallowed: May be poisonous. See *Breathed.*
Long-term exposure: May cause cancer and birth defects. Both skin and lung allergies may develop.
Storage: Store in a regulated, specially marked place in tightly closed containers in a dry, well-ventilated place.

★338★
O-DIANISIDINE
CAS: 119-90-4
Other names: ACETAMINE DIAZO BLACK RD; ACETAMINE DIAZO NAVY RD; AMACEL DEVELOPED NAVY SD; AZOENE FAST BLUE BASE; AZOENO FAST BLUE SALT; AZOFIX BLUE B SALT; AZOGENE FAST BLUE B; AZOGENE FAST; BLUE B SALT; BLUE BASE IRGA B; BLUE BASE NB; BLUE BN BASE; BLUE BN SALT; BLUE SALT NB; BRENTAMINE FAST BLUE B BASE; BRENTAMINE FAST BLUE B SALT; (1,1'-BIPHENYL)-4,4'-DIAMINE,3,3'-DIMETHOXY-; CELLITAZOL B; CELLITAZOL BN; C.I. 24110; C.I. AZOIC DIAZO COMPONENT 48; C.I. AZOIC; DIANISIDINE; SETACYL DIAZO NAVYR; SPECTROKNE BLUE B; 3,3'-DIMETHOXY-1,1-BIPHENYL-4,4'-DIAMINE; 3,3'-DIMETHOXY-4,4'-DIAMINOBIPHENYL; DIAZO COMPONENT 48, FAST BLUE B SALT; CIBACETE DIAZO NAVY BLUE 2B; C.I.DISPERSE BLACK 6; DIACELLITON FAST GREY 6; DIACEL NAVY DC; O-DIANISIDIN (CZECH, GERMAN); HCDB-1018; 0-DIANISIDINA (ITALIAN); O-DIANISIDINE; 0,0'-DIANISIDINE; 3,3'-DIANISIDINE; DIATO BLUE BASE B; DIATO BLUE SALT B; DIAZO FAST BLUE B; 3,3'-DIMETHOXYBENZIDIN (CZECH); 3,3'-DIMETHOSSIBENZODINA (ITALIAN); FAST BLUE B BASE; FAST BLUE BN SALT; FAST BLUE DSC BASE; FAST BLUE DS SALT; FAST BLUE SALT B; FAST BLUE SALT BN; HILTONIL FAST BLUE B BASE; HILTOSAL FAST BLUE B SALT; HINDASOL BLUE B SALT; KAKO BLUE B SALT; KAYAKN BLUE B BASE; KAYAKN FAST FLUE B SALT; LAKE BLUE B BASE; MEISUI TERYL DIAZO BLUE HR; MITSUI BLUE B BASE; MITSUI BLUE B SALT; NAPHTHANIL BLUE B BASE; NATASOL BLUE B SALT; NEUTROGEL NAVY BN; SANYO FAST BLUE SALT B

Danger profile: Carcinogen, mutagen, combustible, Community Right-to-Know List, poisonous gases produced in fire.
Uses: Making dyes.
Appearance: Colorless sand-like crystals, becoming violet with prolonged exposure to heat and light.
Effects of exposure:

Breathed: May cause headaches, dizziness, cyanosis; dust irritates nose severely, causing sneezing.
Eyes: May cause irritation and burns.
Skin: May cause irritation and burns.
Swallowed: Unknown at this time.
Long-term exposure: May cause ovary, breast, bladder, intestine, skin, stomach cancer; damage to the unborn fetus.
Storage: Store in tightly closed containers in a dry, cool, well-ventilated and regulated place. Keep away from oxidizers.

★339★
DIASTASE
CAS: 9000-92-4
Other names: AMYLASE, DIASTASE MALT

Danger profile: Allergen.
Uses: Making bread, malted foods; processing textiles.
Appearance: Yellowish-white powder
Effects of exposure:
Breathed: May cause asthma-like allergy.
Eyes: Contact may cause irritation.
Skin: Contact may cause irritation, especially if moist. Rash and skin allergy may develop.
Long-term exposure: May cause lung allergy with coughing and difficult breathing; skin allergy with itching an rash.
Storage: Store in tightly closed containers in a dry, cool, well-ventilated place.

★340★
DIATOMACEOUS EARTH
CAS: 68855-54-9
Other names: CELITE; CHROMOSORB; DIATOMACEOUS SILICA; DIATOMITE; DICALITE; DIESELGUHR; INFUSORIAL EARTH; SNOWFLOSS CELITE

Danger profile: Exposure limit established.
Uses: Filtering agent; filler in construction materials. In pesticides, paint, varnishes.
Appearance: Gray powder.
Effects of exposure:
Breathed: May cause coughing and shortness of breath. Heavy exposure may cause fibrosis; respiratory crippling; this can be fatal.
Eyes: May cause mechanical irritation, tearing and inflammation of the eyelids.
Long-term exposure: May cause permanent scarring of the lungs with shortness of breath and cough. Symptoms may develop months or years following exposure. Heavy exposure may cause silicosis; respiratory crippling; this can be fatal.
Storage: Store in tightly closed containers in a dry, cool, well-ventilated place.

★341★
DIAZEPAM
CAS: 439-14-5
Other names: ALBORAL; ALISEUM; AMIPROL; ANSIOLIN; ANSIOLISINA; APAURIN; APOZEPAM; ASSIVAL; ATENSINE; ATILEN; CALMOCITENE; CALMPOSE; CERCINE; CEREGULART; 2H-1,4-BENZODIAZEPIN-2-ONE,7-CHLORO-1,3-DIHYDRO-1-METHYL-5-PHENY L-BIALZEPAM; 7-CHLORO-1,3-DIHYDRO-1-METHYL-5-PHENYL-2H-1,4-BENZODIAZEPIN-2-ON E; 7-CHLORO-1-METHYL-5-3H-1,4-BENZODIAZEPIN-2(1H)-ONE; 7-CHLORO-1-METHYL-2-OXO-5-PHENYL-3H-1,4-BENZODIAZEPINE; 7-CHLORO-1-METHYL-5-PHENYL-2H-1,4-BENZODIAZEPIN-2-ONE; 7-CHLORO-1-METHYL-5-PHENYL-3H-1,4-BENZODIAZEPIN-2(1H)-ONE; 7-CHLORO-1-METHYL-5-PHENYL-1,3-DIHYDRO-2H-1,4-BENZODIAZEPIN-2-ON E; CONDITION; DAP; DIACEPAN; DIAPAM; DIAZEMULS; DIAZEPAMU (POLISH); DIAZETARD; DIENPAX; DIPAM; DIPEZONA; DOMALIUM; DUKSEN; DUXEN; E-PAM; ERIDAN; FAUSTAN; FREUDAL; FRUSTAN; GIHITAN; HORIZON; KABIVITRUM; KIATRIUM; LA-III; LEMBROL; LEVIUM; LIBERETAS; METHYL DIAZEPINONE; 1-METHYL-5-PHENYL-7-CHLORO-1,3-DIHYDRO-2H-1,4-BENZODIAZEPIN-2-ON E; MOROSAN; NOAN; NSC-77518; PACITRAN; PARANTEN; PAXATE; PAXEL; PLIDAN; QUETINIL; QUIATRIL; QUIEVITA; RELAMINAL; RELANIUM; RELAX; RENBORIN; RO 5-2807; S.A.R.L.; SAROMET; SEDIPAM; SEDUKSEN; SEDUXEN; SERENACK; SERENAMIN; SERENZIN; SETONIL; SIBAZON; SONACON; STESOLID; STESOLIN; TENSOPAM; TRANIMUL; TRANQDYN; TRANQUIRIT; UMBRIUM; UNISEDIL; USEMPAX AP; VALEO; VALITRAN; VALIUM; VALIUM R; VATRAN; VELIUM; VIVAL; VIVOL; WY-3467; ZIPAN

Danger profile: Mutagen, reproductive hazard, poisonous gases produced in fire.
Uses: Pharmaceutical sedative.
Appearance: Yellow crystalline powder.
Effects of exposure:
   Breathed: Dust may cause drowsiness, lack of coordination, loss of balance; irritability, anxiety, weakness, headache, upset stomach and pains in joints; skin rash, jaundice; may effect white cell count.
   Swallowed: When used as a medical sedative chemical may cause symptoms and side effects listed above.
Long-term exposure: May cause mutations, birth defects, miscarriage and cancer.
Storage: Store in tightly closed containers in a dry, cool, well-ventilated place.

★342★
DIAZINON
CAS: 333-41-5
Other names: ALFA-TOX; BASUDIN; BASUDIN 10 G; BAZUDEN; DAZZEL; O,O-DIAETHYL-O-(2-ISOPROPYL-4-METHYL-PYRIMIDIN-6-YL)-MONOTHIOPHOSPHAT (GERMAN); DIANON; DIATERR-FOS; DIAZAJET; DIAZATOL; DIAZIDE; DIAZINONE; DIAZITOL; DIAZOL; DIMPYLATE; DIPOFENE; DIZINON; DYZOL; ENT 19,507; G 301; G-24480; GARDENTOX; GEIGY 24480; KAYAZINON; KAYAZOL; NCI-CO8673; NECIDOL; NEOCIDOL; NIPSAN; NUCIDOL; PHOSPHORIC ACID, O,O-DIETHYL O-6-METHYL-2-(1-METHYLETHYL)-4-PYRIMIDINYL ESTER; SAROLEX; SPECTRACIDE

Danger profile: Mutagen, exposure limit established, poisonous gases produced in fire.
Uses: Insecticide.
Appearance: Colorless liquid.
Odor: Faint, fruity. The odor is a warning of exposure; however, no smell does not mean you are not being exposed.

Effects of exposure:
   Breathed: Extremely poisonous. May cause headache, dizziness, profuse sweating, blurred vision, nervousness, nausea, vomiting, reduced heart beat, stomach cramps, diarrhea, loss of coordination, slow and weak breathing, fever, loss of consciousness, coma, uncontrollable twitching, loss of reflexes, loss of sphincter control and death. May cause shortness of breath and a dangerous build-up of fluids in the lungs (pulmonary edema), which can cause death. Pulmonary edema is a medical emergency that can be delayed one to two days following exposure.
   Eyes: Rapidly absorbed. Poisoning can happen very rapidly. See *Breathed.*
   Skin: Poisonous. Passes through the unbroken skin; can increase exposure and the severity of the symptoms listed.
   Swallowed: Powerful, deadly poison. See *Breathed.*
Long-term exposure: May damage the developing fetus or cause birth defects; nerve damage including loss of coordination; liver damage. Repeated exposure may cause depression, irritability and personality changes.
Storage: Store in a regulated, specially marked place in tightly closed containers in a dry, well-ventilated place. Keep away from water and oxidizers. Containers may explode in fire. Water may produce toxic gas.
First aid guide: Move victim to fresh air and call emergency medical care; if not breathing, give artificial respiration; if breathing is difficult, give oxygen. In case of contact with material, immediately flush skin or eyes with running water for at least 15 minutes. Speed in removing material from skin is of extreme importance. Remove and isolate contaminated clothing and shoes at the site. Keep victim quiet and maintain normal body temperature. Effects may be delayed; keep victim under observation.

★343★
DIAZOMETHANE
CAS: 334-88-3
Other names: AZIMETHYLENE; DIAZIRINE; DIAZONIUM; METHYLIDE

Danger profile: Carcinogen, mutagen, allergen, explosion hazard, exposure limit established, Community Right-to-Know List, poisonous gases produced in fire and on contact with acid.
Uses: Making other chemicals.
Appearance: Yellow gas at room temperature. Liquid under pressure.
Effects of exposure:
   Breathed: May cause irritation of the eyes, nose, throat and lungs, tears and runny nose, coughing, sore throat, difficult breathing; lethargy, fatigue, and drowsiness; slurred speech, shaking, and dizziness. Does not cause discernible reaction at the time of contact but later, even in minute amounts, produces an inflammatory reaction. Hypersensitivity results, which makes it impossible to work with without attacks of asthma and associated symptoms. Higher exposures could cause shortness of breath, headache, nausea, vomiting, dizziness, diarrhea and a dangerous build-up of fluids in the lungs (pulmonary edema), which can cause death. Pulmonary edema is a medical emergency that can be delayed one to two days following exposure.
   Eyes: May cause irritation, redness, tearing, pain, inflamed eyelids, severe burns and permanent damage.
   Skin: Toxic. May cause painful irritation, redness, and blisters Repeated or prolonged skin contact may cause burns and scaling. Similar symptoms may occur inside the mouth, nose and throat following inhalation.
   Swallowed: May cause irritation of the mouth and throat; stomach and abdominal pain; loss of muscular coordination and mental confusion; convulsions. Also see symptoms for *Breathed.*

Long-term exposure: May cause cancer, birth defects, miscarriage; severe lung damage; asthma-like lung allergy.
Storage: Store in tightly closed containers in a dry, cool, well-ventilated place. Keep away from heat, light, sources of ignition, calcium sulfate, alkali metals and acids. Containers may explode in heat, sunlight or other bright lights.

★344★
DIBENZYLDICHLOROSILANE

CAS: 18414-36-3
Other names: DICHLOROBIS(PHENYL METHYL)SILANE; SILANE, DICHLOROBIS(PHENYLMETHYL)-; SILANE, DIBENZYLDICHLORO-; UN 2434

Danger profile: Corrosive, combustible, poisonous gas produced in fire and on contact with water.
Uses: Making other chemicals; silicones.
Appearance: Colorless liquid.
Effects of exposure:
  Breathed: May cause irritation of the eyes, nose, throat, breathing passages and lungs.
  Eyes: May cause severe burns.
  Skin: May cause severe burns.
  Swallowed: Corrosive and reacts with moisture. May be very harmful.
Storage: Store in tightly closed containers in a dry, cool, well-ventilated place with non-wood floors. Keep away from water, steam and other forms of moisture, and combustible materials.
First aid guide: Move victim to fresh air; call emergency medical care. In case of contact with material, immediately flush skin or eyes with running water for at least 15 minutes. Remove and isolate contaminated clothing and shoes at the site. Keep victim quiet and maintain normal body temperature.

★345★
DIBORANE

CAS: 19287-45-7
Other names: BOROETHANE; BORON HYDRIDE; DIBORANE HEXANHYDRIDE; DIBORON HEXAHYDRIDE; DIBORANE(6)

Danger profile: Fire hazard, exposure limit established, reactive with water, poisonous gas produced in fire.
Uses: Making rubber and other chemicals; fuel for air-breathing engines and rockets; reducing agent; doping agent for semiconductors.
Appearance: Colorless gas.
Odor: Sickly-sweet, repulsive. The odor is a warning of exposure; however, no smell does not mean you are not being exposed.
Effects of exposure:
  Breathed: Poison. May cause irritation of the eyes, nose, throat and lungs, tears and runny nose, coughing, sore throat, difficult breathing; lethargy, fatigue, and drowsiness; slurred speech, shaking, and dizziness. Higher exposures could cause shortness of breath, headache, nausea, vomiting, dizziness, diarrhea and a dangerous build-up of fluids in the lungs (pulmonary edema), which can cause death. Pulmonary edema is a medical emergency that can be delayed one to two days following exposure.
  Eyes: May cause irritation, redness, tearing, pain, inflamed eyelids; burns in corneas and possible loss of sight.
  Skin: Toxic. Passes through the unbroken skin; can increase exposure and the severity of the symptoms listed. May cause painful irritation, redness, burns and blisters. contact with liquid may cause frostbite. Repeated or prolonged skin contact

may cause irritation of the skin and may also cause the legs to feel numb.
  Swallowed: May cause irritation of the mouth and throat; stomach and abdominal pain; loss of muscular coordination and mental confusion; convulsions. Also see symptoms for *Breathed.*
Long-term exposure: Repeated exposure may cause lung damage, bronchitis, cough, phlegm, wheezing and shortness of breath; kidney damage; nervous system damage.
Storage: Store in tightly closed containers in a dry, refrigerated, well-ventilated place. Keep away from heat, sources of ignitions, moisture, air, active metals, oxidizers and halogenated compounds. Chemical may ignite spontaneously in moist air at room temperature. Vapors may travel to sources of ignition and flash back. Containers may explode in fire.
First aid guide: Move victim to fresh air and call emergency medical care; if not breathing, give artificial respiration; if breathing is difficult, give oxygen. In case of contact with material, immediately flush skin or eyes with running water for at least 15 minutes. Remove and isolate contaminated clothing and shoes at the site. Keep victim quiet and maintain normal body temperature. Effects may be delayed; keep victim under observation.

★346★
DIBROMOBENZENE

CAS: 26249-12-7
Other names: BENZENE, DIBROMO-

Danger profile: Combustible, poisonous gas produced in fire.
Uses: For making other chemicals and a solvent for oils.
Appearance: Heavy colorless liquid.
Odor: Pleasant, aromatic. The odor is a warning of exposure; however, no smell does not mean you are not being exposed.
Effects of exposure:
  Breathed: May be poisonous. May cause irritation to the eyes, nose and throat. High levels may cause dizziness, light-headedness and unconsciousness. Possible suffocation.
  Eyes: Contact may cause irritation and burns.
  Skin: May be poisonous if absorbed. Passes through the unbroken skin; can increase exposure and the severity of the symptoms listed. Contact may cause irritation and burns.
Long-term exposure: Similar chemicals cause liver damage.
Storage: Store in tightly closed containers in a dry, cool, well-ventilated place. Keep away from sources of ignition. Containers may explode in fire.
First aid guide: Move victim to fresh air and call emergency medical care; if not breathing, give artificial respiration; if breathing is difficult, give oxygen. In case of contact with material, immediately flush eyes with running water for at least 15 minutes. Wash skin with soap and water. Remove and isolate contaminated clothing and shoes at the site.

★347★
DIBROMOMETHANE

CAS: 74-95-3
Other names: METHYLENE BROMIDE; METHYLENE DIBROMIDE; METHANE, DIBROMO-

Danger profile: Mutagen, poison, combustible, Community Right-to-Know List, poisonous gas produced in fire.
Uses: Solvent; making other chemicals.
Appearance: Colorless liquid.
Odor: Pleasant, sweet, chloroform-like, aromatic. The odor is a warning of exposure; however, no smell does not mean you are not being exposed.
Effects of exposure:

Breathed: May be poisonous. May cause dizziness, light-headedness; anesthetic effects, nausea and drunkenness, unconsciousness; irregualar heartbeat and possible death.
Eyes: Contact may irritate the eyes.
Skin: May be poisonous if absorbed. Contact may irritate the eyes.
Swallowed: Poisonous.
Long-term exposure: May cause liver and kidney damage; dermatitis with dryness, itching and irritation. May cause a buildup of carbon monoxide in the body.
Storage: Store in tightly closed containers in a dry, cool, well ventilated place. Keep away from heat and potassium. Containers may explode in fire.
First aid guide: Move victim to fresh air and call emergency medical care; if not breathing, give artificial respiration; if breathing is difficult, give oxygen. In case of contact with material, immediately flush eyes with running water for at least 15 minutes. Wash skin with soap and water. Remove and isolate contaminated clothing and shoes at the site. Use first aid treatment according to the nature of the injury.

★ 348 ★
DIBUTYLAMINOETHANOL
CAS: 102-81-8
Other names: BU2AE; DBAE; 2-DIBUTYLAMINOETHANOL; 2-(DIBUTYLAMINO) ETHANOL 2-N-DIBUTYLAMINOETHANOL; BETA-N-DIBUTYLAMINOETHYL ALCOHOL; N-N-DIBUTYLETHANOLAMINE; N-N-DI-N-BUTYLAMINOETHANOL; N,N-DIBUTYL N-(2-HYDROXYETHYL)AMINE; ETHANOL, 2-(DIBUTYLAMINO)-

Danger profile: Combustible, exposure limit established, poisonous gases produced in fire.
Uses: Making other chemicals.
Appearance: Colorless liquid.
Odor: Faint. The odor is a warning of exposure; however, no smell does not mean you are not being exposed.
Effects of exposure:
Breathed: May be poisonous. Vapors may cause irritation to the eyes, nose throat and breathing passages; may interfere with nerve enzyme causing symptoms of headache, sweating, nausea and vomiting, diarrhea, muscle twitching and coma; possible death; liver and kidney damage.
Eyes: Contact may cause severe irritation and burns.
Skin: May be poisonous if absorbed. Passes through the unbroken skin; can increase exposure and the severity of the symptoms listed. Contact may cause burns, rash and itching.
Swallowed: Poisonous.
Long-term exposure: May cause dermatitis, weight loss and an increase in clotting time. Repeated exposure may cause liver and kidney damage. Similar and related chemicals may cause skin and lung allergy, skin rash; shortness of breath and a dangerous build-up of fluids in the lungs (pulmonary edema), which can cause death. Pulmonary edema is a medical emergency that can be delayed one to two days following exposure.
Storage: Store in tightly closed containers in a dry, cool, well-ventilated place. Keep away from oxidizing materials, and sources of ignition. Containers may explode in fire.
First aid guide: Move victim to fresh air and call emergency medical care; if not breathing, give artificial respiration; if breathing is difficult, give oxygen. In case of contact with material, immediately flush skin or eyes with running water for at least 15 minutes. Speed in removing material from skin is of extreme importance. Remove and isolate contaminated clothing and shoes at the site. Keep victim quiet and maintain normal body temperature. Effects may be delayed; keep victim under observation.

★ 349 ★
DIBUTYL PHOSPHATE
CAS: 107-66-4
Other names: DIBUTYL ACID O-PHOSPHATE; DIBUTYL ACID PHOSPHATE; DI-N-BUTYL HYDROGEN PHOSPHATE; DIBUTYL PHOSPHORIC ACID; PHOSPHORIC ACID, DIBUTYL ESTER.

Danger profile: Exposure limit established, irritating fumes and poisonous gases produced in fire
Uses: Anti-foaming agent and for making other chemicals.
Appearance: Colorless, amber to brown colorless liquid.
Effects of exposure:
Breathed: May cause irritation to the eyes, nose, throat, and lungs; headache, shortness of breath and coughing.
Eyes: May cause irritation.
Skin: Passes through the unbroken skin; can increase exposure and the severity of the symptoms listed. May cause irritation, dry and cracked skin.
Swallowed: Moderately toxic.
Long-term exposure: May cause skin drying and cracking. Similar chemicals may cause lung damage.
Storage: Store in tightly closed containers in a dry, cool, well-ventilated place. Keep away from oxidizing materials, and heat.

★ 350 ★
DIBUTYL PHTHALATE
CAS: 84-74-2
Other names: O-BENZENEDICARBOXYLIC ACID, DIBUTYL ESTER; BENZENE-O-DICARBOXYLIC ACID DI-N-BUTYL ESTER; 1,2-BENZENEDICARBOXYLIC ACID, DIBUTYL ESTER; DI-N-BUTYL PHTHALATE; N-BUTYLPHTHALATE; PHTHALIC ACID DIBUTYL ESTERCELLUFLEX DPB; DBP; DIBUTYL 1,2-BENZENEDICARBOXYLATE; DI-N-BUTYL PHTHALATE; ELAOL; HEXAPLAS M/B; PALATINOL C; POLYCIZER DBP; PX 104; STAFLEX DBP; WITCIZER 300

Danger profile: Mutagen, reproductive hazard, combustible, exposure limit established, Community Right-to-Know List.
Uses: Insect repellent, lacquer solvent, making flexible plastics.
Appearance: Colorless, oily liquid.
Effects of exposure:
Breathed: May cause irritation of eyes, nasal passages, and upper respiratory system. Also see ***Swallowed.***
Eyes: Contact with vapor or aerosol may cause irritation, tearing.
Skin: Contact may cause irritation.
Swallowed: May cause stomach irritation, light sensitivity, labored breathing, nausea, dizziness, hallucinations, distorted perceptions.
Long-term exposure: May cause damage to the male testes and the developing fetus; changes to the kidneys, bladder and ureter.
Storage: Store in tightly closed containers in a dry, cool, well-ventilated place; refrigeration is preferred. Keep away from heat and sources of ignition, oxidizers such as chlorine, acids and alkalis. Containers may explode in fire.

★351★
1,2-DICHLORETHANE
CAS: 107-06-2
Other names: AETHYLENCHLORID (GERMAN); 1,2-BICHLOROETHANE; BICHLORURE D'ETHYLENE (FRENCH); BORER SOL; BROCIDE; CHLORURE D'ETHYLENE (FRENCH); CLORURO DI ETHENE (ITALIAN); DESTRUXOL BORER-SOL; 1,2-DICHLOORETHAAN (DUTCH); 1,2-DICHLOR-AETHAN (GERMAN); DICHLORE-MULSION; DI-CHLOR-MULSION; DICHLORO-1,2-ETHANE (FRENCH); ALPHA,BETA-DICHLOROETHANE; SYM-DICHLOROETHANE; 1,2-DICHLOROETHANE; DICHLORO-ETHYLENE; 1,2-DICLOROETANO (ITALIAN); DUTCH LIQUID; DUTCH OIL; EDC; ENT 1,656; ETHANE, 1,2-DICHLORO-; ETHANE DICHLORIDE; ETHYLEENDICHLORIDE (DUTCH); ETHYLENE CHLORIDE; ETHYLENE DICHLORIDE; ETHYLENE DICHLORIDE; 1,2-ETHYLENE DI-CHLORIDE; FREON 150; GLYCOL DICHLORIDE; NCI-C00511

Danger profile: Carcinogen, mutagen, reproductive hazard, fire hazard, exposure limit established, Community Right-to-Know List.
Uses: Making vinyl chloride; in gasoline; paint; varnish and finish removers; metal degreasing; soaps and scouring compounds; making other chemicals.
Appearance: Clear liquid.
Odor: Sweet chloroform-like. The odor is a warning of exposure; however, no smell does not mean you are not being exposed.
Effects of exposure:
Breathed: May cause irritation of the nose, throat and lungs; nausea, headaches, dizziness, abdominal pain, nausea, vomiting, liver and kidney damage; loss of consciousness and death. Higher exposures may cause shortness of breath and a dangerous build-up of fluids in the lungs (pulmonary edema), which can cause death. Pulmonary edema is a medical emergency that can be delayed one to two days following exposure.
Eyes: Contact may cause irritation, tearing and inflammation of the eyelids; corneal damage.
Skin: Passes through the unbroken skin; can increase exposure and the severity of the symptoms listed. Contact may cause irritation, redness and rash.
Swallowed: See *Breathed.*
Long-term exposure: May cause liver and kidney damage.
Storage: Store in tightly closed containers in a dry, cool, well-ventilated explosion-proof place. Keep away from oxidizers, acids, chemically active metals such as potassium, sodium, magnesium, and zinc, caustics and dimethylaminoproylamine. Containers may explode in fire.

★352★
DICHLOROACETYLENE
CAS: 7572-29-4
Other names: DCA; DICHLOROETHYNE; ETHYNE, DI-CHLORO- (9CI); ACETYLENE, DICHLORO-

Danger profile: Carcinogen, mutagen, combustible, explosion hazard (explodes before reaching a temperature hot enough to burn); exposure limit established, highly toxic and poisonous gases produced in fire.
Uses: Occurs from thermal decomposition of trichloroethylene.
Appearance: Colorless liquid, but often found in the work place as a vapor.
Effects of exposure:
Breathed: May cause headache, severe nausea, vomiting, facial and jaw pain and paralysis of the face and pain that may occur one to three days following exposure. Higher exposures may cause shortness of breath and a dangerous build-up of fluids in the lungs (pulmonary edema), which can cause death. Pulmonary edema is a medical emergency that can be delayed one to two days following exposure.
Storage: Store in tightly closed containers in a dry, cool, well-ventilated place. Keep away from shock, heat, sources of ignition, oxidizers, acids and acid fumes, potassium, sodium and aluminum powder. Containers may explode in fire.

★353★
1,2-DICHLOROBENZENE
CAS: 95-50-1
Other names: BENZENE, 1,2-DICHLORO-; CHLOROBEN; CHLORODEN; CLOROBEN; DCB; O-DICHLORBENZENE; O-DICHLOR BENZOL; O-DICHLOROBENZENE; DICHLORO-BENZENE, ORTHO, LIQUID; DICHLORICIDE; O-DICHLOROBENZOL; DILANTIN DB; DILATIN DB; DIZENE; DOWTHERM E; NCI-C54944; ODB; ODCB; ORTHODI-CHLOROBENZENE; ORTHODICHLOROBENZOL; SPECIAL TERMITE FLUID; TERMITKIL

Danger profile: Possible carcinogen, mutagen, possible reproductive hazard, combustible, exposure limit established, Community Right-to-Know List, highly poisonous gases produced in fire.
Uses: Insecticide; dry cleaning; degreasing agent.
Appearance: Colorless to pale yellow liquid.
Odor: Pleasant, aromatic.
Effects of exposure:
Breathed: Vapor or dust extremely irritating. Vapor may cause irritation to the throat and mucous membrane; coughing, dizziness, nausea, vomiting; central nervous system depression, lightheadedness and unconsciousness.
Eyes: May cause irritation. Prolonged contact may cause severe burns.
Skin: Passes through the unbroken skin; can increase exposure and the severity of the symptoms listed. May cause irritation. Prolonged contact may cause severe skin rash and burns.
Swallowed: Poisonous.
Long-term exposure: Repeated inhalation of mist or vapors may result in damage to lungs, liver, kidneys; anemia; damage to the testes. There is evidence for an association between exposure to chemical and leukemia. Similar solvents may cause brain or other nerve damage; reduced memory and concentration, personality changes, fatigue, sleep disturbances, reduced coordination and effects on nerves supplying internal organs and the arms and legs.
Storage: Store in tightly closed containers in a dark, dry, cool, outdoor or detached well-ventilated place. Keep away from heat, direct light, sources of ignition, oxidizers, aluminum or aluminum alloys.

★354★
1,4-DICHLOROBENZENE
CAS: 106-46-7
Other names: BENZENE, 1,4-DICHLORO-; DICHLOROCIDE; P-CHLOROPHENYL CHLORIDE; P-DICHLOORBENZEEN (DUTCH); 1,4-DICHLOORBENZEEN (DUTCH); P-DICHLORBENZOL (GERMAN); 1,4-DICHLOR-BENZOL (GERMAN); DI-CHLORICIDE; P-DICHLOROBENZENE; P-DICHLOROBENZOL; DICHLOROBENZENE, PARA, SOLID; 1,4-DICLOROBENZENE (ITALIAN); P-DICLOROBENZENE (ITALIAN); EVOLA; NCI-C54955; PARACIDE; PARA CRYSTALS; PARADI; PARADICHLOROBENZENE; PARADI-CHLOROBENZOL; PARADOW; PARAMOTH; PARANUGGETS; PARAZENE; PDB; PDCB; PERSIA-PERAZOL; SANTOCHLOR

Danger profile: Carcinogen, mutagen, possible reproductive hazard, combustible, exposure limit established, Community Right-to-Know List.
Uses: Fumigant. To control mildew and mold; liquid is used as a deodorant.
Appearance: Colorless to white crystalline (sand-like) material.
Odor: Penetrating, mothball-like. The odor is a warning of exposure; however, no smell does not mean you are not being exposed.
Effects of exposure:
  Breathed: May cause irritation of the upper respiratory tract, headache, dizziness, swelling of the eyes, hands and feet; nausea and vomiting. Higher exposure may cause damage to liver, kidneys and may result in death.
  Eyes: May cause irritation and pain.
  Skin: Passes through the unbroken skin; can increase exposure and the severity of the symptoms listed. Contact may cause irritation and burns. Allergy may develop.
  Swallowed: May be poisonous. See **Breathed.**
Long-term exposure: Repeated exposure may cause damage to the blood cells causing anemia; lungs, liver and kidneys; nervous system causing weakness, trembling and numbness in the legs and arms; skin allergy with itching and rash; jaundice. May damage the developing fetus.
Storage: Store in tightly closed containers in a dry, cool, well-ventilated place. Keep away from heat, sources of ignition and oxidizing materials (violent reaction).

★ 355 ★
DICHLORODIFLUORO-ETHYLENE
CAS: 27156-03-2
Other names: DICHLORODIFLUOROETHYLENE; ETHENE, DICHLORODIFLUORO-; ETHYLENE, DICHLORODIFLUORO-

Danger profile: Reacts with water or vapor to produce corrosive and toxic fumes; poisonous gases produced in fire.
Uses: Research and development.
Appearance: Colorless gas or liquid.
Effects of exposure:
  Breathed: May cause irritation of the eyes nose and throat. High levels may cause dizziness, lightheadedness and unconsciousness. May cause the heart to beat irregularly or to stop, which may result in death.
  Eyes: May cause irritation, tearing an swelling.
  Skin: Contact may cause irritation, rash or burning feeling.
Long-term exposure: May cause liver and kidney damage.
Storage: Store in tightly closed containers in a dry, cool, well-ventilated place. Keep away from water or steam. Containers may explode in fire.
First aid guide: Move victim to fresh air and call emergency medical care; if not breathing, give artificial respiration; if breathing is difficult, give oxygen. In case of contact with material, immediately flush eyes with running water for at least 15 minutes. Wash skin with soap and water. Remove and isolate contaminated clothing and shoes at the site. Use first aid treatment according to the nature of the injury.

★ 356 ★
DICHLORODIFLUOROMETHANE
CAS: 75-71-8
Other names: ALGOFRENE TYPE 2; ARCTON 6; CFC-12; DIFLUORODICHLOROMETHANE; DWUCHLORODWUFLUOROMETAN (POLISH); ELECTRO-CF 12; ESKIMON 12; F 12; FC 12; FLUOROCARBON-12; FREON 12; FREON F-12; FRIGEN 12; GENETRON 12; HALON; HALON 122; ISCEON 122; ISOTRON 12; LEDON 12; METHANE, DICHLORODIFLUORO-; PROPELLANT 12; R 12; REFRIGERANT 12; UCON 12; UCON 12/HALOCARBON 12

Danger profile: Exposure limit established, Community Right-to-Know List, poisonous gases produced in fire.
Uses: Aerosol propellant; refrigerant and air conditioner; making plastics; solvent; freezing of foods by direct contact; for chilling cocktail glasses.
Appearance: Colorless gas or liquid under pressure.
Odor: Ether-like. The odor is a warning of exposure; however, no smell does not mean you are not being exposed.
Effects of exposure:
  Breathed: May cause irritation of the mouth, nose and throat; dizziness, lightheadedness and loss of concentration; tremors. May cause the heart to beat irregularly or to stop, which may result in death. May cause suffocation.
  Eyes: Contact may cause severe eye burns from frostbite.
  Skin: Contact may cause severe burns from frostbite.
Long-term exposure: May cause liver changes.
Storage: Store in tightly closed containers in a dry, cool, well-ventilated place. Keep away from heat and chemically active metals such as sodium, potassium, calcium, powdered aluminum, magnesium and zinc. Containers may explode in fire.
First aid guide: Move victim to fresh air and call emergency medical care; if not breathing, give artificial respiration; if breathing is difficult, give oxygen.

★ 357 ★
1,3-DICHLORO-5,5-DIMETHYL HYDANTOIN
CAS: 118-52-5
Other names: DACTIN; DAKTIN; DANTOIN; DCA; DCDMH; DICHLORANTIN; 1,3-DICHLORO-5,5-DIMETHYLHYDANTOIN; 1,3-DICHLORO-5,5-DIMETHYL-; HYDANTOIN, DICHLORODIMETHYL-; 2,4-IMIDAZOLIDINEDIONE; HALANE; HYDAN; NCI-C03054; OM-CHLOR

Danger profile: Mutagen, exposure limit established, poisonous gases produced in fire.
Uses: Antiseptic; Water treatment; laundry bleach; making pharmaceuticals, drugs and other chemicals.
Appearance: Colorless white powder or liquid.
Odor: Mild, chlorine-like. The odor is a warning of exposure; however, no smell does not mean you are not being exposed.
Effects of exposure:
  Breathed: May cause irritation of the eyes, nose and throat. Higher exposures may cause shortness of breath and a dangerous build-up of fluids in the lungs (pulmonary edema), which can cause death. Pulmonary edema is a medical emergency that can be delayed one to two days following exposure.
  Eyes: May cause irritation.
  Skin: May cause severe irritation.
  Swallowed: Mildly toxic. See **Breathed.**
Long-term exposure: Similar chemicals are central nervous system depressants.
Storage: Store in a refrigerator or tightly closed containers in a dry, cool, well-ventilated place. Keep away from heat, sources of ignition, moisture, acids, easily oxidized materials such as sulfides and ammonium salts, especially xylene. Containers may explode in fire.

★358★
1,1-DICHLOROETHANE
CAS: 75-34-3
Other names: AETHYLIDENCHLORID (GERMAN); CHLORINATED HYDROCHLORIC ETHER; CHLORURE D'ETHYLIDENE (FRENCH); CLORURO DI ETILIDENE (ITALIAN); 1,1-DICHLOORETHAAN (DUTCH); 1,1-DICHLORAETHAN (GERMAN); 1,1-DICHLORETHANE; 1,1-DICLOROETANO (ITALIAN); ETHYLIDENE CHLORIDE; ETHYLIDENE DICHLORIDE; NCI-C04535; ETHANE, 1,1-DICHLORO-; DICHLOROMETHYLETHANE

Danger profile: Dangerous fire hazard, possible reproductive hazard, exposure limit established, highly toxic gases produced in fire.
Uses: Solvent; fumigant.
Appearance: Colorless liquid.
Odor: Chloroform-like. The odor is a warning of exposure; however, no smell does not mean you are not being exposed.
Effects of exposure:
   Breathed: May cause drowsiness, unconsciousness and death; liver and kidney damage.
   Eyes: Contact may cause irritation and burns.
   Skin: Contact may cause irritation and burns.
   Swallowed: Moderately toxic.
Long-term exposure: May cause damage to the developing fetus; liver and kidneys.
Storage: Do not use near welding operation; deadly phosgene gas may be produced. Store in tightly closed containers in a dry, cool, well-ventilated place. Keep away from sources of ignition, sunlight, heat and oxidizers. Containers may explode in fire.
First aid guide: Move victim to fresh air and call emergency medical care; if not breathing, give artificial respiration; if breathing is difficult, give oxygen. In case of contact with material, immediately flush eyes with running water for at least 15 minutes. Wash skin with soap and water. Remove and isolate contaminated clothing and shoes at the site.

★359★
1,2-DICHLOROETHYLENE
CAS: 540-59-0
Other names: ACETYLENE DICHLORIDE; CIS-ACETYLENE DICHLORIDE; TRANS-ACETYLENE DICHLORIDE; 1,2-DICHLOR-AETHEN (GERMAN); SYS-DICHLOROETHYLENE; DICHLORO-1,2-ETHYLENE (FRENCH); SYM-DICHLOROETHYLENE; DIOFORM; ETHENE, 1,2-DICHLORO-; NCI-C56031

Danger profile: Fire hazard, explosion hazard, exposure limit established, Community Right-to-Know List, poisonous gases produced in fire.
Uses: Solvent for organic materials; dye extraction; perfumes; lacquers; thermoplastics; making other chemicals.
Appearance: Colorless liquid.
Odor: Ether or chloroform-like, slightly acrid odor. The odor is a warning of exposure; however, no smell does not mean you are not being exposed.
Effects of exposure:
   Breathed: Poisonous. May cause irritation of the nose, throat and lungs; nausea, vomiting, weakness, dizziness, lightheadedness, tremor, epigastric cramps, central nervous depression, loss of consciousness.
   Eyes: Contact may cause irritation.
   Skin: May be poisonous if absorbed. Contact may cause irritation.
   Swallowed: May be poisonous. Symptoms may vary from slight depression to deep narcosis. Possibly fatal.
Long-term exposure: May cause liver and kidney damage; lung damage and bronchitis; low blood count.

Storage: Store in tightly closed containers in a dry, cool, well-ventilated detached or outside place. Keep away from heat, light, and moisture; copper and copper alloys, oxidizing materials. Vapors may travel to sources of ignition and flash back. Containers may explode in fire.

★360★
DICHLOROETHYL ETHER
CAS: 111-44-4
Other names: BCEE; BIS(BETA-CHLOROETHYL) ETHER; BIS(2-CHLOROETHYL) ETHER; CHLOREX; 1-CHLORO-2-(BETA-CHLOROETHOXY)ETHANE; BIS(2-CHLOROETHYL)ETHER; CHLOROETHYL ETHER; CLOREX; DCEE; 2,2'-DICHLOORETHYLETHER (DUTCH); 2,2'-DICHLORO-DIETHYLETHER; 2,2'-DICHLOR-DIAETHYLAETHER (GERMAN); 2,2'-DICHLORETHYL ETHER; BETA,BETA-DICHLORODIETHYL ETHER; DI-CHLOROETHER; DICHLOROETHYL ETHER; DI(BETA-CHLOROETHYL)ETHER; DI(2-CHLOROETHYL) ETHER; BETA,BETA'-DICHLOROETHYL ETHER; SYM-DICHLOROETHYL ETHER; 2,2'-DICHLOROETHYL ETHER; DICHLOROETHYL OXIDE; 2,2'-DICLOROETILETERE (ITALIAN); DWUCHLORODWUETYLOWY ETER (POLISH); ENT 4,504; ETHANE, 1,1'-OXYBIS 2-CHLORO-; ETHER DICHLORE (FRENCH); 1,1'-OXYBIS(2-CHLORO)ETHANE; OXYDE DE CHLORETHYLE (FRENCH)

Danger profile: Possible carcinogen, mutagen, possible reproductive hazard, combustible, exposure limit established, Community Right-to-Know List, EPA extremely hazardous substance list, poisonous gas produced in fire.
Uses: Making pesticides and textiles.
Appearance: Colorless liquid.
Odor: Chlorinated solvent-like. The odor is a warning of exposure; however, no smell does not mean you are not being exposed.
Effects of exposure:
   Breathed: May be poisonous. May cause irritation of the mucous membrane of the eyes, nose and throat; nausea. Mild narcotic effect. Repeated exposure may cause bronchitis with coughing, phlegm, and shortness of breath. Higher exposures may cause shortness of breath and a dangerous build-up of fluids in the lungs (pulmonary edema), which can cause death. Pulmonary edema is a medical emergency that can be delayed one to two days following exposure.
   Eyes: May cause severe irritation to the mucous membrane with swelling of the eyelids.
   Skin: May be poisonous if absorbed. Passes through the unbroken skin; can increase exposure and the severity of the symptoms listed. May cause severe irritation.
   Swallowed: Poisonous.
Long-term exposure: May cause liver and kidney damage; bronchitis with cough, phlegm and shortness of breath. Has shown to be a cancer causing agent in animals; may have potential for causing cancer and reproductive damage in humans.
Storage: Store in tightly closed containers in a dry, cool, well-ventilated place. Keep away from heat, sources of ignition, oleum, chlorosulfonic acid and oxidizing materials. Containers may explode in fire.
First aid guide: Move victim to fresh air and call emergency medical care; if not breathing, give artificial respiration; if breathing is difficult, give oxygen. In case of contact with material, immediately flush skin or eyes with running water for at least 15 minutes. Speed in removing material from skin is of extreme importance. Remove and isolate contaminated clothing and shoes at the site. Keep victim quiet and maintain normal body temperature. Effects may be delayed; keep victim under observation.

★361★
2,4-DICHLOROPHENOL ACETATE ESTER
CAS: 6341-97-5
Other names: 2,4-DICHLOROPHENOXYACETIC ACID; 2,4-DICHLOROPHENYL ACETATE; PHENOL, 2,4-DICHLORO-ACETATE

Danger profile: Poisonous gases produced in fire.
Uses: Pesticide.
Appearance: White to tan solid.
Effects of exposure:
    Breathed: May cause stomach pain, nausea, vomiting, weakness, muscle pain and twitching. Higher levels may cause damage to the liver, kidneys and nervous system; coma and death.
    Eyes: Contact may cause irritation.
    Skin: Passes through the unbroken skin; can increase exposure and the severity of the symptoms listed. Contact may cause irritation.
    Swallowed: See *Breathed.*
Long-term exposure: May cause damage to the liver and kidneys; nervous system damage causing numbness and weakness in the hands and feet. Repeated contact may cause skin rash.
Storage: Store in tightly closed containers in a dry, cool, well-ventilated place. Containers may explode in fire.

★362★
1,2-DICHLORO-4-PHENYL ISOCYANATE
CAS: 102-36-3
Other names: DICHLOROPHENYL ISOCYANATE, ISOCYANIC ACID,3,4-DICHLOROPHENYL ESTER

Danger profile: Combustible, poisonous gases produced in fire.
Uses: Making other chemicals.
Appearance: White to tan crystalline solid.
Effects of exposure:
    Breathed: May be poisonous. May cause severe irritation of the nose, throat, air passages and lungs.
    Eyes: May cause severe irritation, tearing and inflammation of the eyelids.
    Skin: May be poisonous if absorbed. May cause severe irritation.
    Swallowed: May be poisonous.
Long-term exposure: May cause asthma-like allergy with shortness of breath, wheezing, cough and chest tightness. Repeated skin exposure may cause rash.
Storage: Store in tightly closed containers in a dry, cool, well-ventilated place. Keep away from heat above 140°F, moisture and water, alcohols, bases amines, metals and carboxylic acids. Containers may explode in fire.

★363★
DICHLOROPHENYL TRICHLOROSILANE
CAS: 27137-85-5
Other names: DICHLOROPHENYLTRICHLOROSILANE; (DICHLOROPHENYL)TRICHLOROSILANE; DICHLORO (DICHLOROPHENYL-); SILANE,(DICHLOROPHENYL-); SILANE,(DICHLOROPHENYL)TRICHLORO-; TRICHLORO (DICHLOROPHENYL) SILANE

Danger profile: Combustible, corrosive, releases hydrochloric acid on contact with moisture, Community Right-to-Know List, poisonous gas produced in fire.
Uses: In silicones.
Appearance: Straw colored liquid.
Odor: Pungent. The odor is a warning of exposure; however, no smell does not mean you are not being exposed.

Effects of exposure:
    Breathed: May cause irritation of the eyes, nose and throat; lung irritation with coughing and shortness of breath. Higher exposures may cause shortness of breath and a dangerous build-up of fluids in the lungs (pulmonary edema), which can cause death. Pulmonary edema is a medical emergency that can be delayed one to two days following exposure.
    Eyes: Contact may cause severe burns and permanent damage.
    Skin: Contact may cause severe burns.
    Swallowed: Poisonous. Very irritating corrosive. See *Breathed* and *First aid guide.*
Long-term exposure: May cause lung damage.
Storage: Store in tightly closed containers in a dry, cool, well-ventilated place with non-wood floors. Keep away from moisture and combustible materials.
First aid guide: Move victim to fresh air; call emergency medical care. In case of contact with material, immediately flush skin or eyes with running water for at least 15 minutes. Remove and isolate contaminated clothing and shoes at the site. Keep victim quiet and maintain normal body temperature.

★364★
1,2-DICHLOROPROPANE
CAS: 78-87-5
Other names: BICHLORURE DE PROPYLENE (FRENCH); ALPHA,BETA-DICHLOROPROPANE; DWUCHLOROPROPAN (POLISH); ENT 15,406; NCI-C55141; PROPANE, 1,2-DICHLORO-; PROPYLENE CHLORIDE; ALPHA,BETA-PROPYLENE DICHLORIDE; PROPYLENE DICHLORIDE

Danger profile: Possible carcinogen, mutagen, fire hazard, exposure limit established, Community Right-to-Know List, poisonous gases produced in fire.
Uses: Making other chemicals; in antiknock gasoline; in dry cleaning fluids; solvents for fats, oils, waxes, gums, resins and metal degreasing; scouring compounds; insecticide and soil fumigant for nematodes.
Appearance: Colorless liquid.
Odor: Chloroform-like. The odor is a warning of exposure; however, no smell does not mean you are not being exposed.
Effects of exposure:
    Breathed: Vapor may cause irritation to the nose, throat, eyes and air passages. High exposure may irritate the lungs; dizziness, lightheadedness and unconsciousness; may cause liver, kidney and adrenal gland damage. Narcotic at high vapor concentrations.
    Eyes: Contact may cause severe irritation.
    Skin: Passes through the unbroken skin; can increase exposure and the severity of the symptoms listed. Skin exposure may result in fat extraction and dermatitis.
    Swallowed: Poisonous.
Long-term exposure: Repeated or prolonged skin exposure may cause skin rash. Repeated exposure may cause liver, kidney and brain damage with feeling of drunkenness, headache, nausea and personality changes. Exposure may result in life-threatening heart rhythm changes. There is limited evidence of cancer in animals.
Storage: Store in tightly closed containers in a dry, cool, well-ventilated place or explosion-proof refrigerator. Keep away from heat, direct light, sources of ignition, aluminum, aluminum chloride, o-dichlorobenzene and 1,2-dichloroethane, acids and oxidizing materials. Vapors may travel to sources of ignition and flash back. Containers may explode in fire.

★ 365 ★
1,3-DICHLOROPROPENE
CAS: 542-75-6
Other names: ALPHA-CHLOROALLYL CHLORIDE; 3-CHLORO-ALLYL CHLORIDE; GAMMA-CHLORO ALLYL CHLORIDE; ALPHA,GAMMA-DICHLOROPROPYLENE; DI-CHLOROPROPENE; 1,3-DICHLORO-1-PROPENE; 1,3-DICHLOROPROPENE-1; 1,3-DICHLOROPROPYLENE; NCI-C03985; TELONE; TELONE II SOIL FUMIGANT; VIDDEN D

Danger profile: Carcinogen, fire hazard; exposure limit established, Community Right-to-Know List; poisonous gases produced in fire.
Uses: Making other chemicals; soil fumigant; pesticide.
Appearance: Colorless or straw colored liquid.
Odor: Sweet. The odor is a warning of exposure; however, no smell does not mean you are not being exposed.
Effects of exposure:
   Breathed: May be poisonous. May cause headaches, chest pain, dizziness, unconsciousness. High exposures may cause kidney, liver and lung damage.
   Eyes: Contact may cause severe burns and permanent damage.
   Skin: Passes through the unbroken skin; can increase exposure and the severity of the symptoms listed. Strong irritant. May cause severe irritation and second-degree burns.
   Swallowed: Poison. See *Breathed.*
Long-term exposure: May cause cancer of the stomach, liver and lungs; kidney, liver and lung damage; chronic headache, chest pain and personality changes.
Storage: Store in tightly closed containers in a dry, cool, well-ventilated place. Keep away from heat, sources of ignition, aluminum, magnesium compounds; substances containing fluorine, chlorine, bromine or iodine, alkaline or corrosive materials. Vapors may travel to sources of ignition and flash back. Containers may explode in fire.

★ 366 ★
2,2-DICHLOROPROPIONIC ACID
CAS: 75-99-0
Other names: ALATEX; BASFAPON; BASFAPON B; BASFAPON/BASFAPON N; BH DALAPON; BASINEX; CRISAPON; DALAPON; DALAPON 85; DED-WEED; DEVIPON; ALPHA-DICHLOROPROPIONIC ACID; ALPHA,ALPHA-DICHLOROPROPIONIC ACID; 2,2-DICHLOROPROPIONIC ACID; DOWPON; DOWPON M; GRAMEVIN; KENAPON; LIROPON; PROPANOIC ACID, 2,2-DICHLORO-; PROPROP; RADAPON; REVENGE; S95; UNIPON

Danger profile: Mutagen, corrosive, exposure limit established, poisonous gases produced in fire.
Uses: Herbicide.
Appearance: Although the chemical is a liquid, the commercial herbicide is white to tan powder.
Effects of exposure:
   Breathed: May cause irritation of the eyes, nose throat and lungs; throat pain, loss of appetite, perspiration and nausea.
   Eyes: Corrosive. May cause burns and permanent damage.
   Skin: Corrosive. May cause burns; skin allergy.
   Swallowed: Corrosive. See *First aid guide.*
Long-term exposure: May cause skin allergy. If allergy develops, low future exposures may cause skin rash and itching.
Storage: Store in tightly closed containers in a dry, cool, well-ventilated place with non-wood floors. Keep away from heat and combustible materials.
First aid guide: Move victim to fresh air; call emergency medical care. In case of contact with material, immediately flush skin or eyes with running water for at least 15 minutes. Remove and iso-late contaminated clothing and shoes at the site. Keep victim quiet and maintain normal body temperature.

★ 367 ★
DICOFOL
CAS: 115-32-2
Other names: ACARIN; BENZENEMETHANOL, 4-CHLORO-ALPHA-(4-CHLOROPHENYL)-ALPHA-(TRICHLOROMETHYL)-; CARBOX; CEKUDIFOL; CPCA; DE-COFOL; DICHLOROKELTHANE; DROL; DTMC; ENT 23,648; FW 293; HIFOL; KELTANE; P,P'KELTHANE; KELTHANE; KELTHANE DUST BASE; KELTHANETHANOL; MILBOL; MITIGAN; NCI-C00486; 1,1-BIS(P-CHLOROPHENYL)-2,2,2-TRICHLOROETHANOL; 4-CHLORO-ALPHA-(4-CHLOROPHENYL)-ALPHA-(TRICHLOROMETHYL)BENZENEME THANOL; DI-(P-CHLOROPHENYL)TRICHLOROMETHYLCARBINOL; 4,4'-DICHLORO-ALPHA-(TRICHLOROMETHYL)BENZHYDROL; ENT 23,648; KELTHANETHANOL; NCI-C00486; 2,2,2-TRICHLOOR-1,1-BIS(4-CHLOORFENYL)-ETHANOL (DUTCH); 2,2,2-TRICHLOR-1,1-BIS(4-CHLOR-PHENYL)-AETHANOL (GERMAN); 2,2,2-TRICHLORO-1,1-BIS(4-CHLOROPHENYL)-ETHANOL (FRENCH); 2,2,2-TRICHLORO-1,1-BIS(4-CLORO-FENIL)-ETANOLO (ITALIAN); 2,2,2-TRICHLORO-1,1-DI-(4-CHLOROPHENYL)ETHANOL

Danger profile: Possible carcinogen, mutagen, combustible, Community Right-to-Know List, poisonous gases produced in fire.
Uses: Pesticide (organochlorine).
Appearance: White crystals.
Effects of exposure:
   Breathed: May be poisonous. May cause headaches, irritability, poor appetite or nausea. May effect the nervous system. High exposures may cause weakness, numbness in the arms, legs, face and mouth; muscle twitching, shaking, convulsions, loss of consciousness and death.
   Eyes: May cause irritation
   Skin: Poisonous. Passes through the unbroken skin; can increase exposure and the severity of the symptoms listed. See *Breathed.*
   Swallowed: Poisonous.
Long-term exposure: Closely related to DDT. Evidence of animal cancer; may cause liver cancer. May cause decrease of fertility in females.
Storage: Store in tightly closed containers in a dry, cool, well-ventilated place. Keep away from sources of ignition and food.

★ 368 ★
DICROTOPHOS
CAS: 141-66-2
Other names: BIDRIN; BIDRIN (SHELL); C-709; C-709 (CIBA-GEIGY); CARBICRIN; CIBA 709; CROTONAMIDE, 3-HYDROXY-N,N-DIMETHYL-, CIS-,DIMETHYL PHOSPHATE; CROTONAMIDE, 3-HYDROXY-N-N-DIMETHYL-, DIMETHYL-PHOSPHATE, CIS-; CROTONAMIDE, 3-HYDROXY-N-N-DIMETHYL-, DIMETHYLPHOSPHATE, (E)-; DIAPADRIN; DICROTOFOS (DUTCH); 3-(DIMETHOXYPHOSPHINYLOXY)-N,N-DIMETHYL-CIS-CROTONAMIDE; 3-(DIMETHOXYPHOSPHINYLOXY)-N,NDIMETHYLISOCROTONAMIDE; 3-(DIMETHYLAMINO)-1-METHYL-3-OXO-1-PROPENYL DIMETHYL PHOSPHATE; CIS-2-DIMETHYLCARBAMOYL-1-METHYLVINYL DIMETHYL-PHOSPHATE; O,O-DIMETHYL-O-(2-DIMETHYL-CARBAMOYL-1-METHYL-VINYL)PHOSPHAT (GERMAN); O,O-DIMETHYLO-(N,N-DIMETHYLCARBAMOYL-1-METHYLVINYL) PHOSPHATE; O,O-DIMETHYL-O-(1,4-DIMETHYL-3-OXO- 4-AZA-PENT-1-ENYL)FOSFAAT (DUTCH); O,O-DIMETHYL-O-(1,4-DIMETHYL-3-OXO-4-AZA-PENT-1-ENYL)PHOSPHATE; DIMETHYL PHOSPHATE OF 3-HYDROXY-N,N-DIMETHYL-CIS-CROTONAMIDE; DIMETHYL PHOSPHATE ESTER WITH 3-HYDROXY-N,N-DIMETHYL-CIS-CROTONAMIDE; O,O-DIMETIL-O-(1,4-DIMETIL-3-OXO-4-AZA-PENT-1-ENIL)-FOSFATO (ITALIAN); EKTAFOS; ENT 24,482; 3-HYDROXYDIMETHYL CROTONAMIDE DIMETHYL PHOSPHATE; 3-HYDROXY-N,N-DIMETHYL-CIS-CROTONAMIDE DIMETHYL PHOSPHATE; PHOSPHATEDE DIMETHYLE ET DE 2-DIMETHYLCARBAMOYL 1-METHYL VINYLE (FRENCH); PHOSPHORIC ACID, DIMETHYL ESTER, ESTER WITH CIS-3-HYDROXY-N,N-DIMETHYLCROTONAMIDE; SD 3562; SHELL SD-3562; PHOSPHORIC ACID, DIMETHYL ESTER, ESTER WITH (E)-3-HYDROXY-N,N-DIMETHYLCROTONAMIDE

Danger profile: Mutagen, exposure limit established, EPA extremely hazardous substance list, poisonous gases produced in fire.
Uses: Insecticide (organophosphate). Used on coffee, cotton, citrus and rice pests. Banned for use in some countries; not approved in some others.
Appearance: Brown liquid. Often mixed with water.
Effects of exposure:
Breathed: Extremely poisonous. May cause headache, dizziness, profuse sweating, nausea, vomiting, reduced heart beat, stomach cramps, diarrhea, loss of coordination, slow and weak breathing, fever, loss of consciousness and death. May cause shortness of breath and a dangerous build-up of fluids in the lungs (pulmonary edema), which can cause death. Pulmonary edema is a medical emergency that can be delayed one to two days following exposure.
Eyes: Rapidly absorbed. Poisoning can happen very rapidly. See *Breathed.*
Skin: Poisonous; fatal poisoning can result from contact. Passes through the unbroken skin; can increase exposure and the severity of the symptoms listed.
Swallowed: Powerful, deadly poison. See *Breathed.*
Long-term exposure: May damage the developing fetus or cause birth defects; nerve damage including loss of coordination; liver damage. Repeated exposure may cause depression, irritability and personality changes.
Storage: Store in a regulated, specially marked place in tightly closed containers in a dry, well-ventilated place. Keep away from oxidizers, acids and heat. Containers may explode in fire.

★ 369 ★
DICYCLOPENTADIENE
CAS: 77-73-6
Other names: BICYCLOPENTADIENE; BISCYCLOPENTAD-IENE; 1,3-CYCLOPENTADIENE, DIMER; DICYKLOPENTA-DIEN (CZECH); DIMER CYKLOPENTADIENU (CZECH); 4,7-METHANO-1H-INDENE; 3A,4,7,7A-TETRAHYDRO-4,7-METHANOINDENE; TRICYCLO5.2.1.0 2,6-DECA-3,8-DIENE

Danger profile: Fire hazard, exposure limit established, poisonous gases produced in fire.
Uses: Making insecticides, paints and varnishes; flame retardants for plastics.
Appearance: Colorless crystals.
Odor: Unpleasant. The odor is a warning of exposure; however, no smell does not mean you are not being exposed.
Effects of exposure:
Breathed: May cause irritation of the mucous membranes and respiratory tract; nausea, vomiting, headache, dizziness, loss of balance, convulsions.
Eyes: Contact may cause irritation.
Skin: Contact may cause severe irritation.
Swallowed: Poisonous.
Long-term exposure: May cause damage of the kidneys, lungs with shortness of breath and coughing, and nervous system.
Storage: Store in tightly closed containers in a dry, cool, well-ventilated place. Keep away from heat (with heat, will decompose to cyclopentadine), sources of ignition and oxidizing materials. Containers may explode in fire.
First aid guide: Move victim to fresh air and call emergency medical care; if not breathing, give artificial respiration; if breathing is difficult, give oxygen. In case of contact with material, immediately flush eyes with running water for at least 15 minutes. Wash skin with soap and water. Remove and isolate contaminated clothing and shoes at the site.

★ 370 ★
DIELDRIN
CAS: 60-57-1
Other names: ALVIT; COMPOUND 497; DIELDREX; DIELDRINE (FRENCH); DIELDRITE; 2,7:3,6-DIMETHANONAPTH2,3B OXIRENE,3,4,5,6,9,9-HEXACHLORO-1A,2,2A,3,6,6A,7,7A-OCTAHYDRO-(1A ALPHA,2BETA,2A ALPHA,3BETA,6BETA,6A ALPHA,7BETA,7A ALPHA); ENT 16,225; HEOD; HEXA-CHLOROEPOXYOCTAHYDRO-ENDO,EXO-DIMETHANONAPHTHALENE; ILLOXOL; KILLGERM DE-THLAC INSECTICIDAL LAQUER; NCI-C00124; OCTALOX; OXRALOX; PANORAM; PANORAM D-31; QUINTOX; 1,2,3,4,10,10-HEXACHLORO-6,7-EPOXY-1,4,4A,5,6,7,8,8A-OCTAHYDRO-1 ,4-ENDO-EXO-5,8-DI-METHANONAPHTHALENE; 3,4,5,6,9,9-HEXACHLORO-1A,2,2A,3,6,6A,7,7A-OCTAHYDRO-2,7:3,6-DIM ETHANO

Danger profile: Possible carcinogen, mutagen, reproductive hazard, exposure limit established, poisonous gases produced in fire.
Uses: Insecticide (organochlorine); wood preservative.
Appearance: Colorless to light brown or tan flakes.
Breathed: Poisonous. May cause headaches, irritability, poor appetite or nausea. May effect the nervous system. High exposures may cause weakness, numbness in the arms, legs, face and mouth; muscle twitching, shaking, convulsions, loss of consciousness and death.
Eyes: May cause irritation.
Skin: Poisonous. Passes through the unbroken skin; can increase exposure and the severity of the symptoms listed. May cause irritation and rash. See *Breathed.*
Swallowed: Poisonous. See *Breathed* and *First aid guide.*

Long-term exposure: Closely related to DDT, but more toxic. Evidence of animal cancer; may cause cancer in humans. May cause decrease of fertility in males and females; liver damage. Concentrates in breast milk and may be transferred to infants who are breast feeding. Repeated exposure may cause skin rash, muscle spasms, tremors, loss of appetite, weight loss and fainting. Convulsions may occur weeks or months following exposure.

Storage: Store in tightly closed containers in a dry, cool, well-ventilated place. Keep away from sources of ignition, oxidizers acids, chemically active metals and food. Containers may explode in fire.

First aid guide: Move victim to fresh air and call emergency medical care; if not breathing, give artificial respiration; if breathing is difficult, give oxygen. In case of contact with material, immediately flush skin or eyes with running water for at least 15 minutes. Speed in removing material from skin is of extreme importance. Remove and isolate contaminated clothing and shoes at the site. Keep victim quiet and maintain normal body temperature. Effects may be delayed; keep victim under observation.

★ 371 ★
DIETHANOLAMINE
CAS: 111-42-2
Other names: BIS(2-HYDROXYETHYL)AMINE; BUTANE, 1,2,3,4-DIEPOXY; BUTADIENE DIOXIDE; DEA; DIAETHANOLAMIN (GERMAN); DIETHANOLAMIN (CZECH); N,N-DIETHANOLAMINE; DIETHYLAMINE, 2,2'-DIHYDROXY-; DIETHYLOLAMINE; 2,2'-DIHYDROXYDIETHYLAMINE; DI(2-HYDROXYETHYL)AMINE; DIOLAMINE; ETHANOL, 2,2'-IMINOBIS-; 2,2'-IMINODIETHANOL; 2,2'IMINOBISETHANOL; 2,2'DIHYDROXYETHYLAMINE; NCI-C55174

Danger profile: Exposure limit established, Community Right-to-Know List, poisonous gases produced in fire.
Uses: Liquid detergents, cutting oils, shampoos, cleaners and polishes, textile specialties; absorbent for acid gases; making resins and plastics.
Appearance: Colorless powder or white oily liquid.
Odor: Slight dead fish or ammonia odor. The odor is a warning of exposure; however, no smell does not mean you are not being exposed.
Effects of exposure:
Breathed: May cause irritation of the nose and throat; coughing and smothering sensation, nausea and headache.
Eyes: Contact may cause severe irritation, inflammation of the eyelids and burns.
Skin: Contact may cause irritation and burns.
Swallowed: Moderately toxic.
Long-term exposure: Similar and related chemicals cause allergies of the skin and lungs; brain and nerve damage with loss of memory, personality changes, sleep disorders, reduced coordination and weakness in the arms and legs.
Storage: Store in tightly closed containers in a dry, cool, well-ventilated place. Keep away from sources of ignition and oxidizers. Containers may explode in fire.

★ 372 ★
DIETHOXYPROPENE
CAS: 3054-95-3
Other names: ACROLEIN DIETHYL ACETAL; 3,3-DIETHOXY-1-PROPENE; 1- PROPENE, 3,3-DIETHOXY-

Danger profile: Mutagen, fire hazard, poisonous gas produced in fire.
Uses: Making other chemicals.
Appearance: Colorless liquid.
Effects of exposure:

Breathed: May be poisonous. May cause dizziness and suffocation.
Eyes: May cause irritation and burns.
Skin: May be poisonous if absorbed. May cause irritation and burns.
Long-term exposure: May lead to birth defects, miscarriages or cancer.
Storage: Store in tightly closed containers in a dry, cool, well-ventilated place. Keep away from heat, sources of ignition, oxidizers and acids. Vapors may travel to sources of ignition and flash back. Containers may explode in fire.
First aid guide: Move victim to fresh air and call emergency medical care; if not breathing, give artificial respiration; if breathing is difficult, give oxygen. In case of contact with material, immediately flush eyes with running water for at least 15 minutes. Wash skin with soap and water. Remove and isolate contaminated clothing and shoes at the site.

★ 373 ★
DIETHYLALUMINUM CHLORIDE
CAS: 96-10-6
Other names: ALUMINUM, CHLORODIETHYL-; CHLORODIETHYALUMINUM; CHLORODIETHYLALANE; DIETHYLALUMINUM MONOCHLORIDE

Danger profile: Fire hazard, exposure limit established, poisonous gas produced in fire.
Uses: Making polyolefins and organometallics.
Appearance: Colorless liquid, ignites instantly on contact with air.
Effects of exposure:
Breathed: May cause irritation of the nose, throat, lungs. Higher exposures may cause shortness of breath and a dangerous build-up of fluids in the lungs (pulmonary edema), which can cause death. Pulmonary edema is a medical emergency that can be delayed one to two days following exposure.
Eyes: Contact may cause burns.
Skin: Contact may cause irritation, rash and deep painful burns and subsequent scarring.
Storage: Store in tightly closed containers under a nitrogen or other inert gas blanket. Keep away from air, water, sources of ignition, combustible or reactive materials, strong oxidizing agents and halogenated hydrocarbons. Containers may explode in fire.

★ 374 ★
DIETHYLAMINOETHANOL
CAS: 100-37-8
Other names: DEAE; DIAETHYLAMINOAETHANOL (GERMAN); BETA-DIETHYLAMINOETHANOL; N-DIETHYLAMINOETHANOL; 2-DIETHYLAMINO-; 2-(DIETHYLAMINO) ETHYL ALCOHOL; 2-(DIETHYLAMINO)ETHANOL; 2-N-DIETHYLAMINOETHANOL; BETA-DIETHYLAMINOETHYL ALCOHOL; DIETHYLETHANOLAMINE; N,N-DIETHYLETHANOLAMINE; N,N-DIETHYL-N-(BETA-HYDROXYETHYL)AMINE; 2-HYDROXYTRIETHYLAMINE; N,11-DIETHYL-N-(8-HYDROXY-ETHYL)AMINE; ETHANOL,2-(DIETHYLAMINO)-; N,N-DIETHYL-2-HYDROXYETHYLAMINE

Danger profile: Combustible, exposure limit established, poisonous gases produced in fire.
Uses: Textile softeners; making pharmaceuticals, pesticides and other chemicals; anti-rust compositions; curing resins.
Appearance: Colorless liquid
Odor: Weak, ammonia-like. The odor is a warning of exposure; however, no smell does not mean you are not being exposed.

Effects of exposure:
Breathed: Vapors may cause irritation of the lungs causing perspiration, coughing and shortness of breath; nausea and vomiting.
Eyes: May cause severe burns and permanent damage.
Skin: Passes through the unbroken skin; can increase exposure and the severity of the symptoms listed. May cause irritation, burning sensation and rash.
Swallowed: Moderately toxic. See *Breathed* and *First aid guide.*
Long-term exposure: Skin allergy may develop. Low exposures may cause itching and rash. Similar substances cause lung damage.
Storage: Store in tightly closed containers in a dry, cool, well-ventilated place, preferably a refrigerator. Keep away from heat, moisture, sources of ignition oxidizing agents and acids. Vapors may travel to sources of ignition and flash back. Containers may explode in fire.
First aid guide: Move victim to fresh air and call emergency medical care; if not breathing, give artificial respiration; if breathing is difficult, give oxygen. In case of contact with material, immediately flush skin or eyes with running water for at least 15 minutes. Remove and isolate contaminated clothing and shoes at the site. Keep victim quiet and maintain normal body temperature.

★ 375 ★
DIETHYL ANILINE
CAS: 91-66-7
Other names: BENZENAMINE, N,N-DIETHYL-; DEA; N,N-DIETHYLAMINOBENZENE; DIAETHYLANILIN (GERMAN); N,N-DIETHYLANILIN (CZECH); DIETHYLANILINE; N,N-DIETHYLANILINE; DIETHYLPHENYLAMINE; ANILINE, N,N-DIETHYL-; DIAETHYLANILIN (GERMAN); N,N-DIETHYLANILIN (CZECH); N-PHENYLDIETHYLAMINE

Danger profile: Poisonous, exposure limit established.
Uses: Making other chemicals and dyeing.
Appearance: Colorless to yellow liquid.
Effects of exposure:
Breathed: May be poisonous. May interfere with the ability of the blood to carry oxygen causing headache, dizziness, vertigo, increased pulse and respiration rate, bluish color to the skin lips and nose; difficult breathing and shortness of breath; collapse and death. A single large exposure could lead to liver damage.
Eyes: May cause irritation and burns.
Skin: May be poisonous. Passes through the unbroken skin; can increase exposure and the severity of the symptoms listed. May cause irritation and burns.
Swallowed: May be poisonous.
Long-term exposure: Repeated exposure may cause anemia (low red blood count). May cause liver damage.
Storage: Store in tightly closed containers in a dry, cool, well-ventilated place. Keep away from direct sunlight, sources of ignition, strong acids and strong oxidizers. Containers may explode in fire.
First aid guide: Move victim to fresh air and call emergency medical care; if not breathing, give artificial respiration; if breathing is difficult, give oxygen. In case of contact with material, immediately flush skin or eyes with running water for at least 15 minutes. Speed in removing material from skin is of extreme importance. Remove and isolate contaminated clothing and shoes at the site. Keep victim quiet and maintain normal body temperature. Effects may be delayed; keep victim under observation.

★ 376 ★
DIETHYLENE TRIAMINE
CAS: 111-40-0
Other names: AMINOETHANDIAMINE; 3-AZAPENTANE-1,5-DIAMINE; BIS(2-AMINOETHYL)AMINE; N-(2-AMINOETHYL); DIETHYLENETRIAMINE; 1,2-ETHANEDIAMINE, N-(2-AMINOETHYL)-; ETHYLENEDIAMINE; DETA; BIS(BETA-AMINOETHYL)AMINE; 2,2'-DIAMINODIETHYLAMINE; 1,4,7-TRIAZAHEPTANE

Danger profile: Combustible, corrosive, exposure limit established, reproductive hazard, poisonous gases produced in fire.
Uses: Solvent; fuel component.
Appearance: Colorless to yellow, thick liquid.
Odor: Mild ammonia-like. The odor is a warning of exposure; however, no smell does not mean you are not being exposed.
Effects of exposure:
Breathed: May be poisonous. May cause irritation to the nose and throat; nausea and vomiting.
Eyes: May cause severe irritation and burns.
Skin: Poison. Passes through the unbroken skin; can increase exposure and the severity of the symptoms listed. May cause severe irritation and burns. Possible rash.
Swallowed: Moderately toxic and corrosive. See *Breathed.*
Long-term exposure: Prolonged breathing of vapors may cause asthma. May cause skin allergies with rash and itching and lung allergies with coughing, wheezing and shortness of breath.
Storage: Store in tightly closed containers in a dry, cool, well-ventilated place, or refrigerator. Keep away from heat, sources of ignition, acids halogenated organics, peroxides, nitromethane, cellulouse nitrate, copper and copper alloys. For long-term storage protect from moisture, light and air. Vapors may travel to sources of ignition and flash back. Containers may explode in fire.

★ 377 ★
DI(2-ETHYLHEXYL) PEROXYDICARBONATE
CAS: 16111-62-9
Other names: PEROXYDICARBONIC ACID, BIS(2-ETHYLHEXYL) ESTER; PEROXYDICARBONIC ACID DI(2-ETHYLHEXYL) ESTER

Danger profile: Fire hazard, poisonous gases produced in fire.
Uses: Making other chemicals.
Appearance: Clear liquid.
Effects of exposure:
Breathed: May cause irritation of the nose, throat and breathing passages. High exposures may cause shortness of breath and a dangerous build-up of fluids in the lungs (pulmonary edema), which can cause death. Pulmonary edema is a medical emergency that can be delayed one to two days following exposure.
Eyes: Contact may cause irritation and burns.
Skin: Passes through the unbroken skin; can increase exposure and the severity of the symptoms listed. Contact may cause irritation, skin allergy with itching and rash. Possible burns.
Swallowed: Moderately toxic.
Long-term exposure: May cause skin allergy, rash and itching; lung damage.
Storage: Store in a tightly closed container (temperature at 18°F or below) and prevent loss of temperature control since violent reaction occurs. Keep away from heat, sources of ignition, oxidizers, acids, reducing agents and bases. in tightly closed containers in a dry, cool, well-ventilated place. Containers may explode in heat, from contamination or in fire.
First aid guide: Move victim to fresh air; call emergency medical care. In case of contact with material, immediately flush eyes

with running water for at least 15 minutes. Wash skin with soap and water. Remove and isolate contaminated clothing and shoes at the site. Keep victim quiet and maintain normal body temperature.

★ 378 ★

DIETHYL KETONE

CAS: 96-22-0

Other names: DEK; DIETHYLCETONE (FRENCH); DI-METHYLACETONE; ETHYL KETONE; METACETONE; MET-HACETONE; PENTANONE-3; 3-PENTANONE; PROPIONE

Danger profile: Fire hazard, mutagen, exposure limit established, poisonous gases produced in fire.
Uses: In medicine; making other chemicals.
Appearance: Clear, colorless liquid.
Odor: Nail polish remover-like. The odor is a warning of exposure; however, no smell does not mean you are not being exposed.
Effects of exposure:
  Breathed: May be poisonous. May cause irritation of the nose, throat and breathing passages. Higher exposures may cause dizziness, lightheadedness and loss of consciousness. Possible suffocation.
  Eyes: May cause irritation.
  Skin: May be poisonous if absorbed. Passes through the unbroken skin; can increase exposure and the severity of the symptoms listed. May cause irritation.
  Swallowed: May be poisonous.
Long-term exposure: May cause birth defects, miscarriage or cancer. Similar chemicals cause brain damage with loss of memory, personality changes, fatigue, sleeping disorders, reduced coordination, and nerve damage with numbness and weakness in arms and legs.
Storage: Store in tightly closed containers in a dry, cool, well-ventilated place. Keep away from heat, sources of ignition, and oxidizing materials. Vapors may travel to sources of ignition and flash back. Containers may explode in fire.
First aid guide: Move victim to fresh air and call emergency medical care; if not breathing, give artificial respiration; if breathing is difficult, give oxygen. In case of contact with material, immediately flush eyes with running water for at least 15 minutes. Wash skin with soap and water. Remove and isolate contaminated clothing and shoes at the site.

★ 379 ★

DIETHYL SULPHATE

CAS: 64-67-5

Other names: DIETHYL MONOSULFATE; DIETHYL SULFATE; DS; DIETHYL ESTER SULFURIC ACID; ETHYL SULFATE; SULFURIC ACID, DIETHYL ESTER; DIAETHYLSULFAT (GERMAN)

Danger profile: Carcinogen, mutagen, combustible, Community Right-to-Know List, poisonous gases produced in fire.
Uses: Making other chemicals.
Appearance: Colorless, oily liquid.
Odor: Faint ether-like. The odor is a warning of exposure; however, no smell does not mean you are not being exposed.
Effects of exposure:
  Breathed: Poisonous. May cause irritation of the nose, throat and lungs; cough and shortness of breath. Higher exposures may cause shortness of breath and a dangerous build-up of fluids in the lungs (pulmonary edema), which can cause death. Pulmonary edema is a medical emergency that can be delayed one to two days following exposure.
  Eyes: May cause severe irritation.

  Skin: May be poisonous if absorbed. May cause severe irritation.
  Swallowed: May be poisonous.
Long-term exposure: May cause cancer, birth defects and miscarriage. Similar irritating chemicals cause lung damage.
Storage: Store in tightly closed containers in a dry, cool, well-ventilated place. Keep away from moisture, heat, sources of ignition, iron, potassium tert-butoxide and oxidizing materials. Containers may explode in fire.
First aid guide: Move victim to fresh air and call emergency medical care; if not breathing, give artificial respiration; if breathing is difficult, give oxygen. In case of contact with material, immediately flush skin or eyes with running water for at least 15 minutes. Speed in removing material from skin is of extreme importance. Remove and isolate contaminated clothing and shoes at the site. Keep victim quiet and maintain normal body temperature. Effects may be delayed; keep victim under observation.

★ 380 ★

DIETHYLZINC

CAS: 557-20-0

Other names: ETHYL ZINC; ZINC DIETHYL-; ZINC ETHIDE; ZINC ETHYL

Danger profile: Fire hazard, dangerous explosion hazard, extremely reactive, Community Right-to-Know List, poisonous gases produced in fire.
Uses: Making other chemicals; high energy aircraft and missile fuel.
Appearance: Colorless watery liquid.
Effects of exposure:
  Breathed: Mist or vapor may cause immediate irritation of nose and throat. Excessive or prolonged inhalation of fumes from ignition or decomposition may cause "metal fume fever" (sore throat, headache, fever chills, nausea, vomiting, muscular aches, perspiration, constricting sensation in lungs, weakness, sometimes prostration); symptoms usually last 12-24 hours, with complete recovery in 24-48 hours.
  Eyes: Eyes are immediately and severely irritated on contact with liquid, vapor, or dilute solution; without thorough irrigation, cornea may be permanently damaged.
  Skin: Moisture in skin combines with chemical to cause thermal acid burns at site of contact; pain, nausea, vomiting, cramps, and diarrhea may follow; if untreated, tissue may become ulcerated.
  Swallowed: May be poisonous.
Storage: Store in sealed tubes or cylinders with a dry inert gas blanket. Keep away from air, heat, moisture, all other chemicals including chlorine, hydrazine and oxidizers.
First aid guide: Move victim to fresh air; call emergency medical care. Wipe material from skin immediately; flush skin or eyes with running water for at least 15 minutes. Remove and isolate contaminated clothing and shoes at the site.

★ 381 ★

DIFLUORODIBROMOMETHANE

CAS: 75-61-6

Other names: DIBROMODIFLUOROMETHANE; DIBRO-MODIFLUORO-METHANE; FREON 12-B2; FREON 12B2; HALON 1202; METHANE, DIDIBROMOFLUORO-

Danger profile: Exposure limit established.
Uses: Making dyes, pharmaceuticals, other chemicals; fire-extinguishing agent; processing razor blades, hypodermic needles and scalpels.
Appearance: Colorless liquid or gas.
Odor: Characteristic odor. The odor is a warning of exposure; however, no smell does not mean you are not being exposed.

Effects of exposure:
Breathed: May cause irritation of the nose and throat; dizziness, lightheadedness and loss of consciousness; irregular heart beat and possible death.
Eyes: Contact may cause frostbite.
Skin: Contact may cause frostbite.
Long-term exposure: May cause damage of the liver, kidneys and lungs.
Storage: Store in tightly closed containers in a dry, cool, well-ventilated place. Keep away from heat and chemically active metals.

★ 382 ★
DIFLUOROPHOSPHORIC ACID
CAS: 13779-41-4
Other names: PHOSPHORODIFLUORIDIC ACID; PHOSPHO-RODIFLUORIDIC ACID (ANHYDROUS)

Danger profile: Corrosive, exposure limit established, poisonous gas produced in fire.
Uses: Chemical polishing agent; protective coatings for metal surfaces; making other chemicals.
Appearance: Colorless liquid.
Odor: Sharp, irritating. The odor is a warning of exposure; however, no smell does not mean you are not being exposed.
Effects of exposure:
Breathed: May be poisonous. May cause severe irritation of the eyes, nose throat and upper respiratory tract; nose bleeding, coughing and tightness of the chest. Higher exposures may cause shortness of breath and a dangerous build-up of fluids in the lungs (pulmonary edema), which can cause death. Pulmonary edema is a medical emergency that can be delayed one to two days following exposure.
Eyes: Contact with liquid may cause severe irritation and burns.
Skin: Passes through the unbroken skin; can increase exposure and the severity of the symptoms listed. Contact with liquid may cause severe irritation and burns.
Swallowed: May be poisonous. May cause severe burns of the mouth and stomach.
Long-term exposure: Repeated skin exposure may cause rash. Similar irritating chemicals can cause lung damage.
Storage: Store in tightly closed containers in a dry, cool, well-ventilated place. Keep away from combustable materials.
First aid guide: Move victim to fresh air and call emergency medical care; if not breathing, give artificial respiration; if breathing is difficult, give oxygen. In case of contact with material, immediately flush skin or eyes with running water for at least 15 minutes. Remove and isolate contaminated clothing and shoes at the site. Keep victim quiet and maintain normal body temperature. Effects may be delayed, keep victim under observation.

★ 383 ★
DIGLYCIDYL ETHER
CAS: 2238-07-5
Other names: DGE; DI(2,3-EPOXY)PROPYL ETHER; DIAL-LYL ETHER DIOXIDE; DI (EPOXYPROPYL) ETHER; ETHER, DIGLYCIDYL; ETHER, BIS(2,3-EPOXYPROPYL)-; DI(EPOXY-PROPYL)ETHER; BIS(2-3-EPOXYPROPYL) ETHER; 2-EPOXYPROPYL ETHER; ETHER, DIGLYCIDYL; ETHER, BIS(2,3-EPOXYPROPYL); NSV 54739; OXIRANE, 2,2'-OXYBIS (METHYLENE) BIS-

Danger profile: Carcinogen, mutagen, combustible, exposure limit established, poisonous gas produced in fire.
Uses: In epoxy resin systems.
Appearance: Colorless liquid.

Odor: Strong, irritating. The odor is a warning of exposure; however, no smell does not mean you are not being exposed.
Effects of exposure:
Breathed: May cause irritation of the nose, eyes, throat and lungs. Higher levels may cause dizziness, lightheadedness and loss of consciousness. Death can occur.
Eyes: May cause severe irritation and burns.
Skin: Passes through the unbroken skin; can increase exposure and the severity of the symptoms listed. May cause severe irritation, burns and skin allergy.
Swallowed: Poisonous.
Long-term exposure: Repeated exposure may cause lung, liver and kidney damage; bone marrow depression with lower the blood cell count; cause clouding of the eyes.
Storage: Store in tightly closed containers in a dry, cool, well-ventilated place. Keep away from heat, sources of ignition and strong oxidizers.

★ 384 ★
DIISOPROPYLAMINE
CAS: 108-18-9
Other names: DIISOPROPYL AMINE; 2-PROPANAMINE, N-(1-METHYLETHYL)-; N-(1-METHYETHYL)-2-PROPANAMINE; BIS(ISOPROPYL)AMINE

Danger profile: Mutagen, fire hazard, exposure limit established, poisonous gas produced in fire.
Uses: Making other chemicals.
Appearance: Colorless liquid.
Odor: Ammonia-like. The odor is a warning of exposure; however, no smell does not mean you are not being exposed.
Effects of exposure:
Breathed: Vapors may cause irritation of the lungs, sometimes with nausea and vomiting; can also cause burns to respiratory system. Higher exposures may cause shortness of breath and a dangerous build-up of fluids in the lungs (pulmonary edema), which can cause death. Pulmonary edema is a medical emergency that can be delayed one to two days following exposure.
Eyes: May cause severe irritation, burns and cloudy vision.
Skin: Passes through the unbroken skin; can increase exposure and the severity of the symptoms listed. Contact may cause irritation and skin allergy may develop.
Swallowed: May cause irritation of mouth and stomach. See *Breathed.*
Long-term exposure: May cause skin allergy with rash and itching; lung allergy with shortness of breath, wheezing, coughing and chest tightness. Similar chemicals cause lung damage.
Storage: Store in tightly closed containers in a dry, cool, well-ventilated place. Keep away from heat, sources of ignition acids and oxidizers. May attack some forms of plastic. Vapors may travel to sources of ignition and flash back. Containers may explode in fire.
First aid guide: Move victim to fresh air and call emergency medical care; if not breathing, give artificial respiration; if breathing is difficult, give oxygen. In case of contact with material, immediately flush skin or eyes with running water for at least 15 minutes. Remove and isolate contaminated clothing and shoes at the site. Keep victim quiet and maintain normal body temperature.

★ 385 ★
DIMETHOATE
CAS: 60-51-5
Other names: AC-12682; AMERICAN CYANAMID 12880; BI-58; BIS HCCL 12880; CEKUTHOATE; CYGON; CYGON 4E; CL 12880; CYGON; CYGON INSECTICIDE; DAPHENE; DE-FEND; DE-FEND; DEMOS-L40; DEVIGOS; DIMATE 267; DIMETATE; DIMETHOAAT (DUTCH); DIMETHOAT (GERMAN); DIMETHOATE; DIMETHOATE-267; DIMETHOAT TECHNISCH 95%; DIMETHOGEN; EC8014; O,O-DIMETHYLDITHIOPHOSPHORYLACETIC ACID, N-MONOMETHYLAMIDE SALT; O,O-DIMETHYLS-(2-(METHYLAMINO)-2-OXOETHYL)PHOSPHORODITHIOATE; O,O-DIMETHYL-S-(N-METHYL-CARBAMOYL)-METHYL-DITHIOFOSFAAT (DUTCH); (O,O-DIMETHYL-S-(N-METHYL-CARBAMOYL-METHYL)-DITHIOPHOSPHAT) (GERMAN); O,O-DIMETHYL METHYLCARBAMOYLMETHYLPHOSPHO-RODITHIOATE; O,O-DIMETHYLS-(N-METHYLCARBAMOYLMETHYL)DITHIOPHOSPHATE; O,O-DIMETHYLS-(N-METHYLCARBAMOYLMETHYL)PHOSPHORODITHIOATE; O,O-DIMETHYLS-(N-METHYLCARBAMYLMETHYL)THIOTHIONOPHOSPHATE; O,O-DIMETHYL-S-(N-MONOMETHYL)-CARBAMYL METHYL DITHIOPHOSPHATE; O,O-DIMETHYL-S-(2-OXO-3-AZA-BUTYL)-DITHIOPHOSPHAT (GERMAN); O,O-DIMETIL-S-(N-METIL-CARBAMOIL-METIL)-DITIOFOSFATO (ITALIAN); DIMETON; DIMEVUR; DITHIOPHOSPHATEDE O,O-DIMETHYLE ET DES(-N-METHYLCARBAMOYL-METHYLE) (FRENCH); EI-12880; ENT 24,650; EXPERIMENTAL INSECTICIDE 12,880; FERKETHION; FOSTION MM; NC-262; NCI-C00135; PEI 75; PERFECTHION; PERFEKTHION; PERFEKTION; PHOSPHAMID; PHOSPHAMIDE; PHOSPHO-RODITHIOIC ACID, O,O-DIMETHYL-S-(2-(METHYLAMINO)-2-OXOETHYL) ESTER (9CI); ROGOR; ROXION U.A.; SINORATOX; TRIMETION

Danger profile: Mutagen, reproductive hazard, poison, EPA extremely hazardous substances list, poisonous gas produced in fire.
Uses: Insecticide for flies and mites. Dry formulations not permitted.
Appearance: White to grayish crystalline solid.
Odor: Camphor or moth ball-like. The odor is a warning of exposure; however, no smell does not mean you are not being exposed.
Effects of exposure:
  Breathed: Extremely poisonous. May cause headache, dizziness, profuse sweating, nausea, vomiting, reduced heart beat, stomach cramps, diarrhea, loss of coordination, slow and weak breathing, fever, loss of consciousness and death. May cause shortness of breath and a dangerous build-up of fluids in the lungs (pulmonary edema), which can cause death. Pulmonary edema is a medical emergency that can be delayed one to two days following exposure.
  Eyes: Rapidly absorbed. Poisoning can happen very rapidly. See *Breathed.*
  Skin: Poisonous. Passes through the unbroken skin; can increase exposure and the severity of the symptoms listed.
  Swallowed: Powerful, deadly poison. See *Breathed.*
Long-term exposure: May damage the developing fetus or cause birth defects; nerve damage including loss of coordination; liver damage. Repeated exposure may cause depression, irritability and personality changes.
Storage: Store in a regulated, specially marked place in tightly closed containers in a cold, dry, well-ventilated place. Do not store solid in temperature above 25-30°C; store liquid solutions in temperature above 45°F. Keep away from heat, water, strong alkalis. Containers may explode in fire.

★ 386 ★
DIMETHYL ACETAMIDE
CAS: 127-19-5
Other names: ACETAMIDE N,N-DIMETHYL; ACETDIMETHYLAMIDE; ACETIC ACID, DIMETHYLAMIDE; DIMETHYLACETAMIDE; N,N-DIMETHYLACETAMIDE; N,N-DIMETHYL ACETAMIDE; DIMETHYLACETONE AMIDE; DIMETHYLAMIDE ACETATE; DMA; DMAC; NSC 3138; U-5954

Danger profile: Mutagen, combustible, exposure limit established, poisonous gases produced in fire.
Uses: Solvent for plastics, resins and gums; paint remover; making other chemicals.
Appearance: Colorless liquid.
Odor: Weak, fishy; ammonia-like. The odor is a warning of exposure; however, no smell does not mean you are not being exposed.
Effects of exposure:
  Breathed: May cause nausea, bloating, poor appetite, nausea and/or liver damage and jaundice.
  Eyes: Liquid may cause mild irritation.
  Skin: Passes through the unbroken skin; can increase exposure and the severity of the symptoms listed under *Breathed.* Liquid may cause irritation.
  Swallowed: Mildly toxic. Ingestion causes depression, lethargy, confusion and disorientation, visual and auditory hallucinations, perceptual distortions, delusions, emotional detachment.
Long-term exposure: May cause liver damage; damage to the unborn fetus; brain damage including depression, lethargy, hallucinations, emotionally detached feelings, and personality changes.
Storage: Store in tightly closed containers in a dry, cool, well-ventilated place. Keep away from heat, sources of ignition, halogenated compounds such as carbon tetrachloride when in contact with iron.

★ 387 ★
DIMETHYLAMINE
CAS: 124-40-3
Other names: DIMETHYLAMINE AQUEOUS SOLUTION; N,N-DIMETHYLAMINE; DIMETHYLAMINE ANHYDROUS; DMA; METHANAMINE, N-METHYL-(9CI); DIMETHYLAMINE SOLUTION; N-METHYLMETHANAMINE

Danger profile: Mutagen, fire hazard, exposure limit established, poisonous gas produced in fire.
Uses: Solvent; making other chemicals, rubber, textiles, drugs; dyes flotation agent; gasoline stabilizers; pharmaceuticals; rubber accelerators; electroplating; dehairing agent; missile fuels.
Appearance: Colorless liquefied compressed gas, sometimes solid.
Odor: D Dead fish or ammonia-like. The odor is a warning of exposure; however, no smell does not mean you are not being exposed.
Effects of exposure:
  Breathed: May cause irritation of the eyes, nose and throat; lung irritation, coughing and/or shortness of breath. Higher exposures may cause shortness of breath and a dangerous build-up of fluids in the lungs (pulmonary edema), which can cause death. Pulmonary edema is a medical emergency that can be delayed one to two days following exposure.
  Eyes: Corrosive. Contact may cause severe burns and permanent damage.
  Skin: Contact may cause irritation, redness and severe burns.
  Swallowed: Poisonous.

Long-term exposure: Repeated or high exposure may cause lung damage, bronchitis, liver damage; damage to the developing fetus.

Storage: Store in tightly closed containers in a dry, cool, well-ventilated place. Keep away from heat, sources of ignition, oxidizing materials, mercury, fluorine, maleic anhydride, acrylaldehyde, and chlorine. Vapors may travel to sources of ignition and flash back. Containers may explode in fire.

★ 388 ★
DIMETHYLANILINE
CAS: 121-69-7
Other names: ANILINE, N,N-DIMETHYL-; BENZENAMINE, N,N-DIMETHYL-; (DIMETHYLAMINO)BENZENE; N,N-DIMETHYLANILINE; N,N- DIMETHYLBENZENEAMINE; DI-METHYLPHENYLAMINE; DMA; DWUMETYLOANILINA (POLISH); NCI-C56428; VERSNELLER NL 63/10

Danger profile: Mutagen, combustible, exposure limit established, Community Right-to-Know List, poisonous gases produced in fire.
Uses: Solvent; making dyes, rubber; laboratory chemical.
Appearance: Straw colored to brown liquid.
Odor: Characteristic irritating ammonia-like odor. The odor is a warning of exposure; however, no smell does not mean you are not being exposed.
Effects of exposure:
  Breathed: May be poisonous. May cause nose and throat irritation; headache; blue color to skin, fingertips, lips and nose; weakness, dizziness, labored respiration, increase in pulse and respiration, paralysis, convulsions, collapse and death. Has strong inhalation toxicity, inhibiting nerve center and circulation system.
  Eyes: May cause irritation and burns.
  Skin: May be poisonous if absorbed. Passes through the unbroken skin; can increase exposure and the severity of the symptoms listed. Contact may cause irritation.
  Swallowed: Poisonous.
Storage: Store in tightly closed containers in a dry, cool, well-ventilated place. Keep away from heat, sources of ignition, benzoyl peroxide, diisopropyl peroxdicarbonate. Containers may explode in fire.
First aid guide: Move victim to fresh air and call emergency medical care; if not breathing, give artificial respiration; if breathing is difficult, give oxygen. In case of contact with material, immediately flush skin or eyes with running water for at least 15 minutes. Speed in removing material from skin is of extreme importance. Remove and isolate contaminated clothing and shoes at the site. Keep victim quiet and maintain normal body temperature. Effects may be delayed; keep victim under observation.

★ 389 ★
2,3-DIMETHYLBUTANE
CAS: 79-29-8
Other names: BUTANE, 2,3-DIMETHYL-; DIISOPROPYL; ISOHEXANE

Danger profile: Fire hazard, exposure limit established, poisonous gas produced in fire.
Uses: In high octane fuel; making other chemicals.
Appearance: Colorless liquid.
Effects of exposure:
  Breathed: May cause irritation of the eyes, nose and throat; dizziness, lightheadedness and loss of consciousness.
  Eyes: Contact may cause irritation.
  Skin: Passes through the unbroken skin; can increase exposure and the severity of the symptoms listed. Contact may cause irritation.

Long-term exposure: May cause kidney damage; dermatitis.
Storage: Store in tightly closed containers in a dry, cool, well-ventilated place. Keep away from heat, sources of ignition and oxidizing materials. Containers may explode in fire.

★ 390 ★
DIMETHYLCARBAMOYL CHLORIDE
CAS: 79-44-7
Other names: CARBAMIC CHLORIDE, DIMETHYL-; CAR-BAMOYL CHLORIDE, DIMETHYL-; NSN-DIMETHYL(CARBANDOYLE) CHLORIDE; DDC; DIMETH-YLAMINO CARBONYL CHLORIDE; CHLOROFORMIC ACID DIMETHYLAMIDE; DIMETHYLCARBAMIC ACID CHLORIDE; DIMETHYLCARBAMIC CHLORIDE; DIMETHYLCARBAMID-OYL CHLORIDE; N,N-DIMETHYLCARBAMOYL CHLORIDE; DIMETHYLCARBAMYL CHLORIDE; N,N-DIMETHYLCARBAMYL CHLORIDE; DMCC; TL 389

Danger profile: Carcinogen, mutagen, combustible, corrosive, reacts with water, Community Right-to-Know List, poisonous gas produced in fire.
Uses: M Making pesticides and pharmaceuticals.
Appearance: Clear, colorless liquid.
Effects of exposure:
  Breathed: May cause irritation of the nose, throat and eyes.
  Eyes: May cause irritation, redness, tearing; burns; possible permanent damage.
  Skin: Contact may cause severe irritation and burns.
  Swallowed: Moderately toxic. See **First aid guide.**
Long-term exposure: May cause cancer; reproductive damage; liver and lung damage. Has been shown to cause skin, lung and nose cancer in animals.
Storage: Store in tightly closed containers in a dry, cool, well-ventilated place. Keep away from water and combustable materials.
First aid guide: Move victim to fresh air; call emergency medical care. In case of contact with material, immediately flush skin or eyes with running water for at least 15 minutes. Remove and isolate contaminated clothing and shoes at the site. Keep victim quiet and maintain normal body temperature.

★ 391 ★
2,5-DIMETHYL-2,5-DI-(BENZOYLPEROXY) HEXANE
CAS: 2618-77-1
Other names: BENZENECARBOPEROXOIC ACID, 1,1,4,4-TETRAMETHYL-1,4-BUTANEDIYL ESTER; 2,5-DIMETHYL-2,5-DI-(BENZOYLPEROXY) HEXANE, NOT MORE THAN 82% WITH INERT SOLID; PEROXYBENZOIC ACID,1,1,4,4-TETRAMETHYLTETRAM ETHYLENE ESTER

Danger profile: Combustible, poisonous gases produced in fire.
Uses: Making other chemicals.
Appearance: Usually found diluted with inert solids or water.
Effects of exposure:
  Breathed: May cause irritation of the nose and throat.
  Eyes: May cause irritation and burns.
  Skin: Passes through the unbroken skin; can increase exposure and the severity of the symptoms listed. May cause irritation and burns.
Storage: Store in tightly closed containers in a dry, cool, well-ventilated place. Keep away from any reactive materials. Protect from shock, friction and contamination. Containers may explode in fire.
First aid guide: Move victim to fresh air; call emergency medical care. In case of contact with material, immediately flush eyes with running water for at least 15 minutes. Wash skin with soap and water. Remove and isolate contaminated clothing and

shoes at the site. Keep victim quiet and maintain normal body temperature.

★ 392 ★
DIMETHYLDICHLOROSILANE
CAS: 75-78-5
Other names: DICHLORODIMETHYL SILANE; DIMETHYL-DICHLORSILAN (CZECH); SILANE, DICHLORODIMETHYL-; DICHLOROETHYLSILANE; DIMETHYL-DICHLOROSILAN (CZECH); NCLC50704

Danger profile: Fire hazard, corrosive, water reactive, poisonous gases produced in fire.
Uses: Making silicone products.
Appearance: Light straw colored liquid.
Odor: Sharp, irritating. The odor is a warning of exposure; however, no smell does not mean you are not being exposed.
Effects of exposure:
Breathed: May cause irritation of the eyes, nose and throat; mucous membrane; lungs causing coughing and/or shortness of breath. Higher exposures may cause shortness of breath and a dangerous build-up of fluids in the lungs (pulmonary edema), which can cause death. Pulmonary edema is a medical emergency that can be delayed one to two days following exposure.
Eyes: Corrosive. Contact may cause severe burns and permanent damage.
Skin: Corrosive. Contact may cause severe burns.
Swallowed: May cause severe burns of the mouth and stomach.
Long-term exposure: May cause lung damage.
Storage: Store in tightly closed containers in a dry, cool, well-ventilated place. Keep away from heat, sources of ignition, water or other forms of moisture (reacts with surface moisture to generate hydrogen chloride, which is toxic and corrosive to common metals), oxidizers. Do not store in temperature above 122°F. Vapors may travel to sources of ignition and flash back. Containers may explode in fire.
First aid guide: Move victim to fresh air and call emergency medical care; if not breathing, give artificial respiration; if breathing is difficult, give oxygen. In case of contact with material, immediately flush skin or eyes with running water for at least 15 minutes. Remove and isolate contaminated clothing and shoes at the site. Keep victim quiet and maintain normal body temperature.

★ 393 ★
DIMETHYLDIETHOXYSILANE
CAS: 78-62-6
Other names: SILANE, DIETHOXYDIMETHYL-; DIETHOXY DIMETHYL SILANE; DIMETHYL-DIETHOXYSILAN (CZECH)

Danger profile: Fire hazard, poisonous gas produced in fire
Uses: In water repellents.
Appearance: Clear liquid.
Effects of exposure:
Breathed: May be poisonous. May cause irritation of the eyes, nose and throat; dizziness.
Eyes: May cause irritation, burning, tearing.
Skin: May be poisonous if absorbed. Passes through the unbroken skin; can increase exposure and the severity of the symptoms listed. May cause irritation, burning sensation and rash.
Swallowed: Mildly toxic.
Storage: Store in tightly closed containers in a dry, cool, well-ventilated place. Keep away from heat, sources of ignition and oxidizers. Vapors may travel to sources of ignition and flash back. Containers may explode in fire.

First aid guide: Move victim to fresh air and call emergency medical care; if not breathing, give artificial respiration; if breathing is difficult, give oxygen. In case of contact with material, immediately flush eyes with running water for at least 15 minutes. Wash skin with soap and water. Remove and isolate contaminated clothing and shoes at the site.

★ 394 ★
2,5-DIMETHYL-2,5-DI(TERT-BUTYLPEROXY)HEXANE
CAS: 78-63-7
Other names: 2,5-DIMETHYL-2,5-DI(TERT-BUTYLPEROXY)HEXANE, TECHNICALLY PURE; PEROXIDE (1,1,4,4-TETRAMETHYL-1-1,4-BUTANEDIYL) BIS (1,1- DIMETHYLETHYL); TRIGONOX 101-101/45; VAROX

Danger profile: Combustible, poisonous gas produced in fire.
Uses: Making other chemicals, polyethylene, styrene and polyester resins.
Appearance: Colorless to light yellow liquid.
Odor: Menthol-like. The odor is a warning of exposure; however, no smell does not mean you are not being exposed.
Effects of exposure:
Breathed: May cause irritation of the nose and throat; high levels may cause dizziness, lightheadedness and loss of consciousness.
Eyes: May cause irritation, redness and burns.
Skin: Passes through the unbroken skin; can increase exposure and the severity of the symptoms listed. May cause irritation, redness and burns.
Storage: Store in tightly closed containers in a dry, cool, well-ventilated place. Keep away from heat, sources of ignition and reducing agents. Containers may explode in fire.
First aid guide: Move victim to fresh air; call emergency medical care. In case of contact with material, immediately flush eyes with running water for at least 15 minutes. Wash skin with soap and water. Remove and isolate contaminated clothing and shoes at the site. Keep victim quiet and maintain normal body temperature.

★ 395 ★
DIMETHYLFORMAMIDE
CAS: 68-12-2
Other names: DIMETHYLFORMAMID (GERMAN); DIMETHYL FORMAMIDE; N,N-DIMETHYL FORMAMIDE; N,N-DIMETHYLFORMAMIDE; DIMETILFORMAMIDE (ITALIAN); DIMETYLFORMAMIDU (CZECH); DMF; DMFA; DWUMETYLOFORMAMID (POLISH); N-FORMYLDIMETHYLAMINE; FORMAMIDE, N,N-DIMETHYL-; N,N-DIMETHYLMETHANIDE; NCI-C60913; NSC 5356; U-4224

Danger profile: Mutagen, possible reproductive hazard, combustible, exposure limit established, poisonous gases produced in fire.
Uses: Solvent; making other chemicals.
Appearance: Colorless to very slightly yellow liquid.
Odor: Faint, ammonia-like or fishy. The odor is a warning of exposure; however, no smell does not mean you are not being exposed.
Effects of exposure:
Breathed: May be poisonous. May cause stomach pain, loss of appetite, nausea and vomiting; facial flushing with elevated blood pressure.
Eyes: Contact may cause irritation.
Skin: May be poisonous if absorbed. Passes through the unbroken skin; can increase exposure and the severity of the

symptoms listed. Contact may cause irritation. Repeated exposure may cause drying and cracking.
Swallowed: Exposure and drinking alcohol may cause a reaction with some of the symptoms listed under **Breathed.**
Long-term exposure: May cause damage to the developing fetus; liver damage; drying and cracking skin.
Storage: Store in tightly closed dark containers in a dry, cool, well-ventilated place. Keep away from heat, sources of ignition, oxidizing agents and reactive chemicals, especially carbon tetrachloride and other halogenated compounds. Vapors may travel to sources of ignition and flash back. Containers may explode in fire.
First aid guide: Move victim to fresh air and call emergency medical care; if not breathing, give artificial respiration; if breathing is difficult, give oxygen. In case of contact with material, immediately flush eyes with running water for at least 15 minutes. Wash skin with soap and water. Remove and isolate contaminated clothing and shoes at the site.

★ 396 ★
2,6-DIMETHYLHEPTANONE
CAS: 108-83-8
Other names: DIBK; DIISOBUTILCHETONE (ITALIAN); DI-ISOBUTYLCETONE (FRENCH); DIISOBUTYLKETON (DUTCH, GERMAN); DIISOBUTYL KETONE; S-DIISOPROPYLACETONE; 2,6-DIMETHYL-HEPTAN-4-ON (DUTCH; GERMAN); 2,6-DIMETHYLHEPTAN-4-ONE; 2,6-DIMETHYL-4-HEPTANONE; 2,6-DIMETIL-EPTAN-4-ONE (ITALIAN); ISOBUTYL KETONE; ISOVALERONE; 4-HEPTANONE,2,6-DIMETHYL-; SYM-DIISIPROPYL-ACETONE; VALERONE

Danger profile: Combustible, exposure limit established, poisonous gas produced in fire.
Uses: Solvent for nitrocellulose, rubber, synthetic resins; making other chemicals; lacquers; inks; stains.
Appearance: Colorless oily liquid.
Odor: Mild, sweet, peppermint-like. The odor is a warning of exposure; however, no smell does not mean you are not being exposed.
Effects of exposure:
Breathed: May be poisonous. Vapor may cause irritation of the eyes, nose and throat; headache; coughing, difficult breathing. High concentrations may cause dizziness, lightheadedness and loss of consciousness.
Eyes: May cause irritation, redness and tearing.
Skin: May be poisonous if absorbed. Passes through the unbroken skin; can increase exposure and the severity of the symptoms listed. May cause irritation, burning sensation and rash.
Swallowed: Moderately toxic. May cause irritation of the mouth and throat; nausea and vomiting.
Long-term exposure: May cause skin drying and cracking; liver and kidney damage. Narcotic in high concentrations.
Storage: Store in tightly closed containers in a dry, cool, well-ventilated place. May attack some forms of plastics Keep away from heat, sources of ignition and oxidizing materials. Containers may explode in fire. Vapors may travel to sources of ignition and flash back. Containers may explode in fire.

★ 397 ★
DIMETHYL MERCURY
CAS: 593-74-8
Other names: MERCURY, DIMETHYL

Danger profile: Mutagen, reproductive hazard, extremely toxic, exposure limit established, poisonous gases produced in fire.
Uses: Making other chemicals; laboratory chemical.

Appearance: Colorless liquid.
Effects of exposure:
Breathed: May cause irritation of the nose, throat and bronchial tubes with soreness and cough. High concentrations can cause death.
Eyes: May cause severe irritation.
Skin: Passes through the unbroken skin; can increase exposure and the severity of the symptoms listed. Contact causes redness, blisters and severe burns. Symptoms may be delayed for six to eight hours.
Swallowed: May be poisonous.
Long-term exposure: Extremely toxic chemical that may cause delayed permanent brain damage; damage to the fetus and birth defects.
Storage: Store in tightly closed containers in a dry, cool, well-ventilated place. Keep away from oxidizers. Containers may explode in fire.

★ 398 ★
DIMETHYLPHTHALATE
CAS: 131-11-3
Other names: AVOLIN; 1,2-BENZENEDICARBOXYLIC ACID, DIMETHYL ESTER; DIMETHYL 1,2-BENZENEDICARBOXYLATE; DIMETHYL BENZENEORT-HODICARBOXYLATE; DIMETHYL PHTHALATE; DMP; ENT 262; FERMINE; METHYL PHTHALATE; MIPAX; NTM; PALA-TINOL M; PHTHALSAEUREDIMETHYLESTER (GERMAN); PHTHALIC ACID METHYL ESTER; SOLVANOM; SOLVA-RONE

Danger profile: Mutagen, reproductive hazard, combustible, exposure limit established, EPA extremely hazardous substances list, Community Right-to-Know List, poisonous gases produced in fire.
Uses: Pesticide, insect repellent; plasticizer; solid rocket propellants; lacquers; in plastics; rubber; safety glass; molding powders.
Appearance: Colorless, oily liquid
Odor: Slightly sweet. The odor is a warning of exposure; however, no smell does not mean you are not being exposed.
Effects of exposure:
Breathed: Hot vapor or mist may cause irritation of the eyes, nose and throat.
Eyes: Contact may cause irritation, redness and inflammation of the eyelids. Possible burns.
Skin: May cause irritation and burns.
Swallowed: Moderately toxic. May cause gastrointestinal irritation.
Long-term exposure: May lower fertility in both sexes. May cause birth defects by damaging the fetus.
Storage: Store in tightly closed containers in a dry, cool, well-ventilated place. Keep away from heat, sources of ignition, nitrates, oxidizers, strong acids, and protect from physical damage.

★ 399 ★
DIMETHYL SULFATE
CAS: 77-78-1
Other names: DIMETHYLESTER KYSELINY SIROVE (CZECH); DIMETHYL MONOSULFATE; DIMETHYLSULFAAT (DUTCH); DIMETHYLSULFAT (CZECH); DIMETHYL SUL-PHATE; DIMETILSOLFATO (ITALIAN); DMS; DMS (METHYL SULFATE); DWUMETYLOWY SIARCZAN (POLISH); METH-YLE (SULFATE DE) (FRENCH); METHYL SULFATE; SUL-FATE DE METHYLE (FRENCH); SULFATE DIMETHYLIQUE (FRENCH); SULFURIC ACID, DIMETHYL ESTER

Danger profile: Carcinogen, mutagen, corrosive, combustible, exposure limit established, EPA extremely hazardous substance list, Community Right-to-Know List, poisonous gas produced in fire.
Uses: Making other chemicals.
Appearance: Colorless oily liquid.
Odor: Faint onion-like. The odor is a warning of exposure; however, no smell does not mean you are not being exposed.
Effects of exposure:
Breathed: May be poisonous. Vapor may cause intense irritation to nose, throat, mouth and lungs; mucous membrane. Symptoms of exposure include watering of the eyes, dry painful cough, difficult breathing, headache, nausea, vomiting, muscular weakness, unconsciousness. High exposure can lead to drowsiness, coma and death; buildup of fluid in the lungs (pulmonary edema), and may cause death. Pulmonary edema may be delayed by one to two days following exposure.
Eyes: Severe irritant. Contact may cause, burning and stinging, tearing, irritation and redness and swelling of the eyelids, intense pain; can burn the eyes and lead to blurred vision, chronic irritation and other permanent damage.
Skin: May be poisonous if absorbed. Passes through the unbroken skin; can increase exposure and the severity of the symptoms listed. Severe irritant. Contact may cause burning or stinging feeling, redness, swelling, intense pain, blisters, irritation, rash. Repeated exposure may lead to skin allergy and chronic irritation.
Swallowed: May be poisonous. May cause burning feeling of the mouth and throat, intense thirst, throat swelling, stomach cramps, nausea, vomiting, diarrhea, severe breathing difficulties and unconsciousness. Tissue damage may result. Breathing may stop. Watch for delayed symptoms.
Long-term exposure: May cause nose and throat cancer; possible reproductive damage; bronchitis with coughing, phlegm and shortness of breath; liver, kidney and heart damage; chronic eye damage; hair loss.
Storage: Store in tightly closed containers in a dry, cool, well-ventilated place. Corrodes metal when wet. Keep away from water, sources of ignition, oxidizers and ammonia solutions. Containers may explode in fire.
First aid guide: Move victim to fresh air and call emergency medical care; if not breathing, give artificial respiration; if breathing is difficult, give oxygen. In case of contact with material, immediately flush skin or eyes with running water for at least 15 minutes. Speed in removing material from skin is of extreme importance. Remove and isolate contaminated clothing and shoes at the site. Keep victim quiet and maintain normal body temperature. Effects may be delayed; keep victim under observation.

★ 400 ★
DIMYRISTYL PEROXYDICARBONATE
CAS: 53220-22-7
Other names: DIMYRISTYL PEROXYDICARBONATE, TECHNICALLY PURE; DITETRADECYL PEROXYDICARBONATE; PEROXYDICARBONIC ACID, DITETRADECYL ESTER

Danger profile: Fire hazard, explosion hazard, highly reactive, poisonous gases produced in fire.
Uses: Making polyvinyl chloride polymers; resin curing.
Appearance: Solid or liquid.
Effects of exposure:
Breathed: May cause irritation of the nose, throat and bronchial tubes; cough and phlegm, nosebleeds, sore throat, hoarseness. Higher exposures may cause shortness of breath and a dangerous build-up of fluids in the lungs (pulmonary edema), which can cause death. Pulmonary edema is

a medical emergency that can be delayed one to two days following exposure.
Eyes: May cause severe irritation, redness, tearing and burns.
Skin: Passes through the unbroken skin; can increase exposure and the severity of the symptoms listed. May cause severe irritation and burns.
Long-term exposure: May cause lung damage.
Storage: Store in tightly closed containers in a dry, cool, well-ventilated place. Keep away from heat, sources of ignition, reducing agents and all other chemicals. Store at temperature of 68°F or less. Containers may explode in fire.
First aid guide: Move victim to fresh air; call emergency medical care. In case of contact with material, immediately flush eyes with running water for at least 15 minutes. Wash skin with soap and water. Remove and isolate contaminated clothing and shoes at the site. Keep victim quiet and maintain normal body temperature.

★ 401 ★
DINITOLMIDE
CAS: 148-01-6
Other names: COCCIDINE A; COCCIDOT; DINITOLMID; 2-METHYL-3,5-DINITROBENZAMIDE; 3,5-DINITRO-O-TOLUAMIDE; 3,5-DINITRO-2-METHYLBENZAMIDE; O-TOLUAMIDE, 3,5-DINITRO-; ZOALENE, ZOAMIX

Danger profile: Mutagen, exposure limit established.
Uses: Feed additive for poultry; veterinary drug.
Appearance: Yellowish salt-like crystalline material.
Effects of exposure:
Eyes: May cause irritation.
Skin: May cause irritation.
Swallowed: Moderately toxic.
Long-term exposure: May cause birth defects, miscarriage and cancer; possible liver damage.
Storage: Store in tightly closed containers in a dry, cool, well-ventilated place.

★ 402 ★
DINITROBENZENE
CAS: 25154-54-5
Other names: DINITROBENZINE, SOLUTION; DINITROBENZINE, SOLID; DINITROBENZOL, SOLID

Danger profile: Poison, suspected carcinogen, extremely reactive, extremely explosive, exposure limit established, poisonous gas produced in fire.
Uses: Making dyes and other chemicals; camphor substitute.
Appearance: Pale yellow or colorless solid.
Odor: Slight. The odor is a warning of exposure; however, no smell does not mean you are not being exposed.
Effects of exposure:
Breathed: May cause headache, rapid and difficult breathing, shortness of breath, dizziness, mental confusion, weakness; bluish color of the lips and skin, convulsions, coma and death.
Eyes: May cause irritation, redness, swelling of the eyelids; light causes pain; severe damage.
Skin: Passes through the unbroken skin; can increase exposure and the severity of the symptoms listed. May cause irritation, redness, swelling, blisters. Skin may turn yellow.
Swallowed: Poison. May cause irritation of the mouth, stomach; cramps and diarrhea.
Long-term exposure: May cause liver damage; low blood count; change in vision; hearing loss.
Storage: Store in tightly closed containers in a dry, cool, well-ventilated place. Keep away from heat, sources of ignition, oxi-

dizers, caustics, metals and all other chemicals. Containers may explode in fire.

First aid guide: Move victim to fresh air and call emergency medical care; if not breathing, give artificial respiration; if breathing is difficult, give oxygen. In case of contact with material, immediately flush skin or eyes with running water for at least 15 minutes. Speed in removing material from skin is of extreme importance. Remove and isolate contaminated clothing and shoes at the site. Keep victim quiet and maintain normal body temperature. Effects may be delayed, keep victim under observation.

★ 403 ★
DINITRO-O-CRESOL
CAS: 534-52-1
Other names: ANTINONIN; ANTINONNIN; ARBOROL; BENZENE, CHLORODINITRO-; CAPSINE; CHEMSECT O-CRESOL,4,6-DINITRO-; PHENOL 2,4-DINITRO-6-METHYL-, AND SALTS; 4,6-DINITRO-O-CRESOL AND SALTS; DEGRASSAN; DEKRYSIL; DETAL; DINITROCRESOL; 2,4-DINITRO-O-CRESOL; 4,6-DINITRO-O-CRESOL; 4,6-DINITRO-O-CRESOLO (ITALIAN); DINITRODENDTROXAL; 3,5-DINITRO-2-HYDROXYTOLUENE; 4,6-DINITROKRESOL (DUTCH); 4,6-DINITRO-O-KRESOL (CZECH); DINITROL; DINITROMETHYLCYCLOHEXYLTRIENOL; 2,4-DINITRO-6-METHYLPHENOL; DINOC; DINURANIA; DITROSOL; DN; DNC; DNOC; DN-DRY MIX NO. 2; DNOK (CZECH); DWUNITRO-O-KREZOL (POLISH); EFFUSAN; EFFUSAN 3436; ELGETOL; ELGETOL 30; ELIPOL; EXTRAR; HEDOLIT; HEDOLITE; K III; KIV; KRENITE (OBS.); KRESAMONE; KREZOTOL 50; LE DINITROCRESOL-4,6 (FRENCH); LIPAN; 2-METHYL-4,6-DINITROPHENOL; NITRADOR; NITROFAN; PHENOL, 2-METHYL-4,6-DINITRO- (9CI); PROKARBOL; RAFEX; RAFEX 35; RAPHATOX; SANDOLIN; SANDOLIN A; SELINON; SINOX; TRIFOCIDE; WINTERWASH; ZAHLREICHE BEZEICHNUNGEN (GERMAN)

Danger profile: Mutagen, poison, combustible, exposure limit established, Community Right-to-Know List, EPA extremely hazardous substance list, poisonous gas produced in fire.
Uses: To kill weeds and insects.
Appearance: Yellow sand-like crystals, often used in solution.
Effects of exposure: *Breathed:* May be poisonous. May cause intense thirst, nose irritation, headache, coughing, nausea, stomach pain, restlessness, sweating, and rapid breathing and pulse, anxiety, confusion; yellow face and lips. High levels may cause fever, coma, circulatory collapse and death.
Eyes: May cause burning sensation, irritation, tearing, and inflammation of the eyelids.
Skin: Passes through the unbroken skin; can increase exposure and the severity of the symptoms listed. Symptoms may be delayed. May cause irritation, redness, swelling and blisters.
Swallowed: May be poisonous. May cause burning feeling in the throat and mouth; dizziness, headache, nausea, vomiting, high fever, thirst, fatigue, rapid pulse, convulsions, loss of consciousness. Higher exposures may cause shortness of breath and a dangerous build-up of fluids in the lungs (pulmonary edema), which can cause death. Pulmonary edema is a medical emergency that can be delayed one to two days following exposure.
Long-term exposure: Repeated contact may cause weight loss, unusual thirst, profuse sweating; eye cataracts may develop; may stain the skin and fingernails yellow; liver, kidney and blood cell damage.
Storage: Store in tightly closed containers in a dry, cool, well-ventilated place. Keep away from heat, sources of ignition and oxidizers.
First aid guide: Move victim to fresh air; call emergency medical care. In case of contact with material, immediately flush skin or

eyes with running water for at least 15 minutes. Remove and isolate contaminated clothing and shoes at the site.

★ 404 ★
DINITROPHENOL
CAS: 25550-58-7
Other names: PHENOL, DINITRO-

Danger profile: Explosive, poison, poisonous gases produced in fire.
Uses: In dyes, wood preservatives; to make explosives.
Appearance: Yellow sand-like crystals, often found in solution.
Effects of exposure:
Breathed: May cause intense thirst, nose irritation, headache, coughing, nausea, stomach pain, restlessness, sweating, and rapid breathing and pulse, anxiety, confusion; yellow face and lips. High levels may cause fever, coma and death.
Eyes: May cause burning sensation, irritation, tearing, and inflammation of the eyelids.
Skin: Passes through the unbroken skin; can increase exposure and the severity of the symptoms listed. Symptoms may be delayed. May cause irritation, redness, swelling and blisters.
Swallowed: May cause burning feeling in the throat and mouth; dizziness, headache, nausea, vomiting, high fever, thirst, fatigue, rapid pulse, convulsions, loss of consciousness. Higher exposures may cause shortness of breath and a dangerous build-up of fluids in the lungs (pulmonary edema), which can cause death. Pulmonary edema is a medical emergency that can be delayed one to two days following exposure.
Long-term exposure: May cause damage to the developing fetus; liver and kidney damage; cataracts; underactive thyroid; skin allergy; lung damage, bronchitis with cough and phlegm.
Storage: Store wet in tightly closed containers in a well-ventilated place. Keep away from heat and sources of ignition; shock, oxidizers, metals and metal compounds. Containers may explode in heat and fire.

★ 405 ★
2,4-DINITROTOLUENE
CAS: 121-14-2
Other names: BENZENE, 1-METHYL-2,4-DINITRO-; DINITROTOLUOL; 2,4-DNT; DNT; NCI-CO1865; TOLUENE, 2,4-DINITRO-

Danger profile: Possible carcinogen, mutagen, combustible, highly reactive; explosion hazard; exposure limit established, Community Right-to-Know List; poisonous gas produced in fire.
Uses: Making plastics, dyes, other chemicals and explosives.
Appearance: Yellow to orange-yellow crystalline needles. Often found in a molten state.
Odor: Characteristic. The odor is a warning of exposure; however, no smell does not mean you are not being exposed.
Effects of exposure:
Breathed: May cause interference with the ability of the blood to carry oxygen; headaches, fatigue, dizziness, and blue color to the skin, lips and nose. Prolonged exposure may cause liver damage and low blood count. High exposures may cause nausea, vomiting, shortness of breath, irregular heartbeat, loss of consciousness and death. Symptoms may be delayed for several hours following exposure.
Eyes: Contact or fumes may cause severe burns and permanent damage.
Skin: Passes through the unbroken skin; can increase exposure and the severity of the symptoms listed. Contact or

fumes may cause severe burns. See *Breathed.* Symptoms may be delayed for several hours following exposure.
Swallowed: Poisonous.
Long-term exposure: May cause cancer, liver damage and anemia; damage to the growing fetus.
Storage: Store in tightly closed containers in a dry, cool, well-ventilated place. Keep away from heat, sources of ignition, oxidizers, bases, sodium carbonate, sodium oxide, chemically active metals such as aluminum, tin, zinc, shock (avoid dropping containers). Containers may explode in fire.

★ 406 ★
DIOCTYL PHTHALATE
CAS: 117-84-0
Other names: 1,2-BENZENEDICARBOXYLIC ACID, DI-N-OCTYL ESTER; CELLULEX DOP; DI-N-OCTYL PHTHALATE; DNOP; DINOPOL NOP; PHTHALIC ACID, DIOCTYL ESTER

Danger profile: Carcinogen, possible reproductive hazard, combustible, Community Right-to-Know List, irritating fumes and gas in fire.
Uses: IIn plastics; making plastic products.
Appearance: Colorless, oily liquid.
Effects of exposure:
Breathed: May cause irritation of the nose, throat and bronchial tubes. High levels, if prolonged, may cause death.
Eyes: Contact may cause severe irritation.
Skin: Passes through the unbroken skin; can increase exposure and the severity of the symptoms listed. Contact may cause irritation. Repeated contact may cause dryness, cracking and rash.
Swallowed: Mildly toxic.
Long-term exposure: Possible reproductive hazard. High or repeated exposures may cause liver and kidney damage; skin cracking and rash; lung damage.
Storage: Store in tightly closed containers in a dry, cool, well-ventilated place. Keep away from heat and sources of ignition, oxidizers and alkahalides.

★ 407 ★
DIOXANE
CAS: 123-91-1
Other names: DIETHYLENE DIOXIDE; 1,4-DIETHYLENEDIOXIDE; DIETHYLENE ETHER; DI(ETHYLENE OXIDE); DIOKAN; DIOKSAN (POLISH); DIOSSANO-1,4 (ITALIAN); DIOXAAN-1,4 (DUTCH); 1,4-DIOXACYCLOHEXANE; DIOXAN; DIOXAN-1,4 (GERMAN); P-DIOXAN (CZECH); DIOXANE; DIOXANE-1,4; 1,4-DIOXANE; DIOXANNE (FRENCH); P-DIOXIN, TETRAHYDRO-; DIOXYETHYLENE ETHER; GLYCOL ETHYLENE ETHER; NCI-C03689; TETRAHYDRO-P-DIOXIN; TETRAHYDRO-1,4-DIOXIN; P-DIOXANE; DIETHYLENE-1,4-DIOXIDE

Danger profile: Carcinogen, mutagen, fire hazard, explosion hazard, exposure limit established, Community Right-to-Know List.
Uses: Solvent for many products; lacquers and paints; varnishes; paint and varnish removers; in textile processing, dye baths, stain and printing compositions; cleaning and detergent preparations; cements; cosmetics; deodorants; fumigants; emulsions; polishing compositions.
Appearance: Colorless liquid.
Odor: Pleasant, mild ether-like odor. The odor is a warning of exposure; however, no smell does not mean you are not being exposed.
Effects of exposure:
Breathed: Mildly toxic. May cause irritation of eyes, nose and throat; headache, drowsiness, excitability, drunken behavior,

mental confusion, fatigue. Very high levels of exposure may cause dizziness, lightheadedness and cause victim to pass out, go into coma and die.
Eyes: Contact may cause irritation, pain, burning, stinging, tearing, inflammation of the eyelids. Light may cause pain. Overexposure may cause damage to the cornea.
Skin: Prolonged or repeated exposure may cause rash or burn and absorption of toxic amounts leading to serious injury of liver and kidney. Passes through the unbroken skin; can increase exposure and the severity of the symptoms listed. May remove oil from the skin and cause dryness. Persons with pre-existing skin disorders may be more susceptible to the effects of this substance.
Swallowed: May cause gastro-intestinal irritation.
Long-term exposure: Prolonged or repeated contact may cause cancer; irritation of the skin and dermatitis, and nervous system changes; kidney and liver damage; poor appetite, upset stomach; tenderness of the abdomen.
Storage: Store in a tightly closed container in a cool, well-ventilated place. Keep away from heat and sources of ignition, nitrates, sulfur trioxide, strong oxidizers, strong alkalis and strong acids. Vapors are explosive in heat or flame. Containers may explode in fire.
First aid guide: Move victim to fresh air and call emergency medical care; if not breathing, give artificial respiration; if breathing is difficult, give oxygen. In case of contact with material, immediately flush eyes with running water for at least 15 minutes. Wash skin with soap and water. Remove and isolate contaminated clothing and shoes at the site.

★ 408 ★
DIOXATHION
CAS: 78-34-2
Other names: S,S1-1,4-DIOXANE-2,3-DIY1-0,0,0-TETRAETHYL ESTER; DELNAV; DELNATEX; ENT 22,897; NAVADEL; NCI-C00395; PHOSPHORODITHIOIC ACID, S,S'-1,4-DIOXANE-2,3-DIYL-O,O,O',O'-TETRAETHYL ESTER; PHOSPHORODITHIOIC ACID, O,O-DIETHYL ESTER, S,S-DIESTER WITH P-DIOXANE-2,3-DITHIOL; PHOSPHORODITHIOIC ACID-5-5'-1,4-DIOXANE-2,3-DIYL O,O,O',O'-TETRAETHYL ESTER; BIS(DITHIOPHOSPATE DE O,O-DIETHYLE) DE S,S'-(1,4-DIOXANNE-2,3-DIYLE) (FRENCH)

Danger profile: Mutagen, poison, exposure limit established, poisonous gases produced in fire.
Uses: Insecticide; miticide.
Appearance: Thick reddish-brown liquid.
Effects of exposure:
Breathed: Extremely poisonous. May cause headache, dizziness, profuse sweating, nausea, vomiting, reduced heart beat, stomach cramps, diarrhea, loss of coordination, slow and weak breathing, fever, loss of consciousness and death. May cause shortness of breath and a dangerous build-up of fluids in the lungs (pulmonary edema), which can cause death. Pulmonary edema is a medical emergency that can be delayed one to two days following exposure.
Eyes: Rapidly absorbed. Poisoning can happen very rapidly. See *Breathed.*
Skin: Poisonous. Passes through the unbroken skin; can increase exposure and the severity of the symptoms listed. Fatal poisoning can result from skin contact.
Swallowed: Powerful, deadly poison. See *Breathed.*
Long-term exposure: May cause nerve damage including loss of coordination. Repeated exposure may cause depression, irritability and personality changes.
Storage: Store in a regulated, specially marked place in tightly closed containers in a dry, well-ventilated place. Keep away from water, strong alkalis. Containers may explode in fire.

★ 409 ★
DIPENTENE
CAS: 138-86-3
Other names: ACINTENE DP; ACINTENE DP DIPENTENE; CAJEPUTENE; CINENE; CYCLOHEXENE, 1-METHYL-4-(1-METHYLETHENYL)-; DIPANOL; INACTIVE LIMONENE; KAUTSCHIN; LIMONENE; DL-LIMONENE; P-MENTHA-1,8-DIENE,DL-; 1,8(9)-P-MENTHADIENE; 1-METHYL-4-ISOPROPENYL-1-CYCLOHEXENE; NESOL; DELTA-1,8-TERPODIENE

Danger profile: Combustible, poisonous gas produced in fire.
Uses: Solvent; making rubber, pigments and driers; paints, enamels, lacquers, and varnishes; printing inks; perfumes; flavors; floor waxes and furniture polishes; synthetic resins.
Appearance: Colorless to light yellow liquid.
Odor: Pleasant lemon-like. The odor is a warning of exposure; however, no smell does not mean you are not being exposed.
Effects of exposure:
Breathed: May be poisonous. May cause irritation of the nose and throat; dizziness. Possible suffocation. High exposure may cause kidney damage.
Eyes: Contact may cause irritation.
Skin: May be poisonous if absorbed. Passes through the unbroken skin; can increase exposure and the severity of the symptoms listed. Contact may cause irritation. Skin allergy may develop.
Swallowed: May cause irritation of gastrointestinal tract.
Long-term exposure: May cause skin allergy with itching and rash.
Storage: tore in tightly closed containers in a dry, cool, well-ventilated place. Keep away from heat, sources of ignition and oxidizing materials. Vapors may travel to sources of ignition and flash back. Containers may explode in fire.
First aid guide: Move victim to fresh air and call emergency medical care; if not breathing, give artificial respiration; if breathing is difficult, give oxygen. In case of contact with material, immediately flush eyes with running water for at least 15 minutes. Wash skin with soap and water. Remove and isolate contaminated clothing and shoes at the site.

★ 410 ★
DIPHENYLAMINE
CAS: 122-39-4
Other names: ANILINE, N-PHENYL; ANILINOBENZENE; BENZENE, ANILINO-; BIG DIPPER; C.I. 10355; DFA; N,N-DIPHENYLAMINE; DPA; NO SCALD; N-PHENYLANILINE; N-PHENYLBENZENAMINE; SCALDIP

Danger profile: Combustible, reproductive hazard, exposure limit established, poisonous gas produced in fire.
Uses: Making rubber, plastics, veterinary medicine and pharmaceuticals; solid rocket propellants; pesticides; dyes; making other chemicals; laboratory chemical; apple storage preservative.
Appearance: Colorless to light grey or tan sand-like crystals.
Odor: Pleasant, aromatic floral odor. The odor is a warning of exposure; however, no smell does not mean you are not being exposed.
Effects of exposure:
Breathed: May cause irritation of the mucous membrane; bladder trouble, increased blood pressure and pulse rate.
Eyes: Dust may cause irritation.
Skin: Passes through the unbroken skin; can increase exposure and the severity of the symptoms listed. Contact may cause rash.
Swallowed: Poison. See *Breathed.*
Long-term exposure: May cause liver and kidney damage; bladder problems; increased blood pressure and heart rate. Lim-

ited evidence of cancer and reproductive damage in animals. Similar chemicals cause skin allergy.
Storage: Store in tightly closed containers in a dry, cool, well-ventilated place. Keep away from heat, sources of ignition, hexachloromelamine, and oxidizers.

★ 411 ★
DIPHENYL DICHLOROSILANE
CAS: 80-10-4
Other names: DICHLORODIPHENYLSILANE; DIPHENYLDI-CHLOROSILANE; SILANE, DICHLORODIPHENYL-

Danger profile: Corrosive, combustible, poisonous gases produced in fire.
Uses: Making silicone lubricants.
Appearance: Colorless liquid.
Odor: Sharp, irritating. The odor is a warning of exposure; however, no smell does not mean you are not being exposed.
Effects of exposure:
Breathed: May cause irritation to the eyes, nose and throat. Higher exposures may cause shortness of breath and a dangerous build-up of fluids in the lungs (pulmonary edema), which can cause death. Pulmonary edema is a medical emergency that can be delayed one to two days following exposure.
Eyes: Contact may cause severe burns, redness, tearing, inflammation of the eyelids and permanent damage.
Skin: May cause severe burns.
Swallowed: May cause severe burns of the mouth and stomach.
Long-term exposure: May cause lung problems.
Storage: Store in tightly closed containers in a dry, cool, well-ventilated place. Keep away from moisture (generates hydrogen chloride which is corrosive to common metals), heat, sources of ignition, acids, bases and oxidizers. Vapors may travel to sources of ignition and flash back. Containers may explode in fire.
First aid guide: Move victim to fresh air and call emergency medical care; if not breathing, give artificial respiration; if breathing is difficult, give oxygen. In case of contact with material, immediately flush skin or eyes with running water for at least 15 minutes. Remove and isolate contaminated clothing and shoes at the site. Keep victim quiet and maintain normal body temperature.

★ 412 ★
DIPROPYLENE GLYCOL METHYL ETHER
CAS: 34590-94-8
Other names: ARCOSOLV; DIPROPYLENE GLYCOL MONO-METHYL ETHER; DOWANOL-50B; DOWANOL DPM; PRO-PANOL, OXYBIS-, METHYL ETHER; UCAR SOLVENT 2LM

Danger profile: Combustible, exposure limit established, Community Right-to-Know List, poisonous gas produced in fire.
Uses: Solvent for paints, pastes, inks, dyes; making cosmetics.
Appearance: Colorless liquid.
Odor: Weak. The odor is a warning of exposure; however, no smell does not mean you are not being exposed.
Effects of exposure:
Breathed: May be poisonous. May cause headache, dizziness, lightheadedness and loss of consciousness.
Eyes: Contact may cause irritation.
Skin: May be poisonous if absorbed. Passes through the unbroken skin; can increase exposure and the severity of the symptoms listed. Contact may cause irritation.
Swallowed: May be poisonous.
Long-term exposure: Repeated exposure to high levels may effect the liver. Allergies might develop.

Storage: Store in tightly closed containers in a dry, cool, well-ventilated place. Keep away from heat, sources of ignition and oxidizers. Containers may explode in fire.

★ 413 ★
DIPROPYL KETONE
CAS: 123-19-3
Other names: BUTYRONE; GBL; 4-HEPTANONE; HEPTAN-4-ONE; PROPYL KETONE

Danger profile: Combustible, exposure limit established, poisonous gases produced in fire.
Uses: Solvent for nitrocellulose, oils, resins and polymers; in lacquers; flavoring.
Appearance: Colorless liquid.
Odor: Noticeable. The odor is a warning of exposure; however, no smell does not mean you are not being exposed.
Effects of exposure:
Breathed: May be poisonous. May cause irritation of the eyes, nose and throat; headache, dizziness, lightheadedness and loss of consciousness.
Eyes: May cause irritation.
Skin: May be poisonous if absorbed. Passes through the unbroken skin; can increase exposure and the severity of the symptoms listed. May cause irritation.
Swallowed: May be poisonous.
Long-term exposure: May cause liver and kidney damage; dry skin and rash. Similar solvents cause brain and nerve damage.
Storage: Store in tightly closed containers in a dry, cool, well-ventilated place. Keep away from heat, sources of ignition and oxidizers. Vapors may travel to sources of ignition and flash back. Containers may explode in fire.
First aid guide: Move victim to fresh air and call emergency medical care; if not breathing, give artificial respiration; if breathing is difficult, give oxygen. In case of contact with material, immediately flush eyes with running water for at least 15 minutes. Wash skin with soap and water. Remove and isolate contaminated clothing and shoes at the site.

★ 414 ★
DIQUAT
CAS: 85-00-7
Other names: AQUACIDE; DEIQUAT; DEXTRONE; 9,10-DIHYDRO-8A,10,-DIAZONIAPHENANTHRENE DIBROMIDE; 9,10-DIHYDRO-8A,10A-DIAZONIAPHENANTHRENE(1,1'-ETHYLENE-2,2'-BIPY RIDYLIUM)DIBROMIDE; 5,6-DIHYDRO-DIPYRIDO(1,2A,2,1C)PYRAZINIUM DIBROMIDE; 6,7-DIHYDROPYRIDO(1,2-A, 2',1'-C)PYRAZINEDIUM DIBROMIDE; DIQUAT; DIQUAT DIBROMIDE; 1,1'-ETHYLENE-2,2'-BIPYRIDYLIUMDIBROMIDE; ETHYLENE DIPYRIDYLIUM DIBROMIDE; 1,1-ETHYLENE 2,2-DIPYRIDYLIUM DIBROMIDE; 1,1'-ETHYLENE-2,2'-DIPYRIDYLIUMDIBROMIDE; FB/2; FEGLOX; PREEGLONE; REGLON; REGLONE; REGLOX; WEEDTRINE-D

Danger profile: Mutagen, possible reproductive hazard, exposure limit established, poisonous gas produced in fire.
Uses: Herbicide and plant growth regulator; aquatic weed control; potato haulm destruction; sugarcane flower suppressant.
Appearance: Pale yellow sand-like crystalline substance.
Effects of exposure:
Breathed: May cause irritation of the nose and eyes; sneezing; nose bleed; painful obstruction of the throat.
Eyes: May cause a painful burning feeling, tearing, inflammation of the eyelids; light sensitivity.
Skin: Passes through the unbroken skin; can increase exposure and the severity of the symptoms listed. May cause irri-

tation, inflammation, pain and blisters. The healing of cuts and wounds may be delayed.
Swallowed: Poison. May cause burning feeling in the mouth, throat and stomach; difficult swallowing; nausea, vomiting; painful abdominal cramps, diarrhea (possibly blood-stained).
Long-term exposure: May damage the developing fetus. May cause liver, kidney and lung damage; cracked and dry skin; fingernail damage; cataracts.
Storage: Store in tightly closed containers in a dry, cool, well-ventilated place. Containers may explode in fire.

★ 415 ★
DI-SEC-OCTYL PHTHALATE
CAS: 117-81-7
Other names: BEHP; 1,2-BENZENEDICARBOXYLIC ACID, BIS(2-ETHYLHEXYL) ESTER; BIS(2-ETHYLHEXYL)-1,2-BENZENEDICARB OXYLATE; BIS(2-ETHYLHEXYL)PHTHALATE; BISOFLEX 81; BISOFLEX DOP; COMPOUND 889; DAF 68; DEHP; DI(2-ETHYLHEXYL)ORTHOPHTHALATE; DI(2-ETHYLHEXYL)PHTHALATE; DIOCTYL PHTHALATE; DOP; ERGOPLAST FDO; ETHYLHEXYL PHTHALATE; 2-ETHYLHEXYL PHTHALATE; EVIPLAST 80; EVIPLAST 81; FLEXIMEL; FLEXOL DOP; FLEXOL PLASTICIZER DOP; GOOD-RITE GP 264; HATCOL DOP; HERCOFLEX 260; KO-DAFLEX DOP; MOLLANO; NCI-C52733; NUOPLAZ DOP; OC-TOIL; OCTYL PHTHALATE; PALATINOL AH; PHTHALIC ACID DIOCTYL ESTER; PITTSBURGH PX-138; PLATINOL AH; PLATINOL DOP; RC PLASTICIZER DOP; REOMOL DOP; REOMOL D 79P; SICOL 150; STAFLEX DOP; TRUFLEX DOP; VESTINOL AH; VINICIZER 80; WITCIZER 312

Danger profile: Carcinogen, mutagen, possible reproductive hazard, combustible, exposure limit established, Community Right-to-Know List, poisonous gas produced in fire.
Uses: Plasticizer.
Appearance: Clear, colorless, oily liquid.
Odor: Almost none. The odor is a warning of exposure; however, no smell does not mean you are not being exposed.
Effects of exposure:
Breathed: May cause irritation to the eye, nose and throat; nausea, diarrhea; central nervous system depression.
Eyes: May cause irritation.
Skin: May cause irritation.
Swallowed: May cause irritation, nausea, diarrhea.
Long-term exposure: May cause cancer; birth defect by damaging the fetus; damage to the male testes; kidney and liver damage. Similar chemicals may cause numbness and tingling in the arms and legs.
Storage: Store in tightly closed containers in a dry, cool, well-ventilated place. Keep away from heat, sources of ignition, oxidizing materials. Containers may explode in fire.

★ 416 ★
DISULFIRAM
CAS: 97-77-8
Other names: ABSTENSIL; ABSTINYL; ALCOPHOBIN; ALK-AUBS; ANTABUS; ANTABUSE; ANTADIX; ANTAENYL; AN-AETHAN; ANTAETHYL; ANTAETIL; ANTALCOL; ANTETAN; ANTETHYL; ANTIKOL; AVERSAN; AVERZAN; BIS(DIETH-YLTHIOCARBAMOYL) DISULFIDE; BONIBOL; CONTRALIN; CONTRAPOT; CRONETAL; DICUPRAL; DISETIL; DISULFAN; DISULFURAM; DISULPHURAM; EKAGOM TEDS; EPHOR-RAN; ESPERAL; ETABUS; ETHYLTUADS; ETHYL TUEX; EX-HORAN; EXHORRAN; HOCA; KROTENAL; NCI-C02959; NOCBIN; NOXAL; REFUSAL; RO-SULFIRAM; TATD; TENURID; TENUTEX; TETRAETHYL THIURAM DISULFIDE; TETRAETHYLTHIURAM DISULFIDE; TILLRAM; TIURAM; TTD; TTS; USAF B-33

Danger profile: Mutagen, exposure limit established, poisonous gases produced in fire.
Uses: Fungicide; seed disinfectant; making rubber; prescription drug for alcoholism.
Appearance: White, light gray or brownish sand-like crystalline solid.
Effects of exposure:
Breathed: May be poisonous. Exposure to chemical and alcohol (including some cough medicines and mouth washes) within 24-48 hours of each other may cause reaction, rapid heart beat, vomiting and death.
Skin: May be poisonous if absorbed. Passes through the unbroken skin; can increase exposure and the severity of the symptoms listed. May cause allergy, irritation and skin eruptions.
Swallowed: Poison.
Long-term exposure: May cause damage to the developing fetus; vision, nervous system, kidneys; enlarged thyroid; skin rash.
Storage: Store in tightly closed containers in a dry, cool, well-ventilated place. Keep away from combustible substances, ethylene dibromide and phenols. Containers may explode in fire.

★ 417 ★
DISULFOTON
CAS: 298-04-4
Other names: BAY 19639; BAYER 19639; O,O-DIAETHYL-S-(3-THIA-PENTYL)-DITHIOPHOSPHAT (GERMAN); O,O-DIAETHYL-S-(2-AETHYLTHIO-AETHYL)-DITHIOPHOSPHAT (GERMAN); O,O-DIETHYL S-(2-ETHTHIOETHYL)PHOSPHORODITHIOATE; O,O-DIETHYLS-(2-ETHTHIOETHYL)THIOTHIONOPHOSPHATE; O,O-DIETHYLS-(2-ETHYLMERCAPTOETHYL)DITHIOPHOSPHATE; O,O-DIETHYL-S-(2-ETHYLTHIO-ETHYL)-DITHIOFOSFAAT (DUTCH); O,O-DIETHYL 2-ETHYLTHIOETHYLPHOSPHORODITHIOATE; O,O-DIETHYLS-2-(ETHYLTHIO)ETHYLPHOSPHORODITHIOATE; O,O-DIETIL-S-(2-ETILTIO-ETIL)-DITIOFOSFATO (ITALIAN); DIMAZ; DISULFATON; DI-SYSTON; DISYSTOX; DITHI-ODEMETON; DITHIOPHOSPHATE DE O,O-DIETHYLE ETDE S-(2-ETHYLTHIO-ETHYLE) (FRENCH); DITHIOSYSTOX; ENT 23,437; O,O-ETHYL S-2(ETHYLTHIO)ETHYLPHOSPHORODITHIOATE; S-2-(ETHYLTHIO)ETHYL O,O-DIETHYLESTER OF PHOSPHO-RODITHIOIC ACID; FRUMIN AL; FRUMIN G; M-74; PHO-SPHORODITHIONIC ACID,S-2-(ETHYLTHIO)ETHYL-O,O-DIETHYLESTER; S 276; SOLVIREX; THIODEMETON; THIODEMETRON

Danger profile: Mutagen, exposure limit established, poisonous gases produced in fire.

Uses: Insecticide (organophosphate).
Appearance: Yellow to brown liquid.
Odor: Disagreeable. The odor is a warning of exposure; however, no smell does not mean you are not being exposed.
Effects of exposure:
Breathed: Extremely poisonous. May cause headache, dizziness, profuse sweating, nausea, vomiting, reduced heart beat, stomach cramps, diarrhea, loss of coordination, slow and weak breathing, fever, loss of consciousness and death. May cause shortness of breath and a dangerous build-up of fluids in the lungs (pulmonary edema), which can cause death. Pulmonary edema is a medical emergency that can be delayed one to two days following exposure.
Eyes: May cause irritation and tearing. See *Breathed* for symptoms. Rapidly absorbed; fatal poisoning may occur from skin contact.
Skin: Poisonous. Passes through the unbroken skin; can increase exposure and the severity of the symptoms listed.
Swallowed: Powerful, deadly poison. See *Breathed.*
Long-term exposure: May damage the developing fetus or cause birth defects; nerve damage including loss of coordination; liver damage. Repeated exposure may cause depression, irritability and personality changes.
Storage: Store in a regulated, specially marked place in tightly closed containers in a dry, well-ventilated place. Containers may explode in fire.
First aid guide: Move victim to fresh air and call emergency medical care; if not breathing, give artificial respiration; if breathing is difficult, give oxygen. In case of contact with material, immediately flush skin or eyes with running water for at least 15 minutes. Speed in removing material from skin is of extreme importance. Remove and isolate contaminated clothing and shoes at the site. Keep victim quiet and maintain normal body temperature. Effects may be delayed; keep victim under observation.

★ 418 ★
DIURON
CAS: 330-54-1
Other names: AF 101; CEKIURON; CRISURON; DAILON; DCMU; DIATER; DICHLORFENIDIM; 3-(3,4-DICHLOROPHENOL)-1,1-DIMETHYLUREA; N'-(3,4-DICHLOROPHENYL)-N,N-DIMETHYLUREA; 1-(3,4-DICHLOROPHENYL)-3,3-DIMETHYLUREE (FRENCH); 3-(3,4-DICHLOR-PHENYL)-1,1-DIMETHYL-HARNSTOFF (GER-MAN); 3-(3,4-DICHLOOR-FENYL)-1,1-DIMETHYLUREUM (DUTCH); 3-(3,4-DICLORO-FENYL)-1,1-DIMETIL-UREA (ITALIAN); 1,1-DIMETHYL-3-(3,4-DICHLOROPHENYL)UREA; DI-ON; DIUREX; DIUROL; DIURON 4L; DMU; DREXEL; DURAN; DYNEX; HERBATOX; HW 920; KARMEX; KARMEX DIURON HERBICIDE; KARMEX DW; MARMER; SUP'R FLO; TELVAR DIURON WEED KILLER; UNIDRON; UROX D; USAF P-7; USAF XR-42; VONDURON

Danger profile: Combustible, possible reproductive hazard, exposure limit established, Community Right-to-Know List, poisonous gases produced in fire.
Uses: Weed killer.
Appearance: White sand-like crystalline material.
Effects of exposure:
Breathed: May cause irritation to the nose, throat and mucous membrane.
Eyes: May cause irritation and inflammation of the eyelids.
Skin: May cause irritation.
Swallowed: Moderately toxic.
Long-term exposure: May cause mutations and damage to the developing fetus. Repeated exposure to high levels may lower the red blood count.
Storage: Store in tightly closed containers in a dry, cool, well-ventilated place.

★419★
DIVINYL BENZENE
CAS: 108-57-6
Other names: BENZENE, DIETHENYLBENZENE; M-DIVINYL-; DIVINYLBENZENE; M-DIVINYLBENZEN (CZECH); M-DIVINYLBENZENE; DVB; VINYLSTYRENE; M-VINYLSTYRENE

Danger profile: Combustible; explosion hazard; exposure limit established; poisonous gas produced in fire.
Uses: Making synthetic rubber, drying oils and polyesters.
Appearance: Pale straw colored liquid.
Effects of exposure:
  Breathed: May cause irritation of the nose and throat. Higher exposure may cause dizziness, drowsiness and loss of consciousness.
  Eyes: Contact may cause irritation, redness and inflammation of the eyelids.
  Skin: Passes through the unbroken skin; can increase exposure and the severity of the symptoms listed. Contact may cause irritation; prolonged contact may cause rash.
  Swallowed: May be poisonous.
Long-term exposure: Closely related substances have been shown to cause mutations and damage the developing fetus. May cause skin dryness and rash.
Storage: Store in tightly closed containers in a dry, cool, well-ventilated place. Keep away from heat, sources of ignition, oxidizers and metallic salts.

★420★
DODECYLBENZENESULFONIC ACID
CAS: 27176-87-0
Other names: BENZENESULFONIC ACID, DODECYL-; BENZENE SULFONIC ACID, DODECYL ESTER; DDBSA; DODECYL BENZENESULFONATE

Danger profile: Corrosive, poisonous gases produced in fire.
Uses: Making detergents.
Appearance: Colorless liquid.
Effects of exposure:
  Breathed: May cause irritation to the nose and throat.
  Eyes: May cause irritation, redness, tearing and inflammation of the eyelids.
  Skin: May cause irritation.
  Swallowed: May cause irritation of mouth and stomach; nausea.
Long-term exposure: Repeated contact may cause dryness, itching and rash.
Storage: Store in tightly closed containers in a dry, cool, well-ventilated place. Keep away from combustible materials and metals.
First aid guide: Move victim to fresh air; call emergency medical care. In case of contact with material, immediately flush skin or eyes with running water for at least 15 minutes. Remove and isolate contaminated clothing and shoes at the site. Keep victim quiet and maintain normal body temperature.

★421★
DODECYL TRICHLOROSILANE
CAS: 4484-72-4
Other names: DODECYLTRICHLOROSILANE; DODECYL TRICHLOROSILANE; SILANE, TRICHLORODODECYL-; TRICHLORODODECYLSILANE; SILANE, DODECYLTRICHLORO-

Danger profile: Combustible, corrosive, may react violently with water, poisonous gases produced in fire.
Uses: Making silicones.

Appearance: Colorless liquid.
Odor: Sharp, pungent, irritating. The odor is a warning of exposure; however, no smell does not mean you are not being exposed.
Effects of exposure:
  Breathed: May cause irritation of the lungs with coughing and/or shortness of breath. Higher exposures may cause shortness of breath and a dangerous build-up of fluids in the lungs (pulmonary edema), which can cause death. Pulmonary edema is a medical emergency that can be delayed one to two days following exposure.
  Eyes: May cause severe irritation and permanent damage.
  Skin: May cause irritation, redness and severe burns.
  Swallowed: May cause severe burns of mouth and stomach.
Long-term exposure: Similar chemicals cause lung damage.
Storage: Store in tightly closed containers in a dry, cool, well-ventilated place. Keep away from heat, sources of ignition, moisture (which can generate hydrogen chloride, which is corrosive to common metals).
First aid guide: Move victim to fresh air; call emergency medical care. In case of contact with material, immediately flush skin or eyes with running water for at least 15 minutes. Remove and isolate contaminated clothing and shoes at the site. Keep victim quiet and maintain normal body temperature.

E

★422★
ENDOXAN
CAS: 50-18-0
Other names: ASTA; ASTA B 518; B 518; N,N-BIS-(BETA-CHLORAETHYL)-N',O-PROPYLEN-PHOSPHORSAEURE-ESTER-DIA MID (GERMAN); 2-(BIS(2-CHLOROETHYL)AMINO)-1-OXA-3-AZA-2-PHOSPHOCYCLOHEXANE 2-OXIDE MONOHYDRATE; 1-BIS(2-CHLOROETHYL)AMINO-1-OXO-2-AZA-5-OXAPHOSPHORIDINE MONOHYDRATE; 2-(BIS(2-CHLOROETHYL)AMINO)-2H-1,3, 2-OXAZAPHOSPHORINE 2-OXIDE; (BIS(CHLORO-2-ETHYL)AMINO)-2-TETRAHYDRO-3,4,5,6-OXAZAPHOSPHORINE-1,3, 2-OXIDE-2 HYDRATE; 2-(BIS(2-CHLOROETHYL)AMINO)TETRAHY DRO(2H)-1,3,2-OXAZAPHOSPHORINE 2-OXIDE MONOHYDRATE; N,N-BIS(2-CHLOROETHYL)-N'-(3-HYDROXYPROPYL)PHOSPHORODIAMIDIC ACID INTRA-MOL ESTER HYDRATE; BIS(2-CHLOROETHYL)PHOSPHORAMIDE-CYCLIC PROPANOLA-MIDE ESTER; BIS(2-CHLOROETHYL)PHOSPHORAMIDE CY-CLIC PROPANOLAMIDE ESTER MONOHYDRATE; N,N-BIS(2-CHLOROETHYL)-N',O-PROPYLENEPHOSPHORIC ACID ESTER DIAMIDE; N,N-BIS(BETA-CHLOROETHYL)-N',O-PROPYLENEPHOSPHORIC ACID ESTER AMIDE-MONOHYDRATE; N,N-BIS(BETA-CHLOROETHYL)-N',O-PROPYLENE PHOSPHORIC ACID ESTER DIAMIDEMONO-HYDRATE; N,N-BIS(2-CHLOROETHYL)TETRAHYDRO-2H-1,3,2-OXAPHOSPHORIN-2-AMINE,2 -OXIDE MONOHY-DRATE; N,N-BIS(BETA-CHLOROETHYL)-N',O-TRIM ETHYLENEPHOSPHORIC ACID ESTER DIAMIDE; CB-4564; CLAFEN; CLAPHENE (FRENCH); CP; CPA; CTX; CY; CYCLIC N',O-PROPYLENE ESTER OF N,N-BIS(2-CHLOROETHYL)PHOSPHORODIAMIDIC ACID MONOHY-DRATE; CYCLOPHOSPHAMID; CYCLOPHOSPHAMIDE; CY-CLOPHOSPHAMIDUM; CYCLOPHOSPHAN; CYCLOPHOS-PHANE; CYCLOPHOSPHORAMIDE; CYTOPHOSPHAN; CYTOXAN; 2-(DI(2-CHLOROETHYL)AMINO)-1-OXA-3 AZA-2-PHOSPHACYCLOHEXANE-2-OXIDE MONOHYDRATE; 2-(DI(2-CHLOROETHYL)AMINO)2-OXIDE N,N-DI(2-CHLOROETHYL)AMINO-N,O-PROPYLENE PHOSPHORIC ACID ESTER DIAMIDE MONOHYDRATE; N,N-DI(2-CHLOROETHYL)-N,O-PROPYLENE-PHOSPHORIC ACID ESTER DIAMIDE; ENDOXANA; ENDOXAN-ASTA; ENDOX-ANE; ENDOXAN R; ENDUXAN; ENDOXANAL; GENOXAL; MITOXAN; NCI-C04900; NSC 26271; 2-H-1,3,2-OXAZAPHOSPHORINANE; PHOSPHORODIAMIDIC ACID, N,N-BIS(2-CHLOROETHYL)-N'-(3-HYDROXYPROPYL)-, IN-TRAMOL. ESTER; PROCYTOX; SEMDOXAN; SENDOXAN; SENDUXAN; ZYKLOPHOSPHAMID (GERMAN); 2H-1,3,2-OXAZAPHOSPHORINE,2-BIS(2-CHLOROETHYL)AINOTETRAHYDRO-2- OXIDE

Danger profile: Carcinogen, mutagen, possible reproductive hazard, exposure limit established, poisonous gases produced in fire.
Uses: Medical drug.
Appearance: Colorless to white crystalline powder.
Effects of exposure:
   Breathed: May cause nausea, dizziness, vomiting and interfere with the body's ability to make blood cells leading to increased bleeding, weakness, and increased infections.
   Eyes: Contact may cause irritation and damage.
   Skin: Passes through the unbroken skin; can increase exposure and the severity of the symptoms listed. May cause irritation.

Long-term exposure: May cause leukemia and other cancers; may cause sterility in males and females; hair loss. Repeated exposure (in patients) may cause liver damage and scarring of the lungs.
Storage: Store in tightly closed containers in a dry, cool (below 75°F), well-ventilated place. Keep away from heat above 85°F. Containers may explode in fire.

★423★
ENDRIN
CAS: 72-20-8
Other names: COMPOUND 269; ENDREX; ENDRINE (FRENCH); ENT 17,251; HEXACHLOROEPOXYOCTAHY-DRO-ENDO,ENDO-DIMETHANONAPTHALENE; 1,2,3,4,10,10-HEXACHLORO-6,7-EPOXY-1,4,4A,5,6,7,8,8A-OCTAHYDRO-1 ,4-ENDO-ENDO-1,4,5,8-DIMETHANONAPHTHALENE; HEXADRIN; MENDRIN; NCI-C00157; NENDRIN

Danger profile: Mutagen, possible reproductive hazard, poison, exposure limit established, EPA extremely hazardous substance list; poisonous gas produced in fire.
Uses: Insecticide and rodenticide.
Appearance: White crystals. Colorless to tan in solution.
Effects of exposure:
   Breathed: Poisonous. May cause nausea, dizziness, weakness, confusion; twitching of the arms and legs; severe convulsions and death. Symptoms of poisoning may be delayed from 30 minutes to 10 hours following exposure; convulsions may occur without warning. Headache, dizziness and appetite loss may persist for two to four weeks following exposure.
   Eyes: May cause irritation and burns.
   Skin: Poisonous if absorbed. Passes through the unbroken skin; can increase exposure and the severity of the symptoms listed. May cause irritation and burns.
   Swallowed: Poisonous. May cause stomach pain, nausea, aggressive confusion, twitching of the arms and legs; severe convulsions and death. Symptoms of poisoning may be delayed from 30 minutes to 10 hours following exposure; convulsions may occur without warning. Headache, dizziness and appetite loss may persist for two to four weeks following exposure.
Long-term exposure: May cause birth defects by damaging the fetus; liver and kidney damage; insomnia.
Storage: Store in tightly closed containers in a dry, cool, well-ventilated place. Protect containers against physical damage. Keep away from oxidizers. Containers may explode in fire.
First aid guide: Move victim to fresh air and call emergency medical care; if not breathing, give artificial respiration; if breathing is difficult, give oxygen. In case of contact with material, immediately flush skin or eyes with running water for at least 15 minutes. Speed in removing material from skin is of extreme importance. Remove and isolate contaminated clothing and shoes at the site. Keep victim quiet and maintain normal body temperature. Effects may be delayed; keep victim under observation.

★424★
ENFLURANE
CAS: 13838-16-9
Other names: ANESTHETIC COMPOUND NO. 347; 2-CHLORO-1-(DIFLUOROMETHOXY)-1,1,2-TRIFLUOROETHANE; 2-CHLORO-1,1,2-TRIFLUOROETHYLDIFLUOROMETHYL ETHER; COMPOUND 347; ETHANE,2-CHLORO-1-(DIFLUOROMETHOXY)-1,1,2-TRIFLUORO-; ETHRANE; ETHER,2-CHLORO-1,1,2-TRIFLUOROETHYL DIFLUOROMETHYL; METHYLFLURE-THER; NSC-115944; OHIO 347

Danger profile: Mutagen.
Uses: Anesthetic gas.
Appearance: Clear, colorless liquid; easily turns into gas.
Effects of exposure:
> *Breathed:* May cause lightheadedness and loss of consciousness; decreased urine volume. May impair cardiac performance; cause liver damage; seizures. There is a reported association between exposure and miscarriage and birth defects.
> *Eyes:* May cause irritation.
> *Skin:* May cause irritation.
> *Swallowed:* Mildly toxic. See *Breathed.*

Long-term exposure: Possible carcinogen; miscarriage, birth defects; liver damage, seizures.
Storage: Store in tightly closed containers in a dry, cool, well-ventilated place.

★ 425 ★
EPICHLOROHYDRIN
CAS: 106-89-8
Other names: 2-CHLOROPROPYLENE OXIDE; 3-CHLOROPROPYLENE OXIDE; CHLOROMETHYL OXIRANE; GAMMA-CHLOROPROPYLENE OXIDE; GLYCEROL EPICHLOROHYDRIN; 1-CHLORO-2,3-EPOXYPROPANE; ALPHA-EPICHLOROHYDRIN; ECH; 1,2-EPOXY-3-CHLOROPROPANE; OXIRANE, 2-(CHLOROMETHYL)-

Danger profile: Carcinogen, mutagen, fire hazard, highly reactive, exposure limit established, Community Right-to-Know List, EPA extremely hazardous substance, poisonous gases produced in fire.
Uses: Making epoxy, solvents; resins for paper industry; curing propylene-based rubbers; solvent.
Appearance: Colorless water liquid.
Odor: Irritating, chloroform-like. The odor is a warning of exposure; however, no smell does not mean you are not being exposed.
Effects of exposure:
> *Breathed:* May be poisonous. May cause severe irritation of the eyes, nose, throat, bronchial tubes and lungs. Higher exposures may cause shortness of breath, chemical burns in the lungs and a dangerous build-up of fluids in the lungs (pulmonary edema), which can cause death. Pulmonary edema is a medical emergency that can be delayed one to two days following exposure.
> *Eyes:* Contact may cause irritation, redness and burns; possible permanent damage.
> *Skin:* May be poisonous if absorbed. Passes through the unbroken skin; can increase exposure and the severity of the symptoms listed. Contact may cause blisters, severe pain and burns. Symptoms may be delayed for minutes or hours following contact.
> *Swallowed:* May be poisonous.

Long-term exposure: May cause cancer and decrease fertility in males; lung, liver and kidney damage; temporary sterility.
Storage: Store in tightly closed containers in a dry, cool, well-ventilated place. Keep away from heat, sources of ignition and all other chemicals, especially aniline, trichloroacetylene, potassium tert-butoxide, acids, oxidizers and chemically active metals. Vapors may travel to sources of ignition and flash back. Containers may explode in fire.
First aid guide: Move victim to fresh air and call emergency medical care; if not breathing, give artificial respiration; if breathing is difficult, give oxygen. In case of contact with material, immediately flush skin or eyes with running water for at least 15 minutes. Remove and isolate contaminated clothing and shoes at the site. Keep victim quiet and maintain normal body temperature. Effects may be delayed; keep victim under observation.

★ 426 ★
ETHANE
CAS: 74-84-0
Other names: BIMETHYL; DIMETHYL; ETHYL HYDRIDE; METHYLMETHANE

Danger profile: Fire hazard, acrid smoke and irritating fumes in fire.
Uses: Fuel; making other chemicals; freezing agent.
Appearance: Colorless liquefied compressed gas.
Odor: Mild gasoline-like. The odor is a warning of exposure; however, no smell does not mean you are not being exposed.
Effects of exposure:
> *Breathed:* May cause dizziness, lightheadedness and loss of consciousness. Very high levels can cause death by suffocation.
> *Eyes:* Contact with liquid may cause frostbite.
> *Skin:* Contact with liquid may cause frostbite.

Storage: Store in tightly closed containers in a dry, cool, well-ventilated place. Keep away from heat, sources of ignition, oxidizers, chlorine and dioxygenyl tetrafluoroborate. Vapors may travel to sources of ignition and flash back. Containers may explode in fire.
First aid guide: Move victim to fresh air and call emergency medical care; if not breathing, give artificial respiration; if breathing is difficult, give oxygen. In case of frostbite, thaw frosted parts with water. Keep victim quiet and maintain normal body temperature.

★ 427 ★
ETHANOL
CAS: 64-17-5
Other names: ABSOLUTE ETHANOL; AETHANOL (GERMAN); AETHYLDLKOHOL (GERMAN); ALCOHOL; ALCOHOL, ANHYDROUS; ALCOHOL, DEHYDRATED; ALCOOL ETHYLIQUE (FRENCH); ALCOOL ETILICO (ITALIAN); AL-GRAIN; ALKOHOL (GERMAN); ALKOHOLU ETYLOWEGO (POLISH); ANHYDROL; COLOGNE SPIRIT; COLOGNE SPIRITS; ETAHOLO (ITALIAN); ETHANOL 200 PROOF; ETHYLALCOHOL (DUTCH); ETHYL ALCOHOL ANHYDRO-S; ETHYL HYDRATE; ETHYL HYDROXIDE; ETYLOWY ALKOHOL (POLISH); FERMANTATION ALCOHOL; GRAIN ALCOHOL; JAYSOL S; METHYLCARBINOL; MOLASSES ALCOHOL; NCI-CO3134; POTATO ALCOHOL; SD ALCOHOL 23-HYDROGEN; SPIRITS OF WINE; SPIRIT; TESCOL

Danger profile: Carcinogen, mutagen, fire hazard, exposure limit established, poisonous gas produced in fire.
Uses: Solvent; in alcoholic beverages.
Appearance: Clear, colorless liquid.
Odor: Strong; bitter taste. The odor is a warning of exposure; however, no smell does not mean you are not being exposed.
Effects of exposure:
> *Breathed:* Mildly toxic. May cause irritation of eyes, nose and throat; headache, drowsiness, excitability, drunken behavior, mental confusion, fatigue. Very high levels of exposure may cause dizziness, lightheadedness and cause victim to pass out, go into coma and die.
> *Eyes:* Contact may cause irritation, pain, burning, stinging, tearing, inflammation of the eyelids. Light may cause pain.
> *Skin:* May remove oil from the skin and cause dryness. Persons with pre-existing skin disorders may be more susceptible to the effects of this substance.
> *Swallowed:* May cause gastro-intestinal irritation.

Long-term exposure: Prolonged or repeated contact may cause birth defects; developmental problems; miscarriage; cancer; liver problems; irritation of the skin and dermatitis, and nervous system changes. The primary cause of alcoholism.

Storage: Store in a tightly closed container in a cool, well-ventilated place. Keep away from heat, sources of ignition, acids and oxidizers. Vapors are explosive in heat or flame. Containers may explode in fire.

First aid guide: Move victim to fresh air and call emergency medical care; if not breathing, give artificial respiration; if breathing is difficult, give oxygen. In case of contact with material, immediately flush eyes with running water for at least 15 minutes. Wash skin with soap and water. Remove and isolate contaminated clothing and shoes at the site.

★ 428 ★
ETHANOLAMINE
CAS: 141-43-5

Other names: AETHANOLAMIN (GERMAN); 2-AMINOAETHANOL (GERMAN); 2-AMINOETANOLO (ITALIAN); 2-AMINOETHANOL; BETA-AMINOETHYL ALCOHOL; 1-AMINO-2-HYDROXYETHANE; COLAMINE; ETANOLAMINA (ITALIAN); ETHANOLAMINE SOLUTION; BETA-ETHANOLAMINE; ETHYLOLAMINE; GLYCINOL; 2-HYDROXYETHYLAMINE; BETA-HYDROXYETHYLAMINE; MEA; MONOAETHANOLAMIN (GERMAN); MONOETHANOL-AMINE; OLAMINE; THIOFACO M-50; USAF EK-1597

Danger profile: Mutagen, combustible, corrosive, exposure limit established, poisonous gases produced in fire.
Uses: Making soaps and detergents, ink, dyes and rubber; used in dry cleaning, wool treatment, emulsion paints, polishes, agricultural sprays; pharmaceuticals; corrosion inhibitor.
Appearance: Colorless liquid often mixed in a solution.
Odor: Mild, ammonia-like. The odor is a warning of exposure; however, no smell does not mean you are not being exposed.
Effects of exposure:
Breathed: May cause irritation to the nose, throat and eyes; difficult breathing, coughing. Higher exposures may cause shortness of breath and a dangerous build-up of fluids in the lungs (pulmonary edema), which can cause death. Pulmonary edema is a medical emergency that can be delayed one to two days following exposure.
Eyes: Contact may cause severe burns; possible permanent damage.
Skin: Passes through the unbroken skin; can increase exposure and the severity of the symptoms listed. Contact may cause severe irritation, redness, itching and burns.
Swallowed: May cause burning feeling in mouth, throat and stomach; pain and swelling in throat; nausea and vomiting; stomach cramps, rapid breathing, shock, diarrhea; possible stomach perforation.
Long-term exposure: May cause weight loss; kidney and liver damage; skin irritation and itching. Similar chemicals may cause allergies.
Storage: Store in tightly closed containers in a dry, cool, well-ventilated place. Keep away from heat, sources of ignition, oxidizers, acids, acetic anhydride, acrolein, acrylonitrile, acrolein, cellulose, epichlorohydrin, mesityl oxide, oleum and vinyl acetate. Containers may explode in fire.
First aid guide: Move victim to fresh air; call emergency medical care. In case of contact with material, immediately flush skin or eyes with running water for at least 15 minutes. Remove and isolate contaminated clothing and shoes at the site. Keep victim quiet and maintain normal body temperature.

★ 429 ★
ETHION
CAS: 563-12-2

Other names: AC 3422; BIS(S-(DIETHOXYPHOSPHINOTHIOYL)MERCAPTO)METHANE; BIS (DITHIOPHOSPHATEDE O,O-DIETHYLE) DE S,S'-METHYLENE (FRENCH); BLADAN; DIETHION; EMBATHION; ENT 24,105; ETHIOL; ETHODAN; ETHYL METHYLENE PHOSPHORODITHIOATE; FMC-1240; FOSFONO 50; HYLEMOX; ITOPAZ; KWIT; METHANEDITHIOL, S,S-DIESTER WITH O,O-DIETHYL PHOSPHORODITHIOATE ACID; METHYLEEN-S,S'-BIS(O,O-DIETHYL-DITH IOFOSFAAT) (DUTCH); METHYLENE-S,S'-BIS(O,O-DIAETHYL-DITHIOPHOSPHAT) (GERMAN); S,S'-METHYLENE O,O,O',O'-TETRAETHYL PHOSPHORODITHIOATE; METILEN-S,S'-BIS(O,O-DIETIL-DITIOFOSFATO) (ITALIAN); NIA 1240; NIAGARA 1240; NIALATE; PHOSPHORODITHIOIC ACID, O,O-DIETHYL ESTER, S,S-DIESTER WITH METHANEDITHIOL; PHOSPHOTOX E; RODOCID; RP 8167; SOPRATHION; O,O,O',O'-TETRAAETHYL-BIS(DITHIOPHOSPHAT) (GERMAN); O,O,O,O-TETRAETHYL S,S'-METHYLENEBIS(DITHIOPHOSPHATE); O,O,O',O'-TETRAETHYL S,S'-METHYLENEBISPHOSPHORDITHIOATE; TETRAETHYL S,S'-METHYLENE BIS(PHOSPHOROTHIOLOTHIONATE); O,O,O',O'-TETRAETHYL S,S'-METHYLENE DI(PHOSPHORODITHIOATE); VEGFRUFOSMITE; S,S'-METHYLENE O,O,O',O' TETRAETHYL ESTER PHOSPHORODITHIOIC ACID; VEGFRUFOSMITE

Danger profile: Poison, exposure limit established, EPA extremely hazardous substances list, poisonous gas produced in fire.
Uses: Insecticide and miticide.
Appearance: Colorless to amber liquid. The odor is a warning of exposure; however, no smell does not mean you are not being exposed.
Odor: Disagreeable. The odor is a warning of exposure; however, no smell does not mean you are not being exposed.
Effects of exposure:
Breathed: Extremely poisonous. May cause headache, dizziness, profuse sweating, nausea, vomiting, reduced heart beat, stomach cramps, diarrhea, loss of coordination, slow and weak breathing, fever, loss of consciousness and death. May cause shortness of breath and a dangerous build-up of fluids in the lungs (pulmonary edema), which can cause death. Pulmonary edema is a medical emergency that can be delayed one to two days following exposure.
Eyes: Rapidly absorbed. Poisoning can happen very rapidly. See *Breathed.*
Skin: Poisonous if absorbed. Passes through the unbroken skin; can increase exposure and the severity of the symptoms listed.
Swallowed: Powerful, deadly poison. See *Breathed.*
Long-term exposure: May cause nerve damage including loss of coordination; liver damage. Repeated exposure may cause depression, irritability and personality changes.
Storage: Store in a regulated, specially marked place in tightly-closed containers in a dry, well-ventilated place. Similar chemicals must be kept away from water, strong alkalis. Containers may explode in fire.
First aid guide: Move victim to fresh air and call emergency medical care; if not breathing, give artificial respiration; if breathing is difficult, give oxygen. In case of contact with material, immediately flush skin or eyes with running water for at least 15 minutes. Speed in removing material from skin is of extreme importance. Remove and isolate contaminated clothing and shoes at the site. Keep victim quiet and maintain normal body temperature. Effects may be delayed; keep victim under observation.

★ 430 ★
2-ETHOXYETHANOL
CAS: 110-80-5
Other names: ATHYLENGLYKOL-MONOATHYLATHER (GERMAN); CELLOSOLVE; CELLOSOLVE SOLVENT; DOWANOL EE; 2EE; EKTASOLVE EE; ETHANOL,2-ETHOXY-; ETHER MONOETHYLIQUE DE L'ETHYLENE-GLYCOL (FRENCH); ETHYL CELLOSOLVE; ETHYLENE GLYCOL ETHYL ETHER; ETHYLENE GLYCOL MONOETHYL ETHER; ETOKSYETYLOWY ALKOHOL (POLISH); GLYCOL ETHYL ETHER; GLYCOL MONOETHYL ETHER; HYDROXY ETHER; JEFFERSOL EE; NCI-C54853; OXITOL; POLY-SOLV EE.

Danger profile: Mutagen, combustible, exposure limit established, Community Right-to-Know List, poisonous gases produced in fire.
Uses: Widely used solvent; in brake fluids, aviation and automotive fuels.
Appearance: Colorless liquid.
Odor: Mild, sweetish; nearly odorless. The odor is a warning of exposure; however, no smell does not mean you are not being exposed.
Effects of exposure:
　　Breathed: May be poisonous. May cause irritation of the eyes, nose and throat. Higher levels may cause dizziness, lightheadedness, and loss of consciousness.
　　Eyes: Contact may cause irritation.
　　Skin: May be poisonous if absorbed. Passes through the unbroken skin; can increase exposure and the severity of the symptoms listed. Contact may cause irritation.
　　Swallowed: May be poisonous.
Long-term exposure: May cause damage to the testes; infertility; birth defects by damaging the fetus; low blood count; kidneys.
Storage: Store in tightly closed containers in a dry, cool, well-ventilated place. Keep away from heat, sources of ignition and strong oxidizers. Vapors may travel to sources of ignition and flash back. Containers may explode in fire.

★ 431 ★
2-ETHOXYETHYL ACETATE
CAS: 111-15-9
Other names: ACETATE DE CELLOSOLVE (FRENCH); ACETATE DE L'ETHER MONOETHYLIQUE DE L'ETHYLENE-GLYCOL (FRENCH); ACETATE D'ETHYLGLYCOL (FRENCH); ACETATO DI CELLOSOLVE (ITALIAN); ACETIC ACID, 2-ETHOXYETHYL ESTER; 2-AETHOXY-AETHYLACETAT (GERMAN); AETHYLENGLYKOLAETHERACETAT (GERMAN); CELLOSOLVE ACETATE; CSAC; ETHANOL, 2-ETHOXY-, ACETATE; EKTASOLVE EE ACETATE SOLVENT; ETHOXY ACETATE; 2-ETHOXYETHANOL ACETATE; 2-ETHOXYETHANOL, ESTER WITH ACETIC ACID; 2-ETHOXY-ETHYLACETAAT (DUTCH); ETHOXYETHYL ACETATE; BETA-ETHOXYETHYL ACETATE; 2-ETHOXYETHYLACETATE; 2-ETHOXYETHYLE, ACETATE DE (FRENCH); ETHYL CELLOSOLVE ACETAAT (DUTCH); ETHYLENE GLYCOL ETHYL ETHER ACETATE; ETHYLENE GLYCOL MONOETHYL ETHER ACETATE; ETHYLGLYKOLACETAT (GERMAN); 2-ETOSSIETIL-ACETATO (ITALIAN); GLYCOL MONOETHYL ETHER ACETATE; OCTAN ETOKSYETYLU (POLISH); OXYTOL ACETATE; POLY-SOLV EE ACETATE

Danger profile: Reproductive hazard, exposure limit established, Community Right-to-Know List, poisonous gases produced in fire.
Uses: Widely used solvent; in lacquers, epoxy resins.
Appearance: Colorless liquid.

Odor: Pleasant, mild, ether-like. The odor is a warning of exposure; however, no smell does not mean you are not being exposed.
Effects of exposure:
　　Breathed: May be poisonous. Vapor may cause irritation of the nose, eyes, and throat. High levels may cause dizziness, lightheadedness and loss of consciousness. Very high exposure may cause kidney damage; possible death.
　　Eyes: May cause irritation.
　　Skin: May be poisonous if absorbed. Passes through the unbroken skin; can increase exposure and the severity of the symptoms listed. May cause irritation.
　　Swallowed: May be poisonous.
Long-term exposure: May cause damage to the developing fetus; testes damage and infertility; kidney damage.
Storage: Store in tightly closed containers in a dry, cool, well-ventilated place. Keep away from heat, sources of ignition and oxidizers. Protect from moisture and physical damage. Vapors may travel to sources of ignition and flash back. Containers may explode in fire.

★ 432 ★
ETHYL ACETATE
CAS: 141-78-6
Other names: ACETIC ACID ETHYL ESTER; ACETIC ETHER; ETHYL ACETIC ESTER; ETHYL ETHANOATE; ACETIDIN; ACETOXYETHANE; AETHYLACETAT (GERMAN); ESSIGESTER (GERMAN); ETHYLACETAAT (DUTCH); ETHYL ACETIC ESTER; ETHYLE (ACETATE D') (FRENCH); ETHYL ETHANOATE; ETILE (ACETATO DI) (ITALIAN); FEMA NO. 2414; OCTAN ETYLU (POLISH); VINEGAR NAPHTHA

Danger profile: Mutagen, fire hazard, exposure limit established, poisonous gases produced in fire.
Uses: Solvent; smokeless powders; pharmaceuticals; synthetic fruit flavoring.
Appearance: Colorless liquid.
Odor: Pleasant, fruity. The odor is a warning of exposure; however, no smell does not mean you are not being exposed.
Effects of exposure:
　　Breathed: May be poisonous. May cause irritation of eyes, nose and throat; headache, drowsiness, excitability, drunken behavior, mental confusion, fatigue. Very high levels of exposure may cause dizziness, lightheadedness and cause victim to pass out, go into coma and die.
　　Eyes: Contact may cause irritation, pain, burning, stinging, tearing, inflammation of the eyelids. Light may cause pain.
　　Skin: May be poisonous if absorbed. May remove oil from the skin and cause dryness. Persons with pre-existing skin disorders may be more susceptible to the effects of this substance.
　　Swallowed: May be poisonous. May cause gastro-intestinal irritation.
Long-term exposure: Prolonged or repeated contact may cause irritation of the skin and dermatitis; nervous system changes.
Storage: Store in a tightly closed container in a cool, well-ventilated place. Keep away from heat and sources of ignition, nitrates, strong oxidizers, strong alkalis, oleum and strong acids. Vapors are explosive in heat or flame. Containers may explode in fire.
First aid guide: Move victim to fresh air and call emergency medical care; if not breathing, give artificial respiration; if breathing is difficult, give oxygen. In case of contact with material, immediately flush eyes with running water for at least 15 minutes. Wash skin with soap and water. Remove and isolate contaminated clothing and shoes at the site.

★ 433 ★
ETHYL ACETYLENE
CAS: 107-00-6
Other names: 1-BUTYNE; ETHYL ACETYLENE, INHIBITED; ETHYLETHYNE

Danger profile: Fire hazard, poisonous gas produced in fire.
Uses: Making other chemicals; fuel.
Appearance: Colorless, highly combustible gas; liquid under pressure.
Effects of exposure:
Breathed: May be poisonous. May cause dizziness, lightheadedness and loss of consciousness. Exposure to high levels may cause suffocation and death.
Eyes: Contact may cause irritation and burns. Liquid may cause frostbite.
Skin: Contact may cause irritation and burns. Liquid may cause frostbite.
Storage: Store in tightly closed containers in a dry, cool, well-ventilated place. Keep away from heat, sources of ignition and oxidizers. Vapors may travel to sources of ignition and flash back. Containers may explode violently in fire.
First aid guide: Move victim to fresh air and call emergency medical care; if not breathing, give artificial respiration; if breathing is difficult, give oxygen. In case of frostbite, thaw frosted parts with water. Keep victim quiet and maintain normal body temperature.

★ 434 ★
ETHYL ACRYLATE
CAS: 140-88-5
Other names: ACRYLATE D'ETHYLE (FRENCH); ACRYLIC ACID, ETHYL ESTER; ACRYLSAEUREAETHYLESTER (GERMAN); AETHYLACRYLAT (GERMAN); ETHOXYCARBONYLETHYLENE; ETHYLACRYLAAT (DUTCH); ETHYLAKRYLAT (CZECH); ETHYL PROPENOATE; ETHYL 2-PROPENOATE; ETIL ACRILATO (ITALIAN); ETILACRILATULUI (RUMANIAN); FEMA NO. 2418; NCI-C50384; 2-PROPENOIC ACID, ETHYL ESTER

Danger profile: Carcinogen, mutagen, fire hazard, explosion hazard, Community Right-to-Know List, poisonous gas produced in fire.
Uses: Making acrylic resins, plastics, rubber and denture materials.
Appearance: Colorless liquid.
Odor: Sharp, penetrating. The odor is a warning of exposure; however, no smell does not mean you are not being exposed.
Effects of exposure:
Breathed: May be poisonous. May cause irritation of the eyes, nose, throat and lungs, tears and runny nose, coughing, sore throat, difficult breathing; lethargy, fatigue, and drowsiness; slurred speech, shaking, and dizziness. Higher exposures could cause shortness of breath, headache, nausea, vomiting, dizziness, diarrhea and a dangerous build-up of fluids in the lungs (pulmonary edema), which can cause death. Pulmonary edema is a medical emergency that can be delayed one to two days following exposure.
Eyes: May cause irritation, redness, tearing, pain, inflamed eyelids, and eye injury; burns in corneas and possible loss of sight.
Skin: May be poisonous if absorbed. Passes through the unbroken skin; can increase exposure and the severity of the symptoms listed. May cause painful irritation, redness, and blisters Repeated or prolonged skin contact may cause irritation of the skin and may also cause the legs to feel numb.
Swallowed: May be poisonous. May cause irritation of the mouth and throat; stomach and abdominal pain; loss of muscular coordination and mental confusion; convulsions. Also see symptoms for *Breathed.*
Long-term exposure: May cause skin allergy with itching and skin rash.
Storage: Store in tightly closed containers in a dark, dry, cool, well-ventilated, explosion-proof place. Keep away from heat, sources of ignition, oxidizers, alkalies and moisture. Vapor forms an explosive mixture with air. Containers may explode in fire.
First aid guide: Move victim to fresh air and call emergency medical care; if not breathing, give artificial respiration; if breathing is difficult, give oxygen. In case of contact with material, immediately flush eyes with running water for at least 15 minutes. Wash skin with soap and water. Remove and isolate contaminated clothing and shoes at the site.

★ 435 ★
ETHYL ALUMINUM DICHLORIDE
CAS: 563-43-9
Other names: ALUMINUM DICHLOROETHYL-; ALUMINUM ETHYL DICHLORIDE; DICHLOROETHYLALUMINUM; ETHYLDICHLOROALUMINUM

Danger profile: Fire hazard, explosion hazard, reactive with water and air, exposure limit established, poisonous gases produced in fire.
Uses: Making other chemicals.
Appearance: Colorless to light yellow liquid.
Effects of exposure:
Breathed: May be poisonous. May cause irritation of the nose, throat, breathing passages and lungs. Higher exposures may cause shortness of breath and a dangerous build-up of fluids in the lungs (pulmonary edema), which can cause death. Pulmonary edema is a medical emergency that can be delayed one to two days following exposure. Inhalation of smoke from fire causes metal-fume fever with flu-like symptoms.
Eyes: Contact may cause burns and permanent damage.
Skin: Contact may cause irritation, redness, rash and burns.
Long-term exposure: May cause lung damage and skin rash.
Storage: Store in tightly closed containers under a nitrogen or other inert gas blanket in a dry, cool place. Keep away from water, air, combustible materials and all other chemicals. Reacts with surface moisture to generate hydrogen chloride, which is corrosive to common metals. Ignites spontaneously in air.
First aid guide: Move victim to fresh air; call emergency medical care. Wipe material from skin immediately; flush skin or eyes with running water for at least 15 minutes. Remove and isolate contaminated clothing and shoes at the site.

★ 436 ★
ETHYL ALUMINUM SESQUICHLORIDE
CAS: 12075-68-2
Other names: ALUMINUM, TRICHLOROTRIETHYLDI-; TRICHLOROTRIETHYLDIALUMINUM; TRIETHYLDIALUMINUM TRICHLORIDE

Danger profile: Fire hazard, explosion hazard, exposure limit established, poisonous gaes produced in fire.
Uses: Making other chemicals.
Appearance: Colorless to clear yellow liquid.
Effects of exposure:
Breathed: May be poisonous. May cause irritation of the nose, throat, breathing passages and lungs. Higher exposures may cause shortness of breath and a dangerous build-up of fluids in the lungs (pulmonary edema), which can cause death. Pulmonary edema is a medical emergency that can be delayed one to two days following exposure. Inhalation of

smoke from fire causes metal-fume fever with flu-like symptoms.

Eyes: Contact may cause burns and permanent damage.

Skin: Contact may cause irritation, redness, rash and burns.

Long-term exposure: May cause lung damage and skin rash.

Storage: Store in tightly closed containers under a nitrogen or other inert gas blanket in a dry, cool place. Keep away from air, moisture, carbon tetrachloride, combustible materials and all other chemicals. Reacts with surface moisture to generate hydrogen chloride, which is corrosive to common metals. May ignite spontaneously in air.

First aid guide: Move victim to fresh air; call emergency medical care. Wipe material from skin immediately; flush skin or eyes with running water for at least 15 minutes. Remove and isolate contaminated clothing and shoes at the site.

★437★
ETHYLAMINE
CAS: 75-04-7

Other names: AETHYLAMINE (GERMAN); AMINOETHANE; 1-AMINOETHANE; ETHANAMINE; ETILAMINA (ITALIAN); ETYLOAMINA (POLISH); MONOETHYLAMINE

Danger profile: Fire hazard, exposure limit established, poisonous gas produced in fire.

Uses: Making dyes, rubber, rubber latex and other chemicals; detergents.

Appearance: Colorless liquid or gas.

Odor: Strong, ammonia-like. The odor is a warning of exposure; however, no smell does not mean you are not being exposed.

Effects of exposure:

Breathed: May be poisonous. May cause irritation to the nose, throat and eyes; difficult breathing, coughing. Higher exposures may cause shortness of breath and a dangerous build-up of fluids in the lungs (pulmonary edema), which can cause death. Pulmonary edema is a medical emergency that can be delayed one to two days following exposure.

Eyes: Contact may cause severe burns; possible permanent damage.

Skin: Passes through the unbroken skin; can increase exposure and the severity of the symptoms listed. Contact may cause severe irritation, redness, itching and burns.

Swallowed: May be poisonous. May cause burning feeling in mouth, throat and stomach; pain and swelling in throat; nausea and vomiting; stomach cramps, rapid breathing, shock, diarrhea; possible stomach perforation.

Long-term exposure: May cause weight loss; kidney and liver damage; skin irritation and itching; eye damage.

Storage: Store in tightly closed containers in a dry, cool, well-ventilated place. Keep away from heat, sources of ignition, acids and oxidizers. Will dissolve paint, most plastic materials. Vapors may travel to sources of ignition and flash back. Containers may explode in fire.

First aid guide: Move victim to fresh air and call emergency medical care; if not breathing, give artificial respiration; if breathing is difficult, give oxygen. In case of contact with material, immediately flush skin or eyes with running water for at least 15 minutes. Remove and isolate contaminated clothing and shoes at the site. Keep victim quiet and maintain normal body temperature.

★438★
ETHYL AMYL KETONE
CAS: 541-85-5

Other names: 3-OCTANONE; 3-HEPTANONE, 5-METHYL; AMYL ETHYL KETONE; 3-METHYL-5-HEPTANONE; 5-METHYL-3-HEPTANONE; EAK; ETHYL SEC-AMYL KETONE; ETHYL ISOAMYL KETONE

Danger profile: Combustible, exposure limit established, poisonous gases produced in fire.

Uses: Solvent for resins.

Appearance: Clear liquid.

Odor: Strong, fruity. The odor is a warning of exposure; however, no smell does not mean you are not being exposed.

Effects of exposure:

Breathed: May be poisonous. May cause irritation of the nose and throat; headache and nausea; dizziness, light-headedness and loss of consciousness. High concentrations may be narcotic.

Eyes: Contact may cause irritation.

Skin: May be poisonous if absorbed. Passes through the unbroken skin; can increase exposure and the severity of the symptoms listed. Contact may cause irritation.

Swallowed: May cause mouth, throat and stomach irritation.

Long-term exposure: May cause skin rash.

Storage: Store in tightly closed containers in a dry, cool, well-ventilated place. Keep away from sources of ignition and oxidizers. Vapors may travel to sources of ignition and flash back. Containers may explode in fire.

First aid guide: Move victim to fresh air and call emergency medical care; if not breathing, give artificial respiration; if breathing is difficult, give oxygen. In case of contact with material, immediately flush eyes with running water for at least 15 minutes. Wash skin with soap and water. Remove and isolate contaminated clothing and shoes at the site.

★439★
ETHYLANILINE
CAS: 103-69-5

Other names: AETHYLANILIN (GERMAN); ANILINOETHANOL; ANILINOETHANE; BENZENAMINE, N-ETHYL; BENZENAMINE, N-ETHYL-(9CI); ETHYLPHENYLAMINE; N-ETHYLAMINOBENZENE; N-ETHYLOMINOBENZENE; N-ETHYLANILINE; N-ETHYLBENZENEAMINO; ETHYLPHENYLAMINE

Danger profile: Combustible, allergen, poisonous gases produced in fire.

Uses: Making other chemicals.

Appearance: Yellow-brown oil.

Odor: Fishy. The odor is a warning of exposure; however, no smell does not mean you are not being exposed.

Effects of exposure:

Breathed: May be poisonous. May cause dizziness, headache, weakness, drowsiness, depression, tremors, fatigue, loss of appetite and loss of consciousness. High exposure may cause blue skin and lips; death may result.

Eyes: Contact may cause irritation and burns.

Skin: Passes through the unbroken skin; can increase exposure and the severity of the symptoms listed. Contact may cause irritation and burns. High exposure may result in death.

Swallowed: Poisonous.

Long-term exposure: May cause skin allergy with rash.

Storage: Store in tightly closed containers in a dry, cool, well-ventilated place. Keep away from acid, acid fumes and oxidizing materials. Containers may explode in fire.

First aid guide: Move victim to fresh air and call emergency medical care; if not breathing, give artificial respiration; if breathing is difficult, give oxygen. In case of contact with material, immediately flush skin or eyes with running water for at least 15 minutes. Speed in removing material from skin is of extreme importance. Remove and isolate contaminated clothing and shoes at the site. Keep victim quiet and maintain normal body temperature. Effects may be delayed; keep victim under observation.

★ 440 ★
2-ETHYLANILINE
CAS: 578-54-1
Other names: O-AMINOETHYLBENZENE; ANILINE,O-ETHYL- (8CI); BENZENAMINE,2-ETHYL-; BENZENAMINE,2-ETHYL-(9CI); O-ETHYLANILINE; 2-ETHYL ANILINE; 2-ETHYLBENZENAMINE

Danger profile: Combustible, poison, highly toxic fumes produced in fire.
Uses: Making drugs, pesticides and dyes.
Appearance: Yellow liquid; darkens with age.
Effects of exposure:
  Breathed: May be poisonous. May cause headache, dizziness, nausea and bluish color to the skin and lips. High levels may cause difficult breathing, collapse and possible death.
  Eyes: Contact may cause irritation.
  Skin: May be poisonous if absorbed. Passes through the unbroken skin; can increase exposure and the severity of the symptoms listed. Contact may cause irritation, and the development of allergy.
  Swallowed: Poisonous. See symptoms under Breathed and First aid guide.
Long-term exposure: Skin allergy may develop. May cause liver and kidney damage.
Storage: Store in tightly closed containers in a dry, cool, well-ventilated place. Keep away from acids, oxidizers, chloroformates and acid anhydrides. Containers may explode in fire.
First aid guide: Move victim to fresh air and call emergency medical care; if not breathing, give artificial respiration; if breathing is difficult, give oxygen. In case of contact with material, immediately flush skin or eyes with running water for at least 15 minutes. Speed in removing material from skin is of extreme importance. Remove and isolate contaminated clothing and shoes at the site. Keep victim quiet and maintain normal body temperature. Effects may be delayed; keep victim under observation.

★ 441 ★
ETHYLBENZENE
CAS: 100-41-4
Other names: AETHYLBENZOL (GERMAN); EB; ETHYL-BENZEEN (DUTCH); ETHYLBENZOL; ETILBENZENE (ITALIAN); ETYLOBENZEN (POLISH); NCI-C56393; PHENYLE-THANE

Danger profile: Mutagen, possible reproductive hazard, fire hazard, explosion hazard, exposure limit established, Community Right-to-Know List, poisonous gas produced in fire.
Uses: Making styrene and synthetic polymers; solvent; in automotive and aviation fuels.
Appearance: Colorless liquid.
Odor: Sweet, gasoline-like. The odor is a warning of exposure; however, no smell does not mean you are not being exposed.
Effects of exposure:
  Breathed: May be poisonous. May cause irritation of eyes, nose and throat; headache, drowsiness, excitability, drunken behavior, mental confusion, fatigue. Very high levels of exposure may cause dizziness, lightheadedness; paralysis, breathing problems and cause victim to pass out, go into coma and die. High level exposure may cause liver damage.
  Eyes: Contact may cause irritation, pain, burning, stinging, tearing, inflammation of the eyelids. Light may cause pain.
  Skin: May be poisonous if absorbed. May remove oil from the skin and cause dryness. Persons with pre-existing skin disorders may be more susceptible to the effects of this substance.
  Swallowed: May cause gastro-intestinal irritation. See Breathed.

Long-term exposure: May cause birth defects, miscarriages or cancer. Prolonged or repeated contact may cause irritation of the skin, blistering and dermatitis; brain and nervous system changes.
Storage: Store in a tightly closed container in a cool, well-ventilated place. Keep away from heat and sources of ignition, and strong oxidizers. Vapors are explosive in heat or flame. Containers may explode in fire.
First aid guide: Move victim to fresh air and call emergency medical care; if not breathing, give artificial respiration; if breathing is difficult, give oxygen. In case of contact with material, immediately flush eyes with running water for at least 15 minutes. Wash skin with soap and water. Remove and isolate contaminated clothing and shoes at the site.

★ 442 ★
ETHYL BUTYL KETONE
CAS: 106-35-4
Other names: AETHYLBUTYLKETON (GERMAN); BUTYL ETHYL KETONE; N-BUTYL ETHYL KETONE; EPTAN-3-ONE (ITALIAN); ETHYLBUTYLCETONE (FRENCH); ETHYLBU-TYLKETON (DUTCH); ETILBUTILCHETONE (ITALIAN); FEMA NO. 2545; HEPTAN-3-ON (DUTCH); HEPTAN-3-ON (GERMAN); 3-HEPTANONE; HEPTAH-3-ONE

Danger profile: Combustible, exposure limit established.
Uses: Solvent for various finishes and resins.
Appearance: Colorless liquid.
Odor: Mild, fruity. The odor is a warning of exposure; however, no smell does not mean you are not being exposed.
Effects of exposure:
  Breathed: May be poisonous. High concentrations may cause irritation of the eyes, nose and throat; dizziness, lightheadedness and loss of consciousness.
  Eyes: May cause irritation.
  Skin: May be poisonous if absorbed. Passes through the unbroken skin; can increase exposure and the severity of the symptoms listed. May cause irritation, redness and burning sensation.
  Swallowed: Moderately toxic.
Long-term exposure: May cause skin irritation and dryness.
Storage: Store in tightly closed containers in a dry, cool, well-ventilated place. Keep away from heat, sources of ignition and oxidizing materials. Vapors may travel to sources of ignition and flash back. Containers may explode in fire.

★ 443 ★
ETHYL BUTYRATE
CAS: 105-54-4
Other names: BUTONIC ACID ETHYL ESTER; BUTYRIC ACID, ETHYL ESTER; BUTYRIC ETHER; ETHYL BUTANO-ATE; ETHYL-N-BUTYRATE; FEMA NO. 2427

Danger profile: Fire hazard.
Uses: In flavoring extracts, perfumery; solvent.
Appearance: Colorless liquid.
Odor: Pineapple-like. The odor is a warning of exposure; however, no smell does not mean you are not being exposed.
Effects of exposure:
  Breathed: May be poisonous. May cause irritation of the eyes, nose and throat. Higher levels may cause headache, dizziness and lightheadedness; nausea, vomiting and loss of consciousness.
  Eyes: Contact may cause irritation.
  Skin: May be poisonous if absorbed. Passes through the unbroken skin; can increase exposure and the severity of the symptoms listed. Contact may cause irritation.

Swallowed: Mildly toxic. May cause headache, dizziness, nausea, vomiting and loss of consciousness.
Long-term exposure: Unknown at this time. Similar chemicals may cause lung and liver damage.
Storage: Store in tightly closed containers in a dry, cool, well-ventilated place. Keep away from heat, sources of ignition, oxidizing materials. May attack some forms of plastics. Vapors may travel to sources of ignition and flash back. Containers may explode in fire.
First aid guide: Move victim to fresh air and call emergency medical care; if not breathing, give artificial respiration; if breathing is difficult, give oxygen. In case of contact with material, immediately flush eyes with running water for at least 15 minutes. Wash skin with soap and water. Remove and isolate contaminated clothing and shoes at the site.

★ 444 ★
ETHYL CARBAMATE
CAS: 51-79-6
Other names: A 11032; AETHYLCARBAMAT (GERMAN); AETHYLURETHAN (GERMAN); CARBAMIC ACID, ETHYL ESTER; CARBAMIDSAEURE-AETHYLESTER (GERMAN); ESTANE 5703; ETHYL CARBAMATE; ETHYLURETHAN; ETHYL URETHANE; O-ETHYLURETHANE; LEUCETHANE; LEUCOTHANE; NSC 746; PRACARBAMIN; PRACARB-AMINE; U-COMPOUND; URETAN ETYLOWY (POLISH); URETHAN; URETHANE

Danger profile: Carcinogen, combustible, reproductive hazard, Community Right-to-Know List, poisonous gas produced in fire.
Uses: Making pharmaceuticals, pesticides, fungicides; medical and biomedical research; has been found in many alcoholic beverages sold in the U.S. Chemical formed during processing.
Appearance: Colorless crystalline material or white powder.
Effects of exposure:
Breathed: High exposures may cause dizziness, lightheadedness and loss of consciousness. Very high exposures may cause brain, liver damage, and bone marrow depression (damage to the blood forming organs).
Skin: Passes through the unbroken skin; can increase exposure and the severity of the symptoms listed.
Swallowed: Moderately toxic.
Long-term exposure: May cause cancer, birth defects, fetus damage; depression of bone marrow; degeneration of the brain; liver and central nervous system damage. Mutation data reported. Believed to be a transplacental carcinogen, the chemical has caused cancer in the offsprings of animals exposed during pregnancy.
Storage: Store in tightly closed containers in a dry, cool, well ventilated place. Keep away from sources of ignition, oxidizers, strong acids and bases, phosphorus pentachloride, camphor, menthol, and thymol. Containers may explode in fire.

★ 445 ★
ETHYL CHLORIDE
CAS: 75-00-3
Other names: AETHYLCHLORID (GERMAN); AETHYLIS; AETHYLIS CHLORIDUM; ANODYNON; CHELEN; CHLOO-RETHAAN (DUTCH); CHLORETHYL; CHLORIDUM; CHLOROAETHAN (GERMAN); CHLOROETHANE; CHLO-RURE D'ETHYLE (FRENCH); CHLORYL; CHLORYL ANES-THETIC; CLOROETANO (ITALIAN); CLORURO DI ETILE (ITALIAN); ETHER CHLORATUS; ETHER HYDROCHLORIC; ETHER MURIATIC; ETYLU CHLOREK (POLISH); HYDRO-CHLORIC ETHER; KELENE; MONOCHLORETHANE; MURI-ATIC ETHER; NARCOTILE; NCI-C06224

Danger profile: Dangerous fire hazard, exposure limit established, Community Right-to-Know List, poisonous gases produced in fire.
Uses: Making other chemicals; in refrigeration; solvent for fats, oils, resins and waxes; local anesthetic; laboratory chemical; insecticides.
Appearance: Colorless liquid under pressure or gas.
Odor: Pungent, ether-like. The odor is a warning of exposure; however, no smell does not mean you are not being exposed.
Effects of exposure:
Breathed: May cause irritation of the eyes, nose and throat; headache, dizziness, lightheadedness drunkenness and loss of consciousness; irregular heartbeat, heart stoppage and death. possible lung injury.
Eyes: Contact may cause frostbite.
Skin: Passes through the unbroken skin; can increase exposure and the severity of the symptoms listed. Contact may cause frostbite.
Swallowed: May cause headache, dizziness, lack of coordination, stomach cramps, eventual loss of consciousness. See *Breathed.*
Long-term exposure: May cause liver and kidney damage.
Storage: Store in tightly closed containers in an explosion-proof refrigerator or dry, cool, well-ventilated place. Keep away from heat, sources of ignition, chemically active metals, moisture and oxidizing materials. Containers may explode in fire.
First aid guide: Move victim to fresh air and call emergency medical care; if not breathing, give artificial respiration; if breathing is difficult, give oxygen. In case of contact with material, immediately flush eyes with running water for at least 15 minutes. Wash skin with soap and water. Remove and isolate contaminated clothing and shoes at the site.

★ 446 ★
ETHYL-2-CHLOROPROPIONATE
CAS: 535-13-7
Other names: PROPIONIC ACID, 2-CHLORO-, ETHYL ESTER; PROPANOIC ACID, 2-CHLORO-, ETHYL ESTER (9CI)

Danger profile: Combustible, poisonous gases produced in fire.
Uses: Making other chemicals.
Appearance: Liquid.
Odor: Pleasant. The odor is a warning of exposure; however, no smell does not mean you are not being exposed.
Effects of exposure:
Breathed: May cause irritation of the eyes, nose, throat and breathing passages. Higher exposures may cause shortness of breath and a dangerous build-up of fluids in the lungs (pulmonary edema), which can cause death. Pulmonary edema is a medical emergency that can be delayed one to two days following exposure.
Eyes: Contact may cause irritation and burns.
Skin: Passes through the unbroken skin; can increase exposure and the severity of the symptoms listed. May cause irritation and burns.
Long-term exposure: May cause skin allergy; possible lung damage.
Storage: Store in tightly closed containers in a dry, cool, well-ventilated place. Keep away from heat, sources of ignition, acids, bases, oxidizing and reducing agents. Containers may explode in fire.

★ 447 ★
ETHYLDICHLOROSILANE
CAS: 1789-58-8
Other names: DICHLOROETHYLSILANE; SILANE, DI-CHLOROETHYL-

Danger profile: Fire hazard, corrosive, reacts with water, poisonous gases produced in fire.
Uses: Making silicones.
Appearance: Colorless liquid.
Odor: Sharp, irritating. The odor is a warning of exposure; however, no smell does not mean you are not being exposed.
Effects of exposure:
Breathed: May cause irritation of the lungs with coughing and/or shortness of breath. Higher exposures may cause shortness of breath and a dangerous build-up of fluids in the lungs (pulmonary edema), which can cause death. Pulmonary edema is a medical emergency that can be delayed one to two days following exposure.
Eyes: May cause severe burns and permanent damage.
Skin: Passes through the unbroken skin; can increase exposure and the severity of the symptoms listed. May cause severe burns.
Swallowed: May cause severe burns of the mouth and stomach.
Long-term exposure: Similar chemicals may cause lung damage.
Storage: Store in tightly closed containers in a dry, cool, well-ventilated place. Keep away from heat, sources of ignition, oxidizers and moisture. Reaction with surface moisture will generate hydrogen chloride, which corrodes common metals. Containers may explode in fire.
First aid guide: Move victim to fresh air and call emergency medical care; if not breathing, give artificial respiration; if breathing is difficult, give oxygen. In case of contact with material, immediately flush skin or eyes with running water for at least 15 minutes. Remove and isolate contaminated clothing and shoes at the site. Keep victim quiet and maintain normal body temperature.

★ 448 ★
ETHYLENE
CAS: 74-85-1
Other names: ACETENE; ATHYLEN (GERMAN); BICAR-BURRETTED HYDROGEN; ATHYLEN (GERMAN); DICAR-BURRETTED HYDROGEN; ELAYL; ETHENE; LIQUID ETHYENE; LIQUID ETHELYNE; NCL-C5066; OLEFIANT GAS

Danger profile: Fire hazard, explosion hazard, Community Right-to-Know List, acrid smoke and irritating fumes in fire.
Uses: Making other chemicals; refrigerant; welding and cutting of metals; anesthetic; to accelerate fruit ripening.
Appearance: Colorless gas, or liquid when refrigerated or under pressure.
Effects of exposure:
Breathed: May cause headache, dizziness, fatigue; rapid irregular breathing; nausea, vomiting, confusion, loss of consciousness convulsions and death.
Eyes: Contact with liquid may cause frostbite, pain, stinging, tearing, inflammation of the eyelids and permanent damage.
Skin: Contact with liquid may cause frostbite, intense cold feeling, skin turns white and feels cold and hard.
Swallowed: Liquid may cause headache, confusion, drowsiness, unconsciousness and death.
Storage: Store in tightly closed containers in a dry, cool, well-ventilated place. Keep away from heat , sources of ignition, oxidizing agents, chlorine compounds and computable materials. Vapors may travel to sources of ignition and flash back. Containers may explode in fire.
First aid guide: Move victim to fresh air and call emergency medical care; if not breathing, give artificial respiration; if breathing is difficult, give oxygen. In case of frostbite, thaw frosted parts with water. Keep victim quiet and maintain normal body temperature.

★ 449 ★
ETHYLENEDIAMINE
CAS: 107-15-3
Other names: AETHALDIAMIN (GERMAN); AETHYL-ENEDIAMIN (GERMAN); BETA-AMINOETHYLAMINE; 1,2-ETHYLENEDIAMINE; 1,2-DIAMINOAETHAN (GERMAN); 1,2-DIAMINOETHANE, ANHYDROUS; 1,2-DIAMINO-ETHAAN (DUTCH); DIMETHYLENEDIAMINE; 1,2-ETHANEDIAMINE; ETHYLENE-DIAMINE (FRENCH); NCI-C60402

Danger profile: Mutagen, combustible, corrosive, exposure limit established, EPA extremely hazardous substance list, poisonous gases produced in fire.
Uses: Solvent; fungicide; making resins and other chemicals; textile lubricants; antifreeze solutions.
Appearance: Colorless, thick liquid.
Odor: Ammonia-like. The odor is a warning of exposure; however, no smell does not mean you are not being exposed.
Effects of exposure:
Breathed: May be poisonous. The vapor may cause irritation of the eyes, nose and throat; nausea and vomiting; asthma-like lung allergy may develop. Higher exposures may cause shortness of breath and a dangerous build-up of fluids in the lungs (pulmonary edema), which can cause death. Pulmonary edema is a medical emergency that can be delayed one to two days following exposure.
Eyes: May cause severe pain and permanent injury.
Skin: Passes through the unbroken skin; can increase exposure and the severity of the symptoms listed. May cause irritation, blisters and allergy.
Long-term exposure: May cause lung allergy; kidney, lung and liver damage; skin allergy.
Storage: Store in tightly closed containers in a dry, cool, well-ventilated detached place or in a refrigerator under an inert atmosphere. Outdoor storage is preferred. Keep away from acetic acid, acetic anhydride; acrolein; acrylic acid, acrylonitrile, allyl chloride, carbon disulfide, chlorosulfonic acid, epychlorhydrin, mesityl oxide, silver chloride, hydrochloric and sulfuric acids; oleum; b-propiolactonyl; vinyl acetate. Vapors may travel to sources of ignition and flash back. Containers may explode in fire.
First aid guide: Move victim to fresh air and call emergency medical care; if not breathing, give artificial respiration; if breathing is difficult, give oxygen. In case of contact with material, immediately flush skin or eyes with running water for at least 15 minutes. Remove and isolate contaminated clothing and shoes at the site. Keep victim quiet and maintain normal body temperature.

★ 450 ★
ETHYLENEDIAMINE TETRAACETIC ACID
CAS: 60-00-4
Other names: ACETIC ACID (ETHYLENEDINITRILO)TETRA-; ACIDE ETHYLENEDIAMINETETRACETIQUE (FRENCH); CELON A; CHEELOX; CHEMCOLOX 340; CELON ATH; COMPLEXON II; 3,6-DIAZAOCTANEDIOIC ACID,3,6-BIS(CARBOXYMETHYL)-; EDATHAMIL; EDETIC; EDETIC ACID; EDTA; EDTA ACID; ENDRATE; ETHYLENEDIAMINE-TETRAACETATE; ETHYLENEDIAMINETETRAACETIC ACID; ETHYLENEDIAMINE-N,N,N',N'-TETRAACETIC ACID; ETHYL-ENEDINITRILOTETRAACETIC ACID; GLYCINE, N,N'-1,2-ETHANEDIYLBIS(N-(CARBOXYMETHYL)-9CI); HAVIDOTE; METAQUEST A; NERVANAID B ACID; NULLAPON B ACID; NULLAPON BF ACID; PERMA KLEER 50 ACID; SEQUES-TRENE AA; SEQUESTRIC ACID; SEQUESTROL; TETRINE ACID; TITRIPLEX; TRICON BW; TRILON B; TRILON BW; VERSENE; VERSENE ACID; WARKEELATE ACID

Danger profile: Mutagen, possible reproductive hazard, poisonous gases produced in fire.
Uses: Food additive and preservative; in drugs; detergents, liquid soaps, shampoos, agricultural chemical sprays; metal cleaning and plating; in vegetable oils, pharmaceutical products; blood anticoagulant; in textiles dyeing, scouring, and detergent operations; laboratory chemical; aid in reducing blood cholesterol; in medicine to treat lead poisoning.
Appearance: White sand-like powder.
Effects of exposure:
 Breathed: May cause irritation of the nose, throat and lungs.
 Eyes: May cause irritation and burns.
 Skin: May cause irritation and burns.
 Swallowed: May cause mouth, throat and stomach irritation.
Long-term exposure: Unknown at this time.
Storage: Store in tightly closed containers in a dry, cool, well-ventilated place. Keep away from sources of ignition, oxidizers, bases, and copper and nickel compounds.

★ 451 ★
ETHYLENE DIBROMIDE
CAS: 106-93-4
Other names: AETHYLENBROMID (GERMAN); BROMO-FUME; BROMURO DI ETILE (ITALIAN); CELMIDE; DBE; 1,2-DIBROMAETHAN (GERMAN); 1,2-DIBROMOETANO (ITALIAN); DIBROMOETHANE; ALPHA,BETA-DIBROMOETHANE; SYM-DIBROMOETHANE; 1,2-DIBROMOETHANE; DIBROMURE D'ETHYLENE (FRENCH); 1,2-DIBROOMETHAAN (DUTCH); DOWFUME 40; DOWFUME EDB; DOWFUME W-8; DOWFUME W-85; DWUBROMOETAN (POLISH); EDB; EDB-85; E-D-BEE; ENT 15,349; ETHYLENE BROMIDE; 1,2-ETHYLENE DIBROMIDE; FUMO-GAS; GLYCOL BROMIDE; GLYCOL DIBROMIDE; ISCOBROME D; KOPFUME; NCI-C00522; NEPHIS; PESTMASTER; PESTMASTER EDB-85; SOILBROM-40; SOILBROM-85; SOILBROME-85; SOIL-BROM-90EC; SOILFUME; UNIFUME

Danger profile: Carcinogen, mutagen, possible reproductive hazard, exposure limit established, Community Right-to-Know List, poisonous gases produced in fire.
Uses: In gasoline; fumigant for grains and tree crops; solvent; waterproofing products; making other chemicals.
Appearance: Colorless liquid.
Odor: Sweet. The odor is a warning of exposure; however, no smell does not mean you are not being exposed.
Effects of exposure:
 Breathed: May be poisonous. May cause irritation to the eyes, nose and throat. Higher exposures may cause shortness of breath and a dangerous build-up of fluids in the lungs (pulmonary edema), which can cause death. Pulmonary edema is a medical emergency that can be delayed one to two days following exposure.
 Eyes: Contact may cause irritation and burns.
 Skin: May be poisonous if absorbed. Passes through the unbroken skin; can increase exposure and the severity of the symptoms listed. Contact may cause irritation, blisters, ulcers and burns.
 Swallowed: Poisonous. See symptoms under *Breathed* and *First aid guide.*
Long-term exposure: May cause cancer, damage to the developing fetus and the reproductive systems of both men and women; possible sterility; liver and kidney damage.
Storage: Store in tightly closed containers in a dry, cool, well-ventilated place. Keep away from light, heat, chemically active metals and liquid ammonia. Containers may explode in fire.
First aid guide: Move victim to fresh air and call emergency medical care; if not breathing, give artificial respiration; if breathing is difficult, give oxygen. In case of contact with material, immediately flush skin or eyes with running water for at least 15 minutes. Speed in removing material from skin is of extreme importance. Remove and isolate contaminated clothing and shoes at the site. Keep victim quiet and maintain normal body temperature. Effects may be delayed; keep victim under observation.

★ 452 ★
ETHYLENE GLYCOL
CAS: 107-21-1
Other names: 1,2-DIHYDROXYETHANE; EG; 1,2-ETHANEDIOL; FRIDEX; GLYCOL; GLYCOL ALCOHOL; 2-HYDROXYETHANOL

Danger profile: Mutagen, combustible, exposure limit established, Community Right-to-Know List, poisonous gas produced in fire.
Uses: Coolant and antifreeze; de-icer for aircraft wings and runways; asphalt-emulsion paints; brake fluids; polyester fibers and films; low-freezing dynamite; solvent; extractant for various purposes; solvents mixtures for cellophane; cosmetics; lacquers; alkyd resins; printing inks; wood stains; adhesives; leather dyeing; textile processing; tobacco; ball-point pen inks.
Appearance: Clear, colorless, slightly syrupy liquid.
Odor: The odor is a warning of exposure; however, no smell does not mean you are not being exposed.
Effects of exposure:
 Breathed: Mildly toxic. May cause irritation of eyes, nose and throat; headache, drowsiness, excitability, drunken behavior, mental confusion, fatigue. Very high levels of exposure may cause dizziness, lightheadedness; paralysis, breathing problems and cause victim to pass out, go into coma and die. High level exposure may cause liver damage.
 Eyes: Contact may cause irritation, pain, burning, stinging, tearing, inflammation of the eyelids. Light may cause pain.
 Skin: May remove oil from the skin and cause dryness. Persons with pre-existing skin disorders may be more susceptible to the effects of this substance.
 Swallowed: May cause gastro-intestinal irritation. See *Breathed.*
Long-term exposure: May cause birth defects, miscarriages or cancer. Prolonged or repeated contact may cause irritation of the skin, blistering and dermatitis; brain and nervous system changes.
Storage: Store in a tightly closed resin coated, stainless steel or aluminum containers that are clearly labeled in a cool, well-ventilated place. Keep away from heat moisture, and sources of ignition, oleum, chlorosulfuric acid, sulfuric acid and strong oxidizers. Vapors are explosive in heat or flame. Containers may explode in fire.

★ 453 ★
ETHYLENE GLYCOL DIETHYL ETHER
CAS: 629-14-1
Other names: DIETHOXYETHANE; DIETHYL CELLOSOLVE; ETHANE,1,2-DIETHOXY-; ETHYL GLYME

Danger profile: Fire hazard, Community Right-to-Know List, poisonous gases produced in fire.
Uses: High-boiling inert solvent; specializes solvent and extraction applications.
Appearance: Colorless liquid.
Odor: Mild, pleasant. The odor is a warning of exposure; however, no smell does not mean you are not being exposed.
Effects of exposure:
 Breathed: May be poisonous. May cause irritation of the eyes, nose and throat. High levels may cause dizziness, drowsiness and loss of consciousness.
 Eyes: Contact may cause irritation.

Skin: May be poisonous if absorbed. Passes through the un-broken skin; can increase exposure and the severity of the symptoms listed. Contact may cause irritation.

Swallowed: May cause irritation of the mouth and stomach.

Long-term exposure: May cause kidney damage.

Storage: Store in tightly closed containers in a dry, cool, well-ventilated place. Keep away from heat, sources of ignition, strong acids and oxidizing materials. Vapors may travel to sources of ignition and flash back. Containers may explode in fire.

First aid guide: Move victim to fresh air and call emergency medical care; if not breathing, give artificial respiration; if breathing is difficult, give oxygen. In case of contact with material, immediately flush eyes with running water for at least 15 minutes. Wash skin with soap and water. Remove and isolate contaminated clothing and shoes at the site.

★ 454 ★
ETHYLENE GLYCOL METHYL ETHER
CAS: 109-86-4

Other names: AETHYLENGYKOL-MONOMETHYLAETHER (GERMAN); DOWANOL EM; EGM; EGME; ETHER MONO-METHYLIQUE DE L'ETHYLENE-GLYCIL (FRENCH); ETHYL-ENE GLYCOL MONOMETHYL ETHER; GLYCOL ETHER EM; GLYCOL METHYL ETHER; GLYCOL MONOMETHYL ETHER; JEFFERSOL EM; MECS; 2-METHOXYETHANOL; METHYL CELLOSOLVE; METHYL ETHOXOL; METHYL GLY-COL; METHYL OXITOL; METIL CELLOSOLVE (ITALIAN); METOKSYETYLOWY ALCOHOL (POLISH); POLY-SOLV EM; PRIST

Danger profile: Mutagen, combustible, exposure limit established, Community Right-to-Know List, poisonous gas produced in fire.

Uses: Widely used solvent; lacquers and thinners, dyeing and printing textiles, varnish removers, leather, cleaning solutions; anti-icing component of aviation fuel.

Appearance: Colorless liquid.

Odor: Mild, agreeable. The odor is a warning of exposure; however, no smell does not mean you are not being exposed.

Effects of exposure:

Breathed: Mildly toxic. May cause irritation of eyes, nose and throat; headache, drowsiness, excitability, drunken behavior, mental confusion, fatigue. Very high levels of exposure may cause dizziness, lightheadedness; paralysis, breathing problems and cause victim to pass out, go into coma and die. High level exposure may cause liver damage.

Eyes: Contact may cause irritation, pain, burning, stinging, tearing, inflammation of the eyelids. Light may cause pain.

Skin: May remove oil from the skin and cause dryness. Persons with pre-existing skin disorders may be more susceptible to the effects of this substance.

Swallowed: May cause gastro-intestinal irritation.

Long-term exposure: May cause birth defects, miscarriages or cancer. Prolonged or repeated contact may cause irritation of the skin, blistering and dermatitis; brain and nervous system changes.

Storage: Store in a tightly closed container in a cool, well-ventilated place. Keep away from heat and sources of ignition, and strong oxidizers. Vapors are explosive in heat or flame. Containers may explode in fire.

★ 455 ★
ETHYLENEIMINE
CAS: 151-56-4

Other names: AMINOETHYLENE; AZACYCLOPROPANE; AZIRANE; AZIRIDINE; AZIRINE; 1H-AZIRINE, DIHYDRO-; DI-HYDROAZIRINE; DIHYDRO-1-AZIRINE; DIMETHYLENEI-MINE; DIMETHYLENIMINE; E-1; ETHYLENIMINE; ETHYLI-MENE; TL 337; ETHYLENE IMINE; ETHYLENIMINE; ETHYLIMINE; ETILENIMINA (ITALIAN); TL 337

Danger profile: Carcinogen, mutagen, fire hazard, explosion hazard, exposure limit established, EPA extremely hazardous substance list. Community Right-to-Know List, poisonous gases produced in fire.

Uses: Refining fuel oil and lubricants; protective coatings; pharmaceuticals; adhesives; other chemicals.

Appearance: Clear, colorless liquid.

Odor: Ammonia-like. The odor is a warning of exposure; however, no smell does not mean you are not being exposed.

Effects of exposure:

Breathed: Material gives inadequate warning of overexposure. May cause irritation of the eyes, nose, throat and lungs, tears and runny nose, coughing, sore throat, difficult breathing; lethargy, fatigue, and drowsiness; slurred speech, shaking, and dizziness. Higher exposures could cause shortness of breath, headache, nausea, vomiting, dizziness, diarrhea and a dangerous build-up of fluids in the lungs (pulmonary edema), which can cause death. Pulmonary edema is a medical emergency that can be delayed one to two days following exposure.

Eyes: May cause irritation, redness, tearing, pain, inflamed eyelids, and eye injury; burns in corneas and possible loss of sight.

Skin: Toxic. Material gives inadequate warning of overexposure. Passes through the unbroken skin; can increase exposure and the severity of the symptoms listed. May cause painful irritation, redness, and blisters. Repeated or prolonged skin contact may cause third-degree burns and may also cause the legs to feel numb.

Swallowed: May cause irritation of the mouth and throat; stomach and abdominal pain; loss of muscular coordination and mental confusion; convulsions. May have a corrosive effect on mucous membranes and may cause scarring of esophagus if swallowed. Also see symptoms for *Breathed.*

Long-term exposure: May cause cancer of the lung, liver and lymph glands; damage to the developing fetus; skin allergy; possible lung, liver and kidney damage. May lower white blood cell count.

Storage: Store in tightly closed containers in a dry, cool, well-ventilated place. Contact with silver or aluminum may cause polymerization. Keep away from heat, sources of ignition, acids and oxidizers. Containers may explode in fire.

First aid guide: Move victim to fresh air and call emergency medical care; if not breathing, give artificial respiration; if breathing is difficult, give oxygen. In case of contact with material, immediately flush skin or eyes with running water for at least 15 minutes. Remove and isolate contaminated clothing and shoes at the site. Keep victim quiet and maintain normal body temperature. Effects may be delayed; keep victim under observation.

★ 456 ★
ETHYLENE OXIDE
CAS: 75-21-8
Other names: AETHYLENOXID (GERMAN); ANPROLENE; DIHYDROOXIRENE; DIMETHYLENE OXIDE; E.O.; 1,2-EPOXYAETHAN (GERMAN); EPOXYETHANE (FRENCH); 1,2-EPOXYETHANE; ETHYLEENOXIDE (DUTCH); ETHYLENE (OXYDE D') (FRENCH); ETILENE (OSSIDO DI) (ITALIAN); ETO; ETYLENU TLENEK (POLISH); NCI-C50088; OXACYCLOPROPANE; OXANE; OXIDOETHANE; ALPHA,BETA-OXIDOETHANE; OXIRAAN (DUTCH); OXIRAN; OXIRANE; OXIRENE, DIHYDRO-; 1,2-EPOXY ETHANE

Danger profile: Carcinogen, mutagen, fire hazard, exposure limit established, Community Right-to-Know List, EPA extremely hazardous substances list, poisonous gas produced in fire.
Uses: Making antifreeze, polyesters, laundry detergents and other chemicals; fumigant; rocket propellant; industrial sterilant; fungicide.
Appearance: Colorless gas or liquid.
Odor: Sweet. The odor is a warning of exposure; however, no smell does not mean you are not being exposed.
Effects of exposure:
Breathed: May cause irritation of the eyes, nose, throat and lungs, tears and runny nose, coughing, sore throat, difficult breathing; lethargy, fatigue, and drowsiness; slurred speech, shaking, and dizziness. Exposure to low vapor concentrations often results in delayed nausea and vomiting. Higher exposures could cause shortness of breath, headache, nausea, vomiting, dizziness, diarrhea and a dangerous build-up of fluids in the lungs (pulmonary edema), which can cause death. Pulmonary edema is a medical emergency that can be delayed one to two days following exposure.
Eyes: May cause irritation, redness, tearing, pain, inflamed eyelids, and eye injury; burns in corneas and possible loss of sight.
Skin: Toxic. Passes through the unbroken skin; can increase exposure and the severity of the symptoms listed. May cause painful irritation, redness, and blisters. Repeated or prolonged skin contact may cause irritation of the skin and may also cause the legs to feel numb.
Swallowed: Poison. May cause irritation of the mouth and throat; stomach and abdominal pain; loss of muscular coordination and mental confusion; convulsions. Also see symptoms for *Breathed.*
Long-term exposure: May cause cancer; damage to the developing fetus; gynecological disorders.
Storage: Store in an tightly closed containers in a dry, cool, well-ventilated fire-proof place. Keep away from heat, sources of ignition, acids, alkalis, oxidizers, ammonia, alcohols, air and all other chemicals. Vapors may travel to sources of ignition and flash back. Containers may explode in fire.
First aid guide: Move victim to fresh air and call emergency medical care; if not breathing, give artificial respiration; if breathing is difficult, give oxygen. In case of contact with material, immediately flush skin or eyes with running water for at least 15 minutes. Remove and isolate contaminated clothing and shoes at the site. Keep victim quiet and maintain normal body temperature. Effects may be delayed; keep victim under observation.

★ 457 ★
ETHYLENE THIOUREA
CAS: 96-45-7
Other names: ETHYLENETHIOUREA; ETU; 2-IMIDAZOLIDINETHIONE; 4,5-DIHYDRO-2-MERCAPTO; MERCOZEN; MIDAZOLE

Danger profile: Carcinogen, mutagen, Community Right-to-Know List.

Uses: In insecticides, fungicides, dyes, pharmaceuticals, electroplating, synthetic resins; making neoprene rubber.
Appearance: Colorless to pale green crystals or needles.
Effects of exposure:
Breathed: May cause irritation of the nose, throat and lungs. High exposure may cause sweating, nausea, increased blood pressure. Higher exposures may cause shortness of breath and a dangerous build-up of fluids in the lungs (pulmonary edema), which can cause death. Pulmonary edema is a medical emergency that can be delayed one to two days following exposure.
Eyes: Contact may cause irritation.
Skin: Passes through the unbroken skin; can increase exposure and the severity of the symptoms listed. Contact may cause irritation.
Long-term exposure: May cause cancer; birth defects; thyroid damage. A related chemical (ziram) may cause brain swelling and hemorrhage, muscle weakness and liver and kidney problems.
Storage: Store in tightly closed containers in a dry, cool, well-ventilated place. Keep away from oxidizing agents, acids and acid anhydrides. Containers may explode in fire.

★ 458 ★
ETHYL ETHER
CAS: 60-29-7
Other names: AETHER; ANAESTHETIC ETHER; ANESTHESIA ETHER; ANESTHETIC ETHER; DIAETHYLAETHER (GERMAN); DIETHYL ETHER; DIETHYL OXIDE; DWUETYLOWYETER (POLISH); ETHANE, 1,1'-OXYBIS-; ETERE ETILICO (ITALIAN); ETHER; ETHER, ETHYL; ETHER ETHYLIQUE (FRENCH); ETHOXYETHANE; OXYDE D'ETHYLE (FRENCH); SOLVENT ETHER; SULFURIC ETHER

Danger profile: Mutagen, fire hazard, explosion hazard, exposure limit established, poisonous gas produced in fire.
Uses: Industrial solvent; making other chemicals; smokeless powder; laboratory chemical; has been used as an anesthetic.
Appearance: Colorless liquid.
Odor: Characteristic sweet. The odor is a warning of exposure; however, no smell does not mean you are not being exposed.
Effects of exposure:
Breathed: May cause headache, dizziness, fatigue; rapid irregular breathing; nausea, vomiting, confusion, loss of consciousness convulsions and death.
Eyes: Contact with liquid may cause stinging, tearing, irritation and burns.
Skin: Contact with liquid may cause irritation and burns.
Swallowed: Liquid may cause headache, confusion, drowsiness, unconsciousness and possible death.
Long-term exposure: May cause miscarriage, birth defects; loss of appetite, dizziness, headache, excitement; skin contact may cause drying, cracking and scaling; kidney and/or liver damage.
Storage: Store in tightly closed containers in a dry, cool, well-ventilated place. Keep away from heat, sources of ignition and all other chemicals, especially oxidizers. Vapors may travel to sources of ignition and flash back. Containers may explode in fire.
First aid guide: Move victim to fresh air and call emergency medical care; if not breathing, give artificial respiration; if breathing is difficult, give oxygen. In case of contact with material, immediately flush eyes with running water for at least 15 minutes. Wash skin with soap and water. Remove and isolate contaminated clothing and shoes at the site.

★ 459 ★
ETHYL HEXALDEHYDE
CAS: 123-05-7
Other names: BUTYL ETHYL ACETALDEHYDE; ETHYL BUTYLACETALDEHYDE; 2- ETHYLCAPROALDEHYDE; ALPHA-ETHYLCAPROALDEHYDE; ETHYLHEXALDEHYDE; 2-ETHYLHEXALDEHYDE; 2-ETHYLHEXANAL; 2-ETHYL-1-HEXANAL; HEXANAL, 2-ETHYL

Danger profile: Fire hazard, spontaneously combustible in air, acrid and irritating fumes in fire.
Uses: Making perfumes and other chemicals.
Appearance: White liquid.
Odor: Mild, pleasant. The odor is a warning of exposure; however, no smell does not mean you are not being exposed.
Effects of exposure:
  Breathed: May be poisonous. Vapors may cause irritation of the eyes, throat and breathing passages; coughing and difficult breathing. Higher exposures may cause shortness of breath and a dangerous build-up of fluids in the lungs (pulmonary edema), which can cause death. Pulmonary edema is a medical emergency that can be delayed one to two days following exposure.
  Eyes: Contact may cause severe irritation and burns.
  Skin: May be poisonous if absorbed. Contact may cause severe irritation and burns.
  Swallowed: May cause irritation of the mouth and stomach.
Long-term exposure: Similar chemicals cause lung problems including allergy; skin allergy.
Storage: Store in tightly closed air-tight containers in a dry, cool, well-ventilated place with non-wood floors. May ignite spontaneously when spilled on clothing, paper, or other absorbent materials. Keep away from heat, sources of ignition and oxidizers. Containers may explode in fire.
First aid guide: Move victim to fresh air and call emergency medical care; if not breathing, give artificial respiration; if breathing is difficult, give oxygen. In case of contact with material, immediately flush eyes with running water for at least 15 minutes. Wash skin with soap and water. Remove and isolate contaminated clothing and shoes at the site.

★ 460 ★
ETHYLHEXYL ACETATE
CAS: 103-09-3
Other names: ACETIC ACID, 2-ETHYLHEXYL ESTER; BETA-ETHYLHEXYL ACETATE; 2-ETHYLHEXYL ACETATE; ETHYLHEXYL ETHANOATE; 2-ETHYLHEXYL ETHANOATE; 2-ETHYLHEXANYL ACETATE; OCTYL ACETATE

Effects of exposure:
  Breathed: May be poisonous. May cause irritation of eyes, nose and throat; headache, drowsiness, excitability, drunken behavior, mental confusion, fatigue. Very high levels of exposure may cause dizziness, lightheadedness; paralysis, breathing problems and cause victim to pass out, go into coma and die. High level exposure may cause liver damage.
  Eyes: Contact may cause irritation, pain, burning, stinging, tearing, inflammation of the eyelids. Light may cause pain.
  Skin: May remove oil from the skin and cause dryness. Persons with pre-existing skin disorders may be more susceptible to the effects of this substance.
  Swallowed: May cause gastro-intestinal irritation. See *Breathed.*
Long-term exposure: May cause birth defects, miscarriages or cancer. Prolonged or repeated contact may cause irritation of the skin, blistering and dermatitis; brain and nervous system changes.
Storage: Store in a tightly closed container in a cool, well-ventilated place. Keep away from heat and sources of ignition,

and strong oxidizers. Vapors are explosive in heat or flame. Containers may explode in fire.

★ 461 ★
ETHYLIDENE NORBORNENE
CAS: 16219-75-3
Other names: BICYCLO 221 HEPT-2-ENE, 5-ETHYLIDENE-; ENB; 5-ETHYLIDENEBICYCLO(2.2.1)HEPT-2-ENE; 5-ETHYLIDENE-2-NORBORNENE; 2-NORBORNENE,5-ETHYLIDENE-

Danger profile: Combustible, exposure limit established, poisonous gases produced in fire.
Uses: Making pharmaceuticals, pesticides and specialty resins.
Appearance: Colorless liquid.
Odor: Turpentine-like. The odor is a warning of exposure; however, no smell does not mean you are not being exposed.
Effects of exposure:
  Breathed: May cause irritation of the eyes, nose and throat; dizziness, headache, confusion, respiratory distress; possible severe pneumonia.
  Eyes: May cause irritation and burns.
  Skin: Passes through the unbroken skin; can increase exposure and the severity of the symptoms listed. May cause irritation and burns.
  Swallowed: May cause irritation of entire digestive system.
Long-term exposure: May cause damage to the male reproductive glands; liver, lungs and kidney damage.
Storage: Store in sealed containers in a nitrogen atmosphere. Keep away from air and oxygen. Containers may explode in fire.

★ 462 ★
ETHYL ISOCYANATE
CAS: 109-90-0
Other names: ETHANE, ISOCYANATO-; ETHYL ISOCYANATE; ISOCYANATOETHANE; ISOCYANIC ACID, ETHYL ESTER

Danger profile: Mutagen, fire hazard, poisonous gases produced in fire.
Uses: Making pharmaceuticals and pesticides.
Appearance: Liquid.
Effects of exposure:
  Breathed: May be poisonous. May cause irritation to the nose and throat. Higher exposures may cause shortness of breath and a dangerous build-up of fluids in the lungs (pulmonary edema), which can cause death. Pulmonary edema is a medical emergency that can be delayed one to two days following exposure.
  Eyes: May cause irritation.
  Skin: May be poisonous if absorbed. Passes through the unbroken skin; can increase exposure and the severity of the symptoms listed. May cause irritation.
  Swallowed: May be poisonous. See *Breathed.*
Long-term exposure: May cause birth defects, miscarriages or cancer. Similar chemicals may cause skin allergies and lung problems.
Storage: Store in tightly closed containers in a dry, cool, well-ventilated place. Keep away from water, heat and sources of ignition. Vapors may travel to sources of ignition and flash back. Containers may explode in fire.
First aid guide: Move victim to fresh air and call emergency medical care; if not breathing, give artificial respiration; if breathing is difficult, give oxygen. In case of contact with material, immediately flush skin or eyes with running water for at least 15 minutes. Remove and isolate contaminated clothing and shoes at the site. Keep victim quiet and maintain normal body temperature. Effects may be delayed; keep victim under observation.

★ 463 ★
ETHYLMERCURIC CHLORIDE
CAS: 107-27-7
Other names: CERESAN; CHLOROETHYLMERCURY; EMC; CERESAN; CHLOROETHYLMERCURY; EMC; ETHYLMERCURIC CHLORIDE; ETHYLMERCURY CHLORIDE; GANOZAN; GRANOSAN; MERCURY, CHLOROETHYL-

Danger profile: Mutagen, reproductive hazard, extremely toxic, exposure limit established, Community Right-to-Know List, very toxic gases produced in fire.
Uses: Fungicide.
Appearance: Silvery-white, iridescent leaflets.
Effects of exposure:
Breathed: Poisonous. May cause permanent brain damage with hearing loss, abnormal walk, tremors, personality change. With little or no warning symptoms may occur weeks following exposure. Severe poisoning causes death.
Eyes: Poisonous. May cause irritation and burns.
Skin: Poisonous. Passes through the unbroken skin; can increase exposure and the severity of the symptoms listed. May cause irritation and burns.
Swallowed: Poisonous.
Long-term exposure: May cause mutations, birth defects, miscarriage or cancer; permanent brain damage, personality change, tremors, poor coordination; loss of vision.
Storage: Store in tightly closed containers in a dry, cool, well-ventilated place. Keep away from oxidizers. Containers may explode in fire.

★ 464 ★
ETHYL METHACRYLATE
CAS: 97-63-2
Other names: ETHYL 1-2-METHACRYLATE; ETHYL-ALPHA-METHYLACRYLATE; 1-2-METHACRYLIC ACID, ETHYL ESTER; 2-METHYLE-2-PROPENOIC ACID, ETHYL ESTER; 2-PROPENOIC ACID, 1-METHYL-, ETHYL ESTER; RHOPLEX AC-33

Danger profile: Mutagen, fire hazard, explosion hazard, poisonous gas produced in fire.
Uses: Making other chemicals, plastics, polymers and resins.
Appearance: Colorless liquid.
Odor: Sharp, unpleasant, irritating.
Effects of exposure:
Breathed: May be poisonous. May cause irritation of the eyes, nose, throat and mucous membrane. High levels may cause dizziness, lightheadedness, tiredness, convulsions and loss of consciousness.
Eyes: Contact may cause irritation.
Skin: Contact may cause irritation.
Swallowed: May cause irritation of mouth and stomach.
Long-term exposure: May cause skin and eye irritation; skin allergy; nervous system damage.
Storage: Store in tightly closed containers in a dry, cool, well-ventilated place. Keep away from heat, sources of ignition and oxidizers. Vapors may travel to sources of ignition and flash back. Containers may explode in fire.
First aid guide: Move victim to fresh air and call emergency medical care; if not breathing, give artificial respiration; if breathing is difficult, give oxygen. In case of contact with material, immediately flush eyes with running water for at least 15 minutes. Wash skin with soap and water. Remove and isolate contaminated clothing and shoes at the site.

★ 465 ★
ETHYL PHENYL DICHLOROSILANE
CAS: 1125-27-5
Other names: DICHLOROETHYLPHENYLSILANE; ETHYLPHENYLDICHLOROSILANE; SILANE, DICHLOROETHYLPHENYL-; PHENYLETHYLDICHLOROSILANE

Danger profile: Combustible, corrosive, poisonous gases produced in fire.
Uses: Making silicone products.
Appearance: Colorless liquid; fumes in humid air.
Odor: Sharp, irritating. The odor is a warning of exposure; however, no smell does not mean you are not being exposed.
Effects of exposure:
Breathed: May be poisonous. May cause irritation to the eyes, nose and throat. Higher exposures may cause shortness of breath and a dangerous build-up of fluids in the lungs (pulmonary edema), which can cause death. Pulmonary edema is a medical emergency that can be delayed one to two days following exposure.
Eyes: May cause severe burns and permanent damage.
Skin: Passes through the unbroken skin; can increase exposure and the severity of the symptoms listed. May cause severe burns; possibly second and third degree burns.
Swallowed: Poisonous. May cause sever burns of the mouth and stomach.
Long-term exposure: May cause lung damage.
Storage: Store in tightly closed containers in a dry, cool, well-ventilated place with non-wood floors. Keep away from heat, sources of ignition, water and other forms of moisture, oxidizers and combustible materials. Will react with surface moisture to evolve hydrogen chloride, which is toxic and corrosive to common metals. Containers may explode in fire.
First aid guide: Move victim to fresh air and call emergency medical care; if not breathing, give artificial respiration; if breathing is difficult, give oxygen. In case of contact with material, immediately flush skin or eyes with running water for at least 15 minutes. Speed in removing material from skin is of extreme importance. Remove and isolate contaminated clothing and shoes at the site. Keep victim quiet and maintain normal body temperature.

★ 466 ★
1-ETHYL PIPERIDINE
CAS: 766-09-6
Other names: 1-ETHYLPIPERIDINE; N-AETHYLPIPERIDIN (GERMAN); PIPERIDINE, 1-ETHYL-

Danger profile: Fire hazard, poisonous gases produced in fire.
Uses: Making other chemicals.
Appearance: Liquid.
Effects of exposure:
Breathed: May cause increased blood pressure and heart rate, shortness of breath, weakness. High levels may cause convulsions and death.
Eyes: May cause severe irritation, burns and permanent damage.
Skin: Passes through the unbroken skin; can increase exposure and the severity of the symptoms listed. May cause severe irritation and burns.
Swallowed: May cause severe irritation of the mouth and stomach.
Long-term exposure: May cause lung problems.
Storage: Store in tightly closed containers in a dry, cool, well-ventilated place. Keep away from heat, sources of ignition, acids, and oxidizers. Containers may explode in fire.

★467★
ETHYL SILICATE
CAS: 78-10-4
Other names: ETHYL ORTHOSILICATE; ETYLU KRZEMIAN (POLISH); EXTREMA; SILICATE D'ETHYLE (FRENCH); SILICIC ACID TETRAETHYL ESTER; TEOS; TETRAETHOXYSILANE; TETRAETHYLORTHOSILICATE; TETRAETHYL ORTHOSILICATE; TETRAETHYL SILICATE

Danger profile: Combustible, exposure limit established, poisonous gas produced in fire.
Uses: Hardener for water/weather-resistant concrete, cements, mortars, and stone.
Appearance: Colorless liquid.
Odor: Faint, pleasant. The odor is a warning of exposure; however, no smell does not mean you are not being exposed.
Effects of exposure:
  Breathed: May cause irritation of the nose, eyes, throat and breathing passages with dryness, coughing, nose bleeds, and difficult breathing. Higher exposures may cause shortness of breath and a dangerous build-up of fluids in the lungs (pulmonary edema), which can cause death. Pulmonary edema is a medical emergency that can be delayed one to two days following exposure. Exposure may cause liver, kidneys, lungs and red blood cell damage.
  Eyes: Contact may cause severe irritation and burns.
  Skin: Contact may cause severe irritation and burns.
  Swallowed: May cause nausea, vomiting, and cramps.
Long-term exposure: May cause damage to the liver, kidneys, lungs and red blood cells.
Storage: Store in tightly closed containers in a dry, cool, well-ventilated place. Keep away from heat, sources of ignition and oxidizers. May cause swelling and hardening of some plastics. Vapors may travel to sources of ignition and flash back. Containers may explode in fire.
First aid guide: Move victim to fresh air and call emergency medical care; if not breathing, give artificial respiration; if breathing is difficult, give oxygen. In case of contact with material, immediately flush skin or eyes with running water for at least 15 minutes. Remove and isolate contaminated clothing and shoes at the site. Keep victim quiet and maintain normal body temperature.

★468★
ETHYL TRICHLOROETHYLSILANE
CAS: 115-21-9
Other names: ETHYL SILICON TRICHLORIDE; ETHYL TRICHLOROSILANE; ETHYLTRICHLOROSILANE; SILANE, TRICHLOROETHYL-; SILICANE, TRICHLOROETHYL-; TRICHLOROETHYLSILANE; TRICHLOROETHYLSILICANE

Danger profile: Fire hazard, explosion hazard, corrosive, reacts with water, EPA extremely hazardous substance list; poisonous gases produced in fire.
Uses: Making silicone products.
Appearance: Colorless liquid.
Effects of exposure:
  Breathed: May be poisonous. May cause irritation of the eyes, nose and throat. Higher exposures may cause shortness of breath and a dangerous build-up of fluids in the lungs (pulmonary edema), which can cause death. Pulmonary edema is a medical emergency that can be delayed one to two days following exposure.
  Eyes: Contact may cause burns and permanent damage.
  Skin: Passes through the unbroken skin; can increase exposure and the severity of the symptoms listed. Contact may cause burns.
  Swallowed: May cause burns of the mouth and stomach.
Long-term exposure: May effect the lungs.

Storage: Store in tightly closed containers in a dry, cool, well-ventilated place. Keep away from heat, sources of ignition, water and other forms of moisture and oxidizers. Will react with surface moisture to evolve hydrogen chloride, which is toxic and corrosive to common metals. Vapors may travel to sources of ignition and flash back. Containers may explode in fire.

F

★469★
FENAMIPHOS
CAS: 22224-92-6
Other names: O-AETHYL-O-(3-METHYL-4-METHYLTHIOPHENYL)-ISOPROPYLAMIDO-PHOSPHORS AEURE ESTER (GERMAN); BAY 68138; ENT 27572; ETHYL 3-METHYL-4-(METHYLTHIO)PHENYL(1-METHYLETHYL)PHOSPHORAMIDATE; ETHYL 4-(METHYLTHIO)-M-TOLYLISOPROPYLPHOSPHORAMIDATE; ISOPROPYLAMINO-O-ETHYL-(4-METHYLMER CAPTO-3-METHYLPHENYL)PHOSPHATE; 1-(METHYLETHYL)-ETHYL 3-METHYL-4-(METHYLTHIO)PHENYLPHOSPHORAMIDATE; NEMACUR; NEMACURP; PHENAMIPHOS; PHOSPHORAMIDIC ACID, (1-METHYLETHYL)-, ETHYL (3-METHYL-4-(METHYLTHIO)PHENYL)ESTER; O-AETHYL-O-(3-METHYL-4-METHYLTHIOPHENYL)-ISOPROPYLAMIDO-PHOSPHORS AEURE ESTER (GERMAN); PHOSPHORAMIDIC ACID,ISOPROPYL-, 4-(METHYLTHIO)-M-TOLYL ETHYL ESTER

Danger profile: Combustible, poison, exposure limit established, EPA extremely hazardous substances list, poisonous gas produced in fire.
Uses: Pesticide (organophosphate).
Appearance: White crystalline solid.
Effects of exposure:
  Breathed: Extremely poisonous. May cause headache, dizziness, profuse sweating, nausea, vomiting, reduced heart beat, stomach cramps, diarrhea, loss of coordination, slow and weak breathing, fever, loss of consciousness and death. May cause shortness of breath and a dangerous build-up of fluids in the lungs (pulmonary edema), which can cause death. Pulmonary edema is a medical emergency that can be delayed one to two days following exposure.
  Eyes: Rapidly absorbed. Poisoning can happen very rapidly. See *Breathed.*
  Skin: Poisonous. Passes through the unbroken skin; can increase exposure and the severity of the symptoms listed.
  Swallowed: Powerful, deadly poison. See *Breathed.*
Long-term exposure: May cause nerve damage including loss of coordination; liver damage. Repeated exposure may cause depression, irritability and personality changes.
Storage: Store in a regulated, specially marked place in tightly-closed containers in a dry, well-ventilated place. Keep away from oxidizers. Containers may explode in fire.

★470★
FENSULFOTHION
CAS: 115-90-2
Other names: BAY 25141; BAYER 25141; BAYER S767; CHEMAGRO 25141; DASANIT; O,O-DIETHYLO(P-(METHYLSULFINYL)PHENYL)PHOSPHOROTHIOATE; DMSP; ENTPHOSPHOROTHIOATE; ENT 24,945; PHOSPHOROTHIOIC ACID, O,O-DIETHYLO-(P-(METHYLSULFINYL)PHENYL) ESTER; S 767; TERRACUR P

Danger profile: Poison, exposure limit established, EPA extremely hazardous substances list, poisonous gas produced in fire.
Uses: Insecticide for soil pests. (organophosphate).
Appearance: Yellow or brown liquid.
Effects of exposure:
Breathed: Extremely poisonous. May cause headache, dizziness, profuse sweating, nausea, vomiting, reduced heart beat, stomach cramps, diarrhea, loss of coordination, slow and weak breathing, fever, loss of consciousness and death. May cause shortness of breath and a dangerous build-up of fluids in the lungs (pulmonary edema), which can cause death. Pulmonary edema is a medical emergency that can be delayed one to two days following exposure.
Eyes: Rapidly absorbed. Poisoning can happen very rapidly. See *Breathed.*
Skin: Poisonous. Passes through the unbroken skin; can increase exposure and the severity of the symptoms listed.
Swallowed: Powerful, deadly poison. See *Breathed.*
Long-term exposure: May cause nerve damage including loss of coordination; liver damage. Repeated exposure may cause depression, irritability and personality changes.
Storage: Store in a regulated, specially marked place in tightly closed containers in a dry, well-ventilated place. Keep away from oxidizers. Containers may explode in fire.

★ 471 ★
FENTHION
CAS: 55-38-9
Other names: BAY 29493; BAYCID; BAYER 9007; BAYTEX; O,O-DIMETHYL-O-4(METHYLMERCAPTO)-3-METHYLPHENYL PHOSPHOROTHIOATE; DMTP; ENT 25,540; ENTEX; LEBAYCID; MERCAPTOPHOS; MPP; NCI-C08651; OMS 2; PHOSPHOROTHIOIC ACID,O,O-DIMETHYL-,O-(4-METHYLTHIO)M-TOLYL ESTER; O-(3-METHYL1-4-(METHYLTHIO)PHENYL) ESTER; QUELETOX; S 1752; SPOTTON; TALODEX; TIGUVON

Danger profile: Mutagen, poison, exposure limit established, poisonous gas produced in fire.
Uses: Insecticide; acaricide (organophosphate).
Appearance: Colorless to yellow-brown liquid.
Odor: Garlic-like. The odor is a warning of exposure; however, no smell does mean you are not being exposed.
Effects of exposure:
Breathed: Extremely poisonous. May cause headache, dizziness, profuse sweating, blurred vision, excessive salivation and often secretions, cramps, confusion, nausea, vomiting, stomach cramps, diarrhea, loss of coordination, slow and weak breathing, fever, reduced heart beat, loss of consciousness and death. Symptoms may be delayed one to two days. May cause shortness of breath and a dangerous build-up of fluids in the lungs (pulmonary edema), which can cause death. Pulmonary edema is a medical emergency that can be delayed one to two days following exposure.
Eyes: Rapidly absorbed. Poisoning can happen very rapidly. See *Breathed.*
Skin: Poisonous. Passes through the unbroken skin; can increase exposure and the severity of the symptoms listed.
Swallowed: Powerful, deadly poison. See *Breathed.*
Long-term exposure: May damage the developing fetus or cause birth defects; nerve damage including loss of coordination; liver damage. Repeated exposure may cause depression, irritability and personality changes.
Storage: Store in a regulated, specially marked place in tightly-closed containers in a dry, well-ventilated place. Containers may explode in fire.

★ 472 ★
FERBAM
CAS: 14484-64-1
Other names: AAFERTIS; BERCEMA FERTAM 50; CARBAMATE; CARBAMIC ACID, DIMETHYLDITHIO-, IRON SALT; DIMETHYLCARBAMODITHIOIC ACID, IRON COMPLEX; DIMETHYLCARBAMODITHIOIC ACID, IRON(3) SALT; DIMETHYLDITHIOCARBAMIC ACID, IRON SALT; DIMETHYLDITHIOCARBAMIC ACID, IRON(3) SALT; EISENDIMETHYLDITHIOCARBAMAT (GERMAN); EISEN(III)-TRIS(N,N-DIMETHYLDITHIOCARBAMAT) (GERMAN); ENT 14,689; FERBAM 50; FERBAME; FERBAM, IRON SALT; FERBECK; FERMATE; FERMATE FERBAM FUNGICIDE; FERMOCIDE; FERRADOW; FERRIC DIMETHYLDITHIOCARBAMATE; FUKLASIN; FUKLASIN ULTRA; HEXAFERB; HOKMATE; IRON DIMETHYLDITHIOCARBAMATE; KARBAM BLACK; KNOCKMATE; NIACIDE; SUP'R FLO FERBAM FLOWABLE; TRIFUNGOL; TRIS(DIMETHYLCARBAMODTHIOATO-S,S')IRON; TRIS(DIMETHYLDITHIOCARBAMATO)IRON; TRIS(N,N-DIMETHYLDITHIOCARBAMATO)IRON(111); VANCIDE FE95

Danger profile: Mutagen, possible reproductive hazard, combustible, exposure limit established, poisonous gas produced in fire.
Uses: Fungicide.
Appearance: Black solid.
Effects of exposure:
Breathed: May be poisonous. May cause irritation of the nose and throat.
Eyes: May cause irritation.
Skin: May be poisonous if absorbed. May cause irritation.
Swallowed: May be poisonous.
Long-term exposure: May cause damage to the developing fetus; kidney damage; skin allergy.
Storage: Store in tightly closed containers in a dry, cool, well ventilated place. Keep away from strong oxidizers. Containers may explode in fire.

★ 473 ★
FERRIC AMMONIUM CITRATE
CAS: 1185-57-5
Other names: AMMONIUM FERRIC CITRATE; FERRIC AMMONIUM CITRATE, BROWN; FERRIC AMMONIUM CITRATE, GREEN

Danger profile: Combustible, exposure limit established, poisonous gas produced in fire.
Uses: Medicine; blueprint photography; feed additive.
Appearance: Red-brown flakes or brown-yellow powder.
Odor: Ammonia-like. The odor is a warning of exposure; however, no smell does not mean you are not being exposed.
Effects of exposure:
Breathed: Dust may irritate the nose and throat.
Eyes: Dust may cause irritation and burns.
Skin: May cause mild irritation and burns.
Swallowed: May cause irritation of the mouth and stomach.
Long-term exposure: Large amounts of iron in the body may cause nausea, stomach pain, constipation, black bowel movements. May cause liver damage. May cause a brownish discoloration of the eyes.
Storage: Store in tightly closed containers in a dry, cool, well ventilated place. Keep away from light and sources of ignition.

★474★
FERRIC NITRATE
CAS: 10421-48-4
Other names: FERRIC NITRATE, NONHYDRATE; IRON NITRATE; IRON (III) NITRATE, ANHYDROUS; IRON TRINITRATE; NITRIC ACID, IRON (3) SALT

Danger profile: Mutagen, combustible, exposure limit established, poisonous gas produced in fire.
Uses: Textile dyeing, tanning; laboratory chemical.
Appearance: Green, colorless to pale violet lumpy crystals.
Effects of exposure:
Breathed: Dust may cause irritation of the nose and throat.
Eyes: Dust may cause irritation.
Skin: May cause mild irritation on prolonged contact.
Swallowed: May cause irritation of the mouth and stomach.
Long-term exposure: Large amounts of iron in the body may cause nausea, stomach pain, constipation, black bowel movements. May cause liver damage. Prolonged eye contact may cause a brownish discoloration.
Storage: Store in tightly closed containers in a dry, cool, well ventilated place. Solutions are corrosive to most metals. Keep away from combustible materials such as wood or paper, fuels and powdered aluminum. Containers may explode in fire.
First aid guide: Move victim to fresh air; call emergency medical care. In case of contact with material, immediately flush skin or eyes with running water for at least 15 minutes. Remove and isolate contaminated clothing and shoes at the site.

★475★
FERRIC SULFATE
CAS: 10028-22-5
Other names: IRON TRISULFATE; IRON PERSULFATE; IRON SESQUISULFATE; IRON (III) SULFATE; IRON SULFATE (2:3); IRON (3) SULFATE; IRON TERSULFATE; SULFURIC ACID, IRON (3) SALT (3:2)

Danger profile: Combustible, exposure limit established, poisonous gas produced in fire.
Uses: In pigments; laboratory chemical; metal finishing, pickling and etching; disinfectant; textile dyeing and calico printing; water and sewage treatment.
Appearance: Gray to white powder or yellow lumpy crystals.
Effects of exposure:
Breathed: May cause irritation of the nose and throat.
Eyes: May cause irritation.
Skin: May cause irritation.
Swallowed: May cause irritation of the mouth and stomach.
Long-term exposure: Large amounts of iron in the body may cause nausea, stomach pain, constipation, black bowel movements. May cause liver damage. Prolonged eye contact may cause a brownish discoloration.
Storage: Store in tightly closed containers in a dry, cool, well ventilated place. Keep away from light and moisture. Corrosive to copper, copper alloys, mild steel, and galvanized steel.

★476★
FERROCENE
CAS: 102-54-5
Other names: BISCYCLOPENTADIENYLIRON; BISCYCLOPENTADIENYL IRON; DICYCLOPENTADIENYL IRON; DI-2,4-CYCLOPENTADIEN-1-YL IRON; FERROTSEN; IRON BIS (CYCLOPENTADIENE); IRON DICYCLOPENTADIENYL

Danger profile: Mutagen, combustible, exposure limit established, poisonous gas produced in fire.
Uses: Making rubber, silicone resins, high temperature plastics; oil and gasoline additive.

Appearance: Bright orange, salt-like.
Effects of exposure:
Breathed: May cause irritation of the air passages, lungs; tightness of the chest, possible cough and phlegm.
Eyes: Contact may cause irritation. Eye strain may result if particles are not removed for the eye.
Skin: Contact may cause irritation.
Swallowed: Moderately toxic. See *Breathed.*
Long-term exposure: May cause liver and lung problems; mood changes and irritability.
Storage: Store in tightly closed bottles or fiber drums in a dry, cool, well-ventilated place. Keep away from ammonium perchlorate.

★477★
FERROUS AMMONIUM SULFATE
CAS: 10045-89-3
Other names: AMMONIUM IRON SULFATE; FERROUS AMMONIUM SULFATE HEXAHYDRATE; IRON AMMONIUM SULFATE; MOHR'S SALT; SULFURIC ACID, AMMONIUM IRON (2) SALT (2:2:1)

Danger profile: Combustible, exposure limit established, poisonous gas produced in fire.
Uses: Photography; laboratory chemical; in dosimeters; metallurgy.
Appearance: Pale green or blue-green solid.
Effects of exposure:
Breathed: May cause irritation of the nose and throat.
Eyes: Dust may cause irritation.
Skin: May cause irritation on prolonged contact
Swallowed: May cause irritation of the mouth and stomach
Long-term exposure: May cause liver damage; lung irritation. Large amounts of iron in the body may cause nausea, stomach pain, constipation, black bowel movements. Prolonged eye contact may cause a brownish discoloration.
Storage: Store in tightly closed containers in a dry, cool, well ventilated place.

★478★
FERROUS CHLORIDE
CAS: 7758-94-3
Other names: IRON(II) CHLORIDE (1:2); IRON DICHLORIDE; IRON PROTOCHLORIDE; LAWRENCITE

Danger profile: Mutagen, combustible, corrosive, exposure limit established, poisonous gas produced in fire.
Uses: Mordant in textile dyeing; metallurgy; pharmaceutical preparations; manufacture of other chemicals; sewage treatment.
Appearance: Pale green salt-like powder or crystals.
Effects of exposure:
Breathed: May cause irritation of the nose and throat.
Eyes: May cause severe irritation and burns.
Skin: May cause severe irritation and burns.
Swallowed: May cause serious irritation of the mouth and stomach; stomach pain, constipation and black bowel movements.
Long-term exposure: May cause liver damage. Large amounts of iron in the body may cause nausea, stomach pain, constipation, black bowel movements. Prolonged eye contact may cause a brownish discoloration.
Storage: Store in tightly closed containers in a dry, cool, well ventilated place. Keep away from ethylene oxide, potassium and sodium metals. Protect container against physical damage. Solutions may corrode metals. Containers may explode in fire.
First aid guide: Move victim to fresh air; call emergency medical care. In case of contact with material, immediately flush skin or

eyes with running water for at least 15 minutes. Remove and isolate contaminated clothing and shoes at the site. Keep victim quiet and maintain normal body temperature.

★ 479 ★
FERROUS SULFATE
CAS: 7720-78-7
Other names: COPPERAS; DURETTER; DUROFERON; EXSICCATED FERROUS SULFATE; EXSICCATED FERROUS SULPHATE; FEOSOL; FEOSPAN; FER-IN-SOL; FERROGRADUMET; FERRALYN; FERROSULFAT (GERMAN); FERROSULFATE; FERRO-THERON; FERROUS SULFATE (1:1); FERSOLATE; GREEN VITRIOL IRON MONOSULFATE; IRON PROTOSULFATE; IRON SULFATE (1:1); IRON(II) SULFATE; IRON(2) SULFATE; IRON(2) SULFATE (1:1); IRON VITRIOL; IROSPAN; SLOW-FE; SULFERROUS; SULFURIC ACID IRON SALT (1:1); SULFURIC ACID, IRON(2) SAL (1:1)

Danger profile: Mutagen, exposure limit established, poisonous gas produced in fire.
Uses: Iron oxide pigment; water and sewage treatment; fertilizer; food or feed additive; herbicide; wood preservative; process engraving; making other chemicals.
Appearance: Greenish, bluish-green or yellow fine or lumpy crystals.
Effects of exposure:
  Breathed: May cause irritation to the nose and throat.
  Eyes: Contact may cause irritation. Prolonged contact may discolor the eyes.
  Skin: May cause irritation.
  Swallowed: Poisonous. May cause tiredness, aggression, diarrhea, nausea and vomiting, stomach bleeding and coma.
Long-term exposure: May cause damage to the male reproductive glands and reduce fertility; birth defects, miscarriage, cancer. May cause liver damage. Large amounts of iron in the body may cause nausea, stomach pain, constipation, black bowel movements. Prolonged eye contact may cause a brownish discoloration.
Storage: Store in tightly closed steel, concrete or wood containers in a dry, cool, well ventilated place. Keep away from sources of ignition and alkalis.

★ 480 ★
FLUORIDE
CAS: 16984-48-8
Other names: FLUORIDE(1-); FLUORIDE ION; FLUORIDE ION (1-); PERFLUORIDE

Danger profile: Exposure limit established, poisonous gas produced in fire.
Uses: Widely used in the chemical, steel and aluminum industries; toothpastes and other dentifrices.
Appearance: Found combined with other materials; solid at room temperature.
Effects of exposure: None of the following effects occur at the levels used to treat water for preventing tooth cavities.
  Breathed: May cause irritation of the nose and throat; nausea and headaches. High exposures may cause shortness of breath and a dangerous build-up of fluids in the lungs (pulmonary edema), which can cause death. Pulmonary edema is a medical emergency that can be delayed one to two days following exposure.
  Eyes: May cause irritation and damage.
  Skin: May cause irritation.
  Swallowed: High levels may cause poisoning with weakness, stomach pain, convulsions, nausea, vomiting, loss of appetite, collapse and death.

Long-term exposure: Repeated or high exposures may cause bone and teeth changes; kidney problems; nosebleeds and sinus problems; nausea, vomiting, loss of appetite, diarrhea or constipation.
Storage: Store in tightly closed containers in a dry, cool, well ventilated place. Keep away from water, strong acids, especially nitric acid. Containers may explode in fire.

★ 481 ★
FLUORINE
CAS: 7782-41-4
Other names: BIFLUORIDEN (DUTCH); FLUOR (DUTCH, FRENCH, GERMAN, POLISH); FLUORINE-19; FLUORO (ITALIAN); FLUORURES ACIDE (FRENCH); FLUORURI ACIDI (ITALIAN); SAEURE FLUORIDE (GERMAN)

Danger profile: Highly reactive, dangerous explosion hazard, exposure limit established, water produces poisonous gas, poisonous gas produced in fire.
Uses: Making other chemicals; active constituent of fluoridating compounds used in drinking water, toothpastes and other dentifrices; rocket fuel.
Appearance: Pale yellow gas or yellow liquid.
Odor: Strong, pungent, irritating, choking, intense. The odor is a warning of exposure; however, no smell does mean you are not being exposed.
Effects of exposure:
  Breathed: Vapor extremely irritating. May cause irritation of the nose, eyes, mouth and throat; watering eyes; salivation, coughing, throat pain, difficult breathing, headache, fatigue, nausea, stomach pains. High exposures may cause shortness of breath and a dangerous build-up of fluids in the lungs (pulmonary edema), which can cause death. Pulmonary edema is a medical emergency that can be delayed one to two days following exposure.
  Eyes: May cause intense stinging, irritation, severe burns and irreparable damage and possible loss of sight.
  Skin: May cause irritation, redness, swelling and severe burns.
  Swallowed: May cause irritation; burning feeling in mouth and throat; swallowing causes pain; intense thirst; throat swelling, nausea, vomiting, shock. See *Breathed* for other symptoms.
Long-term exposure: May cause kidney and liver damage; digestive problems; teeth and bone changes.
Storage: Store in tightly closed containers such as cylinders in a dry, cool, well ventilated place with non-wooden floors. Ignites most combustible materials on contact. Keep away from water and all forms of moisture, heat, combustable materials, oxidizable materials, nitric acid. Containers may explode in fire.

★ 482 ★
FLUOROBENZENE
CAS: 462-06-6
Other names: BENZENE, FLUORO; PHENYL FLUORIDE

Danger profile: Dangerous fire hazard, poisonous gas produced in fire.
Uses: Insecticide and larvicide; making plastic or resin polymers.
Appearance: Colorless liquid.
Odor: Aromatic, benzene-like. The odor is a warning of exposure; however, no smell does mean you are not being exposed.
Effects of exposure:
  Breathed: May cause irritation of the eyes, nose, throat and lungs; coughing, transient anesthesia and central nervous system depression; dizziness. High exposures may cause shortness of breath and a dangerous build-up of fluids in the

lungs (pulmonary edema), which can cause death. Pulmonary edema is a medical emergency that can be delayed one to two days following exposure.

Eyes: May cause irritation and burns.

Skin: Passes through the unbroken skin; can increase exposure and the severity of the symptoms listed. May cause irritation and burns; repeated exposure may cause dermatitis.

Swallowed: Mildly toxic. See *Breathed* and *First aid guide.*

Long-term exposure: Chronic inhalation of vapors or mist may result to damage to lungs, liver and kidneys; possible dermatitis with itching and skin rash.

Storage: Store in tightly closed containers in a dry, cool, well ventilated place. Keep away from sources of ignition and oxidizers. Vapors may travel to sources of ignition and flash back. Containers may explode in fire.

First aid guide: Move victim to fresh air and call emergency medical care; if not breathing, give artificial respiration; if breathing is difficult, give oxygen. In case of contact with material, immediately flush eyes with running water for at least 15 minutes. Wash skin with soap and water. Remove and isolate contaminated clothing and shoes at the site.

★ 483 ★

FONOFOS

CAS: 944-22-9

Other names: O-AETHYL-S-PHENYL-AETHYL-DITHIOPHOSPHONAT (GERMAN); DIFONATE; DYFONATE; DYPHONATE; ENT 25,796; O-ETHYL-S-PHENYL ETHYL-PHOSPHONODITHIOATE; O-ETHYLS-PHENYL ETHYLDITHIOPHOSPHONATE; N 2790; STAUFFER N 2790; PHOSPHONODITHIOIC ACID, ETHYL-O-ETHYL S-PHENYL ESTER

Danger profile: Combustible, poison, exposure limit established, EPA extremely hazardous substances list, poisonous gas produced in fire.

Uses: Insecticide; soil fumigant (organophosphate).

Appearance: Clear, colorless liquid.

Odor: Pungent, aromatic. The odor is a warning of exposure; however, no smell does not mean you are not being exposed.

Effects of exposure:

Breathed: Extremely poisonous. May cause headache, dizziness, profuse sweating, nausea, vomiting, reduced heart beat, stomach cramps, diarrhea, loss of coordination, slow and weak breathing, fever, loss of consciousness and death. May cause shortness of breath and a dangerous build-up of fluids in the lungs (pulmonary edema), which can cause death. Pulmonary edema is a medical emergency that can be delayed one to two days following exposure.

Eyes: Rapidly absorbed. Poisoning can happen very rapidly. May cause pain, tearing, twitching eyelids, blurred vision. Also see *Breathed.*

Skin: Poisonous. Passes through the unbroken skin; can increase exposure and the severity of the symptoms listed.

Swallowed: Powerful, deadly poison. May cause nausea, vomiting, loss of appetite, stomach cramps, diarrhea. Also see *Breathed.*

Long-term exposure: May cause nerve damage including loss of coordination; liver damage. Repeated exposure may cause depression, irritability and personality changes; loss of appetite.

Storage: Store in a regulated, specially marked place in tightly-closed containers in a dry, well-ventilated place. Keep away from water, strong alkalis. Containers may explode in fire.

★ 484 ★

FORMALDEHYDE

CAS: 50-00-0

Other names: ALDEHYDE FORMIQUE (FRENCH); ALDEIDE FORMICA (ITALIAN); BFV; FA; FANNOFORM; FORMALDEHYD (CZECH); FORMALDEHYD (POLISH); FORMALIN; FORMALIN 40; FORMALINA (ITALIAN); FORMALINE (GERMAN); FORMALIN-LOESUNGEN (GERMAN); FORMALITH; FORMIC ALDEHYDE; FORMOL; FYDE; IVALON; KARSAN; LYSOFORM; METHANAL; METHYL ALDEHYDE; METHYLENE GLYCOL; METHYLENE OXIDE; MORBICID; NCI-C02799; OPLOSSINGEN (DUTCH); OXOMETHANE; OXYMETHYLENE; PARAFORM; POLYOXYMETHYLENE GLYCOLS; SUPERLYSOFORM; TETRAOXYMETHYLENE; TRIOXANE

Danger profile: Carcinogen, mutagen, combustible, exposure limit established, Community Right-to-Know List, poisonous gas produced in fire.

Uses: Germicide; embalming fluid; in home insulation and various wood products; resins; making other chemicals; fertilizer; dyes; preservative; treatment of textiles; grain treatment.

Appearance: Gas or liquid.

Odor: Strong, pungent and suffocating. The odor is a warning of exposure; however, no smell does not mean you are not being exposed.

Effects of exposure:

Breathed: Vapor may cause intense irritation to nose, throat, mouth and lungs; mucous membrane. Symptoms of exposure include watering of the eyes, coughing, difficult breathing, headache, nausea, vomiting, muscular weakness, unconsciousness. High exposure can lead to drowsiness, coma and death; buildup of fluid in the lungs (pulmonary edema), and may cause death. Pulmonary edema may be delayed by one to two days following exposure.

Eyes: Severe irritant. Contact may cause, burning and stinging, tearing, irritation and redness and swelling of the eyelids, intense pain; can burn the eyes and lead to blurred vision, chronic irritation and other permanent damage. Burns may be delayed for hours.

Skin: Severe irritant. Contact may cause burning or stinging feeling, redness, swelling, intense pain, blisters, irritation, rash. Repeated exposure may lead to skin allergy and chronic irritation. Burns may be delayed for hours.

Swallowed: May cause burning feeling of the mouth and throat, intense thirst, throat swelling, stomach cramps, nausea, vomiting, diarrhea, severe breathing difficulties and unconsciousness. Tissue damage may result. Breathing may stop. Watch for delayed symptoms.

Long-term exposure: May cause cancer; skin allergy; asthma-like allergy; bronchitis.

Storage: Store in tightly closed containers in a dry, cool, well ventilated place. Keep away from heat, sources of ignition and oxidizers. Containers may explode in fire.

First aid guide: Move victim to fresh air and call emergency medical care; if not breathing, give artificial respiration; if breathing is difficult, give oxygen. In case of contact with material, immediately flush skin or eyes with running water for at least 15 minutes. Remove and isolate contaminated clothing and shoes at the site. Keep victim quiet and maintain normal body temperature.

★ 485 ★

FORMAMIDE

CAS: 75-12-7

Other names: CARBAMALDEHYDE; FORMIMIDIC ACID; METHANAMIDE

Danger profile: Mutagen, possible reproductive hazard, combustible, exposure limit established, poisonous gas produced in fire.
Uses: Solvent; making paper; animal glues.
Appearance: Clear, colorless, thick liquid.
Effects of exposure:
  Breathed: May cause irritation of the eyes, nose and throat.
  Eyes: May cause irritation, redness and burns.
  Skin: May cause irritation, rash and burns.
  Swallowed: Moderately toxic. May cause burning feeling in mouth, throat and stomach.
Long-term exposure: May cause damage to the male reproductive glands; birth defects, miscarriage; damage to the fetus; skin irritation and rash.
Storage: Store in tightly closed containers in a dry, cool, well-ventilated place. Keep away from moisture, oxidizing materials and pyridine. Containers may explode in fire.

★ 486 ★
FORMIC ACID
CAS: 64-18-6
Other names: ACIDE FORMIQUE (FRENCH); ACIDO FORMICO (ITALIAN); AMEISENSAEURE (GERMAN); AMINIC ACID; BILORIN; FORMYLIC ACID; HYDROGEN CARBOXYLIC ACID; METHANOIC ACID; MIERENZUUR (DUTCH)

Danger profile: Mutagen, combustible, corrosive, exposure limit established, poisonous gas produced in fire.
Uses: Making textiles, other chemicals, fumigants, insecticides, refrigerants, solvents for perfumes, lacquers, vinyl resins; leather treatment; electroplating; as an antiseptic in brewing; silvering glass.
Appearance: Colorless liquid, may be fuming.
Odor: Strong, pungent, penetrating. The odor is a warning of exposure; however, no smell does not mean you are not being exposed.
Effects of exposure:
  Breathed: Vapor may cause intense irritation to nose, throat, mouth and lungs; mucous membrane. Symptoms of exposure include watering of the eyes, coughing, difficult breathing, headache, nausea, vomiting, muscular weakness, unconsciousness. High exposure can lead to drowsiness, coma and death; buildup of fluid in the lungs (pulmonary edema), and may cause death. Pulmonary edema may be delayed by one to two days following exposure.
  Eyes: Severe irritant. Contact may cause, burning and stinging, tearing, irritation and redness and swelling of the eyelids, intense pain; can burn the eyes and lead to blurred vision, chronic irritation and other permanent damage.
  Skin: Severe irritant. Contact may cause burning or stinging feeling, redness, swelling, intense pain, blisters, irritation, rash. Repeated exposure may lead to skin allergy and chronic irritation.
  Swallowed: May cause burning feeling of the mouth and throat, intense thirst, throat swelling, stomach cramps, nausea, vomiting, diarrhea, severe breathing difficulties and unconsciousness. Tissue damage may result. Breathing may stop. Watch for delayed symptoms.
Long-term exposure: May cause mutations, birth defects, miscarriage; skin allergy; lung injury.
Storage: Store in tightly closed containers in a dry, cool, well ventilated place. Keep away from heat, sources of ignition, strong oxidizers, sulfuric acid, furfuryl alcohol, nitromethane, and caustics.
First aid guide: Move victim to fresh air; call emergency medical care. In case of contact with material, immediately flush skin or eyes with running water for at least 15 minutes. Remove and isolate contaminated clothing and shoes at the site. Keep victim quiet and maintain normal body temperature.

★ 487 ★
FREON 113
CAS: 76-13-1
Other names: ARCTON 63; ARKLONE P; ASAHIFRON 113; CFC-113; DAIFLON S 3; ETHANE, 1,1,2-TRICHLORO-1,2,2-TRIFLUORO-(8CI,9CI); F 113; FC 113; FORANE 113; FREON TF; FREON 113 TR-T; FRIGEN 113; FRIGEN 113A; FRIGEN 113TR; FRIGEN 113TR-N; FLUOROCARBON 113; GENETRON 113; HALOCARBON 113; ISCEON 113; KAISER CHEMICALS 11; KHLADON 113; LEDON 113R 113; R 113 (HALOCARBON); REFRIGERANT 113; TTE; REFRIGERANT 113; REFRIGERANT R 113; TRICHLOROTRIFLUOROETHANE; 1,1,2-TRICHLORO-1,2,2-TRIFLUOROETHANE; 1,1,2-TRIFLUORO-1,2,2-TRICHLOROETHANE; 1,1,2-TRICHLOROTRIFLUOROETHANE; 1,2,2-TRICHLOROTRIFLUOROETHANE; 1,1,2-TRIFLUOROTRICHLOROETHANE; UCON 113; UCON FLUOROCARBON 113; UNCON 113/HALOCARBON 113

Danger profile: Exposure limit established, Community Right-to-Know List, poisonous gas produced in fire.
Uses: In refrigeration and air conditioning equipment; fire extinguishing agents; cleaning fluids and solvents.
Appearance: Colorless gas or liquid.
Odor: Carbon tetrachloride-like at high concentrations. The odor is a warning of exposure; however, no smell does not mean you are not being exposed.
Effects of exposure:
  Breathed: May cause irritation of the eyes, nose, throat and air passages; coughing, headache; increased saliva; reddening of the face; dizziness, lightheadedness, fatigue, loss of consciousness, shock, coma, irregular heartbeat, death. If the victim recovers watch for the possibility of a buildup of fluid in the lungs (pulmonary edema). Pulmonary edema is a medical emergency that can be delayed one to two days following exposure.
  Eyes: Contact may cause irritation, tearing, inflammation of the eyelids; burning sensation. Liquid may cause frostbite.
  Skin: May cause irritation, redness; painful blisters may appear later. Liquid may cause frostbite.
  Swallowed: May cause irritation of the eyes, nose and throat; excitement; headache, drunkenness; gastrointestinal tract, nausea, vomiting, diarreah with possible blood; drowsiness; loss of consciousness; irregular heartbeat and death. See *Breathed.*
Long-term exposure: May accumulate in the brain and kidneys. May cause irregular heartbeat; central nervous system effects. Prolonged exposure can cause narcotic effect or suffocation.
Storage: Store in tightly closed containers in a dry, cool, well ventilated place. Keep away from heat, sources of ignition and chemically active metals. Protect container from physical damage. Containers may explode in fire.

★ 488 ★
FREON 114
CAS: 76-14-2
Other names: ARCTON 33; ARCTON 114; CFC-114; CRYOFLUORAN; CRYOFLUORANE; SYM-DICHLOROTETRAFLUOROETHANE; 1,2-DICHLORO-1,1,2,2-TETRAFLUOROETHANE; DICHLOROTETRAFLUOROETHANE; ETHANE, 1,2-DICHLOROTETRAFLUORO-; ETHANE, 1,2-DICHLORO-1,1,2,2-TETRAFLUORO-F 114; FC 114; FLUORANE 114; FLUOROCARBON 114; FRIGEN 114; FRIGIDERM; GENETRON 114; GENETRON 316; HALOCARBON 114; LEDON 114; PROPELLANT 114; R 114; 1,1,2,2-TETRAFLUORO-1,2-DICHLOROETHANE; UCON 114

Danger profile: Exposure limit established, Community Right-to-Know List, poisonous gas produced in fire.
Uses: In refrigeration and air conditioning equipment; fire extinguishing agents; cleaning fluids and solvents.
Appearance: Clear, water-white liquid or gas.
Odor: Mild, slight, ether-like. The odor is a warning of exposure; however, no smell does not mean you are not being exposed.
Effects of exposure:
　Breathed: May cause irritation of the eyes, nose, throat and air passages; coughing, headache; increased saliva; reddening of the face; dizziness, lightheadedness, fatigue, loss of consciousness, shock, coma, irregular heartbeat, death. If the victim recovers watch for the possibility of a buildup of fluid in the lungs (pulmonary edema). Pulmonary edema is a medical emergency that can be delayed one to two days following exposure.
　Eyes: Contact may cause irritation, tearing, inflammation of the eyelids; burning sensation. Liquid may cause frostbite.
　Skin: May cause irritation, redness; painful blisters may appear later. Liquid may cause frostbite.
　Swallowed: May cause irritation of the eyes, nose and throat; excitement; headache, drunkenness; gastrointestinal tract, nausea, vomiting, diarrhea with possible blood; drowsiness; loss of consciousness; irregular heartbeat and death. See *Breathed.*
Long-term exposure: May accumulate in the brain and kidneys. May cause irregular heartbeat; central nervous system effects. Prolonged exposure can cause narcotic effect or suffocation.
Storage: Store in tightly closed containers in a dry, cool, well ventilated place. Keep away from heat, sources of ignition and chemically active metals. Protect container from physical damage. Containers may explode in fire.

★ 489 ★
FUMARIC ACID
CAS: 110-17-8
Other names: ALLOMALEIC ACID; BOLETIC ACID; 2-BUTENEDIOIC ACID (E); TRANS-BUTENEDIOIC ACID; (E)-BUTENEDIOIC ACID; BUTENEDIOIC ACID, (E)-; 1,2-ETHENEDICARBOXYLIC ACID, TRANS-; TRANS-1,2-ETHYLENEDICARBOXYLIC ACID; 1,2-ETHYLENEDICARBOXYLIC ACID, (E); KYSELINA FUMAROVA (CZECH); LICHENIC ACID; NSC-2752; U-1149; USAF EK-P-583

Danger profile: Mutagen, combustible, poisonous gas produced in fire.
Uses: Food additive; making polyester, alkyd, and phenolic resins, and dyes; paper-size resins; plasticizers; mordant; organic printing inks.
Appearance: White powder.
Effects of exposure:
　Breathed: May cause irritation of the nose, throat and breathing passages.
　Eyes: Prolonged contact may cause irritation.
　Skin: Prolonged contact may cause irritation.
　Swallowed: Mildly toxic.
Long-term exposure: Repeated exposure may effect the liver.
Storage: Store in tightly closed containers in a dry, cool, well ventilated place. Keep away from heat, sources of ignition and oxidizing materials. Containers may explode in fire.

★ 490 ★
FUMARIN
CAS: 117-52-2
Other names: 3-(ALPHA-ACETONYLFURFURYL)-4-HYDROXYCOUMARIN; 2H-1-BENZOPYRAN-2-ONE,3-(1-(2-FURANYL)-3-OXOBUTYL)-4-HYDROXY- (9CI); COUMAFURYL; COUMARIN,3-(ALPHA-ACETONYLFURFURYL)-4-HYDROXY- FOUMARIN; FUMASOL; FURMARIN; 3-(1-FURYL-3-ACETYLETHYL)-4-HYDROXY COUMARIN; KRUMKIL; LURAT; RATAFIN; RAT-A-WAY; TOMARIN

Danger profile: Combustible, poison, poisonous gas produced in fire, poisonous gas produced in fire.
Uses: Rodenticide.
Appearance: White crystalline powder.
Effects of exposure:
　Breathed: May be poisonous. High or repeated exposure may reduce the ability of blood to clot, causing hemorrhage. Early signs may be nosebleeds, bleeding gums, blood in the stool. Massive hemorrhage without warning may occur and cause death.
　Skin: May be poisonous if absorbed.
　Swallowed: May be poisonous.
Long-term exposure: Repeated exposure may reduce the ability of blood to clot, causing hemorrhage, nosebleeds, bleeding gums, bloody stool. Anemia from bleeding may cause organ damage. Similar substances may cause birth defects.
Storage: Store in tightly closed containers in a dry, cool, well ventilated place. Containers may explode in fire.

★ 491 ★
FURFURAL
CAS: 98-01-1
Other names: ARTIFICIAL ANT OIL; FEMA NO. 2489; 2-FURANCARBOXALDEHYDE; FURAL; FURALE; 2-FURALDEHYDE; 2-FURANALDEHYDE; 2-FURANCARBONAL; FURFURALDEHYDE; 2-FURYL-METHANAL; FUROLE; ALPHA-FUROLE; NCI-C56177; PYROMUCIC ALDEHYDE

Danger profile: Mutagen, combustible, poison, exposure limit established, poisonous gas produced in fire.
Uses: Solvent; shoe dyes; making other chemicals, polymers, resins, grinding wheels and brake linings; weed killer; fungicide; flavoring; laboratory chemical.
Appearance: Colorless to light-brown oily liquid.
Odor: Almond-like. The odor is a warning of exposure; however, no smell does mean you are not being exposed.
Effects of exposure:
　Breathed: May be poisonous. May cause irritation of the eyes, nose, throat and lungs, tears and runny nose, coughing, sore throat, difficult breathing; lethargy, fatigue, and drowsiness; slurred speech, shaking, and dizziness. Higher exposures could cause shortness of breath, headache, nausea, vomiting, dizziness, diarrhea and a dangerous build-up of fluids in the lungs (pulmonary edema), which can cause death. Pulmonary edema is a medical emergency that can be delayed one to two days following exposure.
　Eyes: May cause irritation, redness, tearing, pain, inflamed eyelids, and eye injury; burns in corneas and possible loss of sight.
　Skin: May be poisonous if absorbed. Passes through the unbroken skin; can increase exposure and the severity of the symptoms listed. May cause painful irritation, redness, and blisters. Repeated or prolonged skin contact may cause irritation of the skin and may also cause the legs to feel numb.
　Swallowed: May be poisonous. May cause irritation of the mouth and throat; stomach and abdominal pain; loss of muscular coordination and mental confusion; convulsions.

Long-term exposure: May cause birth defects, miscarriage; skin allergy; liver damage; lung problems; loss of sense of taste; headache, tiredness, tremors, watery eyes, throat irritation.

Storage: Store in tightly closed containers in a dry, cool, well ventilated place. Keep away from heat, sources of ignition, acids, alkalis and oxidizing materials. Vapors may travel to sources of ignition and flash back. Containers may explode in fire.

First aid guide: Move victim to fresh air and call emergency medical care; if not breathing, give artificial respiration; if breathing is difficult, give oxygen. In case of contact with material, immediately flush skin or eyes with running water for at least 15 minutes. Remove and isolate contaminated clothing and shoes at the site. Keep victim quiet and maintain normal body temperature.

★ 492 ★
FURFURYL ALCOHOL
CAS: 98-00-0
Other names: 2-HYDROXYMETHYLFURAN; 2-FURYLMETHANOL; 2-FURANCARBINOL; FURFURALCOHOL; 2-FURYL CARBINOL; 2-HYDROXYMETHYL FURAN; A-FURYLCARBINOL; A-FURANCARBINOL A-HYDROXYMETHYLFURAN; NCI-C56224

Danger profile: Mutagen, combustible, exposure limit established, poisonous gas produced in fire.
Uses: Flavoring; making corrosion-resistant sealants and cements, polymers; penetrant; solvent for dyes and resins.
Appearance: Colorless to light yellow liquid; turns red-brown in air.
Odor: Mild alcohol or ether-like. The odor is a warning of exposure; however, no smell does mean you are not being exposed.
Effects of exposure:
  Breathed: May be poisonous. May cause irritation of eyes, nose and throat; headache, drowsiness, excitability, drunken behavior, mental confusion, fatigue. Very high levels of exposure may cause dizziness, lightheadedness and cause victim to pass out, go into coma and die.
  Eyes: Contact may cause irritation, pain, burning, stinging, tearing, inflammation of the eyelids. Liquid may cause inflammation and corneal opacity. Light may cause pain.
  Skin: May be poisonous if absorbed. May remove oil from the skin and cause dryness and irritation. Persons with pre-existing skin disorders may be more susceptible to the effects of this substance.
  Swallowed: Poisonous. May cause gastrointestinal irritation.
Long-term exposure: May cause birth defects, miscarriage. Prolonged or repeated contact may cause headache, nausea, and irritation of mouth and stomach; possible skin rash.
Storage: Store in a tightly closed container in a cool, well-ventilated place. Keep away from heat and sources of ignition, nitrates, strong oxidizers, strong alkalis and strong acids. Vapors are explosive in heat or flame. Containers may explode in fire.
First aid guide: Move victim to fresh air and call emergency medical care; if not breathing, give artificial respiration; if breathing is difficult, give oxygen. In case of contact with material, immediately flush skin or eyes with running water for at least 15 minutes. Speed in removing material from skin is of extreme importance. Remove and isolate contaminated clothing and shoes at the site. Keep victim quiet and maintain normal body temperature. Effects may be delayed; keep victim under observation.

G

★ 493 ★
GALLIUM
CAS: 7440-55-3
Other names: GALLIUM CHLORIDE; GALLIUM, ELEMENTAL; GALLIUM METAL, LIQUID; GALLIUM METAL, SOLID

Danger profile: Mutagen, combustible, corrosive, poisonous gas produced in fire.
Uses: Making semiconductors.
Appearance: Lustrous, silvery liquid or metal, or gray solid.
Effects of exposure:
  Breathed: May cause irritation to the nose, throat, and lungs; coughing, and shortness of breath; metallic taste, nausea, vomiting. High exposures may cause kidney damage; shortness of breath and a dangerous build-up of fluids in the lungs (pulmonary edema), which can cause death. Pulmonary edema is a medical emergency that can be delayed one to two days following exposure.
  Eyes: May cause irritation, redness, shin rash and burns.
  Skin: May cause irritation and burns.
Long-term exposure: May cause birth defects; miscarriage; kidney damage; lung problems. Some gallium compounds may cause nervous system damage.
Storage: Store in tightly closed containers in a dry, cool, well ventilated place. Keep away from sources of ignition, acids and halogens.
First aid guide: Move victim to fresh air; call emergency medical care. In case of contact with material, immediately flush skin or eyes with running water for at least 15 minutes. Remove and isolate contaminated clothing and shoes at the site. Keep victim quiet and maintain normal body temperature.

★ 494 ★
GASOLINE
CAS: 8006-61-9
Other names: BENZIN (GERMAN); CASING HEAD GASOLINE; ESSENCE (FRENCH); MOTOR FUEL; PETROL

Danger profile: Mutagen, fire hazard, explosion hazard, exposure limit established, poisonous gas produced in fire.
Uses: Fuel for internal combustion engines; solvent.
Appearance: Clear liquid. Gasoline with lead may contain colored dyes, usually red, blue, green, or purple.
Odor: Characteristic. The odor is a warning of exposure; however, no smell does mean you are not being exposed.
Effects of exposure:
  Breathed: May be poisonous. May cause irritation of the eyes, nose and throat; dizziness, lightheadedness, rapid breathing, mental confusion, unsteady walk, headache, fatigue, nausea, vomiting, irregular heartbeat, seizures, unconsciousness, coma and death. High exposures during pregnancy may cause damage to the developing fetus.
  Eyes: May cause burning feeling, tearing, irritation, redness and inflammation of the eyelids.
  Skin: May be poisonous if absorbed. Passes through the unbroken skin; can increase exposure and the severity of the symptoms listed above. May cause irritation of the skin, dryness, redness and rash.
  Swallowed: May cause indigestion, dizziness, fatigue, unconsciousness.
Long-term exposure: Gasoline can expose victim to toxic additives such as benzene, ethylene dibromide and tetraethyl lead. High or repeated exposure may cause birth defects; miscarriage; kidney cancer; lung damage, skin rash, kidney damage.

Storage: Store in tightly closed containers in a dry, cool, well ventilated place. Keep away from heat, sources of ignition, direct sunlight and oxidizing materials. Vapors may travel to sources of ignition and flash back. Containers may explode in fire.

First aid guide: Move victim to fresh air and call emergency medical care; if not breathing, give artificial respiration; if breathing is difficult, give oxygen. In case of contact with material, immediately flush eyes with running water for at least 15 minutes. Wash skin with soap and water. Remove and isolate contaminated clothing and shoes at the site.

★ 495 ★
GLUCOSE OXIDASE
CAS: 9001-37-0
Other names: CORYLOPHYLINE; DEOXIN-1; E.C. 1.1.3.4; GLUCOSE AERODEHYDROGENASE; BETA-D-GLUCOSE OXIDASE; MICROCID; NOTATIN; OXIDASE, GLUCOSE; PENIN

Danger profile: Combustible, poisonous gas produced in fire.
Uses: Food preservative for canned foods, beer, etc; stabilizer for vitamins; laboratory chemical.
Appearance: Powder or crystals.
Effects of exposure:
    Breathed: May cause itching, watery eyes, stuffy or runny nose.
Long-term exposure: May cause asthma-like lung allergy with cough, shortness of breath and chest tightness.
Storage: Store in tightly closed containers in a freezer. Containers may explode in fire.

★ 496 ★
GLUTARALDEHYDE
CAS: 111-30-8
Other names: CIDEX; CUDEX; GLUTARAL; GLUTARALDE-HYD (CZECH); GLUTARD DIALDEHYDE; GLUTARIC ACID DIALDEHYSE; GLUTARIC DIALDEHYDE; NCI-C55425; NCL-C55425; 1,5-PENTANEDIAL; 1,5-PENTANEDIONE; POTENTI-ATED ACID GLUTARALDEHYDE; SONACIDE

Danger profile: Mutagen, exposure limit established, poisonous gas produced in fire.
Uses: Leather tanning; sterilization of medical equipment.
Appearance: Clear liquid or colorless crystals.
Odor: Pungent; rotten apples. The odor is a warning of exposure; however, no smell does mean you are not being exposed.
Effects of exposure:
    Breathed: May be poisonous. May cause irritation of the eyes, nose, throat and lungs, tears and runny nose, coughing, sore throat, difficult breathing; lethargy, fatigue, and drowsiness; slurred speech, shaking, and dizziness. Higher exposures could cause shortness of breath, headache, nausea, vomiting, dizziness, diarrhea and a dangerous build-up of fluids in the lungs (pulmonary edema), which can cause death. Pulmonary edema is a medical emergency that can be delayed one to two days following exposure.
    Eyes: May cause irritation, redness, tearing, pain, inflamed eyelids, and eye injury; burns in corneas and possible loss of sight.
    Skin: Toxic. Passes through the unbroken skin in harmful amounts; can increase exposure and the severity of the symptoms listed. May cause painful irritation, redness, and blisters. Repeated or prolonged skin contact may cause irritation of the skin and may also cause the legs to feel numb.
    Swallowed: May be poisonous. May cause irritation of the mouth and throat; stomach and abdominal pain; loss of mus-

cular coordination and mental confusion; convulsions. Also see symptoms for *Breathed.*
Long-term exposure: May cause birth defects, miscarriage; skin allergy; liver and nervous system damage.
Storage: Store in tightly closed containers in a dry, cool, well ventilated place.

★ 497 ★
GLYCERIN
CAS: 56-81-5
Other names: GLYCERINE; GLYCERIN, ANHYDROUS; GLYCERIN, SYNTHETIC; GLYCERINE; GLYCERITOL; GLYC-EROL; GLYCYL ALCOHOL; GROCOLENE; MOON; 1,2,3-PROPANETRIOL; SYNTHETIC GLYCERIN; 90 TECHNICAL GLYCERIN; TRIHYDROXYPROPANE; 1,2,3-TRIHYDROXYPROPANE

Danger profile: Mutagen, combustible, exposure limit established, poisonous gas produced in fire.
Uses: Making alkyd resins, dynamite, drugs, perfumes, cosmetics, baked products, tobacco products, liqueurs; solvent; printer's ink; rubber stamp and copying inks; in cements and mixes; special soaps; lubricant and softener; hydraulic fluid; antifreeze mixtures; to retard bacteria growth.
Appearance: Colorless or pale yellow, oily liquid
Effects of exposure:
    Breathed: May cause irritation of eyes, nose and throat; headache, drowsiness, excitability, drunken behavior, mental confusion, fatigue.
    Eyes: Contact may cause irritation, pain, burning, stinging, tearing, inflammation of the eyelids. Light may cause pain.
    Skin: May remove oil from the skin and cause dryness.
    Swallowed: May cause headache, nausea and vomiting; gastro-intestinal irritation.
Long-term exposure: My cause birth defects, miscarriage; lung problems.
Storage: Store in tightly closed containers in a dry, cool, well ventilated place. Keep away from heat, sources of ignition, hydrogen peroxide, potassium permanganate and oxidizers. Containers may explode in fire.

★ 498 ★
GLYCIDOL
CAS: 556-52-5
Other names: EPIHYDRIN ALCOHOL; 1,2-EPOXY-3-HYDROXY PROPANE; 2,3- EPOXYPROPANOL; GLYCIDE; GLYCIDYL ALCOHOL; 3-HYDROXYPROPYLENE OXIDE; NCI-C5549; OXIRANEMETHOL; OXIRANYMETHANOL

Danger profile: Mutagen, possible reproductive hazard, combustible, exposure limit established, poisonous gases produced in fire.
Uses: Stabilizer for natural oils and vinyl polymers.
Appearance: Colorless liquid.
Effects of exposure:
    Breathed: May cause irritation of the eyes, nose, throat and lungs. High levels may cause dizziness, light headedness, confusion, excitement followed by depression; chemical pneumonia or shortness of breath and a dangerous build-up of fluids in the lungs (pulmonary edema), which can cause death. Pulmonary edema is a medical emergency that can be delayed one to two days following exposure.
    Eyes: May cause severe irritation and burns. Possible permanent damage.
    Skin: Passes through the unbroken skin; can increase exposure and the severity of the symptoms listed above. May cause severe irritation.
    Swallowed: Moderately toxic. See *Breathed.*

Long-term exposure: May cause birth defects, miscarriage, cancer; sterility in males; emphysema; skin allergy; impaired vision from clouding of the eyes.
Storage: Store in tightly closed containers in a dry, cool, well-ventilated place. Keep away from heat, sources of ignition, strong oxidizers, chemically active metals, acids and bases. Containers may explode in fire.

★499★
GLYCIDYL ACRYLATE
CAS: 106-90-1
Other names: ACRYLIC ACID GLYCIDYL ESTER; 2,3-EPOXY-1-PROPANOL ACRYLATE; 2,3-EPOXYPROPYL ACRYLATE; 2,3-EPOXYPROPYL ESTER ACRYLIC ACID; GLYCIDYLACRYLATE; GLYCIDYLESTER KYSELINYAKRYLOVE (CZECH); GLYCIDYL PROPENATE; 1-PROPANOL, 2,3-EPOXY-,ACRYLATE; 2-PROPENOIC ACID, OXIRANYLMETHYL ESTER (9CI)

Danger profile: Mutagen, combustible, poison; poisonous gas produced in fire.
Appearance: Colorless liquid.
Effects of exposure:
Breathed: May be poisonous. May cause irritation of the eyes, nose, throat and lungs, tears and runny nose, coughing, sore throat, difficult breathing; lethargy, fatigue, and drowsiness; slurred speech, shaking, and dizziness. Higher exposures could cause shortness of breath, headache, nausea, vomiting, dizziness, diarrhea and a dangerous build-up of fluids in the lungs (pulmonary edema), which can cause death. Pulmonary edema is a medical emergency that can be delayed one to two days following exposure.
Eyes: May cause irritation, redness, tearing, pain, inflamed eyelids, and eye injury; burns in corneas and possible loss of sight.
Skin: May be poisonous if absorbed. Passes through the unbroken skin; can increase exposure and the severity of the symptoms listed. May cause painful irritation, redness, and blisters. Repeated or prolonged skin contact may cause irritation of the skin and may also cause the legs to feel numb.
Swallowed: May be poisonous. May cause irritation of the mouth and throat; stomach and abdominal pain; loss of muscular coordination and mental confusion; convulsions. Also see symptoms for Breathed.
Long-term exposure: May cause birth defects, miscarriage or cancer.
Storage: Store in tightly closed containers in a dry, cool, well ventilated place. Keep away from heat, sources of ignition and oxidizers. Containers may explode in fire.

★500★
GUM ARABIC
CAS: 9000-01-5
Other names: ACACIA; ACACIA DEALBATA GUM; ACACIA GUM; ACACIA SENEGAL; ACACIA SYRUP; ARABIC GUM; AUSTRALIAN GUM; GUM ACACIA; GUM OVALINE; GUM SENEGAL; INDIAN GUM; NCI-C50748; SENEGAL GUM; STARSOL NO. 1; WATTLE GUM

Danger profile: Mutagen, combustible, poisonous gas produced in fire.
Uses: In confectioneries and foods, pharmaceuticals, cosmetics, adhesives, inks; textile printing.
Appearance: White or yellowish solid. Used in granular form.
Odor: Very slight. The odor is a warning of exposure; however, no smell does mean you are not being exposed.
Effects of exposure:
Breathed: Exposure may lead to lung allergy.

Long-term exposure: May cause birth defects, miscarriage; asthma-like allergy with chronic runny nose; skin rash.
Storage: Store in tightly closed containers in a dry, cool, well ventilated place. Keep away from heat, sources of ignition and oxidizers. Containers may explode in fire.

H

★501★
HALOTHANE
CAS: 151-67-7
Other names: ANESTAN; 2-BROMO-2-CHLORO-1,1,1-TRIFLUORO-; 2-BROMO-2-CHLORO-1,1,1-TRIFLUOROETHANE; CHALOTHANE; ETHANE, 2-BROMO-2-CHLORO-1,1,1-THRIFLUORO-; FLUOTANE; FLUOTHANE; HALOTAN; HALSAN; NARCOTANE; NARCOTANN NESPOFA (RUSSIAN); 1,1,1-TRIFLUORO-2-BROMO-2-CHLOROETHANE; 1,1,1-TRIFLUORO-2-CHLORO-2-BROMOETHANE; 2,2,2-TRIFLUORO-1-CHLORO-1-BROMOETHANE

Danger profile: Mutagen, poisonous gas produced in fire.
Uses: Clinical anesthetic, as a vapor or gas.
Appearance: Clear, colorless liquid.
Odor: Sweetish, pleasant. The odor is a warning of exposure; however, no smell does not mean you are not being exposed.
Effects of exposure:
Breathed: May cause dizziness, lightheadedness, loss of manual dexterity and confusion. High levels may cause respiratory depression, irregular heart beat, loss of consciousness and amnesia.
Eyes: Contact may cause severe irritation.
Skin: May cause irritation.
Swallowed: Poisonous. May cause irritation of the mouth, gastro-intestinal tract, nausea, vomiting, diarrhea with possible blood; drowsiness; loss of consciousness; irregular heartbeat and death. See Breathed.
Long-term exposure: Based on studies of operating room personnel and their families and exposure to anesthetic vapors, there appears to be a risk of cancer, birth defects and miscarriages. May cause liver damage; irregular menstrual vasodilation; possible nerve and brain damage.
Storage: Store in tightly closed containers in a dry, cool, well ventilated place. Keep away from direct sunlight. Chemical may attack some rubbers and plastics. Containers may explode in fire.

★502★
HELIUM
CAS: 7440-59-7
Other names: HELIUM, ELEMENTAL

Danger profile: Asphyxiant.
Uses: Welding gases; growing germanium and silicon crystals; in weather and research balloons; to pressurize rocket fuels; leak detection; chromatography; luminous signs; diving and space vehicle breathing equipment; research.
Appearance: Colorless gas.
Effects of exposure:
Breathed: May cause dizziness, and lightheadedness. Very high levels may cause loss of consciousness and death from suffocation.
Eyes: Contact with liquid may cause frostbite.
Skin: Contact with liquid may cause frostbite.

Storage: Store in tightly closed containers in a dry, cool, well ventilated place. Containers may explode in fire.

★ 503 ★
HEPTACHLOR
CAS: 76-44-8
Other names: AGROCERES; 3-CHLOROCHLORDENE; DRINOX; DRINOX H-34; E 3314; ENT 15,152; EPTACLORO (ITALIAN); 1,4,5,6,7,8,8-EPTACLORO-3A,4,7,7A-TETRAIDRO-4,7-ENDO-METANO-INDE NE (ITALIAN); GPKH; H-34; HEPTACHLOOR (DUTCH); 1,4,5,6,7,8,8-HEPTACHLOOR-3A,4,7,7A-TETRAHYDRO-4,7-ENDO-METHANO- INDEEN (DUTCH); HEPTACHLORE (FRENCH); 3,4,5,6,7,8,8-HEPTACHLORODICYCLOPENTADIENE; 1,4,5,6,7,8,8-HEPTACHLORO-3A,4,7,7A-TETRAHYDRO-4,7-ENDOMETHANOIN DENE; 1,4,5,6,7,8,8A-HEPTACHLORO-3A,4,7,7 A-TETRAHYDRO-4,7-METHANOINDANE; 1,4,5,6,7,8,8-HEPTACHLORO-3A,4,7,7A-TETRAHYDRO-4,7-METHANOINDENE; 1(3A),4,5,6,7,8,8-HEPTACHLORO-3A(1),4,7,7A-TETRAHYDRO-4,7-METHAN OINDENE; 1,4,5,6,7,8,8-HEPTACHLORO-3A,4,7,7A -TETRAHYDRO-4,7-METHANOL-1H-INDENE; 1,4,5,6,7,8,8-HEPTACHLORO-3A,4,7,7,7A-TETRAHYDRO-4,7-METHYLENE INDENE; 1,4,5,6,7,10,10-HEPTACHLORO-4,7,8,9 -TETRAHYDRO-4,7-METHYLENEINDENE; 1,4,5,6,7,10,10-HEPTACHLORO-4,7,8,9-TETRAHYDRO-4,7-ENDOMETHYLENE INDENE; 1,4,5,6,7,8,8-HEPTACHLOR-3A,4,7,7,A-TETRAHYDRO-4,7-ENDO-METHANO- INDEN (GERMAN); HEPTAGRAN; HEPTAMUL; NCI-C00180; RHODIACHLOR; VELSICOL 104; VELSICOL HEPTACHLOR

Danger profile: Carcinogen, mutagen, poison, combustible, exposure limit established, Community Right-to-Know List, poisonous gas produced in fire.
Uses: Pesticide. EPA canceled the registration of insecticides containing heptachlor except for termite control outside of dwelling.
Appearance: White to light tan sand-like material, or a tan, soft, waxy solid.
Odor: Camphor-like. The odor is a warning of exposure; however, no smell does not mean you are not being exposed.
Effects of exposure:
    Breathed: May be poisonous. May cause headache, dizziness, irritability, feeling of anxiety, irritability, muscle twitching. High levels may cause tremors, fits, loss of consciousness, collapse and possible death.
    Eyes: Contact may cause irritation.
    Skin: May be poisonous if absorbed. Passes through the unbroken skin; can increase exposure and the severity of the symptoms listed above. Contact may cause irritation.
    Swallowed: May be poisonous. May cause nausea, vomiting, diarrhea, and irritation of the gastrointestinal tract.
Long-term exposure: May cause cancer; birth defects; liver and kidney damage; brain damage.
Storage: Store in tightly closed containers in a dry, cool, well ventilated place. Protect containers from physical damage. Containers may explode in fire.

★ 504 ★
HEPTANE
CAS: 142-82-5
Other names: DIPROPYL METHANE; DIPROPYLMETHANE; EPTANI (ITALIAN); HEPTAN (POLISH); N-HEPTANE; HEPTANEN (DUTCH); HEPTYL HYDRIDE; DIPROPAL METHANE; NORMAL HEPTANE; SKELLY-SOLVE C

Danger profile: Fire hazard, exposure limit established, poisonous gas produced in fire.

Uses: Industrial solvent; petroleum refining processes; anesthetic; laboratory chemical.
Appearance: Colorless watery liquid.
Odor: Gasoline-like. The odor is a warning of exposure; however, no smell does not mean you are not being exposed.
Effects of exposure:
    Breathed: May cause irritation of the eyes, nose, throat and breathing passages, coughing; depression; dizziness, lightheadedness, rapid breathing, mental confusion, unsteady walk, headache, fatigue, nausea, vomiting, unconsciousness, irregular heart beat, coma and death. High exposures may cause chemical pneumonia and/or a dangerous build-up of fluids in the lungs (pulmonary edema), which can cause death. Pulmonary edema is a medical emergency that can be delayed one to two days following exposure.
    Eyes: May cause burning feeling, tearing, irritation, redness and inflammation of the eyelids.
    Skin: Passes through the unbroken skin; can increase exposure and the severity of the symptoms listed. May cause irritation of the skin, dryness, redness and rash.
    Swallowed: May cause indigestion, headache, dizziness, fatigue, swelling of the abdomen, nausea, vomiting, depression unconsciousness.
Long-term exposure: May cause dermatitis; severe lung irritation. Chemical is often contaminated with n-hexane which may cause nerve damage. Possible brain damage.
Storage: Store in tightly closed containers in a dry, cool, well ventilated place. Keep away from heat, sources of ignition and strong oxidizers. Vapors may travel to sources of ignition and flash back. Containers may explode in fire.
First aid guide: Move victim to fresh air and call emergency medical care; if not breathing, give artificial respiration; if breathing is difficult, give oxygen. In case of contact with material, immediately flush eyes with running water for at least 15 minutes. Wash skin with soap and water. Remove and isolate contaminated clothing and shoes at the site.

★ 505 ★
HEPTANOL
CAS: 543-49-7
Other names: AMYL METHYL CARBINOL; HEPTANOL-2; 2-HYDROXYHEPTANE; METHYL AMYL CARBINOL

Danger profile: Combustible.
Uses: Industrial solvent; synthetic resins; ore processing.
Appearance: Colorless liquid.
Odor: Mild. The odor is a warning of exposure; however, no smell does not mean you are not being exposed.
Effects of exposure:
    Breathed: Mildly toxic. May cause irritation of eyes, nose and throat; headache, drowsiness, excitability, drunken behavior, mental confusion, fatigue. Very high levels of exposure may cause dizziness, lightheadedness and cause victim to pass out, go into coma and die.
    Eyes: Contact may cause severe irritation, pain, burning, stinging, tearing, inflammation of the eyelids. Light may cause pain.
    Skin: May cause irritation; removal of oil from the skin and cause dryness. Persons with pre-existing skin disorders may be more susceptible to the effects of this substance.
    Swallowed: May cause gastro-intestinal irritation. See *Breathed.*
Long-term exposure: Prolonged or repeated contact may cause irritation of the skin and dermatitis, and nervous system changes.
Storage: Store in a tightly closed container in a cool, well-ventilated place. Keep away from heat and sources of ignition, and strong oxidizers. Containers may explode in fire.

★ 506 ★
HEXACHLOROBENZENE
CAS: 118-74-1
Other names: AMATIN; ANTICARIE; BENZENE, HEXA-CHLORO-; BENZENE HEXACHLORIDE; BUNT-CURE; BUNT-NO-MORE; CO-OP HEXA; GRANOX MN; HCB; HEXA C. B.; HEXACHLOROBENZOL (GERMAN); JULIN'S CARBON CHLORIDE; NO BUNT LIQUID; PENTACHLOROPHENYL CHLORIDE; PERCHLOROBENZENE; PHENYL PERCHLO-RYL; SANOCIDE; SMUT-GO; SNIECIOTOX

Danger profile: Carcinogen, mutagen, poison, combustible, suspected reproductive hazard, Community Right-to-Know List, poisonous gas produced in fire.
Uses: Fungicide for treating seeds; wood preservative; making other chemicals.
Appearance: White needles.
Odor: Faint, not unpleasant. The odor is a warning of exposure; however, no smell does not mean you are not being exposed.
Effects of exposure:
  Breathed: May be poisonous. May cause irritation of the nose throat and lungs; headache, nausea, coughing.
  Eyes: May cause irritation, watering and blurred vision.
  Skin: Passes through the unbroken skin; can increase exposure and the severity of the symptoms listed. May cause irritation, redness and possible burns.
  Swallowed: May be poisonous. May cause headache, nausea, vomiting, irritation of the stomach; numbness of hands and arms; tiredness and fatigue. High dose may cause coma.
Long-term exposure: Accumulates in the body. May cause cancer; damage to the developing fetus. High or repeated exposure may cause liver, thyroid, immune system, kidney and nervous system damage; skin rash, skin thickening and pigment changes; dark or red urine.
Storage: Store in tightly closed containers in a dry, cool, well ventilated place. Keep away from heat, sources of ignition and dimethyl formamide. Reacts violently with dimethyl formamide at temperatures above 149°F.
First aid guide: Move victim to fresh air; call emergency medical care. In case of contact with material, immediately flush skin or eyes with running water for at least 15 minutes. Remove and isolate contaminated clothing and shoes at the site.

★ 507 ★
HEXACHLOROBUTADIENE
CAS: 87-68-3
Other names: DOLEN-PUR; GP-40-66:120; 1,3-BUTADIENE,1,1,2,3,4,4-HEXACHLORO-; HCBD; HEXA-CHLORO-1,3-BUTADIENE; PERCHLOROBUTADIENE

Danger profile: Carcinogen, mutagen, combustible, exposure limit established, reproductive hazard, Community Right-to-Know List, poisonous gas produced in fire.
Uses: Solvent; heat-transfer fluid; in transformers; hydraulic fluid; making other chemicals.
Appearance: Clear liquid.
Odor: Mild, faint, turpentine-like. The odor is a warning of exposure; however, no smell not does mean you are not being exposed.
Effects of exposure:
  Breathed: Poisonous. May cause respiratory difficulty and irritation of nose, throat and mucous membrane. May be fatal if inhaled.
  Eyes: May cause irritation and burns.
  Skin: May be poisonous if absorbed. Passes through the unbroken skin; can increase ressource and the severity of the symptoms listed. May cause irritation and burns.
  Swallowed: Poisonous. See *Breathed* and *First aid guide.*

Long-term exposure: May cause cancer; birth defects; liver, kidney damage.
Storage: Store in tightly closed containers in a dry, cool, well ventilated place or a refrigerator. Keep away from heat, sources of ignition, bromine perchlorate and oxidizers. Containers may explode in fire.
First aid guide: Move victim to fresh air and call emergency medical care; if not breathing, give artificial respiration; if breathing is difficult, give oxygen. In case of contact with material, immediately flush skin or eyes with running water for at least 15 minutes. Speed in removing material from skin is of extreme importance. Remove and isolate contaminated clothing and shoes at the site. Keep victim quiet and maintain normal body temperature. Effects may be delayed; keep victim under observation.

★ 508 ★
HEXACHLOROCYCLOPENTADIENE
CAS: 77-47-4
Other names: C-56; GRAPHLOX; HEXACHLORO CYCLO-PENTADIENE; HCCPD; HEX; PCL; PERCHLOROCYCLO-PENTADIENE

Danger profile: Possible reproductive hazard, corrosive, deadly poison, exposure limit established, EPA extremely hazardous substance list, Community Right-to-Know List, poisonous gas produced in fire.
Uses: Making pesticides, flame retardant materials, fungicides, pharmaceuticals, resins, and dyes.
Appearance: Yellow to amber liquid.
Odor: Harsh, pungent. The odor is a warning of exposure; however, no smell does not mean you are not being exposed.
Effects of exposure:
  Breathed: Poison. May cause irritation to the eyes, nose, throat and lungs; tearing, headache, sneezing, salivation, and difficult breathing. Higher exposures may cause shortness of breath and a dangerous build-up of fluids in the lungs (pulmonary edema), which can cause death. Pulmonary edema is a medical emergency that can be delayed one to two days following exposure.
  Eyes: Contact may cause irritation and burns.
  Skin: May be poisonous if absorbed. Passes through the unbroken skin; can increase exposure and the severity of the symptoms listed. Contact may cause irritation, blistering and burns.
  Swallowed: Poison. May cause nausea, vomiting, diarrhea and depression. See *Breathed.*
Long-term exposure: May cause birth defects, liver, kidney and nervous system damage. Possible lung damage.
Storage: Store in tightly closed containers in a dry, cool, well ventilated place. Keep away from sodium, water and other forms of moisture. Protect containers from physical damage. Containers may explode in fire.
First aid guide: Move victim to fresh air and call emergency medical care; if not breathing, give artificial respiration; if breathing is difficult, give oxygen. In case of contact with material, immediately flush skin or eyes with running water for at least 15 minutes. Speed in removing material from skin is of extreme importance. Remove and isolate contaminated clothing and shoes at the site. Keep victim quiet and maintain normal body temperature. Effects may be delayed; keep victim under observation.

★509★
HEXACHLOROETHANE
CAS: 67-72-1
Other names: AVLOTANE; CARBON HEXACHLORIDE; DIS-TOKAL; DISTOPAN; DISTOPIN; EGITOL; ETHANE, HEX-ACHLO-; ETHANE HEXACHLORIDE; ETHANE, 1,1,1,2,2,2,-HEXACHLORO-; ETHYLENE HEXACHLORIDE; FALKITOL; FASCIOLIN; HEXACHLOR-AETHEN (GERMAN); HEXA-CHLOROETHYLENE; MOTTENHEXE; NCI-CO4604; PER-CHLOROETHANE; PHENOHEP

Danger profile: Mutagen, exposure limit established, Community Right-to-Know List, poisonous gas produced in fire.
Uses: Animal medicines; insecticide, camphor substitute; in smoke-making devices; solvent; explosives.
Appearance: White crystals or liquid.
Odor: Mothball-like. The odor is a warning of exposure; however, no smell does not mean you are not being exposed.
Effects of exposure:
  Breathed: May be poisonous. May cause irritation of the eyes, nose, throat and air passages; coughing, headache; increased saliva; reddening of the face; dizziness, lightheadedness, fatigue, loss of consciousness, shock, coma, irregular heartbeat, death. If the victim recovers watch for the possibility of a buildup of fluid in the lungs (pulmonary edema). Pulmonary edema is a medical emergency that can be delayed one to two days following exposure.
  Eyes: Contact may cause imitation, tearing, inflammation of the eyelids; burning sensation.
  Skin: Passes through the unbroken skin; can increase exposure and the severity of the symptoms listed. May cause irritation, redness, burning feeling, swelling; dry skin; rash or blisters.
  Swallowed: Poisonous. Breath may smell of chloroform. May cause irritation of the mouth, gastro-intestinal tract, nausea, vomiting, diarrhea with possible blood; drowsiness; loss of consciousness; irregular heartbeat and death. See *Breathed.*
Long-term exposure: May cause cancer, liver and kidney damage. Possible brain or nerve damage.
Storage: Store in tightly closed containers in a dry, cool, well ventilated place. Keep away from heat, hot iron, zinc, aluminum and alkalis.
First aid guide: Move victim to fresh air; call emergency medical care. In case of contact with material, immediately flush skin or eyes with running water for at least 15 minutes. Remove and isolate contaminated clothing and shoes at the site.

★510★
HEXACHLORONAPHTHALENE
CAS: 1335-87-1
Other names: HALOWAX 1014; NAPTHALENE, HEXA-CHLORO-

Danger profile: Combustible, exposure limit established, Community Right-to-Know List, poisonous gas produced in fire.
Uses: Flame and water proofing; to insulate electrical equipment; additive in lubricants.
Appearance: Light yellow or white wax-like solid.
Odor: Pleasant, aromatic. The odor is a warning of exposure; however, no smell does not mean you are not being exposed.
Effects of exposure:
  Breathed: May cause irritation to the nose, throat and breathing passages.
  Eyes: May cause irritation.
  Skin: Passes through the unbroken skin; can increase exposure and the severity of the symptoms listed. May cause irritation and rash.

Long-term exposure: May cause liver damage; dermatitis with rash. The skin may become more sensitive to light.
Storage: Store in tightly closed containers in a dry, cool, well ventilated place. Keep away from strong oxidizers. Containers may explode in fire.

★511★
HEXACHLOROPHENE
CAS: 70-30-4
Other names: ACIGENA; ALMEDERM; AT 7; B32; BILEVON; BIS(2-HYDROXY-3,5,6-TRICHLOROPHENYL)METHANE; BIS-2,3,5-TRICHLOR-6-HYDROXYFENYLME THAN (CZECH); BIS(3,5,6-TRICHLORO-2-HYDROXYPHENYL)METHANE; COMPOUND G-11; COTOFILM; DERMADEX; 2,2'-DIHYDROXY-3,3',5,5',6,6'-HEXAC HLORODIPHENYLME-THANE; 2,2'-DIHYDROXY-3,5,6,3',5',6'-HEXAC HLO-RODIPHENYLMETHANE; EXOFENE; FOMAC; FOSTRIL; G-11; GAMOPHEN; GAMOPHENE; G-ELEVEN; GERMA-MEDICA; HCP; HEXABALM; 2,2',3,3',5,5'-HEXACHLORO-6,6'-DIHYDROXYDIPHENYLMETHANE; HEXACHLOROFEN (CZECH); HEXACHLOROPHANE; HEXACHLOROPHEN; HEXAFEN; HEXIDE; HEXOPHENE; HEXOSAN; ISOBAC 20; METHANE,BIS(2,3,5-TRICHLORO-6-HYDROXYPHENYL); 2,2'-METHYLENEBIS(3,4,6-TRICHLOROPHENOL); NABAC; NCI-C02653; NEOSEPT; PHISODAN; PHISOHEX; RITOSEPT; SEPTISOL; SEPTOFEN; STERAL; STERASKIN; SURGI-CEN; SURGI-CIN; SUROFENE; TERSASEPTIC; TRICHLORO-PHENE; TURGEX

Danger profile: Poison, combustible, exposure limit established, possible reproductive hazard, Community Right-to-Know List (chlorophenol), poisonous gas produced in fire.
Uses: Antibacterial agent in cosmetics, soaps, shampoos and deodorants; veterinary medicine. The FDA restricts sale of chemical, and products containing high levels are available only through prescription. Permitted in food for human and animal consumption.
Appearance: White powder.
Effects of exposure:
  Breathed: Poisonous. Dust may cause irritation to mucous membranes; stomach problems; loss of appetite, nausea, vomiting, stomach cramps and diarrhea; weakness, convulsions, coma and death. Dehydration may be severe and associated with shock.
  Eyes: Contact may cause irritation.
  Skin: Passes through the unbroken skin; can increase exposure and the severity of the symptoms listed. May cause irritation, burning sensation, rash and skin allergy.
  Swallowed: Poisonous. May cause nausea, vomiting, diarrhea and shock. May damage heart muscle. See *Breathed* and *First aid guide.*
Long-term exposure: There is limited evidence that chemical causes birth defects in animals, and there is a suspected association between exposure to pregnant women and birth defects. Health care personnel who are or may become pregnant should avoid this chemical. May cause brain damage, paralysis, blindness; skin allergy with rash and itching.
Storage: Store in tightly closed containers in a dry, cool, well ventilated place or a refrigerator. Keep away from alkalis.
First aid guide: Move victim to fresh air; call emergency medical care. In case of contact with material, immediately flush skin or eyes with running water for at least 15 minutes. Remove and isolate contaminated clothing and shoes at the site.

★ 512 ★
HEXAFLUOROACETONE
CAS: 684-16-2
Other names: 2-PROPANONE, 1,1,1,2,2,2-HEXAFLUORO-; ACETONE, HEXAFLUORO-; 6FK; NCI-C56440; HFA

Danger profile: Poison, possible reproductive hazard, exposure limit established, poisonous gas produced in fire.
Uses: Making solvents, adhesives, pharmaceuticals and other chemicals.
Appearance: Colorless, non-combustible highly reactive gas.
Effects of exposure:
Breathed: May be poisonous. May cause irritation to the nose, throat and breathing passages. High exposures may cause shortness of breath and a dangerous build-up of fluids in the lungs (pulmonary edema), which can cause death. Pulmonary edema is a medical emergency that can be delayed one to two days following exposure.
Eyes: May cause severe irritation and burns.
Skin: May be poisonous if absorbed. Passes through the unbroken skin; can increase exposure and the severity of the symptoms listed. May cause severe irritation and burns.
Swallowed: Poisonous.
Long-term exposure: May cause damage to the male reproductive glands and affect the production of sperm; kidney, liver, thymus, spleen, lymph nodes and lung injury. May affect the bone marrow and blood cells.
Storage: Store in tightly closed containers in a dry, cool, well ventilated place. Keep away from heat, water, direct sunlight, reducing agents, nitrates and nitric acid.
First aid guide: Move victim to fresh air and call emergency medical care; if not breathing, give artificial respiration; if breathing is difficult, give oxygen. In case of contact with material, immediately flush skin or eyes with running water for at least 15 minutes. Remove and isolate contaminated clothing and shoes at the site. Keep victim quiet and maintain normal body temperature. Effects may delayed; keep victim under observation.

★ 513 ★
HEXAFLUOROETHANE
CAS: 76-16-4
Other names: ETHANE, HEXAFLUORO-; F116; FREON 116; PERFLUOROETHANE

Danger profile: Combustible, poisonous gas produced in fire.
Uses: Coolant, propellant; refrigerant; in dielectric fluids.
Appearance: Colorless gas.
Effects of exposure:
Breathed: May cause headache, dizziness, fatigue; rapid irregular breathing; nausea, vomiting, confusion, abnormal heart rhythms, loss of consciousness, convulsions and death.
Eyes: Contact with liquid may cause frostbite, pain, stinging, tearing, inflammation of the eyelids and permanent damage.
Skin: Contact with liquid may cause frostbite, intense cold feeling, skin turns white and feels cold and hard.
Swallowed: Liquid may cause headache, confusion, drowsiness, unconsciousness and death.
Storage: Store in tightly closed containers in a dry, cool, well ventilated place. Keep away from heat above 125°F, sources of ignition, and metals. Containers may explode in fire.
First aid guide: Move victim to fresh air and call emergency medical care; if not breathing, give artificial respiration; if breathing is difficult, give oxygen.

★ 514 ★
HEXAMETHYLENE DIISOCYANATE
CAS: 822-06-0
Other names: 1,6-DIISOCYANATOHEXANE; HEXAMETHYLENE DI-ISOCYANATE; HEXAMETHYLENE-1,6-DIISOCYANATE; 1,6-HEXAMETHYLENE DIISOCYANATE; 1,6-HEXANEDIOLDIISOCYANATE; HEXANE, 1,6-DIISOCYANATO-; HMDI; TL 78

Danger profile: Poisonous gas produced in fire.
Uses: Making other chemicals and polyurethanes.
Appearance: Colorless liquid.
Odor: Irritating. The odor is a warning of exposure; however, no smell does not mean you are not being exposed.
Effects of exposure:
Breathed: May be poisonous. May cause irritation to the nose, throat and lungs; cough and shortness of breath. High exposures may cause shortness of breath and a dangerous build-up of fluids in the lungs (pulmonary edema), which can cause death. Pulmonary edema is a medical emergency that can be delayed one to two days following exposure. Lung allergy may develop from exposure.
Eyes: Contact may cause severe irritation and burns.
Skin: May be poisonous if absorbed. Passes through the unbroken skin; can increase exposure and the severity of the symptoms listed. Contact may cause severe irritation and burns.
Swallowed: Poisonous.
Long-term exposure: May cause skin or lung allergy; chronic lung problems.
Storage: Store in tightly closed containers in a dry, cool, well ventilated place. Keep away from water and other forms of moisture, strong acids and bases, alcohols, amines, carboxylic acids. Containers may explode in fire.
First aid guide: Move victim to fresh air and call emergency medical care; if not breathing, give artificial respiration; if breathing is difficult, give oxygen. In case of contact with material, immediately flush skin or eyes with running water for at least 15 minutes. Speed in removing material from skin is of extreme importance. Remove and isolate contaminated clothing and shoes at the site. Keep victim quiet and maintain normal body temperature. Effects may be delayed; keep victim under observation.

★ 515 ★
HEXAMETHYL PHOSPHORAMIDE
CAS: 680-31-9
Other names: ENT 50882; HEMPA; HEXAMETAPOL; HEXA-METHYLPHOSPHORIC ACID TRIAMIDE; HEXAMETHYL-PHOSPHORIC TRIAMIDE; N,N,N,N,N,N-HEXAMETHYLPHOSPHORIC TRIAMIDE; HEXAMETHYL-PHOSPHOROTRIAMIDE; HEXAMETHYLPHOSPHOTRIAMIDE; HMPA; HMPT; HPT; MEMPA; PHOSPHORIC TRIS(DIMETHYLAMIDE); PHOSPHO-RYLHEXAMETHYLTRIAMIDE; TRI(DIMETH-YLAMINO)PHOSPHINEOXIDE; TRIS(DIMETH-YLAMINO)PHOSPHINE OXIDE; TRIS(DIMETHYLAMINO)PHOSPHORUS OXIDE; PHOSPHO-RIC TRIAMIDE,HEXAMETHYL-

Danger profile: Carcinogen, mutagen, Community Right-to-Know List, poisonous gas produced in fire.
Uses: Solvent; laboratory chemical; in polyvinyl chloride; sterilizing insects to prevent reproduction.
Appearance: Colorless liquid.
Odor: Spicy. The odor is a warning of exposure; however, no smell does mean you are not being exposed.
Effects of exposure:
Breathed: May cause irritation to the nose, throat and lungs.
Eyes: May cause irritation.

Skin: Passes through the unbroken skin; can increase exposure and the severity of the symptoms listed. May cause irritation.

Long-term exposure: May cause birth defects, miscarriage or cancer; damage to the male reproductive glands and affect sperm production; kidney and lung damage; nose damage with chronic nasal discharge.

Storage: Store in tightly closed containers in a dry, cool, well ventilated place. Keep away from heat, sources of ignition, oxidizers, strong acids and chemically active metals. Containers may explode in fire.

★ 516 ★
HEXANE
CAS: 110-54-3
Other names: ESANI (ITALIAN); HEKSAN (POLISH); N-HEXANE; HEXANEN (DUTCH); NCI-C60571; ESANI (ITALIAN); HEKSAN (POLISH); HEXYL HYDRIDE; NORMAL HEXANE; HEXANEN (DUTCH); SKELLYSOLVE-B

Danger profile: Mutagen, fire hazard, exposure limit established, poisonous gas produced in fire.
Uses: Glue thinner; Solvent, especially for vegetable oils; making other chemicals; in paint and alcohols.
Appearance: Colorless liquid.
Odor: Mild, gasoline-like. The odor is a warning of exposure; however, no smell does not mean you are not being exposed.
Effects of exposure:
Breathed: May be poisonous. May cause irritation of the eyes, nose, throat and respiratory tract; cough, mild depression, dizziness, lightheadedness, rapid breathing, mental confusion, unsteady walk, headache, fatigue, nausea, vomiting, irregular heart beat, unconsciousness, coma and death. High exposures may cause shortness of breath and a dangerous build-up of fluids in the lungs (pulmonary edema), which can cause death. Pulmonary edema is a medical emergency that can be delayed one to two days following exposure.
Eyes: May cause burning feeling, tearing, irritation, redness and inflammation of the eyelids.
Skin: May be poisonous if absorbed. Passes through the unbroken skin; can increase exposure and the severity of the symptoms listed. May cause irritation of the skin, dryness, redness and rash.
Swallowed: May cause indigestion, nausea, vomiting, swelling of the abdomen, headache, depression, dizziness, fatigue, unconsciousness.
Long-term exposure: May cause damage to the developing fetus; damage to the male reproductive glands; nervous system damage; brain damage; dry and cracked skin with possible rash.
Storage: Store in tightly closed containers in a dry, cool, well ventilated place. Keep away from heat, sources of ignition dinitrogen tetraoxide and oxidizers. Vapors may travel to sources of ignition and flash back. Containers may explode in fire.
First aid guide: Move victim to fresh air and call emergency medical care; if not breathing, give artificial respiration; if breathing is difficult, give oxygen. In case of contact with material, immediately flush eyes with running water for at least 15 minutes. Wash skin with soap and water. Remove and isolate contaminated clothing and shoes at the site.

★ 517 ★
HEXENE
CAS: 592-41-6
Other names: BUTYLETHYLENE; BUTYL ETHYLENE; 1-HEXENE; 1-N-HEXENE; HEXYLENE

Danger profile: Fire hazard, poisonous gas produced in fire.
Uses: Making perfumes, dyes, plastic resins; in fuels.
Appearance: Colorless liquid.
Odor: Mild gasoline-like. The odor is a warning of exposure; however, no smell does not mean you are not being exposed.
Effects of exposure:
Breathed: May be poisonous. May cause irritation of the nose and throat; giddiness; loss of coordination, dizziness, nausea and loss of consciousness. Prolonged exposure to high concentrations may cause death.
Eyes: Contact may cause irritation and burns.
Skin: May be poisonous if absorbed. Contact may cause irritation and burns.
Storage: Store in tightly closed containers in a dry, cool, well ventilated place. Keep away from heat, sources of ignition and oxidizing materials. Vapors may travel to sources of ignition and flash back. Containers may explode in fire.
First aid guide: Move victim to fresh air and call emergency medical care; if not breathing, give artificial respiration; if breathing is difficult, give oxygen. In case of contact with material, immediately flush eyes with running water for at least 15 minutes. Wash skin with soap and water. Remove and isolate contaminated clothing and shoes at the site.

★ 518 ★
HEXYLENE GLYCOL
CAS: 107-41-5
Other names: 2,4-DIHYDROXY-2-METHYLPENTANE; DIOLANE; 1,2-HEXANEDIOL; ISOL; 2,4-PENTANEDIOL, 2-METHYL-; PINAKON; A,A,A'-TRIMETHYL; TRIMETHYLENE GLYCOL

Danger profile: Combustible, exposure limit established, poisonous gas produced in fire.
Uses: Making other chemicals; solvent; hydraulic brake fluids; printing inks; textile manufacture; fuel and lubricant additive; prevents ice formation in carburetor; cosmetics.
Appearance: Colorless liquid.
Odor: Mild, sweet, ammonia-like. The odor is a warning of exposure; however, no smell does not mean you are not being exposed.
Effects of exposure:
Breathed: May cause irritation of the eyes (with swelling and blistering of the eyelids), nose and throat. Symptoms may be delayed for several hours. Temporary blindness may occur. High exposures may cause coma and kidney damage.
Eyes: May cause irritation and burns.
Skin: Passes through the unbroken skin; can increase exposure and the severity of the symptoms listed. May cause irritation and burns.
Swallowed: May cause nausea, dizziness, headache, coma and death.
Long-term exposure: May cause liver and kidney damage.
Storage: Store in tightly closed containers in a dry, cool, well ventilated place with non-wooden floors. Keep away from heat, sources of ignition, porous materials such as wood, asbestos, cloth and rusty metal, oxidizing materials.

★ 519 ★
HEXYL TRICHLOROSILANE
CAS: 928-65-4
Other names: HEXYLTRICHLOROSILANE; SILANE, TRICHLOROHEXYL-; SILANE, HEXYLTRICHLORO-

Danger profile: Combustible, corrosive, poisonous and corrosive fumes produced on contact with water, poisonous gas produced in fire.
Uses: Making other chemicals.

Appearance: Colorless liquid. Bubbles in humid air.
Effects of exposure:
Breathed: May be poisonous. May cause irritation of the eyes, nose, throat and lungs; shortness of breath and coughing. High exposures may cause a dangerous build-up of fluids in the lungs (pulmonary edema), which can cause death. Pulmonary edema is a medical emergency that can be delayed one to two days following exposure.
Eyes: May cause severe irritation, burns and permanent damage.
Skin: May cause severe irritation and burns.
Swallowed: May cause irritation of the lips, mouth and throat; pain in swallowing; intense thirst; throat swelling; stomach cramps, nausea and vomiting, convulsions, coma and death.
Long-term exposure: May cause lung problems.
Storage: Store in tightly closed containers in a dry, cool, well ventilated place. Keep away from water and other forms of moisture. Vapors may travel to sources of ignition and flash back. Containers may explode in fire.
First aid guide: Move victim to fresh air and call emergency medical care; if not breathing, give artificial respiration; if breathing is difficult, give oxygen. In case of contact with material, immediately flush skin or eyes with running water for at least 15 minutes. Remove and isolate contaminated clothing and shoes at the site. Keep victim quiet and maintain normal body temperature.

★ 520 ★
HYDRAZINE
CAS: 302-01-2
Other names: DIAMIDE; DIAMINE; DIAMINE, HYDRAZINE BASE; HYDRAZINE, ANHYDROUS; HYDRAZYNA (POLISH)

Danger profile: Carcinogen, mutagen, poison, fire hazard, explosion hazard, highly corrosive, exposure limit established, EPA extremely hazardous substance list, Community Right-to-Know List, poisonous gas produced in fire.
Uses: Rocket propellant; agricultural chemicals; drugs; speeding up chemical reactions; metal plating on glass and plastics; solder fluxes; explosives; photographic chemical; corrosion inhibitors; diving equipment and undersea salvage; reactor cooling water.
Appearance: Colorless, fuming, oily liquid or white solid.
Odor: Ammonia-like. The odor is a warning of exposure; however, no smell does not mean you are not being exposed.
Effects of exposure:
Breathed: May be poisonous. May cause irritation of the throat and lungs; headache, dizziness, excitability, numbness in arms and legs, muscle twitching, convulsions, shallow and slow breathing, shortness of breath, bluish face and lips. High exposures may cause a dangerous build-up of fluids in the lungs (pulmonary edema), which can cause death. Pulmonary edema is a medical emergency that can be delayed one to two days following exposure.
Eyes: May cause irritation and inflammation and blistering of the eyelids. Possible permanent damage.
Skin: May be poisonous if absorbed. Passes through the unbroken skin; can increase exposure and the severity of the symptoms listed. May cause caustic-like burns; tremors, confusion, seizures, slow and shallow breathing and bluish face and lips.
Swallowed: Poisonous. May cause nausea and vomiting. See *Breathed.*
Long-term exposure: May cause cancer; liver, kidney, nervous system; red cell damage and anemia; bronchitis; skin allergy.
Storage: Extremely dangerous chemical and powerful explosive; should not be stored, handled or used without complete and exact instructions from the manufacturer. Store in sealed

containers in a dark, dry, cool, well ventilated place. Keep away from heat, sources of ignition, porous materials, oxidizing materials, strong acids, all other chemicals, especially hydrogen peroxide, and metal oxides. forms an explosive mixture with many chemicals. Containers may explode in fire.

★ 521 ★
HYDRAZINE SULFATE
CAS: 10034-93-2
Other names: DIAMINE SULFATE; HYDRAZINE HYDROGEN SULFATE; HYDRAZINE MONOSULFATE; HYDRAZINIUM SULFATE; HYDRAZONIUM SULFATE; HYDROZINE SULFATE; HS; IDRAZINA SOLFATO (ITALIAN); NSC-150014; SIRAN HYDRAZINU (CZECH)

Danger profile: Carcinogen, mutagen, poison, Community Right-to-Know List, poisonous gas produced in fire.
Uses: Making other chemicals; fungicides and germicides; refining rare metals; in blood tests.
Appearance: White or colorless, sand-like, crystalline material.
Effects of exposure:
Breathed: May be poisonous. May cause irritation of the eyes, nose and throat; dizziness, lightheadedness. High levels may cause trembling, excitation and convulsions.
Eyes: Contact may cause irritation.
Skin: May be poisonous if absorbed. Passes through the unbroken skin; can increase exposure and the severity of the symptoms listed. Contact may cause irritation.
Swallowed: Poison. May cause tiredness and sleep, nausea and vomiting. See *Breathed.*
Long-term exposure: May cause cancer; birth defects, miscarriage; liver and kidney damage; blood cell damage and anemia; brain and nervous system damage; skin allergy.
Storage: Store in tightly closed containers in a dry, cool, well-ventilated place. Keep away from oxidizers and bases. Containers may explode in fire.

★ 522 ★
HYDRIODIC ACID
CAS: 10034-85-2
Other names: ANHYDROUS HYDRIODIC ACID; HYDROGEN IODIDE; HYDROGEN IODIDE, ANHYDROUS; HYDROGEN IODIDE SOLUTION

Danger profile: Corrosive, poison, poisonous gas produced in fire.
Uses: Disinfectant; making drugs and pharmaceuticals; laboratory chemical.
Appearance: Colorless to pale yellow solution.
Effects of exposure:
Breathed: Vapor may cause intense irritation to nose, throat, mouth and lungs; mucous membrane; watering of the eyes, coughing, difficult breathing, headache, nausea, vomiting, muscular weakness, unconsciousness. Fumes may cause windpipe spasms which can be fatal. High exposure can lead to drowsiness, coma and death; buildup of fluid in the lungs (pulmonary edema), and may cause death. Pulmonary edema may be delayed by one to two days following exposure.
Eyes: Severe irritant. Contact may cause, burning and stinging, tearing, irritation and redness and swelling of the eyelids, intense pain; can burn the eyes and lead to blurred vision, chronic irritation and other permanent damage.
Skin: Severe irritant. Contact may cause burning or stinging feeling, redness, swelling, intense pain, blisters, irritation, rash. Repeated exposure may lead to skin allergy and chronic irritation.

Swallowed: May cause burning feeling of the mouth and throat, intense thirst, throat swelling, stomach cramps, nausea, vomiting, diarrhea, severe breathing difficulties and unconsciousness. Tissue damage may result. Breathing may stop. Watch for delayed symptoms.

Long-term exposure: May effect the lungs.

Storage: Store in tightly closed containers in a dry, cool, well ventilated place. Keep away from heat, water, strong acids, chemically active metals (especially phosphorus) and strong oxidizers.

First aid guide: Move victim to fresh air; call emergency medical care. In case of contact with material, immediately flush skin or eyes with running water for at least 15 minutes. Remove and isolate contaminated clothing and shoes at the site. Keep victim quiet and maintain normal body temperature.

★ 523 ★
HYDROCHLORIC ACID
CAS: 7647-01-0
Other names: ACIDE CHLORHYDRIQUE (FRENCH); ACIDO CLORIDRICO (ITALIAN); ANHYDROUS HYDROGEN CHLORIDE; CHLOORWATERSTOF (DUTCH); CHLOROHYDRIC ACID; CHLOROWODOR (POLISH); CHLORWASSERSTOFF (GERMAN); HYDROCHLORIC ACID, ANHYDROUS; HYDROCHLORIDE; HYDROGEN CHLORIDE; MURIATIC ACID; SPIRITS OF SALT

Danger profile: Mutagen, corrosive, poison, exposure limit established, Community Right-to-Know List, EPA extremely hazardous substances list, poisonous gas produced in fire.
Uses: Heavily used in industry. Making other chemicals; adjusting pH in swimming pools; metal processing; food processing; general cleaning, laboratory chemical.
Appearance: Colorless fuming gas or colorless fuming watery liquid.
Odor: Pungent, sharp, irritating. The odor is a warning of exposure; however, no smell does not mean you are not being exposed.
Effects of exposure:
Breathed: Poisonous. Vapor may cause intense irritation to nose, throat, mouth and lungs; mucous membrane. Symptoms of exposure include watering of the eyes, coughing, difficult breathing, headache, nausea, vomiting, muscular weakness, unconsciousness. High exposure can lead to drowsiness, coma and death; buildup of fluid in the lungs (pulmonary edema), and may cause death. Pulmonary edema may be delayed by one to two days following exposure.
Eyes: Severe irritant. Contact may cause, burning and stinging, tearing, irritation and redness and swelling of the eyelids, intense pain; can burn the eyes and lead to blurred vision, chronic irritation and other permanent damage.
Skin: May be poisonous if absorbed. Severe irritant. Contact may cause burning or stinging feeling, redness, swelling, intense pain, blisters, irritation, rash. Repeated exposure may lead to skin allergy and chronic irritation.
Swallowed: May cause burning feeling of the mouth and throat, intense thirst, throat swelling, stomach cramps, nausea, vomiting, diarrhea, severe breathing difficulties and unconsciousness. Tissue damage may result. Breathing may stop. Watch for delayed symptoms.
Long-term exposure: May cause birth defects; lung damage.
Storage: Store in tightly closed containers in a dry, cool, well ventilated place. Keep away from water and other forms of moisture, sulfuric acid and other chemicals. Corrosive to most metals. Containers may explode in fire.

★ 524 ★
HYDROCYANIC ACID
CAS: 74-90-8
Other names: ACIDE CYANHYDRIQUE (FRENCH); ACIDO CIANIDRICO (ITALIAN); AERO LIQUID HCN; BLAUSAEURE (GERMAN); BLAUWZUUR (DUTCH); CYAANWATERSTOF (DUTCH); CYANWASSERSTOFF (GERMAN); CYCLON; CYCLONE B; CYJANOWODOR (POLISH); HCN; HYDROGEN CYANIDE; PRUSSIC ACID; ZACLON DISCOIDS

Danger profile: Combustible, deadly poison, exposure limit established, EPA extremely hazardous substances list, Community Right-to-Know List, poisonous gas produced in fire.
Uses: Manufacture of other chemicals; dyes; rodenticide; pesticide.
Appearance: Colorless or pale blue liquid or gas.
Odor: Characteristic sweetish, bitter almond. The odor is a warning of exposure; however, no smell does not mean you are not being exposed.
Effects of exposure:
Breathed: Poisonous. May cause headache, eye, nose and throat irritation, wheezing, foaming at the mouth, general irritability, nausea; difficult breathing, shortness of breath, vomiting blood, respiratory depression, weakness, chest and abdominal pains, dizziness, diarrhea, tremors, giddiness, convulsions, shock, unconsciousness weak and irregular pulse and death. Reactions may appear hours following overexposure.
Eyes: May cause severe irritation, tearing and chemical burns.
Skin: Poisonous. Passes through the unbroken skin; can increase exposure and the severity of the symptoms listed. May cause severe irritation, redness, stinging and blisters.
Swallowed: May cause severe irritation and chemical burns to the tongue, mouth, throat and stomach as well as many of the symptoms described under *Breathed.*
Long-term exposure: May cause severe nose itch, leading to bleeding, and possibly holes in the nose.
Storage: Store in tightly closed containers in a dry, cool, well ventilated place. Keep away from heat, sources of ignition and acetaldehyde. Containers may explode in fire.
First aid guide: Move victim to fresh air and call emergency medical care; if not breathing, give artificial respiration; if breathing is difficult, give oxygen. In case of contact with material, immediately flush skin or eyes with running water for at least 15 minutes. Keep victim quiet and maintain normal body temperature. Effects may be delayed; keep victim under observation.

★ 525 ★
HYDROFLUORIC ACID
CAS: 7664-39-3
Other names: ANHYDROUS HYDROFLUORIC ACID; FLUORIC ACID; HF-A; HYDROFLUORIC ACID GAS

Danger profile: Mutagen, poison, corrosive, exposure limit established, Community Right-to-Know List, poisonous gas produced in fire.
Uses: Making other chemicals including aluminum and gasoline; fluorocarbons; etching glass.
Appearance: Clear, colorless, fuming, corrosive liquid or gas.
Effects of exposure:
Breathed: May be poisonous. May cause irritation of the nose, eyes, mouth and throat; watering eyes; salivation, coughing, throat pain, difficult breathing, headache, fatigue, nausea, stomach pains. High exposures may cause shortness of breath and a dangerous build-up of fluids in the lungs (pulmonary edema), which can cause death. Pulmonary

edema is a medical emergency that can be delayed one to two days following exposure.

Eyes: May cause intense stinging, irritation, severe burns and irreparable damage and possible loss of sight.

Skin: May be poisonous if absorbed. May cause irritation, redness, swelling and severe burns.

Swallowed: May be poisonous. May cause irritation; burning feeling in mouth and throat; swallowing causes pain; intense thirst; throat swelling, nausea, vomiting, shock. See **Breathed** for other symptoms.

Long-term exposure: May cause lung injury, bronchitis with cough, phlegm and shortness of breath; liver and kidney damage.

Storage: Store in tightly closed containers in a dry, cool, well ventilated place. Keep away from heat, metals, concrete, glass and ceramic. Containers may explode in fire.

First aid guide: Move victim to fresh air and call emergency medical care; if not breathing, give artificial respiration; if breathing is difficult, give oxygen. In case of contact with material, immediately flush skin or eyes with running water for at least 15 minutes. Remove and isolate contaminated clothing and shoes at the site. Keep victim quiet and maintain normal body temperature. Effects may delayed; keep victim under observation.

★526★
HYDROGEN

CAS: 1333-74-0

Other names: HYDROGEN, COMPRESSED; HYDROGEN, REFRIGERATED LIQUID

Danger profile: Fire hazard, explosion hazard, axphyxiant.

Uses: Making other chemicals and vegetable oils; refining petroleum and high-purity metals; welding; instrument-carrying balloons; missile fuel; research chemical.

Appearance: Colorless gas. Shipped and stored as a cryogenic liquid.

Effects of exposure:

Breathed: High levels may cause dizziness and suffocation.

Eyes: Contact may cause frostbite.

Skin: Contact may cause frostbite.

Storage: Store in tightly closed containers in a dry, cool, well ventilated place. Outdoor storage is preferred. Protect containers from physical damage. Keep away from heat, sources of ignition, oxygen, chlorine and all other chemicals. Containers may explode in heat of fire.

First aid guide: Move victim to fresh air and call emergency medical care; if not breathing, give artificial respiration; if breathing is difficult, give oxygen. In case of frostbite, thaw frosted parts with water. Keep victim quiet and maintain normal body temperature.

★527★
HYDROGEN BROMIDE

CAS: 10035-10-6

Other names: ACIDE BROMHYDRIQUE (FRENCH); ACIDO BROMIDRICO (ITALIAN); ANHYDROUS HYDROBROMIC ACID; BROMOWODOR (POLISH); BROMWASSERSTOFF (GERMAN); BROOMWATERSTOF (DUTCH); HYDROBROMIC ACID

Danger profile: Corrosive, exposure limit established, poisonous gas produced in fire and in reaction with water.

Uses: Making other chemicals, pharmaceuticals; solvent.

Appearance: Colorless liquefied compressed gas.

Odor: Sharp, pungent, irritating. The odor is a warning of exposure; however, no smell does not mean you are not being exposed.

Effects of exposure:

Breathed: May cause irritation of the eyes, nose, throat and lungs. Higher exposures may cause shortness of breath and a dangerous build-up of fluids in the lungs (pulmonary edema), which can cause death. Pulmonary edema is a medical emergency that can be delayed one to two days following exposure.

Eyes: Contact may cause severe irritation, burns and permanent damage. Liquid may cause frostbite.

Skin: Contact may cause severe irritation and burns. Liquid may cause frostbite.

Swallowed: May cause burns of the mouth, throat and stomach.

Long-term exposure: May cause chronic indigestion, lung damage and loss of the sense of smell.

Storage: Store in tightly closed containers in a dry, cool, well ventilated place. Keep away from water and other forms of moisture, oxidizers, caustics and metals. Rapidly absorbs moisture forming hydrobromic acid which is corrosive to most metals; contact with metals in the presence of moisture forms combustible hydrogen gas. Containers may explode in fire.

First aid guide: Move victim to fresh air and call emergency medical care; if not breathing, give artificial respiration; if breathing is difficult, give oxygen. In case of contact with material, immediately flush skin or eyes with running water for at least 15 minutes. Remove and isolate contaminated clothing and shoes at the site. Keep victim quiet and maintain normal body temperature. Effects may delayed; keep victim under observation.

★528★
HYDROGEN PEROXIDE

CAS: 7722-84-1

Other names: ALBONE; HYOXYL; HYDROGEN DIOXIDE; HYDROPEROXIDE; INHIBINE; OXYDOL; PERHYDROL; PERONE; PEROXAN; SUPEROXOL; T-STUFF; WASSERSTOSSPEROXIDE (GERMAN); WATERSTOFPEROXYDE (DUTCH)

Danger profile: Mutagen, corrosive, explosion hazard, exposure limit established, poisonous gas produced in fire.

Uses: Bleaching and deodorizing agent for a wide variety of products including textiles, wood pulp and paper, hair, fur, etc; rocket fuel; foam rubber; manufacture of other chemicals; electroplating; antiseptic and skin disinfectant; laboratory chemical; refining and cleaning metals; food and wine processing; seed disinfectant; sewage treatment.

Appearance: Colorless liquid or crystallized solid.

Odor: Slightly sharp. The odor is a warning of exposure; however, no smell does not mean you are not being exposed.

Effects of exposure:

Breathed: Vapor may cause intense irritation to nose, throat, mouth and lungs; mucous membrane. Symptoms of exposure include watering of the eyes, coughing, difficult breathing, headache, nausea, vomiting, muscular weakness, unconsciousness. High exposure can lead to drowsiness, coma and death; buildup of fluid in the lungs (pulmonary edema), and may cause death. Pulmonary edema may be delayed by one to two days following exposure.

Eyes: Severe irritant. Contact may cause, burning and stinging, tearing, irritation and redness and swelling of the eyelids, intense pain; can burn the eyes and lead to blurred vision, chronic irritation and permanent damage.

Skin: Severe irritant. Contact may cause burning or stinging feeling, redness, swelling, intense pain, blisters, irritation, rash. Repeated exposure may lead to skin allergy and chronic irritation.

Swallowed: May cause burning feeling of the mouth and throat, intense thirst, throat swelling, stomach cramps, nausea, vomiting, dirrhea, severe breathing difficulties and unconsciousness; irritation and injury to the mouth and throat

with possible bleeding from the throat and stomach. May produce large quantities of oxygen gas causing enlargement and severe damage to the throat and stomach. Tissue damage may result. Breathing may stop. Watch for delayed symptoms for up to a week following exposure.
Long-term exposure: May cause birth defects, miscarriage and cancer; skin rash; lung damage.
Storage: Highly reactive. Store in covered containers in a dry, cool, dark, well ventilated place with non-wood floors. A powerful oxidizer; keep away from all heat, sources of ignition, other chemicals: alcohols, glycerol, organic materials, radiant heat and sunlight, fuels, powdered metals, iron, copper, chromium, brass, bronze, lead, silver, mangamnese, and their salts. Containers may explode in fire.

★ 529 ★
HYDROGEN SULFIDE
CAS: 7783-06-4
Other names: ACIDE SULFHYDRIQUE (FRENCH); HEPATIC GAS; HYDROGENE SULFURE (FRENCH); IDROGENOSOLFORATO (ITALIAN); SCHWEFELWASSERSTOFF (GERMAN); SIARKOWODOR (POLISH); STINK DAMP; SULFURETED HYDROGEN; SULFUR HYDRIDE; SULFUR HYDROXIDE; ZWAVELWATERSTOF (DUTCH)

Danger profile: Fire hazard, exposure limit established, poisonous gas produced in fire.
Uses: To test and make other chemicals; by-product of chemical reactions such as sewage treatment.
Appearance: Colorless gas.
Odor: Rotten egg-like; but odorless at poisonous concentrations. Initial odor may be irritating, foul or absent and may deaden the sense of smell. The odor is a warning of exposure; however, no smell does not mean you are not being exposed.
Effects of exposure:
   Breathed: Poisonous. Exposure to high levels can cause sudden loss of consciousness and immediate death. Low level exposure may cause irritation of the eyes, nose and throat; eye pain, redness, blurred vision and lumg irritation; nausea, vomiting, dizziness, headache, dry cough, cold sweat, confusion, diarrhea. Higher exposures may cause shortness of breath and a dangerous build-up of fluids in the lungs (pulmonary edema), which can cause death. Pulmonary edema is a medical emergency that can be delayed one to two days following exposure.
   Eyes: May cause severe irritation, watering, redness, swelling of eyelids; permanent injury.
   Skin: May be poisonous if absorbed. May cause slight irritation, inflammation, pain; dark coloration. Liquid may cause freezing and frostbite.
   Swallowed: Unknown at this time.
Long-term exposure: May cause blurred vision and eye irritation, headaches, nausea; loss of appetite; sleep disturbances.
Storage: Store in tightly closed containers in a dry, cool, well ventilated place. Keep away from other combustable materials, oxidizers, and nitric acid. Containers may explode in fire.
First aid guide: Move victim to fresh air and call emergency medical care; if not breathing, give artificial respiration; if breathing is difficult, give oxygen. In case of contact with material, immediately flush skin or eyes with running water for at least 15 minutes. Keep victim quiet and maintain normal body temperature. Effects may be delayed; keep victim under observation.

★ 530 ★
HYDROQUINONE
CAS: 123-31-9
Other names: ARCTUVIN; P-BENZENEDIOL; 1,4-BENZENEDIOL; BENZOHYDROQUINONE; BENZOQUINOL; BLACK AND WHITE BLEACHING CREAM; 1,4-DIHYDROXY-BENZEEN (DUTCH); 1,4-DIHYDROXYBENZEN (CZECH); DI-HYDROXYBENZENE; P-DIHYDROXYBENZENE; 1,4-DIHYDROXYBENZENE; 1,4-DIHYDROXY-BENZOL (GERMAN); 1,4-DIIDROBENZENE (ITALIAN); P-DIOXOBENZENE; ELDOPAQUE; ELDOQUIN; HYDROCHINON (CZECH; POLISH); HYDROQUINOL; HYDROQUINOLE; A-HYDROQUINONE; P-HYDROQUINONE; PARA-HYDROXYPHENOL; IDROCHINONE (ITALIAN); NCI-C55834; QUINOL; B-QUINOL; TECQUINOL; TENOX HQ; USAF EK-356

Danger profile: Mutagen, combustible, exposure limit established, Community Right-to-Know List, poisonous gas produced in fire.
Uses: In photographic developer, paints, varnishes, motor fuels, oils, soil, polymers, medicine; making dyes.
Appearance: White, light tan to gray solid.
Effects of exposure:
   Breathed: May be poisonous. May cause headache, rapid and difficult breathing, shortness of breath, dizziness, mental confusion, weakness; bluish color of the lips and skin, convulsions, coma and death.
   Eyes: May cause irritation, redness, swelling of the eyelids; light causes pain; severe damage.
   Skin: May be poisonous if absorbed. Passes through the unbroken skin; can increase exposure and the severity of the symptoms listed. May cause irritation, redness, swelling, blisters.
   Swallowed: Very toxic poison. May cause irritation of the mouth, stomach; nausea, dizziness; increased respiration rate; cramps and diarrhea; ringing in the ears.
Long-term exposure: May cause birth defects, miscarriage, cancer; skin and eye discoloration; eye clouding and permanent vision damage; skin rash.
Storage: Store in tightly closed containers in a dry, cool, well ventilated place. Keep away from light, oxidizing materials and sodium hydroxide. Containers may explode in fire.
First aid guide: Move victim to fresh air; call emergency medical care. In case of contact with material, immediately flush skin or eyes with running water for at least 15 minutes. Remove and isolate contaminated clothing and shoes at the site.

I

★ 531 ★
INDENE
CAS: 95-13-6
Other names: INDONAPHTHENE

Danger profile: Combustible, exposure limit established, poisonous gas produced in fire.
Uses: Making other chemicals, plastics, varnishes and resins.
Appearance: Colorless liquid.
Effects of exposure:
   Breathed: May cause irritation of the eyes, nose and throat; headache, tight chest and difficult breathing.
   Eyes: May cause irritation, redness, tearing and inflammation of the eyelids.
   Skin: May cause irritation, redness, rash and burns.

Swallowed: May cause burns of the mouth and throat; abdominal pain, diarrhea.
Long-term exposure: May cause liver and kidney damage; skin drying and skin rash.
Storage: Store in tightly closed containers in a dry, cool, well ventilated place. Keep away from light, especially sunlight, heat and acids. Containers may explode in fire.

★ 532 ★
INDIUM
CAS: 7440-74-6

Danger profile: Exposure limit established, reproductive hazard, poisonous gas produced in fire.
Uses: In dental alloys, auto and aircraft bearings and electronic devices.
Appearance: Soft white, slightly-bluish metal.
Effects of exposure:
Breathed: May cause irritation of the nose and throat. High exposures may cause shortness of breath and a dangerous build-up of fluids in the lungs (pulmonary edema), which can cause death. Pulmonary edema is a medical emergency that can be delayed one to two days following exposure.
Eyes: May cause irritation.
Skin: May cause irritation.
Swallowed: See *Breathed.*
Long-term exposure: May cause kidney, lung, heart and liver damage; pain in joints and bones; tooth decay; nervous and gastrointestinal disorder; heart pains and general weakness.
Storage: Store in tightly closed containers in a dry, cool, well ventilated place. Keep away from sulfur, and mineral acids.

★ 533 ★
IODINE
CAS: 7553-56-2
Other names: IODE (FRENCH); IODINE CRYSTALS; IODINE SUBLIMED; IODIO (ITALIAN); JOD (GERMAN; POLISH); JOOD (DUTCH)

Danger profile: Combustible, poison, exposure limit established, poisonous gas produced in fire.
Uses: Making dyes, antiseptics, germicides, special soaps, medicine and iodized salt; process engraving and lithography; laboratory chemical.
Appearance: Bluish-black crystals with a metallic luster.
Odor: Pungent, irritating, chlorine-like. The odor is a warning of exposure; however, no smell does not mean you are not being exposed.
Effects of exposure:
Breathed: Vapor may cause irritation of the eyes, nose, throat, mouth and lungs; mucous membrane. Symptoms of exposure include watering of the eyes, coughing, difficult breathing, headache, nausea, vomiting, muscular weakness, unconsciousness. High exposure can lead to drowsiness, coma and death; buildup of fluid in the lungs (pulmonary edema), and may cause death. Pulmonary edema may be delayed by one to two days following exposure. An allergic reaction can occur with fever, rash and aching.
Eyes: Severe irritant. Contact may cause, burning and stinging, tearing, irritation and redness and swelling of the eyelids, intense pain; can burn the eyes and lead to blurred vision, chronic irritation and other permanent damage.
Skin: Severe irritant. Contact may cause burning or stinging feeling, redness, swelling, intense pain, blisters, irritation, rash. Repeated exposure may lead to skin allergy and chronic irritation.
Swallowed: Poison. May cause burning feeling of the mouth and throat, intense thirst, throat swelling, stomach cramps,

nausea, vomiting, diarrhea, severe breathing difficulties and unconsciousness. Tissue damage may result. Breathing may stop. Watch for delayed symptoms.
Long-term exposure: May cause liver problems; sore mouth, metallic or brassy taste, increased saliva, cough, acne, diarrhea, headache; acne and rash; fever; may inhibit milk production in females. chronic ingestion of large amounts may result in thyroid disease.
Storage: Store in tightly closed containers in a dry, cool, well ventilated place. Keep out of direct sunlight, away from heat and reactive or combustible materials, ammonia, acetylene, acetaldehyde, powdered aluminum and chemically active metals.

★ 534 ★
IODINE MONOCHLORIDE
CAS: 7790-99-0
Other names: IODINE CHLORIDE; PROTOCHLORURE D'IODE (FRENCH); WIJS' CHLORIDE

Danger profile: Combustible, explosion hazard, corrosive, poison, toxic and corrosive fumes produced on contact with water, poisonous gas produced in fire.
Uses: Making other chemicals; laboratory chemical.
Appearance: Black crystals or red-brown liquid.
Odor: Pungent. The odor is a warning of exposure; however, no smell does not mean you are not being exposed.
Effects of exposure:
Breathed: May be poisonous. May cause severe irritation of the nose, throat and lungs. Higher exposures may cause shortness of breath and a dangerous build-up of fluids in the lungs (pulmonary edema), which can cause death. Pulmonary edema is a medical emergency that can be delayed one to two days following exposure.
Eyes: May cause severe irritation and burns
Skin: May cause severe irritation and burns.
Swallowed: Poisonous.
Long-term exposure: May cause cough, phlegm and chronic bronchitis; skin rash.
Storage: Store in tightly closed containers in a dry, cool, well ventilated place. Keep away from water, heat and sources of ignition, organic matter, aluminum foil, cadmium sulfide, lead sulfide, phosphorus, phosphorus trichloride, potassium, rubber, silver sulfide, sodium, zinc sulfide, and other metals. Containers may explode in fire.
First aid guide: Move victim to fresh air and call emergency medical care; if not breathing, give artificial respiration; if breathing is difficult, give oxygen. In case of contact with material, immediately flush skin or eyes with running water for at least 15 minutes. Remove and isolate contaminated clothing and shoes at the site. Keep victim quiet and maintain normal body temperature. Effects may be delayed, keep victim under observation.

★ 535 ★
IODINE PENTAFLUORIDE
CAS: 7783-66-6
Other names: IODINE FLUORIDE; PENTAFLUOROIODINE

Danger profile: Combustible, poison, reproductive hazard, poisonous gas produced in fire or on contact with water or steam.
Uses: Fluorinating agent; in explosives.
Appearance: Colorless liquid.
Effects of exposure:
Breathed: May be poisonous. May cause severe irritation of the nose, throat and lungs. Higher exposures may cause shortness of breath and a dangerous build-up of fluids in the lungs (pulmonary edema), which can cause death. Pulmo-

nary edema is a medical emergency that can be delayed one to two days following exposure.
Eyes: May cause severe irritation and burns
Skin: May be poisonous if absorbed. May cause severe irritation and burns.
Swallowed: Poisonous.
Long-term exposure: May cause cough, phlegm and chronic bronchitis; skin rash.
Storage: Store in tightly closed containers in a dry, cool, well ventilated place with non-wooden floors. Keep away from water, steam, heat and sources of ignition, organic material (i.e., paper, wood, oil), fuels, and all other chemicals, benzene, potassium hydroxide, calcium carbide and metals, especially arsenic, bismuth, phosphorus, silicon, sulfur and tungsten. Attacks glass. Containers may explode in fire.
First aid guide: Move victim to fresh air and call emergency medical care; if not breathing, give artificial respiration; if breathing is difficult, give oxygen. In case of contact with material, immediately flush skin or eyes with running water for at least 15 minutes. Remove and isolate contaminated clothing and shoes at the site. Keep victim quiet and maintain normal body temperature. Effects may be delayed; keep victim under observation.

★ 536 ★
IODOFORM
CAS: 75-47-8
Other names: METHANE, TRIIODO-; NCI-C04568; TRIIODOMETHANE

Danger profile: Mutagen, exposure limit established, poisonous gas produced in fire.
Uses: Disinfectant and antiseptic; making other chemicals.
Appearance: Yellowish powder or crystalline sand-like solid.
Odor: Unpleasant. The odor is a warning of exposure; however, no smell does not mean you are not being exposed.
Effects of exposure:
  Breathed: May cause irritation of the eyes, nose and throat; headaches and nausea; effect the nervous system with symptoms of confusion, irritability, excitement, hallucinations, and poor muscular coordination; general allergic reaction with fever and rash.
  Eyes: Contact may cause irritation and burns.
  Skin: Passes through the unbroken skin; can increase exposure and the severity of the symptoms listed. May cause irritation and burns. Skin allergy may develop.
  Swallowed: Poisonous.
Long-term exposure: May cause birth defects, miscarriage and cancer; generalized allergic reaction with fever, rash; skin allergy; dermatitis, liver and kidney damage.
Storage: Store in tightly closed containers in a dry, cool, well ventilated place or in a refrigerator. Keep away from prolonged exposure to light, steam, mercuric oxide, calomel, silver nitrate, tannin, acetone and lithium.

★ 537 ★
IRON CHLORIDE
CAS: 7705-08-0
Other names: CHLORURE PERRIQUE (FRENCH); FERRIC CHLORIDE; FLORES MARTIS; IRON(III) CHLORIDE; IRON SESQUICHLORIDE; IRON TRICHLORIDE; PERCHLORURE DE FER (FRENCH)

Danger profile: Mutagen, corrosive, exposure limit established, poisonous gas produced in fire and on contact with water or steam.
Uses: Treatment of sewage and industrial wastes; etching agent for engraving, photography, and printed circuitry; dyeing; disin-

fectant; pigment in art materials; medicine; feed additive; water purification.
Appearance: Greenish, or black-brown solid.
Odor: Slightly acidic. The odor is a warning of exposure; however, no smell does mean you are not being exposed.
Effects of exposure:
  Breathed: Dust may cause severe irritation to the nose, throat and lungs, causing coughing, shortness of breath. May cause flu-like symptoms including chills, muscle ache, headache, fever. High exposure may cause nausea, salivation, vomiting, cramps, diarrhea, chest pains, cough, a build-up of fluids in the lungs (pulmonary edema) and possible death. Pulmonary edema may be delayed from one to two days following exposure.
  Skin: May cause severe irritation and thermal and acid burns, especially if skin is wet.
  Eyes: May cause severe irritation, burns, pain and possible blindness.
  Swallowed: May cause severe burns of the mouth, throat and stomach, vomiting, watery or bloody diarrhea. Damage to kidneys and liver, collapse and convulsions can result.
Long-term exposure: May reduce fertility in both males and females; may and cause birth defects, miscarriage or cancer.
Storage: Store in tightly closed containers in a dry, cool, well-ventilated place. Keep away from heat, sources of ignition, water and steam, metals, sulfuric acid, sodium, potassium and allyl chloride.
First aid guide: Move victim to fresh air; call emergency medical care. In case of contact with material, immediately flush skin or eyes with running water for at least 15 minutes. Remove and isolate contaminated clothing and shoes at the site. Keep victim quiet and maintain normal body temperature.

★ 538 ★
IRON OXIDE
CAS: 1309-37-1
Other names: ANCHRED STANDARD; ANHYDROUS IRON OXIDE; ARMENIAN BOLE; BAUXITE RESIDUE; BLACK OXIDE OF IRON; BLENDED RED OXIDES OF IRON; BURNT ISLAND RED; BURNT SIENNA; BURNT UMBER; CALCOTONE RED; CAPUT MORTUUM; C.I.77491; C.I. PIGMENT RED 101 MORTUUM; COLLOIDAL FERRIC OXIDE; DEANOX; ENGLISH RED; FERRIC OXIDE; FERRUGO; INDIAN RED; IRON(III) OXIDE; IRON OXIDE RED; IRON SESQUIOXIDE; JEWELER'S ROUGE; LEVANOX RED 130A; LIGHT RED; MANUFACTURED IRON OXIDES; MARS BROWN; MARS RED; NATURAL IRON OXIDES; NATURAL RED OXIDE; OCHRE; PRUSSIAN BROWN; RADDLE; RED IRON OXIDE; ROUGE; RUBIGO; SIENNA; SPECULAR IRON; STONE RED; SUPRA; SYNTHETIC IRON OXIDE; VENETIAN RED; VITROL RED; VOGEL'S IRON RED; YELLOW FERRIC OXIDE; YELLOW OXIDE OF IRON; ZELAZA THLENK (POLISH)

Danger profile: Spontaneously combustible, exposure limit established, poisonous gas produced in fire.
Uses: Paint pigments; polishing compounds; magnetic inks; coatings for audio and video tapes.
Appearance: Reddish or bluish-black powder.
Effects of exposure:
  Breathed: May cause metal fume fever with flu-like illness; fever, chills, aches, chest tightness, cough, and metallic taste in mouth.
  Eyes: May cause irritation and inflammation of the eyelids.
  Skin: May cause irritation and burns.
Long-term exposure: May cause chronic cough and bronchitis; metallic fume fever; permanent staining of the eyes.
Storage: Store in tightly closed containers in a dry, cool, well ventilated place. Keep away from air, powdered aluminum, car-

bon monoxide; calcium hypochlorite, hydrazine, hydrogen peroxide, ethylene oxide.

First aid guide: Move victim to fresh air; call emergency medical care. In case of contact with material, immediately flush skin or eyes with running water for at least 15 minutes. Remove and isolate contaminated clothing and shoes at the site.

★539★
IRON PENTACARBONYL
CAS: 13463-40-6
Other names: FER PENTACARBONYLE (FRENCH); IRON CARBONYL; IRON, PENTACARBONYL-; PENTACARBON-YLIRON; PENTACARBONYL IRON

Danger profile: Fire hazard, explosion hazard, poison, EPA extremely hazardous substances list, exposure limit established, poisonous gas produced in fire.
Uses: Making other chemicals and finely divided iron used in radio and television high-frequency coils.
Appearance: Oily yellow liquid.
Effects of exposure:
  Breathed: May be poisonous. May cause headache and dizziness; nausea and vomiting; cough, chest pain, shortness of breath and fever may occur 12 to 24 hours later. May cause a dangerous build-up of fluids in the lungs (pulmonary edema), which can cause death. Pulmonary edema is a medical emergency that can be delayed one to two days following exposure.
  Eyes: May cause irritation.
  Skin: May be poisonous if absorbed. Passes through the unbroken skin; can increase exposure and the severity of the symptoms listed. Contact may cause irritation.
  Swallowed: Poisonous.
Long-term exposure: May cause liver, lung and brain damage.
Storage: Store in airtight containers in a dry, cool, well ventilated place. Keep away from heat, flame, sources of ignition, water, oxidizing materials and light and air. Ignites spontaneously in air at about 130°F. Containers may explode in fire.
First aid guide: Move victim to fresh air and call emergency medical care; if not breathing, give artificial respiration; if breathing is difficult, give oxygen. In case of contact with material, immediately flush skin or eyes with running water for at least 15 minutes. Speed in removing material from skin is of extreme importance. Remove and isolate contaminated clothing and shoes at the site. Keep victim quiet and maintain normal body temperature. Effects may be delayed; keep victim under observation.

★540★
ISOAMYL ACETATE
CAS: 123-92-2
Other names: ACETIC ACID, ISOPENTYL ESTER; AMYLACETIC ESTER; BANANA OIL; 3 FEMA NO. 2055; ISOAMYL ETHANOATE; ISOPENTYL ACETATE; ISOPENTYL ALCOHOL ACETATE; 3-METHYLBUTYL ACETATE; 3-METHYL-1-BUTANOL ACETATE; 3-METHYL-1-BUTYL ACETATE; 3-METHYLBUTYL ETHANOATE; PEAR OIL

Danger profile: Fire hazard, exposure limit established, poisonous gas produced in fire.
Uses: Artificial food flavoring; perfumes; solvent; masking undesirable odors.
Appearance: Clear, colorless liquid.
Odor: Banana-like. The odor is a warning of exposure; however, no smell does not mean you are not being exposed.
Effects of exposure:
  Breathed: Mildly toxic. May cause irritation of eyes, nose and throat; headache, drowsiness, excitability, drunken behavior, mental confusion, fatigue. Very high levels of exposure may

cause dizziness, lightheadedness and cause victim to pass out, go into coma and die.
  Eyes: Contact may cause irritation, pain, burning, stinging, tearing, inflammation of the eyelids. Light may cause pain.
  Skin: May remove oil from the skin and cause dryness. Persons with pre-existing skin disorders may be more susceptible to the effects of this substance.
  Swallowed: May cause gastro-intestinal irritation. See *Breathed.*
Long-term exposure: Prolonged or repeated contact may cause irritation of the skin and dermatitis; nervous system changes; lung, kidney and liver damage.
Storage: Store in tightly closed containers in a dry, cool, well ventilated place. Keep away from heat, sources of ignition, strong oxidizers, strong alkalis and strong acids. Vapors may travel to sources of ignition and flash back. Containers may explode in fire.

★541★
ISOAMYL ALCOHOL
CAS: 123-51-3
Other names: ALCOOL AMILICO (ITALIAN); ALCOOL ISOAMYLIQUE (FRENCH); AMYLOWY ALKOHOL (POLISH); 1-BUTANOL, 3-METHYL-; FERMENTATION AMYL ALCOHOL; ISOAMYL ALKOHOL (CZECH); ISOAMYLALKOHOL (GERMAN); ISOAMYLOL; ISOBUTYLCARBINOL; ISOPENTANOL; ISOPENTYL ALCOHOL; 3-METHYL-1-BUTANOL; 2-METHYL-4-BUTANOL; 3-METHYLBUTAN-1-OL; 3-METIL-BUTANOLO (ITALIAN)

Danger profile: Combustible, possible mutagen, exposure limit established, poisonous gas produced in fire.
Uses: Photographic chemicals; making other chemicals; pharmaceutical products; solvent; laboratory chemical.
Appearance: Colorless liquid.
Odor: Pungent; alcohol-like. The odor is a warning of exposure; however, no smell does not mean you are not being exposed.
Effects of exposure:
  Breathed: May be poisonous. May cause irritation to the nose, throat and lungs; shortness of breath and coughing; dizziness, lightheadedness and loss of consciousness.
  Eyes: Contact may cause irritation, inflammation of the eyelids, severe burns and permanent damage.
  Skin: May be poisonous if absorbed. Passes through the unbroken skin; can increase exposure and the severity of the symptoms listed. May cause irritation, drying and cracking.
  Swallowed: Mildly toxic.
Long-term exposure: Long time exposure to high concentrations may be fatal. May cause skin drying and cracking. Possible lung damage.
Storage: Store in tightly closed containers in a dry, cool, well ventilated place. Keep away from heat, sources of ignition and oxidizers such as chlorine and bromine. Vapors may travel to sources of ignition and flash back. Containers may explode in fire.
First aid guide: Move victim to fresh air and call emergency medical care; if not breathing, give artificial respiration; if breathing is difficult, give oxygen. In case of contact with material, immediately flush eyes with running water for at least 15 minutes. Wash skin with soap and water. Remove and isolate contaminated clothing and shoes at the site.

★542★
ISOBUTANE
CAS: 75-28-5
Other names: PROPANE, 2-METHYL-; TRIMETHYLMETHANE

Danger profile: Fire hazard, explosion hazard, poisonous gas produced in fire.
Uses: Making other chemicals; refrigerant; motor fuels; aerosol propellant.
Appearance: Colorless liquefied gas.
Odor: Gasoline-like. The odor is a warning of exposure; however, no smell does not mean you are not being exposed.
Effects of exposure:
Breathed: High levels may cause dizziness, lightheadedness, lack of coordination and loss of consciousness. Very high levels may cause death by suffocation. Irregular heartbeat is rare but is a dangerous complication at anesthetic levels.
Eyes: Contact may cause irritation.
Skin: Contact cause irritation and frostbite.
Storage: Store in tightly closed containers in a dry, cool, well ventilated place. Keep away from heat, sources of ignition, oxygen, and oxidizing materials such as chlorine. Vapors may travel to sources of ignition and flash back. Containers may explode in fire.
First aid guide: Move victim to fresh air and call emergency medical care; if not breathing, give artificial respiration; if breathing is difficult, give oxygen. In case of frostbite, thaw frosted parts with water. Keep victim quiet and maintain normal body temperature.

★ 543 ★
ISOBUTYL ACETATE
CAS: 110-19-0
Other names: ACETATE D'ISOBUTYLE (FRENCH); ACETIC ACID, ISOBUTYL ESTER; ACETIC ACID-2-METHYLPROPYL ESTER; FEMA NO. 2175; 2-METHYLPROPYL ACETATE; 2-METHYL-1-PROPYL ACETATE; BETA-METHYLPROPYL ETHANOATE

Danger profile: Fire hazard, explosion hazard, exposure limit established, poisonous gas produced in fire.
Uses: Solvent; in lacquers, thinners, perfumes; fruit-flavoring agent.
Appearance: Colorless liquid.
Odor: Fruity. The odor is a warning of exposure; however, no smell does not mean you are not being exposed.
Effects of exposure:
Breathed: May be poisonous. May cause irritation of eyes, nose and throat; headache, drowsiness, excitability, drunken behavior, mental confusion, weakness, fatigue. Very high levels of exposure may cause dizziness, lightheadedness and cause victim to pass out, go into coma and die.
Eyes: Contact may cause irritation, pain, burning, stinging, tearing, inflammation of the eyelids. Light may cause pain.
Skin: May be poisonous if absorbed. May remove oil from the skin and cause dryness and cracking. Persons with pre-existing skin disorders may be more susceptible to the effects of this substance.
Swallowed: Mildly toxic. May cause gastro-intestinal irritation.
Long-term exposure: Prolonged or repeated contact may cause irritation of the skin with dryness and cracking; possible nervous system changes with reduced memory, personality changes, fatigue, sleep disturbance.
Storage: Store in a tightly closed container in a cool, well-ventilated place. Keep away from heat and sources of ignition strong acids, alkalies and oxidizers. Vapors may travel to sources of ignition and flash back. Containers may explode in fire.
First aid guide: Move victim to fresh air and call emergency medical care; if not breathing, give artificial respiration; if breathing is difficult, give oxygen. In case of contact with material, immediately flush eyes with running water for at least 15 minutes.

Wash skin with soap and water. Remove and isolate contaminated clothing and shoes at the site.

★ 544 ★
ISOBUTYL ALCOHOL
CAS: 78-83-1
Other names: ALCOOL ISOBUTYLIQUE (FRENCH); FEMA NO. 2179; FERMENTATION BUTYL ALCOHOL; 1-HYDROXYMETHYLPROPANE; IBA; ISOBUTANOL; ISOBUTYLALKOHOL (CZECH); ISOPRYPYLCARBINOL; 2-METHYL-1-PROPANOL; 2-METHYLPROPYL ALCOHOL; 1-PROPANOL, 2-METHYL-

Danger profile: Mutagen, fire hazard, exposure limit established, poisonous gas produced in fire.
Uses: Making other chemicals; solvent in paints and lacquers; paint removers; fruit flavor concentrates.
Appearance: Colorless liquid.
Odor: Mild, sweet and musty. The odor is a warning of exposure; however, no smell does not mean you are not being exposed.
Effects of exposure:
Breathed: May cause irritation of the eyes, nose and throat; headache, nausea, dizziness, stupor and drowsiness.
Eyes: Contact may cause severe irritation and burns.
Skin: Contact may cause irritation, rash, burning feeling; drying and cracking.
Swallowed: Moderately toxic. See *Breathed* and *First aid guide.*
Long-term exposure: May cause birth defects, miscarriage and cancer; drying and cracking of skin. Possible nerve and brain damage.
Storage: Store in tightly closed containers in a dry, cool, well ventilated place. Keep away from heat, sources of ignition, aluminum and strong oxidizers. Containers may explode in fire.
First aid guide: Move victim to fresh air and call emergency medical care; if not breathing, give artificial respiration; if breathing is difficult, give oxygen. In case of contact with material, immediately flush eyes with running water for at least 15 minutes. Wash skin with soap and water. Remove and isolate contaminated clothing and shoes at the site.

★ 545 ★
ISOBUTYRALDEHYDE
CAS: 78-84-2
Other names: FEMA NO. 2220; ISOBUTANAL; ISOBUTYL ALDEHYDE; ISOBUTYALDEHYD (CZECH); ISOBUTYRIC ALDEHYDE; 2-METHYLPROPANAL; 2-METHYL-1-PROPANAL; 2-METHYLPROPIONALDEHYDE; NCI-C60968; PROPANAL, 2-METHYL-; VALINE ALDEHYDE

Danger profile: Fire hazard, Community Right-to-Know List, poisonous gas produced in fire.
Uses: Making vitamins, perfumes, flavorings and many other chemicals.
Appearance: Transparent, colorless liquid.
Odor: Strong, pleasant gasoline-like. The odor is a warning of exposure; however, no smell does not mean you are not being exposed.
Effects of exposure:
Breathed: Vapor may cause intense irritation to nose, throat, mouth and lungs; mucous membrane. Symptoms of exposure include watering of the eyes, coughing, difficult breathing, headache, nausea, vomiting, muscular weakness, unconsciousness. High exposure can lead to drowsiness, coma and death; buildup of fluid in the lungs (pulmonary edema), and may cause death. Pulmonary edema may be delayed by one to two days following exposure.

Eyes: Severe irritant. Contact may cause, burning and stinging, tearing, irritation and redness and swelling of the eyelids, intense pain; can burn the eyes and lead to blurred vision, chronic irritation and other permanent damage.

Skin: Severe irritant. Contact may cause burning or stinging feeling, redness, swelling, intense pain, blisters, irritation, rash. Repeated exposure may lead to skin allergy and chronic irritation.

Swallowed: May cause burning feeling of the mouth and throat, intense thirst, throat swelling, stomach cramps, nausea, vomiting, diarrhea, severe breathing difficulties and unconsciousness. Tissue damage may result. Breathing may stop. Watch for delayed symptoms.

Long-term exposure: May cause lung problems; skin dryness, itching and rash; skin allergy.

Storage: Store in tightly closed containers in a dry, cool, well-ventilated place. Keep away from heat, sources of ignition, oxidizers and reducing materials. Vapors may travel to sources of ignition and flash back. Containers may explode in fire.

First aid guide: Move victim to fresh air and call emergency medical care; if not breathing, give artificial respiration; if breathing is difficult, give oxygen. In case of contact with material, immediately flush eyes with running water for at least 15 minutes. Wash skin with soap and water. Remove and isolate contaminated clothing and shoes at the site.

★546★
ISOBUTYRONITRILE
CAS: 78-82-0
Other names: ISOPROPYL CYANIDE; 2-METHYLPROPIONITRILE

Danger profile: Fire hazard, EPA extremely hazardous substances list, Community Right-to-Know List, poisonous gas produced in fire.

Uses: Making insecticides.

Breathed: Poisonous. May cause headache, eye, nose and throat irritation, wheezing, foaming at the mouth, general irritability, nausea; chest tightness, shortness of breath, vomiting blood, respiratory depression, weakness, chest and abdominal pains, dizziness, diarrhea, tremors, convulsions, shock, unconsciousness weak and irregular pulse and death. Reactions may appear hours following overexposure.

Eyes: May cause severe irritation, tearing and chemical burns.

Skin: Poisonous. Passes through the unbroken skin; can increase exposure and the severity of the symptoms listed. May cause severe irritation, redness, stinging and blisters.

Swallowed: May cause severe irritation and chemical burns to the tongue, mouth, throat and stomach as well as many of the symptoms described under Breathed. Poisonous.

Storage: Store in tightly closed containers in a dry, cool, well-ventilated place. Keep away from heat and sources of ignition. Vapors may travel to sources of ignition and flash back. Containers may explode in fire.

First aid guide: Move victim to fresh air and call emergency medical care; if not breathing, give artificial respiration; if breathing is difficult, give oxygen. In case of contact with material, immediately flush skin or eyes with running water for at least 15 minutes. Remove and isolate contaminated clothing and shoes at the site. Keep victim quiet and maintain normal body temperature. Effects may be delayed; keep victim under observation.

★547★
ISOOCTANE
CAS: 540-84-1
Other names: ISOBUTYLTRIMETHYLMETHANE; PENTANE, 2,2,4-TRIMETHYL-; 2,2,4-TRIMETHYLPENTANE

Danger profile: Mutagen, combustible, poisonous gas produced in fire.

Uses: Solvent; making other chemicals; motor fuel; used to determine the octane number of gasoline.

Appearance: Clear colorless liquid.

Odor: Gasoline-like. The odor is a warning of exposure; however, no smell does not mean you are not being exposed.

Effects of exposure:

Breathed: May be poisonous. May cause irritation of the eyes, nose, throat and lungs; headache, dizziness, nausea, reduced alertness and coordination. Higher levels may cause chemical pneumonia; loss of consciousness and stop breathing.

Eyes: Contact may cause irritation.

Skin: May be poisonous if absorbed. Contact may cause irritation.

Swallowed: May be poisonous.

Long-term exposure: May cause skin dryness and rash.

Storage: Store in tightly closed containers in a dry, cool, well-ventilated place. Keep away from heat, sources of ignition, oxidizers and reducing agents. Vapors may travel to sources of ignition and flash back. Containers may explode in fire.

First aid guide: Move victim to fresh air and call emergency medical care; if not breathing, give artificial respiration; if breathing is difficult, give oxygen. In case of contact with material, immediately flush eyes with running water for at least 15 minutes. Wash skin with soap and water. Remove and isolate contaminated clothing and shoes at the site.

★548★
ISOOCTYL ALCOHOL
CAS: 26952-21-6
Other names: 2-ETHYLHEXANOL; 2-ETHYLHEXYL ALCOHOL; 2-ETHYL-1-HEXANOL; ISOOCTANOL; 6-METHYL-1-HEPTANOL

Danger profile: Combustible, exposure limit established, poisonous gas produced in fire.

Uses: Solvent.

Appearance: Clear liquid.

Odor: Mild; characteristic. The odor is a warning of exposure; however, no smell does not mean you are not being exposed.

Effects of exposure:

Breathed: May be poisonous. May cause irritation of the eyes, nose, throat and lungs.

Eyes: Contact may cause severe irritation, burns and permanent damage.

Skin: May be poisonous if absorbed. Passes through the unbroken skin; can increase exposure and the severity of the symptoms listed. May cause irritation. Prolonged contact may cause burns.

Swallowed: Moderately toxic.

Long-term exposure: May cause lung damage.

Storage: Store in tightly closed containers in a dry, cool, well ventilated place. Keep away from heat, sources of ignition and oxidizers. Vapors may travel to sources of ignition and flash back. Containers may explode in fire.

★ 549 ★
ISOPHORONE
CAS: 78-59-1
Other names: 2-CYCLOHEXEN-1-ONE,3,5,5-TRIMETHYL-; 3,5,5-TRIMETHYL-2-CYCLOHEXENE-1-ONE; ISOACE-TOPHORONE; ISOFORON; ISOFORONE (ITALIAN); IZO-FORON (POLISH); NCI-C55618; 1,1,3- TRIMETHYL-3-CYCLOHEXENE-5-ONE; 3,5,5-TRIMETHYL-2-CYCLOHEXENE-1- ONE; 3,5,5-TRIMETHYL-2-CYCLOHEXEN-1-ONE (GERMAN); 3,5,5- TRIMETIL-2-CICLOESEN-1-ONE (ITALIAN); TRIMETHYLCYCLOHEXE-NONE

Danger profile: Mutagen, combustible, exposure limit established, poisonous gas produced in fire.
Uses: Solvent for resins; making pesticides; lacquers.
Appearance: Colorless or pale liquid.
Odor: Camphor-like. The odor is a warning of exposure; however, no smell does not mean you are not being exposed.
Effects of exposure:
> *Breathed:* May be poisonous. May cause irritation of the eyes, nose and throat; headache, nausea, drunken feeling. Higher levels may cause unconsciousness.
> *Eyes:* May cause severe irritation, burns and permanent damage.
> *Skin:* May be poisonous if absorbed. Passes through the unbroken skin; can increase exposure and the severity of the symptoms listed. May cause irritation, drying and cracking.
> *Swallowed:* May cause irritation of the mouth and stomach.

Long-term exposure: May cause birth defects, miscarriage, cancer; kidney damage; drying and cracking skin; chronic irritation of the nose. Similar chemical solvents may cause nerve and brain damage.
Storage: Store in tightly closed containers in a dry, cool, well ventilated place or in a refrigerator. Keep away from heat, sources of ignition and oxidizing materials. Vapors may travel to sources of ignition and flash back. Containers may explode in fire.

★ 550 ★
ISOPHORONE DIISOCYANATE
CAS: 4098-71-9
Other names: CYCLOHEXANE, 5-ISOCYANATO-1-(ISOCYANATOMETHYL)-1 ,3,3-TRIMETHYL- (9CI); IPDI; 3-ISOCYANATOMETHYL-3,5,5-TRIMETHYLCYCLOHEXYLISOCYANATE; ISOPHORONEDI-AMINE DIISOCYANATE; ISOPHORONEDI-ISOCYANATE; CYCLOHEXANE, 5-ISOCYANATO-1-(ISOCYANATOMETHYL)-1 ,3,3-TRIMETHYL- (9CI); IPDI; 3-ISOCYANATOMETHYL-3,5,5-TRIMETHYLCYCLOHEXYLISOCYANATE; ISOPHORONEDI-AMINE DIISOCYANATE

Danger profile: Combustible, exposure limit established, poisonous gas produced in fire.
Uses: Making paints and varnishes, foams, textile coatings and polyurethanes.
Appearance: Colorless to slightly yellow.
Effects of exposure:
> *Breathed:* Poisonous. May cause an asthma-like allergy. Very high exposures may cause shortness of breath and a dangerous build-up of fluids in the lungs (pulmonary edema), which can cause death. Pulmonary edema is a medical emergency that can be delayed one to two days following exposure.
> *Eyes:* Contact may cause severe irritation, burns and permanent damage.
> *Skin:* May be poisonous if absorbed. Passes through the unbroken skin; can increase exposure and the severity of the symptoms listed. May cause severe irritation and burns.
> *Swallowed:* May be poisonous.

Long-term exposure: May cause lung allergy with shortness of breath and cough; lung damage; skin allergy with rash and itching.
Storage: Store in tightly closed containers in a dry, cool, well ventilated place. Keep away from water, alcohol, phenols, amines, arudes, mercaptans, urethanes and ureas. Containers may explode in fire.
First aid guide: Move victim to fresh air and call emergency medical care; if not breathing, give artificial respiration; if breathing is difficult, give oxygen. In case of contact with material, immediately flush skin or eyes with running water for at least 15 minutes. Speed in removing material from skin is of extreme importance. Remove and isolate contaminated clothing and shoes at the site. Keep victim quiet and maintain normal body temperature. Effects may be delayed; keep victim under observation.

★ 551 ★
ISOPRENE
CAS: 78-79-5
Other names: 1,3-BUTADIENE, 2-METHYL-; BETA-METHYLBIVINYL; 2-METHYLBUTADIENE; 2-METHYL BUTA-DIENE; 2-METHYL-1,3-BUTADIENE

Danger profile: Mutagen, fire hazard, explosion hazard, exposure limit established, poisonous gas produced in fire.
Uses: Making synthetic natural rubber and plastics.
Appearance: Colorless liquid.
Odor: Mild, aromatic. The odor is a warning of exposure; however, no smell does not mean you are not being exposed.
Effects of exposure:
> *Breathed:* May be poisonous. May cause irritation of the eyes, nose and throat; dizziness and possible suffocation.
> *Eyes:* Contact may cause irritation.
> *Skin:* May be poisonous if absorbed. Contact may cause irritation.
> *Swallowed:* May cause irritation of the mouth and throat.

Storage: Store in tightly closed containers in a dry, cool, well ventilated place. Keep away from heat, sources of ignition, nitric acid, sulfuric acid, chlorosulfonic acid and oxidizers. Vapors may travel to sources of ignition and flash back. Containers may explode in fire.
First aid guide: Move victim to fresh air and call emergency medical care; if not breathing, give artificial respiration; if breathing is difficult, give oxygen. In case of contact with material, immediately flush eyes with running water for at least 15 minutes. Wash skin with soap and water. Remove and isolate contaminated clothing and shoes at the site.

★ 552 ★
ISOPROPOXYETHANOL
CAS: 109-59-1
Other names: DOWANOL EIPAT; ETHANOL, ISO-PROPOXY-; ETHYLENE GLYCOL ISOPROPYL ETHER; ETHYLENE GLYCOL, MONOISOPROPYL ETHER; 2-ISOPROPOXYETHANOL; BETA-HYDROXYETHYL ISOPROPYL ETHER; IPE; ISOPROPYLOXITOL; ISOPROPYL CELLOSOLVE; ISOPROPYL GLYCOL; 2-(1- METHYLETHOXY)ETHANOL; MONOISOPROPYL ETHER OF ETHYLENE GLYCOL

Danger profile: Combustible, exposure limit established, Community Right-to-Know List, poisonous gas produced in fire.
Uses: In lacquers and other coatings; solvent for dyes, resins and textiles.

Appearance: Clear liquid.
Effects of exposure:
Breathed: May cause irritation of the eyes, nose, and throat. High exposures may cause shortness of breath and a dangerous build-up of fluids in the lungs (pulmonary edema), which can cause death. Pulmonary edema is a medical emergency that can be delayed one to two days following exposure. Exposure may cause kidney, liver and red blood cell damage; bloody urine.
Eyes: Contact may cause severe irritation, burns and permanent damage.
Skin: Passes through the unbroken skin; can increase exposure and the severity of the symptoms listed. May cause severe irritation.
Swallowed: See *Breathed.*
Long-term exposure: May cause anemia, liver and kidney damage; possible lung damage.
Storage: Store in tightly closed containers in a dry, cool, well ventilated place. Keep away from heat and sources of ignition.

★ 553 ★
ISOPROPYL ACETATE
CAS: 108-21-4
Other names: ACETATE D'ISOPROPYLE (FRENCH); ACETIC ACID ISOPROPYL ETHER; ACETIC ACID-1-METHYLETHYL ESTER (9CI); 2-ACETOXYPROPANE; FEMA NO. 2926; ISOPROPILE (ACETO DI) (ITALIAN); ISOPROPYACETAAT (DUTCH); ISOPROPYLACETAT (GERMAN); ISOPROPYL (ACETATE D') (FRENCH); 2-PROPYL ACETATE

Danger profile: Fire hazard, explosion hazard, exposure limit established, poisonous gas produced in fire.
Uses: Solvent; in lacquers, thinners, printing inks, perfumes; flavoring agent.
Appearance: Colorless liquid.
Odor: Fruity, aromatic. The odor is a warning of exposure; however, no smell does not mean you are not being exposed.
Effects of exposure:
Breathed: May be poisonous. May cause irritation of eyes, nose and throat; headache, drowsiness, excitability, drunken behavior, mental confusion, fatigue. Very high levels of exposure may cause dizziness, lightheadedness and cause victim to pass out, go into coma and die.
Eyes: Contact may cause severe irritation, pain, burning, stinging, tearing, inflammation of the eyelids. Light may cause pain.
Skin: May be poisonous if absorbed. May remove oil from the skin and cause dryness. May cause severe irritation and burns. Persons with pre-existing skin disorders may be more susceptible to the effects of this substance.
Swallowed: May cause gastro-intestinal irritation. See *Breathed.*
Long-term exposure: Health effects may occur below the legal exposure limit. Prolonged or repeated contact may cause liver damage; dry and cracking skin.
Storage: Store in a tightly closed container in a cool, well-ventilated place. Keep away from heat and sources of ignition strong acids, strong alkalis and oxidizers. May dissolve many plastics. Vapors are explosive in heat or flame. Containers may explode in fire.
First aid guide: Move victim to fresh air and call emergency medical care; if not breathing, give artificial respiration; if breathing is difficult, give oxygen. In case of contact with material, immediately flush eyes with running water for at least 15 minutes. Wash skin with soap and water. Remove and isolate contaminated clothing and shoes at the site.

★ 554 ★
ISOPROPYL ALCOHOL
CAS: 67-63-0
Other names: ALCOOL ISOPROPILICO (ITALIAN); ALCOOL ISOPROPYLIQUE (FRENCH); DIMETHYLCARBINOL; ISOHOL; ISOPROPANOL; ISOPROPYLALKOHOL (GERMAN); LUTOSOL; PETROHOL; PROPAN-2-OL; 2-PROPANOL; I-PROPANOL (GERMAN); RUBBING ALCOHOL; SEC-PROPYL ALCOHOL; SPECTAR

Danger profile: Mutagen, fire hazard, exposure limit established, Community Right-to-Know List, poisonous gas produced in fire.
Uses: Solvent; making other chemicals; solvent; deicing agent for liquid fuels; preservative; lotions; rubbing alcohols.
Appearance: Colorless liquid.
Odor: Sweet. The odor is a warning of exposure; however, no smell does not mean you are not being exposed.
Effects of exposure:
Breathed: May be poisonous. May cause irritation of eyes, nose and throat; headache, drowsiness, excitability, clumsiness and drunken behavior, mental confusion, fatigue. Very high levels of exposure may cause dizziness, lightheadedness and cause victim to pass out, go into coma and die.
Eyes: Contact may cause irritation, pain, burning, stinging, tearing, inflammation of the eyelids; burns of the cornea. Light may cause pain.
Skin: May be poisonous if absorbed. May remove oil from the skin and cause dryness. Persons with pre-existing skin disorders may be more susceptible to the effects of this substance.
Swallowed: Poison. May cause gastro-intestinal irritation; drunkenness; nausea.
Long-term exposure: May cause birth defects and miscarriage; dryness and cracking of skin. Similar chemicals may cause brain or nerve damage. There is an increased incidence of sinus cancer in workers in plants manufacturing chemical.
Storage: Store in a tightly closed container in a cool, well-ventilated place. Keep away from heat and sources of ignition, methyl ethyl ketone, hydrogen peroxide, other chemicals, strong oxidizers, strong alkalis and strong acids. Vapors are explosive in heat or flame. Containers may explode in fire.
First aid guide: Move victim to fresh air and call emergency medical care; if not breathing, give artificial respiration; if breathing is difficult, give oxygen. In case of contact with material, immediately flush eyes with running water for at least 15 minutes. Wash skin with soap and water. Remove and isolate contaminated clothing and shoes at the site.

★ 555 ★
ISOPROPYLAMINE
CAS: 75-31-0
Other names: 2-AMINO-PROPAAN (DUTCH); 2-AMONOPROPAN (GERMAN); 2- AMINOPROPANE; ISOPROPILAMINA (ITALIAN); 1-METHYLETHYLAMINE; MONOISOPROPYLAMINE; 2-PROPANAMINE; 2-PROPYLAMINE

Danger profile: Fire hazard, explosion hazard, exposure limit established, poisonous gas produced in fire.
Uses: Solvent; making rubber, pharmaceuticals, dyes, insecticides, bactericides; de-hairing agent; textiles.
Appearance: Colorless, volatile liquid.
Odor: Ammonia-like. The odor is a warning of exposure; however, no smell does not mean you are not being exposed.
Effects of exposure:
Breathed: May be poisous. May cause irritation to the nose, throat and eyes; difficult breathing, coughing. Higher exposures may cause shortness of breath and a dangerous build-up of fluids in the lungs (pulmonary edema), which can cause

death. Pulmonary edema is a medical emergency that can be delayed one to two days following exposure.

Eyes: Contact may cause severe burns; possible permanent damage; blurred vision.

Skin: Passes through the unbroken skin; can increase exposure and the severity of the symptoms listed. Contact may cause severe irritation, redness, itching and burns.

Swallowed: May be poisonous. May cause burning feeling in mouth, throat and stomach; pain and swelling in throat; nausea and vomiting; stomach cramps, rapid breathing, shock, diarrhea; possible stomach perforation.

Long-term exposure: May cause weight loss; kidney and liver damage; skin irritation and itching. Similar chemicals may cause allergies.

Storage: Store in tightly closed containers in a dry, cool, well-ventilated place. Keep away from heat, sources op ignition, perchloryl fluoride, strong acids and oxidizing materials. Vapors may travel to sources of ignition and flash back. Containers may explode in fire.

First aid guide: Move victim to fresh air and call emergency medical care; if not breathing, give artificial respiration; if breathing is difficult, give oxygen. In case of contact with material, immediately flush skin or eyes with running water for at least 15 minutes. Remove and isolate contaminated clothing and shoes at the site. Keep victim quiet and maintain normal body temperature.

★ 556 ★
N-ISOPROPYLANILINE
CAS: 643-28-7
Other names: ANILINE, O-ISOPROPYL-; BENZENAMINE, 2,4-DIMETHOXY- (9CI); BENZENAMINE, N-(1-METHYLETHYL-); O-AMINOISOPROPYLBENZENE; 2-AMINOISOPROPYLBENZENE; BENZENAMINE, 2-(1-METHYLETHYL)-(9CI); O-CUMIDINE; O-ISOPROPYLANILINE; 2-ISOPROPYLANILINE

Danger profile: Combustible, exposure limit established, poisonous gas produced in fire.
Uses: Making other chemicals; dyeing acrylic fibers.
Appearance: Yellowish liquid.
Effects of exposure:
Breathed: May cause headache, lightheadedness and reduce the blood's ability to supply oxygen to the body; lips and other body areas may turn blue. High exposure may cause death.
Eyes: May cause irritation.
Skin: Passes through the unbroken skin; can increase exposure and the severity of the symptoms listed. May cause irritation.
Swallowed: See *Breathed.*
Long-term exposure: May cause headaches, dizziness, loss of appetite; anemia.
Storage: Store in tightly closed containers in a dry, cool, well ventilated place. Keep away from heat and sources of ignition. Containers may explode in fire.

★ 557 ★
ISOPROPYL ETHER
CAS: 108-20-3
Other names: DIISOPROPYLETHER; DIISOPROPYL ETHER; DIISOPROPYL OXIDE; ETHER, ISOPROPYL; ETHER ISOPROPYLIQUE (FRENCH); 2-ISOPROOXYPROPANE; IZOPROPYLOWYETER (POLISH); DI-ISOPROPYL ETHER; 2-ISOPROPOXY PROPANE; DI-ISOPROPYLETHER; PROPANE, 2,2'- OXYBIS-

Danger profile: Fire hazard, explosion hazard, exposure limit established, poisonous gas produced in fire.
Uses: Solvent.
Appearance: Colorless liquid.
Odor: Sharp, sweet, ether-like. The odor is a warning of exposure; however, no smell does not mean you are not being exposed.
Effects of exposure:
Breathed: May be poisonous. May cause irritation of the nose and throat; drowsiness, dizziness, headache and nausea. Higher levels may cause loss of consciousness and even death.
Eyes: Contact may cause irritation.
Skin: May be poisonous if absorbed. Contact may cause irritation. Repeated contact with skin will remove natural oils and may cause dermatitis.
Swallowed: Moderately toxic.
Long-term exposure: May cause dermatitis, with dry, cracking skin.
Storage: Store in tightly closed containers in a dry, cool, well-ventilated detached place. Keep away from direct sunlight, heat, sources of combustion, oxidizers, chlorosulfonic acid, nitric acid, and propionyl chloride. Vapors may travel to sources of ignition and flash back. Containers may explode in fire.
First aid guide: Move victim to fresh air and call emergency medical care; if not breathing, give artificial respiration; if breathing is difficult, give oxygen. In case of contact with material, immediately flush eyes with running water for at least 15 minutes. Wash skin with soap and water. Remove and isolate contaminated clothing and shoes at the site.

★ 558 ★
ISOPROPYL ISOCYANATE
CAS: 1795-48-8
Other names: PROPANE, 2-ISOCYANATO-

Danger profile: Combustible, may react with water, poisonous gas produced in fire.
Uses: Making other chemicals and pesticides.
Appearance: Colorless liquid.
Odor: Strong. The odor is a warning of exposure; however, no smell does not mean you are not being exposed.
Effects of exposure:
Breathed: May be poisonous. May cause severe irritation of the eyes, nose throat and lungs. Lung allergy may develop. High exposures may cause shortness of breath and a dangerous build-up of fluids in the lungs (pulmonary edema), which can cause death. Pulmonary edema is a medical emergency that can be delayed one to two days following exposure.
Eyes: May cause severe irritation.
Skin: May be poisonous if absorbed. May cause severe irritation.
Swallowed: May be poisonous.
Long-term exposure: May cause lung allergy and damage.
Storage: Store in tightly closed containers in a dry, cool, well ventilated place. Keep away from heat and sources of ignition. Vapors may travel to sources of ignition and flash back. Containers may explode in fire.
First aid guide: Move victim to fresh air and call emergency medical care; if not breathing, give artificial respiration; if breathing is difficult, give oxygen. In case of contact with material, immediately flush skin or eyes with running water for at least 15 minutes. Remove and isolate contaminated clothing and shoes at the site. Keep victim quiet and maintain normal body temperature. Effects may be delayed; keep victim under observation.

★ 559 ★
ISOTHIOUREA
CAS: 62-56-6
Other names: CARBAMIDE, THIO-; PSEUDOTHIOUREA; SULOUREA; THIOCARBAMATE; THIOCARBAMIDE; 2-THIOUREA; THIOUREA; THU; TSIZP 34; USAF EK-497

Danger profile: Carcinogen, mutagen, poison; combustible, Community Right-to-Know List, poisonous gas produced in fire.
Uses: In photography and photocopying papers, other chemicals, pesticides, dyes, drugs, hair preparations, rubber, laboratory chemical; mold inhibitor.
Appearance: Colorless crystalline solid.
Effects of exposure:
  Breathed: Poisonous. High exposure of a related chemical may cause shortness of breath and a dangerous build-up of fluids in the lungs (pulmonary edema), which can cause death. Pulmonary edema is a medical emergency that can be delayed one to two days following exposure. It is not known if this chemical will cause the same effect.
  Eyes: May cause irritation.
  Skin: May cause irritation and allergic skin eruptions.
  Swallowed: Poisonous. Bitter taste. May cause hemmorrhage.
Long-term exposure: May cause cancer, mutations, goiter, skin allergy; bone marrow damage; anemia. Possible lung effects.
Storage: Store in tightly closed containers in a dry, cool, well ventilated place. Keep away from nitric acid, acryladehyde and hydrogen peroxide.

K

★ 560 ★
KAOLIN
CAS: 1332-58-7
Other names: ARGILLA; BOLUSALBA; CHINA CLAY; PORCELAIN CLAY; WHITE BO

Danger profile: Combustible, exposure limit established, poisonous gas produced in fire.
Uses: Filler and coatings for paper and rubber, refractories; ceramics; cements; fertilizers; making other chemicals; cosmetics; insecticides; paint.
Appearance: Fine white, yellowish-white or white powder.
Effects of exposure:
  Breathed: May cause irritation of the nose, coughing and sneezing.
  Eyes: May cause irritation, redness, tearing, pain and inflammation of the eyelids.
Storage: Store in tightly closed containers in a dry, cool, well ventilated place. Containers may explode in fire.

★ 561 ★
KEPONE
CAS: 143-50-0
Other names: CHLORDECONE; CIBA 8514; COMPOUND 1189; DECACHLOROKETONE; DECACHLOROOCTAHYDRO-1,3,4-METHENO-2H-CYCLOBUTA(C,D)-PENTALEN-2-ON E; ENT 16,391; GENERAL CHEMICAL 1189; MEREX; NCI-C00191; 1,3,4- METHENO-2H-CYCLOBUTA (CD) PENTALEN-2-ONE,1,1A,3,3A,4,5,5A,5B,6- DECACHLORO-OCTAHYDRO

Danger profile: Carcinogen, mutagen, combustible, reproductive hazard, poisonous gas produced in fire.
Uses: Insecticide; fungicide. Registration suspended by U.S. Environmental Protection Agency.
Appearance: Tan to white crystalline material.
Effects of exposure:
  Breathed: May be poisonous. May cause brain and nerve damage with headache, anxiety and nervousness, tremor, poor coordination, slurred speech, muscle twitching, poor memory, mood changes, muscle weakness.
  Eyes: May cause irritation.
  Skin: May be poisonous if absorbed. Passes through the unbroken skin; can increase exposure and the severity of the symptoms listed. May cause irritation and rash.
  Swallowed: Poison. See **Breathed.**
Long-term exposure: May cause cancer; damage the fetus and cause reproductive damage; liver and kidney damage; brain and nervous system damage with hyperactivity, hyperexcitability, muscle spasms, tremors; low sperm count, sterility, breast enlargement; skin changes.
Storage: Store in tightly closed containers in a dry, cool, well ventilated place. Keep away from strong acids and acid fumes. Containers may explode in fire.

★ 562 ★
KEROSENE
CAS: 8008-20-6
Other names: COAL OIL; KEROSINE; ILLUMINATING OIL; RANGE OIL; A FUEL OIL NO. 1; JET FUEL; JP-1

Danger profile: Mutagen, combustible, exposure limit established, poisonous gas produced in fire.
Uses: Rocket and jet engine fuel; domestic heating; solvent; insecticidal sprays; diesel and tractor fuels.
Appearance: Colorless or pale yellow watery liquid.
Odor: Fuel oil. The odor is a warning of exposure; however, no smell does not mean you are not being exposed.
Effects of exposure:
  Breathed: May be poisonous. May cause irritation of the eyes and nose. If liquid gets into lungs or if victim exposed to high exposures may cause a dangerous build-up of fluids in the lungs (pulmonary edema), which can cause death. Pulmonary edema is a medical emergency that can be delayed one to two days following exposure.
  Eyes: May cause irritation.
  Skin: May be poisonous if absorbed. Passes through the unbroken skin; can increase exposure and the severity of the symptoms listed.
  Swallowed: May cause stomach irritation, hallucinations, coughing, nausea, vomiting and fever.
Long-term exposure: May cause muscular weakness, anemia and changes in the white blood cells.
Storage: Store in tightly closed containers in a dry, cool, well ventilated place, preferably outdoors. Keep away from heat, sources of ignition, combustibles and oxidizing materials. Containers may explode in fire.
First aid guide: Move victim to fresh air and call emergency medical care; if not breathing, give artificial respiration; if breathing is difficult, give oxygen. In case of contact with material, immediately flush eyes with running water for at least 15 minutes. Wash skin with soap and water. Remove and isolate contaminated clothing and shoes at the site.

★ 563 ★
KETENE
CAS: 463-51-4
Other names: CARBOMETHANE; CARBOMETHENE; ETHENOID; ETHENONE; KETEN; KETOETHYLENE

Danger profile: Exposure limit established, poisonous gas produced in fire.
Uses: Making many other commercially important chemicals.
Appearance: Colorless gas.
Odor: Pungent, sharp, disagreeable. The odor is a warning of exposure; however, no smell does not mean you are not being exposed.
Effects of exposure:
Breathed: May cause irritation of the eyes, nose, throat and lungs, tears and runny nose, coughing, sore throat, difficult breathing; lethargy, fatigue, and drowsiness; slurred speech, shaking, and dizziness. Higher exposures could cause shortness of breath, headache, nausea, vomiting, dizziness, diarrhea and a dangerous build-up of fluids in the lungs (pulmonary edema), which can cause death. Pulmonary edema is a medical emergency that can be delayed one to two days following exposure.
Eyes: May cause irritation, redness, tearing, pain, inflamed eyelids, and eye injury; burns in corneas and possible loss of sight.
Skin: Toxic. Passes through the unbroken skin; can increase exposure and the severity of the symptoms listed. May cause painful irritation, redness, and blisters. Repeated or prolonged skin contact may cause irritation of the skin and may also cause the legs to feel numb.
Swallowed: May cause irritation of the mouth and throat; stomach and abdominal pain; loss of muscular coordination and mental confusion; convulsions. Also see symptoms for *Breathed.*
Long-term exposure: May cause permanent lung damage; emphysema; loss of the sense of smell.
Storage: Store in tightly closed containers in a dry, cool, well ventilated place. Keep away from heat, water and organic compounds.

L

★ 564 ★
LEAD
CAS: 7439-92-1
Other names: C.I. 77575; C.I. PIGMENT METAL 4; GLOVER; KS-4; LEAD FLAKE; LEAD S2; LITHARGE; OLOW (POLISH); OMAHA; OMAHA & GRANT; SI; SO

Danger profile: Carcinogen, mutagen, poison, combustible, exposure limit established, reproductive hazard, Community Right-to-Know List, poisonous gas produced in fire.
Uses: Widely used due to properties of softness, high density, resistance to corrosion, etc. Pigment for paint; solder and fusible alloys; storage batteries; gasoline additive; radiation shielding; cable covering; ammunition; foil; bearing alloys.
Appearance: Bluish-white to silvery grey, heavy, soft metal.
Effects of exposure:
Breathed: Poisonous. The effects of exposure may not develop immediately. Dust and fumes may cause wheezing, foaming at the mouth, headache; decreased physical fitness, fatigue, sleep disturbance, aching bones and muscles, constipation, abdominal pains and loss of appetite. These effects may be reversible if exposure ceases. High exposure may lead to convulsions, coma and death.
Eyes: Dust and fumes may cause irritation, redness and inflammation of the eyelids. Passes through the mucous membrane of the eye increasing severity and symptoms listed in *Breathed* and *Swallowed.*
Skin: Dust and fumes may cause irritation and inflammation.

Swallowed: Poisonous. May cause metallic taste, disturbed and rapid respiration and circulation, mental confusion, increased heart beat, irritation of the stomach and intestines, stomach pain, nausea and vomiting, diarrhea, weakness, convulsions, coma and death.
Long-term exposure: May cause birth defects and miscarriage; decrease of fertility in males and females. Lead is a cumulative poison; it collects in the body over a period of time. Long-term exposure may lead to a build-up of lead in the body to a point where symptoms and disability appear; high blood pressure; tiredness; mood changes (depression and irritability); sleeping disturbance; kidney damage; brain damage; weakness and decreased hand grip, paralysis of the wrist joints.
Storage: Store in tightly closed containers in a dry, cool, well-ventilated place. Keep away from potassium and other chemically active metals, oxidizers and excessive heat.

★ 565 ★
LEAD ARSENATE
CAS: 7784-40-9
Other names: ACID LEAD ARSENATE; ACID LEAD ARSENITE; ACID LEAD ORTHOARSENATE; ARSENATE OF LEAD; ARSENIC ACID, LEAD(2); ARSENIC ACID, LEAD SALT; ARSINETTE; DIBASIC LEAD ARSENATE; GYPSINE; LEAD ACID ARSENATE; ORTHO L10 DUST; ORTHO L40 DUST; SALT ARSENATE OF LEAD; SCHULTENITE; SECURITY; SOPRABEL; STANDARD LEAD ARSENATE; TALBOT

Danger profile: Carcinogen, possible reproductive hazard, exposure limit established, Community Right-to-Know List, poisonous gas produced in fire.
Uses: Insecticides; herbicide.
Appearance: White crystals
Effects of exposure:
Breathed: Poisonous. The effects of exposure may not develop immediately. Dust and fumes may cause irritation of the nose, throat and breathing passages; wheezing, foaming at the mouth, headache; decreased physical fitness, fatigue, sleep disturbance, aching bones and muscles, constipation, abdominal pains and loss of appetite; vomiting, muscle cramps. May effect the heart and cause an abnormal EKG. These effects may be reversible if exposure ceases. High exposure may lead to convulsions, coma and death.
Eyes: Dust and fumes may cause irritation, redness and inflammation of the eyelids. Passes through the mucous membrane of the eye increasing severity and symptoms listed in *Breathed* and *Swallowed.*
Skin: Dust and fumes may cause irritation and inflammation.
Swallowed: Poisonous. May cause metallic taste, disturbed and rapid respiration and circulation, mental confusion, increased heart beat, irritation of the stomach and intestines, stomach pain, nausea and vomiting, diarrhea, weakness, convulsions, coma and death.
Long-term exposure: May cause skin, lung and liver cancer; reproductive damage; skin rash; nerve, kidney and brain damage; nose ulcers and sores; high blood pressure. Lead is a cumulative poison; it collects in the body over a period of time. Long-term exposure may lead to a build-up of lead in the body to a point where symptoms and disability appear; kidney damage; decreased hand grip, paralysis of the wrist joints; miscarriage and birth defects. It may take years for the body to get rid of excess lead.
Storage: Store in tightly closed containers in a dry, cool, well-ventilated place. Keep away from potassium and other chemically active metals, oxidizers and excessive heat.
First aid guide: Move victim to fresh air; call emergency medical care. In case of contact with material, immediately flush skin or eyes with running water for at least 15 minutes. Remove and isolate contaminated clothing and shoes at the site.

★566★
LEAD ARSENITE
CAS: 10031-13-7
Other names: LEAD(II) ARSENITE

Danger profile: Carcinogen, reproductive hazard, combustible, exposure limit established, Community Right-to-Know List, poisonous gas produced in fire.
Uses: Insecticide.
Appearance: White powder.
Effects of exposure:
Breathed: Poisonous. The effects of exposure may not develop immediately. Dust and fumes may cause irritation of the nose, throat and breathing passages; wheezing, foaming at the mouth, headache; decreased physical fitness, fatigue, sleep disturbance, aching bones and muscles, constipation, abdominal pains and loss of appetite; vomiting, muscle cramps. May effect the heart and cause an abnormal EKG. These effects may be reversible if exposure ceases. High exposure may lead to convulsions, coma and death.
Eyes: Dust and fumes may cause irritation, redness and inflammation of the eyelids. Passes through the mucous membrane of the eye increasing severity and symptoms listed in *Breathed* and *Swallowed.*
Skin: Dust and fumes may cause irritation and inflammation.
Swallowed: Poisonous. May cause metallic taste, disturbed and rapid respiration and circulation, mental confusion, increased heart beat, irritation of the stomach and intestines, stomach pain, nausea and vomiting, diarrhea, weakness, convulsions, coma and death.
Long-term exposure: May cause skin, lung and liver cancer; reproductive damage; skin rash; nerve, kidney and brain damage; nose ulcers and sores; high blood pressure. Lead is a cumulative poison; it collects in the body over a period of time. Long-term exposure may lead to a build-up of lead in the body to a point where symptoms and disability appear; kidney damage; decreased hand grip, paralysis of the wrist joints; miscarriage and birth defects. It may take years for the body to get rid of excess lead.
Storage: Store in tightly closed containers in a dry, cool, well-ventilated place. Keep away from potassium and other chemically active metals, oxidizers and excessive heat.
First aid guide: Move victim to fresh air; call emergency medical care. In case of contact with material, immediately flush skin or eyes with running water for at least 15 minutes. Remove and isolate contaminated clothing and shoes at the site.

★567★
LEAD CHROMATE
CAS: 7758-97-6
Other names: CROCOITE; CHROMIC ACID, LEAD (2) SALT (1:1); CANARY CHROME YELLOW 40-2250; CHROME GREEN; CHROME YELLOW; C.I. 77600; COLOGNE YELLOW; KING'S YELLOW; LEAD CHROMATE (VI); LEIPZIG YELLOW; LEMON YELLOW; PARIS YELLOW; PLUMBOUS CHROMATE; CROCOITE

Danger profile: Carcinogen, mutagen, combustible, exposure limit established, Community Right-to-Know List, poisonous gas produced in fire.
Uses: Pigment in industrial paints, rubber, plastics, ceramic coatings; laboratory chemical.
Appearance: Yellow or orange-yellow crystalline substance or powder.
Effects of exposure:
Breathed: Poisonous. The effects of exposure may not develop immediately. May cause irritation of the nose, throat and bronchial tubes; wheezing, foaming at the mouth, headache; decreased physical fitness, fatigue, sleep disturbance,

aching bones and muscles, constipation, abdominal pains and loss of appetite. May cause a hole in the inner nose. These effects may be reversible if exposure ceases. High exposure may lead to convulsions, coma and death.
Eyes: Dust and fumes may cause irritation, redness and inflammation of the eyelids. Passes through the mucous membrane of the eye increasing severity and symptoms listed in *Breathed* and *Swallowed.*
Skin: May cause irritation, rash and deep ulcers; allergic skin rash.
Swallowed: Poisonous. May cause metallic taste, disturbed and rapid respiration and circulation, mental confusion, increased heart beat, irritation of the stomach and intestines, stomach pain, nausea and vomiting, diarrhea, weakness, convulsions, coma and death.
Long-term exposure: May cause cancer; allergic skin rash. Lead is a cumulative poison; it accumulates in the body over a period of time. Long-term exposure may lead to a build-up of lead in the body to a point where symptoms and disability appear; kidney damage; weakness and decreased hand grip, paralysis of the wrist joints; loss of appetite, upset stomach, headache, irritability. Similar lead compounds may cause miscarriage and birth defects.
Storage: Store in tightly closed containers in a dry, cool, well-ventilated place. Keep away from oxidizers, potassium and other chemically active metals.

★568★
LEAD CYANIDE
CAS: 592-05-2
Other names: C.I. 77610; C. I. PIGMENT YELLOW 48; CYANURE DE PLOMB (FRENCH); LEAD(II) CYANIDE

Danger profile: Reproductive hazard, exposure limit established, Community Right-to-Know List, poisonous gas produced in fire.
Uses: Metallurgy.
Appearance: White to yellowish powder.
Effects of exposure:
Breathed: May cause irritation of the nose, throat and breathing passages; headache, pounding heart; poor appetite, upset stomach, weakness, irritability and aching joints and muscles. Symptoms may be delayed. High exposure to cyanide may cause death without warning.
Eyes: May cause irritation.
Skin: Passes through the unbroken skin; can increase exposure and the severity of the symptoms listed. May cause irritation.
Swallowed: May be poisonous.
Long-term exposure: May cause birth defects by damaging the fetus; reduced fertility; kidney damage; enlarged thyroid; skin rash. Also see entries for LEAD and CYANIDE entries.
Storage: Store in tightly closed containers in a dry, cool, well-ventilated place. Keep away from magnesium and other chemically active metals, oxidizers and heat. Containers may explode in fire.
First aid guide: Move victim to fresh air; call emergency medical care. In case of contact with material, immediately flush skin or eyes with running water for at least 15 minutes. Remove and isolate contaminated clothing and shoes at the site.

★569★
LEAD FLUOBORATE
CAS: 13814-96-5
Other names: BORATE (1-), TETRAFLUORO-; LEAD (2); LEAD TETRAFLUOROBORATE; TETRAFLUORO BORATE; TETRAFLUORO BORATE (1-) LEAD (2)

Danger profile: Reproductive hazard, combustible, exposure limit established, Community Right-to-Know List, poisonous gas produced in fire.

Uses: Metal finishing.

Appearance: Colorless liquid.

Odor: Faint. The odor is a warning of exposure; however, no smell does not mean you are not being exposed.

Effects of exposure:

Breathed: May be poisonous. Repeated exposure may cause substance to build-up in the body. Low levels may cause headache, tiredness, moodiness, stomach problems, constipation and sleeping disturbances.

Eyes: May cause irritation; possible damage.

Skin: May cause irritation and burns.

Swallowed: May be poisonous. May cause gastrointestinal problems, colic, constipation; weakness and possible paralysis of wrist and ankle muscles. Large amounts may cause local irritation of alimentary tract; pain, leg cramps, muscle weakness, depression, coma, and death may follow in one or two days. See *Breathed* and **First aid guide.**

Long-term exposure: May cause birth defects by damaging the fetus; brain damage; kidney damage; anemia; high blood pressure. Lead buildup in the body may cause symptoms similar to those under *Breathed.*

Storage: Store in tightly closed containers in a dry, cool, well-ventilated place. Keep away from heat, sources of ignition, oxidizers and chemically active metals. Will corrode most metals.

First aid guide: Move victim to fresh air; call emergency medical care. In case of contact with material, immediately flush skin or eyes with running water for at least 15 minutes. Remove and isolate contaminated clothing and shoes at the site.

★ 570 ★
LEAD FLUORIDE
CAS: 7783-46-2

Other names: LEAD DIFLUORIDE; LEAD(II) FLUORIDE; PLOMB FLUORURE (FRENCH); PLUMBOUS FLUORIDE

Danger profile: Reproductive hazard, exposure limit established, Community Right-to-Know List, poisonous gas produced in fire.

Uses: Electronic and optical products.

Appearance: White solid; white to colorless crystals.

Effects of exposure:

Breathed: May be poisonous. Repeated exposure may cause substance to build-up in the body. Low levels may cause headache, tiredness, moodiness, stomach problems, constipation and sleeping disturbances.

Eyes: May cause irritation; possible damage.

Skin: May cause irritation and burns.

Swallowed: May be poisonous. May cause gastrointestinal problems, colic, constipation; weakness and possible paralysis of wrist and ankle muscles. Large amounts may cause local irritation of alimentary tract; pain, leg cramps, muscle weakness, depression, coma, and death may follow in one or two days. See *Breathed* and **First aid guide.**

Long-term exposure: May cause birth defects by damaging the fetus; brain damage; kidney damage; anemia; high blood pressure. Lead buildup in the body may cause symptoms similar to those under *Breathed.*

Storage: Store in tightly closed containers in a dry, cool, well-ventilated place. Keep away from heat, sources of ignition, oxidizers and chemically active metals.

First aid guide: Move victim to fresh air; call emergency medical care. In case of contact with material, immediately flush skin or eyes with running water for at least 15 minutes. Remove and isolate contaminated clothing and shoes at the site.

★ 571 ★
LEAD IODIDE
CAS: 10101-63-0

Danger profile: Reproductive hazard, exposure limit established, Community Right-to-Know List, poisonous gas produced in fire.

Uses: Bronzing; gold pencils; printing; photography; cloud seeding.

Appearance: Bright yellow, heavy powder.

Effects of exposure:

Breathed: Poisonous. The effects of exposure may not develop immediately. Dust and fumes may cause wheezing, foaming at the mouth, headache; decreased physical fitness, fatigue, sleep disturbance, aching bones and muscles, constipation, abdominal pains and loss of appetite. These effects may be reversible if exposure ceases. High exposure may lead to convulsions, coma and death.

Eyes: Dust and fumes may cause irritation, redness and inflammation of the eyelids. Passes through the mucous membrane of the eye increasing severity and symptoms listed in *Breathed* and *Swallowed.*

Skin: Dust and fumes may cause irritation and inflammation.

Swallowed: Poisonous. May cause metallic taste, disturbed and rapid respiration and circulation, mental confusion, increased heart beat, irritation of the stomach and intestines, stomach pain, nausea and vomiting, diarrhea, weakness, convulsions, coma and death.

Long-term exposure: Lead is a cumulative poison; it collects in the body over a period of time. Long-term exposure may lead to a build-up of lead in the body to a point where symptoms and disability appear; kidney damage; decreased hand grip, paralysis of the wrist joints; miscarriage and birth defects.

Storage: Store in tightly closed containers in a dry, cool, well-ventilated place. Keep away from potassium and other chemically active metals and excessive heat.

★ 572 ★
LEAD NITRATE
CAS: 10099-74-8

Other names: LEAD DINITRATE; LEAD (2) NITRATE; LEAD-(II) NITRATE; NITRATE DE PLOMB (FRENCH); NITRIC ACID, LEAD(2) SALT

Danger profile: Mutagen, combustible, exposure limit established, Community Right-to-Know List, poisonous gas produced in fire.

Uses: Dyeing and printing fabrics; making matches fireworks and explosives; in photography; tanning; engraving and lithography.

Appearance: Colorless, white solid.

Effects of exposure:

Breathed: The effects of exposure may not develop immediately. Dust and fumes may cause wheezing, foaming at the mouth, headache; decreased physical fitness, fatigue, sleep disturbance, aching bones and muscles, constipation, abdominal pains and loss of appetite. These effects may be reversible if exposure ceases. High exposure may lead to convulsions, coma and death.

Eyes: May cause irritation, redness and inflammation of the eyelids. Passes through the mucous membrane of the eye increasing severity and symptoms listed in *Breathed* and *Swallowed.*

Skin: Poisonous. Passes through the unbroken skin; can increase exposure and the severity of the symptoms listed. May cause irritation and inflammation.

Swallowed: Poisonous. May cause metallic taste, disturbed and rapid respiration and circulation, mental confusion, increased heart beat, irritation of the stomach and intestines, stomach pain, nausea and vomiting, diarrhea, weakness, convulsions, coma and death.

Long-term exposure: May cause birth defects and miscarriage. Lead is a cumulative poison; it collects in the body over a period of time. Long-term exposure may lead to a build-up of lead in the body to a point where symptoms and disability appear; kidney damage; decreased hand grip, paralysis of the wrist joints.

Storage: Store in tightly closed containers in a dry, cool, well-ventilated place. Keep away from fuels, heat, potassium and other chemically active metals, carbon, lead, combustible materials, oxidizers. Containers may explode in fire.

First aid guide: Move victim to fresh air; call emergency medical care. In case of contact with material, immediately flush skin or eyes with running water for at least 15 minutes. Remove and isolate contaminated clothing and shoes at the site.

★ 573 ★
LEAD(II) OLEATE
CAS: 1120-46-3
Other names: OLEIC ACID LEAD SALT; OLEIC ACID, LEAD(2) SALT (2:1)

Danger profile: Combustible, exposure limit established, Community Right-to-Know List, poisonous gas produced in fire.
Uses: In lacquers and varnishes; drier in paints; high pressure lubricant.
Appearance: White powder or ointment-like granules.
Effects of exposure:
Breathed: May be poisonous. May cause restlessness, trembling hands, headache, nausea and vomiting; insomnia; irritation of the nose, tremor, restlessness, delirium, convulsions.
Eyes: May cause irritation, watering and inflammation of the eyelids.
Skin: May be poisonous if absorbed. May cause irritation with itching and blisters.
Swallowed: May be poisonous.
Storage: Store in tightly closed containers in a dry, cool, well-ventilated place.

★ 574 ★
LEAD OXIDE, BROWN
CAS: 1309-60-0
Other names: BIOXYDE DE PLOMB (FRENCH); C.I.77580; LEAD BROWN; LEAD DIOXIDE; LEAD(IV) OXIDE; LEAD PEROXIDE; LEAD SUPEROXIDE; BIOXYDE DE PLOMB (FRENCH); PEROXYDE DE PLOMB (FRENCH)

Danger profile: Combustible, exposure limit established, Community Right-to-Know List, poisonous gas produced in fire.
Uses: Lead-acid storage batteries; making elastomers; dyeing textiles; matches, fireworks and explosives; laboratory chemical.
Appearance: Brown hexagonal crystals or dark-brown powder.
Effects of exposure:
Breathed: Poisonous. The effects of exposure may not develop immediately. Dust and fumes may cause wheezing, foaming at the mouth, headache; decreased physical fitness, fatigue, sleep disturbance, aching bones and muscles, constipation, abdominal pains and loss of appetite. These effects may be reversible if exposure ceases. High exposure may lead to convulsions, coma and death.
Eyes: Dust and fumes may cause irritation, redness and inflammation of the eyelids. Passes through the mucous membrane of the eye increasing severity and symptoms listed in *Breathed* and *Swallowed.*
Skin: Dust and fumes may cause irritation and inflammation.

Swallowed: Poisonous. May cause metallic taste, disturbed and rapid respiration and circulation, mental confusion, increased heart beat, irritation of the stomach and intestines, stomach pain, nausea and vomiting, diarrhea, weakness, convulsions, coma and death.
Long-term exposure: Lead is a cumulative poison; it collects in the body over a period of time. Long-term exposure may lead to a build-up of lead in the body to a point where symptoms and disability appear; kidney damage; decreased hand grip, paralysis of the wrist joints; miscarriage and birth defects.
Storage: Store in tightly closed containers in a dry, cool, well-ventilated place. Keep away from potassium and chemically active metals, oxidizers and excessive heat.

★ 575 ★
LEAD OXIDE, RED
CAS: 1314-41-6
Other names: C.I. 77578; C.I. PIGMENT RED 105; DILEAD(-II) LEAD (IV) OXIDE; GOLD SATINOBRE; LEAD ORTHO-PLUMBATE; LEAD TETRAOXIDE; MINERAL ORANGE; MINERAL RED; MINIUM; MINIUM NON SETTING RL-95; ORANGE LEAD; PARIS RED; PLUMBOPLUMBIC OXIDE; RED LEAD; RED LEAD OXIDE; SANDIX; SATURN RED; TRILEAD TETROXIDE

Danger profile: Combustible, exposure limit established, Community Right-to-Know List, poisonous gas produced in fire.
Uses: Ceramic glazes, pottery and enameling; varnish; protective paints for metal; packing for pipe joints.
Appearance: Bright red powder.
Effects of exposure:
Breathed: Poisonous. The effects of exposure may not develop immediately. Dust and fumes may cause wheezing, foaming at the mouth, headache; decreased physical fitness, fatigue, sleep disturbance, aching bones and muscles, constipation, abdominal pains and loss of appetite. These effects may be reversible if exposure ceases. High exposure may lead to convulsions, coma and death.
Eyes: Dust and fumes may cause irritation, redness and inflammation of the eyelids. Passes through the mucous membrane of the eye increasing severity and symptoms listed in *Breathed* and *Swallowed.*
Skin: Dust and fumes may cause irritation and inflammation.
Swallowed: Poisonous. May cause metallic taste, disturbed and rapid respiration and circulation, mental confusion, increased heart beat, irritation of the stomach and intestines, stomach pain, nausea and vomiting, diarrhea, weakness, convulsions, coma and death.
Long-term exposure: Lead is a cumulative poison; it collects in the body over a period of time. Long-term exposure may lead to a build-up of lead in the body to a point where symptoms and disability appear; kidney damage; decreased hand grip, paralysis of the wrist joints; miscarriage and birth defects.
Storage: Store in tightly closed containers in a dry, cool, well-ventilated place. Keep away from potassium, combustable materials and excessive heat.

★ 576 ★
LEAD PHOSPHATE
CAS: 7446-27-7
Other names: BLEIPHOSPHAT (GERMAN); C.I. 77622; LEAD ORTHOPHOSPHATE; LEAD PHOSPHATE (3:2); LEAD(2) PHOSPHATE; NORMAL LEAD ORTHOPHOSPHATE; PERLEX PASTE; PHOSPHORIC ACID, LEAD SALT; PHOSPHORIC ACID, LEAD(2) SALT (2:3); PLUMBOUS PHOSPHATE; TRILEAD PHOSPHATE

Danger profile: Carcinogen, reproductive hazard, exposure limit established, Community Right-to-Know List, poisonous gas produced in fire.
Uses: Making plastics.
Appearance: Colorless or white powder.
Effects of exposure:
Breathed: The effects of exposure may not develop immediately. Dust and fumes may cause wheezing, foaming at the mouth, headache; decreased physical fitness, fatigue, sleep disturbance, aching bones and muscles, constipation, abdominal pains and loss of appetite. These effects may be reversible if exposure ceases. High exposure may lead to convulsions, coma and death.
Eyes: Dust and fumes may cause irritation, redness and inflammation of the eyelids. Passes through the mucous membrane of the eye increasing severity and symptoms listed in *Breathed* and *Swallowed*.
Skin: Dust and fumes may cause irritation and inflammation.
Swallowed: Poisonous. May cause metallic taste, disturbed and rapid respiration and circulation, mental confusion, increased heart beat, irritation of the stomach and intestines, stomach pain, nausea and vomiting, diarrhea, weakness, convulsions, coma and death.
Long-term exposure: Lead is a cumulative poison; it collects in the body over a period of time. Long-term exposure may lead to a build-up of lead in the body to a point where symptoms and disability appear; kidney damage; decreased hand grip, paralysis of the wrist joints; miscarriage and birth defects.
Storage: Store in tightly closed containers in a dry, cool, well-ventilated place. Keep away from potassium and excessive heat.

★ 577 ★
LEAD STEARATE
CAS: 7428-48-0
Other names: BLEISTEARAT (GERMAN); OCTADECANOIC ACID, LEAD SALT; STEARIC ACID, LEAD SALT

Danger profile: Combustible, exposure limit established, Community Right-to-Know List, poisonous gas produced in fire.
Uses: Drier in varnishes and lacquers; high-pressure lubricants; component of greases, waxes and paints.
Appearance: White powder.
Effects of exposure:
Breathed: May be poisonous. May cause restlessness, joint and muscle pain; trembling hands, headache, nausea and vomiting; insomnia; irritation of the nose, tremor, restlessness, delirium, convulsions.
Eyes: May cause irritation, watering and inflammation of the eyelids.
Skin: May cause irritation with itching and blisters.
Swallowed: Poisonous. May cause abdominal pain, diarrhea, constipation, loss of appetite, muscular weakness, headache, blue line on gums, metallic taste, nausea, and vomiting. See *Breathed*.
Storage: Store in tightly closed containers in a dry, cool, well-ventilated place. Keep away from oxidizers and chemically active metals.

★ 578 ★
LEAD SUBACETATE
CAS: 1335-32-6
Other names: BASIC LEAD ACETATE; BIS(ACETO)DIHYDROXYTRILEAD; BIS(ACETATO)TETRAHYDROXYTRILEAD; BLA; LEAD ACETATE, BASIC; LEAD, BIS(ACETATO-O)TETRAHYDROXYTRI-; LEAD MONOSUBACETATE; MONOBASIC LEAD ACETATE; SUB-ACETATE LEAD

Danger profile: Mutagen, combustible, exposure limit established, Community Right-to-Know List, poisonous gas produced in fire.
Uses: Decolorizing agent.
Appearance: White powder.
Effects of exposure:
Breathed: Poisonous. The effects of exposure may not develop immediately. Dust and fumes may cause wheezing, foaming at the mouth, headache; decreased physical fitness, fatigue, sleep disturbance, aching bones and muscles, constipation, abdominal pains and loss of appetite. These effects may be reversible if exposure ceases. High exposure may lead to convulsions, coma and death.
Eyes: Dust and fumes may cause irritation, redness and inflammation of the eyelids. Passes through the mucous membrane of the eye increasing severity and symptoms listed in *Breathed* and *Swallowed*.
Skin: Dust and fumes may cause irritation and inflammation.
Swallowed: Poisonous. May cause metallic taste, disturbed and rapid respiration and circulation, mental confusion, increased heart beat, irritation of the stomach and intestines, stomach pain, nausea and vomiting, diarrhea, weakness, convulsions, coma and death.
Long-term exposure: May cause miscarriage or birth defects. Lead is a cumulative poison; it accumulates in the body over a period of time. Long-term exposure may lead to a build-up of lead in the body to a point where symptoms and disability appear; kidney damage; decreased hand grip, paralysis of the wrist joints.
Storage: Store in tightly closed containers in a dry, cool, well-ventilated place. Keep away from potassium and other chemically active chemicals and excessive heat.

★ 579 ★
LEAD SULFATE
CAS: 7446-14-2
Other names: ANGLISITE; BLEISULFAT (GERMAN); C.I. 77630; C.I. PIGMENT WHITE 3; FAST WHITE; FREEMANS WHITE LEAD; LANARKITE; LEAD BOTTOMS; LEAD(II) SULFATE (1:1); MILK WHITE; MULHOUSE WHITE; SULFATE DE PLOMB (FRENCH); SULFURIC ACID, LEAD (2) SALT (1:1); WHITE LEAD

Danger profile: Mutagen, combustible, corrosive, exposure limit established, Community Right-to-Know List, poisonous gas produced in fire.
Uses: Storage batteries; in lithography; rapidly drying oil varnishes; paint pigments.
Appearance: Heavy, white crystalline powder.
Effects of exposure:
Breathed: Poisonous. The effects of exposure may not develop immediately. Dust and fumes may cause wheezing, foaming at the mouth, headache; decreased physical fitness, fatigue, sleep disturbance, aching bones and muscles, constipation, abdominal pains and loss of appetite. These effects may be reversible if exposure ceases. High exposure may lead to convulsions, coma and death.
Eyes: Dust and fumes may cause irritation, redness,inflammation of the eyelids and burns.
Skin: May cause irritation and burns.
Swallowed: Poisonous. May cause metallic taste, disturbed and rapid respiration and circulation, mental confusion, increased heart beat, irritation of the stomach and intestines, stomach pain, nausea and vomiting, diarrhea, weakness, convulsions, coma and death.
Long-term exposure: Lead is a cumulative poison; it collects in the body over a period of time. Long-term exposure may lead to a build-up of lead in the body to a point where symptoms and

disability appear; kidney damage; decreased hand grip, paralysis of the wrist joints; miscarriage and birth defects.
Storage: Store in tightly closed containers in a dry, cool, well-ventilated place. Keep away from potassium and other chemically active metals, oxidizers and heat.
First aid guide: Move victim to fresh air; call emergency medical care. In case of contact with material, immediately flush skin or eyes with running water for at least 15 minutes. Remove and isolate contaminated clothing and shoes at the site. Keep victim quiet and maintain normal body temperature.

★580★
LEAD SULFIDE
CAS: 1314-87-0
Other names: C.I. 77640; GALENA; NATURAL LEAD SULFIDE; PLUMBOUS SULFIDE

Danger profile: Combustible, exposure limit established, Community Right-to-Know List, poisonous gas produced in fire.
Uses: In ceramics and glazes; infrared radiation detector; semiconductors.
Appearance: Silvery, metallic crystalline material or a black powder.
Effects of exposure:
Breathed: Poisonous. The effects of exposure may not develop immediately. Dust and fumes may cause wheezing, foaming at the mouth, headache; decreased physical fitness, fatigue, sleep disturbance, aching bones and muscles, constipation, abdominal pains and loss of appetite. These effects may be reversible if exposure ceases. High exposure may lead to convulsions, coma and death.
Eyes: Dust and fumes may cause irritation, redness and inflammation of the eyelids. Passes through the mucous membrane of the eye increasing severity and symptoms listed in *Breathed* and *Swallowed.*
Skin: Dust and fumes may cause irritation, inflammation, pain and burns.
Swallowed: Poisonous. May cause metallic taste, disturbed and rapid respiration and circulation, mental confusion, increased heart beat, irritation of the stomach and intestines, stomach pain, loss of appetite, nausea and vomiting, diarrhea, weakness, convulsions, coma and death.
Long-term exposure: May cause birth defects or miscarriage. Lead is a cumulative poison; it accumulates in the body over a period of time. Long-term exposure may lead to a build-up of lead in the body to a point where symptoms and disability appear; kidney damage; decreased hand grip, paralysis of the wrist joints.
Storage: Store in tightly closed containers in a dry, cool, well-ventilated place. Keep away from oxidizers, potassium and other chemically active metals and excessive heat.

★581★
LEPTOPHOS
CAS: 21609-90-5
Other names: ABAR; O-(4-BROMO-2,5-DICHLOROPHENYL)O-METHYL PHENYLPHOSPHONOTHIOATE; O-(2,5-DICHLORO-4-BROMOPHENYL) O-METHYL PHENYLTHIOPHOSPHONATE; FOSVEL; K62-105; MBCP; O-METHYL-O-(4- BROMO-2,5-DICHLOROPHENYL)PHENYL THIOPHOSPHONATE; NK 711; PHENYLPHOSPHONOTHIOIC ACID O-(4-BROMO-2,5-BROMO-2,5- DICHLOROPHENYL)O-METHYL ESTER; PHOSPHONOTHIOIC ACID, PHENYL-,O-(4-BROMO-2,5-DICHLOROPHENYL)O-METHYL ESTER; PHOSVEL; V.C.S.; VCS-506; VELSICOL 506; VELSICOL VCS 506

Danger profile: Mutagen, combustible, EPA extremely hazardous substances list; poisonous gas produced in fire.
Uses: In insecticides (organophosphate).
Appearance: White crystalline or colorless amorphous solid; the technical product is a light tan powder.
Effects of exposure:
Breathed: Extremely poisonous. May cause headache, dizziness, profuse sweating, nausea, vomiting, reduced heart beat, stomach cramps, diarrhea, loss of coordination, slow and weak breathing, fever, loss of consciousness and death. May cause shortness of breath and a dangerous build-up of fluids in the lungs (pulmonary edema), which can cause death. Pulmonary edema is a medical emergency that can be delayed one to two days following exposure.
Eyes: Rapidly absorbed. Poisoning can happen very rapidly. See *Breathed.*
Skin: Poisonous. Passes through the unbroken skin; can increase exposure and the severity of the symptoms listed.
Swallowed: Powerful, deadly poison. See *Breathed.*
Long-term exposure: May damage the developing fetus or cause birth defects; nerve damage including loss of coordination; liver damage. Repeated exposure may cause depression, irritability and personality changes.
Storage: Store in a regulated, specially marked place in tightly closed containers in a dry, well-ventilated place. Keep away from water, strong alkalis. Containers may explode in fire.

★582★
LETHANE
CAS: 112-56-1
Other names: 2-(2-BUTOXYETHOXY)ETHYL THIOCYANATE; 2-(2-(BUTOXY)ETHOXY)ETHYLTHIOCYANIC ACID ESTER; 2-(2-BUTOXYETHOXY)ETHYLTHIOKYANAT (CZECH); BUTOXYRHODANODIETHYL ETHER; BETA-BUTOXY-BETA'-THIOCYANODIETHYL ETHER; 2-BUTOXY-2'-THIOCYANODIETHYL ETHER; 1-BUTOXY-ALPHA-(2-THIOCYANOETHOXY)ETHANE; 1-BUTOXY-2-(2-THIOCYANOETHOXY)ETHANE; BUTYL CARBITOL RHODANATE; BUTYL CARBITOL THIOCYANATE; ENT 6; ETHANE, 1-BUTOXY-2-(2-THIOCYANATOETHOXY)-; ETHANOL, 2-(2-BUTOXYETHOXY)-, THIOCYANATE; LETHANE 384; THIOCYANIC ACID,2-(2-BUTOXYETHOXY)ETHYL ESTER

Danger profile: Combustible, poisonous, poisonous gas produced in fire.
Uses: Insecticide.
Appearance: Liquid.
Effects of exposure:
Breathed: May cause rapid, severe poisoning from the formation of cyanide in the body. High exposure may cause drowsiness, twitching, labored breathing, vomiting, blue lips (cyanosis), convulsions, coma and death.
Eyes: May cause irritation.
Skin: May be poisonous if absorbed. Passes through the unbroken skin; can increase exposure and the severity of the symptoms listed. May cause irritation.
Swallowed: May be poisonous.
Long-term exposure: May cause liver and kidney damage.
Storage: Store in tightly closed containers in a dry, cool, well-ventilated place. Containers may explode in fire.

★ 583 ★
LINDANE
CAS: 58-89-9
Other names: AALINDAN; AFICIDE; AGRISOL G-20; AGRO-CIDE; AGRONEXIT; AMEISENATOD; AMEISENMITTEL MERCK; APARASIN; APHTIRIA; APLIDAL; ARBITEX; BBH; BEN-HEX; BENTOX 10; GAMMA-BENZENE HEXACHLO-RIDE; BEXOL; BHC; GAMMA-BHC; CELANEX; CHLORE-SENE; CODECHINE; DBH; DETMOL-EXTRAKT; DETOX 25; DEVORAN; DOL GRANULE; DRILL TOX-SPEZIA AGLUKON; ENT 7,796; ENTOMOXAN; EXAGAMA; FORLIN; GALLOGA-MA GAMACID; GAMAPHEX; GAMENE; GAMMA-HEXACHLORO-CYCLOHEXANE; GAMMA-BHC; GAM-MAHEXA; GAMMAHEXANE; GAMMALIN; GAMMALIN 20; GAMMATERR; GAMMEX; GAMMEXANE; GAMMOPAZ; GEX-ANE; HCCH; HCH; GAMMA-HCH; HECLOTOX; HEXA; HEX-ACHLORAN; GAMMA-HEXACHLORAN; HEXACHLORANE; GAMMA-HEXACHLORANE; GAMMA-HEXACHLOROBENZENE; 1-ALPHA,2-ALPHA,3-BETA,4-ALPHA,5-ALPHA,6-BETA-HEXACHLOROCYCLOHEX ANE; GAMMA-HEXACHLOROCYCLOHEXANE; GAMMA-1,2,3,4,5,6-HEXACHLOROCYCLOHE XANE; HEXACHLORO-CYCLOHEXANE (GAMMA ISOMER) 1,2,3,4,5,6-HEXACHLOR-CYCLOHEXANE; HEXACHLOROCYCLOHEX-ANE, GAMMA-ISOMER; 1,2,3,4,5,6-HEXACHLOROCYCLOHEXANE, GAMMA-ISOMER; HEXA-TOX; HEXAVERM; HEXICIDE; HEXYCLAN; HGI; HORTEX; INEXIT; ISOTOX; JACUTIN; KOKOTINE; KWELL; LENDINE; LENTOX; LIDENAL; LINDAFOR; LINDAGAM; LINDAGRAIN; LINDAGRANOX; GAMMA-LINDANE; LINDAPOUDRE; LINDA-TOX; LINDOSEP; LINTOX; LOREXANE; MILBOL 49; MSZY-COL; NCI-C00204; NEO-SCABICIDOL; NEXEN FB; NEXIT; NEXIT-STARK; NEXOL-E; NICOCHLORAN; NOVIGAM; OM-NITOX; OVADZIAK; OWADZIAK; PEDRACZAK; PFLANZOL; QUELLADA; SANG GAMMA; SILVANOL; SPRITZ-RAPIDIN; SPRUEHPFLANZOL; STREUNEX; TAP 85; TRI-6; VITON

Danger profile: Carcinogen, mutagen, combustible, exposure limit established, Community Right-to-Know List, poisonous gas produced in fire.
Uses: Pesticide.
Appearance: Colorless solid.
Odor: Musty. The odor is a warning of exposure; however, no smell does not mean you are not being exposed.
Effects of exposure:
  Breathed: May cause headache, dizziness; excessive irrita-bility, confusion, anxiety, and excitability; weakness, numb arms and legs; shallow breathing, convulsions, blue skin and lips, coma and death.
  Skin: Poisonous. May cause irritation. Absorption through the unbroken skin is significant and may cause or increase sever-ity of symptoms listed in *Breathed* and *Swallowed*.
  Eyes: May cause irritation and redness. See symptoms in *Breathed* and *Swallowed*.
  Swallowed: Poisonous if absorbed. May cause nausea, vomiting, excessive irritability, convulsions rapid and irregular heartbeat, and death.
Long-term exposure: May cause cancer, damage to the devel-oping fetus and reduce fertility in females; serious drop in blood cell count, or in the white blood cell count; possible liver dam-age.
Storage: Store in tightly closed containers in a dry, cool, well-ventilated place. Keep away from light, powdered metal and strong alkalis. Containers may explode in fire.
First aid guide: Move victim to fresh air and call emergency medical care; if not breathing, give artificial respiration; if breath-ing is difficult, give oxygen. In case of contact with material, im-mediately flush skin or eyes with running water for at least 15 minutes. Speed in removing material from skin is of extreme im-portance. Remove and isolate contaminated clothing and shoes at the site. Keep victim quiet and maintain normal body tempera-ture. Effects may be delayed; keep victim under observation.

★ 584 ★
LIQUIFIED PETROLEUM GAS
CAS: 68476-85-7
Other names: BOTTLED GAS; LIQUIFIED HYDROCARBON GAS; LPG

Danger profile: Fire hazard, exposure limit established, poison-ous gas produced in fire.
Uses: Domestic and industrial fuel; making other chemicals; metal cutting, welding and brazing.
Appearance: Colorless liquefied compressed gas.
Odor: Although odorless, commercial product has skunk-like odorant added as a warning. The odor is a warning of exposure; however, no smell does not mean you are not being exposed.
Effects of exposure:
  Breathed: May cause headache, dizziness, fatigue; rapid ir-regular breathing; nausea, vomiting, confusion, loss of con-sciousness and death from suffocation.
  Eyes: Contact with liquid may cause frostbite, pain, stinging, tearing, inflammation of the eyelids and permanent damage.
  Skin: Contact with liquid may cause frostbite, intense cold feeling, skin turns white and feels cold and hard.
  Swallowed: Liquid may cause headache, confusion, drowsi-ness, unconsciousness and death.
Storage: Store in tightly closed containers in a dry, cool, well-ventilated place. Keep away from heat, sources of ignition and strong oxidizers. Vapors may travel to sources of ignition and flash back. Containers may explode in fire.

★ 585 ★
LITHIUM
CAS: 7439-93-2
Other names: LITHIUM METAL

Danger profile: Mutagen, combustible solid, explosion hazard, poisonous gas produced in fire.
Uses: Making storage batteries; metal alloys.
Appearance: Grayish-white to silvery colored metal.
Effects of exposure:
  Breathed: May be poisonous. May cause irritation of nose, eyes and throat; coughing; difficult breathing; loss of appetite, nausea, muscle twitching, convulsions, coma and death. A dangerous build-up of fluids in the lungs (pulmonary edema) may occur. This can cause death. Pulmonary edema is a medical emergency that can be delayed one to two days fol-lowing exposure.
  Skin: Contact may cause irritation, redness and swelling; burning feeling; burns.
  Eyes: Contact may cause caustic irritation, pain, tearing, burns and serious injury may result.
  Swallowed: May cause burning feeling in the mouth and stomach; mouth and gastric irritation; stomach cramps.
Long-term exposure: May cause lithium poisoning with loss of appetite, nausea, muscle twitching, apathy, shaking, convul-sions, coma and death; allergic skin, lung and other effects; lung irritation.
Storage: Store in tightly closed containers in a dry, cool, well-ventilated place with non-wooden floors. Keep away from water or other forms of moisture, oxidizers, nitric acid and halogenated compounds. May ignite combustible material if they are wet. Containers may explode in fire.

★ 586 ★
LITHIUM ALUMINUM HYDRIDE
CAS: 16853-85-3
Other names: ALUMINUM LITHIUM HYDRIDE; LITHIUM AL-ANATE; LITHIUM ALUMINOHYDRIDE; LITHIUM ALUMINUM TETRAHYDRIDE; LITHIUM TETRAHYDROALUMINATE

Danger profile: Combustible, explosion hazard, exposure limit established, reproductive hazard, poisonous gas produced in fire, dangerous when wet.
Uses: Making pharmaceuticals and drugs, perfumes, polymers and other chemicals; source of hydrogen; propellant.
Appearance: White to gray lumps or powder.
Effects of exposure:
    Breathed: May be poisonous. May cause irritation of the eyes, nose, throat, air passages and lungs; coughing and sneezing. Higher exposures may cause shortness of breath and a dangerous build-up of fluids in the lungs (pulmonary edema), which can cause death. Pulmonary edema is a medical emergency that can be delayed one to two days following exposure.
    Eyes: Contact may cause irritation and burns.
    Skin: Contact may cause irritation and burns.
    Swallowed: May cause burning feeling in the mouth and stomach; mouth and gastric irritation; stomach cramps.
Long-term exposure: May cause lithium poisoning with loss of appetite, nausea, muscle twitching, apathy, shaking, convulsions, coma and death; damage to the nasal septum; possible lung damage.
Storage: Store in tightly closed containers in a dry, cool, well-ventilated place. Keep away from water and all other forms of moisture, air, acids, alcohols, benzoyl peroxide, boron trifluoride etherate, ethers, tetrahydrofuran and oxidizers. Containers may explode in fire.
First aid guide: Move victim to fresh air; call emergency medical care. Wipe material from skin immediately; flush skin or eyes with running water for at least 15 minutes. Remove and isolate contaminated clothing and shoes at the site.

★ 587 ★
LITHIUM CARBONATE
CAS: 554-13-2
Other names: CAMCOLIT; CANDAMIDE; CARBOLITH; CARBONIC ACID, DILITHIUM SALT; CARBONIC ACID LITHIUM SALT; CEGLUTION; CP-15467-61; DILITHIUM CARBONATE; ESKALITH; HYPNOREX; LIMAS; LISKONUM; LITHANE; LITHICARB; LITHINATE; LITHOBID; LITHONATE; LITHOTABS; NSC-16895; PLENUR; PRIADEL; QUILONUM RETARD

Danger profile: Mutagen, combustible, possible reproductive hazard, poisonous gas produced in fire.
Uses: Ceramics and porcelain glazes; pharmaceuticals; luminescent paints, varnishes and dyes; glass ceramics; making aluminum and other chemicals.
Appearance: White crystalline powder.
Effects of exposure:
    Breathed: May cause irritation of nose, eyes and throat; coughing; difficult breathing.
    Skin: Contact with wet skin may cause irritation, redness and swelling; burning feeling.
    Eyes: Contact may cause irritation, pain, tearing; serious injury may result.
    Swallowed: May cause burning feeling in the mouth and stomach; mouth and gastric irritation; stomach cramps; shaking, muscle weakness, increased urine volume.
Long-term exposure: May cause birth defects and developmental abnormalities.

Storage: Store in tightly closed containers in a dry, cool, well-ventilated place. Keep away from fluorine. Containers may explode in fire.

★ 588 ★
LITHIUM HYDRIDE
CAS: 7580-67-8
Other names: HYDRURE DE LITHIUM (FRENCH)

Danger profile: Fire hazard, exposure limit established, EPA extremely hazardous substance list, poisonous gas produced in fire.
Uses: Making other chemicals, desiccant.
Appearance: Gray or blue crystals, or white powder. Darkens rapidly on exposure to light; white translucent crystals.
Effects of exposure:
    Breathed: May be poisonous. May cause irritation of nose, eyes and throat; coughing, sneezing, and burning of nose and throat; difficult breathing.
    Eyes: Contact may cause severe irritation, pain, tearing; serious injury may result; caustic burns, blurred vision, dizziness, weakness, tremors, ringing in the ears.
    Skin: Contact with wet skin may cause severe irritation, redness and swelling; burning feeling; caustic burns.
    Swallowed: May cause burning feeling in the mouth and stomach; mouth and gastric irritation; stomach cramps; severe burns of mouth and stomach; symptoms of central nervous system damage may occur.
Storage: Store in tightly closed containers in a dry, cool, well-ventilated place. Keep away from heat, sources ignition, water and other forms of moisture and liquid oxygen. Dangerous when wet. Powder, whether dry or wet ignites spontaneously in air. Containers may explode in fire.
First aid guide: Move victim to fresh air; call emergency medical care. Wipe material from skin immediately; flush skin or eyes with running water for at least 15 minutes. Remove and isolate contaminated clothing and shoes at the site.

★ 589 ★
LITHIUM HYDROXIDE MONOHYDRATE
CAS: 1310-66-3
Other names: LITHIUM HYDROXIDE

Danger profile: Highly corrosive and reactive, poisonous gas produced in fire.
Uses: Making other chemicals, photographic developer; alkaline storage batteries.
Appearance: Small crystals.
Effects of exposure:
    Breathed: May cause irritation of the eyes, nose, throat, air passages and lungs; sore throat, nose bleeding, coughing and sneezing. Higher exposures may cause shortness of breath and a dangerous build-up of fluids in the lungs (pulmonary edema), which can cause death. Pulmonary edema is a medical emergency that can be delayed one to two days following exposure.
    Eyes: May cause severe irritation, burns, permanent damage and blindness.
    Skin: May cause severe irritation and burns.
    Swallowed: May cause severe attack of membranes in digestive tract and stomach, salivation, loss of appetite, nausea, vomiting, diarrhea.
Long-term exposure: May cause lithium poisoning with loss of appetite, nausea, shaking, muscle twitching, apathy, convulsions, coma and death; chronic nose irritation; possible lung damage.

Storage: Store in tightly closed containers in a dry, cool, well-ventilated place. Keep away from aluminum, zinc, strong acids and oxidizers.

First aid guide: Move victim to fresh air; call emergency medical care. In case of contact with material, immediately flush skin or eyes with running water for at least 15 minutes. Remove and isolate contaminated clothing and shoes at the site. Keep victim quiet and maintain normal body temperature.

★ 590 ★
LITHIUM NITRATE
CAS: 7790-69-4
Other names: NITRIC ACID, LITHIUM SALT

Danger profile: Corrosive, poisonous gas produced in fire.
Uses: Ceramics; fireworks; salt baths; refrigeration systems; rocket propellants.
Appearance: Colorless powder often found in water solution.
Effects of exposure:
 Breathed: Lithium poisoning may develop with loss of appetite, nausea, muscle twitching, apathy, shaking, convulsions, coma and death.
 Eyes: Contact with powder or solution may cause severe irritation and burns.
 Skin: Contact may cause severe irritation and burns.
 Swallowed: May cause burning feeling in the mouth and stomach; mouth and gastric irritation; stomach cramps; nausea, vomiting, diarrhea.
Long-term exposure: May cause lithium poisoning with loss of appetite, nausea, muscle twitching, apathy, shaking, convulsions, coma and death.
Storage: Store in tightly closed containers in a dry, cool, well-ventilated place with non-wooden floors. Keep away from wood, paper and other combustable materials, oil and other fuels, heat and oxidizers. Containers may explode in fire.
First aid guide: Move victim to fresh air; call emergency medical care. In case of contact with material, immediately flush skin or eyes with running water for at least 15 minutes. Remove and isolate contaminated clothing and shoes at the site.

M

★ 591 ★
MAGNESIUM
CAS: 7439-95-4
Other names: MAGNESIO (ITALIAN); MAGNESIUM METAL; MAGNESIUM PELLETS; MAGNESIUM POWDER; MAGNESIUM RIBBONS; MAGNESIUM SCALPINGS; MAGNESIUM SHAVINGS; MAGNESIUM SHEET; MAGNESIUM TURNINGS

Danger profile: Combustible, poison, explosion hazard, exposure limit established, dangerous when wet, poisonous gas produced in fire.
Uses: Making structural parts, die cast automotive parts, precision instruments, optical mirror, missiles.
Appearance: Silvery-white metal.
Effects of exposure:
 Breathed: Dust may cause irritation of the eyes, throat, air passages; cough may develop.
 Eyes: Dust may cause irritation of the eyes.
 Skin: Penetration may cause local irritation, blisters and slow-healing sores and ulcers that may become infected.
 Swallowed: May cause upset stomach.
Long-term exposure: Repeated exposure may cause chemical to be collected in the body and cause upset stomach.

Storage: Store in tightly closed containers in a dry, cool, well-ventilated place. Keep away from sources of ignition, strong oxidizers, strong acids, ethylene oxide, potassium carbonate, chlorine trifluoride. Dangerous when wet. Powders form an explosive mixture in air. Containers may explode in fire.
First aid guide: Move victim to fresh air and call emergency medical care; if not breathing, give artificial respiration; if breathing is difficult, give oxygen. Wipe material from skin immediately; flush skin or eyes with running water for at least 15 minutes. Remove and isolate contaminated clothing and shoes at the site. Keep victim quiet and maintain normal body temperature.

★ 592 ★
MAGNESIUM CHLORATE
CAS: 10326-21-3
Other names: CHLORATE SALT OF MAGNESIUM; CHLORIC ACID, MAGNESIUM; CHLORIC ACID, DE-FOL-ATE; E-Z-OFF; MAGNESIUM SALT; MAGNESIUM DICHLORATE; MAGRON; MC DEFOLIANT; ORTHO MC

Danger profile: Increases intensity of fire, poisonous gas produced in fire.
Uses: Defoliant; drying agent.
Appearance: White powder or crystals.
Effects of exposure:
 Breathed: May cause irritation with sore throat, cough and phlegm; interference with the blood's ability to carry oxygen; headache, dizziness, weakness; bluish lips and skin; breathing difficulties, collapse and possible death.
 Eyes: Contact may cause irritation, burns and permanent damage.
 Skin: Contact may cause irritation and burns.
 Swallowed: May cause stomach irritation. Also see *Breathed.*
Long-term exposure: Possible lung problems; kidney irritation; damage to the heart muscles.
Storage: Store in tightly closed containers in a dry, cool, well-ventilated place. Keep away from organic material, combustable materials, aluminum, arsenic, carbon, copper, phosphorus, sulfur, magnesium oxide, metal sulfides, fuels, and acids.
First aid guide: Move victim to fresh air; call emergency medical care. In case of contact with material, immediately flush skin or eyes with running water for at least 15 minutes. Remove and isolate contaminated clothing and shoes at the site.

★ 593 ★
MAGNESIUM CHLORIDE
CAS: 7786-30-3
Other names: DUS-TOP

Danger profile: Possible mutagen, poisonous gas produced in fire.
Uses: Disinfectants; fire extinguishers, fire proofing wood, ceramics; making various textiles, paper; floor sweeping compounds; making other chemicals.
Effects of exposure:
 Breathed: May cause irritation of the nose, throat and breathing passages; coughing and sneezing.
 Eyes: May cause irritation and watering.
 Skin: Prolonged contact may cause irritation.
 Swallowed: May cause upset stomach and diarrhea.
Long-term exposure: May cause lung problems.
Storage: Store in tightly closed containers in a dry, cool, well-ventilated place. Containers may explode in fire.

★ 594 ★
MAGNESIUM HYDRIDE
CAS: 60616-74-2

Danger profile: Combustible, may ignite spontaneously in air, poisonous gas produced in fire or on contact with water.
Appearance: White crystals or powder.
Effects of exposure:
   Eyes: Contact may cause irritation, burns and permanent damage.
   Skin: Contact may cause irritation and burns.
   Swallowed: May cause stomach irritation. Also see *Breathed.*
Long-term exposure: Possible lung problems; kidney irritation; damage to the heart muscles.
Storage: Store in tightly closed containers in a dry, cool, well-ventilated place. Keep away from water, air, fuels and combustable materials. May catch fire spontaneously in air or in the presence of moisture. Containers may explode in fire.
First aid guide: Move victim to fresh air; call emergency medical care. Wipe material from skin immediately; flush skin or eyes with running water for at least 15 minutes. Remove and isolate contaminated clothing and shoes at the site.

★ 595 ★
MAGNESIUM NITRATE
CAS: 10377-60-3
Other names: NITRIC ACID, MAGNESIUM SALT; MAGNESIUM(II) NITRATE (1:2)

Danger profile: Combustible, increases intensity of fire, poisonous gas produced in fire.
Uses: Fireworks; making other chemicals.
Appearance: Colorless or white powder.
Effects of exposure:
   Breathed: May cause irritation with sore throat, cough and phlegm; interference with the blood's ability to carry oxygen; headache, dizziness, weakness; bluish lips and skin; breathing difficulties, collapse and possible death.
   Eyes: Contact may cause severe irritation, burns and permanent damage.
   Skin: Contact may cause irritation and burns.
   Swallowed: May cause stomach irritation. Also see *Breathed.*
Long-term exposure: Possible lung problems.
Storage: Store in tightly closed containers in a dry, cool, well-ventilated place with non-wooden floors. Keep away from combustible materials, dimethylformamide, fuels, and strong reducing agents.
First aid guide: Move victim to fresh air; call emergency medical care. In case of contact with material, immediately flush skin or eyes with running water for at least 15 minutes. Remove and isolate contaminated clothing and shoes at the site.

★ 596 ★
MAGNESIUM OXIDE
CAS: 1309-48-4
Other names: CALCINED BRUCITE; CALCINED MAGNESIA; CALCINED MAGNESITE; GRANMAG; MAGCAL; MAGLITE; MAGNESIA; MAGNESIA FUME; MAGNESIA USTA; MAGNESIUM OXIDE FUME; MAGNEZU TLENEK (POLISH); MAGOX; PERICLASE; SEAWATER MAGNESIA

Danger profile: Combustible, exposure limit established, poisonous gas produced in fire.
Uses: Making refractories, paper; electrical insulation; pharmaceuticals and cosmetics; cements; fertilizers; removal of sulfur dioxide from stack gases; semiconductors; pharmaceuticals; food and animal feed additive.
Appearance: White fine powder or fume.
Effects of exposure:
   Breathed: May cause irritation of the eyes, nose and metal fume fever; flu-like symptoms of cough, chills, chest pains, weakness, fever and muscle ache for 24 hours.
Storage: Store in tightly closed containers in a dry, cool, well-ventilated place. Keep away from heat, water bromine pentafluoride, chlorine trifluoride, phosphorus pentachloride and oxidizers. Containers may explode in fire.

★ 597 ★
MAGNESIUM PERCHLORATE
CAS: 10034-81-8
Other names: AMMONIUM PERCHLORATE, ANHYDROUS; AMMONIUM PERCHLORATE, HEXAHYDRIDE; ANHYDRONE; PERCHLORIC ACID, MAGNESIUM SALT; DEHYDRITE; PERCHLORATE DE MAGNESIUM (FRENCH)

Danger profile: Explosion hazard, increases intensity of fire, poisonous gas produced in fire.
Uses: As a drying agent for gases; oxidizing agent.
Appearance: White crystals or powder.
Effects of exposure:
   Breathed: May cause irritation of the nose, throat and breathing passages with cough, sore throat an phlegm. High levels may interfere with the blood's ability to carry oxygen; headache, dizziness and bluish color to the lips and skin; possible death from very high exposures.
   Eyes: Contact may cause severe irritation and burns.
   Skin: Contact may cause severe irritation and burns.
   Swallowed: Ingestion of large amounts may be fatal; immediate symptoms include abdominal pains, nausea and vomiting, diarrhea, pallor, blue color to skin, lips, fingernails; shortness of breath, unconsciousness.
Long-term exposure: May cause lung problems.
Storage: Store in tightly closed containers in a dry, cool, well-ventilated place with non-wood floors. Keep away from combustible materials (violent ignition), fuels, powdered metals, ammonia, ethylene oxide, phosphorus, dimethyl sulfoxide, trimethyl phosphite, powdered carbon, potassium permanganate, mineral acids and all other chemicals. A powerful oxidizer and potential explosive. Containers may explode in fire.
First aid guide: Move victim to fresh air; call emergency medical care. In case of contact with material, immediately flush skin or eyes with running water for at least 15 minutes. Remove and isolate contaminated clothing and shoes at the site.

★ 598 ★
MAGNESIUM PEROXIDE
CAS: 14452-57-4
Other names: IXPER 25M

Danger profile: Increases intensity of fires; poisonous gas produced in fire.
Uses: Bleaching and oxidizing agent; medicine (antacid).
Appearance: White powder.
Effects of exposure:
   Breathed: May cause irritation of the nose, throat and breathing passages; sore throat, cough and phlegm.
   Eyes: Contact may cause irritation and burns.
   Skin: contact may cause irritation and burns.
   Swallowed: May cause upset stomach and diarrhea.
Long-term exposure: May cause lung problems.
Storage: Store in tightly closed containers in a dry, cool, well-ventilated place. Keep away from acids and combustible materials.

First aid guide: Move victim to fresh air; call emergency medical care. In case of contact with material, immediately flush skin or eyes with running water for at least 15 minutes. Remove and isolate contaminated clothing and shoes at the site.

★ 599 ★
MAGNESIUM SILICIDE
CAS: 39404-03-0

Danger profile: Fire hazard, poisonous gas produced in fire.
Uses: Making semiconductors.
Appearance: Slate blue powder.
Effects of exposure:
Breathed: May be poisonous. May cause irritation of the nose, throat and breathing passages; sore throat, cough and phlegm.
Eyes: Contact may cause irritation and burns.
Skin: Contact may cause irritation and burns.
Swallowed: May cause upset stomach and diarrhea.
Long-term exposure: May cause lung problems.
Storage: Store in tightly closed containers in a dry, cool, well-ventilated place. Keep away from sources of ignition, water and acids. Ignites spontaneously in air. Containers may explode in fire.
First aid guide: Move victim to fresh air; call emergency medical care. Wipe material from skin immediately; flush skin or eyes with running water for at least 15 minutes. Remove and isolate contaminated clothing and shoes at the site.

★ 600 ★
MAGNESIUM SILICOFLUORIDE
CAS: 18972-56-0
Other names: EULAVA SM; FLUOSILICATE DE MAGNESIUM (FRENCH); MAGNESIUM FLUOSILICATE; MAGNESIUM HEXAFLUOROSILICATE

Danger profile: Combustible, poison, exposure limit established, poisonous gas produced in fire.
Uses: Ceramics; hardeners for concrete; water-proofing; moth proofing; magnesium casting; in laundry or textile operations.
Appearance: White, efflorescent, crystalline powder.
Effects of exposure:
Breathed: May be poisonous. May cause irritation of the nose, throat and breathing passages; sore throat, cough and phlegm.
Eyes: Contact may cause irritation and burns.
Skin: contact may cause irritation and burns.
Swallowed: May be poisonous. May cause upset stomach and diarrhea.
Long-term exposure: May cause lung problems.
Storage: Store in tightly closed containers in a dry, cool, well-ventilated place. Keep away from sources of ignition, oxidizers and acids.
First aid guide: Move victim to fresh air; call emergency medical care. In case of contact with material, immediately flush skin or eyes with running water for at least 15 minutes. Remove and isolate contaminated clothing and shoes at the site.

★ 601 ★
MAGNESIUM SULFATE
CAS: 7487-88-9
Other names: EPSOM SALTS; MAGNESIUM SULFATE (1:1); MAGNESIUM SULPHATE

Danger profile: Possible reproductive hazard, toxic fumes produced in fire.

Uses: Mineral waters; fire proofing; making textiles, paper; ceramics; fertilizers; cosmetics; dietary supplement.
Appearance: Opaque needles or crystalline powder.
Effects of exposure:
Breathed: May cause irritation of the nose and throat; sneezing.
Eyes: May cause irritation and watering.
Skin: Prolonged contact may cause irritation.
Swallowed: May cause diarrhea.
Storage: Store in tightly closed containers in a dry, cool, well-ventilated place.

★ 602 ★
MALATHION
CAS: 121-75-5
Other names: AMERICAN CYANAMID 4,049; BAN-MITE; CALMATHION; COMPOUND 4049; CARBETHOXY MALATHION; CARBAPHOS (RUSSIAN); CARBETOVUR; CARBETOX; CARBOFOS; CARBOPHOS; CELTHION (INDIAN); CHEMATHION; CIMEXAN; COMPOUND 4049; CYTHION; DETMOL MA; DETMOL MA 96%; O,O-DIMETHYL-PHOSPHORODITHIOATE; DURAMITEX; EL 4049; EMMATOS; EMOTOS EXTRA; ENT 17,034; ETHIOLACAR; ETIOL; EXTERMATHION; S-(1,2-BIS(ETHOXYCARBONYL)ETHYLO,O-DIMETHYL)DITHIOPHOSPHATE OF DIETHYL MERCAPTO-SUCCINATE; O,O-DIMETHYLS-(1,2-DICARBETHOXY-ETHYL)PHOSPHORO-DITHIOCITE; FORMAL; FORTHION; FOSFOTHION; FYFANON (DENMARK); HILTHION; HILTHION 25WDP; KARBOFOS; KOP-THION; KYPFOS; MALACIDE; MALAFOR; MALAGRAN; MALAKILL; MALAMAR; MALAMAR 50; MALAPHELE; MALAPHOS; MALASOL; MALASPRAY; MALATHIOZOO; MALATHON; MALATION (POLISH); MALATOL; MALATOX; MALDISON; MALMED; MALPHOS; MALTOX; MERCAPTOTHION; MLT; MOSCARDA; NCI-C00215; PBI CROP SAVER; PRIODERM; SADOFOS; SADOPHOS; SF 60; SIPTOX 1; SUMITOX; TK; TM- 4049; VEGFRU (INDIAN); VEGFRU MALATOX; VETIOL; ZITHIOL

Danger profile: Combustible, poison, mutagen, exposure limit established, poisonous gas produced in fire.
Uses: Insecticide for flies, household insects, human head and body lice, mosquitos; in UK has been used on animal feed, vegetables and fruit including grapes (organophosphate).
Appearance: Yellow to dark brown liquid.
Odor: Skunk-like. The odor is a warning of exposure; however, no smell does not mean you are not being exposed.
Effects of exposure:
Breathed: Extremely poisonous. May cause headache, dizziness, profuse sweating, nausea, vomiting, reduced heart beat, stomach cramps, diarrhea, loss of coordination, slow and weak breathing, fever, loss of consciousness and death. May cause shortness of breath and a dangerous build-up of fluids in the lungs (pulmonary edema), which can cause death. Pulmonary edema is a medical emergency that can be delayed one to two days following exposure. Exposure to fumes from a fire may cause headache, blurred vision, constricted pupils of the eyes, weakness, nausea, cramps, diarrhea, and tightness in the chest. Muscles twitch and convulsions may follow. The symptoms may develop over a period of 8 hours.
Eyes: Rapidly absorbed. Poisoning can happen very rapidly. See *Breathed*.
Skin: Poisonous. Passes through the unbroken skin; can increase exposure and the severity of the symptoms listed. See *Breathed*.
Swallowed: Powerful, deadly poison. May cause difficult breathing, depression of blood pressure and coma. See *Breathed*.

Long-term exposure: May damage the developing fetus or cause birth defects; nerve damage including loss of coordination; liver damage. Repeated exposure may cause depression, irritability and personality changes.

Storage: Store in a regulated, specially marked place in tightly closed containers in a dry, well-ventilated place. Keep away from water, strong alkalis and strong oxidizers. Containers may explode in fire.

First aid guide: Move victim to fresh air and call emergency medical care; if not breathing, give artificial respiration; if breathing is difficult, give oxygen. In case of contact with material, immediately flush skin or eyes with running water for at least 15 minutes. Speed in removing material from skin is of extreme importance. Remove and isolate contaminated clothing and shoes at the site. Keep victim quiet and maintain normal body temperature. Effects may be delayed; keep victim under observation.

★ 603 ★
MALEIC ACID
CAS: 110-16-7
Other names: BUTENEDIOIC ACID, (Z)-; CIS-BUTENEDIOIC ACID; (Z) BUTENEDIOIC ACID; CIS-1,2-ETHYLENEDICARBOXYLIC ACID; 1,2-ETHYLENEDICARBOXYLIC ACID, (Z); MALEINIC ACID; MALENIC ACID; TOXILIC ACID

Danger profile: Combustible, mutagen, poisonous gas produced in fire.
Uses: Making other chemicals and artificial resins, antihistamines; dyeing and finishing cotton, wool and silk; preservative for oils and fats.
Appearance: Colorless crystals or white solid often found in a liquid solution.
Effects of exposure:
Breathed: May cause irritation of the nose and throat with coughing and shortness of breath; dizziness, lightheadedness. High exposures may cause shortness of breath and a dangerous build-up of fluids in the lungs (pulmonary edema), which can cause death. Pulmonary edema is a medical emergency that can be delayed one to two days following exposure.
Eyes: Contact may cause severe irritation, burns and permanent damage.
Skin: Passes through the unbroken skin; can increase exposure and the severity of the symptoms listed. Contact may cause severe irritation.
Swallowed: See *Breathed* and *First aid guide*.
Long-term exposure: May cause lung problems.
Storage: Store in tightly closed containers in a dry, cool, well-ventilated place. Keep away from heat, sources of ignition, combustable materials, oxidizers, amines and alkali metals.
First aid guide: Move victim to fresh air; call emergency medical care. In case of contact with material, immediately flush skin or eyes with running water for at least 15 minutes. Remove and isolate contaminated clothing and shoes at the site. Keep victim quiet and maintain normal body temperature.

★ 604 ★
MALEIC ANHYDRIDE
CAS: 108-31-6
Other names: CIS-BUTENEDIOIC ANHYDRIDE; 2,5-FURANDIONE; MALEIC ACID ANHYDRIDE; 2,5-FURANEDIONE; TOXILIC ANHYDRIDE

Danger profile: Mutagen, combustible, dust vapor may explode, reacts with water and dry chemicals, exposure limit established, Community Right-to-Know List, poisonous gas produced in fire.

Uses: Making other chemicals; polyester resins for permanent press textiles; alkyd coating resins; pesticides; preservative for oils and fats.
Appearance: Colorless solid crystals or white lumps.
Odor: Acrid, strong, choking. The odor is a warning of exposure; however, no smell does not mean you are not being exposed.
Effects of exposure:
Breathed: Dust or vapors may cause irritation to the eyes, nose, throat and lungs; sneezing, shortness of breath and difficult breathing. Lung allergy may develop with asthma-like symptoms. High exposure can lead to drowsiness, coma and death; buildup of fluid in the lungs (pulmonary edema), and may cause death. Pulmonary edema may be delayed by one to two days following exposure.
Eyes: May cause severe irritation, double vision, burns and permanent damage.
Skin: May cause severe irritation, redness, and burns; not immediately painful but deep burns may occur if substance is not removed right away. Skin allergy may develop with itching and rash.
Swallowed: May cause burning feeling of the mouth and throat, intense thirst, throat swelling, stomach cramps, nausea, vomiting, diarrhea, severe breathing difficulties and unconsciousness. Tissue damage may result. Breathing may stop. Watch for delayed symptoms.
Long-term exposure: May lead to birth defects, miscarriage or cancer. May cause skin or lung allergy; lung irritation, bronchitis, cough, phlegm, shortness of breath.
Storage: Store in tightly closed containers in a dry, cool, well-ventilated place. Keep away from water and dry chemical extinguishers.
First aid guide: Move victim to fresh air; call emergency medical care. In case of contact with material, immediately flush skin or eyes with running water for at least 15 minutes. Remove and isolate contaminated clothing and shoes at the site. Keep victim quiet and maintain normal body temperature.

★ 605 ★
MALEIC HYDRAZIDE
CAS: 123-33-1
Other names: BH DOCK KILLER; BOS MH; 1,2-DIHYDROPYRIDAZINE-3,6-DIONE; 1,2-DIHYDRO-3,6-PYRIDAZINEDIONE; EC 300; ENT 18,870; 6-HYDROXY-3(2H)-PYRIDAZINONE; MALEIC ACID HYDRAZIDE; MAZIDE; N,N- MALEOYLHYDRAZINE; REGULOX; ROYAL MH 30; SLO-GRO; STOP-GRO; STUNTMAN; SUPER-DE-SPROUT; VONDALDHYDE; VONDRAX

Danger profile: Mutagen, poisonous fumes produced in fire.
Uses: Herbicide, weed control; treating beet, carrots, onions, potatoes, tobacco plants, beets.
Appearance: Crystals.
Effects of exposure:
Breathed: May cause irritation of the nose and throat.
Eyes: May cause irritation and burns.
Skin: May cause irritation and burns.
Swallowed: Moderately toxic.
Long-term exposure: May cause birth defects, miscarriage, cancer; chronic liver damage and acute central nervous system effects.
Storage: Store in tightly closed containers in a dry, cool, well-ventilated place. Containers may explode in fire.

★ 606 ★
MALONONITRILE
CAS: 109-77-3
Other names: CYANOACETONITRILE; DICYANOMETHANE; MALONIC DINITRILE; METHYLENE CYANIDE; METHANE, DICYANO-; NITRIL KYSELINY MALONOVE (CZECH); PRO-PANEDINITRILE; USAF A-4600

Danger profile: Combustible, poison, EPA extremely hazardous substances list, Community Right-to-Know List, poisonous gases produced in fire.
Uses: Making pharmaceuticals; treating gold.
Appearance: Colorless to white solid.
Effects of exposure:
Breathed: Poison. May cause headache, eye, nose and throat irritation, wheezing, foaming at the mouth, general irritability, nausea; chest tightness, rapid and irregular breathing, shortness of breath, anxiety, confusion, odor of bitter almond on breath, vomiting blood, respiratory depression, weakness, chest and abdominal pains, dizziness, diarrhea, tremors, convulsions, shock, unconsciousness weak and irregular pulse and death. Reactions may appear hours following overexposure.
Eyes: May cause severe irritation, tearing and chemical burns.
Skin: Poisonous. Passes through the unbroken skin; can increase exposure and the severity of the symptoms listed. May cause severe irritation, redness, stinging and blisters.
Swallowed: May be poisonous. May cause severe irritation and chemical burns to the tongue, mouth, throat and stomach as well as many of the symptoms described under Breathing.
Storage: Store in tightly closed containers in a dry, cool, well-ventilated place. Keep away from heat and sources of ignition and strong bases. Containers may explode in fire.
First aid guide: Move victim to fresh air; call emergency medical care. In case of contact with material, immediately flush skin or eyes with running water for at least 15 minutes. Remove and isolate contaminated clothing and shoes at the site.

★ 607 ★
MANEB
CAS: 12427-38-2
Other names: AAMANGAN; AKZO CHEMIE MANAB; BASF-MANEB SPRITZPULVER; BAVISTIN M, COSMIC; CARBAMIC ACID, ETHYLENEBIS(DITHIO-), MANGANESE SALT; CHEM NEB; CHLOROBLE M; CLEANACRES; CR 3029; DELSENE M FLOWABLE; DITHANE M 22 SPECIAL; EBDC; ENT 14,875; 1,2-ETHANEDIYLBIS(CARBAMODITHIOATO)(2-)-MANGANESE; 1,2-ETHANEDIYLBISCARBAMODITHIOIC ACID, MANGANESE COMPLEX; 1,2-ETHANEDIYLBISCARBAMODITHIOIC ACID, MANGANESE(2) SALT(1:1); 1,2-ETHANEDIYLBISMANEB, MANGANESE (2) SALT (1:1); ETHYLENEBISDITHIOCARBAMATE MANGANESE; N,N'-ETHYLENE BIS(DITHIOCARBAMATE MANGANEUX) (FRENCH); ETHYLENEBIS(DITHIOCARBAMATO), MANGANESE; ETHYLENEBIS(DITHIOCARBAMIC ACID), MANGANESE SALT; ETHYLENEBIS(DITHIOCARBAMIC ACID) MANGANOUS SALT; 1,2-ETHYLENEDIYLBIS(CARBAMODITHIOATO)MANGANESE; N,N'-ETILEN-BIS(DITIOCARBAMMATO) DI MANGANESE (ITALIAN); F 10; GRIFFIN MANEX; KYPMAN 80; LONOCOL M; MANAM; MANEB 80; MANEBA; MANEBE (FRENCH); MANEBE 80; MANEBGAN; MANESAN; MANGAAN (II)-(N,N'-ETHYLEEN-BIS(DITHIOCARBAMAAT)) (DUTCH); MANGAN (II)-(N,N'-AETHYLEN-BIS(DITHIOCARBAMATE)) (GERMAN); MANGANESE ETHYLENE-1,2-BISDITHIOCARBAMATE; MANGANESE (II) ETHYLENE DI(DITHIOCARBAMATE); MANGANOUS ETHYLENEBIS(DITHIOCARBAMATE); MANOC; MANZATE; MANZATE D; MANZATE MANEB FUNGICIDE; MANZEB; MANZIN; M-DIPHAR; MEB; MNEBD; MULTI-W, KASCADE; NESPOR; PLANTIFOG 160M; POLYRAM M; REMASAN CHLOROBLE M; RHODIANEHE; SOPRANEBE; SQUADRON AND QUADRANGLE MANEX; SUP R FLO; TERSAN-LSR; TRIMANGOL; TRIMANGOL 80; TRIMANOC; TRITHAC; TUBOTHANE; UNICROP MANEB; VANCIDE; VANCIDE MANEB 80; VASSGRO MANEX

Danger profile: Mutagen, combustible, reproductive hazard, reacts with water, poisonous gas produced in fire, Community Right-to-Know List.
Uses: Pesticide; fungicide for tomatoes, potatoes, tobacco, lettuce and fruits.
Appearance: Yellow powder or crystals.
Effects of exposure:
Breathed: May irritate the eyes, nose and throat. High exposure may cause thyroid problems, nerve system damage, or cause kidney and/or liver problems.
Eyes: May cause irritation and burns.
Skin: May cause irritation, burns and rash.
Long-term exposure: May cause cancer; decreased fertility in men and women, but may be reversible; thyroid problems with possible goiter; nerve damage with loss of muscle strength and tremors; irritation and skin rash; liver effects. Animal carcinogen.
Storage: Store in tightly closed containers in a dry, cool, well-ventilated place. Keep away from water, acids and oxidizing materials. May ignite spontaneously in air.
First aid guide: Move victim to fresh air; call emergency medical care. In case of contact with material, immediately flush skin or eyes with running water for at least 15 minutes. Remove and isolate contaminated clothing and shoes at the site.

★ 608 ★
MANGANESE
CAS: 7439-96-5
Other names: COLLOIDAL MANGANESE; MAGNACAT; MANGAN (POLISH); MANGAN NITRIDOVANY (CZECH); TRONAMANG

Danger profile: Combustible, exposure limit established, Community Right-to-Know List, poisonous gas produced in fire.
Uses: Making and treating steel, aluminum and other metals; dry cell batteries.
Appearance: Reddish-gray or silvery, brittle, metallic solid.
Effects of exposure:
Breathed: Fumes from heated manganese may cause metal fume fever with flu-like symptoms of chills, fever and aching. May last up to 24 hours. Shortness of breath may signal manganese pneumonia with coughing and chest congestion. Liver and kidney damage, and lung allergy may develop.
Eyes: May cause irritation.
Skin: May cause irritation.
Swallowed: See *Breathed*.
Long-term exposure: May cause gradual and permanent brain damage; poor appetite, weakness, tiredness, sleepiness; speech, loss of coordination and balance, tremors, personality changes may occur and symptoms may be identical to Parkinson's disease. May decrease fertility in males; low back pain; lung allergy; kidney and liver damage.
Storage: Store in tightly closed containers in a dry, cool, well-ventilated place. Keep away from water, oxidizers, nitric acid, nitrogen, sulfur dioxide. Containers may explode in fire.

★ 609 ★
MANGANESE DIOXIDE
CAS: 1313-13-9
Other names: BLACK MANGANESE OXIDE; BOG MANGANESE; BRAUNSTEIN (GERMAN); BRUINSTEEN (DUTCH); CEMENT BLACK; C.I. 77728; C.I. PIGMENT BLACK 14; C.I. PIGMENT BROWN 8; MANGAANBIOXYDE (DUTCH); MANGAANDIOXYDE (DUTCH); MANGANDIOXID (GERMAN); MANGANESE BINOXIDE; MANGANESE (BIOSSIDO DI) (ITALIAN); MANGANESE (BIOXYD DE) (FRENCH); MANGANESE BLACK; MANGANESE (DIOSSIDO DI) (ITALIAN); MANGANESE (DIOXYDE DE) (FRENCH); MANGANESE PEROXIDE; MANGANESE SUPEROXIDE; PYROLUSITE BROWN

Danger profile: Combustible, exposure limit established, Community Right-to-Know List, poisonous gas produced in fire.
Uses: Oxidizing agent; dry cell batteries; fireworks, matches, laboratory chemical; decolorizer; textile dyeing.
Appearance: Black crystalline solid or powder.
Effects of exposure:
Breathed: Fumes from heated manganese may cause metal fume fever with flu-like symptoms of chills, fever and aching, which may last up to 24 hours. Shortness of breath with chest congestion and coughing may signal manganese pneumonia. Lung, liver and kidney damage may develop.
Eyes: May cause irritation.
Skin: May cause irritation.
Long-term exposure: May cause gradual and permanent brain damage; poor appetite, weakness, tiredness, sleepiness; speech, loss of coordination and balance, tremors, personality changes may occur and symptoms may be identical to Parkinson's disease. May damage the male testes; lung scarring and allergy; kidney and liver damage.
Storage: Store in tightly closed containers in a dry, cool, well-ventilated place. Keep away from heat, combustible materials, aluminum, hydrogen peroxide, oxidizers and all other chemicals.

★ 610 ★
MANGANESE NITRATE
CAS: 10377-66-9
Other names: MANGANOUS NITRATE; NITRIC ACID, MANGANESE (2) SALT

Danger profile: Mutagen, exposure limit established, Community Right-to-Know List, poisonous gas produced in fire.
Uses: Color agent in porcelain and ceramics; making other chemicals.
Appearance: Colorless or pink crystalline solid.
Effects of exposure:
Breathed: Dust or mist may cause irritation of the nose, throat and breathing passages with cough phlegm. Overexposure may cause the lips and skin to turn blue because the ability of the blood to carry oxygen to body organs has been interfered with. High exposures may cause shortness of breath and a dangerous build-up of fluids in the lungs (pulmonary edema), which can cause death. Pulmonary edema is a medical emergency that can be delayed one to two days following exposure.
Eyes: Contact may cause irritation and burns.
Skin: Contact may cause irritation and burns.
Swallowed: See *Breathed* and *First aid guide*.
Long-term exposure: Repeated exposure may cause gradual brain damage that resembles Parkinson's disease; lung, kidney and liver damage.
Storage: Store in tightly closed containers in a dry, cool, well-ventilated place. Keep away from organic material.
First aid guide: Move victim to fresh air; call emergency medical care. In case of contact with material, immediately flush skin or eyes with running water for at least 15 minutes. Remove and isolate contaminated clothing and shoes at the site.

★ 611 ★
MANGANESE(III) OXIDE
CAS: 1317-34-6
Other names: CASSEL BROWN; C.I. NATURAL BROWN 8; C.I. 77727; COLOGNE EARTH; COLOGNE UMBER; CULLEN EARTH; DIMANGANESE TRIOXIDE; MANGANESE MANGANATE; MANGANESE(3) OXIDE; MANGANESE SISQUIOXIDE; MANGANESE TRIOXIDE; MANGANIC OXIDE; RUBENS BROWN; SOLUBLE VANDYKE BROWN; VANDYKE BROWN; WALNUT STAIN

Danger profile: Exposure limit established, Community Right-to-Know List, poisonous gas produced in fire.
Uses: Paints; artist colors; furniture stains.
Appearance: Black powder.
Effects of exposure:
Breathed: Dust or mist may cause irritation of the nose, throat and breathing passages with cough phlegm. Overexposure may cause the lips and skin to turn blue because the ability of the blood to carry oxygen to body organs has been interfered with. High exposures may cause shortness of breath and a dangerous build-up of fluids in the lungs (pulmonary edema), which can cause death. Pulmonary edema is a medical emergency that can be delayed one to two days following exposure.
Eyes: Contact may cause irritation and burns.
Skin: Contact may cause irritation and burns.
Swallowed: See *Breathed*.
Long-term exposure: Repeated exposure may cause gradual brain damage that resembles Parkinson's disease; lung, kidney and liver damage.
Storage: Store in tightly closed containers in a dry, cool, well-ventilated place. Keep away from organic material.

★612★
MANGANESE TRICARBONYL METHYLCYCLOPENTADIENYL
CAS: 12108-13-3
Other names: AK-33X; ANTIKNOCK-33; CI-2; COMBUSTION IMPROVER-2; MANGANESE, (METHYLCYCLOPENTADIENYL)TRICARBONYL-; METHYLCYCLOPENTADIENYL MANGANESE TRICARBONYL; 2-METHYLCYCLOPENTADIENYL MANGANESETRICARBONYL; MMT

Danger profile: Poisonous, exposure limit established, EPA extremely hazardous substance list; Community Right-to-Know List, poisonous gas produced in fire.
Uses: Antiknock agent for unleaded gasoline. Its use has been prohibited because of its harmful effects on catalytic converters.
Appearance: Straw to dark orange liquid.
Odor: Faint, pleasant, herbaceous. The odor is a warning of exposure; however, no smell does not mean you are not being exposed.
Effects of exposure:
    Breathed: Poisonous. May cause irritation of the eyes, nose and throat; giddiness, thick tongue, nausea, headache. High exposure may cause death.
    Eyes: Poisonous. May cause irritation.
    Skin: Poisonous. Passes through the unbroken skin; can increase exposure and the severity of the symptoms listed. May cause irritation.
    Swallowed: Poisonous.
Long-term exposure: Repeated exposure may cause gradual brain damage that resembles Parkinson's disease; lung, kidney and liver damage.
Storage: Store in tightly closed containers in a dry, cool, well-ventilated place.

★613★
MANNITOL HEXANITRATE
CAS: 15825-70-4
Other names: DILANGIL; HEXANITROL; HYPERTENAIN; INITIATING EXPLOSIVE NITRO MANNITE; MANEXIN; MANICOLE; MANITE; MANNEX; D-MANNITOL HEXANITRATE; MANNITOL HEXANITRATE, CONTAINING, BY WEIGHT, AT LEAST 40% WATER; MANNITRIN; MAXITATE; MEDEMANOL; NITRANITOL; NITRO MANNITE; NITROMANNITE; NITROMANNITOL; SDM NO.5

Danger profile: Combustible, fire hazard, explosive hazard, poisonous gas produced in fire.
Uses: In medicines; making explosive caps.
Appearance: Colorless crystals.
Effects of exposure:
    Breathed: Moderately toxic. May cause a drop in blood pressure; fatigue, weakness, headache, dizziness.
    Eyes: May cause irritation.
    Skin: may cause irritation.
    Swallowed: Moderately toxic. May cause a drop in blood pressure; fatigue, weakness, headache, dizziness.
Long-term exposure: May cause methemoglobinemia with cyanosis.
Storage: Store in tightly closed containers in a dry, cool, well-ventilated place. Keep away from heat, sources of ignition and oxidizers. Shock and heat-sensitive explosive. Containers may explode in fire.

★614★
MERCURIC ACETATE
CAS: 1600-27-7
Other names: ACETIC ACID, MERCURY(2) SALT; BIS(ACETYLOXY)MERCURY; DIACETOXYMERCURY; MERCURIACETATE; MERCURIC DIACETATE; MERCURY ACETATE; MERCURY(2) ACETATE; MERCURY(II) ACETATE; MERCURY DIACETATE; MERCURYL ACETATE

Danger profile: Mutagen, combustible, poison, exposure limit established, possible reproductive hazard, Community Right-to-Know List, EPA extremely hazardous substances list; poisonous gas produced in fire.
Uses: Making other chemicals and pharmaceuticals.
Appearance: White crystalline powder.
Odor: Slight vinegar-like; acetic. The odor is a warning of exposure; however, no smell does not mean you are not being exposed.
Effects of exposure:
    Breathed: May be poisonous. May cause irritation of the nose, throat and air passages. Heating or contact with acid causes mercury vapors and may cause lung problems, lung irritation, bronchitis, coughing and phlegm.
    Eyes: Contact may cause irritation, ulceration and severe burns.
    Skin: Contact may cause irritation, grey skin color; dermatitis. Skin allergy may develop.
    Swallowed: May be poisonous. Ingestion causes pain, vomiting, ulceration of mouth and stomach, kidney failure, metallic taste, pallor, and rapid, weak pulse.
Long-term exposure: May cause birth defects. May cause general symptoms of mercury poisoning, developing rapidly after ingestion but more slowly after low repeated exposures. Mercury poisoning may cause brain damage; irritability, memory loss, personality change, tremors; sore gums; kidney damage; skin rash.
Storage: Store in tightly closed containers in a dry, cool, well-ventilated place. Keep away from light, heat, acids, and oxidizers.
First aid guide: Move victim to fresh air; call emergency medical care. In case of contact with material, immediately flush skin or eyes with running water for at least 15 minutes. Remove and isolate contaminated clothing and shoes at the site.

★615★
MERCURIC BROMIDE
CAS: 7789-47-1
Other names: MERCURIC BROMIDE. SOLID; MERCURY BROMIDE; MERCURY(II) BROMIDE (1:2)

Danger profile: Combustible, poison, exposure limit established, Community Right-to-Know List, poisonous gas produced in fire.
Uses: In medicine.
Appearance: White crystalline sand-like solid.
Effects of exposure:
    Breathed: May be poisonous. May cause irritation of the nose, throat and air passages. Heating or contact with acid releases mercury and bromine vapors and may cause lung problems, lung irritation, bronchitis, coughing and phlegm. Higher exposures may cause shortness of breath and a dangerous build-up of fluids in the lungs (pulmonary edema), which can cause death. Pulmonary edema is a medical emergency that can be delayed one to two days following exposure.
    Eyes: Contact may cause irritation, ulceration, severe burns and permanent damage.
    Skin: Contact may cause irritation and burns; grey skin color; dermatitis. Skin allergy may develop.

Swallowed: May be poisonous. Ingestion causes pain, vomiting, ulceration of mouth and stomach, kidney failure, metallic taste, pallor, and rapid, weak pulse.
Long-term exposure: May cause birth defects. May cause general symptoms of mercury poisoning, developing rapidly after ingestion but more slowly after low repeated exposures. Mercury poisoning may cause brain damage; irritability, memory loss, personality change, tremors; sore gums; kidney damage; skin rash; brown staining of the eyes and peripheral vision effects.
Storage: Store in tightly closed containers in a dry, cool, well-ventilated place. Keep away from light, heat, acids, indium, sodium and potassium.
First aid guide: Move victim to fresh air; call emergency medical care. In case of contact with material, immediately flush skin or eyes with running water for at least 15 minutes. Remove and isolate contaminated clothing and shoes at the site.

★616★
MERCURIC CHLORIDE
CAS: 7487-94-7
Other names: BICHLORIDE OF MERCURY; BICHLORURE DE MERCURE (FRENCH); CALOCHLOR; CHLORID RTUTNATY (CZECH); CHLORURE MERCURIQUE (FRENCH); CLORURO DI MERCURIO (ITALIAN); CORROSIVE MERCURY CHLORIDE; FUNGCHEX; MC; MERCURIC BICHLORIDE; MERCURY BICHLORIDE; MERCURY(II) CHLORIDE; MERCURY PERCHLORIDE; NCI-C60173; QUECKSILBER CHLORID (GERMAN); PERCHLORIDE OF MERCURY; SULEMA (RUSSIAN); SUBLIMAT (CZECH); TL 898

Danger profile: Mutagen, combustible, poison, exposure limit established, Community Right-to-Know List, poisonous gas produced in fire.
Uses: Making other chemicals; disinfectant, fungicide; pesticide for use on turf; laboratory chemical; metallurgy; tanning; sterilant for seed potatoes; insecticide; wood preservative; embalming fluids; paper and textile printing, dry batteries; photography.
Appearance: White solid; crystals or white granules or powder.
Effects of exposure:
Breathed: May be poisonous. May cause irritation of the nose, throat and air passages. Heating or contact with acid causes mercury vapors and may cause lung problems, lung irritation, bronchitis, coughing and phlegm. Higher exposures may cause shortness of breath and a dangerous build-up of fluids in the lungs (pulmonary edema), which can cause death. Pulmonary edema is a medical emergency that can be delayed one to two days following exposure.
Eyes: Contact may cause permanent damage.
Skin: Contact may cause burns.
Swallowed: May be poisonous. Ingestion causes pain, vomiting, ulceration of mouth and stomach, kidney failure, metallic taste, pallor, and rapid, weak pulse.
Long-term exposure: May cause birth defects, damage to the fetus, miscarriage or cancer; reduced fertility in men and women; kidney damage. May cause the general symptoms of mercury poisoning developing rapidly after ingestion but more slowly after low repeated exposures. Mercury poisoning may cause brain damage; irritability, memory loss, personality change, tremors; sore gums; kidney damage; skin rash; brown staining of the eyes and peripheral vision effects.
Storage: Store in tightly closed containers in a dry, cool, well-ventilated place. Keep away from heat, acid, potassium and sodium.
First aid guide: Move victim to fresh air; call emergency medical care. In case of contact with material, immediately flush skin or eyes with running water for at least 15 minutes. Remove and isolate contaminated clothing and shoes at the site.

★617★
MERCURIC CYANIDE
CAS: 592-04-1
Other names: CYANURE DE MERCURE (FRENCH); MERCURY(II) CYANIDE

Danger profile: Combustible, explosion hazard, poison, exposure limit established, Community Right-to-Know List, poisonous gas produced in fire.
Uses: In medicine; antiseptic; germicidal soaps; photography.
Appearance: White or colorless crystals or powder. Darkens on exposure to light.
Effects of exposure:
Breathed: May be poisonous. May cause irritation of the nose, throat and breathing passages. Heating or contact with acid causes mercury and cyanide vapors and may cause lung problems, lung irritation, bronchitis, coughing and phlegm. Acute poisoning has resulted from inhaling low dust concentrations of air; symptoms include tightness and pain in chest, coughing, and difficult breathing; cyanide poisoning can cause anxiety, confusion, dizziness, and shortness of breath, with possible unconsciousness, convulsions, and paralysis; breath may smell like bitter almonds. Overexposure to cyanide may cause kidney damage and/or sudden death.
Eyes: Contact may cause irritation and possible damage.
Skin: Passes through the unbroken skin; can increase exposure and the severity of the symptoms listed. Contact may cause irritation, gray skin color; dermatitis or allergy may develop.
Swallowed: May be poisonous. May cause necrosis, pain, vomiting, and severe purging, plus symptoms listed under *Breathed.*
Long-term exposure: May cause general symptoms of mercury poisoning, developing rapidly after ingestion but more slowly after low repeated exposures. Mercury poisoning may cause brain damage; irritability, memory loss, personality change, tremors; sore gums; kidney damage; skin rash; brown staining of the eyes and peripheral vision effects. Some related mercury compounds may cause decreased fertility in men and women and damage to the developing fetus.
Storage: Store in tightly closed containers in a dry, cool, well-ventilated place. Keep away from heat, acid, fluorine, magnesium, hydrogen cyanide and sodium nitrite. An impact and friction sensitive explosive.
First aid guide: Move victim to fresh air; call emergency medical care. In case of contact with material, immediately flush skin or eyes with running water for at least 15 minutes. Remove and isolate contaminated clothing and shoes at the site.

★618★
MERCURIC IODIDE
CAS: 7774-29-0
Other names: HYDRARGYRUM BIJODATUM (GERMAN); MERCURIC IODIDE, RED; MERCURY BINIODIDE; MERCURY(II) IODIDE; RED MERCURIC IODIDE

Danger profile: Combustible, poison, exposure limit established, Community Right-to-Know List, poisonous gas produced in fire.
Uses: In medicine; laboratory chemical.
Appearance: Scarlet red, heavy powder.
Effects of exposure:
Breathed: Poisoning may result from inhaling dust; symptoms include tightness and pain in chest, coughing, difficult breathing, metallic taste, tremors, irritability, sore gums, memory loss, personality change. All forms of exposure of this compound are hazardous. Acute systemic mercurialism

may be fatal within a few minutes, death by uremic poisoning is usually delayed 5-12 days.
Eyes: May cause burns and permanent damage; ulceration of conjunctiva and cornea.
Skin: Passes through the unbroken skin; can increase exposure and the severity of the symptoms listed. Contact may cause irritation, burns, gray skin color; dermatitis and/or allergy may develop.
Swallowed: May be poisonous. May cause necrosis, pain, vomiting, and severe purging. Also see symptoms under *Breathed.*
Long-term exposure: May cause general symptoms of mercury poisoning, developing rapidly after ingestion but more slowly after low repeated exposures. Mercury poisoning may cause brain damage; irritability, memory loss, personality change, tremors; sore gums; kidney damage; skin rash; brown staining of the eyes and peripheral vision effects. Some related mercury compounds may cause decreased fertility in men and women and damage to the developing fetus.
Storage: Store in tightly closed containers in a dry, cool, well-ventilated place. Keep away from light, heat, acid, chlorine trifluoride, sodium, potassium.

★ 619 ★
MERCURIC NITRATE
CAS: 10045-94-0
Other names: MERCURY(II) NITRATE (1:2); MERCURY NITRATE; MERCURY PERNITRATE; NITRATE MERCURIQUE (FRENCH); NITRIC ACID, MERCURY(II) SALT

Danger profile: Combustible, poison, exposure limit established, Community Right-to-Know List, poisonous gas produced in fire.
Uses: Making felt and other chemicals.
Appearance: White or yellowish powder.
Odor: Sharp odor of nitric acid. The odor is a warning of exposure; however, no smell does not mean you are not being exposed.
Effects of exposure:
  Breathed: May cause irritation of the nose, throat and breathing passages. Heating or contact with acid causes mercury vapors and may cause lung problems, lung irritation, bronchitis, coughing and phlegm. Acute systemic poisoning may be fatal within a few minutes. Acute poisoning has resulted from inhaling dust; symptoms include tightness and pain in chest, coughing, difficult breathing, metallic taste, tremors, irritability, sore gums, memory loss, personality change. All forms of exposure of this compound are hazardous.
  Eyes: Contact may cause irritation and possible damage.
  Skin: Passes through the unbroken skin; can increase exposure and the severity of the symptoms listed. Contact may cause irritation, gray skin color; dermatitis or allergy may develop.
  Swallowed: May be poisonous. May cause necrosis, pain, vomiting, and severe purging, plus symptoms listed under *Breathed.*
Long-term exposure: May cause general symptoms of mercury poisoning, developing rapidly after ingestion but more slowly after low repeated exposures. Mercury poisoning may cause brain damage; irritability, memory loss, personality change, tremors; sore gums; kidney damage; skin rash; brown staining of the eyes and peripheral vision effects. Some related mercury compounds may cause decreased fertility in men and women and damage to the developing fetus.
Storage: Store in tightly closed containers in a dry, cool, well-ventilated place with non-wood floors. Solution will corrode most metals. Contact with wood or paper may cause fire. Keep away from heat, acid, organic and combustable materials, isobutene,

acetylene, ethyl alcohol, phosphine, sulfur, hypophosphoric acid and all other chemicals. Containers may explode in fire.
First aid guide: Move victim to fresh air; call emergency medical care. In case of contact with material, immediately flush skin or eyes with running water for at least 15 minutes. Remove and isolate contaminated clothing and shoes at the site.

★ 620 ★
MERCURIC OXIDE
CAS: 21908-53-2
Other names: C.I. 77760; KANKEREX; MERCURIC OXIDE, RED; MERCURIC OXIDE, YELLOW; OXYDE DE MERCURE (FRENCH); RED OXIDE OF MERCURY; RED PRECIPITATE; SANTAR; YELLOW MERCURIC OXIDE; YELLOW OXIDE OF MERCURY; YELLOW PRECIPITATE; MERCURY OXIDE; SANTAR

Uses: Paint pigments; anti fouling paints; cosmetics and perfumes; pharmaceuticals; ceramics; batteries; polishing compounds; antiseptics; making other chemicals; laboratory chemical; fungicide; as a paint for apple and pear canker; sealant on fruit trees.
Appearance: Heavy red, orange-red or red-yellow crystalline powder.
Effects of exposure:
  Breathed: Poisoning may result from inhaling dust; symptoms include tightness and pain in chest, coughing, difficult breathing, metallic taste, tremors, irritability, sore gums, memory loss, personality change. All forms of exposure of this compound are hazardous. Acute systemic mercurialism may be fatal within a few minutes, death by uremic poisoning is usually delayed 5-12 days.
  Eyes: May cause burns and permanent damage; ulceration of conjunctiva and cornea.
  Skin: Passes through the unbroken skin; can increase exposure and the severity of the symptoms listed. Contact may cause irritation, burns, gray skin color; dermatitis and/or allergy may develop.
  Swallowed: May cause necrosis, pain, vomiting, and severe purging. Also see symptoms under *Breathed.*
Long-term exposure: May cause general symptoms of mercury poisoning, developing rapidly after ingestion but more slowly after low repeated exposures. Mercury poisoning may cause brain damage; irritability, memory loss, personality change, tremors; sore gums; kidney damage; skin rash; brown staining of the eyes and peripheral vision effects. Some related mercury compounds may cause decreased fertility in men and women and damage to the developing fetus.
Storage: Store in tightly closed containers in a dry, cool, well-ventilated place with non-wood floors. Keep away from heat, acid and organic materials.

★ 621 ★
MERCURIC OXYCYANIDE
CAS: 1335-31-5
Other names: MERCURY OXYCYANIDE; MERCURY CYANIDE OXIDE

Danger profile: Combustible, explosion hazard, exposure limit established, Community Right-to-Know List, poisonous gas produced in fire.
Uses: Antiseptic.
Appearance: White crystalline powder.
Effects of exposure:
  Breathed: May be poisonous. May cause irritation of the nose, throat and breathing passages. Heating or contact with acid causes mercury and cyanide vapors and may cause lung problems, lung irritation, bronchitis, coughing and phlegm.

Acute poisoning has resulted from inhaling dust; symptoms include tightness and pain in chest, coughing, and difficult breathing; cyanide poisoning can cause anxiety, confusion, dizziness, and shortness of breath, with possible unconsciousness, convulsions, and paralysis; breath may smell like bitter almonds. Overexposure to cyanide may cause kidney damage and/or sudden death

Eyes: Contact may cause irritation and possible damage.

Skin: Passes through the unbroken skin; can increase exposure and the severity of the symptoms listed. Contact may cause irritation, gray skin color; dermatitis or allergy may develop.

Swallowed: May be poisonous. May cause necrosis, pain, vomiting, and severe purging, plus symptoms listed under *Breathed.*

Long-term exposure: May cause general symptoms of mercury poisoning, developing rapidly after ingestion but more slowly after low repeated exposures. Mercury poisoning may cause brain damage; irritability, memory loss, personality change, tremors; sore gums; kidney damage; skin rash; brown staining of the eyes and peripheral vision effects. Some related mercury compounds may cause decreased fertility in men and women and damage to the developing fetus.

Storage: Store in tightly closed containers in a dry, cool, well-ventilated place. Keep away from heat and acid. An impact and friction sensitive explosive.

First aid guide: Move victim to fresh air; call emergency medical care. In case of contact with material, immediately flush skin or eyes with running water for at least 15 minutes. Remove and isolate contaminated clothing and shoes at the site.

★ 622 ★
MERCURIC SULFATE
CAS: 7783-35-9
Other names: MERCURY(II) SULFATE (1:1); MERCURY BISULFATE; MERCURY PERSULFATE; SULFATE MERCURIQUE (FRENCH); SULFURIC ACID, MERCURY(2) SALT (1:1)

Danger profile: Combustible, poison, exposure limit established, Community Right-to-Know List, poisonous gas produced in fire and on contact with water.
Uses: Making other chemicals; extracting gold and silver from rock; battery electrolyte.
Appearance: White crystalline powder.
Effects of exposure:
Breathed: Poisoning may result from inhaling dust; symptoms include tightness and pain in chest, coughing, difficult breathing, metallic taste, tremors, irritability, sore gums, memory loss, personality change. All forms of exposure of this compound are hazardous. Acute systemic mercurialism may be fatal within a few minutes, death by uremic poisoning is usually delayed 5-12 days.
Eyes: May cause burns and permanent damage; ulceration of conjunctiva and cornea.
Skin: Passes through the unbroken skin; can increase exposure and the severity of the symptoms listed. Contact may cause irritation, burns, gray skin color; dermatitis and/or allergy may develop.
Swallowed: Poisonous. May cause necrosis, pain, vomiting, and severe purging. Also see symptoms under *Breathed.*
Long-term exposure: May cause general symptoms of mercury poisoning, developing rapidly after ingestion but more slowly after low repeated exposures. Mercury poisoning may cause brain damage; irritability, memory loss, personality change, tremors; sore gums; kidney damage; skin rash; brown staining of the eyes and peripheral vision effects. Some related mercury compounds may cause decreased fertility in men and women and damage to the developing fetus.

Storage: Store in tightly closed containers in a dry, cool, well-ventilated place with non-wood floors. Keep away from water and other forms of moisture, heat, acid light and gaseous hydrogen chloride.
First aid guide: Move victim to fresh air; call emergency medical care. In case of contact with material, immediately flush skin or eyes with running water for at least 15 minutes. Remove and isolate contaminated clothing and shoes at the site.

★ 623 ★
MERCURIC THIOCYANATE
CAS: 592-85-8
Other names: BIS(THYOCYANATO)-MERCURY; MERCURIC SULFOCYANATE; MERCURIC SULFO CYANATE, SOLID; MERCURIC SULFOCYANIDE; MERCURY DITHIOCYANATE; MERCURY THIOCYANATE

Danger profile: Combustible, poison; exposure limit established, Community Right-to-Know List, poisonous gas produced in fire.
Uses: Photography; fireworks.
Appearance: White powder.
Effects of exposure:
Breathed: May be poisonous. May cause irritation of the throat and breathing passages; tightness in the chest, difficulty in breathing, coughing, and pain in the chest. High exposure may cause kidney damage.
Eyes: Contact may cause irritation, allergy, and gray skin color. Ulceration of conjunctiva and cornea.
Skin: Contact may cause irritation, allergy, and gray skin color; dermatitis.
Swallowed: May be poisonous. Necrosis, pain, vomiting, severe purging. Patient may die within a few hours from peripheral vascular collapse.
Long-term exposure: Repeated exposure may cause kidney damage. Mercury poisoning may cause irritability, sore gums, tremors, memory loss, personality change; brain damage; bronchitis, lung irritation, cough and phlegm. Closely related mercury compounds may cause decreased fertility of men and women and may cause damage to a developing fetus.
Storage: Store in tightly closed containers in a dry, cool, well-ventilated place. Keep away from heat, direct sunlight and sources of ignition. Very unstable in heat. Containers may explode in fire.

★ 624 ★
MERCUROUS CHLORIDE
CAS: 7546-30-7
Other names: CALOGREEN; CALOMEL; CALOMELANO (ITALIAN); CALOSAN; CHLORURE MERCUREUX (FRENCH); C.I. 77764; CLORURO MERCUROSO (ITALIAN); CLUB ROOT CONTROL; CYCLOSAN; KALOMEL (GERMAN); MERCUROCHLORIDE; MERCURY(I) CHLORIDE; MERCURY MONOCHLORIDE; MERCURY PROTOCHLORIDE; PRECIPITE BLANC; QUECKSILBER(I)-CHLORID (GERMAN); QUECKSILBER CHLORUER (GERMAN); SUBCHLORIDE OF MERCURY; TESCO CLUB ROOT CONTROL

Danger profile: Combustible, exposure limit established, Community Right-to-Know List, poisonous gas produced in fire.
Uses: Fungicide; pesticide on soil to control root maggots, onions, shallots and turf; making pharmaceuticals; ceramic paints; fireworks.
Appearance: White heavy powder or crystals.
Effects of exposure:
Breathed: Poisoning may result from inhaling dust; symptoms include tightness and pain in chest, coughing, difficult breathing, metallic taste, tremors, irritability, sore gums,

memory loss, personality change. All forms of exposure of this compound are hazardous. Acute systemic mercurialism may be fatal within a few minutes, death by uremic poisoning is usually delayed 5-12 days.
Eyes: May cause burns and possible damage.
Skin: Passes through the unbroken skin; can increase exposure and the severity of the symptoms listed. Contact may cause irritation, burns, gray skin color; dermatitis and/or allergy may develop.
Swallowed: May be poisonous. May cause necrosis, pain, vomiting, and severe purging. Also see symptoms under *Breathed.*
Long-term exposure: May cause general symptoms of mercury poisoning, developing rapidly after ingestion but more slowly after low repeated exposures. Mercury poisoning may cause brain damage; irritability, memory loss, personality change, tremors; sore gums; kidney damage; skin rash; brown staining of the eyes and peripheral vision effects. Some related mercury compounds may cause decreased fertility in men and women and damage to the developing fetus.
Storage: Store in tightly closed containers in a dry, cool, well-ventilated place with non-wood floors. Keep away from heat, acid and sunlight.

★ 625 ★
MERCUROUS NITRATE
CAS: 10415-75-5
Other names: MERCURY(I) NITRATE (1:1); NITRATE MERCUREUX (FRENCH); NITRIC ACID, MERCURY(I) SALT

Danger profile: Combustible, poison, exposure limit established, Community Right-to-Know List, poisonous gas produced in fire.
Uses: Laboratory chemical; blackening brass.
Appearance: White solid.
Odor: Slight odor of nitric acid. The odor is a warning of exposure; however, no smell does not mean you are not being exposed.
Effects of exposure:
Breathed: May be poisonous. May cause irritation of the nose, throat and breathing passages. Heating or contact with acid causes mercury vapors and may cause lung problems, lung irritation, bronchitis, coughing and phlegm. Acute systemic poisoning may be fatal within a few minutes. Acute poisoning has resulted from inhaling dust; symptoms include tightness and pain in chest, coughing, difficult breathing, metallic taste, tremors, irritability, sore gums, memory loss, personality change. All forms of exposure of this compound are hazardous.
Eyes: Contact may cause irritation and possible damage.
Skin: Passes through the unbroken skin; can increase exposure and the severity of the symptoms listed. Contact may cause irritation, gray skin color; dermatitis or allergy may develop.
Swallowed: May be poisonous. May cause necrosis, pain, vomiting, and severe purging, plus symptoms listed under *Breathed.*
Long-term exposure: May cause general symptoms of mercury poisoning, developing rapidly after ingestion but more slowly after low repeated exposures. Mercury poisoning may cause brain damage; irritability, memory loss, personality change, tremors; sore gums; kidney damage; skin rash; brown staining of the eyes and peripheral vision effects. Some related mercury compounds may cause decreased fertility in men and women and damage to the developing fetus.
Storage: Store in tightly closed containers in a dry, cool, well-ventilated place with non-wood floors. Solution will corrode most metals. Contact with wood or paper may cause fire. Keep away

from heat, acid, light, carbon and phosphorus. Containers may explode in fire.
First aid guide: Move victim to fresh air; call emergency medical care. In case of contact with material, immediately flush skin or eyes with running water for at least 15 minutes. Remove and isolate contaminated clothing and shoes at the site.

★ 626 ★
MERCUROUS OXIDE
CAS: 15829-53-5
Other names: MERCUROUS OXIDE, BLACK; MERCURY(I) OXIDE QUECKSILBEROXID (GERMAN)

Danger profile: Combustible, poison, exposure limit established, Community Right-to-Know List, poisonous gas produced in fire.
Uses: Making other chemicals.
Appearance: Black to grayish-black powder.
Effects of exposure:
Breathed: May be poisonous. May cause irritation of the nose, throat and breathing passages. Heating or contact with acid causes mercury vapors and may cause lung problems, lung irritation, bronchitis, coughing and phlegm. Acute systemic poisoning may be fatal within a few minutes. Acute poisoning has resulted from inhaling dust; symptoms include tightness and pain in chest, coughing, difficult breathing, metallic taste, tremors, irritability, sore gums, memory loss, personality change. All forms of exposure of this compound are hazardous.
Eyes: Contact may cause irritation and possible damage.
Skin: Passes through the unbroken skin; can increase exposure and the severity of the symptoms listed. Contact may cause irritation, gray skin color; dermatitis or allergy may develop.
Swallowed: May be poisonous. May cause necrosis, pain, vomiting, and severe purging, plus symptoms listed under *Breathed.*
Long-term exposure: May cause general symptoms of mercury poisoning, developing rapidly after ingestion but more slowly after low repeated exposures. Mercury poisoning may cause brain damage; irritability, memory loss, personality change, tremors; sore gums; kidney damage; skin rash; brown staining of the eyes and peripheral vision effects. Some related mercury compounds may cause decreased fertility in men and women and damage to the developing fetus.
Storage: Store in tightly closed containers in a dry, cool, well-ventilated place with non-wood floors. Keep away from heat, acids, light, hydrogen peroxide, chlorine, alkali metals, ethylene, potassium, sodium and sulfur.

★ 627 ★
MERCUROUS SULFATE
CAS: 7783-36-0
Other names: MERCURY(I) SULFATE

Danger profile: Poison, exposure limit established, Community Right-to-Know List, poisonous gas produced in fire.
Uses: Making batteries and other chemicals.
Appearance: White, to yellow crystalline powder
Effects of exposure:
Breathed: May be poisonous. May cause irritation of the nose, throat and breathing passages. Heating or contact with acid causes mercury vapors and may cause lung problems, lung irritation, bronchitis, coughing and phlegm. Acute systemic poisoning may be fatal within a few minutes. Acute poisoning has resulted from inhaling dust; symptoms include tightness and pain in chest, coughing, difficult breathing, metallic taste, tremors, irritability, sore gums, memory loss, per-

sonality change. All forms of exposure of this compound are hazardous.

Eyes: Contact may cause irritation and possible damage.

Skin: Passes through the unbroken skin; can increase exposure and the severity of the symptoms listed. Contact may cause irritation, gray skin color; dermatitis or allergy may develop.

Swallowed: May be poisonous. May cause necrosis, pain, vomiting, and severe purging, plus symptoms listed under ***Breathed***.

Long-term exposure: May cause general symptoms of mercury poisoning, developing rapidly after ingestion but more slowly after low repeated exposures. Mercury poisoning may cause brain damage; irritability, memory loss, personality change, tremors; sore gums; kidney damage; skin rash; brown staining of the eyes and peripheral vision effects. Some related mercury compounds may cause decreased fertility in men and women and damage to the developing fetus.

Storage: Store in tightly closed containers in a dry, cool, well-ventilated place. Keep away from heat and acids.

First aid guide: Move victim to fresh air; call emergency medical care. In case of contact with material, immediately flush skin or eyes with running water for at least 15 minutes. Remove and isolate contaminated clothing and shoes at the site.

★ 628 ★
MERCURY
CAS: 7439-97-6
Other names: COLLOIDAL MERCURY; KWIK (DUTCH); MERCURE (FRENCH); MERCURIO (ITALIAN); MERCURY, METALLIC; METALLIC MERCURY; NCI-C60399; QUECKSILBER (GERMAN); QUICK SILVER; RTEC (POLISH)

Danger profile: Poison, corrosive, mutagen, exposure limit established, Community Right-to-Know List, poisonous gas produced in fire.

Uses: Dental fillings; making other chemicals, electrical equipment and instruments (thermometers, barometers, etc.); mercury vapor lamps; mirror coating; in nuclear power plants as coolant and neutron absorbers.

Appearance: Silvery, heavy, mobile liquid.

Effects of exposure:

Breathed: May cause lung irritation with cough, chest pains shortness of breath, difficult breathing, fever and a dangerous build-up of fluids in the lungs (pulmonary edema), which can cause death. Pulmonary edema is a medical emergency that can be delayed one to two days following exposure.

Eyes: Contact may cause irritation.

Skin: Passes through the unbroken skin; can increase exposure and the severity of the symptoms listed. Contact may cause irritation.

Swallowed: Mercury poisoning can develop with loss of appetite, muscular tremor, nausea, diarrhea; psychic, kidney and cardiovascular disturbances may occur.

Long-term exposure: May cause symptoms of mercury poisoning, developing rapidly after ingestion but more slowly after low repeated exposures by breathing vapors and passing through the skin. Mercury poisoning may cause brain damage; irritability, memory loss, personality change, tremors; sore gums; kidney disease; skin rash; brown staining of the eyes and peripheral vision effects. Exposure may cause spontaneous abortion in women, decreased fertility in men and women and damage to the developing fetus.

Storage: Store in tightly closed containers in a dry, cool, well-ventilated place. Keep away from heat, acids, chlorine dioxide, nitrates, ethylene oxide, chlorine, acetylene, ammonia and nickel.

First aid guide: Move victim to fresh air; call emergency medical care. In case of contact with material, immediately flush skin or eyes with running water for at least 15 minutes. Remove and isolate contaminated clothing and shoes at the site. Keep victim quiet and maintain normal body temperature.

★ 629 ★
MERCURY, ETHYL(P-TOLUENESULFONANILIDATO)-
CAS: 517-16-8
Other names: CERESAN M; CERESAN M-2X; CERESAN M-DB; COMPOUND-1452-F; EMTS; N-ETHYLMERCURI-N-PHENYL-P-TOLUENESU LFONAMIDE; N-(ETHYLMERCURI)-P-TOLUENESULFONANILIDE; N-(ETHYLMERCURI)-P-TOLUENESULPHONANILIDE; ETHYLMERCURY P-TOLUENESULFANILIDE; ETHYLMERCURY-P-TOLUENE SULFONAMIDE; ETHYLMERCURY-P-TOLUENESULFONANILIDE; N-ETHYLMERKURI-P-TOLUENSULFOANILID (CZECH); GRANOSAN M; GRANOSAN MDB; MERCURY,ETHYL(4-METHYL-N-PHENYLBENZENESULFONAMIDATO-N)- (9CI); MERCURY,ETHYL(N-PHENYL-P-TOLUENESULFONAMIDATO)-; MERCURY, ETHYL(N-PHENYL-P-TOLUENESULFONAMIDO)-; MERGON; MERGON D; (N-PHENYL-P-TOLUENESULFONAMIDO)ETHYLMERCURY; SEED DRESSING UNIVERSAL; P-TOLUENESULFONAMIDE,N-(ETHYLMERCURI)-N-PHENYL-; P-TOLUENESULFONANILIDE,N-(ETHYLMERCURI)-; ZAPRAWA NASIENNA UNIVERSAL; ZAPRAWA NASIENNA UNIVERSAL R; ZAPRAWA NASIENNA UNIWERSAL NA

Danger profile: Poison, mutagen, exposure limit established, poisonous gas produced in fire.

Uses: Fungicide; treating seeds and bulbs.

Appearance: Sand-like crystalline solid.

Odor: Garlic-like. The odor is a warning of exposure; however, no smell does not mean you are not being exposed.

Effects of exposure:

Breathed: May cause weakness, clumsiness, tingling feeling of the lips, tongue and fingers; slow loss of peripheral vision; hearing loss, staggered walk, shaking, slurred speech and personality changes. Chemical may cause permanent brain damage without warning; may occur weeks later. High exposure may cause severe poisoning and death.

Eyes: May cause irritation; may increase exposure and the severity of the symptoms listed.

Skin: Passes through the unbroken skin; can increase exposure and the severity of the symptoms listed.

Swallowed: See ***Breathed***.

Long-term exposure: May cause permanent brain damage. Some mercury compounds cause birth defects and skin allergy.

Storage: Store in tightly closed containers in a dry, cool, well-ventilated place. Keep away from oxidizers.

★ 630 ★
MERCURY FULMINATE
CAS: 628-86-4
Other names: FULMINIC ACID, MERCURY(II) SALT; FULMINATE OF MERCURY

Danger profile: Combustible, explosion hazard, exposure limit established, transport forbidden, Community Right-to-Know List, toxic fumes produced in fire.

Uses: Manufacture of detonators and caps for producing explosives.

Appearance: Gray crystalline powder; white solid.

Effects of exposure:

Breathed: May cause a metallic taste in the mouth, salivation, headache, dizziness, staggering and lack of coordination, fatigue; possible tremors.

Eyes: May cause severe irritation, tearing, irritation of the eyelids; lesions may occur.

Skin: Passes through the unbroken skin; can increase exposure and the severity of the symptoms listed. May cause irritation, redness, blisters and ulcerations.

Swallowed: May cause irritation of the mouth, throat; stomach pains, nausea, vomiting; diarrhea. Also see symptoms for *Breathed.*

Storage: Store in tightly closed containers in a cool, well-ventilated place. Keep wet until used. Keep away from heat, sources of ignition, friction, impact, sulfuric acid and oxidizers. Containers may explode in fire.

★ 631 ★
MERCURY(I) IODIDE
CAS: 7783-30-4
Other names: IODURE DE MERCURE (FRENCH); MERCUROUS IODIDE; MERCUROUS IODIDE, SOLID; MERCURY PROTOIODIDE; YELLOW MERCURY IODIDE

Danger profile: Poison, exposure limit established, Community Right-to-Know List, poisonous gas produced in fire.
Uses: An anti-bacterial agent.
Appearance: Yellow crystalline powder.
Effects of exposure:

Breathed: May be poisonous. May cause lung irritation metallic taste; cough, chest pains shortness of breath, difficult breathing, fever, bronchitis, chemical pneumonia; a dangerous build-up of fluids in the lungs (pulmonary edema), which can cause death. Pulmonary edema is a medical emergency that can be delayed up to 15 days following exposure.

Eyes: Contact may cause severe irritation and damage.

Skin: Passes through the unbroken skin; can increase exposure and the severity of the symptoms listed. Contact may cause irritation.

Swallowed: May be poisonous. May cause evil-smelling salivation, inflammation of the mouth, profuse sweating, headache. Mercury poisoning can develop with loss of appetite, muscular tremor, nausea, diarrhea; psychic, kidney and cardiovascular disturbances may occur.

Long-term exposure: May cause symptoms of mercury poisoning, developing rapidly after ingestion but more slowly after low repeated exposures by breathing vapors and passing through the skin. Mercury poisoning may cause brain damage; irritability, memory loss, personality change, tremors; sore gums; kidney disease; skin rash; brown staining of the eyes and peripheral vision effects. Exposure may cause decreased fertility in men and women and damage to the developing fetus.

Storage: Store in tightly closed containers in a dry, cool, well-ventilated place. Keep away from light, heat and acids.

★ 632 ★
MERCURY OXIDE SULFATE
CAS: 1312-03-4
Other names: BASIC MERCURIC SULFATE; MERCURIC BASIC SULFATE; MERCURIC SUBSULFATE; TURPETH MINERAL

Danger profile: Poison, combustible, exposure limit established, Community Right-to-Know List, poisonous gas produced in fire.
Appearance: Lemon-yellow heavy powder.
Effects of exposure:

Breathed: May be poisonous. May cause lung irritation metallic taste; cough, chest pains shortness of breath, difficult breathing, fever, bronchitis, chemical pneumonia; a dangerous build-up of fluids in the lungs (pulmonary edema), which

can cause death. Pulmonary edema is a medical emergency that can be delayed one to two days following exposure.

Eyes: Contact may cause severe irritation and damage.

Skin: Passes through the unbroken skin; can increase exposure and the severity of the symptoms listed. Contact may cause irritation.

Swallowed: May be poisonous. May cause evil-smelling salivation, inflammation of the mouth, profuse sweating, headache. Mercury poisoning can develop with loss of appetite, muscular tremor, nausea, diarrhea; psychic, kidney and cardiovascular disturbances may occur.

Long-term exposure: May cause symptoms of mercury poisoning, developing rapidly after ingestion but more slowly after low repeated exposures by breathing vapors and passing through the skin. Mercury poisoning may cause brain damage; irritability, memory loss, personality change, tremors; sore gums; kidney disease; skin rash; brown staining of the eyes and peripheral vision effects. Exposure may cause decreased fertility in men and women and damage to the developing fetus.

Storage: Store in tightly closed containers in a dry, cool, well-ventilated place. Keep away from heat, acid, magnesium and aluminum powder.

★ 633 ★
METHACRYLIC ACID
CAS: 79-41-4
Other names: ACRYLIC ACID, 2-METHYL-; ALPHA-METHACRYLIC ACID; 2-METHYLPROPIONIC ACID; PROPIONIC ACID, 2-METHYLENE-

Danger profile: Combustible, poison, explosion hazard, corrosive, exposure limit established, poisonous gas produced in fire.
Uses: Making resins, fibers, plastic sheets, moldings and other chemicals.
Appearance: Colorless liquid.
Odor: Unpleasant, repulsive odor. The odor is a warning of exposure; however, no smell does not mean you are not being exposed.
Effects of exposure:

Breathed: May cause irritation to the nose, throat and breathing passages. High levels may cause lung problems.

Eyes: Contact may cause severe irritation, burns and permanent damage.

Skin: Contact may cause irritation and burns.

Long-term exposure: May cause skin rash; kidney, liver and lung problems. Exposure may lead to birth defects, miscarriage or cancer.

Storage: Store in tightly closed containers in a dry, cool, well-ventilated place. Keep away from heat, sources of ignition and oxidizers. Containers may explode in fire.

First aid guide: Move victim to fresh air; call emergency medical care. In case of contact with material, immediately flush skin or eyes with running water for at least 15 minutes. Remove and isolate contaminated clothing and shoes at the site. Keep victim quiet and maintain normal body temperature.

★ 634 ★
METHACRYLONITRILE
CAS: 126-98-7
Other names: 2-CYANOPROPENE-1; ISOPROPENE CYANIDE; ISOPROPENYLNITRILE; ALPHA-METHYLACRYLONITRILE; 2-METHYLPROPENENITRILE; 2-PROPENENITRILE, 2-METHYL-; USAF ST-40

Danger profile: Fire hazard, exposure limit established, EPA extremely hazardous substances list, Community Right-to-Know List; poisonous gas produced in fire.
Uses: Making other chemicals, plastic elastomers and coatings.

Appearance: Colorless liquid.
Effects of exposure:

Breathed: Poison. High concentrations are extremely destructive to tissues of mucous membranes and upper respiratory tract, eyes and skin. May cause headache, eye, nose and throat irritation, wheezing, coughing, foaming at the mouth, general irritability, nausea; chest tightness, shortness of breath, vomiting blood, respiratory depression, weakness, chest and abdominal pains, dizziness, diarrhea, tremors, convulsions, shock, unconsciousness weak and irregular pulse and death. skin and lips may turn blue. Reactions may appear hours following overexposure.
Eyes: May cause severe irritation, tearing and chemical burns.
Skin: Poisonous if absorbed. Passes through the unbroken skin; can increase exposure and the severity of the symptoms listed. May cause severe irritation, redness, stinging and blisters.
Swallowed: May be poisonous. May cause severe irritation and chemical burns to the tongue, mouth, throat and stomach as well as many of the symptoms described under *Breathed.*
Long-term exposure: May cause liver and nervous system damage.
Storage: Store in tightly closed containers in a dry, cool, well-ventilated place. Keep away from heat and all sources of ignition. Vapors may travel to sources of ignition and flash back. Containers may explode in fire.
First aid guide: Move victim to fresh air and call emergency medical care; if not breathing, give artificial respiration; if breathing is difficult, give oxygen. In case of contact with material, immediately flush skin or eyes with running water for at least 15 minutes. Remove and isolate contaminated clothing and shoes at the site. Keep victim quiet and maintain normal body temperature. Effects may be delayed; keep victim under observation.

★ 635 ★
METHALLYL ALCOHOL
CAS: 513-42-8
Other names: ISOPROPENYL CARBINOL; 2-PROPEN-1-OL, 2-METHYL-

Danger profile: Combustible, poisonous gas produced in fire.
Uses: Making other chemicals.
Appearance: Colorless liquid.
Odor: Pungent. The odor is a warning of exposure; however, no smell does not mean you are not being exposed.
Effects of exposure:

Breathed: May be poisonous. May cause irritation of the eyes, nose and throat; dizziness and suffocation.
Eyes: Contact may cause irritation and burns.
Skin: May be poisonous if absorbed. Passes through the unbroken skin; can increase exposure and the severity of the symptoms listed. Contact may cause irritation and burns.
Swallowed: Moderately toxic. See **First aid guide.**
Storage: Store in tightly closed containers in a dry, cool, well-ventilated place. Vapors may travel to sources of ignition and flash back. Containers may explode in fire.
First aid guide: Move victim to fresh air and call emergency medical care; if not breathing, give artificial respiration; if breathing is difficult, give oxygen. In case of contact with material, immediately flush eyes with running water for at least 15 minutes. Wash skin with soap and water. Remove and isolate contaminated clothing and shoes at the site.

★ 636 ★
METHANE
CAS: 74-82-8
Other names: BIOGAS; FIRE DAMP; MARSH GAS; METHYL HYDRIDE; NATURAL GAS

Danger profile: Fire hazard, explosion hazard, reproductive hazard.
Uses: Making other chemical and synthetic proteins; fuel; a source of carbon black.
Appearance: Colorless gas; may be a liquid under pressure or refrigeration.
Odor: Mild, sweet. The odor is a warning of exposure; however, no smell does not mean you are not being exposed.
Breathed: May cause headache, dizziness, fatigue; rapid irregular breathing; nausea, vomiting, confusion, loss of consciousness convulsions and death. High levels may cause suffocation from lack of oxygen.
Eyes: Contact with liquid may cause frostbite, pain, stinging, tearing, inflammation of the eyelids and permanent damage.
Skin: Contact with liquid may cause frostbite, intense cold feeling, skin turns white and feels cold and hard.
Swallowed: Liquid may cause freezing burns of the mouth and throat; headache, confusion, drowsiness, unconsciousness and death.
Storage: Store in tightly closed containers in a dry, cool, well-ventilated place. Keep away from heat, sources of ignition, oxidizers, halogens, air. Vapors may travel to sources of ignition and flash back. Containers may explode in fire.

★ 637 ★
METHANOL
CAS: 67-56-1
Other names: ALCOOL METHYLIQUE (FRENCH); ALCOOL METILICO (ITALIAN); CARBINOL; COLONIAL SPIRIT; COLUMBIAN SPIRIT; METANOLO (ITALIAN); METHYL ALCOHOL; METHYLOL; METHYLALKOHOL (GERMAN); METHYL HYDROXIDE; METYLOWY ALKOHOL (POLISH); MONOHYDROXYMETHANE; PYROXYLIC SPIRIT; WOOD ALCOHOL; WOOD NAPHTHA; WOOD SPIRIT

Danger profile: Mutagen, fire hazard, explosion hazard, poison, exposure limit established, Community Right-to-Know List, poisonous gas produced in fire.
Uses: Making other chemicals synthetic proteins; antifreeze; solvent for various products including shellac, dyes; fuel for utility plants; source of hydrogen for fuel cells; in home heating oils; in furniture finish rejuvenators.
Appearance: Clear, colorless liquid.
Odor: Faintly sweet; slightly alcohol-like. The odor is a warning of exposure; however, no smell does not mean you are not being exposed.
Effects of exposure:

Breathed: Mildly toxic. May cause irritation of eyes, nose and throat; headache, drowsiness, excitability, drunken behavior, mental confusion, fatigue. Very high levels of exposure may cause dizziness, lightheadedness and cause victim to pass out, go into coma and die. May cause blindness.
Eyes: Contact may cause irritation, pain, burning, stinging, tearing, inflammation of the eyelids. Light may cause pain.
Skin: May be poisonous if absorbed. Passes through the unbroken skin; can increase exposure and the severity of the symptoms listed. May remove oil from the skin and cause dryness and cracking. Persons with pre-existing skin disorders may be more susceptible to the effects of this substance.
Swallowed: May be poisonous. May cause gastro-intestinal irritation; blindness. See *Breathed.*

Long-term exposure: Prolonged or repeated contact may cause irritation of the skin and dermatitis with dryness and cracking; liver damage; optic nerve effects, visual field changes; possible blindness; reproductive effects reported. A slow poison which is eliminated very slowly from the body.

Storage: Store in a tightly closed container in a cool, well-ventilated place. Keep away from heat and sources of ignition and oxidizers. Vapors are explosive in heat or flame. Containers may explode in fire.

First aid guide: Move victim to fresh air and call emergency medical care; if not breathing, give artificial respiration; if breathing is difficult, give oxygen. In case of contact with material, immediately flush skin or eyes with running water for at least 15 minutes. Remove and isolate contaminated clothing and shoes at the site. Keep victim quiet and maintain normal body temperature. Effects may be delayed; keep victim under observation.

★ 638 ★
METHIOCARB
CAS: 2032-65-7
Other names: BAY 9026; BAY 37344; 3,5-DIMETHYL-4-(METHYLTHIO)- ,METHYLCARBAMATE; 3,5-DIMETHYL-4-(METHYLTHIO)PHENOLMETHYLCARBAMATE; 3,5-DIMETHYL-4-METHYLTHIOPHENYL N-METHYLCARBAMATE; DRAZA; ENT 25,726; H 321; MER-CAPTODIMETHUR; MESUROL; METHYL CARBAMIC ACID 4-(METHYLTHIO)-3,5-XYLYL ESTER 4-METHYLMERCAPTO-3,5-DIMETHYLPHENYL N-METHYLCARBAMATE; 4-METHYLMERCAPTO-3,5-XYLYLMETHYLCARBAMATE; 4-METHYLTHIO-3,5-DIMETHYLPHENYLMETHYLCARBAMATE; 4-(METHYLTHIO)-3,5-XYLYLMETHYLCARBAMATE; MET-MERCAPTURON; OMS-93; 3,5-XYLENOL,4-(METHYLTHIO)-, METHYLCARBAMATE

Danger profile: Combustible, exposure limit established, EPA extremely hazardous substances list, poisonous gas produced in fire.
Uses: Insecticide (carbamate) used on a wide range of crops; bird and slug repellant.
Appearance: White crystalline powder.
Odor: Mild. The odor is a warning of exposure; however, no smell does not mean you are not being exposed.
Effects of exposure:
Breathed: Poisonous. May cause headache, dizziness, profuse sweating, nausea, vomiting, reduced heart beat, stomach cramps, diarrhea, loss of coordination, slow and weak breathing, fever, loss of consciousness and death. May cause shortness of breath and a dangerous build-up of fluids in the lungs (pulmonary edema), which can cause death. Pulmonary edema is a medical emergency that can be delayed one to two days following exposure.
Eyes: Poisoning can happen very rapidly. Contact may cause irritation and burns. See *Breathed.*
Skin: Poisonous if absorbed. Passes through the unbroken skin; can increase exposure and the severity of the symptoms listed. Contact may cause irritation and burns.
Swallowed: Powerful poison. See symptoms under *Breathed.*
Long-term exposure: May cause nerve damage including loss of coordination; liver damage. Repeated exposure may cause depression, irritability and personality changes.
Storage: Store in a regulated, specially marked place in tightly closed containers in a dry, well-ventilated place. Containers may explode in fire.

★ 639 ★
METHOMYL
CAS: 16752-77-5
Other names: ACETIMIDIC ACID, THIO-N-(METHYLCARBAMOYL)OXY-,METHYL ESTER; ACETIMIDOTHIOIC ACID, METHYL-N-(METHYLCARBAMOYL) ESTER; DUPONT INSETICIDE 1179; ENT 27,341; FRAM FLY KILL; INSECTICIDE 1,179; LANNATE; LANOX 90; LANOX 216; MESOMILE; METHYL N-(METHYLAMINO(CARBONYL)OXY)ETHANIMIDO)THIOATE; METHYL-N-(METHYL(CARBAMOYL)OXY)THIOACETIMIDATE; S-METHYL N-(METHYLCARBAMOYLOXY)THIOACETIMIDATE; 2-METHYLTHIO-PROPIONALDEHYD-O-(METHYLCARBAMOYL)OXIM (GERMAN); METOMIL (ITALIAN); NU-BAIT II; NUDRIN; RENTOKILL; RIDECT; SD 14999; SOREX GOLDEN FLY BAIT; 3-THIABUTAN-2-ONE,O-(METHYLCARBAMOYL)OXIME; WL 18236

Danger profile: Combustible, exposure limit established, EPA extremely hazardous substances list, poisonous gas produced in fire.
Uses: Insecticide for many insects including aphids; on vegetables, hops and fruit (carbamate)
Appearance: White crystalline powder.
Odor: Slightly sulfurous odor. The odor is a warning of exposure; however, no smell does not mean you are not being exposed.
Effects of exposure:
Breathed: Extremely poisonous. May cause headache, dizziness, profuse sweating, nausea, vomiting, reduced heart beat, stomach cramps, diarrhea, loss of coordination, slow and weak breathing, fever, loss of consciousness and death. May cause shortness of breath and a dangerous build-up of fluids in the lungs (pulmonary edema), which can cause death. Pulmonary edema is a medical emergency that can be delayed one to two days following exposure.
Eyes: Rapidly absorbed. Poisoning can happen very rapidly. See *Breathed.*
Skin: Poisonous. Passes through the unbroken skin; can increase exposure and the severity of the symptoms listed.
Swallowed: Powerful, deadly poison. See *Breathed.*
Long-term exposure: May cause birth defects; nerve damage including loss of coordination; liver damage. Repeated exposure may cause depression, irritability and personality changes.
Storage: Store in a regulated, specially marked place in tightly closed containers in a dry, well-ventilated place. Keep away from strong oxidizers. Reacts with certain nitrogen compounds to produce nitrosomethomyl which is potent mutagen, carcinogen and teratogen. Containers may explode in fire.

★ 640 ★
4-METHOXYPHENOL
CAS: 150-76-5
Other names: P-GUAICOL; HYDROQUINONE MONOMETHYL ETHER; 1-HYDROXY-4-METHOXYBENZENE; MEQUINOL; P-METHOXYPHENOL; MME; MONO METHYL ETHER HYDROQUINONE; PHENOL,P-METHOXY; USAF AN-7

Danger profile: Mutagen, combustible, exposure limit established, poisonous gas produced in fire.
Uses: Making other chemicals, pharmaceuticals, plasticizers, dyestuffs; chemical stabilizer.
Appearance: White waxy solid.
Effects of exposure:
Breathed: May cause irritation to the nose, throat and breathing passages.

Eyes: Contact may cause severe irritation, burns and permanent damage. Repeated contact may cause eye discoloration and vision damage.

Skin: Passes through the unbroken skin; can increase exposure and the severity of the symptoms listed. May cause irritation and discoloration. Prolonged contact may cause severe burns and scarring.

Swallowed: Moderately toxic.

Long-term exposure: May cause discoloration of the eyes and skin; possible eye damage.

Storage: Store in tightly closed containers in a dry, cool, well-ventilated place or in a refrigerator. Keep away from heat, sources of ignition, moisture and oxidizers. Containers may explode in fire.

★641★
METHYL ACETATE
CAS: 79-20-9

Other names: ACETATE DE METHYLE (FRENCH); ACETIC ACID, METHYL ESTER; DEVOTON; METHYLACETAAT (DUTCH); METHYLACETAT (GERMAN); METHYL ACETIC ESTER; METHYLE (ACETATE DE) (FRENCH); METHYL ETHANOATE; METHYLESTER KISELINY OCTOVE (CZECH); METILE (ACETATO DI) (ITALIAN); OCTAN METYLU (POLISH); TERETON

Danger profile: Mutagen, fire hazard, explosion hazard, exposure limit established, poisonous gas produced in fire.

Uses: Solvent in paint removers, lacquers; synthetic flavoring; making other chemicals and pharmaceuticals.

Appearance: Colorless liquid.

Odor: Slightly fruity, mild, sweet. The odor is a warning of exposure; however, no smell does not mean you are not being exposed.

Effects of exposure:

Breathed: May be poisonous. May cause irritation of the eyes, nose, throat and lungs; dizziness, lightheadedness; headache, fatigue, and drowsiness; high concentrations can produce central nervous system depression and optic nerve damage. Higher exposures may cause shortness of breath and a dangerous build-up of fluids in the lungs (pulmonary edema), which can cause death. Pulmonary edema is a medical emergency that can be delayed one to two days following exposure. Repeated or high exposure may cause headaches, dizziness, unconsciousness, coma and blindness; possible death.

Eyes: Contact may cause irritation. Overexposure to vapor or liquid may damage optic nerve.

Skin: Contact may cause defatting and cracking.

Swallowed: May be poisonous. Ingestion causes headache, dizziness, drowsiness, fatigue; may cause severe eye damage.

Long-term exposure: Mutation data reported. Repeated exposure may cause blindness; possible lung damage.

Storage: Store in tightly closed containers in a dry, cool, well-ventilated place. Keep away from heat, sources of ignition, oxidizing materials, acids, alkalis and acetate. Vapors may travel to sources of ignition and flash back. Containers may explode in fire.

First aid guide: Move victim to fresh air and call emergency medical care; if not breathing, give artificial respiration; if breathing is difficult, give oxygen. In case of contact with material, immediately flush eyes with running water for at least 15 minutes. Wash skin with soap and water. Remove and isolate contaminated clothing and shoes at the site.

★642★
METHYLACRYLATE
CAS: 96-33-3

Other names: ACRYLATE DE METHYLE (FRENCH); ACRYLIC ACID METHYL ESTER; ACRYLSAEUREMETHYLESTER (GERMAN); CURITHANE 103; METHOXYCARBONYLETHYLENE; METHYL ACRYLATE; METHYLACRYLAAT (DUTCH); METHYL-ACRYLAT (GERMAN); METHYL PROPENATE; METHYL PROPENOATE; METHYL-2-PROPENOATE; METILACRILATO (ITALIAN); PROPENOIC ACID METHYL ESTER; 2-PROPENOIC ACID, METHYL ESTER

Danger profile: Mutagen, fire hazard, explosion hazard, exposure limit established, Community Right-to-Know List, poisonous gas produced in fire.

Uses: Making other chemicals, vitamin B1, adhesives, polymers and surfactants.

Appearance: Clear colorless liquid.

Odor: Sharp, sweet, fruity, fragrant. The odor is a warning of exposure; however, no smell does not mean you are not being exposed.

Effects of exposure:

Breathed: May be poisonous. May cause irritation of the eyes, nose, throat and lungs, tears and runny nose, coughing, sore throat, difficult breathing; lethargy, fatigue, and drowsiness; slurred speech, shaking, and dizziness. Higher exposures could cause shortness of breath, headache, nausea, vomiting, dizziness, diarrhea and a dangerous build-up of fluids in the lungs (pulmonary edema), which can cause death. Pulmonary edema is a medical emergency that can be delayed one to two days following exposure.

Eyes: May cause irritation, redness, tearing, pain, inflamed eyelids, and eye injury; burns in corneas and possible loss of sight.

Skin: Toxic. Passes through the unbroken skin; can increase exposure and the severity of the symptoms listed. May cause painful irritation, redness, and blisters. Repeated or prolonged skin contact may cause irritation of the skin and may also cause the legs to feel numb.

Swallowed: May be poisonous. May cause irritation of the mouth and throat; stomach and abdominal pain; loss of muscular coordination and mental confusion; convulsions. Also see symptoms for *Breathed.*

Long-term exposure: May cause liver and kidney damage; skin allergy; possible lung problems. May lead to birth defects, miscarriage and cancer.

Storage: Store in tightly closed containers in a dry, cool, well-ventilated place. Keep away from heat, sources of ignition, direct sunlight, moisture, and oxidizing materials. Vapors may travel to sources of ignition and flash back. Containers may explode in fire.

First aid guide: Move victim to fresh air and call emergency medical care; if not breathing, give artificial respiration; if breathing is difficult, give oxygen. In case of contact with material, immediately flush eyes with running water for at least 15 minutes. Wash skin with soap and water. Remove and isolate contaminated clothing and shoes at the site.

★643★
METHYLAL
CAS: 109-87-5

Other names: ANESTHENYL; DIMETHYLACETAL FORMALDEHYDE; DIMETHOXYMETHANE; FORMAL; FORMALDEHYDE DIMETHYLACETAL; METHANE, DIMETHOXY-; METHYLENE DIMETHYL ETHER; METYLAL (POLISH); METHYL FORMAL

Danger profile: Fire hazard, explosion hazard, exposure limit established, poisonous gas produced in fire.

Uses: Solvent; perfumes, adhesives and protective coatings; fuel; making other chemicals.
Appearance: Colorless liquid.
Odor: Harsh, chloroform-like. The odor is a warning of exposure; however, no smell does not mean you are not being exposed.
Effects of exposure:
Breathed: May cause irritation to the eyes, nose, throat and lungs; cough, shortness of breath; dizziness, lightheadedness, unconsciousness; depression of central nervous system. Higher exposures may cause shortness of breath and a dangerous build-up of fluids in the lungs (pulmonary edema), which can cause death. Pulmonary edema is a medical emergency that can be delayed one to two days following exposure.
Eyes: Contact may cause irritation.
Skin: Prolonged contact may cause irritation.
Swallowed: May cause depression of central nervous system. See symptoms under *Breathed.*
Long-term exposure: May cause liver, lung, heart and kidney damage; skin irritation.
Storage: Store in tightly closed containers in a dry, cool, well-ventilated place. Keep away from heat, sources of ignition, oxygen and oxidizing materials. Vapors may travel to sources of ignition and flash back. Containers may explode in fire.
First aid guide: Move victim to fresh air and call emergency medical care; if not breathing, give artificial respiration; if breathing is difficult, give oxygen. In case of contact with material, immediately flush eyes with running water for at least 15 minutes. Wash skin with soap and water. Remove and isolate contaminated clothing and shoes at the site.

★ 644 ★
METHYL ALUMINUM SESQUICHLORIDE
CAS: 12542-85-7
Other names: ALUMINUM, TRICHLOROTRIMETHYLDI-; METHYLALUMINUM SESQUICHLORIDE; TRICHLOROTRI-METHYLDIALUMINUM

Danger profile: Combustible, explosive, highly reactive, exposure limit established, poisonous gas produced in fire.
Uses: Making other chemicals.
Appearance: Clear liquid.
Effects of exposure:
Breathed: May be poisonous. Vapors may cause irritation of the nose, throat, breathing passages and lungs; coughing and tightness in the chest. Higher exposures may cause shortness of breath and a dangerous build-up of fluids in the lungs (pulmonary edema), which can cause death. Pulmonary edema is a medical emergency that can be delayed one to two days following exposure.
Eyes: Contact may cause burns and permanent damage.
Skin: Contact may cause irritation, redness, rash and burns.
Swallowed: See *Breathed* and *First aid guide.*
Long-term exposure: May cause skin rash; possible lung problems.
Storage: Store in tightly closed containers in a dry, cool, well-ventilated place. Keep away from air, water and other forms of moisture, combustible or reactive materials. May ignite in presence of air or moisture. Containers may explode in fire.
First aid guide: Move victim to fresh air; call emergency medical care. Wipe material from skin immediately; flush skin or eyes with running water for at least 15 minutes. Remove and isolate contaminated clothing and shoes at the site.

★ 645 ★
METHYLAMINE
CAS: 74-89-5
Other names: AMINOMETHANE; CARBINAMINE; MERCURI-ALIN; METHANAMINE; METHANAMINE (9CI); METHYL-AMINEN (DUTCH); METILAMINE (ITALIAN); METYLOAMINA (POLISH); MONOMETHYLAMINE

Danger profile: Mutagen, fire hazard, severe skin irritant, explosive, exposure limit established, poisonous gas produced in fire.
Uses: Making other chemicals, dyes, pharmaceuticals, insecticides, fungicides; tanning; dyeing textiles; fuel additive; in paint removers; solvent; photographic developer; rocket propellant.
Appearance: Colorless gas or liquid.
Odor: Powerful, ammonia-like, suffocating. The odor is a warning of exposure; however, no smell does not mean you are not being exposed.
Effects of exposure:
Breathed: May be poisonous. May cause irritation of the eyes, nose, throat, breathing passages and lungs. Higher exposures may cause shortness of breath and a dangerous build-up of fluids in the lungs (pulmonary edema), which can cause death. Pulmonary edema is a medical emergency that can be delayed one to two days following exposure.
Eyes: Contact with liquid may cause frostbite, severe irritation and burns.
Skin: Passes through the unbroken skin; can increase exposure and the severity of the symptoms listed. Contact with liquid may cause frostbite, severe irritation and burns.
Swallowed: See *Breathed.*
Long-term exposure: May cause, birth defects, miscarriage or cancer. Repeated exposure may cause bronchitis; cough, shortness of breath and phlegm.
Storage: Store in tightly closed containers in a dry, cool, well-ventilated place with non-wooden floors. Keep away from heat, sources of ignition, nitromethane, mercury, oxidizers, and combustible materials. Vapors may travel to sources of ignition and flash back. Containers may explode in fire.

★ 646 ★
METHYL AMYL KETONE
CAS: 110-43-0
Other names: AMYL-METHYL-CETONE (FRENCH); AMYL METHYL KETONE; N-AMYL METHYL KETONE; BUTYL ACETONE; 2-HEPTANONE; KETONE, METHYL PENTYL; METHYL-AMYL-CETONE (FRENCH); METHYLAMYL KETONE; METHYL (N-AMYL) KETONE; N-AMYL METHYL KETONE; METHYL PENTYL KETONE

Danger profile: Combustible, exposure limit established, poisonous gas produced in fire.
Uses: Lacquer solvent; flavoring agent; making perfumes.
Appearance: Clear, colorless liquid.
Odor: Penetrating, fruity banana-like. The odor is a warning of exposure; however, no smell does not mean you are not being exposed.
Effects of exposure:
Breathed: May be poisonous. May cause dizziness, lightheadedness and unconsciousness. Concentrated vapor may have a narcotic effect.
Eyes: Contact may cause irritation and burns.
Skin: May be poisonous if absorbed. Passes through the unbroken skin; can increase exposure and the severity of the symptoms listed. Prolonged and repeated contact with skin may cause defatting, irritation and burns.
Swallowed: Moderately toxic. May cause gastrointestinal disturbances.

Long-term exposure: May cause liver and kidney damage; skin irritation with dryness and cracking; possible nervous system damage.

Storage: Store in tightly closed containers in a dry, cool, well-ventilated place. Keep away from heat, sources of ignition and oxidizers. Will attack some plastics. Vapors may travel to sources of ignition and flash back. Containers may explode in fire.

First aid guide: Move victim to fresh air and call emergency medical care; if not breathing, give artificial respiration; if breathing is difficult, give oxygen. In case of contact with material, immediately flush eyes with running water for at least 15 minutes. Wash skin with soap and water. Remove and isolate contaminated clothing and shoes at the site.

★647★
METHYLANILINE
CAS: 100-61-8
Other names: ANILINOMETHANE; BENZENENAMINE, N-METHYL- (9CI); (METHYLAMINO)BENZENE; N-METHYLAMINOBENZENE; METHYL ANILINE; N-METHYLANILINE; N-METHYLBENZENAMINE; METHYLPHENYLAMINE; N-METHYLPHENYLAMINE; MONOMETHYL ANILINE; N-MONOMETHYLANILINE; MA; MONOMETHYL ANILINE; N-PHENYLMETHYLAMINE

Danger profile: Combustible, exposure limit established, poisonous gas produced in fire.
Uses: Solvent; making other chemicals.
Appearance: Yellow to light brown oily liquid.
Odor: Weak ammonia-like. The odor is a warning of exposure; however, no smell does not mean you are not being exposed.
Effects of exposure:
   Breathed: May be poisonous. May interfere with the blood's ability to carry oxygen and cause dizziness, headache, weakness, bluish discoloration of the lips, ear lobes, fingernails and nose. Higher exposures may cause shortness of breath, collapse and death.
   Eyes: Contact may cause irritation.
   Skin: May be poisonous if absorbed. Passes through the unbroken skin; can increase exposure and the severity of the symptoms listed.
   Swallowed: May be poisonous. May cause bladder damage and bloody urine. See symptoms under *Breathed.*
Long-term exposure: May cause anemia; liver, bladder and kidney damage.
Storage: Store in tightly closed containers in a dry, cool, well-ventilated place. Keep away from heat, sources of ignition and strong acids. May attack some forms of plastic. Containers may explode in fire.
First aid guide: Move victim to fresh air and call emergency medical care; if not breathing, give artificial respiration; if breathing is difficult, give oxygen. In case of contact with material, immediately flush skin or eyes with running water for at least 15 minutes. Speed in removing material from skin is of extreme importance. Remove and isolate contaminated clothing and shoes at the site. Keep victim quiet and maintain normal body temperature. Effects may be delayed; keep victim under observation.

★648★
METHYL BUTYL KETONE
CAS: 591-78-6
Other names: BUTYL METHYL KETONE; N-BUTYL METHYL KETONE; 2-HEXANONE; HEXANONE-2; KETONE, BUTYL METHYL; MBK; METHYL-N-BUTYL KETONE; MNBK; PROPYLACETONE

Danger profile: Combustible, exposure limit established, poisonous gas produced in fire.
Uses: Solvent.
Appearance: Colorless liquid.
Odor: Odor resembling nail polish remover; acetone-like. The odor is a warning of exposure; however, no smell does not mean you are not being exposed.
Effects of exposure:
   Breathed: May cause irritation of the eyes, nose and throat; dizziness, lightheadedness, and unconsciousness. High concentrations of vapor may result in nerve damage; numbness of the hands and feet.
   Eyes: Contact may cause mild to moderate irritation.
   Skin: Prolonged or repeated contact may cause defatting of the skin with resultant dermatitis; drying and cracking.
   Swallowed: May cause eye effects, headache, nausea and vomiting.
Long-term exposure: May cause damage to the male reproductive system; drying and cracking of the skin; lower the white blood cell count; possible damage to the developing fetus.
Storage: Store in tightly closed containers in a dry, cool, well-ventilated place. Keep away from heat, sources of ignition and oxidizers. Vapors may travel to sources of ignition and flash back. Containers may explode in fire.

★649★
METHYL CELLOSOLVE
CAS: 109-86-4
Other names: AETHYLENGLYKOL-MONOMETHYLAETHER (GERMAN); DOWANOL EM; EGM; EGME; ETHER MONO-METHYLIQUE DE L'ETHYLENE-GLYCOL (FRENCH); ETHYLENE GLYCOL MONOMETHYL ETHER; GLYCOL ETHER EM; GLYCOL MONOMETHYL ETHER; JEFFERSOL EM; MECS; 2-METHOXY-AETHANOL (GERMAN); METHYL GLYCOL; METHLYGLYKOL (GERMAN); 2- METHOXYETHANOL; METHOXYHYDROXYETHANE; POLY-SOLV EM; PRIST

Danger profile: Reproductive hazard, combustible, exposure limit established, Community Right-to-Know List, poisonous gas produced in fire.
Uses: In cleaning compounds; solvent for resins, lacquers, paints, and varnishes; fuel de-icer.
Appearance: Colorless odorless liquid.
Odor: Ether-like. The odor is a warning of exposure; however, no smell does not mean you are not being exposed.
Effects of exposure:
   Breathed: May cause irritation of the eyes, nose, throat and lungs; cough, shortness of breath. Exposure may cause kidney damage. Higher levels of exposures may cause dizziness, lightheadedness, shortness of breath and a dangerous build-up of fluids in the lungs (pulmonary edema), which can cause death. Pulmonary edema is a medical emergency that can be delayed one to two days following exposure.
   Eyes: May cause irritation.
   Skin: Passes through the unbroken skin; can increase exposure and the severity of the symptoms listed. May cause irritation.
   Swallowed: May cause nausea, confusion, weakness, rapid breathing
Long-term exposure: May cause birth defects, miscarriage, cancer; kidney liver and brain damage; headache, drowsiness, tremors, personality changes, gastrointestinal upset, weight loss; anemia; possible lung damage.
Storage: Store in tightly closed containers in a dry, cool, well-ventilated place. Keep away from strong caustics and oxidizers. Containers may explode in fire.

★650★
METHYL CELLOSOLVE ACETATE

CAS: 110-49-6

Other names: ACETATE DE L'ETHER MONOMETHYLIQUE DE L'ETHYLENE-GLYCLO (FRENCH); ACETATE DE METHYLE GLYCOL (FRENCH); ACETO DI METIL CELLOSOLVE (ITALIAN); ETHYLENE GLYCOL METHYL ETHER ACETATE; GLYCOL ETHER EM ACETATE; GLYCOL MONOMETHYL ETHER ACETATE; 2-METHOXYETHYL ACETATE; 2-METHOXY-ETHYL ACETAAT (DUTCH); 2- METHOXYETHYL ACETATE; 2-METHOXYETHYLE, ACETATE DE (FRENCH); METHYL GLYCOL ACETATE; METHYL GLYCOL MONOACETATE; METHYLGLYKOLACETAT (GERMAN); 2-METOSSIETILACETATO (ITALIAN)

Danger profile: Combustible, exposure limit established, Community Right-to-Know List, poisonous gas produced in fire.

Uses: Solvent for oil, grease, ink; in paints, lacquers and adhesives.

Appearance: Colorless liquid.

Odor: Pleasant, sweet, ether-like. The odor is a warning of exposure; however, no smell does mean you are not being exposed.

Effects of exposure:

Breathed: May cause irritation of the eyes, nose, throat and lungs; tearing, cough, shortness of breath. Exposure may cause kidney damage. Higher levels of exposures may cause dizziness, lightheadedness, coughing and shortness of breath.

Eyes: May cause irritation.

Skin: Passes through the unbroken skin; can increase exposure and the severity of the symptoms listed. May cause irritation.

Swallowed: May cause nausea, confusion, weakness, rapid breathing.

Long-term exposure: May cause anemia, kidney, nerve and brain damage; possible lung damage; possible damage to male reproductive organs; possible birth defects and miscarriage.

Storage: Store in tightly closed containers in a dry, cool, well-ventilated place. Keep away from heat, sources of ignition and oxidizers. Vapors may travel to sources of ignition and flash back. Containers may explode in fire.

★651★
METHYL CHLORIDE

CAS: 74-87-3

Other names: ARTIC; CHLOOR-METHAAN (DUTCH); CHLOR-METHAN (GERMAN); CHLOROMETHANE; CHLORURE DE METHYLE (FRENCH); CLOROMETANO (ITALIAN); CLORURO DI METILE (ITALIAN); METHYLCHLORID (GERMAN); METYLU CHLOREK (POLISH); MONOCHLOROMETHANE; METHANE, CHLORO-

Danger profile: Carcinogen, reproductive hazard, mutagen, fire hazard, explosion hazard, exposure limit established, Community Right-to-Know List, poisonous gas produced in fire.

Uses: Making other chemicals; refrigerant; solvent; herbicide; topical anesthetic.

Appearance: Colorless gas.

Odor: Faint, ether-like, sweet. Not noticed at dangerous concentrations. The odor is a warning of exposure; however, no smell does not mean you are not being exposed.

Effects of exposure:

Breathed: May be poisonous. May cause headache, dizziness, fatigue, headache, emotional disturbances (i.e., "drunken behavior"); dimness of vision, double or blurred vision, muscular tremors, blue skin, earlobes and lips (cyanosis), convulsions; rapid irregular breathing; nausea, vomiting, delirium, confusion, loss of consciousness convulsions and death. If the exposure is not fatal, recovery may be slow.

These effects may be delayed for several hours. Very high exposures may cause a dangerous build-up of fluids in the lungs (pulmonary edema), which can cause death. Pulmonary edema is a medical emergency that can be delayed one to two days following exposure.

Eyes: Contact with liquid may cause frostbite, pain, stinging, tearing, inflammation of the eyelids and permanent damage.

Skin: May be poisonous if absorbed. Passes through the unbroken skin; can increase exposure and the severity of the symptoms listed. Contact with liquid may cause frostbite, intense cold feeling, skin turns white and feels cold and hard.

Swallowed: May be poisonous. Liquid may cause headache, confusion, drowsiness, unconsciousness and death.

Long-term exposure: May cause cancer; birth defects, miscarriage; central nervous system damage; damage to the testes with decreased male hormone and sperm production; liver, kidney cardiovascular, and bone marrow damage; brain damage; possible lung damage.

Storage: Store in tightly closed containers in a dry, cool, well-ventilated place. Keep away from direct sunlight, heat, sources of ignition, bromine compounds, zinc, magnesium and alloys, sodium and alloys, potassium and alloys, oxidizers and aluminum chloride, powdered aluminum and other chemically active metals. Vapors may travel to sources of ignition and flash back. Containers may explode in fire.

First aid guide: Move victim to fresh air and call emergency medical care; if not breathing, give artificial respiration; if breathing is difficult, give oxygen. In case of contact with material, immediately flush skin or eyes with running water for at least 15 minutes. Remove and isolate contaminated clothing and shoes at the site. Keep victim quiet and maintain normal body temperature. Effects may be delayed; keep victim under observation.

★652★
METHYL CHLOROFORM

CAS: 71-55-6

Other names: AEROTHENE TT; CHLOROETENE; CHLOROETHENE; CHLOROETHENE NU; CHLOROFORM, METHYL-; CHLOROTHANE NU; CHLOROTHENE; CHLOROTHENE NU; CHLOROTHENE VG; CHLORTEN; INHIBISOL; ETHANE, 1,1,1-TRICHLORO-; METHYLCHLOROFORM; METHYLTRICHLOROMETHANE; NCI-C04626; SOLVENT 111; ALPHA-T; 1,1,1-TCE; 1,1,1-TRICHLOORETHAAN (DUTCH); 1,1,1-TRICHLORAETHAN (GERMAN); TRICHLORO-1,1,1-ETHANE (FRENCH); ALPHA-TRICHLOROETHANE; 1,1,1-TRICHLOROETHANE; 1,1,1-TRICLOROETANO (ITALIAN); TRI-ETHANE

Danger profile: Exposure limit established, Community Right-to-Know List, poisonous gas produced in fire.

Uses: Cleaning solvent, metal degreasing; pesticide; making and finishing textiles; making other chemicals; aerosol propellant.

Appearance: Colorless liquid.

Odor: Chloroform-like, sweetish. The odor is a warning of exposure; however, no smell does mean you are not being exposed.

Effects of exposure:

Breathed: May cause irritation of the eyes, nose, throat and air passages; coughing, headache; increased saliva; reddening of the face; dizziness, lightheadedness, fatigue, loss of consciousness, shock, coma, irregular heartbeat; loss of equilibrium and coordination; high concentration can be fatal due to simple asphyxiation combined with loss of consciousness. If the victim recovers watch for the possibility of a build-up of fluid in the lungs (pulmonary edema). Pulmonary edema is a medical emergency that can be delayed one to two days following exposure.

Eyes: Contact may cause irritation, tearing, inflammation of the eyelids; burning sensation.

Skin: Passes through the unbroken skin; can increase exposure and the severity of the symptoms listed. May cause irritation, redness, burning feeling, swelling; dry skin; rash or blisters.

Swallowed: Poison. Breath may smell of chloroform. May cause irritation of the mouth, gastro-intestinal tract, nausea, vomiting, diarrhea with possible blood; drowsiness; hallucinations or distorted perceptions; irritability, aggression, loss of consciousness; irregular heartbeat and death. Similar to **Breathed.**

Long-term exposure: May cause genetic changes; liver and kidney damage; dermatitis with skin thickening and cracking.

Storage: Store in tightly closed containers in a dry, cool, well-ventilated place. Keep away from heat, moisture, strong caustics, acetone, aluminum, zinc and other chemically active metals, and oxidizers. Keep vapors away from ultraviolet light. Air/vapor mixture of chemical may explode if ignited. Containers may explode in fire.

First aid guide: Move victim to fresh air and call emergency medical care; if not breathing, give artificial respiration; if breathing is difficult, give oxygen. In case of contact with material, immediately flush eyes with running water for at least 15 minutes. Wash skin with soap and water. Remove and isolate contaminated clothing and shoes at the site. Use first aid treatment according to the nature of the injury.

★ 653 ★
METHYL 2-CYANOACRYLATE
CAS: 137-05-3
Other names: ACRYLIC ACID,2-CYANO-, METHYL ESTER; ADHERE; COAPT; ALPHA-CYANOACRYLIC ACID METHYL ESTER; 2-CYANOACRYLIC ACID (ACC); 2-CYANOACRYLIC ACID, METHYL ESTER; CYANOLIT; 2-CYANO-2- PROPENOIC ACID, METHYL ESTER; EASTMAN 910; EASTMAN 910 MONOMER; MECRILAT; MECRYLATE; METHYL ALPHA-ACRYLATE; METHYL CYANOACRYLATE; METHYL ALPHA-CYANOACRYLATE; SUPER GLUE

Danger profile: Combustible, exposure limit established, poisonous gas produced in fire.
Uses: Super Glue liquid adhesive.
Appearance: Thick, clear, colorless liquid.
Odor: Sharp, irritating. The odor is a warning of exposure; however, no smell does mean you are not being exposed.
Effects of exposure:
  Breathed: May cause irritation to the eyes, nose, and throat. Higher exposures may cause shortness of breath and a dangerous build-up of fluids in the lungs (pulmonary edema), which can cause death. Pulmonary edema is a medical emergency that can be delayed one to two days following exposure.
  Eyes: Contact may cause irritation. Can bond instantly to eyelids.
  Skin: Contact may cause irritation; allergy may develop. Can bond instantly to skin.
Long-term exposure: Skin allergy with itching and rash may develop. May cause liver and kidney damage. Repeated breathing of fumes may cause lung problems.
Storage: Store in tightly closed containers in a dry, cool, well-ventilated place. Keep away from water, peroxides and alkaline material. May ignite spontaneously when wet. Containers may explode in fire.

★ 654 ★
METHYL CYCLOHEXANE
CAS: 108-87-2
Other names: CYCLOHEXANE, METHYL-; CYCLOHEXYLMETHANE; HEXAHYDROTOLUENE; METHYLCYCLOHEXANE; METHYLOCYKLOHEKSAN (POLISH); SEXTONE B; TOLUENE HEXAHYDRIDE

Danger profile: Fire hazard, explosion hazard, exposure limit established, poisonous gas produced in fire.
Uses: Solvent; making other chemicals.
Appearance: Colorless liquid.
Odor: Very faint benzene-like. The odor is a warning of exposure; however, no smell does not mean you are not being exposed. Has poor warning properties.
Effects of exposure:
  Breathed: May be poisonous. Vapor or mist is irritating to the eyes, mucous membrane and upper respiratory tract; dizziness, lightheadedness. At higher levels unconsciousness, coma and death may occur.
  Eyes: Vapor may cause slight irritation.
  Skin: May be poisonous if absorbed. Has poor warning properties. Vapor may cause slight irritation. Contact may cause drying and cracking; dermatitis.
  Swallowed: Moderately toxic. See **Breathed** for symptoms.
Long-term exposure: May have a narcotic effect. May cause dermatitis with drying a cracking skin.
Storage: Store in tightly closed containers in a dry, cool, well-ventilated place. Keep away from heat, sources of ignition and strong oxidizers. Vapors may travel to sources of ignition and flash back. Containers may explode in fire.
First aid guide: Move victim to fresh air and call emergency medical care; if not breathing, give artificial respiration; if breathing is difficult, give oxygen. In case of contact with material, immediately flush eyes with running water for at least 15 minutes. Wash skin with soap and water. Remove and isolate contaminated clothing and shoes at the site.

★ 655 ★
METHYL CYCLOHEXANOL
CAS: 25639-42-3
Other names: CYCLOHEXANOL, METHYL-; HEXAHYDROCRESOL; HEXAHYDROMETHYL PHENOL; METHYLCYCLOHEXANOL; METYLOCYKLOHEKSANOL (POLISH); METHYL CYCLOHEXANOL METHYLHEXALIN

Danger profile: Combustable, exposure limit established, poisonous gas produced in fire.
Uses: Solvent for lacquers; textile soaps and detergents; making textiles, soap and silk.
Appearance: Colorless or pale yellow viscous liquid.
Odor: Weak, coconut-like. The odor is a warning of exposure; however, no smell does not mean you are not being exposed.
Effects of exposure:
  Breathed: Prolonged or repeated exposure may cause irritation of the eyes, nose and throat; headaches. High exposure may cause heart, liver, kidney and/or lung damage; possible death.
  Eyes: May cause irritation.
  Skin: Passes through the unbroken skin; can increase exposure and the severity of the symptoms listed. May cause skin rash. Prolonged or repeated exposure may cause irritation of the eyes, nose and throat; headaches. High exposure may cause heart, liver, kidney and/or lung damage; possible death.
  Swallowed: See **Breathed** and **First aid guide.**
Long-term exposure: Prolonged or repeated exposure may cause irritation of the eyes, nose and throat; headaches.

Storage: Store in tightly closed containers in a dry, cool, well-ventilated place. Keep away from heat, sources of ignition and oxidizing agents. Vapors may travel to sources of ignition and flash back. Containers may explode in fire.

First aid guide: Move victim to fresh air and call emergency medical care; if not breathing, give artificial respiration; if breathing is difficult, give oxygen. In case of contact with material, immediately flush eyes with running water for at least 15 minutes. Wash skin with soap and water. Remove and isolate contaminated clothing and shoes at the site.

★ 656 ★
2-METHYLCYCLOHEXANONE
CAS: 583-60-8
Other names: METHYL ANONE; 2-METHYL-CYCLOHEXANON (GERMAN); 2-METHYL- CYCLOHEX-ANON (DUTCH); O-METHYLCYCLOHEXANONE; 2-METILCICLOESANONE (ITALIAN); SEXTON B

Danger profile: Combustible, exposure limit established, poisonous gas produced in fire.
Uses: Solvent in varnish and plastics; making leather.
Appearance: Colorless liquid.
Odor: Peppermint-like. The odor is a warning of exposure; however, no smell does not mean you are not being exposed.
Effects of exposure:
    Breathed: May cause irritation of the eyes, nose and throat; headache, dizziness, lightheadedness.
    Eyes: Contact may cause severe irritation and damage.
    Skin: Passes through the unbroken skin; can increase exposure and the severity of the symptoms listed. May cause irritation, burning sensation and rash.
    Swallowed: Moderately toxic. See *Breathed.*
Long-term exposure: May cause thickening and cracking of the skin; liver, lung and kidney damage.
Storage: Store in tightly closed containers in a dry, cool, well-ventilated place. Keep away from heat, sources of ignition and oxidizers. Vapors may travel to sources of ignition and flash back. Containers may explode in fire.

★ 657 ★
METHYL CYCLOPENTANE
CAS: 96-37-7
Other names: CYCLOPENTANE, METHYL-; METHYLCYCLOPENTANE

Danger profile: Combustible hazard, poisonous gas produced in fire.
Uses: Solvent; making other chemicals.
Appearance: Colorless liquid.
Odor: Gasoline-like. The odor is a warning of exposure; however, no smell does not mean you are not being exposed.
Effects of exposure:
    Breathed: May cause eye, nose and throat irritation; dizziness, lightheadedness, nausea, and vomiting; central nervous system excitement followed by depression; concentrated vapor may cause unconsciousness and collapse. Higher exposures may cause shortness of breath and a dangerous build-up of fluids in the lungs (pulmonary edema), which can cause death. Pulmonary edema is a medical emergency that can be delayed one to two days following exposure.
    Eyes: May cause irritation.
    Skin: Prolonged contact may cause irritation.
    Swallowed: May cause irritation of stomach.
Storage: Store in tightly closed containers in a dry, cool, well-ventilated place. Keep away from heat, sources of ignition and oxidizers. Vapors may travel to sources of ignition and flash back. Containers may explode in fire.

First aid guide: Move victim to fresh air and call emergency medical care; if not breathing, give artificial respiration; if breathing is difficult, give oxygen. In case of contact with material, immediately flush eyes with running water for at least 15 minutes. Wash skin with soap and water. Remove and isolate contaminated clothing and shoes at the site.

★ 658 ★
METHYL DEMETON
CAS: 8022-00-2
Other names: ASHLADE PERSYST; BAY 15203; BAYER 21/116; CAMPBELL'S DSM; CHAFER AZOTOX; DEMETON METHYL; DEMETON-S-METHYL 50; DURATOX; O-2-(ETHYLTHIO)ETHYL O,O-DIMETHYL PHOSPHOROTHIO-ATE; S-2-(ETHYLTHIO)ETHYL O,O-DIMETHYL PHOSPHO-ROTHIOATE; METASYSTOX; METHYL MERCAPTOPHOS; METHYL SYSTOX; PHOSPHOROTHIOC ACID, 0-2 (ETH-YLTHIO)ETHYL O,O-DIMETHYL ESTER, MIXED WITH S-2-(ETHYLTHIO)ETHYL O,O-DIMETHYL PHOSPHOROTHIO-ATE; POWER DSM 580; QUADRANGLE DSM; TRIPART SYSTEMIC INSECTICIDE; VASSGRO DSM

Danger profile: Exposure limit established, poisonous gas produced in fire.
Uses: Insecticide; used on mites, mangold fly, sawfly, leafhoppers and ticks; on wide range of vegetables, fruits, sugar beet (organophosphate).
Appearance: Colorless or yellow liquid.
Odor: Unpleasant. The odor is a warning of exposure; however, no smell does not mean you are not being exposed.
Effects of exposure:
    Breathed: Extremely poisonous. May cause headache, dizziness, profuse sweating, nausea, vomiting, reduced heart beat, stomach cramps, diarrhea, loss of coordination, slow and weak breathing, fever, loss of consciousness and death. May cause shortness of breath and a dangerous build-up of fluids in the lungs (pulmonary edema), which can cause death. Pulmonary edema is a medical emergency that can be delayed one to two days following exposure.
    Eyes: Rapidly absorbed. Poisoning can happen very rapidly. See *Breathed.*
    Skin: Poisonous. Passes through the unbroken skin; can increase exposure and the severity of the symptoms listed.
    Swallowed: Powerful, deadly poison. See *Breathed.*
Long-term exposure: May cause nerve damage including loss of coordination; liver damage. Repeated exposure may cause depression, irritability and personality changes.
Storage: Store in a regulated, specially marked place in tightly closed containers in a dry, well-ventilated place. Keep away from oxidizers. Containers may explode in fire.

★ 659 ★
METHYLDICHLOROSILANE
CAS: 75-54-7
Other names: DICHLOROMETHYLSILANE; METHYL DI-CHLOROSILANE; METHYL-DICHLORSILAN (CZECH); SI-LANE, DICHLOROMETHYL-

Danger profile: Fire hazard, explosion hazard (ignites spontaneously in air), corrosive, poisonous gas produced in fire.
Uses: Making silicone materials.
Appearance: Clear, straw-colored liquid.
Odor: Acrid, sharp, irritating; hydrochloric acid-like. The odor is a warning of exposure; however, no smell does not mean you are not being exposed.
Effects of exposure:
    Breathed: May be poisonous. May cause irritation of lungs and respiratory tract; coughing, difficult breathing, shortness

of breath. Higher exposures may cause shortness of breath and a dangerous build-up of fluids in the lungs (pulmonary edema), which can cause death. Pulmonary edema is a medical emergency that can be delayed one to two days following exposure.

Eyes: Contact may cause severe irritation, inflammation of the eyelids and burns.

Skin: Contact may cause severe irritation and burns.

Swallowed: May cause burns of mouth and stomach.

Long-term exposure: Very irritating; may effect the lungs.

Storage: Store in tightly closed containers in a dry, cool, well-ventilated place. Keep away from heat, sources of ignition, potassium permanganate, lead oxides, copper oxide, silver oxide, and moisture. Reacts with surface moisture to produce hydrogen chloride which is corrosive to common metals. Vapors may travel to sources of ignition and flash back. Containers may explode in fire.

First aid guide: Move victim to fresh air and call emergency medical care; if not breathing, give artificial respiration; if breathing is difficult, give oxygen. In case of contact with material, immediately flush skin or eyes with running water for at least 15 minutes. Remove and isolate contaminated clothing and shoes at the site. Keep victim quiet and maintain normal body temperature.

★ 660 ★
4,4'-METHYLENE BIS (2-CHLOROANILINE)
CAS: 101-14-4

Other names: BIS AMINE; BIS(4-AMINO-3-CHLOROPHENYL)METHANE; BOCA; CL-MDA; CURALIN M; CURENE 442; CYANASET; DACPM; DI(-4-AMINO-3-CHLOROPHENYL)METHANE; DI-(4-AMINO-3-CHLOROFENIL)METANO (ITALIAN); 4,4'-DIAMINO-3,3'-DICHLORODIPHENYLM ETHANE; 3,3'-DICHLOR-4,4'-DIAMINOPHENYLMETHAN (GERMAN); 3,3'-DICHLORO-4,4'-DIAMINODIFENILMETANO (ITALIAN); MBOCA; 4-4'-METHYLENE(BIS)-CHLOROANILINE; METHYLENE 4,4'-BIS(O-CHLOROANILINE); 4,4'-METHYLENE BIS-2-CHLOROBENZENAMINE; 4,4'-METHILENE-BIS-O-CHLOROANILINA (ITALIAN); MOCA

Danger profile: Carcinogen, mutagen, combustible, exposure limit established, Community Right-to-Know List, poisonous gas produced in fire.

Uses: Making epoxy resins and polyurethanes.

Appearance: Tan solid.

Effects of exposure:

Breathed: May cause headache, dizziness, nausea and a bluish color to the skin, and lips. Higher levels may cause difficult breathing, collapse and death.

Eyes: Contact may cause irritation.

Skin: Poisonous. Passes through the unbroken skin; can increase exposure and the severity of the symptoms listed.

Swallowed: Moderately toxic. See symptoms under ***Breathed.***

Long-term exposure: May cause cancer; low blood count; kidney problems and bloody urine.

Storage: Store in tightly closed containers in a dry, cool, well-ventilated place or in a refrigerator under an inert atmosphere. Keep away from chemically active metals such as sodium, potassium magnesium and zinc.

★ 661 ★
METHYLENE BIS(4-CYCLOHEXYLISOCYANATE)
CAS: 5124-30-1

Other names: BIS(4-ISOCYANATOCYCLOHEXYL)METHANE; 1'-METHYLENEBIS 4- ISOCYANATOCYCLOHEXANE; NACCONATE H 12

Danger profile: Exposure limit established, poisonous gas produced in fire.

Uses: Making urethane products.

Appearance: Colorless liquid.

Effects of exposure:

Breathed: Poisonous. May cause irritation of the eyes, nose and throat. Higher exposures may cause shortness of breath and a dangerous build-up of fluids in the lungs (pulmonary edema), which can cause death. Pulmonary edema is a medical emergency that can be delayed one to two days following exposure.

Eyes: May cause serious irritation, burns and permanent damage.

Skin: May cause irritation.

Swallowed: Mildly toxic. See symptoms under ***Breathed.***

Long-term exposure: May cause an asthma-like allergy; difficult breathing, shortness of breath, coughing and wheezing; possible lung damage from repeated exposure.

Storage: Store in tightly closed containers in a dry, cool, well-ventilated place. Keep away from moisture, alcohols, air, amines, strong bases and chemically active metals. Containers may explode in fire.

★ 662 ★
4,4-METHYLENEBIS(N,N-DIMETHYL) ANILINE
CAS: 101-61-1

Other names: ANILINE,4,4'-METHYLENE-BIS (N,N-DI-METHYL)-; P,P'-BIS(DIMETHYLAMINO) DIPHENYLME-THANE; 4,4'-BIS(DIMETHYLAMINO)DISPHENYLMETHANE; BIS(P-DIMETHYLAMINOPHENYL)METHANE; BIS(P-N,N-DIMETHYLAMINO-PHENYL)METHANE; P,P'-BIS(N,N-DIMETHYLAMINOPHENYL)METHANE; P,P-DIMETHYLAMINODIPHENYL-METHANE; METHANE BASE; 4,4'-METHYLENEBIS(N,N-DI-METHYL)BENZENAMINE; MICHLER'S BASE; MICHLER'S HYDRIDE; MICHLER'S METHANE; METHANE BASE; NCI-C01990; TETRA-BASE; TETRAMETHYLDIAMINODIPHENYLMETHANE; 4,4'-TETRAMETHYL-DIAMINODIPHENYL-METHANE; P,P-TETRAMETHYL-DIAMINODIPHENYL-METHANE

Danger profile: Carcinogen, mutagen, Community Right-to-Know List, poisonous gas produced in fire.

Uses: Making dyes; laboratory chemical.

Appearance: Crystalline compound.

Effects of exposure:

Breathed: May cause headache, dizziness, fatigue, and a bluish color to the skin, nose and lips. High levels man cause breathing difficulty, collapse, coma and death.

Eyes: May cause irritation.

Skin: Passes through the unbroken skin; can increase exposure and the severity of the symptoms listed. May cause irritation.

Swallowed: Moderately toxic. See ***Breathed.***

Long-term exposure: Related chemicals may cause skin allergy.

Storage: Store in tightly closed containers in a dry, cool, well-ventilated place. Keep away from strong acids and oxidizers.

★ 663 ★
METHYLENE CHLORIDE
CAS: 75-09-2

Other names: AEROTHENE MM; CHLORURE DE METHY-LENE (FRENCH); DCM; DICHLOROMETHANE; FREON 30; METHANE, DICHLO-; METHANE DICHLORIDE; METHYLENE BICHLORIDE; METHYLENE DICHLORIDE; ME-TYLENU CHLOREK (POLISH); NARKOTIL; NCI-C50102; R-30; SOLAESTHIN; SOLMETHINE

Danger profile: Carcinogen, mutagen, fire hazard, exposure limit established, Community Right-to-Know List, poisonous gas produced in fire.
Uses: Paint removers (strippers); in food; furniture and plastics processing.
Appearance: Colorless liquid.
Odor: Sweet, pleasant, chloroform-like. The odor is a warning of exposure; however, no smell does not mean you are not being exposed.
Effects of exposure:
Breathed: May cause irritation of the eyes, nose, throat and air passages; coughing, headache, nausea, drunkenness; increased saliva; reddening of the face; dizziness, lightheadedness, fatigue, loss of consciousness, shock, coma, irregular heartbeat, death. If the victim recovers watch for the possibility of a buildup of fluid in the lungs (pulmonary edema). Pulmonary edema is a medical emergency that can be delayed one to two days following exposure.
Eyes: Contact may cause irritation, tearing, inflammation of the eyelids; burning sensation, burns.
Skin: Passes through the unbroken skin; can increase exposure and the severity of the symptoms listed. May cause irritation, redness, burning feeling, swelling and burns; dry skin; rash or blisters; altered sleep pattern.
Swallowed: Poison. Breath may smell of chloroform. May cause irritation of the mouth, gastro-intestinal tract, nausea, vomiting, diarrhea with possible blood; drowsiness; loss of consciousness; irregular heartbeat and death. See *Breathed.*
Long-term exposure: May cause cancer; birth defects, miscarriage; liver and brain damage.
Storage: Store in tightly closed containers in a dry, cool, well-ventilated place. Keep away from heat, sources of ignition, alkali metals, lithium, aluminum and other chemically active metals. Air/vapor mistures may explode. Containers may explode in fire.
First aid guide: Move victim to fresh air and call emergency medical care; if not breathing, give artificial respiration; if breathing is difficult, give oxygen. In case of contact with material, immediately flush eyes with running water for at least 15 minutes. Wash skin with soap and water. Remove and isolate contaminated clothing and shoes at the site. Use first aid treatment according to the nature of the injury.

★664★
4,4-METHYLENEDIANILINE
CAS: 101-77-9
Other names: ANILINE, 4,4'-METHYLENEDI-; 4-(4-AMINOBENZYL)ANILINE; BIS-P-AMINOFENYLMETHAN (CZECH); BIS(P-AMINOPHENYL)METHANE; BIS(4-AMINOPHENYL)METHANE; CURITHANE; DDM; P,P'-DIAMINODIFENYLMETHAN (CZECH); 4,4'-DIAMINODIPHENYLMETHAN (GERMAN); DIAMINODIPHE-NYLMETHANE; P,P'-DIAMINODIPHENYLMETHANE; 4,4'-DIAMINODIPHENYLMETHANE; DI-(4-AMINOPHENYL)METHANE; DIANALINEMETHANE; 4,4'-DIPHENYLMETHANEDIAMINE; EPICURE DDM; EPIKURE DDM; HT 972; MDA; METHYLENEBIS(ANILINE); 4,4'-METHYLENEBISANILINE; METHYLENEDIANILINE; 4,4'-METHYLENE DIANILINE; P,P'-METHYLENEDIANILINE; NCI-C54604; TONOX

Danger profile: Carcinogen, poison, combustable, mutagen, Community Right-to-Know List, poisonous gas produced in fire.
Uses: Epoxy hardening agent; making rubber and resins.
Appearance: Light brown sand-like crystalline solid.
Odor: Faint, ammonia-like. The odor is a warning of exposure; however, no smell does not mean you are not being exposed.
Effects of exposure:

Breathed: Poisonous. High exposure may cause serious disease of the liver; fever, abdominal pain, weakness, nausea, vomiting, loss of appetite, fever and chills; jaundice. High dosage may have a narcotic effect.
Eyes: May cause eye irritation.
Skin: Passes slowly through the unbroken skin; can increase exposure and the severity of the symptoms listed.
Swallowed: Poisonous. May cause jaundice and liver problems. See *Breathed.*
Long-term exposure: May cause cancer; toxic hepatitis; liver and kidney damage. May lead to birth defects and miscarriage.
Storage: Store in tightly closed containers in a dry, cool, well-ventilated place. Keep away from heat, sources of ignition and oxidizers.

★665★
METHYL ETHYL ETHER
CAS: 540-67-0
Other names: ETHER, ETHYL METHYL; ETHANE, METHOXY-; ETHYL METHYL ETHER; METHANE, ETHOXY; METHOXYETHANE

Danger profile: Combustible, poisonous gas produced in fire.
Uses: In medicine.
Appearance: Colorless liquid or gas.
Effects of exposure:
Breathed: May be poisonous. May cause dizziness, light-headedness; reduced concentration. High levels may cause unconsciousness and death.
Eyes: May cause irritation and burns.
Skin: May be poisonous if absorbed. May cause irritation and burns.
Long-term exposure: May effect the kidneys.
Storage: Store in tightly closed containers in a dry, cool, well-ventilated place. Keep away from heat, direct sunlight, and sources of ignition. Protect containers from physical damage. Vapors may travel to sources of ignition and flash back. Containers may explode in fire.
First aid guide: Move victim to fresh air and call emergency medical care; if not breathing, give artificial respiration; if breathing is difficult, give oxygen. In case of contact with material, immediately flush eyes with running water for at least 15 minutes. Wash skin with soap and water. Remove and isolate contaminated clothing and shoes at the site.

★666★
METHYL ETHYL KETONE
CAS: 78-93-3
Other names: ACETONE, METHYL-; AETHYLMETHYLKE-TON (GERMAN); BUTANONE; 2-BUTANONE; BUTANONE 2 (FRENCH); ETHYL METHYL CETONE (FRENCH); ETHYL-METHYLKETON (DUTCH); ETHYL METHYL KETONE; MEK; METHYL ACETONE; METHYL KETONE; KETONE, ETHYL METHYL; MEETCO; METILETILCHETONE (ITALIAN); ME-TYLOETYLOKETON (POLISH)

Danger profile: Fire hazard, exposure limit established, reproductive hazard, Community Right-to-Know List, poisonous gas produced in fire.
Uses: Solvent; paint removers; cements and adhesives; making other chemicals, plastics, smokeless powder; cleaning fluids; printing.
Appearance: Clear, colorless liquid.
Odor: Fragrant, mint-like; moderately sharp; pleasant. The odor is a warning of exposure; however, no smell does not mean you are not being exposed.
Effects of exposure:

Breathed: Mildly toxic. May cause irritation of eyes, nose and throat; headache, drowsiness, excitability, drunken behavior, mental confusion, fatigue. Very high levels of exposure may cause dizziness, lightheadedness and cause victim to pass out, go into coma and die.

Eyes: Contact may cause irritation, pain, burning feeling, stinging, tearing, inflammation of the eyelids; burns. Light may cause pain. Possible permantnt damage.

Skin: May be poisonous if absorbed. Passes through the unbroken skin; can increase exposure and the severity of the symptoms listed. May cause irritation, burning feeling ans rash. May remove oil from the skin and cause dryness. Persons with pre-existing skin disorders may be more susceptible to the effects of this substance.

Swallowed: May cause gastro-intestinal irritation. See *Breathed.*

Long-term exposure: Prolonged or repeated contact may cause irritation of the skin and dermatitis, and nervous system changes; nervous disorders; disturbed sleep pattern; birth defects, miscarriage.

Storage: Store in a tightly closed container in a cool, well-ventilated place. Keep away from heat, direct sunlight, sources of ignition, potassium tert-butoxide, chlorosulfonic acid, oleum and strong oxidizers. Vapors are explosive in heat or flame. Containers may explode in fire.

First aid guide: Move victim to fresh air and call emergency medical care; if not breathing, give artificial respiration; if breathing is difficult, give oxygen. In case of contact with material, immediately flush eyes with running water for at least 15 minutes. Wash skin with soap and water. Remove and isolate contaminated clothing and shoes at the site.

★ 667 ★
METHYL ETHYL KETONE PEROXIDE
CAS: 1338-23-4
Other names: 2-BUTANONE, PEROXIDE; HI-POINT 90; LUPERSOL; MEKP; MEK PEROXIDE; METHYLETHYLKE-TONHYDROPEROXIDE; NCI-C55447; QUICKSET EXTRA; SPRAYSET MEKP; THERMACURE

Danger profile: Combustible, explosion hazard, transport forbidden, exposure limit established, poisonous gas produced in fire.

Uses: Making polymers and resins. Added to other chemicals to reduce sensitivity to shock.

Appearance: Colorless liquid.

Effects of exposure:
Breathed: May be extremely irritating. May cause irritation to the nose, throat, breathing passages and lungs.

Eyes: Contact or vapor may cause severe irritation, inflammation of the eyelids, burns, permanent damage and blindness.

Skin: Passes through the unbroken skin; can increase exposure and the severity of the symptoms listed. Contact may cause irritation.

Swallowed: May be poisonous. May cause function of the esophagus; nausea, vomiting; gastrointestinal problems.

Long-term exposure: May damage the liver and kidneys; possible lung damage.

Storage: Store in tightly closed containers in a dry, cool, well-ventilated place, preferably an explosion-proof freezer. Keep away from heat, sources of ignition, sunlight, fuels, shock, and organic compounds and organic materials. Containers may explode in fire.

First aid guide: Move victim to fresh air; call emergency medical care. In case of contact with material, immediately flush eyes with running water for at least 15 minutes. Wash skin with soap and water. Remove and isolate contaminated clothing and

shoes at the site. Keep victim quiet and maintain normal body temperature.

★ 668 ★
METHYL FORMATE
CAS: 107-31-3
Other names: FORMIATE DE METHYLE (FRENCH); FORMIC ACID, METHYL ESTER; METHYLE (FORMIATE DE) (FRENCH); METHYLFORMIAAT (DUTCH); METHYLFOR-MIAT (GERMAN); METHYL METHANOATE; METIL (FORMIA-TO DI) (ITALIAN)

Danger profile: Fire hazard, exposure limit established, poisonous gas produced in fire.

Uses: Insecticide; solvent; making other chemicals; fumigant; larvicide.

Appearance: Colorless liquid.

Odor: Pleasant; agreeable. The odor is a warning of exposure; however, no smell does not mean you are not being exposed.

Effects of exposure:
Breathed: May cause irritation to the eyes, nose and throat; vomiting. Higher exposures may cause irritation of the lungs with shortness of breath and a dangerous build-up of fluids in the lungs (pulmonary edema), which can cause death. Pulmonary edema is a medical emergency that can be delayed one to two days following exposure. High levels may also cause dizziness, lightheadedness, loss of consciousness, coma and death.

Eyes: Contact may cause irritation and inflammation of the eyelids.

Skin: Passes through the unbroken skin; can increase exposure and the severity of the symptoms listed. Prolonged contact may cause irritation.

Swallowed: May cause irritation of mouth and stomach; central nervous system depression, including visual disturbances.

Long-term exposure: Prolonged inhalation can produce narcosis and central nervous symptoms, including some temporary visual disturbance.

Storage: Store in tightly closed containers in a dry, cool, well-ventilated place. Keep away from heat, sources of ignition and strong oxidizers. Protect containers from physical damage. Containers may explode in fire.

★ 669 ★
METHYLHYDRAZINE
CAS: 60-34-4
Other names: HYDRAZINE, METHYL-; HYDRAZOME-THANE; 1-METHYL-HYDRAZINE; METHYL HYDRAZINE; METYLOHYDRAZYNA (POLISH); MMH; MONOMETHYL HY-DRAZINE

Danger profile: Carcinogen, poison, combustible, mutagen, explosion hazard, highly corrosive, exposure limit established, reproductive hazard, Community Right-to-Know List, EPA extremely hazardous substances list, poisonous gas produced in fire.

Uses: Rocket fuel.

Appearance: Clear, colorless liquid.

Odor: Ammonia-like. The odor is a warning of exposure; however, no smell does not mean you are not being exposed.

Effects of exposure:
Breathed: May be poisonous. May cause irritation of the eyes, nose, throat and respiratory tract; respiratory distress; excitability, vomiting, tremors, convulsions and death. High exposures may cause lung irritation, shortness of breath and a dangerous build-up of fluids in the lungs (pulmonary edema), which can cause death. Pulmonary edema is a medi-

cal emergency that can be delayed one to two days following exposure.

Eyes: Contact of liquid may cause severe irritation and burns; can increase exposure and the severity of the symptoms listed. Possible permanent damage.

Skin: May be poisonous if absorbed. Passes through the unbroken skin; can increase exposure and the severity of the symptoms listed. Contact of liquid may cause severe irritation and burns; eye damage.

Swallowed: May be poisonous. May cause severe irritation of mouth and stomach. Also see ***Breathed.***

Long-term exposure: May cause cancer; damage to the liver, kidney and red blood cells; anemia; skin allergy.

Storage: Store in tightly closed containers in a dry, cool, well-ventilated place. Keep away from heat, sunlight, sources of ignition, oxides of iron, hydrogen peroxide, copper, manganese, lead, copper alloys, oxidizers, fuming nitric acid and porous materials. Containers may explode in fire.

First aid guide: Move victim to fresh air and call emergency medical care; if not breathing, give artificial respiration; if breathing is difficult, give oxygen. In case of contact with material, immediately flush skin or eyes with running water for at least 15 minutes. Speed in removing material from skin is of extreme importance. Remove and isolate contaminated clothing and shoes at the site. Keep victim quiet and maintain normal body temperature. Effects may be delayed; keep victim under observation.

★670★
METHYL IODIDE
CAS: 74-88-4
Other names: HALON 10001; IODOMETHANE; IODOMETANO (ITALIAN); IODURE DE METHYLE (FRENCH); JOODMETHAAN (DUTCH); METHANE, IODO-; METHYLJODID (GERMAN); METYLU JODEK (POLISH); MONOIODOMETHANE

Danger profile: Poison, narcotic, combustible, exposure limit established, Community Right-to-Know List, poisonous gas produced in fire.
Uses: Insecticide fumigant; making other chemicals, stain in microscopy.
Appearance: Colorless liquid which turns brown on exposure to light.
Odor: Sweet, ether-like. The odor is a warning of exposure; however, no smell does not mean you are not being exposed.
Effects of exposure:
Breathed: May be poisonous. May cause dizziness, slurred speech, vertigo, visual disturbances, nausea, vomiting, cramps, diarrhea, irritability, loss of muscle control, tiredness, mental problems, coma and death. High exposures may cause kidney damage; lung irritation, coughing, shortness of breath and a dangerous build-up of fluids in the lungs (pulmonary edema), which can cause death. Pulmonary edema is a medical emergency that can be delayed one to two days following exposure.
Eyes: Contact may cause serious irritation and burns.
Skin: May be poisonous if absorbed. Passes through the unbroken skin; can increase exposure and the severity of the symptoms listed. Contact may cause serious irritation, redness and burns. Blistering may occur hours later.
Swallowed: May be poisonous.
Long-term exposure: May cause nervous system damage; brain damage and psychotic behavior; dermatitis with drying a cracking skin.
Storage: Store in tightly closed containers in a dry, cool, well-ventilated place. Keep away from heat, sources of ignition, silver chlorite, sodium, oxygen and strong oxidizers. Containers may explode in fire.

First aid guide: Move victim to fresh air and call emergency medical care; if not breathing, give artificial respiration; if breathing is difficult, give oxygen. In case of contact with material, immediately flush skin or eyes with running water for at least 15 minutes. Speed in removing material from skin is of extreme importance. Remove and isolate contaminated clothing and shoes at the site. Keep victim quiet and maintain normal body temperature. Effects may be delayed; keep victim under observation.

★671★
METHYL ISOAMYL KETONE
CAS: 110-12-3
Other names: 2-HEXANONE, 5-METHYL-; ISOAMYL METHYL KETONE; ISOPENTYL METHYL KETONE; KETONE, METHYL ISOAMYL; 2-METHYL-5-HEXANONE; 5-METHYL-2-HEXANONE; MIAK

Danger profile: Combustible, exposure limit established, poisonous gas produced in fire.
Uses: Solvent; inks, stains; paint and varnish-removers; insect repellent; making other chemicals.
Appearance: Colorless liquid.
Odor: Pleasant, sweet. The odor is a warning of exposure; however, no smell does not mean you are not being exposed.
Effects of exposure:
Breathed: May cause irritation of eyes, nose and throat; headache, drowsiness, excitability, drunken behavior, mental confusion, fatigue. Very high levels of exposure may cause dizziness, lightheadedness and cause victim to pass out, go into coma and die.
Eyes: Contact may cause severe irritation, pain, burning, stinging, tearing, inflammation of the eyelids; possible permanent damage.
Skin: Passes through the unbroken skin; can increase exposure and the severity of the symptoms listed. May remove oil from the skin and cause dryness. Persons with pre-existing skin disorders may be more susceptible to the effects of this substance. If not removed immediately may cause delayed pain and blisters.
Swallowed: May cause irritation of the mouth and stomach; gastro-intestinal irritation. See ***Breathed.***
Long-term exposure: Prolonged or repeated contact may cause irritation of the skin and dermatitis with dryness and cracking; nervous system changes; liver damage.
Storage: Store in a tightly closed container in a cool, well-ventilated place. Keep away from heat and sources of ignition, strong oxidizers, reducing agents and aldehydes. Vapors may travel to sources of ignition and flash back. Vapors may travel to sources of ignition and flash back. Containers may explode in fire.

★672★
METHYL ISOBUTYL CARBINOL
CAS: 108-11-2
Other names: ALCOOL METHYL AMYLIQUE (FRENCH); ISOBUTYLMETHYLCARBINOL; ISOBUTYLMETHYLMETHANOL; MAOH; METHYL AMYL ALCOHOL; METHYLISOBUTYL CARBINOL; 2-METHYL-4-PENTANOL; 4-METHYLPENTANOL-2; 4-METHYL-2-PENTANOL; METILAMIL ALCOHOL (ITALIAN); 4-METILPENTAN-2-OLO (ITALIAN); MIBC; MIC; 3-MIC; 2-PENTANOL, 4-METHYL-

Danger profile: Combustible, exposure limit established, reproductive hazard, poisonous gas produced in fire.
Uses: Brake fluid; solvent.
Appearance: Clear oily liquid.
Odor: Mild alcohol-like. and The odor is a warning of exposure; however, no smell does not mean you are not being exposed.

Effects of exposure:
Breathed: May cause irritation of the eyes, nose and throat. High levels may cause dizziness, lightheadedness, anesthesia, and loss of consciousness.
Eyes: Contact may cause severe irritation and burns.
Skin: Passes through the unbroken skin; can increase exposure and the severity of the symptoms listed. May cause irritation, redness, burning sensation, rash; drying and cracking from prolonged contact.
Long-term exposure: May cause dermatitis with drying and cracking skin.
Storage: Store in tightly closed containers in a dry, cool, well-ventilated place. Keep away from heat, sources of ignition and oxidizing materials. Containers may explode in fire.
First aid guide: Move victim to fresh air and call emergency medical care; if not breathing, give artificial respiration; if breathing is difficult, give oxygen. In case of contact with material, immediately flush eyes with running water for at least 15 minutes. Wash skin with soap and water. Remove and isolate contaminated clothing and shoes at the site.

★ 673 ★
METHYL ISOBUTYL KETONE
CAS: 108-10-1
Other names: FEMA NO. 2731; HEXON (CZECH); HEXONE; ISOBUTYL METHYL KETONE; ISOPROPYLACETONE; METHYL-ISOBUTYL-CETONE (FRENCH); 4-METHYL-2-PENTANON (CZECH); 2-METHYL-4-PENTANONE; 4-METHYL-2-PENTANONE; MIBK; MIK; 2-PENTANONE, 4-METHYL-; SHELL MIBK

Danger profile: Fire hazard, exposure limit established, Community Right-to-Know List, poisonous gas produced in fire.
Uses: Solvent.
Appearance: Colorless liquid.
Odor: Pleasant. The odor is a warning of exposure; however, no smell does not mean you are not being exposed.
Effects of exposure:
Breathed: May cause irritation of eyes, nose and throat; headache, drowsiness, excitability, drunken behavior, mental confusion, fatigue, loss of appetite, nausea, vomiting. Very high levels of exposure may cause dizziness, lightheadedness, depression and cause victim to pass out.
Eyes: Contact may cause irritation.
Skin: May cause severe irritation; remove oil from the skin and cause dryness. Persons with pre-existing skin disorders may be more susceptible to the effects of this substance.
Swallowed: May cause gastro-intestinal irritation. See *Breathed.*
Long-term exposure: Prolonged or repeated contact may cause irritation of the skin and dermatitis, flaking, blisters; liver and kidney damage.
Storage: Store in a tightly closed container in a cool, well-ventilated place. Keep away from heat and sources of ignition, strong oxidizers and potassium-tert-butoxide. Vapors are explosive in heat or flame. Containers may explode in fire.
First aid guide: Move victim to fresh air and call emergency medical care; if not breathing, give artificial respiration; if breathing is difficult, give oxygen. In case of contact with material, immediately flush eyes with running water for at least 15 minutes. Wash skin with soap and water. Remove and isolate contaminated clothing and shoes at the site.

★ 674 ★
METHYL ISOCYANATE
CAS: 624-83-9
Other names: ISOCYANATE DE METHYLE (FRENCH); ISO-CYANATOMETHANE; ISOCYANIC ACID, METHYL ESTER; ISOCYANATE METHANE; METHYLCARBAMYL AMINE; METHYLISOCYANAAT (DUTCH); METHYL ISOCYANAT (GERMAN); METHYL CARBONIMIDE; METIL ISOCIANATO (ITALIAN); MIC; TL 1450

Danger profile: Poison, fire hazard, explosion hazard, exposure limit established, Community Right-to-Know List, poisonous gas produced in fire.
Uses: Making pesticides, polyurethane foams and plastics.
Appearance: Colorless liquid.
Odor: Sharp, unpleasant, causes tears. The odor is a warning of exposure; however, no smell does not mean you are not being exposed.
Effects of exposure:
Breathed: Poisonous; may be fatal if inhaled. May cause severe irritation of the eyes, nose, throat and lungs, tears and runny nose, coughing, sore throat, coughing, dyspnea (difficult or painful breathing, gasping for breath), increased secretions, uncontrollable vomiting, lethargy, fatigue, and drowsiness; slurred speech, shaking, and dizziness; dizziness, diarrhea and a dangerous build-up of fluids in the lungs (pulmonary edema), which can cause death. Pulmonary edema is a medical emergency that can be delayed one to two days following exposure. May cause miscarriage; asthma-like allergy may develop. Most deaths in Bhopal, India, in 1984 have been attributed to various forms of respiratory distress such as massive accumulation of fluid in the lungs or spasmodic constrictions of the bronchial tubes.
Eyes: May cause irritation, redness, tearing, pain, inflamed eyelids, and eye injury; burns and blindness.
Skin: Poisonous if absorbed. Passes through the unbroken skin; can increase exposure and the severity of the symptoms listed. May cause painful irritation, redness, blisters and blindness.
Swallowed: Poisonous. May cause irritation of the mouth and throat; stomach and abdominal pain; loss of muscular coordination and mental confusion; convulsions. Also see symptoms for *Breathed.*
Long-term exposure: May cause birth defects, miscarriage, decrease of fertility in man and women; asthma-like allergy with shortness of breath and coughing an wheezing; lung damage.
Storage: Store in tightly closed containers in a dry, cool, well-ventilated place. Keep away from heat, sources of ignition, water, acid, alkali, amines, iron, steel, zinc, galvanized iron, tin, copper and other chemicals. Use type 304 and 316 stainless steel, nickel, teflon and glass lined equipment; other materials may be unsuitable (and possibly dangerous) to use. Vapors may travel to sources of ignition and flash back. Containers may explode in fire.
First aid guide: Move victim to fresh air and call emergency medical care; if not breathing, give artificial respiration; if breathing is difficult, give oxygen. In case of contact with material, immediately flush skin or eyes with running water for at least 15 minutes. Remove and isolate contaminated clothing and shoes at the site. Keep victim quiet and maintain normal body temperature. Effects may be delayed; keep victim under observation.

★ 675 ★
METHYL ISOPROPYL KETONE
CAS: 563-80-4
Other names: ISOPROPYL METHYL KETONE; 3-METHYL-2-BUTANONE; 3-METHYL BUTAN 2-ONE; MIPK

Danger profile: Combustible, exposure limit established, poisonous gas produced in fire.

Uses: Solvent.

Appearance: Colorless liquid.

Odor: Acetone-like. The odor is a warning of exposure; however, no smell does not mean you are not being exposed.

Effects of exposure:

Breathed: May cause irritation of eyes, nose and throat; headache, drowsiness, excitability, drunken behavior, mental confusion, fatigue. Very high levels of exposure may cause dizziness, lightheadedness and cause victim to pass out, go into coma and die.

Eyes: Contact may cause severe irritation, pain, burning, stinging, tearing, inflammation of the eyelids; possible permanent damage.

Skin: May remove oil from the skin and cause dryness. Persons with pre-existing skin disorders may be more susceptible to the effects of this substance. If not removed immediately may cause delayed pain and blisters.

Swallowed: May cause irritation of the mouth and stomach; gastro-intestinal irritation. See *Breathed.*

Long-term exposure: Prolonged or repeated contact may cause irritation of the skin and dermatitis with dryness and cracking; nervous system changes.

Storage: Store in tightly closed containers in a dry, cool, well-ventilated place. Keep away from heat, sources of ignition and oxidizing materials. Vapors may travel to sources of ignition and flash back. Containers may explode in fire.

First aid guide: Move victim to fresh air and call emergency medical care; if not breathing, give artificial respiration; if breathing is difficult, give oxygen. In case of contact with material, immediately flush eyes with running water for at least 15 minutes. Wash skin with soap and water. Remove and isolate contaminated clothing and shoes at the site.

★ 676 ★
METHYL ISOTHIOCYANATE
CAS: 556-61-6

Other names: EP-161E; ISOTHIOCYANATE DE METHYLE (FRENCH); ISOTHIOCYANIC ACID, METHYL ESTER; ISOTHIOCYANOMETHANE; METHANE, ISOTHIOCYANATO-; METHYL MUSTARD OIL; METHYLSENFOEL (GERMAN); MIC; MIT; MITC; MORTON WP-161-E; TRAPEX; TRAPEXIDE; VORLEX; VORTEX; WN 12

Danger profile: Combustible, poisonous gas produced in fire.

Uses: Pesticide.

Appearance: White sand-like crystalline solid.

Effects of exposure:

Breathed: May be poisonous. May cause severe irritation of the eyes, nose, throat, breathing passages and lungs; coughing, shortness of breath, chest tightness, coughing, wheezing, laryngitis, headache, nausea, and vomiting. Inhalation may be fatal due to spasm, inflammation and edema of the larynx and bronchi, chemical pneumonitis, and a dangerous build-up of fluids in the lungs (pulmonary edema), which can cause death. Pulmonary edema is a medical emergency that can be delayed one to two days following exposure. Asthma-like lung allergy may develop.

Eyes: May cause irritation and permanent damage.

Skin: May be poisonous if absorbed. Passes through the unbroken skin; can increase exposure and the severity of the symptoms listed. May cause severe irritation; skin allergy and rash may develop.

Swallowed: May be poisonous. See *Breathed* and *First aid guide.*

Long-term exposure: May effect the thyroid. May cause skin allergy; lung problems and asthma-like allergy.

Storage: Store in tightly closed containers in a dry, cool, well-ventilated place. Keep away form sources of ignition. Vapors may travel to sources of ignition and flash back. Containers may explode in fire.

First aid guide: Move victim to fresh air and call emergency medical care; if not breathing, give artificial respiration; if breathing is difficult, give oxygen. In case of contact with material, immediately flush skin or eyes with running water for at least 15 minutes. Remove and isolate contaminated clothing and shoes at the site. Keep victim quiet and maintain normal body temperature. Effects may be delayed; keep victim under observation.

★ 677 ★
METHYL MERCAPTAN
CAS: 74-93-1

Other names: MERCAPTAN METHYLIQUE (FRENCH); MERCAPTOMETHANE; METHAANTHIOL (DUTCH); METHANETHIOL; METHANTHIOL (GERMAN); METHVTIOLO (ITALIAN); METHYLMERCAPTLAAN (DUTCH); METHYL SULFHYDRATE; METHYLTHIOALCOHOL; METILMERCAPTANO (ITALIAN); THIOMETHANOL; THIOMETHYL ALCOHOL

Danger profile: Fire hazard, poison, exposure limit established, EPA extremely hazardous substances list, poisonous gas produced in fire.

Uses: Making other chemicals, pesticides, fungicides; jet fuel additives; adds warning odor to natural gas. Common air contaminant.

Appearance: Clear liquid or colorless gas.

Odor: Garlic-like, foul, like rotten cabbage, strong. The odor is a warning of exposure; however, no smell does not mean you are not being exposed.

Effects of exposure:

Breathed: Poisonous. May cause irritation of nose, throat, eyes and respiratory system; headaches, dizziness. Higher exposures may cause shortness of breath and a dangerous build-up of fluids in the lungs (pulmonary edema), which can cause death. Pulmonary edema is a medical emergency that can be delayed one to two days following exposure. High exposures may cause tremors, paralysis, sudden loss of consciousness; breathing stops; death may follow respiratory paralysis.

Eyes: Contact with liquid may cause irritation.

Skin: Contact with liquid may cause irritation and frostbite.

Swallowed: May be poisonous. May cause irritation of mouth and stomach plus symptoms described for inhalation.

Long-term exposure: May cause skin rash; liver and blood cell damage; lung problems.

Storage: Store in tightly closed containers in a dry, cool, well-ventilated place. Keep away from water, strong acids, and oxidizers. Vapors may travel to sources of ignition and flash back. Containers may explode in fire.

First aid guide: Move victim to fresh air and call emergency medical care; if not breathing, give artificial respiration; if breathing is difficult, give oxygen. In case of contact with material, immediately flush skin or eyes with running water for at least 15 minutes. Keep victim quiet and maintain normal body temperature. Effects may be delayed; keep victim under observation.

★ 678 ★
METHYL METHACRYLATE
CAS: 80-62-6
Other names: DIAKON; METAKRYLAN METYLU (POLISH); METHACRYLATE DE METHYLE (FRENCH); METHACRYLIC ACID, METHYL ESTER; METHACRYLSAEUREMETHYL ESTER (GERMAN); METHYLMETHACRYLAAT (DUTCH); METHYL-METHACRYLAT (GERMAN); METHYL METHACRYLATE MONOMER; METHYL ALPHA-METHYLACRYLATE; METHYL 2-METHYLPROPENOATE; METHYL 2-METHYL-2-PROPENOATE; METIL METACRILATO (ITALIAN); MME; "MONOCITE" METHACRYLATE MONOMER; MER; NCI-C50680; 2-PROPENOIC ACID, 2-METHYL-, METHYL ESTER

Danger profile: Fire hazard, exposure limit established, reproductive hazard, Community Right-to-Know List, poisonous gas produced in fire.
Uses: Making plastics, resins; dentures; impregnation of concrete.
Appearance: Colorless liquid.
Odor: Sharp, fruity. The odor is a warning of exposure; however, no smell does not mean you are not being exposed.
Effects of exposure:
Breathed: May be poisonous. May cause irritation of the eyes, nose, throat and lungs, tears and runny nose, coughing, sore throat, difficult breathing; excitement, slurred speech, shaking, and dizziness; loss of appetite, decrease in blood pressure. Higher exposures could cause shortness of breath, headache, nausea, vomiting, dizziness, diarrhea and a dangerous build-up of fluids in the lungs (pulmonary edema), which can cause death. Pulmonary edema is a medical emergency that can be delayed one to two days following exposure.
Eyes: May cause irritation, redness, tearing, pain, inflamed eyelids, and eye injury; burns in corneas and possible loss of sight.
Skin: May cause painful irritation, redness, and blisters Repeated or prolonged skin contact may cause irritation of the skin and may also cause the legs to feel numb.
Swallowed: May be poisonous. May cause irritation of the mouth and throat; stomach and abdominal pain; loss of muscular coordination and mental confusion; convulsions. Also see symptoms for *Breathed.*
Long-term exposure: May cause skin allergy; nervous system damage; possible lung problems.
Storage: Store in tightly closed containers in a dry, cool, well-ventilated place. Keep away from light, heat, sources of ignitions, ionizing radiation, oxidizers, strong alkalis, strong acids, amines, halogens, benzoyl peroxide, dibenzoyl peroxide and reducing agents. Vapors may travel to sources of ignition and flash back. Containers may explode in fire.
First aid guide: Move victim to fresh air and call emergency medical care; if not breathing, give artificial respiration; if breathing is difficult, give oxygen. In case of contact with material, immediately flush eyes with running water for at least 15 minutes. Wash skin with soap and water. Remove and isolate contaminated clothing and shoes at the site.

★ 679 ★
METHYL PARATHION
CAS: 298-00-0
Other names: A-GRO; AZOFOS; AZOPHOS; BAY E-601; BAY 1145; BLADAN-M; CEKUMETHION; DALF; O,O-DIMETHYL 0-4 -NITROPHENYL PHOSPHOROTHIOATE; O,O-DIMETHYL O-P-NITROPHENYLPHOSPHOROTHIOATE; DEVITHION; ENT 17,292; FOLIDOL M; GEARPHOS; ME-PARATHION; MEPATON; MEPOX; METACIDE; METAFOS; METAPHOR; METAPHOS; METHYL-E 605; METHYL FOS-FERNO; METHYL NIRAN; METHYLTHIOPHOS; METRON; NCI- C02971; NITROX; OLEOVOFOTOX; PARAPEST M-50; PARATAF; M-PARATHION; PARATHION METHYL; PARA-THION-METILE (ITALIAN); PATRON M; PARATOX; PENN-CAP-M; PHOSPHOROTHIOIC ACID O,O-DIMETHYL O-(4-NITROPHENYL)ESTER; SINAFID M-48; TEKWAISA; THIOP-HENIT; THYLFAR M-50; TOLL; VOFATOX; WOFATOS; WOFOTOX

Danger profile: Poison, exposure limit established, poisonous gas produced in fire.
Uses: Pesticide; insecticide for cotton and other crops (organophosphate).
Appearance: White to yellow-brown sand-like crystalline solid.
Odor: Garlic-like. The odor is a warning of exposure; however, no smell does not mean you are not being exposed.
Effects of exposure:
Breathed: Extremely poisonous. May cause headache, dizziness, profuse sweating, nausea, vomiting, reduced heart beat, stomach cramps, diarrhea, loss of coordination, slow and weak breathing, fever, loss of consciousness and death. May cause shortness of breath and a dangerous build-up of fluids in the lungs (pulmonary edema), which can cause death. Pulmonary edema is a medical emergency that can be delayed one to two days following exposure. Exposure to fumes from a fire, causes headache, blurred vision, constricted pupils of the eyes, weakness, nausea, cramps, diarrhea, and tightness in the chest. Muscle twitch and convulsions may follow. Symptoms may develop over a period of 8 hours.
Eyes: Rapidly absorbed. Poisoning can happen very rapidly. See *Breathed.*
Skin: Fatal poisoning can result from skin contact. Passes through the unbroken skin; can increase exposure and the severity of the symptoms listed. Exposure to the liquid, causes headache, blurred vision, constricted pupils of the eyes, weakness, nausea, cramps, diarrhea, and tightness in the chest. Muscle twitch and convulsions may follow. Symptoms may develop over a period of 8 hours.
Swallowed: Powerful, deadly poison; headache, sweating, nausea, vomiting, diarrhea, loss of coordination, tremors, twitching and death. See *Breathed.*
Long-term exposure: May cause nerve damage including loss of coordination; liver damage. Repeated exposure may cause depression, irritability and personality changes.
Storage: Store in a regulated, specially marked place in tightly closed containers in a dry, well-ventilated place with non-wood floors. Keep away from heat. Containers may explode in fire.
First aid guide: Move victim to fresh air and call emergency medical care; if not breathing, give artificial respiration; if breathing is difficult, give oxygen. In case of contact with material, immediately flush skin or eyes with running water for at least 15 minutes. Speed in removing material from skin is of extreme importance. Remove and isolate contaminated clothing and shoes at the site. Keep victim quiet and maintain normal body temperature. Effects may be delayed; keep victim under observation.

★680★
METHYLPHENYLDICHLOROSILANE
CAS: 149-74-6
Other names: DICHLOROMETHYLPHENYLSILANE; PHE-NYLMETHYLDICHLOROSILANE; SILANE, DICHLOROMETH-YLPHENYL-

Danger profile: Combustible, corrosive, poisonous gas produced in fire.
Uses: Making silicones.
Appearance: Colorless liquid.
Effects of exposure:
Breathed: May be poisonous. May cause irritation of the eyes, nose, throat and lungs.
Eyes: May cause severe irritation and burns.
Skin: May cause severe irritation and burns.
Long-term exposure: Very irritating substance; capable of causing lung problems.
Storage: Store in tightly closed containers in a dry, cool, well-ventilated place. Keep away from heat, sources of ignition, water, steam and other forms of moisture, and oxidizers. Vapors may travel to sources of ignition and flash back. Containers may explode in fire.
First aid guide: Move victim to fresh air and call emergency medical care; if not breathing, give artificial respiration; if breathing is difficult, give oxygen. In case of contact with material, immediately flush skin or eyes with running water for at least 15 minutes. Remove and isolate contaminated clothing and shoes at the site. Keep victim quiet and maintain normal body temperature.

★681★
METHYL SILICATE
CAS: 681-84-5
Other names: METHYL ESTER OF ORTHO-SILICIC ACID; METHYL ORTHOSILICATE; SILICIC ACID, METHYL ESTER OF ORTHO-; SILICIC ACID, TETRAMETHYL ESTER; TETRA-METHYL ORTHOSILICATE; TETRAMETHOXY SILANE; TL 199

Danger profile: Combustible, exposure limit established, poisonous gas produced in fire.
Uses: Making other chemicals; coating television picture tubes.
Appearance: Clear liquid or sand-like crystalline solid.
Effects of exposure:
Breathed: May be poisonous. Vapor may cause severe eye damage and permanent damage to the cornea. Even if no effects of irritation are noticed victim can be effected for up to 12 hours following exposure. High exposure may cause liver and kidney damage, and pulmonary edema, a dangerous build-up of fluids in the lungs can cause death. Pulmonary edema is a medical emergency that can be delayed one to two days following exposure.
Eyes: May be poisonous if absorbed. May cause severe irritation and permanent blindness. This may happen up to 12 hours following exposure whether or not irritation is noticed at the time of exposure.
Skin: May cause severe irritation.
Swallowed: May be poisonous. See *Breathed.*
Long-term exposure: Repeated exposure, however small, may cause eye damage. May cause liver, kidney and lung damage; kidney damage could be extensive.
Storage: Store in tightly closed containers in a dry, cool, well-ventilated place. Keep away from heat, sources of ignition, water and other forms of moisture, oxidizers, tungsten and molybdenum. Containers may explode in fire.

★682★
METHYL TRICHLOROSILANE
CAS: 75-79-6
Other names: METHYL-TRICHLORSILAN (CZECH); SILANE, TRICHLOROMETHYL-; SILANE,METHYLTRICHLORO-; METHYLCHLOROSILANE; METHYLTRICHLOROSILANE; TRICHLOROMETHYLSILANE

Danger profile: Poison, fire hazard, explosion hazard, corrosive, EPA extremely hazardous substances list, poisonous gas produced in fire.
Uses: Making water repellent silicones, heat resistant paints, electrical insulation and other products.
Appearance: Colorless liquid.
Odor: Sharp, irritating. The odor is a warning of exposure; however, no smell does not mean you are not being exposed.
Effects of exposure:
Breathed: May be poisonous. Vapors may cause irritation of the eyes, nose, mucous membranes and breathing passages. Higher levels cause irritation of the lungs, coughing, shortness of breath and a dangerous build-up of fluids in the lungs (pulmonary edema), which can cause death. Pulmonary edema is a medical emergency that can be delayed one to two days following exposure.
Eyes: May cause severe burns and permanent damage.
Skin: May be poisonous if absorbed. May cause severe irritation and burns.
Swallowed: May be poisonous. May cause severe burns of mouth and stomach.
Long-term exposure: A very irritating substance; may cause lung problems.
Storage: Store in tightly closed containers in a dry, cool, well-ventilated place. Keep away from heat, sources of ignition, water, acids, chemically active metals, alkalies. Reacts with surface moisture to produce hydrogen chloride which is corrosive to metals. May form an explosive mixture with air. Vapors may travel to sources of ignition and flash back. Containers may explode in fire.
First aid guide: Move victim to fresh air and call emergency medical care; if not breathing, give artificial respiration; if breathing is difficult, give oxygen. In case of contact with material, immediately flush skin or eyes with running water for at least 15 minutes. Remove and isolate contaminated clothing and shoes at the site. Keep victim quiet and maintain normal body temperature.

★683★
METRIBUZIN
CAS: 21087-64-9
Other names: 4-AMINO-6-TERT-BUTYL-3-METHYLTHIO-AS-TRIAZIN-5-ONE; 4-AMINO-6-TERT-BUTYL-3-(METHYLTHIO)-1,2,4-TRIAZIN-5-ONE; 4-AMINO-6-(1,1-DIMETHYLETHYL)-3-(METHYLTHIO)-1,2,4-TRIAZIN-5(4H)-ONE; BAY 61597; BAY DIC 1468; BAYER 6159H; BAYER 6443H; BAYER 94337; DIC 1468; LEXONE; SENCOR; SENCORAL; SENCORER; SENCOREX; 1,2,4-TRIAZIN-5-ONE,4-AMINO-6-TERT-BUTYL-3-(METHYLTHIO)-; 1,2,4-TRIAZIN-5(4H)-ONE,4-AMINO-6-(1,1-DIMETHYLETHYL)-3-(ME THYLTHIO)-; AS-TRIAZIN-5(4H)-ONE,4-AMINO-6-TERT-BUTYL-3-(M ETHYLTHIO)-

Danger profile: Poison; exposure limit established, poisonous gas produced in fire.
Uses: Herbicide.
Appearance: White crystalline solid.
Odor: Mild. The odor is a warning of exposure; however, no smell does not mean you are not being exposed.
Effects of exposure:
Breathed: May be poisonous. May cause dizziness, drowsiness and difficult breathing. Related chemicals may cause fa-

tigue, upset stomach, loss of coordination, tremors and weakness.
Eyes: May cause irritation.
Skin: May be poisonous if absorbed. Passes through the unbroken skin; can increase exposure and the severity of the symptoms listed.
Swallowed: May be poisonous. See *Breathed.*
Long-term exposure: High or repeated exposure may cause goiter (enlarged thyroid gland), and may affect function of the thyroid.
Storage: Store in tightly closed containers in a dry, cool, well-ventilated place. Containers may explode in fire.

★ 684 ★
MEXACARBATE
CAS: 315-18-4
Other names: CARBAMATE,4-DIMETHYLAMINO-3,5-XYLYLN-METHYL-; 4-(DIMETHYLAMINE)-3,5-XYLYLN-METHYLCARBAMATE; 4-(DIMETHYLAMINO)-3,5-DIMETHYLPHENOL METHYLCARBAMATE (ESTER); 4-(DIMETHYLAMINO)-3,5-DIMETHYLPHENYL N-METHYLCARBAMATE; 4-(DIMETHYLAMINO)-3,5-XYLENOL,METHYLCARBAMATE (ESTER); 4-DIMETHYLAMINO-3,5-XYLYLMETHYLCARBAMATE; 4-DIMETHYLAMINO-3,5-XYLYL N-METHYLCARBAMATE; 4-(N,N-DIMETHYLAMINO)-3,5-XYLYL N-METHYLCARBAMATE; 5-DIMETHYLPHENOL METHYLCARBAMATE ESTER; DOWCO 139; ENT 25766; METHYLCARBAMIC ACID, 4-(DIMETHYLAMINO)-3,5-XYLYL ESTER; METHYL-4-DIMETHYLAMINO-3,5-XYLYLCARBAMATE; METHYL-4-DIMETHYLAMINO-3,5-XYLYL ESTER OF CARBAMIC ACID; NCI-C00544; OMS-47; PHENOL, 4-(DIMETHYLAMINO)-3,5-DIMETHYL-METHYLCARBAMATE (ESTER); 3,5-XYLENOL, 4-(DIMETHYLAMINO)-,METHYLCARBAMATE; ZACTRAN; ZECTANE; ZECTRAN; ZEXTRAN; DOWCO-139

Danger profile: Combustible, poisonous gas produced in fire.
Uses: Pesticide; insecticide.
Odor: Mild. The odor is a warning of exposure; however, no smell does not mean you are not being exposed.
Effects of exposure:
Breathed: Extremely poisonous. May cause weakness, headache, dizziness, profuse sweating, stomach cramps, metallic taste in mouth, drowsiness, temporary blurring of vision, loss of depth perception, paralysis of extremities, slurring of speech; nausea, vomiting, reduced heart beat, loss of coordination, slow and weak breathing, fever, loss of consciousness and death.
Eyes: Rapidly absorbed. Poisoning can happen very rapidly. See *Breathed.*
Skin: Poisonous if absorbed. Passes through the unbroken skin; can increase exposure and the severity of the symptoms listed.
Swallowed: Powerful, deadly poison. See *Breathed.*
Long-term exposure: May damage the developing fetus or cause birth defects; nerve damage including loss of coordination; liver damage. Repeated exposure may cause depression, irritability and personality changes.
Storage: Store in a regulated, specially marked place in tightly closed containers in a dry, well-ventilated place. Keep away from water, alkalis. Containers may explode in fire.
First aid guide: Move victim to fresh air and call emergency medical care; if not breathing, give artificial respiration; if breathing is difficult, give oxygen. In case of contact with material, immediately flush skin or eyes with running water for at least 15 minutes. Speed in removing material from skin is of extreme importance. Remove and isolate contaminated clothing and shoes at the site. Keep victim quiet and maintain normal body temperature. Effects may be delayed; keep victim under observation.

★ 685 ★
MOLYBDENUM
CAS: 7439-98-7
Other names: MOLYBDATE

Danger profile: Exposure limit established, poisonous gas produced in fire.
Uses: Making iron, steel; in electronic, aircraft and missile parts.
Appearance: Silvery-white metal or grayish-black powder.
Effects of exposure:
Breathed: May cause irritation of the eyes, nose and throat.
Eyes: May cause irritation.
Skin: May cause irritation.
Long-term exposure: May cause head and back aches; pain in the joints. May lead to gout, anemia.
Storage: Keep away from lead dioxide, chlorine trifluoride and strong oxidizers. Containers may explode in fire.

★ 686 ★
MOLYBDENUM TRIOXIDE
CAS: 1313-27-5
Other names: MOLYBDENUM(VI) OXIDE; MOLYBDIC ANHYDRIDE; MOLYBDIC TRIOXIDE

Danger profile: Poison, exposure limit established, EPA extremely hazardous substance list, Community Right-to-Know List, poisonous gas produced in fire.
Uses: Agriculture; making metallic molybdenum; ceramic enamels and glazes; enamels; pigments; laboratory chemical.
Appearance: White or slightly yellow or slightly bluish powder.
Effects of exposure:
Breathed: Dust or mist may cause irritation of the nose, throat, breathing passages; cough, chest tightness.
Eyes: Contact may cause severe irritation.
Skin: Contact may cause severe irritation.
Long-term exposure: May result in pulmonary fibrosis, cough; weight loss, diarrhea, headaches; muscle and joint aches; loss of muscle coordination.
Storage: Store in tightly closed containers in a dry, cool, well-ventilated place. Keep away from strong acids, potassium, sodium, alkalis, bromine pentafluoride, chlorine trifluoride and molten magnesium.

★ 687 ★
MONOCROTOPHOS
CAS: 6923-22-4
Other names: APADRIN; AZODRIN; BILOBRAN; BILOBORN; C 1414; CRISODRIN; CIBA 1414; CRISODIN; CROTONAMIDE, 3-HYDROXY-N-METHYL-, DIMETHYLPHOSPHATE, CIS-; CROTONAMIDE, 3-HYDROXY-N-METHYL- ,DIMETHYLPHOSPHATE, (E)-; 3-(DIMETHOXYPHOSPHINYLOXY)N-METHYL-CIS-CROTONAMIDE; (E)-DIMETHYL 1-METHYL-3-(METHYLAMINO)-3-OXO-1-PROPENYL PHOSPHATE; O,O-DIMETHYL-O-(2-N-METHYLCARBAMOYL-1-METHYL-VINYL)-FOSFAAT (DUTCH); O,O-DIMETHYL-O-(2-N-METHYLCARBAMOYL -1-METHYL)-VINYL-PHOSPHAT (GERMAN); O,O-DIMETHYL-O-(2-N-METHYLCARBAMOYL-1-METHYL-VINYL) PHOSPHATE; DIMETHYL 1-METHYL-2-(METHYLCARBAMOYL)VINYL PHOSPHATE, CIS; DIMETHYL PHOSPHATE ESTER OF 3-HYDROXY-N-METHYL-CIS-CROTONAMIDE; DIMETHYL PHOSPHATE OF 3-HYDROXY-N-METHYL-CIS-CROTONAMINE; O,O-DIMETIL-O-(2-N-METILCARBAMOIL-1-METIL-VINIL)-FOSFATO (ITALIAN); ENT 27,129; GLORE PHOS 36; 3-HYDROXY-N-METHYLCROTONAMIDE DIMETHYL PHOSPHATE; 3-HYDROXY-N-METHYL-CIS-CROTONAMIDE DIMETHYL PHOSPHATE; CIS-1-METHYL-2-METHYL CARBAMOYL VINYL PHOSPHATE; MONOCRON; NUVACRON; PHOSPHATE DE DIMETHYLE ET DE 2-METHYLCARBAMOYL 1-METHYL VINYLE (FRENCH); PHOSPHORIC ACID, DIMETHYL ESTER, ESTER WITH CIS-3-HYDROXY-N-METHYLCROTONAMIDE; PILLARDIN; PLANTDRIN; SD 9129; SHELL SD 9129; SUSVIN; ULVAIR

Danger profile: Poison, combustible, exposure limit established, EPA extremely hazardous substances list, poisonous gas produced in fire.
Uses: May be restricted. Insecticide for mites on cotton, sugar cane, tobacco, potatoes and peanuts (organophosphate).
Appearance: Reddish-brown sand-like crystalline solid.
Odor: Mild. The odor is a warning of exposure; however, no smell does not mean you are not being exposed.
Effects of exposure:
Breathed: Extremely poisonous. May cause headache, dizziness, profuse sweating, nausea, vomiting, reduced heart beat, stomach cramps, diarrhea, loss of coordination, slow and weak breathing, fever, loss of consciousness and death. May cause shortness of breath and a dangerous build-up of fluids in the lungs (pulmonary edema), which can cause death. Pulmonary edema is a medical emergency that can be delayed one to two days following exposure.
Eyes: Rapidly absorbed. Poisoning can happen very rapidly. See *Breathed*.
Skin: Poisonous when absorbed. Quickly passes through the unbroken skin; can increase exposure and the severity of the symptoms listed. Fatal poisoning may result from skin contact even if no irritation occurs.
Swallowed: Powerful, deadly poison. See *Breathed*.
Long-term exposure: May cause nerve damage including loss of coordination; liver damage. Repeated exposure may cause depression, irritability and personality changes; may affect blood cells.
Storage: Store in a regulated, specially marked place in tightly closed containers in a dry, well-ventilated place. Keep away from heat and sources of ignition. Containers may explode in fire.

★ 688 ★
MORPHOLINE
CAS: 110-91-8
Other names: DIETHYLENEIMIDE OXIDE; DIETHYLENE IMIDOXIDE; DIETHYLENE OXIMIDE; N,N-DIMETHYLACETAMIDE; DIETHYLENIMIDE OXIDE; P-ISOXAZINE,TETRAHYDRO-; 1-OXA-4-AZACYCLOHEXANE; 2H-1,4-OXAZINE, TETRAHYDRO-; TETRAHYDRO-1,4-ISOXAZINE; TETRAHYDRO-1,4-OXAZINE; TETRAHYDRO-2H-1,4-OXAZINE; TETRAHYDRO-P-OXAZINE

Danger profile: Mutagen, fire hazard, corrosive, exposure limit established, poisonous gas produced in fire.
Uses: Solvent; making waxes, polishes, cleaners, pharmaceuticals, bactericides, rubber, other chemicals; brightener in detergents; corrosion inhibitor; book paper preservative.
Appearance: Colorless liquid.
Odor: Fishy; ammonia-like. The odor is a warning of exposure; however, no smell does not mean you are not being exposed.
Effects of exposure:
Breathed: May be poisonous. May cause irritation of the eyes, nose, throat, breathing passages and lungs; nausea, headache, difficult breathing, visual disturbances, coughing and respiratory irritation.
Eyes: Contact may cause severe irritation, inflammation of the eyelids and burns.
Skin: Passes through the unbroken skin; can increase exposure and the severity of the symptoms listed. Contact may cause severe irritation and burns.
Swallowed: Moderately toxic.
Long-term exposure: May cause birth defects, miscarriage, cancer; liver and kidney damage.
Storage: Store in tightly closed containers in a dry, cool, well-ventilated place. Keep away from heat, sources of ignition, oxidizers, strong acids, cellulose nitrate and permanganates. Vapors may travel to sources of ignition and flash back. Containers may explode in fire.
First aid guide: Move victim to fresh air and call emergency medical care; if not breathing, give artificial respiration; if breathing is difficult, give oxygen. In case of contact with material, immediately flush skin or eyes with running water for at least 15 minutes. Remove and isolate contaminated clothing and shoes at the site. Keep victim quiet and maintain normal body temperature.

N

★ 689 ★
NALED
CAS: 300-76-5
Other names: ARTHODIBROM; BROMCHLOPHOS; BROMEX; DIBROM; O-(1,2-DIBROM-2,2-DICHLOR-AETHYL)-O,O-DIMETHYL-PHOSPHAT (GERMAN); 1,2-DIBROMO-2,2-DICHLOROETHYL DIMETHYL PHOSPHATE; O-(1,2-DIBROMO-2,2-DICLORO-ETIL)-O,O-DIMETIL-FOSTATO (ITALIAN); O-(1,2-DIBROOM-2,2-DICHLOOR-ETHYL)-O,O-DIMETHYL-FOSFAAT (DUTCH); DIMETHYL1,2-DIBROMO-2,2-DICHLOROETHYLPHOSPHATE; O,O-DIMETHYL-O-(1,2-DIBROMO 2,2-DICHLOROETHYL)PHOSPHATE; O,O-DIMETHYL O-2,2-DICHLORO-1,2-DIBROMOETHYL PHOSPHATE; ENT 24,988; ETHANOL,1,2-DIBROMO-2,2-DICHLORO-, DIMETHYL PHOSPHATE; HIBROM; ORTHO 4355; ORTHODIBROM; ORTHODIBROMO; PHOSPHATE DE O,O-DIMETHLE ET DE O-(1,2-DIBROMO-2,2-DICHLORETHYLE) (FRENCH); RE-4355

Danger profile: Mutagen, reproductive hazard, exposure limit established, poisonous gases produced in fire.
Uses: Insecticide.
Appearance: Colorless solid or straw-colored liquid.
Odor: Strong, slightly pungent. The odor is a warning of exposure; however, no smell does not mean you are not being exposed.
Effects of exposure:
　Breathed: Extremely poisonous. May cause headache, dizziness, profuse sweating, nausea, vomiting, reduced heart beat, stomach cramps, diarrhea, loss of coordination, slow and weak breathing, fever, loss of consciousness and death. May cause shortness of breath and a dangerous build-up of fluids in the lungs (pulmonary edema), which can cause death. Pulmonary edema is a medical emergency that can be delayed one to two days following exposure.
　Eyes: Rapidly absorbed. Poisoning can happen very rapidly. See *Breathed.*
　Skin: Poisonous if absorbed. Passes through the unbroken skin; can increase exposure and the severity of the symptoms listed.
　Swallowed: Powerful, deadly poison. See *Breathed.*
Long-term exposure: May damage the developing fetus or cause birth defects; nerve damage including loss of coordination; liver damage. Repeated exposure may cause depression, irritability and personality changes.
Storage: Store in a regulated, specially marked place in tightly closed containers in a dry, well-ventilated place. Keep away from strong oxidizers. Containers may explode in fire.

★ 690 ★
NAPHTHA
CAS: 8030-30-6
Other names: AROMATIC SOLVENT; BENZIN; COAL TAR NAPHTHA; 160 DEGREE BENZOL; HI-FLASH NAPHTHAYETHYLEN; LIGHT LIGROIN; MINERAL SPIRITS; NAPHTHA DISTILLATE; NAPHTHA, SOLVENT; NAPHTHA PETROLEUM; NAPHTHA SOLVENT; PETROLEUM BENZIN; PETROLEUM DISTILLATES (NAPHTHA); PETROLEUM ETHER; PETROLEUM NAPHTHA; SOLVENT NAPHTHA; VARSOL

Danger profile: Combustible, exposure limit established, poisonous gas produced in fire.
Uses: Solvent, paint thinner; making other chemicals; incorporation in dynamite.
Appearance: Colorless watery liquid.
Odor: Gasoline-like. The odor is a warning of exposure; however, no smell does not mean you are not being exposed.
Effects of exposure:
　Breathed: May be poisonous. May cause irritation of the eyes, nose and throat; dizziness, lightheadedness, rapid breathing, mental confusion, unsteady walk, headache, fatigue, nausea, vomiting; bluish skin, fingernails, earlobes and skin; unconsciousness, coma and death. High exposures may cause shortness of breath and a dangerous build-up of fluids in the lungs (pulmonary edema), which can cause death. Pulmonary edema is a medical emergency that can be delayed one to two days following exposure.
　Eyes: May cause burning feeling, tearing, irritation, redness and inflammation of the eyelids.
　Skin: May be poisonous if absorbed. Passes through the unbroken skin; can increase exposure and the severity of the symptoms listed. May cause irritation of the skin, dryness, redness and rash.
　Swallowed: May cause indigestion, dizziness, fatigue, nausea, sleepiness, unconsciousness. Also see *Breathed.*
Storage: Store in tightly closed containers in a dry, cool, well-ventilated place. Keep away from heat, sources of ignition and

oxidizers. Vapors may travel to sources of ignition and flash back. Containers may explode in fire.
First aid guide: Move victim to fresh air and call emergency medical care; if not breathing, give artificial respiration; if breathing is difficult, give oxygen. In case of contact with material, immediately flush eyes with running water for at least 15 minutes. Wash skin with soap and water. Remove and isolate contaminated clothing and shoes at the site.

★ 691 ★
NAPHTHALENE
CAS: 91-20-3
Other names: CAMPHOR TAR; MIGHTY 150; MOTH BALLS; MOTH FLAKES; NAFTALEN (POLISH); NAPTHALIN; NAPTHALINE; NCI-C52904; TAR CAMPHOR; WHITE TAR

Danger profile: Combustible, exposure limit established, Community Right-to-Know List, poisonous gas produced in fire.
Uses: Moth repellant; making other chemicals, dyes, plastics, resins, explosives and lubricants; fungicide; smokeless powder; cutting fluid; lubricant; synthetic tanning; preservative; antiseptic.
Appearance: White crystalline solid or liquid.
Odor: Mothball-like. The odor is a warning of exposure; however, no smell does not mean you are not being exposed.
Effects of exposure:
　Breathed: May cause irritation of the eyes, nose and throat. Very high levels may cause nausea; liver, kidney and red blood cell damage; mental confusion, convulsions, blue color of the skin and lips with reduced ability of the blood cells to carry oxygen to the body organs; possible death.
　Eyes: May cause irritation, redness, swelling of the eyelids.
　Skin: Passes through the unbroken skin; can increase exposure and the severity of the symptoms listed. May cause irritation, redness and swelling.
　Swallowed: Poison. May cause irritation of the mouth and stomach; cramps and diarrhea, nausea, headache, fever, vomiting, convulsions, coma; anemia, liver and kidney problems.
Long-term exposure: Repeated exposure may cause headache, fatigue, and nausea; eye damage with clouding of the eye lens; skin allergy.
Storage: Store in tightly closed containers in a dry, cool, well-ventilated place. Keep away from heat, sources of ignition, chromium (III) oxide, dinitrogen pentoxide, benzene, ether, carbon tetrachloride, carbon disulfide and oxidizers. May burn rapidly. Containers may explode in fire.
First aid guide: Move victim to fresh air; call emergency medical care. In case of contact with material, immediately flush skin or eyes with running water for at least 15 minutes. Removal of solidified molten material from skin requires medical assistance. Remove and isolate contaminated clothing and shoes at the site.

★ 692 ★
1-NAPHTHYLAMINE
CAS: 134-32-7
Other names: ALFANAFTILAMINA (ITALIAN); ALFA-NAFTYLOAMINA (POLISH); 1-AMINONAFTALEN (CZECH); 1-AMINONAPHTHALENE; C.I. AZOIC DIAZO COMPONENT 114; FAST GARNET B BASE; FAST GARNET BASE B; 1-NAFTILAMINA (SPANISH); ALPHA-NAFTYLAMIN (CZECH); 1-NAFTYLAMINE (DUTCH); 1-NAPHTHALENAMINE; NAPHTHALIDAM; NAPHTHALIDINE; 1-NAPHTHYLAMIN (GERMAN); ALPHA-NAPHTHYLAMINE

Danger profile: Carcinogen, mutagen, combustible, exposure limit established, Community Right-to-Know List, poisonous gas produced in fire.
Uses: Agricultural chemical, weed control; rubber manufacturing; making dyes.
Appearance: White or yellow crystals becoming red or purplish-red on exposure to air.
Odor: Characteristic amine-like. The odor is a warning of exposure; however, no smell does not mean you are not being exposed.
Effects of exposure:
> *Breathed:* May be poisonous. May cause blue color in lips, skin and under finger nails (cyanosis) with less oxygen being carried to body organs.
> *Eyes:* Contact with liquid may cause irritation.
> *Skin:* May be poisonous if absorbed. Passes through the unbroken skin without a sense of irritation or warning; can increase exposure and the severity of the symptoms listed.
> *Swallowed:* May be poisonous. No recognized immediate effects; may increase exposure and the severity of the symptoms listed.

Long-term exposure: May cause bladder or other cancer; changes in the genetic material in a body cell leading to birth defects and miscarriage. Related aromatic amine chemicals may cause liver damage and skin allergies.
Storage: Store in tightly closed containers in a dry, dark, cool, well-ventilated place. Keep away from sources of ignition nitrosyl hydroxide and light. Containers may explode in fire.

★ 693 ★
2-NAPHTHYLAMINE
CAS: 91-59-8
Other names: 2-AMINONAFTHALEN (CZECH); 2-AMINONAPHTHALENE; BETA- NAFTYLOAMINA (POLISH); BETA-NAPHTHYLAMINE; C.I. 37270; FAST SCARLET BASE B; NA; BETA-NAFTILAMINA (ITALIAN); 2-NAFTYLAMINE (DUTCH); USAF CB-22

Danger profile: Carcinogen, mutagen, combustible, Community Right-to-Know List, poisonous and explosive gas produced in fire.
Uses: Making dyes.
Appearance: White to reddish or pink shiny solid in the form of flakes.
Effects of exposure:
> *Breathed:* May cause blue color in lips, skin and under finger nails (cyanosis) with less oxygen being carried to body organs.
> *Eyes:* May cause irritation.
> *Skin:* Passes through the unbroken skin; can increase exposure and the severity of the symptoms listed.
> *Swallowed:* Moderately toxic. Also see *Breathed.*

Long-term exposure: May cause bladder cancer; bladder tumors; liver damage; skin allergy.
Storage: Store in tightly closed containers in a dry, cool, well-ventilated place. Keep away from heat, sources of ignition, light and nitrous acid. Containers may explode in fire.

★ 694 ★
NEON
CAS: 7440-01-9
Other names: NEON, COMPRESSED; NEON REFRIGERATED LIQUID

Danger profile: Asphyxiant.
Uses: In photoelectric bulbs; lasers; research; electronic industry.
Appearance: Colorless gas or liquid under pressure.

Effects of exposure:
> *Breathed:* May cause dizziness and lightheadedness. High levels may cause loss of consciousness, suffocation and death.
> *Eyes:* Contact with liquid may cause frostbite.
> *Skin:* Contact with liquid may cause frostbite.

Long-term exposure: Unknown at this time.
Storage: Store in tightly closed containers in a dry, cool, well-ventilated place. Protect containers from physical damage. Containers may explode in fire.
First aid guide: Move victim to fresh air and call emergency medical care; if not breathing, give artificial respiration; if breathing is difficult, give oxygen.

★ 695 ★
NEOPRENE
CAS: 126-99-8
Other names: 1,3-BUTADIENE, 2-CHLORO-; 2-CHLOOR-1,3-BUTADIEEN (DUTCH); 2-CHLOR-1,3-BUTADIEN (GERMAN); CHLOROBUTADIENE; 2-CHLOROBUTA-1,3-DIENE; 2-CHLORO-1,3-BUTADIENE; CHLOROPREEN (DUTCH); CHLOROPREN (GERMAN); CHLOROPREN (POLISH); CHLOROPRENE; CHLOROPRENE, INHIBITED; CHLOROPRENE, UNINHIBITED; BETA-CHLOROPRENE; 2-CLORO-1,3-BUTADIENE (ITALIAN); CLOROPRENE (ITALIAN)

Danger profile: Mutagen, fire hazard, poison, exposure limit established, Community Right-to-Know List, poisonous gas produced in fire.
Uses: Making synthetic rubber.
Appearance: Colorless liquid.
Odor: Ether-like. The odor is a warning of exposure; however, no smell does not mean you are not being exposed.
Effects of exposure:
> *Breathed:* May be posonous. May cause dizziness, lightheadedness and loss of consciousness. High exposure may cause damage to the developing fetus; spontaneous abortions; affect sperm production; and cause a temporary loss of hair.
> *Eyes:* May cause irritation and burns.
> *Skin:* May be poisonous if absorbed. Passes through the unbroken skin; can increase exposure and the severity of the symptoms listed. May cause irritation and burns.
> *Swallowed:* May be poisonous.

Long-term exposure: May cause damage to the developing fetus; cause spontaneous abortions; affect sperm production; dermatitis, conjunctivitis, anemia, temporary loss of hair, irritability; drop in blood pressure; liver, kidney, lung and other organ effects.
Storage: Store in tightly closed containers in a dark, dry, cool, well-ventilated place below 50°F, or in an explosion-proof freezer. Keep away from light, heat, sources of ignition, fluorine, oxidizers such as peroxides, permanganates, chlorates and perchlorates. Vapors may travel to sources of ignition and flash back. Containers may explode in fire.

★ 696 ★
NICKEL
CAS: 7440-02-0
Other names: C.I. 77775; NI 270; NICKEL 270; NICKEL CATALYST, WET; NICHEL (ITALIAN); NICKEL PARTICLES; NICKEL SPONGE; NI 0901-S; NI 4303T; NP 2; RANEY ALLOY; RANEY NICKEL

Danger profile: Carcinogen, mutagen, fire hazard (powders), exposure limit established, Community Right-to-Know List, poisonous gas produced in fire.

Uses: Making metal alloys, coins, stainless steel, permanent magnets; electroplating; alkaline storage battery;

Appearance: Silvery-white metal.

Effects of exposure:

Breathed: Dusts may cause irritation of the eyes, nose, throat and lungs, tears and runny nose, coughing, sore throat, difficult breathing; lethargy, fatigue, and drowsiness; slurred speech, shaking, and dizziness. Higher exposures could cause pneumonia-like symptoms with headache, nausea, vomiting, dizziness, diarrhea and a dangerous build-up of fluids in the lungs (pulmonary edema), which can cause death. Pulmonary edema is a medical emergency that can be delayed one to two days following exposure.

Eyes: May cause irritation, redness, tearing, pain, inflamed eyelids, and eye injury; burns in corneas and possible loss of sight.

Skin: May cause irritation, redness, and rash. Repeated or prolonged skin contact may cause irritation of the skin and may also cause the legs to feel numb.

Swallowed: May cause irritation of the mouth and throat; stomach and abdominal pain; vomiting, diarrhea, loss of muscular coordination and mental confusion; convulsions. Also see symptoms for *Breathed.*

Long-term exposure: May cause cancer, damage to the developing fetus; asthma or pneumonia-like lung allergy with cough and shortness of breath; nickel dermatitis; metallic taste; metal fume fever.

Storage: Store in tightly closed containers in a dry, cool, well-ventilated place. Keep dust away from heat, sources of ignition, strong acids, fluorine, ammonia, phosphorus, hydrazine, sulfur, selenium, hydrazine and performic acid. Containers may explode in fire.

★ 697 ★
NICKEL AMMONIUM SULFATE

CAS: 15699-18-0

Other names: AMMONIUM NICKEL SULFATE; SULFURIC ACID, AMMONIUM NICKEL (2) SALT (2:2:1)

Danger profile: Carcinogen, exposure limit established, Community Right-to-Know List, poisonous gas produced in fire.

Uses: In electroplating.

Appearance: Dark green-blue powder.

Effects of exposure:

Breathed: May cause irritation of nose and throat. High exposure may cause cough, shortness of breath and a dangerous build-up of fluids in the lungs (pulmonary edema), which can cause death. Pulmonary edema is a medical emergency that can be delayed one to two days following exposure. Lung, heart, liver or kidney damage may result from a single large exposure. Possible lung allergy.

Eyes: May cause irritation.

Skin: May cause irritation, redness and allergy with itching and rash.

Swallowed: May cause vomiting.

Long-term exposure: Repeated exposure may cause lung, heart, kidney or liver damage; skin allergy; lung allergy. Related nickel compounds may cause a reduction of fertility in men and women.

Storage: Store in tightly closed containers in a dry, cool, well-ventilated place. Keep away from strong acids and sulfur. Containers may explode in fire.

★ 698 ★
NICKEL CARBONYL

CAS: 13463-39-3

Other names: NICKEL CARBONYLE (FRENCH); NICHEL TETRACARBONILE (ITALIAN); NICKEL TETRACARBONYL; NICKEL TETRACARBONYLE (FRENCH); NIKKELTETRA-CARBONYL (DUTCH)

Danger profile: Carcinogen, possible reproductive hazard, poison, combustible, explosion hazard, exposure limit established, EPA extremely hazardous substances list; Community Right-to-Know List, poisonous gas produced in fire.

Uses: Making other chemicals and high-purity nickel powder; nickel coatings and films for other metals; refining nickel ore.

Appearance: Colorless to pale yellow liquid.

Odor: Musty. The odor is a warning of exposure; however, no smell does not mean you are not being exposed.

Effects of exposure:

Breathed: May be poisonous. May cause irritation of the eyes, nose, throat and lungs, tears and runny nose, coughing, sore throat, difficult breathing; giddiness, headache, vomiting; lethargy, fatigue, and drowsiness; rapid pulse; slurred speech, shaking, and dizziness. If victim is removed from exposure symptoms may disappear and return 12-36 hours later with blue pallor of skin, fever and cough. Death may occur. High exposures could cause shortness of breath, headache, nausea, vomiting, dizziness, diarrhea and a dangerous build-up of fluids in the lungs (pulmonary edema), which can cause death. Pulmonary edema is a medical emergency that can be delayed one to two days following exposure. Because of delayed effects, medical observation may be advised for up to three days.

Eyes: May cause irritation, redness, tearing, pain, inflamed eyelids, and eye injury; burns in corneas and possible loss of sight.

Skin: May be poisonous if absorbed. Passes through the unbroken skin; can increase exposure and the severity of the symptoms listed. May cause painful irritation, redness, and blisters Repeated or prolonged skin contact may cause irritation of the skin and may also cause the legs to feel numb.

Swallowed: May be poisonous. Ingestion can increase exposure and the severity of the symptoms listed. May cause irritation of the mouth and throat; stomach and abdominal pain; loss of muscular coordination and mental confusion; convulsions. Also see symptoms for *Breathed.*

Long-term exposure: May cause cancer; birth defects by damaging the fetus; lung, liver, heart muscle and kidney damage; skin allergy with rash.

Storage: Store in tightly closed steel cylinder under carbon dioxide atmosphere in a dry, cool, well-ventilated place. Highly reactive. Keep away from heat, sunlight, sources of ignition, all other materials particularly benzene, chloroform, acetone, carbon tetrachloride, air, oxygen, strong oxidizers, strong acids. Containers may explode in fire.

First aid guide: Move victim to fresh air and call emergency medical care; if not breathing, give artificial respiration; if breathing is difficult, give oxygen. In case of contact with material, immediately flush skin or eyes with running water for at least 15 minutes. Speed in removing material from skin is of extreme importance. Remove and isolate contaminated clothing and shoes at the site. Keep victim quiet and maintain normal body temperature. Effects may be delayed; keep victim under observation.

★ 699 ★
NICKEL CHLORIDE

CAS: 7718-54-9

Other names: NICKELOUS CHLORIDE; NICKEL(II) CHLORIDE; NICKEL(II) CHLORIDE (1:2)

Danger profile: Suspected carcinogen, mutagen, poison, exposure limit established, Community Right-to-Know List, poisonous gas produced in fire.
Uses: Electroplating; making ink; laboratory chemical.
Appearance: Golden-yellow powder.
Effects of exposure:
Breathed: Dust may cause irritation of nose, throat and breathing passages; cough and phlegm.
Eyes: May cause severe irritation and burns.
Skin: May cause irritation and skin allergy with redness, itching, and rash.
Swallowed: May cause vomiting.
Long-term exposure: May cause birth defects, miscarriage and cancer; skin and lung allergy; possible lung damage.
Storage: Store in tightly closed containers in a dry, cool, well-ventilated place. Keep away from sulfur, potassium and strong acids. Containers may explode in fire.

★700★
NICKEL CYANIDE
CAS: 557-19-7
Other names: NICKEL CYANIDE, SOLID; NICKEL-(II)CYANIDE

Danger profile: Possible carcinogen, mutagen, poison, exposure limit established, Community Right-to-Know List, poisonous gas produced in fire.
Uses: Metallurgy, electroplating.
Appearance: Yellow-brown powder. May change to light green or apple-green when moisture is absorbed.
Odor: Weak, characteristic almond of cyanide. The odor is a warning of exposure; however, no smell does not mean you are not being exposed.
Effects of exposure:
Breathed: May be poisonous. Dust or fumes may produce bitter taste, salivation, numbness of throat, anxiety, irregular respiration, rapid pulse, convulsions, loss of consciousness, paralysis; pneumonia-like illness; possible death. Higher exposures may cause shortness of breath and a dangerous build-up of fluids in the lungs (pulmonary edema), which can cause death. Pulmonary edema is a medical emergency that can be delayed one to two days following exposure. Lung, heart, liver or kidney damage may result from a single high exposure.
Eyes: Contact may cause irritation.
Skin: Contact may cause irritation.
Swallowed: Poisonous. May cause a bitter taste, salivation, numbness of throat, anxiety, irregular respiration, rapid pulse, convulsions, loss of consciousness, paralysis, death.
Long-term exposure: May cause birth defects, miscarriage, cancer. Repeated exposure may cause lung, heart, liver or kidney damage; asthma-like lung allergy.
Storage: Store in tightly closed containers in a dry, cool, well-ventilated place. Keep away from heat, sources of ignition, magnesium and acids. Containers may explode in fire.
First aid guide: Move victim to fresh air; call emergency medical care. In case of contact with material, immediately flush skin or eyes with running water for at least 15 minutes. Remove and isolate contaminated clothing and shoes at the site.

★701★
NICKEL(II) NITRATE
CAS: 13138-45-9
Other names: NICKEL NITRATE; NICKEL(II) NITRATE (1:2); NITRIC ACID, NICKEL(II) SALT; NITRIC ACID, NICKEL(2) SALT

Danger profile: Carcinogen, mutagen, exposure limit established, Community Right-to-Know List, poisonous gas produced in fire.
Uses: Electroplating; making ceramic colors and other chemicals.
Appearance: Green powder or crystals.
Effects of exposure:
Breathed: May cause irritation of the nose, throat and breathing passages; coughing and phlegm; allergy may develop.
Eyes: Contact may cause irritation and burns.
Skin: Contact may cause irritation and burns.
Long-term exposure: May cause cancer, mutations; decrease of fertility in men; skin and lung allergy; possible lung damage.
Storage: Store in tightly closed containers in a dry, cool, well-ventilated place with non-wood floors. Keep away from strong acids, sulfur, combustibles and organic or easily oxidized materials.

★702★
NICKEL SULFATE
CAS: 7786-81-4
Other names: NICKEL (II) SULFATE; NICKEL(2)SULFATE(1:1); NCI-C60344; NICKELOUS SULFATE; SULFURIC ACID, NICKEL (2) SALT

Danger profile: Carcinogen, mutagen, combustible, exposure limit established, Community Right-to-Know List, poisonous gas produced in fire.
Uses: Making other chemicals; electroplating; dyeing and printing textile; coatings; ceramics.
Appearance: Blue to blue-green solid crystals with a sweet taste.
Effects of exposure:
Breathed: High exposures may cause shortness of breath and a dangerous build-up of fluids in the lungs (pulmonary edema), which can cause death. Pulmonary edema is a medical emergency that can be delayed one to two days following exposure. High exposure may cause lung scarring.
Eyes: Contact may cause irritation.
Skin: Contact may cause irritation.
Long-term exposure: May cause cancer, mutations, infertility in males; scarring of the lungs from repeated exposure; asthma-like lung allergy.
Storage: Store in tightly closed containers in a dry, cool, well-ventilated place with non-wood floors. Keep away from strong acids, wood, organics and other easily combustible materials.

★703★
NICOTINE
CAS: 54-11-5
Other names: EMO-NIB; FLUX MAAY; MACH-NIC; NI-CODUST; NICOFUME; NICOCIDE; PYRIDINE, 3-(1-METHYL-2-PYRROLIDINYL)-; PYRIDINE, (S)-3-(1-METHYL-2-PYRROLIDINYL)-AND SALTS; 3-(1-METHYL-2-PYRROLIDYL) PYRIDINE; TENDUST; XL ALL INSECTICIDE

Danger profile: Mutagen, combustible, exposure limit established, Community Right-to-Know List, poisonous gas produced in fire.
Uses: Insecticide; fumigant; in medicine. Use as insecticide may be restricted.
Appearance: Thick colorless to pale yellow liquid which turns dark brown on exposure to air.
Odor: Fishy when warm. The odor is a warning of exposure; however, no smell does not mean you are not being exposed.
Effects of exposure:

Breathed: May cause irritation of the nose, throat and lungs; burning sensation in mouth and throat; hormone effects. High exposures may cause nausea, headache, weakness, confusion, visual disturbances; increased blood pressure; increased heart rate, tremors, seizures and death.

Eyes: Contact with liquid may cause irritation.

Skin: Passes through the unbroken skin; can increase exposure and the severity of the symptoms listed. Can be absorbed in toxic amounts. Contact may cause irritation.

Swallowed: May cause burning of mouth and stomach, vomiting, excitement, faintness, paralysis of lungs.

Long-term exposure: May cause birth defects by damaging the fetus; hardening of the arteries, increased risk of heart attack, chronic high blood pressure and stroke.

Storage: Store in tightly closed containers in a dry, cool, well-ventilated place. Keep away from strong oxidizers and strong acids. Containers may explode in fire.

First aid guide: Move victim to fresh air and call emergency medical care; if not breathing, give artificial respiration; if breathing is difficult, give oxygen. In case of contact with material, immediately flush skin or eyes with running water for at least 15 minutes. Speed in removing material from skin is of extreme importance. Remove and isolate contaminated clothing and shoes at the site. Keep victim quiet and maintain normal body temperature. Effects may be delayed; keep victim under observation.

★ 704 ★
NITRAPYRIN
CAS: 1929-82-4
Other names: 2-CHLORO-6-(TRICHLOROMETHYL)PYRIDINE; DOWCO-163; NITRAPYRIN; N-SERVE; N-SERVE NITROGEN STABILIZER; PYRIDINE, 2-CHLORO-6-(TRICHLOROMETHYL)-

Danger profile: Possible mutagen, poison, exposure limit established, poisonous gas produced in fire.
Uses: Fertilizer additive; prevents nitrogen loss in soil.
Appearance: White crystalline solid.
Effects of exposure:

Breathed: May be poisonous. May cause irritation of the eyes, nose and throat. High levels may cause dizziness, lightheadedness and loss of consciousness.

Eyes: Contact may cause irritation and permanent damage.

Skin: May be poisonous if absorbed. Passes through the unbroken skin; can increase exposure and the severity of the symptoms listed. Contact may cause irritation and scarring.

Swallowed: May be poisonous.

Long-term exposure: Repeated exposure may cause headache, dizziness, loss of appetite and disturbed sleeping pattern; liver and kidney damage. Evidence of cancer in animals.

Storage: Store in tightly closed, lined containers in a dry, cool, well-ventilated place. Keep away from aluminum, magnesium and their alloys. Containers may explode in fire.

★ 705 ★
NITRIC ACID
CAS: 7697-37-2
Other names: ACIDE NITRIQUE (FRENCH); ACIDO NITRICO (ITALIAN); AQUA FORTIS; AZOTIC ACID; AZOTOWY KWAS (POLISH); HYDROGEN NITRATE; NITAL; NITRIC ACID, RED FUMING; NITRIC ACID, WHITE FUMING; NITRYL HYDROXIDE; NITROUS FUMES; RED FUMING NITRIC ACID; RFNA; SALPETERSAURE (GERMAN); SALPETERZUUROPLOSSINGEN (DUTCH); WHITE FUMING NITRIC ACID; WFNA

Danger profile: Corrosive, poison, exposure limit established, Community Right-to-Know List, poisonous gas produced in fire.

Uses: Making other chemicals, drugs, fertilizers, dyes, and explosives; in metallurgy, photoengraving; urethanes; rubber chemicals; reprocessing nuclear fuel.

Appearance: Colorless, yellow or red fuming liquid when concentrated. May be sold as a dilute solution.

Odor: Acrid; strong; suffocating. The odor is a warning of exposure; however, no smell does mean you are not being exposed.

Effects of exposure:

Breathed: May be poisonous. Vapor may cause intense irritation to nose, throat, mouth and lungs; mucous membrane. Symptoms of exposure include watering of the eyes, coughing, difficult breathing, headache, nausea, vomiting, muscular weakness, unconsciousness. High exposures may cause shortness of breath and a dangerous build-up of fluids in the lungs (pulmonary edema), which can cause death. Pulmonary edema is a medical emergency that can be delayed one to two days following exposure.

Eyes: Severe irritant. Contact may cause, burning and stinging, tearing, irritation and redness and swelling of the eyelids, intense pain; can burn the eyes and lead to blurred vision, chronic irritation and other permanent damage.

Skin: Severe irritant. Contact may cause burning or stinging feeling, redness, swelling, intense pain, blisters, irritation, rash and burns.

Swallowed: May be poisonous. May cause burning feeling of the mouth and throat, intense thirst, throat swelling, stomach cramps, nausea, vomiting, diarrhea, severe breathing difficulties and unconsciousness. Tissue damage may result. Breathing may stop. Watch for delayed symptoms.

Long-term exposure: Fumes may cause tooth erosion; lung damage is possible.

Storage: Store in tightly closed containers in a dry, cool, well-ventilated place. Keep away from water, steam and other forms of moisture; heat, sunlight; aluminum, metallic powders and most metals; carbides, hydrogen sulfide, turpentine, strong bases glass; wood, cloth and other combustible materials, alcohols and amines.

★ 706 ★
NITRIC OXIDE
CAS: 10102-43-9
Other names: BIOXYDE D'AZOTE (FRENCH); NITROGEN MONOXIDE; OXYDE NITRIQUE (FRENCH); STICKMONOXYD (GERMAN)

Danger profile: Explosion hazard, mutagen, poison, exposure limit established, EPA extremely hazardous substances list, poisonous gas produced in fire.

Uses: Making other chemicals; bleaching rayon. Common air pollutant.

Appearance: Colorless gas, blue liquid and solid.

Effects of exposure:

Breathed: May be poisonous. Vapor may cause intense irritation to nose, throat, mouth and lungs; mucous membrane. Symptoms of exposure include watering of the eyes, coughing, difficult breathing, headache, nausea, vomiting, muscular weakness, unconsciousness. High exposures may cause shortness of breath and a dangerous build-up of fluids in the lungs (pulmonary edema), which can cause death. Pulmonary edema is a medical emergency that can be delayed one to two days following exposure.

Eyes: Severe irritant. Contact may cause, burning and stinging, tearing, irritation and redness and swelling of the eyelids, intense pain; can burn the eyes and lead to blurred vision, chronic irritation and other permanent damage. Liquid may cause frostbite.

Skin: Severe irritant. Contact may cause burning or stinging feeling, redness, swelling, intense pain, blisters, irritation, rash. Liquid may cause frostbite.

Swallowed: May cause burning feeling of the mouth and throat, intense thirst, throat swelling, stomach cramps, nausea, vomiting, diarrhea, severe breathing difficulties and unconsciousness. Tissue damage may result. Breathing may stop. Watch for delayed symptoms.

Long-term exposure: Fumes may cause chronic irritation of the respiratory tract with cough, headache, appetite loss, weakness and fatigue; tooth erosion; lung damage is possible.

Storage: Store in tightly closed containers in a dry, cool, well-ventilated place. Keep away from water, steam and other forms of moisture, cumbustable materials, fuels, heat, sunlight, carbon disulfide, methanol, fluorine, nitrogen trichloride, ozone, vinyl chloride, acetic anhydride, aluminum, and all other chemicals. Containers may explode in fire.

First aid guide: Move victim to fresh air and call emergency medical care; if not breathing, give artificial respiration; if breathing is difficult, give oxygen. In case of contact with material, immediately flush skin or eyes with running water for at least 15 minutes. Remove and isolate contaminated clothing and shoes at the site. Keep victim quiet and maintain normal body temperature. Effects maybe delayed; keep victim under observation.

★ 707 ★
P-NITROANILINE
CAS: 100-01-6
Other names: 1-AMINO-4-NITROBENZENE; AZOIC DIAZO COMPONENT 37; BENZENAMINE, 4-NITRO-; DEVELOPER P; FAST RED BASE; FAST RED 2G BASE; 4-NITROANILINE; 4-NITRANBINE; PARA-AMINONITROBENZENE; PNA

Danger profile: Mutagen, combustible, poison, explosion hazard, exposure limit established, poisonous gas produced in fire.
Uses: Making pharmaceuticals, dyes and pesticides.
Appearance: Bright yellow powder.
Odor: Faint, ammonia-like. The odor is a warning of exposure; however, no smell does not mean you are not being exposed.
Effects of exposure:
Breathed: May be poisonous. May cause headache, rapid and difficult breathing, shortness of breath, dizziness, drowsiness, shortness of breath, nausea, mental confusion, weakness; bluish color of the lips and skin, convulsions, collapse, coma and death. A single high exposure may cause liver damage.
Eyes: May cause irritation, redness, swelling of the eyelids; light causes pain; severe damage; possible damage of the cornea.
Skin: May be poisonous if absorbed. Passes through the unbroken skin; can increase exposure and the severity of the symptoms listed. May cause irritation, redness, swelling, blisters.
Swallowed: May be poisonous. May cause irritation of the mouth, stomach; cramps and diarrhea. Also see *Breathed.*
Long-term exposure: May cause birth defects, miscarriage or cancer; liver damage; anemia.
Storage: Store in tightly closed containers in a dry, cool, well-ventilated place. Keep away from heat, sources of ignition, oxidizers, strong bases, reducers and strong acids. Containers may explode in fire.

★ 708 ★
NITROBENZENE
CAS: 98-95-3
Other names: BENZENE, NITRO-; ESSENCE OF MIRBANE; ESSENCE OF MYRBANE; MIRBANE OIL; NCI-C60082; NITROBENZEEN (DUTCH); NITROBENZEN (POLISH); NITROBENZOL; NITROBENZOL,L; NITRO, LIQUID; OIL OF MIRBANE; OIL OF MYRBANE

Danger profile: Mutagen, poison, combustible, exposure limit established, Community Right-to-Know List, EPA extremely hazardous substances list; poisonous gas produced in fire.
Uses: Making shoe polish, dyes, explosives, floor and metal polishes, other chemicals and paints.
Appearance: May be colorless, pale to dark yellow or brown oily liquid.
Odor: Bitter almonds or shoe polish. The odor is a warning of exposure; however, no smell does not mean you are not being exposed.
Effects of exposure:
Breathed: Vapors are highly toxic. May cause headache, rapid and difficult breathing, shortness of breath, dizziness, mental confusion, weakness, nausea, vomiting; bluish color of the lips, nails and skin, convulsions, coma and death. High exposure may cause liver damage.
Eyes: Highly toxic. May cause irritation, redness, swelling of the eyelids; light causes pain; severe damage.
Skin: Highly toxic. Passes through the unbroken skin; can increase exposure and the severity of the symptoms listed. May cause irritation, redness, swelling, blisters; possible skin allergy.
Swallowed: May cause irritation of the mouth, stomach; cramps and diarrhea.
Long-term exposure: May cause damage to the developing fetus; male reproductive organ damage; liver damage; effect blood cell formation.
Storage: Store in tightly closed containers in a dry, cool, well-ventilated place. Keep away from heat, sources of ignition, nitric acid, aluminum chloride, potassium hydroxide and other chemicals. Containers may explode in fire.
First aid guide: Move victim to fresh air and call emergency medical care; if not breathing, give artificial respiration; if breathing is difficult, give oxygen. In case of contact with material, immediately flush skin or eyes with running water for at least 15 minutes. Speed in removing material from skin is of extreme importance. Remove and isolate contaminated clothing and shoes at the site. Keep victim quiet and maintain normal body temperature. Effects may be delayed; keep victim under observation.

★ 709 ★
4-NITROBIPHENYL
CAS: 92-93-3
Other names: BIPHENYL, 4-NITRO-; 1,1'-BIPHENYL, 4-NITRO-; P-NITROBIPHENYL; 4-NITRODIPHENYL; P-NITRODIPHENYL; 1-NITRO- 4-PHENYLBENZENE; P-PHENYL-NITROBENZENE; 4-PHENYL-NITROBENZENE; PNB

Danger profile: Carcinogen, mutagen, combustible, Community Right-to-Know List, poisonous gas produced in fire.
Uses: Making plastics, fungicides; in wood preservatives.
Appearance: Needles from alcohol; yellow plates of needles
Odor: Sweet. The odor is a warning of exposure; however, no smell does not mean you are not being exposed.
Effects of exposure:
Breathed: Similar chemicals have caused liver and nerve damage.
Eyes: May cause irritation.
Skin: Passes through the unbroken skin; can increase exposure and the severity of the symptoms listed. May cause irritation.
Long-term exposure: May cause bladder cancer; liver damage; possible nerve damage.
Storage: Store in tightly closed containers in a dry, cool, well-ventilated place. Keep away from heat and sources of ignition.

★ 710 ★
P-NITROCHLOROBENZENE
CAS: 100-00-5
Other names: 1-CHLOOR-4-NITROBENZEEN (DUTCH); 1-CHLOR-4-NITROBENZOL (GERMAN); P-CHLORONITROBENZENE; 1-CHLORO-4-NITROBENZENE; 4-CHLORONITROBENZENE; 4-CHLORO-1-NITROBENZENE; 1-CLORO-4-NITROBENZENE (ITALIAN); P-NITROCHLOORBENZEEN (DUTCH); P-NITROCHLOROBENZENE; NITROCHLOROBENZENE, PARA-, SOLID; P-NITROCHLOROBENZOL (GERMAN); P-NITROCLOROBENZENE (ITALIAN); PNCB

Danger profile: Mutagen, combustible, poison, exposure limit established, poisonous gas produced in fire.
Uses: Making dyes, rubber, agricultural and other chemicals.
Appearance: Yellow crystals.
Odor: Sweet. The odor is a warning of exposure; however, no smell does not mean you are not being exposed.
Effects of exposure:
  Breathed: May be poisonous. May cause headache, rapid and difficult breathing, shortness of breath, dizziness, mental confusion, weakness; bluish color of the lips and skin, convulsions, coma and death.
  Eyes: May cause irritation, redness, swelling of the eyelids; light causes pain; severe damage.
  Skin: May be poisonous if absorbed. Passes through the unbroken skin; can increase exposure and the severity of the symptoms listed. May cause irritation, redness, swelling, blisters.
  Swallowed: May be poisonous. May cause irritation of the mouth, stomach; cramps and diarrhea.
Long-term exposure: May cause mutations; questionable carcinogen.
Storage: Store in tightly closed containers in a dry, cool, well-ventilated place or refrigerator. Keep away from light, heat (may cause explosion), direct sunlight and sodium methoxide. Containers may explode in fire.

★ 711 ★
NITROFEN
CAS: 1836-75-5
Other names: BENZENE, 2,4-DICHLORO-1-(4-NITROPHENOXY)-; BENZENAMINE, 4- ETHOXY-N-(5-NITRO-2FURANYL)METHYLENE-: 2,4-DECHLOROPHENYL P-NITROPHENYLETHER; 2,4-DICHLORO-4'-NITRODIPHENYLETHER; 2,4-DICHLORO-1-(4-NITROPHENOXY)BENZENE; 2,4-DICHLOROPHENYL P-NITROPHENYL ETHER; 2,4-DICHLOROPHENYL 4-NITROPHENYLETHER; ETHER, 2,4-DICHLOROPHENYL-P-NITROPHENYL; FW 925; MEZOTOX; NCI-C00420; NICLOFEN; NIP; NITOFEN; NITRAPHEN; NITROCHLOR; NITROFENE (FRENCH); NITROPHEN; NITROPHENE; TOK; TOK-2; TOK E; TOK E-25; TOK E-40; TOKKORN; TOK WP-50; TRIZLIN

Danger profile: Carcinogen, mutagen, poison, reproductive hazard, Community Right-to-Know List, poisonous gas produced in fire.
Uses: Broad spectrum herbicide; on cereals and vegetables. Banned in many countries.
Appearance: Dark brown solid.
Effects of exposure:
  Breathed: May be poisonous. May cause intense thirst, nose irritation, headache, coughing, nausea, stomach pain, loss of appetite, restlessness, sweating, fatigue and rapid breathing and pulse, anxiety, confusion; yellow face and lips. High levels may cause fever, coma and death.

Eyes: May cause burning sensation, irritation, tearing, and inflammation of the eyelids.
  Skin: Passes through the unbroken skin; can increase exposure and the severity of the symptoms listed. Symptoms may be delayed. May cause irritation, redness, swelling and blisters.
  Swallowed: May be poisonous. May cause burning feeling in the throat and mouth; dizziness, headache, nausea, vomiting, high fever, thirst, fatigue, rapid pulse, convulsions, loss of consciousness. Higher exposures may cause shortness of breath and a dangerous build-up of fluids in the lungs (pulmonary edema), which can cause death. Pulmonary edema is a medical emergency that can be delayed one to two days following exposure.
Long-term exposure: May cause cancer of the liver and pancreas; liver; nervous system and liver damage. Repeated exposure may cause blood cell damage with a white blood cell count and reduced hemoglobin. Has been shown to be a teratogen in animals with birth defects by damaging the fetus.
Storage: Store in tightly closed containers in a dry, cool, well-ventilated place. Containers may explode in fire.

★ 712 ★
NITROGEN
CAS: 7727-37-9
Other names: NITROGEN, COMPRESSED; NITROGEN GAS; NITROGEN, CRYOGENIC LIQUID; NITROGEN, REFRIGERATED LIQUID

Danger profile: Asphyxiant (gas), explosion hazard (liquid).
Uses: Making other chemicals, explosives; electronics; food refrigeration, freeze drying; food antioxidant; in fertilizers.
Appearance: Colorless gas.
Effects of exposure:
  Breathed: Although low toxic, may cause a shortage of oxygen in a closed space. In high concentrations, it is a simple asphyxiant. May cause headache, dizziness, fatigue; rapid irregular breathing; nausea, vomiting, confusion, loss of consciousness convulsions and death in certain circumstances.
  Eyes: Contact with liquid may cause frostbite, pain, stinging, tearing, inflammation of the eyelids and permanent damage.
  Skin: Contact with liquid may cause frostbite, intense cold feeling, skin turns white and feels cold and hard.
  Swallowed: Liquid may cause headache, confusion, drowsiness, unconsciousness and death.
Long-term exposure: Under certain conditions caisson disease or the "bends" may develop.
Storage: Store in tightly closed containers in a dry, cool, well-ventilated place. Keep away from lithium, neodymium, titanium. Containers may explode in fire.

★ 713 ★
NITROGEN DIOXIDE
CAS: 10102-44-0
Other names: AZOTE (FRENCH); AZOTO (ITALIAN); DINITROGEN TETROXIDE; NITRITO; NITROGEN PEROXIDE; NITROGEN TETROXIDE; NTO; STICKSTOFFDIOXID (GERMAN); STIKSTOFDIOXYDE (DUTCH)

Danger profile: Mutagen, exposure limit established, poisonous gas produced in fire.
Uses: Production of other chemicals, particularly nitric acid; in rocket fuels; polymerization inhibitor for acrylates.
Appearance: Colorless to yellow liquid.
Odor: Pungent, acrid. The odor is a warning of exposure; however, no smell does not mean you are not being exposed.
Effects of exposure:

Breathed: May be poisonous. Vapor may cause intense irritation to nose, throat, mouth and lungs; mucous membrane. Symptoms of exposure include watering of the eyes, coughing, difficult breathing, headache, nausea, vomiting, muscular weakness, unconsciousness. High exposure can lead to drowsiness, coma and death; buildup of fluid in the lungs (pulmonary edema), and may cause death. Pulmonary edema may be delayed by one to two days following exposure.

Eyes: Severe irritant. Contact may cause, burning and stinging, tearing, irritation and redness and swelling of the eyelids, intense pain; can burn the eyes and lead to blurred vision, chronic irritation and other permanent damage; corneal ulceration.

Skin: Severe irritant. Contact may cause burning or stinging feeling, redness, swelling, intense pain, blisters, irritation, rash. Repeated exposure may lead to skin allergy and chronic irritation.

Swallowed: May cause burning feeling of the mouth and throat, intense thirst, throat swelling, stomach cramps, nausea, vomiting, diarrhea, severe breathing difficulties and unconsciousness. Tissue damage may result. Breathing may stop. Watch for delayed symptoms.

Long-term exposure: May cause mutations; erosion of teeth; emphysema; chronic bronchitis; lung damage.

Storage: Store in tightly closed containers in a dry, cool, well-ventilated place. Keep away from combustible materials, fuels, organic and oxidizable material, cylohexane, nitrobenzene, toluene and petroleum. Protect containers from physical damage. Containers may explode in fire.

First aid guide: Move victim to fresh air and call emergency medical care; if not breathing, give artificial respiration; if breathing is difficult, give oxygen. In case of contact with material, immediately flush skin or eyes with running water for at least 15 minutes. Remove and isolate contaminated clothing and shoes at the site. Keep victim quiet and maintain normal body temperature. Effects maybe delayed; keep victim under observation.

★714★
NITROGEN TRIFLUORIDE
CAS: 7783-54-2
Other names: NITROGEN FLUORIDE

Danger profile: Fire hazard, poison, explosion hazard, corrosive, exposure limit established, poisonous gas produced in fire.
Uses: Making other chemicals.
Appearance: Colorless gas.
Odor: Moldy. The odor is a warning of exposure; however, no smell does not mean you are not being exposed.
Effects of exposure:

Breathed: May be poisonous. Vapor may cause intense irritation to nose, throat, mouth and lungs; mucous membrane. Symptoms of exposure include watering of the eyes, coughing, difficult breathing, headache, nausea, vomiting, muscular weakness, unconsciousness. High exposure can lead to drowsiness, coma and death; buildup of fluid in the lungs (pulmonary edema), and may cause death. Pulmonary edema may be delayed by one to two days following exposure.

Eyes: Severe irritant. Contact may cause, burning and stinging, tearing, irritation and redness and swelling of the eyelids, intense pain; can burn the eyes and lead to blurred vision, chronic irritation and other permanent damage.

Skin: Severe irritant. Contact may cause burning or stinging feeling, redness, swelling, intense pain, blisters, irritation, rash. Repeated exposure may lead to skin allergy and chronic irritation.

Swallowed: May cause burning feeling of the mouth and throat, intense thirst, throat swelling, stomach cramps, nausea, vomiting, diarrhea, severe breathing difficulties and un-

consciousness. Tissue damage may result. Breathing may stop. Watch for delayed symptoms.

Long-term exposure: Prolonged exposure may cause erosion of teeth and bones.

Storage: Store in tightly closed containers in a dry, cool, well-ventilated place. Keep away from heat, sources of ignition, charcoal, tetrafluorohydrazine, carbon monoxide, diborane. Protect containers from physical damage. Containers may explode in fire.

First aid guide: Move victim to fresh air and call emergency medical care; if not breathing, give artificial respiration; if breathing is difficult, give oxygen. In case of contact with material, immediately flush skin or eyes with running water for at least 15 minutes. Remove and isolate contaminated clothing and shoes at the site. Keep victim quiet and maintain normal body temperature. Effects may delayed; keep victim under observation.

★715★
NITROGLYCERIN
CAS: 55-63-0
Other names: ANGININE; BLASTING GELATIN; BLASTING OIL; GLONOIN; GLYCERINTRINITRATE (CZECH); GLYCEROLTRINITRAAT (DUTCH); GLYCEROL TRINITRATE; GLYCEROL(TRINITRATE DE) (FRENCH); GLYCEROL, NITRIC ACID TRIESTER; GLYCERYL NITRATE; GLYCERYL TRINITRATE; GTN; KLAVI KORDAL; MYOCON; NG; NIGLYCON; NIONG; NITRIC ACID TRIESTER OF GLYCEROL; NITRINE-TDC; NITROGLICERINA (ITALIAN); NITROGLICERYNA (POLISH); NITROGLYCERINE; NITROGLYCEROL; NITROGLYN; NITROL; NITROLINGUAL; NITROLOWE; NITRONET; NITRONG; NITRO-SPAN; NITROSTAT; NK-843; NTG; PERGLOTTAL; 1,2,3-PROPANETROL, TRINITRATE; 1,2,3-PROPANETRIYL NITRATE; SK-106N; SOUP; TNG; TRINITRIN; TRINITROGLYCERIN; TRINITROGLYCEROL

Danger profile: Mutagen, combustible, dangerous explosion hazard, highly reactive, poison, exposure limit established, Community Right-to-Know List, poisonous gas produced in fire.
Uses: High explosive; production of dynamite and other explosives; medicine (vasodilator); combating fires in oil wells; rocket propellants.
Appearance: Colorless to yellow liquid.
Effects of exposure:

Breathed: May cause intense thirst, nose irritation, headache, coughing, nausea, stomach pain, restlessness, sweating, and rapid breathing and pulse, anxiety, confusion; blue face, fingernails and lips. High levels may cause fever, coma and death.

Eyes: May cause burning sensation, irritation, tearing, and inflammation of the eyelids.

Skin: Passes through the unbroken skin; can increase exposure and the severity of the symptoms listed. Symptoms may be delayed. May cause irritation, redness, swelling and blisters.

Swallowed: May cause burning feeling in the throat and mouth; dizziness, headache, nausea, vomiting, high fever, thirst, fatigue, rapid pulse, convulsions, loss of consciousness; possible death. Drinking alcoholic beverages shortly before or following exposure may aggravate symptoms and cause sudden mental or personality changes; possible loss of control and delirium.

Long-term exposure: May cause mutations. Repeated exposure may result in a developed tolerance; angina and heart attack may occur from a sudden stop to exposure; also, headache may develop when a worker returns to work after being away for an extended time such as a vacation or weekend. Skin irritation may develop. Often found mixed with ethylene glycol dinitrate.
Storage: Extremely explosive and reactive. Should be handled only by those with proper training. Store in tightly closed contain-

ers in a dry, cool, well-ventilated combustible liquid store room. Keep away from heat, sunlight, ultraviolet radiation, flames or any source of ignition, shock of any kind, strong acids, ozone and oxidizing materials. Containers may explode in fire.
First aid guide: Move victim to fresh air and call emergency medical care; if not breathing, give artificial respiration; if breathing is difficult, give oxygen. In case of contact with material, immediately flush eyes with running water for at least 15 minutes. Wash skin with soap and water. Remove and isolate contaminated clothing and shoes at the site.

★ 716 ★
NITROMETHANE
CAS: 75-52-5
Other names: METHANE, NITRO-; NITROCARBOL; NITROMETAN (POLISH)

Danger profile: Fire hazard, exposure limit established, poisonous gas produced in fire.
Uses: Solvent; making other chemicals; rocket fuel; gasoline additive.
Appearance: Colorless oily liquid.
Odor: Strong, disagreeable. The odor is a warning of exposure; however, no smell does not mean you are not being exposed.
Effects of exposure:
  Breathed: May cause irritation of the eyes, nose and throat; dizziness, lightheadedness, rapid breathing, mental confusion, unsteady walk, headache, fatigue, nausea, vomiting, unconsciousness, coma and death.
  Eyes: May cause burning feeling, tearing, irritation, redness and inflammation of the eyelids.
  Skin: Passes through the unbroken skin; can increase exposure and the severity of the symptoms listed. May cause irritation of the skin, dryness, redness and rash.
  Swallowed: May cause indigestion, dizziness, fatigue, loss of appetite, nausea, vomiting, diarrhea, unconsciousness.
Long-term exposure: May cause kidney and liver damage.
Storage: Store in tightly closed containers in a dry, cool, well-ventilated, explosion proof place or refrigerator. Extremely reactive; keep away from heat, sources of ignition and all other chemicals. Protect containers from physical damage and shock. Containers may explode in fire.
First aid guide: Move victim to fresh air and call emergency medical care; if not breathing, give artificial respiration; if breathing is difficult, give oxygen. In case of contact with material, immediately flush eyes with running water for at least 15 minutes. Wash skin with soap and water. Remove and isolate contaminated clothing and shoes at the site.

★ 717 ★
5-NITRO-O-ANISIDINE
CAS: 99-59-2
Other names: 2-AMINO-4-NITROANISOLE; AZOAMINE SCARLET K; BENZENAMINE, C.I. 37130; FAST SCARLET R; 2-METHOXY-5-NITRO-; 2-METHOXY-5- NITROANILINE; NCI-C01934; 3-NITRO-6-METHOXYANILINE; 5-NITRO-2-METHOXYANILINE

Danger profile: Carcinogen, mutagen, Community Right-to-Know List, poisonous gas produced in fire.
Uses: Making dyes and pigments.
Appearance: Orange to red needle-like solid.
Effects of exposure:
  Breathed: High exposure may interfere with the blood's ability to carry oxygen (methemoglobin); skin, fingernails and other body parts may develop a bluish color.
  Skin: Passes through the unbroken skin; can increase exposure and the severity of the symptoms listed.

Long-term exposure: May cause cancer; damage to the male reproductive glands; bone marrow damage; thyroid, liver and other organ problems.
Storage: Store in tightly closed containers in a dry, cool, well-ventilated place. Containers may explode in fire.

★ 718 ★
2-NITROPHENOL
CAS: 88-75-5
Other names: 2-HYDROXYNITROBENZENE; O-NITROPHENOL; ORTHONITROPHENOL; PHENOL, 2-NITRO-; PHENOL, O-NITRO

Danger profile: Combustible, poison, Community Right-to-Know List, poisonous gas produced in fire.
Uses: Making pesticides, dyes and other chemicals.
Appearance: Yellow crystalline solid.
Odor: Peculiar aromatic. The odor is a warning of exposure; however, no smell does not mean you are not being exposed.
Effects of exposure:
  Breathed: May be poisonous. May cause nose and throat irritation, headache, drowsiness, coughing, nausea, stomach pain, restlessness, sweating, and rapid breathing and pulse, anxiety, confusion; blue color in face, ears fingernails and lips. High levels may cause fever, coma and death.
  Eyes: May cause burning sensation, irritation, tearing, and inflammation of the eyelids.
  Skin: May be poisonous if absorbed. Passes through the unbroken skin; can increase exposure and the severity of the symptoms listed. Symptoms may be delayed. May cause irritation, redness, swelling and blisters.
  Swallowed: May be poisonous. May cause burning feeling in the throat and mouth; dizziness, headache, nausea, vomiting, high fever, thirst, fatigue, rapid pulse, convulsions, loss of consciousness.
Long-term exposure: Has produced liver and kidney damage in animals.
Storage: Store in tightly closed containers in a dry, cool, well-ventilated place with non-wood floors. Keep away from heat, sources of ignition, chlorosulfuric acid, potassium hydroxide, and organic material. Containers may explode in fire.

★ 719 ★
3-NITROPHENOL
CAS: 554-84-7
Other names: M-NITROPHENOL; 3-NITROTOLUENE; 3-NITROTOLUOL; 3-METHYLNITROBENZENE

Danger profile: Combustible, poison, poisonous gas produced in fire.
Uses: Making pesticides, dyestuffs, indicator solutions and other chemicals.
Appearance: Yellow crystalline solid.
Odor: Characteristic. The odor is a warning of exposure; however, no smell does mean you are not being exposed.
Effects of exposure:
  Breathed: May be poisonous. May cause nose and throat irritation, headache, coughing, nausea, stomach pain, restlessness, sweating, and rapid breathing and pulse, anxiety, confusion; blue face, ears, fingertips and lips. High levels may cause fever, coma and death.
  Eyes: May cause burning sensation, severe irritation, tearing, and inflammation of the eyelids.
  Skin: May be poisonous if absorbed. Passes through the unbroken skin; can increase exposure and the severity of the symptoms listed. Symptoms may be delayed. May cause irritation, redness, swelling and blisters.

Swallowed: May be poisonous. May cause burning feeling in the throat and mouth; dizziness, headache, nausea, vomiting, high fever, thirst, fatigue, rapid pulse, convulsions, loss of consciousness.
Storage: Store in tightly closed containers in a dry, cool, well-ventilated place with non wood floors. Keep away from sources of ignition and organic materials. Containers may explode in fire.

★ 720 ★
4-NITROPHENOL
CAS: 100-02-7
Other names: 4-HYDROXYNITROBENZENE; NCI-C55992; NIPHEN; 4-NITROFENOL (DUTCH); P-NITROPHENOL; PARANITROFENOL (DUTCH); PARANITROFENOLO (ITALIAN); PARANITROPHENOL (FRENCH; GERMAN); PHENOL, 4-NITRO-; PHENOL, P-NITRO; PNP

Danger profile: Mutagen, combustible, poison, Community Right-to-Know List, poisonous gas produced in fire.
Uses: Fungicide; making other chemicals.
Appearance: Colorless to slightly yellow crystalline material.
Odor: Slight, sweet, characteristic. Burning taste. The odor is a warning of exposure; however, no smell does not mean you are not being exposed.
Effects of exposure:
Breathed: May be poisonous. May cause nose, throat and air passage irritation, headache, coughing, nausea, stomach pain, restlessness, sweating, and rapid breathing and pulse, anxiety, confusion; blue face, lips, fingernails and lips that might lead to heart and brain damage. High levels may cause fever, coma and death.
Eyes: May cause burning sensation, severe irritation, tearing, and inflammation of the eyelids.
Skin: May be poisonous if asorbed. Passes through the un-broken skin; can increase exposure and the severity of the symptoms listed. Symptoms may be delayed. May cause irritation, redness, swelling and blisters.
Swallowed: May be poisonous. May cause burning feeling in the throat and mouth; dizziness, headache, nausea, vomiting, high fever, thirst, fatigue, rapid pulse, convulsions, loss of consciousness.
Long-term exposure: May cause mutations. Related chemicals have cause cataracts.
Storage: Store in tightly closed containers in a dry, cool, well-ventilated place with non-wood floors. Keep away from heat, sources of ignition, strong bases, diethyl phosphite, oxidizers and combustible materials. Containers may explode in fire.

★ 721 ★
2-NITROPROPANE
CAS: 79-46-9
Other names: DIMETHYLNITROMETHANE; ISONITROPROPANE; NIPAR S-20; NIPARS-20 SOLVENT; NIPAR S-30 SOLVENT; NITROISOPROPANE; BETA- NITROPROPANE; 2-NP; PROPANE, 2-NITRO; SEC-NITROPROPANE

Danger profile: Carcinogen, mutagen, fire hazard, exposure limit established, Community Right-to-Know List, poisonous gas produced in fire.
Uses: Solvent; making other chemicals; rocket propellant; gasoline additive.
Appearance: Colorless liquid.
Odor: Mild, fruity. The odor is a warning of exposure; however, no smell does not mean you are not being exposed.
Effects of exposure:
Breathed: May be poisonous. May cause headache, rapid and difficult breathing, shortness of breath, dizziness, nausea, diarrhea, mental confusion, weakness; bluish color of

the lips, ears fingertips and skin, convulsions, coma and death; possible kidney, heart and brain damage. May cause severe liver damage that might result in death.
Eyes: May be poisonous if absorbed. May cause irritation, redness, swelling of the eyelids; light causes pain; severe damage.
Skin: Passes through the unbroken skin; can increase exposure and the severity of the symptoms listed. May cause irritation, redness, swelling, blisters.
Swallowed: May cause irritation of the mouth, stomach; cramps and diarrhea.
Long-term exposure: May cause cancer; damage to the developing fetus. Repeated exposure may cause kidney, liver, heart and brain damage.
Storage: Store in tightly closed containers in a dry, cool, well-ventilated explosion-proof place. Keep away form heat, sources of ignition, oxidizers, strong bases and strong acids. An explosive mixture is formed in air. May attack some forms of plastic. Vapors may travel to sources of ignition and flash back. Containers may explode in fire.

★ 722 ★
N-NITROSODIBUTYLAMINE
CAS: 924-16-3
Other names: 1-BUTANAMINE,N-BUTYL-N-NITROSO-; BUTYLAMINE, N-NITROSODI-; N-BUTYL-N-NITROSO-1-BUTAMINE; DBN; DBNA; DIBUTYLAMINE, N-NITROSO-; DI-N-BUTYLNITROSAMIN (GERMAN); DIBUTYLNITROSAMINE; DI-N-BUTYLNITROSAMINE; N,N-DI-N-BUTYLNITROSAMINE; N,N-DIBUTYLNITROSOAMINE; NDBA; NITROSODIBUTYL-AMINE; N-NITROSO- DI-N-BUTYLAMINE; N-NITROSODI-N-BUTYLAMINE

Danger profile: Carcinogen, mutagen, reproductive hazard, Community Right-to-Know List.
Uses: Research chemical.
Appearance: Pale yellow, oily liquid
Effects of exposure:
Breathed: May cause irritation of the nose, throat and lungs; coughing. Some related chemicals may cause headache, dizziness, stomach cramps and weakness
Long-term exposure: May cause cancer; toxic to the fetus and may cause death; liver damage from repeated exposure.
Storage: Store in tightly closed containers in a dry, cool, well-ventilated place. Keep away from light.

★ 723 ★
O-NITROTOLUENE
CAS: 88-72-2
Other names: 2-METHYLNITROBENZENE; O-METHYLNITROBENZENE; 2-NITROTOLUENE; ONT

Danger profile: Mutagen, combustible, poison, exposure limit established, poisonous gas produced in fire.
Uses: Making other chemicals and synthetic dyes.
Appearance: Yellowish liquid.
Effects of exposure:
Breathed: May be poisonous. May cause headache, flushing of face, rapid and difficult breathing, shortness of breath, dizziness, mental confusion, muscular weakness; bluish color of the lips and skin, nausea, vomiting, increased pulse and respiratory rate, irritability, convulsions, coma and death.
Eyes: May cause irritation, redness, swelling of the eyelids; light causes pain; severe damage.
Skin: May be poisonous if absorbed. Passes through the unbroken skin; can increase exposure and the severity of the

symptoms listed. May cause irritation, redness, swelling, blisters.

Swallowed: May be poisonous. May cause irritation of the mouth, stomach; cramps and diarrhea.

Long-term exposure: May cause mutations; birth defects, miscarriage or cancer. Similar irritating chemicals may cause lung problems.

Storage: Store in tightly closed containers in a dry, cool, well-ventilated place. Keep away from heat, open flame, sources of ignition and alkalis. Containers may explode in fire.

★724★
NITROUS OXIDE
CAS: 10024-97-2
Other names: DINITROGEN MONOXIDE; FACTITIOUS AIR; HYPONITROUS ACID ANHYDRIDE; LAUGHING GAS; NITROGEN OXIDE; STICKDIOXYD (GERMAN)

Danger profile: Mutagen, teratogen, poisonous gas produced in fire.
Uses: Anesthetic in dentistry and surgery; propellant gas in food aerosols; leak detection; foaming agent for whipped cream.
Appearance: Colorless gas.
Odor: Slightly sweet. The odor is a warning of exposure; however, no smell does not mean you are not being exposed.
Effects of exposure:
Breathed: May cause lightheadedness, giddiness, hysteria and sleepy feeling. High levels may cause loss of consciousness. Very high levels may cause suffocation and death.
Eyes: Contact may cause frostbite and burns.
Skin: Contact may cause frostbite and burns.
Swallowed: May cause frostbite and freezing burns of the mouth and throat.
Long-term exposure: May cause birth defects by damaging the fetus; nerve system damage with numbness and weakness in the limbs; possible blood cell damage.
Storage: Store in tightly closed cylinders in a dry, cool, well-ventilated place. Keep away from sources of ignition, fuels, organic peroxides, aluminum, ammonia, carbon monoxide, hydrogen, hydrazine, hydrogen sulfide, sodium and phosphine. An oxidizer, will increase the intensity of any fire. Containers may explode in fire.
First aid guide: Move victim to fresh air; call emergency medical care. Keep victim quiet and maintain normal body temperature.

★725★
NONANE
CAS: 111-84-2
Other names: N-NONANE; SHELLSOL 140

Danger profile: Fire hazard, exposure limit established, poisonous gas produced in fire.
Uses: In gasoline, stoddard solvent; biodegradable detergents; making other chemicals.
Appearance: Colorless liquid.
Odor: Gasoline-like. The odor is a warning of exposure; however, no smell does not mean you are not being exposed.
Effects of exposure:
Breathed: May cause dizziness, lightheadedness, loss of coordination and loss of consciousness. Concentrated vapor may cause severe lung irritation, depression, irritation of respiratory tract, central nervous system excitement followed by depression, shortness of breath and a dangerous build-up of fluids in the lungs (pulmonary edema), which can cause death. Pulmonary edema is a medical emergency that can be delayed one to two days following exposure.
Eyes: May cause irritation.
Skin: Prolonged contact may cause drying and skin cracking.

Swallowed: May cause irritation of mouth and stomach.
Long-term exposure: Drying and cracking of the skin; liver damage.
Storage: Store in tightly closed containers in a dry, cool, well-ventilated place. Keep away from heat, sources of ignition and oxidizing materials. Vapors may travel to sources of ignition and flash back. Containers may explode in fire.
First aid guide: Move victim to fresh air and call emergency medical care; if not breathing, give artificial respiration; if breathing is difficult, give oxygen. In case of contact with material, immediately flush eyes with running water for at least 15 minutes. Wash skin with soap and water. Remove and isolate contaminated clothing and shoes at the site.

★726★
NONYL TRICHLOROSILANE
CAS: 5283-67-0
Other names: NONYLTRICHLOROSILANE; SILANE, TRICHLORONONYL-; SILANE, NONYLTRICHLORO-; TRICHLORONONYLSILANE

Danger profile: Corrosive, poisonous gas produced in fire.
Uses: Making silicones.
Appearance: Clear liquid. Fumes are produced in moist air.
Odor: Irritating. The odor is a warning of exposure; however, no smell does not mean you are not being exposed.
Effects of exposure:
Breathed: Vapors may cause irritation of the eyes, nose, throat and lungs; cough. High exposures may cause shortness of breath and a dangerous build-up of fluids in the lungs (pulmonary edema), which can cause death. Pulmonary edema is a medical emergency that can be delayed one to two days following exposure.
Eyes: May cause severe irritation, burns and permanent damage.
Skin: Contact may cause severe irritation and burns.
Swallowed:
Long-term exposure: Similar irritating substances may cause lung problems.
Storage: Store in tightly closed containers in a dry, cool, well-ventilated place. Keep away from combustible materials. Containers may explode in fire.
First aid guide: Move victim to fresh air; call emergency medical care. In case of contact with material, immediately flush skin or eyes with running water for at least 15 minutes. Remove and isolate contaminated clothing and shoes at the site. Keep victim quiet and maintain normal body temperature.

O

★727★
OCTACHLORONAPHTHALENE
CAS: 2234-13-1
Other names: HALOWAX 1051; NAPHTHALENE, OCTA-CHLORO-; 1,2,3,4,5,6,7,8- OCTACHLORONAPHTHALENE; PERCHLORONAPHTALENE

Danger profile: Exposure limit established, Community Right-to-Know List, poisonous gas produced in fire.
Uses: Lubricants, coatings, cable insulation and other protective coatings.
Appearance: Waxy, pale yellow solid.
Odor: Pleasant, aromatic. The odor is a warning of exposure; however, no smell does not mean you are not being exposed.
Effects of exposure:

Breathed: May cause fatigue, loss of appetite, dark urine and jaundice. Overexposure may damage the liver enough to cause death.
Skin: Passes through the unbroken skin; can increase exposure and the severity of the symptoms listed. May cause dermatitis with rash.
Swallowed: See *Breathed.*
Long-term exposure: May cause liver damage and jaundice; acne-like skin rash. Death may result from repeated lower exposure.
Storage: Store in tightly closed containers in a dry, cool, well-ventilated place. Keep away from heat and strong oxidizers. Containers may explode in fire.

★ 728 ★
OCTAFLUOROCYCLOBUTANE
CAS: 115-25-3
Other names: CYCLOBUTANE; CYCLOOCTAFLUOROBUTANE; FC-C 318; FREON C-318; HALOCARBON C-138; OCTAFLUORO-; PERFLUOROCYCLOBUTANE; PROPELLANT C318; R-C 318

Danger profile: Mutagen, poisonous gas produced in fire.
Uses: Refrigerant; heat transfer; propellent.
Appearance: Colorless gas.
Effects of exposure:
Breathed: May cause lung irritation, coughing, difficult breathing and shortness of breath. High exposures may cause chest tightness, wheezing, and a dangerous build-up of fluids in the lungs (pulmonary edema), which can cause death. Pulmonary edema is a medical emergency that can be delayed one to two days following exposure. Very high levels may cause death.
Eyes: May cause frostbite.
Skin: May cause frostbite.
Long-term exposure: May cause birth defects, miscarriage, cancer; lung problems. Closely related chemicals may cause irregular heart beat and death.
Storage: Store in tightly closed containers in a dry, cool, well-ventilated place. Keep away from heat. Protect cylinders form physical damage. Containers may explode in fire.
First aid guide: Move victim to fresh air and call emergency medical care; if not breathing, give artificial respiration; if breathing is difficult, give oxygen.

★ 729 ★
OCTAMETHYLPYROPHOSPHORAMIDE
CAS: 152-16-9
Other names: ENT 17,291; DIPHOSPHORAMIDE, OCTAMETHYL-; LETHA LAIRE G-59; OMPA; OMPACIDE; OMPATOX; OMPAX; PESTOX; PESTOX 3; PESTOX III; SCHRADAN; SCHRADANE (FRENCH); SYSTAM; SYSTOPHOS; SYTAM

Danger profile: Poison, EPA extremely hazardous substance list, poisonous gas produced in fire.
Uses: An insecticide which is absorbed by the plant and poisons sucking and chewing insects. Banned in the U.S.
Appearance: Viscous liquid.
Effects of exposure:
Breathed: Extremely poisonous. May cause headache, dizziness, profuse sweating, nausea, vomiting, reduced heart beat, stomach cramps, diarrhea, loss of coordination, slow and weak breathing, fever, loss of consciousness and death. May cause shortness of breath and a dangerous build-up of fluids in the lungs (pulmonary edema), which can cause

death. Pulmonary edema is a medical emergency that can be delayed one to two days following exposure.
Eyes: Rapidly absorbed. Poisoning can happen very rapidly. See *Breathed.*
Skin: Poison. Passes through the unbroken skin; can increase exposure and the severity of the symptoms listed.
Swallowed: Poison. May cause lack of appetite, nausea, abdominal cramps, diarrhea, blurred vision, vomiting, shortness of breath. See *Breathed.*
Storage: Store in tightly closed containers in a dry, cool, well-ventilated place. Containers may explode in fire.

★ 730 ★
OCTANE
CAS: 111-65-9
Other names: N-OCTANE; NORMAL OCTANE; NOKTAN (POLISH); OKTANEN (DUTCH); OTTANI (ITALIAN)

Danger profile: Fire hazard, explosion hazard, exposure limit established, poisonous gas produced in fire.
Uses: Solvent; making fuels and solvents; calibrations.
Appearance: Colorless liquid.
Odor: Gasoline-like. The odor is a warning of exposure; however, no smell does not mean you are not being exposed.
Effects of exposure:
Breathed: May be poisonous. May cause irritation of the eyes, nose, throat and breathing passages; dizziness, depression, lightheadedness, rapid breathing, mental confusion, unsteady walk, headache, fatigue, nausea, vomiting, unconsciousness, coma and death. May act as a simple asphyxiant. High exposures may cause shortness of breath and a dangerous build-up of fluids in the lungs (pulmonary edema), which can cause death. Pulmonary edema is a medical emergency that can be delayed one to two days following exposure.
Eyes: May cause burning feeling, tearing, irritation, redness and inflammation of the eyelids.
Skin: May be poisonous if absorbed. Passes through the unbroken skin; can increase exposure and the severity of the symptoms listed. May cause irritation of the skin, dryness, redness cracking and rash.
Swallowed: May cause irritation of the mouth and stomach; indigestion, dizziness, fatigue, unconsciousness.
Long-term exposure: Prolonged contact may result in dry, cracked skin.
Storage: Store in tightly closed containers in a dry, cool, well-ventilated place. Keep away from heat, sources of ignition and strong oxidizers. Vapors may travel to sources of ignition and flash back. Containers may explode in fire.
First aid guide: Move victim to fresh air and call emergency medical care; if not breathing, give artificial respiration; if breathing is difficult, give oxygen. In case of contact with material, immediately flush eyes with running water for at least 15 minutes. Wash skin with soap and water. Remove and isolate contaminated clothing and shoes at the site.

★ 731 ★
OCTYL ALCOHOL
CAS: 111-87-5
Other names: ALCOHOL C-8; ALFOL 8; DYTOL M-83; CAPRYLIC ALCOHOL; HEPTYL CARBINOL; 1-HYDROXYOCTANE; LOROL 20; OCTANOL; N-OCTANOL; 1-OCTANOL; OCTILIN; OCTYL ALCOHOL, NORMAL-PRIMARY; N-OCTYL ALCOHOL; PRIMARY OCTYL ALCOHOL; SIPOL L8

Danger profile: Mutagen, combustible, poisonous gas produced in fire.

Uses: Perfumery; cosmetics; organic synthesis; solvent; manufacture of high-boiling esters; anti-foaming agent; flavoring agent.
Appearance: Colorless liquid.
Odor: Sweet. The odor is a warning of exposure; however, no smell does not mean you are not being exposed.
Effects of exposure:
Breathed: May cause irritation of the eyes, nose and throat.
Eyes: May cause irritation.
Skin: Passes through the unbroken skin; can increase exposure and the severity of the symptoms listed. May cause irritation.
Swallowed: Moderately toxic.
Long-term exposure: May cause birth defects, miscarriage or lead to cancer.
Storage: Store in tightly closed containers in a dry, cool, well-ventilated place. Keep away from heat, sources of ignition and oxidizing materials. Containers may explode in fire.

★ 732 ★
OLEUM
CAS: 8014-95-7
Other names: DISULFURIC ACID; DITHIONIC ACID; FUMING SULFURIC ACID; PYROSULPHURIC ACID; SULFURIC ACID, FUMING; SULFURIC ACID MIXED WITH SULFUR TRIOXIDE

Danger profile: Combustible, extremely corrosive, regulated by OSHA, poisonous gas produced in fire.
Uses: Making dyes, explosives, other chemicals; petroleum refining.
Appearance: Colorless to cloudy, dark brown, fuming, oily liquid.
Odor: Sharp penetrating; choking. The odor is a warning of exposure; however, no smell does not mean you are not being exposed.
Effects of exposure:
Breathed: May be poisonous. Vapor may cause intense irritation to nose, throat, mouth and lungs; mucous membrane. Symptoms of exposure include watering of the eyes, coughing, difficult breathing, headache, nausea, vomiting, muscular weakness, unconsciousness. High exposure can lead to drowsiness, coma and death; buildup of fluid in the lungs (pulmonary edema), and may cause death. Pulmonary edema may be delayed by one to two days following exposure.
Eyes: Severe irritant. Contact may cause, burning and stinging, tearing, irritation and redness and swelling of the eyelids, intense pain; can burn the eyes and lead to blurred vision, chronic irritation and other permanent damage. Liquid may cause severe burns and blindness.
Skin: Severe irritant. Contact may cause burning or stinging feeling, redness, swelling, intense pain, blisters, irritation, rash. Repeated exposure may lead to skin allergy and chronic irritation. Liquid may cause severe third degree burns.
Swallowed: May be poisonous. May cause burning feeling of the mouth and throat, intense thirst, throat swelling, stomach cramps, nausea, vomiting, diarrhea, severe breathing difficulties and unconsciousness. Tissue damage may result. Breathing may stop. Watch for delayed symptoms.
Long-term exposure: May cause bronchitis, emphysema; chronic runny nose, eye irritation, nose bleeds, and ulcers; teeth erosion.
Storage: Store in tightly closed containers in a dry, cool, well-ventilated place. Keep away from cast iron, finely divided metal and other combustable material, carbides, chlorates, fulminates, nitrates, picrates, and organics.
First aid guide: Move victim to fresh air and call emergency medical care; if not breathing, give artificial respiration; if breathing is difficult, give oxygen. In case of contact with material, im-

mediately flush skin or eyes with running water for at least 15 minutes. Speed in removing material from skin is of extreme importance. Remove and isolate contaminated clothing and shoes at the site. Keep victim quiet and maintain normal body temperature.

★ 733 ★
OSMIUM TETROXIDE
CAS: 20816-12-0
Other names: OSMIC ACID; OSMIUM(VIII) OXIDE

Danger profile: Mutagen, exposure limit established, Community Right-to-Know List, poisonous gas produced in fire.
Uses: Microscope tissue staining; photography; in chemical reactions.
Appearance: Colorless to pale yellow solid.
Odor: Strong, unpleasant; chlorine-like; sharp, choking. The odor is a warning of exposure; however, no smell does not mean you are not being exposed. Effects of exposure may occur below the legal exposure limit.
Effects of exposure:
Breathed: Vapor may cause intense irritation to nose, throat, mouth and lungs; mucous membrane. Symptoms of exposure include watering of the eyes, coughing, difficult breathing, headache, nausea, vomiting, muscular weakness, unconsciousness. High exposure can lead to drowsiness, coma and death; buildup of fluid in the lungs (pulmonary edema), and may cause death. Pulmonary edema may be delayed by one to two days following exposure.
Eyes: Contact may cause, burning and stinging, tearing, severe irritation and redness and swelling of the eyelids, intense pain; can burn the eyes and lead to blurred vision, chronic irritation and other permanent damage.
Skin: Passes through the unbroken skin; can increase exposure and the severity of the symptoms listed. Severe irritant. Contact may cause burning or stinging feeling, redness, swelling, intense pain, blisters, irritation, rash. Repeated exposure may lead to skin allergy and chronic irritation.
Swallowed: May cause burning feeling of the mouth and throat, intense thirst, throat swelling, stomach cramps, nausea, vomiting, diarrhea, severe breathing difficulties and unconsciousness. Tissue damage may result. Breathing may stop. Watch for delayed symptoms.
Long-term exposure: Repeated exposure may cause mutations, miscarriage or cancer; kidney and bone marrow damage. May cause lung problems.
Storage: Store in tightly closed containers in a dry, cool, well-ventilated place. Keep away from hydrochloric acid and organic materials. Containers may explode in fire.
First aid guide: Move victim to fresh air and call emergency medical care; if not breathing, give artificial respiration; if breathing is difficult, give oxygen. In case of contact with material, immediately flush skin or eyes with running water for at least 15 minutes. Speed in removing material from skin is of extreme importance. Remove and isolate contaminated clothing and shoes at the site. Keep victim quiet and maintain normal body temperature. Effects may be delayed; keep victim under observation.

★ 734 ★
OXALIC ACID
CAS: 144-62-7
Other names: ACIDE OXALIQUE (FRENCH); ACIDO OSSALICO (ITALIAN); ETHANEDIOIC ACID; KYSELINA STAVELOVA (CZECH); NCI-C55209; OXAALZUUR (DUTCH); OXALSAEURE (GERMAN); OXALIC ACID DIHYDRATE; ETHANE DIOIC ACID

Danger profile: Combustible, poison, corrosive, exposure limit established, poisonous gas produced in fire.

Uses: Has a wide range of uses. Automobile radiator cleanser; general metal and equipment cleaning; leather tanning; laboratory chemical; agent for permanent press resins; bleaching of textiles and wood.

Appearance: Colorless crystals or white powder.

Effects of exposure:

Breathed: Vapor may cause intense irritation to nose, throat, mouth and lungs; mucous membrane. Symptoms of exposure include watering of the eyes, coughing, difficult breathing, headache, nausea, vomiting, muscular weakness, unconsciousness. High exposure can lead to drowsiness, coma and death; buildup of fluid in the lungs (pulmonary edema), and may cause death. Pulmonary edema may be delayed by one to two days following exposure.

Eyes: Severe irritant. Contact may cause, burning and stinging, tearing, irritation and redness and swelling of the eyelids, intense pain; can burn the eyes and lead to blurred vision, chronic irritation and other permanent damage.

Skin: Passes through the unbroken skin; can increase exposure and the severity of the symptoms listed. Contact may cause burning or stinging feeling, redness, swelling, intense pain, blisters, irritation, rash. Repeated exposure may lead to skin allergy and chronic irritation.

Swallowed: May be poisonous. May cause burning feeling and corrosion of the mouth and throat, intense thirst, throat swelling, stomach cramps, nausea, vomiting, diarrhea, severe breathing difficulties and unconsciousness. Tissue damage may result. Breathing may stop. Watch for delayed symptoms.

Long-term exposure: May cause mouth, nose and throat damage; kidney stones; blood vessel damage with pain and blue skin color and gangrene.

Storage: Store in tightly closed containers in a dry, cool, well-ventilated place. Keep away from heat, sources of ignition, furfuryl alcohol, silver and strong oxidizers. Containers may explode in fire.

★735★
OXYGEN
CAS: 7782-44-7
Other names: LIQUID OXYGEN; LOX; OXYGEN, LIQUID

Danger profile: Mutagen, possible reproductive hazard.

Uses: Welding; rocket fuel; in blast furnaces; copper smelting; decompression chambers; spacecraft; medicinal gas for resuscitation; making other chemicals and many industrial processes.

Appearance: Non-toxic colorless gas or light blue liquid under pressure.

Effects of exposure:

Breathed: Pure oxygen at high pressure may cause irritation of the throat and lungs; coughing, nausea, vomiting, dizziness, pain in joints and muscles, tingling in the hands and feet, muscle twitching, loss of vision, palpitations, convulsions, loss of consciousness; cause shortness of breath and a dangerous build-up of fluids in the lungs (pulmonary edema), which can cause death. Pulmonary edema is a medical emergency that can be delayed one to two days following exposure.

Eyes: Contact with liquid oxygen may cause severe burns and frostbite.

Skin: Contact with liquid oxygen may cause severe burns and frostbite.

Swallowed: Liquid may cause severe freezing burns and tissue damage of the mouth and throat.

Long-term exposure: May cause mutations; reproduction problems. Breathing 50-100 percent oxygen at normal pressure

even intermittently over a prolonged period may cause lung damage.

Storage: Store outdoors in tightly closed steel containers in a dry, cool, well-ventilated place. Not combustible but essential to combustion, and greatly increases the intensity and violence of fire. Keep away from heat, sources of ignition, organic and combustible material and other chemicals. Protect cylinders from physical damage. Vapor may explode if ignited in an enclosed area. Containers may explode in fire.

★736★
OXYGEN DIFLUORIDE
CAS: 7783-41-7
Other names: DIFLUORINE MONOXIDE; FLUORINE MONOXIDE; FLUORINE OXIDE; OXYGEN FLUORIDE

Danger profile: Explosion hazard, corrosive, exposure limit established, poisonous gas produced in fire.

Uses: Rocket propellent.

Appearance: Colorless gas.

Odor: Foul. The odor is a warning of exposure; however, no smell does not mean you are not being exposed.

Effects of exposure:

Breathed: May cause irritation of the nose, eyes, mouth and throat; watering eyes; salivation, coughing, throat pain, difficult breathing, headache, fatigue, nausea, stomach pains. High exposures may cause shortness of breath and a dangerous build-up of fluids in the lungs (pulmonary edema), which can cause death. Pulmonary edema is a medical emergency that can be delayed one to two days following exposure.

Eyes: May cause intense stinging, irritation, severe burns and irreparable damage and possible loss of sight. Liquid may cause frostbite.

Skin: May cause severe irritation, redness, swelling and burns. Liquid may cause frostbite.

Swallowed: May cause irritation; burning feeling in mouth and throat; swallowing causes pain; intense thirst; throat swelling, nausea, vomiting, shock. See *Breathed* for other symptoms.

Long-term exposure: May cause chronic lung problems; pulmonary edema; congestion.

Storage: Store in tightly closed containers in a dry, cool, well-ventilated place with non wood floors. Keep away from water, air, adsorbents such as silica gel and alumina, combustibles, fuels, reducing agents and all other chemicals. Forms and explosive mixture with water and combustible gases. Protect cylinders from physical damage. Containers may explode in fire.

First aid guide: Move victim to fresh air and call emergency medical care; if not breathing, give artificial respiration; if breathing is difficult, give oxygen. In case of contact with material, immediately flush skin or eyes with running water for at least 15 minutes. Remove and isolate contaminated clothing and shoes at the site. Keep victim quiet and maintain normal body temperature. Effects maybe delayed; keep victim under observation.

★737★
OZONE
CAS: 10028-15-6
Other names: OZON (POLISH); TRIATOMIC OXYGEN

Danger profile: Mutagen, explosion hazard (liquid), exposure limit established.

Uses: Purifying air and drinking water; industrial waste treatment; deodorizing sewage gases; in bleaching waxes, oils, wet paper and textiles; making other chemicals; bactericide. steroid hormones.

Appearance: Colorless gas; dark blue liquid under pressure.

Odor: Pungent, strong. The odor is a warning of exposure; however, no smell does not mean you are not being exposed.

Effects of exposure:

Breathed: Vapor may cause intense irritation to nose, throat, mouth and lungs; mucous membrane. Symptoms of exposure include watering of the eyes, continual coughing, difficult breathing, headache, nausea, vomiting, muscular weakness, excessive sweating, rapid pulse, unconsciousness. Higher exposure can lead to drowsiness, coma and death; buildup of fluid in the lungs (pulmonary edema), and may cause death. Pulmonary edema may be delayed by one to two days following exposure.

Eyes: Contact with liquid may cause severe chemical burns and frostbite.

Skin: Contact with liquid may cause severe chemical burns and frostbite.

Swallowed: Liquid may cause severe burns and tissue damage of the mouth and throat.

Long-term exposure: May cause mutations; scarring and thickening of small air passages and result in chronic lung problems; possible increased susceptibility to lung disease and infection.

Storage: Store in tightly closed containers in a dry, cool, well-ventilated place with non-wood floors. Keep away from heat, alcohol, ether, sulfur, fats, resins, phosphine vapor, organics, alkenes, aniline, bromine, ether, nitrogen compounds. combustible materials and reducing agents. Liquid ozone is and extreme explosion hazard. High levels may cause rapid deterioration of rubber, paints and metals. Containers may explode in fire.

P

★738★
PANCREATIN
CAS: 8049-47-6

Danger profile: Allergen; poisonous gas produced in fire.
Uses: In medicines; treating textiles and leather.
Appearance: Yellowish to cream-colored powder.
Odor: Strong. The odor is a warning of exposure; however, no smell does not mean you are not being exposed.

Effects of exposure:

Breathed: May cause wheezing, shortness of breath and an asthma-like lung reaction. Hours later fatigue, and possible fever may develop. Possible lung scarring. Similar enzymes may cause nose irritation and nosebleed.

Eyes: Similar enzymes may cause eye irritation.

Skin: May cause skin allergy, itching and rash. Similar enzymes may cause skin irritation and sores.

Swallowed: Similar enzymes may cause irritation of the mouth and tongue.

Long-term exposure: May cause lung problems, asthma-like allergy with tightness of the chest, shortness of breath, wheezing, coughing, chills, fever and possible lung scarring; skin allergy.

Storage: Store in tightly closed containers in a dry, cool, well-ventilated place. Keep away from acids, fuels and alcohols.

★739★
PARAFFIN
CAS: 8002-74-2
Other names: PARAFFIN WAX FUME; PARAFFIN WAX

Danger profile: Combustible, exposure limit established.

Uses: Candles; coating papers; crayons; sealing containers of various kinds; electrical insulation; chewing gum base, candy novelties; photography; polishes.
Appearance: Colorless or white, translucent waxy solid.
Odor: Weak, pleasant. The odor is a warning of exposure; however, no smell does not mean you are not being exposed.

Breathed: May cause irritation of the nose, throat and lungs; headache, nausea, coughing.

Eyes: May cause irritation and watering.

Skin: Passes through the unbroken skin; can increase exposure and the severity of the symptoms listed.

Swallowed: May cause nausea, vomiting, irritation of the stomach; tiredness and fatigue.

Long-term exposure: Many paraffin products contain carcinogens. The chemical itself is a questionable carcinogen.

Storage: Store in tightly closed containers in a dry, cool, well-ventilated place.

★740★
PARAFORMALDEHYDE
CAS: 30525-89-4
Other names: FLO-MORE; FORMAGENE; PARAFORM; PARAFORM 3; TRIFORMOL; TRIOXYMETHYLENE

Danger profile: Mutagen, combustible, may form explosive mixture in air, poisonous gas and combustible vapor produced in fire.
Uses: In fungicides, bactericides, and disinfectants; adhesives; pesticide; contraceptive creams.
Appearance: White crystalline powder.
Odor: Pungent and irritating, formaldehyde-like. The odor is a warning of exposure; however, no smell does not mean you are not being exposed.

Effects of exposure:

Breathed: May cause irritation of the nose, throat and lungs; coughing. Higher exposures may cause shortness of breath and a dangerous build-up of fluids in the lungs (pulmonary edema), which can cause death. Pulmonary edema is a medical emergency that can be delayed one to two days following exposure. May cause lung allergy.

Eyes: May cause severe irritation.

Skin: May cause severe irritation and dermatitis

Swallowed: May cause irritation of the mouth, throat and stomach and may cause death.

Long-term exposure: May cause mutations; kidney damage. See FORMALDEHYDE entry. This chemical gives off formaldehyde gas.

Storage: Store in tightly closed containers in a dry, cool, well-ventilated place. Keep away from liquid oxygen, oxidizers, strong acids, and alkaline materials. Containers may explode in fire.

First aid guide: Move victim to fresh air; call emergency medical care. In case of contact with material, immediately flush skin or eyes with running water for at least 15 minutes. Removal of solidified molten material from skin requires medical assistance. Remove and isolate contaminated clothing and shoes at the site.

★741★
PARALDEHYDE
CAS: 123-63-7
Other names: ACETALDEHYDE, TRIMER; ELALDEHYDE; PARAACETALDEHYDE; PARACETALDEHYDE; PARAL; PARALDEHYD (GERMAN); PARALDEIDE (ITALIAN); PCHO; PORAL; TRIACETALDEHYDE (FRENCH); 2,4,6-TRIMETHYL-1,3,5-TRIOXAAN (DUTCH); 2,4,6-TRIMETHYL-S-TRIOXANE; 2,4,6-TRIMETHYL-1,3,5-TRIOXANE; S-TRIMETHYLTRIOXYMETHYLENE; 2,4,6-TRIMETIL-1,3,5-TRIOSSANO (ITALIAN)

Danger profile: Fire hazard, poisonous gas produced in fire.
Uses: Solvent for cellulose, fats, oils, waxes, gums, resins and leather; making dyes; sedative.
Appearance: Colorless liquid.
Odor: Characteristic, aromatic. The odor is a warning of exposure; however, no smell does not mean you are not being exposed.
Effects of exposure:
  Breathed: May cause irritation, loss of coordination, headache, drowsiness followed by sleep. Higher exposures may cause shortness of breath, coma, weak pulse, shallow respiration, cyanosis, a dangerous build-up of fluids in the lungs (pulmonary edema), which can cause death. Pulmonary edema is a medical emergency that can be delayed one to two days following exposure. High exposure may cause liver and kidney damage.
  Eyes: Contact may cause severe irritation, burns and possible permanent damage.
  Skin: Passes through the unbroken skin; can increase exposure and the severity of the symptoms listed. May cause severe irritation.
  Swallowed: May cause irritation of the digestive tract; loss of coordination, headache, drowsiness followed by sleep; bronchitis. Also see *Breathed.*
Long-term exposure: Repeated exposure may cause liver and kidney damage, fatigue, weight loss, muscular weakness, tremors, changes in personality, speech, mental fatigue, memory loss; poor coordination; skin allergy and rash. Hypnotic agent. Similar chemicals may cause lung problems. After repeated exposures sudden stopping make cause withdrawal symptoms with tremors and hallucinations.
Storage: Store in tightly closed containers in a dry, cool, well-ventilated place. Keep away from heat, sources of ignition, oxidizers, explosives, hydrocyanic acid, organic peroxides, iodides, alkalies and nitric acid. Vapors may travel to sources of ignition and flash back. Containers may explode in fire.
First aid guide: Move victim to fresh air and call emergency medical care; if not breathing, give artificial respiration; if breathing is difficult, give oxygen. In case of contact with material, immediately flush eyes with running water for at least 15 minutes. Wash skin with soap and water. Remove and isolate contaminated clothing and shoes at the site.

★742★
PARAOXON
CAS: 311-45-5
Other names: CHINORTA; DIETHYL-P-NITROPHENYL PHOSPHATE; ENT 16,087; ESTER 25; ETICOL; FOSFAKOL; HC 2072; MINTACO; MINTACOL; MIOTISAL; MIOTISAL A; P-NITROPHENYL DIEYHYLPHOSPHATE; OXYPARATHION; PARAOXONE; PAROXAN; PESTOX 101; PHOSPHACOL; PHOSPHORIC ACID, DIETHYL P-NITROPHENYL ESTER; SOLUGLACIT; TS 219

Danger profile: Mutagen, deadly poison, exposure limit established, poisonous gas produced in fire.
Uses: Pesticide (organophosphate).

Appearance: Reddish-yellow oily liquid.
Odor: Slight. The odor is a warning of exposure; however, no smell does mean you are not being exposed.
Effects of exposure:
  Breathed: Extremely poisonous. May cause headache, dizziness, profuse sweating, nausea, vomiting, reduced heart beat, stomach cramps, diarrhea, loss of coordination, slow and weak breathing, fever, loss of consciousness and death. May cause shortness of breath and a dangerous build-up of fluids in the lungs (pulmonary edema), which can cause death. Pulmonary edema is a medical emergency that can be delayed one to two days following exposure.
  Eyes: Rapidly absorbed. Poisoning can happen very rapidly. See *Breathed.*
  Skin: Poisonous. Passes through the unbroken skin; can increase exposure and the severity of the symptoms listed, or cause fatal poisoning.
  Swallowed: Powerful, deadly poison.
Long-term exposure: May damage the developing fetus or cause birth defects; nerve damage including loss of coordination; liver damage. Repeated exposure may cause depression, irritability and personality changes.
Storage: Store in a regulated, specially marked place in tightly closed containers in a dry, well-ventilated place. Keep away from water, strong alkalis. Containers may explode in fire.

★743★
PARAQUAT
CAS: 4685-14-7
Other names: ACTAR; 4,4'-BIPYRIDINIUM,1,1'-DIMETHYL-, DIHYDRATE; 1,1'-DIMETHYL-4,4'-BIPYRIDINIUM DIHYDRATE DIPYRIDYLDIHYDRATE; CLEANSWEEP; CRISQUAT; DEXTRONE; DEXTRONE-X; DEXURON; DIMETHYL VIOLOGEN; FARMON PDQ; GRAMAZINE; GRAMONOL; GRAMONOL 5; GRAMOXONE 100 S; GRAMOXONE S; GROUNDHOG; METHYL VIOLOGEN (2); NEW WEEDOL; PARAQUAT DICATION; PATHCLEAR; SCYTHE; SOLTAIR; SPEEDWAY; SWEEP; WEEDOL

Danger profile: Mutagen, exposure limit established, poisonous gas produced in fire.
Uses: Herbicide mainly used for weed control in grass and seed crops; on cotton, potatoes, corn, soybeans; aquatic weed control.
Appearance: Yellow granular solid or liquid.
Effects of exposure:
  Breathed: May cause irritation of the nose and eyes; sneezing; nose bleed; painful obstruction of the throat.
  Eyes: May cause a painful burning feeling, tearing, inflammation of the eyelids; light sensitivity.
  Skin: Passes through the unbroken skin; can increase exposure and the severity of the symptoms listed. May cause irritation, inflammation, pain and blisters. The healing of cuts and wounds may be delayed. Fingernail damage possible.
  Swallowed: Poison. May cause burning feeling in the mouth, throat and stomach; headache, cough; difficult swallowing; nausea, vomiting; painful abdominal cramps, diarrhea (possibly blood-stained).
Long-term exposure: May damage the developing fetus. May cause liver, kidney and severe lung damage. Paraquat tends to concentrate in lung tissue and produces fibrosis which reduces capacity of lungs to absorb oxygen. Lung damage may be delayed.
Storage: Store in tightly closed containers in a dry, cool, well-ventilated place. Containers may explode in fire.

★744★
PARAQUAT DICHLORIDE
CAS: 1910-42-5
Other names: CEKUQUAT; CRISQUAT; DEXTRONE; N,N'-DIMETHYL-4,4'-BIPYRIDINIUM DICHLORIDE; 1,1'-DIMETHYL-4,4'-BIPYRIDYNIUM-DICHLORIDE; DIMETHYL VICLOGEN CHLORIDE; ESGRAM; GRAMOZONE S; METHYLVIOLOGEN; METHYL VIOLOGEN (REDUCED); OK 622; ORTHO PARAQUAT CL; PARA-COL; PARAQUAT CHLORIDE; PARAQUAT CL

Danger profile: Mutagen, poison, combustible, exposure limit established, reproductive hazard, EPA extremely hazardous substance list, poisonous gas produced in fire.
Uses: Herbicide.
Appearance: Colorless solid.
Effects of exposure:
Breathed: May cause irritation of the nose and eyes; sneezing; nose bleed; painful obstruction of the throat.
Eyes: May cause a painful burning feeling, tearing, inflammation of the eyelids; light sensitivity.
Skin: Passes through the unbroken skin; can increase exposure and the severity of the symptoms listed. May cause irritation, inflammation, pain and blisters. The healing of cuts and wounds may be delayed. Fingernail damage possible.
Swallowed: Poison. May cause burning feeling in the mouth, throat and stomach; headache, cough; difficult swallowing; nausea, vomiting; painful abdominal cramps, diarrhea (possibly blood-stained).
Long-term exposure: May damage the developing fetus. May cause liver, kidney and severe lung damage. Chemical tends to concentrate in lung tissue and produces fibrosis which reduces capacity of lungs to absorb oxygen. Lung damage may be delayed.
Storage: Store in tightly closed containers in a dry, cool, well-ventilated place. Containers may explode in fire.

★745★
PARATHION
CAS: 56-38-2
Other names: AAT; AATP; AC 3422; ACC 3422; ALKRON; ALLERON; APHAMITE; ARALO; B 404; BAY E-605; BAYER E-605; BLADAN; BLADAN F; COMPOUND 3422; COROTHION; CORTHION; CORTHIONE; DANTHION; DIETHYL P-NITROPHENYL THIONOPHOSPHATE; DNTP; DPP; DREXEL PARATHION 8E; E 605; ECATOX; EKATIN WF & WF ULV; EKATOX; ENT 15,108; ETHLON; ETHYL PARATHION; FOLIDOL; FOLIDOL E605; FOLIDOL E&E 605; FOSFERMO; FOSFERNO; FOSFEX; FOSFIVE; FOSOVA; FOSTERN; FOSTOX; GEARPHOS; GENITHION; KALPHOS; KYPTHION; LETHALAIRE G-54; LIROTHION; MURFOS; NCI-C00226; NIRAN; NIRAN E-4; NITROSTIGMIN (GERMAN); NITROSTIGMINE; NIUIF-100; NOURITHION; OLEOFOS-20; PLEOPARAPHENE; OLEOPARATHION; ORTHOPHOS; PAC; PANTHION; PARADUST; PARAMAR; PARAMAR 50; PARAPHOS; PARATHENE; PARATHION-ETHYL; PARAWET; PESTOX PLUS; PETHION; PHOSKIL; PHOSPHOSTIGMINE; RB; PHOSPHOROTHIOIC ACID, O,O,-DIETHYLO-(P-NITROPHENYL)ESTER; O,O-DIETHYL-O,P-NITROPHENYL PHOSPHOROTHIOATE; RHODIASOL; RHODIATOX; RHODIATROX; SELEPHOS; SNP; SOPRATHION; STATHION; STRATHION; SULPHOS; SUPER RODIATOX; T-47; THIOPHOS; THIOPHOS 3422; TIOFOS; TOX 47; TOXOL (3); VAPOPHOS; VITREX

Danger profile: Mutagen, combustible, deadly poison, exposure limit established, Community Right-to-Know list, EPA extremely hazardous substances list; poisonous gas produced in fire.

Uses: Broad spectrum, contact and stomach insecticide; acaricide. On fruit, vegetables, grasses and nuts. Banned in many countries.
Appearance: Pale yellow or dark brown liquid.
Odor: Garlic-like. The odor is a warning of exposure; however, no smell does not mean you are not being exposed.
Effects of exposure:
Breathed: Extremely poisonous. Mist, dust, or vapor (or ingestion, or absorption through skin) may cause headache, dizziness, constriction of pupils, tightness of chest, nausea, vomiting, abdominal cramps, diarrhea, muscular twitching, profuse sweating, reduced heart beat, stomach cramps, diarrhea, loss of coordination, slow and weak breathing, fever, loss of consciousness and death may follow. May cause a dangerous build-up of fluids in the lungs (pulmonary edema), which can result in death. Pulmonary edema is a medical emergency that can be delayed one to two days following exposure.
Eyes: Rapidly absorbed. Poisoning can happen very rapidly. See *Breathed.*
Skin: Poisonous. Passes through the unbroken skin; can increase exposure and the severity of the symptoms listed.
Swallowed: Powerful, deadly poison. See *Breathed.*
Long-term exposure: May damage the developing fetus or cause birth defects; nerve damage including loss of coordination; liver damage. Repeated exposure may cause depression, irritability and personality changes.
Storage: Store in a regulated, specially marked place in tightly closed containers in a dry, well-ventilated place. Keep away from water, strong alkalis and endrin. Containers may explode in fire.
First aid guide: Move victim to fresh air and call emergency medical care; if not breathing, give artificial respiration; if breathing is difficult, give oxygen. In case of contact with material, immediately flush skin or eyes with running water for at least 15 minutes. Remove and isolate contaminated clothing and shoes at the site. Keep victim quiet and maintain normal body temperature. Effects may be delayed; keep victim under observation.

★746★
PENTABORANE
CAS: 19624-22-7
Other names: DIHYDROPENTABORANE (9); PENTABORANE (9); PENTABORANE UNDECAHYDRIDE; PENTABORON NONAHYDRIDE; STABLE PENTABORANE

Danger profile: Fire hazard, spontaneously combustible in air, poison, exposure limit established, EPA extremely hazardous substances list, poisonous gas produced in fire.
Uses: Fuel for air-breathing engines; in propellant fuels.
Appearance: Colorless gas or liquid.
Odor: Strong, foul, sour milk-like. The odor is a warning of exposure; however, no smell does not mean you are not being exposed.
Effects of exposure:
Breathed: May be poisonous. Low concentrations causes dizziness, blurred vision, nausea, fatigue, light headedness or nervousness; higher concentrations also cause abnormal muscular contractions or twitching of any part of the body, difficult breathing, poor muscular coordination, imperfect articulation of speech, convulsions, and (rarely) coma. High exposures may cause a dangerous build-up of fluids in the lungs (pulmonary edema), which can cause death. Pulmonary edema is a medical emergency that can be delayed one to two days following exposure.
Eyes: May cause severe irritation and inflammation and blistering of the eyelids. Possible permanent damage.
Skin: May be poisonous if absorbed. Passes through the unbroken skin; can increase exposure and the severity of the

symptoms listed. May cause caustic-like burns; tremors, confusion, seizures, slow and shallow breathing and bluish face and lips.
Swallowed: Compound cannot be swallowed, because it is spontaneously combustible in air.
Storage: Store in tightly closed containers in a dry, cool, well-ventilated place. Keep away from heat, sources of ignition, halogenated extinguishing agents, oxygen, dimethyl sulfoxide. Corrosive to rubbers, some synthetics and some greases. Containers may explode in fire. Special fire fighting materials required to fight fire. Do not use water.
First aid guide: Move victim to fresh air and call emergency medical care; if not breathing, give artificial respiration; if breathing is difficult, give oxygen. In case of contact with material, immediately flush skin or eyes with running water for at least 15 minutes. Remove and isolate contaminated clothing and shoes at the site. Keep victim quiet and maintain normal body temperature.

★747★
PENTACHLOROETHANE
CAS: 76-01-7
Other names: ETHANE PENTACHLORIDE; ETHANE, PENTACHLORO-; NCI-C53894; PENTACHLOORETHAAN (DUTCH); PENTACHLORAETHAN (GERMAN); PENTACHLORETHANE (FRENCH); PENTACLOROETANO (ITALIAN); PENTALIN

Danger profile: Mutagen, combustible, explosion hazard, poisonous gas produced in fire.
Uses: As solvent for oil and grease in metal cleaning.
Appearance: Colorless liquid.
Odor: Sweet, chloroform or camphor-like The odor is a warning of exposure; however, no smell does not mean you are not being exposed.
Effects of exposure:
　　Breathed: May be poisonous. May cause irritation of the eyes, nose, throat and air passages; coughing, headache; increased saliva; reddening of the face; dizziness, lightheadedness, fatigue, loss of consciousness, shock, coma, irregular heartbeat, death. If the victim recovers watch for the possibility of a buildup of fluid in the lungs (pulmonary edema). Pulmonary edema is a medical emergency that can be delayed one to two days following exposure.
　　Eyes: Contact may cause irritation, tearing, inflammation of the eyelids; burning sensation.
　　Skin: May be poisonous if absorbed. Passes through the unbroken skin; can increase exposure and the severity of the symptoms listed. May cause irritation, redness, burning feeling, swelling; dry skin; rash or blisters.
　　Swallowed: May be poisonous. Breath may smell of chloroform. May cause irritation of the mouth, gastrointestinal tract, nausea, vomiting, diarrhea with possible blood; drowsiness; loss of consciousness; irregular heartbeat and death. See *Breathed.*
Long-term exposure: May cause liver, lung and/or kidney damage.
Storage: Store in tightly closed containers in a dry, cool, well-ventilated place or a refrigerator. Keep away from alkalis, potassium and reactive metals. Containers may explode in fire.
First aid guide: Move victim to fresh air and call emergency medical care; if not breathing, give artificial respiration; if breathing is difficult, give oxygen. In case of contact with material, immediately flush skin or eyes with running water for at least 15 minutes. Speed in removing material from skin is of extreme importance. Remove and isolate contaminated clothing and shoes at the site. Keep victim quiet and maintain normal body temperature. Effects may be delayed; keep victim under observation.

★748★
PENTACHLORONAPHTHALENE
CAS: 1321-64-8
Other names: HALOWAX 1013

Danger profile: Poison, exposure limit established, poisonous gas produced in fire.
Uses: Making electrical equipment, pesticides and fungicides; water and fire-proofing agent; in lubricants.
Appearance: Waxy white or pale yellow solid.
Odor: Pungent, camphor or mothball-like. Although solid at room temperature, the liquid produces a dangerous vapor. The odor is a warning of exposure; however, no smell does not mean you are not being exposed.
Effects of exposure:
　　Breathed: May cause headache, fatigue, dizziness, loss of appetite. Overexposure may cause severe liver damage and possible death.
　　Skin: Passes through the unbroken skin; can increase exposure and the severity of the symptoms listed. May cause skin rash.
　　Swallowed: See *Breathed.*
Long-term exposure: Toxic to the liver and skin. May cause an acne-like rash; jaundice, liver problems.
Storage: Store in tightly closed containers in a dry, cool, well-ventilated place. Keep away from strong oxidizers.

★749★
PENTACHLOROPHENOL
CAS: 87-86-5
Other names: CHEM-TOL; CHLOROPHEN; CRYPTOGIL OL; DOWCIDE 7; DOWCIDE 7; DOWCIDE EC-7; DOWCIDE G; DOW PENTACHLOROPHENOL DP-2 ANTIMICROBIAL; DUROTOX; EP 30; FUNGIFEN; GLAZD PENTA; GRUNDIER ARBEZOL; LAUXTOL; LAUXTOL A; LIROPREM; NCI-C54933; NCI-C55378; NCI-C56655; PCP; PENCHLOROL; PENTA; PENTACHLOORFENOL (DUTCH); PENTACHLOROFENOL; PENTACLOROFENOLO (ITALIAN); PENTACHLOROPHENATE; 2,3,4,5,6-PENTACHLOROPHENOL; PENTACHLOROPHENOL, DOWICIDE EC-7; PENTACHLOROPHENOL, DP-2; PENTACHLORPHENOL (GERMAN); PENTACHLOROPHENOL, TECHNICAL; PENTACON; PENTA-KIL; PENTASOL; PENWAR; PERATOX; PERMACIDE; PERMAGARD; PERMASAN; PERMATOX DP-2; PERMATOX PENTA; PERMITE; PHENOL, PENTACHLORO-; PRILTOX; SANTOBRITE; SANTOPHEN; SANTOPHEN 20; SINITUHO; TERM-I-TROL; THOMPSON'S WOOD FIX; WEEDONE

Danger profile: Carcinogen, mutagen, poison, exposure limit established, Community Right-to-Know List, poisonous gas produced in fire.
Uses: Making fungicides, bactericides, algicides, herbicides; wood preservative; insecticide for termite control; general herbicide.
Appearance: Needle-like crystals or flakes, light brown or white to dark grey in color.
Odor: Pungent. The odor is a warning of exposure; however, no smell does not mean you are not being exposed.
Effects of exposure:
　　Breathed: May be poisonous. May cause irritation of the nose and throat; poisoning with symptoms of sweating, intense thirst, nose irritation, headache, coughing and sneezing, nausea, stomach pain, restlessness, trouble breathing, rapid breathing and pulse, anxiety, confusion, high fever, chest and abdomen pain. Without treatment, death may happen quickly. High levels may cause fever, coma and death.
　　Eyes: May cause burning sensation, irritation, tearing, and inflammation of the eyelids.

Skin: Passes quickly through the unbroken skin; can increase exposure and the severity of the symptoms listed. Symptoms may be delayed. May cause irritation, redness, swelling and blisters and fatal poisoning.

Swallowed: May be poisonous. May cause loss of appetite, respiratory difficulties, sweating, coma. burning feeling in the throat and mouth; dizziness, headache, nausea, vomiting, high fever, thirst, fatigue, rapid pulse, convulsions, loss of consciousness. High exposures may cause a dangerous build-up of fluids in the lungs (pulmonary edema), which can cause death. Pulmonary edema is a medical emergency that can be delayed one to two days following exposure.

Long-term exposure: May cause damage to the developing fetus and may cause mutations; chronic poisoning with weight loss, fatigue and excessive sweating; liver and kidney damage; bronchitis, cough and phlegm; acne-like skin rash.

Storage: Store in tightly closed containers in a dry, cool, well-ventilated, detached place. Keep away from heat, direct sunlight, sources of ignition and strong oxidizers. Protect containers from physical damage.

★750★
PENTANE
CAS: 109-66-0
Other names: AMYL HYDRIDE; N-PENTANE; NORMAL PENTANE; PENTAN (POLISH); PENTANEN (DUTCH); PENTANI (ITALIAN); SKELLYSOLVE-A

Danger profile: Fire hazard, explosion hazard, exposure limit established, poisonous gas produced in fire.
Uses: Fuel; in gasoline; making other chemicals, plastics, artificial ice and pesticides; low-temperature thermometers; solvent.
Appearance: Colorless liquid.
Odor: Gasoline-like. The odor is a warning of exposure; however, no smell does not mean you are not being exposed. Low toxicity.
Effects of exposure:
Breathed: May cause irritation of the eyes, nose and throat; dizziness, lightheadedness, rapid breathing, mental confusion, unsteady walk, headache, fatigue, nausea, vomiting, unconsciousness, coma and death. High exposures may cause a dangerous build-up of fluids in the lungs (pulmonary edema), which can cause death. Pulmonary edema is a medical emergency that can be delayed one to two days following exposure.
Eyes: May cause burning feeling, tearing, irritation, redness and inflammation of the eyelids.
Skin: Passes through the unbroken skin; can increase exposure and the severity of the symptoms listed. May cause irritation of the skin, dryness, redness and blisters.
Swallowed: May cause indigestion, dizziness, fatigue, unconsciousness.
Long-term exposure: May cause nervous system damage with numbness, and weakness in arms and legs; dry, cracked skin.
Storage: Store in tightly closed containers in a dry, cool, well-ventilated place. Keep away from heat, sources of ignition and strong oxidizers. Vapors may travel to sources of ignition and flash back. Protect containers from shock and physical damage. Containers may explode in fire.
First aid guide: Move victim to fresh air and call emergency medical care; if not breathing, give artificial respiration; if breathing is difficult, give oxygen. In case of contact with material, immediately flush eyes with running water for at least 15 minutes. Wash skin with soap and water. Remove and isolate contaminated clothing and shoes at the site.

★751★
2-PENTANOL
CAS: 6032-29-7
Other names: SEC-AMYL ALCOHOL; ISOAMYL ALCOHOL, SECONDARY; METHYL PROPYL CARBINOL; PENTANOL-2; SEC-PENTYL ALCOHOL

Danger profile: Combustible, poisonous gas produced in fire.
Uses: Solvent for pains and lacquers; making pharmaceuticals.
Appearance: Colorless liquid.
Effects of exposure:
Breathed: Mildly toxic. May cause irritation of eyes, nose and throat; headache, drowsiness, excitability, drunken behavior, mental confusion, fatigue. Very high levels of exposure may cause dizziness, lightheadedness and cause victim to pass out, go into coma and die.
Eyes: Contact may cause irritation, pain, burning, stinging, tearing, inflammation of the eyelids. Light may cause pain.
Skin: May remove oil from the skin and cause dryness. Persons with pre-existing skin disorders may be more susceptible to the effects of this substance.
Swallowed: May cause gastrointestinal irritation. See *Breathed.*
Long-term exposure: Prolonged or repeated contact may cause irritation of the skin and dermatitis, and nervous system changes.
Storage: Store in tightly closed containers in a dry, cool, well-ventilated place. Keep away from heat, sources of ignition, bromine trifluoride and oxidizing materials. Containers may explode in fire.

★752★
2-PENTANONE
CAS: 107-87-9
Other names: ETHYLACETONE; ETHYL ACETONE; METHYL-PROPYL-CETONE (FRENCH); METYLOPROPYLOKETON (POLISH); METHYL PROPYL KETONE; METHYLPROPYL KETONE; METHYL-N-PROPYL KETONE; MPK; PENTAN-2-ONE

Danger profile: Mutagen, fire hazard, exposure limit established, poisonous gas produced in fire.
Uses: Solvent; flavoring agent.
Appearance: Clear, colorless or water-white liquid.
Odor: Strong fruity. The odor is a warning of exposure; however, no smell does not mean you are not being exposed.
Effects of exposure:
Breathed: May cause irritation of the eyes, nose, throat and lungs; coughing, shortness of breath; difficult breathing. Higher exposures may cause dizziness, lightheadedness, and loss of consciousness.
Eyes: May cause severe irritation.
Skin: Passes through the unbroken skin; can increase exposure and the severity of the symptoms listed. May cause irritation, burning sensation and rash. Prolonged contact may cause drying and cracking.
Long-term exposure: May cause birth defects miscarriage, cancer; dermatitis, dry and cracked skin; capable of causing lung problems.
Storage: Store in tightly closed containers in a dry, cool, well-ventilated place. Keep away from heat, flames, sources of ignition and oxidizers. Vapors may travel to sources of ignition and flash back. Containers may explode in fire.

★753★
PERACETIC ACID
CAS: 79-21-0
Other names: ACETIC PEROXIDE; ACETYL HYDROPEROXIDE; ACIDE PERACETIQUE (FRENCH); HYDROPEROXIDE, ACETYL; ETHANEPEROXIC ACID; HYDROPEROXIDE, ACETYL; PAA; PERACETIC ACID SOLUTION; PEROXYACETIC ACID

Danger profile: Combustible, explosion hazard, corrosive, Community Right-to-Know List, EPA extremely hazardous substances list, poisonous gas produced in fire.
Uses: Bleaching textiles, paper, oils, waxes, starch; bactericide and fungicide in food processing; catalyst for epoxy resins; making other chemicals.
Appearance: Colorless liquid.
Odor: Strong, pungent, acrid. The odor is a warning of exposure; however, no smell does not mean you are not being exposed.
Effects of exposure:
Breathed: Vapor may cause intense irritation to nose, throat, mouth and lungs; mucous membrane. Symptoms of exposure include watering of the eyes, coughing, difficult breathing, headache, nausea, vomiting, muscular weakness, unconsciousness. High exposure can lead to drowsiness, coma and death; buildup of fluid in the lungs (pulmonary edema), and may cause death. Pulmonary edema may be delayed by one to two days following exposure.
Eyes: Severe irritant. Contact may cause, burning and stinging, tearing, severe irritation and redness and swelling of the eyelids, intense pain; can burn the eyes and lead to blurred vision, chronic irritation and permanent damage to the surface of the eye.
Skin: Severe irritant. Contact may cause burning or stinging feeling, redness, swelling, intense pain, blisters, irritation, rash. Repeated exposure may lead to skin allergy and chronic irritation.
Swallowed: Poisonous. May cause burning feeling of the mouth and throat, intense thirst, throat swelling, stomach cramps, nausea, vomiting, diarrhea, severe breathing difficulties and unconsciousness. Tissue damage may result. Breathing may stop. Watch for delayed symptoms. Animal data suggest that death may occur from as little as one teaspoon.
Long-term exposure: May cause liver and kidney damage.
Storage: Store in tightly closed containers in a dry, cool, well-ventilated place. Keep away from sources of ignition, organic and combustible materials, olefins, hydrogen peroxide, ether solvents, potassium chloride and other metal chloride solutions, acetic anhydride and reducing substances. Explodes at 230°F, and sensitive to shock. Containers may explode in fire.
First aid guide: Move victim to fresh air; call emergency medical care. In case of contact with material, immediately flush eyes with running water for at least 15 minutes. Wash skin with soap and water. Remove and isolate contaminated clothing and shoes at the site. Keep victim quiet and maintain normal body temperature.

★754★
PERCHLORETHYLENE
CAS: 127-18-4
Other names: ANKILOSTIN; ANTISOL 1; CARBON BICHLORIDE; CARBON DICHLORIDE; CZTEROCHLOROETYLEN (POLISH); DIDAKENE; DOWPER; ENT 1,860; ETHENE, TETRACHLORO-; ETHYLENE TETRACHLORIDE; FEDALUN; NCI-C04580; NEMA; PER; PERAWIN; PERC; PERCHLOORETHYLEEN, PER (DUTCH); PERCHLOR; PERCHLORAETHYLEN, PER (GERMAN); PERCHLORETHYLENE, PER (FRENCH); PERCLENE; PERCLOROETILENE (ITALIAN); PERCOSOLVE; PERK; PERKLONE; PERSEC; TETLEN; TETRACAP; TETRACHLOORETHEEN (DUTCH); TETRACHLORAETHEN (GERMAN); TETRACHLORETHYLENE; TETRACHLOROETHENE; TETRACHLOROETHYLENE; 1,1,2,2,-TETRACHLOROETHYLENE; TETRACLOROETENE (ITALIAN); TETRALENO; TETRALEX; TETRAVEC; TETROGUER; TETROPIL

Danger profile: Possible carcinogen, mutagen, poison, exposure limit established, Community Right-to-Know List, poisonous gas produced in fire.
Uses: Solvent; dry cleaning agent.
Appearance: Colorless liquid.
Odor: Mildly sweet; chloroform-like. The odor is a warning of exposure; however, no smell does not mean you are not being exposed.
Effects of exposure:
Breathed: May cause irritation of the respiratory tract, nausea, headache, abdominal pains, constipation, dizziness, increased fatigue, persperation. Vapor may affect central nervous system and cause anesthesia, hallucinations, symptoms of intoxication, coma and a dangerous build-up of fluids in the lungs (pulmonary edema), which can cause death. Pulmonary edema is a medical emergency that can be delayed one to two days following exposure. At high levels chemical has caused cancer and birth defects in mice.
Eyes: May cause irritation and inflammation of the eyelids.
Skin: Prolonged contact may cause burning feeling, irritation and rash.
Swallowed: May cause irritation of the gastrointestinal tract.
Long-term exposure: May cause cancer, birth defects; central nervous system depression, impaired memory, numbness of arms and legs, impaired vision; kidney and liver damage; dermatitis from repeated contact.
Storage: Store in tightly closed containers in a dry, cool, well-ventilated place. Keep away from beryllium, barium, lithium, sodium hydroxide, and metals. Containers may explode in fire.

★755★
PERCHLORIC ACID
CAS: 7601-90-3

Danger profile: Combustible, explosion hazard, poisonous gas produced in fire.
Uses: Making other chemicals; laboratory chemical; explosives; elecro-polishing; electro-plating
Appearance: Colorless, fuming liquid.
Effects of exposure:
Breathed: Vapor or mist may cause intense irritation and burning sensation to nose, throat, mouth and lungs; mucous membrane. Symptoms of exposure include watering of the eyes, coughing, difficult breathing, headache, nausea, vomiting, muscular weakness, unconsciousness. Prolonged or excessive exposure could cause vomiting and severe coughing. High exposure can lead to drowsiness, coma and death; buildup of fluid in the lungs (pulmonary edema), and may

cause death. Pulmonary edema may be delayed by one to two days following exposure.

Eyes: Severe irritant. Contact may cause, burning and stinging, tearing, irritation and redness and swelling of the eyelids, intense pain; can burn the eyes cause blistering and burns; and lead to blurred vision, chronic irritation and other permanent damage.

Skin: Severe irritant. Contact may cause burning or stinging feeling, redness, swelling, intense pain, blisters, irritation, rash. Repeated exposure may lead to skin allergy and chronic irritation.

Swallowed: May cause burning feeling of the mouth and throat, intense thirst, throat swelling, stomach cramps, nausea, vomiting, diarrhea, severe breathing difficulties and unconsciousness. Tissue damage may result; blistering and burns of the mouth and stomach. Breathing may stop. Watch for delayed symptoms.

Long-term exposure: May cause chronic skin rash in sensitized individuals.

Storage: Store in tightly closed containers in a dry, cool, well-ventilated place. Keep away from combustible materials, organic materials and all other chemicals. Corrosive to most metals. Containers may explode in fire.

★ 756 ★
PERCHLOROMETHYL MERCAPTAN
CAS: 594-42-3
Other names: CLAORSIT; MERCAPTAN METHYLIQUE PERCHLORE (FRENCH); METHANESULFENYL CHLORIDE,TRICHLORO-; TRICHLOROMETHANESULFENYL CHLORIDE; PCM; PERCHLOROMETHYLMERCAPTAN; PERCHLORMETHYLMERKAPTAN (CZECH); PPM; TRICHLOROMETHYL SULFUR CHLORIDE; TRICHLOROMETHYL SULFENYL CHLORIDE; TRICHLOROMETHYLSULFENYL CHLORIDE; TRICHLOROMETHANE PERCHLOROMETHANETHIOL

Danger profile: Poison, exposure limit established, poisonous gas produced in fire.
Uses: Making other chemicals; dye intermediate; fumigant.
Appearance: Pale yellow, oily liquid.
Odor: Foul-smelling, strong, acrid. The odor is a warning of exposure; however, no smell does not mean you are not being exposed.
Effects of exposure:
Breathed: May be poisonous. May cause irritation of the eyes, nose, throat and lungs, tears and runny nose, coughing, sore throat, difficult breathing; lethargy, fatigue, and drowsiness; slurred speech, shaking, and dizziness. Higher exposures could cause shortness of breath, headache, nausea, vomiting, dizziness, diarrhea and a dangerous build-up of fluids in the lungs (pulmonary edema), which can cause death. Pulmonary edema is a medical emergency that can be delayed one to two days following exposure.
Eyes: May cause irritation, redness, tearing, pain, inflamed eyelids, and eye injury; burns in corneas and possible loss of sight.
Skin: May be poisonous if absorbed. Passes through the unbroken skin; can increase exposure and the severity of the symptoms listed. May cause painful irritation, redness, and blisters Repeated or prolonged skin contact may cause irritation of the skin and may also cause the legs to feel numb.
Swallowed: May be poisonous. May cause irritation of the mouth and throat; stomach and abdominal pain; loss of muscular coordination and mental confusion; convulsions. Also see symptoms for *Breathed.*
Storage: Store in tightly closed containers in a dry, cool, well-ventilated place. Reacts with iron and steel producing carbon

tetrachloride. Corrosive to most metals. Containers may explode in fire.
First aid guide: Move victim to fresh air and call emergency medical care; if not breathing, give artificial respiration; if breathing is difficult, give oxygen. In case of contact with material, immediately flush skin or eyes with running water for at least 15 minutes. Speed in removing material from skin is of extreme importance. Remove and isolate contaminated clothing and shoes at the site. Keep victim quiet and maintain normal body temperature. Effects may be delayed; keep victim under observation.

★ 757 ★
PERCHLOROYL FLUORIDE
CAS: 7616-94-6
Other names: CHLORINE FLUORIDE OXIDE; CHLORINE OXYFLUORIDE

Danger profile: Exposure limit established, poisonous gas produced in fire.
Uses: In rocket fuels; making other chemicals.
Appearance: Colorless gas.
Odor: Characteristic, sweet odor. The odor is a warning of exposure; however, no smell does not mean you are not being exposed.
Effects of exposure:
Breathed: Vapor may cause intense irritation to nose, throat, mouth and lungs; mucous membrane. Symptoms of exposure include watering of the eyes, coughing, difficult breathing, headache, nausea, vomiting, muscular weakness, unconsciousness. High exposure can lead to drowsiness, coma and death; buildup of fluid in the lungs (pulmonary edema), and may cause death. Pulmonary edema may be delayed by one to two days following exposure.
Eyes: Severe irritant. Contact may cause, burning and stinging, tearing, irritation and redness and swelling of the eyelids, intense pain; can burn the eyes and lead to blurred vision, chronic irritation and other permanent damage.
Skin: Severe irritant. Passes through the unbroken skin; can increase exposure and the severity of the symptoms listed. Contact may cause burning or stinging feeling, redness, swelling, intense pain, blisters, irritation, rash. Repeated exposure may lead to skin allergy and chronic irritation.
Swallowed: May cause burning feeling of the mouth and throat, intense thirst, throat swelling, stomach cramps, nausea, vomiting, diarrhea, severe breathing difficulties and unconsciousness. Tissue damage may result. Breathing may stop. Watch for delayed symptoms.
Long-term exposure: May destroys red cells causing anemia, loss of appetite, and blue color to skin, lips, fingertips and ears. This is not a permanent injury and recovery is reported to be rapid with little or no after effects.
Storage: Store in tightly closed containers in a dry, cool, well-ventilated place. A powerful oxidizer, keep away from nitrogenous bases, organic materials, carbon black and other chemicals. Containers may explode in fire.

★758★
PHENACETIN
CAS: 62-44-2

Other names: ACETAMIDE, N-(4-ETHOXYPHENYL)-(9CI); 1-ACETAMIDO-4-ETHOXYBENZENE; ACETANILIDE, 4'-ETHOXY-; ACETO-PARA-PHENALIDE; P-ACETOPHENETIDE; ACETO-PARA-PHENETIDIDE; PARA-ACETOPHENETIDIDE; ACETOPHENETIDIN; ACETOPHENE-TIDINE; P-ACETOPHENETIDINE; ACETO-4-PHENETIDINE; ACETOPHENETIN; ACET-P-PHENALIDE; ACETPHENETIDIN; P-ACETPHENETIDIN; ACET-P-PHENETIDIN; ACETYL-PHENETIDIN; N-ACETYL-P-PHENETIDINE; ACHROCIDIN; ANAPAC; APC; ASA COMPOUND; BROMO SELTZER; BUFF-A-COMP; CITRA-FORT; CLISTANOL; CODEMPIRAL; COMMOTIONAL; CONTRADOL; CONTRADOULEUR; CORICIDIN; CORIFORTE; CORYBAN-D; DAPRISAL; DARVON; DARVON COMPOUND; DASIKON; DASIN; DASIN CH; DOLOSTOP; DOLVIRAN; EDRISAL; EMPIRAL; EMPIRIN COMPOUND; EMPRAZIL; EMPRAZIL-C; EPRAGEN; P-ETHOXYACETANILIDE; 4-ETHOXYACETANILIDE; 4'-ETHOXYACETANILIDE; N-PARA-ETHOXYPHENYLACETAMIDE; N-(4-ETHOXYPHENYL)ACETAMIDE; FENACETINA; FENIDINA; FENIA; FENINA; FIORINAL; FORTACYL; GELONIDA; GE-WODIN; HELVAGIT; HJORTON'S POWDER; HOCOPHEN; KAFA; KALMIN; MALEX; MELABON; MELAFORTE; NORGE-SIC; PAMPRIN; PARACETOPHENETIDIN; PARAMETTE; PARATODOL; PERCOBARB; PERTONAL; PHENACET; P-PHENACETIN; PARA-PHENACETIN; PHENACETINE; PHE-NACETINUM; PHENACITIN; PHENACON; PHENAPHEN; PHENAPHEN PLUS; PHENAZETIN; PHENAZETINA; PHENEDINA; P-PHENETIDINE, N-ACETYL-; PHENIDIN; PHENIN; PHENODYNE; PYRAPHEN; PYRROXATE; QUADRONAL; REFORMIN; ROBAXISAL-PH; SALGYDAL; SANALGINE; SARIDON; SERANEX; SINEDAL; SINUBID; SINUTAB; SOMA; STELLACYL; SUPER ANAHIST; SUPRALGIN; SYN-ALGOS-DC; SYNALOGOS; TACOL; TERRACYDIN; TETRACYDIN; THEPHORIN A-C; TREUPEL; VEGANINE; VIDEN; WIGRAINE; XARIL; ZACTIRIN COMPOUND

Danger profile: Carcinogen, poisonous gas produced in fire.
Uses: Medicine (pain killer); veterinary medicines.
Appearance: White crystalline powder.
Effects of exposure:
 Breathed: High exposure lowers the ability of the blood to carry oxygen; bluish color to skin and lips; headache, dizziness, weakness, collapse and possible death.
 Eyes: May cause irritation, rash and itching.
 Skin: May cause allergic reaction with rash and itching.
 Swallowed: May cause gastric irritation, chills, fall in blood pressure, weakness, insomnia.
Long-term exposure: May cause damage to the developing fetus; bladder, urinary or nose cancer; kidney, liver, heart damage; anemia; brownish urine; weight loss, insomnia, shortness of breath, weakness.
Storage: Store in tightly closed containers in a dry, cool, well-ventilated place.

★759★
PHENOL
CAS: 108-95-2

Other names: ACIDE CARBOLIQUE (FRENCH); BAKER'S P AND S LIQUID; BAKER'S P AND S OINTMENT; CARBOLIC ACID; BENZENE,HYDROXY-; MONOHYDROXY BENZENE; CARBOLSAURE (GERMAN); FENOL (DUTCH); FENOL (POLISH); FENOLO (ITALIAN); HYDROXYBENZENE; MONOHYDROXYBENZENE; NCI-C50124; OXYBENZENE; PHENIC ACID; PHENOLE (GERMAN); PHENYL HYDRATE; PHENYL HYDROXIDE; PHENYLIC ACID

Danger profile: Mutagen, combustible exposure limit established, Community Right-to-Know List, EPA extremely hazardous substances list, poisonous gas produced in fire.
Uses: Making pharmaceuticals, other chemicals, plastics, resins, plywood, rubber; refining oils; germicidal paints; laboratory chemical; disinfectant.
Appearance: White to pink crystalline solid or thick liquid. May become red on exposure to light and heat.
Odor: Sickening sweet; tar-like. The odor is a warning of exposure; however, no smell does not mean you are not being exposed.
Effects of exposure:
 Breathed: Vapor may cause intense irritation to nose, throat, mouth and lungs; mucous membrane. Symptoms of exposure include watering of the eyes, coughing, difficult breathing, headache, nausea, vomiting, muscular weakness, unconsciousness. High exposure can lead to drowsiness, coma and death; buildup of fluid in the lungs (pulmonary edema), and may cause death. Pulmonary edema may be delayed by one to two days following exposure.
 Eyes: Severe irritant. Contact may cause, burning and stinging, tearing, irritation and redness and swelling of the eyelids, intense pain; can burn the eyes and lead to blurred vision, chronic irritation and other permanent damage.
 Skin: Severe irritant. Do no handle without good skin protection. Rapidly passes through the unbroken skin; can increase exposure and the severity of the symptoms listed. Contact may cause redness, swelling, intense pain, blisters, irritation, rash. The pain-killing action may cause loss of pain sensation. Repeated exposure may lead to skin allergy and chronic irritation. Death from skin absorption can happen within a half hour to several hours.
 Swallowed: Do not eat, drink or smoke when using phenol. Wash hands thoroughly after use. May cause burning feeling of the lips, mouth and throat, intense thirst, throat swelling, stomach cramps, nausea, vomiting, diarrhea, severe breathing difficulties, gangrene and unconsciousness. Tissue damage may result. Breathing may stop. Watch for delayed symptoms.
Long-term exposure: May cause mutations, birth defects, miscarriage, cancer; damage to the liver, kidneys, pancreas spleen and heart muscle; dermatitis. Many similar chemicals cause lung effects, brain and nerve damage.
Storage: Store in tightly closed containers in a dry, cool, well-ventilated place. Keep away from heat, sources of ignition, light, formaldehyde, peroxydisulfuric acid, peroxymonosulfuric acid, sodium nitrite, calcium hypochlorite and other strong oxidizers. Containers may explode in fire.
First aid guide: Move victim to fresh air and call emergency medical care; if not breathing, give artificial respiration; if breathing is difficult, give oxygen. In case of contact with material, immediately flush skin or eyes with running water for at least 15 minutes. Speed in removing material from skin is of extreme importance. Remove and isolate contaminated clothing and shoes at the site. Keep victim quiet and maintain normal body temperature. Effects may be delayed; keep victim under observation.

★ 760 ★
PHENOTHIAZINE
CAS: 92-84-2
Other names: AFI-TIAZIN; AGRAZINE; ANTIVERM; BIVERM; CONTAVERM; DIBENZOPARATHIAZINE; DIBENZOTHIA-ZINE; DIBENZOTHIAZINE; DIBENZO-1,4-THIAZINE; ENT 38; FEENO; FENOTHIAZINE (DUTCH); FENOTIAZINA (ITALIAN); FENOVERM; FENTIAZIN; HELMETINA; LETHELMIN; NEMA-ZENE; NEMAZINE; ORIMON; PADOPHENE; PENTHAZINE; PHENEGIC; PHENOSAN; 10H-PHENOTHIAZINE; PHENOVERM; PHENOVIS; PHENOXUR; PHENTHIAZINE; RECONOX; SOUFRAMINE; THIODIFENYLAMINE (DUTCH); THIODIPHENYLAMIN (GERMAN); THIODIPHENYLAMINE; TIODIFENILAMINA (ITALIAN); VERMITIN; WURM-THIONAL; XL-50

Danger profile: Combustible, exposure limit established, poisonous gas produced in fire.
Uses: Making pesticides, dyes, other chemicals and drugs; used in veterinary medicine.
Appearance: Greenish-yellow crystalline substance.
Odor: Slight. The odor is a warning of exposure; however, no smell does not mean you are not being exposed.
Effects of exposure:
 Breathed: May be poisonous. May cause irregular heart beat. High levels may cause muscle twitching and shaking.
 Eyes: May cause inflammation of the cornea, which is made more severe by sunlight.
 Skin: May be poisonous if absorbed. Passes through the unbroken skin; can increase exposure and the severity of the symptoms listed. May cause irritation and itching. Exposure to sunlight may cause severe reaction with rash and skin color changes.
 Swallowed: May be poisonous.
Long-term exposure: May cause damage to the developing fetus; hemolytic anemia and severe allergic liver reaction; toxic liver degeneration; blood cell and nervous system damage; skin rash; eye irritation and possible vision effects. Similar chemicals are associated with birth defects.
Storage: Store in tightly closed containers in a dry, cool, well-ventilated place. Keep away from strong acids. Containers may explode in fire.

★ 761 ★
P-PHENYLENEDIAMINE
CAS: 106-50-3
Other names: P-AMINOANILINE; 4-AMINOANILINE; BASF URSOL D; P-BENZENEDIAMINE; 1,4-BENZENEDIAMINE; BENZOFUR D; C.I. 76060; C.I. DEVELOPER 13; C.I. OXIDA-TION BASE 10; DEVELOPER 13; DEVELOPER PF; P-DIAMINOBENZENE; 1,4-DIAMINOBENZENE; DURAFUR BLACK R; FENYLENODWUAMINA (POLISH); FOURAMINE D; FOURRINE D; FOURRINE 1; FUR BLACK 41867; FUR BROWN 41866; FURRO D; FUR YELLOW; FUTRAMINE D; NAKO H; ORSIN; PARA; PARAPHENYLEN-DIAMINE; PELA-GOL D; PELAGOL DR; PELAGOL GREY D; PELTOL D; PPD; RENAL PF; SANTO-FLEX IC; TERTAL D; URSOL D; USAF EK-394; VULKANOX 4020; ZOBA BLACK D

Danger profile: Mutagen, combustible, poison, exposure limit established, Community Right-to-Know List, poisonous gas produced in fire.
Uses: Making dyes, rubber, synthetic fibers; photographic developer; hair and fur dye; laboratory chemical.
Appearance: White to light purple crystals.
Effects of exposure:
 Breathed: Poisonous. Vapor may cause intense irritation to nose, throat, mouth and lungs; mucous membrane. Symptoms of exposure include watering of the eyes, coughing, dif-

ficult breathing, headache, nausea, vomiting, muscular weakness, unconsciousness. High exposure can lead to drowsiness, coma and death; buildup of fluid in the lungs (pulmonary edema), and may cause death. Pulmonary edema may be delayed by one to two days following exposure.
 Eyes: Severe irritant. Contact may cause, burning and stinging, tearing, irritation and redness and swelling of the eyelids, intense pain; can burn the eyes and lead to blurred vision, chronic irritation and other permanent damage.
 Skin: Severe irritant. Contact may cause burning or stinging feeling, redness, swelling, intense pain, blisters, irritation, rash. Repeated exposure may lead to skin allergy and chronic irritation. As a hair dye it has been reportedly caused vertigo, anemia, gastritis, dermatitis, and death.
 Swallowed: Poisonous. May cause burning feeling of the mouth and throat, intense thirst, throat swelling, stomach cramps, nausea, vomiting, diarrhea, severe breathing difficulties and unconsciousness. Tissue damage may result. Breathing may stop. Watch for delayed symptoms.
Long-term exposure: May cause mutations, liver damage; asthma and other respiratory problems; aplastic anemia.
Storage: Store in tightly closed containers in a dry, cool, well-ventilated place. Keep away from heat, sources of ignition and oxidizing materials.

★ 762 ★
PHENYLHYDRAZINE
CAS: 100-63-0
Other names: FENILIDRAZINA (ITALIAN); FENYLHYDRA-ZINE (DUTCH); HYDRAZINE-BENZENE; HYDRAZINOBEN-ZENE; PHENYLHYDRAZIN (GERMAN); PHENYL HYDRA-ZINE; HYDRAZINE, PHENYL-

Danger profile: Carcinogen, mutagen, poison, combustible, exposure limit established, possible reproductive hazard, poisonous gas produced in fire.
Uses: Laboratory chemical; Making other chemicals, dyes and pharmaceuticals.
Appearance: Colorless to pale yellow crystals or oily liquid.
Odor: Weak, aromatic. The odor is a warning of exposure; however, no smell does not mean you are not being exposed.
Effects of exposure:
 Breathed: May be poisonous. May cause headache, rapid and difficult breathing, shortness of breath, dizziness, mental confusion, weakness; bluish color of the lips and skin, convulsions, coma and death.
 Eyes: May cause irritation, redness, swelling of the eyelids; light causes pain; severe damage.
 Skin: Passes through the unbroken skin; can increase exposure and the severity of the symptoms listed. May cause irritation, redness, swelling, blisters.
 Swallowed: May be poisonous. May cause irritation of the mouth, stomach; cramps and diarrhea.
Long-term exposure: May cause cancer; mutations, birth defects, miscarriage; anemia, kidney, bone marrow and liver damage; skin sensitization, severe dermatitis; weakness, gastrointestinal disturbances.
Storage: Store in tightly closed containers under an inert atmosphere in a dry, cool, well-ventilated place or a refrigerator. Keep away from heat, sources of ignition, light, 2-phenylamino-3-phenyloxazirane, lead(IV) oxide, perchloryl fluoride and oxidizers. Containers may explode in fire.
First aid guide: Move victim to fresh air; call emergency medical care. In case of contact with material, immediately flush skin or eyes with running water for at least 15 minutes. Remove and isolate contaminated clothing and shoes at the site.

★763★
PHENYLHYDROXYLAMINE
CAS: 100-65-2
Other names: NCI-C60093; N-PHENYLHYDROXYLAMINE

Danger profile: Combustible, poison, poisonous gas produced in fire.
Appearance: Colorless needles.
Effects of exposure:
  Breathed: May cause headache, rapid and difficult breathing, shortness of breath, dizziness, mental confusion, weakness; bluish color of the lips and skin, convulsions, coma and death.
  Eyes: May cause irritation, redness, swelling of the eyelids; light causes pain; severe damage.
  Skin: Passes through the unbroken skin; can increase exposure and the severity of the symptoms listed. May cause irritation, redness, swelling, blisters.
  Swallowed: May cause irritation of the mouth, stomach; cramps and diarrhea.
Storage: Store in tightly closed containers under an inert atmosphere in a dry, cool, well-ventilated place.

★764★
PHENYL ISOCYANATE
CAS: 103-71-9
Other names: BENZENE, ISOCYANATO-; CARBANIL; ISO-CYANIC ACID, PHENYL ESTER; ISOCYANATO BENZENE; MONDUR P; PHENYL CARBAMIDE; PHENYL CARBONI-MIDE; PHENYLCARBIMIDE

Danger profile: Mutagen, poison, combustible, poisonous gas produced in fire.
Uses: Laboratory chemical for identifying alcohols and amines; making other chemicals.
Appearance: Liquid.
Odor: Irritating, acrid. The odor is a warning of exposure; however, no smell does not mean you are not being exposed.
Effects of exposure:
  Breathed: May be poisonous. May cause irritation of the nose, throat and lungs. Higher exposures may cause shortness of breath and a dangerous build-up of fluids in the lungs (pulmonary edema), which can cause death. Pulmonary edema is a medical emergency that can be delayed one to two days following exposure.
  Eyes: May cause irritation.
  Skin: May be poisonous if absorbed. Passes through the unbroken skin; can increase exposure and the severity of the symptoms listed. May cause irritation and allergic skin rash.
  Swallowed: May be poisonous.
Long-term exposure: May cause liver and kidney damage; allergic skin rash.
Storage: Store in tightly closed containers in a dry, cool, well-ventilated place. Keep away from heat, sources of ignition and oxidizing materials. Containers may explode in fire.
First aid guide: Move victim to fresh air and call emergency medical care; if not breathing, give artificial respiration; if breathing is difficult, give oxygen. In case of contact with material, immediately flush skin or eyes with running water for at least 15 minutes. Speed in removing material from skin is of extreme importance. Remove and isolate contaminated clothing and shoes at the site. Keep victim quiet and maintain normal body temperature. Effects may be delayed; keep victim under observation.

★765★
PHENYLMERCURY ACETATE
CAS: 62-38-4
Other names: ACETATE PHENYLMERCURIQUE (FRENCH); ACETIC ACID, PHENYLMERCURY DERIVITIVE; AGROSAN; ALGIMYCIN; ANTIMUCIN WDR; BUFEN; CEKUSIL; CELMER; CERESAN; CONTRA CREME CYANACIDE; FEMMA; FENYL-MERCURIACETAT (CZECH); FMA; FUNGITOX OR; GALLO-TOX; HL-331; HONG KIEN; HOSTAQUICK; KWIKSAN; LEY-TOSAN; LIQUIPHENE; MERCURY, (ACETO-O)PHENYL-; NORFORMS; NYMERATE; PAMISAN; PHENMAD; PHE-NOMERCURIC ACETATE; PHENOMERCURY ACETATE; PMA; PMAC; PMACETATE; PMAL; PMAS; PURASAN-SC-10; PURATURF 10; QUICKSAN; SANITIZED SPG; SC-110; SEEDTOX; SPOR-KIL; TAG; TAG FUNGICIDE; ZIARNIK

Danger profile: Mutagen, reproductine hazard, poison, combustible, exposure limit established, EPA extremely hazardous substance list, Community Right-to-Know List, poisonous gas produced in fire.
Uses: Fungicide, herbicide; mildewcide for paints; slimicide.
Appearance: White powder or lustrous crystals.
Odor: Slightly vinegar-like. The odor is a warning of exposure; however, no smell does not mean you are not being exposed.
Effects of exposure:
  Breathed: May be poisonous. Inhalation may be fatal as a result of spasm, inflammation and edema of the larynx and bronchi, chemical pneumonitis, and pulmonary edema. Symptoms of exposure may include burning sensation, coughing, wheezing, laryngitis, shortness of breath, headache, nausea, and vomiting. High exposures may cause a dangerous build-up of fluids in the lungs (pulmonary edema), which can cause death. Pulmonary edema is a medical emergency that can be delayed one to two days following exposure.
  Eyes: Extremely destructive. Contact may cause damage and vision loss.
  Skin: May be poisonous if absorbed. Passes through the unbroken skin; can increase exposure and the severity of the symptoms listed. Contact may cause irritation and burns. Prolonged contact may cause blistering.
  Swallowed: May be poisonous. See *Breathed.*
Long-term exposure: May cause mutations, birth defects, miscarriage; kidney damage; prolonged exposure may result in heavy mercury poisoning with shakes, irritability, memory loss, weakness, loss of appetite, personality change and brain damage; skin may turn gray and allergy may develop; possible lung problems.
Storage: Store in tightly closed containers in a dry, cool, well-ventilated place or a refrigerator. Keep away from sources of ignition and oxidizers. Containers may explode in fire.

★766★
PHENYLMERCURY HYDROXIDE
CAS: 100-57-2
Other names: HYDROXYPHENYLMERCURY; MERCURY, HYDROXYPHENYL-; PHENYLMERCURIC HYDROXIDE

Danger profile: Mutagen, reproductive hazard, combustible, exposure limit established, poisonous gas produced in fire.
Uses: Fungicide, germicide, making other chemicals.
Appearance: White or cream colored fine crystalline solid.
Effects of exposure:
  Breathed: May be poisonous. May cause irritation of the throat and breathing passages. Mercury vapors from heated chemical or contact with acid may cause bronchitis, phlegm and lung irritation. High exposures may cause a severe shortness of breath and a dangerous build-up of fluids in the lungs (pulmonary edema), which can cause death. Pulmonary

edema is a medical emergency that can be delayed one to two days following exposure.
Eyes: Contact may cause severe damage with loss of vision.
Skin: Passes through the unbroken skin; can increase exposure and the severity of the symptoms listed. May cause irritation, burns, allergy and gray skin color.
Swallowed: May be poisonous. See **Breathed.**
Long-term exposure: May cause birth defects by damaging the fetus; kidney damage; mercury poisoning with shakes, irritability, sore gums, personality change and brain damage; skin allergy with rash. Similar substances may cause lung damage.
Storage: Store in tightly closed containers in a dry, cool, well-ventilated place. Keep away from sources of ignition and strong oxidizers. Containers may explode in fire.

★767★
N-PHENYL-2-NAPHTHTYLAMINE
CAS: 135-88-6
Other names: ACETO PBN; AGERITE POWDER; AGE RITE POWDER; ANILINONAPHTHALENE; 2-ANILINONAPHTHALENE; ANTIOXIDANT 116; ANTIOXIDANT PBN; N-(2-NAPHTHYL)ANILINE; 2-NAPHTHYLPHENYLAMINE; BETA-NAPHTHYLPHENYLAMINE; NCI-C02915; NEOZON D; NEO-ZONE; NEOZONE D; NILOX PBNA; NONOX D; PBNA; 2-PHENYLAMINONAPHTHALENE; PHENYL-BETA-NAPHTHYLAMINE; PHENYL-2-NAPHTHYLAMINE; N-PHENYL-BETA-NAPHTHYLAMINE; STABILIZATOR AR

Danger profile: Carcinogen, mutagen, combustible, poisonous gas produced in fire.
Uses: Making rubber products and chemicals; in lubricants.
Appearance: Light gray powder with needle-like crystals.
Effects of exposure:
  Breathed: May cause headache, rapid and difficult breathing, shortness of breath, dizziness, mental confusion, weakness; bluish color of the lips and skin, convulsions, coma and death. Very high levels may cause death.
  Eyes: May cause irritation, redness, swelling of the eyelids; light causes pain; severe damage.
  Skin: Passes through the unbroken skin; can increase exposure and the severity of the symptoms listed. May cause irritation, redness, swelling, rash and blisters.
  Swallowed: May cause irritation of the mouth, stomach; cramps and diarrhea.
Long-term exposure: May cause cancer; birth defects, miscarriage; skin irritation and rash; anemia, blue color to skin lips and fingernails.
Storage: Store in tightly closed containers in a dry, cool, well-ventilated place. Keep away from heat and oil.

★768★
O-PHENYLPHENOL
CAS: 90-43-7
Other names: 1,1'-BIPHENYL-2-OL; 2-BIPHENYLPHENOL; 2-BIPHENYLOL; DOWCIDE 1; 2-HYDROXYDIPHENYL; ORTHOXENOL; 2-PHENYLPHENYL; TORSITE; O-XONAL

Danger profile: Mutagen, combustible, exposure limit established, Community Right-to-Know List, poisonous gas produced in fire.
Uses: Making fungicides, rubber chemicals and dyestuffs.
Appearance: White or buff-colored solid in the form of flaky crystals. Looks like powdered soap.
Effects of exposure:
  Breathed: May be poisonous. High exposure may cause kidney damage.

Eyes: Contact may cause severe irritation and possible damage.
Skin: May be poisonous if absorbed. Passes through the unbroken skin; can increase exposure and the severity of the symptoms listed. May cause severe irritation and burns. Skin allergy with rash and itching may develop.
Swallowed: May be poisonous.
Long-term exposure: May cause kidney damage; skin allergy. Similar irritating substances may effect the lungs.
Storage: Store in tightly closed containers in a dry, cool, well-ventilated place. Keep away from heat, sources of ignition and water. Containers may explode in fire.

★769★
PHENYL TRICHLOROSILANE
CAS: 98-13-5
Other names: PHENYLSILICON TRICHLORIDE; SILICON PHENYL TRICHLORORIDE; SILANE, TRICHLOROPHENYL-; TRICHLOROPHENYLSILANE; TRICHLOROPHENYL SILANE

Danger profile: Combustible, poison, corrosive, EPA extremely hazardous substances list, poisonous gas produced in fire.
Uses: Making silicones for water repellents, insulating resins, heat resistant paints; laboratory chemical.
Appearance: Colorless fuming liquid.
Effects of exposure:
  Breathed: May cause irritation of the nose, throat and lungs; chest pain, coughing. Even small amounts may cause coughing, shortness of breath, severe difficulty breathing; possible death. May cause a dangerous build-up of fluids in the lungs (pulmonary edema), which can cause death. Pulmonary edema is a medical emergency that can be delayed one to two days following exposure.
  Eyes: May cause pain, severe irritation, burns and permanent damage.
  Skin: May cause severe irritation and burns.
  Swallowed: May cause severe burns and vomiting.
Long-term exposure: May cause lung damage.
Storage: Store in tightly closed containers in a dry, cool, well-ventilated place. Keep away from water, steam and other forms of moisture, halogens. Containers may explode in fire.

★770★
PHORATE
CAS: 298-02-2
Other names: AC 3911; AMERICAN CYANAMID 3,911; O,O-DIAETHYL-S-(AETHYLTHIO-METHYL)-DITHIOPHOSPHAT (GERMAN); O,O-DIETHYL S-ETHYLMERCAPTOMETHYL DITHIOPHOSPHONATE; O,O-DIETHYL-S-(ETHYLTHIO-METHYL)-DITHIOFOSFAAT (DUTCH); O,O-DIETHYLS-ETHYLTHIOMETHYLDITHIOPHOSPHONATE; O,O-DIETHYLETHYLTHIOMETHYL PHOSPHORODITHIOATE; O,O-DIETHYL S-(ETHYLTHIO)METHYLPHOSPHORODITHIOATE; O,O-DIETHYL S-ETHYLTHIOMETHYLTHIOTHIONOPHOSPHATE; O,O-DIETIL-S-(ETILTIO-METIL)-DITIOF OSFATO (ITALIAN); DITHIOPHOSPHATEDE O,O-DIETHYLE ET D'ETHYLTHIOMETHYLE (FRENCH); EL 3911; ENT 24,042; EXPERIMENTAL INSECTICIDE 3911; FORAAT (DUTCH); GEOMET; GRAMTOX; GRANUTOX; L 11/6; METHANETHIOL, (ETHYLTHIO)-,S-ESTER WITH O,O-DIETHYLPHOSPHORODITHIOATE; PHORAT (GERMAN); PHORATE-10G; RAMPART; TERRATHION GRANULES; THI-MET; TIMET (USSR); VEGFRU; VERGFRU FORATOX

Danger profile: Mutagen, exposure limit established, poisonous gas produced in fire.

Uses: Insecticide; animal feed additive. Used on cotton, maize, sugar beet, fodder crops, carrots, peas and potatoes (organophosphate).

Appearance: Clear liquid.

Effects of exposure:

Breathed: Extremely poisonous. May cause headache, dizziness, profuse sweating, nausea, vomiting, reduced heart beat, stomach cramps, diarrhea, loss of coordination, slow and weak breathing, fever, loss of consciousness and slow. May cause shortness of breath and a dangerous build-up of fluids in the lungs (pulmonary edema), which can cause death. Pulmonary edema is a medical emergency that can be delayed one to two days following exposure.

Eyes: Rapidly absorbed. Poisoning can happen very rapidly. See *Breathed.*

Skin: Poisonous. Quickly passes through the unbroken skin; can increase exposure and the severity of the symptoms listed. Fatal poisoning may result from contact.

Swallowed: Powerful, deadly poison. See *Breathed.*

Long-term exposure: May damage the developing fetus or cause birth defects; damage to the testes; nerve damage including loss of coordination; liver damage. Repeated exposure may cause depression, irritability and personality changes.

Storage: Store in a regulated, specially marked place in tightly closed containers in a dry, well-ventilated place. Keep away from water and strong alkalis. Containers may explode in fire.

★771★
PHOSDRIN
CAS: 7786-34-7

Other names: APAVINPHOS; ALPHA-2-CARBOMETHOXY-1-METHYLVINYL DIMETHYL PHOSPHATE; 1-CARBOMETHOXY-1-PROPEN-2-Y PHOSPHATE; 2-CARBOMETHOXY-1-METHYLVINYLDIMETHYL PHOSPHATE; CMDP; COMPOUND 2046; O,O-DIMETHYL PHOSTENE; 3-((DIMETHYLPHOSPHINYL)OXY)-2-BUTENOIC ACID METHYL ESTHER; DURAPHOS; ENT 22,374; FOSDRIN; GESFID; GESTID; 3-HYDROXYCROTONIC ACID METHYL ESTER DIMETHYL PHOSPHATE; MEVINFOS (DUTCH); MENIPHOS; MEVINPHOS; MENITE; 1-METHOXYCARBONYL-1-PROPEN-2-YLDIMETHYL PHOSPHATE; METHYL 3-(DIMETHOXYPHOSPHINYLOXY)CROTONATE; MEVINOX; OS 2046; CIS-PHOSDRIN; PHOSDRIN 24; PHOSFENE; PHOSPHENE (FRENCH)

Danger profile: Mutagen, combustible, poison, exposure limit established, EPA extremely hazardous substances list, poisonous gas produced in fire.

Uses: Insecticide and acaricide. Against beetles, mites and caterpillars; on fruit trees, strawberries, lettuce and peas.

Appearance: Yellow or pale yellow to orange liquid

Effects of exposure:

Breathed: Extremely poisonous. May cause headache, dizziness, profuse sweating, nausea, vomiting, reduced heart beat, stomach cramps, diarrhea, loss of coordination, slow and weak breathing, fever, loss of consciousness and death. May cause shortness of breath and a dangerous build-up of fluids in the lungs (pulmonary edema), which can cause death. Pulmonary edema is a medical emergency that can be delayed one to two days following exposure.

Eyes: Rapidly absorbed. Poisoning can happen very rapidly. See *Breathed.*

Skin: Poisonous. Passes through the unbroken skin; can increase exposure and the severity of the symptoms listed.

Swallowed: Powerful, deadly poison. See *Breathed.*

Long-term exposure: May damage the developing fetus or cause birth defects; nerve damage including loss of coordina-

tion; liver damage. Repeated exposure may cause depression, irritability and personality changes.

Storage: Store in a regulated, specially marked place in tightly closed containers in a dry, well-ventilated place. Keep away from water, strong alkalis. Containers may explode in fire.

★772★
PHOSGENE
CAS: 75-44-5

Other names: CARBONE (OXYCHLORURE DE) (FRENCH); CARBON OXYCHLORIDE; CARBONIC DICHLORIDE; CARBONYLCHLORID (GERMAN); CARBONYL CHLORIDE; CG; CHLOROFORMYL CHLORIDE; CARBONIO (OSSICLORURO DI) (ITALIAN); DIPHOSGENE; FOSGEEN (DUTCH); FOSGEN (POLISH); FOSGENE (ITALIAN); KOOLSTOFOXYCHLORIDE (DUTCH); NCI-C60219; PHOSGEN (GERMAN)

Danger profile: Combustible, poisonous, exposure limit established, Community Right-to-Know List, EPA extremely hazardous substances list, poisonous gas produced in fire.

Uses: Making other chemicals, polyurethanes, resins, pesticides, herbicides, pharmaceuticals, dyes; military anti-personnel gas.

Appearance: Colorless gas or volatile liquid.

Odor: Like new-mown hay in low concentrations; sharp, pungent in high concentrations. The odor is a warning of exposure; however, no smell does not mean you are not being exposed.

Effects of exposure:

Breathed: May be poisonous. Vapor may cause intense irritation to nose, throat, mouth and lungs; mucous membrane. Symptoms of exposure include watering of the eyes, coughing, difficult breathing, headache, nausea, vomiting, muscular weakness, unconsciousness. High exposure can lead to drowsiness, coma and death; buildup of fluid in the lungs (pulmonary edema), and may cause death. Pulmonary edema may be delayed by one to two days following exposure. Pneumonia may develop up to several days following exposure.

Eyes: Severe irritant. Contact may cause, burning and stinging, tearing, irritation and redness and swelling of the eyelids, intense pain; can burn the eyes and lead to blurred vision, chronic irritation and other permanent damage. Liquid may cause frostbite.

Skin: May be poisonous if absorbed. Severe irritant. Contact may cause burning or stinging feeling, redness, swelling, intense pain, blisters, irritation, rash. Repeated exposure may lead to skin allergy and chronic irritation. Liquid may cause frostbite.

Swallowed: May cause burning feeling of the mouth and throat, intense thirst, throat swelling, stomach cramps, nausea, vomiting, diarrhea, severe breathing difficulties and unconsciousness. Tissue damage may result. Breathing may stop. Watch for delayed symptoms.

Long-term exposure: May cause permanent lung damage.

Storage: Store in tightly closed containers in a dry, cool, well-ventilated place. Keep away from water, steam and other forms of moisture, sunlight, lithium, aluminum, isopropyl alcohol and all other materials. Protect containers from physical damage. Containers may explode in fire.

First aid guide: Move victim to fresh air and call emergency medical care; if not breathing, give artificial respiration; if breathing is difficult, give oxygen. In case of contact with material, immediately flush skin or eyes with running water for at least 15 minutes. Remove and isolate contaminated clothing and shoes at the site. Keep victim quiet and maintain normal body temperature. Effects may delayed; keep victim under observation.

★773★
PHOSMET
CAS: 732-11-6
Other names: APPA; DECEMTHION; DECEMTHION P-6; (O,O-DIMETHYL-PHTHALIMIDIOMETHYL-DITHIOPHOSPHATE); O,O-DIMETHYL S-(N-PHTHALIMIDOMETHYL)DITHIOPHOSPHATE; O,O-DIMETHYL S-PHTHALIMIDOMETHYLPHOSPHORODITHIOATE; ENT 25,705; FTALOPHOS; IMIDAN; KEMOLATE; N-(MERCAPTOMETHYL)PHTHALIMIDE S-(O,O-DIMETHYL PHOSPHORODITHIOATE); PERCOLATE; PHOSPHO-RODITHIOIC ACID, S-((1,3-DIHYDRO-1,3-DIOXO-ISOINDOL-2-YL)METHYL) O,O-DIMETHYL ESTER; PHOSPHO-RODITHIOIC ACID, O,O-DIMETHYL ESTER, S-ESTER WITH N-(MERCAPTOMETHYL)PHTHALIMIDE; PHTHALIMIDE,N-(MERCAPTOMETHYL)-, S-ESTER WITH O,O-DIMETHYL PHOSPHORODITHIOATE; PHTHALIMIDO O,O-DIMETHYL PHOSPHORODITHIOATE; PHTHALIMIDOMETHYL O,O-DIMETHYL PHOSPHORODITHIOATE; PHTHALOPHOS; PMP; PROLATE; R 1504; SMIDAN; STAUFFER R 1504

Danger profile: Mutagen, poison, combustible, EPA extremely hazardous substances list, poisonous gas produced in fire.
Uses: Insecticide (organophosphate). On fruit, citrus, grapes, potatoes; in forestry.
Appearance: White or off-white crystalline solid.
Odor: Offensive. The odor is a warning of exposure; however, no smell does not mean you are not being exposed.
Effects of exposure:
Breathed: Extremely poisonous. May cause headache, dizziness, profuse sweating, nausea, vomiting, reduced heart beat, stomach cramps, diarrhea, loss of coordination, slow and weak breathing, fever, loss of consciousness and death. May cause shortness of breath and a dangerous build-up of fluids in the lungs (pulmonary edema), which can cause death. Pulmonary edema is a medical emergency that can be delayed one to two days following exposure.
Eyes: Rapidly absorbed. Poisoning can happen very rapidly. See *Breathed.*
Skin: Poisonous. Passes through the unbroken skin; can increase exposure and the severity of the symptoms listed.
Swallowed: Powerful, deadly poison. See *Breathed.*
Long-term exposure: May damage the developing fetus or cause birth defects; nerve damage including loss of coordination; liver damage. Repeated exposure may cause depression, irritability and personality changes.
Storage: Store in a regulated, specially marked place in tightly closed containers in a dry, well-ventilated place. Keep away from water, strong alkalis. Containers may explode in fire.

★774★
PHOSPHINE
CAS: 7803-51-2
Other names: CELPHOS; DELICIA; DETIA; FOSFOROWO-DOR (POLISH); GAS-EX-B; HYDROGEN PHOSPHIDE; PHO-SPHOMS TRIHYDRIDE, PHOSPHOROUS HYDRIDE; PHOS-PHORATED HYDROGEN; PHOSPHORWASSERSTOFF (GERMAN); PHOSTOXIN

Danger profile: Mutagen, poison, fire hazard, exposure limit established, poisonous gas produced in fire.
Uses: Making other chemicals, semiconductors; fumigant.
Appearance: Colorless gas.
Odor: Foul, fish-like. The odor is a warning of exposure; however, no smell does not mean you are not being exposed.
Effects of exposure:
Breathed: Poison. Immediately seek medical attention. May cause giddiness, headache, and nausea. More severe expo-sure may cause malaise, weakness, shivering, dizziness; dyspnea; abdominal and back pain, nausea, vomiting; bronze or yellow coloration of skin, jaundice; garlic breath; dark red urine; Delirium, coma and death. High exposures may cause shortness of breath and a dangerous build-up of fluids in the lungs (pulmonary edema), which can cause death. Pulmonary edema is a medical emergency that can be delayed one to two days following exposure. High exposure may cause liver damage.
Eyes: May cause irritation, watering, burning sensation. High levels of exposure may cause permanent damage.
Skin: May be poisonous if absorbed. May cause painful irritation, inflammation, blisters and deep wounds (ulcer).
Swallowed: Poisonous. Immediately seek medical attention. Unlikely route of exposure as most cases of poisoning occur from accidental generation of the gas.
Long-term exposure: May cause mutations, birth defects miscarriage, cancer; liver damage. Possible lung effects.
Storage: Store away from food products, high temperature and strong light. Keep away from air, moisture, heat, sources of ignition, strong oxidizers, strong acids, shock, oxygen halogenated hydrocarbons and all other materials. Vapors may travel to sources of ignition and flash back. Containers may explode in fire.
First aid guide: Move victim to fresh air and call emergency medical care; if not breathing, give artificial respiration; if breathing is difficult, give oxygen. In case of contact with material, immediately flush skin or eyes with running water for at least 15 minutes. Remove and isolate contaminated clothing and shoes at the site. Keep victim quiet and maintain normal body temperature. Effects may be delayed; keep victim under observation.

★775★
PHOSPHORIC ACID
CAS: 7664-38-2
Other names: ACID PHOSPHORIQUE (FRENCH); ACIDO FOSFORICO (ITALIAN); FOSFORZUUROPLOSSINGEN (DUTCH); ORTHOPHOSPHORIC ACID; METAPHOSPHORIC ACID; PHORSAEURELOESUNGEN (GERMANY); WHITE PHOSPHORIC ACID

Danger profile: Combustible, corrosive, poison, exposure limit established, Community Right-to-Know List, poisonous gas produced in fire.
Uses: Making metal products, gelatin, fertilizers; soaps and detergents; rust-proofing; pharmaceuticals; sugar refining; water treatment; animal feeds; electro-polishing; gasoline additive; coatings for metals; making other chemicals; fabric dyeing; yeasts; waxes and polishes; in ceramics; in foods and carbonated beverages; laboratory chemical.
Appearance: Clear colorless syrupy liquid. White crystals below 70°F.
Effects of exposure:
Breathed: Vapor may cause intense irritation to nose, throat, mouth and lungs; mucous membrane. Symptoms of exposure include watering of the eyes, coughing, difficult breathing, headache, nausea, vomiting, muscular weakness, unconsciousness. High exposure can lead to drowsiness, coma and death; buildup of fluid in the lungs (pulmonary edema), and may cause death. Pulmonary edema may be delayed by one to two days following exposure.
Eyes: Severe irritant. Contact may cause, burning and stinging, tearing, irritation and redness and swelling of the eyelids, intense pain; can burn the eyes and lead to blurred vision, chronic irritation and other permanent damage.
Skin: Severe irritant. Contact may cause burning or stinging feeling, redness, swelling, intense pain, blisters, irritation,

rash. Repeated exposure may lead to skin allergy and chronic irritation.

Swallowed: May cause burning feeling of the mouth and throat, intense thirst, throat swelling, stomach cramps, nausea, vomiting, diarrhea, severe breathing difficulties and unconsciousness. Tissue damage may result. Breathing may stop. Watch for delayed symptoms.

Long-term exposure: May cause drying and cracking of the skin; bronchitis with cough, phlegm, shortness of breath.

Storage: Store in tightly closed containers in a dry, cool, well-ventilated place. Keep away from heat and wide temperature changes, water, steam and other forms of moisture, nitromethane, sodium tetraborate, metals and strong caustics. Keep from freezing. Reacts with metal to produce explosive hydrogen gas. Containers may explode in fire.

First aid guide: Move victim to fresh air; call emergency medical care. In case of contact with material, immediately flush skin or eyes with running water for at least 15 minutes. Remove and isolate contaminated clothing and shoes at the site. Keep victim quiet and maintain normal body temperature.

★ 776 ★
PHOSPHORUS
CAS: 7723-14-0

Other names: BONIDE BLUE DEATH RAT KILLER; COMMON SENSE COCKROACH AND RAT PREPARATIONS; FOSFORO BIANCO (ITALIAN); GELBER PHOSPHOR (GERMAN); PHOSPHORE BLANC (FRENCH); PHOSPHORUS ELEMENTAL, WHITE; PHOSPHOROUS YELLOW; RAT-NIP; TETRAFOSFOR (DUTCH); TETRAPHOSPHOR (GERMAN); WEISS PHOSPHOR (GERMAN); WHITE PHOSPHORUS; WP; YELLOW PHOSPHORUS

Danger profile: Fire hazard, explosion hazard, poison, exposure limit established, Community Right-to-Know List, EPA extremely hazardous substances list, poisonous gas produced in fire.

Uses: To kill rodents; chemical fertilizer; explosives; smoke bombs; analytical chemistry.

Appearance: White to yellow, soft, waxy solid.

Odor: Distinctive, disagreeable; pungent, sharp; garlic-like. The odor is a warning of exposure; however, no smell does not mean you are not being exposed.

Effects of exposure:
Breathed: Poisonous. Immediately seek medical attention. may cause severe irritation to the nose, throat and lungs. May cause giddiness, headache, and nausea. More severe exposure may cause malaise, weakness, shivering, dizziness; dyspnea; abdominal and back pain, nausea, vomiting; bronze or yellow coloration of skin, jaundice; garlic breath; dark red urine; delirium, coma and death. High exposure may cause severe or fatal poisoning and/or a dangerous build-up of fluids in the lungs (pulmonary edema), which can cause death. Pulmonary edema is a medical emergency that can be delayed one to two days following exposure.

Eyes: May cause irritation, watering, burning sensation. High levels of exposure may cause permanent damage.

Skin: Passes through the unbroken skin; can increase exposure and the severity of the symptoms listed. May ignite and cause severe burns.

Swallowed: Poisonous.

Long-term exposure: Repeated low exposure may cause bone damage, especially to the jaw bone.

Storage: Store white phosphorus under water under water. Store yellow phosphorus in tightly closed containers in a cool well-ventilated place. Keep away from sources of ignition, air, organic materials, oxidizers, halogens, alkaline materials. Ignites when exposed to air. Containers may explode in fire.

★ 777 ★
PHOSPHORUS OXYCHLORIDE
CAS: 10025-87-3

Other names: PHOSPHORIC CHLORIDE; PHOSPHORUS CHLORIDE OXIDE; PHOSPHORUS OXYTRICHLORIDE; PHOSPHORYL CHLORIDE

Danger profile: Combustible, corrosive, exposure limit established, poisonous gas produced in fire.

Uses: Making other chemicals, semiconductors, gasoline additives, hydraulic fluids, fire-retarding agents and pesticides and other organophosphorus compounds; chlorinating agent.

Appearance: Colorless to pale yellow, strongly fuming, oily liquid.

Odor: Musty and lingering. The odor is a warning of exposure; however, no smell does not mean you are not being exposed.

Effects of exposure:
Breathed: May be poisonous. May cause irritation of the nose, throat and lungs; headache, dizziness, nausea, vomiting; cough, difficult breathing and chest tightness. Higher exposures may cause shortness of breath and a dangerous build-up of fluids in the lungs (pulmonary edema), which can cause death. Pulmonary edema is a medical emergency that can be delayed one to two days following exposure.

Eyes: Contact may cause severe burns and permanent damage.

Skin: Passes through the unbroken skin; can increase exposure and the severity of the symptoms listed. Solid or liquid may cause severe burns.

Swallowed: May be poisonous. May cause nausea, vomiting, jaundice, low blood pressure, depression, delirium, coma, death. Symptoms after ingestion may be delayed for from a few hours to 3 days.

Long-term exposure: May cause kidney and liver damage. Similar irritating substances may cause lung problems.

Storage: Store in tightly closed containers in a dry, cool, well-ventilated place. Keep away from water, steam and other forms of moisture, dimethyl sulfoxide, organic materials, chemically active metal powders, and other chemicals. Corrodes most metals except nickel and lead; rapidly corrodes iron and steel in the presence of moisture. Containers may explode in fire.

First aid guide: Move victim to fresh air and call emergency medical care; if not breathing, give artificial respiration; if breathing is difficult, give oxygen. In case of contact with material, immediately flush skin or eyes with running water for at least 15 minutes. Speed in removing material from skin is of extreme importance. Remove and isolate contaminated clothing and shoes at the site. Keep victim quiet and maintain normal body temperature.

★ 778 ★
PHOSPHORUS PENTACHLORIDE
CAS: 10026-13-8

Other names: FOSFORO (PENTACHLORURO DI) (ITALIAN); FOSFORPENTACHLORIDE (DUTCH); PHOSPHORE(-PENTACHLORURE DE) (FRENCH); PHOSPHORIC CHLORIDE; PHOSPHORPENTACHLORID (GERMAN); PHOSPHORUS PERCHLORIDE; PIECIOCHLOREK FOSFORU (POLISH)

Danger profile: Combustible, corrosive, exposure limit established, EPA extremely hazardous substances list, poisonous gas produced in fire.

Uses: Chlorinating and dehydrating agent; catalyst.

Appearance: Pale yellow, fuming solid.

Odor: Hydrochloric acid-like. The odor is a warning of exposure; however, no smell does not mean you are not being exposed.

Effects of exposure:
Breathed: Poisonous. Immediately seek medical attention. May cause giddiness, headache, and nausea. More severe

exposure may cause malaise, weakness, shivering, dizziness; dyspnea; abdominal and back pain, nausea, vomiting; bronze or yellow coloration of skin, jaundice; garlic breath; dark red urine; delirium, coma and death. High exposures may cause a dangerous build-up of fluids in the lungs (pulmonary edema), which can cause death. Pulmonary edema is a medical emergency that can be delayed one to two days following exposure.

Eyes: May cause severe irritation, watering, burning sensation. High levels of exposure may cause permanent damage.

Skin: May cause painful irritation, inflammation, blisters and deep wounds (ulcer).

Swallowed: May be poisonous. Immediately seek medical attention. Unlikely route of exposure as most cases of poisoning occur from accidental generation of the gas.

Long-term exposure: May cause kidney problems.

Storage: Store in tightly closed containers in a dry, cool, well-ventilated place. Keep away from heat, water, steam and other forms of moisture, carbamates, fluorine, nitrobenzene, phosphorus (III) oxide, aluminum, chlorine dioxide, chlorine, sodium, hydroxylamine and other chemicals. Protect containers from physical damage. Containers may explode in fire.

First aid guide: Move victim to fresh air and call emergency medical care; if not breathing, give artificial respiration; if breathing is difficult, give oxygen. In case of contact with material, immediately flush skin or eyes with running water for at least 15 minutes. Speed in removing material from skin is of extreme importance. Remove and isolate contaminated clothing and shoes at the site. Keep victim quiet and maintain normal body temperature.

★779★
PHOSPHORUS PENTASULFIDE
CAS: 1314-80-3
Other names: PENTASULFURE DE PHOSPHORE (FRENCH); PHOSPHORIC SULFIDE; PHOSPHORUS PERSULFIDE; PHOSPHORUS SULFIDE; SIRNIK FOSFORECNY (CZECH); SULFUR PHOSPHIDE; THIOPHOSPHORIC ANHYDRIDE

Danger profile: Combustible, dangerous when wet, exposure limit established, poisonous gas produced in fire.
Uses: Making lube oil additives, insecticides, fertilizers; ignition compounds, safety matches.
Appearance: Greenish-yellow crystalline solid.
Odor: Rotten egg-like. May paralyze the sense of smell. The odor is a warning of exposure; however, no smell does not mean you are not being exposed.
Effects of exposure:
Breathed: May cause irritation of the nose, coughing and sneezing. Hydrogen sulfide gas formed by reaction with moisture can cause death by respiratory failure. The symptoms may be delayed.
Eyes: May cause severe irritation, redness, tearing, pain and inflammation of the eyelids.
Skin: May cause severe irritation.
Swallowed: Piosonous.
Storage: Store in tightly closed containers in a dry, cool, well-ventilated place. Keep away from air, heat, sources of ignition, water, steam and other forms of moisture, acids, oxidizing materials. Containers may explode in fire.
First aid guide: Move victim to fresh air and call emergency medical care; if not breathing, give artificial respiration; if breathing is difficult, give oxygen. In case of contact with material, immediately flush skin or eyes with running water for at least 15 minutes. Remove and isolate contaminated clothing and shoes at the site. Keep victim quiet and maintain normal body temperature.

★780★
PHOSPHORUS PENTOXIDE
CAS: 1314-56-3
Other names: DIPHOSPHORUS PENTOXIDE; PHOSPHORIC ANHYDRIDE; PHOSPHORUS(V) OXIDE; PHOSPHORUS PENTAOXIDE; PHOSPHORUS OXIDE; POX

Danger profile: Combustible, corrosive, EPA extremely hazardous substances list, poisonous gas produced in fire.
Uses: Making other chemicals; drying agent; sugar refining; laboratory chemicals; fire extinguishing agent.
Appearance: White crystalline solid or powder.
Effects of exposure:
Breathed: May cause irritation of the nose, throat and lungs; coughing, shortness of breath. High exposures may cause shortness of breath and a dangerous build-up of fluids in the lungs (pulmonary edema), which can cause death. Pulmonary edema is a medical emergency that can be delayed one to two days following exposure.
Eyes: May cause severe irritation and permanent damage.
Skin: Contact may cause severe irritation and burns.
Swallowed: May cause severe internal irritation and damage.
Long-term exposure: May cause skin allergy; lung allergy. Possible lung damage.
Storage: Store in tightly closed containers in a dry, cool, well-ventilated place. Keep away from combustibles, water, steam, ammonia, calcium oxide, hydrogen fluoride, chlorine trifluoride, potassium, propargyl alcohol, sodium carbonate, formic acid, metals, sodium hydroxide, methyl hydroperoxide, 3-propynol, iodides, perchloric acid, barium sulfide and hydrogen peroxide. Containers may explode in fire.
First aid guide: Move victim to fresh air and call emergency medical care; if not breathing, give artificial respiration; if breathing is difficult, give oxygen. In case of contact with material, immediately flush skin or eyes with running water for at least 15 minutes. Speed in removing material from skin is of extreme importance. Remove and isolate contaminated clothing and shoes at the site. Keep victim quiet and maintain normal body temperature.

★781★
PHOSPHORUS, RED
CAS: 7723-14-0
Other names: PHOSPHORUS, AMORPHOUS, RED; RED PHOSPHORUS

Danger profile: Combustible, poison, EPA extremely hazardous substance list, Community Right-to-Know List, poisonous gas produced in fire.
Uses: Making other chemicals, phosphor bronzes, semiconductors; coatings, smoke bombs, fireworks, safety matches, fertilizers, rodenticide.
Appearance: Brick red, reddish-brown or violet powder.
Effects of exposure:
Breathed: May effect the liver and kidneys.
Eyes: Contact may cause severe irritation and damage.
Skin: Contact may cause irritation and burns.
Swallowed: Poisonous.
Long-term exposure: May cause liver and kidney damage.
Storage: Store in tightly closed containers in a dry, cool, well-ventilated place. Keep away from heat, direct sunlight, strong oxidizers, alkaline hydroxides and all other chemicals; reacts with alkalis to produce combustible phosphine gas. may have white phosphorus as an impurity. Containers may explode in fire.
First aid guide: Move victim to fresh air; call emergency medical care. In case of contact with material, immediately flush skin or eyes with running water for at least 15 minutes. Removal of solidified molten material from skin requires medical assistance.

Remove and isolate contaminated clothing and shoes at the site.

★782★
PHOSPHORUS TRICHLORIDE
CAS: 7719-12-2
Other names: CHLORIDE OF PHOSPHORUS; FOSFORO (TRICLORURO DI) (ITALIAN); FOSFORTRICHLORIDE (DUTCH); PHOSPHORE (TRICHLORURE DE) (FRENCH); PHOSPHOROUS CHLORIDE; PHOSPHORTRICHLORID (GERMAN); PHOSPHORUS CHLORIDE; TRICHLOROPHOSPHINE; TROJCHLOREK FOSFORU (POLISH)

Danger profile: Combustible, corrosive, exposure limit established, EPA extremely hazardous substances list, poisonous gas produced in fire.
Uses: Making other chemicals, organophosphorus pesticides, dyestuffs; in gasoline additives and textile finishing agents.
Appearance: Colorless, clear to yellow, fuming liquid.
Odor: Pungent, strong, irritating, hydrochloric acid-like. The odor is a warning of exposure; however, no smell does not mean you are not being exposed.
Effects of exposure:
Breathed: Poisonous. Immediately seek medical attention. May cause giddiness, headache, and nausea. More severe exposure may cause weakness, shivering, dizziness; abdominal and back pain, nausea, vomiting; bronze or yellow coloration of skin, jaundice; garlic breath; dark red urine; delirium, coma and death. High exposures may cause a dangerous build-up of fluids in the lungs (pulmonary edema), which can cause death. Pulmonary edema is a medical emergency that can be delayed one to two days following exposure.
Eyes: May cause severe irritation, watering, burning sensation. High levels of exposure may cause permanent damage.
Skin: May cause painful irritation, inflammation, blisters and severe wounds.
Swallowed: Poisonous. Immediately seek medical attention.
Long-term exposure: May cause lung problems.
Storage: Store in tightly closed containers in a dry, cool, well-ventilated place, away from food products. Keep away from water, steam and other forms of moisture, acids, aluminum, chromyl chloride, fluorine, alcohol, sodium peroxide, sodium, tetravinyl lead, potassium, hydroxylamine, lead dioxide and other chemicals. Corrodes most metals, and attacks some forms of rubber, plastics and coatings. Containers may explode in fire.
First aid guide: Move victim to fresh air and call emergency medical care; if not breathing, give artificial respiration; if breathing is difficult, give oxygen. In case of contact with material, immediately flush skin or eyes with running water for at least 15 minutes. Speed in removing material from skin is of extreme importance. Remove and isolate contaminated clothing and shoes at the site. Keep victim quiet and maintain normal body temperature.

★783★
PHTHALIC ANHYDRIDE
CAS: 85-44-9
Other names: ANHYDRIDE PHTALIQUE (FRENCH); ANIDRIDE FTALICA (ITALIAN); 1,2-BENZENEDICARBOXYLIC ANHYDRIDE; 1,2-BENZENEDICARBOXYLIC ACID ANHYDRIDE; 1,3-DIOXOPHTHALAN; ESEN; FTALOWY BEZWODNIK (POLISH); FTAALZUURANHYDRIDE (DUTCH); ISOBENZOFURAN,1,3-DIHYDRO-1,3-DIOXO-; 1,3-ISOBENZOFURANDIONE; NCI-C03601; PAN; PHTHALANDIONE; 1,3-PHTHALANDIONE; PHTHALIC ACID ANHYDRIDE; PHTHALSAEUREANHYDRID (GERMAN); RETARDER AK; RETARDER ESEN; RETARDER PD

Danger profile: Combustible, corrosive, allergen, exposure limit established, Community Right-to-Know List, poisonous gas produced in fire.
Uses: Making plastics, resins, dyes, other chemicals, insecticides, polyesters, drugs; laboratory chemical.
Appearance: White, colorless or pale yellow solid flakes or liquid.
Odor: Characteristic, choking, acrid. The odor is a warning of exposure; however, no smell does not mean you are not being exposed.
Effects of exposure:
Breathed: May cause severe irritation to the nose, throat and lungs; coughing, shortness of breath; excessive discharge and bleeding from the nose. May cause flu-like symptoms including chills, muscle ache, headache, fever. High exposure may cause nausea, salavation, vomiting, cramps, diarrhea, chest pains, shortness of breath and a dangerous build-up of fluids in the lungs (pulmonary edema), which can cause death. Pulmonary edema is a medical emergency that can be delayed one to two days following exposure. Exposure may cause lung allergy to develop.
Eyes: May cause severe irritation, burns, pain and possible permanent damage.
Skin: May cause severe irritation and thermal and acid burns, especially if skin is wet. Exposure may cause skin allergy to develop.
Swallowed: May cause severe burns of the mouth, throat and stomach, vomiting, watery or bloody diarrhea. Damage to kidneys and liver, collapse and convulsions can result. Studies suggest that death may occur from swallowing four to eight ounces.
Long-term exposure: May cause skin and lung allergies to develop; coughing and wheezing, shortness of breath and chest tightness.
Storage: Store in tightly closed containers in a dry, cool, well-ventilated place. Keep away from heat, sources of ignition and strong oxidizers. Molten material should be kept below 300°F and vented to outside of building.
First aid guide: Move victim to fresh air; call emergency medical care. In case of contact with material, immediately flush skin or eyes with running water for at least 15 minutes. Remove and isolate contaminated clothing and shoes at the site. Keep victim quiet and maintain normal body temperature.

★784★
M-PHTHALODINITRILE
CAS: 626-17-5
Other names: 1,3-BENZENEDICARBONITRILE; M-DICYANOBENZENE; 1,3-DICYANOBENZENE; DINITRILE OF ISOPHTHALIC ACID; IPN; ISOPHTHALODINITRILE; ISOPHTHALONITRILE; ISOFTALODINITRIL (CZECH); NITRIL KYSELINY ISOFTALOVE (CZECH); M-PDN

Danger profile: Combustible, exposure limit established, Community Right-to-Know List, poisonous gas produced in fire.
Uses: Making other chemicals, pigments and dyes, high-temperature lubricants, and coatings and insecticides
Appearance: White flaky material; colorless crystals.
Odor: Almond-like. The odor is a warning of exposure; however, no smell does not mean you are not being exposed.
Effects of exposure:
Breathed: High exposure may cause headache, nausea, confusion and loss of consciousness.
Eyes: May cause irritation.
Skin: Passes through the unbroken skin; can increase exposure and the severity of the symptoms listed.
Swallowed: Poisonous.

Long-term exposure: May cause headaches, nausea, loss of memory, tremors of fingers, loss of appetite, facial pallor.
Storage: Store in tightly closed containers in a dry, cool, well-ventilated place. Keep away from sources of ignition and strong oxidizers. Dust is a severe explosion hazard.

★ 785 ★
PICLORAM
CAS: 1918-02-1
Other names: AMDON; AMDON GRAZON; 4-AMINO-3,5,6-TRICHLORO-ROPICOLINIC ACID; 4-AMINO-3,5,6-TRICHLORO-2-PICOLINIC ACID; 4-AMINO-3,5,6-TRICHLORPICOLINSAEURE (GERMAN); ATLADOX; ATLADOX HI; ATCP; BOROLIN; CHLORAMP (RUSSIAN); DIADON; GRAZON; HYDON; K-PIN; NCL-COO237; TORDON; TORDON 10K; TORDON 22K; TORDON 101 MIXTURE; PICOLINIC ACID, 4-AMINO-3,5,6-TRICHLORO-; 3,5,6-TRICHLORO-4-AMINOPICOLINIC ACID

Danger profile: Possible carcinogen, mutagen, combustible, exposure limit established, poisonous gas produced in fire.
Uses: Herbicide. Weed control on crop and non-crop land. Has been used as a forest defoliant and leaves toxic residue in soil. Use has been restricted. Harmful to fish.
Appearance: Colorless powder.
Odor: Chlorine-like. The odor is a warning of exposure; however, no smell does not mean you are not being exposed.
Effects of exposure:
Breathed: May cause irritation of the eyes, nose, throat and respiratory tract.
Eyes: May cause irritation.
Skin: May cause irritation.
Long-term exposure: May cause cancer; damage to the male reproductive glands.
Storage: Store in tightly closed containers in a dry, cool, well-ventilated place. Containers may explode in fire.

★ 786 ★
PICRIC ACID
CAS: 88-89-1
Other names: ACIDE PICRIQUE (FRENCH); ACIDO PICRICO (ITALIAN); CARBAZOTIC ACID; C.I. 10305; 2-HYDROXY-1,3,5-TRINITROBENZENE; LYDDITE; MELINITE; NITROXANTHIC ACID; PERTITE; PHENOL TRINITRATE; PICRONITRIC ACID; PIKRINEZUUR (DUTCH); PIKRINSAEURE (GERMAN); PIKRYNOWY KWAS (POLISH); SHIMOSE; 2,4,6-TRINITROFENOL (DUTCH); 2,4,6-TRINITOPHENOL

Danger profile: Mutagen, fire hazard, explosion hazard, poison, allergen, exposure limit established, Community Right-to-Know List, poisonous gas produced in fire.
Uses: Making dyes, explosives, colored glass, matches; processing leather; electric batteries; etching copper; in textile dyeing; as a drug; laboratory chemical.
Appearance: Yellow crystalline solid or yellow liquid.
Odor: Bitter. The odor is a warning of exposure; however, no smell does not mean you are not being exposed.
Effects of exposure:
Breathed: May cause irritation of the eyes, nose, throat and lungs, tears and runny nose, coughing, sore throat, difficult breathing; lethargy, fatigue, and drowsiness; slurred speech, shaking, and dizziness. Higher exposures could cause shortness of breath, headache, nausea, vomiting, dizziness, diarrhea and a dangerous build-up of fluids in the lungs (pulmonary edema), which can cause death. Pulmonary edema is a medical emergency that can be delayed one to two days following exposure.

Eyes: May cause irritation, redness, tearing, pain, inflamed eyelids, and eye injury; burns in corneas and possible loss of sight.
Skin: Passes through the unbroken skin; can increase exposure and the severity of the symptoms listed. May cause painful irritation, redness, and blisters Repeated or prolonged skin contact may cause irritation of the skin and may also cause the legs to feel numb. Exposure may cause skin allergy to develop.
Swallowed: May cause irritation of the mouth and throat; stomach and abdominal pain; loss of muscular coordination and mental confusion; convulsions. Also see symptoms for *Breathed.*
Long-term exposure: May cause mutations; skin allergies; liver and kidney damage; skin and hair may turn yellow.
Storage: Dangerous explosion hazard. Store in tightly closed containers in a dry, cool, well-ventilated place. Keep away from heat, flame and sources of ignition, bases, aluminum, oxidizing materials. Protect containers from shock and physical damage. Forms unstable salts with various metals concrete, ammonia and bases. These salts are sensitive explosives. Containers may explode in fire.
First aid guide: Move victim to fresh air; call emergency medical care. In case of contact with material, immediately flush skin or eyes with running water for at least 15 minutes. Remove and isolate contaminated clothing and shoes at the site.

★ 787 ★
PIPERAZINE
CAS: 110-85-0
Other names: ANTIREN; 1,4-DIETHYLENEDIAMINE; DISPERMINE; HEXAHYDRO-1,4-DIAZINE; HEXAHYDROPYRAZINE; LUMBRICAL; LUMBUCAL; PIPERAZIDINE; PIPERAZIN (GERMAN); PIPERAZINE, ANHYDROUS; PYRAZINE HEXAHYDRIDE; PYRAZINE, HEXAHYDRO-

Danger profile: Combustible, corrosive, poisonous gas produced in fire.
Uses: Making many products including pharmaceuticals, polymers, dyes, corrosion inhibitors, surfactants; insecticide.
Appearance: White or colorless crystalline solid.
Odor: Mild, fishy, amine-like. The odor is a warning of exposure; however, no smell does not mean you are not being exposed.
Effects of exposure:
Breathed: Dust may cause irritation of the nose and throat. Exposure may cause lung allergy to develop. High exposure may cause irritation of the breathing passages, shortness of breath, weakness, tremors, loss of coordination, epileptic seizures; interfere with the blood's ability to carry oxygen causing headaches, dizziness, bluish color to the skin, lips.
Eyes: Contact may cause severe irritation, burns and permanent damage.
Skin: Passes through the unbroken skin; can increase exposure and the severity of the symptoms listed. Contact may cause irritation, skin allergy and rash.
Swallowed: May cause irritation of mouth and stomach; has been known to cause severe allergic reaction.
Long-term exposure: Repeated contact with skin may cause irritation and skin and allergies to develop; possible lung problems.
Storage: Store in tightly closed containers in a dry, cool, well-ventilated place. Keep away from flame and other sources of ignition and oxidizers.
First aid guide: Move victim to fresh air; call emergency medical care. In case of contact with material, immediately flush skin or eyes with running water for at least 15 minutes. Remove and isolate contaminated clothing and shoes at the site. Keep victim quiet and maintain normal body temperature.

★788★
PIPERAZINE DIHYDROCHLORIDE
CAS: 142-64-3
Other names: DIHYDROCHLORIDE SALT OF DIETHYL-ENEDIAMINE; DHC; DOWZENE; DOWZENE DHC

Danger profile: Combustible, exposure limit established, poisonous gas produced in fire.
Uses: Making fibers, insecticides, rubber products and pharmaceuticals.
Appearance: White salt-like solid.
Effects of exposure:
Breathed: May cause irritation of the respiratory system and asthma-like lung allergy may develop; coughing, shortness of breath, difficult breathing.
Eyes: May cause irritation.
Skin: Passes through the unbroken skin; can increase exposure and the severity of the symptoms listed. May cause irritation; allergy may develop.
Swallowed: May cause burns to the throat and mouth.
Long-term exposure: May cause asthma-like lung allergy with cough, shortness of breath, difficult breathing; skin allergy with itching and rash.
Storage: Store in tightly closed containers in a dry, cool, well-ventilated place.

★789★
PIRIMICARB
CAS: 23103-98-2
Other names: ABOL G; AFICADA; APHOX; APHOX DISPERSIBLE GRAIN; CARBAMIC ACID, DIMETHYL-,2-(DIMETHYLAMINO)-5,6-DIM ETHYL-4-PYRIMIDYL ESTER2-(DIMETHYLAMINO)-5,6-DIMETHYL-4-PYRIMIDINYLDIMETHYLCARBAMAT E; 2-DIMETHYLAMINO-5,6-DIMETHYLPYRIMIDIN-4-YL DIMETHYLCARBAMATE; 5,6-DIMETHYL-2-DIMETHYLAMINO-4-PYRIMIDINYLDIMETHYLCARBAMATE; ENT-27766; FBC PIRIMICARB 50; FERNOS; MYSTIC GREENFLY KILLER; PIRIMOR; PP 062; PYRIMICARBE; PYRIMOR; RAPID

Danger profile: Mutagen, poisonous gas produced in fire.
Uses: Insecticide (carbamate); on apple, pear, cherry, strawberries, lettuce, cucumbers, tomatoes, peppers, potatoes, peas, cereals, ornamental plants.
Appearance: Crystalline material.
Effects of exposure:
Breathed: Extremely poisonous. May cause headache, dizziness, profuse sweating, nausea, vomiting, reduced heart beat, stomach cramps, diarrhea, loss of coordination, slow and weak breathing, fever, loss of consciousness and death. May cause shortness of breath and a dangerous build-up of fluids in the lungs (pulmonary edema), which can cause death. Pulmonary edema is a medical emergency that can be delayed one to two days following exposure.
Eyes: Rapidly absorbed. Poisoning can happen very rapidly. See *Breathed.*
Skin: Poisonous. Passes through the unbroken skin; can increase exposure and the severity of the symptoms listed.
Swallowed: Powerful, deadly poison.
Long-term exposure: May damage the developing fetus or cause birth defects; nerve damage including loss of coordination; liver damage. Repeated exposure may cause depression, irritability and personality changes.
Storage: Store in tightly closed containers in a dry, cool, well-ventilated place. Containers may explode in fire.

★790★
PLATINOUS POTASSIUM CHLORIDE
CAS: 10025-99-7
Other names: PLATINATE(2-),TETRACHLORO- ,DIPOTASSI-UM; PLATINATE(2-),TETRACHLORO- , DIPOTASSIUM (SP-4-1)-; POTASSIUM CHLOROPLATINITE; POTASSIUM PLA-TINOCHLORIDE; POTASSIUM TETRACHLOROPLATINATE(-II)

Danger profile: Mutagen, combustible, corrosive, exposure limit established, poisonous gas produced in fire.
Uses: In photography; making other chemicals.
Appearance: Ruby-red crystalline solid.
Effects of exposure:
Breathed: May cause irritation of the eyes, nose, throat and breathing passages. High levels may cause irritability; possible convulsions or fits.
Eyes: Contact may cause severe irritation and burns.
Skin: Corrosive. Contact may cause irritation and rash may develop.
Swallowed: Poison.
Long-term exposure: May cause mutations; lung scarring; asthma-like allergy with coughing, shortness of breath; allergic skin rash and hives may develop; sores and ulcers may develop on the inner lining of the nose.
Storage: Store in tightly closed containers in a dry, cool, well-ventilated place. Containers may explode in fire.

★791★
PLATINUM
CAS: 7440-06-4
Other names: C.I. 77795; LIQUID BRIGHT PLATINUM; PLATIN (GERMAN); PLATINIC CHLORIDE; PALLADIUM BLACK; PALLADIUM SPONGE

Danger profile: Combustible, exposure limit established.
Uses: Jewelry; electronic equipment; jewelry; catalytic converters; dentistry; electroplating; photography.
Appearance: Silver gray metal or crystalline solids.
Effects of exposure:
Breathed: Metal dust or fumes of salts may cause irritation to the nose, throat breathing passages and lungs; causing coughing, shortness of breath.
Eyes: Dust may cause severe irritation, burning feeling, tearing, scratches and possible damage.
Skin: May cause irritation and dermatitis may result.
Long-term exposure: Salts of platinum may cause may lead to asthma-like allergy, skin allergy.
Storage: Store in tightly closed containers in a dry, cool, well-ventilated place. Keep away from acetate, nitrosyl chloride, arsenic, dioxygen difluoride, ethanol, hydrozine, air, hydrogen peroxide, lithium, ozonides, methyl hydroperoxide, peroxymonosulfuric acid, phosphorus, selenium and tellurium, vanadium dichloride.

★ 792 ★
POLYCHLORINATED BIPHENYL
CAS: 1336-36-3

Other names: AROCLOR; AROCLOR 1221; AROCLOR 1232; AROCLOR 1242; AROCLOR 1248; AROCLOR 1254; AROCLOR 1260; AROCLOR 1262; AROCLOR 1268; ARO-CLOR 2565; AROCLOR 4465; BIPHENYL, POLYCHLORO-; CHLOPHEN; CHLOREXTOL; CHLORINATED BIPHENYL; CHLORINATED DIPHENYL; CHLORINATED DIPHENYLENE; CHLORO BIPHENYL; CHLORO 1,1-BIPHENYL; CLOPHEN; DYKANOL; FENCLOR; INERTEEN; KANECHLOR; KANE-CHLOR 300; KANECHLOR 400; KANECHLOR 500; MON-TAR; NOFLAMOL; PCB; PHENOCHLOR; PHENOCLOR; POLYCHLOROBIPHENYL; PYRALENE; PYRANOL; SANTO-THERM; SANTOTHERM FR; SOVOL; THERMINOL FR-1

Danger profile: Carcinogen, reproductive hazard, exposure limit established, reproductive hazard, Community Right-to-Know List, EPA extremely hazardous substances list; poisonous gas produced in fire.
Uses: In electrical components and systems; insulating and heat exchange fluids; hydraulic fluid; casting processes.
Appearance: A mixture of chemicals that are clear to light yellow oily liquids, or white crystalline powder.
Odor: Weak, practically odorless. The odor is a warning of exposure; however, no smell does not mean you are not being exposed.
Effects of exposure:
   Breathed: May cause irritation of the nose, throat and lungs; headache, nausea, coughing. High exposure may cause liver damage. Serious liver damage may result in coma and death.
   Eyes: May cause irritation and watering.
   Skin: Passes through the unbroken skin; can increase exposure and the severity of the symptoms listed. May cause an acne-like skin rash.
   Swallowed: May cause nausea, vomiting, irritation of the stomach; tiredness and fatigue.
Long-term exposure: May cause cancer; damage the adult reproductive system; acne-like skin rash; liver damage; nervous system damage. May be passed to a nursing child through mother's milk.
Storage: Store in tightly closed containers in a dry, cool, well-ventilated place. Keep away from oxidizers.
First aid guide: In case of contact with material, immediately flush eyes with running water for at least 15 minutes. Wash skin with soap and water. Remove and isolate contaminated clothing and shoes at the site.

★ 793 ★
POTASSIUM
CAS: 7440-09-7
Other names: POTASSIUM, METAL

Danger profile: Dangerous fire hazard, explosion hazard, highly reactive, poisonous gas produced in fire.
Uses: Making other chemicals; heat exchange alloys; laboratory chemical; in fertilizers.
Appearance: Soft silvery-white metal. Quickly tarnishes when freshly cut.
Effects of exposure:
   Breathed: Highly toxic. Dusts or mists may cause severe irritation of the nose, throat and breathing passages with burning feeling; difficult breathing, coughing and sneezing; sore throat. High exposure can lead to a buildup of fluid in the lungs (pulmonary edema), and may cause death. Pulmonary edema may be delayed by one to two days following exposure.
   Eyes: May cause severe irritation with pain in eyes and eyelids; intense watering; burns and damage to the mucous membrane, perforation of the eyes, permanent damage and blindness.
   Skin: Highly toxic. May cause severe irritation, itching, pain, ulceration with slippery feeling; corrosive, serious, deep chemical burns.
   Swallowed: Highly toxic. May cause immediate burning feeling in the mouth, throat and stomach; intense pain in swallowing; throat swelling, vomiting (hemorrhaging may cause vomitus to resemble coffee grounds), watery or bloody diarrhea, shallow respirations; confusion, delirium, coma; swelling of the throat may cause obstruction; perforation of esophagus or stomach, shock, convulsions and collapse and death can result.
Long-term exposure: Prolonged exposure may lead to sores of the inner nose; possible lung problems.
Storage: Store under mineral oil, nitrogen or kerosene in tightly closed containers in a dry, cool, well-ventilated place. Spontaneously combustible in moist air. Reacts violently with many other chemicals and materials. Keep away from heat, air, oxygen, sources of ignition, water, halogenated hydrocarbons, carbon monoxide, silver chloride, silver oxide, carbon tetrachloride and organic and combustible materials. Protect containers from physical damage. Containers may explode in fire.
First aid guide: Move victim to fresh air; call emergency medical care. Wipe material from skin immediately; flush skin or eyes with running water for at least 15 minutes. Remove and isolate contaminated clothing and shoes at the site.

★ 794 ★
POTASSIUM ARSENATE
CAS: 7784-41-0
Other names: ARSENIC ACID, MONOPOTASSIUM SALT; MACQUER'S SALT; MONOPOTASSIUM ARSENATE; MONO-POTASSIUM DIHYDROGEN ARSENATE; POTASSIUM ACID ARSENATE; POTASSIUM ARSENATE, MONOBASIC; PO-TASSIUM DIHYDROGEN ARSENATE; POTASSIUM HYDRO-GEN ARSENATE

Danger profile: Carcinogen, mutagen, poison, exposure limit established, reproductive hazard, Community Right-to-Know List, poisonous gas produced in fire and on contact with acids.
Uses: Making fly paper; insecticide, tanning, paper and textiles industries.
Appearance: Colorless to white powder.
Effects of exposure:
   Breathed: Dust is poisonous. May cause irritation of the nose, throat; coughing, chest pains, and mucous membranes, fever, insomnia, loss of appetite, liver swelling, disturbed heart function. Also see *Swallowed.*
   Eyes: Dust may cause irritation.
   Skin: Passes through the unbroken skin; can increase exposure and the severity of the symptoms listed. Contact may cause irritation; burning and itching; skin color changes.
   Swallowed: Poisonous. May cause burning of the throat and mouth, abdominal pain, vomiting, diarrhea with bleeding, dehydration, jaundice, and collapse.
Long-term exposure: May cause cancer; mutations; thickened skin and color change; ulcer in the nose; nerve damage with numbness in the arms and legs; loss of appetite, nausea, cramps, vomiting
Storage: Store in tightly closed containers in a dry, cool, well-ventilated place. Keep away from acids.
First aid guide: Move victim to fresh air; call emergency medical care. In case of contact with material, immediately flush skin or eyes with running water for at least 15 minutes. Remove and isolate contaminated clothing and shoes at the site.

★795★
POTASSIUM BROMATE
CAS: 7758-01-2
Other names: BROMIC ACID, POTASSIUM SALT

Danger profile: Carcinogen, mutagen, poisonous gas produced in fire.
Uses: Laboratory chemical; oxidizing agent; permanent wave compounds; in flour; food additive.
Appearance: White crystalline powder.
Effects of exposure:
   Breathed: May cause irritation of the nose, throat and bronchial tubes.
   Eyes: Contact may cause irritation. Prolonged contact may cause serious irritation and burns.
   Skin: Passes through the unbroken skin; can increase exposure and the severity of the symptoms listed. Contact may cause irritation. Prolonged contact may cause serious irritation and burns.
   Swallowed: Ingestion may cause vomiting, diarrhea, lowers the ability of the blood to carry oxygen (methemoglobinemia) with bluish color to the skin and lips, collapse and even death; renal injury.
Long-term exposure: May cause cancer; kidney problems; brain damage, personality change; lung problems possible.
Storage: Store in tightly closed containers in a dry, cool, well-ventilated place without wooden floors. Keep away from combustible, organic or other readily oxidizable materials, aluminum, metal sulfides, copper, potassium, selenium. Protect containers from physical damage. Containers may explode in fire.
First aid guide: Move victim to fresh air; call emergency medical care. In case of contact with material, immediately flush skin or eyes with running water for at least 15 minutes. Remove and isolate contaminated clothing and shoes at the site.

★796★
POTASSIUM CARBONATE
CAS: 584-08-7
Other names: CARBONIC ACID, DIPOTASSIUM SALT; KALIUMCARBONAT (GERMAN); K-GRAN; PEARL ASH; POTASH; POTASSIUM CARBONATE (2:1)

Danger profile: Combustible, poisonous gas produced in fire.
Uses: Making other chemicals, pigments, printing inks, soaps, television tubes and other specialized glass products; food additive; laboratory chemical; washing raw wool
Appearance: White granular, translucent powder.
Effects of exposure:
   Breathed: May cause irritation of nose, eyes and throat; coughing; difficult breathing.
   Skin: Contact with wet skin may cause irritation, redness and swelling; burning feeling.
   Eyes: Contact may cause irritation, pain, tearing; serious injury may result.
   Swallowed: May cause burning feeling in the mouth and stomach; mouth and gastric irritation; stomach cramps.
Storage: Store in tightly closed containers in a dry, cool, well-ventilated place. Keep away from magnesium and chlorine trifluoride.

★797★
POTASSIUM CHLORATE
CAS: 3811-04-9
Other names: BERTHOLLET'S SALT; CHLORATE DE POTASSIUM (FRENCH); CHLORATE OF POTASH; CHLORIC ACID, POTASSIUM SALT; FEKABIT; KALIUMCHLORAAT (DUTCH); KALIUMCHLORAT (GERMAN); OXYMURIATE OF POTASH; PEARL ASH; POTASH CHLORATE; POTASSIO (CHLORATODI) (ITALIAN); POTASSIUM (CHLORATE DE) (FRENCH); POTASSIUM OXYMURIATE; POTCRATE; SALT OF TARTER

Danger profile: Combustible, poisonous gas produced in fire.
Uses: Making soap, glass, pottery; in explosives, blasting caps and fireworks, matches; textile printing; disinfectants and bleaching.
Appearance: Transparent colorless crystal or white powder.
Effects of exposure:
   Breathed: May cause irritation of the nose and throat; sneezing and coughing, sore throat; small ulcerations of the nose. Exposure lowers the ability of the blood to carry oxygen with bluish lips and skin, headache, dizziness, collapse and even death.
   Eyes: Contact may cause severe irritation, inflammation of the eyelids, burns and permanent damage.
   Skin: Contact may cause severe irritation and burns. Ulcerations may be delayed.
   Swallowed: May cause stomach pain, nausea, vomiting, diarrhea, bluish face and hands, rapid breathing, dizziness, general weakness and collapse. Also see *Breathed.*
Long-term exposure: May cause nose sores; lung problems are possible; kidney and heart muscle damage; destruction of the red blood cells.
Storage: Store in tightly closed containers in a dry, cool, well-ventilated place with non-wooden floors. Keep away from combustible, organic and other readily oxidizable materials, oxidizers, acids, ammonium salts, sulfur and other chemicals. Protect containers from physical damage. Containers may explode in fire.
First aid guide: Move victim to fresh air; call emergency medical care. In case of contact with material, immediately flush skin or eyes with running water for at least 15 minutes. Remove and isolate contaminated clothing and shoes at the site.

★798★
POTASSIUM CHLORIDE
CAS: 7447-40-7
Other names: CHLORIC DRASELNY (CZECH); CHLOROPOTASSURIL; DIPOTASSIUM DICHLORIDE; EMPLETS POTASSIUM CHLORIDE; ENSEAL; KALITABS; KAOCHLOR; KAON-CL; KAON-CL TABS; KAY CIEL; K-LOR; KLOTRIX; K-PREDNE-DOME; PFIKLOR; POTASSIUM MONOCHLORIDE; POTAVESCENT; REKAWAN; SLOW-K; TRIPOTASSIUM TRICHLORIDE

Danger profile: Mutagen, combustible, poisonous gas produced in fire.
Uses: Fertilizer; in pharmaceutical preparations; photography; plant nutrient; salt substitute; laboratory chemical; food additive.
Appearance: Colorless or white crystals or powder.
Effects of exposure:
   Breathed: May cause irritation of the nose; sneezing and coughing.
   Eyes: May cause irritation, tearing and inflammation of the eyelids.
   Skin: May cause irritation, inflammation and sores.
   Swallowed: Poison. May cause nausea and vomiting; gastrointestinal irritation, fever, rapid breathing, decreased heart beat and possible death.

Long-term exposure: May cause mutations.
Storage: Store in tightly closed containers in a dry, cool, well-ventilated place.

★799★
POTASSIUM CHROMATE
CAS: 7789-00-6
Other names: BIPOTASSIUM CHROMATE; CHROMIC ACID, DIPOTASSIUM SALT; CHROMATE OF POTASSIUM; DI-POTASSIUM CHROMATE; DIPOTASSIUM MONOCHRO-MATE; NEUTRAL POTASSIUM CHROMATE; POTASSIUM CHROMATE (VI); TARAPACAITE

Danger profile: Carcinogen, mutagen, exposure limit established, Community Right-to-Know List, poisonous gas produced in fire.
Uses: Laboratory chemical; making dyes, pigments and inks; textile mordant; enamels.
Appearance: Yellow crystalline solid. Used in solution.
Effects of exposure:
  Breathed: Dust may cause severe irritation to the nose, throat and lungs, causing coughing, shortness of breath. May cause flu-like symptoms including chills, muscle ache, headache, fever. High exposure may cause nausea, salivation, vomiting, cramps, diarrhea, chest pains, cough, a build-up of fluids in the lungs (pulmonary edema) and possible death. Pulmonary edema may be delayed from one to two days following exposure.
  Eyes: May cause severe irritation, inflammation of the eyelids, burns, pain and possible blindness.
  Skin: Passes through the unbroken skin; can increase exposure and the severity of the symptoms listed. May cause severe irritation and thermal and acid burns, especially if skin is wet. Dermatitis or ''chrome sores'' may develop, with slow-healing, hard-rimmed ulcers which leave the infected area vulnerable to infection.
  Swallowed: May cause severe burns of the mouth, throat and stomach, vomiting, vertigo, watery or bloody diarrhea, coma and death. Damage to kidneys and liver, collapse and convulsions can result.
Long-term exposure: May cause lung and throat cancer; mutations; perforation of the nasal septum; liver and kidney damage; dermatitis.
Storage: Store in tightly closed containers in a dry, cool, well-ventilated place with non-wooden floors. Keep away from combustible or organic material.

★800★
POTASSIUM CYANIDE
CAS: 151-50-8
Other names: CYANIDE OF POTASSIUM; CYANURE DE POTASSIUM (FRENCH); HYDROCYANIC ACID, POTASSIUM SALT; KALIUM-CYANID (GERMAN); POTASSIUM CYANIDE SOLUTION

Danger profile: Mutagen, combustible, possible reproductive hazard, poison, corrosive, exposure limit established, Community Right-to-Know List, poisonous gas produced in fire.
Uses: Extraction of gold and silver from ores; electroplating; laboratory chemical; insecticide; hardening steel; fumigant.
Appearance: White solid crystals.
Odor: Faint almond-like. The odor is a warning of exposure; however, no smell does not mean you are not being exposed.
Effects of exposure:
  Breathed: Poisonous. High exposure may cause sudden death. May cause headache, eye, nose and throat irritation, wheezing, foaming at the mouth, general irritability, nausea; chest tightness, shortness of breath, vomiting blood, respira-

tory depression, weakness, chest and abdominal pains, dizziness, diarrhea, tremors, convulsions, shock, pounding heart, unconsciousness, weak and irregular pulse and death. Reactions may appear hours following overexposure.
  Eyes: May cause severe irritation, tearing and chemical burns.
  Skin: Poisonous. Passes through the unbroken skin; can increase exposure and the severity of the symptoms listed. May cause severe irritation, redness, stinging and blisters.
  Swallowed: May cause severe irritation and chemical burns to the tongue, mouth, throat and stomach as well as many of the symptoms described under **Breathed.**
Long-term exposure: May cause enlarged thyroid; nose sores and bleeding; in high doses may decrease fertility in women and men.
Storage: Store in tightly closed containers in a dry, cool, well-ventilated place. Keep away from light, water, steam and other forms of moisture, acids and oxidizing and materials. Contact with even weak acids causes formation of deadly hydrogen cyanide gas. Containers may explode in fire.
First aid guide: Move victim to fresh air and call emergency medical care; if not breathing, give artificial respiration; if breathing is difficult, give oxygen. In case of contact with material, immediately flush skin or eyes with running water for at least 15 minutes. Speed in removing material from skin is of extreme importance. Remove and isolate contaminated clothing and shoes at the site. Keep victim quiet and maintain normal body temperature. Effects may be delayed; keep victim under observation.

★801★
POTASSIUM DICHLOROISOCYANURATE
CAS: 2244-21-5
Other names: 1,3-DICHLORO-S-TRIAZINE-2,4,6(1H,3H,5H)TRIONE POTASSIUM SALT; DICHLOROISO-CYANURIC ACID POTASSIUM SALT; DICHLORO-S-TRIAZINE-2,4,6(1H,3H,5H)-TRIONE POTASSIUM DERIV; DI-CHLOR-S-TRIAZIN-2,4,6(1H,3H,5H)TRIONE POTASSIUM; ISOCYANURIC ACID,DICHLORO-, POTASSIUM SALT; PO-TASSIUM DICHLORO ISOCYANURATE; POTASSIUM DI-CHLORO-S-TRIAZINETRIONE; POTASSIUM TROCLOSENE; 1,3,5-TRIAZINE-2,4,6(1H,3H,5H)-TRIONE, 1,3-DICHLORO-, POTASSIUM SALT; S-TRIAZINE-2,4,6(1H,3H,5H)-TRIONE,DICHLORO-, POTASSIUM DERIV; TROCLOSENE POTASSIUM

Danger profile: Combustible, poisonous gas produced in fire.
Uses: In dish washing compounds, detergents, scouring powders, and household dry bleaches; replacement for calcium hypochlorite.
Appearance: White crystalline powder or granules.
Odor: Chlorine-like. The odor is a warning of exposure; however, no smell does not mean you are not being exposed.
Effects of exposure:
  Breathed: May cause irritation of the nose, throat and breathing passages.
  Eyes: May cause severe irritation.
  Skin: May cause irritation.
  Swallowed: Moderately toxic.
Long-term exposure: Irritant in dry form. May cause weight loss, weakness, lethargy, diarrhea. Autopsy indicates gastrointestinal tract irritation, tissue edema, liver and kidney congestion. Similar irritating substances may cause lung problems.
Storage: Store in tightly closed containers in a dark, dry, cool, well-ventilated place. Keep away from water, steam and other forms of moisture, combustible materials, ammonia, urea, nitrogen compounds, calcium hypochlorite and alkalis. Avoid storage for extended period in summer.
First aid guide: Move victim to fresh air and call emergency medical care; if not breathing, give artificial respiration; if breath-

ing is difficult, give oxygen. In case of contact with material, immediately flush skin or eyes with running water for at least 15 minutes. Remove and isolate contaminated clothing and shoes at the site. Keep victim quiet and maintain normal body temperature. Effects may be delayed; keep victim under observation.

★ 802 ★
POTASSIUM DICHROMATE
CAS: 7778-50-9
Other names: BICHROMATE OF POTASH; CHROMIC ACID, DIPOTASSIUM SALT; DICHROMIC ACID, DIPOTASSIUM SALT; DIPOTASSIUM DICHROMATE; DIPOTASSIUM DICHROMATE; IOPEZITE; KALIUMDICHROMAT (GERMAN); POTASSIUM BICHROMATE; POTASSIUM DICHROMATE (VI); RED CHROMATE OF POTASH

Danger profile: Carcinogen, mutagen, combustible, exposure limit established, Community Right-to-Know List, poisonous gas produced in fire.
Uses: Oxidizing agent; laboratory chemical; electroplating, fireworks, explosives, safety matches; textiles; dyeing and printing; leather tanning; wool processing; wood stains; fly paper; printing industries; perfumes; pigments; alloys; ceramic products.
Appearance: Bright yellowish-red or orange solid crystals.
Effects of exposure:
　Breathed: Dust may cause severe irritation to the nose, throat and lungs, causing coughing, shortness of breath. May cause flu-like symptoms including chills, muscle ache, headache, fever. High exposure may cause nausea, salivation, vomiting, cramps, diarrhea, chest pains, cough, a build-up of fluids in the lungs (pulmonary edema) and possible death. Pulmonary edema may be delayed from one to two days following exposure.
　Eyes: Highly corrosive. May cause severe irritation, burns, pain and possible blindness.
　Skin: Highly corrosive. May cause severe irritation and thermal and acid burns, especially if skin is wet. Passes into the body through the affected area of the skin; can increase exposure and the severity of the symptoms listed. Allergic skin rash may develop.
　Swallowed: Bitter metallic taste. May cause severe burns of the mouth, throat and stomach; violent gastrointestinal problems; muscle cramps, vomiting, watery or bloody diarrhea, coma. Damage to kidneys and liver, collapse and convulsions can result. **Long-term exposure:** May cause throat and lung cancer; lung allergy; irritation of the bronchial tubes, shortness of breath, cough and phlegm; skin allergy with rash; nose bleeding and development of a hole in the bone dividing the nostrils.
Storage: Store in tightly closed containers in a dry, cool, well-ventilated place with non-wood floors. Keep away from organic and oxidizable substances, hydrazine, sulfuric acid, acetone, hydroxylamine, organic mercury, slaked lime, and cyanide.
First aid guide: Move victim to fresh air; call emergency medical care. In case of contact with material, immediately flush skin or eyes with running water for at least 15 minutes. Remove and isolate contaminated clothing and shoes at the site.

★ 803 ★
POTASSIUM FLUORIDE
CAS: 7789-23-3
Other names: FLUORURE DE POTASSIUM (FRENCH); KF; POTASSIUM FLUORIDE SOLUTION; POTASSIUM FLUORURE (FRENCH)

Danger profile: Mutagen, combustible, poison, highly corrosive, exposure limit established, poisonous gas produced in fire.

Uses: Etching glass; preservative; making other chemicals; insecticide; solder flux.
Appearance: White crystalline powder.
Effects of exposure:
　Breathed: Poisonous. May cause irritation of the nose, eyes, mouth and throat; sore throat, watering eyes, salivation, coughing, throat pain, difficult breathing, headache, fatigue, nausea, stomach pains. High exposures may cause shortness of breath and a dangerous build-up of fluids in the lungs (pulmonary edema), which can cause death. Pulmonary edema is a medical emergency that can be delayed one to two days following exposure.
　Eyes: May cause intense stinging, irritation, severe burns and irreparable damage to the cornea and possible loss of sight.
　Skin: May cause irritation, redness, pain, skin gradually turns white and painful swelling and severe burns; shock.
　Swallowed: Poisonous. Sharp salty taste. May cause irritation; burning feeling in mouth and throat; swallowing causes pain; intense thirst; throat swelling, nausea, vomiting, shock. See *Breathed* for other symptoms.
Long-term exposure: It should be noted that these effects do not occur at the levels used in water systems used for cavity prevention. Repeated exposure may cause fluoride to build up in the body; stiffness, brittle bones and eventual crippling; discoloration of the teeth; kidney problems; loss of appetite, nausea, vomiting; nose and sinus problems.
Storage: Store in tightly closed containers in a dry, cool, well-ventilated place. Keep away from acetic substances, especially strong acids.
First aid guide: Move victim to fresh air; call emergency medical care. In case of contact with material, immediately flush eyes with running water for at least 15 minutes. Wash skin with soap and water. Remove and isolate contaminated clothing and shoes at the site.

★ 804 ★
POTASSIUM FLUOROSILICATE
CAS: 16871-90-3
Other names: POTASSIUM FLUOSILICATE; POTASSIUM HEXAFLUOROSILICATE; POTASSIUM SILICOFLUORIDE; SILICATE(2-), HEXAFLUORO-, DIPOTASSIUM

Danger profile: Exposure limit established, poisonous gas produced in fire.
Uses: Vitreous enamel frits; synthetic mica; making ceramics and glass; insecticide.
Appearance: White, fine powder or crystals.
Effects of exposure:
　Breathed: May be poisonous. May cause irritation of the nose, eyes, mouth and throat; congestion, watering eyes; salivation, coughing, throat pain, difficult breathing, headache, fatigue, nausea, stomach pains; pale or bluish face, pain in swallowing, intense thirst. High exposures may cause shortness of breath and a dangerous build-up of fluids in the lungs (pulmonary edema), which can cause death. Pulmonary edema is a medical emergency that can be delayed one to two days following exposure.
　Eyes: May cause intense stinging, irritation, severe burns and irreparable damage and possible loss of sight.
　Skin: May cause irritation, redness, swelling and severe burns.
　Swallowed: May be poisonous. May cause irritation; burning feeling in mouth and throat; swallowing causes pain; intense thirst; throat swelling, nausea, vomiting, diarrhea, shock. See *Breathed* for other symptoms.
Long-term exposure: A strong irritant. Prolonged or repeated exposure to similar chemicals may cause lung problems.

Storage: Store in tightly closed containers in a dry, cool, well-ventilated place. Keep away from hydrofluoric acid.
First aid guide: Move victim to fresh air; call emergency medical care. In case of contact with material, immediately flush skin or eyes with running water for at least 15 minutes. Remove and isolate contaminated clothing and shoes at the site.

★805★
POTASSIUM HEXACHLOROPLATINATE(IV)
CAS: 16921-30-5
Other names: DIPOTASSIUM HEXACHLOROPLATINATE (2-); PLATINIC POTASSIUM CHLORIDE; PLATINATE(2-),HEXACHLORO-, DIPOTASSIUM; POTASSIUM CHLOROPLATINATE; POTASSIUM HEXACHLOROPLATINATE; POTASSIUM PLATINIC CHLORIDE

Danger profile: Possible mutagen, exposure limit established, poisonous gas produced in fire.
Uses: In photography.
Appearance: Yellow crystals, orange-yellow crystals or yellow powder.
Effects of exposure:
  Breathed: May cause irritation of the eyes, nose, throat and breathing passages. High exposures may cause irritability, convulsions and fits. Allergy may develop.
  Eyes: May cause irritation and burning feeling.
  Skin: May cause irritation and skin allergy with hives may develop.
Long-term exposure: May cause severe allergy with hay fever-like symptoms and lung scarring; sores in the nose lining; allergic skin rash.
Storage: Store in tightly closed containers in a dry, cool, well-ventilated place.

★806★
POTASSIUM HYDROXIDE
CAS: 1310-58-3
Other names: CAUSTIC POTASH; HYDROXIDE DE POTASSIUM (FRENCH); KALIUMHYDROXID (GERMAN); KALIUMHYDROXYDE (DUTCH); KOH; LYE; POTASSA; POTASSE CAUSTIQUE (FRENCH); POTASSIO (IDROSSIDO DI) (ITALIAN); POTASSIUM HYDRATE; POTASSIUM (HYDRIXYDE DE) (FRENCH)

Danger profile: Mutagen, combustible, highly corrosive, exposure limit established, poisonous gas produced in fire.
Uses: Making soaps, glass; in alkaline storage batteries; fertilizers; food additive; herbicide; electroplating; paint stripper; making other chemicals and in many chemical manufacturing processes.
Appearance: White solid or clear liquid solution.
Effects of exposure:
  Breathed: May cause irritation of the eyes, nose and throat with burning feeling; difficult breathing, coughing and sneezing. High exposure can lead to a buildup of fluid in the lungs (pulmonary edema), and may cause death. Pulmonary edema may be delayed by one to two days following exposure.
  Eyes: May cause severe irritation with pain in eyes and eyelids; intense watering; irritation of the eyelids; burns and damage to the mucous membrane, perforation of the eyes, permanent damage and blindness.
  Skin: May cause irritation, itching, pain, ulceration with slippery feeling; serious, deep chemical burns.
  Swallowed: May cause immediate burning in the mouth, throat and stomach; intense pain in swallowing, throat swelling, nausea, vomiting (hemorrhaging may cause vomitus to resemble coffee grounds), watery or bloody diarrhea, shallow

respirations; confusion, delirium, coma; swelling of the throat may cause obstruction; perforation of esophagus or stomach, shock, convulsions and collapse and death can result.
Long-term exposure: May cause mutations; sores in the lining of the nose; possible lung problems.
Storage: Store in tightly closed containers in a dry, cool, well-ventilated place with non-wooden floors. Keep away from water, steam and other forms of moisture, metals such as aluminum, tin, lead and zinc, acids, alcohols, combustible materials, explosive products, sugars, organic peroxides and other chemicals. Chemical can absorb water from the air and give off enough heat to ignite combustible materials.
First aid guide: Move victim to fresh air; call emergency medical care. In case of contact with material, immediately flush skin or eyes with running water for at least 15 minutes. Remove and isolate contaminated clothing and shoes at the site. Keep victim quiet and maintain normal body temperature.

★807★
POTASSIUM NITRITE
CAS: 7758-09-0
Other names: NITROUS ACID, POTASSIUM SALT; POTASSIUM NITRITE (1:1)

Danger profile: Mutagen, reproductive hazard, combustible, poisonous gas produced in fire.
Uses: In medications; fertilizers; laboratory chemical; food additive.
Appearance: White to yellow granules, or rod, prism or stick shaped materials.
Effects of exposure:
  Breathed: May cause irritation of the nose, throat, breathing passages and lungs; coughing, phlegm. High exposures may cause headache, weakness, dizziness, bluish color to the lips and skin; shortness of breath and a dangerous build-up of fluids in the lungs (pulmonary edema), which can cause collapse and death. Pulmonary edema is a medical emergency that can be delayed one to two days following exposure.
  Eyes: Contact may cause severe irritation and burns.
  Skin: Contact may cause severe irritation and burns.
  Swallowed: Poisonous.
Long-term exposure: May cause mutations and damage the developing fetus; skin dryness and cracking; possible lung problems.
Storage: Store in tightly closed containers in a dry, cool, well-ventilated place. Keep away from heat, combustable materials, fuels, boron, ammonium sulfate and other ammonium salts, potassium amide, and cyanide salts.
First aid guide: Move victim to fresh air; call emergency medical care. In case of contact with material, immediately flush skin or eyes with running water for at least 15 minutes. Remove and isolate contaminated clothing and shoes at the site.

★808★
POTASSIUM OXIDE
CAS: 12136-45-7

Danger profile: Combustible, corrosive, poisonous gas produced in fire.
Uses: Making other chemicals.
Appearance: Gray crystalline mass.
Effects of exposure:
  Breathed: May cause irritation of the eyes, nose and throat with burning feeling; difficult breathing, coughing and sneezing. High exposure can lead to a buildup of fluid in the lungs (pulmonary edema), and may cause death. Pulmonary

edema may be delayed by one to two days following exposure.

Eyes: May cause severe highly painful irritation; intense watering; burns and damage to the mucous membrane, perforation of the eyes, permanent damage and blindness.

Skin: May cause irritation, itching, pain, ulceration with slippery feeling; serious, deep chemical burns.

Swallowed: May cause intense burning feeling in the mouth, throat and stomach; intense pain in swallowing; throat swelling, vomiting (hemorrhaging may cause vomitus to resemble coffee grounds), watery or bloody diarrhea, shallow respirations; confusion, delirium, coma; swelling of the throat may cause obstruction; perforation of esophagus or stomach, shock, convulsions and collapse and death can result.

Long-term exposure: May cause lung problems.

Storage: Store in tightly closed containers in a dry, cool, well-ventilated place.

First aid guide: Move victim to fresh air; call emergency medical care. In case of contact with material, immediately flush skin or eyes with running water for at least 15 minutes. Remove and isolate contaminated clothing and shoes at the site. Keep victim quiet and maintain normal body temperature.

★ 809 ★
POTASSIUM PERCHLORATE
CAS: 7778-74-7

Other names: ASTRUMAL; IRENAL; IRENAT; PERIODIN; POTASSIUM HYPERCHLORIDE; PERCHLORIC ACID, POTASSIUM SALT (1:1)

Danger profile: Combustible, explosion hazard, reproductive hazard, poisonous gas produced in fire.

Uses: In explosives, flares, solid rocket propellants; oxidizing agent; photography; pyrotechnics an flares; laboratory chemical.

Appearance: Colorless to white crystalline solid.

Effects of exposure:

Breathed: Vapor may cause intense irritation to nose, throat, mouth and lungs; mucous membrane. Symptoms of exposure include watering of the eyes, coughing, difficult breathing, headache, nausea, vomiting, muscular weakness, unconsciousness. High exposure can cause reduced oxygen to the body organs causing blue lips and skin; drowsiness; buildup of fluid in the lungs (pulmonary edema), and may cause death. Pulmonary edema may be delayed by one to two days following exposure.

Eyes: Severe irritant. Contact may cause, burning and stinging, tearing, irritation and redness and swelling of the eyelids, intense pain; can burn the eyes and lead to blurred vision, chronic irritation and other permanent damage.

Skin: Severe irritant. Contact may cause burning or stinging feeling, redness, swelling, intense pain, blisters, irritation, rash. Repeated exposure may lead to skin allergy and chronic irritation.

Swallowed: May cause burning feeling of the mouth and throat, intense thirst, throat swelling, stomach cramps, nausea, vomiting, diarrhea, severe breathing difficulties and unconsciousness. Tissue damage may result. Breathing may stop. Watch for delayed symptoms.

Long-term exposure: May cause reproduction damage; kidney damage; possible lung problems. Has been implicated in aplastic anemia; reduction of oxygen to body organs with blue lips and skin (methemoglobinemia) and kidney injury.

Storage: Store in tightly closed containers in a dry, cool, well-ventilated place. Keep away from heat, oxidizers, powdered metals, ethyl alcohol, combustible materials, organic materials, charcoal, reducing agents and sulfur. Containers may explode in fire.

First aid guide: Move victim to fresh air; call emergency medical care. In case of contact with material, immediately flush skin or eyes with running water for at least 15 minutes. Remove and isolate contaminated clothing and shoes at the site.

★ 810 ★
POTASSIUM PERMANGANATE
CAS: 7722-64-7

Other names: CAIROX; CHAMELEON MINERAL; C.I. 77755; CONDY'S CRYSTALS; KALIUMPERMANGANAAT (DUTCH); KALIUMPERMANGANAT (GERMAN); PERMANGANATE DE POTASSIUM (FRENCH); PERMANGANATE OF POTASH; POTASSIO (PERMANGANATO DI) (ITALIAN); POTASSIUM (PERMANGANATE DE) (FRENCH); PURPLE SALT

Danger profile: Mutagen, combustible, explosion hazard, exposure limit established. Community Right-to-Know List, poisonous gas produced in fire.

Uses: In solution as a disinfectant, deodorizer, bleaching agent; making other chemicals; dye; tanning; laboratory chemicals; medical antiseptic; air and water purification.

Appearance: Dark purple crystalline solid.

Effects of exposure:

Breathed: May cause irritation of the nose, throat and respiratory tract; coughing and difficult breathing. High exposures may cause shortness of breath and a dangerous build-up of fluids in the lungs (pulmonary edema), which can cause death. Pulmonary edema is a medical emergency that can be delayed one to two days following exposure.

Eyes: May cause severe irritation and damage.

Skin: May cause irritation and burns. Skin may harden and stain skin dark brown.

Swallowed: Sweet, astringent taste. Weak solutions may cause burning of the throat, nausea, vomiting and stomach pain; stronger solutions may cause swelling of the throat with a possibility of suffocation, severe distress of the gastrointestinal system; concentrated solutions may cause the above symptoms plus the onset of kidney damage and circulatory collapse.

Long-term exposure: May cause genetic changes in living cells; possible lung problems.

Storage: Store in tightly closed containers in a dry, cool, well-ventilated place with non-wooden floors. Keep away from combustible or oxidizable materials, organic materials, acids, hydrogen peroxide, ethylene glycol, and other chemicals. Chemical attacks most rubbers and fibers. Protect storage containers from physical damage. Containers may explode in fire.

First aid guide: Move victim to fresh air; call emergency medical care. In case of contact with material, immediately flush skin or eyes with running water for at least 15 minutes. Remove and isolate contaminated clothing and shoes at the site.

★ 811 ★
POTASSIUM PERSULFATE
CAS: 7727-21-1

Other names: ANTHION; DIPOTASSIUM PERSULFATE; POTASSIUM PEROXYDISULPHATE; PEROXYDISULFURIC ACID, DISODIUM SALT

Danger profile: Combustible, poisonous gas produced in fire.

Uses: Bleaching; oxidizing agent; in photography; antiseptic; laboratory chemical; pharmaceuticals; making soap, flour and textiles.

Appearance: White or colorless crystalline material.

Effects of exposure:

Breathed: Dust or mist may cause irritation of the nose and throat; sneezing, coughing and sore throat; coughing and sneezing. Prolonged exposure may cause sores on the inner lining of the nose. High exposures may cause shortness of breath and a dangerous build-up of fluids in the lungs (pulmonary edema), which can cause death. Pulmonary edema is a medical emergency that can be delayed one to two days following exposure.

Eyes: May cause severe irritation, burns and permanent damage.

Skin: May cause severe irritation and burns. Rash may develop.

Long-term exposure: Repeated exposures may cause sores to develop on the inner lining of the nose. May cause skin drying and cracking; rash; possible lung problems.

Storage: Store in tightly closed containers in a dry, cool, well-ventilated place. Keep away from heat, moisture, oxidizers, chemically active metals.

First aid guide: Move victim to fresh air; call emergency medical care. In case of contact with material, immediately flush skin or eyes with running water for at least 15 minutes. Remove and isolate contaminated clothing and shoes at the site.

★812★
POTASSIUM SULFIDE
CAS: 1312-73-8
Other names: HEPAR SULFUROUS; POTASSIUM MONO-SULFIDE; POTASSIUM SULFIDE (2:1)

Danger profile: Combustible, corrosive, poisonous gas produced in fire.
Uses: Laboratory chemical; depilatory; pharmaceutical preparations.
Appearance: Brownish-red solid.
Effects of exposure:
Breathed: May cause irritation of the nose, throat and breathing passages; coughing, sneezing and sore throat. Prolonged exposure may cause sores on the inner lining of the nose. High exposures may cause shortness of breath and a dangerous build-up of fluids in the lungs (pulmonary edema), which can cause death. Pulmonary edema is a medical emergency that can be delayed one to two days following exposure.
Eyes: Contact may cause severe irritation, burns and permanent damage.
Skin: Contact may cause severe irritation and burns.
Swallowed: Poisonous. See **First aid guide.**
Long-term exposure: Repeated exposures may cause sores to develop on the inner lining of the nose. May cause lung problems.
Storage: Store in tightly closed containers in a dry, cool, well-ventilated place. Keep away from moisture, sources of ignition, nitrogen oxide, oxidizers, and acids.
First aid guide: Move victim to fresh air; call emergency medical care. In case of contact with material, immediately flush skin or eyes with running water for at least 15 minutes. Removal of solidified molten material from skin requires medical assistance. Remove and isolate contaminated clothing and shoes at the site.

★813★
POTASSIUM ZINC CHROMATE
CAS: 11103-86-9
Other names: BUTTERCUP YELLOW; CHROMATE(1-), HY-DROXYOCTA OXODIZINCATEDI-, POTASSIUM; CHROMIC ACID,POTASSIUM ZINC SALT (2:2:1); CITRON YELLOW; POTASSIUM ZINC CHROMATE HYDROXIDE; ZINC POTAS-SIUM CHROMATE; ZINC YELLOW

Danger profile: Carcinogen, regulated by OSHA, Community Right-to-Know List, poisonous gas produced in fire.

Uses: In metal paints as a rust inhibiting pigment; artist's color.
Appearance: Yellow powder.
Effects of exposure:
Breathed: Higher levels may cause irritation of the nasal passages and throat.
Eyes: May cause irritation.
Skin: Passes through the unbroken skin; can increase exposure and the severity of the symptoms listed. May cause irritation and sores or ulcers. Allergy may develop.
Long-term exposure: May cause cancer; nose sores and bleeding; hole in the nasal septum; skin rash and allergy may develop.
Storage: Store in tightly closed containers in a dry, cool, well-ventilated place. Keep away from combustible and combustable materials and hydrazine. Containers may explode in fire.

★814★
PRAZEPAM
CAS: 2955-38-6
Other names: 2H-1,4-BENZODIAZEPIN-2-ONE,1,3-DIHYDRO-7-CHLORO-1- (CYCLOPROPYLMETHYL)-5-PHENYL-; CENTRAX; 7-CHLORO-1-(CYCLOPROPYLMETHYL)-1,3- DIHYDRO-5-PHENYL-2H-1,4-BENZODIAZEPIN-2-ONE; 7-CHLORO-1-CYCLOPROPYLMETHYL-5-PHENYL-1H-1,4-BENZODIAZEPIN-2(3H) -ONE; DEMETRIN; K-373; LY-SANDIA; VERSTRAN

Danger profile: Combustible, possible reproductive hazard, poisonous gas produced in fire.
Uses: Sedative medicine.
Appearance: Crystalline form.
Effects of exposure:
Breathed: May cause drowsiness, lightheadedness, problems with coordination and balance, difficulty with concentration; irritability, hostility, anxiety, weakness, headache, upset stomach, pains in joints.
Swallowed: May cause drowsiness, lightheadedness, problems with coordination and balance, difficulty with concentration; irritability, hostility, anxiety, weakness, headache, upset stomach, pains in joints.
Long-term exposure: May cause a decrease of fertility in men and women; damage to the developing fetus; personality change with depression, anxiety, irritability; skin rash, jaundice; reduced white blood count.
Storage: Store in tightly closed containers in a dry, cool, well-ventilated place.

★ 815 ★
PREDNISONE
CAS: 53-03-2
Other names: ANCORTONE; BICORTONE; COLISONE; CORTAN; CORTANCYL; DELTA-CORTELAN; CORTIDELT; DELTA-CORTISONE; DELTA-1-CORTISONE; DELTA(SUP1)-CORTISONE; DELTA-CORTONE; COTONE; DACORTIN; DE-CORTANCYL; DECORTIN; DECORTISYL; DELTA-1-DEHYDROCORTISONE; 1-DEHYDROCORTISONE; 1,2-DEHYDROCORTISONE; DEKORTIN; DELTA CORTELAN; DELTACORTISONE; DELTACORTONE; DELTA-DOME; DELTA E; DELTASONE; DELTISON; DELTISONE; DELTRA; DI-ADRESON; 17,21-DIHYDROXYPREGNA-1,4-DIENE-3,1 1,20-TRIONE; ENCORTON; ENCORTONE; ENKORTON; HOSTACORTIN; IN-SONE; JUVASON; LISACORT; META-CORTANDRACIN; METICORTEN; NCI-C04897; NSC 10023; ORASONE; PARACORT; PRECORT; PREGNA-1,4-DIENE-3,11,20-TRIONE, 17,21-HYDROXY-; PREDNICEN-M; PRED-NILONGA; PREDNISON; PREDNIZON; 1,4-PREGNADIENE-17-ALPHA,21-DIOL-3,11,20-TRIONE; RECTODELT; SERVI-SONE; SK-PREDNISONE; DELTA-SONE; SUPERCORTIL; U 6020; ULTRACORTEN; ULTRACORTENE; WOJTAB; ZENA-DRID

Danger profile: Mutagen, combustible, poisonous gas produced in fire.
Uses: Steroid drug.
Appearance: White, crystalline powder. Available in tablet form.
Effects of exposure:
　Breathed: Health effects may not develop after being exposed for days or weeks. May cause an increase in blood pressure, blood sugar, fluid and salt retention, personality changes, increased risk of ulcers, increased risk of infections. Repeated exposure may cause fat deposits on the face, body and liver; thinning of the bones; cataracts. Frequent exposures may interfere with adrenal gland function, with serious risk if exposure is suddenly stopped.
　Skin: Passes through the unbroken skin; can increase exposure and the severity of the symptoms listed.
　Swallowed: See *Breathed.*
Long-term exposure: May cause mutations; damage to the developing fetus; aplastic anemia; stretch marks; increased body hair. See *Breathed.* Sudden stopping of exposure after frequent repeated exposures may cause withdrawal symptoms; fever, weakness, aching, and collapse. Following prolonged exposure, recovery of normal hormone and immune functions may take one to two years. Any withdrawal should be done under medical supervision.
Storage: Store in tightly closed containers in a dry, cool, well-ventilated place.

★ 816 ★
PROPADIENE
CAS: 463-49-0
Other names: ALLENE; 1,2-PROPADIENE

Danger profile: Fire hazard, poisonous gas produced in fire.
Uses: Making other chemicals.
Appearance: Colorless, gas or colorless liquid.
Odor: Slight, sweet, garlic-like. The odor is a warning of exposure; however, no smell does not mean you are not being exposed.
Effects of exposure:
　Breathed: May cause irritation of the eyes, nose and throat. High levels may cause dizziness, lightheadedness and loss of consciousness. Extremely high levels may cause death.
　Eyes: May cause irritation, redness and tearing. Liquid may cause frostbite.
　Skin: May cause irritation. Liquid may cause frostbite.

Storage: Store in tightly closed cylinders in a dry, cool, well-ventilated place. Keep away from heat, strong oxidizers, strong acids and nitrogen oxide. Protect cylinders from physical damage. Vapors may travel to sources of ignition and flash back. Containers may explode in fire.
First aid guide: Move victim to fresh air and call emergency medical care; if not breathing, give artificial respiration; if breathing is difficult, give oxygen. In case of frostbite, thaw frosted parts with water. Keep victim quiet and maintain normal body temperature.

★ 817 ★
PROPANE
CAS: 74-98-6
Other names: DIMETHYLMETHANE; PROPYL HYDRIDE; DI-METHYL METHANE

Danger profile: Fire hazard, asphyxiant, exposure limit established, poisonous gas produced in fire.
Uses: Fuel gas for home, vehicles and industrial use; making other chemicals; solvent; refrigerant; aerosol propellant.
Appearance: Colorless gas or liquid.
Odor: Odorless but may have a foul-smelling odor added. The odor is a warning of exposure; however, no smell does not mean you are not being exposed.
Effects of exposure:
　Breathed: May cause headache, dizziness, fatigue; rapid irregular breathing; nausea, vomiting, confusion and disorientation, loss of consciousness, convulsions and death.
　Eyes: Contact with liquid may cause frostbite, burns, pain, stinging, tearing, inflammation of the eyelids and permanent damage.
　Skin: Contact with liquid may cause frostbite, burns, intense cold feeling, skin turns white and feels cold and hard.
　Swallowed: Liquid may cause headache, confusion, drowsiness, unconsciousness and death.
Storage: Store in tightly closed cylinders in a dry, cool, well-ventilated place. Keep away from heat, sources of ignition and strong oxidizers. Containers may explode in fire.
First aid guide: Move victim to fresh air and call emergency medical care; if not breathing, give artificial respiration; if breathing is difficult, give oxygen. In case of frostbite, thaw frosted parts with water. Keep victim quiet and maintain normal body temperature.

★ 818 ★
PROPANE SULTONE
CAS: 1120-71-4
Other names: 3-HYDROXY-1-PROPANESULPHONIC ACID SULTONE; 3-HYDROXY-1-PROPANESULPHONIC ACID SUL-FONE; 1,2-OXATHROLANE 2,2-DIOXIDE; 1-PROPANESULFONIC ACID-3-HYDROXY-GAMMA-SULTONE; 1,3-PROPANE SULTONE; PROPANESULTONE; 1,2-OXATHIOLANE 2,2-DIOXIDE; 1-PROPANESULFONIC ACID-3-HYDROXY-GAMMA-SULFONE

Danger profile: Carcinogen, mutagen, combustible, Community Right-to-Know List, poisonous gas produced in fire.
Uses: For making other chemicals
Appearance: White crystals or colorless liquid.
Effects of exposure:
　Breathed: May cause irritation of the nose and throat, especially if heated.
　Eyes: May cause irritation.
　Skin: Passes through the unbroken skin; can increase exposure and the severity of the symptoms listed. May cause irritation.
Long-term exposure: May cause cancer; mutations.

Storage: Store in tightly closed containers in a dry, cool, well-ventilated place. Keep away from heat.

★ 819 ★
PROPANOL
CAS: 71-23-8
Other names: ALCOOL PROPILICO (ITALIAN); ALCOOL PROPYLIQUE (FRENCH); ETHYL CARBINOL; 1-HYDROXYPROPANE; OPTAL; OSMOSOL EXTRA; PROPANOL-1; 1-PROPANOL; N-PROPANOL; PROPANOLE (GERMAN); PROPANOLEN (DUTCH); PROPANOLI (ITALIAN); PROPYL ALCOHOL; N-PROPYL ALCOHOL; 1-PROPYL ALCOHOL; N-PROPYL ALKOHOL (GERMAN); PROPYLIC ALCOHOL; PROPYLOWY ALKOHOL (POLISH)

Danger profile: Mutagen, fire hazard, exposure limit established, irritating fumes produced in fire.
Uses: Making other chemicals, textiles, leather products; solvent for waxes, vegetable oils, resins; in printing, polishing compositions; brake fluids; antiseptic.
Appearance: Colorless liquid with a mild alcoholic odor.
Odor: Resembles that of ethyl alcohol. The odor is a warning of exposure; however, no smell does not mean you are not being exposed.
Effects of exposure:
  Breathed: May be poisonous. Vapors may cause irritation of the eyes, nose, mouth and throat. High concentrations, may cause nausea, dizziness, headache, confusion and stupor.
  Eyes: Contact may cause severe irritation and burns.
  Skin: May be poisonous. Passes through the unbroken skin; can increase exposure and the severity of the symptoms listed. May cause irritation, drying and cracking.
  Swallowed: Moderately toxic; may be fatal.
Long-term exposure: May cause mutations. Similar chemicals may cause brain damage. There is limited evidence that chemical causes cancer in animals.
Storage: Store in tightly closed containers in a dry, cool, well-ventilated place. Keep away from heat, sources of ignition and strong oxidizers. Vapors may travel to sources of ignition and flash back. Containers may explode in fire.
First aid guide: Move victim to fresh air and call emergency medical care; if not breathing, give artificial respiration; if breathing is difficult, give oxygen. In case of contact with material, immediately flush eyes with running water for at least 15 minutes. Wash skin with soap and water. Remove and isolate contaminated clothing and shoes at the site.

★ 820 ★
PROPARGYL ALCOHOL
CAS: 107-19-7
Other names: ETHYNYLCARBINOL; ETHYLMETHANOL; METHANOL, ETHYNYL; 1-PROPYNE-3-OL; 2-PROPYN-1-OL; 3-PROPYNOL; 2-PROPYNYL ALCOHOL

Danger profile: Mutagen, fire hazard, poison, explosion hazard, exposure limit established, poisonous gas produced in fire.
Uses: Corrosion inhibitor; laboratory chemical; in solvents; making steel; soil fumigant; making other chemicals.
Appearance: Light straw colored liquid.
Odor: Geranium-like. The odor is a warning of exposure; however, no smell does mean you are not being exposed.
Effects of exposure:
  Breathed: May cause dizziness, lightheadedness and loss of concentration; burning sensation, coughing, wheezing, laryngitis, shortness of breath, headache, nausea, and vomiting. High concentrations are extremely destructive to mucous

membranes, upper respiratory tract; may cause coma and death.
  Eyes: Contact may cause severe irritation, burns and permanent damage.
  Skin: Passes through the unbroken skin; can increase exposure and the severity of the symptoms listed. May cause severe irritation.
  Swallowed: Poisonous.
Long-term exposure: May cause liver and kidney damage; central nervous system depression
Storage: Store in tightly closed containers in a dry, cool, well-ventilated place. Keep away from heat, sources of ignition, alkalis, oxidizing material, mercury(II) sulfate, sulfuric acid, phosphonic anhydride. Containers may explode in fire.
First aid guide: Move victim to fresh air and call emergency medical care; if not breathing, give artificial respiration; if breathing is difficult, give oxygen. In case of contact with material, immediately flush skin or eyes with running water for at least 15 minutes. Remove and isolate contaminated clothing and shoes at the site. Keep victim quiet and maintain normal body temperature. Effects may be delayed; keep victim under observation.

★ 821 ★
PROPENE
CAS: 115-07-1
Other names: METHYLETHENE; METHYLETHYLENE; NCI-C50077; 1-PROPENE (9CI); 1-PROPYLENE; PROPYLENE

Danger profile: Fire hazard, Community Right-to-Know List, poisonous gas produced in fire.
Uses: Making resins, plastics, synthetic rubber, gasoline and many other chemicals.
Appearance: Colorless gas or liquid under pressure; compressed gas.
Odor: Slightly "gassy." The odor is a warning of exposure; however, no smell does not mean you are not being exposed.
Effects of exposure:
  Breathed: Considered a simple asphyxiant. Symptoms are due to lack of oxygen available for breathing and may include dizziness, difficult breathing, bluish color of skin and loss of consciousness. May cause irregular heart beat and liver damage. Irregular heart beat may cause death.
  Eyes: Contact may cause freezing burns.
  Skin: Contact with liquid may cause freezing burns and frostbite.
  Swallowed: Contact with liquid may cause freezing burns of mouth and throat.
Long-term exposure: May cause liver damage.
Storage: Store in tightly closed containers in a dry, cool, well-ventilated place. Containers may explode in fire.

★ 822 ★
PROPIONALDEHYDE
CAS: 123-38-6
Other names: ALDEHYDE PROPIONIQUE (FRENCH); FEMA NO. 2923; METHYLACETALDEHYDE; NCI-C61029; PROPALDEHYDE; PROPANAL; PROPIONIC ALDEHYDE; PROPYL ALDEHYDE; PROPYLIC ALDEHYDE

Danger profile: Mutagen, fire hazard, Community Right-to-Know Lit, poisonous gas produced in fire.
Uses: Making other chemicals, plastic products, rubber chemicals; disinfectant; preservative.
Appearance: Colorless, mobile liquid.
Odor: Pungent, unpleasant, suffocating. The odor is a warning of exposure; however, no smell does not mean you are not being exposed.
Effects of exposure:

Breathed: Vapor may cause intense irritation to nose, throat, mouth and lungs; mucous membrane. Symptoms of exposure include watering of the eyes, coughing, difficult breathing, headache, nausea, vomiting, muscular weakness, unconsciousness. High exposure can lead to drowsiness, coma and death; buildup of fluid in the lungs (pulmonary edema), and may cause death. Pulmonary edema may be delayed by one to two days following exposure.

Eyes: Severe irritant. Contact may cause, burning and stinging, tearing, irritation and redness and swelling of the eyelids, intense pain; can burn the eyes and lead to blurred vision, chronic irritation and other permanent damage.

Skin: Severe irritant. Contact may cause burning or stinging feeling, redness, swelling, intense pain, blisters, irritation, rash. Repeated exposure may lead to skin allergy and chronic irritation.

Swallowed: May cause burning feeling of the mouth and throat, intense thirst, throat swelling, stomach cramps, nausea, vomiting, diarrhea, severe breathing difficulties and unconsciousness. Tissue damage may result. Breathing may stop. Watch for delayed symptoms.

Storage: Store in tightly closed containers in a dry, cool, well-ventilated place. Keep away from heat, sources of ignition, oxidizers and methacrylate. Containers may explode in fire.

First aid guide: Move victim to fresh air and call emergency medical care; if not breathing, give artificial respiration; if breathing is difficult, give oxygen. In case of contact with material, immediately flush eyes with running water for at least 15 minutes. Wash skin with soap and water. Remove and isolate contaminated clothing and shoes at the site.

★ 823 ★
PROPIONIC ACID
CAS: 79-09-4
Other names: ACIDE PROPIONIQUE (FRENCH); CARBONYETHANE; CARBOXYETHANE; ETHANECARBOXYLIC ACID; ETHYLFORMIC ACID; METACETONIC ACID; METHYLACETIC ACID; PROPIONIC ACID GRAIN PRESERVER; PROZOIN; PSEUDOACETIC ACID; SENTRY GRAIN PRESERVER; TENOX P GRAIN PRESERVATIVE

Danger profile: Combustible, corrosive, exposure limit established, poisonous gas produced in fire.
Uses: Food additive and mold inhibitor.
Appearance: Clear, colorless oily liquid.
Odor: Sharp, pungent, rancid, irritating. The odor is a warning of exposure; however, no smell does not mean you are not exposed.
Effects of exposure:
Breathed: Vapors may cause irritation of the eyes, nose, throat and lungs. Lung allergy may develop.
Eyes: Liquid may cause severe irritation, burns and permanent damage.
Skin: Passes through the unbroken skin; can increase exposure and the severity of the symptoms listed. Liquid may cause severe irritation and burns.
Swallowed: Moderately toxic.
Long-term exposure: May cause asthma-like allergy; lung problems.
Storage: Store in tightly closed containers in a dry, cool, well-ventilated place. Keep away from heat, sources of ignition and oxidizers. Corrodes ordinary steel and many other metals.
First aid guide: Move victim to fresh air; call emergency medical care. In case of contact with material, immediately flush skin or eyes with running water for at least 15 minutes. Remove and isolate contaminated clothing and shoes at the site. Keep victim quiet and maintain normal body temperature.

★ 824 ★
PROPOXUR
CAS: 114-26-1
Other names: ARPROCARB; BAY 9010; BAY 39007; BAYER 39007; BAYGON; BIFEX; BLATTANEX; BOYGON; CARBAMIC ACID, METHYL-, O-ISOPROPOXYPHENYL ESTER; CHEMAGRO 9010; DDVP; ENT 25,671; O-IMPC; INVISI-GARD; ISOCARB; O-ISOPROPOXYPHENYL METHYLCARBAMATE; O-ISOPROPOXYPHENYLN-METHYL CARBAMATE; 2-ISOPROPOXYPHENYL METHYLCARBAMATE; 2-ISOPROPOXYPHENYL N-METHYLCARBAMATE; 2-(1-METHYLETHOXY)PHENOLMETHYLCARBAMATE; N-METHYL-2-ISOPROPOXYPHENYLCARBAMATE; OMS-33; PHC; PHENOL,O-ISOPROPOXY-, METHYLCARBAMATE; PHENOL, 2-(1- METHYLETHOXY)-,METHYLCARBAMATE; PILLARGRAN; PROPOGON; PROPOKSURU (POLISH); PROPOXURE; PROPYON; SENDRAN; SUNCIDE; TENDEX; TUGON FLIEGENKUGEL; UNDEN; UNDENE

Danger profile: Mutagen, combustible, exposure limit established, Community Right-to-Know List, poisonous gas produced in fire.
Uses: Insecticide (carbamate). Effective against a wide range of insects including mosquitoes.
Appearance: Colorless crystalline powder.
Odor: Faint, characteristic. The odor is a warning of exposure; however, no smell does not mean you are not being exposed.
Effects of exposure:
Breathed: Extremely poisonous. May cause headache, dizziness, profuse sweating, nausea, vomiting, reduced heart beat, stomach cramps, diarrhea, loss of coordination, slow and weak breathing, fever, loss of consciousness and death. May cause shortness of breath and a dangerous build-up of fluids in the lungs (pulmonary edema), which can cause death. Pulmonary edema is a medical emergency that can be delayed one to two days following exposure.
Eyes: May cause irritation.
Skin: May be poisonous if absorbed. Passes through the unbroken skin; can increase exposure and the severity of the symptoms listed. May cause irritation.
Swallowed: Poisonous.
Long-term exposure: May damage the developing fetus; birth defects; decrease of fertility in men and women.
Storage: Store in a regulated, specially marked place in tightly closed containers in a dry, well-ventilated place. Keep away from water, strong alkalis. Containers may explode in fire.

★ 825 ★
PROPYL ACETATE
CAS: 109-60-4
Other names: ACETATE DE PROPYLE NORMAL (FRENCH); ACETIC ACID, PROPYL ESTER; ACETIC ACID, N-PROPYL ESTER; 1-ACETOXYPROPANE; OCTAN PROPYLU (POLISH); N-PROPYL ACETATE; 1-PROPYL ACETATE; PROPYLACETATE

Danger profile: Fire hazard, exposure limit established, poisonous gas produced in fire.
Uses: In flavoring agents, perfumes; solvent; lacquers; plastics; laboratory chemical; making other chemicals.
Appearance: Colorless liquid.
Odor: Mild fruity. The odor is a warning of exposure; however, no smell does not mean you are not being exposed.
Effects of exposure:
Breathed: Mildly toxic. May cause irritation of eyes, nose and throat; headache, drowsiness, excitability, drunken behavior, mental confusion, fatigue. Very high levels of exposure may

cause dizziness, lightheadedness; loss of consciousness, coma and death.

Eyes: Contact may cause irritation, pain, burning, stinging, tearing, inflammation of the eyelids. Light may cause pain.

Skin: May remove oil from the skin and cause dryness. Persons with pre-existing skin disorders may be more susceptible to the effects of this substance.

Swallowed: May cause gastrointestinal irritation. See **Breathed.**

Long-term exposure: Prolonged or repeated contact may cause irritation of the skin and dermatitis; drying and cracking skin; nervous system changes.

Storage: Store in a tightly closed container in a cool, well-ventilated place. Keep away from heat and sources of ignition, nitrates, strong oxidizers, alkalis and strong acids. Vapors are explosive in heat or flame. Containers may explode in fire.

First aid guide: Move victim to fresh air and call emergency medical care; if not breathing, give artificial respiration; if breathing is difficult, give oxygen. In case of contact with material, immediately flush eyes with running water for at least 15 minutes. Wash skin with soap and water. Remove and isolate contaminated clothing and shoes at the site.

★ 826 ★
PROPYLAMINE
CAS: 107-10-8
Other names: 1-AMINOPROPANE; MONO-N-PROPYLAMINE; MONOPROPYLMINE; PROPANAMINE; 1-PROPANAMINE; N-PROPYLAMINE

Danger profile: Combustible, poisonous gas produced in fire.
Uses: Making textile resins, pharmaceuticals, pesticides and other chemicals; laboratory chemical.
Appearance: Colorless liquid.
Odor: Strong, ammonia-like. The odor is a warning of exposure; however, no smell does not mean you are not being exposed.
Effects of exposure:
Breathed: May cause irritation to the nose, throat and eyes; difficult breathing, coughing. Higher exposures may cause shortness of breath and a dangerous build-up of fluids in the lungs (pulmonary edema), which can cause death. Pulmonary edema is a medical emergency that can be delayed one to two days following exposure.
Eyes: Contact may cause severe burns; possible permanent corneal damage or complete eye destruction.
Skin: Passes through the unbroken skin; can increase exposure and the severity of the symptoms listed. Contact may cause severe irritation, redness, itching and burns.
Swallowed: Corrosive to the gastrointestinal tract. May cause burning feeling in mouth, throat and stomach; pain and swelling in throat; nausea and vomiting; stomach cramps, rapid breathing, shock, diarrhea; possible stomach perforation.

Long-term exposure: May cause weight loss; kidney and liver damage; skin irritation and itching. Similar chemicals may cause allergies.
Storage: Store in tightly closed containers in a dry, cool, well-ventilated, preferably detached, place. Keep away from strong acids, acid chlorides, strong oxidizers, carbon dioxide, triethyl aluminum. Protect containers from physical damage. Vapors may travel to sources of ignition and flash back. Containers may explode in fire.
First aid guide: Move victim to fresh air and call emergency medical care; if not breathing, give artificial respiration; if breathing is difficult, give oxygen. In case of contact with material, immediately flush skin or eyes with running water for at least 15 minutes. Remove and isolate contaminated clothing and shoes at the site. Keep victim quiet and maintain normal body temperature.

★ 827 ★
PROPYLENE CHLOROHYDRIN
CAS: 78-89-7
Other names: 2-CHLORO-1-PROPANOL; 2-CHLOROPROPYL ALCOHOL; PROPYLENECHLOROHYDRIN; 1-PROPANOL, 2-CHLORO-

Danger profile: Fire hazard, poison, poisonous gas produced in fire.
Uses: Making other chemicals.
Appearance: Colorless liquid.
Odor: Mild, non-residual. The odor is a warning of exposure; however, no smell does not mean you are not being exposed.
Effects of exposure:
Breathed: May be poisonous. May cause irritation of the eyes, nose and throat. High exposure may cause dizziness, giddiness, visual disturbances, lightheaded feeling, loss of balance and coordination, drowsiness, nausea, vomiting, coma.
Eyes: Contact may cause severe irritation and burns.
Skin: May be poisonous if absorbed. Passes through the unbroken skin; can increase exposure and the severity of the symptoms listed. Contact may cause severe irritation.
Swallowed: Poisonous.
Storage: Store in tightly closed containers in a dry, cool, well-ventilated place. Keep away from heat, sources of ignition and oxidizers. Containers may explode in fire.
First aid guide: Move victim to fresh air and call emergency medical care; if not breathing, give artificial respiration; if breathing is difficult, give oxygen. In case of contact with material, immediately flush skin or eyes with running water for at least 15 minutes. Speed in removing material from skin is of extreme importance. Remove and isolate contaminated clothing and shoes at the site. Keep victim quiet and maintain normal body temperature. Effects may be delayed; keep victim under observation.

★ 828 ★
PROPYLENE DINITRATE
CAS: 6423-43-4
Other names: PGDN; 1,2-PROPANEDIOL, DINITRATE; PROPYLENE GLYCOL DINITRATE; PROPYLENE GLYCOL 1,2-DINITRATE; 1,2-PROPYLENE GLYCOL DINITRATE

Danger profile: Combustible, explosion hazard, exposure limit established, poisonous gas produced in fire.
Uses: In explosives; fuel.
Appearance: Colorless or red-orange colored liquid.
Odor: Disagreeable. The odor is a warning of exposure; however, no smell does mean you are not being exposed.
Effects of exposure:
Breathed: May cause irritation of the eyes, stuffy nose, loss of coordination and balance, bluish color to the skin and lips; rapid drop in blood pressure, rapid heart beat and death.
Eyes: May cause irritation and inflammation of the eyelids.
Skin: Passes through the unbroken skin; can increase exposure and the severity of the symptoms listed. Contact may cause irritation.
Swallowed: Poisonous.
Long-term exposure: May cause damage to the nervous system; liver and kidney damage.
Storage: Store in tightly closed containers in a dry, cool, well-ventilated place. Keep away form heat, ammonia compounds, amines, oxidizers, combustible materials and reducers. Containers must be handled with extreme care to avoid shock. Containers may explode in fire.

★ 829 ★
PROPYLENE GLYCOL
CAS: 57-55-6
Other names: DOWFROST; 1,2-DIHYDROXYPROPANE; METHYLETHYLENE GLYCOL; METHYL GLYCOL; MONO-PROPYLENE GLYCOL; PG12; PROPANE-1,2-DIOL; 1,2-PROPANEDIOL; 1,2-PROPYLENE GLYCOL; SIRLENE; SOLAR WINTER BAN TRIMETHYL GLYCOL

Danger profile: Mutagen, combustible, poisonous gas produced in fire.
Uses: Making other chemicals, cellophane, textiles; antifreeze; deicing airport runways; solvent; flavorings; perfumes; soft drink syrups; perfumes; hydraulic and brake fluids; coolant in refrigeration systems; bactericide; food additive; cleansing creams; sun tan lotions; pharmaceuticals; tobacco.
Appearance: Colorless, syrup-like liquid.
Effects of exposure:
   Breathed: May cause irritation of nose and throat; headache, drowsiness, excitability, drunken behavior, mental confusion, fatigue. Very high levels of exposure may cause dizziness, lightheadedness and cause victim to pass out, go into coma and die.
   Eyes: Contact may cause irritation, pain, burning, stinging, tearing, inflammation of the eyelids.
   Skin: May cause irritation; remove oil from the skin and cause dryness.
   Swallowed: May cause gastrointestinal irritation, convulsions and fits. See **Breathed.**
Long-term exposure: Prolonged or repeated contact may cause mutations; irritation of the skin.
Storage: Store in a tightly closed container in a cool, well-ventilated place. Keep away from heat and sources of ignition, strong acids and oxidizing materials. Vapors are explosive in heat or flame. Containers may explode in fire.

★ 830 ★
PROPYLENE GLYCOL MONOMETHYL ETHER
CAS: 107-98-2
Other names: DOWANOL 33B; DOWTHERM 209; METHOXY ETHER OF PROPYLENE GLYCOL; 1-METHOXY-2-PROPANOL; POLY-SOLVE MPM; PROPASOL SOLVENT M; PROPYLENE GLYCOL METHYL ETHER; ALPHA-PROPYLENE GLYCOL MONOMETHYL ETHER; PRO-PYLENGLYKOL-MONOMETHYLAETHER (GERMAN)

Danger profile: Fire hazard, exposure limit established, Community Right-to-Know List, poisonous gas produced in fire.
Uses: Solvent. Making dyes, acrylics, inks, stains; solvent-sealing of cellophane; heat transfer fluid.
Appearance: Colorless liquid.
Odor: Mild pleasant; ether-like. The odor is a warning of exposure; however, no smell does mean you are not being exposed.
Effects of exposure:
   Breathed: May cause irritation of the eyes, nose, throat and lungs, tears and runny nose, coughing, sore throat, difficult breathing; lethargy, fatigue, and drowsiness; slurred speech, shaking, and dizziness. Higher exposures could cause shortness of breath, headache, nausea, vomiting, dizziness, diarrhea and a dangerous build-up of fluids in the lungs (pulmonary edema), which can cause death. Pulmonary edema is a medical emergency that can be delayed one to two days following exposure. Very high levels may cause lung, liver and kidney damage.
   Eyes: May cause irritation, redness, tearing, pain, inflamed eyelids, and eye injury; burns in corneas and possible loss of sight.
   Skin: Passes through the unbroken skin; can increase exposure and the severity of the symptoms listed. May cause

painful irritation, redness, and blisters Repeated or prolonged skin contact may cause irritation of the skin and may also cause the legs to feel numb.
   Swallowed: May cause irritation of the mouth and throat; stomach and abdominal pain; loss of muscular coordination and mental confusion; convulsions. Also see symptoms for **Breathed.**
Long-term exposure: May cause lung, liver and kidney damage. A closely related chemical may cause lung allergy, and other glycol ethers are reproductive hazards.
Storage: Store in tightly closed containers in a dry, cool, well-ventilated place. Keep away from heat, sources of ignition and oxidizers. Containers may explode in fire.

★ 831 ★
PROPYLENEIMINE
CAS: 75-55-8
Other names: AZIRIDINE, 2-METHYL- (6CI,8CI,9CI); 2-METHYLAZACLYCLOPROPANE; 2-METHYLAZIRIDINE; 1,2-PROPYLEN IMINE; 2-METHYLETHYLENIMINE; 2-METHYLETHYLEN IMINE; PROPYLEN EIMINE; 1,2-PROPYLENIMINE

Danger profile: Carcinogen, mutagen, combustible, exposure limit established, EPA extremely hazardous substances list, Community Right-to-Know List, poisonous gas produced in fire.
Uses: Making paint, paper, textiles, rubber, pharmaceuticals, and other chemicals.
Appearance: Fuming, colorless oily liquid.
Odor: Strong, fishy, ammonia-like. The odor is a warning of exposure; however, no smell does not mean you are not being exposed.
Effects of exposure:
   Breathed: May cause vomiting, breathing difficulty, and irritation of eyes, nose, and throat. Prolonged exposure tend to redden the whites of the eyes.
   Eyes: May cause severe irritation and burns.
   Skin: Passes through the unbroken skin; can increase exposure and the severity of the symptoms listed. Liquid may cause severe irritation and burns which are slow to heal.
   Swallowed: Poison. May cause burns of mouth and stomach.
Long-term exposure: May cause cancer; mutations; possible lung problems.
Storage: Store in tightly closed containers in a dry, cool, well-ventilated place. Keep away from heat, sources of ignition, acids and acid fumes, and oxidizers. Vapors may travel to sources of ignition and flash back. Containers may explode in fire.
First aid guide: Move victim to fresh air and call emergency medical care; if not breathing, give artificial respiration; if breathing is difficult, give oxygen. In case of contact with material, immediately flush skin or eyes with running water for at least 15 minutes. Remove and isolate contaminated clothing and shoes at the site. Keep victim quiet and maintain normal body temperature. Effects may be delayed; keep victim under observation.

★ 832 ★
PROPYLENE OXIDE
CAS: 75-56-9
Other names: EPOXYPROPANE; 1,2-EPOXYPROPANE; 2,3-EPOXYPROPANE; ETHYLENE OXIDE, METHYL-; METHYL ETHYLENE OXIDE; METHYL OXIRANE; NCI-C50099; OX-IRANE, METHYL-; OXYDE DE PROPYLENE (FRENCH); PROPANE, EPOXY-; 1-PROPEN-3-OL; PROPENE OXIDE; 1,2-PROPYLENE OXIDE

Danger profile: Carcinogen, mutagen, fire hazard, explosion hazard, exposure limit established, EPA extremely hazardous substances list, Community Right-to-Know List, poisonous gas produced in fire.

Uses: Making other chemicals, urethane foams, lubricants, detergents; fumigant; solvent.

Appearance: Clear, colorless liquid.

Odor: Ethereal; sweet, alcoholic; natural gas-like. The odor is a warning of exposure; however, no smell does not mean you are not being exposed.

Effects of exposure:

Breathed: May cause irritation of the eyes, nose, throat and lungs, tears and runny nose, coughing, sore throat, difficult breathing; lethargy, fatigue, and drowsiness; slurred speech, shaking, and dizziness; loss of coordination. Higher exposures could cause shortness of breath, headache, nausea, vomiting, dizziness, diarrhea and a dangerous build-up of fluids in the lungs (pulmonary edema), which can cause death. Pulmonary edema is a medical emergency that can be delayed one to two days following exposure. High or repeated exposure may cause pneumonia.

Eyes: May cause severe irritation, redness, tearing, pain, inflamed eyelids, and eye injury; burns in corneas and possible loss of sight.

Skin: Passes through the unbroken skin; can increase exposure and the severity of the symptoms listed. May cause painful irritation, redness, and blisters. Covered contact may cause burns.

Swallowed: May cause irritation of the mouth and throat; stomach and abdominal pain; loss of muscular coordination and mental confusion; convulsions. Also see symptoms for *Breathed.*

Long-term exposure: May cause cancer; reduce fertility in men and women; lung and liver damage.

Storage: Store in tightly closed containers in a dry, cool, well-ventilated detached place away from direct sun. Keep away from heat, sources of ignition, strong acids, strong bases, peroxides, copper and copper alloys, iron, epoxy resin, oleum, anhydrous metal chlorides. Containers may explode in fire.

First aid guide: Move victim to fresh air and call emergency medical care; if not breathing, give artificial respiration; if breathing is difficult, give oxygen. In case of contact with material, immediately flush eyes with running water for at least 15 minutes. Wash skin with soap and water. Remove and isolate contaminated clothing and shoes at the site.

★ 833 ★
PROPYL ISOCYANATE
CAS: 110-78-1
Other names: 1-ISOCYANATOPROPANE; 1-ISOCYANATO PROPANE; PROPANE,1-ISOCYANATO-; M-PROPYL ISOCYANATE; N-PROPYL ISOCYANATE; 1-PROPYLISOCYANATE; 1-PROPYL ISOCYANATE; ISOCYANIC ACID, PROPYL ESTER; N-PROPYL ISOCYANATE

Danger profile: Fire hazard, poison, poisonous gas produced in fire.

Uses: Making other chemicals and insecticides.

Appearance: Colorless to light yellow liquid.

Odor: Sharp. The odor is a warning of exposure; however, no smell does mean you are not being exposed.

Effects of exposure:

Breathed: Poisonous. May cause irritation of the eyes, nose, throat, breathing passages and lungs; sneezing, coughing, chest tightness and difficult breathing.

Eyes: May cause irritation.

Skin: May be poisonous if absorbed. May pass through the unbroken skin thereby increasing exposure and the severity of the symptoms listed. May cause irritation.

Swallowed: Poisonous.

Long-term exposure: Similar chemicals may cause asthma-like allergy to develop.

Storage: Store in tightly closed containers in a dry, cool, well-ventilated place. Keep away from heat, light, water, steam and other forms of moisture, amines, strong bases and alcohols. Containers may explode in fire.

First aid guide: Move victim to fresh air and call emergency medical care; if not breathing, give artificial respiration; if breathing is difficult, give oxygen. In case of contact with material, immediately flush skin or eyes with running water for at least 15 minutes. Remove and isolate contaminated clothing and shoes at the site. Keep victim quiet and maintain normal body temperature. Effects may be delayed; keep victim under observation.

★ 834 ★
N-PROPYL NITRATE
CAS: 627-13-4
Other names: NITRATE DE PROPYLE NORMAL (FRENCH); NITRIC ACID, PROPYL ESTER; PROPYL NITRATE

Danger profile: Fire hazard, explosion hazard, exposure limit established, poisonous gas produced in fire.

Uses: Used in fuels to promote ignition; making other chemicals and rocket fuels.

Appearance: Colorless to pale yellow liquid.

Odor: Sickly, ether-like. The odor is a warning of exposure; however, no smell does not mean you are not being exposed.

Effects of exposure:

Breathed: Vapor may cause low blood pressure and skin and affect the ability of the blood to carry oxygen, this may cause skin and lips to turn blue.

Storage: Store in tightly closed containers in a dry, cool, well-ventilated place. Keep away from air, combustible materials and oxidizing agents. Containers may explode in fire.

★ 835 ★
PROPYZAMIDE
CAS: 23950-58-5
Other names: BENZAMIDE, 3,5-DICHLORO-N-(1,1-DIMETHYL-2-PROPYNYL)-; CLANEX; 3,5-DICHLORO-N-(1,1-DIMETHYL-2-PROPYNL)BENZAMIDE; N-(1,1-DIMETHYLPROPYNYL)-3,5-DICHLOROBENZAMIDE; KERB; KERB 50W; KERB FLO; KERB GRANULES; MATRIKERB; PRONAMID; PROMAMIDE; PRONAMIDE; RH 315

Danger profile: Possible carcinogen, poisonous gas produced in fire.

Uses: Herbicide for legumes, lettuces, trees, fruit, cabbages, rhubarb, strawberries, turf control of weeds and grass.

Appearance: White to off-white solid.

Effects of exposure:

Breathed: May cause irritation of the nose, throat and bronchial tubes, possibly causing coughing, phlegm and difficult breathing. High exposure may cause liver damage.

Eyes: Contact may cause irritation.

Skin: Passes through the unbroken skin; can increase exposure and the severity of the symptoms listed. Contact may cause irritation. Prolonged exposure may cause rash to develop.

Swallowed: Mildly toxic.

Long-term exposure: May cause cancer; liver damage.

Storage: Store in tightly closed containers in a dry, cool, well-ventilated place.

★ 836 ★
PYRETHRUM
CAS: 8003-34-7
Other names: ANERIN I; ANERIN II; BUHACH; CHRYSAN-THEMUM CINERAREAEFOLIUM; CINERIN I; CINERIN II; DALMATION INSECT FLOWERS; DOOM GARDEN INSECT KILLER; FICAM PLUS; FIRMOTOX; GARDEN INSECT KILL-ER (ENGLAND); GARDEN INSECT SPRAY; INSECT POW-DER; JASNALIN I; JASNALIN II; KERISPRAY; KILLGERM FLYSPRAY; NUVAN FLY KILLER; PYRA FOG; PYRETHRIN I; PYRETHRIN II; PYRETHRINS PYRETHRUM (ACGIH, OSHA); PYRETHRUM INSECTICIDE; RENTOKIL PYRE-THRIN OIL SPRAY; TRIESTE FLOWERS

Danger profile: Allergen, exposure limit established, poisonous gases produced in fire.
Uses: Insecticide.
Appearance: Thick brown liquid or solid.
Effects of exposure:
Breathed: May cause irritation of the eyes, nose, throat and lungs, tears and runny nose, coughing, sore throat, difficult breathing; lethargy, fatigue, and drowsiness; slurred speech, shaking, and dizziness. Higher exposures could cause short-ness of breath, headache, nausea, vomiting, dizziness, diar-rhea and a dangerous build-up of fluids in the lungs (pulmo-nary edema), which can cause death. Pulmonary edema is a medical emergency that can be delayed one to two days following exposure.
Eyes: May cause irritation, redness, tearing, pain, inflamed eyelids, and eye injury; burns in corneas and possible loss of sight.
Skin: Passes through the unbroken skin; can increase expo-sure and the severity of the symptoms listed. May cause painful irritation, redness, and blisters. Even low exposures may cause skin allergy may develop.
Swallowed: May cause irritation of the mouth and throat; stomach and abdominal pain; loss of muscular coordination and mental confusion; convulsions. Also see symptoms for *Breathed.*
Long-term exposure: May cause liver damage; various aller-gies to develop including a generalized allergy with swelling of the air passages, difficult breathing and collapse; asthma-like al-lergy with coughing and tightness of the chest; skin allergy with itching and rash.
Storage: Store in tightly closed containers in a dry, cool, well-ventilated place. Keep away from oxidizers and alkalines.

★ 837 ★
PYRIDINE
CAS: 110-86-1
Other names: AZABENZENE; AZINE; NCL-C55301; PYRIDIN (GERMAN); PRINDINA (ITALIAN); PIRYDYNA (POLISH)

Danger profile: Mutagen, fire hazard, exposure limit estab-lished, Community Right-to-Know List, poisonous gas produced in fire.
Uses: Making vitamins and drugs, alcohols and antifreeze mix-tures; solvent; waterproofing; rubber chemicals; dyeing textiles; fungicides; laboratory chemical.
Appearance: Colorless or yellow liquid.
Odor: Strong, disagreeable, penetrating, sickening. The odor is a warning of exposure; however, no smell does not mean you are not being exposed.
Effects of exposure:
Breathed: May be poisonous. May cause irritation of the eyes, nose, throat and skin. High levels may cause dizziness, lightheadedness, upset stomach, headache, nausea, ner-

vous symptoms, frequent urination and possible coma and death.
Eyes: Contact may cause severe irritation, burns and perma-nent damage.
Skin: May be poisonous if absorbed. Passes through the un-broken skin; can increase exposure and the severity of the symptoms listed. May cause severe irritation.
Long-term exposure: May cause mutations; central nervous system depression, gastrointestinal problems; kidney and liver damage. Repeated exposure, even to low levels, may cause liver and brain damage.
Storage: Store in tightly closed containers in a dry, cool, well-ventilated outdoor or detached place. Keep away from heat, sources of ignition, acids, oxidizers, maleic anhydride, oleum io-dine, chlorosulfonic acid. Containers may explode in fire.
First aid guide: Move victim to fresh air and call emergency medical care; if not breathing, give artificial respiration; if breath-ing is difficult, give oxygen. In case of contact with material, im-mediately flush eyes with running water for at least 15 minutes. Wash skin with soap and water. Remove and isolate contami-nated clothing and shoes at the site.

Q

★ 838 ★
QUINOLINE
CAS: 91-22-5
Other names: 1-AZANAPHTHALENE; 1-BENZAZINE; 1-BENZINE; BENZO(B) PYRIDINE; CHINOLEINE; CHINOLIN (CZECH); CHINOLINE; LEUCOL; LEUCOLINE; LEUKOL; USAF EK-218

Danger profile: Mutagen, combustible, unpredictably violent, Community Right-to-Know List, poisonous gas produced in fire.
Uses: In pharmaceuticals; preserving anatomical specimens; making dyes, paints, and other chemicals; flavoring.
Appearance: Clear, colorless to brown liquid.
Odor: Strong, unpleasant, peculiar. The odor is a warning of ex-posure; however, no smell does not mean you are not being ex-posed.
Effects of exposure:
Breathed: Vapor may cause irritation of the nose, throat and bronchial tubes; headaches, sore throat, nosebleed, cough, phlegm, and difficult breathing. High exposure may cause diz-ziness, nausea, vomiting, trouble breathing and possible death. Higher exposures may cause a dangerous build-up of fluids in the lungs (pulmonary edema), which can cause death. Pulmonary edema is a medical emergency that can be delayed one to two days following exposure.
Eyes: May cause severe irritation. Single high exposure may cause permanent damage.
Skin: Passes through the unbroken skin; can increase expo-sure and the severity of the symptoms listed. May cause se-vere irritation.
Swallowed: May cause irritation of mouth and stomach; vomiting may occur.
Long-term exposure: May cause mutations. Possible lung problems.
Storage: Store in tightly closed containers in a dry, cool, well-ventilated place or refrigerator under an inert atmosphere. Keep away from steam and other forms of moisture, light, strong oxi-dizers, strong acids, hydrogen peroxide, linseed oil, perchro-mates, and nitrogen tetroxide. May attack some forms of plas-tics. Vapors may travel to sources of ignition and flash back. Containers may explode in fire.

First aid guide: Move victim to fresh air and call emergency medical care; if not breathing, give artificial respiration; if breathing is difficult, give oxygen. In case of contact with material, immediately flush skin or eyes with running water for at least 15 minutes. Remove and isolate contaminated clothing and shoes at the site. Keep victim quiet and maintain normal body temperature.

★839★
QUINONE
CAS: 106-51-4
Other names: BENZO-CHINON (GERMAN); 1,4-BENZOQUINONE; BENZOQUINONE; P-BENZOQUINONE; CHINON (DUTCH); CHINON (GERMAN); P-CHINON (GERMAN); CHINONE; CLYCLOHEXADEINEDIONE; 1,4-CYCLOHEXADIENEDIONE; 2,5-CYCLOHEXADIENE-1,4-DIONE; 1,4-DIOXYBENZENE; NCI-C55845; P-QUINONE; USAF P-220

Danger profile: Mutagen, poison, combustible, exposure limit established, Community Right-to-Know List, poisonous gas produced in fire.
Uses: Making and other chemicals; fungicides; laboratory chemical; photography.
Appearance: Yellow sand-like crystals.
Odor: Acrid, penetrating; irritating, chlorine-like. The odor is a warning of exposure; however, no smell does not mean you are not being exposed.
Effects of exposure:
 Breathed: Poisonous; may be fatal if inhaled. Vapor may cause intense irritation to nose, throat, mouth and lungs; mucous membrane. Symptoms of exposure include watering of the eyes, coughing, difficult breathing, headache, nausea, vomiting, muscular weakness, unconsciousness. High exposure can lead to drowsiness, coma and death; buildup of fluid in the lungs (pulmonary edema), and may cause death. Pulmonary edema may be delayed by one to two days following exposure.
 Eyes: Severe irritant. Contact may cause, burning and stinging, tearing, irritation and redness and swelling of the eyelids, intense pain; can burn the eyes and lead to blurred vision, chronic irritation and other permanent damage. The cornea may suffer ulceration and scarring. Chronic eye exposure causes gradual brownish discoloration of the conjunctiva and cornea, small corneal opacities and damage in corneal structure which cause loss of visual acuity.
 Skin: Poisonous; may be fatal if absorbed through the skin. Severe irritant. Contact may cause burning or stinging feeling, redness, swelling, intense pain, blisters, irritation, rash. Repeated exposure may lead to skin allergy and chronic irritation.
 Swallowed: Poisonous; may be fatal. May cause burning feeling of the mouth and throat, intense thirst, throat swelling, stomach cramps, nausea, vomiting, diarrhea, severe breathing difficulties and unconsciousness. Tissue damage may result. Breathing may stop. Watch for delayed symptoms.
Long-term exposure: May cause mutations. Repeated eye exposure to vapors may lead to brown staining, deformed cornea and vision damage.
Storage: Store in tightly closed containers in a dry, cool, well-ventilated place. Keep away from heat, sources of ignition, and strong oxidizers. May attack some forms of plastics, rubber and coatings. Containers may explode in fire.

R

★840★
RESERPINE
CAS: 50-55-5
Other names: AUSTRAPINE; ENT 50,146; ESKASERP; METHYLRESERPATE 3,4,5,- TRIMETHOXYBENZOIC ACID; NCI-C50157; RAU-SED; RAUWOLEAF; RESERPEX; RESERPOID; SERPASIL; SERPASIL APRESOLINE; SERPINE; USAF CB-27; YOHIMBAN-16-CARBOXYLIC ACID, 11,17-DIMETHOXY-18-(3,4,5-TRIMETHOXYBENXOYL)OXY-, METHYL ESTER; YOHIMBAN-16-CARBOXYLIC ACID, 11,17-DIMETHOXY-18-(3,4,5-TRIMETHOXYBENXOYL)OXY-, METHYL ESTER, (3BETA, 16BETA, 17ALPHA, 18BETA, 20ALPHA)-

Danger profile: Carcinogen, mutagen, poisonous gas produced in fire.
Uses: As a medical drug; human and animal food additive.
Appearance: White or pale buff to slightly yellow powder.
Odor: Faint, characteristic. The odor is a warning of exposure; however, no smell does not mean you are not being exposed.
Effects of exposure:
 Breathed: Overdose may cause decreased blood pressure, convulsions and coma.
 Swallowed: Overdose may cause decreased blood pressure, convulsions and coma.
Long-term exposure: May cause cancer; damage to the developing fetus; nightmares, depression, diarrhea, cramps, weight gain.
Storage: Store in tightly closed containers in a dry, cool, well-ventilated place.
First aid guide: In case of contact with material, immediately flush eyes with running water for at least 15 minutes. Wash skin with soap and water. Remove and isolate contaminated clothing and shoes at the site.

★841★
RESORCINOL
CAS: 108-46-3
Other names: BENZENE, M-DIHYDROXY-; M-BENZENEDIOL; 1,3-BENZENEDIOL; C.I. 76505; C.I. DEVELOPER 4; C.I. OXIDATION BASE 31; DEVELOPER O; DEVELOPER R; DEVELOPER RS; M-DIHYDROXYBENZENE; 1,3-DIHYDROXYBENZENE; M-DIOXYBENZENE; DURAFUR DEVELOPER G; FOURAMINE RS; FOURRINE 79; FOURRINE EW; M-HYDROQUINONE; 3-HYDROXYCYCLOHEXADIEN-1-ONE; M-HYDROXYPHENOL; 3-HYDROXYPHENOL; NAKO TGG; NCI-C05970; PELAGOL GREY RS; PELAGOL RS; PHENOL, M-HYDROXY-; RESORCIN; RESORCINE

Danger profile: Mutagen, combustible, exposure limit established, poisonous gas produced in fire.
Uses: Making rubber and neoprene products, adhesives, ultraviolet absorbers, and other chemicals; resins; dyes; pharmaceuticals; wood veneer adhesives; rubber/textile composites; cosmetics.
Appearance: White to pinkish crystalline solid.
Odor: Faint, characteristic. The odor is a warning of exposure; however, no smell does not mean you are not being exposed.
Effects of exposure:
 Breathed: May be poisonous. Vapor may cause intense irritation to nose, throat, mouth and lungs; mucous membrane. Symptoms of exposure include watering of the eyes, coughing, difficult breathing, headache, nausea, vomiting, muscular weakness, unconsciousness. High exposure can lead to drowsiness, coma and death; buildup of fluid in the lungs (pul-

monary edema), and may cause death. Pulmonary edema may be delayed by one to two days following exposure.
Eyes: Severe irritant. Contact may cause, burning and stinging, tearing, irritation and redness and swelling of the eyelids, intense pain; can burn the eyes and lead to blurred vision, chronic irritation and other permanent damage.
Skin: May be poisonous if absorbed. Severe irritant. Contact may cause burning or stinging feeling, redness, swelling, intense pain, blisters, irritation, rash. Repeated exposure may lead to skin allergy and chronic irritation.
Swallowed: Poison. May cause burns of mucous membranes, severe diarrhea, pallor, sweating, weakness, headache, dizziness, ringing in the ears, shock, and severe convulsions; may also cause spleen and kidney problems; burning feeling of the mouth and throat, intense thirst, throat swelling, stomach cramps, nausea, vomiting, severe breathing difficulties and unconsciousness. Tissue damage may result. Breathing may stop. Watch for delayed symptoms.
Long-term exposure: May cause mutations; liver, lung and kidney damage; skin allergy; severe dermatitis; effect the blood's ability to carry oxygen with blue color to skin and lips.
Storage: Store in tightly closed containers in a dry, cool, well-ventilated place. Keep away from sources of ignition, alkalies, oxidizers, ferric salts, methanol, nitric acid, spirit nitrous ether, acetanilide, antipyrine, urethane. Containers may explode in fire.
First aid guide: Move victim to fresh air and call emergency medical care; if not breathing, give artificial respiration; if breathing is difficult, give oxygen. In case of contact with material, immediately flush skin or eyes with running water for at least 15 minutes. Speed in removing material from skin is of extreme importance. Remove and isolate contaminated clothing and shoes at the site. Keep victim quiet and maintain normal body temperature. Effects may be delayed; keep victim under observation.

★ 842 ★
RODINE
CAS: NONE
Other names: BONIDE TOPZOL RAT BAITS AND KILLING SYRUP; RAT-O-CIDE RAT BAIT; RAT'S END; RED SQUILL; ROUGH AND READY RAT BAIT AND RAT PASTE; SCILLIROSIDE GLYCOSIDE; SILMURIN; SQUILL; TOPZOL; URGENEA MARITIMA

Danger profile: Poison, poisonous gas produced in fire.
Uses: Rodenticide.
Appearance: Liquid.
Effects of exposure:
  Breathed: May be poisonous. See **Swallowed**.
  Swallowed: Poison. May cause a bitter taste; nausea, vomiting, abdominal pain, blurred vision; decreased pulse rate; fall in blood pressure; possible convulsions. Heart rhythm disturbances may cause sudden death.
Storage: Store in tightly closed containers in a dry, cool, well-ventilated place.

★ 843 ★
RONNEL
CAS: 299-84-3
Other names: O,O-DIMETHYLO-(2,4,5-TRICHLOROPHENYL) PHOSPHOROTHIOATE; FENCHLOROPHOS; TROLENE

Danger profile: Mutagen, poison, exposure limit established, poisonous gas produced in fire.
Uses: Insecticide (organophosphate).
Appearance: White to tan, waxy solid.
Effects of exposure:
  Breathed: May cause headache, dizziness, profuse sweating, nausea, vomiting, reduced heart beat, stomach cramps,

diarrhea, loss of coordination, slow and weak breathing, fever, loss of consciousness and death. May cause shortness of breath and a dangerous build-up of fluids in the lungs (pulmonary edema), which can cause death. Pulmonary edema is a medical emergency that can be delayed one to two days following exposure.
Eyes: Rapidly absorbed. Poisoning can happen very rapidly. See **Breathed**.
Skin: Severe poisoning may occur from contact. Passes through the unbroken skin; can increase exposure and the severity of the symptoms listed.
Swallowed: May cause rapid poisoning with headache, sweating, salivation, nausea, vomiting, diarrhea, lower limb tremor and muscular cramping, fever, shortness of breath, loss of consciousness and death. See **Breathed**.
Long-term exposure: May damage the developing fetus or cause birth defects; nerve damage including loss of coordination; liver and kidney damage. Repeated exposure may cause depression, irritability and personality changes.
Storage: Store in a regulated, specially marked place in tightly-closed containers in a dry, well-ventilated place. Keep away from water, strong alkalis. Containers may explose in fire.

★ 844 ★
ROTENONE
CAS: 83-79-4
Other names: AROL GORDON DUST; BARBASCO; (1)BENZOPYRANO (3,4-B) FURO (3,3-H) (1) BENZOPYRANO-6(6AH)-ONE,1,2,12,12A-TETRAHYDRO-8,9- DIMETHOXY-2-(1-METHYLETHENYL)-,2R-2ALPHA,6A ALPHA,12A ALPHA; CENOL GARDEN DUST; CHEM FISH; CHEM-MITE; CUBE; CUBE EXTRACT; CUBE-PULVER; CUBE ROOT; CUBOR; CUREX FLEA DUSTER; DACTINOL; DERIL; DERRIN; DERRIS; DRI-KIL; ENT 133; EXTRAX; FISH-TOX; GREEN CROSS WARBLE POWDER; HAIARI; LIQUID DERRIS; MEXIDE; NCI-C55210; NICOULINE; NOXFISH; PARADERIL; POWDER AND ROOT; PRENTOX; PRO-NOX FISH; RO-KO; RONONE; ROTEFIVE; ROTEFOUR; ROTENON; ROTENONA (SPANISH); ROTESSENOL; ROTOCIDE; TUBATOXIN

Danger profile: Mutagen, exposure limit established, reproductive hazard, poisonous gas produced in fire.
Uses: Insecticide; flea powders; fly sprays; moth-proofing agents; fish poison.
Appearance: Colorless to red solid.
Effects of exposure:
  Breathed: May cause irritation of the nose, throat; nausea, vomiting, stomach pain, loss of coordination, numbness, muscle tremors, convulsions; slow and shallow breathing. Breathing may stop causing death unless rescue breathing is started.
  Eyes: Contact may cause severe irritation and permanent damage.
  Skin: May be poisonous if absorbed. May cause irritation and rash.
  Swallowed: Poisonous. See symptoms under **Breathed**.
Long-term exposure: May cause mutations, liver and kidney damage; skin allergy and rash; ulcers in the eyes, nose or mouth; loss of sense of smell. Possible lung problems including allergy. There is limited evidence of cancer in animals and damage to the developing fetus. There is also limited evidence that chemical is passed on to nursing infants through mother's breast milk.
Storage: Store in tightly closed containers in a dry, dark, cool, well-ventilated place. Keep away from light and strong oxidizers.

S

★845★
SACCHARIN
CAS: 81-07-2
Other names: ANHYDRO-O-SULFAMINEBENZOIC ACID; 1,2-BENZISOTHIAZOLIN-3-ONE,1,1-DIOXIDE, AND SALTS; 3-BENZISOTHIAZOLINONE 1,1-DIOXIDE; 1,2-BENZISOTHIAZOL-3(2H)-ONE 1,1-DIOXIDE; O-BENZOIC SULFIMIDE; O-BENZOSULFIMIDE; BENZOICSULPHIMIDE; O-BENZOIC SULPHIMIDE; BENZOSULPHIMIDE; BENZO-2-SULPHIMIDE; BENZO-SULPHIMIDE; O-BENZOYL SULFIMIDE; 1,2-DIHYDRO-2-KETOBENZISOSULFONAZOLE; 1,2-DIHYDRO-2-KETOBENZISOSULPHONAZOLE; 2,3-DIHYDRO-3-OXOBENZISOSULFONAZOLE; 2,3-DIHYDRO-3-OXOBENZISOSULPHONAZOLE; 1,1-DIOXIDE-1,2- BENZOISOTHIAZOL-3(2H)-ONE; GARANTOSE; GLUCID; GLUSIDE; HERMESETAS; 3-HYDROXYBENZISOTHIAZOLE-S,S-DIOXIDE; INSOLUBLE SACCHARIN; KANDISET; NATREEN; SACARINA; SACCHARIMIDE; SACCHARINA; SACCHARIN ACID; SACCHARINE; SACCHARINOL; SACCHARINOSE; SACCHAROL; SAXIN; SUCRE EDULCOR; SUCRETTE; O-SULFOBENZIMIDE; O-SULFOBENZOIC ACID IMIDE; 2-SULPHOBENZOIC IMIDE; SYKOSE; SYNCAL; ZAHARINA

Danger profile: Possible carcinogen, mutagen, Community Right-to-Know List, poisonous gas produced in fire.
Uses: Dietetic sweetener in various food products.
Appearance: White crystalline powder or solid.
Effects of exposure:
Breathed: High levels may cause loss of appetite, nausea, vomiting, diarrhea.
Swallowed: High levels may cause gastric hyperacidity, anorexia, nausea, vomiting, diarrhea.
Long-term exposure: May cause cancer; mutations; general allergic reaction with rash and itching.
Storage: Store in tightly closed containers in a dry, cool, well-ventilated place.

★846★
SELENIUM
CAS: 7782-49-2
Other names: C.I. 77805; ELEMENTAL SELENIUM; SELEN (POLISH); SELENATE; SELENIUM ALLOY; SELENIUM BASE; SELENIUM DUST; SELENIUM ELEMENTAL; SELENIUM HOMOPOLYMER

Danger profile: Combustible, exposure limit established, Community Right-to-Know List, poisonous gas produced in fire.
Uses: Electronic equipment; photocells, solar batteries; xerographic plates; pigments for ceramics, glass, paints and dyes; making rubber; in animal feeds.
Appearance: Gray, black or red metal-like solid.
Effects of exposure:
Breathed: May be poisonous. May cause irritation to the throat, nose and breathing passages. High levels may cause breathing problems, lung irritation; headaches. Higher exposures may cause a dangerous build-up of fluids in the lungs (pulmonary edema), which can cause death. Pulmonary edema is a medical emergency that can be delayed one to two days following exposure.
Eyes: May cause irritation.
Swallowed: May be poisonous. May cause irritation, gastric problems.
Long-term exposure: Repeated overexposure may cause garlic odor on breath, metallic taste; nervousness, fear, irritability and depression; fatigue; dental cavities; upset stomach; vomiting; hair and fingernail loss; nervous system effects; liver problems and damage.
Storage: Store in tightly closed containers in a dry, cool, well-ventilated place. Keep away from strong oxidizers and acids.

★847★
SELENIUM HEXAFLUORIDE
CAS: 7783-79-1
Other names: SELENIUM FLUORIDE

Danger profile: Poison, exposure limit established, Community Right-to-Know List, poisonous gas produced in fire.
Uses: Making other chemicals; gaseous electric insulator.
Appearance: Colorless gas.
Effects of exposure:
Breathed: May be poisonous. May cause irritation of the eyes, nose, throat and breathing passages; breathing difficulties; sneezing, headache, dizziness, nausea and vomiting, cold sweat, diarrhea, weakness and tiredness. High exposures may cause shortness of breath and a dangerous build-up of fluids in the lungs (pulmonary edema), which can cause death. Pulmonary edema is a medical emergency that can be delayed one to two days following exposure. Very high exposures may cause a sudden unconsciousness and death.
Eyes: May cause irritation, tearing, redness and swelling of the eyelids; serious injury. Contact with liquid may cause frostbite.
Skin: May cause irritation, inflammation and possible skin coloration. Contact with liquid may cause frostbite.
Long-term exposure: Repeated overexposure may cause garlic odor on breath, metallic taste; pallor, nervousness, irritability and depression; fatigue; dental cavities; digestive disturbances, upset stomach; vomiting; hair and fingernail loss; liver problems and damage.
Storage: Store in tightly closed containers in a dry, cool, well-ventilated place. Containers may explode in fire.
First aid guide: Move victim to fresh air and call emergency medical care; if not breathing, give artificial respiration; if breathing is difficult, give oxygen. In case of contact with material, immediately flush skin or eyes with running water for at least 15 minutes. Remove and isolate contaminated clothing and shoes at the site. Keep victim quiet and maintain normal body temperature. Effects may delayed; keep victim under observation.

★848★
SELENIUM SULFIDE
CAS: 7446-34-6
Other names: NCI-C50033; SELENIUM MONOSULFIDE; SELENIUM SULPHIDE; SELENSULFID (GERMAN); SULFUR SELENIDE

Danger profile: Carcinogen, mutagen, exposure limit established, Community Right-to-Know List, poisonous gas produced in fire.
Uses: Found in shampoos.
Appearance: Bright orange-yellow tablets or powder.
Odor: The odor is a warning of exposure; however, no smell does mean you are not being exposed.
Effects of exposure:
Breathed: May be poisonous. Dust or mists may cause irritation of the eyes, nose, throat and breathing passages; nosebleeds, coughing, phlegm, headaches. High levels may cause lung irritation and difficult breathing.
Eyes: May cause severe irritation.
Skin: May pass through the unbroken skin; can increase exposure and the severity of the symptoms listed. May cause severe irritation.
Swallowed: Poisonous.

Long-term exposure: May cause cancer of the liver and lungs; repeated overexposure may cause garlic odor on breath, metallic taste; nervousness, fear, irritability and depression; fatigue; dental cavities; upset stomach; vomiting; hair and fingernail loss; liver problems and damage; brain and spinal cord damage with possible disability and death. Related chemicals may cause damage in the developing fetus, and decrease fertility in women.
Storage: Store in tightly closed containers in a dry, cool, well-ventilated place. Keep away from silver oxide, oxidizers and acids.

★ 849 ★
SILANE
CAS: 7803-62-5
Other names: MONOSILANE; SILICANE; SILICON TETRAHYDRIDE

Danger profile: Combustible, explosion hazard, exposure limit established, poisonous gas produced in fire.
Uses: Making solid-state devices and amorphous silicon.
Appearance: Colorless gas.
Odor: Repulsive. The odor is a warning of exposure; however, no smell does not mean you are not being exposed.
Effects of exposure:
  Breathed: May be poisonous. Vapor may cause intense irritation to nose, throat, mouth and lungs; mucous membrane. Symptoms of exposure include watering of the eyes, coughing, difficult breathing, headache, nausea, vomiting, muscular weakness, unconsciousness. High exposure can lead to drowsiness, coma and death; buildup of fluid in the lungs (pulmonary edema), and may cause death. Pulmonary edema may be delayed by one to two days following exposure.
  Eyes: Contact may cause, burning and stinging, tearing, irritation and redness and swelling of the eyelids, intense pain; can burn the eyes and lead to blurred vision, chronic irritation and other permanent damage. Liquid may cause frostbite.
  Skin: Contact may cause burning or stinging feeling, redness, swelling, intense pain, blisters, irritation, rash. Repeated exposure may lead to skin allergy and chronic irritation. Liquid may cause frostbite.
  Swallowed: Highly toxic. May cause burning feeling of the mouth and throat, intense thirst, throat swelling, stomach cramps, nausea, vomiting, diarrhea, severe breathing difficulties and unconsciousness. Tissue damage may result. Breathing may stop. Watch for delayed symptoms.
Storage: Store in tightly closed containers under a blanket of inert gas in a dry, cool, well-ventilated place. Keep away from sources of ignition, oxygen, heat, oxidizers, bromine, chlorine, antimony pentachloride, carbonyl chloride, tin(IV) chloride and other chemicals. May ignite or explode in air or when heated. Containers may explode in fire.
First aid guide: Move victim to fresh air and call emergency medical care; if not breathing, give artificial respiration; if breathing is difficult, give oxygen. In case of frostbite, thaw frosted parts with water. Keep victim quiet and maintain normal body temperature.

★ 850 ★
SILICA, AMORPHOUS, HYDRATED
CAS: 7631-86-9
Other names: SILICA AEROGEL; SILICA GEL; SILICA XEROGEL; SILICIC ACID

Danger profile: Exposure limit established, poisonous gas produced in fire.
Uses: Moisture absorbent; air conditioning; drying gases and liquids; chromatography; in cosmetics, pharmaceuticals, waxes and dietary suppliments. Pure form is non-toxic.

Appearance: A white to gray amorphous powder.
Effects of exposure:
  Breathed: May cause irritation of the nose, coughing and sneezing.
  Eyes: May cause irritation, redness, tearing, pain and inflammation of the eyelids.
Storage: Store in tightly closed containers in a dry, cool, well-ventilated place.

★ 851 ★
SILICA, CRYSTALLINE-CRISTOBALITE
CAS: 14464-46-1
Other names: CALCINED DIATOMITE; CRISTOBALITE

Danger profile: Carcinogen, exposure limit established, poisonous gas produced in fire.
Uses: Making fiber glass; in ceramics; foundry molds, iron and steel castings.
Appearance: Colorless or white crystalline solid.
Effects of exposure:
  Breathed: Very high levels of exposure over a few weeks may cause acute silicosis, a disabling, progressive form of lung scarring; fever, cough, shortness of breath, difficult breathing, reduced lung function, weight loss, crippling and possible death. If silicosis develops, the chances of getting tuberculosis are increased.
Long-term exposure: May cause cancer; silicosis, a progressive and frequently incapacitating disease, coughing, wheezing, fever and impaired pulmonary function; disability or death may occur. The chances of developing tuberculosis is much more likely if person has silicosis and exposed to tuberculosis germs.
Storage: Store in tightly closed containers in a dry, cool, well-ventilated place. Keep away from oxidizers.

★ 852 ★
SILICA, FUSED
CAS: 60676-86-0
Other names: AMORPHOUS FUSED SILICA; FUSED QUARTZ; FUSED SILICA; QUARTZ GLASS; SG-67; SILICA, AMORPHOUS FUSED; SILICA, VITREOUS; SILICON DIOXIDE; SILICONE DIOXIDE; SILICON DIOXIDE (AMORPHOUS); SUPRASIL; VITREOUS QUARTZ

Danger profile: Carcinogen, exposure limit established, poisonous gas produced in fire.
Uses: Making fiber glass, special camera lenses and fibers for reinforcing plastics; in ceramics; iron and steel castings.
Appearance: Microscopic spherical particles.
Effects of exposure:
  Breathed: Very high levels of exposure over a few weeks may cause acute silicosis, a disabling, progressive form of lung scarring; fever, cough, shortness of breath, difficult breathing, reduced lung function, weight loss, crippling and possible death. If silicosis develops, the chances of getting tuberculosis are increased.
Long-term exposure: May cause cancer. Silicosis may develop over many years; disability may occur. The chances of developing tuberculosis is much more likely if person has silicosis and exposd to tuberculosis germs.
Storage: Store in tightly closed containers in a dry, cool, well-ventilated place. Keep away from strong oxidizers, oxygen difluoride, chrome fluoride and manganese trioxide.

★ 853 ★
SILICA, MICA
CAS: 12001-26-2
Other names: AMBER MICA; BIOTITE; LEPIDOLITE; MICA; MICA SILICATE; MUSCOVITE; PHLOGOPITE; ROSCOELITE; SUZORITE MICA; ZINNWALDTE; FLUOROPHLOGOPITE

Danger profile: Exposure limit established, poisonous gas produced in fire.
Uses: Electrical insulation; vacuum tubes; incandescent lamps; lubricant; in exterior paints; cosmetics; glass and ceramic flux; rubber; making roofing shingles, wallpaper, wallboard joint cement and plastics.
Appearance: Transparent, transparent flakes or thin sheets.
Effects of exposure:
   Breathed: May cause lung irritation.
Long-term exposure: May cause lung scarring; shortness of bereath, abnormal chest x-ray; cough; disability may occur; weak; weight lost; clubbing of fingers.
Storage: Store in tightly closed containers in a dry, cool, well-ventilated place.

★ 854 ★
SILICA, QUARTZ
CAS: 14808-60-7
Other names: AGATE; AMETHYST; CHALCEDONY; CHERTS; FLINT; ONYX; PURE QURTZ; QUARTZ; QUAZO PURO (ITALIAN); ROSE QUARTZ; SAND; SILICA, CRYSTALLINE QUARTZ; SILICA FLOUR; SILICIC ANHYDRIDE

Danger profile: Carcinogen, exposure limit established, poisonous gas produced in fire.
Uses: Making fiberglass, electrical insulation; in metal polishes, abrasives, paints; oil drilling muds; chemical filtration.
Appearance: Colorless crystalline solid.
Effects of exposure:
   Breathed: Very high levels of exposure over a few weeks may cause acute silicosis, a disabling, progressive form of lung scarring; fever, cough, shortness of breath, difficult breathing, reduced lung function, weight loss, crippling and possible death. If silicosis develops, the chances of getting tuberculosis are increased.
Long-term exposure: May cause cancer; silicosis, a progressive and frequently incapacitating disease, coughing, wheezing, fever and impaired pulmonary function; diminished chest expansion; disability may occur; liver problems; disability or death may occur. The chances of developing tuberculosis is much more likely if person has silicosis and exposed to tuberculosis germs.
Storage: Store in tightly closed containers in a dry, cool, well-ventilated place. Keep away from oxidizers, chlorine trifluoride, manganese trioxide, and oxygen trifluoride.

★ 855 ★
SILVER
CAS: 7440-22-4
Other names: ARGENTUM; C.I. 77820; SHELL SILVER; SILBER (GERMAN)

Danger profile: Exposure limit established, Community Right-to-Know List, toxic fumes produced in fire.
Uses: Making other chemicals; photographic chemicals; mirrors; electric conductors and equipment; batteries; electroplating; silverware, jewelry; dental, medical and scientific equipment.
Appearance: Shiny, malleable metal.
Effects of exposure:

Eyes: Metal dust or fragments may cause scratches of the eye surface, or become imbedded there and cause gray staining of the eyeball.
Skin: Metal dust or fragments may enter cuts and create permanent marks.
Long-term exposure: May cause kidney damage. All forms of silver accumulate and are excreted very slowly. Blue-grey discoloration of eyes, nose, throat and skin may occur if one gram (1/30 ounce) accumulates in the body. This discoloration is likely to be permanent and disfiguring. Prolonged absorption of fumes or dust may lead to grayish blue discoloration of skin, eyes, nails, mucous membrane and body organs; vision problems.
Storage: Store in tightly closed containers in a dry, cool, well-ventilated place. Keep away from ammonia, acetylene, hydrogen peroxide. Silver dust is combustible.

★ 856 ★
SILVER CYANIDE
CAS: 506-64-9
Other names: CYANURE D'ARGENT (FRENCH); KYANID STRIBRNY (CZECH)

Danger profile: Combustible, poison, exposure limit established, Community Right-to-Know List, poisonous gas produced in fire.
Uses: Silver plating.
Appearance: White or grayish powder that darkens when exposed to light.
Odor: The odor is a warning of exposure; however, no smell does not mean you are not being exposed.
Effects of exposure:
   Breathed: Poisonous. Heated chemical releases deadly cyanide gas; gasping, dizziness, headache, shortness of breath, weakness and loss of consciousness may happen rapidly.
   Eyes: May cause irritation.
   Skin: May be poisonous if absorbed. Passes through the unbroken skin; can increase exposure and the severity of the symptoms listed. May cause irritation.
   Swallowed: Deadly poison.
Long-term exposure: May cause kidney damage. All forms of silver accumulate and are excreted very slowly. Blue-grey discoloration of eyes, nose, throat and skin may occur if one gram (1/30 ounce) accumulates in the body. Dust and fumes may cause permanent blue-grey discoloration of skin, mucous membranes and eyes. This discoloration is likely to be permanent and disfiguring.
Storage: Store in tightly closed containers in a dry, cool, well-ventilated place. Keep away from light, heat, fluorine, acetylene, ammonia, phosphorus tricyanide and hydrogen peroxide.
First aid guide: Move victim to fresh air; call emergency medical care. In case of contact with material, immediately flush skin or eyes with running water for at least 15 minutes. Remove and isolate contaminated clothing and shoes at the site.

★ 857 ★
SILVER NITRATE
CAS: 7761-88-8
Other names: LUNAR CAUSTIC; NITRATE D'ARGENT (FRENCH); NITRIC ACID, SILVER(1) SALT; SILVER(1) NITRATE; SILBERNITRAT

Danger profile: Mutagen, poison, corrosive, exposure limit established, Community Right-to-Know List, poisonous gas produced in fire.
Uses: Photographic film; making other chemicals; indelible inks; silver plating; mirrors; antiseptic and germicide; hair dyeing; laboratory chemical.

Appearance: Colorless to grayish black solid crystals.
Effects of exposure:
Eyes: May cause severe irritation, burns and permanent blindness.
Skin: Passes through the unbroken skin; can increase exposure and the severity of the symptoms listed. Contact may cause severe irritation and burns; ulceration and discoloration of the skin.
Swallowed: May cause violent abdominal pain and other gastroenteric symptoms that may cause death.
Long-term exposure: May cause mutations. All forms of silver accumulate and are excreted very slowly. Blue-grey discoloration of eyes, nose, throat and skin may occur if one gram (1/30 ounce) accumulates in the body. This discoloration is likely to be permanent and disfiguring. Possible kidney damage and lung problems.
Storage: Store in tightly closed black or amber glass bottles in a dry, cool, well-ventilated place. Do not store in plastic containers. Keep away from sunlight, oxidizers, alkalies, oils, fuels and all other combustible materials and chemicals.
First aid guide: Move victim to fresh air and call emergency medical care; if not breathing, give artificial respiration; if breathing is difficult, give oxygen. In case of contact with material, immediately flush skin or eyes with running water for at least 15 minutes. Remove and isolate contaminated clothing and shoes at the site. Keep victim quiet and maintain normal body temperature. Effects may be delayed; keep victim under observation.

★ 858 ★
SILVER PICRATE
CAS: 146-84-9
Other names: PICRIC ACID, SILVER(I) SALT; PHENOL, 2,4,6-TRINITRO-, SILVER(1) SALT.

Danger profile: Carcinogen, combustible, explosion hazard, Community Right-to-Know List, poisonous gas produced in fire, may explode if dried chemical is exposed to heat or shock.
Uses: In antibacterial medications.
Appearance: Yellow powder or crystalline material; turns brown when heated or exposed to light.
Odor: The odor is a warning of exposure; however, no smell does not mean you are not being exposed.
Effects of exposure:
Eyes: Contact may cause severe irritation, burns and permanent damage.
Skin: Passes through the unbroken skin; can increase exposure and the severity of the symptoms listed. May cause severe irritation and burns.
Long-term exposure: May cause cancer or mutations. All forms of silver accumulate and are excreted very slowly. Blue-grey discoloration of eyes, nose, throat and skin may occur if one gram (1/30 ounce) accumulates in the body. This discoloration is likely to be permanent and disfiguring. Possible kidney damage and lung problems.
Storage: Treat as an explosive. Store in tightly closed containers in a dry, cool, well-ventilated place. Keep material wet with water. Keep away from heat, sources of ignition, organic materials. Protect containers from physical damage or sudden impact. Containers may explode if exposed to heat, flame or shock.
First aid guide: Move victim to fresh air; call emergency medical care. In case of contact with material, immediately flush skin or eyes with running water for at least 15 minutes. Remove and isolate contaminated clothing and shoes at the site.

★ 859 ★
SODIUM
CAS: 7440-23-5
Other names: NATRIUM; SODIUM, METAL LIQUID ALLOY; SODIUM METAL

Danger profile: Combustible, explosion hazard, ignites spontaneously in air or oxygen, reacts violently with water producing hydrogen gas, poisonous gas and toxic fumes produced in fire.
Uses: Making other chemicals; laboratory chemical.
Appearance: Soft metal, silvery-white solid; when freshly cut, tarnishes rapidly to gray color.
Effects of exposure:
Breathed: May be poisonous. Vapors may cause irritation of the eyes, nose, throat and lungs; with burning feeling; difficult breathing, coughing and sneezing. High exposure can lead to a buildup of fluid in the lungs (pulmonary edema), and may cause death. Pulmonary edema may be delayed by one to two days following exposure.
Eyes: May cause severe irritation with pain in eyes and eyelids; intense watering; burns and damage to the mucous membrane, perforation of the eyes, permanent damage and blindness.
Skin: May cause irritation, pain and serious, deep chemical burns.
Swallowed: May cause immediate burning feeling in the mouth, throat and stomach; intense pain in swallowing; throat swelling, vomiting (hemorrhaging may cause vomitus to resemble coffee grounds), watery or bloody diarrhea, shallow respirations; confusion, delirium, coma.
Storage: Requires special precautions to avoid contact with moisture, including condensation from other objects and perspiration. Store in an inert atmosphere (such as nitrogen) or under oil (such as kerosene) in tightly closed containers in a dry, cool, well-ventilated fire resistant place. Keep away from water, air, oxygen, heat, and all other chemicals. Protect storage containers from physical damage. May ignite combustible materials if they are damp. Containers may explode in fire.
First aid guide: Move victim to fresh air; call emergency medical care. Wipe material from skin immediately; flush skin or eyes with running water for at least 15 minutes. Remove and isolate contaminated clothing and shoes at the site.

★ 860 ★
SODIUM ALUMINUM HYDRIDE
CAS: 13770-96-2
Other names: ALUMINATE(1-), TETRAHYDRO-, SODIUM; ALUMINUM SODIUM HYDRIDE; SAH 22; SODIUM ALUMINUM TETRAHYDRIDE; SODIUM TETRAHYDROALUMINATE(1-); (T-4) SODIUM; TETRAHYDROALUMINATE (1-) (9CI)

Danger profile: Combustible, dangerous when wet, exposure limit established, poisonous gas produced in fire.
Uses: Making other chemicals.
Appearance: White crystalline powder.
Effects of exposure:
Breathed: May be poisonous. May cause irritation of the eyes, nose and throat.
Eyes: May cause irritation and burns.
Skin: May cause irritation and burns.
Swallowed: May cause severe irritation.
Long-term exposure: May cause lung problems.
Storage: Store in tightly closed containers in a dry, cool, well-ventilated place. Keep away from heat, sources of ignition, water, air, oxidizers, acids, alcohols, tetrahydrofuran and ethers. May ignite and explode spontaneously in moist air. Water may produce a combustible gas. Containers may explode in fire.

First aid guide: Move victim to fresh air; call emergency medical care. Wipe material from skin immediately; flush skin or eyes with running water for at least 15 minutes. Remove and isolate contaminated clothing and shoes at the site.

★ 861 ★
SODIUM ALUMINUM OXIDE
CAS: 11138-49-1
Other names: BETA-ALUMINA; BETA''-ALUMINA; NALCO 680; SODIUM ALUMINATE; SODIUM ALUMINATE, SOLID; SODIUM POLYALUMINATE; ALUMINUM SODIUM OXIDE

Danger profile: Corrosive (solution), poisonous gas produced in fire.
Uses: Making soap, cleaning compounds and milk glass.
Appearance: White granular mass, powder or colorless liquid solution.
Effects of exposure:
  Breathed: May cause irritation of the breathing passages; cough phlegm; difficult breathing, shortness of breath.
  Eyes: May cause irritation, burns and permanent damage.
  Skin: May cause irritation and burns.
Long-term exposure: May cause lung problems.
Storage: Store in tightly closed containers in a dry, cool, well-ventilated place.

★ 862 ★
SODIUM ARSENATE
CAS: 7631-89-2
Other names: ARSENIC ACID, SODIUM SALT; FATSCO ANT POISON; SODIUM METAARSENATE; SODIUM ORTH-OARSENATE; SWEENEY'S ANT-GO

Danger profile: Carcinogen, mutagen, reproductive hazard, exposure limit established, reproductive hazard, poisonous gas produced in fire and on contact with acids.
Uses: In dyeing and printing; making other chemicals; germicide.
Appearance: White or clear, colorless crystalline solid.
Effects of exposure:
  Breathed: May be poisonous. May cause irritation of the nose and throat. Inhalation of large doses can cause nausea, vomiting, muscle cramps, laryngitis, bronchitis.
  Eyes: May cause irritation and sore eyes.
  Skin: May pass through the unbroken skin. Contact may cause burning, itching, skin eruptions, or chronic poisoning.
  Swallowed: Poisonous. May cause constriction in throat and difficulty in swallowing; also causes burning and pain, vomiting, profuse diarrhea, dehydration, cyanosis, coma, convulsions, and death.
Long-term exposure: May cause cancer, birth defects; skin thickening and change of color; nose ulcers and lesions; nervous and respiratory system damage; circulatory system disorders; hearing loss and death.
Storage: Store in tightly closed containers in a dry, cool, well-ventilated place. Keep away from acids, chemically active metals and heat.
First aid guide: Move victim to fresh air; call emergency medical care. In case of contact with material, immediately flush skin or eyes with running water for at least 15 minutes. Remove and isolate contaminated clothing and shoes at the site.

★ 863 ★
SODIUM AZIDE
CAS: 26628-22-8
Other names: AXIUM; AZOTURE DE SODIUM (FRENCH); HYDRAZOIC ACID, SODIUM SALT; KAZOE; NATRIUMAZID (GERMAN); NATRIUMMAZIDE (DUTCH); NCI- C06462; NSC 3072; SODIUM, AZOTURE DE (FRENCH); SODIUM AZO-TURO (ITALIAN); U-3886

Danger profile: Mutagen, poison, exposure limit established, EPA extremely hazardous substances list, poisonous gas produced in fire.
Uses: Making other chemicals, explosives, medicines; air bag inflation.
Appearance: Colorless, crystalline or white solid, or solution.
Effects of exposure:
  Breathed: May be poisonous. May cause irritation of the nose and throat; headache, stuffy nose, blurred vision, dizziness, faint feeling, shortness of breath, pounding heart; decreased blood pressure, collapse. Higher exposures may cause shortness of breath and a dangerous build-up of fluids in the lungs (pulmonary edema), which can cause death. Pulmonary edema is a medical emergency that can be delayed one to two days following exposure. May effect vision and the nervous system.
  Eyes: Dust may cause irritation.
  Skin: May be poisonous if absorbed. Passes through the unbroken skin; can increase exposure and the severity of the symptoms listed. Solid and solution may cause irritation and blistering.
  Swallowed: Poisonous. May cause sleepiness; kidney effects. See *First aid guide.*
Long-term exposure: May cause mutations; brain damage; nervous system effects; vision loss; liver damage.
Storage: Store in tightly closed containers in a dry, cool, well-ventilated place. Keep away from heat, metals such as lead, silver, mercury, brass and copper, sulfuric acid, benzoyl chloride, potassium hydroxide, bromine. Protect containers from friction and contamination. Containers may explode in fire.
First aid guide: Move victim to fresh air and call emergency medical care; if not breathing, give artificial respiration; if breathing is difficult, give oxygen. In case of contact with material, immediately flush skin or eyes with running water for at least 15 minutes. Speed in removing material from skin is of extreme importance. Remove and isolate contaminated clothing and shoes at the site. Keep victim quiet and maintain normal body temperature. Effects may be delayed, keep victim under observation.

★ 864 ★
SODIUM BISULFATE
CAS: 7681-38-1
Other names: NITRE CAKE; SODIUM ACID SULFATE; SODIUM BISULFATE, FUSED; SODIUM BISULFATE, SOLID; SODIUM HYDROGEN SULFATE; SODIUM HYDROGEN SULFATE, SOLID; SODIUM PYROSULFATE; SULFURIC ACID, MONOSODIUM SALT

Danger profile: Mutagen, combustible, corrosive, poisonous gas produced in fire.
Uses: In dyeing; making other chemicals; disinfectant; treating leather; pickling metal.
Appearance: Colorless solid.
Effects of exposure:
  Breathed: Dust or mist may cause irritation of the nose, throat, and air passages. Higher exposures may cause shortness of breath and a dangerous build-up of fluids in the lungs (pulmonary edema), which can cause death. Pulmonary

edema is a medical emergency that can be delayed one to two days following exposure.
Eyes: Contact may cause severe irritation, burns and possible permanent damage.
Skin: Contact may cause severe irritation and burns.
Long-term exposure: May cause mutations. Similar and related chemicals may cause lung problems and lead to erosion of the teeth; emphasema. It is unknown at this time whether chemical may cause these health effects.
Storage: Store in tightly closed containers in a dry, cool, well-ventilated place. Keep away from all forms of moisture and calcium hypochlorite. Containers may explode in fire.
First aid guide: Move victim to fresh air; call emergency medical care. In case of contact with material, immediately flush skin or eyes with running water for at least 15 minutes. Remove and isolate contaminated clothing and shoes at the site. Keep victim quiet and maintain normal body temperature.

★ 865 ★
SODIUM BISULFITE
CAS: 7631-90-5
Other names: BISULFITE DE SODIUM (FRENCH); HYDROGEN SULFITE SODIUM; SODIUM ACID SULFITE; SODIUM BISULFITE, (SOLID); SODIUM BISULFITE, SOLID; SODIUM HYDROGEN SULFITE; SODIUM HYDROGEN SULFITE, SOLID; SODIUM SULHYDRATE; SULFUROUS ACID, MONOSODIUM SALT

Danger profile: Mutagen, corrosive, allergen, exposure limit established, poisonous gas produced in fire.
Uses: Making leather, paper, dyes, and other chemicals; food preservative; textiles; bleaching wool and other products; antiseptic; in brewing (cask cleaning); copper and brass plating; laboratory chemical; dietary supplement.
Appearance: White to yellowish crystalline powder; or clear yellowish solution.
Odor: Rotten eggs or sulfur dioxide when moist. The odor is a warning of exposure; however, no smell does mean you are not being exposed.
Effects of exposure:
Breathed: May cause irritation of the nose and throat; cough, shortness of breath; severe allergic; collapse. May cause lung damage.
Eyes: May cause severe irritation and permanent damage.
Skin: May cause irritation, redness and burns.
Swallowed: May cause stomach irritation. Large doses may cause volent colic, diarrhea, depression and death. Eating food treated with chemical may cause allergic reaction.
Long-term exposure: May cause mutations; asthma-like allergy; lung damage.
Storage: Store in tightly closed containers in a dry, cool, well-ventilated place. Keep away from acids, oxidizers.
First aid guide: Move victim to fresh air; call emergency medical care. In case of contact with material, immediately flush skin or eyes with running water for at least 15 minutes. Remove and isolate contaminated clothing and shoes at the site. Keep victim quiet and maintain normal body temperature.

★ 866 ★
SODIUM BORATE
CAS: 1303-96-4
Other names: ANTIPYONIN; BORASCU; BORATES, TETRA, SODIUM SALT; BORAX; BORAX DECAHYDRATE; BORICIN; GERTLEY BORATE; KAIKIN; NEOBOR; POLYBOR; SODIUM BIBORATE; SODIUM BIBORATE DECAHYDRATE; SODIUM BORATE, ANHYDROUS; SODIUM BORATE DECAHYDRATE; SODIUM DIBORATE; SODIUM METABORATE; SODIUM PYROBORATE; SODIUM PYROBORATE DECAHYDRATE; SODIUM TETRABORATE; SODIUM TETRABORATE DECAHYDRATE

Danger profile: Combustible, exposure limit established, Community Right-to-Know List, poisonous gas produced in fire.
Uses: In soaps, detergents, fertilizers, insecticides, ant poisons; making glazes and enamels; cleaning compounds; heat-resistant glass; herbicides; rust inhibitors; pharmaceuticals; leather; bleaches; paint; flame-retardant fungicide for wood; soldering metals; laboratory chemical; photography chemicals.
Appearance: White, to light gray crystals or solids.
Effects of exposure:
Breathed: Dust may cause severe irritation to the nose, throat and lungs, causing coughing, nosebleeds, shortness of breath. May cause flu-like symptoms including chills, muscle ache, headache, fever. High exposure may cause nausea, salivation, vomiting, cramps, diarrhea, chest pains, cough, a build-up of fluids in the lungs (pulmonary edema) and possible death. Pulmonary edema may be delayed from one to two days following exposure.
Eyes: May cause severe irritation, burns, pain and possible blindness.
Skin: Passes through the unbroken skin; can increase exposure and the severity of the symptoms listed. May cause severe irritation and thermal and acid burns, especially if skin is wet.
Swallowed: May cause severe poisoning with burns of the mouth, throat and stomach, vomiting, watery or bloody diarrhea, shock, coma and death. Damage to kidneys. Ingestion of five to ten grams in young children can cause severe abdominal pain, vomiting, diarrhea, shock, coma, death.
Long-term exposure: May cause chronic stomach pain; nose throat and lung damage, nosebleeds and shortness of breath; skin rash.
Storage: Store in tightly closed containers in a dry, cool, well-ventilated place. Keep away from combustable materials, zirconium, acids and metallic salts.

★ 867 ★
SODIUM CARBONATE (2:1)
CAS: 497-19-8
Other names: CARBONIC ACID, DISODIUM SALT; CRYSTOL CARBONATE; DISODIUM CARBONATE; SODA ASH; TRONA

Danger profile: Poisonous gas produced in fire.
Uses: Smokeless powder; solvent; laboratory chemical; anesthetic. Sodium carbonate has been given the designation of "GRAS" (Generally Recognized As Safe) status by the food and drug administration at levels currently used as a food additive. It occurs often in baked goods as baking soda or baking powder.
Appearance: White or grayish-white crystals or powder.
Effects of exposure:
Breathed: May cause irritation of nose, eyes and throat; coughing; difficult breathing.
Eyes: Contact may cause irritation.
Skin: Contact may cause irritation, burning feeling.

Swallowed: Alkaline tasting. May cause burning feeling in the mouth and stomach; mouth and gastric irritation; stomach cramps.
Storage: Store in tightly closed containers in a dry, cool, well-ventilated place. Keep away from aluminum, sulfuric acid, lithium.

★868★
SODIUM CHLORATE
CAS: 7775-09-9
Other names: ASEX; ATLACIDE; ATRATOL B-HERBATOX; CHLORATE OF SODA; CHLORATE SALT OF SODIUM; CHLORAX; CHLORIC ACID, SODIUM SALT; CHLORSAURE (GERMAN); DE-FOL-ATE; DESOLET; DREXEL DEFOL; DROP LEAF; EVAU-SUPERFALL; GRAIN SORGHUM HARVEST AID; GRANEX OK; HARVEST-AID; KLOREX; KUSA-TOHRUKUSATOL; LOREX; NATRIUMCHLORAAT (DUTCH); NATRIUMCHLORAT (GERMAN); ORTHO C-1 DEFOLIANT & WEED KILLER; OXYCIL; RASIKAL; SHED-A-LEAF; SHED-A-LEAF "L"; SODA CHLORATE; SODIO (CLORATO DI) (ITALIAN); SODIUM (CHLORATE DE) (FRENCH); TRAVEX; TUMBLEAF; UNITED CHEMICAL DEFOLIANT NO. 1; VAL-DROP

Danger profile: Mutagen, combustible, poison, explosion hazard, poisonous gas produced in fire.
Uses: Making paper pulps, dyes, leather, textiles; ore processing; herbicide, weed killer; matches; explosives; flares and fireworks.
Appearance: Colorless to pale yellow solid crystals or powder.
Effects of exposure:
  Breathed: May cause irritation of the nose and throat. May deprive the body of oxygen and cause skin and lips to turn blue. Severe exposure may cause kidney damage.
  Eyes: Contact may cause severe irritation and burns.
  Skin: Contact may cause severe irritation and burns.
  Swallowed: Salty taste. May cause nausea, vomiting, diarrhea, abdominal or gastric pain, dyspnea and other symptoms. The major cause of death from a lethal dose is acute renal failure. Severe exposure may cause kidney damage.
Long-term exposure: May cause damage to the kidneys; blood cells.
Storage: Highly reactive with many products and chemicals. Store in tightly closed containers in a dry, cool, well-ventilated place with non-wood floors. Keep away from sources of ignition, wood and other organic matter, sulfur and metals, ammonium thiosulfate, potassium cyanide, antimony sulfide, arsenic, carbon, charcoal, sulfuric acid, chemical active metals, and thiocyanates. Containers may explode in fire.
First aid guide: Move victim to fresh air; call emergency medical care. In case of contact with material, immediately flush skin or eyes with running water for at least 15 minutes. Remove and isolate contaminated clothing and shoes at the site.

★869★
SODIUM CHLORITE
CAS: 7758-19-2
Other names: CHLOROUS ACID, SODIUM SALT; TEXTILE; TEXTONE

Danger profile: Mutagen, combustible, poison, explosion hazard, corrosive, poisonous gas produced in fire.
Uses: For improving taste and odor of drinkable water; bleaching agent for textiles, paper pulp, shellacs, varnishes, waxes and straw products; laboratory chemical.
Appearance: White crystals, crystalline powder, or liquid.
Effects of exposure:
  Breathed: Vapor may cause intense irritation to nose, throat, mouth and lungs; mucous membrane. Symptoms of expo-

sure include watering of the eyes, nosebleeds, sore throat, coughing, difficult breathing, headache, nausea, vomiting, muscular weakness, unconsciousness. High exposure can lead to drowsiness, coma and death; buildup of fluid in the lungs (pulmonary edema), and may cause death. Pulmonary edema may be delayed by one to two days following exposure.
  Eyes: Severe irritant. Contact may cause, burning and stinging, tearing, irritation and redness and swelling of the eyelids, intense pain; can burn the eyes and lead to blurred vision, chronic irritation and other permanent damage.
  Skin: Severe irritant. Contact may cause burning or stinging feeling, redness, swelling, intense pain, blisters, irritation, rash.
  Swallowed: May cause burning feeling of the mouth and throat, intense thirst, throat swelling, stomach cramps, nausea, vomiting, diarrhea, severe breathing difficulties and unconsciousness. Tissue damage may result. Breathing may stop. Watch for delayed symptoms.
Long-term exposure: May cause mutations; damage to the developing fetus. Possible lung problems.
Storage: Highly reactive. Store in tightly closed containers in a dry, cool, well-ventilated place. Keep away from heat, friction, sources of ignition, acids, oils, organic matter, chemically active metals. Protect the containers from shock. Containers may explode in fire.
First aid guide: Move victim to fresh air; call emergency medical care. In case of contact with material, immediately flush skin or eyes with running water for at least 15 minutes. Remove and isolate contaminated clothing and shoes at the site.

★870★
SODIUM CHLOROPLATINATE
CAS: 1307-82-0
Other names: PLATINIC SODIUM CHLORIDE; PLATINATE(2-),HEXACHLORO-, DISODIUM,TETRAHYDRATE; SODIUM HEXACHLOROPLATINATE (IV); SODIUM PLATINIC CHLORIDE

Danger profile: Exposure limit established, poisonous gas produced in fire.
Uses: Making other chemicals.
Appearance: Yellow crystalline solid.
Effects of exposure:
  Breathed: May cause irritation to the eyes, nose, throat and breathing passages. High exposures may cause irritability and seizures. May cause asthma-like allergy to develop; shortness of breath.
  Eyes: May cause irritation and burning feeling.
Long-term exposure: May cause asthma-like lung allergy; skin allergy. Repeated exposure may result in lung scarring and sores and ulcers on the lining of the nose.
Storage: Store in tightly closed containers in a dry, cool, well-ventilated place. Containers may explode in fire.

★871★
SODIUM CHROMATE
CAS: 7775-11-3
Other names: CHROMATE OF SODA; CHROMIUM DISODIUM OXIDE; CHROMIUM SODIUM OXIDE; DISODIUM CHROMATE; NEUTRAL SODIUM CHROMATE; SODIUM CHROMATE(VI)

Danger profile: Carcinogen, mutagen, poison, combustible, exposure limit established, Community Right-to-Know List, poisonous gas produced in fire.
Uses: Leather tanning; making dyes, inks, pigments; protection of iron against corrosion; wood preservative.

Appearance: Yellow crystalline solid.
Effects of exposure:
Breathed: Dust may cause severe irritation to the nose, throat and lungs, causing coughing, shortness of breath; may ulcerate mucous membranes; continued irritation of the nose may lead to perforation of the nasal septum. May cause flu-like symptoms including chills, muscle ache, headache, fever. High exposure may cause nausea, salivation, vomiting, cramps, diarrhea, chest pains, cough, a build-up of fluids in the lungs (pulmonary edema) and possible death. Pulmonary edema may be delayed from 1-2 days following exposure.
Eyes: May cause severe irritation, inflammation of the eyelids, burns, pain and possible blindness.
Skin: Passes through the unbroken skin; can increase exposure and the severity of the symptoms listed. May cause severe irritation and can cause ulcers; if skin is broken, prolonged contact may cause "chrome sores" (slow-healing, hard-rimmed ulcers), which leave the area vulnerable to infection as a secondary effect.
Swallowed: May cause severe burns of the mouth, throat and stomach, vomiting, watery or bloody diarrhea; severe circulatory collapse and toxic nephritis; may be fatal. Damage to kidneys and liver, collapse and convulsions can result.
Long-term exposure: May cause cancer; mutations; skin allergy; irritation of the bronchial tubes; lung allergy; perforation of the nasal septum.
Storage: Store in tightly closed containers in a dry, cool, well-ventilated place. Keep away from combustibles, organic materials.

★ 872 ★
SODIUM CYANIDE
CAS: 143-33-9
Other names: CIANURO DI SODIO (ITALIAN); CYANIDE OF SODIUM; CYANOBRIK; CYANOGRAN; CYANURE DE SODIUM (FRENCH); CYMAG; HYDROCYANIC ACID, SODIUM SALT; KYANID SODNY (CZECH); SODIUM CYANIDE, SOLID; SODIUM CYANIDE, SOLUTION

Danger profile: Combustible, exposure limit established, Community Right-to-Know List, poisonous gas produced in fire.
Uses: Making paper, leather, food products, dyes, pigments, nylon and other chemicals; electroplating and cleaning metals; metal hardening; insecticide; fumigant.
Appearance: White crystalline powder, flakes or lumps.
Odor: Slight; rotten eggs; almonds; hydrocyanic acid when wet. The odor is a warning of exposure; however, no smell does not mean you are not being exposed.
Effects of exposure: Exposure may cause sudden death.
Breathed: Deadly poison. May cause irritation of the eyes, nose and throat, headache, dizziness, pounding heart and sudden death may occur.
Skin: Deadly poison. Passes through the unbroken skin; can increase exposure and the severity of the symptoms listed. May cause irritation. Strong water solutions, or the solid itself, can be absorbed by the skin and cause deep ulcers which heal slowly.
Swallowed: Deadly poison. May cause hallucinations, muscle weakness and upset stomach. As little as 180 milligrams is a rapidly fatal poison. Non-lethal doses may cause toxic symptoms.
Long-term exposure: May cause loss of appetite, headache, nausea, respiratory irritation; skin rash; enlargement of the thyroid gland and thyroid problems; nose irritation with sores and nose bleed; nervous system damage; changes to blood cell count.
Storage: Store in tightly closed containers in a dry, cool, well-ventilated place. Keep away from acid, oxidizers, nitrates, nitrites, magnesium and fluorine compounds. Reacts with acidic

substances to produce toxic, combustible hydrogen cyanide gas.
First aid guide: Move victim to fresh air and call emergency medical care; if not breathing, give artificial respiration; if breathing is difficult, give oxygen. In case of contact with material, immediately flush skin or eyes with running water for at least 15 minutes. Speed in removing material from skin is of extreme importance. Remove and isolate contaminated clothing and shoes at the site. Keep victim quiet and maintain normal body temperature. Effects may be delayed; keep victim under observation.

★ 873 ★
SODIUM DICHLORO-S-TRIAZINETRIONE
CAS: 2893-78-9
Other names: ACL 60; CDB 63; DICHLOROISOCYANURIC ACID SODIUM SALT; DICHLORO-S-TRIAZINE-2,4,6-(1H,3H,5H)-TRIONE; DIKONIT; DIMANIN C; FICLOR 60 S; ISOCYANURIC ACID, DICHLORO-, SODIUM SALT; OCI 56; SDIC; SIMPLA; SODIUM DICHLOROCYANURATE; SODIUM DICHLORISOCYANURATE; 1-SODIUM-3,5-DICHLORO-S-TRIAZINE-2,4,6-TRIONE; 1-SODIUM-3,5-DICHLORO-1,3,5-TRIAZINE-2,4,6-TRIONE; SODIUM SALT OF DICHLORO-S-TRIAZINETRIONE; S-TRIAZINE-2,4,6(1H, 3H,5H)-TRIONE, DICHLORO-, SODIUM SALT; S-TRIAZINE-2,4,6(1H,3H,5H)-TRIONE, DICHLORO-, SODIUM SALT

Danger profile: Combustible, explosion hazard, poisonous gas produced in fire and on contact with water.
Uses: In bleaches, detergents, disinfectants; sewage treatment; swimming pool chemical. For cleaning, sanitizing, etc.
Appearance: White crystalline solid.
Odor: Bleach-like; chlorine-like. The odor is a warning of exposure; however, no smell does not mean you are not being exposed.
Effects of exposure:
Breathed: May cause irritation of the nose, throat and breathing passages; cough.
Eyes: May cause severe irritation.
Skin: May cause irritation. Irritation may be severe if skin or chemical is wet.
Swallowed: May cause gastrointestinal irritation, stomach ulceration and bleeding; coma and death after very high dose.
Long-term exposure: May cause causes emaciation, weight loss, weakness, lethargy and diarrhea; liver problems; lung congestion.
Storage: Highly reactive. Store in tightly closed containers in a dry, cool, well-ventilated place. Keep away from moisture, organic materials, sawdust, floor sweepings and easily oxidized organics; ammonium compounds, hydrated salts. Contact with water produces chlorine gas. Containers may explode in fire.
First aid guide: Move victim to fresh air and call emergency medical care; if not breathing, give artificial respiration; if breathing is difficult, give oxygen. In case of contact with material, immediately flush skin or eyes with running water for at least 15 minutes. Remove and isolate contaminated clothing and shoes at the site. Keep victim quiet and maintain normal body temperature. Effects may be delayed; keep victim under observation.

★874★
SODIUM DICHROMATE
CAS: 10588-01-9
Other names: BICHROMATE DE SODIUM (FRENCH); BICHROMATE OF SODA; CHROMIC ACID,DISODIUM SALT; CHROMIUM SODIUM OXIDE; DISODIUM DICHROMATE; NATRIUMBICHROMAAT (DUTCH); NATRIUMDICHROMAAT (DUTCH); NATRIUMDICHROMAT (GERMAN); SODIO (DICROMATO DI) (ITALIAN); SODIUM BICHROMATE; SODIUM CHROMATE; SODIUM DICHROMATE(VI); SODIUM-(DICHROMATE DE) (FRENCH)

Danger profile: Carcinogen, mutagen, poison, combustible, exposure limit established, Community Right-to-Know List, poisonous gas produced in fire.
Uses: Making other chemicals, pigments; wood preservative; in electroplating; corrosion inhibitor; leather tanning; weed and vegetation killer.
Appearance: Red to orange crystals.
Effects of exposure:
Breathed: Dust may cause severe irritation to the nose, throat and lungs, causing coughing, shortness of breath; respiratory irritation sometimes resembling asthma; nasal septal perforation may occur. May cause flu-like symptoms including chills, muscle ache, headache, fever. High exposure may cause nausea, salivation, vomiting, cramps, diarrhea, chest pains, cough, a build-up of fluids in the lungs (pulmonary edema) and possible death. Pulmonary edema may be delayed from 1-2 days following exposure.
Eyes: Contact may cause severe irritation, burns, pain and possible blindness.
Skin: Passes through the unbroken skin; can increase exposure and the severity of the symptoms listed. May cause severe irritation and thermal and acid burns, especially if skin is wet.
Swallowed: May cause severe burns of the mouth, throat and stomach, vomiting, watery or bloody diarrhea; Damage to kidneys stomach and liver; collapse and convulsions can result.
Long-term exposure: May cause cancer; exposure may cause skin allergy and dermatitis; irritation of the bronchial tubes with cough and phlegm; lung allergy; kidney damage; perforation of the nasal septum.
Storage: Store in tightly closed containers in a dry, cool, well-ventilated place with non-wood floors. Keep away from heat, easily oxidized materials, finely divided combustibles such as sawdust, sulfur, aluminum, acetic acid, hydrazine, alcohols, acetic anhydride, sulfuric acid, hydroxylamine, trinitrotoluene.
First aid guide: Move victim to fresh air; call emergency medical care. In case of contact with material, immediately flush skin or eyes with running water for at least 15 minutes. Remove and isolate contaminated clothing and shoes at the site.

★875★
SODIUM DODECYLBENZENE SULFONATE
CAS: 25155-30-0
Other names: AA-9; ABESON NAM; BENZENE SULFONIC ACID, DODECLY-, SODIUM SALT; BIO-SOFT D-40; BIO-SOFT D-60; BIO-SOFT D-62; BIO-SOFT D-35X; CALSOFT F-90; CALSOFT L-40; CALSOFT L-60; CONCO AAS-35; CONCO AAS-40; CONCO AAS-65; CONCO AAS-90; CONOCO C-50; CONOCO C-60; CONOCO SD 40; DETERGENT HD-90; DODECYL BENZENE SODIUM SULFONATE; DODECYLBENZENESULPHONATE, SODIUM SALT; DODECYLBENZENSULFONAN SODNY (CZECH); MERCOL 25; MERCOL 30; NACCANOL NR; NACCANOL SW; NACCONOL 40F; NACCONOL 90F; NACCONOL 35SL; NECCANOL SW; PILOT HD-90; PILOT SF-40; PILOT SF-60; PILOT SF-96; PILOT SF-40B; PILOT SF-40FG; PILOT SP-60; RICHONATE 1850; RICHONATE 45B; RICHONATE 60B; SANTOMERSE 3; SANTOMERSE NO. 1; SANTOMERSE NO.85; SODIUM LAURYLBENZENESULFONATE; SOLAR 40; SOLAR 90; SOL SODOWA KWASU LAURYLOBENZENOSULFONOWEGO (POLISH); SULFAPOL; SULFAPOLU (POLISH); SULFRAMIN 85; SULFRAMIN 40 FLAKES; SULFRAMIN 90 FLAKES; SULFRAMIN 40 GRANULAR; SULFRAMIN 40RA; SULFRAMIN 1238 SLURRY; SULFRAMIN 1250 SLURRY; P-1',1',4',4'-TETRAMETHYLOKTYLBENZENSULFONAN SODNY (CZECH); ULTRAWET K; ULTRAWET 60K; ULTRAWET KX; ULTRAWET SK

Danger profile: Poisonous gas produced in fire.
Uses: Antiseptic, detergent; making pharmaceuticals and pesticides.
Appearance: White to light yellow flakes, granules or powder.
Effects of exposure:
Breathed: Dust may cause irritation of the nose, throat, bronchial tubes and lungs; cough, difficult breathing.
Eyes: May cause severe irritation and burning feeling.
Skin: May cause irritation and burning feeling.
Swallowed: Moderately toxic. May cause vomiting.
Long-term exposure: Similar chemicals may cause skin and lung allergies; lung problems. It is not know for certain whether this chemical causes the same problems.
Storage: Store in tightly closed containers in a dry, cool, well-ventilated place. Containers may explode in fire.

★876★
SODIUM FLUORIDE
CAS: 7681-49-4
Other names: ALCOA SODIUM FLUORIDE; ANTIBULIT; CAV-TROL; CHEMIFLOUR; CREDO; DISODIUM DIFLUORIDE; FDA 0101; F1-TABS; FLORIDINE; FLUORIDENT; FLUORID SODNY (CZECH); FLUORIGARD; FLUORINEED; FLUORINSE; FLUORITAB; FLOUR-O-KOTE; FLUORURE DE SODIUM (FRENCH); FLURA-GEL; FLURCARD; FUNGOL B; GELUTION; IRADICAV; KARIDIUM; KARIGEL; KARI-RINSE; LEA-COV; LEMOFLUR; LURIDE; NAFEEN; NATRIUM FLUORIDE; NCI C55221; NUFLOUR; OSSALIN; OSSIN; PEDIAFLOR; PEDIDENT; PENNWHITE; PERGANTENE; PHOSFLUR; POINT TWO; RESCUE SQUAD; ROACH SALT; SODIUM HYDROFLUORIDE; SODIUM MONOFLUORIDE; SOFLO; STAY-FLO; STUDAFLOUR; SUPER-DENT; T-FLUORIDE; THERA-FLUR-N; TRISODIUM TRIFLUORIDE; VILLIAUMITE

Danger profile: Combustible, corrosive, poison, exposure limit established, reproductive hazard, poisonous gas produced in fire.
Uses: Widely used chemical. Making other chemicals and glass; treatment of municipal water; chemical cleaning; wood preservative; insecticide; fungicide; rodenticide; electroplating; vitre-

ous enamels; preservative for adhesives; toothpastes; disinfectant; dentistry; in ultraviolet and infrared radiation detecting systems.

Appearance: Clear crystals; white crystalline powder or solution. The insecticide is dyed blue.

Effects of exposure:

Breathed: May be poisonous. May cause irritation of the nose, eyes, mouth and throat; watering eyes; salivation, coughing, throat pain, difficult breathing, headache, fatigue, nausea, stomach pains. High exposures may cause shortness of breath and a dangerous build-up of fluids in the lungs (pulmonary edema), which can cause death. Pulmonary edema is a medical emergency that can be delayed one to two days following exposure.

Eyes: May cause intense stinging, irritation, severe burns and irreparable damage and possible loss of sight.

Skin: May cause irritation, redness, swelling and severe burns; skin rash and ulcers.

Swallowed: May be poisonous. May cause irritation; burning feeling in mouth and throat; swallowing causes abdominal pain; intense thirst; throat swelling, nausea, vomiting, shock; diarrhea, convulsions, collapse, disturbed color vision, acute toxic nephritis; possible death. See *Breathed* for other symptoms.

Long-term exposure: May cause infertility in women; skin rash and ulcers; kidney damage; increase bone density, stimulate new bone growth or cause calcium deposits in ligaments. This may become a problem at levels of 20 to 50 mg/m3 or higher. Mottling of tooth enamel may also occur. These effects are not expected to occur when chemical is properly used in water for dental cavity prevention.

Storage: Store in tightly closed containers in a dry, cool, well-ventilated place. Keep away from acids, chemically active metals, alkalis. Protect containers from physical damage. Containers may explode in fire.

First aid guide: Move victim to fresh air; call emergency medical care. In case of contact with material, immediately flush eyes with running water for at least 15 minutes. Wash skin with soap and water. Remove and isolate contaminated clothing and shoes at the site.

★ 877 ★
SODIUM FLUOROSILICATE
CAS: 16893-85-9
Other names: DESTRUXOL APPLEX; DISODIUM HEXA-FLUOROSILICATE; DISODIUM HEXAFLUOROSILICATE (2-); DISODIUM SILICOFLUORIDE; ENS-ZEM WEEVIL BAIT; FLUOSILICATE DE SODIUM; ORTHO EARWIG BAIT; ORTHO WEEVIL BAIT; PRODAN; PSC CO-OP WEEVIL BAIT; SAFSAN; SALUFER; SILICON SODIUM FLUORIDE; SODIUM HEXAFLUOROSILICATE; SODIUM HEXAFLUOSILICATE; SODIUM SILICOFLUORIDE; SODIUM SILICON FLUORIDE; SUPER PRODAN

Danger profile: Poison, exposure limit established, poisonous gas produced in fire.
Uses: Fluoridation; opalescent glass; vitreous enamel; insecticide and rodenticide; glue, leather and wood preservative; moth repellent; making other chemicals.
Appearance: White powder or granules.
Effects of exposure:

Breathed: Poisonous. May cause irritation of the nose, eyes, mouth and throat; watering eyes; salivation, coughing, throat pain, difficult breathing, headache, fatigue, nausea, stomach pains. High exposures may cause shortness of breath and a dangerous build-up of fluids in the lungs (pulmonary edema), which can cause death. Pulmonary edema is a medical emer-

gency that can be delayed one to two days following exposure.

Eyes: May cause intense stinging, severe irritation, burns and irreparable damage and possible loss of sight.

Skin: May cause irritation, redness, burning feeling, rash sometime followed by the formation of ulcers.

Swallowed: Poisonous. May cause irritation; burning feeling in mouth and throat; swallowing causes pain; intense thirst; throat swelling, nausea, vomiting, diarrhea, dehydration and shock. In sever cases, convulsions, shock, and cyanosis are followed by death in two to four hours.

Storage: Store in tightly closed containers in a dry, cool, well-ventilated place. Keep away from acids. Containers may explode in fire.

First aid guide: Move victim to fresh air; call emergency medical care. In case of contact with material, immediately flush skin or eyes with running water for at least 15 minutes. Remove and isolate contaminated clothing and shoes at the site.

★ 878 ★
SODIUM HYDRIDE
CAS: 7646-69-7
Other names: NAH 80; SODIUM MONOHYDRIDE

Danger profile: Fire hazard, explosion hazard, highly reactive, poisonous gas produced in fire.
Uses: Used in chemical reactions; descaling metals.
Appearance: Gray-white or brownish-white powder.
Effects of exposure:

Breathed: May be poisonous. May cause irritation of he lungs; coughing, difficult breathing. Higher exposures may cause shortness of breath and a dangerous build-up of fluids in the lungs (pulmonary edema), which can cause death. Pulmonary edema is a medical emergency that can be delayed one to two days following exposure.

Eyes: Contact may cause severe irritation, burns and possible loss of vision.

Skin: May cause severe irritation. If skin is moist may cause burns.

Swallowed: Moisture of body converts compound to caustic soda, which irritates all tissues.

Long-term exposure: May cause a breakdown of red blood cells; kidney damage; lung problems.

Storage: Powder ignites spontaneously in air. Store in tightly closed containers in a dry, cool, well-ventilated place. Keep away from heat, air, sources of ignition, oxidizers, water, sulfur, sulfur dioxide, dimethylformamide, acetylene. Containers may explode in fire.

First aid guide: Move victim to fresh air; call emergency medical care. Wipe material from skin immediately; flush skin or eyes with running water for at least 15 minutes. Remove and isolate contaminated clothing and shoes at the site.

★879★
SODIUM HYDROXIDE
CAS: 1310-73-2
Other names: CAUSTIC SODA; CAUSTIC SODA, BEAD; CAUSTIC SODA, DRY; CAUSTIC SODA, FLAKE; CAUSTIC SODA, GRANULAR; CAUSTIC SODA, SOLID; HYDROXYDE DE SODIUM (FRENCH); LEWIS-RED DEVIL LYE; LYE; LYE SOLUTION; NATRIUMHYDROXID (GERMAN); NATRIUMHYDROXYDE (DUTCH); SODA LYE; SODIO(IDROSSIDO DI) (ITALIAN); SODIUM HYDRATE; SODIUM HYDRATE; SODIUM HYDROXIDE, BEAD; SODIUM HYDROXIDE, DRY; SODIUM HYDROXIDE, FLAKE; SODIUM HYDROXIDE, GRANULAR; SODIUM HYDROXIDE, SOLID; SODIUM(HYDROXYDE DE) (FRENCH); WHITE CAUSTIC; SODIUM HYDROXIDE CAUSTIC SODA SOLUTION; LYE SOLUTION; SODA LYE; SODIUM HYDRATE SOLUTION; SODIUM HYDROXIDE LIQUID; SODIUM HYDROXIDE SOLUTION; WHITE CAUSTIC; WHITE CAUSTIC, SOLUTION

Danger profile: Possible mutagen, combustible, corrosive, exposure limit established, poisonous gas produced in fire.
Uses: Among the ten highest volume chemicals produced in the U.S. In a water solution chemicals has a wide range of industrial uses. Making other chemicals, mercerized cotton, rayon, cellophane, aluminum, paper, detergents, soaps, textiles, vegetable oils; processing vegetables and fruit; laboratory chemical; electroplating; etching; food additive; metal cleaner.
Appearance: A white solid in the form of flakes, pellets, cakes, chips or sticks. Also available as a clear, colorless water solution.
Effects of exposure:
  Breathed: Dust or mist may cause irritation of the eyes, nose and throat with burning feeling; difficult breathing, coughing and sneezing; possible lung damage. High exposure can lead to a buildup of fluid in the lungs (pulmonary edema), and may cause death. Pulmonary edema may be delayed by one to two days following exposure.
  Eyes: May cause severe irritation with pain in eyes and eyelids; intense watering; burns and damage to the mucous membrane, perforation of the eyes, permanent damage and blindness.
  Skin: Contact may cause irritation, itching, pain, ulceration with slippery feeling; serious, deep chemical burns.
  Swallowed: May cause immediate burning feeling in the mouth, throat and stomach; intense pain in swallowing; throat swelling, vomiting (hemorrhaging may cause vomitus to resemble coffee grounds), watery or bloody diarrhea, shallow respirations; confusion, delirium, coma; swelling of the throat may cause obstruction; perforation of esophagus or stomach, shock, convulsions and collapse and death can result.
Long-term exposure: May cause mutations; lung problems and permanent damage.
Storage: Dangerous material. Store in tightly closed containers in a dry, cool, well-ventilated place. Keep away from heat, combustible materials, water, acids, metals, combustible liquids, nitro compounds, organic halogen compounds and all other chemicals.
First aid guide: Move victim to fresh air; call emergency medical care. In case of contact with material, immediately flush skin or eyes with running water for at least 15 minutes. Remove and isolate contaminated clothing and shoes at the site. Keep victim quiet and maintain normal body temperature.

★880★
SODIUM HYPOCHLORITE
CAS: 7681-52-9
Other names: ANTIFORMIN; B-K LIQUID; CARREL-DAKIN SOLUTION; CHLOROS; CHLOROX; CLOROX; DAKIN'S SOLUTION; HYCLORITE; HYPOCHLOROUS ACID, SODIUM SALT; MILTON; SURCHLOR

Danger profile: Mutagen, poisonous gas produced in fire.
Uses: Industrial bleaching product; water purification; disinfectant; medicine; fungicides; swimming pool chemical; laundry bleach; germicide.
Appearance: White crystalline solid. Sodium hypochlorite is usually found dissolved in water as a clear pale green or yellowish solution (if strong) or a clear colorless liquid (if weak).
Odor: Bleach or chlorine-like. The odor is a warning of exposure; however, no smell does not mean you are not being exposed.
Effects of exposure:
  Breathed: Dust may cause severe irritation to the nose, throat and lungs, causing coughing, shortness of breath. May cause flu-like symptoms including chills, muscle ache, headache, fever. High exposure may cause nausea, salivation, vomiting, cramps, diarrhea, chest pains, cough, a build-up of fluids in the lungs (pulmonary edema) and possible death. Pulmonary edema may be delayed from one to two days following exposure.
  Eyes: May cause severe irritation, burns, pain and possible blindness.
  Skin: May cause severe irritation and thermal and chemical burns, especially if skin is wet.
  Swallowed: May cause severe burns of the mouth, throat and stomach, vomiting, watery or bloody diarrhea. Damage to kidneys and liver, collapse and convulsions can result.
Long-term exposure: May cause mutations; irritation and skin rash; lung problems.
Storage: Strong oxidizer. Store in tightly closed containers in a dry, cool, well-ventilated place with non-wood floors. Keep away from combustible materials, formic acid, amines, ethylene imine, cellulose, ammonia and other ammonium compounds, hydrochloric acid. Long storage without decomposition is impossible.
First aid guide: Move victim to fresh air; call emergency medical care. In case of contact with material, immediately flush skin or eyes with running water for at least 15 minutes. Remove and isolate contaminated clothing and shoes at the site. Keep victim quiet and maintain normal body temperature.

★881★
SODIUM METABISULFITE
CAS: 7681-57-4
Other names: DISODIUM PYROSULFITE; PYROSULFUROUS ACID, DISODIUM SALT; SODIUM PYROSULFITE

Danger profile: Mutagen, exposure limit established, poisonous gas produced in fire.
Uses: In foods and pharmaceuticals, as preservative; laboratory chemical.
Appearance: White powder or crystals.
Odor: Rotten egg-like. The odor is a warning of exposure; however, no smell does not mean you are not being exposed.
Effects of exposure:
  Breathed: May cause irritation of the nose, throat, sinuses and lungs; cough and difficult breathing; possible lung allergy.
  Eyes: May cause irritation.
  Skin: May cause irritation.
Long-term exposure: May cause mutations, asthma-like allergy; sinus and lung problems.
Storage: Store in tightly closed glass, earthenware, metal or plastic containers in a dry, cool, well-ventilated place. Keep

away from water and combustible materials. Forms sulfur dioxide on contact with water.

★882★
SODIUM MONOXIDE
CAS: 1313-59-3
Other names: CALCINED SODA; DISODIUM MONOXIDE; DISODIUM OXIDE; SODIUM MONOXIDE, SOLID; SODIUM OXIDE

Danger profile: Corrosive; poisonous gas produced in fire.
Uses: Making other chemicals; dehydrating agent.
Appearance: White-gray crystals.
Effects of exposure:
 Breathed: May cause irritation of the eyes, nose and throat with burning feeling; difficult breathing, coughing and sneezing. High exposure can lead to a buildup of fluid in the lungs (pulmonary edema), and may cause death. Pulmonary edema may be delayed by one to two days following exposure.
 Eyes: May cause severe irritation with pain in eyes and eyelids; intense watering; burns and damage to the mucous membrane, perforation of the eyes, permanent damage and blindness.
 Skin: May cause irritation, itching, pain, ulceration with slippery feeling; serious, deep chemical burns.
 Swallowed: May cause immediate burning feeling in the mouth, throat and stomach; intense pain in swallowing; throat swelling, vomiting (hemorrhaging may cause vomitus to resemble coffee grounds), watery or bloody diarrhea, shallow respirations; confusion, delirium, coma; swelling of the throat may cause obstruction; perforation of esophagus or stomach, shock, convulsions and collapse and death can result.
Long-term exposure: May cause lung problems.
Storage: Store in tightly closed containers in a dry, cool, well-ventilated place. Keep away from heat, water, nitric acid, 2,4-dinitrotoluene. Containers may explode in fire.
First aid guide: Move victim to fresh air; call emergency medical care. In case of contact with material, immediately flush skin or eyes with running water for at least 15 minutes. Remove and isolate contaminated clothing and shoes at the site. Keep victim quiet and maintain normal body temperature.

★883★
SODIUM NITRATE
CAS: 7631-99-4
Other names: CHILE SALTPETER; CUBIC NITER; NITRATINE; NITRATE DE SODIUM (FRENCH); NITRIC ACID, SODIUM SALT; SODA NITER; SODIUM NITRATE (1:1)

Danger profile: Mutagen, combustible, poisonous gas produced in fire.
Uses: Rocket fuel; fertilizer; soldering flux; making glass; fireworks; laboratory chemical; refrigerant; matches; dynamites; black powders; dyes; pharmaceuticals; an aphrodisiac; food preservative; enamel for pottery; in tobacco.
Appearance: Colorless crystalline solid.
Effects of exposure:
 Breathed: Vapors or dust may be irritating.
 Eyes: May cause irritation.
 Skin: May cause irritation.
 Swallowed: Moderately toxic. May cause dizziness, abdominal cramps, vomiting, bloody diarrhea, weakness, convulsions, and collapse.
Long-term exposure: May cause mutations. Small repeated doses may cause headache and mental impairment.
Storage: Dangerous and powerful oxidizer. Store in tightly closed containers in a dry, cool, well-ventilated place with non-

wood floors. Keep away from combustible and organic materials, especially wood; sodium hypophosphite, antimony, cyanide compounds, acetic anhydride, boron phosphite, aluminum powder, bitumen, magnesium.
First aid guide: Move victim to fresh air; call emergency medical care. In case of contact with material, immediately flush skin or eyes with running water for at least 15 minutes. Remove and isolate contaminated clothing and shoes at the site.

★884★
SODIUM PERCHLORATE
CAS: 7601-89-0
Other names: IRENAT; NATRIUMPERCHLORAAT (DUTCH); NATRIUMPERCHLORAT (GERMAN); PERCHLORATE DE SODIUM (FRENCH); PERCHLORIC ACID, SODIUM SALT; SODIO (PERCLORATO DI) (ITALIAN); SODIUM (PERCHLORATE DE) (FRENCH)

Danger profile: Mutagen, combustible, poisonous gas produced in fire.
Uses: Explosives; jet fuels; making other chemicals; laboratory chemical.
Appearance: Colorless to white crystalline solid.
Effects of exposure:
 Breathed: Dust may cause irritation of the nose, throat and lungs. May deprive the body of oxygen and cause skin and lips to turn blue. Severe exposure may cause kidney damage.
 Eyes: Contact may cause severe irritation and burns.
 Skin: Contact may cause severe irritation and burns.
 Swallowed: May cause nausea, vomiting, diarrhea, abdominal or gastric pain, dyspnea and other symptoms. The major cause of death from a lethal dose is acute renal failure. Severe exposure may cause kidney damage.
Long-term exposure: May cause kidney damage. Possible lung, liver and other body organ damage.
Storage: Powerful oxidizer. Store in tightly closed containers in a dry, cool, well-ventilated place. Keep away from water, combustible and organic materials, ethanolamine, acetone, ethylene glycol, acids, powdered metals, ammonium nitrate, formamide, hydrazine, charcoal, sulfuric acid. Containers may explode in fire.
First aid guide: Move victim to fresh air; call emergency medical care. In case of contact with material, immediately flush skin or eyes with running water for at least 15 minutes. Remove and isolate contaminated clothing and shoes at the site.

★885★
SODIUM PEROXIDE
CAS: 1313-60-6
Other names: DISODIUM DIOXIDE; DISODIUM PEROXIDE; FLOCOOL 180; SODIUM DIOXIDE; SODIUM OXIDE; SODIUM SUPEROXIDE; SOLOZONE

Danger profile: Explosion hazard, corrosive, reacts with water, poisonous gas produced in fire.
Uses: Bleaching agent; deodorant; germicidal soap; antiseptics; fungicide; water treatment; pharmaceuticals; oxygen generation for diving bells, submarines, and other underwater apparatus; textile dyeing and printing; laboratory chemical.
Appearance: Yellowish-white powder.
Effects of exposure:
 Breathed: May cause irritation of the eyes, nose and throat with burning feeling; difficult breathing, coughing and sneezing. High exposure can lead to a buildup of fluid in the lungs (pulmonary edema), and may cause death. Pulmonary edema may be delayed by one to two days following exposure.

Eyes: May cause severe irritation with pain in eyes and eyelids; intense watering; burns and damage to the mucous membrane, perforation of the eyes, permanent damage and blindness.

Skin: may cause irritation, itching, pain, ulceration with slippery feeling; serious, deep chemical burns.

Swallowed: May cause immediate burning feeling in the mouth, throat and stomach; intense pain in swallowing; throat swelling, vomiting (hemorrhaging may cause vomitus to resemble coffee grounds), watery or bloody diarrhea, shallow respirations; confusion, delirium, coma; swelling of the throat may cause obstruction; perforation of esophagus or stomach, shock, convulsions and collapse, and death can result.

Long-term exposure: May cause skin dryness and cracking; rash; lung problems.

Storage: Store in tightly closed containers in a dry, cool, well-ventilated place with non-wood floors. Keep away from strong acids, water, powdered metals, fuels, organic and combustible materials such as wood, paper, oil, etc. Containers may explode in fire.

First aid guide: Move victim to fresh air and call emergency medical care; if not breathing, give artificial respiration; if breathing is difficult, give oxygen. In case of contact with material, immediately flush skin and eyes with running water for at least 15 minutes. Remove and isolate contaminated clothing and shoes at the site. Keep victim quiet and maintain normal body temperature.

★ 886 ★
SODIUM PHOSPHATE
CAS: 7601-54-9

Other names: DRI-TRI; EMULSIPHOS 440/660; NUTRIFOS STP; PHOSPHORIC ACID, TRISODIUM SALT; OAKITE; TRIBASIC SODIUM PHOSPHATE; TRINATRIUMPHOSPHAT (GERMAN); TRISODIUM ORTHOPHOSPHATE; TRISODIUM PHOSPHATE, TRIBASIC; TRISODIUM'-O-PHOSPHATE; SODIUM PHOSPHATE, ANHYDROUS; TROMETE; TSP

Danger profile: Mutagen, poisonous gas produced in fire.
Uses: In household cleaners; water softeners; boiler water compounds; detergent; metal cleaner; textiles; manufacture of paper; laundering; tanning; sugar purification; photographic developers; paint removers; industrial cleaners; dietary supplement; buffer; emulsifier; food additive.
Appearance: Colorless to white crystals or crystalline powder.
Effects of exposure:
Breathed: May cause irritation of the eyes, nose and throat; difficult breathing.
Eyes: May cause severe irritation, pain, tearing, burns and serious damage.
Skin: Contact with wet skin may cause severe irritation, redness, burning feeling and rash.
Swallowed: May cause burning feeling in the mouth and stomach; mouth and gastric irritation; stomach cramps.
Long-term exposure: May cause mutations; skin rash and itching.
Storage: Strong caustic. Store in tightly closed containers in a dry, cool, well-ventilated place.

★ 887 ★
SODIUM PHOSPHATE, DIBASIC
CAS: 7558-79-4
Other names: DIBASIC SODIUM PHOSPHATE; DISODIUM HYDROGEN PHOSPHATE; DISODIUM MONOHYDROGEN PHOSPHATE; DISODIUM ORTHOPHOSPHATE; DISODIUM PHOSPHATE; DISODIUM PHOSPHORIC ACID; DSP; EXSICCATED SODIUM PHOSPHATE; NATRIUMPHOSPHAT (GERMAN); PHOSPHORIC ACID, DISODIUM SALT; SODA PHOSPHATE, DIBASIC; SODIUM HYDROGEN PHOSPHATE; SODIUM MONOHYDROGEN PHOSPHATE (2:1:1)

Danger profile: Poisonous gas produced in fire.
Uses: In fertilizers, pharmaceuticals, textiles, detergents; fire proofing wood, paper; ceramic glazes; leather tanning; dietary supplement; food processing; cheese; laboratory chemical.
Appearance: Colorless, translucent crystals or white powder.
Effects of exposure:
Breathed: May cause irritation of the eyes, nose and throat.
Eyes: May cause irritation.
Skin: May cause irritation, burning feeling and rash.
Swallowed: Mildly toxic.
Long-term exposure: May cause skin rash and itching.
Storage: Store in tightly closed containers in a dry, cool, well-ventilated place. Keep away from alkaloids, antipyrine, chloral hydrate, lead acetate, pyrogallol and resorcinol.

★ 888 ★
SODIUM SILICATE
CAS: 6834-92-0
Other names: B-W; CRYSTAMET; DISODIUM METASILICATE; DISODIUM MONOSILICATE; METSO 20; METSO PENTABEAD 20; ORTHOSIL; SODIUM METASILICATE, ANHYDROUS; WATER GLASS

Danger profile: Caustic, poisonous gas produced in fire.
Uses: In cosmetics, soaps and detergents; adhesives; water treatment; making abrasive wheels, textiles and paper products; pigments; flame retardant; waterproofing cements
Appearance: White powders, or clear or cloudy syrupy liquid.
Effects of exposure:
Breathed: May cause irritation of nose, eyes and throat; coughing; difficult breathing.
Skin: Contact with wet skin may cause severe irritation, redness and swelling; burning feeling.
Eyes: Contact may cause severe irritation, pain, tearing; serious injury may result.
Swallowed: Poison. May cause burning feeling in the mouth and stomach; mouth and gastric irritation; stomach cramps.
Storage: A caustic material. Store in tightly closed containers in a dry, cool, well-ventilated place. Keep away from fluorine compounds.

★ 889 ★
SODIUM SULFATE
CAS: 7757-82-6
Other names: DISODIUM SULFATE; NATRIUMSULFAT (GERMAN); SALT CAKE; SODIUM SULFATE (2:1); SODIUM SULFATE, ANHYDROUS; SODIUM SULPHATE; SULFURIC ACID DISODIUM SALT; THENARDITE; TRONA

Danger profile: Poisonous gas produced in fire.
Uses: Among the 50 highest volume chemicals produced in the U.S. Making paper, textiles, dyes, cardboard, glass; food additive; in detergents, ceramic glazes; leather tanning, drugs; laboratory chemical.
Appearance: White crystals.
Effects of exposure:

Breathed: May cause irritation of the nose; coughing and sneezing.
Eyes: May cause irritation, redness and tearing.
Skin: Prolonged contact may cause irritation.
Swallowed: May cause diarrhea.
Long-term exposure: Experimental reproductive hazard.
Storage: Store in tightly closed containers in a dry, cool, well-ventilated place. Keep away from aluminum.

★ 890 ★
SODIUM THIOCYANATE
CAS: 540-72-7
Other names: HAIMASED; NATRIUMRHODANID (GERMAN); SCYAN; SODIUM ISOTHIOCYANATE; SODIUM SULFOCYANATE; SODIUM RHODANIDE; SODIUM THIOCYANIDE; RHODANATE; USAF EK-T-434

Danger profile: Poison, poisonous gas produced in fire.
Uses: Dyeing; textile printing; electroplating; making other chemicals; laboratory chemical.
Appearance: Colorless crystals or white powder.
Effects of exposure:
Breathed: May cause irritation of the nose and throat; coughing and sneezing.
Eyes: May cause irritation.
Skin: Prolonged contact with skin may produce various skin eruptions, dizziness, cramps, nausea, and mild to severe disturbance of the nervous system.
Swallowed: May cause diarrhea, vomiting and convulsions. Large doses may cause vomiting, extreme cerebral excitement, convulsions, and death in 10-48 hours.
Long-term exposure: Chronic poisoning can cause flu-like symptoms, skin rashes, weakness, fatigue, vertigo, nausea, vomiting, diarrhea, confusion.
Storage: Store in tightly closed containers in a dry, cool, well-ventilated place.

★ 891 ★
SODIUM THIOSULFATE
CAS: 7772-98-7
Other names: ANTICHLOR; AMEDOX; CHLORINE CONTROL; CHLORINE CURE; DECLOR-IT; DISODIUM THIOSULFATE; HYPO; S-HYDRIL; SODIUM HYPOSULFITE; SODIUM OXIDE SULFIDE; SODIUM THIOSULFATE, ANHYDROUS; SODIUM THIOSULPHATE; SODOTHIOL; SULFOTHIORIUE; THIOSULFURIC ACID, DISODIUM SALT

Danger profile: Poisonous gas produced in fire.
Uses: Photography; water treatment; mordant; bleaching; dyeing; antidote for cyanide poisoning; laboratory chemical.
Appearance: Colorless crystals or powder.
Effects of exposure:
Breathed: May cause irritation of the nose; coughing and sneezing.
Eyes: May cause irritation, redness and tearing.
Skin: Prolonged contact may cause irritation.
Swallowed: May cause diarrhea.
Storage: Store in tightly closed containers in a dry, cool, well-ventilated place. Keep away from sodium nitrite.

★ 892 ★
STANNIC CHLORIDE PENTAHYDRATE
CAS: 10026-06-9
Other names: STANNIC CHLORIDE, HYDRATED; STANNIC CHLORIDE, PENTAHYDRATE; TETRACHLOROSTANNANE PENTAHYDRATE; TIN(IV) CHLORIDE, PENTAHYDRATE (1:4:5)

Danger profile: Corrosive, exposure limit established, poisonous gas produced in fire.
Uses: Treating silk to add fabric weight; fixing specific textile dyes.
Appearance: White or yellow powder.
Odor: Faint, hydrochloric acid. The odor is a warning of exposure; however, no smell does mean you are not being exposed.
Effects of exposure:
Breathed: Mist or dust may cause irritation of the throat and bronchial tubes; cough, difficult breathing. Heated chemical can release toxic chlorine gas and may cause shortness of breath and a dangerous build-up of fluids in the lungs (pulmonary edema), which can cause death. Pulmonary edema is a medical emergency that can be delayed one to two days following exposure.
Eyes: May cause severe irritation and burns.
Skin: May cause severe irritation, redness and burns.
Long-term exposure: May cause lung problems; iron deficiency anemia.
Storage: Store in tightly closed containers in a dry, dark, cool, well-ventilated place. Keep away from combustible materials, chlorine, turpentine, ethylene oxide, alkyl nitrates. May attack some plastics, rubbers and coatings.

★ 893 ★
STANNOUS CHLORIDE
CAS: 7772-99-8
Other names: C.I. 77864; NCI-C02722; STANNOCHLOR; STANNICHLOR; STANNOUS CHLORIDE, SOLID; TIN CHLORIDE; TIN(II) CHLORIDE (1:2); TIN DICHLORIDE; TIN PROTOCHLORIDE

Danger profile: Mutagen, exposure limit established, poisonous gas produced in fire.
Uses: Making other chemicals, chemicals preservatives and glass; food additives; dyes; textiles; silvering mirrors.
Appearance: White or colorless crystalline solid.
Effects of exposure:
Breathed: Mist or dust may cause irritation of the throat, breathing passages; cough, difficult breathing.
Eyes: May cause severe irritation and burns.
Skin: May cause severe irritation and burns.
Swallowed: May cause pain in mouth and throat, retching, vomiting, diarrhea.
Long-term exposure: May cause mutations, lung problems; iron deficiency anemia.
Storage: Store in tightly closed containers in a dry, cool, well-ventilated place. Keep away from heat, sources of ignition, water, steam and other forms of moisture, combustible materials, bromine trifluoride, sodium, calcium acetylide, metal nitrates, nitrates, potassium, hydrazine hydrate, sodium peroxide, ethylene oxide, hydrogen peroxide.
First aid guide: Move victim to fresh air; call emergency medical care. In case of contact with material, immediately flush skin or eyes with running water for at least 15 minutes. Remove and isolate contaminated clothing and shoes at the site. Keep victim quiet and maintain normal body temperature.

★ 894 ★
STANNOUS FLUORIDE
CAS: 7783-47-3
Other names: FLUORISTAN; TIN BIFLUORIDE; TIN DIFLUORIDE; TIN FLUORIDE

Danger profile: Exposure limit established, poisonous gas produced in fire.
Uses: Ingredient in certain toothpastes.

Appearance: White crystalline powder.
Effects of exposure:
> *Breathed:* May cause irritation of the throat and breathing passages. Heated chemical releases toxic fluorine gas and may cause shortness of breath and a dangerous build-up of fluids in the lungs (pulmonary edema), which can cause death. Pulmonary edema is a medical emergency that can be delayed one to two days following exposure.
> *Eyes:* Contact may cause severe irritation and burns.
> *Skin:* Contact may cause severe irritation and burns.

Long-term exposure: May cause lung problems and iron deficiency anemia.
Storage: Store in tightly closed containers in a dry, cool, well-ventilated place. Keep away from chlorine and turpentine.

★ 895 ★
STIBINE
CAS: 7803-52-3
Other names: ANTIMIMONWASSERSTOFFES (GERMAN); ANTIMONY HYDRIDE; ANTIMONY TRIHYDRIDE; ANTYMONOWODOR (POLISH); HYDROGEN ANTIMONIDE

Danger profile: Fire hazard, poison, exposure limit established, Community Right-to-Know List, poisonous gas produced in fire.
Uses: Fumigating agent.
Appearance: Colorless gas.
Odor: Unpleasant, rotten egg-like. The odor is a warning of exposure; however, no smell does not mean you are not being exposed.
Effects of exposure:
> *Breathed:* Poisonous Immediately seek medical attention. May cause giddiness, headache, and nausea. More severe exposure may cause malaise, weakness, shivering, dizziness; dyspnea; abdominal and back pain, nausea, vomiting; bronze or yellow coloration of skin, jaundice; garlic breath; dark red urine; accumulation of fluid in the lungs, delirium, coma and death. Symptoms may be delayed for one to two days following exposure.
> *Eyes:* May cause irritation, watering, burning sensation. High levels of exposure may cause permanent damage. Liquid may cause frostbite.
> *Skin:* May be poisonous if absorbed. May cause painful irritation, inflammation, blisters and deep wounds (ulcer). Liquid may cauise frostbite.
> *Swallowed:* Poisonous. Immediately seek medical attention. Unlikely route of exposure as most cases of poisoning occur from accidental generation of the gas.

Long-term exposure: May cause lung, liver and kidney problems.
Storage: Store away from food products, high temperature and strong light. Keep away from strong oxidizers, open flame, shock, chlorine, nitric acid, mixtures of potassium and ammonia. Arsine can be formed when inorganic arsenic reacts with freshly formed (nascent) hydrogen gas. Store in tightly closed containers in a dry, cool, well-ventilated place. Keep away from heat, sources of ignition, oxidizers, acids, ammonia, ozone, halogenated hydrocarbons. Containers may explode in fire.
First aid guide: Move victim to fresh air and call emergency medical care; if not breathing, give artificial respiration; if breathing is difficult, give oxygen. In case of contact with material, immediately flush skin or eyes with running water for at least 15 minutes. Remove and isolate contaminated clothing and shoes at the site. Keep victim quiet and maintain normal body temperature. Effects may be delayed; keep victim under observation.

★ 896 ★
STODDARD SOLVENT
CAS: 8052-41-3
Other names: DRY CLEANING SAFETY SOLVENT; MINERAL SPIRITS; NAPHTHA SAFETY SOLVENT; VARNOLINE; VARSOL; WHITE SPIRITS

Danger profile: Combustible, exposure limit established, poisonous gas produced in fire.
Uses: Dry cleaning; spot and stain removal; degreasing; cleaning in mechanical shops, herbicide.
Appearance: Colorless watery liquid.
Odor: Kerosene or gasoline-like. The odor is a warning of exposure; however, no smell does not mean you are not being exposed.
Effects of exposure:
> *Breathed:* May cause irritation of the eyes, nose and throat; dizziness, lightheadedness, rapid breathing, mental confusion, unsteady walk, headache, fatigue, nausea, vomiting, unconsciousness, coma and death. Highly concentrated vapors may cause intoxication.
> *Eyes:* May cause burning feeling, tearing, irritation, redness and inflammation of the eyelids.
> *Skin:* Passes through the unbroken skin; can increase exposure and the severity of the symptoms listed above. May cause severe irritation of the skin, dryness, redness and rash; skin ulcers.
> *Swallowed:* May cause indigestion, dizziness, fatigue, unconsciousness.

Long-term exposure: May cause skin drying and cracking; liver damage; bone marrow damage, destruction of the blood cells and aplastic anemia.
Storage: Store in tightly closed containers in a dry, cool, well-ventilated place. Keep away from heat, sources of ignition, oxidizing materials. Attacks plastics, rubber and coatings. Containers may explode in fire.

★ 897 ★
STRONTIUM CHROMATE
CAS: 7789-06-2
Other names: CHROMIC ACID, STRONTIUM SALT (1:1); C.I. PIGMENT YELLOW 32; DEEP LEMON YELLOW; STRONTIUM CHROMATE (1:1); STRONTIUM CHROMATE 12170; STRONTIUM CHROMATE A; STRONTIUM CHROMATE (VI); STRONTIUM CHROMATE X-2396; STRONTIUM YELLOW

Danger profile: Carcinogen, mutagen, exposure limit established, Community Right-to-Know List, poisonous gas produced in fire.
Uses: Metal protective coatings to prevent corrosion; color resins; fireworks; in electroplating baths.
Appearance: Yellow powder.
Effects of exposure:
> *Breathed:* Dust may cause irritation of the mouth, nose and breathing passages; ulceration of the nasal septum.
> *Eyes:* May cause severe irritation and burns.
> *Skin:* May cause irritation and burns.
> *Swallowed:* May cause dizziness, intense thirst, abdominal pain, vomiting, shock, reduced ability to make and excrete urine.

Long-term exposure: Repeated exposure may cause chemical to accumulate in the body, and cause persistent effects. May cause lung cancer; mutations; skin, lung and liver damage; nervous system effects.
Storage: Store in tightly closed containers in a dry, cool, well-ventilated, well-marked place. Keep away from strong oxidizers.

★ 898 ★
STRONTIUM NITRATE
CAS: 10042-76-9
Other names: STRONTIUM(II) NITRATE (1:2); NITRATE DE STRONTIUM (FRENCH); NITRIC ACID, STRONTIUM SALT

Danger profile: Combustible, poisonous gas produced in fire.
Uses: Fireworks, marine signals, railroad flares; matches.
Appearance: White powder.
Effects of exposure:
> *Breathed:* Dust may cause irritation of the mouth, nose and breathing passages; ulceration of the nasal septum.
> *Eyes:* May cause severe irritation and burns.
> *Skin:* May cause irritation and burns.
> *Swallowed:* Moderately toxic.

Long-term exposure: Repeated exposure may cause chemical to accumulate in the body, and cause persistent effects. May cause heart muscle, skin, lung, kidney and liver damage; nervous system effects.
Storage: Powerful oxidizer. Store in tightly closed containers in a dry, cool, well-ventilated place. Keep away from heat, fuels, combustible, organic or other readily oxidizable materials.
First aid guide: Move victim to fresh air; call emergency medical care. In case of contact with material, immediately flush skin or eyes with running water for at least 15 minutes. Remove and isolate contaminated clothing and shoes at the site.

★ 899 ★
STYRENE
CAS: 100-42-5
Other names: ANNAMENE; BENZENE, VINYL-; CINNAMENE; CINNAMENOL; CINNAMOL; DIAREX HF 77; ETHYLBENZENE; ETHYLENE, PHENYL-; NCI-C02200; PHENYLETHENE; PHENYLETHYLENE; STIROLO (ITALIAN); STYREEN (DUTCH); STYREN (CZECH); STYRENE MONOMER; STYRENE MONOMER, INHIBITED; STYROL (GERMAN); STYROLE; STYROLENE; STYRON; STYROPOL; STYROPOR; VINYLBENZEN (CZECH); VINYLBENZENE; VINYLBENZOL

Danger profile: Possible carcinogen, mutagen, fire hazard, explosion hazard, exposure limit established, Community Right-to-Know List, poisonous gas produced in fire.
Uses: Solvent; making plastics, synthetic rubber, protective coatings, resins, polyesters; making other chemicals.
Appearance: Colorless to light yellow watery liquid.
Odor: Sweet at low concentrations; characteristic pungent; sharp; disagreeable. The odor is a warning of exposure; however, no smell does not mean you are not being exposed.
Effects of exposure:
> *Breathed:* May cause irritation of the eyes, nose, throat and lungs, tears and runny nose, coughing, sore throat, difficult breathing; decrease coordination and dexterity; lethargy, fatigue, and drowsiness; slurred speech, shaking, and dizziness. Very high levels may cause brain and liver damage; possible death.
> *Eyes:* May cause irritation, redness, tearing, pain, inflamed eyelids, severe itching, and eye injury; burns in corneas and possible loss of sight.
> *Skin:* Passes through the unbroken skin; can increase exposure and the severity of the symptoms listed. May cause painful irritation, redness, and blisters. Repeated or prolonged skin contact may cause irritation of the skin and may also cause the legs to feel numb.
> *Swallowed:* May cause irritation of the mouth and throat; stomach and abdominal pain; loss of muscular coordination and mental confusion; convulsions. Also see symptoms for *Breathed.*

Long-term exposure: May cause cancer; damage to the fetus; liver damage; decrease fertility in women; headache and upset stomach. Repeated low levels of exposure may effect balance, reflexes, memory, concentration and ability to learn. Styrene has been found to produce lung tumors in mice and cause changes in the genetic material of laboratory organisms. Some human studies revealed conflicting results.
Storage: Store outside in tightly closed containers in a dry, cool, well-ventilated place. Keep away from heat, sources of ignition, oxidizing agents, oleum, peroxides, acids, and aluminum chlorides. Protect containers from physical damage. Vapors may travel to sources of ignition and flash back. Containers may explode in fire.

★ 900 ★
SULFALLATE
CAS: 95-06-7
Other names: CARBAMIC ACID, DIETHYLDITHIO-,2-CHLOROALLYL ESTER; CARBAMODITHIOIC ACID, DIETHYL-,2-CHLORO-2-PROPENYL ESTER (9CI); CDEC; CHLORALLYL DIETHYLDITHIOCARBAMATE; 2-CHLOROALLYL DIETHYLDITHIOCARBAMATE; 2-CHLOROALLYL N,N-DIETHYLDITHIOCARBAMATE; 2-CHLORO-2-PROPENYL DIETHYLCARBAMODITHIOATE; CP 4572; CP 4,742; DIETHYLDITHIOCARBAMIC ACID 2-CHLOROALLYL ESTER; NCI-C00453; 2-PROPENE-1-THIOL, 2-CHLORO-, DIETHYLDITHIOCARBAMATE; THIOALLATE; VEGADEX; VEGADEX SUPER; VEGEDEX

Danger profile: Carcinogen, mutagen, combustible, poisonous gas produced in fire.
Uses: Herbicide; control of weeds and grass.
Appearance: Yellowish-orange to red liquid.
Effects of exposure:
> *Breathed:* May be poisonous. May cause irritation of the nose, throat and breathing passages. High exposure may cause headache, dizziness, fatigue, sleepiness, upset stomach, muscle weakness and collapse; personality change; kidney damage.
> *Eyes:* Contact may cause irritation.
> *Skin:* May be poisonous if absorbed. Passes through the unbroken skin; can increase exposure and the severity of the symptoms listed. Contact may cause irritation and rash.
> *Swallowed:* Moderately toxic.

Long-term exposure: May cause cancer; kidney damage. related chemicals may cause skin allergy.
Storage: Store in tightly closed containers in a dry, cool, well-ventilated place. Keep away from heat, sources of ignition, Containers may explode in fire.

★ 901 ★
SULFAMIC ACID
CAS: 5329-14-6
Other names: AMIDOSULFONIC ACID; AMIDOSULFURIC ACID; AMINOSULFONIC ACID; KYSELINA AMIDOSULFONOVA (CZECH); KYSELINA SULFAMINOVA (CZECH); SULFAMIDIC ACID; SULPHAMIC ACID

Danger profile: Combustible, corrosive, poisonous gas produced in fire.
Uses: Stabilizer or swimming pool chemicals; metal cleaning; making paper.
Appearance: White crystals.
Effects of exposure:
> *Breathed:* Vapor may cause irritation to the nose, mouth and breathing passages; cough. High exposures may cause shortness of breath and a dangerous build-up of fluids in the lungs (pulmonary edema), which can cause death. Pulmo-

nary edema is a medical emergency that can be delayed one to two days following exposure.
Eyes: Contact may cause severe irritation and burns.
Skin: Contact may cause severe irritation and burns.
Long-term exposure: Similar and related chemicals may cause skin allergy and lung problems. It is unknown at this time if this chemical causes these problems.
Storage: Store in tightly closed containers in a dry, cool, well-ventilated place. Keep away from acids, chlorine and metal nitrates.
First aid guide: Move victim to fresh air; call emergency medical care. In case of contact with material, immediately flush skin or eyes with running water for at least 15 minutes. Remove and isolate contaminated clothing and shoes at the site. Keep victim quiet and maintain normal body temperature.

★ 902 ★
SULFOTEP
CAS: 3689-24-5
Other names: ASP 47; BAYER-E-393; BIS-O,O-DIETHYLPHOSPHOROTHIONIC ANHYDRIDE; BLADAFUME; BLADAFUN; DITHIO; DITHIODIPHOSPHORIC ACID, TETRA-ETHYL ESTER; DITHIOFOS; DITHION; DITHIONE; DITHIOPHOS; DI(THIOPHOSPHORIC) ACID, TETRAETHYL ESTER; DITHIOPYROPHOSPHATE DE TETRAETHYLE (FRENCH); DITHIOTEP; E393; ENT 16,273; ETHYL THIOPY-ROPHOSPHATE; LETHALAIRE G-57; PIROFOS; PLANT DITHIO AEROSOL; PLANTFUME 103 SMOKE GENERATOR; PYROPHOSPHORODITHIOIC ACID, TETRAETHYL ESTER; PYROPHOSPHORODITHIOIC ACID,O,O,O,O-TETRAETHYL ESTER; SULFATEP; SULFOTEPP; TEDP; TEDTP; O,O,O,O-TETRAETHYL-DITHIO-DIFOSFAAT (DUTCH); TETRAETH-YLDITHIONOPYROPHOSPHATE; TETRAETHYL DITHIOPY-ROPHOSPHATE; O,O,O,O-TETRAETHYLDITHIOPYROPHOSPHATE; TETRAETH-YLDITHIO PYROPHOSPHATE, LIQUID; O,O,O,O-TETRAETIL-DI TIO-PIROFOSFATO (ITALIAN); THIOTEPP

Danger profile: Mutagen, combustible, exposure limit established, poisonous gas produced in fire.
Uses: Insecticide (organophosphate).
Appearance: Yellow liquid.
Odor: Garlic-like. The odor is a warning of exposure; however, no smell does not mean you are not being exposed.
Effects of exposure:
　　Breathed: Extremely poisonous. May cause headache, dizziness, profuse sweating, nausea, vomiting, reduced heart beat, stomach cramps, diarrhea, loss of coordination, slow and weak breathing, fever, loss of consciousness and death. May cause shortness of breath and a dangerous build-up of fluids in the lungs (pulmonary edema), which can cause death. Pulmonary edema is a medical emergency that can be delayed one to two days following exposure.
　　Eyes: Rapidly absorbed. Poisoning can happen very rapidly. See symptoms listed. May cause irritation.
　　Skin: Poisonous. Passes through the unbroken skin; can increase exposure and the severity of the symptoms listed. May cause irritation.
　　Swallowed: Poisonous. Ingestion of liquid or inhalation of mist causes nausea, vomiting, mental confusion, abdominal pain, sweating, giddiness, apprehension, and restlessness; later, muscular twitching of the eyelids and tongue begin, then other muscles of face and neck become involved; generalized twitching and muscle weakness may occur; pulmonary edema, tremor, and convulsions may advance to coma. See **Breathed.**
Long-term exposure: May damage the developing fetus or cause birth defects; nerve damage including loss of coordina-

tion; liver damage. Repeated exposure may cause depression, irritability and personality changes.
Storage: Store in tightly closed containers in a dry, cool, well-ventilated place. Corrosive to most metals when moisture is present. Keep away from oxidizers. Containers may explode in fire.

★ 903 ★
SULFUR
CAS: 7704-34-9
Other names: BENSULFOID; BRIMSTONE; COLLOIDAL-S; COLLOIDAL SULFUR; COLLOKIT; COLSUL; COROSUL D; COROSUL S; COSAN; COSAN 80; CRYSTEX; FLOUR SUL-PHUR; FLOWERS OF SULPHUR; GROUND VOCLE SUL-PHUR; HEXASUL; KOLOFOG; KOLOSPRAY; KUMULUS; MAGNETIC 70; MAGNETIC 90; MAGNETIC 95; MICROFLO-TOX; PRECIPITATED SULFUR; SOFRIL; SPERLOX-S; SPER-SUL; SPERSUL THIOVIT; SUBLIMED SULFUR; SULFIDAL; SULFORON; SULFUR, SOLID; SULKOL; SUPER COSAN; SULPHUR; SULSOL; TECHNETIUM TC 99M SULFUR COL-LOID; TESULOID; THIOLUX; THIOVIT

Danger profile: Combustible, poisonous gas produced in fire.
Uses: In fungicides, pharmaceuticals, gunpowder; making other chemicals.
Appearance: Yellow powder or crystalline solid.
Effects of exposure:
　　Breathed: May cause irritation of the nose, throat, and lungs.
　　Eyes: May cause irritation, inflammation of the eyelids and eye damage.
　　Skin: May cause irritation; allergy may develop.
　　Swallowed: Poisonous. See **First aid guide.**
Long-term exposure: May cause skin allergy; bronchitis; cough, irritation of the respiratory tract; difficult breathing; dust may cause permanent eye damage.
Storage: Store in tightly closed containers in a dry, cool, well-ventilated place. Keep away from water, steam, sources of ignition, oxidizers, chemically active metals, aluminum, sodium, charcoal, phosphorus, metal nitrates, metal carbides, metal halogens. Containers may explode in fire.
First aid guide: Move victim to fresh air; call emergency medical care. In case of contact with material, immediately flush skin or eyes with running water for at least 15 minutes. Removal of solidified molten material from skin requires medical assistance. Remove and isolate contaminated clothing and shoes at the site.

★ 904 ★
SULFUR DIOXIDE
CAS: 7446-09-5
Other names: BISULFITE; FERMENICIDE; SCHWEFELD-DIOXYD (GERMAN); SIARKI DWUTLENEK (POLISH); SUL-FUROUS ACID ANHYDRIDE; SULFUROUS ANHYDRIDE; SULFUROUS OXIDE

Danger profile: Mutagen, combustible, poison, exposure limit established, EPA extremely hazardous substances list, poisonous gas produced in fire.
Uses: Food preservative; fumigant; bleach; making other chemical. A common air contaminant; generated in various manufacturing processes including the burning of various fuels, making paper.
Appearance: Colorless gas or liquid under pressure.
Odor: Sharp; suffocating and irritating odor; like burning sulfur. The odor is a warning of exposure; however, no smell does not mean you are not being exposed.
Effects of exposure:

Breathed: May be poisonous. Vapor may cause intense irritation to nose, throat, mouth and lungs; mucous membrane; with severe choking. Symptoms of exposure include watering of the eyes, coughing, difficult breathing, headache, nausea, vomiting, muscular weakness, unconsciousness. High exposure can lead to drowsiness, coma and death; buildup of fluid in the lungs (pulmonary edema), and may cause death. Pulmonary edema may be delayed by one to two days following exposure.

Eyes: Severe irritant. Contact may cause, burning and stinging, tearing, irritation and redness and swelling of the eyelids, intense pain; can burn the eyes and lead to blurred vision, chronic irritation and other permanent damage. Liquid may cause frostbite.

Skin: Severe irritant. Contact may cause burning or stinging feeling, redness, swelling, intense pain, blisters, irritation, rash. Repeated exposure may lead to skin allergy and chronic irritation. Liquid may cause frostbite.

Swallowed: May cause burning feeling of the mouth and throat, intense thirst, throat swelling, stomach cramps, nausea, vomiting, diarrhea, severe breathing difficulties and unconsciousness. Tissue damage may result. Breathing may stop. Watch for delayed symptoms.

Long-term exposure: May cause mutations; chronic irritation of the eyes, nose and throat; loss of smell; lung damage.

Storage: Store in tightly closed containers in a dry, cool, well-ventilated, fireproof, outdoor place. Keep away from water and chemically active metals. Keep temperature below 130°F; protect containers from damage. Corrodes aluminum. Containers may explode in fire.

First aid guide: Move victim to fresh air and call emergency medical care; if not breathing, give artificial respiration; if breathing is difficult, give oxygen. In case of contact with material, immediately flush skin or eyes with running water for at least 15 minutes. Remove and isolate contaminated clothing and shoes at the site. Keep victim quiet and maintain normal body temperature.

★ 905 ★
SULFURIC ACID

CAS: 7664-93-9

Other names: BOU; DIPPING ACID; HYDROGEN SULFATE; NORDHAUSEN ACID; OIL OF VITRIOL; SPIRIT OF SULFUR; SULPHURIC ACID; VITRIOL BROWN OIL

Danger profile: Combustible, corrosive, violent reaction with water, exposure limit established, Community Right-to-Know List, poisonous gas produced in fire.

Uses: Fertilizers; dyes, pigments; petroleum refining; etching metals; electroplating; rayon, film; explosives; laboratory chemical; making iron, steel and other chemicals.

Appearance: Colorless oily liquid.

Odor: Odorless unless hot, then choking. The odor is a warning of exposure; however, no smell does not mean you are not being exposed.

Effects of exposure:

Breathed: Vapor may cause intense irritation to nose, throat, mouth and lungs; mucous membrane. Symptoms of exposure include watering of the eyes, coughing, difficult breathing, headache, nausea, vomiting, muscular weakness, unconsciousness. High exposure can lead to drowsiness, coma and death; buildup of fluid in the lungs (pulmonary edema), and may cause death. Pulmonary edema may be delayed by one to two days following exposure. Inhalation of concentrated vapors may cause rapid unconsciousness with serious damage to lung tissue; shock and collapse; severe exposure may cause chemical pneumonitis.

Eyes: Severe irritant. Contact may cause, burning and stinging, tearing, irritation and redness and swelling of the eyelids,

intense pain; can burn the eyes and lead to blurred vision, chronic irritation and permanent damage.

Skin: Severe irritant. Contact may cause burning or stinging feeling, redness, swelling, intense pain, blisters, irritation, rash. Repeated exposure may lead to skin allergy and chronic irritation.

Swallowed: May cause burning feeling of the mouth and throat, intense thirst, throat swelling, stomach cramps, nausea, vomiting, diarrhea, severe breathing difficulties and unconsciousness. Tissue damage may result. Breathing may stop. Watch for delayed symptoms.

Long-term exposure: After long exposure workers lose sensitivity to irritating action.

Storage: Violent reaction with water. Do not store in unlined, acid resistant metal drums. Store in tightly closed containers in a dry, cool, well-ventilated place with non-wood floors. Keep away from water and steam, chlorates, chromates, carbides, fulminates, nitrates, picrates, powdered metals. Containers may explode in fire.

First aid guide: Move victim to fresh air and call emergency medical care; if not breathing, give artificial respiration; if breathing is difficult, give oxygen. In case of contact with material, immediately flush skin or eyes with running water for at least 15 minutes. Speed in removing material from skin is of extreme importance. Remove and isolate contaminated clothing and shoes at the site. Keep victim quiet and maintain normal body temperature.

★ 906 ★
SULFUROUS ACID

CAS: 7782-99-2

Other names: SCHWEFLIGE SAURE (GERMAN); SULFUR DIOXIDE SOLUTION; SULFUROUS ACID SOLUTION

Danger profile: Corrosive, poisonous gas produced in fire.

Uses: Bleach, antiseptic, making paper, wine and other chemicals.

Appearance: Colorless liquid.

Odor: Strong, irritating. The odor is a warning of exposure; however, no smell does not mean you are not being exposed.

Effects of exposure:

Breathed: Vapor may cause intense irritation to nose, throat, mouth and lungs; mucous membrane. Symptoms of exposure include watering of the eyes, dryness, coughing, difficult breathing, headache, nausea, vomiting, muscular weakness, unconsciousness. High exposure can burn the lungs; drowsiness, coma and death; buildup of fluid in the lungs (pulmonary edema), and may cause death. Pulmonary edema may be delayed by one to two days following exposure.

Eyes: Severe irritant. Contact may cause, burning and stinging, tearing, irritation and redness and swelling of the eyelids, intense pain; can burn the eyes and lead to blurred vision, chronic irritation and other permanent damage.

Skin: Severe irritant. Passes through the unbroken skin; can increase exposure and the severity of the symptoms listed. Contact may cause burning or stinging feeling, redness, swelling, intense pain, blisters, irritation, rash. Repeated exposure may lead to skin allergy and chronic irritation.

Swallowed: May cause burning feeling of the mouth and throat, intense thirst, throat swelling, stomach cramps, nausea, vomiting, diarrhea, severe breathing difficulties and unconsciousness. Tissue damage may result. Breathing may stop. Watch for delayed symptoms.

Long-term exposure: May cause chronic irritation of the eyes, nose and throat; lung damage.

Storage: Store in tightly closed containers in a dry, cool, well-ventilated place.

First aid guide: Move victim to fresh air; call emergency medical care. In case of contact with material, immediately flush skin or

eyes with running water for at least 15 minutes. Remove and isolate contaminated clothing and shoes at the site. Keep victim quiet and maintain normal body temperature.

★907★
SULFUR TETRAFLUORIDE
CAS: 7783-60-0
Other names: SULFUR FLUORIDE; TETRAFLUROSULFURANE

Danger profile: Exposure limit established, EPA extremely hazardous substances list, poisonous gas produced in fire.
Uses: Fluorinating agent; making water and oil repellents; and lubricity improvers; making pesticides.
Appearance: Colorless gas.
Effects of exposure:
Breathed: May cause irritation of the nose, throat and lungs. Higher exposures may cause shortness of breath and a dangerous build-up of fluids in the lungs (pulmonary edema), which can cause death. Pulmonary edema is a medical emergency that can be delayed one to two days following exposure.
Eyes: May cause extreme irritation, frostbite and burns.
Skin: May cause extreme irritation, frostbite and burns.
Swallowed: A powerful irritant. May cause extreme irritation and burns of the mouth and stomach.
Long-term exposure: May cause brittle bones or stiff ligaments; kidney damage.
Storage: Store in tightly closed containers in a dry, cool, well-ventilated place. Keep away from water, steam and other forms of moisture, dioxygen difluoride and acids. Containers may explode in fire.
First aid guide: Move victim to fresh air and call emergency medical care; if not breathing, give artificial respiration; if breathing is difficult, give oxygen. In case of contact with material, immediately flush skin or eyes with running water for at least 15 minutes. Remove and isolate contaminated clothing and shoes at the site. Keep victim quiet and maintain normal body temperature. Effects may delayed; keep victim under observation.

★908★
SULFUR TRIOXIDE
CAS: 7446-11-9
Other names: SULFAN; SULFURIC ANHYDRIDE; SULFURIC OXIDE; SULFUR TRIOXIDE, STABILIZED

Danger profile: EPA extremely hazardous substances list, poisonous gas produced in fire.
Uses: Making other chemicals; solar energy collectors.
Appearance: Solid.
Effects of exposure:
Breathed: May be poisonous. Vapor may cause intense irritation to nose, throat, mouth and lungs; mucous membrane. Symptoms of exposure include watering of the eyes, coughing, difficult breathing, headache, nausea, vomiting, muscular weakness, unconsciousness. High exposure can lead to drowsiness, coma and death; buildup of fluid in the lungs (pulmonary edema), and may cause death. Pulmonary edema may be delayed by one to two days following exposure.
Eyes: Severe irritant. May cause coughing, choking and severe discomfort in a concentration of 1ppm. Contact may cause, burning and stinging, tearing, irritation and redness and swelling of the eyelids, intense pain; can burn the eyes and lead to blurred vision, chronic irritation and other permanent damage.
Skin: Severe irritant. Contact may cause burning or stinging feeling, redness, swelling, intense pain, blisters, irritation,

rash. Repeated exposure may lead to skin allergy and chronic irritation.
Swallowed: May be poisonous. May cause burning feeling of the mouth and throat, intense thirst, throat swelling, stomach cramps, nausea, vomiting, diarrhea, severe breathing difficulties and unconsciousness. Tissue damage may result. Breathing may stop. Watch for delayed symptoms.
Long-term exposure: May cause lung problems.
Storage: Oxidizing agent. Store in tightly closed containers in a dry, cool, well-ventilated place with non-wood floors. Keep away from steam, water and other forms of moisture, sulfuric acid, combustible and organic materials, pyridine, metal oxides, acetonitrile, iodine, metal oxides and other chemicals.
First aid guide: Move victim to fresh air and call emergency medical care; if not breathing, give artificial respiration; if breathing is difficult, give oxygen. In case of contact with material, immediately flush skin or eyes with running water for at least 15 minutes. Speed in removing material from skin is of extreme importance. Remove and isolate contaminated clothing and shoes at the site. Keep victim quiet and maintain normal body temperature.

★909★
SULFURYL FLUORIDE
CAS: 2699-79-8
Other names: FLUORURE DE SULFURYLE (FRENCH); SULFURIC OXYFLUORIDE; VIKANE; VIKANE FUMIGANT

Danger profile: Exposure limit established, poisonous gas produced in fire.
Uses: Insecticide; fumigant.
Appearance: Colorless gas.
Effects of exposure:
Breathed: May be poisonous. May cause irritation to the nose, throat and breathing passages; nosebleeds, cough and phlegm. High exposures may cause nausea, vomiting, seizures; shortness of breath and a dangerous build-up of fluids in the lungs (pulmonary edema), which can cause death. Pulmonary edema is a medical emergency that can be delayed one to two days following exposure. May be narcotic in high concentrations.
Eyes: Contact may cause irritation and burns.
Skin: Contact may cause irritation and burns.
Swallowed: May be poisonous.
Long-term exposure: May cause lung, kidney damage; accumulation of fluoride in the body may effect and damage bones and teeth.
Storage: Store in tightly closed containers in a dry, cool, well-ventilated place. Keep away from water, steam and other forms of moisture. Containers may explode in fire.
First aid guide: Move victim to fresh air and call emergency medical care; if not breathing, give artificial respiration; if breathing is difficult, give oxygen. In case of contact with material, immediately flush skin or eyes with running water for at least 15 minutes. Remove and isolate contaminated clothing and shoes at the site. Keep victim quiet and maintain normal body temperature. Effects may delayed; keep victim under observation.

★910★
SULPROFOS
CAS: 35400-43-2
Other names: BAY-NTN-9306; BOLSTAR; O-ETHYLO-(4-(METHYLTHIO)PHENYL) S-PROPYLPHOSPHORODITHIOATE; HELOTHION; PHOSPHORODITHIOIC ACID, O-ETHYLO-(4-(METHYLTHIO)PHENYL)S-PROPYL ESTER

Danger profile: Exposure limit established, poisonous gas produced in fire.
Uses: Insecticide.
Appearance: Tan colored liquid.
Effects of exposure:

Breathed: May be poisonous. May cause headache, dizziness, profuse sweating, nausea, vomiting, reduced heart beat, stomach cramps, diarrhea, loss of coordination, slow and weak breathing, fever, loss of consciousness and death. May cause shortness of breath and a dangerous build-up of fluids in the lungs (pulmonary edema), which can cause death. Pulmonary edema is a medical emergency that can be delayed one to two days following exposure.

Eyes: Rapidly absorbed. Poisoning can happen very rapidly. See *Breathed.*

Skin: Poisonous if absorbed. Passes through the unbroken skin; can increase exposure and the severity of the symptoms listed above.

Swallowed: Poisonous.

Long-term exposure: May cause depression, anxiety, irritability, personality changes; nerve damage.
Storage: Store in tightly closed containers in a dry, cool, well-ventilated place. Containers may explode in fire.

★911★
SWAT
CAS: 122-10-1
Other names: BOMYL; DIMETHYL 1,3-BIS(CARBOMETHOXY)-1-PROPEN-2-YL PHOSPHATE; DI-METHYL 1,3-DI(CARBOMETHOXY)-1-PROPEN-2-YL PHOSPHATE; ENT 24,833; FLY BAIT GRITS; GC 3707; GENERAL CHEMICALS 3707; GLUTONIC ACID, 3-HYDROXY-,DIMETHYL ESTER, DIMETHYL PHOSPHATE; 3- HYDROXY-GLUTACONIC ACID, DIMETHYL ESTER, DIMETHYL PHOSPHATE; PHOSPHORIC ACID, DIMETHYL ESTER WITH DI-METHYL 3- HYDROXYGLUTACONATE

Danger profile: Poison, poisonous gas produced in fire.
Uses: Insecticide (organophosphate).
Appearance: Liquid.
Effects of exposure:

Breathed: Poisonous. May cause headache, dizziness, profuse sweating, nausea, vomiting, reduced heart beat, stomach cramps, diarrhea, loss of coordination, slow and weak breathing, fever, loss of consciousness and death. May cause shortness of breath and a dangerous build-up of fluids in the lungs (pulmonary edema), which can cause death. Pulmonary edema is a medical emergency that can be delayed one to two days following exposure.

Eyes: Poisoning can happen very rapidly. See *Breathed.*
Skin: Poisonous. Passes through the unbroken skin; can increase exposure and the severity of the symptoms listed above.

Swallowed: Powerful poison. See *Breathed.*

Long-term exposure: Similar organophosphate chemicals may cause nerve damage.
Storage: Store in specially marked place in tightly-closed glass container in a dry, well-ventilated place. Corrosive to iron, steel and brass. Containers may explode in fire.

T

★912★
2,4,5-T ACID
CAS: 93-76-5
Other names: ACETIC ACID, (2,4,8-TRICHLOROPHYNOXY)-; ACIDE 2,4,5-TRICHLOROPHENOXYACETIQUE (FRENCH); ACIDO (2,4,5-TRICLORO-FENOSSI)-ACETICO (ITALIAN); BCF-BUSHKILLER; BRUSH-OFF 445 LOW VOLATILE BRUSH KILLER; BRUSH RHAP; BRUSHTOX; DACAMINE; DE-BROUSSAILLANT CONCENTRE; DEBROUSSAILLANT SUPER CONCENTRE; DECAMINE 4T; DED-WEED BRUSH KILLER; DED-WEED LV-6 BRUSH KIL; T-5 BRUSH KIL; DINOXOL; ENVERT-T; ESTERCIDE T-2 AND T-245; ESTERON; ESTERON 245; ESTERON BRUSH KILLER; FENCE RIDER; FORRON; FORSTU 46; FORTEX; FRUITONE A; IN-VERTON 245; LINE RIDER; PHORTOX; REDDON; REDDOX; SPONTOX; SUPER D WEEDONE; 2,4,5-T; TIPPON; TOR-MONA; TRANSAMINE; TRIBUTON; (2,4,5-TRICHLOOR-FENOXY)-AZIJNZUUR (DUTCH); 2,4,5-TRICHLOROPHENOXYACETIC ACID; (2,4,5-TRICHLOR-PHENOXY)-ESSIGSAEURE (GERMAN); TRINOXOL; TRIOX-ON; TRIOXONE; U 46; VEON; VEON 245; VERTON 2T; VISKO RHAP LOW VOLATILE ESTER; WEEDAR; WEEDONE

Danger profile: Carcinogen, mutagen, exposure limit established, reproductive hazard, poisonous gas produced in fire.
Uses: Weed control; herbicide.
Appearance: Colorless to tan solid.
Effects of exposure:

Breathed: May be poisonous. May cause irritation of the nose, throat and breathing passages; weakness, loss of appetite, nausea. High levels may cause muscle spasms, fatigue and seizures or fits; possible liver and kidney levels
Eyes: Contact may cause irritation.
Skin: May be poisonous if absorbed. Passes through the unbroken skin; can increase exposure and the severity of the symptoms listed. Contact may cause itching, redness, swelling and rash; acne-like sores, loss of skin; coloration in small patches.

Swallowed: Poisonous. May cause fatigue, nausea, vomiting, diarrhea, lowered blood pressure, convulsions, coma, cardiac arrest and death; mouth ulcers.

Long-term exposure: May cause cancer, mutations, birth defects; decrease of fertility in men and women; liver and kidney effects; gastrointestinal tract ulcers, nerve disorders resulting in difficulty controlling muscles. Animal studies also indicate the possibility of an increased susceptibility to infection. Chemical may contain the chemical TCDD as an impurity.
Storage: Store in tightly closed containers in a dry, cool, well-ventilated place. Keep away from heat, direct sunlight, oxidizers, strong bases, acid and acid fumes. Corrosive to common metals. Containers may explode in fire.
First aid guide: Move victim to fresh air and call emergency medical care; if not breathing, give artificial respiration; if breathing is difficult, give oxygen. In case of contact with material, immediately flush skin or eyes with running water for at least 15 minutes. Speed in removing material from skin is of extreme importance. Remove and isolate contaminated clothing and shoes at the site. Keep victim quiet and maintain normal body temperature. Effects may be delayed; keep victim under observation.

★913★
TALC
CAS: 14807-96-6
Other names: AGALITE; ALPINE TALC; ASBESTINE; C.I. 77718; DESERTALC 57; EMTAL 596; FIBRENE C 400; FRENCH CHALK; HYDROUS MAGNESIUM SILICATE; LO MICRON TALC 1; METRO TALC; MISTRON; MISTRON STAR; MISTRON SUPER FROST; MISTRON VAPOR; MP-12-50; MP 25-38; NCI- C06008; NON-ASBESTIFORM TALC; NON-FIBROUS TALC; NYTAL; OOS; OXO; PURETALC USP; SIERRA C-400; SNOWGOOSE; SEAWHITE; STEATITE TALC; SUPREME DENSE; TALC (NON-ASBESTOS FORM); TALCUM

Danger profile: Exposure limit established.
Uses: In a wide range of products including baby and cosmetic powders such as bath and face powder, eye shadows, liquid powders, skin fresheners, foot powders, face creams, drugs, ceramics, paints, plasters, putty. Similar in chemical composition to asbestos.
Appearance: Grayish-white powder.
Effects of exposure:
Breathed: May cause irritation of the nose, coughing and sneezing. When carelessly used and inhaled by babies, talc has caused coughing, vomiting and pneumonia. Frequently talc may be contaminated with asbestos or silica.
Eyes: Contact may cause irritation, redness, tearing, pain and inflammation of the eyelids. Serious eye damage may develop.
Long-term exposure: Repeated high exposures may cause lung scarring; cough, shortness of breath; abnormal x-ray; talc pneumoconiosis. Questionable carcinogen; however, talc-based powders have been linked to ovarian cancer. Talc containing asbestos fibers are confirmed carcinogens.
Storage: Store in tightly closed containers in a dry, cool, well-ventilated place.

★914★
2,3,6-TBA
CAS: 50-31-7
Other names: ACIDE TRICHLOROBENZOIQUE (FRENCH); BENZAC; BENZAC-1281; BENZBOR; BENZOIC ACID, 2,3,6-TRICHLORO-; CAMBILINE; CAMPBELL'S CAMTOX; FEN-ALL; HC 12811; NCI-C60242; POCKET TOUCHWEEDER; R262; T-2; TBA; TCB; TCBA; 2,3,6-TCBA; TRIBAC; TRIBEN; TRYSBEN; TRYSBEN 200; ZOBAR

Danger profile: Poison, poisonous gas produced in fire.
Uses: Herbicide; on cereals; for grass seed control. Products containing 2,3,6-TBA were canceled in the U.S.
Appearance: Liquid.
Effects of exposure:
Breathed: May be poisonous. May cause irritation of the nose, throat and breathing passages; nosebleeds, coughing; chest tightness; nausea, increased saliva, tremor and depression. High levels may effect the kidneys.
Eyes: May cause irritation, redness and inflammation of the eyelids.
Skin: May cause irritation. Prolonged contact may cause rash, redness and swelling.
Swallowed: May be poisonous.
Long-term exposure: May cause lung problems.
Storage: Store in tightly closed containers in a dry, cool, well-ventilated place.

★915★
TCDD
CAS: 1746-01-6
Other names: DIOKSYNY (POLISH); DIBENZO(B,E) 1,4-DIOXIN,2,3,7,8- TETRACHLORO-; DIOXINE; DIOXIN; NCI-C03714; TCDBD; 2,3,7,8-TCDD; 2,3,7,8-TETRACHLORODIBENZO-P-DIOXIN; TETRADIOXIN

Danger profile: Carcinogen, mutagen, poison; reproductive hazard, poisonous gas produced in fire, among the most toxic synthetic chemicals.
Uses: Occurs as an impurity in the manufacture of fungicides, herbicides and other chemicals such as polychlorinated biphenols (PCBs); incineration of chemical wastes. May last in soils for years. Most toxic of a group of more than 70 chemicals known as dioxins.
Appearance: Colorless, crystalline, needle-like solid.
Effects of exposure:
Breathed: May be poisonous. May cause headaches, weakness, numbness and tingling in the arms and legs; liver and nervous system damage. Symptoms may slowly develop over a period of days. Death may result a lethal dose by weeks.
Eyes: May cause irritation.
Skin: Passes through the unbroken skin; can increase exposure and the severity of the symptoms listed. May cause chloracne, an acne-like skin rash. Symptoms may slowly develop over a period of days.
Swallowed: May be poisonous. May cause headaches, weakness, stomach upset, numbness and pain in the legs; liver and nervous system damage. Symptoms may slowly develop over a period of days. See *Breathed.*
Long-term exposure: Low levels may cause cancer and birth defects; chloracne; liver and neurological damage. A blood abnormality may occur which may include light sensitive skin, blisters, dark skin coloration, excessive hair growth and dark urine.
Storage: Use the same precautions required for radioactive work. Store in sealed containers in a dry, cool, well-ventilated place.

★916★
TELLURIUM HEXAFLUORIDE
CAS: 7783-80-4
Other names: TELLURIUM FLUORIDE

Danger profile: Poison, exposure limit established, EPA extremely hazardous substances list, poisonous gas produced in fire
Appearance: Colorless gas.
Odor: Repulsive. The odor is a warning of exposure; however, no smell does not mean you are not being exposed.
Effects of exposure:
Breathed: May be poisonous. Vapor may cause intense irritation to nose, throat, mouth and lungs; mucous membrane. Symptoms of exposure include watering of the eyes, coughing, difficult breathing, headache, nausea, vomiting, muscular weakness, unconsciousness. High exposure can lead to drowsiness, coma and death; buildup of fluid in the lungs (pulmonary edema), and may cause death. Pulmonary edema may be delayed by one to two days following exposure.
Eyes: Severe irritant. Contact may cause, burning and stinging, tearing, irritation and redness and swelling of the eyelids, intense pain; can burn the eyes and lead to blurred vision, chronic irritation and other permanent damage. Liquid may cause frostbite.
Skin: May be poisonous if absorbed. Severe irritant. Contact may cause burning or stinging feeling, redness, swelling, intense pain, blisters, irritation, rash. Repeated exposure may

lead to skin allergy and chronic irritation. Liquid may cause frostbite.

Swallowed: May be poisonous. May cause burning feeling of the mouth and throat, intense thirst, throat swelling, stomach cramps, nausea, vomiting, diarrhea, severe breathing difficulties and unconsciousness. Tissue damage may result. Breathing may stop. Watch for delayed symptoms.

Storage: Store in tightly closed containers in a dry, cool, well-ventilated place. Containers may explode in fire.

First aid guide: Move victim to fresh air and call emergency medical care; if not breathing, give artificial respiration; if breathing is difficult, give oxygen. In case of contact with material, immediately flush skin or eyes with running water for at least 15 minutes. Remove and isolate contaminated clothing and shoes at the site. Keep victim quiet and maintain normal body temperature. Effects may delayed; keep victim under observation.

★917★
TEMEPHOS
CAS: 3383-96-8

Other names: ABAT; ABATE; ABATHION; AC 52160; AMERICAN CYANAMID AC-52,160; BIOTHION; BITHION; CL 52160; DIFENTHOS; O,O-DIMETHYLPHOSPHOROTHIOATE O,O-DIESTER WITH 4,4'-THIODIPHENOL; ECOPRO; EI 52160; ENT 27,165; EXPERIMENTAL INSECTICIDE 52160; NIMITEX; NIMITOX; PHENOL,4,4'-THIODI-, O,O-DIESTER WITH O,O-DIMETHYLPHOSPHOROTHIOATE; PHOSPHOROTHIOIC ACID, O,O'-(THIODI-4,1-PHENYLENE)O,O,O',O'-TETRAMETHYL ESTER; SWEBATE; TEMEFOS; TEMOPHOS; TETRAMETHYL-O,O'-THIODI-P-PHENYLENE PHOSPHOROTHIOATE; O,O,O',O'-TETRAMETHYLO,O'-THIODI-P-PHENYLENEPHOSPHOROTHIOATE; O,O'-(THIODI-4,1-PHENYLENE)BIS(O,O-DIMETHYL PHOSPHOROTHIOATE); O,O'-(THIODI-P-PHENYLENE)O,O,O',O'-TETRAMETHYLBIS(PHOSPHOROTHIOAT E)

Danger profile: Exposure limit established, poisonous gas produced in fire.

Uses: Insecticide (organophosphate).

Appearance: White crystalline solid or a brown liquid.

Effects of exposure:

Breathed: May cause headache, dizziness, profuse sweating, nausea, vomiting, reduced heart beat, stomach cramps, diarrhea, loss of coordination, slow and weak breathing, fever, loss of consciousness and death. May cause shortness of breath and a dangerous build-up of fluids in the lungs (pulmonary edema), which can cause death. Pulmonary edema is a medical emergency that can be delayed one to two days following exposure.

Eyes: Rapidly absorbed. Poisoning can happen very rapidly. See ***Breathed.***

Skin: May cause severe poisoning. Passes through the unbroken skin; can increase exposure and the severity of the symptoms listed. Mild skin irritant.

Swallowed: May cause headache, sweating, nausea, vomiting, diarrhea, loss of coordination and death. human volunteers have tolerated an oral dosage of 256 mg/man per day for five days or 64mg/man/day without clinical symptoms or side effects, and without detectable effects on red blood cells or plasma cholinesterase.

Long-term exposure: May cause nerve damage including loss of coordination; liver damage. Repeated exposure may cause depression, irritability and personality changes.

Storage: Store in a regulated, specially marked place in tightly closed containers in a dry, well-ventilated place. Keep away from strong alkalis. Containers may explode in fire.

★918★
TEPP
CAS: 107-49-3

Other names: BIS-O,O-DIETHYLPHOSPHORIC ANHYDRIDE; BLADAN; BLADON; DIPHOSPHORIC ACID, TETRAETHYL ESTER; ENT 18,771; ETHYL PYROPHOSPHATE, TETRA-; FOSVEX; GRISOL; HEPT; HEXAMITE; KILLAX; KILMITE 40; LETHALAIRE G-52; LIROHEX; MORTOPAL; MOTOPAL; NIFOS; NIFOS T; NIFROST; PHOSPHORIC ACID, TETRAETHYL ESTER; COMMERCIAL 40%; PYROPHOSPHATE DE TETRAETHYLE (FRENCH); TEP; O,O,O,O-TETRAAETHYL-DIPHOSPHAT, BIS(O,O-DIAETHYLPHOSPHORSAEURE)-ANHYDRID (GERMAN); O,O,O,O-TETRAETHYL-DIFOSFAAT (DUTCH); O,O,O,O-TETRAETIL-PIROFOSFATO (ITALIAN); TETRAETHYL PYROFOSFAAT (BELGIAN); TETRAETHYL-PYROPHOSPHATE; TETRAETHYL PYROPHOSPHATE, LIQUID; TETRASTIGMINE; TETRON; TETRON-100; VAPOTONE

Danger profile: Poison, exposure limit established, EPA extremely hazardous substances list, poisonous gas produced in fire.

Uses: Insecticide for aphids and mites; rodenticide (organophosphate).

Appearance: Colorless to amber liquid.

Odor: Aromatic; faint fruity. The odor is a warning of exposure; however, no smell does not mean you are not being exposed.

Effects of exposure:

Breathed: Poisonous. May cause headache, dizziness, profuse sweating, nausea, vomiting, reduced heart beat, stomach cramps, diarrhea, loss of coordination, slow and weak breathing, fever, loss of consciousness and death. May cause shortness of breath and a dangerous build-up of fluids in the lungs (pulmonary edema), which can cause death. Pulmonary edema is a medical emergency that can be delayed one to two days following exposure.

Eyes: Rapidly absorbed. Poisoning can happen very rapidly. See ***Breathed.***

Skin: Poisonous. Passes through the unbroken skin; can increase exposure and the severity of the symptoms listed. Fatal poisoning can occur form skin contact; there may be no feeling of irritation.

Swallowed: Powerful, deadly poison. May cause excitement, nausea, vomiting, mental confusion, abdominal pain, sweating, giddiness, apprehension, and restlessness; later, muscular twitching of eyelids and tongue begin, then other muscles of face and neck become involved; pulmonary edema, ataxia, tremor, and convulsions may advance to coma.

Long-term exposure: May cause nerve damage including loss of coordination; liver damage. Repeated exposure may cause depression, irritability and personality changes.

Storage: Store in a regulated, specially marked place in tightly closed containers in a dry, well-ventilated place. Keep away from oxidizers. Corrosive to aluminum; slowly corrosive to copper, brass zinc and tin. Containers may explode in fire.

★919★
TESTOSTERONE
CAS: 58-22-0
Other names: ANDROLIN; ANDRONAQ; ANDROST-4-EN-3-ONE,17-HYDROXY-, (17-BETA)-; ANDROST-4-EN-17(BETA)-OL-3-ONE; ANDROST-4-EN-3-ONE, 17-BETA-HYDROXY-; ANDRUSOL; CRISTERONE T; GENO-CRISTAUX GREMY; HOMOSTERON; HOMOSTERONE; 17-BETA-HYDROXYANDROST-4-EN-3-ONE; 17-BETA-HYDROXY-4-ANDROSTEN-3-ONE; 7-BETA-HYDROXYANDROST-4-EN-3-ONE; MALESTRONE (AMPS); MERTESTATE; NEO-TESTIS; ORETON; ORETON-F; ORQUISTERON; PERANDREN; PERCUTACRINE ANDROGENIQUE; PRIMOTEST; PRIMOTESTON; SUSTANONE; SYNANDROL F; TESLEN; TESTANDRONE; TESTICULOSTERONE; TESTOBASE; TESTOPROPON; TESTOSTEROID; TESTOSTERON; TESTOSTERONE HYDRATE; TESTOSTOSTERONE; TESTOVIRON SCHERING; TESTOVIRON T; TESTRONE; TESTRYL; VIRORMONE; VIROSTERONE

Danger profile: Carcinogen, mutagen, reproductive hazard, poisonous gas produced in fire.
Uses: In medicines; biochemical research.
Appearance: White to cream-white crystalline solid.
Effects of exposure:
　Breathed: May cause acne, nausea; fluid retention.
Long-term exposure: May cause cancer; mutations; development of male features, baldness, increased body hair, deep voice, smaller breasts and menstrual cycle changes in women. Lowered fertility and sperm count, painful breast development and enlarged prostate and excessive production of red blood cells in men. Stunted growth, premature puberty in boys and abnormal genitals in girls.
Storage: Store in tightly closed containers in a dry, cool, well-ventilated place.

★920★
TETRABUTYLSTANNINE
CAS: 1461-25-2
Other names: TETRA-N-BUTYLCIN (CZECH); TETRABUTYLTIN

Danger profile: Combustible, poison, exposure limit established, acrid smoke and irritating fumes produced in fire.
Uses: Making heat resistant paint; improving bonding of paints, rubber, and plastics to metal surfaces.
Appearance: Colorless to light yellow liquid.
Effects of exposure:
　Breathed: May cause irritation of the eyes; headache, nausea; sneezing and coughing.
　Eyes: May cause irritation, itching, tearing and swelling of the eyelids.
　Skin: Poisonous. May cause irritation with itching, redness, swelling; blisters and skin ulcerations.
　Swallowed: May cause irritation of the mouth, throat and stomach; pain; headache, nausea and vomiting, dizziness, convulsions, unconsciousness, coma; possible permanent paralysis of the limbs, death.
Long-term exposure: May cause lung tissue and liver damage.
Storage: Store in tightly closed containers in a dry, cool, well-ventilated place. Exercise caution; little is known about organometallic compounds; may be highly reactive and dangerous. Containers may explode in fire.

★921★
1,1,1,2-TETRACHLORO-2,2-DIFLUOROETHANE
CAS: 76-11-9
Other names: 1,1-DIFLUOROPERCHLOROETHANE; 2,2-DIFLUORO-1,1,1,2- TETRACHLORETHANE; ETHANE, 1,1,1,2-TETRACHLORO-2,2-DIFLUORO-; REFRIGERANT 112A; HALOCARBON 112A

Danger profile: Exposure limit established, poisonous gas produced in fire.
Uses: Refrigerant; solvent; making plastics; in dry cleaning.
Appearance: Colorless liquid, solid or gas.
Odor: Slight, ether-like. The odor is a warning of exposure; however, no smell does not mean you are not being exposed.
Effects of exposure:
　Breathed: May cause irritation of the eyes, nose, throat and air passages; coughing, headache; increased saliva; shortness of breath; reddening of the face; dizziness, lightheadedness, fatigue, loss of consciousness, shock, coma, irregular heartbeat, death. If the victim recovers watch for the possibility of a buildup of fluid in the lungs (pulmonary edema). Pulmonary edema is a medical emergency that can be delayed one to two days following exposure. Very high exposures may cause unconsciousness and death.
　Eyes: Contact may cause irritation, tearing, inflammation of the eyelids; burning sensation.
　Skin: May cause irritation, redness, burning feeling, swelling; dry skin; rash or blisters.
　Swallowed: Breath may smell of chloroform. May cause irritation of the mouth, gastrointestinal tract, nausea, vomiting, diarrhea with possible blood; drowsiness; loss of consciousness; irregular heartbeat and death. See *Breathed.*
Long-term exposure: May cause liver damage; reduce the number of white blood cells; irregular heart beat; lung effects.
Storage: Store in tightly closed containers in a dry, cool, well-ventilated place. Keep away from heat, chemically active metals, powdered aluminum, magnesium or zinc. Containers may explode in fire.

★922★
1,1,2,2-TETRACHLORO-1,2-DIFLUOROETHANE
CAS: 76-12-0
Other names: 1,2-DIFLUORO-1,1,2,2-TETRACHLOROETHANE; ETHANE,1,1,2,2-TETRACHLORO-1,2-DIFLUORO-; F-112; FREON 112; GENETRON 112; HALOCARBON 112; REFRIGERANT 111; TETRACHLORO-1,2- DIFLUOROETHANE; SYM-TETRACHLORO-1,2-DIFLUOROETHANE

Danger profile: Exposure limit established, poisonous gas produced in fire.
Uses: Refrigerant; in dry cleaning; making plastics.
Appearance: Colorless liquid, white solid or refrigerant gas.
Odor: Slight, ether-like. The odor is a warning of exposure; however, no smell does not mean you are not being exposed.
Effects of exposure:
　Breathed: May cause irritation to the lungs; coughing, shortness of breath; dizziness, lightheadedness, loss of consciousness; irregular heartbeat. High exposures may cause shortness of breath and a dangerous build-up of fluids in the lungs (pulmonary edema), which can cause death. Pulmonary edema is a medical emergency that can be delayed one to two days following exposure. Very high exposures may cause death.
　Eyes: Contact may cause irritation, redness, tearing and inflamed eyelids. Liquid may cause frostbite.
　Skin: Contact may cause irritation, redness, burning sensation and rash. Liquid may cause frostbite.
　Swallowed: Moderately toxic.

Long-term exposure: May cause liver damage; reduced white blood cell count.
Storage: Store in tightly closed containers in a dry, cool, well-ventilated place. Keep away from heat, chemically active metals. Containers may explode in fire.

★ 923 ★
1,1,2,2-TETRACHLOROETHANE
CAS: 79-34-5
Other names: ACETYLENE TETRACHLORIDE; BONO-FORM; CELLON; 1,1,2,2-CZTEROCHLOROETAN (POLISH); 1,1-DICHLORO-2,2-DICHLOROETHANE; ETHANE,1,1,2,2-TETRACHLORO-; NCI-C03554; TCE; 1,1,2,2-TETRACHLOORETHAAN (DUTCH); 1,1,2,2-TETRACHLORAETHAN (GERMAN); TETRACHLORETHANE; 1,1,2,2-TETRACHLORETHANE (FRENCH); SYM-TETRACHLOROETHANE; 1,1,2,2-TETRACHLORO-; 1,1,2,2-TETRACLOROETANO (ITALIAN); TETRACHLORURE D'ACETYLENE (FRENCH); WESTRON

Danger profile: Carcinogen, mutagen, poison, exposure limit established, Community Right-to-Know List, poisonous gas produced in fire.
Uses: Making other chemicals, insecticides, paints, rust removers and varnishes. Restricted or forbidden to be used in certain countries.
Appearance: Colorless or pale yellow liquid.
Odor: Sickly sweet, like chloroform. The odor is a warning of exposure; however, no smell does not mean you are not being exposed.
Effects of exposure:
  Breathed: May cause irritation of the eyes, nose, throat and air passages; coughing, headache; increased saliva; reddening of the face; dizziness, lightheadedness, fatigue, loss of consciousness, shock, coma, irregular heartbeat, death. If the victim recovers watch for the possibility of a buildup of fluid in the lungs (pulmonary edema). Pulmonary edema is a medical emergency that can be delayed one to two days following exposure. Exposure may cause enough liver and kidney damage to result in death.
  Eyes: Contact may cause severe irritation, tearing, inflammation of the eyelids; burning sensation and permanent damage.
  Skin: Passes through the unbroken skin; can increase exposure and the severity of the symptoms listed. May cause severe irritation, redness, burning feeling, swelling; dry skin; rash or blisters.
  Swallowed: Poison. Breath may smell of chloroform. May cause irritation of the mouth, gastrointestinal tract, nausea, vomiting, abdominal pain, diarrhea with possible blood; tremor of fingers, drowsiness; headache; irritability; nervousness; hallucinations; distorted perceptions; loss of consciousness; irregular heartbeat and death.
Long-term exposure: May cause cancer; mutations; liver, kidney, blood forming organ and nerve damage.
Storage: Severe industrial hazard. Store in tightly closed containers in a dry, cool, well-ventilated place. Keep away from heat and sources of ignition, chemically active metals, potassium hydroxide and acids. Containers may explode in fire.

★ 924 ★
TETRACHLORONAPHTHALENE
CAS: 1335-88-2
Other names: HALOWAX; NAPHTHALENE, TETRA-CHLORO-; NIBREN WAX; SEEKAY WAX; 1,2,3,4-TETRACHLORO-1,2,3,4-TETRAHYDRONAPHTHALENE

Danger profile: Probable poison, exposure limit established, poisonous gas produced in fire.
Uses: Making waterproofing materials, flame retardants, metal and electrical products.
Appearance: Colorless to pale yellow waxy solid.
Odor: Sweet, aromatic. The odor is a warning of exposure; however, no smell does not mean you are not being exposed.
Effects of exposure:
  Breathed: High exposure may cause liver damage and result in death.
  Eyes: Contact may cause irritation.
  Skin: Passes through the unbroken skin; can increase exposure and the severity of the symptoms listed. May cause irritation, burning sensation and rash.
  Swallowed: May be poisonous.
Long-term exposure: May cause chloracne, and acne-form skin rash; skin may burn and be affected more easily by the sun; liver damage; weight loss.
Storage: Store in tightly closed containers in a dry, cool, well-ventilated place. Keep away from heat and oxidizers.

★ 925 ★
TETRACYCLINE
CAS: 60-54-8
Other names: ABRAMYCIN; ABRICYCLINE; ACHROMYCIN; AGROMICINA; AMBRAMICINA; AMBRAMYCIN; AMYCIN; BIO-TETRA; BRISTACICLIN ALPHA; BRISTACICLINA; BR-ISTACYCLINE; CEFRACYCLINE SUSPENSION; CRISEOCI-CLINA; CYCLOMYCIN; DEMOCRACIN; DESCHLOROBIOMY-CIN; HOSTACYCLIN; LIQUAMYCIN; 6-METHYL-1,11-DIOXY-2-NAPHTHACENECA RBOXAMIDE; 2-NAPHTHACENECARBOXAMIDE; NEOCYCLINE; OLETETRIN; OMEGAMYCIN; ORLYCYCLINE; PANMYCIN; PENTAHYDROXY-6-METHYL-1,1,1-DIOXO-; POLYCYCLINE; PUROCYCLINA; ROBITET; SANCLOMYCIN; SIGMAMYCIN; STECLIN; T-125; TETRABON; TETRACYCLINE I; TETRACY-CLINE II; SK-TETRACYCLINE; TETRACYN; TETRADECIN; TETRAVERINE; TSIKLOMITSIN

Danger profile: Mutagen, possible reproductive hazard, poison, poisonous gas produced in fire.
Uses: Antibiotic medication for human and animal infections.
Appearance: Yellow crystalline powder.
Effects of exposure:
  Breathed: May cause sneezing, nose itching and upset stomach.
  Skin: May cause allergic skin reaction to occur.
  Swallowed: Moderately toxic. May cause sleepiness, constipation, and a decrease in urine volume.
Long-term exposure: May cause mutations; damage to the fetus; liver and kidney damage; skin allergy.
Storage: Store in tightly closed containers in a dry, cool, well-ventilated place. Keep away from sunlight.

★ 926 ★
TETRAETHYLENEPENTAMINE
CAS: 112-57-2
Other names: AMINO ETHYL-1,2-ETHANEDIAMINE, 1,4,7,10,13,- PENTAAZATRIDECANE; D.E.H. 26; 1,2-ETHANEDIAMINE, N-(2-AMINOETHYL)-N'-(2-(2-AMINOETHYL)ETHYL)-; 1,2-ETHANEDIAMINE, N-(2-AMINOETHYL)-N'-(2-AMINOETHYL) AMINOETHYL-; 1,4,7,10,13-PENTAAZATRIDECANE

Danger profile: Mutagen, combustible, corrosive, poisonous gas produced in fire.
Uses: Solvent for resins and dyes; making synthetic rubber and oil additives; dispersant in motor oils.

Appearance: Thick yellow liquid.
Odor: Ammonia-like; disagreeable; penetrating. The odor is a warning of exposure; however, no smell does not mean you are not being exposed.
Effects of exposure:
Breathed: May cause irritation to the nose, throat and breathing passages; nausea; nosebleeds. Chemical is a sensitizer, and prolonged contact may cause asthma. High exposures may cause shortness of breath and a dangerous build-up of fluids in the lungs (pulmonary edema), which can cause death. Pulmonary edema is a medical emergency that can be delayed one to two days following exposure.
Eyes: May cause severe irritation and burns.
Skin: Passes through the unbroken skin; can increase exposure and the severity of the symptoms listed. May cause severe irritation and burns.
Swallowed: May cause burns of mouth, throat and stomach.
Long-term exposure: May cause mutations. Repeated skin contact may cause dermatitis and skin allergy. Repeated exposure may cause asthma; brain, liver, kidney and/or heart muscle damage. Related chemicals may cause lung allergy.
Storage: Store in tightly closed containers in a dry, cool, well-ventilated place. Keep away from heat, sources of ignition and oxidizers. May attack some forms of plastics. Containers may explode in fire.
First aid guide: Move victim to fresh air; call emergency medical care. In case of contact with material, immediately flush skin or eyes with running water for at least 15 minutes. Remove and isolate contaminated clothing and shoes at the site. Keep victim quiet and maintain normal body temperature.

★927★
TETRAETHYL LEAD
CAS: 78-00-2
Other names: CZTEROETHLEK OLOWIU (POLISH); LEAD TETRAETHYL; MOTOR FUEL ANTI-KNOCK COMPOUND; NCI-C54988; PLUMBANE, TETRAETHYL-; TEL; TETRAETHYLPLUMBANE

Danger profile: Possible carcinogen, possible reproductive hazard, poison; combustible, explosion hazard, exposure limit established, Community Right-to-Know List, EPA extremely hazardous substances list; poisonous gas produced in fire.
Uses: Gasoline additive, antiknock agent; in aviation gasoline; common air contaminant.
Appearance: Colorless oily liquid; generally dyed red.
Odor: Sweetish, slightly musty; fruity, pleasant. The odor is a warning of exposure; however, no smell does not mean you are not being exposed.
Effects of exposure:
Breathed: May be poisonous. May cause restlessness, trembling hands, headache, nausea and vomiting; insomnia; irritation of the nose, tremor, restlessness, delirium, convulsions. A large degree of absorption may cause excitability, headaches, delirium, coma and death. High exposure may cause kidney or brain damage. Symptoms may be delayed.
Eyes: May cause irritation, watering and inflammation of the eyelids.
Skin: May be poisonous if absorbed. Passes through the unbroken skin; can increase exposure and the severity of the symptoms listed. May cause irritation with itching and blisters. A large degree of absorption may cause insomnia, headaches, excitability, delirium, coma and death. Symptoms may be delayed.
Swallowed: May be poisonous.
Long-term exposure: Prolonged exposure to low levels may cause symptoms listed. May cause cancer; reproductive damage; increased blood pressure; kidney and brain damage. May

cause a decrease in fertility in men and women; damage the testes and reduce sperm production.
Storage: Highly reactive. Store in tightly closed containers in a dry, cool, well-ventilated place. Keep away from air, heat, sunlight, sources of ignition, oxidizers, chemically active metals. Protect containers from physical damage.
First aid guide: Move victim to fresh air and call emergency medical care; if not breathing, give artificial respiration; if breathing is difficult, give oxygen. In case of contact with material, immediately flush skin or eyes with running water for at least 15 minutes. Speed in removing material from skin is of extreme importance. Remove and isolate contaminated clothing and shoes at the site. Keep victim quiet and maintain normal body temperature. Effects may be delayed, keep victim under observation.

★928★
TETRAETHYL TIN
CAS: 597-64-8
Other names: TETRAETHYLSTANNANE; TET; TETRAETHYLTIN; TIN, TETRAETHYL-; STANNANE, TETRAETHYL-

Danger profile: Exposure limit established, poisonous gas produced in fire.
Appearance: Colorless liquid.
Effects of exposure:
Breathed: May cause irritation of the eyes; headache, nausea; sneezing and coughing.
Eyes: May cause irritation, itching, tearing and swelling of the eyelids.
Skin: May cause irritation with itching, redness, swelling; blisters and skin ulcerations.
Swallowed: May cause irritation of the mouth, throat and stomach; pain; headache, nausea and vomiting, dizziness, convulsions, unconsciousness, coma; possible permanent paralysis of the limbs, death.
Storage: Store in tightly closed containers in a dry, cool, well-ventilated place. Exercise caution; little is known about organo-metallic compounds; may be highly reactive and dangerous. Containers may explode in fire.

★929★
TETRAFLUOROETHYLENE
CAS: 116-14-3
Other names: ETHYLENE, TETRAFLUORO- (INHIBITED); FLUROPLAST 4; 1,1,2,2- TETRAFLUOROETHYLENE; TETRAFLUOROETHYLENE, INHIBITED; TFE

Danger profile: Fire hazard, explosion hazard, poisonous gas produced in fire.
Uses: Making other chemicals.
Appearance: Colorless compressed gas.
Odor: Faint. The odor is a warning of exposure; however, no smell does not mean you are not being exposed.
Effects of exposure:
Breathed: May be poisonous. May cause irritation of the nose, throat and breathing passages; headache, dizziness, fatigue; rapid irregular breathing; nausea, vomiting, sweating, weakness, confusion, loss of consciousness convulsions and death. High levels of exposure may suffocation and/or damage the liver and kidneys.
Eyes: Contact with liquid may cause frostbite, pain, stinging, tearing, inflammation of the eyelids and permanent damage.
Skin: Contact with liquid may cause frostbite, intense cold feeling, skin turns white and feels cold and hard.
Long-term exposure: May cause lung irritation; liver and kidney damage.
Storage: Highly reactive. Store in tightly closed cylinders in a dry, cool, well-ventilated detached place or explosion-proof

freezer. Keep away from heat, sunlight, oxidizers, other combustible materials. Vapors may travel to sources of ignition and flash back. Containers may explode in fire.

First aid guide: Move victim to fresh air and call emergency medical care; if not breathing, give artificial respiration; if breathing is difficult, give oxygen. In case of frostbite, thaw frosted parts with water. Keep victim quiet and maintain normal body temperature.

★930★
TETRAFLUOROMETHANE
CAS: 75-73-0
Other names: ARCTON O; CARBON FLUORIDE; F 14; FC 14; FREON 14; HALOCARBON 14; HALON 14; METHANE, TETRAFLUORO-; PERFLUOROMETHANE; R 14; CARBON TETRAFLUORIDE

Danger profile: Poisonous gas produced in fire.
Uses: Refrigerant; gaseous insulator.
Appearance: Colorless compressed gas.
Effects of exposure:
  Breathed: May cause headache, dizziness, fatigue; rapid irregular breathing; nausea, vomiting, confusion, loss of consciousness convulsions and death. High concentrations are narcotic. Similar chemicals may cause irregular heart beat.
  Eyes: Contact with liquid may cause frostbite, pain, stinging, tearing, inflammation of the eyelids and permanent damage.
  Skin: Contact with liquid may cause frostbite, intense cold feeling, skin turns white and feels cold and hard.
Long-term exposure: Unknown at this time.
Storage: Store in tightly closed steel cylinders in a dry, cool, well-ventilated place. Keep away from heat, sources of ignition, aluminum, powdered metals. Containers may explode in fire.
First aid guide: Move victim to fresh air and call emergency medical care; if not breathing, give artificial respiration; if breathing is difficult, give oxygen.

★931★
TETRAHYDROFURAN
CAS: 109-99-9
Other names: BUTANE, 1,4-EPOXY-; BUTYLENE OXIDE; CYCLOTETRAMETHYLENE OXIDE; DIETHYLENE OXIDE; 1,4-EPOXYBUTANE; FURANIDINE; FURAN, TETRAHYDRO-; HYDROFURAN; NCI-C60560; OXACYCLOPENTANE; OXOLANE; TETRAHYDROFURAAN (DUTCH); TETRAHYDROFURANNE (FRENCH); TETRAIDROFURANO (ITALIAN); TETRAMETHYLENE OXIDE; THF

Danger profile: Possible mutagen, fire hazard, exposure limit established, poisonous gas produced in fire.
Uses: Solvent for natural and synthetic resins, particularly vinyl, in topcoating solutions, polymer coating, cellophane, protective coatings, adhesives, magnetic tapes, printing inks, etc., chemical intermediate and monomer.
Appearance: Colorless liquid.
Odor: Ether-like; faint, fruity; similar to acetone. The odor is a warning of exposure; however, no smell does not mean you are not being exposed.
Effects of exposure:
  Breathed: Vapors may cause irritation of the eyes, nose, throat and lungs; dizziness, headache, nausea, cough, shortness of breath and a dangerous build-up of fluids in the lungs (pulmonary edema), which can cause death. Pulmonary edema is a medical emergency that can be delayed one to two days following exposure. High exposures are narcotic and may cause liver and kidney damage. Very high exposure may cause loss of consciousness and death.
  Eyes: Contact may cause severe irritation and damage.

Skin: Passes through the unbroken skin; can increase exposure and the severity of the symptoms listed. Contact may cause defatting of skin, drying and severe irritation. Prolonged contact may cause blisters.
  Swallowed: Moderately toxic.
Long-term exposure: May cause mutations, liver and kidney damage; skin drying and cracking; possible skin rash.
Storage: Vapors are explosive. Store in tightly closed containers in a dry, cool, well-ventilated place. Keep away from heat, sources of ignition, potassium hydroxide, sodium hydroxide, borane and strong oxidizers. Containers may explode in fire.
First aid guide: Move victim to fresh air and call emergency medical care; if not breathing, give artificial respiration; if breathing is difficult, give oxygen. In case of contact with material, immediately flush eyes with running water for at least 15 minutes. Wash skin with soap and water. Remove and isolate contaminated clothing and shoes at the site.

★932★
TETRAHYDROPHTHALIC ACID ANHYDRIDE
CAS: 85-43-8
Other names: ANHYDRID KYSELINY TETRAHYDROFTALOVE (CZECH); 1,3-ISOBENZOFURANDIONE,3A,4,7,7A-TETRAHYDRO-; MALEIC ANHYDRIDE ADDUCT OF BUTADIENE; MEMTETRAHYDROPHTHALIC ANHYDRIDE; MEMTETRAHYDRO PHTALIC ANHYDRIDE; PHTHALIC ANHYDRIDE, 1,2,3,6-TETRAHYDRO-; TETRAHYDROFTALANHYDRID (CZECH); TETRAHYDROPHTHALIC ANHYDRIDE; DELTA(SUP 4)-TETRAHYDROPHTHALIC ANHYDRIDE; 1,2,3,6-TETRAHYDROPHTHALIC ANHYDRIDE; THPA

Danger profile: Combustible, corrosive, acrid smoke and irritating fumes produced in fire.
Uses: Curing epoxy resins; making electrical equipment, polyesters, alkyd resins and plasticizer.
Appearance: White crystalline powder.
Effects of exposure:
  Breathed: Vapors may cause irritation of the nose, throat and breathing passages. High exposures may cause shortness of breath and a dangerous build-up of fluids in the lungs (pulmonary edema), which can cause death. Pulmonary edema is a medical emergency that can be delayed one to two days following exposure.
  Eyes: Contact may cause severe irritation and burns.
  Skin: Passes through the unbroken skin; can increase exposure and the severity of the symptoms listed. Contact may cause severe irritation and burns.
  Swallowed: Mildly toxic.
Long-term exposure: Similar and related chemicals may cause brain and nerve damage; skin and lung allergy.
Storage: Store in tightly closed containers in a dry, cool, well-ventilated place. Keep away from water, sources of ignition and oxidizing materials.

★933★
TETRAMETHYL LEAD
CAS: 75-74-1
Other names: PLUMBANE, TETRAMETHYL-; LEAD, TETRAMETHYL-; TETRAMETHYLPLUMBANE; TML

Danger profile: Reproductive hazard, fire hazard, explosion hazard, poison, exposure limit established, EPA extremely hazardous substance list, poisonous gas produced in fire.
Uses: In gasoline supplements and aviation gasoline; octane amplifier and "antiknock" agent.
Appearance: Colorless liquid or dyed red, orange, or blue.

Odor: Faint, fruity, musty. The odor is a warning of exposure; however, no smell does not mean you are not being exposed.
Effects of exposure:
Breathed: May cause severe or fatal poisoning. Symptoms may be delayed; restlessness, trembling hands, headache, nausea and vomiting; insomnia, hallucinations, irritation of the nose, tremor, delirium, convulsions and death. Overexposure may cause kidney damage.
Eyes: May cause irritation, watering and inflammation of the eyelids.
Skin: Passes through the unbroken skin; can increase exposure and the severity of the symptoms listed. May cause irritation with itching and blisters. Severe or fatal poisoning may result. Symptoms may be delayed; restlessness, trembling hands, headache, nausea and vomiting; insomnia, hallucinations, irritation of the nose, tremor, delirium, convulsions and death. Overexposure may cause kidney damage.
Swallowed: Poisonous.
Long-term exposure: Lead compounds may cause reproductive damage, reduced fertility and changes in menstrual cycle. May effect vision; cause high blood pressure; nervous system and brain damage.
Storage: Highly reactive; explosion and fire hazard. Store in tightly closed containers in a dry, cool, well-ventilated place. Keep away from heat, sources of ignition, chemically active metals and oxidizers. Containers may explode in fire.
First aid guide: Move victim to fresh air and call emergency medical care; if not breathing, give artificial respiration; if breathing is difficult, give oxygen. In case of contact with material, immediately flush skin or eyes with running water for at least 15 minutes. Speed in removing material from skin is of extreme importance. Remove and isolate contaminated clothing and shoes at the site. Keep victim quiet and maintain normal body temperature. Effects may be delayed, keep victim under observation.

★934★
TETRAMETHYL SILANE
CAS: 75-76-3
Other names: SILANE, TETRAMETHYL

Danger profile: Fire hazard, poisonous gas produced in fire.
Uses: Making aviation fuel; laboratory chemical
Appearance: Clear, colorless liquid.
Effects of exposure:
Breathed: May cause irritation of the eyes, nose, throat and lungs.
Eyes: May cause severe burns.
Skin: May cause severe burns.
Long-term exposure: May effect the lungs.
Storage: Store in tightly closed containers in a dry, cool, well-ventilated place. Keep away from heat and sources of ignition. Vapors may travel to sources of ignition and flash back. Containers may explode in fire.
First aid guide: Move victim to fresh air and call emergency medical care; if not breathing, give artificial respiration; if breathing is difficult, give oxygen. In case of contact with material, immediately flush skin or eyes with running water for at least 15 minutes. Remove and isolate contaminated clothing and shoes at the site. Keep victim quiet and maintain normal body temperature.

★935★
TETRAMETHYLSUCCINONITRILE
CAS: 3333-52-6
Other names: TMSN; TETRAMETHYL SUCCINONITRILE

Danger profile: Poison, exposure limit established, Community Right-to-Know List, poisonous gas produced in fire.

Uses: From the manufacture of sponge rubber.
Appearance: Colorless crystalline solid.
Effects of exposure:
Breathed: Poison. May cause headache, eye, nose and throat irritation, wheezing, foaming at the mouth, general irritability, nausea; chest tightness, shortness of breath, vomiting blood, respiratory depression, weakness, chest and abdominal pains, dizziness, diarrhea, tremors, convulsions, shock, unconsciousness weak and irregular pulse and death. Reactions may appear hours following overexposure.
Eyes: May cause severe irritation, tearing and chemical burns.
Skin: Poison. Passes through the unbroken skin; can increase exposure and the severity of the symptoms listed. May cause severe irritation, redness, stinging and blisters.
Swallowed: May cause severe irritation and chemical burns to the tongue, mouth, throat and stomach as well as many of the symptoms described under **Breathed.**
Long-term exposure: Reproductive effects have been reported.
Storage: Store in tightly closed containers in a dry, cool, well-ventilated place. Containers may explode in fire.

★936★
TETRANITROMETHANE
CAS: 509-14-8
Other names: METHANE, TETRANITRO-; NCI-C55947; TNM

Danger profile: Fire hazard, explosion hazard, poison, exposure limit established, EPA extremely hazardous substances list, poisonous gas produced in fire.
Uses: Rocket fuel; laboratory chemical; diesel fuel booster; in TNT.
Appearance: Colorless to pale yellow liquid or solid.
Odor: Pungent. The odor is a warning of exposure; however, no smell does not mean you are not being exposed.
Effects of exposure:
Breathed: May be poisonous. Central nervous system depressant. May cause irritation of the mucous membrane, upper respiratory tract, nose and eyes; tears, headache, rapid and difficult breathing, shortness of breath, dizziness, mental confusion, weakness; bluish color of the lips and skin, convulsions, coma and death. May cause a dangerous build-up of fluids in the lungs (pulmonary edema), which can cause death. Pulmonary edema is a medical emergency that can be delayed one to two days following exposure.
Eyes: May cause irritation, redness, swelling of the eyelids; light causes pain; severe damage.
Skin: Passes through the unbroken skin; can increase exposure and the severity of the symptoms listed. May cause irritation, redness, swelling, blisters. Absorption into the body leads to the formation of methemoglobin which may lead to cyanosis. Onset may be delayed two to four hours or longer.
Swallowed: May cause irritation of the mouth, stomach; cramps and diarrhea.
Long-term exposure: May cause liver damage.
Storage: Explosion hazard. Store in tightly close containers in a dry, cool, well-ventilated place. Keep away from heat, sources of ignition, fuels and combustable materials. Protect containers from shock. Containers may explode in fire.
First aid guide: Move victim to fresh air and call emergency medical care; if not breathing, give artificial respiration; if breathing is difficult, give oxygen. In case of contact with material, immediately flush skin and eyes with running water for at least 15 minutes. Remove and isolate contaminated clothing and shoes at the site. Keep victim quiet and maintain normal body temperature.

★ 937 ★
TETRASODIUM PYROPHOSPHATE
CAS: 7722-88-5
Other names: ANHYDROUS TETRASODIUM PYROPHOS-PHATE; DIPHOSPHORIC ACID, TETRASODIUM SALT; NATRIUMPYROPHOSPHAT; PHOSPHOTEX; PYROPHOS-PHATE; PYROPHOSPHORIC ACID, TETRASODIUM SALT-; SODIUM PYROPHOSPHATE; TETRANATRIUMPY-ROPHOSPHAT (GERMAN); TETRASODIUM DIPHOSPHATE; TETRASODIUM PYROPHOSPHATE, ANHYDROUS; TSPP; VICTOR TSPP

Danger profile: Poison, exposure limit established, poisonous gas produced in fire.
Uses: Water softener; in soaps, detergents, cleaning compounds; metal cleaner; de-inking news print; making synthetic rubber; textile dyeing; wool processing; food additive.
Appearance: Colorless or white crystalline solid.
Effects of exposure:
Breathed: Dust may cause irritation of the eyes, nose, throat and respiratory passages.
Eyes: Contact may cause irritation.
Skin: Contact may cause irritation.
Swallowed: Poisonous.
Long-term exposure: May effect the lungs.
Storage: Store in tightly closed containers in a dry, cool, well-ventilated place. Keep away from magnesium.

★ 938 ★
TETRYL
CAS: 479-45-8
Other names: N-METHYL-N-2,4,6-TETRANITROANILINE; NITRAMINE; PICRYLNITROMETHYLAMINE; TETRALITE; 2,4,6-TETRYL; N,2,4,5- TETRANITRO-N-METHYLANILINE; TRINITROPHENYLMETHYLNITRAMINE; 2,4,6-TRINITROPHENYLMETHYLNITRAMINE; 2,4,6-TRINITROPHENYL-N- METHYLNITRAMINE

Danger profile: Possible mutagen, explosion hazard, exposure limit established, poisonous gas produced in fire.
Uses: Detonating agent for high explosives; pH indicator.
Appearance: Colorless to yellow crystalline solid.
Effects of exposure:
Breathed: May cause irritation of the eyes, nose, throat and lungs, tears and runny nose, coughing, sore throat, difficult breathing; lethargy, fatigue, and drowsiness; slurred speech, shaking, and dizziness. Higher exposures could cause shortness of breath, headache, nausea, vomiting, dizziness, diarrhea and a dangerous build-up of fluids in the lungs (pulmonary edema), which can cause death. Pulmonary edema is a medical emergency that can be delayed one to two days following exposure.
Eyes: May cause irritation, redness, tearing, pain, inflamed eyelids, and eye injury; burns in corneas and possible loss of sight.
Skin: Passes through the unbroken skin; can increase exposure and the severity of the symptoms listed. May cause painful irritation, redness, and blisters. Repeated or prolonged skin contact may cause irritation of the skin and may also cause the legs to feel numb.
Swallowed: May cause irritation of the mouth and throat; stomach and abdominal pain; loss of muscular coordination and mental confusion; convulsions. Also see symptoms for *Breathed.*
Long-term exposure: May cause mutations; dermatitis; kidney damage; anemia; insomnia; gastrointestinal effects.
Storage: Powerful oxidant. Store in tightly closed containers in a dry, cool, well-ventilated place. Keep away from heat, sources

of ignition, hydrazine. Protect containers from shock and friction. Containers may explode in fire.

★ 939 ★
THALLIUM
CAS: 7440-28-0

Danger profile: Poison, exposure limit established, Community Right-to-Know List, highly poisonous gas produced in fire.
Uses: Rodenticide; fungicide; making semiconductors; photoelectric applications; making low-melting glass products; lenses, prisms and thermometers.
Appearance: Bluish-white, soft metal.
Effects of exposure:
Breathed: May cause fatigue, weakness, irritability, confusion, tremor, delirium, hallucinations, convulsions, coma and death. Symptoms are often delayed hours or days.
Eyes: May cause irritation and other symptoms listed.
Skin: Passes through the unbroken skin; can increase exposure and the severity of the symptoms listed. May cause tremor, delirium, hallucinations, convulsions, coma and death. Symptoms are often delayed hours or days.
Swallowed: Extremely toxic. May cause nerve changes, sweating, changes in vision and other effects.
Long-term exposure: May cause fatigue, loss of appetite, insomnia, pains and swelling in the arms and legs, tremors; brain damage; hair color discoloration and loss; vision loss.
Storage: Store in tightly closed containers in a dry, cool, well-ventilated place. Keep away from fluorine, acids and oxidizers.

★ 940 ★
THIOACETAMIDE
CAS: 62-55-5
Other names: ACETAMIDE, THIO-(8CI); ACETIMIDIC ACID, THIO-(7CI); ACETOTHIOAMIDE; ETHANETHIOAMIDE; TAA; THIACETAMIDE; USAF CB-21; USAF EK-1719

Danger profile: Carcinogen, mutagen, poison, combustible, reproductive hazard, Community Right-to-Know List, poisonous gas produced in fire.
Uses: Hydrogen sulfite substitute.
Appearance: Colorless leaf-like material; crystals from benzene.
Odor: Slight, mercaptan-like. The odor is a warning of exposure; however, no smell does not mean you are not being exposed.
Effects of exposure:
Breathed: May cause irritation of the nose, throat and breathing passages. Higher exposures may cause shortness of breath and a dangerous build-up of fluids in the lungs (pulmonary edema), which can cause death. Pulmonary edema is a medical emergency that can be delayed one to two days following exposure. High exposure may cause liver damage and death.
Eyes: May cause slight eye irritation.
Skin: Passes through the unbroken skin; can increase exposure and the severity of the symptoms listed. May cause skin eruptions. High exposure may cause liver damage and death.
Swallowed: May cause drowsiness, fatigue, nausea, acidosis.
Long-term exposure: May cause cancer; mutations; liver damage. Reproductive damage reported.
Storage: Store in tightly closed containers in a dry, cool, well-ventilated place. Keep away from strong oxidizers, acids, bases, water and mineral acids.

★941★
4,4'-THIOBIS(6-TERT-BUTYL-M-CRESOL)
CAS: 96-69-5
Other names: BIS(3-TERT-BUTYL-4-HYDROXY-6-METHYLPHENYL) SULFIDE; BIS(4-HYDROXY-5-TERT-BUTYL-2-METHYLPHENYL) SULFIDE; M-CRESOL,4,4'-THIOBIS(6-TERT-BUTYL-); DISPERSE MB-61; SANTOX; SANTONOX; SANTOWHITE CRYSTALS; THIOALKOFEN BM4; 4,4'-THIOBIS(3-METHYL-6-TERT-BUTYLPHENOL); 1,1'-THIOBIS(2-METHYL-4-HYDROXY-5-TERT-BUTYLBENZENE); USAF B-15; YOSHINOX S

Danger profile: Mutagen, poison, exposure limit established, poisonous gas produced in fire.
Uses: Making plastic and rubber products.
Appearance: Light gray to tan powder.
Effects of exposure:
   Breathed: May cause poisoning.
   Swallowed: May cause poisoning, gastroenteritis.
Long-term exposure: May cause mutations; liver damage. Retarded weight gain and enlarged livers in rat experiments.
Storage: Store in tightly closed containers in a dry, cool, well-ventilated place.

★942★
THIOGLYCOLIC ACID
CAS: 68-11-1
Other names: ACETIC ACID, MERCAPTO-; ACIDE THIOGLYCOLIQUE (FRENCH); ALPHA-MERCAPTOACETIC ACID; GLYCOLIC ACID, THIO-; GLYCOLIC ACID, 2-THIO-; MERCAPTOACETIC ACID; 2-MERCAPTOACETIC ACID; THIOGLYCOLIC ACID; 2-THIOGLYCOLIC ACID; THIOGLYCOLLIC ACID; THIOVANIC ACID; USAF CB-35

Danger profile: Combustible, corrosive, exposure limit established, poisonous gas produced in fire.
Uses: Making other chemicals, permanent wave solutions, hair removers, pharmaceuticals; vinyl stabilizer.
Appearance: Colorless liquid.
Odor: Unpleasant, strong. The odor is a warning of exposure; however, no smell does not mean you are not being exposed.
Effects of exposure:
   Breathed: May cause irritation to the nose and throat; weakness, gasping, convulsions.
   Eyes: May cause severe irritation, burns and permanent damage.
   Skin: Passes through the unbroken skin; can increase exposure and the severity of the symptoms listed. May cause severe irritation and burns.
   Swallowed: Poisonous and corrosive to living tissue.
Long-term exposure: May cause skin rash.
Storage: Store in tightly closed containers in a dry, cool, well-ventilated place.
First aid guide: Move victim to fresh air; call emergency medical care. In case of contact with material, immediately flush skin or eyes with running water for at least 15 minutes. Remove and isolate contaminated clothing and shoes at the site. Keep victim quiet and maintain normal body temperature.

★943★
THIRAM
CAS: 137-26-8
Other names: AATACK; ACETO TETD; ARASAN; AULES; BIS((DIMETHYLAMINO)CARBONOTHIOYL) DISULPHIDE; BIS(DIMETHYL-THIOCARBAMOYL) DISULFID (GERMAN); DISOLFURO DI TETRAMETILTIOURAME (ITALIAN); ALPHA, ALPHA'-DITHIOBIS(DIMETHYLTHIO)FORMAMIDE; CHIPCO THIRAM 75; CYURAM DS; DISOLFURO DI TETRAMETIL-TIOURAME (ITALIAN); DISULFURE DE TETRAMETH-YLTHIOURAME (FRENCH); DURALYN; EBRAGOM TB; EKA-GOM TB; ENT 987; FALITIRAM; FERNIDE; FERNACOL; FERNASAN; FERNIDE; FLO PRO T SEED PROTECTANT; HCDB; HERMAL; HERMAT TMT; HERYL; HEXATHIR; KRE-GASAN; MERCURAM; N,N'-(DITHIODICARBOROTHIOYL)BIS(N-METHYL-METHANAMINE); METHYL THIRAM; METHYL THIURAMDI-SULFIDE; METHYL TUADS; NOBECUTAN; NOMERSAN; NORMERSAN; NOWERGAN; NSC 1771; PANORAM 75; PEZIFILM; POLYRAM ULTRA; POMARSOL; POMASOL; PU-RALIN; REZIFILM; ROYAL TMTD; SADOPLON; SPOTRETE; SQ 1489; TERSAN; TETRAMETHYL THIURAM DISULFIDE; TETRAMETHYLDIURANE SULPHITE; TETRAMETHYLE-NETHIURAM DISULPHIDE; TETRAMETHYLTHIOCARBAM-OYLDISULPHIDE; TETRAMETHYLTHIOPEROXYDICAR-BONIC DIAMIDE; TETRAMETHYL THIORAM DISULFIDE (DUTCH); TETRAMETHYL-THIRAM DISULFID (GERMAN); TETRAMETHYLTHIURAM BISULFIDE; TETRAMETHYLTHI-URAM DISULFIDE; N,N,N',N'-TETRAMETHYL THIURAM DI-SULFIDE; TETRAMETHYL THIURANE DISULFIDE; TETRA-METHYLTHIURAM DISULFIDE; THILLATE; THIMER; THIOSAN; THIOTOX; THIRAMAD; THIRAME (FRENCH); TI-URAM (POLISH); THIURAD; THIURAMYL; THYLATE; TIURA-MYL; TMTD; TTD; TUADS; TUEX; TULISAN; USAF B-30; USAF EK-2089; USAF P-5; VANCIDA TM-95; VANCIDE TM; VUAGT-1-4; VULCAFOR TMTD; VULKACIT MTIC; VULKACIT THIURAM; VULKACIT THIURAM/C; VULKAUT THIRAM

Danger profile: Mutagen, exposure limit established, reproductive hazard, poisonous gas produced in fire.
Uses: In making rubber; control of bacteria growth; fungicide, insecticide; seed disinfectant; additive to lubrication oils; animal repellent.
Appearance: Colorless to cream solid (some commercial products are dyed blue).
Effects of exposure:
   Breathed: Dust may cause respiratory irritation; nose, throat and breathing passages. High exposure may cause headache, fatigue, nerve effects, kidney and/or liver damage.
   Eyes: Liquid may cause irritation.
   Skin: Liquid may cause irritation and allergic eczema and skin rash with itching in sensitive individuals.
   Swallowed: Poisonous. May cause nausea, vomiting, diarrhea, all of which may be persistent; paralysis may develop.
Long-term exposure: May cause mutations; kidney and/or liver damage; skin allergy; personality change; lung allergy. Occupational exposures to 0.03 Mg/m3 over a 5 year period has caused mild irritation of the nose and throat. Prolonged contact has caused eye irritation, tearing, increased sensitivity to light, reduced night vision and blurred vision. Thiram has caused birth defects in laboratory animals. Whether it has this effect in humans is not known.
Storage: Store in tightly closed containers in a dry, cool, well-ventilated place. Keep away from heat, sources of ignition, oxidizers, strong acids, oxidizable materials.
First aid guide: Move victim to fresh air and call emergency medical care; if not breathing, give artificial respiration; if breathing is difficult, give oxygen. In case of contact with material, immediately flush skin or eyes with running water for at least 15 minutes. Speed in removing material from skin is of extreme im-

portance. Remove and isolate contaminated clothing and shoes at the site. Keep victim quiet and maintain normal body temperature. Effects may be delayed; keep victim under observation.

★ 944 ★
THORIUM NITRATE
CAS: 13823-29-5
Other names: NITRIC ACID, THORIUM SALT; NITRIC ACID, THORIUM(4) SALT (8CI,9CI); THORIUM(4) NITRATE; THORIUM(IV) NITRATE; THORIUM(IV) NITRATE TETRAHYDRATE; THORIUM TETRANITRATE

Danger profile: Carcinogen, combustible, radioactive, poisonous gas produced in fire.
Uses: Making other chemicals; laboratory chemical.
Appearance: White solid.
Effects of exposure:
    Breathed: A primary route of exposure. May cause irritation. See **Long-term exposure.**
    Eyes: May cause irritation.
    Skin: Passes through the unbroken skin; can increase exposure and the severity of the symptoms listed. May cause irritation.
    Swallowed: Moderately toxic.
Long-term exposure: May contain radioactive contaminants. Repeated exposure may result in an accumulation of chemical in various body systems. May cause lung scarring.
Storage: Oxidizing material; when in contact with readily combustible material, causes violent combustion. Store in tightly closed containers in a dry, cool, well-ventilated place. Keep away from combustible, organic and readily oxidizable materials. Containers may explode in fire.
First aid guide: Use first aid treatment according to the nature of the injury. In case of contact with material, immediately flush skin or eyes with running water for at least 15 minutes. If not affecting injury, remove and isolate suspected contaminated clothing and shoes; wrap victim in sheet or blanket before transporting. If there is no injury, remove and isolate suspected contaminated clothing and shoes; assist person to shower with soap and water and notify Radiation Authority of action. Advise medical personnel that injured persons may be contaminated with low-level radioactive material and may have chemical burns.

★ 945 ★
TIN
CAS: 7440-31-5
Other names: SILVER MATT POWDER; TIN (ALPHA); TIN FLAKE; TIN POWDER; ZINN (GERMAN)

Danger profile: Exposure limit established, poisonous gas produced in fire.
Uses: Soldering alloys; tin plate; babbitt and similar metal types; pewter; bronze; corrosion-resistant coatings; in electroplating; hot-dipped coatings; organ pipes, dental amalgams, die casting, white, type and casting metal; making other chemicals; in coating copper cooking utensils and lead sheet, or lining lead pipe for distilled water, beer, carbonated beverages.
Appearance: Soft white, silvery metal.
Effects of exposure:
    Breathed: Dust and fumes may cause irritation of the throat and breathing passages; lung effects, coughing.
    Eyes: Dust may cause irritation.
Long-term exposure: Dust or fumes is known to cause a pneumoconiosis (stannosis); changes in chest x-ray; reduced lung function. Tin may contain contaminants of arsenic or lead. May contribute to iron deficiency by interfering with the body's ability to absorb iron from food and vitamins.

Storage: Store in tightly closed containers in a dry, cool, well-ventilated place. Keep away from chlorine, bromine, copper nitrate, turpentine, potassium dioxide.

★ 946 ★
TIN(IV) CHLORIDE (1:4)
CAS: 7646-78-8
Other names: ETAIN (TETRACHLORURE D') (FRENCH); LIBAVIUS FUMING SPIRIT; STAGNO (TETRACLORURO DI) (ITALIAN); STANNIC CHLORIDE; STANNIC CHLORIDE, ANHYDROUS; TIN CHLORIDE, FUMING; TIN PERCHLORIDE; TIN TETRACHLORIDE; TIN TETRACHLORIDE, ANHYDROUS; TINTETRACHLORIDE (DUTCH); ZINNTETRACHLORID (GERMAN)

Danger profile: Corrosive, exposure limit established, poisonous gas produced in fire.
Uses: Electro-conductive and electroluminescent coatings; mordant in dyeing textiles; in perfumes, resins; making ceramic coatings, blue print and other sensitized papers; producing sugar; bacteria and fungi control in soaps.
Appearance: Colorless fuming liquid.
Effects of exposure:
    Breathed: May cause irritation of the nose, throat and breathing passages; coughing, difficult breathing. Higher exposures may cause shortness of breath and a dangerous build-up of fluids in the lungs (pulmonary edema), which can cause death. Pulmonary edema is a medical emergency that can be delayed one to two days following exposure.
    Eyes: Contact may cause severe irritation and burns.
    Skin: Contact may cause severe irritation and burns.
    Swallowed: Highly corrosive; hydrochloric acid produced on contact with moisture.
Long-term exposure: May contribute to iron deficiency anemia, because tin in the body blocks the body's ability to absorb iron from food or vitamins. Possible lung problems.
Storage: Store in tightly closed containers in a dry, cool, well-ventilated place. Keep away from water, heat, sunlight, combustible material, alcohols, amines, chlorine, turpentine, ethylene oxide, alkyl nitrate, potassium and sodium.

★ 947 ★
TITANIUM
CAS: 7440-32-6
Other names: CONTIMET 30; C.P. TITANIUM; IMI 115; NCI-C04251; OREMET; T 40; TITANIUM 50A; TITANIUM ALLOY; TITANIUM METAL POWDER, DRY; TITANIUM SPONGE GRANULES; TITANIUM SPONGE POWDERS; VT 1

Danger profile: Combustible, spontaneously combustible in air, poisonous gas produced in fire.
Uses: Structural material in jet engines, marine equipment, chemical equipment and surgical instruments.
Appearance: Silvery-white metal or dark gray powder.
Effects of exposure:
    Breathed: May cause irritation of the nose, throat and air passages; coughing, phlegm and sneezing.
    Eyes: May cause irritation, redness, tearing, pain and inflammation of the eyelids.
Long-term exposure: There is limited evidence that chemical causes cancer; damage to the fetus, chronic bronchitis and emphysema.
Storage: Dust may ignite spontaneously in air. Store in tightly closed containers in a dry, cool, well-ventilated place. Keep away from sources of ignition, all forms of water, carbon dioxide, metal carbonates, trichloroethylene, trichlorotrifluorethane, halocarbons, halogens, aluminum.

★ 948 ★
TITANIUM CHLORIDE
CAS: 7550-32-6
Other names: TAC 121; TAC 131; TITAQNIUM(III) CHLORIDE; TITANIUM TRICHLORIDE; TITANIUM TRICHLORIDE, PYROPHORIC; TITANOUS CHLORIDE; TRICHLORO TRITANIUM

Danger profile: Corrosive, spontaneously combustible in air, poisonous gas produced in fire.
Uses: Making other chemicals; laundry agent.
Appearance: Colorless to light yellow liquid.
Effects of exposure:
Breathed: Dust may cause severe irritation to the nose, throat and lungs, causing coughing, shortness of breath. May cause flu-like symptoms including chills, muscle ache, headache, fever. High exposure may cause nausea, salivation, vomiting, cramps, diarrhea, chest pains, cough, a build-up of fluids in the lungs (pulmonary edema) and possible death. Pulmonary edema may be delayed from 1-2 days following exposure.
Skin: May cause severe irritation and thermal and acid burns, especially if skin is wet. Wipe off with dry cloth before washing with water.
Eyes: May cause severe irritation, burns, pain and possible blindness.
Swallowed: May cause severe burns of the mouth, throat and stomach, vomiting, watery or bloody diarrhea. Damage to kidneys and liver, collapse and convulsions can result.
Long-term exposure: There is limited evidence that titanium causes cancer in lab animals. Possible lung effects; chronic bronchitis and emphysema.
Storage: May ignite spontaneously in air. Store in sealed containers in a dry, cool, well-ventilated place. Keep away from water, heat, sources of ignition, potassium, hydrogen fluoride. Containers may explode in fire.

★ 949 ★
TITANIUM DIOXIDE
CAS: 13463-67-7
Other names: A-FIL CREAM; ATLAS WHITE TITANIUM; ANATASE; AUSTIOX; BAYERITIAN; BAYERTITAN; BROOKITE; CALCOTONE WHITE T; C.I. 77891; C.I. PIGMENT WHITE 6; COSMETIC WHITE C47-5175; C-WEISS 7 (GERMAN); FLAMENCO; HOMBITAN; HORSE HEAD A-410; KH 360; KRONOS TITANIUM DIOXIDE; LEVANOX WHITE RKB; NCI-C04240; RAYOX; RUNA RH20; RUTILE; TIOFINE; TIOXIDE; TITANDIOXID (SWEDEN); TITANIUM OXIDE; TRONOX UNITANE 0-110; ZOPAQUE; 1700 WHITE

Danger profile: Exposure limit established, Community Right-to-Know List, exposure limit established, poisonous gas produced in fire.
Uses: Pigment in paints, paper, inks, rubber, plastics, cosmetics, floor coverings, glassware and ceramics; processing synthetic fibers; welding rods; radioactive decontamination of skin. A common air contaminant and nuisance dust.
Appearance: White powder or colorless crystals.
Effects of exposure:
Breathed: May cause irritation of the nose, throat and breathing passages; coughing and sneezing.
Eyes: May cause irritation, redness, tearing, pain and inflammation of the eyelids.
Skin: May cause irritation.
Long-term exposure: May cause lung fibrosis and scarring; bronchitis, cough and phlegm; emphysema.
Storage: Store in tightly closed containers in a dry, cool, well-ventilated place. Keep away from high temperatures and metals.

★ 950 ★
O-TOLIDINE
CAS: 119-93-7
Other names: BIANISIDINE; 1,1'-BIPHENYL-4,4'-DIAMINE,3,3'-DIMETHYL-; 4,4'-BI-O-TOLUIDINE; C.I. 37230; C.I. AZOIC DIAZO COMPONENT 113; 4,4'-DIAMINO-3,3'-DIMETHYLBIPHENYL; 4,4'-DIAMINO-3,3'-DIMETHYLDIPHENYL; DIAMINODITOLYL; 3,3'-DIMETHYLBENZIDIN; 3,3'-DIMETHYLBENZIDINE; 3,3'-DIMETHYL-4,4'-BIPHENYLDIAMINE; 3,3'-DIMETHYLBIPHENYL-4,4'-DIAMINE; 3,3'-DIMETHYL-4,4'-DIPHENYLDIAMINE; 3,3'-DIMETHYLDIPHENYL-4,4'-DIAMINE; 4,4'-DI-O-TOLUIDINE; DMB; FAST DARK BLUE BASE R; 2-METHYLANILINE O-TOLIDIN; 2-TOLIDIN (GERMAN); 2-TOLIDINA (ITALIAN); TOLIDINE; O,O'-TOLIDINE; 2-TOLIDINE; 3,3'-TOLIDINE; O-TOLUIDINE

Danger profile: Carcinogen, mutagen, combustible, exposure limit established, Community Right-to-Know List, poisonous gas produced in fire.
Uses: Making dyes and resins; chemical testing procedures, including for free chlorine in swimming pools.
Appearance: Glistening plates, white to reddish.
Effects of exposure:
Breathed: May cause headache, rapid and difficult breathing, shortness of breath, dizziness, mental confusion, weakness; bluish color of the lips and skin, convulsions, coma and death.
Eyes: May cause irritation, redness, swelling of the eyelids; light causes pain; severe damage.
Skin: Passes through the unbroken skin; can increase exposure and the severity of the symptoms listed. May cause irritation, redness, swelling, blisters.
Swallowed: Moderately toxic. May cause irritation of the mouth, stomach; cramps and diarrhea.
Long-term exposure: May cause cancer; kidney and bladder effects. Possible cause of mutations.
Storage: Store in tightly closed containers in a dry, cool, well-ventilated place. Keep away from heat, sources of ignition and direct light.

★ 951 ★
TOLUENE
CAS: 108-88-3
Other names: ANTISAL LA; BENZENE, METHYL-; METHACIDE; METHANE, PHENYL-; METHYLBENZENE; METHYLBENZOL; NCI-C07272; PHENYLMETHANE; TOLUEEN (DUTCH); TOLUEN (CZECH); TOLUOL; TOLUOLO; TOLUSOL

Danger profile: Mutagen, fire hazard, exposure limit established, Community Right-to-Know List, poisonous gas produced in fire.
Uses: Solvent for paints, in aviation gasoline and high-octane blending stock; making other chemicals, explosives (TNT), detergents, saccharin, medicines, dyes, perfumes; adhesive solvent in plastic models and toys. A common air contaminant.
Appearance: Colorless liquid.
Odor: Strong, pleasant, benzene-like. The odor is a warning of exposure; however, no smell does not mean you are not being exposed.
Effects of exposure:
Breathed: May cause irritation of the eyes, nose, throat and upper respiratory tract; dizziness, lightheadedness, rapid breathing, mental confusion, unsteady walk, headache, fatigue, nausea, vomiting, unconsciousness, coma, respiratory arrest and death. High exposures may cause shortness of breath and a dangerous build-up of fluids in the lungs (pulmonary edema), which can cause death. Pulmonary edema is

a medical emergency that can be delayed one to two days following exposure. Levels below 200 ppm may produce headache, tiredness and nausea. From 200 to 750 ppm symptoms may include insomnia, irritability, dizziness, some loss of memory, loss of appetite, a feeling of drunkenness and disturbed menstruation. Levels up to 1,500 ppm may cause heart palpitations and loss of coordination.

Eyes: May cause burning feeling, tearing, severe irritation, redness and inflammation of the eyelids.

Skin: Passes through the unbroken skin; can increase exposure and the severity of the symptoms listed. May cause irritation of the skin, dryness, redness and rash.

Swallowed: May cause indigestion, vomiting, diarrhea, dizziness, fatigue, depressed respiration and unconsciousness.

Long-term exposure: May cause mutations, damage to the fetus; bone marrow damage, lower blood cell count; skin rash; headaches, loss of appetite. Possible brain damage. Blood effects and anemia have been reported but are probably due to contamination by benzene.

Storage: Store outdoors or in a detached building in tightly closed containers in a dry, cool, well-ventilated place. If indoors, store in a standard combustible liquid storage room or cabinet. Keep away from heat, sources of ignition, dinitrogen tetroxide, nitric acid and oxidizers. Protect containers from physical damage. Containers may explode in fire.

First aid guide: Move victim to fresh air and call emergency medical care; if not breathing, give artificial respiration; if breathing is difficult, give oxygen. In case of contact with material, immediately flush eyes with running water for at least 15 minutes. Wash skin with soap and water. Remove and isolate contaminated clothing and shoes at the site.

★ **952** ★
TOLUENE-2,4-DIISOCYANATE
CAS: 584-84-9

Other names: BENZENE,2,4-DIISOCYANATOMETHYL-; BENZENE,2,4-DIISOCYANATO- 1-METHYL-; CRESORCINOL DIISOCYANATE; DESMODUR T80; DI-ISOCYANATE DE TOLUYLENE (FRENCH); DI-ISO-CYANATOLUENE; 2,4-DIISOCYANATO-1-METHYLBENZENE (9CI); 2,4-DIISOCYANATOTOLUENE; DIISOCYANAT-TOLUOL (GERMAN); HYLENE T; HYLENE TCPA; HYLENE TLC; HYLENE TM; HYLENE TM-65; HYLENE TRF; ISOCYANIC ACID, METHYLPHENYLENE ESTER; ISOCYANIC ACID, 4-METHYL-M-PHENYLENE ESTER; 4-METHYL-PHENYLENE DIISOCYANATE; 4-METHYL-PHENYLENE ISOCYANATE; MONDUR TDS; NACCONATE IOO; NCI-C50533; NIAX TDI; NIAX TDI-P; TDI; 2,4-TDI; TDI-80; TOLUEEN-DIISOCYANAAT (DUTCH); TOLUEN-DISOCIANATO (ITALIAN); TOLUENE DIISOCYANATE; TOLUENE DIISOCYANATE; 2,4-TOLUENEDIISOCYANATE; TOLUILENODWUIZOCYJANIAN (POLISH); TOLUYLENE-2,4-DIISOCYANATE; TOLYENE 2,4-DIISOCYANATE; TOLYLENE-2,4-DIISOCYANATE 2,4-TOLYLENEDIISOCYANAT E; 2,4-TOLYLENE DIISOCYANATE; TOLUENE DI-ISOCYANATE; TULUYLENDIISOCYANAT (GERMAN)

Danger profile: Carcinogen, combustible, poison, regulated by OSHA, Community Right-to-Know List, EPA extremely hazardous substances list; poisonous gas produced in fire.

Uses: Making poyurethane foams and coatings. A common air contaminant.

Appearance: Colorless, yellow, or dark liquid or solid.

Odor: Sweet, fruity, pungent. The odor is a warning of exposure; however, no smell does not mean you are not being exposed.

Effects of exposure:

Breathed: May be poisonous. May cause irritation of the eyes, nose, throat and lungs, tears and runny nose, coughing, sore throat, difficult breathing; lethargy, fatigue, and drowsiness; slurred speech, shaking, and dizziness. Night cough is common. Potent sensitizer and lung irritant. Higher exposures could cause shortness of breath, headache, nausea, vomiting, dizziness, diarrhea and a dangerous build-up of fluids in the lungs (pulmonary edema), which can cause death. Pulmonary edema is a medical emergency that can be delayed one to two days following exposure.

Eyes: May cause severe irritation, redness, tearing, pain, inflamed eyelids, and eye injury; burns in corneas and possible loss of sight.

Skin: Toxic. Passes through the unbroken skin; can increase exposure and the severity of the symptoms listed. May cause painful irritation, redness, and blisters Repeated or prolonged skin contact may cause irritation of the skin and may also cause the legs to feel numb.

Swallowed: May be poisonous. May cause irritation of the mouth and throat; stomach and abdominal pain; loss of muscular coordination and mental confusion; convulsions. Also see symptoms for *Breathed.*

Long-term exposure: May cause cancer; asthma-like lung allergy; shortness of breath; chronic lung disease; psychological effects. Limited evidence of temporary impotence in men. Possible mutagen.

Storage: Store in sealed containers in a dry, cool, well-ventilated place. Do not store in polyethylene containers. Keep away from water, heat, light, sources of ignition, amines, alcohols, acyl chlorides, acids, and bases. Containers may explode in fire.

★ **953** ★
TOLUENE-2,6-DIISOCYANATE
CAS: 91-08-7

Other names: 2,6-DIISOCYANATO-1-METHYLBENZENE; 2,6- DIISOCYANATOTOLUENE; HYLENE TM; 2-METHYL-M-THENYLENE ESTER, ISOCYANIC ACID; 2-METHYL-M-PHENYLENE ISOCYANATE; NIAX TDI; 2,6- TDI; 2,6-TOLUENE DIISOCYANATE; TOLYENE-2,6-DIISOCYANATE

Danger profile: Poison, Community Right-to-Know List, EPA hazardous substances list, poisonous gas produced in fire.

Uses: Making polyurethane foams, coatings.

Appearance: Water-white to pale yellow liquid.

Effects of exposure:

Breathed: May be poisonous. May cause irritation of the eyes, nose, throat and lungs, tears and runny nose, coughing, sore throat, difficult breathing; lethargy, fatigue, and drowsiness; slurred speech, shaking, and dizziness. Higher exposures could cause shortness of breath, headache, nausea, vomiting, dizziness, diarrhea and a dangerous build-up of fluids in the lungs (pulmonary edema), which can cause death. Pulmonary edema is a medical emergency that can be delayed one to two days following exposure.

Eyes: May cause irritation, redness, tearing, pain, inflamed eyelids, and eye injury; burns in corneas and possible loss of sight.

Skin: Toxic. Passes through the unbroken skin; can increase exposure and the severity of the symptoms listed. May cause painful irritation, redness, and blisters Repeated or prolonged skin contact may cause irritation of the skin and may also cause the legs to feel numb.

Swallowed: May be poisonous. May cause irritation of the mouth and throat; stomach and abdominal pain; loss of muscular coordination and mental confusion; convulsions. Also see symptoms for *Breathed.*

Long-term exposure: Suspected carcinogen. May cause lung problems.

Storage: Store in tightly closed containers in a dry, cool, well-ventilated place. Containers may explode in fire.

★ 954 ★
TOLUENE SULPHONIC ACID
CAS: 25231-46-3
Other names: BENZENESULFONIC ACID, METHYL-; TOLUENE SULFONIC ACID; TOLUENE SULFONIC ACID, LIQUID; TOLUENESULPHONIC ACID; TOLUENE SULPHONIC ACID, LIQUID

Danger profile: Corrosive, poisonous gas produced in fire.
Uses: Making soaps, dyes, insecticides, pharmaceuticals and other chemicals.
Appearance: Colorless liquid or crystalline material.
Effects of exposure:
  Breathed: May cause irritation of the eyes, nose, throat and lungs; burning feeling, dryness and coughing; shortness of breath and a dangerous build-up of fluids in the lungs (pulmonary edema), which can cause death. Pulmonary edema is a medical emergency that can be delayed one to two days following exposure.
  Eyes: May cause severe irritation and burns.
  Skin: May cause severe irritation and burns.
Long-term exposure: May cause chronic irritation of the eyes, nose and throat; burning and dryness. Possible lung effects.
Storage: Store in tightly closed containers in a dry, cool, well-ventilated place. Keep away from combustible materials.
First aid guide: Move victim to fresh air; call emergency medical care. In case of contact with material, immediately flush skin or eyes with running water for at least 15 minutes. Remove and isolate contaminated clothing and shoes at the site. Keep victim quiet and maintain normal body temperature.

★ 955 ★
M-TOLUIDINE
CAS: 108-44-1
Other names: 3-AMINO-1-METHYLBENZENE; 3-AMINOPHENYLMETHANE; 3-AMINOTOLUEN (CZECH); M-AMINOTOLUENE; 3-AMINOTOLUENE; ANILINE, 3-METHYL-; BENZENEAMINE, 3-METHYL-; M-METHYLANILINE; 3-METHYLANILIN E; M-METHYLBENZENAMINE; 3-METHYLBENZENAMINE; M-TOLUIDIN (CZECH); 3-TOLUIDINE; M-TOLYLAMINE

Danger profile: Combustible, exposure limit established, poisonous gas produced in fire.
Uses: Making dyes and other chemicals.
Appearance: Colorless liquid.
Odor: Aniline-like, aromatic. The odor is a warning of exposure; however, no smell does not mean you are not being exposed.
Effects of exposure:
  Breathed: May be poisonous. May lower the ability of the blood to carry oxygen, causing bluish color of the lips, skin, nails, headache, dizziness, nausea, vomiting, coma and even death.
  Eyes: Contact may cause irritation and possible permanent damage.
  Skin: May be poisonous if absorbed. Passes through the unbroken skin; can increase exposure and the severity of the symptoms listed. Contact may cause irritation, burning sensation and rash.
  Swallowed: May be poisonous. See symptoms under *Breathed.*
Long-term exposure: May cause kidney and bladder damage with pain and bloody urine. Repeated inhalation of low concentrations may cause pallor, low-grade secondary anemia, fatigue, and loss of appetite.
Storage: Store outside or in detached building in tightly closed containers in a dry, cool, well-ventilated place. Keep away from heat, sources of ignition, acids, oxidizers. Containers may explode in fire.

First aid guide: Move victim to fresh air and call emergency medical care; if not breathing, give artificial respiration; if breathing is difficult, give oxygen. In case of contact with material, immediately flush skin or eyes with running water for at least 15 minutes. Speed in removing material from skin is of extreme importance. Remove and isolate contaminated clothing and shoes at the site. Keep victim quiet and maintain normal body temperature. Effects may be delayed; keep victim under observation.

★ 956 ★
O-TOLUIDINE
CAS: 95-53-4
Other names: 1-AMINO-2-METHYLBENZENE; 2-AMINO-1-METHYLBENZENE; O-AMINOTOLUENE; ORTHO-AMINOTOLUENE; 2-AMINOTOLUENE; ANILINE, 2-METHYL-; BENZENAMINE,2-METHYL- (9CI); C.I. 37077; 1-METHYL-2-AMINOBENZENE; O-METHYLANILINE; 1-METHYL-1,2-AMINO-BENZENE; 2-METHYL-1-AMINOBENZENE; O-METHYLANILINE; 2-METHYLANILINE; O-METHYLBENZENAMINE; 2-METHYLBENZENAMINE; O-TOLUIDIN (CZECH); 2-TOLUIDINE; O-TOLUIDYNA (POLISH); O-TOLYLAMINE

Danger profile: Carcinogen, mutagen, poison, combustible, exposure limit established, Community Right-to-Know List, poisonous gas produced in fire.
Uses: Making other chemicals, dyes; medical testing.
Appearance: Colorless to pale yellow liquid; turns reddish-brown on exposure to air and light
Odor: Weak, pleasant, aromatic, aniline-like. The odor is a warning of exposure; however, no smell does not mean you are not being exposed.
Effects of exposure:
  Breathed: May be poisonous. May cause headache, rapid and difficult breathing, shortness of breath, dizziness, mental confusion, weakness; bluish color (cyanosis) of the lips, nails and skin, convulsions, coma and death.
  Eyes: May cause severe irritation, redness, swelling of the eyelids; light causes pain; severe damage.
  Skin: May be poisonous if absorbed. Passes through the unbroken skin; can increase exposure and the severity of the symptoms listed. May cause severe irritation, redness, swelling, blisters.
  Swallowed: May be poisonous. May cause irritation of the mouth, stomach; cramps and diarrhea.
Long-term exposure: May cause cancer; mutations; pallor, low grade secondary anemia, fatigue, loss of appetite; kidney and bladder damage with bloody urine; skin drying and cracking.
Storage: Store in sealed containers in a dry, cool, well-ventilated outside or detached place. Keep away from light, air, heat, sources of ignition, oxidizers and nitric acid. Containers may explode in fire.
First aid guide: Move victim to fresh air and call emergency medical care; if not breathing, give artificial respiration; if breathing is difficult, give oxygen. In case of contact with material, immediately flush skin or eyes with running water for at least 15 minutes. Speed in removing material from skin is of extreme importance. Remove and isolate contaminated clothing and shoes at the site. Keep victim quiet and maintain normal body temperature. Effects may be delayed; keep victim under observation.

★957★
O-TOLUIDINE HYDROCHLORIDE
CAS: 636-21-5
Other names: 1-AMINO-2-METHYLBENZENE HYDROCHLO-RIDE; 2-AMINO-1- METHYLBENZENE HYDROCHLORIDE; 2-AMINOTOLUENE HYDROCHLORIDE; O-AMINOTOLUENE HYDROCHLORIDE; BENZENAMINE, 2-METHYL-, HYDRO-CHLORIDE; O-METHYLANILINE; O-METHYLANILINE HY-DROCHLORIDE; NCI-C02335; 2-TOLUIDINE HYDROCHLO-RIDE; O-TOLYAMINE HYDROCHLORIDE

Danger profile: Carcinogen, combustible, Community Right-to-Know List, poisonous gas produced in fire.
Uses: Making dyes.
Appearance: Colorless crystalline solid.
Effects of exposure:
　Breathed: May cause methemoglobinemia, a condition that causes interference with the blood's ability to carry oxygen, with the skin, fingernails and lips to turn blue.
　Eyes: May cause irritation.
　Skin: Passes through the unbroken skin; can increase exposure and the severity of the symptoms listed. May cause irritation; methemoglobinemia, a condition that causes interference with the blood's ability to carry oxygen, with the skin, fingernails and lips to turn blue.
　Swallowed: Moderately toxic.
Long-term exposure: May cause cancer; irritation of the kidney and bladder with frequent urination, pain and blood; kidney, bladder and spleen damage, anemia.
Storage: Store in tightly closed containers in a dry, cool, well-ventilated place. Keep away from light. Containers may explode in fire.

★958★
TOXAPHENE
CAS: 8001-35-2
Other names: AGRICIDE MAGGOT KILLER (F); ALLTEX; ALLTOX; ATTAC-2; ATTAC 6; ATTAC 6-3; CAMPHECHLOR; CAMPHOCHLOR; CAMPHOCLOR; CAMPHOFENE HUILEUX; CHEM-PHENE; CHLORINATED CAMPHENE; CHLOROCAM-PHENE; CLOR CHEM T-590; COMPOUND 3956; CRES-TOXO; CRISTOXO 90; ENT 9,735; ESTONOX; FASCO-TERPENE; GENIPHENE; GY-PHENE; HERCULES 3956; HERCULES TOXAPHENE; KAMFOCHLOR; M 5055; MELI-PAX; MOTOX; NCI-C00259; OCTACHLOROCAMPHENE; PCC; PHENACIDE; PHENATOX; POLYCHLORCAMPHENE; POLYCHLORINATED CAMPHENE; POLYCHLOROCAM-PHENE; STROBANE-T; STROBANE-T-90; SYNTHETIC 3956; TOXADUST; TOXAFEEN (DUTCH); TOXAKIL; TOXAPHEN (GERMAN); TOXASPRAY; TOXON 63; TOXYPHEN; VERTAC 90%; VERTAC TOXAPHENE 90

Danger profile: Carcinogen, reproductive hazard, poison, combustible, exposure limit established, reproductive hazard, Community Right-to-Know List, poisonous gas produced in fire.
Uses: Insecticide, primarily for foliage of cotton, small grains and vegetables; corn, fruit, soy beans; livestock pest control. May be banned for use in some countries, and for some specific uses in others (organochlorine).
Appearance: Yellow or amber-colored waxy solid. Usually dissolved in liquid.
Odor: Mild turpentine-like. The odor is a warning of exposure; however, no smell does not mean you are not being exposed.
Effects of exposure:
　Breathed: May be poisonous. May cause irritation of the nose and throat; headache, dizziness; excessive irritability, confusion and excitability; weakness, numb arms and legs; shallow breathing, convulsions, blue skin and lips, coma and death. High exposures may cause shortness of breath and

a dangerous build-up of fluids in the lungs (pulmonary edema), which can cause death. Pulmonary edema is a medical emergency that can be delayed one to two days following exposure. High exposure may effect the nervous system with tremors, weakness, dizziness, increased saliva, nausea, vomiting, convulsions and loss of consciousness, and cause liver and kidney damage.
　Eyes: May cause irritation and redness.
　Skin: Passes through the unbroken skin; can increase exposure and the severity of the symptoms listed. May cause irritation.
　Swallowed: Poisonous. May cause nausea, vomiting, increased salivation, leg and back muscle spasms, excessive irritability, hyperexcitability, tremors, shivering, convulsions, rapid and irregular heartbeat, and death. Lethal doses cause respiratory failure.
Long-term exposure: May cause cancer; possible damage to the fetus; kidney and liver damage; low blood count; brain damage.
Storage: Store in sealed containers in a dry, cool, well-ventilated place. Keep away from oxidizers. Containers may explode in fire.
First aid guide: Move victim to fresh air and call emergency medical care; if not breathing, give artificial respiration; if breathing is difficult, give oxygen. In case of contact with material, immediately flush skin or eyes with running water for at least 15 minutes. Speed in removing material from skin is of extreme importance. Remove and isolate contaminated clothing and shoes at the site. Keep victim quiet and maintain normal body temperature. Effects may be delayed; keep victim under observation.

★959★
TRIBUTYL ALUMINUM
CAS: 1116-70-7
Other names: TNBA; TRI-N-BUTYL ALUMINUM

Danger profile: Fire hazard, explosion hazard, exposure limit established, poisonous gas produced in fire.
Uses: Making other chemicals.
Appearance: Colorless liquid.
Effects of exposure:
　Breathed: May be poisonous. Fumes may cause irritation of the nose, throat, breathing passages and lungs; headache, nausea, vomiting, chills, cough and shortness of breath.
　Eyes: Contact may cause severe burns.
　Skin: Passes through the unbroken skin; can increase exposure and the severity of the symptoms listed. Contact may cause severe burns.
Long-term exposure: May cause lung problems.
Storage: Highly reactive. May ignite spontaneously in air. Store in tightly closed containers in a dry, cool, well-ventilated place. Keep away from heat, sources of ignition, combustible materials, air, water, alcohols, acid, oxidizing agents and halogenated compounds. Containers may explode in fire.
First aid guide: Move victim to fresh air; call emergency medical care. Wipe material from skin immediately; flush skin or eyes with running water for at least 15 minutes. Remove and isolate contaminated clothing and shoes at the site.

★960★
TRIBUTYL PHOSPHATE
CAS: 126-73-8
Other names: BUTYL PHOSPHATE, TRI-; CELLUPHOS 4; PHOSPHORIC ACID TRIBUTYL ESTER; TBP; TRIBUTILFOS-FATO (ITALIAN); TRIBUTYLE (PHOSPHATE DE) (FRENCH); TRIBUTYLFOSFAAT (DUTCH); TRIBUTYLPHOSPHAT (GER-MAN); TRI-N-BUTYL PHOSPHATE

Danger profile: Exposure limit established, poisonous gas produced in fire.
Uses: Solvent; plasticizer; making pigments; anti-foam agent.
Appearance: Colorless to pale yellow liquid.
Effects of exposure:
Breathed: May cause irritation of the nose, throat, and breathing passages; headaches, weakness, muscle twitching, nausea, collapse; possible death. High exposures may cause shortness of breath and a dangerous build-up of fluids in the lungs (pulmonary edema), which can cause death. Pulmonary edema is a medical emergency that can be delayed one to two days following exposure.
Eyes: May cause severe irritation and burns.
Skin: May cause severe irritation and burns.
Swallowed: Moderately toxic.
Long-term exposure: May cause stimulation of the central nervous system; lung problems.
Storage: Store in tightly closed containers in a dry, cool, well-ventilated place. Keep away from heat, sources of ignition and strong oxidizers.

★961★
TRICHLOROACETIC ACID
CAS: 76-03-9
Other names: ACETIC ACID, TRICHLORO-; ACETO-CAUSTIN; ACIDE TRICHLORACETIQUE (FRENCH); ACIDOTRICLOROACETICO (ITALIAN); AMCHEM GRASS KILLER; DOW SODIUM TCA INHIBITED; KONESTA; SODIUM TCA; TCA; TRICHLOORAZIJNZUUR (DUTCH); TRICHLORACETIC ACID; TRICHLORESSIGSAEURE (GERMAN); TRICHLOROACETIC ACID SOLUTION; TRICHLOROETHANOIC ACID; VARITOX

Danger profile: Poison, corrosive, exposure limit established, poisonous gas produced in fire.
Uses: Making medicines, drugs, herbicides, pesticides and other chemicals; laboratory chemical.
Appearance: Colorless crystalline solid; colorless solution in water.
Odor: Acrid, pungent. The odor is a warning of exposure; however, no smell does not mean you are not being exposed.
Effects of exposure:
Breathed: May be poisonous. Vapor may cause intense irritation to nose, throat, mouth and lungs with a persistent cough; mucous membrane. Symptoms of exposure include watering of the eyes, coughing, choking, difficult breathing, headache, dizziness, nausea, vomiting, muscular weakness, unconsciousness. Disturbances of the digestive tract may also be noticed; these should only be significant at levels above the exposure limit. High exposure can lead to drowsiness, coma and death; buildup of fluid in the lungs (pulmonary edema), and may cause death. Pulmonary edema may be delayed by one to two days following exposure.
Eyes: Severe irritant. Contact may cause, burning and stinging, tearing, irritation and redness and swelling of the eyelids, intense pain; can burn the eyes and lead to blurred vision, chronic irritation and other permanent damage.
Skin: May be poisonous. Severe irritant. Contact may cause burning or stinging feeling, redness, swelling, intense pain, blisters, irritation, rash. Repeated exposure may lead to skin allergy and chronic irritation.
Swallowed: May be poisonous. May cause burning feeling of the mouth and throat, intense thirst, throat swelling, stomach cramps, nausea, vomiting, diarrhea, severe breathing difficulties and unconsciousness. Tissue damage may result. Breathing may stop. Watch for delayed symptoms.
Long-term exposure: May cause lung problems. Mutation data reported. Possible carcinogen.

Storage: Store in tightly closed containers in a dry, cool, well-ventilated place. Keep away from heat, sun, sources of ignition and combustable materials.
First aid guide: Move victim to fresh air and call emergency medical care; if not breathing, give artificial respiration; if breathing is difficult, give oxygen. In case of contact with material, immediately flush skin or eyes with running water for at least 15 minutes. Remove and isolate contaminated clothing and shoes at the site. Keep victim quiet and maintain normal body temperature. Effects may be delayed, keep victim under observation.

★962★
1,2,4-TRICHLOROBENZENE
CAS: 120-82-1
Other names: BENZENE, 1,2,4-TRICHLORO-; UNSYM-TRICHLOROBENZENE; TROJCHLOROBENZEN (POLISH)

Danger profile: Possible reproductive hazard, poison, combustible, exposure limit established, Community Right-to-Know List, poisonous gas produced in fire.
Uses: Making insecticides, fungicides and other chemicals; heat transfer fluids.
Appearance: Clear, colorless to slightly yellow liquid.
Odor: Pleasant, characteristic chlorobenzene odor. The odor is a warning of exposure; however, no smell does not mean you are not being exposed.
Effects of exposure:
Breathed: May be poisonous. Exposures to high concentrations are potentially hazardous to the lungs, kidneys and liver. Prolonged exposures or short exposure to high concentrations are potentially hazardous to the lungs, kidneys and liver.
Eyes: Prolonged or repeated exposure to the eyes is likely to result in moderate pain and irritation.
Skin: Passes through the unbroken skin; can increase exposure and the severity of the symptoms listed. Prolonged or repeated contact may result in irritation, burns and possible systemic effects.
Swallowed: May be poisonous. May cause kidney and liver damage.
Long-term exposure: May cause damage to the fetus. Repeated exposures may result in damage to the lungs, kidneys and liver.
Storage: Store in tightly closed containers in a dry, dark, cool, well-ventilated place. Keep away from sources of ignition, acids, oxidizers.

★963★
1,1,2-TRICHLOROETHANE
CAS: 79-00-5
Other names: ETHANE TRICHLORIDE; ETHANE,1,1,2-TRICHLORO-; NCI-C04579; BETA-T; 1,1,2-TRICHLORETHANE; BETA-TRICHLOROETHANE; TROJCHLOROETAN(1,1,2) (POLISH); VINYL TRICHLORIDE

Danger profile: Suspected carcinogen, poison, combustible, exposure limit established, Community Right-to-Know List, poisonous gas produced in fire.
Uses: Solvent; making adhesives and other chemicals.
Appearance: Colorless liquid.
Odor: Sweet, chloroform-like. The odor is a warning of exposure; however, no smell does not mean you are not being exposed.
Effects of exposure:
Breathed: May cause irritation of the eyes, nose, throat, air passages and lungs; coughing, headache; increased saliva; reddening of the face; dizziness, lightheadedness, fatigue, loss of consciousness, shock, coma, irregular heartbeat, death. If the victim recovers watch for the possibility of a

buildup of fluid in the lungs (pulmonary edema). Pulmonary edema is a medical emergency that can be delayed one to two days following exposure. High concentrations may cause death by respiratory failure or irregular heart beat. May cause liver and/or kidney damage.

Eyes: Contact may cause irritation, tearing, inflammation of the eyelids; burning sensation.

Skin: Passes through the unbroken skin; can increase exposure and the severity of the symptoms listed. May cause irritation, redness, burning feeling, swelling; dry skin; rash or blisters. Vapor may produce superficial burns or defatting type dermatitis.

Swallowed: Poison. Breath may smell of chloroform. May cause severe irritation of the mouth, gastrointestinal tract, nausea, vomiting, diarrhea with possible blood; drowsiness; loss of consciousness; irregular heartbeat and death. May cause liver and/or kidney damage. See **Breathed.**

Long-term exposure: My cause liver and kidney damage; skin cracking. Similar chemical solvents may cause brain and nerve damage. Has caused cancer in laboratory animals, and mutagen data has been reported.

Storage: Store in tightly closed containers in a dry, cool, dark, well-ventilated place. Keep away from heat, sources of combustion, oxidizers, chemically active metals. Do not store in aluminum containers. Containers may explode in fire.

★ 964 ★
TRICHLOROETHYLENE
CAS: 79-01-6
Other names: ACETYLENE TRICHLORIDE; ALGYLEN; ANAMENTH; BENZINOL; BLACOSOLV; CECOLENE; CHLORYLEA; CHORYLEN; CIRCOSOLV; CRAWHASPOL; DENSINFLUAT; DOW-TRI; DUKERON; ETHINYL TRICHLORIDE; ETHYLENE TRICHLORIDE; FLECK-FLIP; FLUATE; GERMALGENE; LANADIN; LETHURIN; NARCOGEN; NARKOSOID; NCI-C04546; NIALK; PERM-A-CHLOR; PETZINOL; TCE; THRETHYLENE; TRI; TRIAD; TRIASOL; TRICHLOORETHEEN (DUTCH); TRICHLORAETHEN (GERMAN); TRICHLORORAN; TRICHLORETHENE (FRENCH); TRICHLOROETHYLENE TRI (FRENCH); TRICHLOROETHENE; TRI-CLENE; TRICHLORETHENE (ITALIAN); TRICLOROETILENE (ITALIAN); 1,1,2-TRICHLOROETHYLENE; TRILENE; TRIMAR; TRI-PLUS; VESTROL; VITRAN; WESTROSOL

Danger profile: Suspected carcinogen, mutagen, combustible, exposure limit established, Community Right-to-Know List, poisonous gas produced in fire.

Uses: Solvent for oils, fats, waxes, degreasing and dry cleaning; solvent dyeing; refrigerant; heat exchange liquid; fumigant; in paints and adhesives; textile processing; making other chemicals. Common air contaminant.

Appearance: Colorless, watery liquid.

Odor: Sweet, chloroform-like; ethereal. The odor is a warning of exposure; however, no smell does not mean you are not being exposed.

Effects of exposure:
Breathed: May cause irritation of the eyes, nose, throat and air passages; coughing, headache; increased saliva; reddening of the face; dizziness, lightheadedness, giddiness, nervousness, increased sensitivity to alcohol, fatigue, loss of consciousness, shock, coma, irregular heartbeat, death. If the victim recovers watch for the possibility of a buildup of fluid in the lungs (pulmonary edema). Pulmonary edema is a medical emergency that can be delayed one to two days following exposure. High levels may cause liver and kidney damage, lung irritation, brain damage and death. Very high levels may cause irregular heart beat and death.

Eyes: Contact may cause irritation, tearing, inflammation of the eyelids; burning sensation.

Skin: Passes through the unbroken skin; can increase exposure and the severity of the symptoms listed. May cause irritation, redness, burning feeling, swelling; dry skin; rash or burns.

Swallowed: Poison. Breath may smell of chloroform. May cause irritation of the mouth, gastrointestinal tract, nausea, vomiting, diarrhea with possible blood; drowsiness; loss of consciousness; irregular heartbeat and death. See **Breathed.**

Long-term exposure: May cause cancer; birth defects; skin allergy; liver and kidney damage; memory loss, intolerance of alcohol. Repeated exposure may cause paralysis of the fingers and facial nerves. May become additive.

Storage: Store in sealed steel or plastic cans, or dark glass bottles in a dry, cool, well-ventilated place. Keep away from heat, sources of strong flame, high intensity ultraviolet light, alkali, aluminum and other chemically active metals, liquid oxygen, hydrochloric acid, metal powders and shavings. Containers may explode in fire.

First aid guide: Move victim to fresh air and call emergency medical care; if not breathing, give artificial respiration; if breathing is difficult, give oxygen. In case of contact with material, immediately flush eyes with running water for at least 15 minutes. Wash skin with soap and water. Remove and isolate contaminated clothing and shoes at the site. Use first aid treatment according to the nature of the injury.

★ 965 ★
TRICHLOROFLUOROMETHANE
CAS: 75-69-4
Other names: ALGOFRENE TYPE 1; ARCTRON 9; CFC-11; ELECTRO-CF 11; ESKIMON 11; F-11; FC-11; FLUOROCARBON 11; FLUOROTRICHLOROMETHANE; FREON 11; FREON MF; GENETRON 11; HALOCARBON 11; ISCEON 131; ISOTRON 11; LEDON 11; MONOFLUROTRICHLOROMETHANE; NCI-C04637; METHANE, TRICHLOROFLUORO-; REFRIGERANT 11; TRICHLOROMONOFLUOROMETHANE; UCON REFRIGERANT 11

Danger profile: Exposure limit established, Community Right-to-Know List, poisonous gas produced in fire.

Uses: Solvent; fire extinguisher; making foams and other chemicals.

Appearance: Colorless liquid or gas.

Effects of exposure:
Breathed: May cause headache, dizziness, drowsiness, fatigue; rapid irregular breathing; nausea, vomiting, confusion, loss of consciousness convulsions and death. Overexposure may cause lightheadedness, irregular heart beat, loss of consciousness and death.

Eyes: Contact with liquid may cause frostbite, pain, stinging, tearing, inflammation of the eyelids and permanent damage.

Skin: Contact with liquid may cause frostbite, intense cold feeling, skin turns white and feels cold and hard.

Swallowed: Liquid may cause headache, confusion, drowsiness, unconsciousness and death.

Long-term exposure: May cause lung problems; skin drying and rash.

Storage: Store in tightly closed containers in a freezer. Keep away from heat, aluminum, barium and lithium.

★966★
TRICHLOROISOCYANURIC ACID
CAS: 87-90-1
Other names: ACL 85; CBD 90; FICHLOR 91; FI CLOR 91; ISOCYANURIC CHLORIDE; KYSELINA TRICHLOISOKY-ANUROVA (CZECH); NSC-405124; SYMCLOSEN; SYMCLO-SENE; TRICHLORINATED ISOCYANURIC ACID; TRI-CHLOROCYANURIC ACID; TRICHLOROISOCYANIC ACID; N,N',N''-TRICHLOROISOCYANURIC ACID; 1,3,5-TRICHLOROISOCYANURIC ACID; NSC-405124; 1,3,5-TRIAZINE-2,4,6(1H,3H,5H)-TRIONE, 1,3,5-TRICHLORO-; TRI-CHLORO-S-TRIAZINETRIONE; 1,3,5-TRICHLORO-S-TRIAZINE-2,4,6(1H,3H,5H)-TRIONE; 1,3,5-TRICHLORO-2,4,6-TRIOXOHEXAHYDRO-S-TRIAZINE; 1,3,5,TRICHLORO-1,2,5-TRIAZINE-2,4,6(1H,3H,5H)-TRIONE; TRICHLORO-S-TRIAZINE-2,4,6(1H,3H,5H)-TRIONE; TRICHLORO-S-TRIAZINETRIONE, DRY

Danger profile: Poisonous gas produced in fire.
Uses: In household bleaches and detergents; swimming pool chemical.
Appearance: White powder or crystalline solid.
Odor: Chlorine-like. The odor is a warning of exposure; however, no smell does not mean you are not being exposed. Inhalation causes sneezing and coughing.
Effects of exposure:
Breathed: Dust may cause irritation to the nose, throat and breathing passages.
Eyes: May cause severe irritation.
Skin: Passes through the unbroken skin; can increase exposure and the severity of the symptoms listed. May cause severe irritation with redness and itching.
Swallowed: May cause burns of the mouth and stomach; ulceration, stomach bleeding.
Long-term exposure: May cause skin irritation, rash; loss of appetite, emaciation, fatigue, weakness and delayed death. Possible lung effects. Autopsy shows inflammation of gastrointestinal tract, liver discoloration, and kidney problems.
Storage: Strong oxidizer. Store in tightly closed containers in a dry, cool, well-ventilated place with non-wood floors. Keep away from water, heat, sources of ignition, fuels and combustible materials such as paper, wood and oil.
First aid guide: Move victim to fresh air and call emergency medical care; if not breathing, give artificial respiration; if breathing is difficult, give oxygen. In case of contact with material, immediately flush skin or eyes with running water for at least 15 minutes. Remove and isolate contaminated clothing and shoes at the site. Keep victim quiet and maintain normal body temperature. Effects may be delayed; keep victim under observation.

★967★
2,4,6-TRICHLOROPHENOL
CAS: 88-06-2
Other names: DOWICIDE 2S; NCI-C02904; OMAL; PHENA-CHLOR; PHENOL, 2,4,6-TRICHLORO-; 2,4,6-TRICHLORFENOL (CZECH)

Danger profile: Carcinogen, Community Right-to-Know List, poisonous gas produced in fire.
Uses: Preservative; in insecticides; treating mildew; germicide.
Appearance: Colorless to yellow solid.
Odor: Strong. The odor is a warning of exposure; however, no smell does not mean you are not being exposed.
Effects of exposure:
Breathed: May cause irritation of the eyes and throat. High levels of exposure may cause weakness, breathing difficulties, tremors, convulsions, coma and death.
Eyes: May cause severe irritation.
Skin: May cause severe irritation.

Swallowed: May be poisonous.
Long-term exposure: May cause cancer. A related chemical may cause liver and kidney damage; it is unknown at this time of this chemical has the same effect. May be contaminated with 2,3,7,8-tetrachlorodibenzo-p-dioxin; this chemical may cause skin rash and damage to the nervous system.
Storage: Store in tightly closed containers in a dry, cool, well-ventilated place or a refrigerator. Keep away from acid chlorides, acid anhydrides, oxidizers.

★968★
TRIETHANOLAMINE
DODECYLBENZENESULFONATE
CAS: 27323-41-7
Other names: 2-2',2''-NITRILOTRIS-DODECYLBENZENESULFONATE (SALT)

Danger profile: Combustible, poisonous gas produced in fire.
Uses: In household detergent products.
Appearance: Yellowish-brown liquid.
Effects of exposure:
Breathed: High levels may cause irritation of the throat and lungs.
Eyes: Contact may cause irritation.
Skin: Contact may cause irritation.
Long-term exposure: May cause skin drying, chronic irritation and rash.
Storage: Store in tightly closed containers in a dry, cool, well-ventilated place. Keep away from open flames and other sources of ignition.

★969★
TRIETHYLALUMINIUM
CAS: 97-93-8
Other names: ALUMINUM TRIETHYL; ATE; TEA

Danger profile: Fire hazard, spontaneously combustible, exposure limit established, poisonous gas produced in fire.
Uses: Making other chemicals, fuels; aluminum plating.
Appearance: Colorless liquid.
Effects of exposure:
Breathed: May be poisonous. May cause irritation of the nose and respiratory tract; sneezing. Exposure to smoke from fire causes metal-fume fever with flu-like symptoms.
Eyes: May cause burns, great pain, swelling of eyelids and permanent damage.
Skin: May cause third degree burns, shock, loss of consciousness and possible death.
Swallowed: Extremely destructive to living tissue. Unlikely due to handling restrictions.
Long-term exposure: May cause lung problems.
Storage: Violent explosion in presence of water. Store in sealed containers in a dry, cool, well-ventilated place. Keep away from heat, moisture, sources of ignition, combustible materials, oxidizing agents, alcohols, halogenated hydrocarbons, triethyl borine. Containers may explode in fire.

★970★
TRIETHYLAMINE
CAS: 121-44-8
Other names: (DIETHYLAMINO)ETHANE; N,N-DIETHYLETHANEAMINE; ETHANAMINE,N,N-DIETHYL-; TEA; TEN; TRIAETHYLAMIN (GERMAN); TRIETILAMINA (ITALIAN)

Danger profile: Fire hazard, exposure limit established, poisonous gas produced in fire.
Uses: Making other chemicals, rubber; wetting agent; corrosion inhibitor; propellant; waterproofing agent.
Appearance: Colorless watery liquid.
Odor: Ammonia-like; fishy. The odor is a warning of exposure; however, no smell does not mean you are not being exposed.
Effects of exposure:
   Breathed: May be poisonous. May cause irritation to the nose, throat, eyes and lungs; difficult breathing, coughing, choking. Higher exposures may cause shortness of breath and a dangerous build-up of fluids in the lungs (pulmonary edema), which can cause death. Pulmonary edema is a medical emergency that can be delayed one to two days following exposure. Exposure may cause liver, kidney and/or heart damage.
   Eyes: Contact may cause severe burns; possible permanent damage.
   Skin: Passes through the unbroken skin; can increase exposure and the severity of the symptoms listed. Contact may cause severe irritation, redness, itching and burns.
   Swallowed: May be poisonous. May cause burning feeling in mouth, throat and stomach; pain and swelling in throat; nausea and vomiting; stomach cramps, rapid breathing, shock, diarrhea; possible stomach perforation.
Long-term exposure: May cause weight loss; heart, kidney and liver damage; skin irritation and itching. Similar chemicals may cause allergies. Mutation and reproductive data reported.
Storage: Store in tightly closed containers in a dry, cool, well-ventilated place or explosion-proof refrigerator. Keep away from heat, sources of ignition, acids, and oxidizers. Vapors may travel to sources of ignition and flash back. Vapor may explode if ignited in an enclosed area. Containers may explode in fire.
First aid guide: Move victim to fresh air and call emergency medical care; if not breathing, give artificial respiration; if breathing is difficult, give oxygen. In case of contact with material, immediately flush skin or eyes with running water for at least 15 minutes. Remove and isolate contaminated clothing and shoes at the site. Keep victim quiet and maintain normal body temperature.

★971★
TRIETHYLENE GLYCOL
CAS: 112-27-6
Other names: DI-BETA-HYDROXYETHOXYETHANE; 3,6-DIOXAOCTANE-1,8-DIOL; 2,2'-(1,2-ETHANEDIYLBIS(OXY)) BISEHANOL; ETHANOL,2,2'-(ETHYLENEDIOXY)DI-; 2,2'-ETHYLENEDIOXYDIETHANOL; 2,2'-ETHYLENEDIOXYETHANOL; ETHYLENE GLYCOL-BIS-(2-HYDROXYETHYL ETHER); ETHYLENE GLYCOL DIHYDROXYDIETHYL ETHER; GLYCOL BIS(HYDROXYETHYL) ETHER; TEG; TRIGEN; TRIGLYCOL

Danger profile: Community Right-to-Know List, poisonous gas produced in fire.
Uses: Solvent; in vinyl, polyester and polyurethane resins, printing inks
Appearance: Colorless liquid.
Odor: Very mild, sweet. The odor is a warning of exposure; however, no smell does not mean you are not being exposed.
Effects of exposure:
   Breathed: Mildly toxic. May cause irritation of eyes, nose and throat; headache, drowsiness, excitability, drunken behavior, mental confusion, fatigue. Very high levels of exposure may cause dizziness, lightheadedness and cause victim to pass out, go into coma and die.
   Eyes: Contact may cause irritation, pain, burning, stinging, tearing, inflammation of the eyelids. Light may cause pain.

   Skin: May remove oil from the skin and cause dryness. Persons with pre-existing skin disorders may be more susceptible to the effects of this substance.
   Swallowed: May cause gastrointestinal irritation. See *Breathed.*
Long-term exposure: Prolonged or repeated contact may cause irritation of the skin and dermatitis, and nervous system changes.
Storage: Store in a tightly closed container in a cool, well-ventilated place. Keep away from heat and sources of ignition, oxidizers. Vapors are explosive in heat or flame. Containers may explode in fire.

★972★
TRIETHYLLEAD FLUOROACETATE
CAS: 562-95-8
Other names: ACETIC ACID, FLUORO-,TRIETHYLLEAD SALT; FLUOROACETIC ACID, TRIETHYLLEAD SALT

Danger profile: Exposure limit established, Community Right-to-Know List, poisonous gas produced in fire.
Effects of exposure:
   Breathed: May cause restlessness, trembling hands, headache, nausea and vomiting; insomnia; irritation of the nose, tremors, restlessness, delirium, convulsions.
   Eyes: May cause irritation, watering and inflammation of the eyelids.
   Skin: May cause irritation with itching and blisters.
   Swallowed: May be poisonous.
Storage: Store in tightly closed containers in a dry, cool, well-ventilated place.

★973★
TRIFLUOROBROMOMETHANE
CAS: 75-63-8
Other names: BROMOFLUOROFORM; BROMOTRIFLUOROMETHANE; F-13B1; FREON 13B1; HALOCARBON 13B1; HALON 1301; METHANE, BROMOTRIFLUORO; MONOBROMOTRIFLUORMETHANE; REFRIGERANT 13B1; TRIFLUOROMONOBROMOMETHANE

Danger profile: Exposure limit established, Community Right-to-Know List, poisonous gas produced in fire.
Uses: In refrigeration systems; for extinguishing fires.
Appearance: Colorless gas.
Odor: Slight, ether-like. The odor is a warning of exposure; however, no smell does not mean you are not being exposed.
Effects of exposure:
   Breathed: May cause lightheadedness, dizziness, numbness and loss of concentration. High concentrations may cause irregular heartbeat and death. Possible suffocation.
   Eyes: Contact with liquid may cause frostbite.
   Skin: Contact with liquid may cause frostbite.
Long-term exposure: Related chemicals may cause liver damage. It is unknown at this time if this chemical has the same effects.
Storage: Noncombustible gas. Store in tightly closed cylinders in a dry, cool, well-ventilated place. Keep away from heat, aluminum and other chemically active metals. Containers may explode in fire.

★974★
TRIFLUOROETHANE
CAS: 27987-06-0
Other names: ETHANE, TRIFLUORO-

Danger profile: Combustible, poisonous gas produced in fire.

Uses: Making other chemicals.
Appearance: Colorless gas, or liquid under pressure.
Effects of exposure:
  Breathed: May cause headache, dizziness, fatigue; rapid irregular breathing; nausea, vomiting, confusion, loss of consciousness, convulsions and death. May cause suffocation.
  Eyes: Contact with liquid may cause frostbite, pain, stinging, tearing, inflammation of the eyelids and permanent damage.
  Skin: Contact with liquid may cause frostbite, intense cold feeling, skin turns white and feels cold and hard.
  Swallowed: Liquid may cause headache, confusion, drowsiness, unconsciousness and death.
Storage: Store in tightly closed containers in a dry, cool, well-ventilated place. Keep away from heat and sources of ignition. Vapors may travel to sources of ignition and flash back. Vapor may explode if ignited in an enclosed area. Containers may explode in fire.
First aid guide: Move victim to fresh air and call emergency medical care; if not breathing, give artificial respiration; if breathing is difficult, give oxygen. In case of frostbite, thaw frosted parts with water. Keep victim quiet and maintain normal body temperature.

★975★
TRIFLUOROMETHANE
CAS: 75-46-7
Other names: ARCTON; FLUROFORM; FLUORYL; FREON 23; FREON-23; GENETRON- 23; HALOCARBON 23; METHYL TRIFLUORIDE; R 23

Danger profile: Poisonous gas produced in fire.
Uses: Refrigerant; making other chemicals, urethane foams.
Appearance: Colorless gas.
Effects of exposure:
  Breathed: May cause irritation. High exposures are narcotic and may cause dizziness or suffocation.
  Eyes: Contact with liquid may cause frostbite, pain, stinging, tearing, inflammation of the eyelids and permanent damage.
  Skin: Contact with liquid may cause frostbite, intense cold feeling, skin turns white and feels cold and hard.
  Swallowed: Liquid may cause headache, confusion, drowsiness, unconsciousness.
Storage: Store in tightly closed containers in a dry, cool, well-ventilated place. Containers may explode in fire.

★976★
TRIFLURALIN
CAS: 1582-09-8
Other names: AGREFLAN; AGRIFLAN 24; BENZENAMINE, 2,6-DINITRO-N,N-DIPROPYL-4-(TRIFLUOROMETHYL-); CRISALIN; DIGERMIN; 2,6-DINITRO-N,N-DIPROPYL-4-(TRIFLUO ROMETHYL)BENZENAMINE; 2,6-DINITRO-N,N-DI-N-PROPYL-ALPHA,ALPHA,ALPHA-TRIFLURO-P-TOLUIDINE; 4-(DI-N-PROPYLAMINO)-3,5-DINITRO-1-TRIFLUOROMETHYLBENZENE; N,N-DI-N-PROPYL-2,6-DINITRO-4-TRIFLUOROMETHYLANILINE; 2,6-DINITRO-4-TRIFLUORMETHYL-N,N-DIPROPYLANILIN (GERMAN); N,N-DIPROPYL-4-TRIFLUOROMETHYL-2,6-DINITROANILINE; ELANCOLAN; IPERSAN; L-36352; LILLY 36,352; MARKSMAN 2, TRIGARD; NCI-C00442; NITRAN; OLITREF; SINFLOWAN; SU SEGURO CARPIDOR; P-TOLUIDINE,ALPHA,ALPHA,ALPHA-TRIFLUO RO-2,6-DINITRO-N,N-DIPROPYL-; TREFANOCIDE; TREFICON; TREFLAN; TREFLANOCIDE ELANCOLAN; TRIFLUORALIN; TRIFLURALINE; TRISTAR; ALPHA,ALPHA,ALPHA-TRIFLUORO-2,6-DINITRO-N,N-DIPROPYL-P-TOLUIDINE; TRIFUREX; TRIKEPIN; TRIM

Danger profile: Possible mutagen, Community Right-to-Know List, poisonous gas produced in fire.
Uses: To control weeds and grasses; on a wide range of crops including broccoli, sprouts, carrots, lettuce, strawberries, sugar beet, cotton, vineyards.
Appearance: Yellow to yellow-orange crystalline solid.
Effects of exposure:
  Breathed: May be poisonous. May cause irritation of the nose, throat and breathing passages. High exposure may cause liver and kidney problems, or cause anemia.
  Eyes: Contact may cause irritation.
  Skin: Passes through the unbroken skin; can increase exposure and the severity of the symptoms listed. Contact may cause irritation and rash.
  Swallowed: May be poisonous.
Long-term exposure: Repeated exposure may cause liver and kidney problems, or cause anemia. Rash may develop that can be made worse by sunlight. Questionable carcinogen; animal data reported. Possible mutagen; human data reported. May cause damage to the fetus.
Storage: Store in tightly closed containers in a dry, well-ventilated place in temperatures above 40°F. Containers may explode in fire.

★977★
TRIISOBUTYL ALUMINIUM
CAS: 100-99-2
Other names: ALUMINUM, TRIISOBUTYL-; ALUMINUM, TRIS(2-METHYLPROPYL)- (9CI); TIBA; TIBAL; TRIISOBUTYLALANE

Danger profile: Fire hazard, explosion hazard, exposure limit established, poisonous gas produced in fire.
Uses: Making other chemicals, olefins and fuels.
Appearance: Colorless liquid.
Effects of exposure:
  Breathed: May be poisonous. May cause irritation of the nose, throat and lungs; metal fume fever with flu-like symptoms; sneezing.
  Eyes: Flames may cause burns, great pain, swelling of eyelids and permanent damage.
  Skin: May cause third degree burns, shock, loss of consciousness and possible death.
Long-term exposure: May cause lung problems.
Storage: Highly reactive. May ignite spontaneously in air. Store in sealed containers in a dry, cool, well-ventilated place with non-wood floors. Keep away from heat, sources of ignition, air, water, combustible materials, acid, alcohols, oxidizing agents, halogenated compounds. Containers may explode in fire.
First aid guide: Move victim to fresh air; call emergency medical care. Wipe material from skin immediately; flush skin or eyes with running water for at least 15 minutes. Remove and isolate contaminated clothing and shoes at the site.

★ 978 ★
TRIMELLITIC ANHYDRIDE
CAS: 552-30-7
Other names: ANHYDRIDE-ETHOMID HT POLYMER; AN-HYDRO TRIMELLIC ACID; 1,2,4-BENZENETRICARBOXYLIC ACID ANHYDRIDE; 1,2,4-BENZENETRICARBOXYLIC ACID, CYCLIC 1,2-ANHYDRIDE; 1,2,4-BENZENETRICARBOXYLIC ANHYDRIDE; 4-CARBOXYPHTHALIC ANHYDRIDE; 1,3-DIHYDRO-1,3-DIOXO-5-ISOBENZOFURANCARBOXYLIC ACID; 1,3-DIOXO-5-PHTHALANCARBOXYLIC ACID; DIPHE-NYLMETHANE-4,4'-DIISOCYANATE-TRIMELLIC ANHY-DRIDE-ETHANOMID HT POLYMER; 5-PHTHALANACARBOXYLIC ACID, 1,3-DIOXO-; TMA; TMAN; TRIMELLITIC ACID ANHYDRIDE; TRIMELLIC ACID 1,2-ANHYDRIDE; TRIMELLITIC ACID CYCLIC 1,2-ANHYDRIDE

Danger profile: Exposure limit established, poisonous gas produced in fire.
Uses: Making plastics, resins, other chemicals, dyes; gaskets; automotive upholstery.
Appearance: Colorless to white solid.
Effects of exposure:
  Breathed: Dust may cause severe irritation to the nose, throat breathing passages and lungs; coughing, wheezing, shortness of breath, running nose. May cause flu-like symptoms including chills, muscle ache, headache, fever. High exposure may cause nausea, salivation, vomiting, cramps, diarrhea, chest pains, cough, a build-up of fluids in the lungs (pulmonary edema) and possible death. Pulmonary edema may be delayed from one to two days following exposure.
  Skin: May cause severe irritation and thermal and acid burns, especially if skin is wet.
  Eyes: May cause severe irritation, burns, pain and possible blindness.
  Swallowed: May cause severe burns of the mouth, throat and stomach, vomiting, watery or bloody diarrhea. Damage to kidneys and liver, collapse and convulsions may result.
Long-term exposure: May cause asthma-like allergy; lung disease, low blood count.
Storage: Store in tightly closed containers in a dry, cool, well-ventilated place. Containers may explode in fire.

★ 979 ★
TRIMETHYL ALUMINUM
CAS: 75-24-1
Other names: ALUMINUM TRIMETHYL; TRIMETHYLALANE

Danger profile: Fire hazard, explosion hazard, exposure limit established, poisonous gas produced in fire.
Uses: Making other chemicals.
Appearance: Colorless liquid.
Odor: The odor is a warning of exposure; however, no smell does not mean you are not being exposed.
Effects of exposure:
  Breathed: May be poisonous. May cause irritation of the nose, throat and lungs; metal fume fever with flu-like symptoms, headache, nausea, vomiting, chills, coughing, sneezing, difficult breathing.
  Eyes: Flames may cause burns, great pain, swelling of eyelids and permanent damage.
  Skin: May cause third degree burns, shock, loss of consciousness and possible death.
  Swallowed: Unlikely due to handling restrictions.
Long-term exposure: May cause lung problems.
Storage: Highly reactive chemical. May spontaneously ignite in air. Store in sealed containers in a dry, cool, well-ventilated place with non-wood floors. Keep away from heat, sources of ignition, air, water, combustible materials, acid, alcohols, oxidizing agents, halogenated compounds. Containers may explode in fire.
First aid guide: Move victim to fresh air; call emergency medical care. Wipe material from skin immediately; flush skin or eyes with running water for at least 15 minutes. Remove and isolate contaminated clothing and shoes at the site.

★ 980 ★
TRIMETHYLAMINE
CAS: 75-50-3
Other names: N,N-DIMETHYLMETHANAMINE; TMA

Danger profile: Fire hazard, exposure limit established, poisonous gas produced in fire.
Uses: Making other chemicals; attracting insects; warning smell in natural gas.
Appearance: Colorless liquefied compressed gas.
Odor: Pungent, fishy, ammonia-like. The odor is a warning of exposure; however, no smell does not mean you are not being exposed.
Effects of exposure:
  Breathed: May be poisonous. May cause irritation to the nose, throat and eyes; difficult breathing, coughing. Higher exposures may cause shortness of breath and a dangerous build-up of fluids in the lungs (pulmonary edema), which can cause death. Pulmonary edema is a medical emergency that can be delayed one to two days following exposure.
  Eyes: Contact may cause severe burns; possible permanent damage. Liquid may cause frostbite.
  Skin: Passes through the unbroken skin; can increase exposure and the severity of the symptoms listed. Contact may cause severe irritation, redness, itching and burns. Prolonged contact may cause burns and serious injury. Liquid may cause frostbite.
  Swallowed: May cause burning feeling in mouth, throat and stomach; pain and swelling in throat; nausea and vomiting; stomach cramps, rapid breathing, shock, diarrhea; possible stomach perforation.
Long-term exposure: May cause weight loss; kidney and liver damage; skin irritation and itching. Similar chemicals may cause allergies and lung problems.
Storage: Store in tightly closed containers in a dry, cool, well-ventilated place. Keep away from heat, sources of ignition, ethylene oxide, and strong oxidizers. Vapors may travel to sources of ignition and flash back. Containers may explode in fire.

★ 981 ★
TRIMETHYLBENZENE
CAS: 25551-13-7
Other names: BENZENE, TRIMETHYL- (MIXED ISOMERS)

Danger profile: Exposure limit established, poisonous gas produced in fire.
Uses: Making other chemicals.
Appearance: Colorless liquid.
Odor: Peculiar. The odor is a warning of exposure; however, no smell does not mean you are not being exposed.
Effects of exposure:
  Breathed: May be poisonous. May cause dizziness, light-headedness, and loss of consciousness. Bronchitis may develop with cough and difficult breathing.
  Eyes: May cause irritation and burns.
  Skin: May be poisonous if absorbed. May cause irritation and burns.
  Swallowed: Mildly toxic.
Long-term exposure: May cause liver problems; bronchitis, cough, difficult breathing; headaches, tiredness and nervousness; blood clotting effects. Lung problems are possible.

Storage: Store in tightly closed containers in a dry, cool, well-ventilated place. Keep away from heat, sources of ignition and oxidizers. Vapors may travel to sources of ignition and flash back. Vapor may explode if ignited in an enclosed area. Containers may explode in fire.
First aid guide: Move victim to fresh air and call emergency medical care; if not breathing, give artificial respiration; if breathing is difficult, give oxygen. In case of contact with material, immediately flush eyes with running water for at least 15 minutes. Wash skin with soap and water. Remove and isolate contaminated clothing and shoes at the site.

★ 982 ★
1,2,3-TRIMETHYL BENZENE
CAS: 526-73-8
Other names: HEMIMELLITENE; 1,2,3-TRIMETHYL BENZOL

Danger profile: Exposure limit established, poisonous gas produced in fire.
Uses: Making other chemicals.
Appearance: Colorless liquid.
Odor: Peculiar. The odor is a warning of exposure; however, no smell does not mean you are not being exposed.
Effects of exposure:
 Breathed: May be poisonous. May cause dizziness, light-headedness, and loss of consciousness. Bronchitis may develop with cough and difficult breathing.
 Eyes: May cause irritation and burns.
 Skin: May be poisonous if absorbed. May cause irritation and burns.
 Swallowed: Mildly toxic.
Long-term exposure: May cause liver problems; bronchitis, cough, difficult breathing; headaches, tiredness and nervousness; blood clotting effects. Lung problems are possible.
Storage: Store in tightly closed containers in a dry, cool, well-ventilated place. Keep away from heat, sources of ignition and oxidizers. Vapors may travel to sources of ignition and flash back. Vapor may explode if ignited in an enclosed area. Containers may explode in fire.

★ 983 ★
1,2,4-TRIMETHYL BENZENE
CAS: 95-63-6
Other names: ASYMMETRICAL TRIMETHYL BENZENE; PSI-CUMENE; PSEUDOCUMENE; PSEUDOCUMOL; 1,2,5-TRIMETHYL BENZENE

Danger profile: Exposure limit established, Community Right-to-Know List, poisonous gas produced in fire.
Uses: Making other chemicals.
Appearance: Colorless liquid.
Odor: Peculiar. The odor is a warning of exposure; however, no smell does not mean you are not being exposed.
Effects of exposure:
 Breathed: May be poisonous. May cause dizziness, light-headedness, and loss of consciousness. Bronchitis may develop with cough and difficult breathing.
 Eyes: May cause irritation and burns.
 Skin: May be poisonous if absorbed. May cause irritation and burns.
 Swallowed: Mildly toxic.
Long-term exposure: May cause liver problems; central nervous system depression; bronchitis, cough, difficult breathing; headaches, tiredness and nervousness; blood clotting effects; anemia. Lung problems are possible.
Storage: Store in tightly closed containers in a dry, cool, well-ventilated place. Keep away from heat, sources of ignition and oxidizers. Vapors may travel to sources of ignition and flash

back. Vapor may explode if ignited in an enclosed area. Containers may explode in fire.

★ 984 ★
1,3,5-TRIMETHYL BENZENE
CAS: 108-67-8
Other names: FLEET-X; MESITLENE; SYM-TRIMETHYL BENZENE; TRIMETHYL BENZENE; TRIMETHYL BENZOL

Danger profile: Combustible, exposure limit established, poisonous gas produced in fire.
Uses: Making other chemicals.
Appearance: Colorless liquid.
Odor: Peculiar. The odor is a warning of exposure; however, no smell does not mean you are not being exposed.
Effects of exposure:
 Breathed: May be poisonous. May cause dizziness, light-headedness, and loss of consciousness; tiredness. Bronchitis may develop with cough and difficult breathing.
 Eyes: May cause irritation and burns.
 Skin: May be poisonous if absorbed. May cause irritation and burns.
 Swallowed: Mildly toxic.
Long-term exposure: May cause liver problems; bronchitis, cough, difficult breathing; headaches, tiredness and nervousness; blood clotting effects. Lung problems are possible. Mutation data reported.
Storage: Store in tightly closed containers in a dry, cool, well-ventilated place. Keep away from heat, sources of ignition nitric acid and oxidizers. Vapors may travel to sources of ignition and flash back. Vapor may explode if ignited in an enclosed area. Containers may explode in fire.

★ 985 ★
TRIMETHYL CHLOROSILANE
CAS: 75-77-4
Other names: CHLOROTRIMETHYL SILANE; CHLOROTRI-METHYLSILICANE; SILANE, CHLOROTRIMETHYL-; SILANE TRIMETHYLCHLORO-; SILICANE, CHLOROTRIMETHYL-; TL 1163; TRIMETHYLCHLOROSILANE

Danger profile: Fire hazard, explosion hazard, corrosive, poisonous gas produced in fire.
Uses: Making silicones, silicone lubricants; water repellency chemical.
Appearance: Colorless liquid.
Odor: Sharp, irritating, acrid; hydrochloric acid-like. The odor is a warning of exposure; however, no smell does not mean you are not being exposed.
Effects of exposure:
 Breathed: May be poisonous. Vapor may cause irritation of the lungs; cough, difficult breathing. Higher exposures may cause shortness of breath and a dangerous build-up of fluids in the lungs (pulmonary edema), which can cause death. Pulmonary edema is a medical emergency that can be delayed one to two days following exposure.
 Eyes: Contact may cause severe burns and permanent damage.
 Skin: Contact may cause severe burns.
 Swallowed: May cause severe burns of the mouth and stomach.
Long-term exposure: May cause lung problems to develop.
Storage: Very reactive. Store in tightly closed containers in a dry, cool, well-ventilated place. Keep away from heat, sources of ignition, water, steam and other forms of moisture. Reacts with surface moisture to evolve hydrogen chloride, which will corrode common metals and form combustible hydrogen gas. Vapors may travel to sources of ignition and flash back. Vapor

may explode if ignited in an enclosed area. Containers may explode in fire.

First aid guide: Move victim to fresh air and call emergency medical care; if not breathing, give artificial respiration; if breathing is difficult, give oxygen. In case of contact with material, immediately flush skin or eyes with running water for at least 15 minutes. Remove and isolate contaminated clothing and shoes at the site. Keep victim quiet and maintain normal body temperature.

★ 986 ★
TRIMETHYL PHOSPHITE
CAS: 121-45-9
Other names: FOSFORYN TROJMETYLOWY (CZECH); METHYL PHOSPHITE; PHOSPHOROUS ACID,TRIMETHYL ESTER; PHTHALIC ACID, BIS(2-METHOXYETHYL)ESTER; TMP; TRIMETHOXYPHOSPHINE; TRIMETHYLPHOSPHITE

Danger profile: Exposure limit established, poisonous gas produced in fire.
Uses: Making pesticides, insecticides and flame retardants.
Appearance: Colorless liquid.
Odor: Pungent, pyridine-like. The odor is a warning of exposure; however, no smell does not mean you are not being exposed.
Effects of exposure:
  Breathed: May be poisonous. May cause lung irritation; burning sensation, coughing, wheezing, laryngitis, shortness of breath, headache, nausea, and vomiting; emphysema.
  Eyes: Contact or vapor may cause severe irritation and permanent damage.
  Skin: May be poisonous if absorbed. Passes through the unbroken skin; can increase exposure and the severity of the symptoms listed. May cause severe irritation and skin allergy.
  Swallowed: Moderately toxic.
Long-term exposure: May cause emphysema; skin allergy; liver and kidney damage.
Storage: Store in tightly closed containers in a dry, cool, well-ventilated place. Keep away from heat, sources of ignition, magnesium diperchlorate. Vapors may travel to sources of ignition and flash back. Vapor may explode if ignited in an enclosed area. Containers may explode in fire.
First aid guide: Move victim to fresh air and call emergency medical care; if not breathing, give artificial respiration; if breathing is difficult, give oxygen. In case of contact with material, immediately flush eyes with running water for at least 15 minutes. Wash skin with soap and water. Remove and isolate contaminated clothing and shoes at the site.

★ 987 ★
1,3,5-TRINITROBENZENE
CAS: 99-35-4
Other names: BENZENE, 1,3,5-TRINITRO-; TNB; TRINITRO-BENZEEN (DUTCH); TRINITROBENZENE; TRINITROBEN-ZENE, DRY; TRINITROBENZOL (GERMAN); SYM-TRINITROBENZENE

Danger profile: Explosion hazard, poisonous gas and explosion produced in fire.
Uses: Explosive materials.
Appearance: Yellow crystals.
Effects of exposure:
  Breathed: May cause headache, rapid and difficult breathing, shortness of breath, dizziness, mental confusion, weakness; bluish color of the lips and skin, convulsions, coma and death.
  Eyes: May cause irritation, redness, swelling of the eyelids; light causes pain; severe damage.

Skin: Passes through the unbroken skin; can increase exposure and the severity of the symptoms listed. May cause irritation, redness, swelling, blisters.
Swallowed: May cause irritation of the mouth, stomach; cramps and diarrhea.
Long-term exposure: In experimental animals: liver damage, central nervous system damage, methemoglobin formation.
Storage: High explosive with more destructive power than TNT. Store in tightly closed containers in a cool, well-ventilated place. Keep material wet with water or treat as an explosive. Keep away from heat, sources of ignition and reducing material. Protect containers from shock. Containers may explode in fire.

★ 988 ★
TRINITROTOLUENE
CAS: 118-96-7
Other names: ENTSUFON; NCI-C56155; TNT; ALPHA-TNT; TNT-TOLITE (FRENCH); TOLIT; TOLITE; TOLUENE, 2,4,6-TRINITRO,-(WET); 2,4,6-TRINITROTOLUEEN (DUTCH); SYM-TRINITROTOLUENE; 2,4,6-TRINITROTOLUENE; TRINI-TROTOLUENE, WET; S-TRINITROTOLUOL; SYM-TRINITROTOLUOL; 2,4,6-TRINITROTOLUOL (GERMAN); TRITOL; TROJNITROTOLUOL (POLISH); TROTYL; TROTYL OIL

Danger profile: Explosion hazard, combustible, exposure limit established, poisonous gas and explosion produced in fire.
Uses: Explosive; making dyes and photographic chemicals.
Appearance: Colorless to pale yellow crystalline solid.
  Breathed: May cause headache, rapid and difficult breathing, shortness of breath, dizziness, mental confusion, weakness; bluish color of the lips and skin, convulsions, coma and death.
  Eyes: May cause irritation, redness, swelling of the eyelids; light causes pain; severe damage.
  Skin: Passes through the unbroken skin; can increase exposure and the severity of the symptoms listed. May cause irritation, redness, swelling, blisters.
  Swallowed: May cause irritation of the mouth, stomach; cramps and diarrhea; distorted perceptions, hallucinations; gastrointestinal problems.
Long-term exposure: May cause liver damage, headache, weakness, anemia. Possible mutagen. Data on aplastic anemia and mutations reported.
Storage: Powerful high explosive. Store in tightly closed containers in a cool, well-ventilated place. Keep material wet with water or treat as an explosive. Keep away from heat, sources of ignition, metal, nitric acid and reducing materials. Protect containers from shock. Containers may explode in fire.
First aid guide: Move victim to fresh air; call emergency medical care. In case of contact with material, immediately flush skin or eyes with running water for at least 15 minutes. Remove and isolate contaminated clothing and shoes at the site.

★ 989 ★
TRIPHENYL AMINE
CAS: 603-34-9
Other names: N,N-DIPHENYLANILINE; N,N-DIPHENYLBENZENAMINE

Danger profile: Exposure limit established, poisonous gas produced in fire.
Uses: Making photographic film.
Appearance: Colorless crystalline solid.
Effects of exposure:
  Skin: Contact may cause irritation.
  Swallowed: Moderately toxic.

Long-term exposure: Related chemicals may cause skin and lung allergies.
Storage: Store in tightly closed containers in a dry, cool, well-ventilated place. Keep away from aldehydes, ketones, nitrates, oxidizers, oxygen and peroxides.

★ 990 ★
TRIPOLI
CAS: 1317-95-9
Other names: FINELY GROUND SILICA; SILICA, CRYSTAL-LINE-TRIPOLI; SILICA FLOUR; SILICA, TRIPOLI

Danger profile: Carcinogen, exposure limit established, poisonous gas produced in fire.
Uses: Abrasives; in metal polishes; paint filler.
Appearance: Fine white or gray sand-like material.
Effects of exposure:
  Breathed: High levels may cause silicosis; cough, difficult breathing; increased possibility of tuberculosis developing. Silicosis may cause death.
Long-term exposure: May cause silicosis; increased susceptibility to tuberculosis; disability.
Storage: Store in tightly closed containers in a dry, cool, well-ventilated place.

★ 991 ★
TRIPROPYLALUMINUM
CAS: 102-67-0
Other names: ALUMINUM, TRIPROPYL-; TNPA; TRIPROPYL ALUMINUM

Danger profile: Fire hazard, explosion hazard, exposure limit established, reacts violently with water, poisonous gas produced in fire.
Uses: Making other chemicals.
Appearance: Colorless liquid.
Effects of exposure:
  Breathed: May be poisonous. May cause irritation of the nose, throat and lungs; metal fume fever with flu-like symptoms; headache, nausea, chills, vomiting, etc.
  Eyes: Contact may cause immediate severe burns and permanent damage.
  Skin: Contact may cause sever burns.
Long-term exposure: May cause metal fume fever. Possible lung problems.
Storage: May ignite spontaneously in air. Water reactive. Store in tightly closed containers in a dry, cool, well-ventilated place. Keep away from air, water, acid, alcohols, oxidizing agents and halogenated compounds.
First aid guide: Move victim to fresh air; call emergency medical care. Wipe material from skin immediately; flush skin or eyes with running water for at least 15 minutes. Remove and isolate contaminated clothing and shoes at the site.

★ 992 ★
TRIS
CAS: 126-72-7
Other names: ANFRAM 3PB; APEX 462-5; BROMKAL P 67-6HP; 2,3DIBROMO-1- PROPANOL, PHISPHATE (3:1); FIRE-MASTER T23P-LV; FLACAVON R; FLAMMEX AP; FYROL HB32; NCI-C03270; PHOSPHORIC ACID, TRIS(2,3- DIBRO-MOPROPYL)ESTER; 1-PROPANOL, 2,3-DIBROMO-, PHOS-PHATE (3:1); TDBP (CZECH); TRIS(2,3-DIBROMOPROPYL) PHOSPHATE; USAF DO-41; ZETIFEX ZN

Danger profile: Carcinogen, mutagen, reproductive hazard, Community Right-to-Know List, poisonous gas produced in fire.

Uses: Use restricted. Once used as flame retardant for plastics and synthetic fibers; to control flammability of cloth children's sleep wear.
Appearance: Pale yellow liquid.
Effects of exposure:
  Breathed: May cause irritation of the nose and throat.
  Eyes: May cause severe irritation.
  Skin: Passes through the unbroken skin; can increase exposure and the severity of the symptoms listed. May cause severe irritation.
  Swallowed: Moderately toxic. Can be chewed or sucked off sleep wear by infants.
Long-term exposure: May cause cancer; mutations; injury to the fetus; testicular atrophy and sterility.
Storage: Store in tightly closed containers in a dry, cool, well-ventilated place.

★ 993 ★
TRITHION
CAS: 786-19-6
Other names: ACARITHION; AKARITHION; CARBOFE-NOTHION (DUTCH); CARBOPHENOTHION; S-((P-CHLOROPHENYLTHIO)METHYL)O,O-DIETHYL PHOSPHO-RODITHIOATE; S-(4-CHLOROPHENYLTHIOMETHYL)DIETHYL PHOSPHO-ROTHIOLOTHIONATE; DAGADIP; O,O-DIAETHYL-S-((4-CHLOR-PHENYL-THIO)-METHYL)DITHIOPHOSPHAT (GERMAN); O,O-DIETHYL-S-((4-CHLOOR-FENYL-THIO) -METHYL)-DITHIOFOSFAAT (DUTCH); O,O-DIETHYL-S-P-CHLORFENYLTHIOMETHYLESTER KYSELINY DITHIOFOS-FORECNE (CZECH); O,O-DIETHYL P-CHLOROPHENYLMERCAPTOMETHYLDITHIOPHOSPHATE; O,O-DIETHYL S-P-CHLOROPHENYLTHIOMETHYLDITHIOPHOSPHATE; O,O-DIETHYL S-(4-CHLOROPHENYLTHIOMETHYL) DITHIOPHOSPHATE; O,O-DIETHYL S-(P-CHLOROPHENYLTHIOMETHYL) PHOSPHORODITHIOATE; O,O-DIETHYL DITHIOPHOSPHORIC ACID P-CHLOROPHENYLTHIOMETHYL ESTER; O,O-DIETIL-S-((4-CLORO-FENIL-TIO)-METILE)-DITIOFOSFATO (ITALIAN); DITHIOPHOSPHATE DE O,O-DIETHYLE ET DE (4-CHLORO-PHENYL) THIOMETHYLE (FRENCH); ENDYL; ENT 23,708; GARRATHION; LETHOX; NEPHOCARP; OLEOAKARITHION; PHOSPHORODITHIOIC ACID, S-(((P-CHLOROPHENYL)THIO)METHYL) O,O-DIETHYL ESTER; R-1303; STAUFFER R-1,303; TRITHION MITICIDE

Danger profile: Poisonous gas produced in fire.
Uses: Insecticide (organophosphate).
Appearance: Off-white to amber liquid.
Odor: Mild sulfur; rotten egg-like. The odor is a warning of exposure; however, no smell does not mean you are not being exposed.
Effects of exposure:
  Breathed: Extremely poisonous. May cause headache, dizziness, profuse sweating, nausea, vomiting, reduced heart beat, stomach cramps, diarrhea, loss of coordination, slow and weak breathing, fever, loss of consciousness and death. May cause shortness of breath and a dangerous build-up of fluids in the lungs (pulmonary edema), which can cause death. Pulmonary edema is a medical emergency that can be delayed one to two days following exposure.
  Eyes: Rapidly absorbed. Poisoning can happen very rapidly. See **Breathed.**
  Skin: Poisonous. Passes through the unbroken skin; can increase exposure and the severity of the symptoms listed.
  Swallowed: Powerful, deadly poison. See **Breathed.**
Long-term exposure: May damage the developing fetus or cause birth defects; nerve damage including loss of coordina-

tion; liver damage. Repeated exposure may cause depression, irritability and personality changes.
Storage: Store in tightly closed containers in a dry, cool, well-ventilated place. Containers may explode in fire.

★994★
TUNGSTEN
CAS: 7440-33-7
Other names: WOLFRAM

Danger profile: Flammmable (dust), exposure limit established.
Uses: Making electronic tubes and lamp bulbs.
Appearance: Steel gray to tin-white metal or fine powder.
Effects of exposure:
Eyes: May cause irritation.
Skin: May cause irritation.
Swallowed: May be toxic.
Long-term exposure: Some tungsten compounds may cause skin and lung problems.
Storage: Store in tightly closed containers in a dry, cool, well-ventilated place. Keep powder away from sources of ignition, especially flame, air and oxidants such as chlorine, iodine, potassium, bromine and fluorine compounds. Containers may explode in fire.

★995★
TUNGSTEN HEXAFLUORIDE
CAS: 7783-82-6
Other names: TUNGSTEN FLUORIDE

Danger profile: Corrosive, exposure limit established, poisonous gas produced in fire.
Uses: Application of tungsten coatings to other surfaces; fluorinating agent.
Appearance: Colorless gas or light yellow liquid.
Effects of exposure:
Breathed: May be poisonous. May cause irritation of the nose, throat and breathing passages. Tungsten hexafluoride is toxic to humans although the effects may be due to the fluoride content which may cause nausea, vomiting, abdominal pain, convulsions and kidney damage.
Eyes: Contact may cause irritation. Contact with liquid may cause frostbite.
Skin: May be poisonous if absorbed. Contact may cause irritation. Contact with liquid may cause frostbite.
Swallowed: See *Breathed.*
Long-term exposure: The fluoride content in chemical may cause chronic stomach pain, nausea, diarrhea; teeth and bone changes.
Storage: Store in tightly closed containers in a dry, cool, well-ventilated place. Keep away from water. Containers may explode in fire.
First aid guide: Move victim to fresh air and call emergency medical care; if not breathing, give artificial respiration; if breathing is difficult, give oxygen. In case of contact with material, immediately flush skin or eyes with running water for at least 15 minutes. Remove and isolate contaminated clothing and shoes at the site. Keep victim quiet and maintain normal body temperature. Effects may delayed; keep victim under observation.

★996★
TURPENTINE
CAS: 8006-64-2
Other names: GUM SPIRITS; GUM TURPENTINE; OIL OF TURPENTINE; SPIRITS OF TURPENTINE; SULFATE WOOD TURPENTINE; STEAM DISTILLED TURPENTINE; TEREBENTHINE (FRENCH); TERPENTIN OEL (GERMAN); TURPENTINE STEAM DISTILLED; TURPS; WOOD TURPENTINE

Danger profile: Fire hazard, exposure limit established, poisonous gas produced in fire.
Uses: Solvent for paints and other products; making other chemicals, inks and many other products; in household cleaning products. A common air contaminant.
Appearance: Colorless liquid.
Odor: Aromatic, characteristic, penetrating. The odor is a warning of exposure; however, no smell does not mean you are not being exposed.
Effects of exposure:
Breathed: May be poisonous. May cause irritation of the eyes, nose and throat. Higher levels may cause headache, dizziness, lightheadedness, rapid breathing, mental confusion, ringing in the ears, unsteady walk, headache, fatigue, nausea, vomiting; still higher levels may cause kidney damage, unconsciousness, coma and death. If liquid is taken into lungs, causes severe pneumonitis.
Eyes: May cause burning feeling, tearing, irritation, redness and inflammation of the eyelids.
Skin: May be poisonous if absorbed. Passes through the unbroken skin; can increase exposure and the severity of the symptoms listed. May cause irritation of the skin, dryness, redness and rash. Skin allergy may develop.
Swallowed: May cause irritation of the entire digestive system; indigestion, dizziness, fatigue, unconsciousness. May cause kidney damage.
Long-term exposure: May cause skin rash, and allergy may develop; kidney and bladder damage; nervous system damage.
Storage: Store in tightly closed containers in a dry, cool, well-ventilated place. Keep away from sources of ignition, chlorine and oxidizers. Vapors may travel to sources of ignition and flash back. Containers may explode in fire.
First aid guide: Move victim to fresh air and call emergency medical care; if not breathing, give artificial respiration; if breathing is difficult, give oxygen. In case of contact with material, immediately flush eyes with running water for at least 15 minutes. Wash skin with soap and water. Remove and isolate contaminated clothing and shoes at the site.

U

★997★
UREA
CAS: 57-13-6
Other names: B-I-K; CARBAMIDE; CARBAMIDE ACID; CARBAMIDE RESIN; CARBONYL DIAMINE; CARBAMIMIDIC ACID; CARBONYL DIAMIDE; CARBONYLDIAMIDE; ISOUREA; NCI-C02119; PRESPERSION, 75 UREA; PSEUDOUREA; SUPERCEL 3000; UREAPHIL; UREOPHIL; UREVERT; VARIOFORM II

Danger profile: Possible mutagen, poisonous gas produced in fire.
Uses: In fertilizers and animal feeds; making resins and plastics; stabilizer in explosives and in medicines.
Appearance: White crystals or powder.

Odor: Similar to ammonia. The odor is a warning of exposure; however, no smell does not mean you are not being exposed.

Effects of exposure:

Breathed: Dust may cause difficult breathing especially if the victim has asthma.

Eyes: Contact may cause irritation.

Skin: Contact with liquid may cause burning and stinging and mild irritation.

Swallowed: Moderately toxic. Some toxic effects have been found in animals with impaired liver function.

Long-term exposure: Possible human mutagen. May effect the fetus of an exposed woman.

Storage: Store in tightly closed containers in a dry, cool, well-ventilated place. Keep away from high temperatures, gallium perchlorate and strong oxidizing agents.

V

★ 998 ★
N-VALERALDEHYDE
CAS: 110-62-3

Other names: AMYL ALDEHYDE; BUTYL FORMAL; PENTANAL; N-PENTANAL; VALERIC ALDEHYDE; VALERAL; VALERIANIC ALDEHYDE; VALERIC ACID ALDEHYDE; N-VALERIC ALDEHYDE; VALERALDEHYDE

Danger profile: Carcinogen, combustible, exposure limit established, poisonous gas produced in fire.

Uses: Food flavorings; making rubber products.

Appearance: Colorless watery liquid.

Odor: Fruity. The odor is a warning of exposure; however, no smell does not mean you are not being exposed.

Effects of exposure:

Breathed: May be poisonus. Vapor may cause eye irritation. High levels may cause dizziness, lightheadedness and suffocation.

Eyes: Contact with liquid may cause severe irritation.

Skin: May be poisonus if absorbed. Passes through the unbroken skin; can increase exposure and the severity of the symptoms listed. Contact with liquid may cause severe irritation.

Swallowed: Moderately toxic.

Storage: Store in tightly closed containers in a dry, cool, well-ventilated place. Keep away from heat, sources of ignition. Vapors may travel to sources of ignition and flash back. Containers may explode in fire.

★ 999 ★
VANADIUM
CAS: 7440-62-2

Danger profile: Exposure limit established, Community Right-to-Know List, poisonous gas produced in fire.

Uses: Making x-ray equipment, synthetic rubber, alloy steels and other chemicals.

Appearance: A gray or bright, white, soft metal.

Effects of exposure:

Breathed: May be poisonous. Dust may cause irritation to the nose, throat and lungs, causing cough, phlegm and shortness of breath. Higher exposure may cause flu-like symptoms including chills, muscle ache, headache, fever. High exposure may cause nausea, salivation, vomiting, cramps, diarrhea, chest pains, cough, a build-up of fluids in the lungs

(pulmonary edema) and possible death. Pulmonary edema may be delayed from one to two days following exposure.

Eyes: May cause irritation, burns, pain and possible damage.

Skin: May cause irritation and burns, especially if skin is wet.

Swallowed: May be poisonous. May cause burns of the mouth, throat and stomach, vomiting, watery or bloody diarrhea.

Long-term exposure: May effect the lungs; green coating on tongue.

Storage: Store in tightly closed containers in a dry, cool, well-ventilated place. Keep away from heat, sources of ignition, and oxidizers.

★ 1000 ★
VANADIUM PENTOXIDE
CAS: 1314-62-1

Other names: VANADIC ANHYDRIDE; VANADIUM OXIDE; VANADIUM(V) OXIDE FUME; VANADIUM PENTOXIDE DUST; VANADIUM PENTOXIDE (FUME)

Danger profile: Exposure limit established, EPA extremely hazardous substances list, poisonous gas produced in fire.

Uses: Making alloys and other chemicals; photographic developer; dye for textiles; ceramic coloring material.

Appearance: Yellow to red crystal powder, dark gray flakes or fume (finely divided particulate dispersed in air).

Effects of exposure:

Breathed: May be poisonous. Dust or fume may cause severe irritation to the nose, throat and lungs, causing coughing, sore throat, phlegm, shortness of breath. May cause flu-like symptoms including chills, muscle ache, headache, fever. High exposure may cause nausea, salivation, vomiting, cramps, diarrhea, chest pains, cough, a build-up of fluids in the lungs (pulmonary edema) and possible death. Pulmonary edema may be delayed from one to two days following exposure.

Eyes: May cause severe irritation, burns, pain, possible damage and blindness.

Skin: May cause severe irritation and thermal and acid burns, especially if skin is wet.

Swallowed: May be poisonous. May cause severe burns of the mouth, throat and stomach, vomiting, watery or bloody diarrhea. Damage to kidneys.

Long-term exposure: May cause emphysema; chronic bronchitis; metal taste; discoloration (green coating) of the tongue; skin allergy.

Storage: Store in tightly closed containers in a dry, cool, well-ventilated place. Keep away from heat, lithium, chlorine trifluoride, peroxyformic acid. Containers may explode in fire.

First aid guide: Move victim to fresh air; call emergency medical care. In case of contact with material, immediately flush skin or eyes with running water for at least 15 minutes. Remove and isolate contaminated clothing and shoes at the site.

★ 1001 ★
VANADIUM SESQUIOXIDE
CAS: 1314-34-7

Other names: VANADIC OXIDE; VANADIUM OXIDE; VANADIUM TRIOXIDE

Danger profile: Exposure limit established, poisonous gas produced in fire.

Uses: Making other chemicals.

Appearance: Black crystalline powder. Exposure to air may cause gradual change in color to indigo blue.

Effects of exposure:

Breathed: May be poisonous. Dust may cause severe irritation to the eyes, nose and throat, causing coughing, shortness of breath. May cause flu-like symptoms including chills, muscle ache, headache, fever. High exposure may cause nausea, salivation, vomiting, cramps, diarrhea, chest pains, cough, pneumonia and/or a build-up of fluids in the lungs (pulmonary edema) and possible death. Pulmonary edema may be delayed from one to two days following exposure.

Skin: May cause irritation and thermal and acid burns, especially if skin is wet.

Eyes: May cause irritation, burns, pain and possible damage.

Swallowed: Poisonous. May cause severe burns of the mouth, throat and stomach, vomiting, watery or bloody diarrhea.

Long-term exposure: May cause skin irritation and rash; possible chronic lung problems; skin discoloration (green coating) of the tongue.

Storage: Store in tightly closed air tight containers in a dry, cool, well-ventilated place. Keep away from heat and air. Exposure to air may cause conversion to vanadium tetroxide; color changes to indigo blue. Containers may explode in fire.

★ 1002 ★
VANADIUM TETRACHLORIDE
CAS: 7632-51-1
Other names: VANADIUM CHLORIDE

Danger profile: Corrosive, explosion hazard, exposure limit established, poisonous gas produced in fire.
Uses: Making other chemicals; in textile dyeing.
Appearance: Thick reddish-brown liquid that fumes in moist air.
Effects of exposure:
　Breathed: May be poisonous. May cause severe irritation to the nose, throat and lungs, causing coughing, shortness of breath. May cause flu-like symptoms including chills, muscle ache, headache, fever. High exposure may cause nausea, salivation, vomiting, cramps, diarrhea, chest pains, cough, a build-up of fluids in the lungs (pulmonary edema) and possible death. Pulmonary edema may be delayed from 1-2 days following exposure.
　Eyes: May cause severe irritation, burns, pain and possible blindness.
　Skin: Passes through the unbroken skin; can increase exposure and the severity of the symptoms listed. May cause severe irritation and thermal and acid burns, especially if skin is wet.
　Swallowed: Poisonous. May cause severe burns of the mouth, throat and stomach, vomiting, watery or bloody diarrhea. Damage to kidneys may result.
Long-term exposure: May cause lung problems; discoloration (green coating) of the tongue; kidney problems.
Storage: Highly corrosive and reactive; an oxidizer. Store in tightly closed containers in a dry, cool, well-ventilated place. Keep away from heat, water, fuels, and combustible material. Containers may explode in fire.
First aid guide: Move victim to fresh air and call emergency medical care; if not breathing, give artificial respiration; if breathing is difficult, give oxygen. In case of contact with material, immediately flush skin or eyes with running water for at least 15 minutes. Speed in removing material from skin is of extreme importance. Remove and isolate contaminated clothing and shoes at the site. Keep victim quiet and maintain normal body temperature.

★ 1003 ★
VANADYL SULFATE
CAS: 27774-13-6
Other names: C.I. 77940; OXYSULFATOVANADIUM; VANADIUM, OXYSULFATO (2-)- O-

Danger profile: Exposure limit established, poisonous gas produced in fire.
Uses: In textile dyeing; coloring glass and ceramics; making other chemicals.
Appearance: Pale blue powder.
Effects of exposure:
　Breathed: May be poisonous. Dust may cause irritation of the nose, throat and lungs. Higher exposures may cause pneumonia and/or a dangerous build-up of fluids in the lungs (pulmonary edema), which can cause death. Pulmonary edema is a medical emergency that can be delayed one to two days following exposure.
　Eyes: Contact may cause irritation.
　Skin: May be poisonous if absorbed. Passes through the unbroken skin; can increase exposure and the severity of the symptoms listed.
　Swallowed: May be poisonous. May cause irritation of the mouth and stomach.
Long-term exposure: May cause anemia, kidney disorders, discoloration (green coating) of the tongue; blindness; skin irritation, eczema and rash.
Storage: Store in tightly closed containers in a dry, cool, well-ventilated place. Containers may explode in fire.
First aid guide: Move victim to fresh air and call emergency medical care; if not breathing, give artificial respiration; if breathing is difficult, give oxygen. In case of contact with material, immediately flush skin or eyes with running water for at least 15 minutes. Speed in removing material from skin is of extreme importance. Remove and isolate contaminated clothing and shoes at the site. Keep victim quiet and maintain normal body temperature. Effects may be delayed; keep victim under observation.

★ 1004 ★
VINYL ACETATE
CAS: 108-05-4
Other names: ACETATE DE VINYLE (FRENCH); ACETIC ACID, ETHENYL ESTER; ACETIC ACID, ETHENYL ESTER; ACETIC ACID, VINYL ESTER; 1-ACETOXYETHYLENE; ETHENYL ACETATE; ETHENYLETHANOATE; OCTAN WINYLU (POLISH); VAC; VINILE (ACETATO DI) (ITALIAN); VINYLACETAAT (DUTCH); VINYLACETAT (GERMAN); VINYL ACETATE H.Q.; VINYL A MONOMER; VINYLE (ACETATE DE) (FRENCH); VYAC; ZESET T

Danger profile: Fire hazard, explosion hazard, exposure limit established, Community Right-to-Know List, EPA extremely hazardous substances list, poisonous gas produced in fire, storage hazard.
Uses: In latex paints; paper coating; adhesives; textile finishes; chewing gum base; safety glass interlayers; making other chemicals.
Appearance: Clear, colorless watery liquid.
Odor: Initially sweet, pleasant, fruity, but becoming sharp and irritating. The ability to recognize the odor may be lost after about two hours of exposure. The odor is a warning of exposure; however, no smell does not mean you are not being exposed.
Effects of exposure:
　Breathed: May be poisonous. May cause irritation of eyes, nose and throat; sore throat, hoarseness, cough, headache, drowsiness, excitability, drunken behavior, mental confusion, fatigue. Very high levels of exposure may cause dizziness, lightheadedness and cause victim to pass out, go into coma and die.
　Eyes: Contact may cause irritation, pain, burning, stinging, tearing, inflammation of the eyelids. Light may cause pain.
　Skin: May be poisonous if absorbed. May cause irritation dryness and rash; removes oil from the skin and cause dryness.

Persons with pre-existing skin disorders may be more susceptible to the effects of this substance.

Swallowed: Moderately toxic. May cause gastrointestinal irritation. See ***Breathed.***

Long-term exposure: Prolonged or repeated contact may cause irritation of the skin, blisters, burns and dermatitis; eye irritation; lung irritation and/or damage; nervous system changes; heart and liver problems. Human mutation data reported.

Storage: Highly reactive; a storage hazard. Store in sealed containers in a dry, cool, well-ventilated place. Keep away from heat, air, water, direct sunlight, sources of ignition, 2-aminoethanol, peroxides, aldehydes, strong mineral acids, ozone, oxygen, 2-amino ethanol, ethylene diamine, ethylene amine, chlorosulfonic acid and strong oxidizers. Vapors may travel to sources of ignition and flash back. Containers may explode in fire.

First aid guide: Move victim to fresh air and call emergency medical care; if not breathing, give artificial respiration; if breathing is difficult, give oxygen. In case of contact with material, immediately flush eyes with running water for at least 15 minutes. Wash skin with soap and water. Remove and isolate contaminated clothing and shoes at the site.

★ 1005 ★
VINYL BROMIDE

CAS: 593-60-2

Other names: BROMOETHENE; BROMOETHYLENE; BROMOMETHANE; BROMURE DE VINYLE (FRENCH); ETHENE, BROMO-; MONOBROMOETHYLENE; VINILE (BROMURO DI) (ITALIAN); VINYLBROMID (GERMAN); VINYLE (BROMURE DE) (FRENCH)

Danger profile: Fire hazard, carcinogen, exposure limit established, Community Right-to-Know List, poisonous gas produced in fire.

Uses: Making other chemicals and plastics; flame-retarding agent for fibers.

Appearance: Colorless gas.

Effects of exposure:

Breathed: High levels may cause dizziness and lightheadedness.

Eyes: May cause irritation.

Skin: Passes through the unbroken skin; can increase exposure and the severity of the symptoms listed.

Long-term exposure: May cause cancer; tumors. May cause mutations.

Storage: Store in tightly closed containers in a dry, cool, well-ventilated place. Keep away from heat, direct sunlight, sources of ignition and oxidizers.

First aid guide: Move victim to fresh air; call emergency medical care. In case of contact with material, immediately flush skin or eyes with running water for at least 15 minutes. Remove and isolate contaminated clothing and shoes at the site. Keep victim quiet and maintain normal body temperature.

★ 1006 ★
VINYL CHLORIDE

CAS: 75-01-4

Other names: CHLOROETHYLENE; VINYL CHLOROIDE; CHLOROETHENE; CHLOROETHYLENE; CHLORURE DE VINYLE (FRENCH); CHLORO DI VINYLE (ITALIAN); ETHYLENE MONOCHLORIDE; MONOCHLOROETHENE; MONOCHLOROETHYLENE; TROVIDUER; VINYL CHLORIDE MONOMER; VINYL C MONOMER; WINYLU CHLORED (POLISH); VC; VCM; VCL

Danger profile: Carcinogen, mutagen, fire hazard, exposure limit established, reproductive hazard, Community Right-to-Know List, poisonous gas produced in fire.

Uses: Making plastics and other chemicals; refrigerant; adhesives for plastics.

Appearance: Colorless gas or liquid by refrigeration or pressurization.

Odor: Pleasant, faintly sweet, ether-like. The odor is a warning of exposure; however, no smell does not mean you are not being exposed.

Breathed: May be poisonous. May cause irritation of the eyes, nose, throat, air passages and lungs; coughing, headache; increased saliva; reddening of the face; dizziness, lightheadedness, feeling of intoxication, fatigue, loss of consciousness, shock, coma, irregular heartbeat, death. If the victim recovers watch for the possibility of a buildup of fluid in the lungs (pulmonary edema). Pulmonary edema is a medical emergency that can be delayed one to two days following exposure.

Eyes: Contact may cause severe irritation, tearing, inflammation of the eyelids; burning sensation.

Skin: Passes through the unbroken skin; can increase exposure and the severity of the symptoms listed. May cause frostbite, redness, burning feeling, swelling; dry skin; rash, burns and blisters.

Swallowed: Poisonous. Breath may smell of chloroform. May cause irritation of the mouth, gastrointestinal tract, nausea, vomiting, diarrhea with possible blood; drowsiness; loss of consciousness; irregular heartbeat and death. See ***Breathed.***

Long-term exposure: May cause liver cancer; damage to the fetus; liver, spleen and kidney damage; stomach problems; nervous system damage; skin allergy. Liver and spleen damage may occur. May cause club-like swelling and shortening of finger tips. Skin may become thickened and stiff with coarse, whitish patches. Bones and joints of arms and legs may suffer blood vessel damage (Raynaud's syndrome), causing the hands or feet to turn numb, blue or pale with mild cold exposure. Not all symptoms disappear after exposure stops. Spontaneous abortions have occurred among female spouses of workers who have been occupationally exposed to this chemical. Increased rates of birth defects have been reported in areas where production plants of this chemical are located.

Storage: Very dangerous material; obtain specific storage instructions from supplier. Store in sealed containers in a dry, cool, well-ventilated, regulated and marked area. Long contact with air may result in formation of unstable and explosive peroxides. Corrosive to iron at high temperature in presence of water. Keep away from heat, sources of ignition and oxidizing materials. Vapors may travel to sources of ignition and flash back. Containers may explode in fire.

First aid guide: Move victim to fresh air and call emergency medical care; if not breathing, give artificial respiration; if breathing is difficult, give oxygen. In case of frostbite, thaw frosted parts with water. Keep victim quiet and maintain normal body temperature.

★ 1007 ★
VINYL CYCLOHEXENE DIOXIDE
CAS: 106-87-6
Other names: CHISSONOX 206; EP-206; 1,2-EPOXY-4-(EPOXYETHYL)CYCLOHEXANE; 1-EPOXYETHYL-3,4-EPOXYCYCLOHEXANE; 3-(EPOXYETHYL)-7-OXABICYCLO(4.1.0)HEPTANE; 3-(1,2-EPOXYETHYL)-7-OXABICYCLO(4.1.0)HEPTANE; 4-(1,2-EPOXYETHYL)-7-OXABICYCLO(4.1.0)HEPTANE; 4-(EPOXYETHYL)-7-OXABICYCLO(4.1.0)HEPTANE; ERLA-2270; ERLA-2271; 1-ETHYLENEOXY-3,4-EPOXYCYCLOHEXANE; NCI-C60139; 7-OXABICYCLO (4.1.0)HEPTANE, 3-OXIRANYL-; 3-OXIRANYL-7-OXABICYCLO(4.1.0)HEPTENE; UCET TEXTILE FINISH 11-74 (OBS.); UNOX EPOXIDE 206; VINYL CYCLOHEXENE DIEPOXIDE; 4-VINYLCYCLOHEXENE DIEPOXIDE; 4-VINYL-1-CYCLOHEXENE DIEPOXIDE; 4-VINYL-1,2-CYCLOHEXENE DIEPOXIDE; 1-VINYL-3-CYCLOHEXENE DIOXIDE; 4-VINLYCYCLOHEXENE DIOXIDE; 4-VINYL-1-CYCLOHEXENE DIOXIDE

Danger profile: Carcinogen, mutagen, poison, exposure limit established, poisonous gas produced in fire.
Uses: Making polymers, plastics and other chemicals.
Appearance: Colorless liquid.
Effects of exposure:
   Breathed: May cause irritation of the eyes, nose, throat and lungs; congestion.
   Eyes: May cause severe irritation.
   Skin: Moderately toxic. Passes through the unbroken skin; can increase exposure and the severity of the symptoms listed. May cause severe irritation. Prolonged contact may cause burns and blisters. Allergy may develop.
   Swallowed: Moderately toxic. May cause nausea, vomiting.
Long-term exposure: May cause cancer, mutations; male reproductive system damage; skin allergy; change in the production of white blood cells; damage to the thymus.
Storage: Store in tightly closed containers in a dry, cool, well-ventilated place. Keep away from heat, flame and sources of ignition.

★ 1008 ★
VINYL FLUORIDE
CAS: 75-02-5
Other names: FLUOROETHENE; FLUORETHYLENE; MONOFLUROETHYLENE

Danger profile: Fire hazard, poison, exposure limit established, poisonous gas produced in fire.
Uses: Making polymers, plastics, resins and other chemicals.
Appearance: Colorless gas or liquid under pressure.
Effects of exposure:
   Breathed: May cause headache, dizziness, fatigue; rapid irregular breathing; nausea, vomiting, confusion, loss of consciousness convulsions and death.
   Eyes: Contact with liquid may cause frostbite, pain, stinging, tearing, inflammation of the eyelids and permanent damage.
   Skin: Contact with liquid may cause frostbite, intense cold feeling, skin turns white and feels cold and hard.
   Swallowed: Liquid may cause headache, confusion, drowsiness, unconsciousness and death.
Storage: Store in tightly closed cylinders in a dry, cool, well-ventilated place. Keep away from heat, flame and sources of ignition. Containers may explode in fire.
First aid guide: Move victim to fresh air and call emergency medical care; if not breathing, give artificial respiration; if breathing is difficult, give oxygen. In case of frostbite, thaw frosted parts with water. Keep victim quiet and maintain normal body temperature.

★ 1009 ★
VINYLIDENE FLUORIDE
CAS: 75-38-7
Other names: 1,1-DIFLUOROETHYLENE; HALOCARBON 1132A; NCI-C60208; VDF

Danger profile: Suspected carcinogen, fire hazard, explosion hazard, poisonous gas produced in fire.
Uses: Making plastics and other chemicals.
Appearance: Colorless gas.
Odor: Faint, ether-like. The odor is a warning of exposure; however, no smell does not mean you are not being exposed.
Effects of exposure:
   Breathed: Mildly toxic. May cause headache, dizziness, fatigue; rapid irregular breathing; nausea, vomiting, confusion, loss of consciousness convulsions and death.
   Eyes: Contact with liquid may cause frostbite, pain, stinging, tearing, inflammation of the eyelids and permanent damage.
   Skin: Contact with liquid may cause frostbite, intense cold feeling, skin turns white and feels cold and hard.
   Swallowed: Liquid may cause headache, confusion, drowsiness, unconsciousness and death.
Long-term exposure: May cause cancer and mutations.
Storage: Store in tightly closed containers in a dry, cool, well-ventilated place. Keep away from heat, flame, hydrogen chloride, sources of ignition and oxidizers. Vapor may explode if ignited in an enclosed area. Vapors may travel to sources of ignition and flash back. Containers may explode in fire.
First aid guide: Move victim to fresh air and call emergency medical care; if not breathing, give artificial respiration; if breathing is difficult, give oxygen. In case of frostbite, thaw frosted parts with water. Keep victim quiet and maintain normal body temperature.

★ 1010 ★
VINYLIDINE CHLORIDE
CAS: 75-35-4
Other names: CHLORURE DE VINYLIDENE (FRENCH); 1,1-DCE; 1,1-DICHLOROETHENE; 1,1-DICHLOROETHYLENE; ETHENE,1,1-DICHLORO-; NCI-C54262; SCONATEX; VDC; VINYLIDENE CHLORIDE (II); VINYLIDENE DICHLORIDE

Danger profile: Carcinogen, fire hazard, explosion hazard, exposure limit established, reproductive hazard, Community Right-to-Know List, poisonous gas produced in fire.
Uses: Making plastics and other chemicals.
Appearance: Colorless liquid.
Odor: Sweet. The odor is a warning of exposure; however, no smell does not mean you are not being exposed.
Effects of exposure:
   Breathed: May cause irritation of the eyes, nose, throat and air passages; coughing, headache; increased saliva; reddening of the face; dizziness, drunkenness, lightheadedness, fatigue, loss of consciousness, shock, coma, irregular heartbeat, death. If the victim recovers watch for the possibility of a buildup of fluid in the lungs (pulmonary edema). Pulmonary edema is a medical emergency that can be delayed one to two days following exposure.
   Eyes: Contact may cause irritation, tearing, inflammation of the eyelids; burning sensation.
   Skin: Passes through the unbroken skin; can increase exposure and the severity of the symptoms listed. May cause irritation, redness, burning feeling, swelling; dry skin; rash or blisters.
   Swallowed: Poison. Breath may smell of chloroform. May cause irritation of the mouth, gastrointestinal tract, nausea, vomiting, diarrhea with possible blood; drowsiness; loss of consciousness; irregular heartbeat and death. See *Breathed.*

Long-term exposure: May cause cancer, damage to the fetus; reproductive damage in males; bronchitis; liver, kidney and/or lung damage.

Storage: Store in tightly closed containers in a dry, cool, well-ventilated place. Keep away from heat, flames and sources of ignition, copper, aluminum, strong acids and oxidizers. Containers may explode in fire.

★ 1011 ★
VINYL TOLUENE
CAS: 25013-15-4
Other names: BENZENE, ETHENYLMETHYL-; METHYL STYRENE; NCI-C56406; TOLUENE, VINYL-(MIXED ISOMERS), INHIBITED; VINYLTOLUENE

Danger profile: Combustible, exposure limit established, poisonous gas produced in fire.
Uses: Solvent; making other chemicals.
Appearance: Colorless liquid.
Odor: Strong, disagreeable. The odor is a warning of exposure; however, no smell does not mean you are not being exposed.
Effects of exposure:
 Breathed: May be poisonous. May cause irritation to the eyes, nose and throat. High concentrations produce anesthetic effect; with dizziness, lightheadedness and possible loss of consciousness.
 Eyes: Contact may cause irritation.
 Skin: May be poisonous if absorbed. Contact may cause irritation.
 Swallowed: Moderately toxic.
Long-term exposure: May cause liver, kidney and nervous system damage. Mutation data reported.
Storage: Store in tightly closed containers in a dry, cool, well-ventilated place. Keep away from heat, flame, sources of ignition, strong acids and oxidizers. Vapor may explode if ignited in an enclosed area. Vapors may travel to sources of ignition and flash back. Containers may explode in fire.
First aid guide: Move victim to fresh air and call emergency medical care; if not breathing, give artificial respiration; if breathing is difficult, give oxygen. In case of contact with material, immediately flush eyes with running water for at least 15 minutes. Wash skin with soap and water. Remove and isolate contaminated clothing and shoes at the site.

W

★ 1012 ★
WARFARIN
CAS: 81-81-2
Other names: 3-(ALPHA-ACETONYLBENZYL)-4-HYDROXYCOUMARIN; ARAB RAT DETH; ATROMBINE-K; 2H-1-BENZOPYRAN-2-ONE,4-HYDROXY-3-(3-OXO-1-PHE-NYLBUTYL)-; BRUMIN; COMPOUND 42; D-CON; CORAX; COUMADIN; COUMARIN,3-(ALPHA-ACETONYLBENZYL)-4-HYDROXY-; COUMAFENE; DETHMORE; EASTERN STATES DUOCIDE; KUMANDER; LIQUA-TOX; MOUSE PAK; PROTHROMADIN; RAT-A-WAY; RAT-B-GON; RAT-O-CIDE; RAT-GARD; RAT & MICE BAIT; RATRON; RATS-NO-MORE; RO-DETH; ROUGH & READY MOUSE MIX; SOLFARIN; SPRAY-TROL BRANCH RODEN-TROL; TWIN LIGHT RAT AWAY; VAMPIRINIP; ZOOCOUMARING (RUSSIAN)

Danger profile: Reproductive hazard, poison, exposure limit established, EPA extremely hazardous substances list, poisonous gas produced in fire.

Uses: Rodent control; oral anti-coagulant.
Appearance: Colorless solid.
Effects of exposure:
 Breathed: May be poisonous. High exposure may reduce the ability of blood to clot; causing hemorrhage; easy bruising, pallor, bleeding nose, gums, bruises, bloody urine and/or feces; internal bleeding (hematoma) around the joints and buttocks. Symptoms may be delayed a few days.
 Skin: May be poisonous if absorbed. Passes through the unbroken skin; can increase exposure and the severity of the symptoms listed.
 Swallowed: May be poisonous. See symptoms under *Breathed.*
Long-term exposure: Repeated exposures may cause a decrease of fertility in females; damage to the fetus; anemia; kidney and liver damage; reduction of the blood's ability to clot; hemorrhage; easy bruising, pallor, bleeding nose, gums, bruises, bloody urine and/or feces; internal bleeding (hematoma) around the joints and buttocks.
Storage: Store in tightly closed containers in a dry, cool, well-ventilated place. Keep away from oxidizers, strong acids, strong bases.

X

★ 1013 ★
XYLENE
CAS: 1330-20-7
Other names: BENZENE, DIMETHYL; DIMETHYLBENZENE; KSYLEN (POLISH); METHYL TOLUENE; NCI-C55232; VIOLET 3; XILOLI (ITALIAN); XYLENEN (DUTCH); XYLOL; XYLOLE (GERMAN)

Danger profile: Fire hazard, exposure limit established, Community Right-to-Know List, poisonous gas produced in fire.
Uses: Aviation gasoline; protective coatings; solvent for alkyd resins, lacquers, enamels, rubber cements; synthesis of organic chemicals.
Appearance: Clear, colorless or light-colored liquid. May form crystals at temperatures below 57°F.
Odor: Pleasant, benzene-like. The odor is a warning of exposure; however, no smell does not mean you are not being exposed.
Effects of exposure:
 Breathed: May cause irritation of the eyes, nose and throat; dizziness, lightheadedness, rapid breathing, excitement, mental confusion, unsteady walk, headache, fatigue, drowsiness, nausea, vomiting, unconsciousness, slurred speech, coma and death. High levels may cause liver and/or kidney damage and/or lung congestion.
 Eyes: May cause burning feeling, tearing, irritation, redness and inflammation of the eyelids.
 Skin: Passes through the unbroken skin; can increase exposure and the severity of the symptoms listed. Vapor or liquid may cause irritation of the skin, dryness, redness and rash.
 Swallowed: May cause immediate burning sensation in the mouth and throat; indigestion, sharp stomach pains, nausea, vomiting, dizziness, fatigue, unconsciousness. Drinking alcohol may make the effects worse.
Long-term exposure: May cause liver and kidney damage; intestinal tract disturbances; central nervous system depression; bone marrow damage; dermatitis, dry and cracking skin; eye damage, redness and swelling; brain effects. Many of these effects are reversible and may disappear when the chemical is removed. May cause reproductive effects.

Storage: Store in tightly closed containers in a dry, cool, well-ventilated place. Keep away from heat, flames, sources of ignition and oxidizers such as chlorine and permanganate. Protect containers from physical damage. Vapor may explode if ignited in an enclosed area. Vapors may travel to sources of ignition and flash back. Containers may explode in fire.

First aid guide: Move victim to fresh air and call emergency medical care; if not breathing, give artificial respiration; if breathing is difficult, give oxygen. In case of contact with material, immediately flush eyes with running water for at least 15 minutes. Wash skin with soap and water. Remove and isolate contaminated clothing and shoes at the site.

★ 1014 ★
M-XYLENE
CAS: 108-38-3
Other names: M-DIMETHYLBENZENE; 1,3-DIMETHYLBENZENE; META-XYLENE; 1,3-XYLENE; M-XYLOL

Danger profile: Fire hazard, explosion hazard, exposure limit established, Community Right-to-Know List, poisonous gas produced in fire.
Uses: Solvent; making dyes, insecticides and other chemicals; in aviation fuels. A common air contaminant.
Appearance: Clear, colorless liquid.
Odor: Strong, pleasant. The odor is a warning of exposure; however, no smell does not mean you are not being exposed.
Effects of exposure:
 Breathed: May cause irritation of the eyes, nose and throat; dizziness, lightheadedness, rapid breathing, mental confusion, unsteady walk, headache, fatigue, nausea, vomiting, unconsciousness, coma and death.
 Eyes: May cause burning feeling, tearing, irritation, redness and inflammation of the eyelids.
 Skin: Passes through the unbroken skin; can increase exposure and the severity of the symptoms listed. May cause irritation of the skin, dryness, redness and rash.
 Swallowed: May cause indigestion, dizziness, fatigue, unconsciousness. Drinking alcohol may worsen effects.
Long-term exposure: May cause liver and kidney damage; bone marrow damage; dermatitis, dry and cracking skin; eye damage, redness and swelling; brain effects. May cause reproductive effects.
Storage: Store in tightly closed containers in a dry, cool, well-ventilated place. Keep away from heat, flames, sources of ignition and oxidizers. Vapor may explode if ignited in an enclosed area. Vapors may travel to sources of ignition and flash back. Containers may explode in fire.

★ 1015 ★
M-XYLENE-ALPHA,ALPHA'-DIAMINE
CAS: 1477-55-0
Other names: 1,3-BIS-AMINOMETHYLBENZEN (CZECH); ALPHA,ALPHA'-DIAMINO-M-XYLENE; METHYLAMINE, M-PHENYLENEBIS-; MXDA; M-PHENYLENEBIS(METHYLAMINE); M-XYLYLENDIAMIN (CZECH); M-XYLENE-A,A'-DIAMINE

Danger profile: Exposure limit established, poisonous gas produced in fire.
Uses: Making plastics, synthetic fibers for textiles, resins and rubber; curing agent for epoxy resins.
Appearance: Colorless liquid.
Effects of exposure:
 Breathed: May be poisonous. May cause irritation of the nose and throat.
 Eyes: Contact may cause severe irritation and burns.

 Skin: May be poisonous. Passes through the unbroken skin; can increase exposure and the severity of the symptoms listed. Contact may cause severe irritation and burns.
 Swallowed: May be poisonous.
Long-term exposure: Skin and lung allergies may develop. Possible kidney and liver damage.
Storage: Store in tightly closed containers in a dry, cool, well-ventilated place. Keep away from strong acids and oxidizers. Containers may explode in heat of fire.

★ 1016 ★
O-XYLENE
CAS: 95-47-6
Other names: O-DIMETHYLBENZENE; 1,2-DIMETHYLBENZENE; O-METHYLTOLUENE; 1,2-XYLENE; ORTHO-XYLENE; O-XYLOL

Danger profile: Fire hazard, explosion hazard, exposure limit established, Community Right-to-Know List, poisonous gas produced in fire.
Uses: Making other chemicals, pharmaceuticals, vitamins, dyes, insecticides, motor fuels. A common air contaminant.
Appearance: Colorless liquid.
Odor: Strong, pleasant. The odor is a warning of exposure; however, no smell does not mean you are not being exposed.
Effects of exposure:
 Breathed: May cause irritation of the eyes, nose and throat; dizziness, lightheadedness, rapid breathing, mental confusion, unsteady walk, headache, fatigue, nausea, vomiting, unconsciousness, coma and death.
 Eyes: May cause burning feeling, tearing, irritation, redness and inflammation of the eyelids.
 Skin: Passes through the unbroken skin; can increase exposure and the severity of the symptoms listed. May cause irritation of the skin, dryness, redness and rash.
 Swallowed: May cause indigestion, dizziness, fatigue, unconsciousness. Drinking alcohol may worsen effects.
Long-term exposure: May cause liver and kidney damage; bone marrow damage; dermatitis, dry and cracking skin; eye damage, redness and swelling; brain effects. May cause reproductive effects.
Storage: Store in tightly closed containers in a dry, cool, well-ventilated place. Keep away from heat, flames, sources of ignition and oxidizers. Vapor may explode if ignited in an enclosed area. Vapor may explode if ignited in an enclosed area. Vapors may travel to sources of ignition and flash back. Containers may explode in fire.

★ 1017 ★
P-XYLENE
CAS: 106-42-3
Other names: CHROMAR; P-DIMETHYLBENZENE; 1,4-DIMETHYLBENZENE; P- METHYLTOLUENE; SCINTILLAR; PARA-XYLENE; 1,4-XYLENE; P-XYLOL

Danger profile: Fire hazard, exposure limit established, Community Right-to-Know List, poisonous gas produced in fire.
Uses: Making polyester fibers and resins, vitamins, pharmaceuticals, insecticides.
Appearance: Colorless liquid. One of the highest volume chemicals produced in the U.S.
Odor: Strong, pleasant. The odor is a warning of exposure; however, no smell does not mean you are not being exposed.
Effects of exposure:
 Breathed: May cause irritation of the eyes, nose and throat; dizziness, lightheadedness, rapid breathing, mental confu-

sion, unsteady walk, headache, fatigue, nausea, vomiting, unconsciousness, coma and death.
Eyes: May cause burning feeling, tearing, irritation, redness and inflammation of the eyelids.
Skin: Passes through the unbroken skin; can increase exposure and the severity of the symptoms listed. May cause irritation of the skin, dryness, redness and rash.
Swallowed: May cause indigestion, dizziness, fatigue, unconsciousness. Drinking alcohol may worsen effects.
Long-term exposure: May cause liver and kidney damage; bone marrow damage; dermatitis, dry and cracking skin; eye damage, redness and swelling; brain effects. May cause reproductive effects.
Storage: Store in tightly closed containers in a dry, cool, well-ventilated place. Keep away from heat, flames, sources of ignition, acetic acid and oxidizers. Vapor may explode if ignited in an enclosed area. Vapors may travel to sources of ignition and flash back. Containers may explode in fire.

★ 1018 ★
XYLENOL
CAS: 1300-71-6
Other names: DIMETHYLPHENOL; PHENOL, DIMETHYL-; XILENOLI (ITALIAN); XYLENOLEN (DUTCH)

Danger profile: Combustible, poisonous gas produced in fire.
Uses: Making disinfectants, solvents, pharmaceuticals, insecticides, fungicides, plasticizers and other chemicals; additives to lubricants and gasolines; dyestuffs.
Appearance: White solid needle-like crystals.
Odor: Sweet, tarry. The odor is a warning of exposure; however, no smell does not mean you are not being exposed.
Effects of exposure:
Breathed: May cause irritation of the eyes, nose, throat and lungs. Readily absorbed through the mucous membranes producing toxic symptoms of weakness, dizziness, headache, difficult breathing and twitching.
Eyes: Contact may cause severe irritation, burns and permanent damage.
Skin: Toxic. Passes through the unbroken skin; can increase exposure and the severity of the symptoms listed. Contact may cause severe burns, temporary prickling, intense burning, then numbness and local anesthesia. Affected areas initially show white discolartion, wrinkling and softening, then become red, then brown or black. This is a sign of gangrene. Extensive burns may permit absorption of chemical to produce toxic symptoms described under *Breathed.*
Swallowed: Toxic. May cause irritation of mouth and stomach, nausea, abdominal pain, weakness, dizziness, headache, difficult breathing, and twitching.
Long-term exposure: May lead to vomiting, headaches, dizziness and fainting. May cause liver and kidney damage. Possible lung damage.
Storage: Store in tightly closed containers in a dry, cool, well-ventilated place. Keep away from flames and sources of ignition and oxidizers. Containers may explode in heat of fire.
First aid guide: Move victim to fresh air and call emergency medical care; if not breathing, give artificial respiration; if breathing is difficult, give oxygen. In case of contact with material, immediately flush skin or eyes with running water for at least 15 minutes. Speed in removing material from skin is of extreme importance. Remove and isolate contaminated clothing and shoes at the site. Keep victim quiet and maintain normal body temperature. Effects may be delayed; keep victim under observation.

★ 1019 ★
2,3-XYLENOL
CAS: 526-75-0
Other names: 2,3-DIMETHYLPHENOL; O-XYLENOL

Danger profile: Combustible, poisonous gas produced in fire.
Uses: Making disinfectants, solvents, pharmaceuticals, insecticides, fungicides, plasticizers and other chemicals.
Appearance: White solid needle-like crystals.
Odor: Sweet, tarry. The odor is a warning of exposure; however, no smell does not mean you are not being exposed.
Effects of exposure:
Breathed: May cause irritation of the eyes, nose, throat and lungs. Readily absorbed through the mucous membranes producing toxic symptoms of weakness, dizziness, headache, difficult breathing and twitching.
Eyes: Contact may cause severe irritation, burns and permanent damage.
Skin: Toxic. Passes through the unbroken skin; can increase exposure and the severity of the symptoms listed. Contact may cause severe burns, temporary prickling, intense burning, then numbness and local anesthesia. Affected areas initially show white discolartion, wrinkling and softening, then become red, then brown or black. This is a sign of gangrene. Extensive burns may permit absorption of chemical to produce toxic symptoms described under *Breathed.*
Swallowed: Toxic. May cause irritation of mouth and stomach, nausea, abdominal pain, weakness, dizziness, headache, difficult breathing, and twitching.
Long-term exposure: May lead to vomiting, headaches, dizziness and fainting. May cause liver and kidney damage. Possible lung damage.
Storage: Store in tightly closed containers in a dry, cool, well-ventilated place. Keep away from flames and sources of ignition and oxidizers. Containers may explode in heat of fire.

★ 1020 ★
2,4-XYLENOL
CAS: 105-67-9
Other names: 2,4-DIMETHYLPHENOL; 4,6-DIMETHYLPHENOL; 1-HYDROXY-2,4- DIMETHYLBENZENE; M-XYLENOL

Danger profile: Combustible, poisonous gas produced in fire.
Uses: Making disinfectants, solvents, pharmaceuticals, insecticides, fungicides, plasticizers and other chemicals; additives to lubricants and gasolines; dyestuffs.
Appearance: White solid needle-like crystals.
Odor: Sweet, tarry. The odor is a warning of exposure; however, no smell does not mean you are not being exposed.
Effects of exposure:
Breathed: May cause irritation of the eyes, nose, throat and lungs; readily absorbed through the mucous membranes producing toxic symptoms of weakness, dizziness, headache, difficult breathing and twitching.
Eyes: Contact may cause severe irritation, burns and permanent damage.
Skin: Toxic. Passes through the unbroken skin; can increase exposure and the severity of the symptoms listed. Contact may cause severe burns, temporary prickling, intense burning, then numbness and local anesthesia. Affected areas initially show white discolartion, wrinkling and softening, then become red, then brown or black. This is a sign of gangrene. Extensive burns may permit absorption of chemical to produce toxic symptoms described under *Breathed.*
Swallowed: Toxic. May cause irritation of mouth and stomach, nausea, abdominal pain, weakness, dizziness, headache, difficult breathing, and twitching.

Long-term exposure: May lead to vomiting, headaches, dizziness and fainting. May cause liver and kidney damage. Possible lung damage.

Storage: Store in tightly closed containers in a dry, cool, well-ventilated place. Keep away from flames and sources of ignition and oxidizers. Containers may explode in heat of fire.

★ 1021 ★
2,5-XYLENOL
CAS: 95-87-4
Other names: 2,5-DIMETHYLPHENOL; 3,6-DIMETHYLPHENOL; 2,5-DMP; 6- METHYL-M-CRESOL; P-XYLENOL; 1,2,5-XYLENOL

Danger profile: Combustible, poisonous gas produced in fire.
Uses: Making disinfectants, solvents, pharmaceuticals, insecticides, fungicides, plasticizers and other chemicals; additives to lubricants and gasolines; dyestuffs.
Appearance: White solid needle-like crystals.
Odor: Sweet, tarry. The odor is a warning of exposure; however, no smell does not mean you are not being exposed.
Effects of exposure:
> **Breathed:** May cause irritation of the eyes, nose, throat and lungs. Readily absorbed through the mucous membranes producing toxic symptoms of weakness, dizziness, headache, difficult breathing and twitching.
> **Eyes:** Contact may cause severe irritation, burns and permanent damage.
> **Skin:** Toxic. Passes through the unbroken skin; can increase exposure and the severity of the symptoms listed. Contact may cause severe burns, temporary prickling, intense burning, then numbness and local anesthesia. Affected areas initially show white discolartion, wrinkling and softening, then become red, then brown or black. This is a sign of gangrene. Extensive burns may permit absorption of chemical to produce toxic symptoms described under **Breathed.**
> **Swallowed:** Toxic. May cause irritation of mouth and stomach, nausea, abdominal pain, weakness, dizziness, headache, difficult breathing, and twitching.

Long-term exposure: May lead to vomiting, headaches, dizziness and fainting. May cause liver and kidney damage. Possible lung damage.
Storage: Store in tightly closed containers in a dry, cool, well-ventilated place. Keep away from flames and sources of ignition and oxidizers. Containers may explode in heat of fire.

★ 1022 ★
2,6-XYLENOL
CAS: 546-26-1
Other names: 2,6-DIMETHYLPHENOL; 2,6-DMP

Danger profile: Combustible, poisonous gas produced in fire.
Uses: Making disinfectants, solvents, pharmaceuticals, insecticides, fungicides, plasticizers and other chemicals.
Appearance: White solid needle-like crystals.
Odor: Sweet, tarry. The odor is a warning of exposure; however, no smell does not mean you are not being exposed.
Effects of exposure:
> **Breathed:** May cause irritation of the eyes, nose, throat and lungs. Readily absorbed through the mucous membranes producing toxic symptoms of weakness, dizziness, headache, difficult breathing and twitching.
> **Eyes:** Contact may cause severe irritation, burns and permanent damage.
> **Skin:** Toxic. Passes through the unbroken skin; can increase exposure and the severity of the symptoms listed. Contact may cause severe burns, temporary prickling, intense burning, then numbness and local anesthesia. Affected areas ini-

tially show white discolartion, wrinkling and softening, then become red, then brown or black. This is a sign of gangrene. Extensive burns may permit absorption of chemical to produce toxic symptoms described under **Breathed.**
> **Swallowed:** Toxic. May cause irritation of mouth and stomach, nausea, abdominal pain, weakness, dizziness, headache, difficult breathing, and twitching.

Long-term exposure: May lead to vomiting, headaches, dizziness and fainting. May cause liver and kidney damage. Possible lung damage.
Storage: Store in tightly closed containers in a dry, cool, well-ventilated place. Keep away from flames and sources of ignition and oxidizers. Containers may explode in fire.

★ 1023 ★
3,4-XYLENOL
CAS: 95-65-8
Other names: 3,4-DIMETHYLPHENOL; 3,4-DMP

Danger profile: Combustible, poisonous gas produced in fire.
Uses: Making disinfectants, solvents, pharmaceuticals, insecticides, fungicides, plasticizers and other chemicals.
Appearance: White solid needle-like crystals.
Odor: Sweet, tarry. The odor is a warning of exposure; however, no smell does not mean you are not being exposed.
Effects of exposure:
> **Breathed:** May cause irritation of the eyes, nose, throat and lungs. Readily absorbed through the mucous membranes producing toxic symptoms of weakness, dizziness, headache, difficult breathing and twitching.
> **Eyes:** Contact may cause severe irritation, burns and permanent damage.
> **Skin:** Toxic. Passes through the unbroken skin; can increase exposure and the severity of the symptoms listed. Contact may cause severe burns, temporary prickling, intense burning, then numbness and local anesthesia. Affected areas initially show white discolartion, wrinkling and softening, then become red, then brown or black. This is a sign of gangrene. Extensive burns may permit absorption of chemical to produce toxic symptoms described under **Breathed.**
> **Swallowed:** Toxic. May cause irritation of mouth and stomach, nausea, abdominal pain, weakness, dizziness, headache, difficult breathing, and twitching.

Long-term exposure: May lead to vomiting, headaches, dizziness and fainting. May cause liver and kidney damage. Possible lung damage.
Storage: Store in tightly closed containers in a dry, cool, well-ventilated place. Keep away from flames and sources of ignition and oxidizers. Containers may explode in heat of fire.

★ 1024 ★
3,5-XYLENOL
CAS: 108-68-9
Other names: 3,5-DIMETHYLPHENOL; 3,5-DMP; 1,3,5-XYLENOL

Danger profile: Combustible, poisonous gas produced in fire.
Uses: Making disinfectants, solvents, pharmaceuticals, insecticides, fungicides, plasticizers and other chemicals.
Appearance: White solid needle-like crystals.
Odor: Sweet, tarry. The odor is a warning of exposure; however, no smell does not mean you are not being exposed.
Effects of exposure:
> **Breathed:** May cause irritation of the eyes, nose, throat and lungs. Readily absorbed through the mucous membranes producing toxic symptoms of weakness, dizziness, headache, difficult breathing and twitching.

Eyes: Contact may cause severe irritation, burns and permanent damage.

Skin: Toxic. Passes through the unbroken skin; can increase exposure and the severity of the symptoms listed. Contact may cause severe burns, temporary prickling, intense burning, then numbness and local anesthesia. Affected areas initially show white discolartion, wrinkling and softening, then become red, then brown or black. This is a sign of gangrene. Extensive burns may permit absorption of chemical to produce toxic symptoms described under **Breathed.**

Swallowed: Poison. May cause irritation of mouth and stomach, nausea, abdominal pain, weakness, dizziness, headache, difficult breathing, and twitching.

Long-term exposure: May lead to vomiting, headaches, dizziness and fainting. May cause liver and kidney damage. Possible lung damage.

Storage: Store in tightly closed containers in a dry, cool, well-ventilated place. Keep away from flames and sources of ignition and oxidizers. Containers may explode in heat of fire.

★ 1025 ★
XYLIDINE
CAS: 1300-73-8

Other names: ACID LEATHER BROWN 2G; ACID ORANGE 24; AMINODIMETHYLBENZENE; BENZENAMINE, AR,AR-DIMETHYL-; 11460 BROWN; DIMETHYLANILINE; DIMETHYLPHENYLAMINE; RESORCINE BROWN J; RESORCINE BROWN R; XILIDINE (ITALIAN); XYLIDINEN (DUTCH)

Danger profile: Carcinogen, poison, combustible, exposure limit established, poisonous gas produced in fire.
Uses: Making dyes, pharmaceuticals and other chemicals.
Appearance: Pale yellow to brown liquid.
Odor: Weak, ammonia-like. The odor is a warning of exposure; however, no smell does not mean you are not being exposed.
Effects of exposure:
Breathed: Poisonous. Highly toxic, severe exposure may occur before many or any of the following symptoms appear. May interfere with the blood's capacity for carrying oxygen, causing headaches, dizziness, nausea and cyanosis (a bluish color the skin, fingernails and lips). High levels may cause difficult breathing, weakness, drowsiness, tremors, collapse, liver damage and possible death.
Skin: May be poisonous if absorbed. Passes through the unbroken skin; can increase exposure and the severity of the symptoms listed. Severe and possibly fatal effects can result without warning signs through skin absorption.
Swallowed: Poisonous.
Long-term exposure: May cause liver and blood damage; central nervous system depression. Possible kidney and lung effects.
Storage: Store in tightly closed containers in a dry, cool, well-ventilated place. Keep away from heat, flame, sources of ignition, bleaches containing hypochlorite and oxidizing materials. Containers may explode in fire.
First aid guide: Move victim to fresh air and call emergency medical care; if not breathing, give artificial respiration; if breathing is difficult, give oxygen. In case of contact with material, immediately flush skin or eyes with running water for at least 15 minutes. Speed in removing material from skin is of extreme importance. Remove and isolate contaminated clothing and shoes at the site. Keep victim quiet and maintain normal body temperature. Effects may be delayed; keep victim under observation.

★ 1026 ★
2,4-XYLIDINE
CAS: 95-68-1

Other names: 1-AMINO-2,4-DIMETHYLBENZENE; 4-AMINO-1,3-DIMETHYLBENZENE; 4-AMINO-3-METHYLTOLUENE; 4-AMINO-1,3-XYLENE; 2,4-DIMETHYLANILINE; 2,4-DIMETHYLBENZENAMINE; 2,4-DIMETHYLPHENYLAMINE; 2-METHYL-P-TOUIDINE; 4-METHYL-O-TOLUIDINE; M-XYLIDINE; M-4-XYLIDINE

Danger profile: Carcinogen, possible mutagen, poison, poisonous gas produced in fire.
Uses: Making dyes, pharmaceuticals and organic compounds.
Appearance: Liquid.
Odor: No desc given The odor is a warning of exposure; however, no smell does not mean you are not being exposed.
Effects of exposure:
Breathed: May be poisonous. May cause irritation of the eyes, nose and throat; dizziness, lightheadedness, rapid breathing, mental confusion, unsteady walk, headache, fatigue, nausea, vomiting, unconsciousness, coma and death.
Eyes: May cause burning feeling, tearing, irritation, redness and inflammation of the eyelids.
Skin: May be poisonous. Passes through the unbroken skin; can increase exposure and the severity of the symptoms listed. May cause irritation of the skin, dryness, redness and rash.
Swallowed: Poison. May cause indigestion, dizziness, fatigue, unconsciousness.
Long-term exposure: May cause cancer. Mutation data reported.
Storage: Store in tightly closed containers in a dry, cool, well-ventilated place. Containers may explode in fire.

Z

★ 1027 ★
ZEOLITE
CAS: 1318-02-1

Danger profile: Long-term health effects.
Uses: In water softening, detergents; in drying agents.
Appearance: Crystalline solid.
Effects of exposure:
Breathed: No short-term effects are listed at this time. Health effects are gradual over months or years.
Long-term exposure: May cause lung scarring, with difficulty in breathing.
Storage: Store in tightly closed containers in a dry, cool, well-ventilated place. Containers may explode in fire.

★ 1028 ★
ZINC
CAS: 7440-66-6

Other names: ASAREO L15; BLUE POWDER; C.I. 77945; C.I.PIGMENT BLACK 16; C.I. PIGMENT METAL 6; EMANAY ZINC DUST; GRANULAR ZINC; JASAD; MERRILLITE; PASCO; ZINC DUST; ZINC POWDER

Danger profile: Combustible, Community Right-to-Know List, poisonous gas produced in fire.
Uses: Galvanized coatings on iron and steel sheet metal; making paint, dyes and brass metal alloys; in casting, printing, batter-

ies and electrical equipment; laboratory chemical; nutritional trace element.
Appearance: Soft white metal with a bluish tinge.
Effects of exposure:
Breathed: Fumes may result in sweet taste and cause throat dryness, cough, weakness, generalized aching, chills, fever, nausea, vomiting. Pure zinc in the form of powder, dust or fume is relatively non-toxic to humans. See **Long-term exposure.**
Eyes: Dust particles may cause irritation; metal fragments may cause eye scratches.
Skin: Passes through the unbroken skin; can increase exposure and the severity of the symptoms listed. May cause irritation.
Swallowed: One third ounce of zinc metal taken over two days has caused sluggishness, lightheadedness, a staggered gait and difficulty in writing. See *Breathed.*
Long-term exposure: Although chronic health effects are not known at this time, difficulties arise from oxidation of zinc (zinc oxide) fumes prior to inhalation, or the presence of impurities such as lead, arsenic, cadmium, antimony. Some of these impurities cause cancer and are poisonous.
Storage: Dry powder may ignite spontaneously in air. Store in tightly closed containers in a dry, cool, well-ventilated place. Keep away from heat, flame, sources of ignition, water, acids, chromic anhydride, halocarbons, manganese chloride, chlorates, chlorinated rubbers, chlorine, magnesium, alkali hydroxides, oxidants and other substances that react with chemically active metals.

★ **1029** ★
ZINC BORATE
CAS: 1332-07-6
Other names: BORAX 2335; BORIC ACID, ZINC SALT; ZB 112; ZB 237; ZN 100

Danger profile: Community Right-to-Know List, poisonous gas produced in fire.
Uses: Medicine; fire proofing textiles; fungicide; mildewcide; making ceramics.
Appearance: White powder.
Effects of exposure:
Breathed: May cause irritation of the nose and throat. High exposures may cause poisoning.
Eyes: May cause irritation.
Skin: May cause irritation.
Swallowed: May cause gastrointestinal disturbances, convulsions, central nervous system depression, skin eruptions, shock and death.
Long-term exposure: May cause poisoning.
Storage: Store in tightly closed containers in a dry, cool, well-ventilated place.

★ **1030** ★
ZINC BROMIDE
CAS: 7699-45-8
Other names: ZINC DIBROMIDE

Danger profile: Community Right-to-Know List, poisonous gas produced in fire.
Uses: Making photographic emulsions, rayon. In shielding viewing windows for nuclear reactors.
Appearance: White granular powder.
Effects of exposure:
Breathed: May cause irritation to the nose, throat and air passages.
Eyes: Contact may cause severe irritation and burns.

Skin: Passes through the unbroken skin; can increase exposure and the severity of the symptoms listed. Contact may cause severe irritation and burns.
Swallowed: May cause irritation or corrosion of the mouth, throat and digestive tract; violent vomiting, severe stomach pain, diarrhea, shock and collapse. Scars may form on the throat and stomach. If a large amount is swallowed and not thrown up, drowsiness and other symptoms of bromide poisoning may occur. Long lasting kidney irritation may occur. Less than an ounce may cause death.
Long-term exposure: May cause bromide poisoning leading to depression, loss of appetite, fatigue; acne-like skin rash. Possible lung effects.
Storage: Store in tightly closed containers in a dry, cool, well-ventilated place. Keep away from high heat, sodium and potassium. Protect containers from physical damage.

★ **1031** ★
ZINC CARBONATE
CAS: 3486-35-9
Other names: CARBONIC ACID, ZINC SALT (1:1)

Danger profile: Community Right-to-Know List, poisonous gas produced in fire.
Uses: In ceramics, cosmetics and lotions; pharmaceutical ointments, dusting powders; medicine, topical antiseptic.
Appearance: Colorless to white powder.
Effects of exposure:
Breathed: Dust or fumes may cause dry throat, cough and chest discomfort; fever and sweating.
Eyes: Contact may cause severe irritation and burns.
Skin: Passes through the unbroken skin; can increase exposure and the severity of the symptoms listed. Contact may cause severe irritation and burns.
Swallowed: May cause nausea and vomiting.
Storage: Store in tightly closed containers in a dry, cool, well-ventilated place.

★ **1032** ★
ZINC CHLORATE
CAS: 10361-95-2
Other names: CHLORIC ACID, ZINC SALT

Danger profile: Explosion hazard, Community Right-to-Know List, poisonous gas produced in fire.
Uses: Making other chemicals, adhesives and explosives.
Appearance: Colorless to yellow powder.
Effects of exposure:
Breathed: May interfere with the blood's ability to carry oxygen; headaches, dizziness, fatigue, bluish color to the skin, lips and fingernails; collapse and possible death. High exposure may also cause kidney and red cell damage. Symptoms may be delayed; and, in the event of kidney damage, transfusion may be required.
Eyes: Contact may cause irritation.
Skin: Passes through the unbroken skin; can increase exposure and the severity of the symptoms listed. Contact may cause irritation.
Long-term exposure: May cause kidney damage.
Storage: Highly reactive; a powerful oxidizer. Store in tightly closed containers in a dry, cool, well-ventilated place with non-wood floors. Keep away from combustible materials, aluminum, arsenic, charcoal, copper, sulfur, phosphorus, manganese oxide, copper(II)sulfide and other metal sulfides, sulfuric acid and organic compounds. Containers may explode in fire.
First aid guide: Move victim to fresh air; call emergency medical care. In case of contact with material, immediately flush skin or

eyes with running water for at least 15 minutes. Remove and isolate contaminated clothing and shoes at the site.

★ 1033 ★
ZINC CHLORIDE
CAS: 7646-85-7
Other names: BUTTER OF ZINC; CHLORURE DE ZINC (FRENCH); TINNING GLUX; ZINC BUTTER; ZINC CHLORIDE FUME; ZINC (CHLORURE DE) (FRENCH); ZINC MURIATE SOLUTION; ZINCO (CLORURO DI) (ITALIAN); ZINC DICHLORIDE; ZINKCHLORID (GERMAN); ZINKCHLORIDE (DUTCH)

Danger profile: Mutagen, corrosive, exposure limit established, Community Right-to-Know List, poisonous gas produced in fire.
Uses: Making other chemicals and textiles; fire proofing agent; preserving wood; in soldering fluxes and smoke screens; burnishing and polishing compounds for steel; electroplating; antiseptics; deodorants; dyeing textile ; adhesives; dental cements; glass etching; dentifrices; embalming and taxidermists fluids; medicine.
Appearance: White crystals or fumes; or, colorless solution.
Odor: Fumes are acrid. The odor is a warning of exposure; however, no smell does not mean you are not being exposed.
Effects of exposure:
Breathed: May cause severe irritation to the nose, throat and lungs, causing coughing, shortness of breath. May cause flu-like symptoms including chills, muscle ache, headache, fever. High exposure may cause nausea, salivation, vomiting, cramps, diarrhea, chest pains, cough, a build-up of fluids in the lungs (pulmonary edema) and possible death. Pulmonary edema may be delayed from 1-2 days following exposure.
Eyes: May cause severe burning irritation, extreme pain, chemical burns, redness and swelling. Permanent damage and blindness amy result.
Skin: Passes through the unbroken skin; can increase exposure and the severity of the symptoms listed. May cause severe irritation and thermal and acid burns, especially if skin is wet or broken.
Swallowed: May cause intoxication, severe irritation of stomach, corrosive burns of the mouth and stomach, nausea, vomiting, swelling of the throat, watery or bloody diarrhea. Damage to kidneys and liver, collapse and convulsions can result.
Long-term exposure: May cause mutations; permanent lung damage. Cancer and reproductive data reported.
Storage: Store in tightly closed containers in a dry, cool, well-ventilated place. Keep away from potassium.
First aid guide: Move victim to fresh air; call emergency medical care. In case of contact with material, immediately flush skin or eyes with running water for at least 15 minutes. Remove and isolate contaminated clothing and shoes at the site. Keep victim quiet and maintain normal body temperature.

★ 1034 ★
ZINC CHROMATE
CAS: 13530-65-9
Other names: BASIC ZINC CHROMATE; BASIC ZINC CHROMATE X-2259; BUTTERCUP YELLOW; CHROMIC ACID, ZINC SALT; CHROMIUM ZINC OXIDE; C.I. 77955; C.I. PIGMENT YELLOW 36; CITRON YELLOW; C.P. ZINC YELLOW X-883; PRIMROSE YELLOW; PURE ZINC CHROME; PURE ZINC YELLOW; ZINC CHROMATE C; ZINC CHROMATE O; ZINC CHROMATE T; ZINC CHROMATE(VI) HYDROXIDE; ZINC CHROMATE Z; ZINC CHROME; ZINC CHROME (ANTI-CORROSION); ZINC CHROME YELLOW; ZINC CHROMIUM OXIDE; ZINC HYDROXYCHROMATE; ZINC TETRAOXYCHROMATE; ZINC TETRAOXYCHROMATE 76A; ZINC TETRAOXYCHROMATE 780B; ZINC TETROXYCHROMATE; ZINC YELLOW; ZINC YELLOW 1; ZINC YELLOW 1425; ZINC YELLOW 40-9015; ZINC YELLOW AZ-16; ZINC YELLOW AZ-18; ZINC YELLOW KSH; ZINC YELLOW 386N; ZINC YELLOWS

Danger profile: Carcinogen, combustible, exposure limit established, Community Right-to-Know List, poisonous gas produced in fire.
Uses: In rust inhibiting states, artists colors, varnishes, lacquers; making linoleum.
Appearance: Yellow crystalline powder.
Effects of exposure:
Breathed: High levels may cause irritation of the nose, throat and breathing passages; nasal septum irritation.
Eyes: Contact may cause irritation and inflammation of the eyelids.
Skin: Passes through the unbroken skin; can increase exposure and the severity of the symptoms listed. May cause irritation, rash and skin ulcers. Possible skin allergy.
Swallowed: May cause burns of the mouth and stomach; corrosion of the alimentary tract. Possible circulatory collapse and death.
Long-term exposure: May cause cancer; liver and kidney damage; skin allergy; nose bleeding and sores leading to a hole in the nasal septum. Mutation data reported.
Storage: Store in tightly closed containers in a dry, cool, well-ventilated place.

★ 1035 ★
ZINC CYANIDE
CAS: 557-21-1
Other names: CYANURE DE ZINC (FRENCH); ZINC DICYANIDE

Danger profile: Exposure limit established, Community Right-to-Know List, poisonous gas produced in fire.
Uses: Metal plating; insecticide; laboratory chemical.
Appearance: Colorless white powder.
Odor: Almond-like. The odor is a warning of exposure; however, no smell does not mean you are not being exposed.
Effects of exposure:
Breathed: Poisonous. May cause cyanide poisoning; headache, weakness, nausea, vomiting; gasping for air, collapse and death.
Eyes: May cause burns.
Skin: Passes through the unbroken skin; can increase exposure and the severity of the symptoms listed. May cause irritation.
Swallowed: Poisonous. May cause a bitter or acrid taste followed by constriction or numbness of the throat; salivation and nausea; possible vomiting; anxiety, confusion, vertigo, giddiness and stiffness in the lower jaw; respiration changes; loss of consciousness, convulsions, death from respiratory arrest.

Long-term exposure: May cause thyroid effects and enlargement (goiter).

Storage: Store in a dry, cool, well-ventilated place. Keep away from acids and acid mists, magnesium and oxidizers.

First aid guide: Move victim to fresh air; call emergency medical care. In case of contact with material, immediately flush skin or eyes with running water for at least 15 minutes. Remove and isolate contaminated clothing and shoes at the site.

★ 1036 ★
ZINC DITHIONITE
CAS: 7779-86-4
Other names: DITTHIONOUS ACID, ZINC SALT (1:1); ZINC HYDROSULFITE

Danger profile: Combustible, Community Right-to-Know List, poisonous gas produced in fire.
Uses: Making paper; treating beet and cane sugar juices; bleaching textiles, vegetables, straw and hemp; animal glue.
Appearance: White powder.
Effects of exposure:
  Breathed: May cause lung effects and allergy.
  Eyes: Contact may cause irritation and burns.
  Skin: Contact may cause irritation and burns.
Storage: Water, heat or air may cause ignition. Store in tightly closed containers in a dry, cool, well-ventilated place. Keep away from air, heat, flames and sources of ignition, water, acids and oxidizers. Containers may explode in fire.
First aid guide: Move victim to fresh air; call emergency medical care. In case of contact with material, immediately flush skin or eyes with running water for at least 15 minutes. Removal of solidified molten material from skin requires medical assistance. Remove and isolate contaminated clothing and shoes at the site.

★ 1037 ★
ZINC FLUORIDE
CAS: 7783-49-5
Other names: ZINC FLUORURE (FRENCH)

Danger profile: Exposure limit established, Community Right-to-Know List, poisonous gas produced in fire.
Uses: In ceramic glazes and enamels; wood preservative; galvanizing and electroplating; fluoridation agent.
Appearance: Colorless to white powder.
Effects of exposure:
  Breathed: Irritation of nasal passages, throat and lungs; dryness, and nose bleed. High exposures may cause shortness of breath and a dangerous build-up of fluids in the lungs (pulmonary edema), which can cause death. Pulmonary edema is a medical emergency that can be delayed one to two days following exposure.
  Eyes: Contact may cause severe irritation and burns.
  Skin: Passes through the unbroken skin; can increase exposure and the severity of the symptoms listed. May cause severe irritation and burns. Excessive exposure may cause rash.
  Swallowed: Salty or soapy taste; salivation, nausea, burning abdominal pain and cramps; vomiting, diarrhea, dehydration and thirst.
Long-term exposure: May cause loss of appetite, nausea, constipation or diarrhea. Repeated fluoride exposure may cause stiffness of body joints and brittle bones leading to eventual crippling.
Storage: Store in tightly closed containers in a dry, cool, well-ventilated place. Keep away from potassium and acids.

★ 1038 ★
ZINC FORMATE
CAS: 557-41-5
Other names: FORMIC ACID, ZINC SALT; ZINC DIFORMATE

Danger profile: Community Right-to-Know List, poisonous gas produced in fire.
Uses: Making other chemicals and textiles; waterproofing agent; antiseptic.
Appearance: Colorless powder. Turns white from absorbed moisture.
Effects of exposure:
  Breathed: May cause irritation of the nasal passages and throat.
  Eyes: May cause severe irritation, inflammation of the eyelids and damage; corneal opacity.
  Skin: Passes through the unbroken skin; can increase exposure and the severity of the symptoms listed. May cause severe irritation and burns.
  Swallowed: May cause nausea and vomiting.
Storage: Store in tightly closed containers in a dry, cool, well-ventilated place.

★ 1039 ★
ZINC NITRATE
CAS: 7779-88-6
Other names: NITRATE DE ZINC (FRENCH); NITRIC ACID, ZINC SALT

Danger profile: Community Right-to-Know List, poisonous gas produced in fire.
Uses: Making other chemicals; latex coagulant; laboratory chemical; fixing dyes.
Appearance: Colorless crystals or needles.
Effects of exposure:
  Breathed: Dust may irritate nose and throat. High exposure may cause headaches, nausea, weakness and confusion.
  Eyes: Contact may cause irritation and burns. May be delayed.
  Skin: Passes through the unbroken skin; can increase exposure and the severity of the symptoms listed. Contact may cause irritation and burns.
  Swallowed: May cause irritation; corrosion of the alimentary tract; nausea and vomiting.
Storage: Oxidizer. Does not burn but may increase intensity of fire. Store in tightly closed containers in a dry, cool, well-ventilated place with non-wood floors. Keep away from fuels, combustible materials, carbon, copper, phosphorus, sulfur and metal sulfides. Containers may explode in fire.
First aid guide: Move victim to fresh air; call emergency medical care. In case of contact with material, immediately flush skin or eyes with running water for at least 15 minutes. Remove and isolate contaminated clothing and shoes at the site.

★ 1040 ★
ZINC OXIDE
CAS: 1314-13-2

Other names: AMALOX; AKRO-ZINC BAR 85; AKRO ZINC BAR 90; AMALOX; AZO-33; AZO-55; AZO-66; AZO-77; AZO-DOX-55; CALAMINE; CHINESE WHITE; C.I. 77947; C.I.PIGMENT WHITE 4; CYNKU TLENEK (POLISH); EMANAY ZINC OXIDE; EMAR; FELLING ZINC OXIDE; FLOWERS OF ZINC; GREEN SEAL-8; HUBBUCK'S WHITE; KADOX-25; K-ZINC; OZIDE; OZLO; PASCO; PERMANENT WHITE; PHILOSOPHER'S WOOL; PROTOX TYPE 166; PROTOX TYPE 167; PROTOX TYPE 168; PROTOX TYPE 169; PROTOX TYPE 267; PROTOX TYPE 268; RED-SEAL-9; SNOW WHITE; WHITE FLOWER OF ZINC; WHITE SEAL-7; ZINCITE; ZINCOID; ZINC OXIDE FUME; ZINC WHITE

Danger profile: Exposure limit established, Community Right-to-Know List, poisonous gas produced in fire.
Uses: Pigment in paints, rubber products, lacquers, varnishes, ceramics, cosmetics; fungicide; seed disinfectant; food additive; photocopying; pharmaceuticals; dentistry.
Appearance: White to yellowish-white powder. Fume is white; produced by exposure of zinc compounds to high temperatures, as in welding.
Effects of exposure:
   Breathed: May cause metal fume fever, a flu-like illness with chills, fever, headache, tightness in the chest, cough, metallic taste in the mouth; muscle and joint pain; upset stomach; mouth dryness and thirst; nausea and vomiting; back pain; pulmonary changes, occasionally bronchitis or pneumonia. Symptoms may delayed for several hours following exposure and lasts one to two days.
   Eyes: May cause irritation.
   Skin: Passes through the unbroken skin; can increase exposure and the severity of the symptoms listed. May cause irritation that can result in rash.
Long-term exposure: May cause liver effects; symptoms of ulcers; skin irritation.
Storage: Store in tightly closed containers in a dry, cool, well-ventilated place. Keep away from high heat, magnesium, linseed oil and chlorinated rubber.

★ 1041 ★
ZINC PERMANGANATE
CAS: 23414-72-4
Other names: PERMANGANIC ACID, ZINC SALT

Danger profile: Carcinogen, exposure limit established, Community Right-to-Know List, poisonous gas produced in fire.
Uses: Oxidizing agent; medical antiseptic; astringent.
Appearance: Violet-brown to black powder.
Effects of exposure:
   Breathed: May cause irritation to the nasal passages and throat.
   Eyes: Contact may cause irritation and burns.
   Skin: Passes through the unbroken skin; can increase exposure and the severity of the symptoms listed. Contact may cause irritation and burns.
Storage: Oxidizing agent; will increase the intensity of a fire. Store in tightly closed containers in a dry, cool, well-ventilated place. Keep away from combustible materials, reducing agents, acetic acid, acetic anhydride. Protect containers from shock. Containers may explode in fire.
First aid guide: Move victim to fresh air; call emergency medical care. In case of contact with material, immediately flush skin or eyes with running water for at least 15 minutes. Remove and isolate contaminated clothing and shoes at the site.

★ 1042 ★
ZINC PEROXIDE
CAS: 1314-22-3
Other names: ZINC SUPEROXIDE

Danger profile: Community Right-to-Know List, poisonous gas produced in fire.
Uses: Making pharmaceuticals, cosmetics, rubber and elastomers; in high-temperature oxidation.
Appearance: White to yellowish-white powder.
Effects of exposure:
   Breathed: May be poisonous. May cause irritation of the throat and air passages.
   Eyes: Contact may cause severe irritation and burns.
   Skin: Passes through the unbroken skin; can increase exposure and the severity of the symptoms listed. Contact may cause severe irritation and burns.
Storage: Store in tightly closed containers in a dry, cool, well-ventilated place. Keep away from heat, flames, sources of ignition, water, flammable and combustible materials, reducing agents, chemically active metals such as aluminum and zinc. Containers may explode in fire.
First aid guide: Move victim to fresh air and call emergency medical care; if not breathing, give artificial respiration; if breathing is difficult, give oxygen. In case of contact with material, immediately flush skin and eyes with running water for at least 15 minutes. Remove and isolate contaminated clothing and shoes at the site. Keep victim quiet and maintain normal body temperature.

★ 1043 ★
ZINC SULFATE
CAS: 7733-02-0
Other names: BONAZEN; BUFOPTO ZINC SULFATE; OPTHAL-ZIN; SULFATE DE ZINC (FRENCH); SULFURIC ACID, ZINC SALT (1:1); VERAZINC; WHITE COPPERAS; WHITE VITRIOL; ZINC VITRIOL; ZINKOSITE

Danger profile: Mutagen, Community Right-to-Know List, poisonous gas produced in fire.
Uses: Making rayon; dietary supplement; fixing dyes; wood preservative; laboratory chemical.
Appearance: Colorless crystalline powder.
Effects of exposure:
   Breathed: Dust may cause irritation of the nose and throat. High doses may cause spontaneous abortions or stillbirths.
   Eyes: May cause severe irritation and burns.
   Skin: May cause severe irritation and burns.
   Swallowed: May cause irritation and corrosion of the alimentary tract; gastrointestinal changes; increased pulse, decreased blood pressure rate, diarrhea; a dangerous build-up of fluids in the lungs (pulmonary edema), which can cause death. Pulmonary edema is a medical emergency that can be delayed one to two days following exposure.
Long-term exposure: May cause mutation and damage to the fetus.
Storage: Store in tightly closed containers in a dry, cool, well-ventilated place. Keep away from flames and sources of ignition.

★ 1044 ★
ZINC SULFIDE
CAS: 1314-98-3
Other names: ALBALITH; IRTRAN 2; ZINC BLENDE; ZINC MONOSULFIDE; SPALERITE; WURTZITE

Danger profile: Community Right-to-Know List, poisonous gas produced in fire, or on contact with moisture or acid.

Uses: Making white or opaque glass, dyes, rubber, plastic luminous paints; phosphor in x-ray and television screens; fungicide.
Appearance: White or grayish-white or yellowish powder.
Effects of exposure:
Breathed: Dust may cause irritation of the mouth, nose and throat. When heated chemical emits zinc oxide fumes, causing metal fume fever with chills, aching, nausea, fever, dry throat, cough and weakness.
Eyes: May cause irritation.
Skin: May cause irritation.
Swallowed: May cause irritation and corrosion of the throat, stomach and intestines and possible death.
Storage: Store in tightly closed containers in a dry, cool, well-ventilated place. Keep away from heat, flames, sources of ignition, water, acids and iodine monochloride.

★ 1045 ★
ZINEB
CAS: 12122-67-7
Other names: ASPOR; ASPORUM; BERCEMA; BLIGHTOX; BLITEX; BLIZENE; CARBADINE; CHEM ZINEB; CINEB; CRITTOX; CYNKOTOX; DAISEN; DIPHER; DITHANE Z; DI-TIAMINA; ENT 14,874; 1,2-ETHANEDIYLBIS(CARBAMODITHIOATO)ZINC; (1,2-ETHANEDIYLBIS(CARBAMODITHIOATO))(2-)ZINC; 1,2-ETHANEDIYLBIS(CARBAMODITHIOATO)(2-)-S,S'-ZINC; 1,2-ETHANEDIYLBISCARBAMODITHIOIC ACID, ZINC COMPLEX; 1,2-ETHANEDIYLBISCARBAMOTHIOIC ACID, ZINC SALT; ETHYLENEBIS(DITHIOCARBAMATO)ZINC; ETHYLENEBIS(DITHIOCARBAMIC ACID),ZINC SALT; ETHYL ZIMATE; HEXATHANE; KUPRATSIN; KYPZIN; LIROTAN; LONACOL; MICIDE; MILTOX; MILTOX SPECIAL; NOVOSIR N; NOVOZIN N 50; NOVOZIR; NOVOZIR N; NOVOZIR N 50; PAMOSOL 2 FORTE; PARZATE; PARZATE ZINEB; PEROSIN; PEROSIN 75B; PEROZIN; PEROZINE; POLYRAM Z; SPERLOX-Z; THIODOW; TIEZENE; TRITOFTOROL; TSINEB (RUSSIAN); Z-78; ZEBENIDE; ZEBTOX; ZIDAN; ZIMATE; ZINC ETHYLENEBIS (DITHIOCARBAMATE); ZINC ETHYLENE-1,2-BISDITHIOCARBAMATE; ZINC, (ETHYLENEBIS(DITHIOCARBAMATO))-; ZINK-(N,N'-AETHYLEN-BIS(DITHIOCARBAMAT)) (GERMAN); ZINOSAN

Danger profile: Community Right-to-Know List, poisonous gas produced in fire.
Uses: Insecticide; fungicide.
Appearance: Light colored powder.
Effects of exposure:
Breathed: May be poisonous. May cause irritation to the nasal passages and throat.
Eyes: May cause irritation.
Skin: May cause irritation and rash.
Swallowed: May be poisonous.
Long-term exposure: May cause cancer; reduce fertility in women; skin rash. Has caused cancer in animals exposed during pregnancy. Human mutation data reported.
Storage: Store in tightly closed containers in a dry, cool, well-ventilated place. Keep away from heat; decomposition produces very toxic fumes. Containers may explode in heat of fire.

★ 1046 ★
ZINOPHOS
CAS: 297-97-2
Other names: AC 18133; AMERICAN CYANAMID 18133; CL 18133; CYNEM; O,O-DIETHYL O-PARAZINYL PHOSPHOROTHIOATE; PHOSPHOROTHIOIC ACID, O,O-DIETHYL O-PYRAZINYL ESTER; EN 18133; ENT 25,580; EXPERIMENTAL NEMATOCIDE 18,133; NEMAFOS; NEMAPHOS; THIONAZIN

Danger profile: Poison, EPA extremely hazardous substances list, poisonous gas produced in fire.
Uses: Insecticide (organophosphate).
Appearance: Amber liquid.
Effects of exposure:
Breathed: Poisonous. May cause headache, dizziness, profuse sweating, nausea, vomiting, reduced heart beat, stomach cramps, diarrhea, loss of coordination, slow and weak breathing, fever, loss of consciousness and death. May cause shortness of breath and a dangerous build-up of fluids in the lungs (pulmonary edema), which can cause death. Pulmonary edema is a medical emergency that can be delayed one to two days following exposure.
Eyes: Rapidly absorbed. Poisoning can happen very rapidly. See *Breathed.*
Skin: Poisonous. Passes through the unbroken skin; can increase exposure and the severity of the symptoms listed.
Swallowed: Deadly poison. See *Breathed.*
Long-term exposure: May damage the developing fetus or cause birth defects; nerve damage including loss of coordination; liver damage. Repeated exposure may cause depression, irritability and personality changes.
Storage: Store in a regulated, specially marked place in tightly closed containers in a dry, well-ventilated place. Keep away from water, strong alkalis. Containers may explode in fire.

★ 1047 ★
ZIRCONIUM
CAS: 7440-67-7
Other names: ZIRCAT; ZIRCONIUM METAL; ZIRCONIUM SHAVINGS; ZIRCONIUM SHEETS; ZIRCONIUM TURNINGS

Danger profile: Fire hazard (powder), explosion hazard (powder), exposure limit established, poisonous gas produced in fire, slightly radioactive.
Uses: In corrosion resistant alloys, photoflash bulbs, explosives, fireworks, vacuum tubes and steel; to coat nuclear fuel rods.
Appearance: Grayish, hard, lustrous powder or liquid suspension.
Effects of exposure:
Skin: Contact may cause an allergic skin reaction.
Long-term exposure: May cause allergic skin reaction.
Storage: Dust or powder may ignite spontaneously. Store in tightly closed containers in a dry, cool, well-ventilated place. Keep away from heat, flames, sources of ignition, flammable materials, alkali materials, chromates, molybidates, sulfates, borax, lead, potassium, oxidizers and other chemicals. Containers may explode in fire.

Part 2: Resource Listings

Hotlines and Clearinghouses

Hotlines and clearinghouses are arranged by subject of inquiry. Business hours of the hotlines are provided when available. "800" numbers are toll free.

Air Quality

★ 1048 ★ Air Resources Information Clearinghouse
Center for Environmental Information, Inc.
99 Court St.
Rochester, NY 14604-1824
Phone: (716)546-3796
Contact: Elizabeth Thorndike, Pres.
Provides information relating to air pollution, including air toxins, indoor air pollution, acid rain, and ozone depletion and global warming. Maintains ACIDOC, a bilingual database on acidic disposition.

★ 1049 ★ Air RISC Hotline
Office of Air Quality Planning and Standards (MD-13)
U.S. Environmental Protection Agency (EPA)
Air Risk Information Support Center
Research Triangle Park, NC 27711
Hotline: (919)541-0888
Provides state and local air pollution agencies and EPA regional offices with health, exposure, and risk information on asbestos and toxic air pollutants and assists in interpreting that data. Monday through Thursday, 8 a.m.-5 p.m., EST.

★ 1050 ★ Best Available Control Technology/Lowest Achievable Emission Rate (BACT/LAER)
Office of Air Quality Planning and Standards (MD-13)
U.S. Environmental Protection Agency (EPA)
Emission Standards and Engineering Division
Research Triangle Park, NC 27711
Phone: (919)541-5432
Contact: Robert Blaszczak, Exec. Officer
Assists federal, state, and local agencies in exchanging information about BACT and LAER determinations as established under the Clean Air Act. Monday through Friday, 8:30 a.m.-4:30 p.m., EST.

★ 1051 ★ Control Technology Center for Air Toxics Hotline
Office of Air Quality Planning and Standards
U.S. Environmental Protection Agency (EPA)
Research Triangle Park, NC 27711
Phone: (919)541-0800
Provides engineering guidance on air pollution technology to state and local agencies and regional EPA offices that implement air pollution programs. Monday through Friday, 8 a.m.-5 p.m., EST.

Art and Craft Materials

★ 1052 ★ Art and Craft Materials Institute
715 Boyleston St.
Boston, MA 02116
Phone: (617)266-6800
Provides information on those art and craft products that the Institute has certified and approved for safety.

★ 1053 ★ Art Hazards Information Center
Center for Safety in the Arts
5 Beekman St.
New York, NY 10038
Phone: (212)227-6220
Responds to questions relating to paints, clays, make-up, and other chemicals used in visual and performing arts and crafts and refers callers to doctors specializing in toxic-related illnesses.

Asbestos

★ 1054 ★ Asbestos and Small Business Ombudsman Hotline
U.S. Environmental Protection Agency (EPA)
401 M St., N.W., A-149C
Washington, DC 20460
Phone: (703)557-1938
Fax: (703)557-1462
Hotline: 800-368-5888
Contact: Karen Brown, Ombudsman
Assists small businesses in complying with environmental laws and EPA regulations; investigates with the EPA problems encountered by small-quantity generators of hazardous waste. Monday through Friday, 8 a.m.-4:30 p.m., EST. There are Small Business Ombudsmen in each EPA regional office.

★ 1055 ★ Asbestos Hotline
Office of Toxic Substances (TS-799)
U.S. Environmental Protection Agency (EPA)
401 M St., S.W.
Washington, DC 20460
Fax: (202)260-5603
Hotline: 800-835-6700 (for asbestos in schools only); (202)260-1404
Provides technical information concerning asbestos abatement in schools and public and commercial buildings. Answers inquiries about hazards from asbestos and compliance under the Asbestos Hazard Emergency Response Act (AHERA; 1986). Monday through Friday, 8:30 a.m.-5 p.m., EST.

★ 1056 ★ Asbestos Technical Information and Referral Hotline
U.S. Environmental Protection Agency (EPA)
401 M St., S.W.
Washington, DC 20460
Phone: (202)260-1404
Hotline: 800-334-8571
Provides technical information on asbestos, including bulk identification sampling, analysis, and regulations.

Emergency Planning and Response

★ 1057 ★ American Trucking Association (ATA)
2200 Mill Rd.
Alexandria, VA 22314
Hotline: 800-ATA-LINE
Provides emergency procedure information for handling hazardous spills resulting from accidents involving trucks.

★ 1058 ★ Bureau of Explosives Hotline
Association of American Railroads Hazardous Materials Systems
American Railroads Bldg.
50 F St., N.W.
Washington, DC 20001
Phone: (202)639-2222
Hotline: (202)639-2132
24-hour hotline provides emergency information and assistance for hazardous materials incidents involving railroads.

★ 1059 ★ Center for Disease Control Chemical Spill Hotline
U.S. Department of Health and Human Services
Atlanta, GA
Phone: (404)633-5313

★ 1060 ★ Chemical Emergency Preparedness, Emergency Planning and Community Right-to-Know Information Hotline
Office of Solid Waste and Emergency Response
U.S. Environmental Protection Agency (EPA)
401 M St., S.W., (OS-120)
Washington, DC 20460
Phone: (703)920-9877; (202)260-2449
Hotline: 800-535-0202
Provides communities with information on preparing for chemical accidents and on the Toxics Release Inventory (TRI) reporting

requirements of SARA Title III. Monday through Friday, 8:30 a.m.-7:30 p.m., EST.

★ 1061 ★ Chemical Transportation Emergency Center Hotline (CHEMTREC)
2501 M St., N.W.
Washington, DC 20037
Phone: (202)260-1100
Hotline: 800-424-9300
24-hour hotline provides immediate information on controlling hazardous material spills, leaks, fires, accidents, and other emergencies involving identified substances and assists in the identification of unknown substances. Notifies manufacturers and shippers of accidents involving their products.

★ 1062 ★ DOT Hotline
Office of Hazardous Materials Transportation Standards
Department of Transportation (DOT)
400 Seventh St., S.W.
Washington, DC 20590
Phone: (202)366-4488
Provides information on proper procedures and protection when transporting hazardous materials and toxic chemicals in all modes except bulk transportation by water.

★ 1063 ★ EPA Region I Emergency Response Office
Phone: (617)860-4300
Hotline: (617)223-7265
24-hour hotline serves Connecticut, Maine, Massachusetts, New Hampshire, Rhode Island, and Vermont.

★ 1064 ★ EPA Region II Emergency Response Office
Phone: (201)321-6658
Hotline: (201)548-8730
24-hour hotline serves New Jersey, New York, Puerto Rico, and the Virgin Islands

★ 1065 ★ EPA Region III Emergency Response Office
Hotline: (215)597-9898
24-hour hotline serves Delaware, District of Columbia, Maryland, Pennsylvania, Virginia, and West Virginia.

★ 1066 ★ EPA Region IV Emergency Response Office
Phone: (404)347-3931
Hotline: (404)347-4062
24-hour hotline serves Alabama, Florida, Georgia, Kentucky, Mississippi, North Carolina, South Carolina, and Tennessee.

★ 1067 ★ EPA Region IX Emergency Response Office
Phone: (415)744-2386
Hotline: (415)744-2000
24-hour hotline serves Arizona, California, Hawaii, Nevada, American Samoa, Guam, and the Commonwealth of the Northern Mariana Islands.

★ 1068 ★ EPA Region V Emergency Response Office
Phone: (312)353-2316
Hotline: (312)353-2318
24-hour hotline serves Illinois, Indiana, Michigan, Minnesota, Ohio, and Wisconsin.

★ 1069 ★ EPA Region VI Emergency Response Office
Phone: (214)655-2270
Hotline: (214)655-2222
24-hour hotline serves Arkansas, Louisiana, New Mexico, Oklahoma, and Texas.

★ 1070 ★ EPA Region VII Emergency Response Office
Phone: (913)236-3888
Hotline: (913)236-3778
24-hour hotline serves Iowa, Kansas, Missouri, and Nebraska.

★ 1071 ★ EPA Region VIII Emergency Response Office
Phone: (303)294-7528
Hotline: (303)293-1788
24-hour hotline serves Colorado, Montana, North Dakota, South Dakota, Utah, and Wyoming.

★ 1072 ★ EPA Region X Emergency Response Office
Phone: (206)442-1263
Hotline: (206)442-1263
24-hour hotline serves Alaska, Idaho, Oregon, and Washington.

★ 1073 ★ National Response Center for Oil and Hazardous Spills
U.S. Coast Guard
2100 Second St., S.W., Rm. 2611
Washington, DC 20593
Phone: (202)260-2675
Hotline: 800-424-8802
Report chemical, oil, and other hazardous material spills to this center, which is open 24 hours. The Center provides hazard assessments, movement forecasting, and other technical information to the on-scene coordinators of the National Response Team.

★ 1074 ★ U.S. Army Corps of Engineers
Hotline: (202)272-0001
Provides emergency information and assistance in dealing with toxic chemical spills on inland waterways.

General Information

★ 1075 ★ Center for Environmental Research Information
Office of Research and Development
Research Information Unit (MS-G-72)
U.S. Environmental Protection Agency (EPA)
26 W. Martin Luther King Dr.
Cincinnati, OH 45268
Phone: (513)569-7391
Contact: Dorothy Williams, Exec. Officer
Distributes information on current EPA research projects.

★ 1076 ★ Environment Canada
Phone: (819)953-7156
Hotline: 800-567-1999 (within Canada)

★ 1077 ★ EPA Superfund/RCRA Hotline
U.S. Environmental Protection Agency (EPA)
401 M St., S.W.
Washington, DC 20460
Phone: (202)260-3000; (703)920-9810
Hotline: 800-424-9346
Provides information on the Comprehensive Environmental Response, Compensation and Liability Act (Superfund) and the Resource Conservation and Recovery Act (RCRA). Maintains the Electronic Information Exchange System (EIES), a computerized information network accessible by personal computers with appropriate software. Monday through Friday 8:30 a.m.-7:30 p.m., EST.

★ 1078 ★ Information Center
National Technical Information Service (NTIS)
5285 Port Royal Rd.
Springfield, VA 22161
Phone: 800-336-4700
Fax: (703)321-8547
National clearinghouse for all documents published by federal and foreign research and development agencies, including those of the EPA and OSHA.

★ 1079 ★ Pollution Prevention Information Clearinghouse (PPIC)
Pollution Prevention Office
U.S. Environmental Protection Agency (EPA)
401 M St., S.W.
Washington, DC 20460
Hotline: 800-424-9346
Provides information on government and industry pollution prevention programs, project funding opportunities, and conferences and seminars and assists in document searches and ordering.

★ 1080 ★ Public Information Center
U.S. Environmental Protection Agency (EPA)
401 M. St., S.W.
Washington, DC 20460
Phone: (202)260-7751
Helps locate the proper EPA department for emergency information and response guidance. Offers general, nontechnical literature on various environmental topics.

★ 1081 ★ Public Information Reference Unit
U.S. Environmental Protection Agency (EPA)
401 M St., S.W., Rm. 2904
Washington, DC 20460
Phone: (202)260-5926
Dockets, comprising records of and information on the EPA's legal decisions on a variety of subjects, including air and water quality and toxic substances, are available to the public through this office.

Hazardous Substances

★ 1082 ★ Hazardous Materials Information Hotline
Center for Hazardous Materials Research
University of Pittsburgh Applied Science Center
320 William Pitt Way
Pittsburgh, PA 15238
Phone: (412)826-5320
Fax: (412)826-5552
Hotline: 800-334-CHMR
Responds to inquiries concerning hazardous substances found in household products and their proper disposal and provides communities with information and guidance on hazardous waste management problems.

★ 1083 ★ Substance Identification Service (SIS)
American Chemical Society (ACS)
Chemical Abstracts Service (CAS)
1155 16th St., N.W.
Washington, DC 20036
Hotline: 800-848-6538
Identifies hazardous chemicals by name and Chemical Abstract Service number (CAS).

★ 1084 ★ Toxicology Information Response Center
Oak Ridge National Laboratory
Bldg. 2001
P.O. Box 2008
Oak Ridge, TN 37831-6050
Phone: (615)576-1746
Fax: (615)574-9888
Contact: Kimberly Slusher, Exec. Officer
International center collects, analyzes, and disseminates toxic-related information on chemicals, including pesticides, food additives, industrial chemicals, heavy metals, environmental pollutants, and pharmaceuticals. The Center is on-line to more than 400 computerized databases, including DIALOG, MEDLARS, STN International, ITIS, and DROLS, and performs searches for outside users for a fee.

★ 1085 ★ Toxic Substances Control Act (TSCA) Assistance Information Service
Office of Pesticides and Toxic Substances (TS-799)
U.S. Environmental Protection Agency (EPA)
401 M St., S.W.
Washington, DC 20460
Phone: (202)260-1404
Fax: (202)260-5603
Hotline: 800-424-9065; (800)835-6700 (for asbestos in schools only)
Provides technical information, including publications and audiovisual material, on toxic substances, including asbestos. Monday through Friday, 8:30 a.m.-5 p.m., EST.

★ 1086 ★ Toxic Substances Information Hotline
Natural Resources Defense Council (NRDC)
40 W. 20th St.
New York, NY 10011
Phone: (212)727-2700
Hotline: 800-648-6732
Provides information on household chemicals.

Hazardous Waste

★ 1087 ★ Alternative Treatment Technology Information Center
Office of Environmental Engineering and Technology Demonstration
U.S. Environmental Protection Agency (EPA)
401 M St., S.W.
Washington, DC 20460
Phone: (202)260-7161
Hotline: 800-424-9346
Contact: Myles Morris, Exec. Officer
Provides EPA staff with technical information on hazardous waste for Superfund cleanup activities.

★ 1088 ★ Citizens' Clearinghouse for Hazardous Waste, Inc. (CCHW)
P.O. Box 6806
Falls Church, VA 22040
Phone: (703)237-CCHW
Contact: Lois Marie Gibbs, Exec. Dir.
Provides information and guidance for dealing with toxic substances and conducts research on chemicals to determine dangerous levels of usage. **Publications:** *Everyone's Backyard*, a bimonthly newsletter; more than 60 books, pamphlets, and technical fact sheets.

★ 1089 ★ Hazardous Waste Ombudsman
U.S. Environmental Protection Agency (EPA)
401 M St., S.W.
Washington, DC 20460
Phone: (202)260-9361
Assists those with problems concerning hazardous waste issues. There is a Hazardous Waste Ombudsman in each of the EPA's Regional Offices.

Health and Safety

★ 1090 ★ Cancer Hotline
American Cancer Society (ACS)
1599 Clifton Rd., N.E.
Atlanta, GA 30329
Hotline: 800-ACS-2345
Answers questions about carcinogens present in products and the environment.

★ 1091 ★ Cancer Information Service Hotline
National Cancer Institute
National Institutes of Health
Office of Cancer Communication
NCI Bldg. 31, Rm. 10A1B
Bethesda, MD
Hotline: 800-4-CANCER
Provides information about cancer and related diseases, treatments, low-cost clinics and support groups, and medical consultation and referrals. Monday through Friday, 6 a.m.-10 p.m.; Saturday, 10 a.m.-6 p.m., EST.

★ 1092 ★ Centers for Disease Control Public Inquiry Office
U.S. Department of Health and Human Services
Atlanta, GA
Phone: (404)639-3535
Responds to inquiries concerning Centers for Disease Control programs and specific questions dealing with environmental health problems, including environmental, chemical, and radiation emergencies.

★ 1093 ★ Chemical Health and Safety Division
American Chemical Society
1155 16th St., N.W.
Washington, DC 20036
Hotline: (202)872-4401
Provides information about harmful effects of hazardous chemicals and appropriate safety precautions to take when handling them.

★ 1094 ★ Chemical Referral Center
Chemical Manufacturers Association (CMA)
2501 M St., N.W.
Washington, DC 20037
Phone: (202)887-1318
Hotline: 800-CMA-8200
Provides non-emergency information to consumers about chemical manufacturers who provide health and safety information about their products.

★ 1095 ★ Lungline
National Jewish Center for Immunology and Respiratory Medicine
1400 Jackson St.
Denver, CO 80206
Phone: (303)355-LUNG
Hotline: 800-222-LUNG
Provides information about lung and immunologic diseases from exposure to

asbestos and other toxic chemicals. Monday through Friday, 8 a.m.-5 p.m., MST.

★ 1096 ★ National Health Information Clearinghouse
Office of Disease Prevention and Health Promotion
U.S. Department of Health and Human Services
P.O. Box 1133
Washington, DC 20013-1133
Phone: (301)565-4167
Hotline: 800-336-4797
Assists the general public and professionals in locating appropriate health resources to resolve health questions. Does not provide medical advice, diagnosis, or physician referrals. Monday through Friday, 9 a.m.-5 p.m., EST.

★ 1097 ★ National Injury Information Clearinghouse
U.S. Consumer Product Safety Commission (CPSC)
Westwood Tower Bldg., Rm. 625
5401 Westbard Ave.
Bethesda, MD
Hotline: 800-638-CPSC; teletype for the hearing impaired: (800)492-8104, (800)6388270
Compiles and publishes statistics relating to injuries from consumer products, including those caused by the misuse of products containing potentially toxic substances.

★ 1098 ★ Technical Information and Assistance
Occupational Safety and Health Administration (OSHA)
National Institute for Occupational Safety and Health (NIOSH)
9467 Columbia Pkwy.
Cincinnati, OH 45226
Phone: (513)533-8287
Hotline: 800-356-4674
Distributes OSHA technical reports, studies, position papers, and other technical literature pertaining to hazardous substances.

Pesticides

★ 1099 ★ Chevron/Ortho Emergency Information Center
Chevron Corporation
Hotline: (415)233-3737 (collect calls accepted)
24-hour hotline provides information and assistance in emergencies involving Ortho or Chevron brand products, including pesticides, herbicides, fertilizers, and automobile fluids.

★ 1100 ★ National Coalition Against the Misuse of Pesticides
701 E St., S.E.
Washington, DC 20003
Phone: (202)543-5450
Provides information on the effects of pesticides on health and suggests alternative nontoxic products for use in gardens and on farms.

★ 1101 ★ National Pesticide Information Retrieval Service
Purdue University
Entomology Hall
West Lafayette, IN 47907
Phone: (317)494-6614
Contact: James H. White, Exec. Dir.
Provides current information on products registered with the EPA, tolerance data for pesticides and hazardous chemicals, state

registration information, and descriptions of studies on pesticides and hazardous chemicals.

★ 1102 ★ **National Pesticide Information Retrieval System**
Purdue University
Entomology Hall
West Lafayette, IN 47907
Phone: (317)494-6616
Fax: (317)494-0535
Contact: James H. White, Exec. Dir.
Membership organization sponsored by the Office of Pesticide Programs of the U.S. Environmental Protection Agency that maintains the National Pesticide Information Retrieval Service (NPIRS), which provides current information on products registered with the EPA and tolerance data for pesticides and hazardous chemicals. It also provides state registration information, registration guideline information, and descriptions of studies on pesticides and hazardous chemicals.

★ 1103 ★ **National Pesticides Telecommunication Network**
Texas Tech University Health Science Center
School of Medicine
Department of Preventive Medicine and Community Health
Lubbock, TX 79430
Phone: (806)743-3091
Fax: 800-743-3094
Hotline: 800-858-PEST; for medical and government personnel: (800)858-7377.
Contact: Dr. Anthony Way, Exec. Dir.
24-hour hotline provides information on pesticide-related health problems, toxicity, minor cleanup problems, and residues. Also provides specific information on safety practices, health and environmental effects, and cleanup and disposal procedures. Operates in the contiguous United States, Puerto Rico, and the Virgin Islands.

★ 1104 ★ **Pesticide Action Network North American Regional Center (PAN/NA)**
965 Mission St., Ste. 514
San Francisco, CA 94103
Phone: (415)541-9140
Fax: (415)541-9253
Contact: Monica Moore, Exec. Dir.
Links pesticide action groups in North America with counterpart citizens movements on other continents, including in Nairobi, Kenya; Dakar, Senegal; Penang, Malaysia; London, England; and Palmira, Colombia. **Publications:** *Global Pesticide Campaigner,* a quarterly newsletter; pamphlets, reports, manuals, and booklets.

★ 1105 ★ **Pesticides Hotline**
U.S. Environmental Protection Agency (EPA)
401 M St., S.W.
Washington, DC 20460
Phone: (202)260-2902
Responds to inquiries concerning pesticides, including exposure to them and treatments and safe disposal.

Poisons

★ 1106 ★ **National Animal Poison Information Network (NAPINET)**
National Animal Poison Control Center
University of Illinois
Champaign-Urbana, IL
Hotline: 800-548-2423; (217)333-3611
In order to sustain the operation of this national 24-hour hotline, a small fee may be charged per inquiry.

★ 1107 ★ **National Poison Control Center Hotline**
Georgetown University Hospital
Washington, DC
Hotline: (202)625-3333 (collect calls accepted)
Provides emergency first aid advice for accidental swallowing of chemicals, poisons, or drugs and information on acute and longterm effects of toxic chemical exposure and on specific product ingredients.

Radon

★ 1108 ★ **National Radon Hotline**
U.S. Environmental Protection Agency (EPA)
401 M St., S.W.
Washington, DC 20460
Phone: (202)260-9605
Hotline: 800-SOS-RADON
Responds to inquiries concerning radon and other sources of radiation in homes and commercial and industrial establishments.

Water Quality

★ 1109 ★ **Safe Drinking Water Hotline**
U.S. Environmental Protection Agency (EPA)
401 M St., S.W.
Washington, DC 20460
Phone: (202)260-5533
Hotline: 800-426-4761; (800)426-9607
Provides information about regulations and programs under the Safe Drinking Water Act (SDWA). Monday through Friday, 8:30 a.m.4:30 p.m., EST.

Poison Control Centers

According to the American Association of Poison Control Centers, certified regional poison control centers should make available to the public these services: 1) 24-hour emergency treatment advice on a variety of topics, from common household ingestion to contact with hazardous materials in the workplace; 2) referrals for specialized care when necessary; 3) telephone follow-up of all exposures requiring treatment, whether at home or in a medical facility; and 4) community education in poison prevention and professional education in the treatment of poisoning.

Alabama

★ 1110 ★ **Alabama Poison Control Center**
809 University Blvd., E.
Tuscaloosa, AL 35401
Phone: (205)345-0600
Hotline: 800-462-0800 (within Alabama)
Certified by the American Association of Poison Control Centers.

★ 1111 ★ **Children's Hospital of Alabama Poison Control Center**
1600 Seventh Ave., S.
Birmingham, AL 35233-1711
Phone: (205)939-9201; (205)933-4050
Hotline: 800-292-6678 (within Alabama)
Certified by the American Association of Poison Control Centers.

Alaska

★ 1112 ★ **Anchorage Poison Center**
Providence Hospital
P.O. Box 196604
Anchorage, AK 99519-0604
Hotline: 800-478-3193 (within Alaska)

Arizona

★ 1113 ★ **Arizona Poison and Drug Information Center**
University of Arizona
Arizona Health Science Center, Rm. 3204K
Tucson, AZ 85724
Phone: (602)626-6016
Hotline: 800-362-0101 (within Arizona)
Certified by the American Association of Poison Control Centers.

★ 1114 ★ **Samaritan Regional Poison Center**
Good Samaritan Medical Center
1130 E. McDowell Rd., Ste. A-5
Phoenix, AZ 85006
Phone: (602)253-3334
Certified by the American Association of Poison Control Centers.

Arkansas

★ 1115 ★ **Statewide Poison Control Drug Information Center**
University of Arkansas Medical Sciences
43091 W. Markham St.
Little Rock, AR 72205
Hotline: 800-482-8948 (within Arkansas)

California

★ 1116 ★ **Fresno Regional Poison Control Center of Fresno Community Hospital and Medical Center**
2823 Fresno St.
P.O. Box 1232
Fresno, CA 93715
Phone: (209)445-1222
Hotline: 800-346-5922 (within California)
Certified by the American Association of Poison Control Centers.

★ 1117 ★ **Los Angeles County Medical Association Regional Poison Center**
1925 Wilshire Blvd.
Los Angeles, CA 90057
Phone: (213)484-5151
Hotline: 800-825-2722; (800)777-6476
Certified by the American Association of Poison Control Centers.

★ 1118 ★ **San Diego Regional Poison Center**
UCSD Medical Center
225 Dickinson St.
San Diego, CA 92103
Phone: (619)543-6000
Hotline: 800-876-4766
Certified by the American Association of Poison Control Centers.

★ 1119 ★ **San Francisco Bay Area Regional Poison Control Center**
San Francisco General Hospital
1001 Potrero Ave., Rm. 1E86
San Francisco, CA 94110
Phone: (415)476-6600
Hotline: 800-523-2222 (within area codes 415 and 707)
Certified by the American Association of Poison Control Centers.

★ 1120 ★ **Santa Clara Valley Medical Center Regional Poison Center**
751 S. Bascom Ave.
San Jose, CA 95128
Hotline: 800-662-9886 (within California)

★ 1121 ★ **UCDMC Regional Poison Control Center**
2315 Stockton Blvd.
Sacramento, CA 95817
Phone: (916)453-3414
Hotline: 800-342-9293 (within California)
Certified by the American Association of Poison Control Centers.

Colorado

★ 1122 ★ **Mid-Plains Poison Control Center**
Children's Memorial Hospital
8301 Dodge St.
Omaha, NE 68114
Phone: (402)390-5400
Hotline: 800-642-9999; (800)228-9515 (within CO, IA, KS, MO, SD, WY)

★ 1123 ★ **Rocky Mountain Poison and Drug Center**
645 Bannock St.
Denver, CO 80204-4507
Phone: (303)629-1123
Hotline: 800-332-3073; (800)525-5042 (within Montana); (800)442-2702 (within Wyoming)
Certified by the American Association of Poison Control Centers.

District of Columbia

★ 1124 ★ **National Capital Poison Center**
Georgetown University Hospital
3600 Reservoir Rd., N.W.
Washington, DC 20007
Phone: (202)625-3333; (202)784-4660
Certified by the American Association of Poison Control Centers.

Florida

★ 1125 ★ Florida Poison Information Center at the Tampa General Hospital
P.O. Box 1289
Tampa, FL 33601
Phone: (813)253-4444
Hotline: 800-282-3171 (within Florida)
Certified by the American Association of Poison Control Centers.

Georgia

★ 1126 ★ Georgia Poison Control Center
Grady Memorial Hospital
80 Butler St., S.E.
P.O. Box 26066
Atlanta, GA 30335-3801
Phone: (404)589-4400; (404)525-3323
Hotline: 800-282-5846 (within Georgia)
Certified by the American Association of Poison Control Centers.

Hawaii

★ 1127 ★ Hawaii Poison Center
Kapiolani-Children's Medical Center
1319 Punahou St.
Honolulu, HI 96826
Hotline: 800-362-3585 (within Hawaii)

Idaho

★ 1128 ★ Idaho Poison Control Center
St. Alphonsus Regional Medical Center
1055 N. Curtis Rd.
Boise, ID 83704
Hotline: 800-632-8000 (within Idaho)

★ 1129 ★ Spokane Poison Center
715 S. Cowley
Spokane, WA 99202
Hotline: 800-572-5842; (800)541-5624 (within northern Idaho and western Wyoming)

Illinois

★ 1130 ★ Central and Southern Illinois Regional Poison Resource Center
St. John's Hospital
800 E. Carpenter St.
Springfield, IL 62769
Hotline: 800-252-2022 (within Illinois)

★ 1131 ★ Chicago and Northeastern Illinois Regional Poison Control Center
Rush-Presbyterian-St. Luke's Medical Center
1753 W. Congress Pkwy.
Chicago, IL 60612
Hotline: 800-942-5969 (within Illinois)

Indiana

★ 1132 ★ Indiana Poison Center
Methodist Hospital of Indiana
1701 N. Senate Blvd.
P.O. Box 1367
Indianapolis, IN 46206
Phone: (317)929-2323
Hotline: 800-382-9097 (within Indiana)
Certified by the American Association of Poison Control Centers.

Iowa

★ 1133 ★ McKennen Hospital Poison Center
800 E. 21st St.
P.O. Box 5045
Sioux Falls, SD 57117-5045
Hotline: 800-952-0123; (800)843-0505 (within Iowa, Minnesota, Nebraska)

★ 1134 ★ Mid-Plains Poison Control Center
Children's Memorial Hospital
8301 Dodge St.
Omaha, NE 68114
Phone: (402)390-5400
Hotline: 800-642-9999; (800)228-9515 (within CO, IA, KS, MO, SD, WY)

★ 1135 ★ University of Iowa Hospitals and Clinics Poison Control Center
Iowa City, IA 52242
Hotline: 800-272-6477 (within Iowa)

★ 1136 ★ Variety Club Poison and Drug Information Center
Iowa Methodist Medical Center
1200 Pleasant St.
Des Moines, IA 50309
Hotline: 800-362-2327 (within Iowa)

Kansas

★ 1137 ★ Mid-America Poison Center
University of Kansas Medical Center
39th and Rainbow Blvd.
Kansas City, KS 66103
Hotline: 800-332-6633 (within Kansas)

★ 1138 ★ Mid-Plains Poison Control Center
Children's Memorial Hospital
8301 Dodge St.
Omaha, NE 68114
Phone: (402)390-5400
Hotline: 800-642-9999; (800)228-9515 (within CO, IA, KS, MO, SD, WY)

Kentucky

★ 1139 ★ Kentucky Regional Poison Center of Kosair Children's Hospital
P.O. Box 35070
Louisville, KY 40232-5070
Phone: (502)589-8222
Hotline: 800-722-5725 (within Kentucky)
Certified by the American Association of Poison Control Centers.

Maine

★ 1140 ★ Maine Poison Control Center at Maine Medical Center
22 Bramhall St.
Portland, ME 04102
Hotline: 800-442-6305 (within Maine)

Maryland

★ 1141 ★ Maryland Poison Center
University of Maryland School of Pharmacy
20 N. Pine St.
Baltimore, MD 21201
Phone: (301)528-7701
Hotline: 800-492-2414 (within Maryland)
Certified by the American Association of Poison Control Centers.

★ 1142 ★ National Capital Poison Center
Georgetown University Hospital
3600 Reservoir Rd., N.W.
Washington, DC 20007
Phone: (202)625-3333; (202)784-4660
Certified by the American Association of Poison Control Centers.

Massachusetts

★ 1143 ★ Massachusetts Poison Control System
300 Longwood Ave.
Boston, MA 02115
Phone: (617)232-2120
Hotline: 800-682-9211 (within Massachusetts)
Certified by the American Association of Poison Control Centers.

Michigan

★ 1144 ★ Blodgett Regional Poison Center
1840 Wealthy, S.E.
Grand Rapids, MI 49506
Hotline: 800-632-2727; (800)356-3232 (within neighboring states)
Certified by the American Association of Poison Control Centers.

★ 1145 ★ Children's Hospital of Michigan Poison Control Center
3901 Beaubien Blvd.
Detroit, MI 48201
Phone: (313)745-5711
Hotline: 800-462-6642 (within Michigan)
Certified by the American Association of Poison Control Centers.

★ 1146 ★ Great Lakes Poison Center
Bronson Methodist Hospital
252 E. Lovell St.
Kalamazoo, MI 49001
Hotline: 800-442-4112 (within area code 616)

Minnesota

★1147★ Hennepin Regional Poison Center
Hennepin County Medical Center
701 Park Ave.
Minneapolis, MN 55415
Phone: (612)347-3141; (612)337-7474
Certified by the American Association of Poison Control Centers.

★1148★ McKennen Hospital Poison Center
800 E. 21st St.
P.O. Box 5045
Sioux Falls, SD 57117-5045
Hotline: 800-952-0123; (800)843-0505 (within Iowa, Minnesota, Nebraska)

★1149★ Minnesota Regional Poison Center
St. Paul-Ramsey Medical Center
640 Jackson St.
St. Paul, MN 55101
Phone: (612)221-2113
Hotline: 800-222-1222 (within Minnesota)
Certified by the American Association of Poison Control Centers.

Missouri

★1150★ Cardinal Glennon Children's Hospital Regional Poison Center
1465 S. Grand Blvd.
St. Louis, MO 63104
Phone: (314)772-5200; (314)577-5336
Hotline: 800-392-9111 (within Missouri); (800)366-8888
Certified by the American Association of Poison Control Centers.

★1151★ Mid-Plains Poison Control Center
Children's Memorial Hospital
8301 Dodge St.
Omaha, NE 68114
Phone: (402)390-5400
Hotline: 800-642-9999; (800)228-9515 (within CO, IA, KS, MO, SD, WY)

Montana

★1152★ Rocky Mountain Poison and Drug Center
645 Bannock St.
Denver, CO 80204-4507
Phone: (303)629-1123
Hotline: 800-332-3073; (800)525-5042 (within Montana); (800)442-2702 (within Wyoming)
Certified by the American Association of Poison Control Centers.

Nebraska

★1153★ McKennen Hospital Poison Center
800 E. 21st St.
P.O. Box 5045
Sioux Falls, SD 57117-5045
Hotline: 800-952-0123; (800)843-0505 (within Iowa, Minnesota, Nebraska)

★1154★ Mid-Plains Poison Control Center
Children's Memorial Hospital
8301 Dodge St.
Omaha, NE 68114
Phone: (402)390-5400
Hotline: 800-642-9999; (800)228-9515 (within CO, IA, KS, MO, SD, WY)

★1155★ The Poison Center
8301 Dodge St.
Omaha, NE 68114
Phone: (402)390-5400
Hotline: 800-642-9999; (800)228-9515 (outside Nebraska)
Certified by the American Association of Poison Control Centers.

New Hampshire

★1156★ New Hampshire Poison Information Center
2 Maynard St.
Hanover, NH 03756
Hotline: 800-562-8236 (within New Hampshire)

New Jersey

★1157★ New Jersey Poison Information and Education System
201 Lyons Ave.
Newark, NJ 07112
Phone: (201)923-0764
Hotline: 800-962-1253 (within New Jersey)
Certified by the American Association of Poison Control Centers.

New Mexico

★1158★ New Mexico Poison and Drug Information Center
University of New Mexico
Albuquerque, NM 87131
Phone: (505)843-2551
Hotline: 800-432-6866 (within New Mexico)
Certified by the American Association of Poison Control Centers.

New York

★1159★ Central New York Poison Control Center
Upstate Medical Center
750 E. Adams St.
Syracuse, NY 13210
Hotline: 800-252-5655 (within New York)

★1160★ Long Island Regional Poison Control Center
Nassau County Medical Center
2201 Hempstead Tnpk.
East Meadow, NY 11556
Phone: (516)542-2323
Certified by the American Association of Poison Control Centers.

★1161★ New York City Poison Control Center
455 First Ave., Rm. 123
New York, NY 10016
Phone: (212)340-4494; (202)POI-SONS
Certified by the American Association of Poison Control Centers.

North Carolina

★1162★ Duke University Poison Control Center
Duke University Medical Center
Durham, NC 27710
Hotline: 800-672-1697 (within North Carolina)

★1163★ Triad Poison Center
Moses H. Cone Memorial Hospital
1200 N. Elm St.
Greensboro, NC 27401-1020
Hotline: 800-722-2222 (within North Carolina)

★1164★ Western North Carolina Poison Control Center
Memorial Mission Hospital
509 Biltmore Ave.
Asheville, NC 28801
Hotline: 800-542-4225

North Dakota

★1165★ North Dakota Poison Information Center
St. Luke's Hospitals
Fifth St. N. and Mills Ave.
Fargo, ND 58122
Hotline: 800-732-2200 (within North Dakota)

Ohio

★1166★ Akron Regional Poison Control Center
Children's Hospital Medical Center of Akron
281 Locust St.
Akron, OH 44308
Hotline: 800-362-9922 (within Ohio)

★1167★ Central Ohio Poison Control Center
Children's Hospital
700 Children's Dr.
Columbus, OH 43205
Phone: (614)228-1323; (614)228-2272
Hotline: 800-682-7625 (within Ohio)
Certified by the American Association of Poison Control Centers.

★1168★ Lorain Community Hospital Poison Control Center
3700 Kolbe Rd.
Lorain, OH 44053
Hotline: 800-821-8972 (within Ohio)

★ 1169 ★ Regional Poison Control System and Cincinnati Drug and Poison Information Center
University of Cincinnati Medical Center
231 Bethesda Ave., M.L. No. 144
Cincinnati, OH 45267-0144
Phone: (513)558-5111
Hotline: 800-872-5111 (within the region)
Certified by the American Association of Poison Control Centers.

★ 1170 ★ Western Ohio Regional Poison and Drug Information Center
Children's Medical Center
1 Children's Plaza
Dayton, OH 45404-1815
Hotline: 800-762-0727 (within Ohio)

Oklahoma

★ 1171 ★ Oklahoma Poison Control Center
Oklahoma Children's Memorial Hospital
940 N.E. Tenth St.
P.O. Box 26307
Oklahoma City, OK 73126
Hotline: 800-522-4611 (within Oklahoma)

Oregon

★ 1172 ★ Oregon Poison Center
Oregon Health Science University
3181 S.W. Sam Jackson Park Rd.
Portland, OR 97201
Phone: (503)279-8968
Hotline: 800-452-7165 (within Oregon)
Certified by the American Association of Poison Control Centers.

Pennsylvania

★ 1173 ★ Delaware Valley Regional Poison Control Center
1 Children's Center
34th and Civic Center Blvd.
Philadelphia, PA 19104
Phone: (215)386-2100
Certified by the American Association of Poison Control Centers. **Publication:** *The Poison Control Center Newsletter.*

★ 1174 ★ Northwest Regional Poison Center
Saint Vincent Health Center
232 W. 25th St.
Erie, PA 16544
Hotline: 800-822-3232 (within Pennsylvania)

★ 1175 ★ Pittsburgh Poison Center
3705 Fifth Ave. at DeSoto St.
Pittsburgh, PA 15213
Phone: (412)681-6669
Certified by the American Association of Poison Control Centers.

Rhode Island

★ 1176 ★ Rhode Island Poison Center
Rhode Island Hospital
593 Eddy St.
Providence, RI 02902
Phone: (401)277-5727
Certified by the American Association of Poison Control Centers.

South Dakota

★ 1177 ★ McKennen Hospital Poison Center
800 E. 21st St.
P.O. Box 5045
Sioux Falls, SD 57117-5045
Hotline: 800-952-0123; (800)843-0505 (within Iowa, Minnesota, Nebraska)

★ 1178 ★ Mid-Plains Poison Control Center
Children's Memorial Hospital
8301 Dodge St.
Omaha, NE 68114
Phone: (402)390-5400
Hotline: 800-642-9999; (800)228-9515 (within CO, IA, KS, MO, SD, WY)

★ 1179 ★ St. Luke's Midland Poison Control Center
305 S. State St.
Aberdeen, SD 57401
Hotline: 800-592-1889

Texas

★ 1180 ★ North Texas Poison Center
P.O. Box 35926
Dallas, TX 75235
Phone: (214)590-5000
Hotline: 800-441-0040 (within Texas)
Certified by the American Association of Poison Control Centers.

★ 1181 ★ Texas State Poison Center
University of Texas Medical Branch
Galveston STR, TX 77650-2780
Phone: (409)765-1420; (713)654-1701; (512)478-4490
Hotline: 800-392-8548 (within Texas)
Certified by the American Association of Poison Control Centers.

Utah

★ 1182 ★ Intermountain Regional Poison Control Center
50 N. Medical Dr., Bldg. 428
Salt Lake City, UT 84132
Phone: (801)581-2151
Hotline: 800-456-7707 (within Utah)
Certified by the American Association of Poison Control Centers.

Virginia

★ 1183 ★ Blue Ridge Poison Center
University of Virginia
Blue Ridge Hospital, Box 67
Charlottesville, VA 22901
Hotline: 800-451-1428

★ 1184 ★ National Capital Poison Center
Georgetown University Hospital
3600 Reservoir Rd., N.W.
Washington, DC 20007
Phone: (202)625-3333; (202)784-4660
Certified by the American Association of Poison Control Centers.

★ 1185 ★ Tidewater Poison Center
150 Kingsley Ln.
Norfolk, VA 23505
Hotline: 800-522-6337 (within Virginia)

Washington

★ 1186 ★ Central Washington Poison Center
Yakima Valley Memorial Hospital
2811 Tieton Dr.
Yakima, WA 98902
Hotline: 800-572-9176 (within Washington)

★ 1187 ★ Mary Bridge Poison Center
Mary Bridge Children's Health Center
1317 S. K St.
P.O. Box 5299
Tacoma, WA 98405-0987
Hotline: 800-542-6319 (within Washington)

★ 1188 ★ Seattle Poison Center
Children's Hospital and Medical Center
4800 Sand Point Way, N.E.
P.O. Box C5371
Seattle, WA 98105-0371
Hotline: 800-732-6985 (within Washington)

★ 1189 ★ Spokane Poison Center
715 S. Cowley
Spokane, WA 99202
Hotline: 800-572-5842; (800)541-5624 (within northern Idaho and western Wyoming)

West Virginia

★ 1190 ★ West Virginia Poison Center
West Virginia University Health Science Center/Charleston Division
3110 McCorkle Ave., S.E.
Charleston, WV 25304
Phone: (304)348-4211
Hotline: 800-642-3625 (within West Virginia)
Certified by the American Association of Poison Control Centers.

Wyoming

★ 1191 ★ Mid-Plains Poison Control Center
Children's Memorial Hospital
8301 Dodge St.
Omaha, NE 68114
Phone: (402)390-5400
Hotline: 800-642-9999; (800)228-9515 (within CO, IA, KS, MO, SD, WY)

★ 1192 ★ Rocky Mountain Poison and Drug Center
645 Bannock St.
Denver, CO 80204-4507
Phone: (303)629-1123
Hotline: 800-332-3073; (800)525-5042 (within Montana); (800)442-2702 (within Wyoming)
Certified by the American Association of Poison Control Centers.

★ 1193 ★ Spokane Poison Center
715 S. Cowley
Spokane, WA 99202
Hotline: 800-572-5842; (800)541-5624 (within northern Idaho and western Wyoming)

Federal Government Organizations

★ 1194 ★ Agency for Toxic Substances and Disease Registry
Centers for Disease Control
U.S. Department of Health and Human Services
1600 Clifton Rd., N.E.
Atlanta, GA 30333
Phone: (404)452-4100; (404)639-2886
Carries out health-related responsibilities of the Comprehensive Environmental Response, Compensation, and Liability Act (CERCLA; 1980) and its various amendments such as the Superfund Act. Its programs are designed to protect the public and workers from exposure and adverse health effects of hazardous substances in storage sites or released in fires, explosions, or transportation accidents.

★ 1195 ★ Alberta Environmental Protection Service
Oxbridge Place
9820 106th St.
Edmonton, AB, Canada T5K 2J6
Phone: (403)427-5872

★ 1196 ★ Asbestos Action Program
Office of Pesticides and Toxic Substances (TS-794)
U.S. Environmental Protection Agency (EPA)
Federal Asbestos Task Force
401 M St., S.W.
Washington, DC 20460
Phone: (202)382-3862
Coordinates public programs administered by the federal government concerning potential health hazards from exposure to asbestos and other fibrous materials.

★ 1197 ★ The Asbestos List
Freedom of Information Division
U.S. Consumer Product Safety Commission (CPSC)
5401 Westbard Ave.
Washington, DC 20207
Phone: 800-492-8363; (301)492-6580
Hotline: 800-638-CPSC; teletype for the hearing impaired: (800)492-8104, (800)6388270
Provides a free listing of asbestos-detection organizations that have been certified by the EPA.

★ 1198 ★ British Columbia Ministry of Environment
Environmental Protection Division
810 Blanshard St., Second Floor
Victoria, BC, Canada V8V 1X5
Phone: (604)387-9965

★ 1199 ★ Canadian Department of the Environment, Conservation and Protection
Ottawa, ON, Canada K1A 0H3
Phone: (819)997-1575
Contact: Robert de Cotret, Minister

★ 1200 ★ Canadian Department of the Environment, Conservation and Protection
Atlantic Region Office
45 Alderney Dr., Fifth Floor
Dartmouth, NH B2Y 2N6
Phone: (902)426-3593
Contact: K. Hamilton, Acting Dir.
Includes New Brunswick, Newfoundland, Nova Scotia, and Prince Edward Island.

★ 1201 ★ Canadian Department of the Environment, Conservation and Protection
Ontario Region Office
25 St. Clair Ave., E.
Toronto, ON, Canada M4T 1M2
Phone: (416)973-1055
Contact: S. Llewellyn, Dir.

★ 1202 ★ Canadian Department of the Environment, Conservation and Protection
Pacific and Yukon Region Office
Kapilano 100-Park Royal
West Vancouver, BC, Canada V7T 1A2
Phone: (604)666-0064
Contact: B. A. Heskin, Dir.
Includes British Columbia and Yukon Territory.

★ 1203 ★ Canadian Department of the Environment, Conservation and Protection
Quebec Region Office
1179, rue de Bluery
Montreal, PQ, Canada H3B 3H9
Phone: (514)283-0178
Contact: Raymond Perrier, Dir.

★ 1204 ★ Canadian Department of the Environment, Conservation and Protection
Western and Northern Region Office
Twin Atria No. 2, Second Floor
Edmonton, AB, Canada T6B 2X3
Phone: (403)468-8074
Contact: Dr. Robert K. Lane, Dir.
Includes Alberta, Manitoba, Northwest Territories, and Saskatchewan.

★ 1205 ★ Chemical Emergency Preparedness and Prevention
Office of Solid Waste and Emergency Response
U.S. Environmental Protection Agency (EPA)
401 M St., S.W., Rm. M3103
Washington, DC 20460
Phone: (202)260-8600

★ 1206 ★ Emergency Response Division
Office of Emergency and Remedial Response/Superfund
U.S. Environmental Protection Agency (EPA)
401 M St., S.W., Rm. 2710
Washington, DC 20460
Phone: (202)260-8720

★ 1207 ★ Environmental Assistance Division
Office of Toxic Substances
U.S. Environmental Protection Agency (EPA)
401 M St., S.W., Rm. 543
Washington, DC 20460
Phone: (202)260-7024

★ 1208 ★ Environmental Fate and Effects Division
Office of Pesticide Programs
U.S. Environmental Protection Agency (EPA)
Crystal Mall Bldg. 2, Tenth Floor
1921 Jefferson Davis Hwy.
Arlington, VA 22202
Phone: (703)557-7695

★ 1209 ★ Federal Insecticide, Fungicide, and Rodenticide Act Scientific Advisory Panel (FIFRA)
Office of Pesticide Programs (H7509C)
U.S. Environmental Protection Agency (EPA)
401 M St., S.W.
Washington, DC 20460
Phone: (703)557-4369
Reviews the impact of pesticide regulations on health and the environment and makes recommendations to the Office of Pesticide Programs.

★ 1210 ★ Health and Environmental Review Division
Office of Toxic Substances
U.S. Environmental Protection Agency (EPA)
401 M St., S.W., Rm. 617
Washington, DC 20460
Phone: (202)260-4241

★ 1211 ★ Health Effects Division
Office of Pesticide Programs
U.S. Environmental Protection Agency (EPA)
Crystal Mall Bldg. 2, Rm. 821
1921 Jefferson Davis Hwy.
Arlington, VA 22202
Phone: (703)557-7351

★ 1212 ★ Interagency Toxic Substances Data Committee (ITSDC)
Office of Program Integration and Information
U.S. Environmental Protection Agency (EPA)
Office of Toxic Substances
401 M St., S.W.
Washington, DC 20460
Coordinates the planning and activities concerning chemical data and information projects of the major federal producers and users of chemical data.

★ 1213 ★ Manitoba Environmental Approvals
139 Tuxedo Ave., Bldg. 2
Winnipeg, MB, Canada R3N 0H6
Phone: (204)945-7071
Important contact numbers: Manitoba Clean Environment Commission (includes air quality), (204)945-7120; Environmental Operations, Dangerous Goods (includes hazardous waste management), (204)945-7039.

★ 1214 ★ National Health Information Center
Office of Disease Prevention and Health Promotion (ODPHP)
U.S. Department of Health and Human Services
P.O. Box 1133
Washington, DC 20013-1133
Phone: (301)565-4167; 800-336-4797
Operates the National Health Information Clearinghouse (NHIC), a health information referral service available to both health professionals and the general public. *Publications: Healthfinders,* a series of resource lists on current health concerns; *Health Information Resources in the Federal Government,* a directory; *Locating Funds for Health Promotion Projects,* a guidebook; a database of 1,000 healthrelated organizations, available on-line through DIRLINE, part of the National Library of Medicine's MEDLARS system.

★ 1215 ★ National Injury Information Clearinghouse (NIIC)
U.S. Consumer Product Safety Commission (CPSC)
Westwood Tower Bldg.
5401 Westbard Ave., Rm. 625
Bethesda, MD
Phone: 800-492-8363; (301)492-6580
Hotline: 800-638-CPSC; teletype for the hearing-impaired: (800)492-8104, (800)6388270
Compiles and publishes statistics relating to injuries from consumer products, including those caused by the misuse of products containing potentially toxic substances.

★ 1216 ★ National Institute for Occupational Safety and Health (NIOSH)
Centers for Disease Control
U.S. Department of Health and Human Services (DHHS)
1600 Clifton Rd., N.E.
Mail Stop D-32
Atlanta, GA 30333
Phone: (404)329-3691
Conducts research on various safety and health problems, including those related to the use of toxic substances in the workplace, and recommends standards for adoption by the Occupational Safety and Health Administration (OSHA). It may undertake investigations in the workplace: gathering testimonies from employers, measuring employee exposure to potentially hazardous substances, and requiring employers to provide medical examinations and tests to determine the incidence of illness

caused by these substances among the employees.

★ 1217 ★ National Institute for Occupational Safety and Health (NIOSH)
Western Region Office
Federal Office Bldg.
1961 Stout St., Rm. 1185
Denver, CO 80294
Phone: (303)844-6166
Includes Arizona, California, Hawaii, Nevada, American Samoa, Guam, and the Pacific Trust Territories

★ 1218 ★ National Institute for Occupational Safety and Health Educational Resource Center (NIOSH/ERC)
Division of Training and Development
4676 Columbia Pkwy., C-11
Cincinnati, OH 45226-1998
Phone: (513)533-8225
Fax: (513)533-8560
Contact: Marsha Striley, Training Registrar

★ 1219 ★ National Institute for Occupational Safety and Health Educational Resource Center (NIOSH/ERC)
Harvard University
Harvard School of Public Health
Harvard Educational Resource Center
677 Huntington Ave., LL-23
Boston, MA 02115
Phone: (617)432-1171
Fax: (617)432-1969
Contact: Daryl J. Bichel, Dir., Continuing Education

★ 1220 ★ National Institute for Occupational Safety and Health Educational Resource Center (NIOSH/ERC)
Johns Hopkins University
Department of Environmental Health Sciences
The Johns Hopkins Educational Resource Center
615 N. Wolfe St., Rm. 6001
Baltimore, MD 21205
Phone: (301)955-2609
Contact: Dr. Jacqueline Corn, Dir., Continuing Education

★ 1221 ★ National Institute for Occupational Safety and Health Educational Resource Center (NIOSH/ERC)
Midwest Center for Occupational Health and Safety
Minnesota Educational Resource Center
640 Jackson St.
St. Paul, MN 55101
Phone: (612)221-3992
Contact: Jeanne F. Ayers, Dir.

★ 1222 ★ National Institute for Occupational Safety and Health Educational Resource Center (NIOSH/ERC)
Southwest Center for Occupational Health and Safety
P.O. Box 20186
Houston, TX 77225
Phone: (713)792-4648
Contact: Kay Garcia, Dir., Continuing Education

★ 1223 ★ National Institute for Occupational Safety and Health Educational Resource Center (NIOSH/ERC)
University of Alabama at Birmingham
School of Public Health
Deep South Center for Occupational Health and Safety
University Station
Birmingham, AL 35294
Phone: (205)934-7178
Contact: Elizabeth H. Murray, Dir.

★ 1224 ★ National Institute for Occupational Safety and Health Educational Resource Center (NIOSH/ERC)
University of California
Center for Occupational and Environmental Health
Northern California Educational Resource Center
2521 Channing Way
Berkeley, CA 94720
Phone: (415)642-5507
Contact: Marion Gillen, Coord., Continuing Education

★ 1225 ★ National Institute for Occupational Safety and Health Educational Resource Center (NIOSH/ERC)
University of Cincinnati (ML 56)
Department of Environmental Health
University of Cincinnati Educational Resource Center
3223 Eden Ave.
Cincinnati, OH 45267-0056
Phone: (513)558-1730
Contact: Judy L. Jarrel, Dir.

★ 1226 ★ National Institute for Occupational Safety and Health Educational Resource Center (NIOSH/ERC)
University of Illinois at Chicago
School of Public Health
Occupational Health and Safety Center
Illinois Educational Resource Center
2121 W. Taylor St.
Chicago, IL 60612
Phone: (312)413-0459
Contact: Louis Rowitz

★ 1227 ★ National Institute for Occupational Safety and Health Educational Resource Center (NIOSH/ERC)
University of Medicine and Dentistry of New Jersey
Robert Wood Johnson Medical School
New York/New Jersey Educational Resource Center
45 Knightsbridge Rd., Brookwood II
Piscataway, NJ 08854
Phone: (908)463-5062
Fax: (908)463-5133
Contact: Lee Lausten, Deputy Dir.

★ 1228 ★ National Institute for Occupational Safety and Health Educational Resource Center (NIOSH/ERC)
University of Michigan
Center for Occupational Health and Safety Engineering
Michigan Educational Resource Center
1205 Beal Ave.
176 IOE Bldg.
Ann Arbor, MI 48109-2117
Phone: (313)936-0148
Contact: Randy Rabourn, Dir., Continuing Education

★ 1229 ★ National Institute for Occupational Safety and Health Educational Resource Center (NIOSH/ERC)
University of North Carolina
North Carolina Educational Resource Center
109 Conner Dr., Ste. 1101
Chapel Hill, NC 27514
Phone: (919)962-2101
Contact: Larry D. Hyde, Dir., Continuing Education

★ 1230 ★ National Institute for Occupational Safety and Health Educational Resource Center (NIOSH/ERC)
University of Southern California
Institute of Safety and Systems Management
Southern California Educational Resource Center
927 W. 35th Pl., Rm. 102
Los Angeles, CA 90007
Phone: (213)740-3998
Contact: Ramona Cayuela, Dir., Continuing Education

★ 1231 ★ National Institute for Occupational Safety and Health Educational Resource Center (NIOSH/ERC)
University of Utah
Rocky Mountain Center for Occupational and Environ. Health
Bldg. 512
Salt Lake City, UT 84112
Phone: (801)581-5710
Contact: Connie Crandall, Dir., Continuing Education

★ 1232 ★ National Institute for Occupational Safety and Health Educational Resource Center (NIOSH/ERC)
University of Washington
School of Public Health, SC-34
Department of Environmental Health
Northwest Center for Occupational Health and Safety
Seattle, WA 98195
Phone: (206)543-1069
Contact: Sharon Morris, Dir.

★ 1233 ★ National Pesticide Information Retrieval Service (NPIRS)
Pesticide and Hazardous Chemical Center
Purdue University
Entomology Hall
West Lafayette, IN 47907
Phone: (317)494-6616

★ 1234 ★ National Pesticides Hazard Assessment Program
Oregon State University
Department of Agriculture Chemistry
Corvallis, OR 97331
Phone: (503)757-5086
Contact: Dr. Sheldon Wagner, M.D.
Provides basic and clinical information on the toxicology of pesticides.

★ 1235 ★ National Pesticides Telecommunication Network (NPTN)
Texas Tech University Health Science Center
School of Medicine
Department of Preventive Medicine
Lubbock, TX 79430
Phone: (806)743-3091
Fax: 800-743-3094
Hotline: 800-858-PEST
Provides information on pesticide-related health problems, toxicity, minor cleanup problems, residue and safety practices, health and environmental effects, and cleanup and disposal procedures.

★ 1236 ★ National Technical Information Service (NTIS)
5285 Port Royal Rd.
Springfield, VA 22161
Phone: 800-336-4700
Fax: (703)321-8547
National clearinghouse for all documents published by federal and foreign research and development agencies, including those of the EPA and OSHA.

★ 1237 ★ National Toxicology Program
National Institute for Environmental Health Sciences (NIEHS)
U.S. Department of Health and Human Services (DHHS)
P.O. Box 12233
Research Triangle Park, NC 27709
Phone: (919)541-3345
Primary federal agency for biomedical research on the effects of chemical, physical, and biological environmental agents on human health and well being. It seeks to identify potentially hazardous environmental agents and develop biological test systems that can be used to predict toxicity from exposure to environmental factors.

★ 1238 ★ New Brunswick Department of the Environment
P.O. Box 6000
364 Argyle St.
Fredericton, NB, Canada E3B 5H1
Phone: (506)453-2861

★ 1239 ★ Newfoundland Department of Environment and Lands
P.O. Box 8700
Confederation Bldg., Fourth Floor, West Block
St. John's, NF, Canada A1B 4J6
Phone: (709)576-2555

★ 1240 ★ Nova Scotia Department of the Environment
P.O. Box 2107
Halifax, NS, Canada B3J 3B7
Phone: (902)424-5300

★ 1241 ★ Office of Consumer Affairs
U.S. Food and Drug Administration (FDA)
U.S. Department of Health and Human Services (DHHS)
5600 Fischers Lane
Rockville, MD 20857
Phone: (301)443-3170
Enforces the Federal Food, Drug and Cosmetic Act and related laws to insure the purity and safety of foods, drugs, cosmetics, and therapeutic devices and the truthful, informative labeling of such products. Food and color additives, antibiotic drugs, and insulin and most prescription drugs are subject to pre-marketing approval by the Agency. It licenses the production of vaccines, serums, and other biologic drugs and regulates blood banks. The FDA enforces radiation safety standards for products such as X-ray equipment, color televisions, lasers, sunlamps, and microwave ovens. The FDA has regional offices throughout the U.S.

★ 1242 ★ Office of Pesticide Programs
Document Management Section (H7502C)
U.S. Environmental Protection Agency (EPA)
ISB/PMSD
401 M St. S.W.
Washington, DC 20460
Phone: (202)260-4474
Provides free publications on pesticide use and safety and document numbers for Pesticide

Fact Sheets available through the National Technical Information Service (NTIS).

★ 1243 ★ Office of Research and Development
Center for Environmental Research Information (CERI)
U.S. Environmental Protection Agency (EPA)
26 W. Martin Luther King Dr.
Cincinnati, OH 45268
Phone: (513)569-7391

★ 1244 ★ Ontario Ministry of the Environment
135 St. Clair Ave., W., 15th Floor
Toronto, ON, Canada M4V 1P5
Phone: (416)323-4360
Important contact numbers: Air Quality and Meteorology Branch, (416)326-1700; Waste Management Branch, (416)323-5200.

★ 1245 ★ Pesticides and Toxic Substances Department
Office of Pesticide Programs
U.S. Environmental Protection Agency (EPA)
Crystal Mall Bldg. 2, Rm. 1115
1921 Jefferson Davis Hwy.
Arlington, VA 22202
Phone: (703)557-7090

★ 1246 ★ Pesticides and Toxic Substances Department
Office of Toxic Substances
U.S. Environmental Protection Agency (EPA)
401 M St., S.W., Rm. 539
Washington, DC 20460
Phone: (202)260-3810

★ 1247 ★ Pesticides and Toxic Substances Enforcement Division
Office of Civil Enforcement
U.S. Environmental Protection Agency (EPA)
401 M St., S.W., Rm. 113
Washington, DC 20460
Phone: (202)260-8690

★ 1248 ★ Pollution Prevention Information Clearinghouse (PPIC)
Pollution Prevention Office
U.S. Environmental Protection Agency (EPA)
401 M St., S.W.
Washington, DC 20460
Hotline: 800-424-9346
Provides information on government and industry pollution prevention programs, project funding opportunities, upcoming events, and conferences and seminars.

★ 1249 ★ Public Information Center
Information Access Branch
U.S. Environmental Protection Agency (EPA)
401 M St. S.W., PM-211B
Washington, DC 20460
Phone: (202)260-7751
Distributes general, nontechnical information about the EPA and various environmental topics to the public and to other government agencies at all levels. Refers technical information requests to the appropriate EPA office or other government agencies. 8 a.m.-5:30 p.m. EST, Monday through Friday.

★ 1250 ★ Quebec Direction de l'Expertise scientifique
3900, rue Marly
Sainte-Foy, PQ, Canada G1X 4E4
Phone: (418)644-3420
Important contact number: Gestion dechets et lieux contamines (hazardous waste management), (418)643-3794.

★ 1251 ★ Radon Division
Office of Radiation Programs
U.S. Environmental Protection Agency (EPA)
401 M St., S.W.
Washington, DC 20460
Phone: (202)260-9622

★ 1252 ★ Saskatchewan Department of Environmental and Public Safety
Air and Land Protection Branch
3085 Albert St.
Regina, SK, Canada S4S 0B1
Phone: (306)787-6195

★ 1253 ★ Superfund Division
Office of Civil Enforcement
U.S. Environmental Protection Agency (EPA)
401 M St., S.W., Rm. 3105-G
Washington, DC 20460
Phone: (202)260-3050

★ 1254 ★ Title III Reporting Center
U.S. Environmental Protection Agency (EPA)
470/490 l'Enfant Plaza, Ste. 7103
Washington, DC 20022
Phone: (202)260-1501
A reading room which makes available to the public copies of the reporting forms submitted to the Toxic Release Inventory (TRI). The TRI contains information on the estimated annual releases of toxic chemicals into the environment, the names and locations of the releasing facilities, and the amounts of specific substances that are released into the air, water, or soil.

★ 1255 ★ Toxic Substances Control Act (TSCA) Assistance Information Service
Office of Pesticides and Toxic Substances
U.S. Environmental Protection Agency (EPA)
401 M St., S.W.
Washington, DC 20460
Phone: (202)260-1404
Fax: (202)260-5603
Hotline: 800-424-9065; (800)835-6700 (for asbestos in schools only)
Provides general and technical information, including publications and audiovisual material on toxic substances and regulations, to the chemical industry, labor and trade organizations, environmental groups, and the public.

★ 1256 ★ U.S. Consumer Product Safety Commission (CPSC)
Bethesda, MD
Phone: 800-492-8363; (301)492-6580
Hotline: 800-638-CPSC; teletype for the hearing-impaired: (800)492-8104, (800)6388270
Contact: Jacqueline Jones-Smith, Chair
Regulates consumer products against unreasonable risks of injury in or around the home, schools, and recreational areas; evaluates the comparative safety of consumer products; develops uniform safety standards and minimizes conflicting state and local regulations; promotes research and investigation into the causes and prevention of product-related deaths, illnesses, and injuries. The following products are generally exempted from the Commissions's authority: foods, drugs, cosmetics, medical devices, motor vehicles, boats, airplanes, tobacco, firearms, alcohol, and pesticides.

★ 1257 ★ U.S. Consumer Product Safety Commission (CPSC)
Albuquerque Resident Post
Albuquerque Technical Vocational Institute
604 Bueno Vista, S.E.
Albuquerque, NM 87106
Phone: (505)766-2108
Hotline: 800-638-CPSC

★ 1258 ★ U.S. Consumer Product Safety Commission (CPSC)
Atlanta Satellite Office
730 Peachtree St., N.E., Ste. 871
Atlanta, GA 30365
Phone: (404)347-2231
Hotline: 800-638-CPSC

★ 1259 ★ U.S. Consumer Product Safety Commission (CPSC)
Baltimore Resident Post
303 E. Fayette St., Lower Level
Baltimore, MD 21202
Phone: (301)922-0622
Hotline: 800-638-CPSC

★ 1260 ★ U.S. Consumer Product Safety Commission (CPSC)
Boston Resident Post
Federal Office Bldg.
10 Causeway St., Rm. 224
Boston, MA 02222
Phone: (617)565-5915
Hotline: 800-638-CPSC

★ 1261 ★ U.S. Consumer Product Safety Commission (CPSC)
Bridgeport Resident Post
U.S. Courthouse and Federal Bldg.
915 Lafayette Blvd., Rm. 314-B
Bridgeport, CT 06604
Phone: (203)579-5915
Hotline: 800-638-CPSC

★ 1262 ★ U.S. Consumer Product Safety Commission (CPSC)
Buffalo Resident Post
111 W. Huron St.
P.O. Box 3001
Buffalo, NY 14202
Phone: (716)846-4116
Hotline: 800-638-CPSC

★ 1263 ★ U.S. Consumer Product Safety Commission (CPSC)
Central Regional Center
230 S. Dearborn St., Rm. 2945
Chicago, IL 60604
Phone: (312)353-8260
Hotline: 800-638-CPSC

★ 1264 ★ U.S. Consumer Product Safety Commission (CPSC)
Charlotte Resident Post
222 S. Church St., Ste. 408
Charlotte, NC 28202
Phone: (704)371-1615
Hotline: 800-638-CPSC

★ 1265 ★ U.S. Consumer Product Safety Commission (CPSC)
Cincinnati Resident Post
Federal Office Bldg., Rm. 3527
550 Main St.
Cincinnati, OH 45202
Phone: (513)684-2872
Hotline: 800-638-CPSC

★ 1266 ★ U.S. Consumer Product Safety Commission (CPSC)
Cleveland Resident Post
1 Playhouse Square
1375 Euclid Ave., Rm. 606
Cleveland, OH 44115
Phone: (216)522-3886
Hotline: 800-638-CPSC

★ 1267 ★ U.S. Consumer Product Safety Commission (CPSC)
Dallas Satellite Office
1100 Commerce St., Rm. 1C-10
Dallas, TX 75242
Phone: (214)767-0841
Hotline: 800-638-CPSC

★ 1268 ★ U.S. Consumer Product Safety Commission (CPSC)
Denver Resident Post
U.S. Customs House
721 19th St., Rm. 579
Denver, CO 80202
Phone: (303)844-2904
Hotline: 800-638-CPSC

★ 1269 ★ U.S. Consumer Product Safety Commission (CPSC)
Detroit Resident Post
McNamara Federal Bldg.
477 Michigan Ave., Rm. M-24
Detroit, MI 48226
Phone: (313)226-4040
Hotline: 800-638-CPSC

★ 1270 ★ U.S. Consumer Product Safety Commission (CPSC)
Eastern Regional Center
6 World Trade Center
Vesey St., Third Floor
New York, NY 10048
Phone: (212)264-1125
Hotline: 800-638-CPSC

★ 1271 ★ U.S. Consumer Product Safety Commission (CPSC)
Honolulu Resident Post
P.O. Box 50052
Honolulu, HI 96850
Phone: (808)546-7523
Hotline: 800-638-CPSC

★ 1272 ★ U.S. Consumer Product Safety Commission (CPSC)
Houston Resident Post
405 Main St., Rm. 605
Houston, TX 77002
Phone: (713)226-2814
Hotline: 800-638-CPSC

★ 1273 ★ U.S. Consumer Product Safety Commission (CPSC)
Indianapolis Resident Post
Corporate Square West
Bldg. 10, Ste. 1000
5610 Crawfordsville Rd.
Indianapolis, IN 46224
Hotline: 800-638-CPSC

★ 1274 ★ U.S. Consumer Product Safety Commission (CPSC)
Kansas City Resident Post
1009 Cherry St.
Kansas City, MO 64106
Phone: (816)374-2034
Hotline: 800-638-CPSC

★ **1275** ★ **U.S. Consumer Product Safety Commission (CPSC)**
Little Rock Resident Post
Arkansas Department of Health
4815 W. Markham
Little Rock, AR 72201
Phone: (501)378-6631
Hotline: 800-638-CPSC

★ **1276** ★ **U.S. Consumer Product Safety Commission (CPSC)**
Los Angeles Resident Post
4221 Wilshire Blvd., Ste. 220
Los Angeles, CA 90010
Phone: (213)251-7476
Hotline: 800-638-CPSC

★ **1277** ★ **U.S. Consumer Product Safety Commission (CPSC)**
Miami Resident Post
299 E. Broward Blvd., Rm. 202-F
Ft. Lauderdale, FL 33301
Phone: (305)527-7161
Hotline: 800-638-CPSC

★ **1278** ★ **U.S. Consumer Product Safety Commission (CPSC)**
Milwaukee Resident Post
Federal Bldg., Rm. 607
517 E. Wisconsin Ave.
Milwaukee, WI 53202
Phone: (414)291-1468
Hotline: 800-638-CPSC

★ **1279** ★ **U.S. Consumer Product Safety Commission (CPSC)**
New Orleans Resident Post
4298 Elysian Fields Ave.
New Orleans, LA 70122
Phone: (504)589-3742
Hotline: 800-638-CPSC

★ **1280** ★ **U.S. Consumer Product Safety Commission (CPSC)**
Orlando Resident Post
c/o Food and Drug Administration
7200 Lake Ellenor Dr.
Orlando, FL 32809
Phone: (305)420-6261
Hotline: 800-638-CPSC

★ **1281** ★ **U.S. Consumer Product Safety Commission (CPSC)**
Philadelphia Resident Post
Customs House Bldg., Rm. 209
Philadelphia, PA 19106
Phone: (215)597-9105
Hotline: 800-638-CPSC

★ **1282** ★ **U.S. Consumer Product Safety Commission (CPSC)**
Phoenix Resident Post
U.S. Dept. of Health and Human Svcs., Environmental Health
3738 N. 16th St., Ste. A
Phoenix, AZ 85016-5981
Phone: (602)241-2397
Hotline: 800-638-CPSC

★ **1283** ★ **U.S. Consumer Product Safety Commission (CPSC)**
Pittsburgh Resident Post
Federal Bldg., Rm. 2318
1000 Liberty Ave.
Pittsburgh, PA 15222
Phone: (412)644-6582
Hotline: 800-638-CPSC

★ **1284** ★ **U.S. Consumer Product Safety Commission (CPSC)**
Portland Resident Post
U.S. Courthouse, Rm. 207
620 S.W. Main St.
Portland, OR 97205
Phone: (503)221-3056
Hotline: 800-638-CPSC

★ **1285** ★ **U.S. Consumer Product Safety Commission (CPSC)**
San Juan Resident Post
U.S. Courthouse and Federal Bldg.
Carlos Chardon Ave., Rm. 300
Hato Rey, PR 00918
Phone: (809)753-4403
Hotline: 800-638-CPSC

★ **1286** ★ **U.S. Consumer Product Safety Commission (CPSC)**
Seattle Resident Post
6046 Federal Office Bldg.
909 First Ave.
Seattle, WA 98174
Phone: (206)442-5276
Hotline: 800-638-CPSC

★ **1287** ★ **U.S. Consumer Product Safety Commission (CPSC)**
St. Louis Resident Post
808 N. Collins
St. Louis, MO 63102
Phone: (314)425-6281
Hotline: 800-638-CPSC

★ **1288** ★ **U.S. Consumer Product Safety Commission (CPSC)**
Trenton Resident Post
City Hall, First Floor Annex
319 E. State St.
Trenton, NJ 08608
Phone: (609)989-2062
Hotline: 800-638-CPSC

★ **1289** ★ **U.S. Consumer Product Safety Commission (CPSC)**
Tulsa Resident Post
333 W. Fourth St., Rm. 3097
Tulsa, OK 74103
Phone: (918)581-7606
Hotline: 800-638-CPSC

★ **1290** ★ **U.S. Consumer Product Safety Commission (CPSC)**
Twin Cities Resident Post
Federal Courts Bldg., Ste. 128
316 N. Robert St.
St. Paul, MN 55101
Phone: (612)725-7781
Hotline: 800-638-CPSC

★ **1291** ★ **U.S. Consumer Product Safety Commission (CPSC)**
Washington Resident Post
11820 Coakley Circle
Rockville, MD 20852
Phone: (301)443-1152
Hotline: 800-638-CPSC

★ **1292** ★ **U.S. Consumer Product Safety Commission (CPSC)**
Western Regional Center
555 Battery St., Rm. 415
San Francisco, CA 94111
Phone: (415)556-1816
Hotline: 800-638-CPSC

★ **1293** ★ **U.S. Environmental Protection Agency (EPA)**
401 M St., S.W.
Washington, DC 20460
Phone: (202)260-2080; (202)260-7751
Fax: (202)260-7883
Contact: William K. Reilly, Admin.
The Agency's mission is to control and abate pollution in the areas of air, water, solid waste, pesticides, radiation, and toxic substances. The EPA coordinates and supports research and anti-pollution activities by state and local governments, private and public groups, individuals, and educational institutions. *Publication: EPA Journal*, a monthly.

★ **1294** ★ **U.S. Environmental Protection Agency (EPA)**
Region I
John F. Kennedy Federal Bldg., Rm. 2203
Boston, MA 02203
Phone: (617)565-3715; (617)565-3424
Contact: Julie D. Belaga, Regional Admin.
Includes Connecticut, Maine, Massachusetts, New Hampshire, Rhode Island, and Vermont.

★ **1295** ★ **U.S. Environmental Protection Agency (EPA)**
Region II
Jacob K. Javitz Federal Bldg.
26 Federal Plaza
New York, NY 10278
Phone: (212)264-2515
Contact: Constantine Sidamon-Eristoff, Regional Admin.
Includes New Jersey, New York, Puerto Rico, and the Virgin Islands.

★ **1296** ★ **U.S. Environmental Protection Agency (EPA)**
Region III
841 Chestnut Bldg.
Philadelphia, PA 19107
Phone: (215)597-9800; (215)597-9370
Contact: Edwin B. Erickson, Regional Admin.
Includes Delaware, District of Columbia, Maryland, Pennsylvania, Virginia, and West Virginia.

★ **1297** ★ **U.S. Environmental Protection Agency (EPA)**
Region IV
345 Cortland St., N.E.
Atlanta, GA 30365
Phone: 800-282-0289; 800-241-1754; (404)347-3004
Contact: Greer C. Tidwell, Regional Admin.
Includes Alabama, Florida, Georgia, Kentucky, Mississippi, North Carolina, South Carolina, and Tennessee.

★ **1298** ★ **U.S. Environmental Protection Agency (EPA)**
Region V
230 S. Dearborn St.
Chicago, IL 60604
Phone: 800-572-2515; 800-621-8431; (312)353-2072
Contact: Valdas V. Adamkus, Regional Admin.
Includes Illinois, Indiana, Michigan, Minnesota, Ohio, and Wisconsin.

★ **1299** ★ **U.S. Environmental Protection Agency (EPA)**
Region VI
First Interstate Bank Tower at Fountain Place
1445 Ross Ave., Ste. 1200
Dallas, TX 74202
Phone: (214)655-2200
Contact: Robert E. Layton, Jr., Regional Admin.

Includes Arkansas, Louisiana, New Mexico, Oklahoma, and Texas.

★ 1300 ★ U.S. Environmental Protection Agency (EPA)
Region VII
726 Minnesota Ave.
Kansas City, KS 66101
Phone: (913)551-7000; (913)551-7003
Contact: Morris Kay, Regional Admin.
Includes Iowa, Kansas, Missouri, and Nebraska.

★ 1301 ★ U.S. Environmental Protection Agency (EPA)
Region VIII
999 18th St., Ste. 500
Denver, CO 80202-2405
Phone: 800-759-4372; (303)293-1692
Contact: James J. Scherer, Regional Admin.
Includes Colorado, Montana, North Dakota, South Dakota, Utah, and Wyoming.

★ 1302 ★ U.S. Environmental Protection Agency (EPA)
Region IX
75 Hawthorne St.
San Francisco, CA 94105
Phone: (415)744-1600; (415)744-1015
Contact: Daniel W. McGovern, Regional Admin.
Includes Arizona, California, Hawaii, Nevada, American Samoa, Guam, and the Commonwealth of the Northern Mariana Islands.

★ 1303 ★ U.S. Environmental Protection Agency (EPA)
Region X
1200 Sixth Ave.
Seattle, WA 98101
Phone: (206)442-5810; (206)442-1466
Contact: Dana A. Rasmussan, Regional Admin.
Includes Alaska, Idaho, Oregon, and Washington.

★ 1304 ★ U.S. Occupational Safety and Health Administration (OSHA)
Region I
133 Portland St., First Floor
Boston, MA 02114
Phone: (617)565-7164
Contact: John B. Miles, Regional Admin.
Includes Connecticut, Maine, Massachusetts, New Hampshire, Rhode Island, and Vermont.

★ 1305 ★ U.S. Occupational Safety and Health Administration (OSHA)
Region II
201 Varick St., Rm. 670
New York, NY 10014
Phone: (212)337-2378
Contact: James Stanley, Regional Admin.
Includes New Jersey, New York, Puerto Rico, and the Virgin Islands.

★ 1306 ★ U.S. Occupational Safety and Health Administration (OSHA)
Region III
3535 Market St.
Philadelphia, PA 19104
Phone: (215)596-1201
Contact: Linda R. Ancu, Regional Admin.
Includes Delaware, District of Columbia, Maryland, Pennsylvania, Virginia, and West Virginia.

★ 1307 ★ U.S. Occupational Safety and Health Administration (OSHA)
Region IV
1375 Peachtree St., N.E., Ste. 587
Atlanta, GA 30367
Phone: (404)347-3573
Contact: R. Davis Layne, Regional Admin.
Includes Alabama, Florida, Georgia, Kentucky, Mississippi, North Carolina, South Carolina, and Tennessee.

★ 1308 ★ U.S. Occupational Safety and Health Administration (OSHA)
Region V
230 S. Dearborn St., Rm. 3244
Chicago, IL 60604
Phone: (312)353-2220
Contact: Michael Connors, Regional Admin.
Includes Illinois, Indiana, Michigan, Minnesota, Ohio, and Wisconsin.

★ 1309 ★ U.S. Occupational Safety and Health Administration (OSHA)
Region VI
525 Griffin St., Rm. 602
Dallas, TX 75202
Phone: (214)767-4731
Contact: Gilbert J. Saulter, Regional Admin.
Includes Arkansas, Louisiana, New Mexico, Oklahoma, and Texas.

★ 1310 ★ U.S. Occupational Safety and Health Administration (OSHA)
Region VII
911 Walnut St.
Kansas City, MO 64106
Phone: (816)426-5861
Includes Iowa, Kansas, Missouri, and Nebraska.

★ 1311 ★ U.S. Occupational Safety and Health Administration (OSHA)
Region VIII
1961 Stout St.
Denver, CO 80294
Phone: (303)844-3061
Contact: Byron R. Chadwick, Regional Admin.
Includes Colorado, Montana, North Dakota, South Dakota, Utah, and Wyoming.

★ 1312 ★ U.S. Occupational Safety and Health Administration (OSHA)
Region IX
71 Stevenson St., Rm. 415
San Francisco, CA 94105
Phone: (415)744-6670
Contact: Frank Strasheim, Regional Admin.
Includes Arizona, California, Hawaii, Nevada, American Samoa, Guam, and the Trust Territories of the Pacific.

★ 1313 ★ U.S. Occupational Safety and Health Administration
Region X (OSHA)
1111 Third Ave., Ste. 715
Seattle, WA 98174
Phone: (206)442-5930
Contact: James W. Lake, Regional Admin.
Includes Alaska, Idaho, Oregon, and Washington.

★ 1314 ★ U.S. Occupational Safety and Health Administration (OSHA)
U.S. Department of Labor
200 Constitution Ave., N.W.
Washington, DC 20210
Phone: (202)523-8151
Contact: Gerard F. Scannell, Asst. Sec.
OSHA develops and promulgates safety and health standards in the workplace; develops and issues regulations; conducts investigations and inspections to determine the status of compliance with safety and health standards and regulations; and issues citations and proposes penalties for noncompliance with safety and health standards and regulations. Establishes minimum permissible exposure limits to control exposure of hazardous substances in the workplace and sets standards for communicating the presence of potentially hazardous substances to employees. Not subject to the jurisdiction of OSHA are state and municipal employees and workers covered by other laws applying to mine safety, atomic energy, and railroads.

★ 1315 ★ Water Department
Office of Drinking Water
U.S. Environmental Protection Agency (EPA)
401 M St., S.W., Rm. 1011
Washington, DC 20460
Phone: (202)260-5543

State Government Organizations

American Samoa

★ 1316 ★ American Samoa Environmental Quality Commission
P.O. Box 366
Pago Pago, American Samoa 96799
Phone: (684)663-2304
Fax: (684)633-5801

Guam

★ 1317 ★ Guam Environmental Protection Agency
130 Rojas St., Unit D-107
Harmon, Guam 96911
Phone: (671)646-8863

Northern Mariana Islands

★ 1318 ★ Northern Mariana Islands Public Health and Environmental Service
Environmental Quality Division
P.O. Box 1304
Saipan, Northern Mariana Islands 96950
Phone: (607)234-6114
Fax: (607)234-1003

Virgin Islands

★ 1319 ★ Virgin Islands Department of Conservation and Cultural Affairs
Division of Natural Resources Management
111 Watergut Homes
14 F Bldg.
Christiansted, St. Croix, Virgin Islands of the United States 00820-5065
Phone: (809)773-0565

★ 1320 ★ Virgin Islands Department of Labor
Box 890
Christiansted, St. Croix, Virgin Islands of the United States 00820
Phone: (809)773-1994

★ 1321 ★ Virgin Islands Department of Planning and Natural Resource
Division of Environmental Protection
Nisky Center
P.O. Box 4399
Charlotte Amalie, St. Thomas, Virgin Islands of the United States 00801
Phone: (809)774-3320

Alabama

★ 1322 ★ Alabama Department of Agriculture and Industries
Agriculture Chemistry/Plant Industry Division
Pesticide Information and Registration
P.O. Box 3336
Montgomery, AL 36193
Phone: (205)242-2656

★ 1323 ★ Alabama Department of Environmental Management
1751 William L. Dickinson Dr.
Montgomery, AL 36109
Phone: (205)271-7700
Fax: (205)271-7950
Important contact numbers: Air Division, (205)2717861; Hazardous Waste Branch, Land Division, (205)271-7735; Radiological Health Branch (radon information), (205)261-5313; Solid Waste Division, (205)271-7761; Superfund, (205)271-7730; Water Quality Management, (205)271-7823.

Alaska

★ 1324 ★ Alaska Department of Environmental Conservation
Division of Environmental Quality
3220 Hospital Dr.
P.O. Box 0
Juneau, AK 99811-1800
Phone: (907)465-2606
Important contact numbers: Air and Hazardous Waste Management, (907)465-2666; Pesticide Information and Registration, (907)465-2609; Solid and Hazardous Waste Management (includes recycling), (907)465-2671; Spill Planning and Prevention, (907)4652630; Water Quality Management, (907)465-2653.

★ 1325 ★ Alaska Department of Health and Social Services
Division of Public Health
P.O. Box H
Juneau, AK 99811-0610
Phone: (907)465-3090
Important contact number: Radon Information, (907)465-3019.

★ 1326 ★ Alaska Department of Labor
P.O. Box 21149
Juneau, AK 99802-1149
Phone: (907)465-2700

★ 1327 ★ Alaska Department of Natural Resources
400 Willoughby Ave.
Juneau, AK 99801
Phone: (907)465-2400

Arizona

★ 1328 ★ Arizona Commission of Agriculture and Horticulture
Agriculture Chemical and Environmental Services Division
1688 W. Adams, Ste. 103
Phoenix, AZ 85007
Phone: (602)542-4373
Important contact number: Pesticide Information and Registration, (602)833-5442.

★ 1329 ★ Arizona Department of Environmental Quality
2005 N. Central Ave., Rm. 701
Phoenix, AZ 85004
Phone: (602)257-2300
Important contact numbers: Office of Air Quality, (602)257-2308; Emergency Response/ Community Right-to-Know, (602)257-2338; Hazardous Waste Management, (602)257-2215; Office of Waste Programs (includes water quality), (602)257-2318; Radon Information, (602)255-4845.

★ 1330 ★ Arizona Emergency Response Commission
Division of Emergency Service
5636 E. McDowell Rd.
Phoenix, AZ 85008-3495
Phone: (602)231-6245
Important contact number: Emergency Response/Community Right-to-Know, (602)231-6326.

★ 1331 ★ Arizona Energy Office
Community Energy Planning
1700 W. Washington St.
Phoenix, AZ 85007
Phone: (602)542-3633

★ 1332 ★ Arizona Industrial Commission
800 W. Washington St.
Phoenix, AZ 85007
Phone: (602)542-5795

★ 1333 ★ Arizona Land Department
1616 W. Adams St.
Phoenix, AZ 85007
Phone: (602)255-4621

★ 1334 ★ Arizona Radiation Regulatory Agency
4814 S. 40th St.
Phoenix, AZ 85040
Phone: (602)255-4845

Arkansas

★ 1335 ★ Arkansas Department of Health
Division of Radiation Control and Emergency Management
4815 W. Markham St.
Little Rock, AR 72205-3867
Phone: (501)661-2301

★ 1336 ★ Arkansas Department of Labor
10421 W. Markham St.
Little Rock, AR 72205
Phone: (501)682-4500
Important contact number: Emergency Response/Community Right-to-Know, Section 313, (501)682-4534.

★ 1337 ★ Arkansas Department of Pollution Control and Ecology
Hazardous Waste Division
8001 National Dr.
P.O. Box 8913
Little Rock, AR 72219-8913
Phone: (501)562-7444; (501)562-6533

★ 1338 ★ Arkansas State Plant Board
Division of Feed, Fertilizer and Pesticides
Pesticide Information and Registration
1 Natural Resources Dr.
P.O. Box 1069
Little Rock, AR 72203
Phone: (501)225-1598

California

★ 1339 ★ California Department of Conservation
Division of Recycling
1416 Ninth St., Rm. 1320
Sacramento, CA 95814
Phone: (916)323-3743

★ 1340 ★ California Department of Food and Agriculture
Environmental Protection and Worker Safety
Division of Pest Management
1220 N St.
Sacramento, CA 95814
Phone: (916)445-8614

★ 1341 ★ California Department of Health Services
400 P St., Rm. 1253
Sacramento, CA 95814
Phone: (916)323-2913
Important contact numbers: Environmental Health Division (includes radon), (916)322-2308; Indoor Quality Program (air quality), (916)540-2134; Office of Public and Government Liaison, (916)322-0476; Toxic Substances Control Program (hazardous and solid waste), (916)323-9723.

★ 1342 ★ California Department of Industrial Relations
395 Oyster Point Blvd., Third Floor, Wing C
San Francisco, CA 94080
Phone: (415)737-2960

★ 1343 ★ California Department of Toxic Substances Control
P.O. Box 806
Sacramento, CA 95812-0806
Phone: (916)324-7193

★ 1344 ★ California Environmental Protection Agency
1102 Q St.
P.O. Box 2815
Sacramento, CA 95812
Phone: (916)322-5840
Important contact numbers: Air Resources Board, (916)322-2990; Emergency Response/Community Right-to-Know, Section 313, (916)324-8124; Waste Management Board, (916)322-3330.

★ 1345 ★ California State Water Resources Control Board
Division of Clean Water Programs
P.O. Box 944212
Sacramento, CA 94244-2120
Phone: (916)739-4333

Colorado

★ 1346 ★ Colorado Department of Agriculture
Division of Pesticides
700 Kipling St., Ste. 400
Lakewood, CO 80215-5894
Phone: (303)239-4140

★ 1347 ★ Colorado Department of Health
4210 E. 11th Ave.
Denver, CO 80220
Phone: (303)331-4513
Important contact numbers: Air Pollution Control Division, (303)331-8500; Emergency Response/Community Right-toKnow, Section 313, (303)331-4858, after hours number, (303)3776326; Hazardous Materials and Waste Management Division, Superfund Unit (includes recycling), (303)331-4830; Radiation Control Division, (303)331-8700; Water Quality Division, (303)331-4534.

Connecticut

★ 1348 ★ Connecticut Department of Environmental Protection
State Office Bldg.
165 Capitol Ave.
Hartford, CT 06106
Phone: (203)566-5599
Important contact numbers: Bureau of Air Management, (203)566-2506; Emergency Response/Community Right-to-Know, (203)566-4856; Hazardous Materials Management Unit (includes pesticides), (203)566-4924; Solid Waste Division (includes recycling), (203)566-5847; Spills and Emergency Cleanups, Superfund (203)566-4633; Waste Management Bureau, (203)566-8476; Water Compliance Unit, (203)566-3245.

★ 1349 ★ Connecticut Department of Health Service
Health Systems Regulation Bureau
150 Washington St.
Hartford, CT 06106
Phone: (203)566-3207
Important contact numbers: Radon Program, (203)5663122; Toxic Hazards Section, (203)566-8167.

★ 1350 ★ Connecticut Department of Labor
200 Folly Brook Blvd.
Wethersfield, CT 06109
Phone: (203)566-5123

★ 1351 ★ Connecticut Resource Recovery Authority
Professional Bldg.
179 Allyn St., Ste. 603
Hartford, CT 06103
Phone: (203)549-6390

Delaware

★ 1352 ★ Delaware Department of Agriculture
Pesticide Compliance Office
2302 S. Dupont Hwy.
Dover, DE 19901
Phone: (302)736-4811

★ 1353 ★ Delaware Department of Natural Resources and Environmental Control
89 Kings Highway
Richardson and Robbins Bldg.
P.O. Box 1401
Dover, DE 19903
Phone: (302)736-4403
Important contact numbers: Air Resources Section, (302)736-4791; Division of Public Health (radon information), (800)544-INFO; Emergency Response/Community Right-to-Know, (302)739-4764, in Delaware, (800)662-8802; Hazardous Waste Management Branch, (302)739-3689; Office of Radiation Control, (302)736-4731; Solid Waste Management Branch (includes recycling), (302)736-3689; Water Resources Division, (302)736-4860.

★ 1354 ★ Delaware Department of Public Safety
Emergency Planning and Operations Division
P.O. Box 527
Delaware City, DE 19706
Phone: (302)834-4531
Hotline: 800-292-9588

★ 1355 ★ Delaware Solid Waste Authority
Solid Waste Management
P.O. Box 455
Dover, DE 19903
Phone: (302)736-5361

District of Columbia

★ 1356 ★ District of Columbia Department of Consumer and Regulatory Affairs
Housing and Environmental Regulations Administration
614 H St., N.W., Rm. 505
Washington, DC 20001
Phone: (202)727-7395
Important contact numbers: Air and Water Quality Branch, (202)767-7370; Pesticides and Hazardous Waste Management Branch, (202)404-1167; Radiation Information, (202)727-7728.

★ 1357 ★ District of Columbia Department of Public Works
Environment Policy Division
2000 14th St., N.W., Sixth Floor
Washington, DC 20009
Phone: (202)939-8115

★ 1358 ★ District of Columbia Office of Emergency Preparedness
Emergency Response Commission
2000 14th St., N.W., Eight Floor
Washington, DC 20009
Phone: (202)727-6161

Florida

★ 1359 ★ Florida Department of Agriculture and Consumer Services
Bureau of Pesticides
Mayo Bldg., Rm. 213
Tallahassee, FL 32301
Phone: (904)487-2130

★ 1360 ★ Florida Department of Community Affairs
Emergency Management Division
2740 Centerview Dr.
Tallahassee, FL 32399-2149
Phone: (904)487-4918
Hotline: 800-635-7179 (in Florida)

★ 1361 ★ Florida Department of Environmental Regulation
Twin Towers Office Bldg.
2600 Blairstone Rd.
Tallahassee, FL 32399-2400
Phone: (904)488-4805
Important contact numbers: Division of Air Resources Management, (904)488-1344; Division of Waste Management (solid waste), (904)488-0190; Division of Water Facilities, (904)487-1855; Solid and Hazardous Waste Section (includes recycling), (904)4880300.

★ 1362 ★ Florida Department of Health and Rehabilitative Services
1317 Winewood Blvd.
Tallahassee, FL 32399-0700
Phone: (904)488-4854
Hotline: 800-543-8273 (in Florida)
Important contact number: Office of Radiation

Control, (904)488-1525, in Florida, (800)543-8273.

Georgia

★ 1363 ★ Georgia Department of Agriculture
Entomology and Pesticides Division
19 Martin Luther King Dr., S.W.
Agriculture Bldg.
Capitol Sq.
Atlanta, GA 30334
Phone: (404)656-4958

★ 1364 ★ Georgia Department of Human Resources
Radiological Health Section
47 Trinity Ave., S.W.
Atlanta, GA 30334-1202
Phone: (404)894-6644

★ 1365 ★ Georgia Department of Natural Resources
Environmental Protection Division
Floyd Towers East, Ste. 1154
205 Butler St., S.E.
Atlanta, GA 30334
Phone: 800-334-2373; (404)656-4713
Important contact numbers: Air Protection Branch, (404)656-6900; Emergency Response/ Community Right-to-Know, Section 313, (404)656-6905; Emergency Release Notification, (800)241-4113; Industrial and Hazardous Waste Management Program, (404)656-7404; Land Protection Branch (includes solid waste management and recycling), (404)656-2833; Water Protection Branch, (404)656-4708.

Hawaii

★ 1366 ★ Hawaii Department of Agriculture
Division of Plant Industry
Pesticide Information and Registration
350 Capitol Hill Ave.
P.O. Box 22159
Honolulu, HI 96822-0159
Phone: (808)548-7124

★ 1367 ★ Hawaii Department of Health
Environmental Protection and Health Services Division
591 Ala Moana Blvd.
P.O. Box 3378
Honolulu, HI 96813
Phone: (808)548-6455
Important contact numbers: Clean Air Branch, (808)543-8200; Hazard Evaluation and Emergency Response, (808)5438248; Hazardous Waste Programs (includes recycling), (808)548-6410; Radon Information, (808)548-4383; Solid Waste Management and Water Quality, (808)548-4139.

★ 1368 ★ Hawaii Department of Labor and Industrial Relations
830 Punchbowl St.
Honolulu, HI 96813
Phone: (808)548-3150

★ 1369 ★ Hawaii Department of Land and Natural Resources
1151 Punchbowl St.
P.O. Box 621
Honolulu, HI 96809
Phone: (808)548-6550

Idaho

★ 1370 ★ Idaho Department of Agriculture
Division of Plant Industries
Bureau of Pesticides
P.O. Box 790
Boise, ID 83701
Phone: (208)334-3243

★ 1371 ★ Idaho Department of Health and Welfare
Division of Environmental Quality
450 W. State St.
Tower Bldg., Third Floor
Boise, ID 83720
Phone: (208)334-5840
Important contact numbers: Air Quality Bureau, (208)334-5898; Bureau of Hazardous Materials (includes solid waste management and recycling), (208)334-5879; Bureau of Preventive Medicine (radon information), (208)334-5927; Bureau of Water Quality, (208)334-4250; Emergency Response/ Community Right-to-Know, (800)632-8000.

Illinois

★ 1372 ★ Illinois Department of Agriculture
Plant Industry and Consumer Services Division
Bureau of Plant Safety and Apiary Protection
State Fair Ground
P.O. Box 19281
Springfield, IL 62794-9281
Phone: (217)785-2427

★ 1373 ★ Illinois Department of Energy and Natural Resources
Hazardous Waste Research and Information Center
1 E. Hazelwood Dr.
Champaign, IL 61820
Phone: (217)333-8941
Fax: (217)333-8944

★ 1374 ★ Illinois Department of Energy and Natural Resources
Office of Solid Waste and Renewable Resources
Hazardous Waste Management
325 W. Adams, Rm. 300
Springfield, IL 62704
Phone: (217)785-2800

★ 1375 ★ Illinois Department of Nuclear Safety
Radioecology Division
1301 Knotts St.
Springfield, IL 62703
Phone: 800-225-1245; (217)785-9900
Important contact number: Nuclear Safety Laboratory (radon information), (217)786-7126.

★ 1376 ★ Illinois Department of Public Health
Office of Health Regulation
535 W. Jefferson
Springfield, IL 62761
Phone: (217)782-2913
Important contact number: Pesticide Information and Registration, (217)782-4674.

★ 1377 ★ Illinois Emergency Service and Disaster Agency
Emergency Response Commission
110 E. Adams St.
Springfield, IL 62706
Phone: (217)782-4694

★ 1378 ★ Illinois Environmental Protection Agency
2200 Churchill Rd.
P.O. Box 19276
Springfield, IL 62794-9276
Phone: (217)782-3397
Fax: (217)782-9039
Important contact numbers: Division of Air Pollution Control, (217)782-7326; Division of Land Pollution Control (includes hazardous and solid waste management), (217)782-6760; Emergency Response/Community Right-to-Know, Section 313, (217)7823637; Water Quality, (217)782-5544.

Indiana

★ 1379 ★ Indiana Department of Environmental Management
105 S. Meridian St.
P.O. Box 6015
Indianapolis, IN 46206-6015
Phone: (317)232-8603
Important contact numbers: Office of Air Management, (317)232-5586; Office of Solid and Hazardous Waste, Hazardous Waste Management Branch, (317)232-4458, Solid Waste Management Branch, (317)232-4473; Office of Water Management, (317)232-8476; Pesticide Information and Registration, State Chemist Office, (317)494-1492.

★ 1380 ★ Indiana Department of Labor and Industrial Relations
1013 State Office Bldg.
100 N. Senate Ave.
Indianapolis, IN 46204-2287
Phone: (317)232-2665

★ 1381 ★ Indiana Emergency Response Commission
5500 W. Bradbury Ave.
Indianapolis, IN 46241
Phone: (317)243-5176

★ 1382 ★ Indiana State Board of Health
Division of Industrial Hygiene and Radiological Health
1330 W. Michigan St.
P.O. Box 1964
Indianapolis, IN 46206-1964
Phone: (317)633-0692
Hotline: 800-272-9723 (in Indiana)
Important contact numbers: Air Pollution Control Division, (317)232-8217; Radon Information, (317)633-0153.

Iowa

★ 1383 ★ Iowa Department of Agriculture and Land Stewardship
Pesticide Control Bureau
Henry A. Wallace State Office Bldg.
E. Ninth St. and Grand Ave.
Des Moines, IA 50319
Phone: (515)281-8591

★ 1384 ★ Iowa Department of Natural Resources
Environmental Protection Division
Air Quality and Solid Waste Protection Bureau
Henry A. Wallace State Office Bldg.
900 E. Grand Ave.
Des Moines, IA 50319-0034
Phone: (515)281-8852
Important contact numbers: Air Quality, (515)2815145; RECRA Hazardous Waste issues, (913)236-2888 or (800)223-0425; Recycling Coordinator, (515)281-8176; Solid Waste Protection Bureau, (515)281-8693; Superfund Unit, (515)281-4968; Water Quality Section, (515)281-8869.

★ 1385 ★ Iowa Department of Public Health
Radiological Health Unit
Lucas State Office Bldg.
Des Moines, IA 50319-0075
Phone: (515)281-3478

★ 1386 ★ Iowa Division of Labor Services
1000 E. Grand Ave.
Des Moines, IA 50319
Phone: (515)281-3447

Kansas

★ 1387 ★ Kansas Department of Health and Environment
Landon State Office Bldg.
901 S.W. Jackson St.
Topeka, KS 66612
Phone: (913)296-1529
Important contact numbers: Bureau of Air and Waste Management (includes recycling), (913)296-1593; Emergency Response Commission, Community Right-to-Know Program, (913)296-1690; Emergency Release Notification, (913)296-3176; Radiation Control Program, Bureau of Environmental Health Services, (913)296-5600; Water Bureau, (913)296-5500.

★ 1388 ★ Kansas State Board of Agriculture
Plant Health Division
Pesticide Information and Registration
901 S. Kansas Ave.
Topeka, KS 66612-1280
Phone: (913)296-5395

Kentucky

★ 1389 ★ Kentucky Department of Agriculture
Consumer Safety Office
Division of Pesticides
Capitol Plaza Tower
Frankfort, KY 40601
Phone: (502)564-7274

★ 1390 ★ Kentucky Department of Military Affairs
Disaster and Emergency Services
Emergency Response Commission
Boone National Guard Center
EOC Bldg.
Frankfort, KY 40601-6168
Phone: (502)564-8680
Important contact number: Emergency Response/Community Right-to-Know, (502)564-8660.

★ 1391 ★ Kentucky Human Resources Cabinet
Department for Health Services
Division of Radiation and Product Safety, Radiation Control Branch
275 E. Main St.
Frankfort, KY 40621
Phone: (502)564-3700

★ 1392 ★ Kentucky Labor Cabinet
U.S. Hwy. 127, S.
Frankfort, KY 40601
Phone: (502)564-3070

★ 1393 ★ Kentucky Natural Resources and Environmental Protection Cabinet
Department for Environmental Protection
Fort Boone Plaza
18 Reilly Rd., Bldg. 2
Frankfort, KY 40601
Phone: (502)564-2150
Important contact numbers: Community Affairs, (502)564-3350; Division of Air Quality, (502)564-3382; Division of Waste Management (includes recycling), (502)564-6716; Division of Water Compliance and Enforcement, (502)564-3410.

Louisiana

★ 1394 ★ Louisiana Department of Agriculture and Forestry
Office of Agriculture and Environmental Sciences
Pesticide Commission
P.O. Box 44153
Baton Rouge, LA 70804
Phone: (504)925-3763

★ 1395 ★ Louisiana Department of Environmental Quality
438 Main St.
P.O. Box 44066
Baton Rouge, LA 70804-4307
Phone: (504)342-1266
Important contact numbers: Air Quality and Radiation Protection Division, (504)342-1206; Emergency Response Coordinator, (504)342-8617; Nuclear Energy Division (radon information), (504)925-4518; Solid and Hazardous Waste Division (includes recycling), (504)765-0355; Water Resources Division, (504)342-6363.

Maine

★ 1396 ★ Maine Department of Agriculture
Board of Pesticide Control
State Office Bldg., Section 28
Augusta, ME 04333
Phone: (207)289-2731

★ 1397 ★ Maine Department of Environmental Protection
State Office Bldg., Station 17
Augusta, ME 04333
Phone: (207)289-2811
Important contact numbers: Bureau of Air Quality, (207)289-2437; Bureau of Water Quality, (207)289-3355; Division of Response Services, Bureau of Hazardous Materials and Solid Waste Control, Superfund, (207)289-2651; Emergency Response Commission, (207)289-4080, in Maine (800)452-8735.

★ 1398 ★ Maine Department of Human Services
Bureau of Health
Indoor Air Program
State House, Station 11
Augusta, ME 04333
Phone: (207)289-3205

★ 1399 ★ Maine Waste Management Agency
Office of Waste Reduction and Recycling
286 Water St.
State House, Station 154
Augusta, ME 04333
Phone: (207)289-5300

Maryland

★ 1400 ★ Maryland Department of Agriculture
Pesticide Applicators Law Section
50 Harry S Truman Pkwy.
Annapolis, MD 21401
Phone: (301)841-5710

★ 1401 ★ Maryland Department of the Environment
2500 Broening Hwy.
Baltimore, MD 21224
Phone: (301)631-3084
Important contact numbers: Air Management Administration (301)631-3255; Hazardous and Solid Waste Management Administration, (301)631-3304; Office of Waste Minimization and Recycling, (301)631-3315; Radiological Health, (301)631-3300; Toxics Information Center, (301)631-3778; Water Management Administration, (301)631-3567.

★ 1402 ★ Maryland Division of Labor and Industry
Department of Licensing and Regulation
501 St. Paul Place, 15th Floor
Baltimore, MD 21202-2272
Phone: (301)333-4179

★ 1403 ★ Maryland Emergency Management Agency
2 Sudbrook Ln., E.
Pikesville, MD 21208
Phone: (301)486-4422

Massachusetts

★ 1404 ★ Massachusetts Department of Environmental Affairs
100 Cambridge St., 20th Floor
Boston, MA 02202
Phone: (617)727-9800
Fax: (617)727-2754

Important contact number: Water Resource Issues, (617)727-3267.

★ 1405 ★ Massachusetts Department of Environmental Protection
1 Winter St., Third Floor
Boston, MA 02108
Phone: (617)292-5856
Important contact numbers: Air Quality Control Division, (617)292-5593; Hazardous and Solid Waste Division, (617)292-5589; Public Affairs Office, (617)292-5515; Title III Emergency Response Commission, (617)292-5993; Water Pollution Control Division, (617)292-5646.

★ 1406 ★ Massachusetts Department of Food and Agriculture
Pesticides Division
100 Cambridge St., 21st Floor
Boston, MA 02202
Phone: (617)727-7712

★ 1407 ★ Massachusetts Department of Public Health
150 Tremont St., 11th Floor
Boston, MA 02111
Phone: (617)727-0201
Important contact numbers: Lead Poisoning Prevention Division, (617)522-3700; Radiation Control Division, (617)727-6124.

Michigan

★ 1408 ★ Michigan Department of Agriculture
Pesticide and Plant Pest Management Division
Ottawa Bldg., North Tower, Fourth Floor
611 W. Ottawa St.
P.O. Box 30017
Lansing, MI 48909
Phone: (517)373-1087

★ 1409 ★ Michigan Department of Labor
309 N. Washington Sq.
P.O. Box 30015
Lansing, MI 48909
Phone: (517)373-9600

★ 1410 ★ Michigan Department of Natural Resources
Environmental Protection Bureau
Mason Bldg.
Box 30028
Lansing, MI 48909
Phone: (517)373-1214
Important contact numbers: Air Quality Division, (517)373-7023; Recycling and Recovery Unit, (517)373-0540; Waste Management Division, Hazardous and Solid Waste Management, (517)373-2730, Title III Notification, (517)373-8481; Superfund Section, (517)335-3393; Water Management Division, (517)373-2329.

★ 1411 ★ Michigan Department of Public Health
Bureau of Environmental and Occupational Health
3423 N. Logan
P.O. Box 30195
Lansing, MI 48909
Phone: (517)335-8022
Important contact number: Division of Radiological Health, (517)335-8200.

Minnesota

★ 1412 ★ Minnesota Department of Agriculture
Division of Agronomy Services
90 W. Plato Blvd.
St. Paul, MN 55107
Phone: (612)296-1161

★ 1413 ★ Minnesota Department of Health
Environmental Health Division
717 Delaware St., S.E.
Box 9441
Minneapolis, MN 55440
Phone: 800-652-9747; (612)623-5033
Important contact number: Radiation Control Section, (612)627-5062.

★ 1414 ★ Minnesota Department of Labor and Industry
443 Lafayette Rd.
St. Paul, MN 55155
Phone: (612)296-2342

★ 1415 ★ Minnesota Department of Public Safety
Emergency Response Commission
Transportation Bldg., Rm. 211
St. Paul, MN 55155
Phone: (612)643-3002

★ 1416 ★ Minnesota Pollution Control Agency
520 Lafayette Rd., N.
St. Paul, MN 55155-4001
Phone: (612)296-7283
Important contact numbers: Division of Air Quality, (612)296-7331; Hazardous Waste Division (includes solid waste), (612)296-8502; Division of Water Quality, (612)296-7202; Recycling and Recovery Unit, (612)296-6300; Site Response Section/Superfund, (612)296-7290.

Mississippi

★ 1417 ★ Mississippi Department of Agriculture and Commerce
Division of Plant Industry
P.O. Box 5207
Mississippi State, MS 39762
Phone: (601)325-3390

★ 1418 ★ Mississippi Department of Health
Division of Radiological Health
3150 Lawson St.
P.O. Box 1700
Jackson, MS 39215-1700
Phone: (601)354-6657

★ 1419 ★ Mississippi Department of Natural Resources
Pollution Control Bureau
Southport Mall
P.O. Box 10385
Jackson, MS 39209
Phone: (601)961-5100
Important contact numbers: Air Quality Branch, (601)961-5104; Groundwater Protection Branch (includes recycling), (601)961-5119; Hazardous Waste Division, (601)961-5062; Public Information Office, (601)961-5015; Solid Waste Management Section, (601)961-5171; Water Quality Branch, (601)961-5102.

★ 1420 ★ Mississippi Emergency Management Agency

Emergency Response Commission
Fondren Station
P.O. Box 4501
Jackson, MS 39296-4501
Phone: (601)960-9973

Missouri

★ 1421 ★ Missouri Department of Agriculture

Bureau of Pesticide Control
P.O. Box 630
Jefferson City, MO 65102
Phone: (314)751-5503

★ 1422 ★ Missouri Department of Health

Bureau of Radiological Health
1730 E. Elm St.
P.O. Box 570
Jefferson City, MO 65102
Phone: (314)751-6083
Hotline: 800-669-7236 (in Missouri)

★ 1423 ★ Missouri Department of Natural Resources

Division of Environmental Quality
205 Jefferson St.
P.O. Box 176
Jefferson City, MO 65102
Phone: (314)751-4810
Hotline: 800-334-6946 (in Missouri)
Important contact numbers: Air Pollution Control Program, (314)751-4817; Emergency Response Commission, (314)7517929; Solid Waste Planning Unit (includes recycling), (314)7511492; Superfund Section, Waste Management Program (hazardous and solid waste), (314)751-3176; Water Pollution Control Program, (314)751-1300.

Montana

★ 1424 ★ Montana Department of Agriculture

Environmental Management Division
Agriculture and Livestock Bldg.
Capitol Station, Rm. 317
Sixth and Roberts Sts.
Helena, MT 59620-0205
Phone: (406)444-2944

★ 1425 ★ Montana Department of Health and Environmental Sciences

Environmental Sciences Division
Cogswell Bldg.
Capitol Station
Helena, MT 59620
Phone: (406)444-3948
Important contact numbers: Air Quality Bureau, (406)444-3454; Emergency Response Commission, (406)444-6911; Occupational Health Bureau (radon information), (406)444-3671; Superfund Program, Solid and Hazardous Waste Bureau (includes recycling), (406)444-2821; Water Quality Bureau, (406)444-2406.

Nebraska

★ 1426 ★ Nebraska Department of Agriculture

Plant Industry Division
301 Centennial Mall S.
P.O. Box 94947
Lincoln, NE 68509
Phone: (402)471-2394
Important contact number: Pesticide Information and Registration, (402)471-2341.

★ 1427 ★ Nebraska Department of Environmental Control

301 Centennial Mall S.
State House Station
P.O. Box 94877
Lincoln, NE 68509-8922
Phone: (402)471-2186
Important contact numbers: Air Quality Division, (402)471-2189; Emergency Response Commission, (402)471-2186, after hours, (402)471-4545; Superfund Section, (402)471-4217; Water Quality Division, (402)471-4239.

★ 1428 ★ Nebraska Department of Health

301 Centennial Mall, S.
P.O. Box 95007
Lincoln, NE 68509-5007
Phone: (402)471-4047
Important contact numbers: Division of Radiological Health, (402)471-2168; Environmental Health and Housing Surveillance (water quality), (402)471-2541.

Nevada

★ 1429 ★ Nevada Department of Agriculture

Plant Industry Division
350 Capitol Hill Ave.
P.O. Box 11100
Reno, NV 89510-1100
Phone: (702)688-1180

★ 1430 ★ Nevada Department of Conservation and Natural Resources

Division of Environmental Protection
Capitol Complex
201 S. Fall St., Rm. 221
Carson City, NV 89710
Phone: (702)687-4670
Important contact numbers: Bureau of Air Quality, (702)687-5065; Division of Emergency Management, (702)687-4240, Emergency Release Notification, (702)687-5300; Office of Community Services (includes recycling), (702)687-4908; Waste Management Bureau (includes hazardous and solid waste), (702)687-5872.

★ 1431 ★ Nevada Department of Human Resources

Health Division
Bureau of Radiological Health
505 E. King St., Rm. 203
Carson City, NV 89710
Phone: (702)885-5394

★ 1432 ★ Nevada Department of Industrial Relations

Division of Occupational Safety and Health
Capitol Complex
1370 S. Curry St.
Carson City, NV 89710
Phone: (702)885-5240

New Hampshire

★ 1433 ★ New Hampshire Department of Agriculture

Division of Pesticides Control
10 Ferry St.
Concord Center
Caller Box 2042
Concord, NH 03301
Phone: (603)271-3550

★ 1434 ★ New Hampshire Department of Environmental Services

Public Health Services Division
Health and Welfare Bldg.
6 Hazen Dr.
Concord, NH 03301-6509
Phone: (603)271-4583
Important contact numbers: Air Resources Agency, (603)271-1370; Radiological Health Program, (603)271-4588; Solid Waste Compliance and Enforcement Bureau, (603)271-4662; State Recycling Coordinator, (603)271-2902; Superfund Site Management Bureau (hazardous waste materials), (603)271-2908; Waste Management Division, (603)271-2906; Water Supply and Pollution Control Division, (603)271-3503.

★ 1435 ★ New Hampshire Emergency Management Agency

State Office Park S.
107 Pleasant St.
Concord, NH 03301
Phone: (603)271-2231

New Jersey

★ 1436 ★ New Jersey Department of Environmental Protection

Environmental Management and Control
401 E. State St, CN-402
Trenton, NJ 08625
Phone: (609)292-8058
Important contact numbers: Air Pollution Control Program, (609)292-6704; Division of Hazardous Waste Management, (609)292-1250; Division of Solid Waste, (609)292-8879; Division of Water Resources, (609)292-1637; Emergency Response Commission, SARA Title III, Section 313, (609)292-6714, emergency number, (609)2927172; Office of Recycling, (609)292-0331; Pesticide Control Program, (609)530-4070; Radon Information, (609)987-6402, in New Jersey, (800)648-0394.

New Mexico

★ 1437 ★ New Mexico Department of Health and Environment

Environmental Improvement Division
1190 St. Francis Dr.
P.O. Box 968
Santa Fe, NM 87504-0968
Phone: (505)827-2850
Important contact numbers: Air Quality Bureau, (505)827-0031; Hazardous and Radioactive Materials Bureau, (505)827-2922; Radiation Licensing and Registration Section, (505)827-2948; Superfund, Hazardous Waste Department, (505)8272775; Surface Water Quality Bureau, (505)827-2793; Waste

Management Section (includes radon information), (505)827-2786.

★ 1438 ★ New Mexico Department of Public Safety
Emergency Response Commission
P.O. Box 1628
Santa Fe, NM 87504-1628
Phone: (505)827-9222

★ 1439 ★ New Mexico Environmental Improvement Division
Health and Environmental Department
1190 St. Francis Dr., N2200
Santa Fe, NM 87503-0968
Phone: (505)827-2850

★ 1440 ★ New Mexico State Department of Agriculture
Division of Agriculture and Environmental Sciences
Pesticides Management Bureau
New Mexico State University
Department of Agriculture
P.O. Box 3005-3AQ1
Las Cruces, NM 88003-0005
Phone: (505)646-2133

New York

★ 1441 ★ New York Department of Environmental Conservation
50 Wolf Rd.
Albany, NY 12233
Phone: (518)457-3446
Hotline: 800-631-0666 (in New York)
Important contact numbers: Bureau of Hazardous Waste Remediation, (518)457-5861; Bureau of Pesticides, (518)457-7482; Bureau of Waste Reduction and Recycling, (518)457-7337; Division of Air Quality, (518)457-7230; Division of Water, (518)457-6674; Emergency Response Commission, Bureau of Spill Response, Section 313, (518)457-4107; Solid Waste Management Division, (518)457-6603.

★ 1442 ★ New York Department of Health
Environmental Health Center
Empire State Plaza
Corning Tower
Albany, NY 12237-0001
Phone: (518)458-6400
Hotline: 800-458-1158 (in New York)
Important contact number: Radon Information, (518)458-6450.

★ 1443 ★ New York Department of Labor
1 Main St.
Brooklyn, NY 11201
Phone: (518)457-3518

North Carolina

★ 1444 ★ North Carolina Department of Agriculture
Food and Drug Protection Division
Pesticide Office
State Agriculture Bldg.
P.O. Box 27647
Raleigh, NC 27611
Phone: (919)733-3556

★ 1445 ★ North Carolina Department of Environment, Health and Natural Resources
Division of Environmental Management
512 N. Salisbury St.
P.O. Box 27687
Raleigh, NC 27611-7687
Phone: (919)733-7015
Important contact numbers: Air Quality Section, (919)733-3340; Community Assistance Division, (919)733-2850; Emergency Response Commission, (919)733-3867, for general information on North Carolina, (800)451-1403.

★ 1446 ★ North Carolina Department of Environment, Health and Natural Resources
Division of Solid Waste Management
P.O. Box 27687
Raleigh, NC 27611-7687
Phone: (919)733-4996
Fax: (919)733-4810
Important contact numbers: CERCLA Program Unit, Solid and Hazardous Waste Management Branch, (919)733-2178; Radiation Protection Section, (919)733-4283; Solid Waste Management Branch (includes recycling), (919)733-0692.

★ 1447 ★ North Carolina Department of Labor
4 W. Edenton St.
Raleigh, NC 27603
Phone: (919)733-7166

North Dakota

★ 1448 ★ North Dakota Department of Agriculture
Plant Industries Division
Pesticide/Noxious Weed Division
600 East Blvd., Sixth Floor
Bismarck, ND 58505-0020
Phone: (701)224-4756

★ 1449 ★ North Dakota Department of Health and Consolidated Laboratories
Environmental Health Section
1200 Missouri Ave.
P.O. Box 5520
Bismarck, ND 58505-5520
Phone: (701)224-2374
Important contact numbers: Division of Environmental Engineering (includes air quality and radon information), (701)2242348; Division of Waste Management and Special Studies (includes Superfund and recycling), (701)224-2366; Division of Water Supply and Pollution Control, (701)224-2354.

Ohio

★ 1450 ★ Ohio Department of Agriculture
Division of Plant Industry
8995 E. Main St.
Reynoldsburg, OH 43068
Phone: (614)866-6361

★ 1451 ★ Ohio Department of Health
Division of Technical Environmental Health Services
Radiological Health Program
1224 Kinnear Rd.
Columbus, OH 43212
Phone: (614)644-2727
Hotline: 800-523-4439 (in Ohio)

★ 1452 ★ Ohio Department of Natural Resources
Office of Litter Prevention and Recycling
Fountain Square, Bldg. E1
Columbus, OH 43224
Phone: (614)265-6333

★ 1453 ★ Ohio Environmental Protection Agency
1800 Watermark Dr.
P.O. Box 1049
Columbus, OH 43266-0149
Phone: (614)644-2782
Important contact numbers: Division of Air Pollution Control (includes Superfund), (614)644-2270; Division of Solid and Hazardous Waste Management, (614)644-2917; Division of Water Quality, (614)644-2856.

Oklahoma

★ 1454 ★ Oklahoma Department of Health
Environmental Health Services
1000 N.E. Tenth St.
P.O. Box 53551
Oklahoma City, OK 73152
Phone: (405)271-4200
Important contact numbers: Air Quality Services, (405)271-5220; Hazardous Waste Management Service (includes recycling), (405)271-5338; Radiation and Special Hazards Services, (405)271-5221.

★ 1455 ★ Oklahoma Office of Civil Defense
Emergency Response Commission
P.O. Box 53365
Oklahoma City, OK 73152
Phone: (405)521-2481

★ 1456 ★ Oklahoma State Department of Agriculture
Plant Industry Division
Pest Management Section
2800 N. Lincoln Blvd.
Oklahoma City, OK 73105
Phone: (405)521-3864

★ 1457 ★ Oklahoma State Department of Health
Department of Pollution Control
1000 N.E. Tenth St.
P.O. Box 53554
Oklahoma City, OK 73152
Phone: (405)271-4677

★ 1458 ★ Oklahoma Water Resources Board
Water Quality Division
1000 N.E. Tenth St.
P.O. Box 53585
Oklahoma City, OK 73151
Phone: (405)271-2500

Oregon

★ 1459 ★ Oregon Department of Agriculture
Plant Division
635 Capitol St., N.E.
Salem, OR 97301-0110
Phone: (503)378-3776

★ 1460 ★ Oregon Department of Environmental Quality
Executive Bldg., Ninth Floor
811 S.W. Sixth Ave.
P.O. Box 1760
Portland, OR 97204
Phone: 800-452-4011; (503)229-5395
Important contact numbers: Air Quality Control Division, (503)229-5397; Environmental Cleanup Division (Superfund), (503)229-5254; Hazardous Waste Section, (503)229-6434; Solid Waste Section (includes recycling), (503)229-5913; Water Quality Control Division, (503)229-5324.

★ 1461 ★ Oregon Department of Human Resources
Health Division
Radiation Control Section
1400 S.W. Fifth Ave.
P.O. Box 231
Portland, OR 97207
Phone: (503)229-5797

★ 1462 ★ Oregon Emergency Response Commission
c/o State Fire Marshall
3000 Market St. Plaza, Ste. 534
Salem, OR 97310
Phone: (503)378-2885

★ 1463 ★ Oregon Occupational Safety and Health Division
Oregon Department of Insurance and Finance
21 Labor and Industries Bldg., Rm. 160
Salem, OR 97310
Phone: (503)378-3304

Pennsylvania

★ 1464 ★ Pennsylvania Department of Agriculture
Bureau of Plant Industry
Agronomic Services
2301 N. Cameron St.
Harrisburg, PA 17110-9408
Phone: (717)787-4843

★ 1465 ★ Pennsylvania Department of Environmental Resources
Fulton Bldg., Ninth Floor
P.O. Box 2063
Harrisburg, PA 17120
Phone: (717)787-2814
Important contact numbers: Bureau of Air Quality, (717)787-9702; Bureau of Right-to-Know, Section 313, (717)783-2071; Bureau of Waste Management, (717)787-2388; Bureau of Water Quality Management, (717)787-2666; Radon Hotline, (800)23-RADON; Radon Program Office, Bureau of Radiation Protection, (717)787-2480; Recycling and Waste Reduction Section, (717)787-7382.

Puerto Rico

★ 1466 ★ Puerto Rico Department of Agriculture
Analysis and Registration of Agricultural Materials
Laboratory Division
P.O. Box 10163
Santurce, PR 00908
Phone: (809)796-1715

★ 1467 ★ Puerto Rico Department of Labor and Human Resources
Prudencio Rivera Martinez Bldg.
505 Munoz Rivera Ave.
Hato Rey, PR 00918
Phone: (809)854-2119

★ 1468 ★ Puerto Rico Environmental Quality Board
Sernades Juncos Station
P.O. Box 11488
Santurce, PR 00910-1488
Phone: (809)767-8181; (809)767-8056
Fax: (809)767-2483
Important contact numbers: Air Quality Program, (809)722-5140; Radon Information Office, (809)767-3563; SERC Commissioner, Title III, Section 313, (809)722-0077; Solid Waste Authority, (809)765-7584; Water Quality Program, (809)722-5140.

Rhode Island

★ 1469 ★ Rhode Island Department of Environmental Management
9 Hayes St.
Providence, RI 02908-5003
Phone: (401)277-2771
Important contact numbers: Division of Agriculture and Marketing (pesticides), (401)277-2781; Division of Air and Hazardous Materials (includes toxic release inventory), (401)2772797; Division of Water Resources, (401)277-2234; Emergency Release Notification, (401)277-3070; Recycling Coordinator, (401)277-3434; Superfund Program, Solid Waste Management Corporation, (401)2772979.

★ 1470 ★ Rhode Island Department of Health
Division of Occupational Health and Radiation Control
206 Cannon Bldg.
75 Davis St.
Providence, RI 02908
Phone: (401)277-3601
Important contact number: Radon Information, (401)277-2438.

South Carolina

★ 1471 ★ South Carolina Department of Health and Environmental Control
Division of Environmental Quality Control
2600 Bull St., Ste. 415
Columbia, SC 29201
Phone: (803)734-5360
Important contact numbers: Bureau of Air Quality Control, (803)758-5406; Bureau of Radiological Health, (803)7344700; Bureau of Solid and Hazardous Waste Management (includes Superfund Title III and recycling), (803)734-5164; Bureau of Water Pollution Control, (803)734-5300; Department of Fertilizer and Pest Control, (803)656-0394.

★ 1472 ★ South Carolina Department of Labor
3600 Forest Dr.
P.O. Box 11329
Columbia, SC 29211-1329
Phone: (803)734-9594

South Dakota

★ 1473 ★ Governor's Office of Energy Policy
Alternative Energy Program
2171/2 W. Missouri
Pierre, SD 57501
Phone: (605)733-3603

★ 1474 ★ South Dakota Department of Agriculture
Division of Regulatory Services
Feed, Fertilizer, and Pesticide Program
Anderson Bldg.
445 E. Capitol
Pierre, SD 57501
Phone: (605)773-3724

★ 1475 ★ South Dakota Department of Water and Natural Resources
Division of Environmental Regulations
Joe Foss Bldg.
523 E. Capitol Ave.
Pierre, SD 57501-3181
Phone: (605)773-3351
Important contact numbers: Division of Land and Water Quality, (605)773-3296; Office of Air Quality and Solid Waste (includes hazardous waste and radon information), (605)773-3151; Office of Waste Management, (605)773-3153; Superfund, Section 313, (605)773-3296.

Tennessee

★ 1476 ★ Tennessee Department of Agriculture
Plant Industries Division
Melrose Station
P.O. Box 40627
Nashville, TN 37204
Phone: (615)360-0130

★ 1477 ★ Tennessee Department of Environment and Conservation
Environmental Bureau
344 Cordell Hull Bldg.
Nashville, TN 37247-0101
Phone: (615)741-3657
Important contact numbers: Division of Air Pollution Control, (615)741-3931; Division of Radiological Health, (615)7417812; Division of Solid Waste Management, (615)741-3424; Division of Superfund, (615)741-6287; Division of Water Pollution Control, (615)741-2275.

★ 1478 ★ Tennessee Department of Labor
501 Union Bldg., Second Floor, Ste. A
Nashville, TN 37219
Phone: (615)741-2582

★1479★ **Tennessee Emergency Management Agency**
Emergency Response Commission
3041 Sidco Dr.
Nashville, TN 37204
Phone: (615)252-3300
Hotline: 800-262-3300 (in Tennessee); (800)258-3300 (out of state)

Texas

★1480★ **Texas Air Control Board**
6330 Highway 290 E.
Austin, TX 78723
Phone: (512)451-5711

★1481★ **Texas Department of Agriculture**
Division of Pesticide Enforcement
P.O. Box 12847
Austin, TX 78711
Phone: (512)463-7550

★1482★ **Texas Department of Health**
Environmental and Consumer Health Protection
1100 W. 49th St.
Austin, TX 78756
Phone: (512)458-7541
Important contact numbers: Radiation Control Bureau, (512)835-7000; Solid Waste Management Division (includes recycling), (512)458-7271.

★1483★ **Texas Water Commission**
1700 N. Congress
Capitol Station
P.O. Box 13087
Austin, TX 78711-3087
Phone: (512)463-8028
Important contact numbers: Emergency Response Unit, (512)463-8527; Emergency Release Notification, (512)458-7410; Hazardous and Solid Waste Division, (512)463-7760; Water Quality Division, (512)463-8412.

Utah

★1484★ **Utah Department of Agriculture**
Division of Plant Industries
350 N. Redwood Rd.
Salt Lake City, UT 84116
Phone: (801)538-7180

★1485★ **Utah Department of Environmental Quality**
288 N. 1460 West St.
P.O. Box 16690
Salt Lake City, UT 84114-4880
Phone: (801)538-6121
Important contact numbers: Division of Air Quality, (801)538-6108; Division of Radiation Control, (801)538-6734; Division of Solid and Hazardous Waste (includes recycling), (801)538-6170; Division of Water Pollution Control, (801)538-6146.

★1486★ **Utah Occupational Safety and Health Department**
160 E. 300, S.
P.O. Box 5800
Salt Lake City, UT 84110-5800
Phone: (801)530-6900

Vermont

★1487★ **Vermont Agency of Natural Resources**
Department of Environmental Conservation
103 S. Main St.
Waterbury, VT 05676
Phone: (802)244-8755
Important contact numbers: Air Pollution Control Division, (802)244-8731; Hazardous Materials Management Division, (802)244-8702; Recycling and Resource Conservation Section, (802)244-8731; Solid Waste Programs, (802)244-8731; Water Quality Division, (802)244-6951.

★1488★ **Vermont Department of Agriculture**
Plant Industry Division
116 State St.
State Office Bldg.
Montpelier, VT 05602
Phone: (802)828-2435
Important contact number: Pesticide Information and Registration, (802)828-2431.

★1489★ **Vermont Department of Health**
60 Main St.
P.O. Box 70
Burlington, VT 05402
Phone: (802)863-7281
Important contact number: Occupational and Radiological Health Division, (802)828-2886.

★1490★ **Vermont Department of Labor and Industry**
120 State St.
Montpelier, VT 05602
Phone: (802)828-2765

★1491★ **Vermont Environmental Board**
State Office Bldg.
Montpelier, VT 05602
Phone: (802)828-3309

Virginia

★1492★ **Virginia Council on the Environment**
Ninth Street Office Bldg., Rm. 903
Richmond, VA 23219
Phone: (804)786-4500
Hotline: 800-552-2075 (in Virginia)
Important contact numbers: Department of Air Pollution Control, (804)786-7913; Department of Waste Management (includes recycling), (804)225-2667; Emergency Response/Community Right-to-Know, (804)225-2513.

★1493★ **Virginia Department of Agriculture and Consumer Service**
Office of Pesticide Management
P.O. Box 1163
Richmond, VA 23209
Phone: (804)786-3798; (804)786-3162

★1494★ **Virginia Department of Health**
Division of Health Hazards Control
109 Governor St., N.
Richmond, VA 23219
Phone: (804)786-5932
Hotline: 800-468-0138 (in Virginia)
Important contact number: Bureau of Radiological Health, (804)786-5932, in Virginia, (800)468-0138; Toxic Substances Information Bureau, (804)786-1763.

★1495★ **Virginia Department of Labor and Industry**
P.O. Box 12064
Richmond, VA 23241-0064
Phone: (804)786-2376

★1496★ **Virginia Water Control Board**
2107 Hamilton St.
P.O. Box 11143
Richmond, VA 23230
Phone: (804)257-6384

Washington

★1497★ **Washington Department of Agriculture**
Pesticide Management Division
406 General Administration Bldg., GT-12
Olympia, WA 98504-0641
Phone: (206)753-5064

★1498★ **Washington Department of Community Development**
Emergency Management Division
Ninth and Columbia Bldg., GH-51
Olympia, WA 98504-4151
Phone: (206)459-9191; (206)753-5625
Hotline: 800-633-7585 (in Washington)

★1499★ **Washington Department of Ecology**
Mail Stop PV-11
Olympia, WA 98504-8711
Phone: 800-633-7585; (206)459-6839
Important contact number: Air Quality Program, (206)459-6256; Hazardous Substance Information Office, (206)4387252; Solid and Hazardous Waste Program, (206)459-6316; Water Quality Program, (206)438-7090.

★1500★ **Washington Department of Labor and Industries**
General Administration Bldg., Rm. 334-AX-31
Olympia, WA 98504-0631
Phone: (206)753-6307

★1501★ **Washington Department of Social and Health Services**
Environmental Protection Section
Office of Radiation Protection
Thurston Airdustrial Center
Bldg. 5, LE-13
Olympia, WA 98504
Phone: (206)586-3303

West Virginia

★1502★ **West Virginia Bureau of Public Health**
Environmental Health Office
Industrial Hygiene Division
1800 Washington St., E.
State Capitol Complex, Bldg. No. 3
Charleston, WV 25305
Phone: (304)348-3526
Important contact numbers: Environmental Engineering Division (water quality), (304)348-2981; Radiological Health Program, (304)348-3427.

★ 1503 ★ West Virginia Department of Agriculture
Plant Protection Division
State Capitol Bldg.
Charleston, WV 25305
Phone: (304)348-2212

★ 1504 ★ West Virginia Department of Commerce, Labor and Environmental Resources
State Capitol Bldg.
Charleston, WV 25305
Phone: (304)348-3381
Important contact numbers: Air Pollution Control Commission, (304)348-2275; Division of Water Resources, (304)3482771; Waste Management Section, (304)348-5929.

★ 1505 ★ West Virginia Office of Emergency Services
Emergency Response Commission
State Capital Bldg. 1, Rm. EB-80
Charleston, WV 25305
Phone: (304)348-5380

Wisconsin

★ 1506 ★ Wisconsin Department of Agriculture Trade and Consumer Protection
Agriculture Resources Management Division
Bureau of Plant Industry
801 W. Badger Rd.
P.O. Box 8911 ·
Madison, WI 53708
Phone: (608)266-7131

★ 1507 ★ Wisconsin Department of Health and Social Services
Division of Health
Section of Radiation Programs
5708 Odana Rd.
Madison, WI 53719
Phone: (608)272-6421

★ 1508 ★ Wisconsin Department of Natural Resources
Division of Environmental Equality
P.O. Box 7921
Madison, WI 53707
Phone: (608)266-1099
Important contact numbers: Bureau of Air Management, (608)266-0603; Bureau of Solid and Hazardous Waste Management (includes recycling), (608)266-1327; Bureau of Waste Water Management, (608)266-3910; Emergency Response/Community Right-toKnow, Section 313, (608)266-9255.

★ 1509 ★ Wisconsin Emergency Response Commission
Division of Emergency Government
4802 Sheboygan Ave.
P.O. Box 7865
Madison, WI 53707
Phone: (608)266-3232

Wyoming

★ 1510 ★ Wyoming Department of Agriculture
Weed and Pest Division
2219 Carey Ave.
Cheyenne, WY 82002-0100
Phone: (307)777-6590

★ 1511 ★ Wyoming Department of Environmental Quality
Herschler Bldg.
122 W. 25th St.
Cheyenne, WY 82002
Phone: (307)777-7938
Important contact numbers: Division of Air Quality, (307)777-7391; Division of Solid Waste Management (includes recycling), (307)777-7752; Division of Water Quality, (307)7777781; Hazardous Spill Response, (307)777-7096.

★ 1512 ★ Wyoming Department of Occupational Health and Safety
Herschler Bldg., Second Floor E.
122 E. 25 St.
Cheyenne, WY 82002
Phone: (307)777-7786

★ 1513 ★ Wyoming Division of Health
Division of Health and Medical Services
Radiological Health Services
Hathaway Bldg., Fourth Floor
Cheyenne, WY 82002-0710
Phone: (307)777-6015

★ 1514 ★ Wyoming Emergency Management Agency
Emergency Response Commission
P.O. Box 1709
Cheyenne, WY 82003
Phone: (307)777-7566

Organizations

★ 1515 ★ Academy of Hazard Control Management (AHCM)
8009 Carita Ct.
Bethesda, MD 20817
Phone: (301)984-8969
Contact: Harold W. Gordon, Exec. Dir.
Promotes professional development and the exchange of information within the field of hazard control management. **Publication:** *Hazard Control Monitor.*

★ 1516 ★ Agriculture Resources Center Pesticide Education Project (ARC/PESTED)
115 W. Main St.
Carrboro, NC 27510
Phone: (919)967-1886; (919)967-3054
Contact: Allen Splat, Dir.
Promotes the safe use of pesticides and nontoxic alternatives, participates in statewide and regional studies of pest controls, and serves as a center for reporting pesticide problems. **Publications:** *PESTed News,* a periodic newsletter; studies and position papers.

★ 1517 ★ Agroecology Program
University of California—Santa Cruz
Santa Cruz, CA 95064
Phone: (408)429-4140
Provides research and information about organic farming and sustainable agriculture.

★ 1518 ★ Air and Waste Management Association (AandWMA)
P.O. Box 2861
Pittsburgh, PA 15230
Phone: (412)232-3444
Fax: (412)232-3450
Contact: Martin E. Rivers, Exec. V. Pres.
Nonprofit educational and technical organization that promotes a clean environment through involvement in the technological, economic, and social issues of air and waste management. **Publication:** *Journal of the Air and Waste Management Association.*

★ 1519 ★ Air Pollution Health Effects Laboratory
University of California—Irvine
Department of Community and Environmental Health
Irvine, CA 92717
Phone: (714)856-5860
Fax: (714)856-4763
Contact: Robert H. Phalen, Dir.
Studies health effects of air pollutants on the lungs, including inhalation exposure to acidic aerosol mixtures.

★ 1520 ★ Air Pollution Research Laboratory
New Jersey Institute of Technology
323 Martin Luther King Blvd.
Newark, NJ 07102
Phone: (201)596-3587
Fax: (201)643-3934
Contact: J. Bozzelli, Co-director
Conducts research in the fields of organic, analytic, and physical chemistry as they pertain to pollution of the atmosphere, and analyzes major mutagens on airborne particulates and volatile organics.

★ 1521 ★ Air Pollution Research Laboratory
University of Florida
408 Black Hall
Gainesville, FL 32611
Phone: (904)392-0845
Fax: (904)392-3076
Contact: Dale Lundgren, Dir.
Studies the causes, effects, evaluation, and control of air pollution and the resulting problems of occupational health.

★ 1522 ★ Alabama Conservancy
2717 Seventh Ave. S., Ste. 201
Birmingham, AL 35233
Phone: (205)322-3126
Contact: Kyle G. Crider, Exec. Dir.
Promotes the conservation of natural resources through environmental preservation. Clearinghouse for information on nontoxic alternatives, renewable agriculture, organic gardening, and pesticides and a source for reporting pesticide problems. **Publications:** *ACT,* a monthly newsletter; *Alabama Recycler-Herald; TacNotes.*

★ 1523 ★ Alabama Wildlife Federation (AWF)
P.O. Box 1109
Montgomery, AL 36102
Phone: (205)832-9453
Contact: Douglas Schofield, Exec. Dir.
Concerned with pesticide problems affecting agriculture and wildlife, including the use of hazardous substances to control the boll weevil. **Publication:** *Alabama Wildlife Magazine.*

★ 1524 ★ Alaska Center for the Environment
519 W. Eighth St., Ste. 201
Anchorage, AK 99501
Phone: (907)274-3621
Fax: (907)274-4145
Contact: Sue Libenson, Exec. Dir.
Nonprofit membership public interest advocacy group dedicated to the conservation of Alaska's natural resources. It is the leading Alaskan environmental organization working on hazardous materials and waste reduction

programs. **Publication:** *Alaska Center for the Environment—Center News,* a newsletter.

★ 1525 ★ Alaska Health Project
417 W. Eighth Ave.
Anchorage, AK 99501
Phone: (907)276-2864
Concerned with health issues affecting workers and the general public, including toxic effects of the exposure to pesticides, right-to-know laws, and the improvement of state and local pesticide regulations. **Publication:** *Alaska Health Bulletin.*

★ 1526 ★ Alaska Public Interest Research Group (AKPIRG)
P.O. Box 1093
Anchorage, AK 99510
Phone: (907)278-3661
Contact: Steve Conn, Dir.

★ 1527 ★ Alaska Survival
P.O. Box 344
Talkeetna, AK 99676
Contact: Becky Long, Information Coord.
Seeks to preserve natural resources through environmental conservation. Originally organized to halt the Alaska Railroad from spraying herbicides on its rights-of-way, it has extended its lobbying efforts to controlling the use of pesticides in the area. Provides information on pesticides and is a source for reporting pesticide problems.

★ 1528 ★ Alliance for Responsible Chlorofluorocarbon Policy (ARCFCP)
1901 N. Fort Myer Dr., Ste. 1204
Arlington, VA 22209
Phone: (703)841-9363
Contact: Kevin J. Fay, Exec. Dir.
Trade association of 500 users and producers of chlorofluorocarbons (CFC) that seeks to ensure scientifically sound and economically and socially effective government policies pertaining to the use of CFCs. CFCs are a family of compounds containing carbon, chlorine, fluorine, and sometimes hydrogen that are used primarily as refrigerants, specialty solvents, and agents for forming plastics.

★ 1529 ★ American Association for Aerosol Research (AAAR)
4330 E. West Hwy., Ste. 117
Bethesda, MD 20814
Phone: (301)718-6508
Promotes aerosol research in areas including industrial processes, air pollution, and industrial hygiene. Composed of scientists and engineers associated with universities, technical institutes, and private firms; government representatives; and interested firms. **Publications:** *Aerosol*

Science and Technology, a newsletter published eight times a year; a quarterly newsletter of association activities; a periodic directory.

★ 1530 ★ American Association of Poison Control Centers (AAPCC)
Arizona Poison and Drug Information Center
University of Arizona
Health Science Center, Rm. 3204K
1501 N. Campbell
Tucson, AZ 85721
Phone: (602)626-1587
Fax: (602)626-4063
Contact: Theodore Tong, Exec. Sec.
Aids in the procurement of information on the contents and potential toxicity of substances that may cause accidental poisonings and on the proper management of poisonings. Establishes standards for poison information and control centers and maintains a national database on poisoning.

★ 1531 ★ American Chemical Society (ACS)
Department of Government Relations and
   Science Policy
1155 16th St., N.W.
Washington, DC 20036
Phone: 800-227-5558; (202)872-4395
Contact: John K. Crum, Exec. Dir.
Helps interpret technical data and refers citizens to local scientists. ***Publications:*** Chemistry journals and information pamphlets on a variety of hazardous materials and pollution subjects, available upon request in writing or by calling (202)872-8725.

★ 1532 ★ American College of Toxicology (ACT)
9650 Rockville Pike
Bethesda, MD 20814
Phone: (301)571-1840
Fax: (301)530-7133
Contact: Alexandra Ventura, Exec. Dir.
Professionals who have a common interest in toxicology address related issues and disseminate information. ***Publications:*** *Journal of American College of Toxicology,* bimonthly; *Acute Toxicity Data,* bimonthly; scientific papers, studies, and reports.

★ 1533 ★ American Conference of Governmental and Industrial Hygienists (ACGIH)
6500 Glenway Ave., Bldg. D-7
Cincinnati, OH 45211-4438
Phone: (513)661-7881
Fax: (513)661-7195
Contact: William D. Kelly, Exec. Sec.
Professional society of persons conducting research in occupational safety and health or responsible for implementing industrial hygiene programs in governmental and industrial organizations. Establishes exposure limits for toxic chemicals used in the workplace. ***Publications:*** *Applied Occupational and Environmental Hygiene,* a monthly journal; proceedings, manuals, guides, and research studies.

★ 1534 ★ American Council on Science and Health (ACSH)
1995 Broadway, 18th Floor
New York, NY 10023-5860
Phone: (212)362-7044
Fax: (212)362-4919
Contact: Elizabeth Whelan, Pres.
Nonprofit organization provides consumers with up-to-date, scientifically sound information on the relationship between chemicals, foods, nutrition, lifestyle, and the environment and human health. ***Publications:*** *Priorities,* a quarterly magazine for consumers; *Inside ACSH* and *Media Updates,* quarterly newsletters; research reports on environmental and public health issues.

★ 1535 ★ American Defender Network (ADN)
P.O. Box 911
Lake Zurich, IL 60047
Phone: (815)455-3243; (312)438-7516
Contact: Lorens Tronet, Exec. Dir.
Seeks increased control on the use of pesticides in food, lawn care, and aquatic weed management programs through local, state, and federal legislation.

★ 1536 ★ American Federation of State, County and Municipal Employees (AFSCME)
Department of Research
1625 L St., N.W.
Washington, DC 20036-5687
Phone: (202)429-1000
Fax: (202)429-1293
Contact: Carlos Eduardo Siqueira
Researches occupational health and safety practices and produces reports and guidelines on the use of hazardous materials and toxic substances in the workplace that are available to union and non-union members.

★ 1537 ★ American Industrial Health Council (AICH)
1330 Connecticut Ave., N.W., Ste. 300
Washington, DC 20036-1790
Phone: (202)659-0060
Contact: Ronald A. Lang, Pres.
Coalition of commercial firms and trade associations that promotes the implementation of scientific methods to identify potential industrial carcinogens and other health hazards. Seeks to establish basis for the review, risk assessment, and regulation of substances that may pose significant chronic health risks in the workplace. ***Publications:*** *American Industrial Health Council—Quarterly,* a newsletter; annual report.

★ 1538 ★ American Industrial Hygiene Association (AIHA)
345 White Pond Rd.
P.O. Box 8390
Akron, OH 44320
Phone: (216)873-2442
Fax: (216)873-1642
Contact: O. Gordon Banks, Exec. Dir.
Professional society promotes the study and control of environmental factors affecting the health and well-being of factory workers, including procedures for monitoring the exposure to toxic materials in the work place. ***Publications:*** *Journal,* a monthly journal; annual directory; books, guides, and manuals related to industrial hygiene.

★ 1539 ★ American Petroleum Institute (API)
Health and Environmental Sciences
   Department
1220 L St., N.W.
Washington, DC 20005
Phone: (202)682-8000
Fax: (202)682-8030
Contact: Charles J. DiBona, Pres.
Sponsors research and information programs on the manufacture, handling, transportation, and use of petroleum products in the fields of occupational health, product safety, environmental biology, environmental technology, and community health. Reports and publications are free to members and available for a fee to nonmembers.

★ 1540 ★ Americans for Safe Food
Center for Science in the Public Interest
   (CSPI)
1501 16th St, N.W.
Washington, DC 20036
Phone: (202)332-9110
Nonprofit organization researches issues of environmental and toxic substances concerns and provides information on pesticide residues in food.

★ 1541 ★ Appropriate Technology Transfer for Rural Areas (ATTRA)
P.O. Box 3657
Fayetteville, AR 72702
Phone: 800-346-9140; (501)442-9824
Contact: Julie P. Rogers, Public Information
   Specialist
Provides free information and technical assistance to farmers, county extension agents, agricultural support groups, and professional agriculturists on a broad range of agricultural topics, from non-toxic control methods for specific crops to the direct marketing of "clean beef." Funded by the U.S. Fish and Wildlife Service of the Department of the Interior and managed by the National Center for Appropriate Technology. ***Publication:*** *ATTRAnews,* a monthly newsletter.

★ 1542 ★ Arizona Public Interest Research Group
PIRG Organizing Project
P.O. Box 32474
Tucson, AZ 85741
Phone: (602)326-2732
Contact: Brian Fagin, Dir.

★ 1543 ★ Arizona Toxics Information
P.O. Box 1896
Bisbee, AZ 85603
Phone: (602)432-7340
Clearinghouse for information on pesticides and a source for reporting pesticide problems.

★ 1544 ★ Arkansas Public Interest Research Group
PIRG Organizing Project
39 E. Center
Fayetteville, AR 72701
Phone: (501)521-3939
Contact: Larry Froelich, Dir.

★ 1545 ★ The Art and Craft Materials Institute, Inc. (ACMI)
100 Boyleston St., Ste. 1050
Boston, MA 02116
Phone: (617)426-6400
Fax: (617)426-6639
Contact: Deborah M. Fanning, Exec. V. Pres.
Trade association fosters the safe use of arts and crafts materials. Conducts certification program to ensure the nontoxicity and quality of products and their proper labeling and works with the American Standards Institute to develop industry-wide standards for children's art products. ***Publication:*** *Institute Items,* a monthly newsletter.

★ 1546 ★ Asbestos Information Association/North America (AIA/NA)
1745 Jefferson Davis Hwy., Ste. 509
Arlington, VA 22202
Phone: (703)979-1150
Fax: (703)979-1152
Contact: B. J. Pigg, Pres.
Nonprofit trade organization acts as a clearinghouse for medical and technical information on asbestos-related diseases and ecological considerations. Works with government agencies in developing and

implementing industry-wide standards for exposure to asbestos dust emissions into community air and water and exchanges information on methods and techniques of asbestos dust control. *Publications: Notes and News,* a monthly newsletter; informational and technical materials.

★ 1547 ★ **Asbestos Litigation Group (ALG)**
c/o Ness, Motley, Loadholt, Richardson, and Poole
P.O. Box 1137
Charleston, SC 29402
Phone: (803)577-6747
Contact: Ronald Motley, Pres.
Represents litigants in asbestos-related disease cases throughout the United States.

★ 1548 ★ **Asbestos Victims of America (AVA)**
P.O. Box 559
Capitola, CA 95010
Phone: (408)476-3646
Contact: Heather A. Maurer, CEO
Support network and information center for asbestos victims and their families helps solve related medical, emotional, and financial problems; provides referrals to doctors, attorneys, and other professionals; provides information on consumer and construction products that contain asbestos and on alternative products; and lobbies legislative and regulatory bodies on issues concerning asbestos. *Publications: Asbestos Victims of America Advisor,* a semiannual newspaper; *AVA Resource Directory,* periodical; bulletins, fact sheets, and pamphlets.

★ 1549 ★ **Association of American Pesticide Control Officials (AAPCO)**
P.O. Box 1249
Hardwick, VT 05843
Phone: (802)472-6956
Contact: Phillip H. Gray, Sec.
Promotes uniform laws, regulations, and enforcement policies controlling the sale, use, and distribution of pesticides and chemicals. *Publication:* An annual directory.

★ 1550 ★ **Association of Battery Recyclers (ABR)**
Sanders Lead Co., Corp.
Sanders Rd.
P.O. Drawer 707
Troy, AL 36081
Phone: (205)566-1563
Association of recyclers of lead, oxide manufacturers, industrial equipment suppliers, consulting companies, and battery producers that investigates Office of Safety and Health Administration and U.S. Environmental Protection Agency compliance methods related to lead recycling and its use in manufacturing. Provides information relative to safety and environmental controls and conducts industry-wide research in monitoring, administrative, and protection procedures related to batteries.

★ 1551 ★ **Association of Local Air Pollution Control Officials (ALAPCO)**
444 N. Capitol St., N.W., Ste. 306
Washington, DC 20001
Phone: (202)624-7864
Fax: (202)624-7862
Contact: S. William Becker, Exec. Dir.
Publications: Association of Local Air Pollution Control Officials Washington Update, a monthly newsletter; position papers.

★ 1552 ★ **Association of New Jersey Environmental Commissions (ANJEC)**
300 Mendham Rd.
P.O. Box 157
Mendham, NJ 07945
Phone: (201)539-7547
Fax: (201)539-7713
Contact: Sally Dudley, Exec. Dir.
Statewide nonprofit organization educates, informs, and assists municipal environmental commissions and interested citizens in preserving and protecting New Jersey's environment. *Publication: ANJEC Report,* a quarterly booklet.

★ 1553 ★ **Association of Occupational and Environmental Clinics**
142 Cold Springs St.
New Haven, CT 06511
Phone: (203)785-5885
Responds to workers' occupational health questions and offers suggestions for occupational health clinics and doctors. Spanish is spoken in this office.

★ 1554 ★ **Association of State and Interstate Water Pollution Control Administrators (ASIWPCA)**
444 N. Capitol St., N.W., Ste. 330
Washington, DC 20001
Phone: (202)624-7782
Contact: Robbi J. Savage, Exec. Dir.
Comprised of chief water pollution control administrators from 50 states, territories, and intergovernmental agencies who are responsible for the prevention, abatement, and control of water pollution. Promotes the coordination of state agency programs with those of the U.S. Environmental Protection Agency, Congress, and other federal agencies, conducts research, and maintains a speakers bureau. *Publications: Association of State and Interstate Water Control Administrators—Position Statements,* an annual; annual report and membership directory.

★ 1555 ★ **Association of State and Territorial Solid Waste Management Officials (ASTSWMO)**
444 N. Capitol St., N.W., Ste. 388
Washington, DC 20001
Phone: (202)624-5828
Fax: (202)624-7875
Contact: Thomas J. Kennedy, Exec. Dir.
Works in coordination with the U.S. Environmental Protection Agency to develop programs in solid waste management. Represents state directors in promoting uniform laws and regulations at the federal and local levels, disseminates technical and management information, and conducts studies and training programs. *Publications: Washington Update,* a newsletter issued 10 times a year; surveys, proceedings, and directories.

★ 1556 ★ **Association of University Environmental Health Sciences Centers**
University of Cincinnati
Institute of Environmental Health
3223 Eden Ave.
Cincinnati, OH 45267-0056
Phone: (513)558-5701
Fax: (513)558-1756
Contact: Roy E. Albert, MD, Pres.
Coordinating association of 14 university-based environmental health science centers supported by the National Institute of Environmental Health Sciences serves as a forum for the exchange of information among the centers and for the collaboration of projects.

★ 1557 ★ **Audubon Naturalist Society of the Central Atlantic States**
8940 Jones Mill Rd.
Chevy Chase, MD 20815
Phone: (301)652-5964
Contact: Neal Fitzpatrick, Coord.
Works for pesticide right-to-know legislation and integrated pest management (IPM) education, research, and implementation at the local and state levels; focuses on proper disposal methods of household toxic substances. *Publication: Audubon Naturalist News,* a monthly newspaper.

★ 1558 ★ **Battery Council International (BCI)**
111 E. Wacker Dr., Ste. 600
Chicago, IL 60601
Phone: (312)644-6610
Fax: (312)565-4658
Contact: Edward M. Craft, Exec. Dir.
Not-for-profit trade association of the international leadacid battery industry focuses on lead battery recycling and legislative and environmental issues. *Publications: BCI News,* a periodic newsletter; directories, proceedings, and data books.

★ 1559 ★ **Bio-Integral Resource Center (BIRC)**
P.O. Box 7414
1307 Acton St.
Berkeley, CA 94707
Phone: (415)524-2567
Fax: (415)524-1758
Contact: Sheila Daar, Exec. Dir.
Provides research and information on organic farming, natural alternatives to pesticides, integrated pest management (IPM), and least-toxic methods of pest control. *Publication: Common Sense Pest Control Quarterly,* a newsletter.

★ 1560 ★ **Black Lung Association (BLA)**
Box 872
Craborchard, WV 25827
Phone: (304)252-9654
Contact: Bill Bailey, Exec. Officer
Nonprofit organization promotes safer working conditions for coal miners and just compensation for miners disabled from black lung (pneumoconiosis) and other chronic respiratory and pulmonary diseases caused by coal dust. *Publications: Black Lung Bulletin,* a biweekly newsletter; *Federal Black Lung Journal,* a bimonthly; a biennial directory; a monthly newsletter.

★ 1561 ★ **British Columbia Public Interest Research Group (BCPIRG)**
Simon Fraser University, Rm. TC-304
Burnaby, BC, Canada V5A 1S6
Phone: (604)291-4360
Contact: Phil Lyons

★ 1562 ★ **Brown Lung Association (BLA)**
P.O. Box 7583
Greenville, SC 29610
Phone: (803)269-8048
Contact: Lemar Case, Pres.
Provides education programs on brown lung disease (byssinosis) and obtains fair compensation and safety conditions for workers in cotton mills.

★ 1563 ★ **California Certified Organic Farmers**
P.O. Box 8136
Santa Cruz, CA 95061
Phone: (408)423-2263
Provides research and information about organic farming and sustainable agriculture.

★ 1564 ★ California Coalition for Alternatives to Pesticides (CCAP)
494 H St., Ste. C
Arcata, CA 95521
Phone: (707)822-8497; (707)822-8640
Contact: Jerry Rhodes, Coord.
Coalition of 20 organizations in northern California active in pesticide issues, especially as they relate to crop spraying and forest management. Lobbies for controls and regulations and assists groups in developing relief through litigation. **Publication:** *CCAP Newsletter,* issued nine times a year.

★ 1565 ★ Californians for Alternatives to Toxics (CATS)
494 H St., Ste. C
Arcata, CA 95521
Phone: (707)822-8497; (707)822-8640
Contact: Nancy Correll, Coord.
Educational organization compiles research and provides information on non-chemical forest, park, roadside, and urban pest management. **Publication:** *Drift Dodger,* a newsletter.

★ 1566 ★ California Public Health Foundation
2001 Addison St., Ste. 210
Berkeley, CA 94704-1103
Phone: (415)644-8200
Fax: (415)644-9319
Contact: Joseph M. Hafey, Exec. Dir.
Conducts research on a broad range of public and occupational health issues, including toxic chemicals, air and industrial hygiene, and cancer prevention.

★ 1567 ★ California Public Interest Research Group
1147 S. Robertson Blvd., Ste. 203
Los Angeles, CA 90035
Phone: (213)278-9244
Contact: Deb Bruns, Dir.

★ 1568 ★ Carrying Capacity Network (CCN)
1325 G St., N.W., Ste. 1003
Washington, DC 20005-3104
Phone: (202)879-3044
Contact: Stephen Mabley, Network Coord.
Nonprofit activist network disseminates information on carrying capacity issues such as environmental protection, growth control, and resources conservation. "Carrying capacity" refers to the number of individuals who can be supported without degrading the physical, ecological, cultural, and social environment, that is, without reducing the ability of the environment to sustain the desired quality of life over the long term. **Publication:** *Clearinghouse Bulletin,* a periodic newsletter.

★ 1569 ★ Cause for Concern (CC)
R.D. 1, Box 570
Stewartsville, NJ 08886
Phone: (201)479-4110
Contact: Adele T. McIntosh, Exec. Officer
Environmental education organization concerned with toxic household products. **Publication:** A quarterly newsletter.

★ 1570 ★ Center for Air Environment Studies (CAES)
Environmental Resources Research Institute
Pennsylvania State University
226 Fenske Laboratory
University Park, PA 16802
Phone: (814)865-1415
Contact: Archie J. McDonnell, Dir.
A unit of the Environmental Resources Research Institute, an interdisciplinary research program concerned with the use and management of environmental resources. **Publication:** *Air Pollution Titles,* a bibliographic journal.

★ 1571 ★ Center for Biomedical Toxicological Research and Hazardous Waste Management
Florida State University
Bellamy Bldg.
Tallahassee, FL 32306
Phone: (904)644-5524
Fax: (904)574-6704
Contact: Roy C. Herndon, Dir.
Provides a research base for studies in the areas of environmental toxicology, health effects assessment, occupational health, and effects of toxic organic substances, heavy metals, and inorganic pollutants on humans and the environment.

★ 1572 ★ Center for Emergency Response Planning (CERP)
Workplace Health Fund
815 16th St., N.W.
Washington, DC 20006
Phone: (202)842-7834
Consortium, with industrial departments of the unions of the American Federation of Labor-Congress of Industrial Organizations (AFL-CIO), involved in planning, researching, and disseminating information about issues concerning workers' safety, including general information on chemicals in the workplace.

★ 1573 ★ Center for Environmental Management (CEM)
Tufts University
Curtis Hall
474 Boston Ave.
Medford, MA 02155
Phone: (617)381-3486
Fax: (617)381-3084
Contact: William R. Moomaw, Dir.
Independent international center concerned with critical environmental problems, particularly in the areas of toxic substances and hazardous and solid wastes. Conducts research in a number of areas, analyzes policy, and develops educational programs to meet the needs of society, industry, and government. **Publications:** *CEM Report,* a quarterly newsletter.

★ 1574 ★ Center for Environmental Research
IIT Research Institute
10 W. 35th St.
Chicago, IL 60616
Phone: (312)567-4250
Fax: (313)567-4577
Contact: Demetrios Moschandreas, Dir.
Performs basic, applied, and policy research in a number of environmental areas, including toxic and odorous air emissions, toxicology, asbestos abatement, and environment assessment. **Publications:** Maintains the Hazardous Waste Management computer bulletin board.

★ 1575 ★ Center for Environmental Toxicology
Michigan State University
C231 Holden Hall
East Lansing, MI 48824
Phone: (517)353-6469
Fax: (517)335-4603
Contact: Lawrence J. Fischer, Dir.
Conducts research in a broad range of toxicological matters that effect the environment, including analytical, behavioral, biochemical, and food toxicology, human epidemiology, the fate of chemicals in the environment, and hazardous waste management.

★ 1576 ★ Center for Hazardous Materials Research (CHMR)
University of Pittsburgh Applied Research Center
320 William Pitt Way
Pittsburgh, PA 15238
Phone: (412)826-5320
Fax: (412)826-5552
Hotline: 800-334-CHMR
Contact: Edgar Berkey, Pres.
Develops and implements practical solutions to technical, environmental, institutional, public health, and public policy problems involving hazardous materials and waste. Offers technical assistance to industry and government on pollution prevention, environmental compliance, and recycling and waste minimization and provides independent third-party reviews, forensic investigations, and technical services to communities concerning hazardous waste sites, hazardous materials handling and emergency response, pesticide applications, and other hazardous materials and environmental problems. **Publications:** *The Minimizer,* a quarterly newsletter; manuals, reports and information packets; fact sheets on strategies for pollution prevention in selected industries.

★ 1577 ★ Center for Safety in the Arts
5 Beekman St.
New York, NY 10038
Phone: (212)227-6220
Contact: Michael McCann, Exec. Dir.
Nonprofit organization researches health hazards from toxins found in materials used in the performing and visual arts and suggests suitable precautions. **Publications:** *Art Hazards News,* a monthly newsletter; books, articles, and data sheets.

★ 1578 ★ Central Wisconsin Citizens Pesticide Control Committee
3216 Welsby Ave.
Stevens Point, WI 54481
Phone: (715)344-5446
Clearinghouse for information on pesticides and a source for reporting pesticide problems.

★ 1579 ★ Chemical Manufacturers Association (CMA)
Chemical Referral Center
2501 M St., N.W.
Washington, DC 20037
Phone: (202)887-1255
Fax: (202)887-1237
Hotline: 800-CMA-8200; (202)887-1315 (call collect in Alaska)
Contact: Charles W. Van Vleck, Sec. and V. Pres.
Research organization for chemical companies provides nonemergency health and safety information on chemicals.

★ 1580 ★ Chemical Specialties Manufacturers Association (CSMA)
1913 Eye St., N.W.
Washington, DC 20006
Phone: (202)872-8110
Fax: (202)872-8114
Contact: Ralph Engel, Pres.
Trade association of manufacturers, marketers, formulators, and suppliers of household, industrial, and personal care products acts as a clearinghouse for legislative and scientific developments that affect potentially toxic products used in the home and businesses. **Publications:** *Chemical Times and Trends,* a

quarterly; *Legislative Reporter* and *Executive Watch*, weekly newsletters; directories and reports of laws and regulations.

★ 1581 ★ **The Chlorine Institute (CI)**
2001 L St., N.W., Ste. 506
Washington, DC 20036
Phone: (202)775-2790
Fax: (202)223-7225
Contact: Robert G. Smerko, Pres.
Comprised of producers of gaseous and liquid processing of chlorine and caustic soda. Promotes the safe handling of chlorine, caustic soda, and caustic potash and advises on safe plant operations and transportation and the uses of chlorine and the materials that are used in its manufacture. *Publications:* Pamphlets, lists, manuals, related technical material; audiovisual aids.

★ 1582 ★ **Chlorobenzene Producers Association (CPA)**
1330 Connecticut Ave., N.W.
Washington, DC 20036
Phone: (202)659-0060
Contact: Alan W. Rautio, Exec. Dir.
Chlorobenzene producers who seek to coordinate industry efforts regarding their product. Chlorobenzenes are colorless, flammable, volatile, toxic liquids used as solvents in organic processes. CPA is concerned with health and environmental issues involving these chemicals in response to EPA activities under the Toxic Substances Control Act (TSCA).

★ 1583 ★ **Citizens' Clearinghouse for Hazardous Waste, Inc. (CCHW)**
Appalachia Office
P.O. Box 639
Floyd, VA 24091
Phone: (703)745-3400
Contact: Pete Castelli, Coord.

★ 1584 ★ **Citizens' Clearinghouse for Hazardous Waste, Inc. (CCHW)**
Appalachia Office
P.O. Box 42
Wendel, PA 15691
Phone: (412)864-0845
Contact: Diana Steck, Coord.
Nonprofit grassroots environmental crisis organization of 50 regional groups, 250 state groups, and 6,000 local organizations concerned with the likelihood of adverse physical effects of human contact with toxic chemicals and other hazardous waste. Provides information and guidance for dealing with toxic problems and conducts research on chemicals to determine their dangerous levels of usage. *Publications:* *Everyone's Backyard,* a bimonthly newsletter; over 60 books, pamphlets, and technical fact sheets.

★ 1585 ★ **Citizens' Clearinghouse for Hazardous Waste, Inc. (CCHW)**
Midwest Office
744 Eastgate Dr.
Spencerville, OH 45887
Phone: (419)647-6824
Contact: Sally Teets, Coord.

★ 1586 ★ **Citizens' Clearinghouse for Hazardous Waste, Inc. (CCHW)**
South Office
447 Moreland Ave.
Atlanta, GA 30307
Phone: (404)522-4827
Contact: Hubert Dixon, Coord.

★ 1587 ★ **Citizens' Clearinghouse for Hazardous Waste, Inc. (CCHW)**
West Office
P.O. Box 33124
Riverside, CA 92519
Phone: (714)681-9913
Contact: Penny Newman, Coord.

★ 1588 ★ **Citizens' Environmental Laboratory**
National Toxics Campaign
1168 Commonwealth Ave.
Boston, MA 02134
Phone: (617)232-0327
Testing laboratory serving member organizations of the National Toxics Campaign that provides laboratory services on a wide range of environmental and waste samples.

★ 1589 ★ **Citizens for a Better Environment (CBE)**
407 S. Dearborn, Ste. 1775
Chicago, IL 60605
Phone: (312)939-1530
Contact: William Davis, Exec. Dir.
Nonprofit organization based in the upper Midwest concerned with protecting human health by reducing toxic air and water pollution from solid and hazardous wastes and pesticides. Watchdogs public agencies and industry, performs litigation and scientific research, and provides technical information to the public. *Publications:* *Environmental Review, Midwest Edition,* a quarterly journal; fact sheets on consumer concerns of local interest.

★ 1590 ★ **Citizens for a Better Environment (CBE)**
Green Bay Office
1222 Walnut St.
Green Bay, WI 54301
Phone: (414)432-1909
Contact: Susan Mudd, State Dir.

★ 1591 ★ **Citizens for a Better Environment (CBE)**
Madison Office
222 E. Hamilton St.
Madison, WI 53703
Phone: (608)251-2804

★ 1592 ★ **Citizens for a Better Environment (CBE)**
Milwaukee Office
647 W. Virginia St., Ste. 303
Milwaukee, WI 53204
Phone: (414)271-7280

★ 1593 ★ **Citizens for a Better Environment (CBE)**
Minnesota Office
3255 Hennepin Ave., S.
Minneapolis, MN 55408
Phone: (612)824-8637

★ 1594 ★ **Citizens for a Better Environment (CBE)**
Western Office
942 Market St., Ste. 505
San Francisco, CA 94102
Phone: (415)788-0690
Fax: (415)788-0423
Nonprofit organization working to protect human health by reducing toxic pollution in air and water from solid and hazardous wastes and pesticides, mainly in urban areas in the upper midwestern states and northern California. Watchdogs public agencies and industry, performs litigation and scientific research, and provides technical information and organizing

skills to the public. *Publications:* *Environmental Review, Western Edition,* a quarterly journal; fact-sheets on consumer concerns of local interest.

★ 1595 ★ **Citizens for Alternatives to Chemical Contamination (CACC)**
9496 School St.
Lake, MI 48632
Phone: (517)544-3318
Contact: Ann Hunt, Exec. Dir.
Statewide organization concerned with pesticides and federal and state legislation. Serves as a clearinghouse for pesticide information and conducts an education program on safe pest management. *Publication:* *CACC Newsletter,* a quarterly.

★ 1596 ★ **Citizens for Environmental Protection (CEP)**
27427 M-60 W.
Cassopolis, MI 49031
Phone: (616)445-8769; (616)683-5793
Contact: Merrill Clark, Chair
Focuses on water contamination from pesticides and hazardous substances from landfills in southwestern Michigan counties, where chemicals are widely used in the orchards. Affiliated with the Michigan Pesticide Network.

★ 1597 ★ **Citizens for Environmental Quality (CEQ)**
Rte. 4, Box 152
St. Maries, ID 83861
Phone: (208)245-2368
Contact: Janice Masterjohn, V. Pres.
Concerned principally with local environmental issues, including the control and disposal of hazardous waste, the elimination of noxious weeds, forest management, and implementing integrated pest management (IPM) policies by the state.

★ 1598 ★ **Clean Water Action (CWA)**
Clean Water Fund
317 Pennsylvania Ave., S.E., Ste. 200
Washington, DC 20003
Phone: (202)638-1196
Contact: David Zwick, Dir.
National nonprofit citizens' action group concerned with making water safe from toxic chemicals, the protection of natural resources, and the promotion of alternative waste treatment technologies for recycling. *Publication:* *Clean Water Action News,* a quarterly newsletter.

★ 1599 ★ **Clean Water Fund**
1320 18th St., N.W.
Washington, DC 20036
Phone: (202)457-0336
Fax: (202)457-0287
Nonprofit research and educational organization sponsors programs that promote public interest in issues relating to clean, safe water, the control of toxic chemicals, and the protection of natural resources. *Publication:* *Household Alternatives Chart.*

★ 1600 ★ **Coalition Against Toxics (CAT)**
223 Park Ave.
Atco, NJ 08004
Phone: (609)767-1110; (609)596-0757
Contact: Wynne Falkowski, Chair
Organizes education programs and lobbying activities focusing on the control and elimination of toxic pesticides from farm and residential use in southern New Jersey. Seeks right-to-know legislation and pesticide notification regulations for residential applications.

★ 1601 ★ Colorado Coalition for Alternatives to Pesticides (CCAP)
777 Grant St., Ste. 203
Denver, CO 80203-3518
Focuses its efforts primarily in the area of mosquito and weed control and promotes pesticide alternatives such as integrated pest management (IPM) and manual eradication.

★ 1602 ★ Colorado Environmental Coalition (CEC)
777 Grant St., Ste. 606
Denver, CO 80203-3518
Phone: (303)837-8701
Contact: Elizabeth Otto, Adm.
Statewide network of organizations active in hazardous waste reduction, nontoxic alternatives, pesticide control, and other environmental concerns. *Publication: Colorado Environmental Report;* a newsletter.

★ 1603 ★ Colorado Pesticide Network
c/o Colorado Chapter of the Sierra Club
777 Grant St., No. 606
Denver, CO 80203
Phone: (303)771-0595; (303)688-4431
Promotes the safe use of pesticides and acts as a center for reporting pesticide problems.

★ 1604 ★ Colorado Public Interest Research Group (COPIRG)
1724 Gilpin St.
Denver, CO 80218
Phone: (303)355-1861
Contact: Rich McClintock, Dir.

★ 1605 ★ Committee for Sustainable Agriculture
P.O. Box 1300
Colfax, CA 95713
Phone: (916)346-2777
Provides research and information about organic farming and sustainable agriculture.

★ 1606 ★ Concern, Inc.
1794 Columbia Rd., N.W.
Washington, DC 20009
Phone: (202)328-8160
Fax: (202)328-8161
Contact: Susan Boyd, Exec. Dir.
Nonprofit organization provides environmental information to those concerned with the environment and public education and public policy development. Develops, publishes, and distributes community action guides, each providing a comprehensive explanation of a specific environmental issue, extensive resources, guidelines for action, and pertinent state and federal legislation. *Publications: Alternatives to Pesticides; Drinking Water; Ground Water; Farmland: A Community Issue; Household Waste: Issues and Opportunities; Waste: Choices for Communities.*

★ 1607 ★ Connecticut Citizen Action Group (CCAG)
2074 Park St.
Hartford, CT 06106
Phone: (203)523-9232
Activist group concerned with such issues as toxic air regulations, pesticides, ground water contamination, and toxic substances in landfills. *Publication:* A newsletter.

★ 1608 ★ Connecticut Fund for the Environment (CFE)
1032 Chapel St.
New Haven, CT 06510
Phone: (203)787-0646
Contact: Suzanne Y. Mattei, Dir.
Seeks to protect the natural resources of Connecticut by working on a broad range of issues, including toxic and solid waste management, air and land pollution, and preservation of wetlands. Active in promoting the safe use of pesticides and is a source for reporting pesticide problems. *Publication: Newsletter,* quarterly; fact sheets on pesticides; booklets.

★ 1609 ★ Connecticut Public Interest Research Group (CONNPIRG)
219 Park Rd., Second Floor
West Hartford, CT 06119
Phone: (203)523-5873
Contact: Jim Leahy, Dir.

★ 1610 ★ Consumer Pesticide Project (CPP)
425 Mississippi St.
San Francisco, CA 94107
Phone: (415)826-6314
Fax: (415)826-6314
Contact: Craig Merrlees, Dir.
Activist organization of twelve state groups works with farmers, consumers, and environmental organizations to promote the practical reduction of dangerous pesticides. *Publication: A Practical Strategy to Reduce Dangerous Pesticides in Our Environment,* an organizing kit.

★ 1611 ★ Co-op America
2100 M St., N.W., Ste. 403
Washington, DC 20063
Phone: (202)872-5307
Contact: Alisa Gravitz, Exec. Dir.
Encourages the purchase of environmentally sound products and services. *Publications: Co-op America Alternative Catalog* and *Co-op America Quarterly.*

★ 1612 ★ Coordinating Committee on Toxics and Drugs (CCTD)
825 West End Ave., Ste. 70
New York, NY 10025
Phone: (212)663-6378
Focuses on the export of hazardous products, primarily pesticides, from the U.S. to developing countries.

★ 1613 ★ Cosmetic, Toiletry and Fragrance Association (CTFA)
1110 Vermont Ave., N.W., Ste. 800
Washington, DC 20005
Phone: (202)331-1770
Fax: (202)331-1969
Contact: E. Edward Kavanaugh, Pres.
Trade association provides scientific, legal, regulatory, and legislative services for its members and publishes material that seeks to prevent the use of toxic substances in products. *Publications: Cosmetic Ingredient Review; CTFA Newsletter,* biweekly; periodic newsletters and directories.

★ 1614 ★ C.U.R.E. Formaldehyde Poisoning Association
9255 Lynnwood Rd.
Waconia, MN 55387-9583
Phone: (612)448-5441
Contact: Connie Smrecek, Exec. Dir.
Seeks to educate health and legal professionals concerning problems caused by formaldehyde. C.U.R.E. is an acronym for "Citizens United to Reduce Emissions of." *Publication: The Environmental Guardian,* a quarterly newsletter.

★ 1615 ★ Delmarva Rural Ministries
26 Wyoming Ave.
Dover, DE 19901
Phone: (302)678-2000
Contact: Debra Singletary, Exec. Dir.
Focuses on educating the public and farm workers about the dangers of toxic substances used in agriculture and present in the environment. Helps farm workers contaminated by pesticides primarily in Delaware, Maryland, and Virginia.

★ 1616 ★ Department of Environmental Health
University of Cincinnati
3223 Eden Ave.
Cincinnati, OH 45267-0056
Phone: (513)558-5701
Fax: (513)558-1756
Contact: Roy E. Albert, Dir.
Conducts research in a number of environmental health sciences fields, including areas related to heavy metals, metabolism and toxicology of environmental chemicals and drugs, and environmental carcinogenesis, mutagenesis, pulmonary metabolism, and toxicology.

★ 1617 ★ Department of Toxicology
North Carolina State University
P.O. Box 7633
Raleigh, NC 27695
Phone: (919)737-2274
Fax: (919)737-7169
Contact: Ernest Hodgson
Conducts studies on the biological and general aspects of environmental health sciences, with emphasis on pesticides.

★ 1618 ★ Division of Occupational and Environmental Medicine
University of California—Davis
Davis, CA 95616
Phone: (916)752-4256
Contact: Mark B. Schenker, Division Chief
Conducts research into the epidemiology of occupational and environmental health problems, focusing on agricultural health problems, occupational cancer, reproductive hazards, and respiratory diseases.

★ 1619 ★ East Michigan Environmental Action Council (EMEAC)
21220 W. 14 Mile Rd.
Birmingham, MI 48010
Phone: (313)258-5188
Contact: Elizabeth Harris, Exec. Dir.
Citizens' action group focuses on educational programs that urge the use of integrated pest management (IPM) for homes and farms. *Publication: Consumer Pesticide Information Guide.*

★ 1620 ★ Ecology Center of Ann Arbor
417 Detroit St.
Ann Arbor, MI 48104
Phone: (313)761-3186
Contact: Tara Ward, Coord.
Nonprofit organization encourages community involvement in environmental issues affecting the area. Promotes the safe use of pesticides and acts as a center for reporting pesticide problems. *Publication: Ecology Reports,* a newsletter issued 10 times a year.

★ 1621 ★ Enterprise for Education, Inc.
1320-A Third St., Ste. 202
Santa Monica, CA 90401
Phone: (213)394-9864
Fax: (213)394-3539
Private corporation prepares materials about hazardous household wastes, global warming, energy, and health for distribution to elementary and secondary schools. *Publications: Hazardous Waste from Homes: A Student Guide* and *The Greenhouse Effect and Global Warming.*

★ 1622 ★ Environmental Action Coalition (EAC)
625 Broadway, Second Floor
New York, NY 10012-2611
Phone: (212)677-1601
Contact: Nancy Wolf, Exec. Dir.
Nonprofit corporation focuses on recycling and waste reduction, water conservation, use of non-polluting energy, environmental education, and urban forestry, primarily in the New York City area. Serves as a clearinghouse for environmental services in urban areas nationwide. *Publication: Cycle,* a membership newsletter.

★ 1623 ★ Environmental Action Foundation (EAF)
Waste and Toxic Substance Project
1525 New Hampshire Ave., N.W.
Washington, DC 20036
Phone: (202)745-4970
Contact: Ruth Caplan, Dir.
Organization focuses on the proper manufacture and disposal of pesticides and other toxic substances. Assists grassroots organizations, works for strong state and federal environmental laws, and, through their Toxics Accountability Project, offers victims of toxic exposure legal assistance. *Publications: Environmental Action,* a bimonthly magazine; fact packets on hazardous waste issues.

★ 1624 ★ Environmental and Occupational Health Science Institute (EOHSI)
University of Medicine and Dentistry of New Jersey
Robert Wood Johnson Medical School
Brookwood II
45 Knightsbridge Rd.
Piscataway, NJ 08854
Phone: (908)463-5353
Fax: (908)463-5133
Hotline: 800-843-0054 (in New Jersey)
Provides the public with scientific information on environmental and occupational health issues. *Publications: INFOletter,* a newsletter; INFOsheets on selected topics.

★ 1625 ★ Environmental Defense Fund (EDF)
257 Park Ave., S., 16th Floor
New York, NY 10010-7304
Phone: (212)505-2100
Fax: (212)505-2375
Contact: Frederick D. Krupp, Exec. Dir.
Public interest organization of attorneys, scientists, economists, and interested individuals that seeks to protect and improve environmental quality and public health. It works toward reform of public policy in many fields of hazardous substances, including toxic chemical regulations, toxicology, radiation, air and water quality, energy, agriculture, wildlife conservation, ozone depletion, and the greenhouse effect. It also initiates litigation in environment-related matters and conducts public service and educational campaigns about protecting the environment. *Publication: EDF Letter,* a bimonthly newsletter.

★ 1626 ★ Environmental Defense Fund (EDF)
California Office
5655 College Ave.
Oakland, CA 94618
Phone: (415)658-8008

★ 1627 ★ Environmental Defense Fund (EDF)
Capital Office
1616 P St., N.W.
Washington, DC 20036
Phone: (202)387-3500

★ 1628 ★ Environmental Defense Fund (EDF)
North Carolina Office
128 E. Hargett St.
Raleigh, NC 27601
Phone: (919)821-7793

★ 1629 ★ Environmental Defense Fund (EDF)
Rocky Mountain Office
1405 Arapahoe Ave.
Boulder, CO 80302
Phone: (303)440-4901

★ 1630 ★ Environmental Defense Fund (EDF)
Texas Office
1800 Guadalupe
Austin, TX 78701
Phone: (512)478-5161

★ 1631 ★ Environmental Defense Fund (EDF)
Virginia Office
1108 E. Main St.
Richmond, VA 23219
Phone: (804)780-1297

★ 1632 ★ Environmental Hazards Management Institute (EHMI)
10 Newmarket Rd.
P.O. Box 932
Durham, NH 03824
Phone: (603)868-1496
Fax: (603)868-1547
Contact: Alan John Borner, Exec. Dir.
Provides information on environmental management to private citizens, industry, nonprofit agencies, government, and academia. *Publication: Re:Source,* a newsletter; *HazMat World Magazine;* EMHI Educational/ Informational Wheels.

★ 1633 ★ Environmental Health Network (EHN)
2536 Leimert Blvd.
Oakland, CA 94602
Phone: (415)530-1724
Contact: Susan Del Solar, Pres.
Group of individuals who have been exposed to toxic chemicals promote public awareness of health problems caused by chemicals and environmental pollutants and educate people exposed to toxic chemicals about their legal rights and the public services available to them. *Publications: Reactor,* a newsletter published four to five times a year; *Environmental Illness Guide.*

★ 1634 ★ Environmental Health Sciences Center
Oregon State University
317 Weniger Hall
Corvallis, OR 97331-6504
Phone: (503)737-3608
Fax: (503)737-0481
Contact: Donald J. Reed, Dir.
Conducts interdisciplinary research to assess the impact of environmental chemicals on human health, including toxicology of naturally occurring toxins and environmental chemicals and carcinogenesis of environmental chemicals.

★ 1635 ★ Environmental Health Sciences Research Laboratory
Tulane University
F. Edward Hebert Research Center
Belle Chasse, LA 70037
Phone: (504)394-2233
Contact: Jamal Y. Shamas, Research Coord.
Conducts research in a number of environmental and pollution areas, including occupational health and safety, toxicology, toxic effects of pesticides, and industrial hygiene.

★ 1636 ★ Environmental Law Institute (ELI)
1616 P St., N.W.
Washington, DC 20036
Phone: (202)328-5150
Fax: (202)328-5002
Contact: J. William Futrell, Pres.
Nonprofit research and educational organization serves as a national forum for professionals of all disciplines in the environmental field. Conducts and sponsors research on environmental law and policy; sponsors workshops, seminars, and conferences; and maintains an extensive publishing program and an environmental law information clearinghouse. *Publications: The Environmental Forum,* a bimonthly journal; loose-leaf services, newsletters, and journals pertaining to environmental law.

★ 1637 ★ Environmental Research Foundation
P.O. Box 3541
Princeton, NJ 08543-3541
Phone: (202)328-1119
Contact: Robin Lee Zeff, Assoc. Dir.
Nonprofit organization conducts research on hazardous materials and waste and other environmental issues. *Publication: Rachel's Hazardous Waste News,* a newsletter.

★ 1638 ★ Environmental Resources Research Institute (ERRI)
Pennsylvania State University
Land and Water Research Bldg., Rm. 103
University Park, PA 16802
Phone: (814)865-1415
Contact: Archie J. McDonnell, Dir.
Interdisciplinary research program concerned with the use and management of environmental resources, comprising the Center for Air Environment Studies, Office of Hazardous and Toxic Waste Management, Pennsylvania Center for Water Resources Research, Center for Land Resources Research, and Office for Remote Sensing of Earth Resources. *Publication:* Environmental Resources Research Institute Newsletter.

★ 1639 ★ Environmental Toxicology Center
University of Wisconsin—Madison
309 Infirmary
Madison, WI 53706
Phone: (608)263-4580
Contact: Colin R. Jefcoate, Dir.
Studies toxicology and problems related to the presence of potentially hazardous chemicals in the environment, such as pesticides, heavy metals, toxins carried by food, and chlorinated hydrocarbons. *Publication:* Environmental Toxicology Center Newsletter.

★ 1640 ★ Farmworker Justice Fund (FJF)
P.O. Box 53396
2001 S St., N.W., Ste. 210
Washington, DC 20009
Phone: (202)462-8192
Contact: Mike Hancock, Exec. Dir.
National advocacy group tracks federal and state legislation, administrative, and judicial

matters concerning pesticide reform, immigration, and occupational health concerns of migrant farm workers and their families. **Publications:** *Newsletter,* a quarterly; books and monographs; *Toxic Chemicals: The Interface Between Law and Science; The Occupational Health of Migrant and Seasonal Farmworkers in the United States.*

★ 1641 ★ **Farmworkers Legal Assistance Project (FLAP)**
P.O. Box 4823
Lafayette, LA 70502
Phone: 800-252-3129; (318)237-4320
Contact: Stanley Halpin, Dir.
Informs farmworkers in Louisiana of their legal rights and remedies for exposure to harmful agriculture pesticides and other toxic chemicals.

★ 1642 ★ **Feed and Fertilizer Laboratory**
Louisiana State University
Baton Rogue, LA 70893
Phone: (504)388-2755
Contact: Hershel F. Morris, Dir.
Conducts research in the fields of fertilizers, pesticides, and pesticide residues.

★ 1643 ★ **Florida Consumer Action Network (FL CAN)**
4601 W. Kennedy Blvd., No. 226
Tampa, FL 33609
Phone: (813)286-1226
Contact: Monte Belote, Adm. Dir.
Statewide nonprofit citizen lobbying group seeks to give consumers a greater voice in the legislative and regulatory processes, particularly as they affect Florida's groundwater contamination from pesticides and other toxic substances. **Publication:** *Consumer Action News.*

★ 1644 ★ **Florida Public Interest Research Group (FPIRG)**
308 E. Park Ave., Ste. 213
Tallahassee, FL 32301
Phone: (904)224-5304
Contact: Ann Whitfield, Dir.

★ 1645 ★ **Floridians for a Safe Environment**
R.R. 1, Box 1213-Y
Tavares, FL 32778
Phone: (904)343-9127
Contact: Ann Williams, Pres.
Activist group organized by residents to prevent the use of the pesticide Temik (aldicarb) in their residential and agricultural area. **Publication:** *Floridians for a Safe Environment,* a newsletter.

★ 1646 ★ **Formaldehyde Institute, Inc.**
1330 Connecticut Ave., N.W., Ste. 300
Washington, DC 20036
Phone: (202)659-0060
Fax: (202)659-1699
Contact: John F. Murray, Pres.
Conducts health and toxicity studies on formaldehyde and products containing formaldehyde.

★ 1647 ★ **Friends of the Earth (FOE)**
218 D St., S.E.
Washington, DC 20003
Phone: (202)544-2600
Fax: (202)543-4710
Contact: Michael S. Clark, Exec. Officer
Member organization that merged with the Environmental Policy Institute and the Oceanic Society in 1990. Lobbies the federal government on behalf of grassroots citizens groups, provides guidance for lobbying to state and local governments, and assists in community organizing. **Publications:** *Friends of*

the Earth, a newsletter; *The Community Plume,* a newsletter; *Not Man Apart,* a bimonthly magazine.

★ 1648 ★ **Friends of the Earth (FOE)**
Northwest Office
4512 University Way, N.E.
Seattle, WA 98105
Phone: (206)633-1661
Fax: (206)633-1935
Contact: David E. Ortman, Northwest Representative

★ 1649 ★ **Friends of the Mountain**
Rte. 2 Tate City
Clayton, GA 30525
Phone: (404)896-3421
Contact: Audry Cornelius, Sec.
Conservation group works on a number of forest-related issues, including the use of herbicides in forest management. It lobbies the U.S. Forest Service and local governments against using the pesticides Velpar and Garlon in their surroundings. **Publication:** *The Friends of the Mountain,* a newsletter.

★ 1650 ★ **Friends of Wildlife**
P.O. Box 958
Boca Grande, FL 33921
Phone: (813)964-0110
Contact: Dolores Heimann, Dir.
Activist group seeks to prevent the use of such pesticides as Propoxur in aerial spraying and to promote safer alternatives.

★ 1651 ★ **Georgia Environmental Project**
136 Marietta St., N.W., Ste. 238
Atlanta, GA 30303
Contact: Patrick Kessing
Working to stop the heavy use of pesticides to combat the cotton boll weevil.

★ 1652 ★ **Golden Empire Health Planning Center**
909 12th South St., Ste. 203
Sacramento, CA 95814
Phone: (916)448-1198
Sacramento county-based nonprofit organization conducts research and sponsors educational programs on hazardous substances, waste disposal, pesticides, and other environmental issues. **Publications:** *Household Hazardous Waste: Solving the Disposal Dilemma* and *Making the Switch: Alternatives to Using Toxic Chemicals in the Home.*

★ 1653 ★ **Graphic Arts Technical Foundation (GATF)**
Graphic Communications Industries, Inc.
4615 Forbes Ave.
Pittsburgh, PA 15213-3796
Phone: (412)621-6941
Fax: (412)621-3049
Contact: Gary A. Jones, Mgr., Environmental Information
Nonprofit scientific, technical, and educational organization serves the graphic arts industry. Conducts studies on processes and equipment, including those relating to exposure to toxics and hazardous wastes. **Publication:** *Environmental Control Report,* a newsletter; *Gatfworld Magazine,* a bimonthly magazine reporting activities of the Foundation; product catalogs, textbooks, proceedings, special reports.

★ 1654 ★ **Grass Roots the Organic Way**
38 Llangollen Ln.
Newton Square, PA 19073
Phone: (215)353-2838

Promotes the safe use of pesticides and acts as a center for reporting pesticide problems.

★ 1655 ★ **Greenpeace U.S.A.**
1436 U St., N.W.
Washington, DC 20009
Phone: (202)462-1177
Fax: (202)462-4507
Contact: Peter Bahouth, Exec. Dir.
International organization dedicated to the protection of the natural environment through direct action, education, and legislative lobbying. There are five regional Greenpeace offices in the U.S.; Greenpeace International has offices in nineteen countries. **Publication:** *Greenpeace Magazine,* a quarterly.

★ 1656 ★ **Hawaii Coalition Against the Misuse of Pesticides (HCAMP)**
R.R. 1, Box 22A
Wailuka, HI 96793
Phone: (808)242-7296
Contact: Hazel Cunningham, Coord.
Focuses on the general regulations of pesticides in Hawaii, especially in the pineapple industry, where the use of pesticides has affected ground water.

★ 1657 ★ **Hazardous Materials Control Research Institute (HMCRI)**
7237-A Hanover Pkwy.
Greenbelt, MD 20770-3602
Phone: (301)982-9500
Fax: (301)220-3870
Contact: Harold Bernard, Exec. Dir.
Organization of corporations, scientists, engineers, academics, government administrators, and interested individuals concerned with the safe management of hazardous materials and the prevention, control, and disposal of waste. Disseminates information about technical advancements and institutional requirements in hazardous waste disposal; conducts training programs on the control and management of toxic and hazardous materials; encourages the minimization of hazardous materials released into the environment; and fosters the development of reasonable standards, test procedures, and monitoring and reporting requirements. **Publications:** *Focus,* a monthly newsletter; *Hazardous Materials Control Directory,* biennial; *Hazardous Materials Control Magazine,* bimonthly; *Journal on Hazardous Waste and Hazardous Materials,* annual.

★ 1658 ★ **Hazardous Substances Management Program**
Center for Technology, Policy and Industrial Development
Massachusetts Institute of Technology
Technology, Business and Environment Program
77 Massachusetts Ave., E40-241
Cambridge, MA 02139
Phone: (617)253-0902
Fax: (617)253-7140
Contact: John R. Ehrenfeld, Dir.
Interdisciplinary program designed to advance the study of techniques necessary to manage the environment and to understand public policy matters affecting control of the environment. Seeks to discover the effects of hazardous substances upon human health, to understand the transport and fate of chemicals in soil and water, and to determine how incineration destroys and creates hazardous substances. **Publications:** *News from the Hazardous Substances Management Program,* a quarterly newsletter; articles and working papers.

★ 1659 ★ Hazardous Substances Research Center (HSRC)
EPA Regions VII and VIII
Kansas State University
133 Ward Hall
Manhattan, KS 66506-2502
Phone: (913)532-6519
Fax: (913)532-6952
Contact: Larry E. Erickson, Dir.
Consortium of the EPA and regional universities provides short-term and long-term research relating to key hazardous substance problems in the region, including those affecting the agriculture, forestry, mining, mineral processing, and related industries. Disseminates information through training programs and the media. **Publications:** *HazTech Transfer,* a quarterly newsletter; proceedings and bibliographies.

★ 1660 ★ Hazardous Waste and Toxic Substance Research and Management Center
Clarkson University
Rowley Laboratories
Potsdam, NY 13699
Phone: (315)268-6400
Contact: Thomas L. Theis, Dir.
Conducts research in toxicology, environmental health, air pollution, assessments of exposure, and other environmental topics.

★ 1661 ★ Hazardous Waste Federation (HWF)
c/o Department 3220
Div. 3314
P.O. Box 5800
Albuquerque, NM 87185
Phone: (505)845-8889
Contact: Gordon J. Smith, Chair
Seeks to increase public awareness of problems related to hazardous waste management and such issues as air pollution, groundwater and water pollution, radioactive materials, leaking storage tanks, and transportation of toxic substances.

★ 1662 ★ Health Effects Institute—Asbestos Research (HEI-AR)
141 Portland St., Ste. 7100
Cambridge, MA 02139
Phone: (617)225-0866
Fax: (617)225-2211
Contact: Andrew Sivak, Exec. Dir.
Independent nonprofit organization formed to support research to determine airborne asbestos exposure levels prevalent in buildings, to characterize peak exposures and their significance, and to evaluate the effectiveness of asbestos management and abatement strategies. **Publication:** *Asbestos in Public and Commercial Buildings: A Literature Review and Synthesis of Current Knowledge,* a 387-page research report, published 1991.

★ 1663 ★ Health Effects of Lawn Care Pesticides Task Force (HELP)
2290 Clematis Dr.
Sarasota, FL 34239
Phone: (813)954-2291
Contact: Ann Mason, Dir.
Seeks to promote a healthier residential and urban environment through public awareness of potential health problems associated with the use of pesticides. Investigates alternatives to pest control procedures and works with local and state agencies to reduce the amount of toxic chemicals used in pesticides.

★ 1664 ★ Household Hazardous Waste Project (HHWP)
Southwest Missouri State University
Office of Continuing Education
901 S. National
Box 108
Springfield, MO 65804
Phone: (417)836-5777
Contact: Sondra Goodman, Dir.
Community education program assists the public in making informed decisions about the safe use, storage, and disposal of hazardous products commonly found around the home. **Publication:** *Guide to Hazardous Products around the Home.*

★ 1665 ★ Human Ecology Action League (HEAL)
P.O. Box 49126
Atlanta, GA 30359-1126
Phone: (404)248-1898
Contact: Kenneth Dominy, Pres.
Represents individuals who have illnesses caused by sensitivity to natural and synthetic chemical substances in the environment. **Publications:** *Human Ecologist,* a quarterly journal; directories.

★ 1666 ★ Humboldt Herbicide Task Force (HHTF)
P.O. Box 4536
Arcata, CA 95521
Phone: (707)822-8497
Contact: Lynn Duggins
Local group concerned with the use of herbicides in forestry and insecticides on California's apple crop. Seeks bans on toxic aerial sprays and encourages the use of integrated pest management (IPM).

★ 1667 ★ Illinois Public Interest Research Group (ILPIRG)
202 S. State St., Ste. 1400
Chicago, IL 60604
Phone: (312)341-0814
Contact: Diane Brown, Dir.

★ 1668 ★ Indiana Division of the Izaak Walton League (IWL)
1802 Chapman Rd.
Box 141
Huntertown, IN 46748
Phone: (219)637-6264
Contact: Jane Dustin, Chair, Water Quality
State branch of the Izaak Walton League concerned with a number of pesticide-related issues, including the state-wide registration of pesticides and monitoring Monsanto Corporation's training program for Alachlor use.

★ 1669 ★ Indiana Public Interest Research Group (INPIRG)
Activities Desk
Indiana Memorial Union
Bloomington, IN 47405
Phone: (812)335-7575

★ 1670 ★ INFORM, Inc.
381 Park Ave., S.
New York, NY 10016
Phone: (212)689-4040
Contact: Joanna D. Underwood, Pres.
Nonprofit environmental research and educational organization identifies practical ways to protect and conserve natural resources. Clarifies corporate social responsibilities and recommends specific actions for change. Researches toxic waste reduction in the organic chemical industry, municipal solid waste reduction, clean fuel alternatives, and other topics dealing with conservation on a national

and regional level. **Publication:** *INFORM Reports,* a quarterly newsletter.

★ 1671 ★ Institute for Environmental Studies
Center for Solid Waste Management and Research
University of Illinois at Urbana-Champaign
1101 W. Peabody Dr.
Urbana, IL 61801
Phone: (217)333-4178
Contact: Roger A. Minear, Dir.
Administrates an interdisciplinary scientific research program examining the sociological, biological, and physical aspects of waste reduction, recycling, incineration, and landfilling. **Publication:** *Solid Waste Management Newsletter.*

★ 1672 ★ Institute of Analytical and Environmental Chemistry
University of New Haven
300 Orange Ave.
West Haven, CT 06516
Phone: (203)932-7171
Performs water and soil testing for lead, heavy metals, pesticides, and other environmental pollutants.

★ 1673 ★ Institute of Chemical Toxicology
Wayne State University
2727 Second Ave.
Detroit, MI 48201
Phone: (313)577-0100
Contact: Raymond F. Novak, Dir.
Studies the chemical hazards of the environment and the industrial workplace, including toxicology of air pollutants, solvents, and heavy metals; studies carcinogenesis and the metabolic and hematologic effects resulting from exposure to toxic agents.

★ 1674 ★ Institute of Environmental and Industrial Health
University of Michigan
School of Public Health
Ann Arbor, MI 48109
Phone: (313)764-3188
Fax: (313)763-5455
Contact: Robert Gray, Dir.
Conducts research in the fields of environmental and industrial health and safety; environmental toxicology and epidemiology; toxic contamination of the environment; hazardous materials evaluation, control, and environmental impacts; and measurement, evaluation, and control of radiation in the workplace.

★ 1675 ★ Institute of Environmental Medicine
New York University
550 First Ave.
New York, NY 10016
Phone: (212)340-5280
Contact: Arthur Upton, Dir.
Conducts research in the areas of toxicology, chemical carcinogenesis, respiratory diseases and aerosol physiology, environmental hazards to which industrial and community populations are exposed, and industrial and environmental health hazards and means for their control.

★ 1676 ★ Institute of Environmental Toxicology and Chemistry
Western Washington University
Huxley College
Environmental Studies Center
Mail stop 9079
Bellingham, WA 98225
Phone: (206)647-6136
Fax: (206)647-7284
Contact: David B. Thorud, Dir.
Studies the effects of toxic substances on terrestrial and aquatic organisms.

★ 1677 ★ Institute of Toxicology and Environmental Health
University of California—Davis
Davis, CA 95616
Phone: (916)752-1340
Fax: (916)752-5300
Contact: James W. Overstreet, Dir.
Coordinates interdisciplinary research on biomedical and toxicological problems related to exposure to chemical, physical, and biological toxic agents.

★ 1678 ★ Instituto Laboral
2947 16th St.
San Francisco, CA 94130
Phone: (415)431-7522
Provides information to Hispanic workers about their legal rights concerning occupational safety and health matters, such as injuries from hazardous chemicals in the workplace, and will suggest legal clinics and attorneys. Spanish is spoken in this organization.

★ 1679 ★ International Hazardous Materials Association (IHMA)
640 E. Wilmington Ave.
Salt Lake City, UT 84106
Phone: (801)466-3500
Fax: (801)466-9616
Association of firemen and other first responders that sponsors annual conferences and seminars about the prevention of chemical accidents and effective actions to take when responding to toxic spills and other accidents. **Publications:** A quarterly newsletter.

★ 1680 ★ Iowa Audubon Council
R.R. 2
Ames, IA 50010
Phone: (515)232-3807
Contact: Cindy Hildebrand
Composed of members of the National Audubon Society, the council is concerned with a variety of environmental issues, including water contamination by agricultural chemicals, roadside pesticide use, pesticide storage and disposal, and chemical trespass. **Publication:** *Legislative News,* a periodic newsletter.

★ 1681 ★ Iowa Environmental Coalition (IEC)
3500 Kingman Blvd.
Des Moines, IA 50010
Phone: (515)292-4014
Contact: Sherry Drugula, Office Mgr.
Lobbies for legislation on pesticides and other agricultural chemicals. **Publication:** A bimonthly newsletter.

★ 1682 ★ Iowa Public Interest Research Group
Iowa State University
Memorial Union, Rm. 306
Ames, IA 50011
Phone: (515)294-8094
Contact: Jim Dubert, Dir.

★ 1683 ★ Izaak Walton League of America (IWLA)
1401 Wilson Blvd., Level B
Arlington, VA 22209
Phone: (703)528-1818
Fax: (703)528-1836
Contact: Jack Lorenz, Exec. Dir.
Organization works to educate the public to conserve the natural resources of the country. Sponsors educational programs on a number of environmental issues, including wetlands, acid rain, clean streams, and outdoor ethics, and is active in joining litigation to outlaw harmful pesticides that endanger nature. **Publications:** *League Leader,* a bimonthly newsletter; *Outdoor America,* a quarterly magazine; *Outdoor Ethics,* a quarterly newsletter; booklets and proceedings.

★ 1684 ★ Kansans for Safe Pest Control
R.R. 5, Box 163
Lawrence, KS 66046
Phone: (913)748-0950

★ 1685 ★ Kansas Organic Producers (KOP)
P.O. Box 153
Beattie, KS 66406
Phone: (913)353-2414
Contact: Judy Nickelson, Sec.
Encourages the use of nontoxic alternatives to pesticides and sustainable agriculture. **Publication:** A quarterly newsletter.

★ 1686 ★ Laboratory for Pest Control Application Technology
Ohio State University
Ohio Agriculture Research and Development Center
Wooster, OH 44691
Phone: (216)263-3726
Fax: (216)263-3713
Contact: Franklin R. Hall, Dir.
Studies pest control agents, application technologies, biological responses of the pesticide delivery system, and other issues pertaining to pesticides and their effects on crops and exposure to humans. Develops pesticides risk/benefit computer models to assist regulatory decision making.

★ 1687 ★ Labor Occupational Health Program (LOHP)
University of California at Berkeley
Institute of Industrial Relations
2521 Channing Way
Berkeley, CA 94720
Phone: (415)642-5507
Contact: Patricia Quinlan
Education arm of the United Farm Workers union conducts workshops on pesticides and other farm health and safety concerns for members of the union, other unions, and community organizations. Spanish is spoken in this office. **Publications:** *Fruits of Your Labor: A Guide to Pesticides Hazards for Field Workers; Health and Safety Issues Commonly Faced by Farm Workers: Answers to 67 Most Frequently Asked Questions; Monitor,* a quarterly newsletter.

★ 1688 ★ Labor Occupational Safety and Health Program
University of California at Los Angeles
1001 Gayley Ave., Second Floor
Los Angeles, CA 90024
Phone: (213)825-7012; (213)825-3877
Provides materials on hazardous chemicals in the workplace and other safety information and conducts training programs.

★ 1689 ★ League of Women Voters of the United States
1720 M St., N.W.
Washington, DC 20036
Phone: (202)429-1965
Fax: (202)429-0854
Nonpartisan political organization promotes the active involvement of citizens in government, influences public policy through education and advocacy, and sponsors studies and reports on many issues, including farming, pesticides, environmental protection, and the preservation of natural resources. **Publication:** *America's Growing Dilemma: Pesticides in Food and Water.*

★ 1690 ★ Los Angeles Committee on Occupational Safety and Health
2501 S. Hill St.
Los Angeles, CA 90007
Phone: (213)749-6161
Provides information to workers on industrial health and safety and on hazardous chemicals in the workplace.

★ 1691 ★ Maine Audubon Society
Gisland Farm
118 U.S. Route One
Falmouth, ME 04105
Phone: (207)781-2330

★ 1692 ★ Maine Public Interest Research Group
University of Southern Maine
96 Falmouth
Portland, ME 04103
Phone: (207)874-6597

★ 1693 ★ Manasota-88
5314 Bay State Rd.
Palmetto, FL 34221
Phone: (813)722-7413
Contact: Gloria Rains, Chair
Statewide organization seeks to protect the natural resources of local areas and promotes the safe use of pesticides. Concerned with the broad use of pesticides for lawn care, conducts studies and testifies on the impact of pesticide spraying for mosquito control as well as household and agriculture pest control, and functions as a source for reporting pesticide problems. **Publication:** *Newsletter,* monthly.

★ 1694 ★ Maricopa County Organizing Project (MCOP)
Pesticide Project
5040 S. Central Ave., Ste. C-1
Phoenix, AZ 85040
Phone: (602)268-6099
Contact: Francisca Cavazos, Program Coord.
Citizens' activist group addresses issues concerning the regulation of hazardous chemicals and pesticides in Maricopa County and across the state. Active in pressing for more state control over safe drinking water standards, the use of pesticides, and related occupational health standards in farming and farm labor camps. **Publication:** *Cuidado,* a bilingual pesticide flyer for farm laborers.

★ 1695 ★ Marin Conservation League (MCL)
35 Mitchell Blvd., Ste. 11
San Rafael, CA 94903
Phone: (415)472-6170
Contact: Karin Urquhart, Exec. Dir.
Seeks the reduction of toxic and hazardous materials in the San Rafael area by sponsoring forums on hazardous waste management and the use of integrated pest management (IPM) systems. **Publication:** A monthly newsletter.

★ **1696** ★ **Maryland Public Interest Research Group**
3110 S. Campus Dining Hall
College Park, MD 20742
Phone: (301)314-8353
Contact: Dan Pontious, Dir.

★ **1697** ★ **Massachusetts Audubon Society**
South Great Rd.
Lincoln, MA 01773
Phone: (617)259-9500
Fax: (617)259-8899
Hotline: 800-541-3443
Contact: Gerard A. Bertrand, Dir.
Active in programs to preserve the environment. Promotes the safe use of pesticides and acts as a center for reporting pesticide problems. *Publication: Sanctuary,* a monthly magazine.

★ **1698** ★ **Massachusetts Public Interest Research Group (MASSPIRG)**
29 Temple Place
Boston, MA 02111-1305
Phone: (617)292-4800
Contact: Janet Domenitz, Dir.
Part of the U.S. Public Interest Research Group, which coordinates the activities of public interest research and advocacy groups in 26 states and Canadian provinces. Serves as lobbying office for reforms on consumer, environmental, and energy issues. Its current efforts include support for laws protecting consumers from unsafe products and reducing the use of toxic chemicals. *Publication: U.S. PIRG Citizen Agenda,* a quarterly newsletter; reports and studies; *Masscitizen,* a newsletter.

★ **1699** ★ **McHenry County Defenders**
132 Cass St.
Woodstock, IL 60098
Contact: Jerry Paulson, Exec. Dir.
Seeks to stop the spraying of Carbaryl in McHenry county to control the gypsy moth population. Promotes the use of integrated pest management (IPM) programs as alternatives to toxics and issues fact sheets on chemical alternatives. *Publication:* A newsletter.

★ **1700** ★ **Michigan Environmental Education Association (MEEA)**
414 Chesley Dr.
Lansing, MI 48917
Phone: (517)351-8888; (517)321-0142
Contact: David Chapman, Project Coord.
Concerned with a broad range of environmental issues, including the misuse of pesticides in the home, in industry, and on farms. *Publications: Hazardous Waste and the Consumer Connection,* a newsletter.

★ **1701** ★ **Michigan Pesticide Network (MPN)**
c/o Ecology Center for Ann Arbor
417 Detroit St.
Ann Arbor, MI 48104
Phone: (313)445-8769
Monitors legislation in Michigan affecting pesticide management, chemical notification laws, and other programs concerning the use of alternatives to pesticides and toxic chemicals.

★ **1702** ★ **Minnesota Herbicide Coalition (MHC)**
5913 Ewing Ave., S.
Minneapolis, MN 55410
Phone: (612)926-8780
Contact: Bette Kent
Volunteer network concerned with a variety of pesticide issues, including mosquito controls in wetlands, forest spraying, the misuse of pesticides in commercial lawn care products,
and the use of nontoxic alternatives to pesticides.

★ **1703** ★ **Minnesota Public Interest Research Group (MPIRG)**
2512 Delaware St., S.E.
Minneapolis, MN 55414
Phone: (612)627-4035
Contact: Heather Cusick, Dir.

★ **1704** ★ **Missoulians for a Clean Environment**
P.O. Box 2885
Missoula, MT 59801
Promotes conservation of natural resources through environmental preservation. Provides information about pesticides and serves as a source for reporting pesticide problems.

★ **1705** ★ **Missouri Public Interest Research Group**
(MOPIRG)
4069 1/2 Shenandoah Ave.
St. Louis, MO 63110
Phone: (314)772-7710
Contact: Beth Zilbert, Dir.

★ **1706** ★ **Montana Public Interest Research Group (MONTPIRG)**
363 Corbin Hall
Missoula, MT 59812
Phone: (406)243-2907
Contact: Brad Martin, Dir.

★ **1707** ★ **National Asbestos Council, Inc. (NAC)**
1777 Northeast Expressway, Ste. 150
Atlanta, GA 30329
Phone: (404)633-2622
Fax: (404)633-5714
Contact: James H. Stark, Exec. Dir.
An organization of individuals and companies concerned with the management and control of asbestos in buildings. Develops and disseminates information about asbestos and provides technical assistance, training, informational conferences, publications, and a referral service to members concerning health issues related to asbestos. *Publications: Council Currents,* a bimonthly newsletter; *NAC Journal,* a quarterly magazine; annual directory and reports.

★ **1708** ★ **National Association for Environmental Management (NAEM)**
4400 Jenifer St., N.W., Ste. 310
Washington, DC 20015
Phone: (202)966-0019
Contact: Jay McCrensky, Exec. Dir.
An organization of environmental managers that seeks to promote professionalism in its members and to increase awareness of environmental strides in the private sector. *Publication: NAEM Network News,* a quarterly newsletter.

★ **1709** ★ **National Association for Solvent Recyclers (NASR)**
1333 New Hampshire Ave., N.W., Ste. 1100
Washington, DC 20036
Phone: (202)463-6956
Fax: (202)775-4163
Contact: Brenda Pulley, Dir.
Trade association of firms engaged in recycling and reclamation of used industrial solvents for industrial fuel.

★ **1710** ★ **National Center for Environmental Health Strategies (NCEHS)**
1100 Rural Ave.
Voorhees, NJ 08043
Phone: (609)429-5358
Contact: Mary Lamielle, Pres. and Dir.
Clearinghouse provides technical, referral, support, and advocacy services for those with environmentally or occupationally induced chemical sensitivity disorders. Conducts research and educational programs and compiles statistics on indoor and outdoor pollutants, pesticides, less-toxic products, natural foods, and such disability resources as alternative employment, workplace accommodations, and workers compensation. *Publications: The Delicate Balance,* a quarterly newsletter; *Chemical Sensitivity: A Report to the New Jersey Department of Health;* reports, bibliographies, books.

★ **1711** ★ **National Clean Air Coalition (NCAC)**
1400 16th St., N.W.
Washington, DC 20036
Phone: (202)797-5436
Contact: Richard Ayres, Chair
A network of state and local organizations concerned with environment, health, labor, and other resources that are threatened by air pollutants. Lobbies for national clean air legislation and seeks the effective implementation of legislation through local and state air programs.

★ **1712** ★ **National Coalition Against the Misuse of Pesticides (NCAMP)**
701 E St., S.E., Ste. 200
Washington, DC 20003
Phone: (202)543-5450
Fax: (202)543-4791
Contact: Jay Feldman, National Coord.
National network committed to pesticide safety and the adoption of alternative pest management strategies that reduce or eliminate a dependency on toxic chemicals. Clearinghouse for a broad range of urban and rural issues, from farm worker safety to lawn care pesticide use. Concerned with federal legislative, regulatory, and enforcement activities. *Publications: Pesticides and You,* a newsletter issued five times a year; *NCAMP Technical Report,* monthly data sheets; *The Network Guide,* a periodic directory; *Legislative Alerts,* periodic; *Pesticide Review: Health and Environmental Effects and Alternatives,* periodic; publications list.

★ **1713** ★ **National Coalition of Hispanic Health and Human Services Organizations**
1030 15th St., N.W., Ste. 1053
Washington, DC 20005
Phone: (202)371-2100
Responds to workers' occupational health questions and refers them to occupational health clinics and doctors. Spanish is spoken in this office.

★ **1714** ★ **National Institute for Chemical Studies (NICS)**
2300 MacCorkle Ave., S.E.
Charleston, WV 25304
Phone: (304)346-6264
Fax: (304)346-6349
Contact: Paul L. Hill, Pres.
Serves as liaison between chemical industry representatives, residents living near chemical plants, and the general public principally in the Kenawha Valley of West Virginia.

★ 1715 ★ National Lawyers Guild
Labor Law Network
1613 Smith Tower
Seattle, WA 98104
Phone: (206)623-0900
Provides information to workers about their legal rights concerning occupational safety and health matters, such as injuries from hazardous chemicals in the workplace, and will refer them to legal clinics and attorneys.

★ 1716 ★ National Lawyers Guild
Toxics Law Network
760 N. First St., Ste. 1
San Jose, CA 95112
Phone: (408)287-7744
Provides information to workers about their legal rights concerning occupational safety and health matters, such as injuries from hazardous chemicals in the workplace, and will suggest legal clinics and attorneys.

★ 1717 ★ National Pesticide
Telecommunication Network (NPTN)
Texas Tech University
Health Science Center
School of Medicine
Department of Preventive Medicine and Community Health
Thompson Hall, Rm. S-129
Lubbock, TX 79430
Phone: (806)743-3091
Fax: (806)743-3094
Hotline: 800-858-7378
Contact: Anthony Way, Exec. Dir.
Clearinghouse for pesticides information, including on toxicology, symptomatic reviews, health and environmental effects, safety and cleanup, and disposal. Service is available to anyone in the contiguous United States, Puerto Rico, and the Virgin Islands.

★ 1718 ★ National Safety Council (NSC)
444 N. Michigan Ave.
Chicago, IL 60611
Phone: (312)527-4800
Fax: (312)527-9381
Contact: T. C. Gilchrest, Pres.
Large nongovernmental organization promotes accident reduction in the home and industrial environments by providing a forum for the exchange of ideas, techniques, and experience. Their Safety Training Institute and home study courses for supervisors seek to reduce the number of accidents due to the misuse of hazardous and toxic substances. **Publications:** *Family Safety and Health,* a quarterly magazine; a wide range of newsletters focusing on safety elements in specific industries.

★ 1719 ★ National Toxics Campaign (NTC)
1168 Commonwealth Ave.
Boston, MA 02134
Phone: (617)232-0327
Fax: (617)232-3945
Contact: John O'Connor, Exec. Dir.
Political action coalition of 80 national labor, environmental, and citizens' groups that lobbies for stronger laws against chemical contamination and conducts grassroots petition drives. **Publications:** *Toxic Times,* a quarterly newsletter; *Touching Bases,* a newsletter of the Military Toxics Network, a project of the National Toxics Campaign Fund.

★ 1720 ★ National Toxics Campaign (NTC)
Dallas Office
11520 N. Central Expressway, No. 133
Dallas, TX 75243
Phone: (214)343-6090

★ 1721 ★ National Toxics Campaign (NTC)
Houston Office
3400 Montrose St., Ste. 225
Houston, TX 77006
Phone: (713)529-8038
Fax: (713)529-8759

★ 1722 ★ National Toxics Campaign (NTC)
Louisiana Office
8841 Blue Bonnet Rd., Ste. C
Baton Rouge, LA
Phone: (504)769-7939

★ 1723 ★ National Toxics Campaign (NTC)
North Carolina Office
530 N. Blount
Raleigh, NC 27604
Phone: (919)828-5539

★ 1724 ★ National Toxics Campaign (NTC)
Northeast Regional Office
R.R. 1, Box 2020
Litchfield, ME 04350
Phone: (207)268-4071

★ 1725 ★ National Toxics Campaign (NTC)
Oklahoma Office
3000 United Founders Blvd., Ste. 125
Oklahoma City, OK 73112
Phone: (405)843-3249

★ 1726 ★ National Toxics Campaign (NTC)
Richmond Office
P.O. Box 1447
Richmond, CA 94801
Phone: (415)232-3427

★ 1727 ★ National Toxics Campaign (NTC)
Rocky Mountain Office
3570 E. 12th St., Ste. 206
Denver, CO 80206
Phone: (303)333-9714

★ 1728 ★ National Toxics Campaign (NTC)
Sacramento Office
1330 21st St., Ste. 102
Sacramento, CA 95814
Phone: (916)446-3350
Fax: (916)446-3394

★ 1729 ★ National Toxics Campaign (NTC)
San Francisco Office
425 Mississippi St.
San Francisco, CA 94107
Phone: (415)826-6314
Fax: (415)826-6314
The San Francisco Consumer Pesticide Project provides information on pesticide residue in food.

★ 1730 ★ National Toxics Campaign (NTC)
Southern Regional Office
P.O. Drawer 1526
Livingston, AL
Phone: (205)652-9854

★ 1731 ★ National Wildlife Federation (NWF)
Environmental Quality Division
1412 16th St., N.W.
Washington, DC 20036-2266
Phone: (202)797-6800
Contact: Jay D. Hair, Pres.
Federation of 6,500 state and local conservation organizations and associate members that develops educational programs, publications, and research activities to promote the wise use of natural resources. Its pesticide efforts, a part of its farranging conservation activities, are focused on both the national and international implications of the use of pesticides. The NWF is active in the Pesticide

Action Network and in the United Nations Food and Agriculture Organization Code of Conduct for the Use and Distribution of Pesticides. **Publications:** *International Wildlife,* a bimonthly magazine; *National Wildlife,* a monthly magazine; newsletters and directories.

★ 1732 ★ Natural Resources Council of
Maine (NRCM)
271 State St.
Augusta, ME 04330
Phone: (207)622-3101
Nonprofit organization focuses on educational and community action activities concerning pesticide issues, including the reduction of chemical insecticides in Maine's forests. **Publications:** *Maine Environment* and *Growth Management News.*

★ 1733 ★ Natural Resources Defense
Council (NRDC)
40 W. 20th St.
New York, NY 10011
Phone: (212)727-2700
Fax: (212)727-1773
Contact: John A. Adams, Exec. Dir.
Broad-based nonprofit activist group conducts research and public education and develops public policy on environmental and ecological issues. Concerns include toxic substances, air and water pollution, protection of wildlife, wetlands, and wilderness, land use, and coastal protection. Engages in litigation that may set precedents or preserve natural resources and monitors federal departments and regulatory agencies concerned with the environment and the control of toxic substances. **Publications:** *NRDC Newsline,* a newsletter published five times a year; *NRDC—tlc, Truly Loving Care,* a quarterly newsletter for children; *The Amicus Journal,* a quarterly journal; brochures, books, and reports.

★ 1734 ★ Natural Resources Defense
Council (NRDC)
1350 New York Ave., N.W., Ste. 300
Washington, DC 20005

★ 1735 ★ Natural Resources Defense
Council (NRDC)
California Office
25 Kearney St.
San Francisco, CA 94108

★ 1736 ★ New Alchemy Institute (NAI)
237 Hatchville Rd.
East Falmouth, MA 02536
Phone: (508)564-6301
Fax: (508)457-9680
Contact: Greg Watson, Exec. Dir.
Develops food processing systems that work with and emulate natural processes; conducts research; provides educational programs for household and small farm economics on resource-efficient housing and landscape design; seeks to reduce toxins in the home and garden; encourages integrated pest management (IPM) and solar greenhouse agriculture. **Publications:** *New Alchemy Institute Catalogue: Books and Products for Ecological Living,* periodic; *New Alchemy Quarterly.*

★ 1737 ★ New Jersey Coalition for
Alternatives to Pesticides (NJ CAP)
P.O. Box 627
Boonton, NJ 07005
Phone: (201)334-7975
Contact: Susan Shaw
Focuses on state-level legislative reforms concerning pesticide use in insect control

programs, ground water contamination, and farm worker issues.

**★ 1738 ★ New Jersey Environmental
Federation**
Clean Water Fund
808 Belmar Plaza
Belmar, NJ 07719
Phone: (908)846-4224
Educational and research organization that promotes public interest in issues related to clean water, toxic substances, and natural resources. Works to promote the effectiveness of local environmental groups, promotes the safe use of pesticides, and acts as a clearinghouse for reporting pesticide problems.

**★ 1739 ★ New Jersey Public Interest
Research Group (NJPIRG)**
103 Bayard St.
New Brunswick, NJ 08901
Phone: (201)247-4606
Contact: Ken Ward, Dir.

**★ 1740 ★ New Mexico Public Interest
Research Group (NMPIRG)**
Box 66 SUB
University of New Mexico
Albuquerque, NM 87131
Phone: (505)277-2757
Contact: Claudia Lucas, Dir.

**★ 1741 ★ New York Coalition for
Alternatives to Pesticides (NY CAP)**
Westside Station
P.O. Box 310
Buffalo, NY 14213
Contact: Karen Murphy, Coord.
Statewide organization of individuals and groups seeking to reduce the exposure to pesticides and encourage the use of safe alternatives lobbies for safe alternatives and pesticide right-toknow laws and notification regulations. **Publication:** *NY CAP News,* a quarterly newsletter.

**★ 1742 ★ New York Coalition for
Alternatives to Pesticides**
33 Central Ave.
Albany, NY 12210
Phone: (518)426-8246

**★ 1743 ★ New York Committee on
Occupational Safety and Health**
275 Seventh Ave.
New York, NY 10001
Phone: (212)672-3900
Provides information to workers on industrial health and safety and on hazardous chemicals in the workplace.

**★ 1744 ★ New York Public Interest
Research Group (NYPIRG)**
9 Murray St.
New York, NY 10007
Phone: (212)349-6460
Contact: Jay Halfon, Dir.

**★ 1745 ★ New York State Center for
Hazardous Waste Management**
State University of New York at Buffalo
207 Jarvis Hall
Buffalo, NY 14260
Phone: (716)636-3446
Contact: Ralph R. Rumer, Exec. Dir.
Conducts research and development aimed at reducing the generation of hazardous waste, treating hazardous waste so that it can be safely disposed, and cleaning up inactive hazardous waste disposal sites. Sponsors projects, conferences, roundtables, and workshops and

conducts an extensive publishing program. **Publications:** *Waste Management Research Report,* published three times a year and coordinated with the Waste Management Institute of the State University of New York at Stony Brook and the Cornell Waste Management Institute; reports and studies.

**★ 1746 ★ North American Radon
Association (NARA)**
8445 River Birch
Roswell, GA 30075
Phone: (404)993-5033
Contact: Stewart M. Huey, Adm. Dir.
An organization of scientists, educators, mitigation professionals and individuals that seeks to eliminate health problems caused by radon and to encourage legislation, establish standards, and improve the methods of detection and removal of radon by-products. **Publication:** *NARA Newsletter,* a bimonthly.

**★ 1747 ★ North Georgia Citizens Opposed
to Paraquat Spraying (COPS)**
P.O. Box 818
Sautee, GA 30571
Phone: (404)896-3421
Contact: Donald Cornelius, Coord.
Seeks to make the public aware of toxic substances in northern Georgia and lobbies to halt the spraying of the herbicide paraquat in national forests and on rights-of-way.

**★ 1748 ★ Northwest Coalition for
Alternatives to Pesticides (NCAP)**
P.O. Box 1393
Eugene, OR 97440
Phone: (503)344-5044
Contact: Norma Grier, Exec. Dir.
Nonprofit organization composed of individuals and groups primarily from the Northwest that works to reduce pesticide use through policy reform and education about pesticide hazards and alternatives. Provides documented information, advice, and referrals to groups that want to affect the use of pesticides in their own communities, and serves as a clearinghouse for reporting pesticide problems. **Publications:** *Journal of Pesticide Reform,* a quarterly; *On the Trail of Pesticides;* factsheets and information packets.

★ 1749 ★ Occupational Lung Disease Center
Tulane University
School of Medicine
1700 Perdido St.
New Orleans, LA 70112
Phone: (504)588-5265
Fax: (504)588-5035
Contact: Hans Weill, Dir.
Studies occupational lung diseases in industries whose workers may have been exposed to airborne contaminants.

**★ 1750 ★ Ohio Coalition Against the Misuse
of Pesticides**
2128 Halcyon Rd.
Beachwood, OH 44122
Phone: (216)382-4341

**★ 1751 ★ Ohio Public Interest Research
Group (OPIRG)**
2084 1/2 N. High St.
Columbus, OH 43201
Phone: (614)299-7474
Contact: John Rumpler, Campaign Dir.

★ 1752 ★ OMB Watch
1731 Connecticut Ave., N.W.
Washington, DC 20009-1146
Phone: (202)234-8494
Fax: (202)234-8584
Contact: Gary D. Bass, Executive Dir.
Nonprofit research, educational, and advocacy organization that monitors the activities of the White House Office of Management and Budget (OMB) that affect nonprofit, public interest, and community groups. OMB Watch sponsors with the Unison Institute RTK NET, a bulletin board network and database of right-to-know information and resources. **Publications:** *Government Information Insider,* a bimonthly magazine about government information and regulation issues; *OMB Watcher,* a bimonthly roundup of news about OMB activities.

**★ 1753 ★ Ontario Public Interest Research
Group (OPIRG)**
455 Spadina Ave.
Toronto, ON, Canada M5S 2G8
Phone: (416)598-1576
Contact: Lisa Kristensen, Province Coord.

**★ 1754 ★ Oregon Public Interest Research
Group (ORPIRG)**
1536 S.E. 11th St.
Portland, OR 97214
Phone: (503)231-4181
Contact: Joel Ario, Dir.

★ 1755 ★ Panhandle Environmental League
P.O. Box 963
Sandpoint, ID 83864
Phone: (208)262-1020
Contact: Jane Fritz
Promotes alternatives to pesticides and herbicides in area developments, forest lands, and water bodies and seeks to eliminate chemical contamination from a local treatment facility. Affiliated with the Idaho Conservation League.

**★ 1756 ★ Pennsylvania Environmental
Council (PEC)**
1211 Chestnut St., Ste. 900
Philadelphia, PA 19107
Phone: (215)563-0250
Fax: (215)563-0528
Hotline: 800-322-9214
Contact: Joanne R. Denworth, Exec. Dir.
Activists group concerned with environmental issues in Pennsylvania and EPA Region II. Resolves disputes through PennACCORD, which facilitates policy negotiation and mediation, and sponsors citizens events and professional and public education programs. **Publications:** *PEC Forum,* a monthly newsletter; *Legislative Updates;* special bulletins.

**★ 1757 ★ Pennsylvania Environmental
Council (PEC)**
Harrisburg Office
c/o Killian and Gephart
216-218 Pine St.
Harrisburg, PA 17108
Phone: (717)232-1851
Fax: (717)238-0592
Hotline: 800-322-9214
Contact: Joanne R. Denworth, Exec. Dir.

★ 1758 ★ Pennsylvania Environmental Council (PEC)
Western Pennsylvania Office
Benedum Trees Bldg.
223 Fourth Ave., Ste. 503
Pittsburgh, PA 15222
Phone: (412)471-1770
Fax: (412)471-1661
Hotline: 800-322-9214
Contact: Joanne R. Denworth, Exec. Dir.

★ 1759 ★ Pennsylvania Public Interest Research Group (PENNPIRG)
3507 Lancaster Ave.
Philadelphia, PA 19104
Phone: (215)387-8144
Contact: David Yeaworth, Dir.

★ 1760 ★ Pesticide Action Network North American Regional Center (PAN/NA)
965 Mission St., Ste. 514
San Francisco, CA 94103
Phone: (415)541-9140
Fax: (415)541-9253
Contact: Monica Moore, Exec. Dir.
Nonprofit, tax-exempt member organization of Pesticide Action Network, a group of non-governmental coalitions, organizations, and individuals in more than 50 countries involved with pesticide issues. Serves as a networking tool for linking pesticide action groups in North America with counterpart citizen's movements in other countries. PAN also maintains Regional Centers in Nairobi, Kenya, Dakar, Senegal, Penang, Malaysia, London, England, and Palmira, Columbia. **Publications:** *Global Pesticide Campaigner,* a quarterly newsletter; pamphlets, reports, manuals, and booklets. Material is maintained on a computerized bibliographic database available on a fee basis. Also subscribes to DIALOG and will perform searches on pesticides on a fee basis.

★ 1761 ★ Pesticide Laboratory
University of Puerto Rico
Puerto Rico Agricultural Experiment Station
P.O. Box 21360
Rio Piedras, PR 00928
Phone: (809)767-9705
Fax: (809)758-5158
Contact: Nilsa M. Acin-Diaz, Dir.
Studies the behavior of pesticides in the soil and the fate of pesticides in the environment and develops methodologies for determining pesticide residues.

★ 1762 ★ Pesticide Research Center
Michigan State University
East Lansing, MI 48824-1311
Phone: (517)353-9430
Contact: Robert M. Hollingworth, Dir.
Conducts research into all aspects of pesticide science, including integrated pest management (IPM) and alternatives to pesticides.

★ 1763 ★ Pesticide Research Laboratory
University of Florida
Gainesville, FL 32611
Phone: (904)392-1978
Fax: (904)392-1988
Studies pesticide residue in food, water, and soil.

★ 1764 ★ Pesticide Research Laboratory and Graduate Study Center
Pennsylvania State University
Department of Entomology
University Park, PA 16802
Phone: (814)863-0844
Fax: (814)865-3048
Contact: James L. Frazier, Dir.

Studies the environmental fate of pesticides and the toxicology and physiology of insects.

★ 1765 ★ Pesticide Residue Research Laboratory
North Carolina State University
3709 Hillsborough St.
Raleigh, NC 27607
Phone: (919)737-3391
Contact: T. J. Sheets, Dir.
Studies pesticide residues in food and feed crops and the behavior of pesticides in soils, air, and water.

★ 1766 ★ Pesticide Residue, Toxic Waste, and Basic Research Analytical Laboratory
University of Miami
12500 S.W. 152nd St., Bldg. B
Miami, FL 33177-1411
Phone: (305)232-8202
Fax: (305)232-7461
Contact: Jon B. Mann, Laboratory Mgr.
Studies pesticides, air pollutants, chemicals, and carcinogens.

★ 1767 ★ Pesticide Watch
1147 S. Robertson Blvd., Ste. 203
Los Angeles, CA 90035
Phone: (213)278-9244
Clearinghouse for information on pesticides and a source for reporting pesticide problems.

★ 1768 ★ Protect Our Environment From Sprayed Toxins (PEST)
One Cathance Ln.
Cooper, ME 04638
Phone: (207)454-8029
Contact: Bo Yerxa, Sec.
Statewide organization addresses issues concerning aerial spraying of spruce budworm insecticides, forest management, the abuse and misuse of right-of-way and agricultural pesticides, and toxic abuses in ground water contamination, aerial drift, and occupational health. **Publication:** *The Gadfly,* a quarterly newsletter.

★ 1769 ★ Public Citizen
2000 P St., N.W., Ste. 610
P.O. Box 19404
Washington, DC 20036
Phone: (202)293-9142
Contact: Joan Claybrook, Pres.
Nonprofit research, lobbying, and litigation organization that seeks consumer rights in the marketplace for safe products, a healthy environment and workplace, clean and safe energy resources, and corporate and government accountability. **Publication:** *Public Citizen,* a newsletter available for a contribution of $20.00 or more.

★ 1770 ★ Public Citizen
Critical Mass Energy Project
215 Pennsylvania Ave., S.E.
Washington, DC 20003
Phone: (202)546-4996
Produces research, studies, and reports concerning the safety, economic, and waste problems posed by nuclear power as well as the availability of safer, cheaper, and cleaner alternatives. **Publications:** A variety of reports on nuclear power issues, including worker exposure and nuclear waste recycling.

★ 1771 ★ Public Interest Research Group in Michigan (PIRGIM)
212 S. Fourth Ave., Ste. 207
Ann Arbor, MI 48104
Phone: (313)662-6597
Contact: Andy Buchsbaum, Dir.

★ 1772 ★ Public Voice for Food and Health Policy
1001 Connecticut Ave., N.W., Ste. 522
Washington, DC 20036
Phone: (202)659-5930
Contact: Ellen Haas, Exec. Dir.
Nonprofit, tax-exempt organization advances citizen interest in national food and health policy making. **Publications:** *Advocacy Update,* a newsletter; *Information Service for Decision Makers,* a compilation of working materials.

★ 1773 ★ Rachel Carson Council
8940 Jones Mill Rd.
Chevy Chase, MD 20815
Phone: (301)652-1877
Contact: Shirley A. Briggs, Exec. Dir.
Seeks to further the philosophy of scientist and author Rachel Carson by developing an awareness of the problems of environmental contamination and by serving as a information clearinghouse on the health and environmental effects of pesticides and other poisons. **Publication:** Semiannual Report, pamphlets, and booklets.

★ 1774 ★ Radioactive Waste Campaign (RWC)
7 West St.
Warwick, NY 10990
Phone: (914)986-1115
Contact: Patrick J. Malloy, Exec. Dir.
Supports grassroots organizations by providing technical, organizing, and legal information about nuclear and toxic materials waste disposal problems. Also provides information on waste technologies, resources and organizing, and other citizens' groups activities. **Publications:** *RWC Reports,* a quarterly newsletter; facts sheets and books.

★ 1775 ★ Renew America
1400 16th St., N.W., Ste. 710
Washington, DC 20036
Phone: (202)232-2252
Contact: Tina Hobson, Exec. Dir.
Nonprofit, tax-exempt clearinghouse for solutions to environmental problems that fosters the expansion of successful environmental programs and encourages cooperation among environmental interests. **Publication:** *Searching for Success—Environmental Success Index,* an annual directory of award-winning environmental programs.

★ 1776 ★ Residents Against Spraying of Pesticides (RASP)
1008 W. Kensington Ln.
Los Angeles, CA 90026
Contact: Patty Prickett, Dir.

★ 1777 ★ Risk and Systems Analysis for the Control of Toxics Program (RSACT)
University of California—Los Angeles
Department of Mechanical, Aerospace, and Nuclear Engineering
48-121 Engineering IV
Los Angeles, CA 90024
Phone: (213)825-2045
Fax: (213)825-0761
Contact: William Kastenberg, Dir.
Studies the control of hazardous substances from their source to human exposure.

★ 1778 ★ Rocky Mountain Center for Occupational and Environmental Health
University of Utah
Bldg. 512
Salt Lake City, UT 84112
Phone: (801)581-8719
Fax: (801)581-7224
Contact: Royce Moser, Jr., Managing Dir.
Studies occupational and environmental health and safety with emphasis on the neurobehavioral effects of toxic exposures and asbestos-related health problems.

★ 1779 ★ Rodale Institute
222 Main St.
Emmaus, PA 18098
Phone: (215)967-5171
Provides research and information about organic farming and sustainable agriculture.

★ 1780 ★ Safe Alternatives for Fruitfly Eradication (SAFE)
P.O. Box 63302
Los Angeles, CA 90063
Phone: (818)359-2491

★ 1781 ★ Safe Alternatives for Our Forest Environment (SAFE)
P.O. Box 297
Hayfork, CA 96041
Phone: (916)628-4474
Contact: Jan Mountjoy, Sec.
Advocates the use of nontoxic alternatives to pesticides by the U.S. Forest Service and the timber industry in the Hayfork area.
Publications: *SAFE News,* a monthly newsletter.

★ 1782 ★ Safe Water Coalition (SWC)
150 Woodland Ave.
San Anselmo, CA 94960
Phone: (415)453-0158
Contact: Shirley Graves, Chair
Regional group that seeks to inform its members and the public about the toxic effects of fluoride in public water systems and in dental hygiene products. A clearinghouse on fluoride research. ***Publication:*** *National Fluoridation News,* a quarterly.

★ 1783 ★ San Francisco Household Hazardous Waste Program
City Hall, Rm. 271
San Francisco, CA 94102
Phone: (415)554-6194
Large municipal program educates the public on the best methods of disposing of hazardous household waste and encourages the use of nontoxic alternatives. ***Publication:*** *Garbage Cans and Can't,* a pamphlet.

★ 1784 ★ Sierra Club
730 Polk St.
San Francisco, CA 94109
Phone: (415)776-2211
Fax: (415)776-0350
Contact: Sue Morrow, Pres.
Grassroots organization of 12 regional groups and 57 state groups that promotes the protection and conservation of the natural resources of the U.S. and the world. It's extensive network of local committees, campaigns, and lobbying programs extend to all areas of environmental concerns: toxic pollutants, clean air and water, forestry, hazardous waste, wilderness and coastal protection, energy conservation, and land use.
Publications: The Sierra Club has an extensive publishing program that focuses on a range of environmental issues, from toxics and pollution to ecology, recycling, and the protection of public lands. A *Sierra Club Sourcebook* and current book catalogs are available free upon request by writing to the Sierra Club at the above address. Some useful publications are: *Art: Toxics on the Home Front, Hazardous Materials/Water Resources Newsletter,* and *Pesticide Alert: A Guide to Pesticides in Fruits and Vegetables.*

★ 1785 ★ Sierra Club
Alabama Chapter
P.O. Box 55591
Birmingham, AL 35255
Phone: (205)592-9233

★ 1786 ★ Sierra Club
Alaska Chapter
P.O. Box 103-441
Anchorage, AK 99501
Phone: (907)276-8768

★ 1787 ★ Sierra Club
Alaska Field Office
241 E. Fifth Ave., Ste. 205
Anchorage, AK 99501
Phone: (907)276-4048
Contact: Jack Hession, Alaska Rep.

★ 1788 ★ Sierra Club
Angeles Chapter
3550 W. Sixth St., No. 321
Los Angeles, CA 90020
Phone: (213)387-4287

★ 1789 ★ Sierra Club
Appalachian Field Office
P.O. Box 667
Annapolis, MD 21404-0667
Phone: (301)268-7411
Contact: Joy Oakes, Appalachian Rep.
The Appalachian Field Office serves Delaware, Maryland, North Carolina, South Carolina, Tennessee, Washington, DC, Virginia, and West Virginia.

★ 1790 ★ Sierra Club
Arizona Branch Office
3201 N. 16th St., Ste. 6-A
Phoenix, AZ 85016
Phone: (602)277-8079
Contact: Rob Smith, Associate Rep.

★ 1791 ★ Sierra Club
Arizona Chapter
c/o Dawson Henderson
Rt. 4, Box 886
Flagstaff, AZ 86001
Phone: (602)774-1571

★ 1792 ★ Sierra Club
Arkansas Chapter
c/o George Oleson
1669 Carolyn Dr.
Fayetteville, AR 72701
Phone: (501)521-9794

★ 1793 ★ Sierra Club
Austin Regional Group Pesticide Committee
2502 Albata Ave.
Austin, TX 78757
Phone: (512)459-8063
Promotes the safe use of pesticides and serves as a center for reports of pesticide problems.

★ 1794 ★ Sierra Club
Colorado Chapter
777 Grant St., Ste. 606
Denver, CO 80203-3518
Phone: (303)861-8819
Contact: Angela Medberry, Chair
Publication: *Peak and Prairie,* a newsletter.

★ 1795 ★ Sierra Club
Connecticut Chapter
118 Oak St.
Hartford, CT 06106
Phone: (203)527-9788

★ 1796 ★ Sierra Club
Delaware Chapter
1116 C West St.
Annapolis, MD 21403
Phone: (301)268-3935

★ 1797 ★ Sierra Club
District of Columbia Chapter
1116 C West St.
Annapolis, MD 21403
Phone: (301)268-3935

★ 1798 ★ Sierra Club
Eastern Canada Chapter
2316 Queen St. East
Toronto, ON, Canada M4E 1G8
Phone: (416)698-8446

★ 1799 ★ Sierra Club
Florida Branch Office
1201 N. Federal Hwy., Rm. 250H
North Palm Beach, FL 33408
Phone: (407)775-3846
Contact: Theresa Woody, Associate Rep.

★ 1800 ★ Sierra Club
Florida Chapter
c/o Beth Ann Brick
462 Fernwood
Key Biscayne, FL 33149
Phone: (305)361-1292

★ 1801 ★ Sierra Club
Georgia Chapter
P.O. Box 467151
Atlanta, GA 30346
Phone: (404)921-5389

★ 1802 ★ Sierra Club
Hawaii Chapter
P.O. Box 2577
Honolulu, HI 96803
Phone: (808)538-6616

★ 1803 ★ Sierra Club
Idaho Chapter
c/o Edwina Allen
1408 Joyce St.
Boise, ID 83706
Phone: (208)344-4565

★ 1804 ★ Sierra Club
Illinois Chapter
506 S. Wabash, No. 525
Chicago, IL 60605
Phone: (312)431-0158

★ 1805 ★ Sierra Club
Indiana Chapter
P.O. Box 40275
Indianapolis, IN 46240
Phone: (317)253-2687

★ 1806 ★ Sierra Club
Iowa Chapter
Thoreau Center
3500 Kingman Blvd.
Des Moines, IA 50311
Phone: (515)277-8868
Contact: Sherry Dragula, Office Mgr.

★ 1807 ★ Sierra Club
Kansas Chapter
c/o Michael Moore
P.O. Box 47319
Wichita, KS 67201
Phone: (316)683-8492

★ 1808 ★ Sierra Club
Kentucky Chapter
c/o David Stawicki
2004 Writt Ct.
Lexington, KY 40505
Phone: (606)259-1922

★ 1809 ★ Sierra Club
Kern-Kaweah Chapter
1013 Hogan Way
Bakersfield, CA 93309
Phone: (805)834-8035

★ 1810 ★ Sierra Club
Loma Prieta Chapter
2448 Watson Ct.
Palo Alto, CA 94303
Phone: (415)494-9901

★ 1811 ★ Sierra Club
Los Padres Chapter
c/o David Sheehan
1411 Ramona Dr.
Newbury Park, CA 91320
Phone: (805)498-1736

★ 1812 ★ Sierra Club
Louisiana Chapter
P.O. Box 19469
Bienville St.
New Orleans, LA 70119
Phone: (504)482-9566

★ 1813 ★ Sierra Club
Maine Chapter
3 Joy St.
Boston, MA 02108
Phone: (617)227-5339

★ 1814 ★ Sierra Club
Maryland Chapter
1116 C West St.
Annapolis, MD 21403
Phone: (301)268-3935

★ 1815 ★ Sierra Club
Massachusetts Chapter
3 Joy St.
Boston, MA 02108
Phone: (617)227-5339

★ 1816 ★ Sierra Club
Michigan Chapter
115 W. Allegan, No. 10-B
Cap Hall
Lansing, MI 48933
Phone: (517)484-2372

★ 1817 ★ Sierra Club
Midwest Field Office
214 N. Henry St., Ste. 203
Madison, WI 53703
Phone: (608)257-4994
Contact: Jane Elder, Midwest Rep.
The Midwest Field Office serves Illinois, Indiana, Iowa, Kentucky, Michigan, Minnesota, Missouri, Ohio, and Wisconsin.

★ 1818 ★ Sierra Club
Mother Lode Chapter
P.O. Box 1335
Sacramento, CA 95812
Phone: (916)444-2180

★ 1819 ★ Sierra Club
North Carolina Chapter
c/o Bill Thomas
P.O. Box 272
Cedar Mountain, NC 28718
Phone: (704)885-8229

★ 1820 ★ Sierra Club
North Dakota Chapter
P.O. Box 1624
Rapid City, SD 57709
Phone: (605)348-1351

★ 1821 ★ Sierra Club
Northeast Field Office
360 Broadway
Saratoga Springs, NY 12866
Phone: (518)587-9166
Fax: (518)587-9171
Contact: Chris Ballantyne, Northeast Rep.
The Northeast Field Office serves Connecticut, Maine, Massachusetts, New Hampshire, New Jersey, New York, Pennsylvania, Rhode Island, and Vermont.

★ 1822 ★ Sierra Club
Northern California/Nevada Field Office
5428 College Ave.
Oakland, CA 94618
Phone: (415)654-7847
Contact: Sally Kabisch, Northern California/ Nevada Rep.

★ 1823 ★ Sierra Club
Northern Plains Field Office
23 N. Scott, Rm. 25
Sheridan, WY 82801
Phone: (307)672-0425
Contact: Larry Mehlhaff, Northern Plains Rep.
The Northern Plains Field Office serves Montana, Nebraska, North Dakota, South Dakota, and Wyoming.

★ 1824 ★ Sierra Club
Northwest Field Office
1516 Melrose Ave.
Seattle, WA 98122
Phone: (206)621-1696
Contact: Bill Arthur, Northwest Rep.
The Northwest Field Office serves Hawaii, Idaho, Oregon, and Washington.

★ 1825 ★ Sierra Club
Ohio Chapter
145 N. High St.
Columbus, OH 43215
Phone: (614)461-0734

★ 1826 ★ Sierra Club
Oklahoma Chapter
c/o Karin Derichsweiler
312 Keith
Norman, OK 73069
Phone: (405)360-6445

★ 1827 ★ Sierra Club
Oregon Chapter
c/o John Albrecht
3550 Willamette
Eugene, OR 97405
Phone: (503)343-5902

★ 1828 ★ Sierra Club
Pennsylvania Chapter
600 N. Second St.
P.O. Box 663
Harrisburg, PA 17108
Phone: (717)232-0101

★ 1829 ★ Sierra Club
Redwood Chapter
P.O. Box 466
Santa Rosa, CA 95402
Phone: (707)544-7651

★ 1830 ★ Sierra Club
Rhode Island Chapter
3 Joy St.
Boston, MA 02108
Phone: (617)227-5339

★ 1831 ★ Sierra Club
Sacramento Office
1014 Ninth St., Ste. 201
Sacramento, CA 95814
Phone: (916)444-6906

★ 1832 ★ Sierra Club
San Diego Chapter
3820 Ray St.
San Diego, CA 92104
Phone: (619)299-1743

★ 1833 ★ Sierra Club
San Francisco Bay Chapter
6014 College Ave.
Oakland, CA 94618
Phone: (415)653-6127

★ 1834 ★ Sierra Club
San Gorgonio Chapter
568 N. Mountain View Ave., No. 130
San Bernardino, CA 92401
Phone: (714)381-5015

★ 1835 ★ Sierra Club
Santa Lucia Chapter
P.O. Box 15755
San Luis Obispo, CA 93406
Phone: (805)549-5059

★ 1836 ★ Sierra Club
South Carolina Chapter
P.O. Box 12112
Columbia, SC 29211
Phone: (803)256-8487

★ 1837 ★ Sierra Club
South Dakota Chapter
P.O. Box 1624
Rapid City, SD 57709
Phone: (605)348-1351

★ 1838 ★ Sierra Club
Southeast Field Office
P.O. Box 11248
Knoxville, TN 37939-1248
Phone: (615)588-1892
Contact: Jim Price, Southeast Representative
The Southeast Field Office serves Alabama, Florida, Georgia, Louisiana, and Mississippi.

★ 1839 ★ Sierra Club
Southern California/Nevada Field Office
3550 W. 6th St., Ste. 323
Los Angeles, CA 90020
Phone: (213)387-6528
Contact: Bob Hattoy, Southern California/ Nevada Rep.

★ 1840 ★ Sierra Club
Southern Plains Field Office
6220 Gaston, Ste. 609
Dallas, TX 75214
Phone: (214)824-5930
Contact: Beth Johnson, Southern Plains Rep.
The Southern Plains Field Office serves Arkansas, Kansas, Oklahoma, and Texas.

★ 1841 ★ Sierra Club
Southwest Field Office
1240 Pine St.
Boulder, CO 80302
Phone: (303)449-5595
Contact: Maggie Fox, Southwest Rep.
The Southwest Field Office serves Arizona, Colorado, New Mexico, and Utah.

★ 1842 ★ Sierra Club
Tehipite Chapter
P.O. Box 5396
Fresno, CA 93755
Phone: (209)233-1820

★ 1843 ★ Sierra Club
Tennessee Chapter
c/o Richard Mochow
871 Kensington Pl.
Memphis, TN 38107
Phone: (901)274-1510

★ 1844 ★ Sierra Club
Texas Chapter
P.O. Box 1931
Austin, TX 78767
Phone: (512)477-1729

★ 1845 ★ Sierra Club
Utah Branch Office
177 E. 900 South, No. 200
Salt Lake City, UT 84111
Phone: (801)355-0509
Contact: Lawson LeGate, Associate Rep.

★ 1846 ★ Sierra Club
Utah Chapter
177 East 900 South, No. 102
Salt Lake City, UT 84111
Phone: (801)363-9621

★ 1847 ★ Sierra Club
Ventana Chapter
P.O. Box 5667
Carmel, CA 93921
Phone: (408)624-8032

★ 1848 ★ Sierra Club
Vermont Chapter
3 Joy St.
Boston, MA 02108
Phone: (617)227-5339

★ 1849 ★ Sierra Club
Virginia Chapter
P.O. Box 14648
Richmond, VA 23221
Phone: (703)978-4782

★ 1850 ★ Sierra Club
Washington Chapter
1516 Melrose
Seattle, WA 98119
Phone: (206)621-1696

★ 1851 ★ Sierra Club
Washington, DC, Office
408 C St., N.E.
Washington, DC 20002
Phone: (202)547-1141
Fax: (202)547-6009
Hotline: (202)547-5550

★ 1852 ★ Sierra Club
Western Canada Chapter
620 View St., No. 314
Victoria, BC, Canada V8W 1J6
Phone: (604)386-5255

★ 1853 ★ Sierra Club
West Virginia Chapter
P.O. Box 4142
Morgantown, WV 26505
Phone: (304)274-1130

★ 1854 ★ Sierra Club
Wisconsin Chapter
111 King St.
Madison, WI 53703
Phone: (608)256-0565

★ 1855 ★ Sierra Club
Wyoming Chapter
c/o Meredith Taylor
Rt. 31, Box 807
Dubois, WY 82513
Phone: (307)455-2161

★ 1856 ★ Sierra Club Legal Defense Fund (SCLDF)
Hawaii Office
212 Merchant St., Ste. 202
Honolulu, HI 96813
Phone: (808)599-2436

★ 1857 ★ Sierra Club Legal Defense Fund (SCLDF)
Rocky Mountain Office
1600 Broadway, Ste. 1600
Denver, CO 80202
Phone: (303)863-9898

★ 1858 ★ Sierra Club Legal Defense Fund (SCLDF)
San Francisco Office
2044 Fillmore
San Francisco, CA 94115
Phone: (415)567-6100
Fax: (415)567-7740
The legal arm of the Sierra Club uses existing legal avenues to protect the natural environment of the U.S. and seeks to develop a realistic and enforceable body of environmental law through the implementation of existing statutes, regulations, and common law principles. **Publication:** *Sierra Club Legal Defense Fund—In Brief: A Quarterly Newsletter on Environmental Law.*

★ 1859 ★ Sierra Club Legal Defense Fund (SCLDF)
Seattle Office
216 First Ave., S., Ste. 303
Seattle, WA 98104
Phone: (206)343-7340

★ 1860 ★ Sierra Club Legal Defense Fund (SCLDF)
Southeast Alaska Office
325 Fourth St.
Juneau, AK 99801
Phone: (907)586-2751

★ 1861 ★ Sierra Club Legal Defense Fund (SCLDF)
Washington, DC, Office
1531 P St. N.W., Ste. 200
Washington, DC 20005
Phone: (202)667-4500

★ 1862 ★ Soap and Detergent Association (SDA)
475 Park Ave., S.
New York, NY 10016
Phone: (212)725-1262
Trade association of manufacturers of soaps, detergents, fatty acids, and glycerine that conducts environmental and human safety research and is a clearinghouse for information on ingredients. **Publications:** *Cleanliness Facts*, a bimonthly newsletter; *Detergents—In Depth Symposium Proceedings,* a periodic.

★ 1863 ★ Society of Toxicology (SOT)
1101 14th St., N.W., Ste. 1100
Washington, DC 20005
Phone: (202)371-1393
Fax: (202)371-1090
Contact: Joan Walsh Cassedy, Exec. Sec.
Professional association of people concerned with the effects of chemicals on human and animal health and the environment and who have conducted and published original investigations in some phase of toxicology. **Publications:** *Fundamental and Applied Toxicology,* a technical journal published eight times a year; *Society of Toxicology—Membership Directory,* annual; *Society of Toxicology—Newsletter,* bimonthly; *Toxicology and Applied Pharmacology,* a professional journal, 15 issues a year.

★ 1864 ★ Society of Toxicology of Canada (STC)
c/o D. J. Ecobichon
Department of Pharmacology and Therapeutics
McGill University
Montreal, PQ, Canada H3G 1Y6
Contact: D. J. Ecobichon, Dir.
Organization of professional toxicologists. **Publication:** *STC Newsletter.*

★ 1865 ★ Spill Control Association of America (SCAA)
400 Renaissance Ctr., Ste. 1900
Detroit, MI 48243-1509
Phone: (313)567-0500; (313)849-2649
Fax: (313)849-2685
Contact: Clayton W. Evans, Exec. Dir.
Nonprofit trade organization that represents the environmental services and pollution emergency response industry throughout the United States and individuals in private and governmental capacities involved with spill cleanup and associated companies. Provides information on oil and hazardous material emergency response and the remediation industry's practices. **Publications:** *Spill Briefs,* a monthly newsletter; *Newsletter,* a biweekly; technical reports, papers, training material.

★ 1866 ★ Sustainable Agriculture Program
University of California—Davis
Agronomy Extension
Davis, CA 95616
Phone: (916)752-8667; (916)752-7645
Contact: Mark Van Horn, Program Dir.
Conducts research and provides information on organic farming and sustainable agriculture.

★ 1867 ★ Toxicology Information Response Center (TIRC)
Oak Ridge National Laboratory
Bldg. 2001
P.O. Box 2008
Oak Ridge, TN 37831-6050
Phone: (615)576-1746
Fax: (615)574-9888
Contact: Kimberly Slusher, Exec. Officer
Nonprofit information center sponsored by the Toxicology Information Program of the U.S. Department of Health and Human Services Library of Medicine. International center for collecting, analyzing, and disseminating toxic-related information on chemicals, including pesticides, food additives, industrial chemicals, heavy metals, environmental pollutants, and pharmaceuticals. The Center is online to more than 400 computerized databases, including DIALOG, MEDLARS, STN International, ITIS, and DROLS. It performs searches for outside users for a fee.

★ 1868 ★ Toxics Chemicals Laboratory
Cornell University
New York State College of Agriculture
Tower Rd.
Ithaca, NY 14853-7401
Phone: (607)255-4538
Contact: Donald J. Lisk, Dir.
Conducts pesticide and toxic chemical metabolite studies of soils, fruits, plants, animals, fish, and other farm products as well as on forage crops and resultant products.

★ 1869 ★ Toxics Use Reduction Institute
University of Lowell
1 University Ave.
Lowell, MA 01854-9985
Phone: (508)934-3250
Fax: (508)452-5711
Contact: Ken Geiser, Dir.
Studies technologies for reducing the use of toxic chemicals.

★ 1870 ★ Toxic Waste Investigative Group, Inc. (TWIG)
P.O. Box 6173
Phoenix, AZ 85005
Phone: (602)272-6997
Contact: Pamela Swift, Chair
Activist and lobbying group that acts as a forum for communities and individuals with hazardous chemical and pesticide problems. It particularly concerns itself with crop dusting and the illegal disposal of hazardous chemicals in the Phoenix area.

★ 1871 ★ University Center for Environmental and Hazardous Materials Studies
Virginia Polytechnic Institute and State University
1020 Derring Hall
Blacksburg, VA 24061-0415
Phone: (703)951-5538
Contact: John Cairns, Jr., Dir.
Conducts research on a wide range of environmental issues, including hazard evaluation of toxic chemicals.

★ 1872 ★ U.S. Public Interest Research Group (USPIRG)
215 Pennsylvania Ave., S.E.
Washington, DC 20003
Phone: (202)546-9707
Contact: Gene Karpinski, Exec. Dir.
Coordinates activities of public interest research groups in 26 states and Canadian provinces. Serves as the national lobbying office for state PIRGs around the country for reforms on consumer, environmental, energy, and governmental issues. Conducts independent research for national environmental and consumer protection and maintains an information bank on pesticides. Contributions to U.S. PIRG are not tax deductible as charitable contributions. **Publication:** *U.S. PIRG Citizen Agenda,* a quarterly newsletter; research reports and studies.

★ 1873 ★ U.S. Public Interest Research Group (USPIRG)
Field Office
113 1/2 W. Franklin St.
Chapel Hill, NC 27514
Phone: (919)933-9994
Contact: Ed Johnson, Campaign Dir.

★ 1874 ★ U.S. Public Interest Research Group (USPIRG)
Field Office
1206 San Antonio
Austin, TX 78701-1887
Phone: (512)479-8481
Contact: Ross Crow, Campaign Dir.

★ 1875 ★ U.S. Public Interest Research Group (USPIRG)
Field Office
321 14th Ave., S.E., Third Floor
Minneapolis, MN 55414
Phone: (612)378-9393
Contact: John Miyasato, Campaign Dir.

★ 1876 ★ Vermont Public Interest Research Group (VPIRG)
43 State St.
Montpelier, VT 05602
Phone: (802)223-5221
Contact: John Gilroy, Dir.

★ 1877 ★ Washington Public Interest Research Group (WASHPIRG)
340 15th Ave., E.
Seattle, WA 98112
Phone: (206)322-9064
Contact: Rick Bunch, Dir.

★ 1878 ★ Washington Toxics Coalition
4516 University Way, N.E., Ste. 6
Seattle, WA 98105
Phone: (206)632-1545
Nonprofit organization of individuals and grassroots citizen groups active on issues such as pesticide reform, toxic household products, and hazardous waste management. Promotes the safe use of pesticides and serves as a clearinghouse for reporting pesticide problems. **Publications:** *WTC News,* a quarterly newsletter; a series of fact sheets on alternatives to toxic household products.

★ 1879 ★ Waste Management Research and Education Institute (WMREI)
Energy Environment and Resources Center (EERC)
University of Tennessee
327 S. Stadium Hall
Knoxville, TN 37996-0710
Phone: (615)974-4251
Contact: E. William Colglazier, Dir.
Researches various national and regional waste management issues, including those concerned with chemical, low- and highlevel radioactive waste, and solid waste, and investigates innovative public policy approaches for resolving waste management conflicts. **Publications:** *Forum for Applied Research and Public Policy,* a quarterly; a quarterly newsletter.

★ 1880 ★ Waste Watch Center (WWC)
Dana Duxbury and Associates
16 Haverhill St.
Andover, MA 01810
Phone: (508)470-3044
Fax: (508)470-3384
Contact: Dana Duxbury, Editor-in-chief
Private consulting group active in establishing the toxicity of commonly used consumer products and in recommending nontoxic alternatives. **Publications:** *Household Hazardous Waste Management News,* a newsletter; a series of booklets in the $5.00-$6.50 range, identifying experts in various fields of waste disposal, including household batteries, car batteries, paint collection and recycling, used oil collection, and farm pesticide collection programs.

★ 1881 ★ Water Pollution Control Federation (WPCF)
601 Wythe St.
Alexandria, VA 22314
Phone: (703)684-2400
Fax: (703)684-2492
Contact: Quincalee Brown, Exec. Dir.
Organization of technical societies concerned with water pollution, issues under the Clean Water Act, and the control of household hazardous wastes. **Publications:** *Household Hazardous Waste,* a public relations brochure available for imprinting with the names of local organizations; *The Bench Sheet,* a bimonthly newsletter; *Highlights,* a monthly newsletter; *Safety and Health Bulletin,* a quarterly; technical journals, bulletins, and fact sheets; and training materials and public relations material.

★ 1882 ★ White Lung Association (WLA)
P.O. Box 1483
Baltimore, MD 21203-1483
Phone: (301)243-5864
Contact: James Fite, Exec. Dir.
National nonprofit organization for those with conditions associated with exposure to asbestos. Educates the public about the dangers of asbestos and serves as a clearinghouse providing information on legal and medical assistance to asbestos victims. Conducts investigations of places with suspected asbestos contamination and monitors the removal of asbestos from workplaces. **Publication:** *Asbestos Watch,* a semiannual newsletter.

★ 1883 ★ White Lung Association (WLA)
Alabama Office
P.O. Box 190097
Tillman's Corner, AL 36619
Phone: (205)661-6532

★ 1884 ★ White Lung Association (WLA)
Florida Office
4940 S.W. 100th Ct.
Miami, FL 33165
Phone: (305)595-3495

★ 1885 ★ White Lung Association (WLA)
Illinois Office
Branch Box 1522
Edwardsville, IL 62026
Phone: (618)656-5957

★ 1886 ★ White Lung Association (WLA)
Northern California Office
924 El Camino Real
South San Francisco, CA 94080
Phone: (415)952-6840

★ 1887 ★ White Lung Association (WLA)
South Carolina Office
P.O. Box 7583
Greenville, SC 29610
Phone: (803)269-8048

★ 1888 ★ White Lung Association (WLA)
Southern California Office
P.O. Box 5089
San Pedro, CA 90731
Phone: (213)832-5864

★ 1889 ★ Wisconsin Public Interest Research Group (WISPIRG)
306 N. Brooks
Madison, WI 53715
Phone: (608)251-1918
Contact: Michael Leffel, Dir.

★ 1890 ★ **Working Group on Community Right-to-Know**
218 D St., S.E.
Washington, DC 20003
Phone: (202)544-2600
Coalition of public interest and environmental groups that provides information on the Emergency Planning and Community Rightto-Know Act and compiles documents on Title III.

★ 1891 ★ **Workplace Health Fund**
815 16th St., N.W., Ste. 301
Washington, DC 20006
Phone: (202)842-7833

Provides information to workers on industrial health and safety and on hazardous chemicals in the workplace.

★ 1892 ★ **World Wildlife Fund and The Conservation Foundation (WWF)**
1250 24th St., N.W.
Washington, DC 20037
Phone: (202)293-4800
Fax: (202)293-9211
Contact: Kathryn S. Fuller, Pres.
Leading U.S. organization devoted to saving endangered wildlife and habitats around the world and to protecting the biological resources on which human well-being depends. It supports individuals, public and private conservation institutions, and governments who carry out scientifically designed conservation programs that have immediate and long-term conservation benefits and programs to reduce global warming through the reduction of air pollution. It is part of the international WWF network, which includes national organizations in 26 countries. **Publications:** *World Wildlife Fund Letter*, an infrequent newsletter with single-issue features; *Focus*, a bimonthly newsletter; an annual catalog and specialized books and educational materials.

Books

★ 1893 ★ *Acceptable Common Names and Chemical Names for the Ingredient Statements on Pesticide Labels, 4th ed.*
Office of Pest Programs, Hazards Evaluation Division
U.S. Environmental Protection Agency
Washington, DC
Frequency: 1979. EPA-540/9-77-017.

★ 1894 ★ *Acceptable Risk?: Making Decisions in a Toxic Environment*
Lee Clark
University of California Press, Berkeley
Berkeley, CA
Frequency: 1989. *Price:* $30.00.

★ 1895 ★ *Access EPA: Clearinghouses and Hotlines*
Headquarters Library, Information Management and Services Division
U.S. Environmental Protection Agency, National Technical Information Service
Springfield, VA
Frequency: 1990. *Price:* $15.00 paper, $8.00 microfiche. EPA/IMSD/90-009; NTIS PB90-237082/XAB.

★ 1896 ★ *Access EPA: Library and Information Services*
Headquarters Library, Information Management and Services Division
U.S. Environmental Protection Agency, National Technical Information Service
Springfield, VA
Frequency: 1990. *Price:* $23.00 paper, $8.00 microfiche. EPA/IMSD/90-008; NTIS PB90-237074/XAB.

★ 1897 ★ *Access EPA: Major EPA Dockets*
Public Information Reference Unit, Information Management and Services Division
U.S. Environmental Protection Agency, National Technical Information Service
Springfield, VA
Frequency: 1990. *Price:* $15.00 paper, $8.00 microfiche. EPA/IMSD/90-006; NTIS PB90-237066/XAB.

★ 1898 ★ *Access EPA: State Environmental Libraries*
Headquarters Library, Information Management and Services Division
U.S. Environmental Protection Agency, National Technical Information Service
Springfield, VA
Frequency: 1990. *Price:* $17.00 paper, $8.00 microfiche. EPA/IMSD/90-004; NTIS PB90-237058/XAB.

★ 1899 ★ *Agent Orange and Vietnam: An Annotated Bibliography*
Scarecrow Press
Metuchen, NJ
Frequency: 1988. 401 pp.

★ 1900 ★ *Agricultural Chemicals*
W. T. Thomson
Thomson Publications
Fresno, CA
Vol. 1: *Insecticides*, 1989, 288 pp.; Vol. 2: *Herbicides*, 1986, 330 pp.; Vol. 3: *Fumigants, Growth Regulators, Repellents, Rodenticides*, 1991, 210 pp. *Price:* $16.50 each.

★ 1901 ★ *Agrochemicals Handbook*
Royal Society of Chemistry
Cambridge, Cambs., England
Frequency: Annual with semiannual updates.

★ 1902 ★ *Air and Waste Management Association Directory and Resource Book*
Air and Waste Management Association
Pittsburgh, PA
Frequency: Annual.

★ 1903 ★ *Air and Waste Management Association Government Agencies Directory*
Harold M. Englund, ed.
Air and Waste Management Association
Pittsburgh, PA
Frequency: Annual. *Price:* $8.00.

★ 1904 ★ *Air Pollution Control Directory*
American Business Directories, Inc.
Omaha, NE
Frequency: Annual. *Price:* $85.00.

★ 1905 ★ *All about OSHA, revised ed.*
Occupational Safety and Health Administration
U.S. Department of Labor
Washington, DC
Frequency: 1991. *Price:* Free. 48 pp. OSHA 2056; GPO 1991-282-150/45318.

★ 1906 ★ *American Academy of Environmental Medicine Directory*
American Academy of Environmental Medicine
Denver, CO
Frequency: Annual.

★ 1907 ★ *America the Poisoned*
Lewis Regenstein
Acropolis Books
Washington, DC
Frequency: 1982. *Price:* $16.95. 414 pp.

★ 1908 ★ *Artist Beware: The Hazards and Precautions in Working with Art and Craft Material*
Michael McCann
Watson-Guptill Publications
New York, NY
Frequency: 1979. *Price:* $16.95. 378 pp. Exhaustive guide to toxic substances used in arts and crafts.

★ 1909 ★ *Asbestos Abatement Resource Guide, 2nd ed.*
Bureau of National Affairs
Washington, DC
Frequency: 1989. *Price:* $250.00.

★ 1910 ★ *Asbestos: Directory of Unpublished Studies, 2nd ed.*
Sandro Amaducci
Elsevier Science Publishing
Frequency: 1987. *Price:* $54.00. 222 pp.

★ 1911 ★ *Asbestos Hazard Management: Guidebook to Abatement*
Robert P. Ouellette, Ramesh Talreja, et al.
Technomic Publishing
Lancaster, PA
Frequency: 1987. *Price:* $45.00 paperback. 403 pp.

★ 1912 ★ *Asbestos in Buildings*
Safe Building Alliance
Metropolitan Square
655 15th St., N.W., Suite 1200
Washington, DC 20005
Price: Free.

★ 1913 ★ *Asbestos in Public and Commercial Buildings: A Literature Review and Synthesis of Current Knowledge*
Health Effects Institute—Asbestos Research
Cambridge, MA
Phone: (617)225-0866
Fax: (617)225-2211
Frequency: 1991. *Price:* Free. 387 pages. Report prepared under a congressional mandate to investigate exposure to asbestos and evaluate the effectiveness of asbestos management; summarizes the state of knowledge reflected in scientific articles, reports, and unpublished data.

★ 1914 ★ *Asbestos in the Home*
U.S. Consumer Product Safety Commission
U.S. Government Printing Office
Washington, DC
Hotline: 800-638-2772
Frequency: 1989. *Price:* $1.00. S/N 052-011-00246-1.

★ 1915 ★ *Asbestos Management Sourcebook*
Environmental Publications
Lutherville, MD
Frequency: Annual. *Price:* $20.00.

★ 1916 ★ *Asbestos: The Hazardous Fiber*
Melvin A. Benardes, ed.
CRC Press
Boca Raton, FL
Frequency: 1990. *Price:* $49.95. 490 pp. Provides information on the history of asbestos, as well as the current problem with it and protective and disposal methods one should follow when dealing with it.

★ 1917 ★ *ATSDR Fact Sheet*
Agency for Toxic Substances and Disease Registry
Atlanta, GA
Phone: (404)639-0727
Price: Free.

★ 1918 ★ *Best's Safety Directory*
Kathleen M. Guindon, ed.
A. M. Best
Oldwick, NJ
2 vols. *Frequency:* Annual published in November. *Price:* $33.00. Primarily a buyers' guide to sources for 5,200 safety, industrial hygiene, and industrial pollution control products and services; contains Occupational Safety and Health Administration (OSHA) standards.

★ 1919 ★ *The Bhopal Syndrome: Pesticides, Environment and Health*
David Weir
Sierra Club Books
San Francisco, CA
Frequency: 1988. *Price:* $8.95. 224 pp. The incident at Bhopal, India, and other industrial chemical accidents that have either occurred or are waiting to happen.

★ 1920 ★ *The Bhopal Tragedy: Language, Logic, and Politics in the Production of a Hazard*
William Bogard
Westview Press
Boulder, CO
Frequency: 1989. *Price:* $28.50 paperback. 154 pp.

★ 1921 ★ *A Blueprint for Pesticide Policy: Changing the Way We Safeguard, Grow and Market Food*
Public Voice for Food and Health Policy
Washington, DC
Frequency: 1989. *Price:* $10.00. Examines the economic factors involved in pesticide use.

★ 1922 ★ *Borrowed Earth, Borrowed Time: Healing America's Chemical Wounds*
Glenn E. Schweitzer
Plenum Publishing
New York, NY
Frequency: 1991.

★ 1923 ★ *Breaking the Pesticide Habit, 2nd ed.*
Terry Gip
International Alliance for Sustainable Agriculture
Minneapolis, MN
Frequency: 1988. *Price:* $26.95 hardbound, $19.95 paperback. 372 pp. Covers the 12 most toxic pesticides.

★ 1924 ★ *Calculated Risks: Understanding the Toxicity and Human Health Risks of Chemicals in Our Environment*
Joseph V. Rodricks
Cambridge University Press
New York, NY
Frequency: 1992. *Price:* $22.95.

★ 1925 ★ *A Catalog of Hazardous and Solid Waste Publications, 4th ed.*
Office of Solid Waste and Emergency Response
U.S. Environmental Protection Agency
Washington, DC
Phone: 800-424-9346
Frequency: 1990. *Price:* Free.

★ 1926 ★ *A Catalog of Superfund Program Publications*
Office of Emergency and Remedial Response
U.S. Environmental Protection Agency
Washington, DC 20460
Frequency: 1990. *Price:* Free. EPA/540/890/015.

★ 1927 ★ *CEP Annual Software Directory*
American Institute of Chemical Engineers
345 E. 47th St.
New York, NY 10017
Frequency: Annual supplement to the October issue of *Chemical Engineering Progress* magazine. Listings of software available for air pollution, waste management, and other Occupational Safety and Health Administration (OSHA) and U.S. Environmental Protection Agency regulated programs and material safety data sheets.

★ 1928 ★ *CERCLA Regulations and Keyword Index*
McCoy and Associates
Frequency: 1991. *Price:* $40.00 paperback. 401 pp. Purchase through Hazardous Materials Control Resources Institute, Greenbelt, MD.

★ 1929 ★ *Chemical Deception: The Toxic Threat to Health and the Environment*
Marc Lappe
Sierra Club Books
San Francisco, CA
Frequency: 1991. *Price:* $27.00. 360 pp.

★ 1930 ★ *Chemical Hazard Communication Guidebook: OSHA, EPA and DOT Requirements*
Andrew B. Waldo and Richard Hinds
Executive Enterprise Publications
New York, NY
Frequency: 1988. *Price:* $75.00.

★ 1931 ★ *Chemical Hazards in the Workplace: Industrial Toxicology Introduction*
Ronald M. Scott
Lewis Publishers
Boca Raton, FL
Frequency: 1989. *Price:* $49.95. 171 pp.

★ 1932 ★ *Chemical Hazards Response Information System (CHRIS)*
U.S. Coast Guard
U.S. Government Printing Office
Washington, DC
Frequency: 1978. An invaluable guide for handling hazardous materials waste and spills.

★ 1933 ★ *Chemical Plant Wastes: A Bibliography*
Mary Vance
Vance Bibliographies
Monticello, IL
Frequency: 1989. *Price:* $5.00. 20 pp. List of journal articles dealing with wastes from chemical plants. Public Administration Series, Bibliography No. P2749.

★ 1934 ★ *Chemical Risk: A Primer*
American Chemical Society
Washington, DC
Frequency: 1984. *Price:* Free. 12 pp. Overview of toxic substances in the environment.

★ 1935 ★ *Chemical Risk: Personal Decisions*
American Chemical Society
Washington, DC
Frequency: 1989. *Price:* Free. 16 pp.

★ 1936 ★ *Chemicals, the Press and the Public: A Journalist's Guide to Reporting on Chemicals in the Community*
Environmental Health Center
National Safety Council
Chicago, IL
Frequency: 1989. *Price:* $9.95.

★ 1937 ★ *A Citizen's Guide to Promoting Toxic Waste Reduction*
Lauren Kenworthy and Eric Schaeffer
Inform, Inc.
New York, NY
Frequency: 1990. *Price:* $17.50. 122 pp.

★ 1938 ★ *A Citizen's Guide to Radon: What It Is and What to Do about It*
U.S. Environmental Protection Agency and Centers for Disease Control
U.S. Government Printing Office
Washington, DC
Frequency: 1986. *Price:* Free. For a copy contact EPA Public Information Center, (202)260-7751. EPA-OPA-86-004; GPO No. 617-003/84298.

★ 1939 ★ *The Citizen's Toxics Protection Manual*
National Toxics Campaign
Boston, MA
Price: $35.00. 650 pp. How-to manual for organizing a toxics campaign in your local community.

★ 1940 ★ *Clean and Green: The Complete Guide to Nontoxic and Environmentally Safe Housekeeping*
Annie Berthold-Bond
Ceres Press
Woodstock, NY
Frequency: 1990. 161 pp.

★ 1941 ★ *Clean Water Deskbook*
Environmental Law Reporter staff
Environmental Law Institute
Washington, DC
Frequency: 1989. *Price:* $85.00.

★ 1942 ★ *Clearer, Cleaner, Safer, Greener: A Blueprint for Detoxifying Your Environment*
Gary Null
Villard Books
New York, NY
Frequency: 1990. *Price:* $18.95. 293 pp.

★ 1943 ★ *Clinical Toxicology of Commercial Products, 5th ed.*
Robert Gosselin, Roger P. Smith, and Harold C. Hodge
Williams and Wilkins
Baltimore, MD
Frequency: 1984. *Price:* $129.95. 2012 pp. Lists more than 1,000 commercial products by trade name and their label contents.

★ 1944 ★ Common-Sense Pest Control
William Olkowski, Sheila Daar, and Helga Olkowski
Taunton Press
Newtown, CT
Frequency: 1991. **Price:** $39.95. 715 pp. Extensive and encyclopedic reference work on least-toxic solutions for the home, garden, pets, and community; presents steps necessary for integrated pest management (IPM).

★ 1945 ★ Common Synonyms for Chemicals Listed under Section 313 of the Emergency Planning and Community Right-to-Know Act
U.S. Environmental Protection Agency
Office of Pesticides and Toxic Substances
Washington, DC
Frequency: 1991. **Price:** Free. 114 pp. EPA 5604-91-005.

★ 1946 ★ Community Right-to-Know Compliance Handbook
Thomas Balf
Business and Legal Reports
Madison, CT
Frequency: 1988. **Price:** $160.00.

★ 1947 ★ Community Right-to-Know Deskbook
Environmental Law Reporter staff
Environmental Law Institute
Washington, DC
Frequency: 1988. **Price:** $85.00.

★ 1948 ★ Compendium of Hazardous Chemicals in Schools and Colleges
Forum for Scientific Excellence
J. B. Lippincott
Philadelphia, PA
Frequency: 1990.

★ 1949 ★ The Complete Book of Home Environmental Hazards
Roberta Altman
Facts on File
New York, NY
Frequency: 1991. **Price:** $12.95.

★ 1950 ★ Complete Guide to Pest Control with and without Chemicals
George W. Ware
Thomson Publishing
Fresno, CA

★ 1951 ★ Concern's Community Action Guides on Environmental Topics
Concern, Inc.
Washington, DC
Frequency: 1991.

★ 1952 ★ Concise Manual of Chemical and Environmental Safety in Schools and Colleges
Forum for Scientific Excellence
J. B. Lippincott
Philadelphia, PA
Vol. 1: *Basic Principles*, 1989; Vol. 2: *Hazardous Chemical Classes*, 1989; Vol. 3: *Chemical Interactions*, 1990; Vol. 4: *Safe Chemical Storage*, 1990; Vol. 5: *Safe Disposal of Chemicals*, 1990. **Price:** $17.95 each. Contents of all five manuals are included in the *Handbook of Chemical and Environmental Safety in Schools and Colleges*.

★ 1953 ★ Congressional Directory: Environment 1989
John T. Grupenhoff and Betty Farkey, eds.
Science and Health Communications Group
Bethesda, MD
Frequency: Biennial published in April of odd years. **Price:** $87.50. Lists members of the U.S. Congress and staffs of committees dealing with environmental matters.

★ 1954 ★ A Consumer's Dictionary of Cosmetic Ingredients, 3rd ed.
Ruth Winter
Crown
New York, NY
Frequency: 1989. **Price:** $10.95. 327 pp.

★ 1955 ★ Controlling Asbestos in Buildings: An Economic Investigation; An RFF Study
Donald N. Dewees
Resources for the Future/Johns Hopkins University Press
Baltimore, MD
Frequency: 1986. **Price:** $9.95.

★ 1956 ★ Control of Radon in Houses
National Council on Radiation Protection and Measurements
NCRP Publications
Bethesda, MD
Frequency: 1989. **Price:** $14.00. NCRP Report No. 103.

★ 1957 ★ Co-op America's Business and Organizational Member Directory
Co-op America
Washington, DC
Frequency: Annual. **Price:** $2.00.

★ 1958 ★ CRC Handbook of Environmental Control
Conrad P. Straub, ed.
CRC Press
Boca Raton, FL
Frequency: 1988. **Price:** $49.95.

★ 1959 ★ Criteria Pollutant Point Source Directory
George Crocker, ed.
North American Water Office
Lake Elma, MN
Frequency: Biennial published in even years. **Price:** $33.00. Lists more than 1,600 utilities, refineries, and other facilities that emit more than 1,000 tons of air pollutants a year.

★ 1960 ★ Dangerous Properties of Industrial Materials, 7th ed.
Irving N. Sax and Richard J. Lewis, Sr.
Van Nostrand Reinhold
New York, NY
3 vols. **Frequency:** 1988. **Price:** $395.00. 4,000 pp. Contains technical profiles of more than 20,000 chemicals, including their exposure limits, toxic reviews, and hazard ratings.

★ 1961 ★ Dan's Practical Guide to Least Toxic Home Pest Control
Dan Stein
Hulogosi Communications
Eugene, OR
Frequency: 1991. **Price:** $8.95. 87 pp.

★ 1962 ★ A Database of Safer Substitutes for Hazardous Household Products
Philip Dickey
Washington Toxics Coalition
Seattle, WA
2 vols. **Frequency:** 1990.

★ 1963 ★ Defusing the Toxics Threat: Controlling Pesticides and Industrial Waste
Sandra Postel
Worldwatch Institute
Washington, DC
Frequency: 1987. **Price:** $4.00. 69 pp. Discussion of what countries other than the United States are doing to reduce pesticides and other toxic substances in industry and agriculture. Worldwatch Paper No. 79.

★ 1964 ★ Dictionary of Environmental Science and Technology
Andrew Porteous
John Wiley & Sons
New York, NY
Frequency: 1991. **Price:** $88.00 hardbound, $33.00 paperback. 416 pp.

★ 1965 ★ Dictionary of Toxicology
Ernest Hodgson, Richard Mailman, and Janice Chambers
Van Nostrand Reinhold
New York, NY
Frequency: 1988. **Price:** $82.95. 470 pp.

★ 1966 ★ Diet for a Poisoned Planet: How to Choose Safe Foods for You and Your Family
David Steinman
Harmony Books
New York, NY
Frequency: 1990. **Price:** $12.95. 392 pp. Reviews 500 foods and beverages and tells which ones are safe and which contain residues of pesticides, hormones, and other chemicals.

★ 1967 ★ Dioxin, Agent Orange: The Facts
Michael Grough
Plenum Press
New York, NY
Frequency: 1986. **Price:** $19.95.

★ 1968 ★ Directory of Commercial Hazardous Waste Treatment and Recycling Facilities
Office of Solid Waste
U.S. Environmental Protection Agency
Washington, DC
Frequency: Irregular, latest edition published 1987. **Price:** $23.00.

★ 1969 ★ Directory of Computerized Data Files
National Technical Information Service
Springfield, VA
Frequency: 1989. **Price:** $59.00. Lists 2,400 numeric and textual data files from 50 federal agencies, including coverage in environment, health, and energy. NTIS PB89-191761FAA.

★ 1970 ★ Directory of Environmental Groups in New England
Office of Public Affairs, Region I
U.S. Environmental Protection Agency
Boston, MA
Frequency: Irregular, latest edition published 1990. **Price:** Free. Federal and state agencies and nonprofit organizations with environmental concerns.

★ 1971 ★ Directory of Environmental Information Sources, 3rd ed.
Thomas F. P. Sullivan, ed.
Government Institutes
Rockville, MD
Frequency: 1990. **Price:** $65.00. 298 pp. Lists more than 1,400 federal and state agencies, trade associations, organizations, periodicals, databases, and other sources for information on the environment.

★ 1972 ★ Directory of EPA/State Contacts by Specialty
U.S. Environmental Protection Agency
Frequency: 1990. **Price:** $17.00. 73 pp. To order, write the National Technical Information

Service, U.S. Department of Commerce, Springfield, VA 22161. EPA-540-8-90-002; NTIS PB90-249749.

★ 1973 ★ *Directory of Federal Contacts on Environmental Protection*
Patricia Murray, ed.
Naval Energy and Environmental Support
 Activity
Port Hueneme, CA
Frequency: Biennial published in even years. *Price:* Free. Lists federal agencies engaged in some aspect of environmental protection, including military departments.

★ 1974 ★ *Directory of Federal Laboratories and Technology Resources, 1990-1991: A Guide to Services, Facilities, and Expertise*
National Technical Information Service
Springfield, VA
Frequency: 1990. *Price:* $59.95. Provides comprehensive information on more than 1,100 government research centers, testing facilities, and laboratories that will share their expertise, specialized services, and facilities with private businesses and the academic community. Includes a listing of U.S. Government laboratories' technology transfer offices. NTIS PB90-104480FAA.

★ 1975 ★ *Directory of Occupational Health and Safety Software Version 3.0*
Computers in Occupational Users Group
American College of Occupational Medicine
Charlottesville, VA
Frequency: 1990. *Price:* $25.00.

★ 1976 ★ *The Dose Makes the Poison, 2nd ed.*
M. Alice Ottoboni
Van Nostrand Reinhold
New York, NY
Frequency: 1991. 244 pp. A plain-language guide to toxicology.

★ 1977 ★ *The Effects of Pesticides on Human Health*
Scott R. Baker and Chris F. Wilkinson, eds.
Princeton Scientific Publishing
Princeton, NJ
Frequency: 1988. Proceedings of a workshop.

★ 1978 ★ *Elements of Toxicology and Chemical Risk Assessment: A Handbook for Nonscientists, Attorneys, and Decision Makers*
Environ Corp.
Washington, DC
Frequency: 1986.

★ 1979 ★ *Emergency Response Directory for Hazardous Materials Accidents*
Pamela Lawrence, ed.
Odin Press
New York, NY
Frequency: Biennial published in odd years. *Price:* $34.00. Lists more than 1,000 federal, state, and local agencies, chemical manufacturers and transporters, hotlines, strike teams, disaster centers, burn centers, and other organizations concerned with the containment and cleanup of chemical spills and other hazardous materials accidents.

★ 1980 ★ *Emergency Response Guidebook*
Research and Special Programs
 Administration, Office of Hazardous
 Materials Transportation
U.S. Dept. of Transportation
Washington, DC
Frequency: Annual. *Price:* $5.00. The standard

manual for first responders to hazardous materials incidents, including fire fighters, police, and emergency services personnel. Reprints are available from commercial publishers. DOT P 5800.5.

★ 1981 ★ *The Encyclopedia of Environmental Issues*
William Ashworth, ed.
Facts on File
New York, NY
Frequency: 1989.

★ 1982 ★ *Environmental Compliance Management Sourcebook*
Environmental Publications
Lutherville, MD
Frequency: 1990. *Price:* $20.00. Directory of information resources related to health, exposure, and risk and assessment of air toxics. For a free copy contact the U.S. Environmental Protection Agency Risk Information Support Center at (919)541-0888. EPA 450-3-88-015.

★ 1983 ★ *The Environmental Dictionary*
James J. King, ed.
Executive Enterprises Publications
New York, NY
Frequency: 1989.

★ 1984 ★ *The Environmental Directory*
Terrell K. Higashi, ed.
Northwest Technical Publishing
Kent, WA
Frequency: Annual. *Price:* $40.00. Regional directories of environmental consultants and suppliers.

★ 1985 ★ *Environmental Hazards: Air Pollution, A Reference Handbook*
E. Willard Miller and Ruby M. Miller
ABC-CLIO
Santa Barbara, CA
Frequency: 1989. *Price:* $39.50. 250 pp.

★ 1986 ★ *Environmental Hazards Management Institute (EHMI) Wheels*
Durham, NH
The series of three wheels (*Household Hazardous Waste Wheel, Water Sense Wheel,* and *Recycling Wheel*) are quick reference tools available in bulk to organizations for fundraising and/or promotion.

★ 1987 ★ *Environmental Hazards: Radioactive Materials and Wastes, A Reference Handbook*
E. Willard Miller and Ruby M. Miller
ABC-CLIO
Santa Barbara, CA
Frequency: 1990. *Price:* $39.50. 298 pp.

★ 1988 ★ *Environmental Hazards: Toxic Waste and Hazardous Material*
E. Willard Miller and Ruby M. Miller
ABC-CLIO
Santa Barbara, CA
Frequency: 1991. *Price:* $39.50. 286 pp. Concise overview of hazardous waste, waste site evaluation, and corresponding legislative history.

★ 1989 ★ *Environmental Information— Environmental Services Directory*
Cary L. Perket, ed.
Environmental Information
Bloomington, MN
Frequency: Annual. *Price:* $275.00. Describes waste-handling facilities, spill response firms, consultants, laboratories, recovery services, asbestos services, and other companies that

service spills and the containment and disposal of hazardous materials.

★ 1990 ★ *Environmental Telephone Directory*
Government Institutes
Rockville, MD
Frequency: Published every 18 months, the latest edition in December 1989. *Price:* $55.00.

★ 1991 ★ *EPA Guides to Pollution Prevention*
U.S. Environmental Protection Agency
Washington, DC
Titles include *The Printed Circuit Board Manufacturing Industry, The Fabricated Metal Products Industry, The Commercial Printing Industry, The Pesticide Formulating Industry, The Paint Manufacturing Industry,* and *Selected Hospital Waste Streams.*

★ 1992 ★ *EPA HQ Telephone Directory*
U.S. Government Printing Office
Washington, DC
Frequency: Periodical. *Price:* $12.00.

★ 1993 ★ *EPA Information Resources Directory*
U.S. Environmental Protection Agency, Office
 of Information Resources Management
National Technical Information Service
Springfield, VA
Frequency: 1989. *Price:* $29.50. Provides contacts for federal agencies and commissions, local, national, and international interest groups, scientific, professional, and advocacy groups, and trade associations concerned with the environment. NTIS PB90-132192KOF.

★ 1994 ★ *Evaluation of Chemical Products against Selected Pests of Greenhouse and Outdoor Ornamental Crops, 1977-1982*
Department of Agriculture
Agriculture Research Service
Washington, DC
Frequency: 1985. Production Research Report No. 183.

★ 1995 ★ *The Facts on File Dictionary of Environmental Science*
L. Harold Stevenson and Bruce C. Wyman, eds.
Facts on File
New York, NY
Frequency: 1991. *Price:* $24.95. 288 pp.

★ 1996 ★ *Farm Chemicals Handbook*
Charlotte Sine, ed.
Meister Publishing
Willoughby, OH
Frequency: Annual.

★ 1997 ★ *Fighting Toxics: A Manual for Protecting Your Family, Community and Workplace*
Gary Cohen and John O'Connor, eds.
Island Press
Washington, DC
Frequency: 1990. *Price:* $19.95 paperback. 325 pp.

★ 1998 ★ *First Aid Manual for Chemical Accidents, 2nd ed.*
Shirley Conibear, ed.
Van Nostrand Reinhold
New York, NY
Frequency: 1989. *Price:* $26.95.

★ 1999 ★ *Fundamentals of Pesticides: A Self-Instruction Guide, 2nd ed.*
George W. Ware
Thompson Publications
Fresno, CA
Frequency: 1986. *Price:* $11.95.

★ 2000 ★ *Gale Environmental Sourcebook: A Guide to Organizations, Agencies, and Publications, 1st ed.*
Karen Hill and Annette Piccirelli, eds.
Gale Research
Detroit, MI
Frequency: 1992. *Price:* $75.00. 688 pp.

★ 2001 ★ *Glossary of Environmental Terms and Acronym List*
Office of Public Affairs
U.S. Environmental Protection Agency
Washington, DC
Phone: (202)260-4361
Frequency: 1989. *Price:* Free. EPA OPA-87-017.

★ 2002 ★ *Goldfrank's Toxicological Emergencies, 4th ed.*
Lewis R. Goldfrank, ed.
Appleton and Lange
East Norwalk, CT
Frequency: 1990. *Price:* $135.00. 1012 pp.

★ 2003 ★ *The Green Book: The Environmental Resource Directory, New England*
The Green Book
P.O. Box 23
Belmont, MA 02178
Frequency: Annual. *Price:* $89.95. 224 pp. Directory of environmental companies, organizations, and government officials in EPA Region I.

★ 2004 ★ *The Green Consumer: A Guide for the Environmentally Aware*
John Elkington and Joel Makower
Penguin Books
New York, NY
Frequency: 1990. *Price:* $9.95.

★ 2005 ★ *Green Lifestyle Handbook: 1001 Ways You Can Heal the Earth*
Jeremy Rifkin, ed.
Henry Holt
New York, NY
Frequency: 1990. *Price:* $10.95. 202 pp.

★ 2006 ★ *The Green Pages: Environmental Information and Savings Guide for Vancouver and Lower Mainland*
Green Pages International
306-1450 Chestnut St.
Vancouver, BC, Canada V6J 3K3
Frequency: 1991. *Price:* $20.00. 204 pp. Environmental information and discount coupons for environmental products and services in British Columbia.

★ 2007 ★ *The Green Pages: Your Everyday Shopping Guide to Environmentally Safe Products*
Steven J. Bennett
Random House
New York, NY
Frequency: 1990. *Price:* $8.95.

★ 2008 ★ *Groundwater Chemicals Desk Reference*
John H. Montgomery and Linda M. Welkam
Lewis Publishers
Boca Raton, FL
Frequency: 1989. *Price:* $79.95. 600 pp.

★ 2009 ★ *Guidance for Controlling Asbestos-Containing Materials in Buildings*
Office of Pesticides and Toxic Substances
U.S. Environmental Protection Agency
Washington, DC
Frequency: 1985. EPA 560/5-85-024.

★ 2010 ★ *Guide to EPA Hotlines, Clearinghouses, Libraries, and Dockets*
Office of Public Affairs
U.S. Environmental Protection Agency
Washington, DC
Phone: (202)260-7751
Frequency: 1990. *Price:* Free. OPA 007-89.

★ 2011 ★ *Guide to EPA Libraries and Information Services*
Information Management and Services Division
U.S. Environmental Protection Agency
Washington, DC
Phone: (202)260-5922
Frequency: 1989. *Price:* Free. EPA/IMSD/89-008.

★ 2012 ★ *Guide to Hazardous Products around the House*
Household Hazardous Waste Project
Southwest Missouri State University
Springfield, MO
Frequency: 1989. *Price:* $9.95. 178 pp. A-to-Z listing of hazardous substances found in the home includes information on how to dispose of hazardous products, how to recycle, and how to read labels. Also contains information on safer alternatives to chemicals and a directory of Missouri resources.

★ 2013 ★ *Guide to Key Environmental Statistics in the U.S. Government*
Janet N. Abramovitz, Douglas S. Baker, and Daniel B. Tunstall
World Resources Institute
Washington, DC
Frequency: 1990. *Price:* $15.00.

★ 2014 ★ *Guide to State Environmental Programs*
Deborah Hitchcock Jessup, ed.
BNA Books
Washington, DC
Frequency: Irregular, latest edition published 1990. *Price:* $48.00.

★ 2015 ★ *Handbook of Chemical and Environmental Safety in Schools and Colleges*
Forum for Scientific Excellence
J. B. Lippincott
Philadelphia, PA
Frequency: 1990. (Includes the five volumes of the *Concise Manual of Chemical and Environmental Safety in Schools and Colleges* and the Pocket Guides produced by the Forum for Scientific Excellence.)

★ 2016 ★ *Handbook of Environmental Definition*
C. C. Lee
Hazardous Materials Control Resources Institute
Greenbelt, MD
Frequency: 1992. *Price:* $59.00 paperback. 500 pp.

★ 2017 ★ *Handbook of Radon in Buildings: Detection, Safety, and Control*
Mueller Associates, SYSCON Corp., and Brookhaven National Laboratory, compilers
Hemisphere Publishing
New York, NY
Frequency: 1989. *Price:* $49.50.

★ 2018 ★ *Handbook of Toxic and Hazardous Chemicals and Carcinogens, 2nd ed.*
Marshall Sittig
Noyes
Park Ridge, NJ
Frequency: 1985. *Price:* $96.00. 950 pp.

★ 2019 ★ *Hawley's Condensed Chemical Dictionary, 11th ed.*
N. Irving Sax and Richard J. Lewis, eds.
Van Nostrand Reinhold
New York, NY
Frequency: 1987. *Price:* $52.95.

★ 2020 ★ *Hazard Communication: A Compliance Kit*
Occupational Safety and Health Administration
U.S. Government Printing Office
Washington, DC
Frequency: Annual.

★ 2021 ★ *The Hazard Communication Standard*
John W. Wells
Fairmont Press
Lilburn, GA
Frequency: 1989. *Price:* $85.00. 149 pp.

★ 2022 ★ *Hazardous and Toxic Materials: Safe Handling and Disposal, 2nd ed.*
Howard H. Fawcett
John Wiley & Sons
New York, NY
Frequency: 1988. *Price:* $63.00. 514 pp.

★ 2023 ★ *Hazardous Chemicals Desk Reference, 2nd ed.*
Richard J. Lewis, Sr.
Van Nostrand Reinhold
New York, NY
Frequency: 1990. *Price:* $84.95. 1,100 pp.

★ 2024 ★ *Hazardous Chemicals on File*
Craig T. Norback and Judith C. Norback, eds.
Facts on File
New York, NY
3 vols. *Frequency:* 1988. *Price:* $250.00. 756 pp. Written for the general public, this loose-leaf reference describes more than 325 chemicals, their health hazards, storage and disposal methods, and steps to take for personal protection when dealing with them.

★ 2025 ★ *Hazardous Materials Control Buyer's Guide and Source Book, 1992*
Hazardous Materials Control Research Institute
Greenbelt, MD
Frequency: Annual. *Price:* Free to members, $65.00 for nonmembers. 384 pp. Lists more than 3,000 companies offering services and products in the hazardous materials management industry.

★ 2026 ★ *Hazardous Materials Emergency Response Pocket Handbook*
Paul N. Cheremisinoff
Technomic Publishing
Lancaster, PA
Frequency: 1989. *Price:* $39.00.

★ 2027 ★ Hazardous Materials Spills Handbook
Gary F. Bennett, Frank S. Feates, and Ira Wilder, eds.
McGraw-Hill
New York, NY
Frequency: 1982. Discusses handling hazardous chemical spills and their legal and political aspects.

★ 2028 ★ Hazardous Materials Storage and Handling Handbook
Defense Logistics Agency, Directories of Supply Operations, Depot Operations Division
Department of Defense
Alexandria, VA
Frequency: Annual. 118 pp. DLAH 4145.6.

★ 2029 ★ Hazardous Substances: A Reference
Melvin Berger
Enslow Publishers
Hillside, NJ
Frequency: 1986. **Price:** $17.95. 128 pp. Nontechnical guide to commonly used hazardous substances in laboratories and factories, with bibliographies.

★ 2030 ★ Hazardous Substances in Our Environment: A Citizen's Guide to Understanding Health Risks and Reducing Exposure
Office of Policy Planning and Evaluation
U.S. Environmental Protection Agency
Washington, DC
Frequency: 1990. **Price:** Free. 125 pp. EPA 230/09/90/081.

★ 2031 ★ Hazardous Waste and Hazardous Materials Transportation Directory
Howard Roth, ed.
K-III Press
New York, NY
Frequency: Annual. Directory of suppliers in the hazardous materials field.

★ 2032 ★ Hazardous Waste Bibliography
Office of Solid Waste and Emergency Response
U.S. Environmental Protection Agency
Washington, DC
Frequency: 1987. 42 pages. OSWER Directive 9380.102.

★ 2033 ★ Hazardous Waste from Homes
John Lord
Enterprise for Education
Santa Monica, CA
Frequency: 1988. **Price:** $3.00. 36 pp.

★ 2034 ★ Health Detective's Handbook: A Guide to the Investigation of Environmental Health Hazards by Nonprofessionals
Marvin S. Ligator, et al.
Johns Hopkins University Press
Baltimore, MD
Frequency: 1985. **Price:** $70.80 hardbound, $12.95 paperback. 272 pp.

★ 2035 ★ Health Hazards Manual for Artists, 3rd ed.
Michael McCann
Lyons & Buford
New York, NY
Frequency: 1985. **Price:** $9.95. 104 pp.

★ 2036 ★ Health Hotlines: 800 Numbers from DIRLINE
Specialized Information Services
National Library of Medicine
8600 Rockville Pike
Bethesda, MD
Phone: (301)496-1131
Frequency: 1990. **Price:** Free. 56 pp.

★ 2037 ★ The Healthy Home: An Attic to Basement Guide to Toxin-Free Living
Linda Mason Hunter
Rodale Press
Emmaus, PA
Frequency: 1989. **Price:** $21.95. 313 pp.

★ 2038 ★ Herbicide Handbook, 6th ed.
N. E. Humberg
Weed Science Society of America
Champaign, IL
Frequency: 1989. **Price:** $25.00. 301 pp.

★ 2039 ★ Household Pollutants Guide
Albert J. Fritsch, ed.
Center for Science in the Public Interest/ Anchor Books
Garden City, NY
Frequency: 1978. **Price:** $3.50. 309 pp.

★ 2040 ★ Household Waste: Issues and Opportunities
Susan Boyd, Burks Lapham, and Cynthia McGrath, eds.
Concern, Inc.
Washington, DC
Frequency: 1989. **Price:** Free. 30 pages. Primer on how to reduce, recycle, and dispose of hazardous waste in the household. Contains information on state and federal regulations and suggested courses for community action.

★ 2041 ★ How to Comply with the OSHA Hazard Communication Standard: A Complete Guide to Compliance with OSHA Worker Right-to-Know Regulations
Environmental Resources Center
Van Nostrand Reinhold
New York, NY
Frequency: 1989. **Price:** $24.95.

★ 2042 ★ Indoor Radon and Its Hazards
D. Bodansky, M. A. Robkin, and D. R. Stadler, eds.
University of Washington Press
Seattle, WA
Frequency: 1987.

★ 2043 ★ Information Resources in Toxicology
Philip Wexler
Elsevier Science Publishing
New York, NY
Frequency: 1988. **Price:** $96.00. 510 pp.

★ 2044 ★ International Directory for Sources of Environmental Information
U.S. National Focal Point for UNEP/ INFOTERRA
U.S. Environmental Protection Agency
Frequency: Triennial with supplements. **Price:** $200.00. World-wide directory of organizations, government agencies, universities, research centers, and other sources of environmental information.

★ 2045 ★ Is Your Drinking Water Safe?
Office of Water
U.S. Environmental Protection Agency
Washington, DC

Frequency: 1989. **Price:** Free. 26 pp. EPA 570989-005.

★ 2046 ★ Keeping Your Company Green
Stefan Bechtel and editors at Rodale Press
Rodale Press
Emmaus, PA
Frequency: 1990. **Price:** Free. 92 pp. Compendium of simple steps companies can take to help the environment.

★ 2047 ★ Lawn Care Pesticides: Risks Remain Uncertain while Prohibited Safety Claims Continue
Government Accounting Office
Washington, DC
Frequency: 1990. Report to the U.S. Senate, Chairman of the Subcommittee on Toxic Substances, Environmental Oversight, and Research and Development of the Committee on Environment and Public Works. GAO/ RCED-90-134.

★ 2048 ★ Lead and Your Drinking Water
Office of Water
U.S. Environmental Protection Agency
Washington, DC
Frequency: 1987. **Price:** Free. 8 pp. OPA-87006.

★ 2049 ★ Lead in the Environment
National Science Foundation
U.S. Government Printing Office
Washington, DC
Frequency: 1977. No. 038-000-00338-1.

★ 2050 ★ Making Darkrooms Safer
National Press Photographers Association
3200 Croasdaile Dr.
Durham, NC 27705
Frequency: 1989. **Price:** $6.50. 70 pp. Booklet on the safe handling of hazardous photochemicals.

★ 2051 ★ Managing Asbestos in Place—A Building Owner's Guide to Operations and Maintenance Programs for Asbestos-Containing Materials
Office of Pesticides and Toxic Substances
U.S. Environmental Protection Agency
Washington, DC
Frequency: 1990. **Price:** $3.25. 20T-2003.

★ 2052 ★ Merck Index: An Encyclopedia of Chemicals and Drugs, 11th ed.
Susan Budavari, et al., eds.
Merck and Co.
Rahway, NJ
Frequency: 1989. **Price:** $35.00.

★ 2053 ★ MSDS Pocket Dictionary: What Does an MSDS Mean, Term Used on MSDS
Joseph O. Accrocco, ed.
Genium Publishing
Schenectady, NY
Frequency: 1988. **Price:** $3.75.

★ 2054 ★ New Jersey Department of Health Hazardous Substance Fact Sheets
Department of Health Right-to-Know Program
Trenton, NJ
Phone: (609)984-2202
Frequency: 1985-90. Series of more than 560 chemicalspecific fact sheets available in printed and electronic form via CD-ROM and online access.

★ 2055 ★ *1991/1992 Safe Home/Business Resource Guide*
Lloyd Publishing
New Canaan, CT
Frequency: 1991. *Price:* $22.50. Listings for alternative building products and personal care items, services, consultants, and other resources.

★ 2056 ★ *1992 Earth Journal Environmental Almanac and Resource Directory*
Editors of Buzzworm Magazine
Buzzworm Books
Boulder, CO
Frequency: 1992. *Price:* $7.95. 352 pp.

★ 2057 ★ *The 1992 Environmental Almanac*
World Resources Institute
Houghton-Mifflin
Boston, MA
Frequency: 1991. *Price:* $10.00. 606 pp.

★ 2058 ★ *NIOSH Pocket Guide to Chemical Hazards*
U.S. Dept. of Health and Human Services, National Institute for Occupational Safety and Health
U.S. Government Printing Office
Washington, DC
Frequency: Annual. *Price:* $7.00. 245 pp. Convenient source of information on general industrial hygiene and medical monitoring practices. NIOSH No. 90-117.

★ 2059 ★ *Nontoxic and Natural Guide for Consumers: How to Avoid Dangerous Everyday Products and Buy or Make Safe Ones*
Deborah Lynn Dadd
Jeremy P. Tarcher
Los Angeles, CA
Frequency: 1984. *Price:* $9.95. 287 pp. Chemical composition and label warnings of 1,000 products in 300 categories are discussed, as well as safe alternatives to hazardous chemicals and do-it-yourself hints.

★ 2060 ★ *The Nontoxic Home: Protecting Yourself and Your Family from Everyday Toxics and Health Hazards*
Deborah Lynn Dadd
Jeremy P. Tarcher
Los Angeles, CA
Frequency: 1986. *Price:* $9.95. 213 pp. Guide to safe alternatives to hazardous substances commonly used in household products.

★ 2061 ★ *Not in Our Backyards! Community Action for Health and the Environment*
Nicholas Freudenberg
Monthly Review Press
New York, NY
Frequency: 1984. *Price:* $26.00 hardbound, $10.00 paperback. 320 pp. What actions communities can take when hazardous waste sites are threatened to be installed nearby; with case histories.

★ 2062 ★ *Occupational Health Resource Guide*
Resource Center of Environmental and Occupational Health Sciences Institute
New Jersey Department of Health, University of Medicine and Dentistry of New Jersey, Robert Wood Johnson Medical School
Phone: (908)463-5353
Frequency: 1989. *Price:* $15.00. 256 pp. Exhaustive consolidation of information in the field of occupational health; a tool for those with occupational health responsibilities in management, labor organizations, government, and citizen activist groups.

★ 2063 ★ *On the Way to Market: Roadblocks to Reducing Pesticide Use in Produce*
Public Voice for Food and Health Policy
Washington, DC
Frequency: 1991. *Price:* $10.00. 39 pp. Report on the impact of pesticide use and organic farming on produce marketing.

★ 2064 ★ *Organic Chemicals in Drinking Water*
Association of State Drinking Water Administrators
Arlington, VA
Frequency: 1990. *Price:* Free. 8 pp.

★ 2065 ★ *OSHA Handbook for Small Businesses, revised ed.*
Occupational Safety and Health Administration
U.S. Department of Labor
Frequency: 1990. *Price:* Free. 56 pp. Assists small business employers in meeting the legal requirements imposed by the Occupational Safety and Health Administration (OSHA) Act of 1970 and achieving in-compliance status voluntarily prior to an inspection; also contains information on how to comply with rightto-know aspects for reporting hazardous materials in the workplace. OSHA 2209.

★ 2066 ★ *An Ounce of Toxic Pollution Prevention: Rating States' Toxics Use Reduction Laws*
William Ryan and Richard Schrader
National Environmental Law Center (Boston) and Center for Policy Alternatives (Washington, DC)
Frequency: 1984. *Price:* Free. 62 pp.

★ 2067 ★ *Pennsylvania Classroom Guide to Safety in the Arts*
Debra L. Oltman
Pennsylvania Department of Education
Harrisburg, PA
Frequency: 1990. *Price:* Free. 139 pp.

★ 2068 ★ *Pesticide Alert: A Guide to Pesticides in Fruit and Vegetables*
Lawrie Mott and Karen Snyder
National Resources Defense Council/Sierra Club Books
San Francisco, CA
Frequency: 1987. *Price:* $6.95. 128 pp.

★ 2069 ★ *Pesticide Book, 3rd ed.*
George W. Ware
Thomson Publishing
Frequency: 1989. *Price:* $27.00. 340 pp.

★ 2070 ★ *Pesticide Directory: A Guide to Producers and Products, Regulators, Researchers and Associations in the United States*
W. T. Thomson and Lori Thomson Harvey, eds.
Thomson Publications
Fresno, CA
Frequency: Annual. *Price:* $75.00.

★ 2071 ★ *Pesticide Handbook*
Peter Hurst, et al.
Unwin Hyman
New York, NY
Frequency: 1989. *Price:* $24.95. 128 pp.

★ 2072 ★ *Pesticide Information Pamphlet*
American Chemical Society
Washington, DC
Frequency: 1987. *Price:* Free. 12 pp.

★ 2073 ★ *Pesticide Policy and the Consumer: How We Safeguard, Grow and Market Our Food*
Public Voice for Food and Health Policy
Washington, DC
Frequency: 1989. *Price:* $2.50.

★ 2074 ★ *Pesticides: Helpful or Harmful?*
Leonard T. Flynn
American Council on Science and Health
New York, NY
Frequency: 1988. *Price:* $3.00. 64 pp.

★ 2075 ★ *Pocket Guide to MSDSs and Labels, Your Keys to Chemical Safety*
Business and Legal Reports
Madison, CT
Frequency: 1990. *Price:* $3.95. 60 pp.

★ 2076 ★ *Preventing Lead Poisoning in Young Children: A Statement by the Centers for Disease Control*
Center for Environmental Health, Chronic Disease Division
U.S Department of Health and Human Services, Centers for Disease Control
Atlanta, GA
Frequency: 1985.

★ 2077 ★ *Proctor and Hughes' Chemical Hazards of the Workplace, 3rd ed.*
Gloria Hathaway, Nick H. Proctor, James P. Hughes, and Michael Fischman
Van Nostrand Reinhold
New York, NY
Frequency: 1990. *Price:* $56.95. 900 pp. Introduction to the toxicology of more than 600 chemicals most likely to be encountered in the workplace.

★ 2078 ★ *Publications That Can Work for You!*
Technology Clearinghouse, Toxic Substances Control Program
California Department of Health Services
P.O. Box 942732
Sacramento, CA 94234-7320
Phone: (916)324-1807
Frequency: 1990. Listing of free reports and fact sheets available to businesses, individuals, and government agencies located in California.

★ 2079 ★ *Radon in Buildings*
B. Pathak
Canadian Centre for Occupational Health and Safety
Hamilton, ON, Canada
Frequency: 1989. *Price:* $3.50 U.S., $2.50 Canada. P89-3E.

★ 2080 ★ *Radon in Drinking Water*
Richard Cothern and Paul Rebers, eds.
Lewis Publishers
Boca Raton, FL
Frequency: 1990. *Price:* $59.95.

★ 2081 ★ *Radon in Ground Water*
Barbara Graves, ed.
Lewis Publishers/National Water Well Association
Boca Raton, FL
Frequency: 1987. *Price:* $55.00.

★ 2082 ★ *Rapid Guide to Hazardous Chemicals in the Workplace, 2nd ed.*
Richard J. Lewis, Sr.
Van Nostrand Reinhold
New York, NY
Frequency: 1990. *Price:* $33.95. 303 pp.

★ 2083 ★ Recognition and Management of Pesticide Poisonings, 4th ed.
Donald Morgan
U.S. Environmental Protection Agency
Frequency: 1989. *Price:* $8.00. EPA-540/9-88-001.

★ 2084 ★ Regulating Pesticides in Foods: The Delaney Paradox
Committee on Scientific and Regulatory Issues Underlying Pesticide Use Patterns and Agricultural Innovations
Board on Agriculture, National Research Council
Washington, DC
Frequency: 1987. *Price:* $19.95. 288 pp. Examines the contradictions between the U.S. Food and Drug Administration (FDA) and the U.S. Environmental Protection Agency pesticide regulations.

★ 2085 ★ Report of the Royal Commission on Matters of Health and Safety Arising from the Use of Asbestos in Ontario
Ontario Ministry of Government Services
Toronto, ON, Canada
3 vols. *Frequency:* 1984. Order from the Duplications Services Branch, Mail Order Service, 880 Bay St., Toronto, Ontario, Canada.

★ 2086 ★ Residues in Fresh Produce, 1989
Pest Enforcement Branch
California Department of Food and Agriculture
1220 N St.
Sacramento, CA 95814
Frequency: 1990. *Price:* Free. 60 pp.

★ 2087 ★ Resources for Organic Pest Control
Rodale Press
Emmaus, PA
Frequency: Biennial published in January of odd years. *Price:* Free. Lists manufacturers and suppliers of natural pesticides and beneficial organisms and the names of organizations providing information on natural pesticides.

★ 2088 ★ Safety at Home: A Guide to the Hazards of Lawn and Garden Pesticides and Safer Ways to Manage Pests
National Coalition Against the Misuse of Pesticides
Washington, DC
Frequency: 1991. *Price:* $11.00. 75 pp.

★ 2089 ★ Safety Catalog, 1990
Coastal Video Communications
Virginia Beach, VA
Frequency: Annual. *Price:* Free. Lists video programs on safety, industrial hygiene, and environmental training.

★ 2090 ★ SARA Title III: Regulations and Keyword Index
McCoy and Associates
Frequency: 1991. *Price:* $30.00 paperback. 309 pp. Purchase through Hazardous Materials Control Resources Institute, Greenbelt, MD.

★ 2091 ★ School Science Laboratories: A Guide to Some Hazardous Substances
Council of State Science Supervisors, U.S. Consumer Products Safety Commission
U.S. Government Printing Office
Washington, DC
Frequency: 1984. 60 pp. Identifies potentially hazardous chemicals that are used in school laboratories with guidelines for their proper storage, handling, use, and removal if necessary.

★ 2092 ★ Searching for Success— Environmental Success Index
John Jester
Renew America
Washington, DC
Frequency: 1991. *Price:* $25.00. 154 pp. Describes award-winning programs selected by the National Environmental Awards Council that have effectively protected, restored, or enhanced the environment in 20 categories.

★ 2093 ★ Shopping for a Better Environment: A Brand Name Guide to Environmentally Responsible Shopping
Laurence Tasaday with Katherine Stevenson
Meadowbrook Press
Deephaven, MN
Frequency: 1991. *Price:* $10.00. 341 pp. Discusses the environmental problems of a wide range of consumer products and provides the names of nontoxic and environmentally sound alternatives.

★ 2094 ★ Shopping for a Better World
Council on Economic Priorities
New York, NY
Frequency: Annual. Guide to socially responsible supermarket shopping.

★ 2095 ★ Silent Spring, 25th Anniversary Edition
Rachel Carson
Houghton Mifflin
Boston, MA
Frequency: 1987. *Price:* $17.95 hardbound, $8.95 paperback. 448 pp. First published in 1962, this work awakened the public to the indiscriminate misuse of pesticides and their harmful effect on the environment.

★ 2096 ★ Snake Venom Poisoning Wall Chart
American Association of Poison Control Center's Scientific Review Subcommittee and the Arizona Poison and Drug Information Center
Tucson, AZ
Frequency: 1990. *Price:* $8.50. 20'' x 24'' chart detailing assessment, laboratory testing, and first-aid and medical treatment of bites by venomous snakes in the United States. To order, write: Wall Chart, Arizona Poison Information Center, Arizona Health Sciences Center, Rm. 3204K, 1501 N. Campbell Ave., Tucson, AZ 85724.

★ 2097 ★ The Soil Chemistry of Hazardous Materials
James Dragun
Hazardous Materials Control Resources Institute
Greenbelt, MD
Frequency: 1988. *Price:* $65.00. 458 pp.

★ 2098 ★ Some Publicly Available Sources of Computerized Information on Environmental Health and Toxicology
Kathryn S. Deck and Sandra E. Bonzo
Center for Environmental Health and Injury Control, U.S. Centers for Disease Control
Atlanta, GA
Phone: (404)488-4588
Frequency: 1990. *Price:* Free.

★ 2099 ★ Sourcebook on Asbestos Diseases: Medical, Legal, and Engineering Aspects
George A. Peters and Barbara J. Peters
Garland Publishing
New York, NY
2 vols. *Frequency:* 1980-86. Covers legal and medical aspects of asbestos-related diseases, including texts of cases covering liability for asbestos-related diseases; contains lists of manufacturers with trade names.

★ 2100 ★ Superfund Deskbook
Environmental Law Reporter staff
Environmental Law Institute
Washington, DC
Frequency: 1989. *Price:* $85.00.

★ 2101 ★ Suspect Chemicals Sourcebook, 4th ed.
Synthetic Organic Chemical Manufacturers Association, ed.
Roytech Publications
Burlingham, CA
Frequency: 1985. Written for the professional who must comply with the Occupational Safety and Health Administration (OSHA) Chemical Hazard Communication Standard, this reference has indexes to 6,000 chemicals.

★ 2102 ★ Suspended, Canceled, and Restricted Pesticides
Office of Pesticides and Toxic Substances
U.S. Environmental Protection Agency
Washington, DC
Frequency: 1990.

★ 2103 ★ Title III Fact Sheet
U.S. Environmental Protection Agency
Washington, DC
Hotline: 800-535-0202
Frequency: 1988.

★ 2104 ★ Toxic and Hazardous Materials: A Sourcebook and Guide to Information Sources
James Webster, ed.
Greenwood
Westport, CT
Frequency: 1987. *Price:* $449.95. 431 pp. Guide to 1,600 information sources, including indexes, databases, associations, agencies, and research organizations.

★ 2105 ★ Toxicants in Consumer Products
Susan M. Ridgley
Household Hazardous Waste Disposal Project, Municipality of Metropolitan Seattle
Seattle, WA
Frequency: 1982. *Price:* Free. 228 pp. Metro Toxicant Program Report No. 1B.

★ 2106 ★ Toxic Chemicals, Health, and the Environment
Lester B. Lave and Arthur C. Upton, eds.
Johns Hopkins University Press
Baltimore, MD
Frequency: 1987. *Price:* $45.00 hardbound, $17.50 paperback. 336 pp. Collection of papers from a forum on toxic substances, stressing better communications about toxic effects on health.

★ 2107 ★ The Toxics Directory: References and Resources on the Health Effects of Toxic Substances
Hanafi Russell, ed.
Hazard Identification and Risk Assessment Branch, California Department of Health Services
Berkeley, CA
Frequency: 1990. *Price:* $5.50. Directory is available through the Publications Section, California Department of General Services, P.O. Box 1015, North Highlands, CA 95660, tel.: (916)973-3700. Pub. No. 7540-958-1300-3.

★ 2108 ★ *Toxic Substances Control Guide*
Mary Worobec and Girard Ordway
Bureau of National Affairs
Washington, DC
Frequency: 1988. *Price:* $35.00.

★ 2109 ★ *Toxic Terror*
Elizabeth M. Whelen
Jameson Books
Ottawa, IL
Frequency: 1985. *Price:* $18.95. 348 pp. Case studies of significant events concerning toxic substances.

★ 2110 ★ *Toxic Truth and Consequences: The Magnitude of and the Problems Resulting from America's Use of Toxic Chemicals*
Geoffrey Lomax, Marc Osten, and William Ryan
National Environmental Law Center (Boston) and U.S. Public Interest Research Group (Washington, DC)
Frequency: 1991. *Price:* Free. 36 pp.

★ 2111 ★ *Toxins in the Home and Safe Alternatives*
Maureen R. McClelland
New Alchemy Institute
East Falmouth, MA
Frequency: 1991. *Price:* $4.50. 20 pp. Discussion of hazardous and safe building materials and household products.

★ 2112 ★ *Tracking Hazardous Substances at Industrial Facilities: Engineering Mass Balance versus Materials*
National Research Council
National Academy Press

Frequency: 1990. *Price:* $18.00. 100 pp.

★ 2113 ★ *Understanding Toxic Substances: An Introduction to Chemical Hazards in the Workplace*
Jon Rosenberg, ed.
Hazard Evaluation System and Information Service, California Occupational Health Program
2151 Berkeley Way
Berkeley, CA 94704
Frequency: 1986. *Price:* Free. 38 pp.

★ 2114 ★ *Volatile Organic Chemicals: Are VOCs in Your Drinking Water?*
Office of Water
U.S. Environmental Protection Agency
Washington, DC
Frequency: 1989. *Price:* Free.

★ 2115 ★ *What Chemicals Each Industry Uses, A Source Book for Citizens*
Environmental Research Foundation
New York, NY
Frequency: 1988. *Price:* $25.00. 288 pp. Provides information on what kinds of chemicals are stored on-site and used by different industries; organized by Standard Industrial Classifications (SIC) and Chemical Abstract Service Numbers (CAS).

★ 2116 ★ *Winning the Right to Know: A Handbook for Toxics Activists*
Caron Chess
Delaware Valley Toxics Coalition
Frequency: 1983. *Price:* $7.95. 132 pp.

★ 2117 ★ *World Directory of Environmental Organizations*
Thaddeus C. Trzyna, ed.
California Institute of Public Affairs
Sacramento, CA
Frequency: Irregular, latest edition published 1989. *Price:* $35.00.

★ 2118 ★ *World Directory of Pesticide Control Organisations*
George Ekstrom and Hamish Kidd, eds.
Royal Society of Chemistry/CRC Press
Boca Raton, FL
Frequency: 1988. *Price:* $69.95.

★ 2119 ★ *World Environment Directory*
WED Division
Business Publishers
Silver Springs, MD
Frequency: Irregular, latest edition published 1991. *Price:* $225.00.

★ 2120 ★ *Worldwatch Paper 79: Defusing the Toxic Threat; Controlling Pesticides and Industrial Waste*
Sandra Postel
Worldwatch Institute
Washington, DC
Frequency: 1987. *Price:* $4.00 paperback. 36 pp.

★ 2121 ★ *Your Resource Guide to Environmental Organizations*
John Seredich, ed.
Smiling Dolphin Press
Irvine, CA
Frequency: 1991. *Price:* $15.95 paperback. 514 pp.

Magazines

★ 2122 ★ Abstracts on Health Effects of Environmental Pollutants
Biological Abstracts
2100 Arch St.
Philadelphia, PA 19103
Phone: (215)587-4800

★ 2123 ★ Air Pollution Consultant
McCoy and Associates, Inc.
13701 W. Jewel Ave., Ste. 202
Denver, CO 80228
Phone: (303)987-0333
Fax: (303)989-7917
Frequency: Bimonthly. Professional journal provides regulatory and technical information for the consultant.

★ 2124 ★ American Industrial Hygiene Association Journal
American Industrial Hygiene Association
345 White Pond Dr.
P.O. Box 8390
Akron, OH 44320
Phone: (216)873-AIHA
Frequency: Monthly. **Price:** $75/year. Professional journal contains scientific articles and technical reports relating to industrial hygiene, health in the workplace, detection and exposure limits of hazardous substances, human and animal toxicology, and product safety.

★ 2125 ★ Amicus Journal
Natural Resources Defense Council, Inc.
40 W. 20th St.
New York, NY 10011
Phone: (212)727-4412
Fax: (212)727-1773
Contact: Peter Borrelli, ed.
Frequency: Quarterly. **Price:** Free with membership; $10/year, nonmembers and libraries. Journal on national and international environmental policy.

★ 2126 ★ Applied Occupational and Environmental Hygiene
American Conference of Governmental
  Industrial Hygienists
6500 Glenway Ave., Bldg. D-7
Cincinnati, OH 45211-4438
Phone: (513)661-7881
Fax: (513)661-7195
Professional journal for governmental and university managers concerned with the administrative and technical aspects of health and safety in the workplace.

★ 2127 ★ Asbestos Issues
Mediacom, Inc.
760 Whalers Way, Ste. 100, Bldg. A
Fort Collins, CO 80525
Phone: (303)229-0029
Fax: (303)455-3773
Contact: Julie Larson Bricher, ed.
Frequency: Monthly.

★ 2128 ★ Buzzworm, The Environmental Journal
Buzzworm, Inc.
2305 Canyon Blvd., Ste. 206
Boulder, CO 80302
Phone: (303)442-1969
Fax: (303)442-4875
Contact: Joseph E. Daniel, ed.
Frequency: Bimonthly. **Price:** $18/year. Consumer magazine focusses on environmental issues, including the conservation of nature.

★ 2129 ★ Canadian Environmental Protection
Baum Publications Ltd.
831 Helmcken St.
Vancouver, BC, Canada V6Z 1B1
Phone: (604)689-2804
Fax: (604)682-8347
Contact: Len Webster, ed.

★ 2130 ★ Comments on Toxicology
Gordon and Breach Science Publishers
P.O. Box 786
Cooper Station
New York, NY 10276
Phone: (212)206-8900
Fax: (212)645-2459
Contact: A. Wallace Hayes and William O. Berndt, eds.
Frequency: Bimonthly.

★ 2131 ★ Co-op America Quarterly
Co-op America
2100 M St., N.W., Ste. 403
Washington, DC 20063
Phone: 800-424-2667
Price: Free with membership. Magazine covers strategies for making environmentally sound choices involving work, housing, sustainable agriculture, health, and community.

★ 2132 ★ ECON: The Environmental Magazine
PTN Publishing Inc.
445 Broad Hollow Rd., Ste. 21
Melville, NY 11747
Phone: (516)845-2700
Fax: (516)845-7109
Contact: Judith N. Hogan, ed.

★ 2133 ★ Ecotoxicology and Environmental Safety
Academic Press, Inc.
1250 Sixth Ave.
San Diego, CA 92101
Phone: (619)699-6825
Fax: (619)699-6800
Contact: Frederick Coulston and Friedhelm Korte, eds.
Frequency: Bimonthly. **Price:** $200/year, U.S. and Canada; $251/year, elsewhere. Reports research on the biologic and toxic effects of natural and synthetic chemical pollutants on animal, plant, and microbial systems.

★ 2134 ★ E Magazine
Earth Action Network, Inc.
P.O. Box 5098
Westport, CT 06881
Phone: (203)854-5559
Fax: (203)866-0602
Contact: Douglas Moss, ed.
Frequency: Bimonthly. **Price:** $20/year; $36/two years; $50/three years. Magazine provides news and commentary on a broad range of consumer environmental issues such as waste reduction, pesticides and other hazardous substances, nutrition, farming, wildlife, and current trends.

★ 2135 ★ Environmental Action Magazine
Environmental Action Foundation
1525 New Hampshire Ave., N.W.
Washington, DC 20036
Phone: (202)745-4870
Frequency: Bimonthly. **Price:** Included in membership.

★ 2136 ★ Environmental Business Journal
Environmental Business Publishing, Inc.
4452 Park Blvd., Ste. 306
San Diego, CA 92116
Phone: (619)295-7685
Fax: (619)295-5743
Contact: Grant Ferrier, editor-in-chief
Frequency: Monthly. **Price:** $395/year, U.S. and Canada; $445, elsewhere. Reports on strategic issues of environmental concern, including company profiles, business and financial statistics on companies in the environmental field, and current developments.

★ 2137 ★ Environmental Claims Journal
Executive Enterprises Publications, Inc.
22 W. 21st St.
New York, NY 10010
Phone: (212)645-7880
Fax: (212)645-8689
Summary of litigation claims and settlements in cases involving such environmental issues as toxic substances and wastes.

★ 2138 ★ *The Environmental Forum*
Environmental Law Institute
1616 P St. N.W., Ste. 200
Washington, DC 20036
Phone: (202)328-5150
Fax: (202)328-5002
Contact: Steve Dujack, ed.
Frequency: Bimonthly. **Price:** Included in membership; $60/year, nonmembers; $50/year, nonprofit, academic, and government bodies; $25/year, student rate. Professional journal provides in-depth analysis of environmental issues.

★ 2139 ★ *Environmental Health Perspectives*
U.S. Government Printing Office
Superintendent of Documents
Washington, DC 20402
Phone: (202)783-3238

★ 2140 ★ *Environmental Impact Assessment Review*
Elsevier Science Publishing Co.
655 Avenue of the Americas
New York, NY 10010
Phone: (212)989-5800
Fax: (212)633-3965

★ 2141 ★ *Environmental Periodicals Bibliography*
International Academy at Santa Barbara
800 Garden St., Ste. D
Santa Barbara, CA 93101-1552
Phone: (805)965-5010
Fax: (805)965-6071
Contact: Miriam Flacks, ed.
Frequency: Bimonthly. Professional journal provides tables of contents and subject and author indexes of articles from more than 380 environmental periodicals.

★ 2142 ★ *Environmental Protection Agency Journal*
Office of Public Affairs
U.S. Environmental Protection Agency
401 M St., S.W., A-107
Waterside Mall
Washington, DC 20460
Phone: (202)382-4359
Fax: (202)252-0231
Contact: John Heritage, ed.
Frequency: Monthly. **Price:** $11/year, U.S.; $13.75/year, elsewhere. Magazine contains articles on regulations and other environmental subjects and news about activities in the EPA. To order, write the Superintendent of Documents, U.S. Government Printing Office, Washington, DC 20040.

★ 2143 ★ *Environmental Review*
Citizens for a Better Environment
407 S. Dearborn, Ste. 1775
Chicago, IL 60605
Phone: (312)939-1530
Frequency: Quarterly. **Price:** Included in membership. Journal with in-depth background articles on emerging pollution issues in Illinois, Minnesota, and Wisconsin.

★ 2144 ★ *Environmental Science and Technology*
American Chemical Society
1155 16th St., N.W.
Washington, DC 20036
Phone: (202)872-4600
Contact: Russell F. Christman, ed.
Frequency: Monthly. **Price:** $36/year, members; $67/year, nonmembers. Scientific journal contains research articles on environmental chemistry, hazardous materials, and scientific aspects of environmental management.

★ 2145 ★ *Environmental Toxicology and Chemistry*
Society of Environmental Toxicology and Chemistry
1101 14th St., N.W., Ste. 1110
Washington, DC 20005
Phone: (202)371-1275
Fax: (202)371-1090

★ 2146 ★ *Environment Today*
Enterprise Communications Inc.
1905 Powers Ferry Rd., Ste. 120
Marietta, GA 30067
Phone: (404)988-9558
Fax: (404)859-9166
Contact: Paul Harris, ed.

★ 2147 ★ *Florida Environments*
Florida Environments Publishing Co.
215 N. Main St.
P.O. Box 1617
High Springs, FL 32643-1617
Phone: (904)454-2007
Fax: (904)454-3113
Frequency: Monthly. **Price:** $14.95/year.

★ 2148 ★ *Food Additives and Contaminants*
Taylor and Francis
79 Madison Ave.
New York, NY 10016
Phone: (212)725-1999; 800-821-8312
Frequency: Bimonthly. **Price:** $250/year. Professional journal publishes research papers and reviews of manmade additives and contaminants in the food chain.

★ 2149 ★ *Fundamental and Applied Toxicology: An Official Journal of the Society of Toxicology*
Society of Toxicology
1101 14th St., N.W., Ste. 1110
Washington, DC 20005
Phone: (202)371-1393
Fax: (202)371-1090
Contact: Bernard A. Schwetz, ed.
Frequency: 8x/year. **Price:** Included in membership; $200/year, nonmembers in U.S. and Canada; $236/year, elsewhere. Scientific journal contains articles assessing the risk to human and animal health of exposure to toxic agents.

★ 2150 ★ *Garbage*
Old House Journal Corp.
435 Ninth St.
Brooklyn, NY 11215
Phone: (718)788-1700
Fax: (718)788-9051
Contact: Patricia Poor, ed.
Frequency: Bimonthly. **Price:** $21/year. Consumer magazine covers all aspects of living an environmentally sound lifestyle.

★ 2151 ★ *GATFWORLD*
Graphic Arts Technical Foundation
4615 Forbes Ave.
Pittsburgh, PA 15213-3796
Phone: (412)621-6941
Fax: (412)621-3049
Contact: Thomas M. Destree, Managing ed.
Frequency: Bimonthly. **Price:** Included in membership. Magazine covers activities of the Foundation and other environmental concerns of the graphic arts industry.

★ 2152 ★ *Greenpeace Magazine*
Greenpeace USA
1436 U St., N.W.
Washington, DC 20009
Phone: (202)462-1177
Contact: Andre Carothers, editor-in-chief
Frequency: Bimonthly. **Price:** Included in membership in Greenpeace USA. Covers world-wide environmental issues and the activities of Greenpeace.

★ 2153 ★ *Hazard Monthly*
Research Alternatives, Inc.
966 Hungerford Dr., Ste. 1
Rockville, MD 20850
Phone: (301)424-2803
Fax: (301)738-1026
Contact: K. C. Chartrand, ed.
Frequency: Monthly. **Price:** $29/year; $39/year, institutions and libraries. Tabloid discusses management of natural and manmade disasters.

★ 2154 ★ *Hazardous Materials Control*
Hazardous Materials Control Research Institute
7237 Hanover Parkway
Greenbelt, MD 20770
Phone: (301)982-9500
Fax: (301)220-3870
Contact: Christopher Hoezel, ed.
Frequency: Bimonthly. **Price:** Free with membership; $18/year, $5 single issue. Professional magazine concerned with the management and disposal of hazardous wastes and other items of interest to the hazardous materials management industry.

★ 2155 ★ *HazMat World*
Tower-Borner Publishing Inc.
800 Roosevelt Rd., Bldg. C, Ste. 206
Glen Ellyn, IL 60137-5851
Phone: (708)858-1888
Fax: (708)858-1957
Contact: Jim Bishop, ed.
Frequency: Monthly. **Price:** Free upon request. Professional journal covers hazardous materials and hazardous waste management issues, including regulatory and enforcement trends, emergency response training, insurance and liability, and case histories.

★ 2156 ★ *Industrial Hygiene News*
Rimbach Publishing, Inc.
8650 Babcock Blvd.
Pittsburgh, PA 15237
Phone: 800-245-3182; (412)364-5366
Frequency: Bimonthly. **Price:** Free. Magazine provides product information, book reviews, and buying guides for industrial hygiene and toxicology managers.

★ 2157 ★ *Industrial Safety and Hygiene News*
Chilton Company
Chilton Way
Radnor, PA 19089
Phone: (215)964-4055
Contact: David Johnson, ed.
Frequency: Monthly. **Price:** Free upon request. Tabloid reports on new products, research developments, and industry news for people involved with industrial safety and hygiene, security, hazardous materials, and similar areas.

★ 2158 ★ *Journal of American College of Toxicology*
American College of Toxicology
Mary Ann Liebert, Inc.
1651 Third Ave.
New York, NY 10128
Phone: (212)289-2300
Frequency: Bimonthly. *Price:* $157/year.

★ 2159 ★ *Journal of Environmental Science and Health. Part B: Pesticides, Food Contaminants, and Agricultural Wastes*
Marcel Dekker, Inc.
270 Madison Ave.
New York, NY 10016
Phone: 800-228-1160; (212)696-9000
Fax: (212)685-4540
Frequency: Bimonthly. *Price:* $360/year.

★ 2160 ★ *Journal of Environmental Science and Health. Part C: Environmental Carcinogenesis Reviews*
Marcel Dekker, Inc.
270 Madison Ave.
New York, NY 10016
Phone: 800-228-1160; (212)696-9000
Fax: (212)685-4540
Frequency: Twice a year. *Price:* $89.50/year, individuals; $185/year, institutions.

★ 2161 ★ *Journal of Pesticide Reform*
Northwest Coalition for Alternatives to Pesticides
P.O. Box 1393
Eugene, OR 97440
Phone: (503)344-5044
Contact: Caroline Cox, ed.

★ 2162 ★ *Journal of the Air and Waste Management Association*
Air and Waste Management Association
Three Gateway Center
Four West
Pittsburgh, PA 15222
Phone: (412)232-3444
Fax: (412)232-3450
Contact: Harold M. Englund, ed.
Frequency: Monthly. *Price:* $90/year, nonprofit libraries and institutions; $200/year, all others. Magazine covers technical aspects of air pollution and activities of the association.

★ 2163 ★ *Journal of Toxicology and Environmental Health*
Hemisphere Publishing Co.
1101 Vermont Ave., N.W., Ste. 200
Washington, DC 20005
Phone: (202)289-2174
Fax: (202)289-3665
Frequency: Monthly. *Price:* $172/year, individuals; $598/year, institutions. Medical

research journal publishes papers on environmental factors affecting health.

★ 2164 ★ *Journal on Hazardous Waste and Hazardous Materials*
Hazardous Materials Control Research Institute
7237 Hanover Pkwy.
Greenbelt, MD 20770
Phone: (301)587-9390
Frequency: Quarterly. *Price:* $99/year.

★ 2165 ★ *NAC Journal*
National Asbestos Council, Inc.
1777 Northeast Expressway, Ste. 150
Atlanta, GA 30329
Phone: (404)633-2622
Fax: (404)633-5714
Contact: Karen S. Weber, ed.
Frequency: Quarterly. *Price:* Free with membership; $30/year, U.S.; $35/year, Canada; $50/year, elsewhere. Professional journal contains technical articles concerning asbestos and other environmental health hazards to occupants of buildings, industrial sites, and other facilities.

★ 2166 ★ *Occupational Hazards*
Penton Publishing Co.
1100 Superior Ave.
Cleveland, OH 44114
Phone: (216)696-7000
Fax: (216)696-8765
Contact: Stephen G. Minter, ed.
Frequency: Monthly. *Price:* Free upon request. Magazine covers industrial safety, occupational hygiene and health, first aid, housekeeping, medical care, training, and similar fields of interest to health and safety managers in industrial and government installations.

★ 2167 ★ *Occupational Health and Safety News*
Stevens Publishing Corp.
P.O. Box 2573
Waco, TX 76702
Phone: (817)776-9000
Fax: (817)776-9018
Contact: Greg Densmore, ed.
Frequency: Monthly. *Price:* Free. Magazine carries news about occupational safety and health, legal developments, and technical information for occupational physicians, managers, nurses, and hygienists.

★ 2168 ★ *Pest Control Technology*
Gie, Inc.
4012 Bridge Ave.
Cleveland, OH 44113
Phone: (216)961-4130
Frequency: Monthly. *Price:* $30/year. Trade

magazine of the National Pest Control Association provides technical articles about pests and pesticides.

★ 2169 ★ *Public Citizen*
Public Citizen Foundation
2000 P St., N.W.
Washington, DC 20036
Phone: (202)293-9142
Contact: Ana Radelet, ed.
Price: Free with membership. Presents articles of interest to environmental activists in all fields, including the use of pesticides and the dangers of hazardous materials in commonly used products.

★ 2170 ★ *Texas Environmental News*
AES Marketing, Inc.
760 Whalers Way, Ste. 100-A
Fort Collins, CO 80525
Frequency: Monthly. *Price:* $18/year.

★ 2171 ★ *Toxicity Assessment*
John Wiley & Sons, Inc.
605 Third Ave.
New York, NY 10016
Phone: (212)850-6289

★ 2172 ★ *Toxicology and Applied Pharmacology*
Society of Toxicology
1101 14th St., N.W., Ste. 1110
Washington, DC 20005
Phone: (202)371-1393
Fax: (202)371-1090
Frequency: 15x/year. *Price:* Included in membership; $610/year, nonmembers in U.S. and Canada; $729, elsewhere.

★ 2173 ★ *Toxic Substances Journal*
Hemisphere Publishing Corp.
1101 Vermont Ave., N.W.
Washington, DC 20005
Phone: 800-821-8312; (202)289-2174
Fax: (202)289-3665
Contact: George S. Dominguez, ed.
Frequency: Quarterly. *Price:* $60/year; $120/year, institutions. Professional journal covers legislation, testing, and guidelines relating to toxic substances.

★ 2174 ★ *Toxic Substances Report*
Friends of the Everglades
202 Park St.
Miami Springs, FL 33166
Phone: (305)888-1230

Newsletters

★ 2175 ★ ACS Washington Alert
American Chemical Society
1155 16th St., N.W.
Washington, DC 20036
Phone: 800-227-5558
Contact: Keith Bilton, ed.
Frequency: Biweekly. **Price:** $42/year, members; $129/year, nonmembers. Covers government regulatory and legislative activities concerning such chemical issues as hazardous waste, air and water quality, toxic substances, and biotechnology.

★ 2176 ★ Advocacy Update
Public Voice for Food and Health Policy
1001 Connecticut Ave., N.W., Ste. 522
Washington, DC 20036
Phone: (202)659-5930
Contact: Jodie Silverman, ed.
Frequency: 11x/year. **Price:** Included in membership; $100/year, nonmembers. Covers health, nutrition, and food safety policy issues.

★ 2177 ★ Agrarian Advocate
California Action Network
P.O. Box 464
Davis, CA 95617
Phone: (916)756-8518
Contact: Judith Redmond, ed.
Frequency: Quarterly. **Price:** Included in membership. Educates rural residents of California on issues such as pesticide safety, organic farming practices, and ground water quality.

★ 2178 ★ Air Pollution Titles: A Guide to Current Air Pollution Literature
Center for Air Environment Studies,
Environmental Resources Research
Institute
Pennsylvania State University
226 Fenske Laboratory
University Park, PA 16802
Phone: (814)865-1415
Contact: Elizabeth J. Carroll, ed.
Frequency: Bimonthly. **Price:** $120/year, U.S.; $130/year, elsewhere; microfiche: $65/year, U.S. and Canada; $70/year, elsewhere. Spiral-bound guide to current air pollution literature covers more than 1,000 American and foreign journals and technical reports on research projects conducted or sponsored by the U.S. Government; includes bibliographies.

★ 2179 ★ Air Toxics Report
Business Publishers, Inc.
951 Pershing Dr.
Silver Springs, VA 20910
Phone: 800-274-0122; (301)587-6300
Fax: (301)587-1081
Frequency: Monthly. **Price:** $250/year.

Reports on changes in U.S. Government regulations, compliance, violations, legal actions, and administrative developments concerning the Clean Air Act, Toxic Substances Control Act (TSCA), Resources Conservation and Recovery Act (RCRA), Occupational Safety and Health Administration (OSHA), and FIFRA, for transportation, storage, and manufacturing professionals. Available online through Predicasts, Inc.

★ 2180 ★ Air/Water Pollution Report
Business Publishers, Inc.
951 Pershing Dr.
Silver Springs, MD 20910-4464
Phone: 800-274-0122; (301)587-6300
Fax: (301)587-1081
Frequency: Weekly. **Price:** $507.50/year. Focuses on legislation, regulations, amendments, and other provisions of the Clean Air Act, Clean Water Act, Safe Drinking Water Act, Superfund, hazardous waste management, toxics in air and water, and waste water funding. Available online through NewsNet and Predicasts, Inc.

★ 2181 ★ ANJEC Report
Association of the New Jersey Environmental
Control Commissions
300 Mendham Rd.
P.O. Box 157
Mendham, NJ 07945
Phone: (201)539-7547
Fax: (201)539-7713
Contact: Susan Hanna, ed.
Frequency: Quarterly. **Price:** Included in membership. Contains articles on current environmental subjects concerning New Jersey.

★ 2182 ★ Arizona Poison and Drug Information Center Newsletter
Arizona Poison and Drug Information Center
University of Arizona, College of Pharmacy
Tucson, AZ 85721
Phone: (602)626-6016
Price: Free upon request. Covers issues concerning toxic substances, pesticides, and air and water quality in the Southwest.

★ 2183 ★ Art Hazards News
Center for Safety in the Arts
5 Beekman St., Ste. 1030
New York, NY 10038
Phone: (212)227-6220
Contact: Michael McCann, ed.
Frequency: 10x/year. **Price:** $18.50/year, U.S.; $21.50/year, Canada; $22.50/year, elsewhere. Concerned with regulatory and legislative developments and health and safety hazards in the arts and crafts materials fields,

including dance, theater, and museum conservation.

★ 2184 ★ Asbestos Abatement Report
Bureau of National Affairs, Inc. (BNA)
1350 Connecticut Ave., N.W., Ste. 1000
Washington, DC 20036
Phone: (202)452-7889
Fax: (202)862-0990
Contact: Ken Doyle, ed.
Frequency: Biweekly. **Price:** $327/year, U.S. and Canada; $349, elsewhere. Covers issues relating to the removal or containment of asbestos, such as legislation, regulations, enforcement, litigation, liability insurance, and costs.

★ 2185 ★ Asbestos Control Report
Business Publishers, Inc.
951 Pershing Dr.
Silver Springs, MD 20910-4464
Phone: 800-274-0122; (301)587-6300
Fax: (301)587-1081
Contact: Bryan Lee, ed.
Frequency: Biweekly. **Price:** $355/year. Covers activities and developments at the federal and state levels on asbestos control techniques, work site health and safety, medical research findings, waste disposal, contracts, liability insurance concerns, and pending litigation and judicial decisions under the Asbestos Hazard Emergency Response Act of 1986. Available online through Newsnet and Predicasts, Inc.

★ 2186 ★ Asbestos Litigation Reporter
Andrews Publications, Inc.
1646 West Chester Pike
P.O. Box 1000
Westtown, PA 19395
Phone: 800-345-1101; (215)399-6600
Fax: (215)399-6610
Contact: Harry G. Armstrong, Exec. ed.
Frequency: Biweekly. **Price:** $900/year. Focuses on the most recent developments in lawsuits alleging personal injuries from exposure to asbestos and on suits involving the recovery of costs for removing asbestos building material.

★ 2187 ★ Asbestos Property Litigation Reporter
Andrews Publications, Inc.
1646 West Chester Pike
P.O. Box 1000
Westtown, PA 19395
Phone: 800-345-1101; (215)399-6600
Fax: (215)399-6610
Contact: Harry G. Armstrong, Executive ed.
Frequency: Biweekly. **Price:** $600/year. Provides information on lawsuits concerning the

recovery of costs for removing or encapsulating asbestos-containing building materials.

★ 2188 ★ Asbestos Victims of America Advisor
Asbestos Victims of America
P.O. Box 550
Capitola, CA 95010
Phone: (408)476-3646
Contact: H. R. Maurer, ed.
Frequency: Quarterly. **Price:** Included in membership. Covers the hazards of asbestos and asbestos-related diseases and provides information on asbestos abatement and emergency damage control.

★ 2189 ★ Asbestos Watch
White Lung Association
P.O. Box 1483
Baltimore, MD 21203-1483
Phone: (301)243-5864
Contact: D.G. Soutter, ed.
Frequency: Monthly. **Price:** Included in membership. Provides information for victims of asbestos, including the course of litigation and insurance claims for victims.

★ 2190 ★ BNA's National Environment Watch
Bureau of National Affairs, Inc. (BNA)
1231 25th St., N.W.
Washington, DC 20037
Phone: 800-372-1033; (202)452-4200
Fax: (202)822-8092
Frequency: Weekly. **Price:** $450/year.

★ 2191 ★ The Book of Chemical Lists
Business Legal Reports, Inc.
39 Academy St.
Madison, CT 06443-1513
Phone: 800-727-5257
Fax: (203)245-2559
Frequency: Annual. **Price:** $159.95/year. Loose-leaf service provides information on chemicals regulated by the federal and state governments.

★ 2192 ★ California Environmental Law Handbook, 4th ed.
Government Institutes, Inc.
966 Hungerford Dr., No. 24
Rockville, MD 20850
Phone: (301)251-9250
Fax: (301)251-0638
Contact: Charlene Ikonomou, Managing ed.
Frequency: Irregular. **Price:** $64/year. Looseleaf service.

★ 2193 ★ Canadian Occupational Health and Safety News
Corpus Information Services
Southern Business Information and Communications Group, Inc.
1450 Don Mills Rd.
Don Mills, ON, Canada M3B 2X7
Phone: (416)445-6641
Contact: Tov Rasmussen and Angela Stelmakowich, eds.
Frequency: Weekly. **Price:** $307/year. Provides information on new legal requirements, approaches to reducing hazards, and training and education alternatives for the workplace.

★ 2194 ★ CBE Environmental Review, Midwest Edition
Citizens for a Better Environment
33 E. Congress St., Ste. 523
Chicago, IL 60605
Phone: (312)939-1530
Contact: Sharon McGowan, ed.
Frequency: Quarterly. **Price:** Free with

membership; $25/year, nonmembers. Covers such environmental health issues as toxins in the home, pesticides, and air and water pollution.

★ 2195 ★ CBE Environmental Review, Western Edition
Citizens for a Better Environment
924 Market St., Ste. 505
San Francisco, CA 94102
Phone: (415)788-0690
Contact: Hannah Creighton, ed.
Frequency: Quarterly. **Price:** Free with membership; $25/year, nonmembers. Covers such environmental health issues as toxins in the home, pesticides, and air and water pollution.

★ 2196 ★ CEM Report
Center for Environmental Management
Tufts University
Curtis Hall
474 Boston Ave.
Medford, MA 02155
Phone: (617)381-3486
Contact: Kate Philbin, ed.
Frequency: Quarterly. **Price:** Free upon request. Provides news items and articles on the activities of the Center.

★ 2197 ★ Center News
Alaska Center for the Environment
519 W. Eighth St., Ste. 201
Anchorage, AK 99501
Phone: (907)274-3621
Fax: (907)274-4145
Contact: Cliff Eames, Issues Dir.
Frequency: 5x/year. **Price:** $30/year; included in membership. Discusses hazardous and toxic wastes—particularly oil and gas industry wastes—on the regional and national level, Alaska land issues, and alternative energy policy.

★ 2198 ★ Chemical Newsletter
National Safety Council
44 N. Michigan Ave.
Chicago, IL 60611
Phone: (312)527-4800
Fax: (312)527-9391
Frequency: Bimonthly. **Price:** Included in membership; $9.98/year, nonmembers. Concerned with occupational safety in the chemical industry and the overall safety of chemical products; includes research news and calendar of events.

★ 2199 ★ Chemical Regulation Reporter
Bureau of National Affairs, Inc. (BNA)
1231 25th St. N.W.
Washington, DC 20037
Phone: 800-372-1033; (202)452-4430
Fax: (202)822-8092
Frequency: Weekly. **Price:** $1,326/year. Looseleaf reference service concerning chemical regulations provides information on congressional hearings and legislation, administrative regulations and rulings, and litigation decisions, opinions, and settlements. Available online through HRIN, LEXIS, and WESTLAW.

★ 2200 ★ Chemical Regulatory Cross-reference
J. J. Keller Associates, Inc.
3003 W. Breezewood Lane
P.O. Box 368
Neenah, WI 54957
Phone: 800-327-6868
Contact: Linda Wereley, ed.

Price: $59.95/year. Loose-leaf reference service concerned with chemical regulations.

★ 2201 ★ Chemtox Communique
The Chemtox System
Division of Resource Consultants, Inc.
7121 Cross Roads Blvd.
P.O. Box 1848
Brentwood, TN 37027
Phone: (615)373-5040; (609)737-9009
Fax: (615)370-4339
Contact: Richard P. Pohanish, ed.
Frequency: Quarterly. **Price:** Free to Chemtox users; $50/year, nonusers in the U.S.; $60/year, elsewhere. Features general news of U.S. Environmental Protection Agency, Occupational Safety and Health Administration (OSHA), and the Chemtox System.

★ 2202 ★ Clean Water Action News
Clean Water Action
317 Pennsylvania Ave., S.E., Ste. 200
Washington, DC 20003
Phone: (202)638-1196
Contact: David Zwick, ed.
Frequency: Quarterly. **Price:** $24/year, individuals; $40/year, institutions. Discusses pending and recent legislation at the national and state levels on issues affecting water quality.

★ 2203 ★ Clearinghouse Bulletin
Carrying Capacity Network
1325 G St., N.W., Ste. 1003
Washington, DC 20005
Phone: (202)879-3044
Contact: Stephen M. Mabley, ed.
Frequency: Monthly. **Price:** Free with membership; $25/year, nonmembers. Provides information on activities of environmental activist groups and updates of government policies and legislation.

★ 2204 ★ Common Sense Pest Control Quarterly
The Bio-Integral Resources Center
P.O. Box 7414
Berkeley, CA 94707
Phone: (415)524-2567
Contact: S. Daar, H. Olkowski, and W. Olkowski, eds.
Price: $30/year, individuals; $50/year, institutions in U.S. and Canada; $40/year, individuals elsewhere; $60/year, institutions elsewhere. Evaluates alternative pest-control strategies.

★ 2205 ★ Community and Worker Right-to-Know News
Thompson Publishing Group
1725 K St., N.W., Ste. 200
Washington, DC 20006
Phone: 800-424-2959
Fax: (301)543-2921
Contact: Lucy Caldwell-Stair, ed.
Frequency: 24x/year. **Price:** $349/year. Looseleaf reference covers federal, state, local, and industry activity concerning community right-to-know and emergency response issues.

★ 2206 ★ The Community Plume
Friends of the Earth
218 D St., S.E.
Washington, DC 20003
Phone: (202)544-2600
Fax: (202)543-4710
Contact: Fred Miller, ed.
Frequency: Quarterly. **Price:** Free with membership. Reports on accidents, legislative developments, and other Title III concerns for members of Local Emergency Planning

Committees (LEPC), fire fighters, first responders, and other citizens involved in toxic chemical incidents.

★ 2207 ★ Community Right-to-Know Compliance Handbook
Business Legal Reports, Inc.
39 Academy St.
Madison, CT 06443-1513
Phone: 800-727-5257
Fax: (203)245-2559
Price: $179.95/year. Loose-leaf reference guide concerned with compliance with SARA Title III regulations.

★ 2208 ★ Community Right-to-Know Manual: A Guide to SARA Title III
Thompson Publishing Group
1725 K St., N.W., Ste. 200
Washington, DC 20006
Phone: 800-424-2959
Fax: (301)543-2921
Contact: Lucy Caldwell-Stair, ed.
Frequency: Monthly. **Price:** $274/year. Looseleaf reference.

★ 2209 ★ Compost Patch
Compost Patch, Inc.
306 Coleridge Ave.
Altoona, PA 16602
Phone: (814)946-9291
Contact: Charles Leiden, ed. and dir.
Price: Free with membership. Reviews new publications in the environmental field and describes new organizations and programs.

★ 2210 ★ The Conscious Consumer
The New Consumer Institute
P.O. Box 51
Wauconda, IL 60084
Phone: (708)526-0522
Contact: John F. Wasik, ed.
Frequency: Bimonthly. **Price:** $29.95/year. Reports on environmental issues affecting consumer products and services.

★ 2211 ★ Conservation Foundation Letter
The Conservation Foundation, Inc.
1250 24th St., N.W.
Washington, DC 20037
Phone: (202)293-4800
Fax: (202)293-9211
Contact: Jonathan Adams, ed.
Frequency: Bimonthly. **Price:** Free with membership. Covers a wide variety of environmental concerns, including pesticides, solar energy, and technology.

★ 2212 ★ Consolidated Chemical Regulation Guidebook: Federal/State/International Requirements
Executive Enterprises Publications Co., Inc.
22 W. 21st St.
New York, NY 10010-6904
Phone: (212)645-7880
Fax: (212)645-8689
Contact: Andrew Wald, ed.
Frequency: 3x/year. **Price:** $495/year. Looseleaf service provides regulatory information on more than 11,000 chemicals.

★ 2213 ★ Consumer News
U.S. Office of Consumer Affairs
1620 L St., N.W., Ste. 700
Washington, DC 20036
Phone: (202)634-4310
Contact: Theresa Michel, Dir. of Public Affairs
Frequency: Monthly. **Price:** Free upon request. Summarizes the administration's consumer activities and related areas of activities.

★ 2214 ★ Council Currents
National Asbestos Council, Inc.
1777 Northeast Expressway, Ste. 150
Atlanta, GA 30329
Phone: (404)633-2622
Fax: (404)633-5714
Contact: Laura Tener, ed.
Frequency: Bimonthly. **Price:** Free with membership; $15/year, nonmembers in the U.S.; $18/year, Canada; $25/year, elsewhere. Covers Council news and timely issues concerning asbestos in the environment.

★ 2215 ★ Cycle
Environmental Action Coalition (EAC)
625 Broadway, Second Floor
New York, NY 10012
Phone: (212)677-1601
Contact: Jennie Tichenor, ed.
Frequency: Quarterly. **Price:** Free with membership. Covers EAC programs, primarily in the New York City area.

★ 2216 ★ DA Hazardous Waste Hotline
Deuel and Associates, Inc.
7208 Jefferson St., N.E.
Albuquerque, NM 87109
Phone: (505)345-8732
Contact: N. Carol Evans, ed.
Frequency: Monthly. **Price:** $43/year, U.S. and Canada. Reports on national and regional hazardous waste issues, including developments in regulatory policy and treatment technology and standards.

★ 2217 ★ Dangerous Properties of Industrial Materials Report
Van Nostrand Reinhold
115 Fifth Ave.
New York, NY 10003
Phone: (212)254-3232
Fax: (212)673-1239
Frequency: Bimonthly. **Price:** $195/year, U.S.; $235/year, elsewhere. Chemical and environmental review of hazardous industrial materials.

★ 2218 ★ The Delicate Balance
National Center for Environmental Health Strategies
1100 Rural Ave.
Voorhees, NJ 08043
Phone: (609)429-5358
Contact: Mary Lamielle, ed.
Frequency: Quarterly. **Price:** Free with membership. Concerned with coping with environmental and occupational health problems caused by chemical sensitivity.

★ 2219 ★ ECB Newsletter
Environmental Conservation Board
Graphic Communications Industries, Inc.
4615 Forbes Ave.
Pittsburgh, PA 15213
Phone: (412)621-6941
Contact: Gary A. Jones, ed.
Frequency: Bimonthly. **Price:** Included in membership. Covers environmental and occupational safety and health regulations as they affect the printing industry.

★ 2220 ★ Econews
Northcoast Environmental Center
879 Ninth St.
Arcata, CA 95521
Phone: (707)822-6918
Contact: Sid Dominitz, ed.
Frequency: Monthly, except January. **Price:** Included in membership. Concerned with such issues as water resources, forestry, wildlife, and nuclear energy affecting Northwest California. Available online through ECONET.

★ 2221 ★ EDF Letter
Environmental Defense Fund
257 Park Ave., S., 16th Floor
New York, NY 10010
Phone: (212)505-2100
Contact: Norma H. Watson, ed.
Frequency: Monthly. **Price:** Free with membership; free single copy to nonmembers; $20/year, nonmembers. Reports on activities of the EDF and on national and global environment problems in the areas of toxic chemical use and regulation, air quality, water resources management, and wildlife protection.

★ 2222 ★ EI Digest—Industrial and Hazardous Waste Management
Environmental Information, Ltd.
4801 W. 81st St., Ste. 119
Minneapolis, MN 55437
Phone: (612)831-2473
Contact: Cary L. Perket, Sr., ed.
Frequency: Monthly. **Price:** $400/year. Booklets with information on a broad spectrum of hazardous waste issues, including company and facility profiles and analyses of regulations and state programs.

★ 2223 ★ EI Environmental Services Directory
Environmental Information, Ltd.
4801 W. 81st St., Ste. 119
Minneapolis, MN 55437
Phone: (612)831-2473
Frequency: Annual. **Price:** National edition, $290/year; regional editions, $75/year each. Directory of environmental services organized by state containing state regulatory information, hazardous waste facilities, transporters, spill response firms, consultants, laboratories, and special services and suppliers.

★ 2224 ★ ENFO
Environmental Information Center
Florida Conservation Foundation, Inc.
1251-B Miller Ave.
Winter Park, FL 32789
Phone: (305)644-5377
Contact: Gerald Grow, ed.
Frequency: Bimonthly. **Price:** Included in membership. Covers a wide variety of environmental subjects as they affect the state of Florida, such as toxic industrial wastes, wildlife, water quality, and land use.

★ 2225 ★ The Environmental Audit Handbook
Business Legal Reports, Inc.
39 Academy St.
Madison, CT 06443-1513
Phone: 800-727-5257
Fax: (203)245-2559
Price: $159.95/year. Loose-leaf reference concerned with how to do an environmental audit program.

★ 2226 ★ Environmental Compliance Manual
Business Legal Reports, Inc.
39 Academy St.
Madison, CT 06443
Phone: 800-727-5257
Fax: (203)245-2559
Contact: Faith Gavin Kuhn, ed.
Frequency: Bimonthly. **Price:** $595/year. Loose-leaf reference guide concerned with complying with the environmental rules and regulations established by the U.S. Government and 17 states.

★ 2227 ★ *Environmental Control News for Southern Industry: Southern Pollution Control Newsletter*
E. F. Williams Associates, Inc.
3637 Park Ave., Ste. 224
Memphis, TN 38111
Phone: (901)458-4696
Contact: E. F. Williams III and A. H. Williams, eds.
Frequency: Monthly. *Price:* $29.95/year, U.S. and Canada; $38/year, elsewhere. Reports regulatory, legislative, and technical developments of environmental concerns in the South, plus activities of environmental groups.

★ 2228 ★ *Environmental Control Report*
Graphic Arts Technical Foundation, Inc.
(GATF)
4615 Forbes Ave.
Pittsburgh, PA 15213
Phone: (412)621-6941
Contact: Gary A. Jones, ed.
Frequency: 3-4x/year, consolidated into the GATF's magazine. *Price:* Included in membership; $7.50/year, nonmembers. Covers environmental control regulation and legislation in the printing industry, including air and water pollution, toxic substances, and hazardous waste.

★ 2229 ★ *The Environmental Guardian*
C.U.R.E. Formaldehyde Poisoning Association, Inc.
Winonia, MN 55387-9583
Phone: (612)442-4665
Contact: Judith Ulseth, ed.
Frequency: Quarterly. *Price:* Included in membership. Provides information concerning formaldehyde, including products, state coordination officials, and medical and legal options for victims of formaldehyde poisoning.

★ 2230 ★ *Environmental Health Choices*
Environmental Health Service
New Jersey State Department of Health, Division of Occupational and Environmental Health
CN-360, Rm. 706
Trenton, NJ 08625-0360
Phone: (609)633-2043
Contact: D. L. Kiel, B. Bishop, and L. A. Pyrch, eds.
Frequency: 2x/year. *Price:* Free upon request. Provides public health information on environmental issues to citizens of New Jersey. Available electronically through the Environmental Health Bulletin Board of the Public Health Network.

★ 2231 ★ *Environmental Health News*
University of Washington
School of Public Health
Department of Environmental Health
461 Health Science Bldg.
Seattle, WA 98195
Phone: (206)543-6955
Fax: (206)543-8123
Contact: David Kalman, ed.
Frequency: 3x/year. *Price:* Free upon request. Covers industrial hygiene issues, hazardous and toxic materials safety concerns, and environmental problems in the workplace.

★ 2232 ★ *Environmental Law Deskbook*
The Environmental Reporter staff
Environmental Law Institute
1616 P St., N.W., Ste. 200
Washington, DC 20036
Phone: (202)328-5150
Fax: (202)328-5002
Frequency: Irregular. *Price:* $85/year.

Compendium of 16 major environmental statutes encompassing resource protection, pollution control, and administrative procedures.

★ 2233 ★ *Environmental Law Handbook, 10th ed.*
Government Institutes, Inc.
966 Hungerford Dr., No. 24
Rockville, MD 20850
Phone: (301)251-9250
Fax: (301)251-0638
Frequency: Irregular. *Price:* $59.95/year.

★ 2234 ★ *Environmental Law Reporter*
Environmental Law Institute
1616 P St., N.W., Ste. 200
Washington, DC 20036
Phone: (202)328-5150
Fax: (202)328-5002
Contact: Barry Breen, editor-in-chief
Frequency: Monthly, plus updates 3/x month. *Price:* $860/year. Loose-leaf reference reports legislation, litigation, regulations, and other developments concerning environmental issues.

★ 2235 ★ *Environmental Liability Monitor*
Business Publishers, Inc.
951 Pershing Dr.
Silver Springs, MD 20910-4464
Phone: 800-274-0122; (301)587-6300
Fax: (301)587-1081
Frequency: Monthly. *Price:* $383/year. Reports court cases, regulations, and liability problems in environmental contamination and real estate transactions.

★ 2236 ★ *Environmental Manager's Compliance Advisor*
Business Legal Reports, Inc.
39 Academy St.
Madison, CT 06443
Phone: 800-727-5257
Fax: (203)245-2559
Contact: Bruce W. Perry, ed.
Frequency: Biweekly. *Price:* $239.50/year. News roundup of governmental, technical, legislative, and corporate development matters for the environmental professional.

★ 2237 ★ *Environmental Report*
Food Marketing Institute
1750 K St., N.W., Ste. 700
Washington, DC 20006-2394
Phone: 800-433-8200; (202)452-8444
Fax: (202)429-4529
Frequency: Bimonthly. *Price:* Free to members; $100/year, nonmembers. Provides updates on environmental issues that affect the food distribution industry, including food additives and preservatives, environmental trends, solid waste management, consumer education, new technologies, and pertinent legislation.

★ 2238 ★ *Environmental Resources Research Institute Newsletter*
Environmental Resources Research Institute
Pennsylvania State University
Land and Water Research Bldg., Rm. 125
University Park, PA 16802
Phone: (814)865-1415
Contact: Lonnie Balaban, ed.
Frequency: Quarterly. *Price:* Free upon request.

★ 2239 ★ *Environmental Shopping Update*
Pennsylvania Resources Council, Inc.
P.O. Box 88
Media, PA 19063
Phone: (215)565-9131
Contact: Ruth Becker, ed.
Frequency: 3x/year. *Price:* $5/year. Promotes elimination of toxic substances, recycling and waste reduction, litter control, and preservation of natural beauty.

★ 2240 ★ *Environmental Toxicology Center Newsletter*
Environmental Toxicology Center
University of Wisconsin at Madison
309 Infirmary
Madison, WI 53706
Phone: (608)263-4580
Reports on activities of the Center and studies done on effects of hazardous substances.

★ 2241 ★ *Environment Pollution and Control: An Abstract Newsletter*
National Technical Information Service (NTIS)
U.S. Department of Commerce
5285 Port Royal Rd.
Springfield, VA 22161
Phone: (703)487-4630
Frequency: Weekly. *Price:* $135/year, U.S., Canada, and Mexico; $185, elsewhere. Provides abstracts of reports published by NTIS and other agencies on pesticide pollution and control, air and water pollution control, radiation, and environmental health and safety; summarizes environmental impact statements and includes information on how to obtain them.

★ 2242 ★ *Environment Reporter*
Bureau of National Affairs, Inc. (BNA)
1231 25th St., N.W.
Washington, DC 20037
Phone: 800-372-1033; (202)452-4430
Fax: (202)823-8092
Contact: Wallis E. McClain, Jr., editor
Frequency: Weekly. *Price:* $2,079/year. Looseleaf reference concerned with state and federal environmental litigation and regulations. Partially available online through HRIN, LEXIS, and WESTLAW.

★ 2243 ★ *Environment Week*
King Communications Group, Inc.
627 National Press Bldg.
Washington, DC 20045
Phone: (202)662-9720
Frequency: Weekly. Covers a wide variety of environmental issues, including hydrocarbon contamination, hazardous materials, toxic waste, acid rain, and federal and state regulatory actions. Available online through NewsNet and Predicasts, Inc.

★ 2244 ★ *Environ Report*
Environ Corp.
1000 Potomac St., N.W.
Washington, DC 20007
Phone: (202)337-7444
Frequency: Quarterly. *Price:* Free upon request.

★ 2245 ★ *EPA Environmental News*
U.S. Environmental Protection Agency, Region I
Public Affairs Office
JFK Bldg.
Boston, MA 02203
Phone: (617)565-3424
Contact: Brooke Chamberlain-Cook and Curt Spaulding, eds.
Price: Free upon request. Covers federal programs in toxic substances, pesticide control,

air and water pollution, waste management, safe drinking water, and other environmental concerns.

★ 2246 ★ *Everyone's Backyard*
Citizen's Clearinghouse for Hazardous Waste
P.O. Box 6806
Falls Church, VA 22040
Phone: (703)237-2249
Contact: Brian Lipsett, ed.
Frequency: Bimonthly. **Price:** Free with membership; $25/year, nonmembers. Reports on activities of environmental activist groups.

★ 2247 ★ *Family Safety and Health*
National Safety Council
444 N. Michigan Ave.
Chicago, IL 60611
Phone: (312)527-4800
Fax: (312)527-9381
Frequency: Quarterly. **Price:** $7.25/year with membership; $9.06/year, nonmembers. Newsletter devoted to off-the-job safe and healthy living.

★ 2248 ★ *FDA Surveillance Index for Pesticides*
National Technical Information Service (NTIS)
U.S. Department of Commerce
5285 Port Royal Road
Springfield, VA 22161
Phone: (703)487-4630
Frequency: Monthly. **Price:** $50/copy. Booklets provide information on the health risks of individual pesticides from a dietary exposure viewpoint; risks evaluated by using U.S. Food and Drug Administration (FDA) monitoring, usage estimates, and chemical, biological, and toxicology data.

★ 2249 ★ *Focus*
Hazardous Materials Control Research Institute
7237 Hanover Pkwy.
Greenbelt, MD 20770-3602
Phone: (301)982-9500
Fax: (301)220-3870
Contact: Patricia Segato, Managing ed.
Frequency: Monthly. **Price:** Free with membership. Summarizes news about the U.S. Environmental Protection Agency, hazardous waste, and environmental litigation.

★ 2250 ★ *Food Chemical News*
Food Chemical News, Inc.
1101 Pennsylvania Ave., S.E.
Washington, DC 20003
Phone: (202)544-1980
Fax: (202)546-3890
Contact: Louis Rothchild, Jr., ed.
Frequency: Weekly. **Price:** $750/year. Provides in-depth and timely coverage of administrative and regulatory developments in federal agencies affecting food and food processing. Available online through Predicasts, Inc.

★ 2251 ★ *Food News for Consumers*
Food Safety and Inspection Service
U.S. Department of Agriculture
FSIS/ILA, Rm. 1165 South
Washington, DC 20250
Phone: (202)447-9351
Contact: Mary Ann Parmley, ed.
Frequency: Quarterly. **Price:** $5/year. Covers all aspects of using food in the home, from nutrition to residual pesticides.

★ 2252 ★ *Friends of the Earth*
Friends of the Earth
218 D St., S.E.
Washington, DC 20003
Phone: (202)544-2600
Fax: (202)543-4710
Contact: Frances Raskin, ed.
Frequency: 10x/year. **Price:** Included in membership. Reports on activities of Friends of the Earth and world-wide events of environmental concern.

★ 2253 ★ *From the State Capitals: Waste Disposal and Pollution Control*
Wakeman/Walworth, Inc.
300 N. Washington St., Ste. 204
Alexandria, VA 22314
Phone: (703)540-8606
Fax: (703)549-1372
Contact: Keyes Walworth, ed.
Frequency: Monthly. **Price:** $118/year. Reports on developments in the pollution control industry and news affecting the transportation, storage, and clean-up of hazardous wastes. Also monitors pesticide policies on the state and municipal levels.

★ 2254 ★ *Global Pesticide Campaigner*
Pesticide Action Network
North American Regional Center
965 Mission St., Ste. 514
San Francisco, CA 94103
Phone: (415)541-9140
Fax: (415)541-9253
Contact: Bill Hall, ed.
Frequency: Quarterly. **Price:** $25/year, individuals and nonprofit organizations; $15/year, low income individuals; $50/year, government and corporate institutions. The journal of the Pesticide Action Network North American Regional Center and PAN International's Dirty Dozen Campaign. Focuses on international pesticide issues and developments related to the Dirty Dozen Campaign to ban world wide the most dangerous pesticides. Each thematic issue examines ecological, health, and economic aspects of pesticide problems, as well as the development and implementation of alternatives.

★ 2255 ★ *GREEN-keeping*
Green-keeping, Inc.
Box 110 Annandale Rd.
Annandale-on-Hudson, NY 12504
Phone: (914)246-6948
Contact: Annie Berthold-Bond, ed.
Frequency: Bimonthly. **Price:** $22.50/year. Covers hazardous substances found in common household products and their non-toxic alternatives.

★ 2256 ★ *Green MarketAlert*
The Bridge Group
MarketAlert Publications
345 Wood Creek Rd.
Bethlehem, CT 06751
Phone: (203)266-7209
Fax: (203)266-5049
Contact: Carl Frankel, ed.
Frequency: Monthly. **Price:** $295/year. Analyzes the impact upon businesses of green consumerism, including company plans and programs, trends, and pending regulations. Available online through Predicasts, Inc.

★ 2257 ★ *Green Marketing Report*
Business Publishers, Inc.
951 Pershing Dr.
Silver Springs, MD 20910-4464
Phone: 800-274-0122; (301)587-6300
Fax: (301)587-1081
Frequency: Monthly. Covers the advertising and marketing of products based upon their environmental benefits and claims. Includes federal, state, and local laws affecting environmental claims of products. Also available online through Predicasts, Inc.

★ 2258 ★ *Handbook of Hazardous Waste Regulation. Volume I: Comprehensive Introduction to RCRA Compliance*
Business Legal Reports, Inc.
39 Academy St.
Madison, CT 06443-1513
Phone: 800-727-5257
Fax: (203)245-2559
Price: $109.95/year. Loose-leaf reference provides an introduction to Resources Conservation and Recovery Act (RCRA) compliance, with hazardous waste regulations and contingency plans.

★ 2259 ★ *Handbook of Hazardous Waste Regulation. Volume II: How to Protect Employees during Environmental Incident Response*
Business Legal Reports, Inc.
39 Academy St.
Madison, CT 06443-1513
Phone: 800-727-5257
Fax: (203)245-2559
Price: $69.95. Loose-leaf reference provides steps on how to handle spills and other toxic accidents; includes safety procedures, site safety plans, and information on safety equipment for various situations.

★ 2260 ★ *Hazard Communication Guide*
J. J. Keller Associates, Inc.
3003 W. Breezewood Lane
P.O. Box 368
Neenah, WI 54957

Contact: Linda Wereley, ed.
Price: $125/year. Guide to SARA Title III regulations.

★ 2261 ★ *Hazard Communication Handbook: A RTK Compliance Guide*
Clark Boardman Company
375 Hudson St.
New York, NY 10014
Phone: 800-221-9428
Fax: (212)924-0460
Contact: Craig A. Moyer and Michael A. Francis, eds.
Frequency: Annual. **Price:** $85/year. Information on how to comply with right-to-know regulations.

★ 2262 ★ *Hazardous Chemicals on File*
Facts on File, Inc.
460 Park Ave. S.
New York, NY 10016
Phone: (212)683-2244
Contact: Craig T. Norback, ed.
Frequency: Irregular. Loose-leaf reference features profiles of more than 320 hazardous chemicals.

★ 2263 ★ *Hazardous Materials Intelligence Report*
World Information Systems
P.O. Box 535
Cambridge, MA 02238
Phone: 800-666-4430; (617)491-5100
Fax: (617)492-3312
Frequency: Weekly. **Price:** $325/year. Reviews federal, state, and local regulations and programs related to hazardous waste and hazardous materials management; covers litigation, cleanups of chemical spills and waste sites, and new technical developments. Available online through NewsNet.

★ 2264 ★ *Hazardous Materials Newsletter*
John R. Cashman, Publisher
P.O. Box 204
Barre, VT 05641
Phone: (802)479-2307
Contact: John R. Cashman, ed.
Frequency: Bimonthly. **Price:** $41/year, U.S., Canada, Mexico; $45/year, elsewhere. Newsletter aimed at hazardous materials emergency personnel.

★ 2265 ★ *Hazardous Materials Transportation*
Washington Business Information, Inc.
1117 N. 19th St., Ste. 200
Arlington, VA 22209-1798
Phone: (703)247-3424
Fax: (703)247-3421
Contact: Dee Richards, ed.
Frequency: Semimonthly. **Price:** $307/year. Provides shippers with information on the safe transportation of hazardous materials, including the latest rules and regulations. Available online through DIOGENES.

★ 2266 ★ *Hazardous Materials Transportation Report*
J. J. Keller Associates
P.O. Box 368
Neenah, WI 54957
Phone: 800-558-5011
Contact: Thomas Ziebell, ed.
Frequency: Monthly. **Price:** $125/year. Looseleaf reference reports changes in transportation regulations.

★ 2267 ★ *Hazardous Materials Transportation Reporter*
Bureau of National Affairs, Inc. (BNA)
1231 25th St, N.W.
Washington, DC 20037
Phone: 800-372-1033
Fax: (202)822-8092
Contact: Bernard A. Chabel, ed.
Frequency: Monthly. **Price:** $466/year. Looseleaf reference reports on transport rules and regulations and administrative decisions governing the shipment of hazardous materials by rail, air, ship, highway, and pipeline.

★ 2268 ★ *Hazardous Materials Water Resources Newsletter*
Sierra Club
2439 Crestline Dr.
Olympia, WA 98502
Contact: Doris Cellarius, ed.
Frequency: Quarterly. **Price:** $10/year.

★ 2269 ★ *Hazardous Substances and Public Health*
Agency for Toxic Substances and Disease Registry (E-22)
U.S. Department of Health and Human Services
1600 Clifton Rd., N.E.
Mailstop E-33
Atlanta, GA 30333
Phone: (404)639-0736
Fax: (404)639-0746
Contact: Teresa L. Ramsey, ed.
Frequency: Irregular. **Price:** Free upon request. Provides summaries of federal and state regulations related to public health issues concerning chemicals in the environment, including new research findings, education and training programs, and product information.

★ 2270 ★ *Hazardous Waste Litigation Reporter*
Andrews Publications
1646 West Chester Pike
P.O. Box 1000
Westtown, PA 19395
Phone: 800-345-1101; (215)399-6600
Fax: (215)399-6610
Contact: Harry G. Armstrong, Exec. ed.
Frequency: 24x/year. **Price:** $850/year. Provides full texts of all documents, motions, briefs, and opinions in federal, state, and private party lawsuits against hazardous waste generators, transporters, and past and present facility owners and operators.

★ 2271 ★ *Hazardous Waste News*
Business Publishers, Inc.
951 Pershing Dr.
Silver Springs, MD 20910-4464
Phone: 800-274-0122; (301)587-6300
Fax: (301)587-1081
Contact: Robert W. Thompson, ed.
Frequency: Weekly. **Price:** $489.50/year. Provides information on Superfund and Resources Conservation and Recovery Act (RCRA) developments, disposal and recycling, regulations, and current court cases affecting compliance with government standards. Available online through NewsNet and Predicasts, Inc.

★ 2272 ★ *Hazardous Waste Operations and Emergency Response Communications Manual*
J. J. Keller Associates, Inc.
3003 W. Breezewood Lane
P.O. Box 368
Neenah, WI 54957
Phone: 800-327-6868
Contact: Linda Wereley, ed.

★ 2273 ★ *Hazardous Waste Quarterly*
Hazardous Waste Advisement Program, Division of Hazardous Waste Management
New Jersey Department of Environmental Protection
401 E. State St., Fifth Floor
Trenton, NJ 08625
Phone: (609)292-8341
Contact: Bill DeStefano, ed.
Frequency: Quarterly. **Price:** Free upon request. Reports on environmental concerns in New Jersey and EPA Region II.

★ 2274 ★ *Hazardous Waste Report*
Aspen Publishing, Inc.
200 Orchard Ridge Dr.
Gaithersburg, MD 20878
Phone: 800-638-8437; (301)417-7500
Fax: (301)417-7550
Contact: Sue Darcey, ed.
Frequency: 24x/year. **Price:** $440/year.

Reports on current developments in the hazardous waste field, including those concerning U.S. Environmental Protection Agency policies, the Resources Conservation and Recovery Act (RCRA), Superfund, state requirements, congressional activity, and environmental litigation and enforcement.

★ 2275 ★ *Hazchem Alert: The Global Communique on Hazardous Chemicals News and Information*
Van Nostrand Reinhold
115 Fifth Ave.
New York, NY 10003
Phone: (212)254-3232
Fax: (212)673-1239
Contact: Judith A. Douville, ed.
Frequency: Biweekly. **Price:** $260/year, U.S.; $315, elsewhere. Reports developments in toxicology, industrial hygiene, occupational safety and health, government regulations, and similar issues.

★ 2276 ★ *Hazmet Transport News*
Business Publishers, Inc.
951 Pershing Dr.
Silver Springs, MD 20910-4464
Phone: 800-274-0122; (301)587-6300
Fax: (301)587-1081
Contact: Melanie Scott, ed.
Frequency: Biweekly. **Price:** $305.50/year. Formerly titled *Toxic Materials Transport*. Monitors federal, state, and local regulations affecting the routing of hazardous and radioactive materials by rail, truck, barge, air, and pipeline. Covers uniform hazardous waste manifest requirements under the Resources Conservation and Recovery Act (RCRA), the U.S. Environmental Protection Agency, the Department of Transportation, including the Bureau of Motor Carrier Safety, and the Office of Hazardous Materials Transport. Available online through NewsNet and Predicasts, Inc.

★ 2277 ★ *Haznet*
Capitol Reports, Inc.
921 11th St., Ste. 701
Sacramento, CA 95814
Phone: (916)441-4427
Fax: (916)441-4560
Contact: Margaret K. Akins, ed.
Frequency: Monthly. **Price:** $18/year. Covers hazardous waste issues in California.

★ 2278 ★ *HazTECH News*
HazTECH Publications, Inc.
14120 Huckleberry Lane
Silver Springs, MD 20906
Phone: (301)871-3289
Contact: Cathy Dombrowski, ed.
Frequency: Biweekly. **Price:** $333/year. Describes new technologies for handling hazardous waste products.

★ 2279 ★ *HazTech Transfer*
Hazardous Substance Research Center, EPA Regions VII and VIII
Kansas State University
133 Ward Hall
Manhattan, KS 66506
Phone: (913)532-6026
Fax: (913)532-6952
Contact: Mike Dorcey and Steven Galitzer, eds.
Frequency: Quarterly. **Price:** Free upon request. Covers research on hazardous substances and the activities of the Center, which heads a consortium of universities in EPA Regions VII and VIII.

★ 2280 ★ Home, Yard and Garden Pest Newsletter
Agriculture Newsletter Service
University of Illinois
116 Mumford Hall
1301 W. Gregory Dr.
Urbana, IL 61801
Frequency: 20x/year. *Price:* $15/year. Covers current pest controls, the storage and disposal of pesticides, and application methods and equipment.

★ 2281 ★ Household Hazardous Waste Management News
Waste Watch Center
16 Haverhill St.
Andover, NH
Phone: (508)470-3044
Fax: (508)470-3384
Contact: Dana Duxbury, editor-in-chief
Frequency: Quarterly. *Price:* Free upon request. Covers matters concerning waste collection in the U.S. Environmental Protection Agency regions and world wide.

★ 2282 ★ How to Meet OSHA's Safety and Health Guidelines
Business Legal Reports, Inc.
39 Academy St.
Madison, CT 06443
Phone: 800-727-5257
Fax: (203)245-2559
Contact: Faith Gavin Kuhn, ed.
Price: $74.95. Loose-leaf service provides sample policies, programs, and sources of help to comply with (OSHA) safety program guidelines.

★ 2283 ★ Indoor Pollution Law Report
Leader Publications
New York Law Publishing Co.
111 Eighth Ave.
New York, NY 10011
Phone: (212)463-5709
Contact: Lawrence S. Kirsch, ed.
Frequency: Monthly. *Price:* $155/year. Alerts attorneys to developments in litigation cases involving radon, asbestos, and formaldehyde.

★ 2284 ★ Industrial Health and Hazards Update
Merton Allen Associates
P.O. Box 15640
Plantation, FL 33318-5640
Phone: (305)473-9560
Fax: (305)473-0544
Contact: Merton Allen, Pres.
Frequency: Monthly. Publishes abstracts from reports on industrial health and hazards; includes information on regulations, legislation, litigation, and toxicity of industrial materials and products. Available online through NewsNet and Predicasts, Inc.

★ 2285 ★ INFOletter
Public Education and Risk Communication Division
Environmental and Occupational Health Science Institute
UMDNJ-Robert Wood Johnson Medical School
Brookwood II
45 Knightsbridge Rd.
Piscataway, NJ 08854
Phone: (201)463-5353
Contact: Cindy Rovins, ed.
Frequency: Bimonthly. *Price:* $24/year. Covers environmental and occupational health issues.

★ 2286 ★ Information Service for Decision Makers
Public Voice for Food and Health Policy
1001 Connecticut Ave., Ste. 522
Washington, DC 20036
Phone: (202)659-5930
Contact: Jodie Silverman, Dir. of Communications
Frequency: Monthly. *Price:* $350/year. Compilation of topical working materials highlighting consumer positions on key food and health policies.

★ 2287 ★ INFORM Reports
Inform, Inc.
381 Park Ave. S.
New York, NY 10016
Phone: (212)689-4040
Contact: Sibyl R. Golden, ed.
Frequency: Quarterly. *Price:* $35/year. Covers news of INFORM, Inc. concerning research in the areas of chemical hazards prevention, toxic wastes, and hazardous materials in household and building products. Also articles of interest on municipal solid waste management, urban air quality, and agricultural water conservation.

★ 2288 ★ Inside EPA
Inside Washington Publishers
P.O. Box 7167
Ben Franklin Station
Washington, DC 20044
Phone: 800-424-9068; (703)892-8500
Fax: (703)685-2606
Contact: Julie Edelson, ed.
Frequency: Weekly. *Price:* $730/year. Reports on activities within the U.S. Environmental Protection Agency and environmental developments in Congress, the courts, and regulatory agencies.

★ 2289 ★ Institute Items
The Art and Craft Materials Institute, Inc.
100 Boylston St., Ste. 1050
Boston, MA 02116
Phone: (617)426-6400
Fax: (617)426-6639
Contact: Laurie Doyle, ed.
Frequency: Monthly. *Price:* Free to members; $20/year, nonmembers. Covers safety and health hazards in arts and crafts materials.

★ 2290 ★ Integrated Risk Information System
National Technical Information Service (NTIS)
U.S. Department of Commerce
5285 Port Royal Rd.
Springfield, VA 22161
Phone: (703)487-4630
Frequency: Quarterly. *Price:* Available on a diskette data file only; $130/year, IBM PC/AT; $230/year, IBM PC. Covers the effects of chemicals on human health, and includes information on reference doses and carcinogen assessments.

★ 2291 ★ The IPM Practitioner
Bio-Integral Resource Center
P.O. Box 7414
Berkeley, CA 94707
Phone: (415)524-2567
Contact: S. Daar, H. Olkowski, and W. Olkowski, eds.
Frequency: 10x/year. *Price:* Included in membership. Publishes information on all aspects of environmentally sound pest control. Investigates the least-toxic methods of controlling pests in agriculture, urban landscapes, greenhouses, and other settings.

★ 2292 ★ Job Safety and Health
Bureau of National Affairs, Inc. (BNA)
1231 25th St., N.W.
Washington, DC 20037
Phone: (202)452-4430
Fax: (202)823-8092
Frequency: Biweekly. *Price:* $528/year. Looseleaf reference covers Occupational Safety and Health Administration (OSHA) rules and regulations.

★ 2293 ★ The Lawrence Review of Natural Products
Facts and Comparisons
111 W. Port Plaza, Ste. 423
St. Louis, MO 63146-3098
Phone: 800-223-0554; (314)878-2515
Fax: (314)878-5563
Contact: Bernie R. Olin, ed.
Frequency: Monthly. *Price:* $24.95/year, U.S. and Canada. Provides reference information on the origins, uses, and toxicity of natural products.

★ 2294 ★ Louisiana Industry Environmental Alert
Environmental Compliance Reporter, Inc.
3154-B College Dr., Ste. 522
Baton Rouge, LA 70808
Phone: 800-729-1964
Contact: Natale Nogosek, ed.
Frequency: 13x/year. *Price:* $235/year. Covers activities of the Louisiana Department of Environmental Quality, Department of Natural Resources, and the Hazardous Materials Unit of the State Police, including proposed regulations and enforcement activities, and those activities of the U.S. Environmental Protection Agency that are of significance to Louisiana. Available online through Predicasts, Inc.

★ 2295 ★ The Manager's Clean Air Act Compliance Manual
Business Legal Reports, Inc.
39 Academy St.
Madison, CT 06443-1513
Phone: 800-727-5257
Fax: (203)245-2559
Frequency: Annual updates. *Price:* $249/year with updating service; $149/year, without updating service. Looseleaf service analyzing the new Clean Air Act of 1990.

★ 2296 ★ Masscitizen
Massachusetts Public Interest Research Group (MASSPIRG)
29 Temple Place
Boston, MA 02111-1305
Phone: (617)292-4800
Contact: Kathleen Traphagen, ed.
Frequency: Quarterly. *Price:* Included in membership. Reports on environmental and other interests of the Northeast and on MASSPIRG activities.

★ 2297 ★ The Minimizer
Center for Hazardous Materials Research
University of Pittsburgh Applied Science Center
320 William Pitt Way
Pittsburgh, PA 15238
Phone: 800-334-CHMR; (412)826-5320
Contact: Robin A. Day, ed.
Frequency: Quarterly. *Price:* Free upon request. Covers hazardous waste matters, particularly those that affect Pennsylvania small quantity generators.

★ 2298 ★ Multinational Environmental Outlook
Business Publishers, Inc.
951 Pershing Dr.
Silver Springs, MD 20910-4464
Phone: 800-274-0122; (301)587-6300
Fax: (301)587-1081
Contact: Hiram Reisner, ed.
Frequency: Biweekly. **Price:** $355.50/year.
Reports on international developments in air and water pollution control, toxic substances, waste management, and other environmental concerns and their potential effect on operations and plans of multinational corporations. Available online through NewsNet and Predicasts, Inc.

★ 2299 ★ NAEM Network News
National Association for Environmental Management
4400 Jenifer St., N.W., Ste. 310
Washington, DC 20015
Phone: (202)966-0019
Contact: Alan J. Borner, editor-in-chief
Frequency: Quarterly. **Price:** Free with membership. Covers issues of interest to environmental managers in the private sector.

★ 2300 ★ NATICH Newsletter
National Air Toxics Information Clearinghouse
U.S. Environmental Protection Agency (EPA)
Pollution Assessment Branch (MD-13)
Research Triangle Park, NC 27711
Phone: (919)541-5348
Frequency: Bimonthly. **Price:** Free.

★ 2301 ★ National Coalition Against the Misuse of Pesticides-Technical Report
National Coalition Against the Misuse of Pesticides
701 E St., S.E.
Washington, DC 20003
Phone: (202)543-5450
Contact: Catherine Karr, ed.
Frequency: Monthly. Tracks congressional legislation and administrative rulings concerning pesticides, covers litigation and settlements, and reports pesticide poisoning incidents. Available online through the publisher.

★ 2302 ★ NCOSH Safety and Health News
North Carolina Safety and Health Project
P.O. Box 2514
Durham, NC 27715
Phone: (919)286-9249
Contact: Mark Schulz, ed.
Frequency: Quarterly. **Price:** $25/year. Covers North Carolina activities concerning toxic substances, right-toknow campaigns, worker's compensation, and medical clinics and surveys.

★ 2303 ★ New Jersey Industry Environmental Alert
Environmental Compliance Reporter, Inc.
3154-B College Dr., Ste. 522
Baton Rouge, LA 70808
Phone: 800-729-1964
Contact: Frederick Muckerman, Managing ed.
Frequency: 12x/year. **Price:** $255/year.
Covers activities of the New Jersey Department of Environmental Protection, New Jersey environmental legislation, and proposed rules and policy changes. Available online through Predicasts, Inc.

★ 2304 ★ News and Notes
Asbestos Information Association/North America (AIA/NA)
1745 Jefferson Davis Hwy., Ste. 509
Arlington, VA 22202
Phone: (703)979-1150
Fax: (703)979-1152
Contact: B. J. Pigg, Pres.
Frequency: Monthly. **Price:** Free with membership. Provides updates of asbestos management issues.

★ 2305 ★ News from the Hazardous Substances Management Program
Center for Technology, Policy and Industrial Development
Massachusetts Institute of Technology
77 Massachusetts Ave., Rm. E40-241
Cambridge, MA 021139
Phone: (617)253-0902
Fax: (617)253-7140
Contact: Jennifer Nash, ed.
Frequency: Quarterly. **Price:** Available upon request. Reports on activities of the Management Program and highlights corporate programs and research activities.

★ 2306 ★ NPTN Now!
National Pesticide Telecommunications Network
Department of Preventive Medicine, Texas Tech Health Science Center
Texas Tech University
S-129 Thompson Hall
Lubbock, TX 79430
Phone: (806)743-3091
Hotline: 800-858-7378
Frequency: Quarterly. **Price:** Free upon request. News about the NPTN and articles on topics of similar interest such as pesticides, rodenticides, herbicides, and animal poisoning.

★ 2307 ★ NRDC Newsline
Natural Resources Defense Council
40 W. 20th St.
New York, NY 10011
Phone: (212)727-2700
Fax: (212)727-1773
Contact: Paul Allen, ed.
Frequency: 5x/year. **Price:** Free with membership. Covers news about the NRDC, federal and local environmental issues, legislative news, and court decisions.

★ 2308 ★ NRDC-tlc
Natural Resources Defense Council
40 W. 20th St.
New York, NY 10011
Phone: (212)727-2700
Fax: (212)727-1773
Frequency: Quarterly. **Price:** $10/year, members; $15/year, nonmembers. Covers environmental and health issues concerning families, with emphasis on children.

★ 2309 ★ NTIS Newsline
National Technical Information Service
U.S. Department of Commerce
5285 Port Royal Rd., Rm. 2004-S
Springfield, VA 22161
Phone: (703)487-4812
Frequency: Quarterly. **Price:** Free. Provides current information about NTIS products and services.

★ 2310 ★ Occupational Health and Safety Letter
Business Publishers, Inc.
951 Pershing Dr.
Silver Springs, MD 20910-4464
Phone: 800-274-0122; (301)587-6300
Fax: (301)587-1081
Frequency: 24x/year. **Price:** $155/year.

★ 2311 ★ Occupational Safety and Health Reporter
Bureau of National Affairs, Inc. (BNA)
1231 25th St., N.W.
Washington, DC 20037
Phone: (202)452-4430
Fax: (202)823-8092
Frequency: Weekly. **Price:** $808/year.
Looseleaf reference reports OSHA litigation, rules and regulations.

★ 2312 ★ Online—The RTK NET Newsletter
Focus Project/OMB Watch
1731 Connecticut Ave., N.W.
Washington, DC 20009-1146
Phone: (202)234-8494
Fax: (202)234-8584
Contact: Rich Puchalsky, Research Coord.
Frequency: Quarterly. News for users of the RTK NET, a bulletin board network of right-to-know information.

★ 2313 ★ OSHA Compliance Advisor
Business Legal Reports, Inc.
39 Academy St.
Madison, CT 06443
Phone: 800-727-5257
Fax: (203)245-2559
Contact: John F. Brady, ed.
Frequency: 24x/year. **Price:** $239.50/year.
Loose-leaf service for corporate managers.

★ 2314 ★ OSHA Compliance Encyclopedia
Business Legal Reports, Inc.
39 Academy St.
Madison, CT 06443
Phone: 800-727-5257
Fax: (203)245-2559
Contact: Faith Gavin Kuhn, ed.
Frequency: Biweekly newsletter updates.
Price: $424/year, for Encyclopedia and newsletters; $329.95 for Encyclopedia only; $239.50 for newsletter only. Covers Occupational Safety and Health Administration (OSHA) rules and regulations and training requirements.

★ 2315 ★ PEC Forum
Pennsylvania Environmental Council, Inc.
1211 Chestnut St., Ste. 900
Philadelphia, PA 19107
Phone: 800-322-9214
Fax: (215)563-0528
Contact: Joanne R. Denworth, editor-in-chief
Frequency: Quarterly. **Price:** $30/year. News on the PEC and Pennsylvania environmental issues.

★ 2316 ★ Pennsylvania Industry Environmental Alert
Environmental Compliance Reporter, Inc.
3154-B College Dr., Ste. 522
Baton Rouge, LA 70808
Phone: 800-729-1964
Contact: Frederick Muckerman, Managing ed.
Frequency: 12x/year. **Price:** $255/year.
Covers activities of the Pennsylvania Department of Environmental Resources, including proposed regulations, changes in policy, and enforcement activities, the Pennsylvania environmental legislation, and the U.S. Environmental Protection Agency, when of

significance to Pennsylvania. Available online through Predicasts, Inc.

★ 2317 ★ Pesticide and Toxic Chemical News
Food Chemical News, Inc.
1101 Pennsylvania Ave., S.E.
Washington, DC 20003
Phone: (202)544-1980
Fax: (202)546-3890
Contact: Cathy Cooper, ed.
Frequency: Weekly. **Price:** $615/year. Reports on hazardous wastes, pesticides, toxic substances, and general issues of chemical regulatory and legislative developments. Covers U.S. Environmental Protection Agency and other national and international organizations. Also available online through Predicasts, Inc.

★ 2318 ★ Pesticide Progress
Interstate Professional Applicator's
    Association
P.O. Box 1377
Milton, WA 98354
Phone: (206)848-3407
Contact: Mary Ellen Smith, ed.
Frequency: Quarterly. **Price:** Included in membership. Provides information on the application of horticultural pesticides.

★ 2319 ★ Pesticides and You
National Coalition Against the Misuse of
    Pesticides (NCAMP)
701 E St., S.E.
Washington, DC 20003
Phone: (202)543-5450
Contact: Susan Cooper, ed.
Frequency: 5x/year. **Price:** $20/year; included in membership. Concerned with pesticide hazards and safety and the activities of the NCAMP. Promotes alternatives to pesticide use, monitors relevant government administrative and congressional activities, and seeks to advance the awareness of public health, environmental, and economic problems caused by pesticide abuse.

★ 2320 ★ Pest Management and Crop Development Bulletin
Agriculture Newsletter Service
University of Illinois
116 Mumford Hall
1301 W. Gregory Dr.
Urbana, IL 61801
Frequency: 25x/year: weekly, April through August, plus five additional issues. **Price:** $20/year. Reports on current insect and plant diseases, gives advice on control methods, and covers new developments in pesticide application techniques.

★ 2321 ★ Poison Pen Notes
Delaware Valley Regional Poison Control
    Center
One Children's Center
34th Street and Civic Center Blvd.
Philadelphia, PA 19104
Phone: (215)590-2003
Contact: Patricia A. Farina and Marcie Gelber, eds.
Price: Free with membership. Provides information on poison prevention and news about the Poison Control Center.

★ 2322 ★ Priorities
American Council on Science and Health
1995 Broadway, 16th Floor
New York, NY 10023
Phone: (212)362-7044
Fax: (212)362-4919
Contact: Elizabeth Whelen, ed.
Frequency: Quarterly. **Price:** $25/year. Articles and reports on a wide variety of topics essential to the public health, including exposure to hazardous substances in commonly used products.

★ 2323 ★ Product Safety and Liability Reporter
Bureau of National Affairs, Inc. (BNA)
1231 25th St., N.W.
Washington, DC 20037
Phone: (202)452-4430
Fax: (202)822-8092
Frequency: Weekly. **Price:** $736/year. Looseleaf reference concerned with litigation in the products liability field and cases involving the Consumer Products Safety Commission.

★ 2324 ★ Rachel's Hazardous Waste News
Environmental Research Foundation
P.O. Box 3541
Princeton, NJ 08543-6994
Phone: (609)683-0707
Contact: Peter Montague, ed.
Frequency: Weekly. **Price:** $40/year, individuals and citizen groups; $80/year, government agencies; $15/year, students and seniors; $400/year, businesses and professionals. Focuses on technology, government regulations, and corporate strategies in environmental and hazardous waste issues such as toxins, regulations, and health and environment. Available online.

★ 2325 ★ Radon Research Notes
Oak Ridge National Laboratory
Bldg. 2001
P.O. Box 2008
Oak Ridge, TN 37831-6050
Phone: (615)574-7759
Contact: Gloria M. Caton, ed.
Frequency: Quarterly. **Price:** Free.

★ 2326 ★ Regulated Chemicals Directory
Chapman Hall
29 W. 34th St.
New York, NY 10001-2291
Phone: (212)244-3336
Fax: (212)563-2269
Contact: Patricia L. Dsida, ed.
Frequency: Three quarterly updates. **Price:** $375/year. Loose-leaf reference presents summaries of hazards and regulations concerning toxic substances; contains indexes by Chemical Abstract Service (CAS) number and chemical name and synonyms. Available on diskettes through ChemADVISOR.

★ 2327 ★ Report to the Consumer
Honor Publications
1275 Carson Woods Rd.
Fortuna, CA 95540
Phone: (707)725-4995
Contact: Ida Honorof, ed.
Frequency: 24x/year. **Price:** $8/year. Discusses the effects on human health of herbicides, drugs, food additives, and other environmental pollutants.

★ 2328 ★ Re:SOURCE
Environmental Hazards Management Institute
10 Newmarket Rd.
P.O. Box 70
Durham, NH 03824
Phone: (603)868-1946
Fax: (603)868-1547
Contact: Alan John Borner, ed.
Frequency: Quarterly. **Price:** Free upon request. Covers environmental management and education issues among businesses, municipalities, and nonprofit groups.

★ 2329 ★ Right-to-Know Compliance Manual
J. J. Keller Associates, Inc.
3003 W. Breezewood Lane
P.O. Box 368
Neenah, WI 54957
Phone: 800-327-6868
Contact: Linda Wereley, ed.
Price: $179.95/year. How to comply with SARA Title III requirements.

★ 2330 ★ RWC Report
Radioactive Waste Campaign
7 West St.
Warwick, NY 10990
Phone: (914)986-1115
Contact: Jean Fazzino, ed.
Frequency: Quarterly. **Price:** Included in membership; $15/year, nonmembers. Contains articles on the environmental and social impact of the generation, transportation, storage, and management of radioactive waste.

★ 2331 ★ Safe Home Digest
Lloyd Publishing Inc.
24 East Ave., Ste. 1300
New Canaan, CT 06840
Phone: (203)966-2099
Contact: Richard Thiel, ed.
Frequency: Monthly. **Price:** $27.96/year. Reports on hazardous substances and nontoxic alternatives for consumers.

★ 2332 ★ School Asbestos Alert
Andrews Publications, Inc.
P.O. Box 1000
1646 West Chester Pike
Westtown, PA 19395
Phone: 800-345-1101; (215)399-6600
Fax: (215)399-6610
Frequency: Monthly. **Price:** $180/year. Covers current legal and regulatory developments, construction, and medical problems relating to exposure to asbestos in schools; also monitors U.S. Environmental Protection Agency and state regulations.

★ 2333 ★ Solid Waste Management Newsletter
Center for Solid Waste Management and
    Research, Office of Technology Transfer
Institute for Environmental Studies
University of Illinois at Urbana-Champaign
1101 W. Peabody Dr.
Urbana, IL 61801
Phone: (217)333-4178
Frequency: Monthly. **Price:** Free upon request. Reports on Illinois solid waste legislation, regulatory hearings and rulings, and news of the Center.

★ 2334 ★ Spill Briefs
Spill Control Association of America
400 Renaissance Center, Ste. 1990
Detroit, MI 48243-1509
Phone: (313)567-0500
Contact: Marc K. Shaye and Clayton W. Evans, eds.
Frequency: Monthly. **Price:** Free with

membership. Covers material relevant to oil and hazardous substances spill control technology and news about the Association.

★ 2335 ★ *State Environment Report*
Business Publishers, Inc.
951 Pershing Dr.
Silver Springs, MD 20910-4464
Phone: 800-274-0122; (301)587-6300
Fax: (301)587-1081
Contact: Phil Zahodiakin, ed.
Frequency: Weekly. **Price:** $429.50/year.
Formerly titled *State Regulation Report*. Covers state laws, court cases and regulations of toxic substances, air, water, and hazardous waste, including pesticide certification and enforcement, right-to-know laws, environmental liability, and environmental and occupational health. Available online through NewsNet and Predicasts, Inc.

★ 2336 ★ *STC Newsletter*
Society of Toxicology of Canada
c/o Dr. D.J. Ecobichon
Department of Pharmacology Therapeutics
McGill University
Montreal, PQ, Canada H3G 1Y6
Phone: (514)398-3604
Contact: D. J. Ecobichon, ed.
Frequency: 3x/year. **Price:** Included in membership. Discusses current issues in the field of toxicology.

★ 2337 ★ *Summary of California Environmental Laws*
Hazard Identification and Risk Assessment Branch
California Department of Health Services
2151 Berkeley Way
Berkeley, CA 94704
Phone: (415)540-3063
Price: Free.

★ 2338 ★ *Superfund*
Pasha Publications, Inc.
1401 Wilson Blvd., Ste. 900
Arlington, VA 22209-9970
Phone: 800-424-2908; (703)528-1244
Fax: (703)528-1253
Contact: Bowman Cox, ed.
Frequency: 25x/year. **Price:** $295/year, U.S. and Canada; $310/year, elsewhere. Reports on developments in hazardous waste cleanup policy required under the federal and various state Superfund laws, remedial projects, new technology, liability, litigations, and settlements. Available online through NewsNet and Predicasts.

★ 2339 ★ *The Synergist*
American Industrial Hygiene Association (AIHA)
345 White Pond Dr.
P.O. Box 8390
Akron, OH 44320
Phone: (216)373-2442
Fax: (216)873-1642
Frequency: Monthly. **Price:** Included in membership. Reports news of the AIHA and developments affecting industrial health and safety, the use of hazardous materials in the workplace, and the impact of industrial accidents on workers.

★ 2340 ★ *Synergos*
Great Lakes and Mid-Atlantic Hazardous Substance Research Institute
Michigan State University
Holden Hall, Rm. C231
East Lansing, MI 48824-1206
Phone: (517)353-9718
Fax: (517)355-4603
Contact: Karen Vigmostad, ed.
Frequency: 2/x year. **Price:** Free upon request. Focuses on topics concerning site remediation and provides news about projects in the universities that compose the Institute.

★ 2341 ★ *Technical Transfer*
Office of Research and Development
Center for Environmental Research Information
26 W. Martin Luther King Dr.
Cincinnati, OH 45268
Frequency: Monthly. **Price:** Free upon request. Reports on technical developments in the environmental field.

★ 2342 ★ *Texas Industry Environmental Alert*
Environmental Compliance Reporter, Inc.
3154-B College Dr., Ste. 522
Baton Rouge, LA 70808
Phone: 800-729-1964
Contact: Duggan Flanakin, ed.
Frequency: 22x/year. **Price:** $295/year. Covers activities of the Texas Water Commission, Texas Air Control Board, Railroad Commission, and the Texas Department of Health, including proposed regulations, changes in policy, and enforcement activities. Also includes Texas environmental legislation and U.S. Environmental Protection Agency actions of significance to Texas. Available online through Predicasts, Inc.

★ 2343 ★ *Touching Bases*
Military Toxics Network
National Toxics Campaign Fund
Pacific Studies Center
222B View St.
Mountain View, CA 94041
Phone: (415)969-1545
Fax: (415)968-1126
Contact: Lenny Siegel, ed.
Frequency: Quarterly. **Price:** $25/year. Deals with toxic wastes at U.S. military bases and of defense contractors.

★ 2344 ★ *Toxic Chemicals Litigation Reporter*
Andrews Publications, Inc.
1646 West Chester Pike
P.O. Box 1000
Westtown, PA 19395
Phone: 800-345-1101; (215)399-6600
Fax: (215)399-6610
Contact: Harry G. Armstrong, Exec. ed.
Frequency: 24x/year. **Price:** $850/year. Reports on lawsuits alleging personal injuries or property damage from exposure to toxic chemicals such as Agent Orange, lead paint, dioxin, pesticides, herbicides, and polychlorinated biphenyls (PCBs); provides full text of selected documents.

★ 2345 ★ *Toxic Materials News*
Business Publishers, Inc.
951 Pershing Dr.
Silver Springs, MD 20910-4464
Phone: 800-274-0122; (301)587-6300
Fax: (301)587-1081
Contact: Charles Knebl, ed.
Frequency: Weekly. **Price:** $489.50/year, U.S. and Canada. Monitors developments in toxic

substances control that affects business development and occupational health. standards and regulations governing the manufacture, handling, transport, distribution, and disposal of toxic chemicals and pesticides. It focuses on emergency response, right-to-know, U.S. Environmental Protection Agency and Occupational Safety and Health Administration (OSHA) regulations, the Toxic Substances Control Act (TSCA), sections of the Clean Air Act, and pesticides. Available online through NewsNet and Predicasts, Inc.

★ 2346 ★ *Toxics Law Reporter*
Bureau of National Affairs, Inc. (BNA)
1231 25th St., N.W.
Washington, DC 20037
Phone: 800-372-1033; (202)452-4200
Fax: (202)822-8092
Contact: William Harris Frank, ed.
Frequency: Weekly. **Price:** $1,077/year. Looseleaf service covers legal developments concerning toxic tort and hazardous waste lawsuits and related insurance issues, Superfund and Resources Conservation and Recovery Act (RCRA) litigation, and related cases, regulations, and legislation. Available online through HRIN and LEXIS.

★ 2347 ★ *Toxic Times*
National Toxics Campaign
1168 Commonwealth Ave.
Boston, MA 02134
Phone: (617)232-0327
Contact: Michael Stein, ed.
Frequency: Quarterly. **Price:** $15/year. Reports on citizen-based programs on toxins and environmental problems.

★ 2348 ★ *TSCA Chemicals in Progress Bulletin*
Toxic Substances Control Act Assistance Office
U.S. Environmental Protection Agency (EPA)
401 M St., S.W.
Washington, DC 20460
Phone: (202)554-1404
Price: Free. Reports activities of the TSCA office.

★ 2349 ★ *UCLA Hazardous Substances Bulletin*
Engineering Research Center for Hazardous Substances Control
School of Engineering and Applied Sciences
University of California, Los Angeles
6722 Boelter Hall
405 Hilgard Ave.
Los Angeles, CA 90024-1600
Phone: (213)206-3071
Fax: (213)206-3906
Frequency: Quarterly. **Price:** Free upon request. Focuses on topics concerning the prevention, handling, and disposal of hazardous substances and news about the Engineering Research Center.

★ 2350 ★ *U.S. PIRG Citizen Agenda*
U.S. Public Interest Research Group (USPIRG)
215 Pennsylvania Ave., S.E.
Washington, DC 20003
Phone: (202)546-9707
Frequency: Quarterly. **Price:** Free with membership. Covers activities of PIRG organizations.

★ 2351 ★ *Washington Environmental Protection Report*
Callahan Publications
P.O. Box 3751
Washington, DC 20007
Phone: (703)356-1925
Fax: (703)356-9614
Contact: Vincent F. Callahan, Jr., ed.
Frequency: 24x/year. *Price:* $175/year. Provides information on environmental issues, including legislation, new regulatory developments, and contracting opportunities with the U.S. Environmental Protection Agency and other federal agencies.

★ 2352 ★ *Waste Management Research Report*
College of Environmental Science and Forestry
State University of New York
123 Bray Hall
1 Forestry Dr.
Syracuse, NY 13210
Phone: (315)470-6644
Contact: Louise W. Laughton, ed.
Frequency: 3x/year. *Price:* Free upon request. Each issue focuses on one major area of hazardous waste management of interest to New York state and highlights research news about the environmental research institutions in the state.

★ 2353 ★ *Water Connection*
New England Interstate Water Pollution Control Commission
85 Merrimac St.
Boston, MA 02114
Phone: (617)437-1524
Contact: Ellen Frye, ed.
Frequency: Quarterly. *Price:* Free upon request. Covers regional water quality, air quality, and hazardous waste management issues.

★ 2354 ★ *WTC News*
Washington Toxics Coalition
4516 University Way, N.E.
Seattle, WA 98105
Phone: (206)632-1545
Frequency: Quarterly *Price:* Included in membership. Articles, news from local groups, and information on statewide issues concerning hazardous substances, waste, and environmental protection, primarily in Washington state.

Publishers

★ 2355 ★ ABC-CLIO, Inc.
130 Cremona Dr.
P.O. Box 1911
Santa Barbara, CA 93117
Phone: 800-422-2546; (805)968-1911
Fax: (805)685-9685
Publishes books on such matters as air pollution, hazardous materials, toxic and radioactive waste, space exploration, world hunger, and human rights.

★ 2356 ★ Academic Press, Inc.
1250 Sixth Ave.
San Diego, CA 92101-4311
Phone: 800-321-5058; (619)699-6825
Fax: (619)699-6800
Publishes technical and scientific journals.

★ 2357 ★ American Chemical Society
1155 16th St., N.W.
Washington, DC 20036
Phone: 800-227-5558; (202)872-4600
Fax: (202)872-4615
Publishes professional and reference books, journals, and technical monographs related to chemistry and chemicals, industrial processes, industrial hygiene, and related issues.

★ 2358 ★ American Conference of Governmental Industrial Hygienists
6500 Glenway Ave., Bldg. D-7
Cincinnati, OH 45211-4438
Phone: (513)661-7881
Publishes reports and studies concerning hazardous substances in the workplace.

★ 2359 ★ American Council on Science and Health
1995 Broadway, 18th Floor
New York, NY 10023-5860
Phone: (212)362-7044
Fax: (212)362-4919
Nonprofit organization publishes studies and reports on scientific issues related to the relationship between chemicals, food, and the environment and on health and lifestyle factors.

★ 2360 ★ Andrews Publications, Inc.
1646 West Chester Pike
P.O. Box 1000
Westtown, PA 19395
Phone: 800-345-1101; (215)399-6600
Fax: (215)399-6610
Private company publishes newsletters that report litigation in the field of asbestos abatement, U.S. Environmental Protection Agency (EPA) regulations, and hazardous substances.

★ 2361 ★ Aspen Publishing, Inc.
200 Orchard Ridge Dr.
Gaithersburg, MD 20878
Phone: 800-638-8437
Fax: (301)417-7550
Publishes professional references concerning hazardous waste regulatory matters, including those involving the U.S. Environmental Protection Agency (EPA).

★ 2362 ★ Beacon Press
25 Beacon St.
Boston, MA 02108
Phone: (617)742-2110
Fax: (617)367-3237
Publishes books of environmental concern.

★ 2363 ★ Bio-Integral Resource Center
1307 Acton St.
P.O. Box 7414
Berkeley, CA 94707
Phone: (415)524-2567
Fax: (415)524-1758
Publishes research studies and information about organic farming and natural alternatives to pesticides.

★ 2364 ★ BNA Books
1231 25th St., N.W.
Washington, DC 20037
Phone: (202)452-4276
Fax: (202)452-9186
Division of the Bureau of National Affairs publishes legal and regulatory books on many subjects, including the environment and occupational safety and health.

★ 2365 ★ Bureau of National Affairs, Inc.
1350 Connecticut Ave., N.W., Ste. 1000
Washington, DC 20036
Phone: (202)452-7889
Fax: (202)862-0990
Publishes loose-leaf references for attorneys and professionals in such fields as environmental regulation, litigation, and asbestos and other toxic chemicals.

★ 2366 ★ Business and Legal Reports, Inc.
39 Academy St.
Madison, CT 06443
Phone: 800-727-5257; (203)245-7448
Fax: (203)245-2559
Publishes newsletters and loose-leaf references for environmental and occupational safety and health compliance and community right-to-know matters.

★ 2367 ★ Business Publishers, Inc.
951 Pershing Dr.
Silver Springs, MD 20910-4464
Phone: 800-274-0122; (301)587-6300
Fax: (301)587-1081
Publishes professional reference newsletters and directories in the fields of environmental regulation and litigation.

★ 2368 ★ Buzzworm, Inc.
2305 Canyon Blvd., Ste. 206
Boulder, CO 80302
Phone: (303)442-1969
Fax: (303)442-4875
Publishes *Buzzworm* magazine and other generalinterest materials on the environment.

★ 2369 ★ California Department of General Services, Publications Section
P.O. Box 1015
North Highlands, CA 95660
Phone: (916)973-3700
Produces guidebooks, directories, manuals, and other materials pertaining to environmental and occupational health and safety concerns in California.

★ 2370 ★ Canadian Centre for Occupational Health and Safety
250 Main St., E.
Hamilton, ON, Canada L8N 1H6
Publishes references and other material of environmental and occupational health and safety concerns in Canada.

★ 2371 ★ Center for Environmental Information, Inc.
46 Prince St.
Rochester, NY 14607-1016
Phone: (716)271-3550
Nonprofit clearinghouse for environmental information serving northwestern New York.

★ 2372 ★ Chemical Manufacturers Association
2501 M St., N.W.
Washington, DC 20037
Phone: (202)887-1255
Publishes guidebooks, manuals, and technical references concerning the manufacture, transport, handling, and disposal of hazardous chemicals.

★ 2373 ★ Chemical Publishing Co., Inc.
80 Eighth Ave.
New York, NY 10011
Phone: (212)255-1950
Publishes technical and college textbooks.

★ 2374 ★ Chilton Company
Chilton Way
Radnor, PA 19089
Phone: (215)964-4055
Publishes a controlled circulation tabloid on occupational safety and health.

★ 2375 ★ Concern, Inc.
1794 Columbia Rd., N.W.
Washington, DC 20009
Phone: (202)328-8160
Fax: (202)328-8161
Nonprofit organization publishes a variety of guides on specific environmental issues and provides information on community-action and public education and policy development.

★ 2376 ★ CRC Press, Inc.
2000 Corporate Blvd., N.W.
Boca Raton, FL 33431
Phone: 800-272-7737; (407)994-0555
Fax: (407)997-0949
Subsidiary of Yearbook Medical Publishers, Inc., publishes medical, scientific, and technical reference books.

★ 2377 ★ Earth Action Network, Inc.
P.O. Box 5098
Westport, CT 06881
Phone: (203)854-5559
Fax: (203)866-0602
Publishes *E Magazine,* a consumer magazine concerning the environment.

★ 2378 ★ Earthworks Press
1400 Shattuck Ave.
Box 25
Berkeley, CA 94709
Phone: (415)528-4616
Publishes books on environmental subjects.

★ 2379 ★ Elsevier Science Publishing Co., Inc.
655 Avenue of the Americas
New York, NY 10010
Phone: (212)989-5800
Fax: (212)633-3965
Publishes technical and scientific journals, textbooks, and reference books.

★ 2380 ★ Enslow Publishers
Box 777
Bloy St. and Ramsey Ave.
Hillside, NJ 07205
Phone: (201)964-4116

★ 2381 ★ Environmental Business Publishing, Inc.
827 Washington St.
San Diego, CA 92103
Phone: (619)295-7685
Fax: (619)295-5743

★ 2382 ★ Environmental Compliance Reporter, Inc.
Mark St. J. Kouhig, Publisher
3154-B College Dr., Ste. 522
Baton Rouge, LA 70808
Phone: 800-729-1964
Publishes state environmental compliance newsletters for New Jersey, Louisiana, Texas, and Pennsylvania.

★ 2383 ★ Environmental Information, Ltd.
4801 W. 81st St., Ste. 119
Minneapolis, MN 55437
Phone: (612)831-2473
Publishes a directory of 5,300 environmental products and services.

★ 2384 ★ Environmental Law Institute
1616 P St., N.W., Ste. 200
Washington, DC 20036
Phone: (202)328-5150
Nonprofit center for research on environmental law and policy publishes reference books on a variety of environmental and community right-to-know topics.

★ 2385 ★ Environmental Publications Associates, Ltd.
17 Jefryn Blvd., W.
Deer Park, NY 11729
Phone: (516)667-8896
Publishes introductory material on environmental sciences and works pertaining to Long Island, New York.

★ 2386 ★ Environmental Publications, Inc.
1400 Front Ave.
Lutherville, MD 21903
Phone: (301)828-6618
Publishes books and references on environmental subjects.

★ 2387 ★ EPA Public Information Center
U.S. Environmental Protection Agency
401 M St., S.W., PM-211B
Washington, DC 20460
Phone: (202)260-7751
Publishes handbooks, pamphlets, directories, and other materials pertaining to the activities of the EPA.

★ 2388 ★ EPA TSCA Assistance Information Service
Toxic Substances Control Act Assistance Office
401 M St., S.W., TS-799
Washington, DC 20460
Phone: (202)260-3795
Publishes guidebooks, references, directories, and other material relating to pesticides and toxic substances and other pollution matters.

★ 2389 ★ Executive Enterprises Publications Co., Inc.
22 W. 21st St.
New York, NY 10010-6904
Phone: 800-332-1105; 800-332-8804
Fax: (212)645-8689
Publishes books and periodicals for corporate managers in the fields of environmental control, employee safety, employee relations, equal employment, and other employee relations matters.

★ 2390 ★ Facts on File, Inc.
460 Park Ave. S.
New York, NY 10016
Phone: 800-322-8755; (212)683-2244
Publishes library reference books and trade books on a variety of social science subjects, current events, and technical subjects.

★ 2391 ★ Fairmont Press, Inc.
700 Indian Trail
Lilburn, GA 30247
Phone: (404)925-9388
Publishes reference, professional, and how-to-do books on the environment, safety, and energy.

★ 2392 ★ Food Chemical News, Inc.
1101 Pennsylvania Ave., S.E.
Washington, DC 20003
Phone: (202)544-1980
Fax: (202)546-3890
Provides information services on federal and state regulatory matters concerning pesticides and other toxic chemicals.

★ 2393 ★ Gale Research Inc.
835 Penobscot Bldg.
Detroit, MI 48226
Phone: 800-347-GALE; (313)961-2242
Publishes reference books on a variety of subjects, including the environment.

★ 2394 ★ Garland Publishing, Inc.
136 Madison Ave.
New York, NY 10016
Phone: 800-627-6273; (212)686-7492
Publishes reference books on medicine, law, science and technology, and business.

★ 2395 ★ Genium Publishing Corp.
1145 Catalyn St.
Schenectady, NY 12303
Phone: (518)377-8854
Fax: (518)377-1891
Publishes technical information products dealing with plant safety and emergencies.

★ 2396 ★ Gordon and Breach Science Publishers
P.O. Box 786
Cooper Station
New York, NY 10276
Phone: (212)206-8900
Fax: (212)645-2459
Publishes technical and scientific journals.

★ 2397 ★ Government Institutes, Inc.
966 Hungerford Dr., Ste. 24
Rockville, MD 20850
Phone: (301)251-3801
Fax: (301)251-0638
Publishes handbooks and directories in the fields of government and environmental law and regulations and hazardous wastes.

★ 2398 ★ The Green Book
9 Old Middlesex Rd.
P.O. Box 23
Belmont, MA 02178-9957
Phone: (617)935-4800
Publishes an environmental directory for New England.

★ 2399 ★ Green-keeping, Inc.
Box 110 Annandale Rd.
Annandale-on-Hudson, NY 12504
Phone: (914)246-6948
Publishes books and newsletters on hazardous materials in commonly used products.

★ 2400 ★ Greenwood Press Inc.
88 Post Rd. W.
P.O. Box 5007
Westport, CT 06881
Phone: 800-225-5800; (203)226-3571
Publishes technical reference books.

★ 2401 ★ Hazardous Materials Control Resources Institute
7237 Hanover Pkwy.
Greenbelt, MD 20970-3602
Phone: (301)982-9500
Fax: (301)220-3870
Membership organization publishes periodicals, training manuals, and conference proceedings and markets publications of other publishers that produce text, references, and loose-leaf services for the hazardous materials industry.

★ 2402 ★ Hemisphere Publishing Corp.
1101 Vermont Ave., N.W.
Washington, DC 20005
Phone: 800-821-8312; (202)289-2174
Fax: (202)289-3665
Publishes technical and scientific journals and newsletters on toxic substances.

★ 2403 ★ **Hulogosi Communications, Inc.**
P.O. Box 1188
Eugene, OR 97440
Publishes buyers guides to environmental products and services.

★ 2404 ★ **INFORM, Inc.**
381 Park Ave. S.
New York, NY 10016
Phone: (212)689-4040
Nonprofit environmental research organization publishes newsletters, study reports, research publications, and citizens' guides for reducing toxic wastes and identifying toxic substances in household products.

★ 2405 ★ **International Alliance for Sustainable Agriculture**
1701 University Ave., S.E.
Minneapolis, MN 55414
Phone: (612)331-1099
Publishes in the field of organic farming, including about the use of nontoxic alternatives to pesticides.

★ 2406 ★ **Island Press**
1718 Connecticut Ave., N.W., Ste. 300
Washington, DC 20009
Phone: 800-828-1302; (202)232-7933
Fax: (202)983-6414
Publishes books and other material for conservationists and environmental activists.

★ 2407 ★ **J. B. Lippincott Co.**
227 E. Washington Square
Philadelphia, PA 19106-3780
Phone: 800-638-3030; (215)238-4200
Fax: (215)238-4227
Publishes professional medical books, nursing, critical care, internal medicine, and allied medical and health care fields.

★ 2408 ★ **Jeremy P. Tarcher, Inc.**
9110 Sunset Blvd.
Los Angeles, CA 90069
Phone: 800-255-3362; (213)273-3274
Fax: (213)273-5732

★ 2409 ★ **J. J. Keller and Associates, Inc.**
3003 W. Breezewood Lane
P.O. Box 368
Neenah, WI 54957-0368
Phone: 800-558-5011; (415)722-2848
Publishes newsletters, loose-leaf references, and training materials on compliance with U.S. Environmental Protection Agency (EPA) and Occupational Safety and Health Administration (OSHA) regulations in the workplace.

★ 2410 ★ **Krieger Publishing Company**
P.O. Box 9542
Melbourne, FL 32902-9542
Phone: (407)724-9542; (407)727-7270
Fax: (407)951-3671
Publishes environmental books about marine waste management and pollution control and transportation of hazardous materials.

★ 2411 ★ **Labelmaster**
5724 N. Pulaski Rd.
Chicago, IL 60646
Phone: (312)478-0900
Publishes training materials, labels, employee safety and health aids, and other material pertaining to the handling, transportation, and use of hazardous substances.

★ 2412 ★ **Lewis Publishers, Inc.**
2000 Corporate Blvd., N.W.
Boca Raton, FL 33431
Phone: 800-272-7737; (407)994-0555
Fax: (407)997-0949
Publishes references and college textbooks on hazardous materials, industrial hygiene, occupational health and safety, toxicology, and environmental sciences.

★ 2413 ★ **Marcel Dekker, Inc.**
270 Madison Ave.
New York, NY 10016
Phone: 800-228-1160; (212)696-9000
Fax: (212)685-4540
Publishes scientific journals and reference books in the fields of chemistry, pesticides, and toxicology.

★ 2414 ★ **McDonald Publishers of London**
238A High St.
Uxbridge UB8 1UA, England
Publishes directories of pesticides and chemicals.

★ 2415 ★ **Meadowbrook Press, Inc.**
18318 Minnetonka Blvd.
Deephaven, MN 55391
Phone: 800-338-2232; (612)473-5400
Fax: (612)475-0736

★ 2416 ★ **Meister Publishing Company**
37841 Euclid Ave.
Willoughby, OH 44094
Phone: (216)942-2000
Publishes books on farming and farm chemicals.

★ 2417 ★ **Merck and Company, Inc.**
P.O. Box 2000
Rahway, NJ 07065
Phone: (201)855-4560
Publishes medical manuals.

★ 2418 ★ **The MIT Press**
55 Hayward St.
Cambridge, MA 02142
Phone: 800-356-0343; (617)253-5646
Fax: (617)258-6779
Publishes scholarly and professional books, nonfiction and trade books, and reference books.

★ 2419 ★ **Monthly Review Press**
122 W. 27th St.
New York, NY 10001
Phone: (212)691-2555
Publishes community action books on various subjects, including the environment, ecology, and hazardous substances.

★ 2420 ★ **National Coalition Against the Misuse of Pesticides**
530 Seventh St., S.E.
Washington, DC 20003
Phone: (202)543-5450
Publishes pamphlets, reports, and studies on a wide range of pesticide issues, including pesticide alternatives, chemicals in pesticides, and sustainable agriculture.

★ 2421 ★ **National Institute for Occupational Safety and Health**
U.S. Department of Health and Human Services
Public Health Service, Centers for Disease Control
4676 Columbia Pkwy.
Cincinnati, OH 45226-1998
Publishes reports on health hazards, hazard evaluations, industry studies, and technical topics concerning occupational safety and health. NIOSH publications must be ordered through the U.S. Government Printing Office or the National Technical Information Service.

★ 2422 ★ **National Safety Council**
444 N. Michigan Ave.
Chicago, IL 60611
Phone: 800-621-7619; (312)527-4800
Fax: (312)527-9381
Publishes a variety of pamphlets and newsletters for specific industries, each focusing on safety elements in the industry, exposure to hazardous substances, and other elements of workers' safety.

★ 2423 ★ **National Technical Information Service**
U.S. Department of Commerce
5285 Port Royal Rd.
Springfield, VA 22161
Phone: (703)487-4650
Publishes the results of both U.S. and foreign governmentsponsored research and development activities, of which the NTIS provides weekly awareness bulletins; access to more than 1.5 million technical reports of government agencies.

★ 2424 ★ **New Jersey State Department of Health**
Department of Occupational and Environmental Health
Environmental Health Section
CN 360, Seventh Floor
TRENTON, NJ 08625-0360
Phone: (609)633-2043
Publishes technical fact sheets and material safety data sheets (MSDS) reports, pesticides pamphlets, and information on hazardous substances and environmental guidelines.

★ 2425 ★ **Noyes Data Corp.**
Mill Rd. at Grand Ave.
Park Ridge, NJ 07656
Phone: (201)391-8484
Publishes environmental and hazardous substances reference books.

★ 2426 ★ **Nuclear Information and Resource Service**
1424 16th St., N.W., Ste. 601
Washington, DC 20036
Phone: (202)328-0002
A national clearinghouse and networking center for information about nuclear power issues, including radioactive waste, nuclear accidents, and health.

★ 2427 ★ **Old House Journal Corp.**
435 Ninth St.
Brooklyn, NY 11215
Phone: (718)788-1700
Fax: (718)788-9051
Publishes *Garbage,* a consumer magazine for people interested in preserving the environment.

★ 2428 ★ **Penton Publishing Co.**
1100 Superior Ave.
Cleveland, OH 44114
Phone: (216)696-7000
Fax: (216)696-8765
Publishes a controlled circulation tabloid in the field of occupational safety and health.

★ 2429 ★ **Perry-Wagner Publishing Co., Inc.**
3524 S. Wakefield St., No. A-1
Arlington, VA 22206-1729
Phone: (703)820-5083
Publishes technical books and other material on

hazardous waste and allied environmental topics.

★ 2430 ★ Pesticide Action Network/North America
965 Mission St., Ste. 514
San Francisco, CA 94103
Phone: (415)514-9140
Fax: (415)541-9253
Publishes books, newsletters, posters, and other materials in the field of pesticides and nontoxic alternatives.

★ 2431 ★ Plenum Publishing Corp.
233 Spring St.
New York, NY 10013
Phone: 800-221-9369; (212)620-8000
Publishes college textbooks and references in chemistry, physics, medicine, and the social sciences.

★ 2432 ★ Princeton Scientific Publishing Co., Inc.
P.O. Box 2155
Princeton, NJ 08543
Phone: (609)683-4750
Publishes scientific journals, including on pesticides and toxicology, and workshop proceedings.

★ 2433 ★ Raven Press
1140 Avenue of the Americas
New York, NY 10036
Phone: (212)930-9500
Fax: (212)869-3495
Scientific publisher with works on pesticide exposure in animals.

★ 2434 ★ Rodale Press, Inc.
33 E. Minor St.
Emmaus, PA 18049
Phone: 800-441-7761; (215)967-5171
Fax: (215)967-3044
Publishes books and magazines on organic gardening, nontoxic pesticides, the environment, health, nutrition, and related subjects.

★ 2435 ★ The Royal Society of Chemistry
University of Nottingham
Nottingham, Notts. NG7 2RD, England
Publishes handbooks on agricultural chemicals and pesticides.

★ 2436 ★ Roytech Publications, Inc.
840 Hinkley Rd.
Burlingame, CA 94010
Phone: (415)697-0541
Fax: (415)697-6255
Publishes chemical reference books.

★ 2437 ★ Sierra Club Books
730 Polk St.
San Francisco, CA 94109
Phone: (415)776-2211
Publishes general titles on the environment and ecology, natural history and science, sports and the outdoors, photography, and travel, as well as calendars.

★ 2438 ★ Smiling Dolphin Press
4 Segura
Irvine, CA 92715
Phone: (714)733-1065

★ 2439 ★ Specialty Technical Publishers, Inc.
2034 W. 12th Ave., Ste. 101
Vancouver, BC, Canada V6J 2G2
Phone: (604)684-5425

★ 2440 ★ Springer-Verlag New York, Inc.
175 Fifth Ave., 19th Floor
New York, NY 10010
Phone: 800-SPRINGER; (212)460-1500
Fax: (212)473-6272
Publishes scientific, medical, and technical reference books and journals.

★ 2441 ★ Stevens Publishing Corp.
P.O. Box 2573
Waco, TX 76702
Phone: (817)776-9000
Fax: (817)776-9018
Publishes a controlled circulation magazine for occupational safety and health managers.

★ 2442 ★ Taunton Press, Inc.
63 S. Main St.
P.O. Box 5506
Newtown, CT 06470-5500
Phone: 800-243-7252; (203)426-8171

★ 2443 ★ Technomic Publishing Co.
851 New Holland St.
P.O. Box 3535
Lancaster, PA 17604
Phone: 800-233-9936; (717)291-5609
Publishes environmental reference books.

★ 2444 ★ Thomson Publications
P.O. Box 9335
Fresno, CA 93791
Phone: (209)435-2163
Publishes technical reference books on farm chemicals, pesticides, and nontoxic alternatives to pesticides.

★ 2445 ★ Tower-Borner Publishing Inc.
800 Roosevelt Rd.
Bldg. C, Ste. 206
Glen Ellyn, IL 60137-5851
Phone: (708)858-1888
Fax: (708)858-1957
Publishes journals on hazardous materials and waste management.

★ 2446 ★ Unwin Hyman, Inc.
10 E. 53 St.
New York, NY 10022
Phone: (212)207-7626
Publishes college textbooks and reference books on earth and life sciences.

★ 2447 ★ U.S. Environmental Directories, Inc.
P.O. Box 65156
St. Paul, MN 55165
Publishes a directory of nongovernmental environmental and conservation organizations.

★ 2448 ★ U.S. General Accounting Office
Document Handling and Information Service
P.O. Box 6015
Gaithersburg, MD 20877
Phone: (202)275-6241
Publishes documents issued by the GAO that directly or indirectly discuss some aspects of environmental protection. The first five copies of each GAO report are free.

★ 2449 ★ U.S. Government Printing Office
Superintendent of Documents
Washington, DC 20402-9325
Phone: (202)783-3238
Handles the publishing and sale of publications issued by federal departments and agencies.

★ 2450 ★ Vance Bibliographies
112 N. Charter St.
P.O. Box 229
Monticello, IL 61856
Phone: (217)762-3831

★ 2451 ★ Van Nostrand Reinhold
115 Fifth Ave.
New York, NY 10003
Phone: (212)254-3232
Fax: (212)254-9499
Publishes technical and reference books on hazardous chemicals and their properties.

★ 2452 ★ Watson-Guptill Publications
1515 Broadway
New York, NY 10036
Phone: 800-451-1741; (212)764-7300
Fax: (212)536-5236
Publishes books on fine arts and crafts, art instruction, environmental design, and safety in art and allied fields.

★ 2453 ★ W. B. Saunders Co.
The Curtis Center
West Independence Square
Philadelphia, PA 19106
Phone: 800-545-2522; (215)238-7800
Fax: (215)238-7883
Publishes books, journals, dictionaries, and clinical studies in medicine and allied health and veterinary medicine.

★ 2454 ★ Weed Society of America
309 W. Clark St.
Champaign, IL 61820
Phone: (217)356-3182
Publishes books and other materials on pesticides and leasttoxic methods of pest control.

★ 2455 ★ Williams and Wilkins
428 E. Preston St.
Baltimore, MD 21202
Phone: 800-638-0672; (301)528-4000
Fax: (301)528-8597
Publishes medical and science books and journals.

★ 2456 ★ Worldwatch Institute
1776 Massachusetts Ave., N.W.
Washington, DC 20036

Online Databases

★ 2457 ★ AGRICOLA
U.S. National Agriculture Library
10301 Baltimore Blvd.
Beltsville, MD 20705
Phone: (301)344-3813
Fax: (301)344-3675
Contact: Gary K. McCone, Head of Database Adm. Branch
Complete bibliographic information for all monographs and serials received at the U.S. National Agriculture Library from agricultural literature. Covers pesticides, fertilizers, and other agricultural subjects. Available online through BRS Information Technologies, DIALOG, Knowledge Index, and DIMDI; on CD-ROM through OCLC Search, SilverPlatter, and Abt Books, Inc.

★ 2458 ★ Agrochemicals Handbook
Royal Society of Chemistry
Information Services
Thomas Graham House
Science Park
Milton Rd.
Cambridge, Cambs. CB4 4WF, England
Phone: (022)3 420 066
Provides chemical, physical, analytical, agricultural, toxicological, and environmental data on more than 750 active ingredients in agrochemical products used worldwide, including pesticides, herbicides, fungicides, insecticides, and rodenticides. Available online through DATA-STAR, DIALOG, Knowledge Index; on CDROM through The Pesticides Disc; on diskette through the producer.

★ 2459 ★ Asbestos Control Report
Business Publishers, Inc.
951 Pershing Dr.
Silver Springs, MD 20910-4464
Phone: 800-274-0122; (301)587-6300
Fax: (301)587-1081
Contact: Kimberlee Brown, Marketing Mgr.
Reports news and developments about the asbestos industry, including technical information on control techniques, work site health and safety, federal, state, and local standards and regulations, waste disposal, insurance requirements, and contracts available and awarded. Available online through Predicasts, Inc., and NewsNet, Inc.

★ 2460 ★ Asbestos Information System (AIS)
U.S. Environmental Protection Agency
Office of Pesticides and Toxic Substances, PM-218-B
401 M St., S.W.
Washington, DC 20460
Phone: (202)260-8844
Contains information about asbestos, including

chemical use, exposure data, and environmental releases. Available online through MEDLARS.

★ 2461 ★ BAKER Materials Safety Data Sheets
J. T. Baker Co.
222 Red School Ln.
Phillipsburg, NJ 08865
Phone: (201)859-2151
Contact: Michael Stehlin, Marketing Mgr.
Contains the complete texts of 1,688 Occupational Safety and Health Administration (OSHA) formatted material safety data sheets (MSDS). Available online through Chemical Information Systems (CIS).

★ 2462 ★ BIOSIS Previews—Biological Abstracts
BIOSIS
2100 Arch St.
Philadelphia, PA 19103-1399
Phone: 800-523-4806
Provides comprehensive worldwide coverage of research in life sciences, with extensive coverage of toxicology. Covers 9,000 primary journals as well as symposia, reviews, preliminary reports, and secondary sources. Available online through BRS, DATA-STAR, DIALOG, ORBIT, STN, and TOXLINE; on CD-ROM through SilverPlatter.

★ 2463 ★ BNA Daily News
Bureau of National Affairs, Inc.
BNA ONLINE
1231 25th St., N.W.
Washington, DC 20037
Phone: 800-862-4636; (202)452-4132
Fax: (202)822-8092
Contains the complete text of 15 BNA publications that provide news on federal government and private sector activities in the areas of the environment, hazardous materials, pollution control, solid waste, recycling, Superfund, and toxic litigation. Available online through DIALOG and LEXIS/NEXIS.

★ 2464 ★ BNA Toxics Law Daily
Bureau of National Affairs, Inc.
BNA ONLINE
1231 25th St., N.W.
Washington, DC 20037
Phone: 800-862-4636; (202)452-4132
Fax: (202)822-8092
Covers torts involving toxins and hazardous waste litigation cases and related insurance issues. Available online through HRIN, LEXIS/NEXIS, and WESTLAW.

★ 2465 ★ CCINFOdisc/CCINFOline
Canadian Centre for Occupational Health and Safety
250 Main St., E.
Hamilton, ON, Canada
Phone: 800-263-8276; (416)572-2981
Fax: (416)572-2206
Subscription service of occupational health and safety information is fully operable in both English and French. Available online through CCINFOline; on CD-ROM through CCINFOdisc.

★ 2466 ★ CHEM-BANK
SilverPlatter Information, Inc.
37 Walnut St.
Wellesley Hills, MA 02181
Phone: 800-343-0064; (617)239-0306
Fax: (617)235-1715
Contact: Elizabeth Morley, Communications Mgr.
Contains four databases relating to potentially hazardous chemicals: Registry of Toxic Effects of Chemical Substances; Oil and Hazardous Materials Technical Assistance Data System; Chemical Hazard Response Information System; and Toxic Substances Control Act (TSCA) Chemical Substances Inventory. Available on CD-ROM through SilverPlatter, Abt Books Inc., and the Faxon Company.

★ 2467 ★ Chemical Carcinogenicity Research Information System (CCRIS)
Chemical Information Systems, Inc. (CIS)
7215 York Rd.
Baltimore, MD 21212
Phone: 800-CIS-USER; (410)821-5980
Fax: (301)296-0712
Contains individual assay results and test conditions for 1,451 chemicals in the areas of carcinogenicity, cocarcinogenicity, mutagenicity, and tumor promotion. Available online through the producer.

★ 2468 ★ Chemical Collection System/ Request Tracking (CCS/RTS)
U.S. Environmental Protection Agency (EPA)
Office of Pesticides and Toxic Substances, TS-793
401 M St., S.W.
Washington, DC 20460
Phone: (202)260-3619
Provides information on properties of a number of chemicals, such as environmental and health effects and test and analysis methods. Available through the EPA.

★ 2469 ★ Chemical Evaluation Search and Retrieval System (CESARS)
Michigan State Department of Natural Resources
Surface Water Quality Division
Great Lakes and Environmental Assessment Section
Knapp's Office Centre
P.O. Box 30028
Lansing, MI 48909
Phone: (517)373-2190; (517)335-3308
Provides detailed, evaluated, fully referenced profiles of more than 194 chemicals of environmental importance to the Great Lakes Basin. Available online through CCINFOline and Chemical Information Systems (CIS); on CD-ROM through CCINFOdisc and TOMES Plus.

★ 2470 ★ Chemical Exposure
Science Applications International Corp.
800 Oak Ridge Turnpike
P.O. Box 2501
Oak Ridge, TN 37831
Phone: (615)482-9031; (615)481-2348
Contact: Marialice Wilson
Provides information on more than 1820 chemicals that have been identified in the tissues and body fluids of humans and feral and food animals. Available online through DIALOG.

★ 2471 ★ Chemical Hazard Response Information System (CHRIS)
U.S. Coast Guard
Office of Marine Safety, Security and Environmental Protection
2100 Second St., S.W., G-MER-2
Washington, DC 20590
Phone: 800-424-8802; (202)267-2611
Fax: (202)426-7881
Provides emergency response and chemical handling information for more than 1,000 chemical substances that would be involved in chemical transport accidents, particularly by water. Contains information on labeling, physical and chemical properties, health and fire hazards and hazard classifications, chemical reactivity, and water pollution. Includes safety procedures for preventing emergency situations. Available online through Chemical Information Systems (CIS); on CD-ROM through CHEM-BANK, as part of TOMES Plus System; on magnetic tape through TOMES Plus System.

★ 2472 ★ Chemical Incident Reporting Advisor
OXKO Corporation
175 Admiral Cochrane Dr., North Lobby
P.O. Box 6674
Annapolis, MD 21401
Phone: (301)266-1671
Fax: (301)266-6572
Provides information on when a chemical release incident must be reported and which report items are significant. Available on diskette through the producer.

★ 2473 ★ Chemical Regulation Reporter
Bureau of National Affairs, Inc.
BNA ONLINE
1231 25th St., N.W.
Washington, DC 20037
Phone: 800-862-4636; (202)452-4132
Fax: (202)822-8092
Reports legislative and regulatory developments affecting the production and use of pesticides and toxic substances and new and existing chemical regulations, including the control of chemicals in the air, water, land, the workplace, and during transport. The online version contains the Current Reports section only. Available online through HRIN, LEXIS, and WESTLAW.

★ 2474 ★ Chemical Regulations and Guidelines System
Network Management Inc./CRC Systems, Inc.
11242 Waples Mill Rd.
Fairfax, VA 22030
Phone: (703)359-9400
Fax: (703)273-2719
Contact: Michael Weaver, Tech. Info. Spec.
Indexes U.S. federal regulatory material relating to the control of chemical substances, providing references to each document in the cycle. Covers statutes, regulations, rules, guidelines, standards, and support documents. Available online through DIALOG.

★ 2475 ★ Chemical Right-to-Know
Bureau of National Affairs, Inc.
BNA ONLINE
1231 25th St., N.W.
Washington, DC 20037
Phone: 800-862-4636; (202)452-4132
Fax: (202)822-8092
Covers state and federal right-to-know laws for chemicals. Available online through HRIN, LEXIS, and DIALOG.

★ 2476 ★ Chemical Safety NewsBase (CSNB)
Royal Society of Chemistry
Information Services
Thomas Graham House
Science Park
Milton Rd.
Cambridge, Cambs. CB4 4WF, England
Phone: (022)3 420 066
Fax: (022)3 423 623
Contact: Alison Hibbard
Provides abstracts and indexes of international journal articles on health and safety in chemical and allied industries. Available online through DATA-STAR, DIALOG, ORBIT Search Service, and STN International.

★ 2477 ★ Chemical Substance Control
Bureau of National Affairs, Inc.
BNA ONLINE
1231 25th St., N.W.
Washington, DC 20037
Phone: 800-862-4636; (202)452-4132
Fax: (202)822-8092
Reports on government and private sector activities regarding regulatory compliance and management of toxic and hazardous chemicals and pesticides, from premanufacture through disposal. Available online through HRIN, LEXIS, and DIALOG.

★ 2478 ★ CHEMINFO/INFOCHIM
Canadian Centre for Occupational Health and Safety
250 Main St., E.
Hamilton, ON, Canada L8N 1H6
Phone: 800-263-8276; (416)572-2981
Fax: (416)572-2206
Provides information on pure chemicals, natural substances, and mixtures which result from industrial processes. Available online through CCINFOline and the producer; on CD-ROM through CCINFOdisc and the producer.

★ 2479 ★ CHEMTOX Database
Resources Consultants, Inc.
7121 CrossRoads Blvd.
P.O. 1848
Brentwood, TN 37024-1848
Phone: 800-338-2815; (615)373-5040
Fax: (615)370-4339
Contact: Richard P. Pohanish, Vice Pres.
Covers chemicals regulated by government agencies. Provides information on more than 6,000 chemical substances that are hazardous and common to the environment and the workplace. Available on diskette through the producer.

★ 2480 ★ CHEMTREC Hazard Information Transmission
Chemical Manufacturers Association (CMA)
2501 M St., N.W.
Washington, DC 20037
Phone: (202)887-1255
Provides chemical profiles and emergency response information to emergency medical services and other groups which respond to chemical emergencies; data provided by members of the CMA. Available through the producer.

★ 2481 ★ Clinical Toxicology of Commercial Products (CTCP)
Chemical Information Systems, Inc. (CIS)
7215 York Rd.
Baltimore, MD 21212
Phone: 800-CIS-USER; (410)321-8440
Fax: (410)296-0712
Information on more than 1,500 ingredients of 23,000 commercial products; based upon the print version of the same title by Gosselin, Smith, and Hodge. Available online through the producer.

★ 2482 ★ Current Contents Search
Institute for Scientific Information, Inc.
3501 Market St.
Philadelphia, PA 19104
Phone: 800-523-1857; (215)386-0100
Fax: (215)386-6362
Provides indexes and tables of contents pages from the latest issues of more than 6,500 leading scientific journals in the fields of agriculture, biological and environmental sciences, and physical, chemical, and earth sciences. Available online through BRS Information Technology, DIALOG, DIMDI, and Petro-Link Society of Petroleum Engineers; on diskette through the producer.

★ 2483 ★ Directory of Information Resources Online (DIRLINE)
National Library of Medicine
Specialized Information Services
8600 Rockville Pike
Bethesda, MD 20894
Phone: 800-638-8480; (301)496-6193
Fax: (301)480-3537
Contains telephone numbers, addresses, and descriptions of services and facilities for 15,000 organizations in mainly the health and biomedical fields, including those concerned with poisoning, genetic diseases, AIDS, and biotechnology. Available through the National Library of Medicine computer system (MEDLARS).

★ 2484 ★ Doane Pesticide Profile
Doane Marketing Research, Inc.
555 N. New Ballas
St. Louis, MO 63141
Phone: (314)993-4949
Fax: (314)993-7033
Contact: David M. Tugend, Mgr. of Client Services
Contains annual measurements of U.S. growers' purchases of pesticides for major and minor crops. Available online through Telmar Group, Inc.

★ 2485 ★ Electronic Information Exchange System (EIES)
U.S. Environmental Protection Agency (EPA)
Office of Research and Development, RD-618
401 M St., S.W.
Washington, DC 20460
Phone: 800-424-9346; (703)506-1025
Hotline: (703)506-1025
Provides information about government and industry pollution prevention programs, grant and project funding opportunities, and upcoming events, conferences, and seminars.

★ 2486 ★ ENFLEX DATA
ERM Computer Services, Inc.
855 Springdale Rd.
Exton, PA 19341
Phone: 800-544-3118; (215)524-3600
Contact: George Esry, National Sales Mgr.
Provides the information necessary to monitor hazardous chemicals in the workplace and to file the reports required by the U.S. Environmental Protection Agency (EPA) and Occupational Safety and Health Administration (OSHA); also includes information to prepare material safety data sheets (MSDS) and Section 313 reports. Available on CD-ROM through the producer.

★ 2487 ★ ENFLEX INFO
ERM Computer Services, Inc.
855 Springdale Rd.
Exton, PA 19341
Phone: 800-544-3118; (215)524-3600
Contact: George Esry, National Sales Mgr.
Provides complete text of all U.S. federal and state environmental regulations, including those covering hazardous materials, their transportation, and their health and safety requirements. Available on CD-ROM through the producer.

★ 2488 ★ Environmental Bibliography
International Academy at Santa Barbara
Environmental Studies Institute
800 Garden St., Ste. D
Santa Barbara, CA 93101-1553
Phone: (805)965-5010
Contact: Joanne St. John, Vice President
Provides citations to current international periodical literature dealing with environmental topics such as chemical waste, air and water pollution, water treatment, energy conservation, noise abatement, and soil mechanics. Available online through DIALOG.

★ 2489 ★ Environmental Fate (ENVIROFATE)
U.S. Environmental Protection Agency (EPA)
Office of Pesticides and Toxic Substances, MS-TS799
401 M St., S.W.
Washington, DC 20460
Phone: (202)260-1404
Provides information on the environmental fate or behavior of 800 chemicals which are released into the environment. Available online through Chemical Information Systems (CIS).

★ 2490 ★ Environmental Health News (EHN)
Occupational Health Services, Inc.
11 W. 42nd St., 12th Floor
New York, NY 10036
Phone: 800-445-MSDS; (212)789-3535
Fax: (212)268-9276
Provides news alerts on late-breaking events related to hazardous substances, including court decisions, regulatory changes, and medical and scientific news. Available online through HRIN and the producer.

★ 2491 ★ Environmental Law Reporter
Environmental Law Institute
1616 P St., N.W., Ste. 200
Washington, DC 20036
Phone: (202)328-5150
Provides news, analysis, commentary, primary documents, and other materials dealing with environmental law, statutes, regulations, litigation, and Superfund cases. Available online through WESTLAW.

★ 2492 ★ Environmental Mutagen Information Center (EMIC)
U.S. Department of Energy
Oak Ridge National Laboratory
Environmental Mutagen Information Center
Bldg. 9224
P.O. Box Y
Oak Ridge, TN 37830
Phone: (615)574-7871
Provides information on chemical, biological, and physical agents tested for mutagenicity. Available online through the producer.

★ 2493 ★ Environment Library
Online Computer Library Center, Inc. (OCLC)
6565 Frantz Rd.
Dublin, OH 43017
Phone: 800-848-5878; (614)764-6000
Fax: (614)764-6096
Provides full bibliographic and cataloging information for English- and foreign-language materials published in all formats on environmental issues. Material selected from the OCLC Online Union Catalog (OLUC) by a combination of Library of Congress and Dewey Decimal classification numbers. Available on CD-ROM through the producer.

★ 2494 ★ Environment Reporter
Bureau of National Affairs, Inc.
BNA ONLINE
1231 25th St., N.W.
Washington, DC 20037
Phone: 800-862-4636; (202)452-4132
Fax: (202)822-8092
Reports legislative, regulatory, judicial, industrial, and technical activities and developments concerning the environment. The online version contains only the Current Reports section. Available online through HRIN, LEXIS, and WESTLAW.

★ 2495 ★ Environment Week
King Communications Group, Inc.
627 National Press Bldg.
Washington, DC 20045
Phone: (202)638-4260
Fax: (202)662-9744
Contact: JoAnn Wood, Asst. Publisher
Provides news on current environmental issues from the U.S. Environmental Protection Agency (EPA), Congress, the Department of Energy, and other government agencies. Available online through NewsNet, Inc.

★ 2496 ★ Federal Register Search System (FRSS)
Chemical Information Systems, Inc. (CIS)
7215 York Rd.
Baltimore, MD 21212
Phone: 800-CIS-USER; (410)821-5980
Fax: (301)296-0712
Contains regulations, rules, standards, and guidelines involving chemical substances accumulated by the U.S. Environmental Protection Agency (EPA) from 1978 through November, 1983. The EPA ceased funding the updates to the Federal Register but the citations (162,492 of them) remain useful information. Available online through the producer.

★ 2497 ★ FIREDOC
U.S. National Institute of Standards and Technology
Center for Fire Research
Bldg. 224, Rm. A252
Gaithersburg, MD 20899
Phone: (301)975-6862
Fax: (301)975-4052
Contact: Nora H. Jason, Project Leader
Contains citations to fire research publications and articles, including combustion toxicology, extinguishment, and suppression. Available online through the producer.

★ 2498 ★ From the State Capitals: Waste Disposal and Pollution Control
Wakeman/Walworth, Inc.
300 N. Washington St., Ste. 204
Alexandria, VA 22314
Phone: (703)549-8606
Fax: (703)549-1372
Contact: Keyes Walworth, ed.
Reports state and municipal environmental laws, regulations, and developments concerning pesticide policies, the transportation, storage, and clean-up of hazardous wastes, resource recovery, and insurance protection. Available online through WESTLAW.

★ 2499 ★ Genetic Toxicity (GENETOX)
U.S. Environmental Protection Agency (EPA)
Office of Pesticides and Toxic Substances, TS-799
401 M St., S.W.
Washington, DC 20460
Phone: (202)260-1404
Provides mutagenicity information on more than 4,300 chemicals tested on 38 biological systems. Available online through Chemical Information Systems (CIS).

★ 2500 ★ Hazard Communication Compliance Manual Database
Bureau of National Affairs, Inc.
1231 25th St., N.W.
Washington, DC 20037
Phone: 800-862-4636; (202)452-4132
Fax: (202)822-8092
Information on how to establish a hazard right-to-know program in the workplace. Available on diskette through the producer.

★ 2501 ★ HAZARDLINE
Occupational Health Services, Inc. (OHS)
11 W. 42nd St., 12th Floor
New York, NY 10036
Phone: 800-445-MSDS; (212)789-3535
Provides full-text information on more than 10,000 hazardous substances used in the workplace. Information includes physical and chemical profiles, standards and regulations, and emergency response procedures. Available online through BRS Information Technologies, Inc., HRIN, and through the producer.

★ 2502 ★ Hazardous Chemicals Information and Disposal (HAZINF)
University of Alberta
Department of Chemistry
Edmonton, AB, Canada T6G 2G2
Phone: (403)492-3254
Provides instructions for safe handling and disposal of more than 220 hazardous chemicals that are likely to occur in the laboratory, compiled from *Hazardous Chemicals Information and Disposal Guide*. Available online through Chemical Information Systems (CIS).

★ 2503 ★ Hazardous Materials Intelligence Report
World Information Systems
P.O. Box 535
Cambridge, MA 02238
Phone: (617)491-5100
Fax: (617)492-3312
Provides information on federal, state, and local legislation, regulations, and programs related to hazardous waste and hazardous materials management. Includes research news, book reviews, contract opportunities, and calendar of events. Available online through NewsNet, Inc.

★ 2504 ★ Hazardous Substances Data Bank (HSDB)
National Library of Medicine
Toxicology Information Program
8600 Rockville Pike
Bethesda, MD 20894
Phone: (301)496-1131
Fax: (301)480-3537
Contact: Jeanne C. Goshorn, Chief, Bioinformation Services
Contains toxicological information on more than 4,200 chemicals augmented by data for such aspects as emergency handling procedures, environmental fate, human exposure, detection methods, and regulatory requirements. Available online through DATA-STAR, DIMDI, ORBIT, and MEDLARS (as part of TOXNET); on CD-ROM as part of TOMES Plus System and SilverPlatter.

★ 2505 ★ Hazardous Waste News
Business Publishers, Inc.
951 Pershing Dr.
Silver Springs, MD 20910-4464
Phone: 800-274-0122; (301)587-6300
Fax: (301)587-1081
Contact: Kimberlee Brown, Marketing Mgr.
Covers legislative, regulatory, and judicial decisions at the federal and state levels relating to hazardous waste management. Available online through NewsNet, Inc., and Predicasts, Inc.

★ 2506 ★ Industrial Health and Hazards Update
Merton Allen Associates
P.O. Box 15640
Plantation, FL 33318-5640
Phone: (305)473-9560
Contact: Merton Allen, Pres.
Contains abstracts from reports on industrial health and hazards, covering such topics as regulations, legislation, litigation, plant surveys, and toxicity of industrial materials and products. Available online through NewsNet, Inc, and Predicasts, Inc.

★ 2507 ★ Information System for Hazardous Organics in Water (ISHOW)
U.S. Environmental Protection Agency (EPA)
Office of Pesticides and Toxic Substances, TS-799
401 M St., S.W.
Washington, DC 20460
Phone: (202)260-1404
Provides information on more than 5,000 hazardous chemicals found in water, including the chemical's melting and boiling points, water solubility, and vapor pressure. Available online through Chemical Information Systems (CIS).

★ 2508 ★ Integrated Risk Information System (IRIS)
U.S. Environmental Protection Agency (EPA)
Office of Research and Development
Research Triangle Park, NC 27711
Provides information on how levels of exposure to some 400 hazardous chemicals effect human health. Covers levels of exposure to hazardous chemicals below which no adverse health effects are expected to occur in various segments of the human population. Available online through Chemical Information Systems (CIS) and TOXNET; on CD-ROM through TOMES Plus.

★ 2509 ★ International Register of Potentially Toxic Chemicals (IRPTC)
United Nations Environment Programme (UNEP)
Palais des Nations
CH-1211 Geneva, Switzerland
Phone: (022)798 5850
Fax: (022)733 2673
Provides information on the physical and chemical properties of chemicals that are potentially toxic to humans as well as hazardous to the environment. Available online as part of Environmental Chemicals Data and Information Network.

★ 2510 ★ Material Safety Data Sheets (MSDS)
Occupational Health Services, Inc. (OHS)
11 W. 42nd St., 12th Floor
New York, NY 10036
Phone: 800-445-MSDS; (212)789-3535
Fax: (212)789-3646
Provides more than 60,000 Occupational Safety and Health Administration (OSHA) formatted MSDS. Fifteen thousand available online through National Pesticides Information Retrieval System; on CD-ROM through the producer; on microfiche through the producer.

★ 2511 ★ National Air Toxics Information Clearinghouse (NATICH)
U.S. Environmental Protection Agency (EPA)
Pollutant Assessment Branch
Office of Air Quality Planning and Standards (MD-13)
Research Triangle Park, NC 27711
Phone: (919)541-0850
Assists federal, state, and local agencies in exchanging information about air toxics and the development of air toxic programs; contains all of the air pollutant information collected from federal, state, and local agencies and is organized according to agency, pollutant, and emission source.

★ 2512 ★ National Safety Council Library Database
National Safety Council
444 N. Michigan Ave.
Chicago, IL 60611
Phone: (312)527-4800
Fax: (312)527-7381
Contact: Robert J. Marecek, Library Mgr.
Furnishes citations to books, journals, and other material relating to health and safety, including accidents and injuries from the misuse of hazardous materials. Available through the producer.

★ 2513 ★ New Jersey Department of Health Hazardous Substance Fact Sheets
New Jersey Department of Health
Right-to-Know Program
Trenton, NJ
Provides more than 560 chemical-specific fact sheets. Available online through RACHEL; on CD-ROM through CCINFOdisc.

★ 2514 ★ NIOSH Database
National Institute of Occupational Safety and Health
U.S. Public Health Service
Standards Development and Technology Transfer Division
4676 Columbia Pkwy.
Cincinnati, OH 45226
Phone: (513)684-8326
Covers all aspects of occupational safety and health, including hazardous agents and waste, unsafe environment in the workplace, toxicology, chemistry, and control technology. Available through the producer.

★ 2515 ★ NIOSH Technical Information Center Database (NIOSHTIC)
U.S. National Institute for Occupational Safety and Health
Standards Development and Technology Transfer Division
Technical Information Branch
4676 Columbia Pkwy.
Cincinnati, OH 45226
Phone: 800-35-NIOSH; (513)533-8326
Contact: Vivian Morgan, Chief, Tech. Info. Branch
Provides citations and abstracts of published and unpublished technical literature dealing with occupational safety and health worldwide, including works on industrial hygiene, chemistry, and toxicology. Available online through CCINFOline, DIALOG, and ORBIT Search Service; on CD-ROM through CCINFOdisc and OSH-ROM.

★ 2516 ★ NPIRS Pesticide and Hazardous Chemical Databases
National Pesticide Information Retrieval System
Purdue University
Entomology Hall, Rm. 1158
West Lafayette, IN 47907
Phone: (317)494-6616
Fax: (317)494-0535
Contact: Ed Ramsey, NPIRS User Service Mgr.
Provides information on more than 60,000 pesticide products and hazardous chemicals registered with the U.S. Environmental Protection Agency (EPA) and 36 state agencies. Available online through the producer; on CD-ROM through SilverPlatter and the Faxon Company.

★ 2517 ★ Oil and Hazardous Materials Technical Assistance Data System (OHM/TADS)
U.S. Environmental Protection Agency (EPA)
Emergency Response Division
Office of Solid Waste and Emergency Response (OS-210)
401 M St., S.W.
Washington, DC 20460
Phone: (202)260-2190
Fax: (202)260-2155
Contact: John Cunningham, Project Officer
Provides emergency spill response personnel with immediate information to identify 1,402 substances designated as oils or hazardous materials by the EPA from their physical, chemical, and toxicological properties, and assists to formulate correct reaction. Available online through Chemical Information Systems (CIS); on CD-ROM as part of CHEM-BANK and as part of TOMES Plus System.

★ 2518 ★ OPTS Regulation Tracking System (OPTS/RTS)
U.S. Environmental Protection Agency (EPA)
Office of Pesticides and Toxic Substances (TS-793)
401 M St., S.W.
Washington, DC 20460
Phone: (202)260-3619
Provides histories of various pesticides and toxics regulations and information about compliance with them. Available through the producer.

★ 2519 ★ OSH-ROM
SilverPlatter Information, Inc.
37 Walnut St.
Wellesley Hills, MA 02181
Phone: 800-343-0064; (617)239-0306
Fax: (617)235-1715
Contact: Elizabeth Morley, Communications Mgr.
Contains references to the world's literature dealing with all aspects of occupational health and safety. Available on CD-ROM through the producer.

★ 2520 ★ Pest Control Literature Documentation (PESTDOC)
Derwent Publications Ltd.
Rochdale House
128 Theobalds Rd.
London WC1X 8RP, England
Phone: 800-451-3451; (071)242 5823
Fax: (071)405 3630
Contact: Brian Gore, Marketing Mgr.
Provides references and abstracts from scientific journals on pesticides and agricultural chemicals. Available online through ORBIT Search Service.

★ 2521 ★ The Pesticides Disc
Royal Society of Chemistry
Information Services
Thomas Graham House
Science Park
Milton Rd.
Cambridge, Cambs. CB4 4WF, England
Phone: (022)342 0066
Fax: (022)342 3623
Contact: Alison Hibberd, Sales Exec.
Provides technical and commercial data on pesticides, directories of agricultural chemicals and products, and information on worldwide pesticide control organizations. Available on CD-ROM through Pergamon Press; on diskette through the producer.

★ 2522 ★ Poisindex System
Micromedex, Inc.
600 Grant St.
Denver, CO 80203-3527
Phone: 800-525-9083; (303)831-1400
Fax: (303)837-1717
Contact: Allen C. Howerton, Vice Pres., Sales and Marketing
Provides toxicity information on more than 650,000 commercial products, plants, and drugs. Available online as part of MEDIS; on DC-ROM as part of the Computerized Clinical Information System (CCIS).

★ 2523 ★ PressNet Environmental Reports (PER)
PressNet Systems, Inc.
400 E. Pratt St.
Baltimore, MD 21202
Phone: 800-666-3236
Contains summaries of articles on environmental issues published in newspapers around the U.S. Available online through Chemical Information Systems (CIS).

★ 2524 ★ Registry of Toxic Effects of Chemical Substances (RTECS)
National Institute for Occupational Safety and Health
U.S. Public Health Service
Standards Development and Technology Transfer Division
4676 Columbia Pkwy.
Cincinnati, OH 45226
Phone: (513)684-8326
Provides test results for more than 108,500 chemical substances in four categories: substance identification; toxicity/biomedical effects; toxicology and carcinogenicity review; and exposure standards and regulations. Available online through CCINFOline, Chemical Information Systems (CIS), DIALOG, DIMDI, BRS, MEDLARS (as part of TOXNET), and the producer; on CD-ROM through CCINFOdisc, CHEM-BANK, and TOMES Plus System.

★ 2525 ★ Regscan 4.0
Regulation Scanning
30 W. Third St.
Williamsport, PA 17701
Phone: 800-326-9303
Fax: (717)323-8082
Provides safety, environmental, and transportation regulations appearing in the Code of Federal Regulations (CFR), including 29 CFR (Occupational Safety and Health Administration; OSHA), 40 CFR (U.S. Environmental Protection Agency; EPA), and 49 CFR (Department of Transportation; DOT). Available on diskette through the producer.

★ 2526 ★ Regulated Materials Database (REGMAT)
Chemical Information Systems, Inc. (CIS)
7215 York Rd.
Baltimore, MD 21212
Phone: 800-CIS-USER; (410)821-5980
Fax: (301)296-0712
Summarizes the U.S. federal regulations as they relate to chemical substances under the Toxic Substances Control Act, Safe Drinking Water Act, Hazardous Materials Transportation Act, and the Occupational Safety and Health Act. Available online through the producer.

★ 2527 ★ REPROTOX
Reproductive Toxicology Center
Columbia Hospital for Women
2440 M St., N.W., Ste. 217
Washington, DC 20037-1404
Phone: 800-535-1400; (202)293-5137
Contact: Anthony R. Scialli, Dir.
Provides current information and summaries of articles dealing with industrial and environmental chemicals and their effects upon human fertility, pregnancy, and fetal development. Available online through the producer.

★ 2528 ★ SARA!
OSHA-Soft, Inc.
P.O. Box 668
Amherst Station
Amherst, MA 03031
Phone: 800-446-3427; (603)672-7230
Contact: Peter E. Bragdon, Pres.
Provides reports on hazardous chemicals in the workplace required by Title III of the U.S. Environmental Protection Agency (EPA) Superfund Amendments and Reauthorization Act (SARA). Available on diskette through the producer.

★ 2529 ★ Standard Pesticide File (SPF)
Derwent Publications Ltd
Rochdale House
128 Theobalds Rd.
London WC1X 8RP, England
Phone: 800-451-3451; (071)242 5823
Fax: (703)790-1426
Contact: Brian Gore, Marketing Mgr.
Lists more than 3,900 known pesticides and agricultural chemicals. Available online through ORBIT Search Service.

★ 2530 ★ State Environment Report
Business Publishers, Inc.
951 Pershing Dr.
Silver Springs, MD 20910-4464
Phone: 800-274-0122; (301)587-6300
Fax: (301)587-1081
Contact: Kimberlee Brown, Marketing Mgr.
Provides information on toxic substances control and hazardous waste management at the state and local levels, with particular attention given to programs relating to toxic substances used in the workplace. Available online through NewsNet, Inc., and Predicasts, Inc.

★ 2531 ★ Superfund
Pasha Publications, Inc.
1401 Wilson Blvd., Ste. 900
Arlington, VA 22209
Phone: 800-424-2908; (703)528-1244
Fax: (703)528-1253
Reports news and developments in the cleanup of hazardous wastes required under federal and state Superfund laws. Available online through NewsNet, Inc., and Predicasts, Inc.

★ 2532 ★ Suspect Chemicals Sourcebook (SCS)
Roytech Publications, Inc.
840 Hinckley Rd., Ste. 147
Burlingame, CA 94010
Phone: (415)697-0541
Fax: (415)697-6255
Contact: Kenneth B. Clansky, Pres.
Provides references to federal regulations and data pertaining to the manufacture, sales, storage, use, and transportation of more than 5,000 industrial chemicals. Covers more than 50 federal advisory and regulatory programs and selected state and international regulations. Includes data on hazardous substances, health and safety regulations, and air and water quality criteria. Available online through Chemical Information Systems (CIS).

★ 2533 ★ TOMES Plus System
Micromedex, Inc.
600 Grant Ave.
Denver, CO 80203-3527
Phone: 800-525-9083; (303)831-1400
Fax: (303)837-1717
Contact: Allen C. Howerton, Vice Pres.
Contains toxicological, environmental, and industrial medicine information on the hazardous effects related to chemical exposure in manufacturing and transportation, and gives the proper response in chemical emergency situations. Contains several proprietary files as well as files of major international journals, standard reference sources, professional specialists, chemical manufacturers, government agencies, and poison centers. Available on CD-ROM through the producer.

★ 2534 ★ Toxic Chemical Release Inventory (TRI)
U.S. Environmental Protection Agency (EPA)
401 M St., S.W.
Washington, DC 20406
Phone: (202)260-3596
Contains the annual estimated releases of more than 300 toxic chemicals into the environment by reporting facilities, based upon information collected by the EPA. Available online through TOXNET; on CD-ROM through the U.S. Government Printing Office.

★ 2535 ★ Toxic Materials News
Business Publishers, Inc.
951 Pershing Dr.
Silver Springs, MD 20910-4464
Phone: 800-274-0122; (301)587-6300
Fax: (301)587-1081
Contact: Kimberlee Brown, Marketing Mgr.
Monitors developments in legislation, regulations, and litigation concerning toxic substances. Available online through Predicasts, Inc., and NewsNet, Inc.

★ 2536 ★ Toxic Materials Transport
Business Publishers, Inc.
951 Pershing Dr.
Silver Springs, MD 20910-4464
Phone: 800-274-0122; (301)587-6300
Fax: (301)587-1081
Contact: Kimberlee Brown, Marketing Mgr.
Covers current news and trends relating to hazardous materials shipments by rail, trucks, barge, air, and pipeline. Available online through Predicasts, Inc., and NewsNet, Inc.

★ 2537 ★ Toxics Law Reporter
Bureau of National Affairs, Inc.
1231 25th St., N.W.
Washington, DC 20037
Phone: 800-862-4636; (202)452-4132
Fax: (202)822-8092
Covers toxic tort and hazardous waste litigation cases, news, and related liability issues. Available online through HRIN and LEXIS.

★ 2538 ★ Toxic Substances Control Act (TSCA) Chemical Substances Inventory
U.S. Environmental Protection Agency (EPA)
Office of Pesticides and Toxic Substances (TS-799)
401 M St., S.W.
Washington, DC 20460
Phone: (202)260-1404
Provides information about chemical substances manufactured, imported, or processed in the United States for commercial purposes. Available online through DIALOG and CHEMLINE.

★ 2539 ★ TOXLINE (NLM)
National Library of Medicine
Toxicology Information Program
8600 Rockville Pike
Bethesda, MD 20894
Phone: (301)496-1131
Fax: (301)480-3537
Contact: Jeanne C. Goshorn, Chief, Bioinformation Services
Provides citations to published sources of information on the pharmacological, biochemical, physiological, and toxicological effects of drugs and other chemicals. TOXLINE has four files: TOXLINE (covers 1981 to the present), TOXLINE65 (covers 1965 through 1980), TOXLIT (Toxicology Literature from Special Sources; covers 1981 to the present), and TOXLIT65 (covers 1965 through 1980). Available online through BRS Information Technologies, DATASTAR, DIMDI, DIALOG, and MEDLARS; on CD-ROM through SilverPlatter.

★ 2540 ★ TSCA Plant and Production Data (TSCAPP)
Chemical Information Systems, Inc. (CIS)
7215 York Rd.
Baltimore, MD 21212
Phone: 800-CIS-USER; (410)821-5980
Fax: (301)296-0712
Contains 127,000 production citations for 53,000 unique substances, representing the non-confidential portion of reports received by the U.S. Environmental Protection Agency (EPA) under the Toxic Substances Control Act (TSCA). Available online through the producer.

★ 2541 ★ WESTLAW Environmental Law Library
West Publishing Company
50 W. Kellogg Blvd.
P.O. Box 64526
St. Paul, MN 55164-0526
Phone: 800-WES-TLAW; (612)228-2500
Provides the complete texts of U.S. federal court decisions, statutes, regulations, administrative law publications, specialized files, texts and periodicals dealing with environmental law. Available online through the producer.

Online Database Producers

★ 2542 ★ **American Chemical Society**
ACS
Distribution Department 127
West End Station
P.O. Box 57136
Washington, DC 20037
Phone: 800-227-5558

★ 2543 ★ **Aries Systems Corp.**
ARIES
79 Boxford St.
North Andover, MA 01845-3219
Phone: (508)689-9334

★ 2544 ★ **Bibliographic Services**
BLAISE-LINE
Marketing and Support Group
2 Sheraton St.
London W1V 4BH, England
Phone: (01)323 7077

★ 2545 ★ **BIOSIS Connection**
2100 Arch St.
Philadelphia, PA 19103-1399
Phone: 800-523-4806

★ 2546 ★ **BRS Information Technologies**
BRS/COLLEAGUE
8000 Westpark Dr.
McLean, VA 22102
Phone: 800-955-0906

★ 2547 ★ **Burrelle's Information Services**
75 E. Northfield Rd.
Livingston, NJ 07039
Phone: 800-631-1160

★ 2548 ★ **C.A.B. International**
CABI
Wallingford, Oxon. 0X10 8DE, England

★ 2549 ★ **Cambridge Scientific Abstracts**
7200 Wisconsin Ave.
Bethesda, MD 20814
Phone: 800-843-7751

★ 2550 ★ **Canadian Centre for Occupational Health and Safety**
CCOHS
Inquiries Service
250 Main St., E.
Hamilton, ON, Canada L8N 1H6
Phone: 800-263-8276; (416)572-2981
Fax: (416)572-2206

★ 2551 ★ **CD Plus**
951 Amsterdam Ave., Ste. 2C
New York, NY 10025

★ 2552 ★ **Chadwyck-Healey**
Electronic Publishing Division
1101 King St.
Alexandria, VA 22314
Phone: (703)683-4890

★ 2553 ★ **Chemical Information Systems, Inc.**
CIS
Fein-Marquart Associates
7215 York Rd.
Baltimore, MD 21212
Phone: 800-CIS-USER; (410)321-8440
Fax: (410)296-0712

★ 2554 ★ **Chemical Manufacturers Association**
CHEMTREC HITS
2501 M St., N.W.
Washington, DC 20037
Phone: (202)887-1255

★ 2555 ★ **Commerce Clearing House, Inc.**
ELSS (Electronic Legislative Search System)
4025 W. Peterson Ave.
Chicago, IL 60646
Phone: (312)940-4600

★ 2556 ★ **Compact Publishing, Co.**
P.O. Box 40310
Washington, DC 20016
Phone: (202)224-4770

★ 2557 ★ **Congressional Information Service**
4520 East-West Hwy., Ste. 800
Bethesda, MD 20814-3389
Phone: 800-638-8380

★ 2558 ★ **Cornell University**
Media Services Distribution Center
8 Research Park
Ithaca, NY 14850
Phone: (607)255-2091

★ 2559 ★ **Counterpoint Publishing**
20 Williams St., Ste. G-70
P.O. Box 9135
Wellesley, MA 02181
Phone: (617)235-4667

★ 2560 ★ **Data-Star Marketing, Inc.**
485 Devon Park Dr., Ste. 110
Wayne, PA 19087
Phone: 800-221-7754; (215)687-6777

★ 2561 ★ **DataTimes**
818 N.W. 63rd St.
Oklahoma City, OK 73116
Phone: 800-642-2525

★ 2562 ★ **DIALOG Information Services, Inc.**
3460 Hillview Ave.
Palo Alto, CA 94304
Phone: 800-334-2564; 800-982-5838

★ 2563 ★ **Digital Diagnostics, Inc.**
601 University Ave.
Sacramento, CA 95825
Phone: (916)456-5931

★ 2564 ★ **Earthnet**
P.O. Box 330072
Kahului, Maui, HI 96733
Phone: (808)872-6090

★ 2565 ★ **EBSCO Electronic Information**
P.O. Box 13787
Torrance, CA 90503
Phone: 800-888-3272

★ 2566 ★ **Econet**
3228 Sacramento
San Francisco, CA 94114

★ 2567 ★ **Environmental Research Foundation**
RACHEL
P.O. Box 3541
Princeton, NJ 08543-3541
Phone: (609)683-0707

★ 2568 ★ **Environmental Technical Information System**
ETIS-CELDS
CELDS Office
Urbana, IL 61801
Phone: (217)333-3675

★ 2569 ★ **The Fire Service Software Co.**
FIRSTsystem
134 Middle Neck Rd., Ste. 210
Great Neck, NY 11021
Phone: (516)829-5858

★ 2570 ★ **Global Education Motivators**
GEMNET
Chestnut Hill College
Germantown and Northwestern Ave.
Chestnut Hill, PA 19118-2695
Phone: (215)248-1150

★ 2571 ★ **Greenpeace Action**
ENVIRONET
Bldg. E
Fort Mason, CA 94123
Phone: (415)474-6767

★ 2572 ★ Hazox, Inc.
P.O. Box 637
Chadds Ford, PA 19317
Phone: (215)388-2030

★ 2573 ★ The Human Resource Information Network
HRIN
Executive Telecom System, Inc.
9585 Valparaiso Court
Indianapolis, IN 46268
Phone: 800-421-8884; (317)872-2045
Fax: (317)872-2059

★ 2574 ★ The H.W. Wilson Company
WILSONDISC/WILSONLINE
950 University Ave.
Bronx, NY 10452
Phone: 800-622-4002

★ 2575 ★ Information Access Company
IAC
362 Lakeside Dr.
Foster City, CA 94404
Phone: 800-227-8431

★ 2576 ★ Information Handling Service
IHS
15 Inverness Way E.
P.O. Box 1154
Englewood, CO 80150
Phone: 800-241-7824

★ 2577 ★ Institute for Scientific Information
ISI
3501 Market St.
Philadelphia, PA 19104
Phone: 800-523-1850

★ 2578 ★ IST-Informatheque, Inc.
2, Complexe Desjardins
Montreal, PQ, Canada H5B 1B3
Phone: (514)284-1111

★ 2579 ★ J. B. Lippincott Company
Lippincott Information Services
E. Washington Square
Philadelphia, PA 19105
Phone: 800-523-2945

★ 2580 ★ Longman/Microinfo World Research Database
Omega Park
P.O. Box 3
Alton, Hants. 0420-86848, England

★ 2581 ★ Maxwell Electronic Publishing
124 Mount Auburn St.
Cambridge, MA 02138
Phone: 800-342-1338

★ 2582 ★ McGraw-Hill, Inc.
11 W. 19th St.
New York, NY 10011
Phone: (212)512-2000

★ 2583 ★ Mead Data Central, Inc.
LEXIS/NEXIS
0443 Springboro Pike
P.O. Box 933
Dayton, OH 45401
Phone: 800-346-9759; 800-543-6862

★ 2584 ★ Micromedex, Inc.
600 Grant St.
Denver, CO 80203-3527
Phone: 800-525-9083; (303)831-1400
Fax: (303)837-1717

★ 2585 ★ National Air Toxics Information Clearinghouse
NATICH (EPA)
U.S. Environmental Protection Agency
Pollutant Assessment Branch (MD-12)
Research Triangle Park, NC 27711
Phone: (919)541-0850

★ 2586 ★ National Groundwater Information Center
NGWIC
500 W. Wilson Bridge Rd.
Worthington, OH 43085
Phone: (614)846-9355

★ 2587 ★ National Information Services Corp.
NISC
Wyman Towers
3100 St. Paul St., Ste. 6
Baltimore, MD 21218
Phone: (301)243-0797

★ 2588 ★ National Library of Medicine
MEDLARS Management Section
Bldg. 38-A, Rm. 4N421
Bethesda, MD
Phone: 800-638-8480

★ 2589 ★ National Library of Medicine
TOXNET
MEDLARS Management Section
Bldg. 38-A, Rm. 4N421
Bethesda, MD
Phone: 800-638-8480

★ 2590 ★ National Pesticide Information Retrieval System
NPIRS
Purdue University
Entomology Hall
West Lafayette, IN 47907
Phone: (317)494-6614

★ 2591 ★ National Safety Council
CAMEO
444 N. Michigan Ave.
Chicago, IL 60611
Phone: (312)527-4800

★ 2592 ★ National Technical Information Service
NTIS
5285 Port Royal Rd.
Springfield, VA 22161
Phone: (703)487-4650

★ 2593 ★ NEWSNET, Inc.
945 Haverford Rd.
Bryn Mawr, PA 19010
Phone: 800-952-0122; (215)527-8030

★ 2594 ★ Nimbus Information Systems
P.O. Box 7427
Charlottesville, VA 22906
Phone: 800-782-0778

★ 2595 ★ Occupational Health Services, Inc.
OHS
11 W. 42nd St., 12th Floor
New York, NY 10036
Phone: 800-445-MSDS; (212)789-3535
Fax: (212)789-3646
Contact: Richard Cohen, Marketing Dir.

★ 2596 ★ Online Computer Library Center, Inc.
EPIC
6565 Frantz Rd.
Dublin, OH 43017-0702
Phone: 800-848-8286; 800-848-5878

★ 2597 ★ Online Computer Library Center, Inc.
OCLC
6565 Frantz Rd.
Dublin, OH 43017-0702
Phone: (614)764-6000

★ 2598 ★ Online Products Corporation
20251 Century Blvd.
Germantown, MD 20874
Phone: 800-922-9204

★ 2599 ★ ORBIT Search Service
8000 Westpark Dr.
McLean, VA 22102
Phone: 800-456-7248; (703)442-0900

★ 2600 ★ OXKO Corporation
175 Admiral Cochrane Dr., North Lobby
P.O. Box 6674
Annapolis, MD 21401
Phone: (301)266-1671
Fax: (301)266-6572

★ 2601 ★ Predicasts, Inc.
University Circle Research Center
11001 Cedar Ave.
Cleveland, OH 44106
Phone: 800-321-6388; (216)795-3000
Fax: (216)229-9944

★ 2602 ★ The Public Health Foundation
PHNET (Public Health Network)
1220 L St., N.W., Ste. 350
Washington, DC 20006
Phone: (202)898-5600

★ 2603 ★ Questel
16 I St., N.W., Ste. 818
Washington, DC 20006
Phone: 800-424-9600

★ 2604 ★ Reproductive Toxicology Center
REPROTOX
2425 L St., N.W.
Washington, DC 20037
Phone: (202)293-5137

★ 2605 ★ Resources Consultants, Inc.
RCI
7121 CrossRoads Blvd.
P.O. Box 1848
Brentwood, TN 37024-1848
Phone: (615)373-5040
Fax: (615)370-4339
Contact: Richard P. Pohanish, Vice Pres.

★ 2606 ★ R. R. Bowker Co.
121 Chanlon Rd.
New Providence, NJ 07974
Phone: 800-521-8110

★ 2607 ★ SilverPlatter Information, Inc.
One Newton Executive Park
Newton Lower Falls, MA 02162-1449
Phone: 800-343-0064; (617)969-5554

★ 2608 ★ STN International
Chemical Abstracts Service
2540 Olentangy Rd.
P.O. Box 3012
Columbus, OH 43210
Phone: 800-848-6533; 800-848-6538; 800-848-6535

★ 2609 ★ Timeplace, Inc.
EDNET
460 Totten Pond Rd.
Waltham, MA 02154
Phone: 800-544-4023

★ 2610 ★ **Toxicology Information Response
Center**
TIRC
Oak Ridge National Laboratory
Bldg. 2001
P.O. Box 2008
Oak Ridge, TN 37831-6050
Phone: (615)576-1746
Fax: (615)574-9888
Contact: Kimberly Slusher, Exec. Officer

★ 2611 ★ **University Microfilms International**
UMI
300 N. Zeeb Rd.
Ann Arbor, MI 48106
Phone: 800-521-0600

★ 2612 ★ **U.S. Government Printing Office**
USGPO
Superintendent of Documents
Washington, DC 20402-9325
Phone: (202)783-3238

★ 2613 ★ **West Publishing Company**
WESTLAW
50 W. Kellogg Blvd.
P.O. Box 64526
St. Paul, MN 55164-0526
Phone: 800-WES-TLAW; 800-328-0109

★ 2614 ★ **WRI Publications**
Hampden Station
P.O. Box 4852
Baltimore, MD 21211
Phone: (301)338-6983

Indexes

Chemical Name Index

Bold numbers denote main chemical entries; roman numerals denote text.

403

Bold numbers denote main chemical entries; roman numerals denote text.

ALLYLEHYDE 21
ALLYL EHYDE 21
ALLYL 2,3-EPOXYPROPYL ETHER 34
ALLYLGLYCIDAETHER (GERMAN) 34
ALLYL GLYCIDYL ETHER **34**
ALLYLIC ALCOHOL 31
ALLYL IODIDE **35**
ALLYL ISORHODANIDE 36
ALLYL ISOSULFOCYANATE 36
ALLYL ISOTHIOCYANATE **36**
ALLYL MUSTARD OIL 36
1-ALLYLOXY-2,3-EPOXY-PROPAAN
  (DUTCH) 34
1-ALLYLOXY-2,3-EPOXYPROPAN (GER-
  MAN) 34
1-ALLYLOXY-2,3-EPOXY-PROPANE 34
1-(ALLYLOXY)-2,3-EPOXYPROPANE 34
ALLYL PROPYL DISULFIDE **37**
ALLYLSENEVOL 36
ALLYLSENFOEL (GERMAN) 36
ALLYLSILICONE TRICHLORIDE 38
ALLYLTRICHLOROSILANE 38
ALLYL TRICHLOROSILANE **38**
ALLYL TRICHOROSILANE, STABILIZED 38
ALMEDERM 511
ALMITE 43
ALMOND ARTIFICIAL ESSENTIAL OIL 142
ALON 43
ALON C 43
ALOXITE 43
ALPHANAPHTHYL THIOUREA 108
ALPHANAPHTYL THIOUREE
  (FRENCH) 108
AL-PHOS 45
ALPINE TALC 913
ALRATO 108
ALTOX 29
ALUM 47
ALUMIGEL 41
ALUMINA 43
ALPHA-ALUMINA 43
BETA-ALUMINA 861, 43
BETA''-ALUMINA 861
GAMMA-ALUMINA 43
ALUMINA FIBRE 39
ALUMINA HYDRATE 41
ALUMINA HYDRATED 41
ALUMINATE(1-), TETRAHYDRO-, SODI-
  UM 860
ALUMINA TRIHYDRATE 41
ALPHA-ALUMINA TRIHYDRATE 41
ALUMINIC ACID 41
ALUMINITE 37 43
ALUMINIUM 39
ALUMINIUM BRONZE 39
ALUMINIUMCHLORID (GERMAN) 40
ALUMINIUM FLAKE 39
ALUMINIUM HYDROXIDE 41
ALUMINOPHOSPHORIC ACID 44
ALUMINUM **39**, lxiii, cii
ALUMINUM 27 39
ALUMINUM ALUM 47
ALUMINUM CHLORIDE **40**, xciii
ALUMINUM CHLORIDE (1:3) 40
ALUMINUM CHLORIDE ANHYDROUS 40
ALUMINUM, CHLORODIETHYL- 373
ALUMINUM DEHYDRATED 39
ALUMINUM DICHLOROETHYL- 435
ALUMINUM ETHYL DICHLORIDE 435
ALUMINUM FOSFIDE (DUTCH) 45
ALUMINUM HYDRATE 41
ALUMINUM HYDROXIDE **41**
ALUMINUM (III) HYDROXIDE 41
ALUMINUM HYDROXIDE GEL 41
ALUMINUM LITHIUM HYDRIDE 586
ALUMINUM, METALLIC, POWDER 39
ALUMINUM MONOPHOSPHIDE 45
ALUMINUM NITRATE **42**

ALUMINUM (III) NITRATE (1:3) 42
ALUMINUM NITRATE NONHYDRATE 42
ALUMINUM OXIDE **43**
ALUMINUM OXIDE (2:3) 43
ALUMINUM OXIDE-3H2O 41
ALPHA-ALUMINUM OXIDE 43
BETA-ALUMINUM OXIDE 43
GAMMA-ALUMINUM OXIDE 43
ALUMINUM OXIDE HYDRATE 41
ALUMINUM OXIDE TRIHYDRATE 41
ALUMINUM PHOSPHATE **44**
ALUMINUM PHOSPHIDE **45**
ALUMINUM PHOSPHITE 45
ALUMINUM POWDER 39
ALUMINUM SALT OF NITRIC ACID 42
ALUMINUM SESQUIOXIDE 43
ALUMINUM SILICATE, MONTMORILLON-
  ITE 140
ALUMINUM SODIUM FLUORIDE **46**
ALUMINUM SODIUM HYDRIDE 860
ALUMINUM SODIUM OXIDE 861
ALUMINUM SULFATE **47**
ALUMINUM SULFATE (2:3) 47
ALUMINUM TRICHLORIDE 40
ALUMINUM, TRICHLOROTRIETHYLDI- 436
ALUMINUM, TRICHLOROTRIMETHYLDI-
  644
ALUMINUM TRIETHYL 969
ALUMINUM TRIHYDRAT 41
ALUMINUM TRIHYDROXIDE 41
ALUMINUM, TRIISOBUTYL- 977
ALUMINUM TRIMETHYL 979
ALUMINUM TRINITRATE 42
ALUMINUM TRIOXIDE 43
ALUMINUM, TRIPROPYL- 991
ALUMINUM, TRIS(2-METHYLPROPYL)-
  (9CI) 977
ALUMINUM TRISULFATE 47
ALUMITE 43
ALUMITE (OXIDE) 43
ALUNDUM 43
ALUNDUM 600 43
ALUPHOS 44
ALUSAL 41
ALVIT 370
ALZODEF 222
AMACEL DEVELOPED NAVY SD 338
AMALOX 1040
AMARTHOL FAST RED TR BASE 281
AMARTHOL FAST RED TR SALT 281
AMATIN 506
AMAZE lxii
AMBER MICA 853
AMBEROL ST 140F 41
AMBEST (GERMAN) 117
AMBIBEN 246
AMBOFEN 247
AMBRAMICINA 925
AMBRAMYCIN 925
AMBUSH 28
AMCHEM GRASS KILLER 961
AMCHLOR 63
AMCIDE 81
AMDON 785
AMDON GRAZON 785
AMEDOX 891
AMEISENATOD 146, 583
AMEISENMITTEL MERCK 146, 583
AMEISENSAEURE (GERMAN) 486
AMERCIDE 233
AMERICAN CYANAMID 3,911 770
AMERICAN CYANAMID 4,049 602
AMERICAN CYANAMID 12880 385
AMERICAN CYANAMID 18133 1046
AMERICAN CYANAMID AC-52,160 917
AMEROL 56
AMETHYST 854
AM-FOL 57

AMIANTHUS 117
AMIBEN 246
AMIBIN 246
AMICIDE 81
AMID KYSELINY OCTOVE 4
AMIDOCYANAGEN 315
AMIDOCYANOGEN 315
AMIDOSULFONIC ACID 901
AMIDOSULFURIC ACID 901
AMIDOX 330
AMINE lvii, lxii
AMINE DERIVATIVES lvii
AMINE (9CI), O-
  METHOXYPHENYLAMINE 96
AMINIC ACID 486
AMINO xxxv
2-AMINOAETHANOL (GERMAN) 428
4-AMINOANILINE 761
P-AMINOANILINE 761
4-AMINOANISOLE 97
O-AMINOANISOLE 96
ORTHO-ORTHO-AMINOANISOLE 96
P-AMINOANISOLE 97
AMINOAZOTOLUENE 48
3,2-AMINO-5-AZOTOLUENE 48
O-AMINOAZOTOLUENE 48
O-AMINOAZOTOLUOL (GERMAN) 48
2-AMINO-5-AZOTOULENE **48**
AMINOBENZENE 95
4-(4-AMINOBENZYL)ANILINE 664
4-AMINOBIPHENYL 51
P-AMINOBIPHENYL 51
1-AMINO-BUTAAN (DUTCH) 197
1-AMINOBUTAN (GERMAN) 197
1-AMINOBUTANE 197
4-AMINO-6-TERT-BUTYL-3-(METHYLTHIO)-
  1,2,4-TRIAZIN-5-ONE 683
4-AMINO-6-TERT-BUTYL-3-METHYLTHIO-
  AS-TRIAZIN-5-ONE 683
1-AMINO-2-CHLORO-6-
  METHYLBENZENE 264
1-AMINO-3-CHLORO-2-
  METHYLBENZENE 264
1-AMINO-3-CHLORO-4-
  METHYLBENZENE 265
1-AMINO-3-CHLORO-6-
  METHLBENZENE 266
2-AMINO-4-CHLOROPHENOL **49**
2-AMINO-3-CHLOROTOLUENE 280
2-AMINO-4-CHLOROTOLUENE 266
2-AMINO-6-CHLOROTOLUENE 264
4-AMINO-2-CHLOROTOLUENE 265
2-AMINO-5-CHLOROTOLUENE HYDRO-
  CHLORIDE 281
AMINOCYCLOHEXANE 323
3-AMINO-2,5-DICHLOROBENZOIC
  ACID 246
3-AMINO-2,6- DICHLOROBENZOIC
  ACID 246
2-AMINO-5-DIETHYLAMINOPENTANE **50**
4-AMINODIFENIL (SPANISH) 51
AMINODIMETHYLBENZENE 1025
1-AMINO-2,4-DIMETHYLBENZENE 1026
4-AMINO-1,3-DIMETHYLBENZENE 1026
4-AMINO-6-(1,1-DIMETHYLETHYL)-3-
  (METHYLTHIO)-1,2,4-TRIAZIN-5(4H)
  -ONE 683
4-AMINODIPHENYL **51**
2-AMINOETANOLO (ITALIAN) 428
AMINOETHANDIAMINE 376
AMINOETHANE 437
1-AMINOETHANE 437
2-AMINOETHANOL 428
N-(2-AMINOETHYL) 376
BETA-AMINOETHYL ALCOHOL 428
BETA-AMINOETHYLAMINE 449
O-AMINOETHYLBENZENE 440
AMINOETHYLENE 455

AMINO ETHYL-1,2-ETHANEDIAMINE, 1,4,7,10,13,- PENTAAZATRIDECANE 926
AMINOHEXAHYDROBENZENE 323
1-AMINO-2-HYDROXYETHANE 428
2-AMINOISOBUTANE 198
2-AMINOISOPROPYLBENZENE 556
O-AMINOISOPROPYLBENZENE 556
AMINOMETHANE 645
1-AMINO-4-METHOXYBENZENE 97
1-AMINO-2-METHYL-9,10-ANTHRACENEDIONE 52
1-AMINO-2-METHYL-ANTHRAQUINONE **52**
1-AMINO-2-METHYLBENZENE 956
2-AMINO-1-METHYLBENZENE 956
3-AMINO-1-METHYLBENZENE 955
1-AMINO-2-METHYLBENZENE HYDRO-CHLORIDE 957
2-AMINO-1- METHYLBENZENE HYDRO-CHLORIDE 957
2-AMINO-2-METHYLPROPANE 198
4-AMINO-3-METHYLTOLUENE 1026
1-AMINONAFTALEN (CZECH) 692
2-AMINONAFTHALEN (CZECH) 693
1-AMINONAPHTHALENE 692
2-AMINONAPHTHALENE 693
4-AMINO-2-NITROANILINE **53**
2-AMINO-4-NITROANISOLE 717
1-AMINO-4-NITROBENZENE 707
AMINOPHEN 95
AMINOPHENOL **54**
2-AMINOPHENOL (95-55-6) 54
3-AMINOPHENOL (591-27-5) 54
4-AMINOPHENOL (123-30-8) 54
M-AMINOPHENOL (591-27-5) 54
O-AMINOPHENOL (95-55-6) 54
P-AMINOPHENOL (123-30-8) 54
3-AMINOPHENYLMETHANE 955
2-AMINO-PROPAAN (DUTCH) 555
1-AMINOPROPANE 826
2- AMINOPROPANE 555
AMINOPROPYLPIPERAZINE 165
2-AMINOPYRIDINE **55**
AMINO-2 PYRIDINE 55
ALPHA-AMINOPYRIDINE 55
O-AMINOPYRIDINE 55
3-AMINO-S-TRIAZOLE 56
AMINOSULFONIC ACID 901
3-AMINOTOLUEN (CZECH) 955
3-AMINOTOLUENE 955
M-AMINOTOLUENE 955
2-AMINOTOLUENE 956
O-AMINOTOLUENE 956
2-AMINOTOLUENE HYDROCHLORIDE 957
O-AMINOTOLUENE HYDROCHLORIDE 957
AMINOTRIAZOLE 56
2-AMINOTRIAZOLE 56
2-AMINO-1,3,4-TRIAZOLE 56
3-AMINOTRIAZOLE 56
3-AMINO-1,2,4-TRIAZOLE 56
3-AMINO-1H-1,2,4-TRIAZOLE 56
AMINO TRIAZOLE WEEDKILLER 90 56
4-AMINO-3,5,6-TRICHLORO-2-PICOLINIC ACID 785
4-AMINO-3,5,6-TRICHLORO-ROPICOLINIC ACID 785
4-AMINO-3,5,6-TRICHLORPICOLINSAEURE (GERMAN) 785
4-AMINO-1,3-XYLENE 1026
AMIPROL 341
AMIRAL lxii
AMITOL 56
AMITRAZ lxiii
AMITRIL 56
AMITROL **56**

AMITROL 90 56
AMITROL-T 56
AMITROLE 56
AMIZOL 56
AMMAT 81
AMMATE 81
AMMATE HERBICIDE 81
AMMONERIC ̈ 63
AMMONIA **57**, xxxii, xxxiii, cii, cviii, cix
AMMONIA ANHYDROUS 57
AMMONIA AQUEOUS 70
AMMONIAC (FRENCH) 57
AMMONIACA (ITALIAN) 57
AMMONIA FLUOSILICATE 69
AMMONIA GAS 57
AMMONIALE (GERMAN) 57
AMMONIA SOLUTION 70
AMMONIA WATER 70
AMMONIO (BICROMATO DI) (ITALIAN) 67
AMMONIO (DICROMATO DI) (ITALIAN) 67
AMMONIUM ACETATE **58**
AMMONIUM AMIDOSULPHATE 81
AMMONIUM AMINOFORMATE 61
AMMONIUM AMINOSULFONATE 81
AMMONIUM BICARBONATE **59**
AMMONIUM BICARBONATE (1:1) 59
AMMONIUMBICHROMAAT (DUTCH) 67
AMMONIUM BICHROMATE 67
AMMONIUM BIFLUORIDE **60**
AMMONIUM BIPHOSPHATE 78
AMMONIUM CARBAMATE **61**
AMMONIUMCARBONAT (GERMAN) 62
AMMONIUM CARBONATE **62**, 59, lxxvii
AMMONIUMCHLORID (GERMAN) 63
AMMONIUM CHLORIDE **63**
AMMONIUM CHLOROPLATINATE **64**
AMMONIUM CHROMATE **65**
AMMONIUM CHROMATE(IV) 65
AMMONIUM CHROMIUM OXIDE 65
AMMONIUM CITRATE **66**
AMMONIUM CITRATE, DIBASIC 66
AMMONIUMDICHROMAAT (DUTCH) 67
AMMONIUMDICHROMAT (GERMAN) 67
AMMONIUM DICHROMATE **67**
AMMONIUM DICHROMATE(VI) 67
AMMONIUM (DICHROMATE D') (FRENCH) 67
AMMONIUM DIHYDROGEN PHOS-PHATE 78
AMMONIUM D-TARTRATE 85
AMMONIUM FERRIC CITRATE 473
AMMONIUM FLUORIDE **68**
AMMONIUM FLUORURE (FRENCH) 68
AMMONIUM FLUOSILICATE 69
AMMONIUM HEXACHLOROPLATINATE (IV) 64
AMMONIUM HEXAFLUOROSILICATE **69**
AMMONIUM HYDROFLUORIDE 60
AMMONIUM HYDROGEN BIFLUORIDE 60
AMMONIUM HYDROGEN CARBONATE 59
AMMONIUM HYDROGEN DIFLUORIDE 60
AMMONIUM HYDROGEN FLUORIDE 60
AMMONIUM HYDROGEN PHOSPHATE 79
AMMONIUM HYDROSULFIDE 82
AMMONIUM HYDROXIDE **70**, 57
AMMONIUM IRON SULFATE 477
AMMONIUM METAVANADATE **71**
AMMONIUM MOLYBDATE **72**
AMMONIUM MONOSULFIDE 83
AMMONIUM MURIATE 63
AMMONIUM NICKEL SULFATE 697
AMMONIUM NITRATE **73**
(1:1) AMMONIUM NITRATE 73
AMMONIUM (1) NITRATE 73

AMMONIUM OXALATE **74**
AMMONIUM OXALATE HYDRATE 74
AMMONIUM PARAMOLYBDATE 72
AMMONIUM PERCHLORATE **75**, xciii
AMMONIUM PERCHLORATE, ANHY-DROUS 597
AMMONIUM PERCHLORATE, HEXAHY-DRIDE 597
AMMONIUM PERMANGANATE **76**
AMMONIUM PEROXYDISULFATE 77
AMMONIUM PERSULFATE **77**
AMMONIUM PHOSPHATE **78**
AMMONIUM PHOSPHATE DIBASIC **79**
AMMONIUM PICRATE **80**
AMMONIUM PICRATE, DRY 80
AMMONIUM PICRATE, WET 80
AMMONIUM PICRATE WETTED WITH LESS THAN 10% WATER 80
AMMONIUM PICRATE WETTED WITH MORE THAN 10% WATER 80
AMMONIUM PICRONITRATE 80
AMMONIUM PLATINIC CHLORIDE 64
AMMONIUM RHODANATE 87
AMMONIUM RHODANIDE 87
AMMONIUM RHODANTATE 87
AMMONIUM SALT OF NITRIC ACID 73
AMMONIUM SALZ DER AMIDOSULFONS-AURE (GERMAN) 81
AMMONIUM SILICOFLUORIDE 69
AMMONIUM SULFAMATE **81**
AMMONIUM SULFIDE 82, 83
AMMONIUM SULFIDE, SOLID **82**
AMMONIUM SULFIDE, SOLUTION **83**
AMMONIUM SULFITE **84**
AMMONIUM SULFOCYANATE 87
AMMONIUM SULFOCYANIDE 87
AMMONIUM SULPHAMATE 81
AMMONIUM TARTRATE **85**
AMMONIUM TETRACHLOROPLATI-NATE **86**
AMMONIUMTHIOCYANATE 87
AMMONIUM THIOCYANATE **87**
AMMONIUM VANADATE 71
AMOBEN 246
AMONIAK (POLISH) 57
2-AMONO-6- CHLOROTOLUENE 264
2-AMONOPROPAN (GERMAN) 555
AMORPHOUS FUSED SILICA 852
AMOSITE lxxi
AMOSITE (CAS 12172-73-5) 117
AMOSITE (OBS) 117
AMOXONE 330
AMPHENICOL 247
AMPHIBOLE 117
AMPHICOL 247
AMPHOJEL 41
AMS 81
AMSECLOR 247
AMTHIO 87
AMYCIN 925
AMYL ACETATE 88, lxxvii
N-AMYL ACETATE **88**
SEC-AMYL ACETATE **89**
AMYLACETIC ESTER 540
AMYL ACETIC ESTER 88
AMYL ACETIC ETHER 88
AMYL ALCOHOL **90**
N-AMYL ALCOHOL 90
SEC-AMYL ALCOHOL 751
AMYL ALDEHYDE 998
N-AMYLALKOHOL (CZECH) 90
AMYLASE, DIASTASE MALT 339
AMYLAZETAT (GERMAN) 88
AMYL ETHYL KETONE 438

Bold numbers denote main chemical entries; roman numerals denote text.

AMYL HYDRIDE 750
AMYL METHYL CARBINOL 505
AMYL-METHYL-CETONE (FRENCH) 646
AMYL METHYL KETONE 646
N-AMYL METHYL KETONE 646
AMYL NITRATE **91**
N-AMYL NITRATE 91
AMYL NITRITE **92**, 93
N-AMYL NITRITE **93**
AMYLOL 90
AMYLOWY ALKOHOL (POLISH) 541
AMYLTRICHLOROSILANE **94**
AMYL TRICHLOROSILANE 94
ANAC 110 297
ANADOMIS GREEN 288
ANAESTHETIC ETHER 458
ANAETHAN 416
ANAMENTH 964
ANAPAC 758
ANATASE 949
ANCHRED STANDARD 538
ANCORTONE 815
ANDROLIN 919
ANDRONAQ 919
ANDROST-4-EN-3-ONE,17-HYDROXY-, (17-
    BETA)- 919
ANDROST-4-EN-3-ONE, 17-BETA-HYDROXY-
    919
ANDROST-4-EN-17(BETA)-OL-3-ONE 919
ANDRUSOL 919
ANERIN I 836
ANERIN II 836
ANESTAN 501
ANESTHENYL 643
ANESTHESIA ETHER 458
ANESTHETIC COMPOUND NO. 347 424
ANESTHETIC ETHER 458
ANFRAM 3PB 992
ANGININE 715
ANGLISITE 579
ANHYDRIDE ACETIQUE (FRENCH) 6
ANHYDRIDE ARSENIEUX (FRENCH) 111
ANHYDRIDE ARSENIQUE (FRENCH) 112
ANHYDRIDE CARBONIQUE (FRENCH) 237
ANHYDRIDE CHROMIQUE (FRENCH) 290
ANHYDRIDE-ETHOMID HT POLYMER 978
ANHYDRIDE PHTALIQUE (FRENCH) 783
ANHYDRID KYSELINY TETRAHYDROF-
    TALOVE (CZECH) 932
ANHYDROL 427
ANHYDRONE 597
ANHYDRO-O-SULFAMINEBENZOIC
    ACID 845
ANHYDRO TRIMELLIC ACID 978
ANHYDROUS AMMONIA 57
ANHYDROUS BORIC ACID 174
ANHYDROUS HYDRIODIC ACID 522
ANHYDROUS HYDROBROMIC ACID 527
ANHYDROUS HYDROFLUORIC ACID 525
ANHYDROUS HYDROGEN CHLORIDE 523
ANHYDROUS IRON OXIDE 538
ANHYDROUS TETRASODIUM PYROPHOS-
    PHATE 937
ANIDRIDE ACETICA (ITALIAN) 6
ANIDRIDE CHROMICA (ITALIAN) 290
ANIDRIDE CROMIQUE (FRENCH) 288
ANIDRIDE FTALICA (ITALIAN) 783
ANILAZINE lxii
ANILIN (CZECH) 95
ANILINA (ITALIAN, POLISH) 95
ANILINE **95**, cii
ANILINE, N,N-DIETHYL- 375
ANILINE, N,N-DIMETHYL- 388
ANILINE,O-ETHYL- (8CI) 440
ANILINE, HEXAHYDRO- 323
ANILINE, O-ISOPROPYL- 556
ANILINE, 2-METHYL- 956
ANILINE, 3-METHYL- 955

ANILINE,4,4'-METHYLENE-BIS (N,N-DI-
    METHYL)- 662
ANILINE, 4,4'-METHYLENEDI- 664
ANILINE OIL 95
ANILINE OIL, LIQUID 95
ANILINE, N-PHENYL 410
PREGNA-1,4-DIENE-3,11,20-TRIONE, 17,21-
    HYDROXY- 815
ANILINOBENZENE 51, 410
ANILINOETHANE 439
ANILINOETHANOL 439
ANILINOMETHANE 647
ANILINONAPHTHALENE 767
2-ANILINONAPHTHALENE 767
ANIMAL CARBON 25
4-ANISIDINE 97
O-ANISIDINE **96**
P-ANISIDINE **97**
ANKILOSTIN 754
ANNAMENE 899
ANODYNON 445
ANOFEX 332
ANOL 319
ANONE 320
ANPROLENE 456
2-ANSIDINE 96
ANSIOLIN 341
ANSIOLISINA 341
ANTABUS 416
ANTABUSE 416
ANTADIX 416
ANTAENYL 416
ANTAETHYL 416
ANTAETIL 416
ANTALCOL 416
ANTETAN 416
ANTETHYL 416
ANTHION 811
ANTHIUM DIOXCIDE 252
ANTHON lxii
ANTHOPHYLITE (CAS 77536-67-5) 117
ANTHOPHYLLITE (17068-78-9) 117
ANTHRACEN (GERMAN) 98
ANTHRACENE **98**, cii
9,10-ANTHRACENEDIONE, 1-AMINO-2-
    METHYL- 52
ANTHRACIN 98
ANTHROPHYLITE lxxi
ANTHROPHYLLITE lxxi
ANTIBULIT 876
ANTICARIE 506
ANTICHLOR 891
ANTIFORMIN 880
ANTIFREEZE lvi, lix, lx
ANTIKNOCK-33 612
ANTIKOL 416
ANTIMIMONWASSERSTOFFES (GER-
    MAN) 895
ANTIMOINE FLUORURE (FRENCH) 106
ANTIMOINE (TRICHLORURE D')
    (FRENCH) 105
ANTIMONATE (2-), BIS .MU.-2,3-
    DIHYDROXYBUTANEDIOATA (4-)-
    01,02:03,04DI-, DIPOTASSIUM, TRIHY-
    DRATE, STEREOISOMER 103
ANTIMONIC CHLORIDE 101
ANTIMONIO (PENTACLORURO DI) (ITAL-
    IAN) 101
ANTIMONIO (TRICHLORURO DI) (ITAL-
    IAN) 105
ANTIMONIOUS OXIDE 107
ANTIMONIUS CHLORIDE 105
ANTIMONOUS FLUORIDE 106
ANTIMONPENTACHLORID (GERMAN) 101
ANTIMONY **99**, cii
ANTIMONY BLACK 99
ANTIMONY BUTTER 105
ANTIMONY (III) CHLORIDE 105

ANTIMONY (V) CHLORIDE, ANTIMONY PER-
    CHLORIDE 101
ANTIMONY FLUORIDE 102
ANTIMONY (III) FLUORIDE (1:3) 106
ANTIMONY (V) FLUORIDE 102
ANTIMONY HYDRIDE 895
ANTIMONY LACTATE **100**
ANTIMONY LACTATE SOLID 100
ANTIMONY (III) OXIDE 107
ANTIMONY(3) OXIDE 107
ANTIMONYL POTASSIUM TARTRATE 103
ANTIMONY PENTACHLORIDE **101**
ANTIMONY PENTAFLUORIDE **102**
ANTIMONY (V) PENTAFLUORIDE 102
ANTIMONY PEROXIDE 107
ANTIMONY POTASSIUM TARTRATE **103**
ANTIMONY POWDER 99
ANTIMONY, REGULUS 99
ANTIMONY SESQUIOXIDE 107
ANTIMONY TRIBROMIDE **104**
ANTIMONY TRICHLORIDE **105**
ANTIMONY TRIFLUORIDE **106**
ANTIMONY TRIHYDRIDE 895
ANTIMONY TRIOXIDE **107**
ANTIMONY WHITE 107
ANTIMOONPENTACHLORIDE (DUTCH) 101
ANTIMOONTRICHLRIDE (DUTCH) 105
ANTIMUCIN WDR 765
ANTINONIN 403
ANTINONNIN 403
ANTIOXIDANT 29 163
ANTIOXIDANT 116 767
ANTIOXIDANT DBPC 163
ANTIOXIDANT PBN 767
ANTIPYONIN 866
ANTIREN 787
ANTISAL LA 951
ANTISOL 1 754
ANTIVERM 760
ANTOX 107
ANTRACINE 8 163
ANTU **108**
ANTURAT 108
ANTYMON (POLISH) 99
ANTYMONOWODOR (POLISH) 895
ANYVIM 95
AO4K 163
AO 29 163
AO A1 39
AP 50 107
APADRIN 687
APARASIN 146, 583
APAURIN 341
APAVINPHOS 771
APC 758
APEX 462-5 992
APHAMITE 745
APHOX 789
APHOX DISPERSIBLE GRAIN 789
APHTIRIA 146, 583
APLIDAL 146, 583
APOZEPAM 341
APPA 773
AQUA AMMONIA 57
AQUA AMMONIA, AQUEOUS AMMONIA 70
AQUACAT 295
AQUACIDE 414
AQUA FORTIS 705
AQUA-KLEEN 330
AQUALIN 21
AQUALINE 21
AQUAMYCETIN 247
AQUATHOL lxii
AR2 39
ARABIC GUM 500
ARAB RAT DETH 1012
ARAGONITE 219
ARALO 745

ARASAN 943
ARBITEX 146, 583
ARBOROL 403
ARCOSOLV 412
ARCTON 975
ARCTON O 930
ARCTON 4 261
ARCTON 6 356
ARCTON 33 488
ARCTON 63 487
ARCTON 114 488
ARCTRON 9 965
ARCTUVIN 530
ARGENTUM 855
ARGILLA 560
ARILATE 139
ARKLONE P 487
ARKOTINE 332
ARMENIAN BOLE 538
AROCLOR 792
AROCLOR 1221 792
AROCLOR 1232 792
AROCLOR 1242 792
AROCLOR 1248 792
AROCLOR 1254 792
AROCLOR 1260 792
AROCLOR 1262 792
AROCLOR 1268 792
AROCLOR 2565 792
AROCLOR 4465 792
AROL GORDON DUST 844
AROMATIC HYDROCARBONS xxvii, xlv
AROMATIC SOLVENT 690
ARPROCARB 824
ARRHENAL lxii
ARSEN (GERMAN, POLISH) 109
ARSENATE OF LEAD 565
ARSENIC **109**, xl, xlii, lxxxix
ARSENIC-75 109
ARSENIC ACID **110**
ARSENIC ACID ANHYDRIDE 112
ARSENIC ACID, LEAD(2) 565
ARSENIC ACID, LEAD SALT 565
ARSENIC ACID, MONOPOTASSIUM
  SALT 794
ARSENIC ACID, SODIUM SALT 862
ARSENICALS 109
ARSENIC ANHYDRIDE 112
ARSENIC BLACK 109
ARSENIC BLANC (FRENCH) 111
ARSENIC COMPOUNDS cx
ARSENIC HYDRIDE 116
ARSENIC OXIDE **111**
ARSENIC (III) OXIDE 111
ARSENIC (V) OXIDE 112
ARSENIC PENTOXIDE **112**
ARSENIC SESQUIOXIDE 111
ARSENIC SESQUISULFIDE 115
ARSENIC, SOLID 109
ARSENIC SULFIDE 115
ARSENIC TRICHLORIDE **113**
ARSENIC TRIHYDRIDE 116
ARSENIC TRIOXIDE **114**, lxiii
ARSENIC TRIOXIDE, SOLID 111
ARSENIC TRISULFIDE **115**
ARSENICUM ALBUM 111
ARSENIC YELLOW 115
ARSENIGEN SAURE (GERMAN) 111
ARSENIOUS ACID 111
ARSENIOUS OXIDE 111
ARSENIOUS TRIOXIDE 111
ARSENITE 111
ARSENIURETTED HYDROGEN 116
ARSENOLITE 111

ARSENOUS ACID 111
ARSENOUS ACID ANHYDRIDE 111, 114, 119
ARSENOUS ANHYDRIDE 111
ARSENOUS CHLORIDE 113
ARSENOUS HYDRIDE (POLISH) 116
ARSENOUS OXIDE 111
ARSENOUS OXIDE ANHYDRIDE 111
ARSENOUS TRICHLORIDE 113
ARSENTRIOXIDE 111
ARSENWASSERSTOFF (GERMAN) 116
ARSINE **116**
ARSINETTE 565
ARSINYL lxii
ARSODENT 111
ARSONIC ACID, COPPER(2)SALT (1:1)
  (9CI) 299
ARTHODIBROM 689
ARTIC 651
ARTIFICIAL ALMOND OIL 142
ARTIFICIAL ANT OIL 491
ARTIFICIAL BARITE 137
ARTIFICIAL HEAVY SPAR 137
ARTISIL ORANGE 3RP 52
AR-TOLUENOL 303
ARWOOD COPPER 297
ASA 20
ASA COMPOUND 758
A.S.A. EMPRIN 20
ASAHIFRON 113 487
ASAREO L15 1028
ASATARD 20
ASBESTINE 913
ASBESTOS **117**, lxxi, lxxii, lxxiii, lxxxix, cii, cviii
7-45 ASBESTOS (CAS 12001-29-5) 117
ASBESTOSE (GERMAN) 117
ASBESTOS FIBER 117
ASBESTOS FIBRE 117
ASBESTOS, WHITE (CAS 12001-29-5) 117
ASCARITE 117
ASEX 868
ASHLADE PERSYST 658
ASP 47 902
ASPHALT **118**
ASPHALTUM 118
ASPIRIN 20
ASPIRINE 20
ASPON-CHLORDANE 248
ASPOR 1045
ASPORUM 1045
ASPRO 20
ASSIVAL 341
ASTA 422
ASTA B 518 422
ASTERIC 20
ASTRUMAL 809
ASUNTHOL 301
ASUNTOL 301
ASYMMETRICAL TRIMETHYL BEN-
  ZENE 983
AT 56
AT 7 511
ATA 56
ATCP 785
ATE 969
ATENSINE 341
ATHYLEN (GERMAN) 448
ATHYLENGLYKOL-MONOATHYLATHER
  (GERMAN) 430
ATILEN 341
ATLACIDE 868
ATLADOX 785
ATLADOX HI 785

ATLAS WHITE TITANIUM 949
AT LIQUID 56
ATOMIT 219
ATRATOL B-HERBATOX 868
ATRAZINE **119**, lxii, lxv
ATROMBINE-K 1012
ATTAC-2 958
ATTAC 6 958
ATTAC 6-3 958
AULES 943
AUSTIOX 949
AUSTRACIL 247
AUSTRACOL 247
AUSTRALIAN GUM 500
AUSTRAPINE 840
AV00 39
AV000 39
AVADEX 336
AVERSAN 416
AVERZAN 416
AVIBEST C (CAS 12001-29-5) 117
AVLOTANE 509
AVOLIN 398
AWPA NO. 1 302
AXIUM 863
AZABENZENE 837
AZACYCLOPROPANE 455
1-AZANAPHTHALENE 838
AZANIL RED SALT TRD 281
3-AZAPENTANE-1,5-DIAMINE 376
AZAPLANT 56
AZBILLEN ASBESTOS (17068-78-9) 117
AZBOLEN ASBESTOS (CAS 77536-67-5) 117
AZETYLAMINOFLUOREN (GERMAN) 10
AZIDES xciii
AZIJNZUUR (DUTCH) 5
AZIJNZUURANHYDRIDE (DUTCH) 6
AZIMETHYLENE 343
AZINE 837
AZINPHOS METHYL lxiii
AZIRANE 455
AZIRIDINE 455
AZIRIDINE, 2-METHYL- (6CI,8CI,9CI) 831
AZIRINE 455
1H-AZIRINE, DIHYDRO- 455
AZO-33 1040
AZO-55 1040
AZO-66 1040
AZO-77 1040
AZOAMINE SCARLET K 717
AZO COMPOUNDS xciii
AZODOX-55 1040
AZODRIN 687
AZOENE FAST BLUE BASE 338
AZOENE FAST RED TR SALT 281
AZOENO FAST BLUE SALT 338
AZOFIX BLUE B SALT 338
AZOFOS 679
AZOGENE FAST 338
AZOGENE FAST BLUE B 338
AZOGENE FAST RED TR 281
AZOIC DIAZO COMPONENT 11 BASE 281
AZOIC DIAZO COMPONENT 37 707
AZOICDIAZO COMPONENT 46 264
AZOLAN 56
AZOLE 56
AZOPHOS 679
AZOTE (FRENCH) 713
AZOTIC ACID 705
AZOTO (ITALIAN) 713
AZOTOWY KWAS (POLISH) 705
AZOTOX 332
AZOTURE DE SODIUM (FRENCH) 863

Bold numbers denote main chemical entries; roman numerals denote text.

Bold numbers denote main chemical entries; roman numerals denote text.

BIS(4-GLYCIDYLOXYPHE-NYL)DIMETHYAMETHANE 171
2,2-BIS(P-GLYCIDYLOXYPHENYL)PROPANE 171
BIS HCCL 12880 385
BIS(4-HYDROXY-5-TERT-BUTYL-2-METHYLPHENYL) SULFIDE 941
BIS(2-HYDROXYETHYL)AMINE 371
BIS(4-HYDROXYPHENYL)DIMETHYLMETHANE DIGLYCIDYL ETHER 171
2,2-BIS(4-HYDROXYPHE-NYL)PROPANE,DIGLYCIDYL ETHER 171
BIS(2-HYDROXY-3,5,6-TRICHLOROPHENYL)METHANE 511
BIS(4-ISOCYANATOCYCLOHEX-YL)METHANE 661
BIS(ISOPROPYL)AMINE 384
BIS(1-METHYLETHYL) CARBAMOTHIOIC ACID, S-(2,3-DICHLORO-2-PROPENYL)ESTER 336
BISOFLEX 81 415
BISOFLEX DOP 415
BISPHENOL A DIGLYCIDYL ETHER **171**
BIS(THYOCYANATO)-MERCURY 623
BIS-2,3,5-TRICHLOR-6-HYDROXYFENYLME THAN (CZECH) 511
BIS(3,5,6-TRICHLORO-2-HYDROXYPHENYL)METHANE 511
BIS(TRICHLOROPHENYL)ETHER 250
BISULFITE 904
BISULFITE DE SODIUM (FRENCH) 865
BITHION 917
BITHIONOL lxiii
BITUMEN 118
BIVERM 760
BIVINYL 182
B-K LIQUID 880
B-K POWDER 225
BLA 578
BLACK AND WHITE BLEACHING CREAM 530
BLACK MANGANESE OXIDE 609
BLACK OXIDE OF IRON 538
BLACOSOLV 964
BLADAFUME 902
BLADAFUN 902
BLADAN 429, 745, 918
BLADAN F 745
BLADAN-M 679
BLADEX **172**
BLADEX 80WP 172
BLADON 918
BLANC FIXE 137
BLASTING GELATIN 715
BLASTING OIL 715
BLATTANEX 824
BLAUSAEURE (GERMAN) 524
BLAUWZUUR (DUTCH) 524
BLEACHING POWDER 225
BLEIPHOSPHAT (GERMAN) 576
BLEISTEARAT (GERMAN) 577
BLEISULFAT (GERMAN) 579
BLENDED RED OXIDES OF IRON 538
BLIGHTOX 1045
BLITEX 1045
BLIZENE 1045
BLUE ASBESTOS (CAS 12001-28-4) 117
BLUE BASE IRGA B 338
BLUE BASE NB 338
BLUE BN BASE 338

BLUE B SALT 338
BLUE BN SALT 338
BLUE COPPER 313
BLUE OIL 95
BLUE POWDER 1028
BLUE SALT NB 338
BLUE STONE 313
BLUE VITRIOL 313
BNM 139
BOCA 660
BOEHMITE 43
BOG MANGANESE 609
BOLETIC ACID 489
BOLSTAR 910
BOLUSALBA 560
BOMYL 911
BONAZEN 1043
BONIBOL 416
BONIDE BLUE DEATH RAT KILLER 776
BONIDE TOPZOL RAT BAITS AND KILLING SYRUP 842
BONOFORM 923
BORACIC ACID 173
BORANE, TRIBROMO- 175
BORANE, TRIFLUORO- 176
BORASCU 866
BORATES, TETRA, SODIUM SALT 866
BORATE (1-), TETRAFLUORO- 569
BORAX 866
BORAX 2335 1029
BORAX DECAHYDRATE 866
BORDERMASTER lxii
BOREA 177
BORER SOL 351
BORIC ACID **173**
BORIC ACID, ZINC SALT 1029
BORIC ANHYDRIDE 174
BORICIN 866
BORIC OXIDE 174
BORNANE, 2-OXO- 231
2-BORNANONE 231
BOROCIL EXTRA 177
BOROETHANE 345
BOROFAX 173
BOROLIN 785
BORON BROMIDE 175
BORON FLUORIDE 176
BORON HYDRIDE 333, 345
BORON OXIDE **174**
BORON SEQUIOXIDE 174
BORON TRIBROMIDE **175**
BORON TRIBROMIDE 6 175
BORON TRIFLUORIDE **176**
BORON TRIOXIDE 174
BORSAURE (GERMAN) 173
BOSAN SUPRA 332
BOS MH 605
BOTTLED GAS 584
BOU 905
BOVIDERMOL 332
BOYGON 824
BP 150
BPZ-250 154
B-QUINOL 530
BRAKE FLUID lvi
BRAUNSTEIN (GERMAN) 609
BRAVO 278, lxii
BRAVO-W-75 278
BRENTAMINE FAST BLUE B BASE 338
BRENTAMINE FAST BLUE B SALT 338
BRENTAMINE FAST RED TR SALT 281
BRICK OIL 302
BRIGHTENERS xxvii
BRIMESTONE lxii

BRIMSTONE 903
BRISTACICLIN ALPHA 925
BRISTACICLINA 925
BRISTACYCLINE 925
BRITISH ALUMINUM AF 260 41
BROCIDE 351
BROCKMANN, ALUMINUM OXIDE 43
BRODAN 282
BROM (GERMAN) 178
BROMACIL **177**
BROMACIL 1.5 177
ALPHA BROMACIL 80 WP 177
BROMALLYLENE 32
BROMAX 177
BROMAZIL 177
BROMCHLOPHOS 689
BROME (FRENCH) 178
BROMELLITE 162
BROMEX 689
BROMIC ACID, BARIUM SALT 124
BROMIC ACID, POTASSIUM SALT 795
BROMINE **178**
BROMINE FLUORIDE 179, 180
BROMINE PENTAFLUORIDE **179**
BROMINE TRIFLUORIDE **180**
BROMKAL P 67-6HP 992
BROMO (ITALIAN) 178
1-BROMOBUTANE 199
5-BROMO-3-SEC-BUTYL-6-METHYLURACIL 177
BROMOCHLOROMETHANE 259
2-BROMO-2-CHLORO-1,1,1-TRIFLUORO-501
2-BROMO-2-CHLORO-1,1,1-TRIFLUOROETHANE 501
O-(4-BROMO-2,5-DICHLOROPHENYL)O-METHYL PHENYLPHOSPHONOTHIO-ATE 581
BROMOETHENE 1005
BROMOETHYLENE 1005
BROMOFLUOROFORM 973
BROMOFORM **181**
BROMOFORME (FRENCH) 181
BROMOFORMIO (ITALIAN) 181
BROMOFUME 451
BROMOMETHANE 1005
(BROMOMETHYL)BENZENE 155
5-BROMO-6-METHYL-3-(1-METHYLPROPYL)-2,4(1H,3H)-PYRIMIDINEDIONE 177
P-(BROMOMETHYL)NITROBENZENE 155
BROMOPHENYLMETHANE 155
3-BROMOPROPENE 32
3-BROMOPROPYLENE 32
BROMO SELTZER 758
ALPHA-BROMOTOLUENE 155
BROMOTRIFLUOROMETHANE 973
BROMOWODOR (POLISH) 527
BROMOXYNIL lxiii
BROMOXYNIL BUTYRATE lxiii
BROMURE DE VINYLE (FRENCH) 1005
BROMURO DI ETILE (ITALIAN) 451
BROMWASSERSTOFF (GERMAN) 527
BRONZE POWDER 297
BROOKITE 949
BROOM (DUTCH) 178
BROOMWATERSTOF (DUTCH) 527
BRUINSTEEN (DUTCH) 609
BRUMIN 1012
BRUSH BUSTER lxii
BRUSH-OFF 445 LOW VOLATILE BRUSH KILLER 912
BRUSH-RHAP 330
BRUSH RHAP 912
BRUSHTOX 912

Bold numbers denote main chemical entries; roman numerals denote text.

Bold numbers denote main chemical entries; roman numerals denote text.

CARBIMIDE 315
CARBINAMINE 645
CARBINOL 637
CARBOFENOTHION (DUTCH) 993
CARBOFOS 602
CARBOFURAN **235**, lxiii
CARBOLIC ACID 759
CARBOLITH 587
CARBOLSAURE (GERMAN) 759
CARBOMETHANE 563
CARBOMETHENE 563
2-CARBOMETHOXY-1-
    METHYLVINYLDIMETHYL PHOS-
    PHATE 771
ALPHA-2-CARBOMETHOXY-1-METHYLVINYL
    DIMETHYL PHOSPHATE 771
1-CARBOMETHOXY-1-PROPEN-2-Y PHOS-
    PHATE 771
CARBONA 241, xxxiii
CARBON BICHLORIDE 754
CARBON BISULFIDE 238
CARBON BISULPHIDE 238
CARBON BLACK **236**, xxxv, xxxix
CARBON BROMIDE 240
CARBON CHLORIDE 241
CARBON DICHLORIDE 754
CARBON DIOXIDE **237**, xcvii
CARBON DISULFIDE **238**, xlvii, ciii
CARBON DISULPHIDE 238
CARBONE (OXYCHLORURE DE)
    (FRENCH) 772
CARBONE (OXYDE DE) (FRENCH) 239
CARBONE (SUFURE DE) (FRENCH) 238
CARBON FLUORIDE 930
CARBON HEXACHLORIDE 509
CARBONIC ACID, AMMONIUM SALT 62
CARBONIC ACID, CALCIUM SALT (1:1) 219
CARBONIC ACID, DIAMMONIUM SALT 62
CARBONIC ACID, DILITHIUM SALT 587
CARBONIC ACID, DIPOTASSIUM SALT 796
CARBONIC ACID, DISODIUM SALT 867
CARBONIC ACID GAS 237
CARBONIC ACID LITHIUM SALT 587
CARBONIC ACID, MONOAMMONIUM
    SALT 59
CARBONIC ACID, ZINC SALT (1:1) 1031
CARBONIC ANHYDRIDE 237
CARBONIC DICHLORIDE 772
CARBONIC OXIDE 239
CARBONIO (OSSICLORURO DI) (ITAL-
    IAN) 772
CARBONIO (OSSIDO DI) (ITALIAN) 239
CARBONIO (SOLFURO DI) (ITALIAN) 238
CARBON MONOXIDE **239**
CARBON MONOXIDE MONOSULFIDE 242
CARBON NAPTHA 143
CARBON NITRIDE 316
CARBON OIL 143
CARBON OXIDE (CO) 239
CARBON OXIDE SULFIDE 242
CARBON OXYCHLORIDE 772
CARBON OXYGEN SULFIDE 242
CARBON OXYSULFIDE 242
CARBONS lxix
CARBON SULFIDE 238
CARBON TET 241
CARBON TETRABROMIDE **240**
CARBON TETRACHLORIDE **241**, xxvii, xxx-
    iii, xliii, lxiii, cviii
CARBON TETRAFLUORIDE 930
CARBONYETHANE 823
CARBONYLCHLORID (GERMAN) 772
CARBONYL CHLORIDE 772
CARBONYLDIAMIDE 997
CARBONYL DIAMIDE 997
CARBONYL DIAMINE 997
CARBONYL SULFIDE **242**
CARBONYLSULFID-(32)S 242

CARBOPHENOTHION 993
CARBOPHOS 602
CARBORAFFIN 25
CARBORAFINE 25
CARBOSIP 5G 235
CARBOSPOL 36
CARBOX 367
CARBOXIDE 224
CARBOXYBENZENE 151
CARBOXYETHANE 823
4-CARBOXYPHTHALIC ANHYDRIDE 978
CARDONA lxii
CARPIN 20
CARPOLIN 234
CARREL-DAKIN SOLUTION 880
CARYOLYSIN 166
CASALIS GREEN 288
CASING HEAD GASOLINE 494
CASORON lxii
CASSEL BROWN 611
CASTRIX 304
CATALIN CAO-3 163
CATAPAL S 43
CATECHIN 243
CATECHOL **243**
CATILAN 247
CAUSTIC POTASH 806
CAUSTIC SODA 879
CAUSTIC SODA, BEAD 879
CAUSTIC SODA, DRY 879
CAUSTIC SODA, FLAKE 879
CAUSTIC SODA, GRANULAR 879
CAUSTIC SODA, SOLID 879
CAV-TROL 876
CB 259
CB-4564 422
CBD 90 966
CBM 259
CCC 222
CCH 225
CCS 203 193
CCS 301 194
CD 68 248
CDA 101 297
CDA 102 297
CDA 110 297
CDA 122 297
CDB 63 873
CDDP 293
CDEC 900
CECOLENE 964
CEFRACYCLINE SUSPENSION 925
CEGLUTION 587
CEKIURON 418
CEKUDIFOL 367
CEKUMETHION 679
CEKUQUAT 744
CEKUSIL 765
CEKUTHOATE 385
CELANEX 146, 583
CELITE 340
CELLITAZOL B 338
CELLITAZOL BN 338
CELLITON ORANGE R 52
CELLON 923
CELLOSOLVE 430
CELLOSOLVE ACETATE 431
CELLOSOLVE SOLVENT 430
CELLULEX DOP 406
CELLULOSIC RESINS xxxvii
CELLUPHOS 4 960
CELMER 765
CELMIDE 451
CELON A 450
CELON ATH 450
CELPHOS 774
CELPHOS (INDIAN) 45
CELTHION (INDIAN) 602

CEMENT BLACK 609
CENOL GARDEN DUST 844
CENTRAX 814
CERCINE 341
CEREGULART 341
CERESAN 463, 765
CERESAN M 629
CERESAN M-2X 629
CERESAN M-DB 629
CERUBIDIN 331
CESIUM HYDRATE 244
CESIUM HYDROXIDE **244**
CESIUM HYDROXIDE DIMER 244
CFC-11 965
CFC-12 356
CFC-113 487
CFC-114 488
CFC-115 269
CFV 249
CG 772
CGA 26351 249
CHA 323
CHAFER AZOTOX 658
CHALCEDONY 854
CHALK 219
CHALOTHANE 501
CHAMELEON MINERAL 810
CHANNEL BLACK 236
CHARCOAL, ACTIVATED 25
CHEELOX 450
CHELEN 445
CHEMAGRO 9010 824
CHEMAGRO 25141 470
CHEMANOX 11 163
CHEMATHION 602
CHEMCOLOX 340 450
CHEMETRON FIRE SHIELD 107
CHEM FISH 844
CHEMICAL 109 108
CHEMICETIN 247
CHEMICETINA 247
CHEMIFLOUR 876
CHEM-MITE 844
CHEM NEB 607
CHEM-PHENE 958
CHEMSECT O-CRESOL,4,6-DINITRO- 403
CHEM-TOL 749
CHEM ZINEB 1045
CHERTS 854
CHILE SALTPETER 883
CHINA CLAY 560
CHINESE WHITE 1040
CHINOLEINE 838
CHINOLIN (CZECH) 838
CHINOLINE 838
CHINON (DUTCH) 839
CHINON (GERMAN) 839
P-CHINON (GERMAN) 839
CHINONE 839
CHINORTA 742
CHIPCO THIRAM 75 943
CHIPCO TURF HERBICIDE "D" 330
CHISSONOX 206 1007
CHLOFENVINPHOS 249
CHLOMIN 247
CHLOMYCOL 247
CHLOOR (DUTCH) 251
CHLOORBENZEEN (DUTCH) 258
2-CHLOOR-1,3-BUTADIEEN (DUTCH) 695
CHLOORDAAN (DUTCH) 248
O-2-CHLOOR-1-(2,4-DICHLOOR-FENYL)-
    VINYL-O,O-DIETHYLFOSFAAT
    (DUTCH) 249
2-CHLOOR-4-DIMETHYLAMINO-6-METHYL-
    PYRIMIDINE (DUTCH) 304
CHLOORETHAAN (DUTCH) 445
CHLOOR-METHAAN (DUTCH) 651

1-CHLOOR-4-NITROBENZEEN (DUTCH) 710
CHLOORPIKRINE (DUTCH) 274
CHLOORWATERSTOF (DUTCH) 523
CHLOPHEN 792
CHLOR (GERMAN) 251
CHLORACETIC ACID 255
CHLORAL **245**
CHLORAL, ANHYDROUS, INHIBITED. 245
CHLORALLYL DIETHYLDITHIOCARBA-MATE 900
CHLORALLYLENE 33
CHLORAMBED 246
CHLORAMBEN **246**
CHLORAMBENE 246
CHLORAMEX 247
CHLORAMINE 166
CHLORAMP (RUSSIAN) 785
CHLORAMPHENICOL **247**
D-CHLORAMPHENICOL 247
D-THREO-CHLORAMPHENICOL 247
D-(-)-THREO-CHLORAMPHENICOL 247
CHLORAMSAAR 247
CHLORANIL lxiii
CHLORASOL 247
CHLORA-TABS 247
CHLORATE DE CALCIUM (FRENCH) 220
CHLORATE DE POTASSIUM (FRENCH) 797
CHLORATE OF POTASH 797
CHLORATE OF SODA 868
CHLORATE SALT OF MAGNESIUM 592
CHLORATE SALT OF SODIUM 868
CHLORAX 868
CHLORBENZEN 258
CHLORBENZILATE lxiii
2-CHLOR-1,3-BUTADIEN (GERMAN) 695
CHLORDAN 248
GAMMA-CHLORDAN 248
CHLORDANE **248**, lxiii
CHLORDECONE 561
O-2-CHLOR-1-(2,4-DICHLOR-PHENYL)-VINYL-O,O-DIAETHYLPHOSPHAT (GERMAN) 249
CHLORDIMEFORM lxiii
CHLORDIMETHYLETHER (CZECH) 267
CHLORE (FRENCH) 251
CHLORESENE 146, 583
CHLORETHAZINE 166
CHLORETHYL 445
CHLOREX 360
CHLOREXTOL 792
P-CHLORFENOL (CZECH) 272
CHLORFENVINFOS **249**
CHLORFENVINPHOS 249, lxiii
CHLORHYDRATE DE 4-CHLOROORTHOTOLUIDINE (FRENCH) 281
CHLORIC ACID, BARIUM SALTCHLORIC ACID, BARIUM SALT 126
CHLORIC ACID, CALCIUM SALT 220
CHLORIC ACID, DE-FOL-ATE 592
CHLORIC ACID, MAGNESIUM 592
CHLORIC ACID, POTASSIUM SALT 797
CHLORIC ACID, SODIUM SALT 868
CHLORIC ACID, ZINC SALT 1032
CHLORIC DRASELNY (CZECH) 798
CHLORICOL 247
CHLORID AMONNY (CZECH) 63
CHLORID ANTIMONITY 105
CHLORIDE OF LIME 225
CHLORIDE OF PHOSPHORUS 782
CHLORID RTUTNATY (CZECH) 616
CHLORIDUM 445
CHLORINATED BIPHENYL 792

CHLORINATED CAMPHENE 958
CHLORINATED DIPHENYL 792
CHLORINATED DIPHENYLENE 792
CHLORINATED DIPHENYL OXIDE **250**
CHLORINATED HYDROCARBONS xxvii, xl, cx
CHLORINATED HYDROCHLORIC ETHER 358
CHLORINATED LIME 225
CHLORINATED PHENOLS xl, xli
CHLORINDAN 248
CHLORINE **251**, xxxiii, ciii, cix
CHLORINE CONTROL 891
CHLORINE CURE 891
CHLORINE DIOXIDE **252**
CHLORINE FLUORIDE 253
CHLORINE FLUORIDE OXIDE 757
CHLORINE MOLECULAR 251
CHLORINE OXIDE 252
CHLORINE(IV) OXIDE 252
CHLORINE OXYFLUORIDE 757
CHLORINE PEROXIDE 252
CHLORINE TRIFLUORIDE **253**
CHLOR KIL 248
CHLOR-METHAN (GERMAN) 651
CHLORMETHINE 166
1-CHLOR-4-NITROBENZOL (GERMAN) 710
CHLOROACETALDEHYDE **254**
2-CHLOROACETALDEHYDE 254
CHLOROACETALDEHYDE (40% AQUE-OUS) 254
CHLOROACETALDEHYDE MONOMER 254
CHLOROACETIC ACID **255**
CHLOROACETIC ACID CHLORIDE 256
CHLOROACETIC CHLORIDE 256
CHLOROACETYL CHLORIDE **256**
CHLOROAETHAN (GERMAN) 445
3-CHLORO-ALLYL CHLORIDE 365
ALPHA-CHLOROALLYL CHLORIDE 365
GAMMA-CHLORO ALLYL CHLORIDE 365
2-CHLOROALLYL DIETHYLDITHIOCARBA-MATE 900
2-CHLOROALLYL N,N-DIETHYLDITHIOCARBAMATE 900
CHLOROALONIL 278
P-CHLORO-O-AMINOPHENOL 49
2-CHLORO-4-AMINOTOLUENE 265
3-CHLORO-2-AMINOTOLUENE 280
4-CHLORO-2-AMINOTOLUENE 266
5-CHLORO-2-AMINOTOLUENE HYDRO-CHLORIDE 281
CHLOROANILINE **257**
CHLOROBEN 353
CHLOROBENZAL 141
ALPHA-CHLOROBENZALDEHYDE 153
CHLOROBENZEN (POLISH) 258
CHLOROBENZENE **258**, ciii
CHLOROBENZOL 258
CHLORO BIPHENYL 792
CHLORO 1,1-BIPHENYL 792
CHLOROBLE M 607
CHLOROBROMOMETHANE **259**
CHLOROBUTADIENE 695
2-CHLOROBUTA-1,3-DIENE 695
2-CHLORO-1,3-BUTADIENE 695
1-CHLOROBUTANE 200
CHLOROCAMPHENE 958
CHLOROCAPS 247
3-CHLOROCHLORDENE 503
1-CHLORO-2-(BETA-CHLOROETHOXY)ETHANE 360
1-CHLORO-2-(BETA-CHLOROETHYLTHIO)ETHANE 167

CHLORO(CHLORO-METHOXY)METHANE 168, 169
4-CHLORO-ALPHA-(4-CHLOROPHENYL)-ALPHA-(TRICHLOROMETHYL)BENZENEME THANOL 367
CHLOROCID 247
CHLOROCIDE 247
CHLOROCIDIN C 247
CHLOROCIDIN C TETRAN 247
CHLOROCOL 247
2-CHLORO-4-(1-CYANO-1-METHYLETHYLAMINO)-6-ETHYLAMINO-1,3,5-TRIAZINE 172
4-CHLORO-2-CYCLOPENTYLPHENOL xxxi
7-CHLORO-1-(CYCLOPROPYLMETHYL)-1,3-DIHYDRO-5-PHENYL-2H-1,4-BENZODIAZEPIN-2-ONE 814
7-CHLORO-1-CYCLOPROPYLMETHYL-5-PHENYL-1H-1,4-BENZODIAZEPIN-2(3H)-ONE 814
CHLORODANE 248
CHLORODEN 353
2-CHLORO-1-(2,4-DICHLOROPHENYL)VINYL DIETHYL PHOSPHATE 249
BETA-2-CHLORO-1-(2',4'-DICHLOROPHENYL) VINYL DIETHYL-PHOSPHATE 249
CHLORODIETHYALUMINUM 373
CHLORODIETHYLALANE 373
CHLORODIFLUOROETHANE **260**
1-CHLORO-1,1-DIFLUOROETHANE 260
CHLORODIFLUOROMETHANE **261**
2-CHLORO-1-(DIFLUOROMETHOXY)-1,1,2-TRIFLUOROETHANE 424
7-CHLORO-1,3-DIHYDRO-1-METHYL-5-PHENYL-2H-1,4-BENZODIAZEPIN-2-ONE 341
CHLORO DI VINYLE (ITALIAN) 1006
1-CHLORO-2,3-EPOXYPROPANE 425
CHLOROETENE 652
2-CHLOROETHANAL 254
2-CHLORO-1-ETHANAL 254
CHLOROETHANE 445
CHLOROETHANOIC ACID 255
CHLOROETHENE 652, 1006
CHLOROETHENE NU 652
2-CHLORO-4-ETHYLAMINO-6-(1-CYANO-1-METHYL)ETHYLAMINO-S-TRIAZINE 172
2-((4-CHLORO-6-(ETHYLAMINO)-S- TRIAZIN-2-YL)AMINO)-2-METHYLPROPIONITRILE
2-(4-CHLORO-6-(ETHYLAMINO)-1,3,5-TRIAZIN-2-YL)AM-2-METHYL- 172
CHLOROETHYLENE 1006
CHLOROETHYL ETHER 360
ALPHA-CHLOROETHYLIDENE FLUO-RIDE 260
CHLOROETHYLMERCURY 463
CHLOROFENVINPHOS 249
CHLOROFLUOROCARBONS lxxvi, lxix, lxx
CHLOROFORM **262**, xliii, ciii
CHLOROFORME (FRENCH) 262
CHLOROFORMIC ACID DIMETHYLA-MIDE 390
CHLOROFORM, METHYL- 652
CHLOROFORMYL CHLORIDE 772
CHLOROHYDRIC ACID 523
3-CHLORO-7-HYDROXY-4-METHYL-COUMARIN-O,O-DIETHYLPHOSPHOROTHIONAT E 301
3-CHLORO-7-HYDROXY-4-METHYL-COUMARIN O,O-DIETHYL PHOSPHO-ROTHIOATE 301

Bold numbers denote main chemical entries; roman numerals denote text.

Bold numbers denote main chemical entries; roman numerals denote text.

Bold numbers denote main chemical entries; roman numerals denote text.

DIAMINE 520
CIS-DIAMINEDICHLOROPLATINUM 293
DIAMINE, HYDRAZINE BASE 520
DIAMINE SULFATE 521
1,2-DIAMINOAETHAN (GERMAN) 449
2,4-DIAMINOANISOLE SULFATE **337**
2,4-DIAMINOANISOLE SULPHATE 337
2,4-DIAMINO-ANISOL SULPHATE 337
1,4-DIAMINOBENZENE 761
P-DIAMINOBENZENE 761
4,4'-DIAMINOBIPHENYL 149
4,4'-DIAMINO-1,1'-BIPHENYL 149
P,P'-DIAMINOBIPHENYL 149
DI-(4-AMINO-3-CHLOROFENIL)METANO
   (ITALIAN) 660
DI(-4-AMINO-3-
   CHLOROPHENYL)METHANE 660
4,4'-DIAMINO-3,3'-DICHLORODIPHENYLM
   ETHANE 660
2,2'-DIAMINODIETHYLAMINE 376
P,P'-DIAMINODIFENYLMETHAN
   (CZECH) 664
4,4'-DIAMINO-3,3'-
   DIMETHYLBIPHENYL 950
4,4'-DIAMINO-3,3'-
   DIMETHYLDIPHENYL 950
4,4'-DIAMINODIPHENYL 149
P-DIAMINODIPHENYL 149
4,4'-DIAMINODIPHENYL ETHER ciii
4,4'-DIAMINODIPHENYLMETHAN (GER-
   MAN) 664
DIAMINODIPHENYLMETHANE 664
4,4'-DIAMINODIPHENYLMETHANE 664
P,P'-DIAMINODIPHENYLMETHANE 664
DIAMINODITOLYL 950
1,2-DIAMINO-ETHAAN (DUTCH) 449
1,2-DIAMINOETHANE, ANHYDROUS 449
1,2-DIAMINOETHANE COPPER COM-
   PLEX 314
2,4-DIAMINO-1-METHOXYBENZENE 337
1,4-DIAMINO-2-NITROBENZENE 53
DI-(4-AMINOPHENYL)METHANE 664
ALPHA,ALPHA'-DIAMINO-M-XYLENE 1015
CIS-DIAMMINEDICHLOROPLATINUM 293
DIAMMONIUM CARBONATE 62
DIAMMONIUM CHROMATE 65
DIAMMONIUM CITRATE 66
DIAMMONIUM HEXACHLOROPLATINATE (2-
) 64
DIAMMONIUM HEXAFLUOROSILICATE 69
DIAMMONIUM MOLYBDATE 72
DIAMMONIUM OXALATE 74
DIAMMONIUM SULFITE 84
DIAMMONIUM TARTRATE 85
DIANALINEMETHANE 664
P,P'-DIANILINE 149
O-DIANISIDIN (CZECH, GERMAN) 338
0-DIANISIDINA (ITALIAN) 338
DIANISIDINE 338
O-DIANISIDINE **338**, 338
0,0'-DIANISIDINE 338
3,3'-DIANISIDINE 338
DIANON 342
DIANTIMONY TRIOXIDE 107
DIAPADRIN 368
DIAPAM 341
DIAREX HF 77 899
DIARSENIC PENTOXIDE 112
DIARSENIC TRIOXIDE 111, 114, 119
DIARSENIC TRISULFIDE 115
DIASPORE 43
DIASTASE **339**
DIATER 418
DIATERR-FOS 342

DIATO BLUE BASE B 338
DIATO BLUE SALT B 338
DIATOMACEOUS EARTH **340**
DIATOMACEOUS SILICA 340
DIATOMITE 340
DIAZAJET 342
3,6-DIAZAOCTANEDIOIC ACID,3,6-
   BIS(CARBOXYMETHYL)- 450
DIAZATOL 342
DIAZEMULS 341
DIAZEPAM **341**
DIAZEPAMU (POLISH) 341
DIAZETARD 341
DIAZIDE 342
DIAZINON **342**, lxii, lxiv
DIAZINONE 342
DIAZIRINE 343
DIAZITOL 342
DIAZO COMPONENT 48, FAST BLUE B
   SALT 338
DIAZO COMPOUNDS xciii
DIAZO FAST BLUE B 338
DIAZO FAST RED TR 281
DIAZO FAST RED TRA 281
DIAZOL 342
DIAZOMETHANE **343**
DIAZONIUM 343
DIBASIC LEAD ARSENATE 565
DIBASIC SODIUM PHOSPHATE 887
DIBENZO(B,E) 1,4-DIOXIN,2,3,7,8- TETRA-
   CHLORO- 915
1,2,5,6-DIBENZONAPHTHALENE 292
DIBENZOPARATHIAZINE 760
DIBENZOTHIAZINE 760
DIBENZO-1,4-THIAZINE 760
DIBENZOYLPEROXID (GERMAN) 154
DIBENZOYL PEROXIDE 154
DIBENZYLDICHLOROSILANE **344**
DIBK 396
DIBORANE **345**
DIBORANE(6) 345
DIBORANE HEXANHYDRIDE 345
DIBORON HEXAHYDRIDE 345
DIBOVAN 332
DIBROM 689
1,2-DIBROMAETHAN (GERMAN) 451
O-(1,2-DIBROM-2,2-DICHLOR-AETHYL)-O,O-
   DIMETHYL-PHOSPHAT (GERMAN) 689
DIBROMOBENZENE **346**
DIBROMOCHLOROPROPANE lxxxix
1,2-DIBROMO-2,2-DICHLOROETHYL DI-
   METHYL PHOSPHATE 689
1,2-DIBROMO-3-DICHLOROPROPANE xliii
O-(1,2-DIBROMO-2,2-DICLORO-ETIL)-O,O-
   DIMETIL-FOSFATO (ITALIAN) 689
DIBROMODIFLUOROMETHANE 381
DIBROMODIFLUORO-METHANE 381
1,2-DIBROMOETANO (ITALIAN) 451
DIBROMOETHANE 451
1,2-DIBROMOETHANE 451
ALPHA,BETA-DIBROMOETHANE 451
SYM-DIBROMOETHANE 451
DIBROMOMETHANE **347**
2,3DIBROMO-1- PROPANOL, PHISPHATE
   (3:1) 992
DIBROMURE D'ETHYLENE (FRENCH) 451
DIBRON lxii
O-(1,2-DIBROOM-2,2-DICHLOOR-ETHYL)-
   O,O-DIMETHYL-FOSFAAT (DUTCH) 689
1,2-DIBROOMETHAAN (DUTCH) 451
DIBUTYL ACID PHOSPHATE 349
DIBUTYL ACID O-PHOSPHATE 349
DIBUTYLAMINE, N-NITROSO- 722
DIBUTYLAMINOETHANOL **348**

2-DIBUTYLAMINOETHANOL 348
N-N-DI-N-BUTYLAMINOETHANOL 348
2-(DIBUTYLAMINO) ETHANOL 2-N-
   DIBUTYLAMINOETHANOL 348
BETA-N-DIBUTYLAMINOETHYL ALCO-
   HOL 348
DIBUTYL 1,2-
   BENZENEDICARBOXYLATE 350
2,6-DI-TERT-BUTYL-P-CRESOL 163
N-N-DIBUTYLETHANOLAMINE 348
DIBUTYL ETHER 202
DI-N-BUTYL ETHER 202
DI-N-BUTYL HYDROGEN PHOSPHATE 349
N,N-DIBUTYL N-(2-
   HYDROXYETHYL)AMINE 348
2,6-DI-T-BUTYL-4-METHYLPHENOL 163
DI-N-BUTYLNITROSAMIN (GERMAN) 722
DIBUTYLNITROSAMINE 722
N,N-DIBUTYLNITROSOAMINE 722
N,N-DI-N-BUTYLNITROSAMINE 722
DI-N-BUTYLNITROSAMINE 722
DIBUTYL OXIDE 202
DIBUTYL PHOSPHATE **349**
DIBUTYL PHOSPHORIC ACID 349
DIBUTYL PHTHALATE **350**, liii, ciii
DI-N-BUTYL PHTHALATE 350
DIC 1468 683
DICALITE 340
DICAMBA lxii, lxv
DICARBAM 234
DICARBURRETTED HYDROGEN 448
DICHLOBENIL lxii
P-DICHLOORBENZEEN (DUTCH) 354
1,4-DICHLOORBENZEEN (DUTCH) 354
1,1-DICHLOORETHAAN (DUTCH) 358
1,2-DICHLOORETHAAN (DUTCH) 351
2,2'-DICHLOORETHYLETHER (DUTCH) 360
(2,4-DICHLOOR-FENOXY)-AZIJNZUUR
   (DUTCH) 330
3-(3,4-DICHLOOR-FENYL)-1,1-
   DIMETHYLUREUM (DUTCH) 418
1,1-DICHLORAETHAN (GERMAN) 358
1,2-DICHLOR-AETHAN (GERMAN) 351
1,2-DICHLOR-AETHEN (GERMAN) 359
S-(2,3-DICHLOR-ALLYL)-N,N-DIISOPROPYL-
   MONOTHIOCARBAMAAT (DUTCH) 336
DICHLORAMINE 166
DICHLORANTIN 357
O-DICHLORBENZENE 353
1,4-DICHLOR-BENZOL (GERMAN) 354
O-DICHLOR BENZOL 353
P-DICHLORBENZOL (GERMAN) 354
2,2'-DICHLOR-DIAETHYLAETHER (GER-
   MAN) 360
3,3'-DICHLOR-4,4'-
   DIAMINOPHENYLMETHAN (GER-
   MAN) 660
DICHLORDIMETHYLAETHER (GER-
   MAN) 168, 169
DICHLOREMULSION 351
DICHLOREN (GERMAN) 166
1,1-DICHLORETHANE 358
1,2-DICHLORETHANE **351**
2,2'-DICHLORETHYL ETHER 360
BETA,BETA-DICHLOR-ETHYL-
   SULPHIDE 167
DICHLORFENIDIM 418
DI-CHLORICIDE 354
DICHLORINE 251
DI-CHLOR-MULSION 351
D-(-)-THREO-2-DICHLOROACETAMIDO-1-P-
   NITROPHENYL-1,3-PROPANEDIOL 247
DICHLOROACETYLENE **352**

Bold numbers denote main chemical entries; roman numerals denote text.

Bold numbers denote main chemical entries; roman numerals denote text.

N,N-DIMETHYLCARBAMOYL CHLO-
RIDE 390
CIS-2-DIMETHYLCARBAMOYL-1-
METHYLVINYL DIMETHYLPHOS-
PHATE 368
DIMETHYLCARBAMYL CHLORIDE 390
N,N-DIMETHYLCARBAMYL CHLORIDE 390
NSN-DIMETHYL(CARBANDOYLE) CHLO-
RIDE 390
DIMETHYLCARBINOL 554
DIMETHYLCHLOROETHER 267
2,5-DIMETHYL-2,5-DI-(BENZOYLPEROXY)
HEXANE **391**
2,5-DIMETHYL-2,5-DI-(BENZOYLPEROXY)
HEXANE, NOT MORE THAN 82% WITH
INERT SOLID 391
DIMETHYL1,2-DIBROMO-2,2-
DICHLOROETHYLPHOSPHATE 689
O,O-DIMETHYL-O-(1,2-DIBROMO-2,2-
DICHLOROETHYL)PHOSPHATE 689
2,5-DIMETHYL-2,5-DI(TERT-
BUTYLPEROXY)HEXANE **394**
2,5-DIMETHYL-2,5-DI(TERT-
BUTYLPEROXY)HEXANE, TECHNICALLY
PURE 394
O,O-DIMETHYLS-(1,2-DICARBETHOXY-
ETHYL)PHOSPHORO-DITHIOCITE 602
DIMETHYL 1,3-DI(CARBOMETHOXY)-1-
PROPEN-2-YL PHOSPHATE 911
O,O-DIMETHYL O-2,2-DICHLORO-1,2-
DIBROMOETHYL PHOSPHATE 689
DIMETHYL-1,1'-DICHLOROETHER 168, 169
1,1-DIMETHYL-3-(3,4-
DICHLOROPHENYL)UREA 418
DIMETHYL-DICHLOROSILAN (CZECH) 392
DIMETHYLDICHLOROSILANE **392**
DIMETHYL-DICHLORSILAN (CZECH) 392
DIMETHYL-DIETHOXYSILAN (CZECH) 393
DIMETHYLDIETHOXYSILANE **393**
2,2-DIMETHYL-2,2-
DIHYDROBENZOFURANYL-7 N-
METHYLCARBAMATE 235
DIMETHYL DIKETONE 184
5,6-DIMETHYL-2-DIMETHYLAMINO-4-
PYRIMIDINYLDIMETHYLCARBA-
MATE 789
O,O-DIMETHYL-O-(2-DIMETHYL-
CARBAMOYL-1-METHYL-
VINYL)PHOSPHAT (GERMAN) 368
O,O-DIMETHYLO-(N,N-
DIMETHYLCARBAMOYL-1-METHYLVINYL)
PHOSPHATE 368
O,O-DIMETHYL-O-(1,4-DIMETHYL-3-OXO- 4-
AZA-PENT-1-ENYL)FOSFAAT
(DUTCH) 368
O,O-DIMETHYL-O-(1,4-DIMETHYL-3-OXO-4-
AZA-PENT-1-ENYL)PHOSPHATE 368
3,3'-DIMETHYLDIPHENYL-4,4'-
DIAMINE 950
3,3'-DIMETHYL-4,4'-
DIPHENYLDIAMINE 950
DIMETHYLDITHIOCARBAMIC ACID, IRON
SALT 472
DIMETHYLDITHIOCARBAMIC ACID, IRON(3)
SALT 472
O,O-DIMETHYLDITHIOPHOSPHORYLACETIC
ACID, N-MONOMETHYLAMIDE SALT 385
DIMETHYLENEDIAMINE 449
DIMETHYLENEIMINE 455
3,3-DIMETHYLENENORCAMPHENE 230
DIMETHYLENE OXIDE 456
DIMETHYLENIMINE 455
DIMETHYLESTER KYSELINY SIROVE
(CZECH) 399

1,1-DIMETHYLETHANOL 195
1,1-DIMETHYLETHYLAMINE 198
DIMETHYLFORMALDEHYDE 7
DIMETHYLFORMAMID (GERMAN) 395
DIMETHYLFORMAMIDE **395**
DIMETHYL FORMAMIDE 395
N,N-DIMETHYLFORMAMIDE 395
N,N-DIMETHYL FORMAMIDE 395
DIMETHYLFORMEHYDE 7
DIMETHYL GLYOXAL 184
2,6-DIMETHYL-HEPTAN-4-ON (DUTCH; GER-
MAN) 396
2,6-DIMETHYLHEPTANONE **396**
2,6-DIMETHYLHEPTAN-4-ONE 396
2,6-DIMETHYL-4-HEPTANONE 396
DIMETHYLKETAL 7
DIMETHYL KETONE 7
DIMETHYL MERCURY **397**
N,N-DIMETHYLMETHANAMINE 980
DIMETHYLMETHANE 817
DIMETHYL METHANE 817
N,N-DIMETHYLMETHANIDE 395
2,2-DIMETHYL-4-(N-
METHYLAMINOCARBOXYLATO)-1,3-
BENXODIOXOLE 138
O,O-DIMETHYLS-(2-(METHYLAMINO)-2-
OXOETH-
YL)PHOSPHORODITHIOATE 385
(O,O-DIMETHYL-S-(N-METHYL-
CARBAMOYL-METHYL)-
DITHIOPHOSPHAT) (GERMAN) 385
O,O-DIMETHYL METHYLCARBAMOYLMETH-
YLPHOSPHORODITHIOATE 385
O,O-DIMETHYL-S-(N-METHYL-
CARBAMOYL)-METHYL-DITHIOFOSFAAT
(DUTCH) 385
O,O-DIMETHYLS-(N-
METHYLCARBAMOYLMETH-
YL)DITHIOPHOSPHATE 385
O,O-DIMETHYLS-(N-
METHYLCARBAMOYLMETH-
YL)PHOSPHORODITHIOATE 385
O,O-DIMETHYLS-(N-
METHYLCARBAMYLMETH-
YL)THIOTHIONOPHOSPHATE 385
O,O-DIMETHYL-O-(2-N-
METHYLCARBAMOYL-1-METHYL-VINYL)-
FOSFAAT (DUTCH) 687
O,O-DIMETHYL-O-(2-N-
METHYLCARBAMOYL -1-METHYL)-VINYL-
PHOSPHAT (GERMAN) 687
O,O-DIMETHYL-O-(2-N-
METHYLCARBAMOYL-1-METHYL-VINYL)
PHOSPHATE 687
2,2-DIMETHYL-3-METHYLENE- 230
2-2-DIMETHYL-3-METHYLENE NORBO-
RANE 230
3,3-DIMETHYL-2-METHYLENE NORCAM-
PHONE 230
O,O-DIMETHYL-O-4(METHYLMERCAPTO)-3-
METHYLPHENYL PHOSPHOROTHIO-
ATE 471
(E)-DIMETHYL 1-METHYL-3-
(METHYLAMINO)-3-OXO-1-PROPENYL
PHOSPHATE 687
DIMETHYL 1-METHYL-2-
(METHYLCARBAMOYL)VINYL PHOS-
PHATE, CIS 687
3,5-DIMETHYL-4-(METHYLTHIO)- ,METHYL-
CARBAMATE 638
3,5-DIMETHYL-4-
(METHYLTHIO)
PHENYLMETHYLCARBAMATE 638

3,5-DIMETHYL-4-METHYLTHIOPHENYL N-
METHYLCARBAMATE 638
DIMETHYL MONOSULFATE 399
DIMETHYLNITROMETHANE 721
O,O-DIMETHYL 0-4 -NITROPHENYL PHO-
SPHOROTHIOATE 679
O,O-DIMETHYL-S-(N-MONOMETHYL)-
CARBAMYL METHYL DITHIOPHOS-
PHATE 385
O,O-DIMETHYL O-P-
NITROPHENYLPHOSPHOROTHIO-
ATE 679
O,O-DIMETHYL-S-(2-OXO-3-AZA-BUTYL)-
DITHIOPHOSPHAT (GERMAN) 385
DIMETHYLPHENOL 1018
2,3-DIMETHYLPHENOL 1019
2,4-DIMETHYLPHENOL 1020
2,5-DIMETHYLPHENOL 1021
2,6-DIMETHYLPHENOL 1022
3,4-DIMETHYLPHENOL 1023
3,5-DIMETHYLPHENOL 1024
3,6-DIMETHYLPHENOL 1021
4,6-DIMETHYLPHENOL 1020
5-DIMETHYLPHENOL METHYLCARBAMATE
ESTER 684
DIMETHYLPHENYLAMINE 388, 1025
2,4-DIMETHYLPHENYLAMINE 1026
DIMETHYL PHOSPHATE ESTER OF 3-
HYDROXY-N-METHYL-CIS-
CROTONAMIDE 687
DIMETHYL PHOSPHATE ESTER WITH 3-
HYDROXY-N,N-DIMETHYL-CIS-
CROTONAMIDE 368
DIMETHYL PHOSPHATE OF 3-HYDROXY-
N,N-DIMETHYL-CIS-CROTONAMIDE 368
DIMETHYL PHOSPHATE OF 3-HYDROXY-N-
METHYL-CIS-CROTONAMINE 687
3-((DIMETHYLPHOSPHINYL)OXY)-2-
BUTENOIC ACID METHYL ESTHER 771
O,O-DIMETHYL-
PHOSPHORODITHIOATE 602
O,O-DIMETHYLPHOSPHOROTHIOATE O,O-
DIESTER WITH 4,4'-THIODIPHENOL 917
O,O-DIMETHYL PHOSTENE 771
DIMETHYLPHTHALATE **398**
DIMETHYL PHTHALATE 398, civ
(O,O-DIMETHYL-PHTHALIMIDIOMETHYL-
DITHIOPHOSPHATE) 773
O,O-DIMETHYL S-(N-
PHTHALIMIDOMETH-
YL)DITHIOPHOSPHATE 773
O,O-DIMETHYL S-
PHTHALIMIDOMETHYLPHOSPHO-
RODITHIOATE 773
N-(1,1-DIMETHYLPROPYNYL)-3,5-
DICHLOROBENZAMIDE 835
DIMETHYLSULFAAT (DUTCH) 399
DIMETHYLSULFAT (CZECH) 399
DIMETHYL SULFATE **399**, civ
DIMETHYL SULPHATE 399
O,O-DIMETHYLO-(2,4,5-
TRICHLOROPHENYL) PHOSPHOROTHIO-
ATE 843
DIMETHYL VICLOGEN CHLORIDE 744
DIMETHYL VIOLOGEN 743
2,6-DIMETIL-EPTAN-4-ONE (ITALIAN) 396
DIMETILFORMAMIDE (ITALIAN) 395
O,O-DIMETIL-O-(1,4-DIMETIL-3-OXO-4-AZA-
PENT-1-ENIL)-FOSFATO (ITALIAN) 368
O,O-DIMETIL-S-(N-METIL-CARBAMOIL-
METIL)-DITIOFOSFATO (ITALIAN) 385
O,O-DIMETIL-O-(2-N-METILCARBAMOIL-1-
METIL-VINIL)-FOSFATO (ITALIAN) 687
DIMETILSOLFATO (ITALIAN) 399

Bold numbers denote main chemical entries; roman numerals denote text.

DITTHIONOUS ACID, ZINC SALT (1:1)   1036
DIUREX   418
DIUROL   56, 418
DIURON   **418**
DIURON 4L   418
DIVINYL   182
M-DIVINYL-   419
M-DIVINYLBENZEN (CZECH)   419
DIVINYL BENZENE   **419**
DIVINYLBENZENE   419
M-DIVINYLBENZENE   419
DIZENE   353
DIZINON   342
DKC 1347   265
DL-LIMONENE   409
DM   331
DMA   386, 387, 388
DMA-4   330
DMAC   386
D-MANNITOL HEXANITRATE   613
DMB   950
DMCC   390
DMDT   lxii
DMF   395
DMFA   395
DMP   398
2,5-DMP   1021
2,6-DMP   1022
3,4-DMP   1023
3,5-DMP   1024
DMS   399
DMS (METHYL SULFATE)   399
DMSP   470
DMTP   471
DMU   418
DN   403
DNC   403
DN-DRY MIX NO. 2   403
DNOC   403
DNOK (CZECH)   403
DNOP   406
DNT   405
2,4-DNT   405
DNTP   745
DOA   170
DOCTAMICINA   247
DODAT   332
DODECYL BENZENE SODIUM SULFO-
   NATE   875
DODECYL BENZENESULFONATE   420
DODECYLBENZENESULFONIC ACID   **420**
DODECYLBENZENSULPHONATE, SODIUM
   SALT   875
DODECYLBENZENSULFONAN SODNY
   (CZECH)   875
DODECYLTRICHLOROSILANE   421
DODECYL TRICHLOROSILANE   421
DOLEN-PUR   507
DOL GRANULE   146, 583
DOLOSTOP   758
DOLVIRAN   758
DOMALIUM   341
DOMATOL   56
DOMOLITE   219
DOOM GARDEN INSECT KILLER   836
DOP   415
DORMONE   330
DOTMENT 324   43
DOTMENT 358   43
DOWANOL 33B   830
DOWANOL-50B   412
DOWANOL DPM   412
DOW ANOL EB   186
DOWANOL EE   430

DOWANOL EIPAT   552
DOWANOL EM   454, 649
DOWCHLOR   248
DOWCIDE 1   768
DOWCIDE 7   749
DOWCO 132   307
DOWCO 139   684
DOWCO-139   684
DOWCO-163   704
DOWCO 179   282
DOWFROST   829
DOWFUME 40   451
DOWFUME EDB   451
DOWFUME W-8   451
DOWFUME W-85   451
DOWICIDE 2S   967
DOWICIDE 7   749
DOWICIDE EC-7   749
DOWICIDE G   749
DOW PENTACHLOROPHENOL DP-2 ANTIMI-
   CROBIAL   749
DOWPER   754
DOWPON   366
DOWPON M   366
DOW SODIUM TCA INHIBITED   961
DOWTHERM 209   830
DOWTHERM E   353
DOW-TRI   964
DOWZENE   788
DOWZENE DHC   788
DOXCIDE 50   252
DPA   410
2,4-D PARAQUAT   lxiii
DPP   745
DRACYLIC ACID   151
DRAZA   638
DRC 1339   265
DREXEL   418
DREXEL DEFOL   868
DREXEL PARATHION 8E   745
DRI-KIL   844
DRILL TOX-SPEZIA AGLUKON   583
DRILL TOX-SPEZIAL AGLUKON   146
DRINOX   29, 503
DRINOX H-34   503
DRI-TRI   886
DROL   367
DROP LEAF   868
DRY CLEANING SAFETY SOLVENT   896
DRY ICE (SOLID)   237
DS   379
DSMA   lxii, lxv
DSP   887
DTMC   367
DUKERON   964
DUKSEN   341
DUPONT 1991   139
DUPONT HERBICIDE 976   177
DUPONT INSETICIDE 1179   639
DURAFUR BLACK R   761
DURAFUR BROWN   53
DURAFUR BROWN 2R   53
DURAFUR DEVELOPER C   243
DURAFUR DEVELOPER G   841
DURALYN   943
DURAMAX   20
DURAMITEX   602
DURAN   418
DURANOL ORANGE G   52
DURAPHOS   771
DURATOX   658
DURETTER   479
DUROFERON   479
DUROTOX   749

DURSBAN   282, lxii
DURSBAN 4   282
DURSBAN 5G   282
DURSBAN F   282
DUSICNAN BARNATY (CZECH)   132
DUS-TOP   593
DUTCH LIQUID   351
DUTCH OIL   351
DUXEN   341
DVB   419
DW3418   172
DWUBROMOETAN (POLISH)   451
DWUCHLORODWUETYLOWY ETER (POL-
   ISH)   360
DWUCHLORODWUFLUOROMETAN (POL-
   ISH)   356
2,4-DWUCHLOROFENOKSYOCTOWY KWAS
   (POLISH)   330
DWUCHLOROPROPAN (POLISH)   364
DWUETYLOWYETER (POLISH)   458
DWUMETYLOANILINA (POLISH)   388
DWUMETYLOFORMAMID (POLISH)   395
DWUMETYLOWY SIARCZAN (POLISH)   399
DWUNITRO-O-KREZOL (POLISH)   403
DYCARB   138
DYE GS   53
DYFONATE   483
DYKANOL   792
DYKOL   332
DYLOX   lxii
DYMID   lxii
DYNEX   418
DYPHONATE   483
DYRENE   lxii
DYTOL M-83   731
DYZOL   342
E-1   455
E393   902
E 605   745
E-1059   334
E 3314   503
EAK   438
EARTHCIDE   lxii
EASTERN STATES DUOCIDE   1012
EASTMAN 910   653
EASTMAN 910 MONOMER   653
EB   441
EBDC   607
EBRAGOM TB   943
EC 300   605
E.C. 1.1.3.4   495
EC8014   385
ECATOX   745
ECH   425
ECM   20
ECONOCHLOR   247
ECOPRO   917
ECOTRIN   20
EDATHAMIL   450
EDB   451, lxiii
EDB-85   451
E-D-BEE   451
EDC   351
EDETIC   450
EDETIC ACID   450
EDRISAL   758
EDTA   450
EDTA ACID   450
2EE   430
EEREX   177
EEREX GRANULAR WEED KILLER   177
EEREX WATER SOLUBLE GRANULAR
   WEED KILLER   177
EF 121   282

Bold numbers denote main chemical entries; roman numerals denote text.

EF737 172
EFFUSAN 403
EFFUSAN 3436 403
EG 452
EGITOL 509
EGM 454, 649
EGME 454, 649
EHYDE ACETIQUE (FRENCH) 2
EHYDE ACRYLIQUE (FRENCH) 21
EHYDE BUTYRIQUE (FRENCH) 196
EI-12880 385
EI 52160 917
EIDE ACETICA (ITALIAN) 2
EIDE ACRILICA (ITALIAN) 21
EIDE BUTIRRICA (ITALIAN) 196
EISENDIMETHYLDITHIOCARBAMAT (GER-
  MAN) 472
EISEN(III)-TRIS(N,N-
  DIMETHYLDITHIOCARBAMAT) (GER-
  MAN) 472
EKAGOM TB 943
EKAGOM TEDS 416
EKATIN WF & WF ULV 745
EKATOX 745
EKTAFOS 368
EKTASOLVE EB 186
EKTASOLVE EB ACETATE 187
EKTASOLVE EE 430
EKTASOLVE EE ACETATE SOLVENT 431
EL 3911 770
EL 4049 602
ELALDEHYDE 741
ELANCOLAN 976
ELAOL 350
ELAYL 448
ELDOPAQUE 530
ELDOQUIN 530
ELECTRO-CF 11 965
ELECTRO-CF 12 356
ELECTRO-CF 22 261
ELEMENTAL SELENIUM 846
11460 BROWN 1025
ELGETOL 403
ELGETOL 30 403
ELIPOL 403
ELMASIL 56
EMANAY ATOMIZED ALUMINUM POW-
  DER 39
EMANAY ZINC DUST 1028
EMANAY ZINC OXIDE 1040
EMAR 1040
EMBACETIN 247
EMBATHION 429
EMBICHIN 166
EMC 463
EMERALD GREEN 298
EMERY 43
EMETIQUE (FRENCH) 103
EMETREN 247
EMISOL 56
EMMATOS 602
EMO-NIB 703
EMOTOS EXTRA 602
EMPIRAL 758
EMPIRIN 20
EMPIRIN COMPOUND 758
EMPLETS POTASSIUM CHLORIDE 798
EMPRAZIL 758
EMPRAZIL-C 758
EMTAL 596 913
EMTS 629
EMULSAMINE BK 330
EMULSAMINE E-3 330
EMULSIPHOS 440/660 886
EN 18133 1046
ENAMEL WHITE 137
ENB 461
ENCORTON 815

ENCORTONE 815
ENDOTHALL lxii, lxv
ENDOXAN **422**
ENDOXANA 422
ENDOXANAL 422
ENDOXAN-ASTA 422
ENDOXANE 422
ENDOXAN R 422
ENDRATE 450
ENDREX 423
ENDRIN **423**, lxiii
ENDRINE (FRENCH) 423
ENDUXAN 422
ENDYDOL 20
ENDYL 993
ENFLURANE **424**
ENGLISH RED 538
ENICOL 247
ENKORTON 815
ENSEAL 798
ENS-ZEM WEEVIL BAIT 877
ENT 6 582
ENT 38 760
ENT 54 24
ENT 133 844
ENT 262 398
ENT 884 298
ENT 987 943
ENT 1,506 332
ENT 1,656 351
ENT 1,860 754
ENT 4,504 360
ENT 4705 241
ENT 7,796 146, 583
ENT 8,601 144
ENT 9,232 145
ENT 9,735 958
ENT 9,932 248
ENT 14,689 472
ENT 14,874 1045
ENT 14,875 607
ENT 15,108 745
ENT 15,152 503
ENT 15,349 451
ENT 15,406 364
ENT 15949 29
ENT 16,087 742
ENT 16,225 370
ENT 16,273 902
ENT 16,391 561
ENT 17,034 602
ENT 17,251 423
ENT 17,291 729
ENT 17,292 679
ENT 17295 334
ENT 17,957 301
ENT 18,771 918
ENT 18,870 605
ENT 19,507 342
ENT 22,374 771
ENT 22,897 408
ENT 23,437 417
ENT 23,648 367
ENT 23,708 993
ENT 23,969 234
ENT 24,042 770
ENT 24,105 429
ENT 24,482 368
ENT 24,650 385
ENT 24,833 911
ENT 24,945 470
ENT 24969 249
ENT 24,984 46
ENT 24,988 689
ENT-25294 166
ENT 25445 56
ENT 25,540 471
ENT 25,552-X 248

ENT 25,580 1046
ENT 25,602-X 307
ENT 25,671 824
ENT 25,705 773
ENT 25,726 638
ENT 25766 684
ENT 25,796 483
ENT 26,538 233
ENT-27093 28
ENT 27,129 687
ENT 27,164 235
ENT 27,165 917
ENT 27311 282
ENT 27,341 639
ENT 27572 469
ENT-27766 789
ENT 50,146 840
ENT 50,434 103
ENT 50882 515
ENTEROMYCETIN 247
ENTEROSARINE 20
ENTEX 471
ENTOMOXAN 146, 583
ENTPHOSPHOROTHIOATE 470
ENTROPHEN 20
ENTSUFON 988
ENVERT 171 330
ENVERT DT 330
ENVERT-T 912
ENVOY 172
E.O. 456
EP 30 749
EP-161E 676
EP-206 1007
E-PAM 341
EPHORRAN 416
EPICHLOROHYDRIN **425**, civ
ALPHA-EPICHLOROHYDRIN 425
EPICLEAR 154
EPICURE DDM 664
EPIHYDRIN ALCOHOL 498
EPIKURE DDM 664
EPI-REZ 508 171
EPI-REZ 510 171
EPN lxiii
EPON 828 171
EPOXIDE A 171
EPOXY cix
1,2-EPOXYAETHAN (GERMAN) 456
1,2-EPOXYBUTANE 201
1,4-EPOXYBUTANE 931
1,2-EPOXY-3-CHLOROPROPANE 425
EPOXY COMPOUNDS xciii
1,2-EPOXY-4-
  (EPOXYETHYL)CYCLOHEXANE 1007
EPOXYETHANE (FRENCH) 456
1,2-EPOXYETHANE 456
1,2-EPOXY ETHANE 456
1-EPOXYETHYL-3,4-
  EPOXYCYCLOHEXANE 1007
3-(EPOXYETHYL)-7-
  OXABICYCLO(4.1.0)HEPTANE 1007
3-(1,2-EPOXYETHYL)-7-
  OXABICYCLO(4.1.0)HEPTANE 1007
4-(EPOXYETHYL)-7-
  OXABICYCLO(4.1.0)HEPTANE 1007
4-(1,2-EPOXYETHYL)-7-
  OXABICYCLO(4.1.0)HEPTANE 1007
1,2-EPOXY-3-HYDROXY PROPANE 498
1,2-EPOXYPROPANE 832
2,3-EPOXYPROPANE 832
2,3- EPOXYPROPANOL 498
2,3-EPOXY-1-PROPANOL ACRYLATE 499
2,3-EPOXYPROPYL ACRYLATE 499
2,3-EPOXYPROPYL BUTYL ETHER 203
2,3-EPOXYPROPYL ESTER ACRYLIC
  ACID 499
2-EPOXYPROPYL ETHER 383

EPRAGEN 758
EPOXYPROPANE 832
EPSOM SALTS 601
EPTACLORO (ITALIAN) 503
1,4,5,6,7,8,8-EPTACLORO-3A,4,7,7A-
TETRAIDRO-4,7-ENDO-METANO-INDE NE
(ITALIAN) 503
EPTANI (ITALIAN) 504
EPTAN-3-ONE (ITALIAN) 442
ERADEX 282
ERBAPLAST 247
ERGOPLAST FDO 415
ERIDAN 341
ERL-27774 171
ERLA-2270 1007
ERLA-2271 1007
ERTILEN 247
ERYTHRENE 182
ESAIDRO-1,3,5-TRINITRO-1,3,5-TRIAZINA
(ITALIAN) 326
ESANI (ITALIAN) 516
ESEN 783
ESGRAM 744
ESKALITH 587
ESKASERP 840
ESKIMON 11 965
ESKIMON 12 356
ESKIMON 22 261
ESOPHOTRAST 137
ESPERAL 416
ESSENCE (FRENCH) 494
ESSENCE OF MIRBANE 708
ESSENCE OF MYRBANE 708
ESSIGESTER (GERMAN) 432
ESSIGSAEURE (GERMAN) 5
ESSIGSAEUREANHYDRID (GERMAN) 6
ESSOFUNGICIDE 406 233
ESTANE 5703 444
ESTER 25 742
ESTERCIDE T-2 AND T-245 912
ESTERON 330, 912
ESTERONE FOUR 330
ESTERON 44 WEED KILLER 330
ESTERON 76 BE 330
ESTERON 99 330
ESTERON 99 CONCENTRATE 330
ESTERON 245 912
ESTERON BRUSH KILLER 912
ESTERS xliv, xlvi
ESTONATE 332
ESTONE 330
ESTONOX 958
ETABUS 416
ETAHOLO (ITALIAN) 427
ETAIN (TETRACHLORURE D')
(FRENCH) 946
ETANOLAMINA (ITALIAN) 428
ETERE ETILICO (ITALIAN) 458
ETHANAL 2
ETHANAL OXIME 3
ETHANAMIDE 4
ETHANAMINE 437
ETHANAMINE,2-CHLORO-N-(2-
CHLOROETHYL)-N-METHYL-(9CI) 166
ETHANAMINE,N,N-DIETHYL- 970
ETHANE **426**
ETHANE, 2- BROMO-2-CHLORO-1,1,1-
THRIFLUORO- 501
ETHANE, 1-BUTOXY-2-(2-
THIOCYANATOETHOXY)- 582
ETHANECARBOXYLIC ACID 823
ETHANE, 1-CHLORO-1-DIFLUORO- 260

ETHANE,2-CHLORO-1-
(DIFLUOROMETHOXY)-1,1,2-TRIFLUORO-
424
ETHANE, CHLOROPENTAFLUORO- 269
1,2-ETHANEDIAMINE 449
1,2-ETHANEDIAMINE, N-(2-AMINOETHYL)-
376
1,2-ETHANEDIAMINE, N-(2-AMINOETHYL)-
N'-(2-AMINOETHYL) AMINOETHYL- 926
1,2-ETHANEDIAMINE, N-(2-AMINOETHYL)-
N'-(2-(2-AMINOETHYL)ETHYL)- 926
ETHANE DICHLORIDE 351
ETHANE, 1,1-DICHLORO- 358
ETHANE, 1,2-DICHLORO- 351
ETHANE, 1,2-DICHLOROTETRAFLUORO-
488
ETHANE, 1,2-DICHLORO-1,1,2,2-
TETRAFLUORO-F 114 488
ETHANE, 1,2-DIETHYL- 453
ETHANEDINITRILE 316
ETHANEDIOIC ACID 734
ETHANE DIOIC ACID 734
ETHANEDIOIC ACID DIAMMONIUM
SALT 74
1,2-ETHANEDIOL 452
1,2-
ETHANEDIYLBIS(CARBAMODITHIOA-
TO)(2-)-MANGANESE 607
1,2-
ETHANEDIYLBIS(CARBAMODITHIOA-
TO)ZINC 1045
(1,2-
ETHANEDIYLBIS(CARBAMODITHIOA-
TO))(2-)ZINC 1045
1,2-
ETHANEDIYLBIS(CARBAMODITHIOA-
TO)(2-)-S,S'-ZINC 1045
1,2-ETHANEDIYLBISCARBAMODITHIOIC
ACID, MANGANESE COMPLEX 607
1,2-ETHANEDIYLBISCARBAMODITHIOIC
ACID, MANGANESE(2) SALT(1:1) 607
1,2-ETHANEDIYLBISCARBAMODITHIOIC
ACID, ZINC COMPLEX 1045
1,2-ETHANEDIYLBISCARBAMOTHIOIC ACID,
ZINC SALT 1045
1,2-ETHANEDIYLBISMANEB, MANGANESE
(2) SALT (1:1) 607
2,2'-(1,2-ETHANEDIYLBIS(OXY)) BISE-
HANOL 971
ETHANE, HEXACHLO- 509
ETHANE HEXACHLORIDE 509
ETHANE, 1,1,1,2,2,2,-HEXACHLORO- 509
ETHANE, HEXAFLUORO- 513
ETHANE, ISOCYANATO- 462
ETHANE, METHOXY- 665
ETHANENITRILE 9
ETHANE, 1,1'-OXYBIS- 458
ETHANE, 1,1'-OXYBIS 2-CHLORO- 360
ETHANE PENTACHLORIDE 747
ETHANE, PENTACHLORO- 747
ETHANEPEROXIC ACID 753
ETHANE,1,1,2,2-TETRABROMO- 17
ETHANE,1,1,2,2-TETRACHLORO- 923
ETHANE, 1,1,1,2-TETRACHLORO-2,2-
DIFLUORO- 921
ETHANE,1,1,2,2-TETRACHLORO-1,2-
DIFLUORO- 922
ETHANETHIOAMIDE 940
ETHANE TRICHLORIDE 963
ETHANE, 1,1,1-TRICHLORO- 652
ETHANE,1,1,2-TRICHLORO- 963
ETHANE, 1,1,2-TRICHLORO-1,2,2-
TRIFLUORO-(8CI,9CI) 487
ETHANE, TRIFLUORO- 974

ETHANOIC ACID 5
ETHANOIC ANHYDRATE 6
ETHANOIC ANHYDRIDE 6
ETHANOL **427**
ETHANOL 200 PROOF 427
ETHANOLAMINE **428**
BETA-ETHANOLAMINE 428
ETHANOLAMINE SOLUTION 428
ETHANOL, 2-BUTOXY- 186
ETHANOL, 2-BUTOXY-, ACETATE 187
ETHANOL, 2-(2-BUTOXYETHOXY)-, THIOCY-
ANATE 582
ETHANOL,1,2-DIBROMO-2,2-DICHLORO-, DI-
METHYL PHOSPHATE 689
ETHANOL, 2-(DIBUTYLAMINO)- 348
ETHANOL,2-(DIETHYLAMINO)- 374
ETHANOL,2-ETHOXY- 430
ETHANOL, 2-ETHOXY-, ACETATE 431
ETHANOL,2,2'-(ETHYLENEDIOXY)DI- 971
ETHANOL, 2,2'-IMINOBIS- 371
ETHANOL, 2,2'-IMINOBIS- 552
ETHANOL, ISOPROPOXY- 552
ETHANOYL BROMIDE 12
ETHANOYL CHLORIDE 13
ETHENE 448
ETHENE, BROMO- 1005
1,2-ETHENEDICARBOXYLIC ACID, TRANS-
489
ETHENE, 1,2-DICHLORO- 359
ETHENE,1,1-DICHLORO- 1010
ETHENE, TRANS- 1,2-DICHLORO- 16
ETHENE, DICHLORODIFLUORO- 355
ETHENE, TETRACHLORO- 754
ETHENOID 563
ETHENONE 563
ETHENYL ACETATE 1004
ETHENYLETHANOATE 1004
ETHER 458
ETHER, ALLYL 2,3-EPOXYPROPYL 34
ETHER, BIS(2,3-EPOXYPROPYL) 383
ETHER, BIS(2,3-EPOXYPROPYL)- 383
ETHER, BUTYL 2,3-EPOXYPROPYL 203
ETHER, BUTYL GLYCIDYL 203
ETHER BUTYLIQUE (FRENCH) 202
ETHER CHLORATUS 445
ETHER,2-CHLORO-1,1,2-TRIFLUOROETHYL
DIFLUOROMETHYL 424
ETHER DICHLORE (FRENCH) 360
ETHER, 2,4-DICHLOROPHENYL-P-
NITROPHENYL 711
ETHER, DIGLYCIDYL 383
ETHER, DIMETHYL CHLOR O 267
ETHER, ETHYL 458
ETHER ETHYLIQUE (FRENCH) 458
ETHER, ETHYL METHYL 665
ETHER, HEXACHLOROPHENYL 250
ETHER HYDROCHLORIC 445
ETHER, ISOPROPYL 557
ETHER ISOPROPYLIQUE (FRENCH) 557
ETHER METHYLIQUE MONOCHLORE
(FRENCH) 267
ETHER MONOETHYLIQUE DE L'ETHYLENE-
GLYCOL (FRENCH) 430
ETHER MONOMETHYLIQUE DE
L'ETHYLENE-GLYCIL (FRENCH) 454
ETHER MONOMETHYLIQUE DE
L'ETHYLENE-GLYCOL (FRENCH) 649
ETHER MURIATIC 445
ETHINE 15
ETHINYL TRICHLORIDE 964
ETHIOL 429
ETHIOLACAR 602
ETHION **429**
ETHLON 745
ETHODAN 429

Bold numbers denote main chemical entries; roman numerals denote text.

ETHYLMERCURY P-
TOLUENESULFANILIDE 629
ETHYLMERCURY-P-TOLUENE SULFON-
AMIDE 629
ETHYLMERCURY-P-
TOLUENESULFONANILIDE 629
N-ETHYLMERKURI-P-TOLUENSULFOANILID
(CZECH) 629
ETHYL METHACRYLATE **464**
ETHYL 1-2-METHACRYLATE 464
ETHYLMETHANOL 820
ETHYL-ALPHA-METHYLACRYLATE 464
ETHYLMETHYL CARBINOL 194
ETHYL METHYL CARBINOL 194
ETHYL METHYL CETONE (FRENCH) 666
ETHYL METHYLENE PHOSPHORODITHIO-
ATE 429
ETHYL METHYL ETHER 665
ETHYLMETHYLKETON (DUTCH) 666
ETHYL METHYL KETONE 666
ETHYL 3-METHYL-4-
(METHYLTHIO)PHENYL(1-
METHYLETH-
YL)PHOSPHORAMIDATE 469
ETHYL 4-(METHYLTHIO)-M-
TOLYLISOPROPYLPHOSPHORAMI-
DATE 469
ETHYL NITRILE 9
ETHYLOLAMINE 428
O-ETHYLO-(4-(METHYLTHIO)PHENYL) S-
PROPYLPHOSPHORODITHIOATE 910
N-ETHYLOMINOBENZENE 439
ETHYL ORTHOSILICATE 467
2-ETHYLOXIRANE 201
ETHYL PARATHION 745, lxiii
ETHYLPHENYLAMINE 439
ETHYLPHENYLDICHLOROSILANE 465
O-ETHYLS-PHENYL ETH-
YLDITHIOPHOSPHONATE 483
O-ETHYL-S-PHENYL ETHYLPHOSPHO-
NODITHIOATE 483
ETHYL PHENYL DICHLOROSILANE **465**
1-ETHYLPIPERIDINE 466
1-ETHYL PIPERIDINE **466**
ETHYL PROPENOATE 434
ETHYL 2-PROPENOATE 434
ETHYL PYROPHOSPHATE, TETRA- 918
ETHYL SILICATE **467**
ETHYL SILICON TRICHLORIDE 468
ETHYL SULFATE 379
S-2-(ETHYLTHIO)ETHYL O,O-
DIETHYLESTER OF PHOSPHO-
RODITHIOIC ACID 417
O-2-(ETHYLTHIO)ETHYL O,O-DIMETHYL
PHOSPHOROTHIOATE 658
S-2-(ETHYLTHIO)ETHYL O,O-DIMETHYL
PHOSPHOROTHIOATE 658
ETHYL THIOPYROPHOSPHATE 902
ETHYL TRICHLOROETHYLSILANE **468**
ETHYLTRICHLOROSILANE 468
ETHYL TRICHLOROSILANE 468
ETHYLTUADS 416
ETHYL TUEX 416
ETHYLURETHAN 444
ETHYL URETHANE 444
O-ETHYLURETHANE 444
ETHYL ZIMATE 1045
ETHYL ZINC 380
ETHYNE 15
ETHYNE, DICHLORO- (9CI) 352
ETHYNYLCARBINOL 820
ETICOL 742
ETIL ACRILATO (ITALIAN) 434
ETILACRILATULUI (RUMANIAN) 434

ETILAMINA (ITALIAN) 437
ETILBENZENE (ITALIAN) 441
ETILBUTILCHETONE (ITALIAN) 442
ETILE (ACETATO DI) (ITALIAN) 432
N,N'-ETILEN-BIS(DITIOCARBAMMATO) DI
MANGANESE (ITALIAN) 607
ETILENE (OSSIDO DI) (ITALIAN) 456
ETILENIMINA (ITALIAN) 455
ETIOL 602
ETO 456
ETOKSYETYLOWY ALKOHOL (POL-
ISH) 430
2-ETOSSIETIL-ACETATO (ITALIAN) 431
ETU 457
ETYLENU TLENEK (POLISH) 456
ETYLOAMINA (POLISH) 437
ETYLOBENZEN (POLISH) 441
ETYLOWY ALKOHOL (POLISH) 427
ETYLU CHLOREK (POLISH) 445
ETYLU KRZEMIAN (POLISH) 467
EULAVA SM 600
EVAU-SUPERFALL 868
EVIPLAST 80 415
EVIPLAST 81 415
EVOLA 354
EWEISS 137
EXAGAMA 146, 583
EXHAUST GAS 239
EXHORAN 416
EXHORRAN 416
EXITELITE 107
EXOFENE 511
EXOLON XW 60 43
EXOTHERM 278
EXPERIMENTAL INSECTICIDE 3911 770
EXPERIMENTAL INSECTICIDE 7744 234
EXPERIMENTAL INSECTICIDE 12,880 385
EXPERIMENTAL INSECTICIDE 52160 917
EXPERIMENTAL NEMATOCIDE
18,133 1046
EXPLOSIVE D 80
EXSICCATED FERROUS SULFATE 479
EXSICCATED FERROUS SULPHATE 479
EXSICCATED SODIUM PHOSPHATE 887
EXTERMATHION 602
EXTRAR 403
EXTRAX 844
EXTREMA 467
E-Z-OFF 592
E-Z-PAQUE 137
F1-TABS 876
F 10 607
F-11 965
F 12 356
F-13B1 973
F 14 930
F 22 261
F-112 922
F 113 487
F-115 269
F116 513
F 360 (ALUMINA) 43
F 1991 139
FA 484
FAA 10
2-FAA 10
FACTITIOUS AIR 724
FALITIRAM 943
FALKITOL 509
FANNOFORM 484
FARMCOCCID 294
FARMICETINA 247
FARMON PDQ 743
FASCIOLIN 241, 509

FASCO-TERPENE 958
FASERTON 43
FASERTONERDE 43
FAST BLUE B BASE 338
FAST BLUE BN SALT 338
FAST BLUE DS SALT 338
FAST BLUE DSC BASE 338
FAST BLUE SALT B 338
FAST BLUE SALT BN 338
FAST CORINTH BASE B 149
FAST DARK BLUE BASE R 950
FAST GARNET B BASE 692
FAST GARNET BASE B 692
FAST OIL YELLOW 48
FAST RED BASE 707
FAST RED 2G BASE 707
FAST RED 5CT SALT 281
FAST RED SALT TR 281
FAST RED SALT TRA 281
FAST RED SALT TRN 281
FAST RED TR SALT 281
FAST SCARLET BASE B 693
FAST SCARLET R 717
FAST SCARLET TR BASE 264
FAST WHITE 579
FAST YELLOW AT 48
FATSCO ANT POISON 862
FAUSTAN 341
FB/2 414
FBC PIRIMICARB 50 789
FC-11 965
FC 12 356
FC 14 930
FC 113 487
FC 114 488
FC-C 318 728
FDA 0101 876
FD&C BLUE NO. 1 lxxvii
FD&C NO. 17 lxxvii
FD&C RED NO. 4 lxxvi, lxxvii
FD&C YELLOW NO. 5 lxxvii
FD&C YELLOW NO. 10 lxxvii
FEDAL-UN 754
FEENO 760
FEGLOX 414
FEKABIT 797
FELLING ZINC OXIDE 1040
FEMA NO. 2003 2
FEMA NO. 2006 5
FEMA NO. 2011 26
FEMA NO. 2034 36
3 FEMA NO. 2055 540
FEMA NO. 2174 189
FEMA NO. 2175 543
FEMA NO. 2178 193
FEMA NO. 2179 544
FEMA NO. 2219 196, 212
FEMA NO. 2220 545
FEMA NO. 2221 213
FEMA NO. 2229 230
FEMA NO. 2370 184
FEMA NO. 2414 432
FEMA NO. 2418 434
FEMA NO. 2427 443
FEMA NO. 2489 491
FEMA NO. 2545 442
FEMA NO. 2731 673
FEMA NO. 2923 822
FEMA NO. 2926 553
FEMMA 765
FENACETINA 758
FEN-ALL 914
FENAMINE 56
FENAMIPHOS **469**

Bold numbers denote main chemical entries; roman numerals denote text.

Bold numbers denote main chemical entries; roman numerals denote text.

GENOXAL　422
GENUINE RARIS GREEN　298
GEOMET　770
GERMAIN'S　234
GERMALGENE　964
GERMA-MEDICA　511
GERTLEY BORATE　866
GESAFID　332
GESAPON　332
GESAREX　332
GESAROL　332
GESFID　771
GESTID　771
GEWODIN　758
GEXANE　146, 583
GHA 331　41
GHA 332　41
GIBBSITE　43
GIHITAN　341
GK (OXIDE)　43
GLACIAL ACETIC ACID　5
GLACIAL ACRYLIC ACID　23
GLAZD PENTA　749
GLOBAL CRAWLING INSECT BAIT　282
GLOBENICOL　247
GLONOIN　715
GLORE PHOS 36　687
GLOROUS　247
GLOVER　564
GLUCID　845
GLUCINIUM　158
GLUCINUM　158
GLUCOSE AERODEHYDROGENASE　495
GLUCOSE OXIDASE　**495**
BETA-D-GLUCOSE OXIDASE　495
GLUSIDE　845
GLUTARAL　496
GLUTARALDEHYD (CZECH)　496
GLUTARALDEHYDE　**496**
GLUTARD DIALDEHYDE　496
GLUTARIC ACID DIALDEHYSE　496
GLUTARIC DIALDEHYDE　496
GLUTONIC ACID, 3-HYDROXY-,DIMETHYL
　ESTER, DIMETHYL PHOSPHATE　911
GLYCERIN　**497**
GLYCERIN, ANHYDROUS　497
GLYCERIN, SYNTHETIC　497
GLYCERINE　497
GLYCERINTRINITRATE (CZECH)　715
GLYCERITOL　497
GLYCEROL　497
GLYCEROL EPICHLOROHYDRIN　425
GLYCEROL, NITRIC ACID TRIESTER　715
GLYCEROLTRINITRAAT (DUTCH)　715
GLYCEROL TRINITRATE　715
GLYCEROL(TRINITRATE DE)
　(FRENCH)　715
GLYCERYL NITRATE　715
GLYCERYL TRINITRATE　715
GLYCIDE　498
GLYCIDOL　**498**
GLYCIDYLACRYLATE　499
GLYCIDYL ACRYLATE　**499**
GLYCIDYL ALCOHOL　498
GLYCIDYLBUTYLETHER　203
GLYCIDYL BUTYL ETHER　203
GLYCIDYLESTER KYSELINYAKRYLOVE
　(CZECH)　499
GLYCIDYL PROPENATE　499
GLYCINE, N,N'-1,2-ETHANEDIYLBIS(N-
　(CARBOXYMETHYL)-9CI)　450
GLYCINOL　428
GLYCOL　452
GLYCOL ALCOHOL　452
GLYCOL BIS(HYDROXYETHYL)
　ETHER　971
GLYCOL BROMIDE　451
GLYCOL BUTYL ETHER　186

GLYCOL DIBROMIDE　451
GLYCOL DICHLORIDE　351
GLYCOL ETHER　lvi
GLYCOL ETHER EB　186
GLYCOL ETHER EB ACETATE　186
GLYCOL ETHER EM　454, 649
GLYCOL ETHER EM ACETATE　650
GLYCOL ETHERS　cx
GLYCOL ETHYLENE ETHER　407
GLYCOL ETHYL ETHER　430
GLYCOLIC ACID, THIO-　942
GLYCOLIC ACID, 2-THIO-　942
GLYCOL METHYL ETHER　454
GLYCOL MONOBUTYL ETHER　186
GLYCOL MONOBUTYL ETHER ACE-
　TATE　187
GLYCOL MONOETHYL ETHER　430
GLYCOL MONOETHYL ETHER ACE-
　TATE　431
GLYCOL MONOMETHYL ETHER　454, 649
GLYCOL MONOMETHYL ETHER ACE-
　TATE　650
GLYCOLS　xxvii
GLYCYL ALCOHOL　497
GLYODEX 3722　233
GLYPHOSATE　lxii, lxv
GOLD BRONZE　297
GOLD SATINOBRE　575
GOOD-RITE GP 264　415
GP-40-66:120　507
GPKH　503
GRAIN ALCOHOL　427
GRAIN DUST　xxxvi
GRAIN SORGHUM HARVEST AID　868
GRAMAZINE　743
GRAMEVIN　366
GRAMONOL　743
GRAMONOL 5　743
GRAMOXONE 100 S　743
GRAMOXONE S　743
GRAMOZONE S　744
GRAMTOX　770
GRANEX OK　868
GRANOSAN　463
GRANOSAN M　629
GRANOSAN MDB　629
GRANOX MN　506
GRANULAR ZINC　1028
GRANUTOX　770
GRAPHLOX　508
GRAZON　785
GREEN CHROMIC OXIDE　288
GREEN CINNABAR　288
GREEN CROSS WARBLE POWDER　844
GREEN GA　288
GREEN OIL　98
GREEN ROUGE　288
GREEN SEAL-8　1040
GREEN VITRIOL IRON MONOSUL-
　FATE　479
GREY ARSENIC　109
GRIFFIN MANEX　607
GRISOL　918
GROCOLENE　497
GROUNDHOG　743
GROUND VOCLE SULPHUR　903
GRUNDIER ARBEZOL　749
GTN　715
P-GUAICOL　640
GUESAROL　332
GUM ACACIA　500
GUM ARABIC　**500**
GUM CAMPHOR　231
GUM OVALINE　500
GUM SENEGAL　500
GUM SPIRITS　996
GUM TURPENTINE　996

GY-PHENE　958
GYPSINE　565
GYRON　332
H-34　503
H 46　41
H 321　638
HAIARI　844
HAIMASED　890
HALANE　357
HALIZAN　lxii
HALOCARBON 11　965
HALOCARBON 13B1　973
HALOCARBON 14　930
HALOCARBON 23　975
HALOCARBON 112　922
HALOCARBON 112A　921
HALOCARBON 113　487
HALOCARBON 114　488
HALOCARBON 1132A　1009
HALOCARBON C-138　728
HALOGENATED HYDROCARBONS　cx
HALOMYCETIN　247
HALON　356
HALON 14　930
HALON 104　241
HALON 122　356
HALON 1011　259
HALON 1202　381
HALON 1301　973
HALON 10001　670
HALOTAN　501
HALOTHANE　**501**
HALOWAX　924
HALOWAX 1013　748
HALOWAX 1014　510
HALOWAX 1051　727
HALSAN　501
HARTSHORN　62
HARVEST-AID　868
HATCOL DOP　415
HAVERO-EXTRA　332
HAVIDOTE　450
HC 2072　742
HC 12811　914
HCB　506
HCBD　507
HCCH　144, 146, 583
HCCPD　508
HCDB　943
HCDB-1018　338
HCH　146, 583
ALPHA-HCH　145
GAMMA-HCH　146, 583
HCN　524
HCP　511
HCS 3260　248
HEAVY OIL　302
HECLOTOX　146, 583
HEDOLIT　403
HEDOLITE　403
HEDONAL　330
HEDONAL (THE HERBICIDE)　330
HEKSAN (POLISH)　318, 516
HEKSOGEN (POLISH)　326
HELICON　20
HELIUM　**502**
HELIUM, ELEMENTAL　502
HELMETINA　760
HELOTHION　910
HELVAGIT　758
HEMIMELLITENE　982
HEMPA　515
HEOD　370
HEPAR SULFUROUS　812
HEPATIC GAS　529
HEPT　918
HEPTACHLOOR (DUTCH)　503

Bold numbers denote main chemical entries; roman numerals denote text.

M-HYDROXYPHENOL 841
O-HYDROXYPHENOL 243
HYDROXYPHENYLMERCURY 766
1-HYDROXYPROPANE 819
3-HYDROXY-1-PROPANESULPHONIC ACID
SULFONE 818
3-HYDROXY-1-PROPANESULPHONIC ACID
SULTONE 818
3-HYDROXYPROPENE 31
3-HYDROXYPROPYLENE OXIDE 498
6-HYDROXY- 3(2H)-PYRIDAZINONE 605
HYDROXYTOLUENE 163
HYDROXYTOLUOLE (GERMAN) 303
2-HYDROXY-,TRIANHYDRIDE WITH ANTI-
MONIC ACID 100
2-HYDROXYTRIETHYLAMINE 374
2-HYDROXY-1,3,5-TRINITROBENZENE 786
HYDROZINE SULFATE 521
HYDRURE DE LITHIUM (FRENCH) 588
HYLEMOX 429
HYLENE T 952
HYLENE TCPA 952
HYLENE TLC 952
HYLENE TM 952, 953
HYLENE TM-65 952
HYLENE TRF 952
HYOXYL 528
HYPALOX II 43
HYPERIZ 309
HYPERTENAIN 613
HYPNORÈX 587
HYPO 891
HYPOCHLORIC ACID cix
HYPOCHLORIC ACID, BARIUM SALT 131
HYPOCHLOROUS ACID, CALCIUM 225
HYPOCHLOROUS ACID, SODIUM
SALT 880
HYPONITROUS ACID ANHYDRIDE 724
HYVAR 177
HYVAREX 177
HYVAR X 177
HYVAR X BROMACIL 177
HYVAR X WEED KILLER 177
I 337A 247
IBA 544
IDRAZINA SOLFATO (ITALIAN) 521
IDROCHINONE (ITALIAN) 530
IDROGENOSOLFORATO (ITALIAN) 529
IDROPEROSSIDO DI CUMENE (ITAL-
IAN) 309
IDROPEROSSIDO DI CUMOLO (ITAL-
IAN) 309
4-IDROSSI-4-METIL-PENTAN-2-ONE (ITAL-
IAN) 335
IKURIN 81
ILLOXOL 370
ILLUMINATING OIL 562
IMI 115 947
IMIDAN 773
2,4-IMIDAZOLIDINEDIONE 357
2-IMIDAZOLIDINETHIONE 457
2,2'IMINOBISETHANOL 371
2,2'-IMINODIETHANOL 371
IMPERIAL GREEN 298
IMPRUVOL 163
IMVITE I.G.B.A. 140
INACTIVE LIMONENE 409
INCIDOL 154
INCRACIDE E 51 313
INDENE **531**
INDIAN GUM 500
INDIAN RED 538
INDIUM **532**
INDONAPHTHENE 531

INERTEEN 792
INEXIT 146, 583
INFUSORIAL EARTH 340
INHIBINE 528
INHIBISOL 652
INITIATING EXPLOSIVE NITRO MAN-
NITE 613
INSECTICIDE 1,179 639
INSECT POWDER 836
INSOLUBLE SACCHARIN 845
IN-SONE 815
INTRAMYCETIN 247
INVERTON 245 912
INVISI-GARD 824
IODE (FRENCH) 533
IODINE **533**, xxiii
IODINE CHLORIDE 534
IODINE CRYSTALS 533
IODINE FLUORIDE 535
IODINE MONOCHLORIDE **534**
IODINE PENTAFLUORIDE **535**
IODINE SUBLIMED 533
IODIO (ITALIAN) 533
IODOFORM **536**
IODOMETANO (ITALIAN) 670
IODOMETHANE 670
3-IODOPROPENE 35
3-IODO-1-PROPENE 35
1-IODO-2-PROPENE 35
3-IODOPROPYLENE 35
IODURE DE MERCURE (FRENCH) 631
IODURE DE METHYLE (FRENCH) 670
IONOL 163
IONOL CP 163
IOPEZITE 802
IPANER 330
IPDI 550
IPE 552
IPERSAN 976
IPN 784
I-PROPANOL (GERMAN) 554
IRADICAV 876
IRENAL 809
IRENAT 809, 884
IRON AMMONIUM SULFATE 477
IRON BIS (CYCLOPENTADIENE) 476
IRON CARBONYL 539
IRON CHLORIDE **537**
IRON(III) CHLORIDE 537
IRON(II) CHLORIDE (1:2) 478
IRON DICHLORIDE 478
IRON DICYCLOPENTADIENYL 476
IRON DIMETHYLDITHIOCARBAMATE 472
IRON NITRATE 474
IRON (III) NITRATE, ANHYDROUS 474
IRON OXIDE **538**, xxxv, xxxix
IRON OXIDE RED 538
IRON(III) OXIDE 538
IRON PENTACARBONYL **539**
IRON, PENTACARBONYL- 539
IRON PERSULFATE 475
IRON PROTOCHLORIDE 478
IRON PROTOSULFATE 479
IRON SESQUICHLORIDE 537
IRON SESQUIOXIDE 538
IRON SESQUISULFATE 475
IRON SULFATE (1:1) 479
IRON SULFATE (2:3) 475
IRON(II) SULFATE 479
IRON(2) SULFATE 479
IRON(2) SULFATE (1:1) 479
IRON (III) SULFATE 475
IRON (3) SULFATE 475
IRON TERSULFATE 475

IRON TRICHLORIDE 537
IRON TRINITRATE 474
IRON TRISULFATE 475
IRON VITRIOL 479
IROSPAN 479
IRTRAN 2 1044
ISAZOPHOS lxiv
ISCEON 22 261
ISCEON 113 487
ISCEON 122 356
ISCEON 131 965
ISCOBROME D 451
ISICETIN 247
ISMICETINA 247
ISOACETOPHORONE 549
ISOAMYL ACETATE **540**
ISOAMYL ALCOHOL **541**
ISOAMYL ALCOHOL, SECONDARY 751
ISOAMYLALKOHOL (GERMAN) 541
ISOAMYL ALKOHOL (CZECH) 541
ISOAMYL ETHANOATE 540
ISOAMYL METHYL KETONE 671
ISOAMYLOL 541
ISOBAC 20 511
ISOBENZOFURAN,1,3-DIHYDRO-1,3-DIOXO-
783
1,3-ISOBENZOFURANDIONE 783
1,3-ISOBENZOFURANDIONE,3A,4,7,7A-
TETRAHYDRO- 932
ISOBUTANAL 545
ISOBUTANE **542**, xxxii
ISOBUTANOL 544
ISOBUTYALDEHYD (CZECH) 545
ISOBUTYL ACETATE **543**
ISOBUTYL ALCOHOL **544**
ISOBUTYL ALDEHYDE 545
ISOBUTYLALKOHOL (CZECH) 544
ISOBUTYLCARBINOL 541
ISOBUTYL KETONE 396
ISOBUTYLMETHYLCARBINOL 672
ISOBUTYL METHYL KETONE 673
ISOBUTYLMETHYLMETHANOL 672
ISOBUTYLTRIMETHYLMETHANE 547
ISOBUTYRALDEHYDE **545**
ISOBUTYRIC ALDEHYDE 545
ISOBUTYRONITRILE **546**
ISOCARB 824
ISOCYANATE DE METHYLE (FRENCH) 674
ISOCYANATE METHANE 674
ISOCYANATES cviii
ISOCYANATO BENZENE 764
ISOCYANATOETHANE 462
ISO-CYANATOMETHANE 674
3-ISOCYANATOMETHYL-3,5,5-
TRIMETHYLCYCLOHEXYLISOCYA-
NATE 550
1-ISOCYANATOPROPANE 833
1-ISOCYANATO PROPANE 833
ISOCYANIC ACID 324
ISOCYANIC ACID, CYCLOHEXYL
ESTER 324
ISOCYANIC ACID, ETHYL ESTER 462
ISOCYANIC ACID, METHYL ESTER 674
ISOCYANIC ACID, METHYLPHENYLENE
ESTER 952
ISOCYANIC ACID, 4-METHYL-M-
PHENYLENE ESTER 952
ISOCYANIC ACID, PHENYL ESTER 764
ISOCYANIC ACID, PROPYL ESTER 833
ISOCYANURIC ACID,DICHLORO-, POTASSI-
UM SALT 801
ISOCYANURIC ACID, DICHLORO-, SODIUM
SALT 873
ISOCYANURIC CHLORIDE 966

Bold numbers denote main chemical entries; roman numerals denote text.

Chemical Name Index

Bold numbers denote main chemical entries; roman numerals denote text.

MANICOLE 613
MANITE 613
MANNEX 613
MANNITOL HEXANITRATE **613**
MANNITOL HEXANITRATE, CONTAINING,
  BY WEIGHT, AT LEAST 40% WATER 613
MANNITRIN 613
MANOC 607
MANUFACTURED IRON OXIDES 538
MANZATE 607
MANZATE D 607
MANZATE MANEB FUNGICIDE 607
MANZEB 607
MANZIN 607
MAOH 672
MARBLE 219
MARKSMAN 2, TRIGARD 976
MARLATE lxii
MARMER 418
MARS BROWN 538
MARSH GAS 636
MARS RED 538
MARTINAL 41
MARTOXIN 43
MASTIPHEN 247
MATRICARIA CAMPHOR 231
MATRIKERB 835
MAXATASE 120
MAXITATE 613
MAZIDE 605
MBA 166
MBC 139
MBCP 581
MBK 648
MBOCA 660
MC 616
MC 6897 138
MCA 255
MCB 258
MC DEFOLIANT 592
M2 COPPER 297
MCPA lxii, lxv
MCPP lxv
MCPP/MECOPROP lxii
MDA 664
MEA 428
MEASURIN 20
MEB 607
MECHLORETHAMINE 166
MECOPEX lxii
MECRILAT 653
MECRYLATE 653
MECS 454, 649
MEDEMANOL 613
MEDIAMYCETINE 247
MEETCO 666
MEISUI TERYL DIAZO BLUE HR 338
MEK 666
MEKP 667
MEK PEROXIDE 667
MELABON 758
MELAFORTE 758
MELDANE 301
MELDONE 301
MELINITE 786
MELIPAX 958
MEMPA 515
MEMTETRAHYDRO PHTALIC ANHY-
  DRIDE 932
MEMTETRAHYDROPHTHALIC ANHY-
  DRIDE 932
MENDRIN 423
MENEB lxii, lxvi
MENIPHOS 771

MENITE 771
1,8(9)-P-MENTHADIENE 409
P-MENTHA-1,8-DIENE,DL- 409
ME-PARATHION 679
MEPATON 679
MEPOX 679
MEQUINOL 640
MER 678
MERCAPTAN METHYLIQUE (FRENCH) 677
MERCAPTAN METHYLIQUE PERCHLORE
  (FRENCH) 756
MERCAPTOACETIC ACID 942
2-MERCAPTOACETIC ACID 942
ALPHA-MERCAPTOACETIC ACID 942
1-MERCAPTOBUTANE 205
MERCAPTODIMETHUR 638
MERCAPTOMETHANE 677
N-(MERCAPTOMETHYL)PHTHALIMIDE S-
  (O,O-DIMETHYL PHOSPHORODITHIO-
  ATE) 773
MERCAPTOPHOS 334, 471
MERCAPTOTHION 602
MERCOL 25 875
MERCOL 30 875
MERCOZEN 457
MERCURAM 943
MERCURE (FRENCH) 628
MERCURIACETATE 614
MERCURIALIN 645
MERCURIC ACETATE **614**
MERCURIC BASIC SULFATE 632
MERCURIC BICHLORIDE 616
MERCURIC BROMIDE **615**
MERCURIC BROMIDE. SOLID 615
MERCURIC CHLORIDE **616**
MERCURIC CYANIDE **617**
MERCURIC DIACETATE 614
MERCURIC IODIDE **618**
MERCURIC IODIDE, RED 618
MERCURIC NITRATE **619**
MERCURIC OXIDE **620**
MERCURIC OXIDE, RED 620
MERCURIC OXIDE, YELLOW 620
MERCURIC OXYCYANIDE **621**
MERCURIC SUBSULFATE 632
MERCURIC SULFATE **622**
MERCURIC SULFOCYANATE 623
MERCURIC SULFO CYANATE, SOLID 623
MERCURIC SULFOCYANIDE 623
MERCURIC THIOCYANATE **623**
MERCURIO (ITALIAN) 628
MERCUROCHLORIDE 624
MERCUROUS CHLORIDE **624**
MERCUROUS IODIDE 631
MERCUROUS IODIDE, SOLID 631
MERCUROUS NITRATE **625**
MERCUROUS OXIDE **626**
MERCUROUS OXIDE, BLACK 626
MERCUROUS SULFATE **627**
MERCURY **628**, xxvi, xl, lxiii, cv, cviii, cx
MERCURY ACETATE 614
MERCURY(II) ACETATE 614
MERCURY(2) ACETATE 614
MERCURY, (ACETO-O)PHENYL- 765
MERCURY BICHLORIDE 616
MERCURY BINIODIDE 618
MERCURY BISULFATE 622
MERCURY BROMIDE 615
MERCURY(II) BROMIDE (1:2) 615
MERCURY(I) CHLORIDE 624
MERCURY(II) CHLORIDE 616
MERCURY, CHLOROETHYL- 463
MERCURY(II) CYANIDE 617
MERCURY CYANIDE OXIDE 621

MERCURY DIACETATE 614
MERCURY, DIMETHYL 397
MERCURY DITHIOCYANATE 623
MERCURY,ETHYL(4-METHYL-N-
  PHENYLBENZENESULFONAMIDATO-N)-
  (9CI) 629
MERCURY,ETHYL(N-PHENYL-P-
  TOLUENESULFONAMIDATO)- 629
MERCURY, ETHYL(N-PHENYL-P-
  TOLUENESULFONAMIDO)- 629
MERCURY, ETHYL(P-
  TOLUENESULFONANILIDATO)- **629**
MERCURY FULMINATE **630**
MERCURY, HYDROXYPHENYL- 766
MERCURY(I) IODIDE **631**
MERCURY(II) IODIDE 618
MERCURYL ACETATE 614
MERCURY, METALLIC 628
MERCURY MONOCHLORIDE 624
MERCURY NITRATE 619
MERCURY(I) NITRATE (1:1) 625
MERCURY(II) NITRATE (1:2) 619
MERCURY OXIDE 620
MERCURY(I) OXIDE QUECKSILBEROXID
  (GERMAN) 626
MERCURY OXIDE SULFATE **632**
MERCURY OXYCYANIDE 621
MERCURY PERCHLORIDE 616
MERCURY PERNITRATE 619
MERCURY PERSULFATE 622
MERCURY PROTOCHLORIDE 624
MERCURY PROTOIODIDE 631
MERCURY(I) SULFATE 627
MERCURY(II) SULFATE (1:1) 622
MERCURY THIOCYANATE 623
MEREX 561
MERGON 629
MERGON D 629
MERPAN 233
MERRILLITE 1028
MERTESTATE 919
MESITLENE 984
MESOMILE 639
MESUROL 638
METAARSENIC ACID 110
METACETONE 378
METACETONIC ACID 823
METACIDE 679
METACORTANDRACIN 815
METAFOS 679
METAKRYLAN METYLU (POLISH) 678
METALDEHYDE lxii, lxiii
METAL DUST liii
METALLIC ARSENIC 109
METALLIC MERCURY 628
METAL PIGMENTS cviii
METANA 39
METANA ALUMINUM PASTE 39
METANE, BROMOCHLORO- 259
METANOLO (ITALIAN) 637
METAPHOR 679
METAPHOS 679
METAPHOSPHORIC ACID 775
METAQUEST A 450
METASON lxii
METASYSTOX 658
META-XYLENE 1014
METHAANTHIOL (DUTCH) 677
METHACETONE 378
METHACIDE 951
METHACRYLATE DE BUTYLE
  (FRENCH) 206
METHACRYLATE DE METHYLE
  (FRENCH) 678

Bold numbers denote main chemical entries; roman numerals denote text.

Bold numbers denote main chemical entries; roman numerals denote text.

MOLLANO 415
MOLYBDATE 685
MOLYBDENUM **685**
MOLYBDENUM(VI) OXIDE 686
MOLYBDENUM TRIOXIDE **686**
MOLYBDIC ACID, DIAMMONIUM SALT 72
MOLYBDIC ANHYDRIDE 686
MOLYBDIC TRIOXIDE 686
MONDUR P 764
MONDUR TDS 952
MONOAETHANOLAMIN (GERMAN) 428
MONOAMMONIUM CARBONATE 59
MONOAMMONIUM PHOSPHATE 78
MONOAMMONIUM SALT OF SULFAMIC
　ACID 81
MONOAMMONIUM SULFAMATE 81
MONOBASIC CHROMIUM SULFATE 291
MONOBASIC CHROMIUM SULPHATE 291
MONOBASIC LEAD ACETATE 578
MONOBROMOETHYLENE 1005
MONOBROMOTRIFLUORMETHANE 973
MONOBUTYLAMINE 197
MONO-N-BUTYLAMINE 197
MONOBUTYL GLYCOL ETHER 186
MONOCHLOORAZIJNZUUR (DUTCH) 255
MONOCHLOORBENZEEN (DUTCH) 258
MONOCHLORBENZENE 258
MONOCHLORBENZOL (GERMAN) 258
MONOCHLORESSIGSAEURE (GER-
　MAN) 255
MONOCHLORETHANE 445
MONOCHLOROACETALDEHYDE 254
MONOCHLOROACETIC ACID 255
MONOCHLOROACETYL CHLORIDE 256
MONOCHLOROBENZENE 258
MONOCHLORODIFLUOROMETHANE 261
MONOCHLOROETHENE 1006
MONOCHLOROETHYLENE 1006
MONOCHLOROMETHANE 651
MONO-CHLORO-MONO-BROMO-
　METHANE 259
MONOCHLOROPENTAFLUOROE-
　THANE 269
MONOCHLOROSULFURIC ACID 277
MONOCHROMIUM OXIDE 290
MONOCHROMIUM TRIOXIDE 290
"MONOCITE" METHACRYLATE MONO-
　MER 678
MONOCRON 687
MONOCROTOPHOS **687**
MONOCYANOGEN 316
MONOETHANOLAMINE 428
MONOETHYLAMINE 437
MONOFLUOROETHYLENE 1008
MONOFLUROTRICHLOROMETHANE 965
MONOHYDROXYBENZENE 759
MONOHYDROXY BENZENE 759
MONOHYDROXYMETHANE 637
MONOIODOMETHANE 670
MONOISOPROPYLAMINE 555
MONOISOPROPYL ETHER OF ETHYLENE
　GLYCOL 552
MONOMETHYLAMINE 645
MONOMETHYL ANILINE 647
N-MONOMETHYLANILINE 647
MONO METHYL ETHER HYDROQUI-
　NONE 640
MONOMETHYL HYDRAZINE 669
MONOPOTASSIUM ARSENATE 794
MONOPOTASSIUM DIHYDROGEN ARSE-
　NATE 794
MONO-N-PROPYLAMINE 826
MONOPROPYLENE GLYCOL 829
MONOPROPYLMINE 826

MONOSAN 330
MONOSILANE 849
MONTAR 792
MONTMORILLONITE 140
MONTREL 307
MOON 497
MORBICID 484
MOROSAN 341
MORPHOLINE **688**
MORTON WP-161-E 676
MORTOPAL 918
MOSCARDA 602
MOSS GREEN 298
MOTH BALLS 691
MOTH FLAKES 691
MOTOPAL 918
MOTOR BENZOL 143
MOTOR FUEL 494
MOTOR FUEL ANTI-KNOCK COM-
　POUND 927
MOTOX 958
MOTTENHEXE 509
MOUNTAIN GREEN 298
MOUSE PAK 1012
MOXONE 330
MP-12-50 913
MP 25-38 913
MPK 752
MPP 471
MSMA lxii, lxv
MSZYCOL 146, 583
MULHOUSE WHITE 579
MULTAMAT 138
MULTIMET 138
MULTI-W, KASCADE 607
MURFOS 745
MURIATIC ACID 523
MURIATIC ETHER 445
MURPHY SUPER ROOT GUARD 282
MUSCATOX 301
MUSCOVITE 853
MUSTARD GAS 167
MUSTARD HD 167
MUSTARD OIL 36
MUSTARD VAPOR 167
MUSTARGEN 166
MUSTINE 166
MUTHMANN'S LIQUID 17
MUTOXIN 332
MXDA 1015
MYCHEL 247
MYCINOL 247
MYCROLYSIN 274
MYOCON 715
MYSORITE (CAS 12172-73-5) 117
MYSTIC GREENFLY KILLER 789
N 2790 483
NA 693
NABAC 511
NACCANOL NR 875
NACCANOL SW 875
NACCONATE H 12 661
NACCONATE IOO 952
NACCONOL 35SL 875
NACCONOL 40F 875
NACCONOL 90F 875
NADONE 320
NAFEEN 876
NAFTALEN (POLISH) 691
1-NAFTILAMINA (SPANISH) 692
BETA-NAFTILAMINA (ITALIAN) 693
1-NAFTIL-TIOUREA (ITALIAN) 108
ALPHA-NAFTYLAMIN (CZECH) 692
1-NAFTYLAMINE (DUTCH) 692

2-NAFTYLAMINE (DUTCH) 693
ALPHA-NAFTYL-N-METHYLKARBAMAT
　(CZECH) 234
BETA- NAFTYLOAMINA (POLISH) 693
1-NAFTYLTHIOUREUM (DUTCH) 108
NAH 80 878
NAKO H 761
NAKO TGG 841
NALCO 680 861
NALED **689**, lxii
NALKIL 177
NAPHTENATE DE COBALT (FRENCH) 296
NAPHTHA **690**, lvi, cx
2-NAPHTHACENECARBOXAMIDE 925
5,12-NAPHTHACENEDIONE,8-ACETYL-10-(3-
　AMINO-2,3,6-TRIDEOXY-ALPHA-L-LYXO-
　HEXOPYRANOSYL)OXY-7,8,9,10-
　TETRAHYDRO-6,8,11-TRIHYDROXY-1-
　METHOXY-, (8S-CIS)- 331
NAPHTHA DISTILLATE 690
ALPHA-NAPHTHALAMINE ci
1-NAPHTHALENAMINE 692
NAPHTHALENE **691**, cv, cix, cx
NAPHTHALENE, OCTACHLORO- 727
NAPHTHALENE OIL 302
NAPHTHALENE, TETRACHLORO- 924
METHYLCARBAMATE 1-
　NAPHTHALENOL 234
METHYLCARBAMATE 1-NAPHTHALENOL,
　METHYCARBAMATE 234
NAPHTHALIDAM 692
NAPHTHALIDINE 692
NAPHTHANIL BLUE B BASE 338
NAPHTHA PETROLEUM 690
NAPHTHA SAFETY SOLVENT 896
NAPHTHA SOLVENT 690
NAPHTHA, SOLVENT 690
NAPHTHENATE DE COBALT
　(FRENCH) 296
NAPHTHENIC ACID xli, cx
NAPHTHENIC ACID, COBALT SALT 296
1-NAPHTHOL 234
1-NAPHTHOL N-METHYLCARBAMATE 234
ALPHA-NAPHTHOTHIOUREA 108
1-NAPHTHYLAMIN (GERMAN) 692
1-NAPHTHYLAMINE **692**
2-NAPHTHYLAMINE **693**
ALPHA-NAPHTHYLAMINE 692
BETA-NAPHTHYLAMINE 693
N-(2-NAPHTHYL)ANILINE 767
1-NAPHTHYLMETHYLCARBAMATE 234
1-NAPHTHYL N-METHYLCARBAMATE 234
1-NAPHTHYL N-METHYL-CARBAMATE 234
ALPHA-NAPHTHYL N-
　METHYLCARBAMATE 234
2-NAPHTHYLPHENYLAMINE 767
BETA-NAPHTHYLPHENYLAMINE 767
ALPHA-NAPHTHYLTHIOCARBAMIDE 108
1-NAPHTHYL-THIOHARNSTOFF (GER-
　MAN) 108
1-NAPHTHYL THIOUREA 108
1-(1-NAPHTHYL)-2-THIOUREA 108
ALPHA-NAPHTHYLTHIOUREA 108
N-(1-NAPHTHYL)-2-THIOUREA 108
1-NAPHTHYL-THIOUREE (FRENCH) 108
NAPHTOX 108
NAPTHALENE, HEXACHLORO- 510
NAPTHALIN 691
NAPTHALINE 691
NARCOGEN 964
NARCOTANE 501
NARCOTANN NE-SPOFA (RUSSIAN) 501
NARCOTILE 445
NARCYLEN 15

Bold numbers denote main chemical entries; roman numerals denote text.

NEO-TESTIS 919
NEOZON D 767
NEOZONE 767
NEOZONE D 767
NEPHIS 451
NEPHOCARP 993
NERVANAID B ACID 450
NESOL 409
NESPOR 607
NETAGRONE 330
NETAGRONE 600 330
NEURONILCA 20
NEUTRAL AMMONIUM CHROMATE 65
NEUTRAL AMMONIUM FLUORIDE 68
NEUTRAL POTASSIUM CHROMATE 799
NEUTRAL SODIUM CHROMATE 871
NEUTRAL VERDIGRIS 310
NEUTROGEL NAVY BN 338
NEUTROSEL RED TRVA 281
NEW GREEN 298
NEW WEEDOL 743
NEX 235
NEXEN FB 146, 583
NEXIT 146, 583
NEXIT-STARK 146, 583
NEXOL-E 146, 583
NG 715
NI 270 696
NI 0901-S 696
NI 4303T 696
NIA 10242 235
NIA 1240 429
NIACIDE 472
NIAGARA 1240 429
NIAGARA 10242 235
NIALATE 429
NIALK 964
NIAX TDI 952, 953
NIAX TDI-P 952
NIBREN WAX 924
NICHEL (ITALIAN) 696
NICHEL TETRACARBONILE (ITALIAN) 698
NICKEL **696**, li, cv, cviii
NICKEL 270 696
NICKEL AMMONIUM SULFATE **697**
NICKEL CARBONYL **698**
NICKEL CARBONYLE (FRENCH) 698
NICKEL CATALYST, WET 696
NICKEL CHLORIDE **699**
NICKEL(II) CHLORIDE 699
NICKEL(II) CHLORIDE (1:2) 699
NICKEL CYANIDE **700**
NICKEL(II)CYANIDE 700
NICKEL CYANIDE, SOLID 700
NICKEL NITRATE 701
NICKEL(II) NITRATE **701**
NICKEL(II) NITRATE (1:2) 701
NICKELOUS CHLORIDE 699
NICKELOUS SULFATE 702
NICKEL PARTICLES 696
NICKEL SPONGE 696
NICKEL SULFATE **702**
NICKEL (II) SULFATE 702
NICKEL(2)SULFATE(1:1) 702
NICKEL TETRACARBONYL 698
NICKEL TETRACARBONYLE
(FRENCH) 698
NICLOFEN 711
NICOCHLORAN 146, 583
NICOCIDE 703
NICODUST 703
NICOFUME 703
NICOTINE **703**, lxvii
NICOULINE 844

NIFOS 918
NIFOS T 918
NIFROST 918
NIGLYCON 715
NIKKELTETRACARBONYL (DUTCH) 698
NILOX PBNA 767
NIMITEX 917
NIMITOX 917
NIOMIL 138
NIONG 715
NIP 711
NIPAR S-20 721
NIPARS-20 SOLVENT 721
NIPAR S-30 SOLVENT 721
NIPHEN 720
NIPSAN 342
NIRAN 248, 745
NIRAN E-4 745
NITAL 705
NITOFEN 711
NITRADOR 403
NITRAM 42, 73
NITRAMINE 938
NITRAMYL 93
NITRAN 976
4-NITRANBINE 707
NITRANITOL 613
NITRAPHEN 711
NITRAPYRIN **704**
NITRATE D'AMYLE (FRENCH) 91
NITRATE D'ARGENT (FRENCH) 857
NITRATE DE BARYUM (FRENCH) 132
NITRATE DE PLOMB (FRENCH) 572
NITRATE DE PROPYLE NORMAL
(FRENCH) 834
NITRATE DE SODIUM (FRENCH) 883
NITRATE DE STRONTIUM (FRENCH) 898
NITRATE DE ZINC (FRENCH) 1039
NITRATE ESTERS xciii
NITRATE MERCUREUX (FRENCH) 625
NITRATE MERCURIQUE (FRENCH) 619
NITRATINE 883
NITRE CAKE 864
NITRIC ACID **705**, cv, cix
NITRIC ACID, ALUMINUM (3) 42
NITRIC ACID, ALUMINUM SALT 42
NITRIC ACID, AMMONIUM SALT 73
NITRIC ACID, BARIUM SALT 132
NITRIC ACID, BERYLLIUM SALT 161
NITRIC ACID, CALCIUM SALT 226
NITRIC ACID, CHROMIUM (3) SALT 287
NITRIC ACID, COPPER (2) SALT 311
NITRIC ACID, IRON (3) SALT 474
NITRIC ACID, LEAD(2) SALT 572
NITRIC ACID, LITHIUM SALT 590
NITRIC ACID, MAGNESIUM SALT 595
NITRIC ACID, MANGANESE (2) SALT 610
NITRIC ACID, MERCURY(I) SALT 625
NITRIC ACID, MERCURY(II) SALT 619
NITRIC ACID, NICKEL(II) SALT 701
NITRIC ACID, NICKEL(2) SALT 701
NITRIC ACID, PENTYL ESTER 91
NITRIC ACID, PROPYL ESTER 834
NITRIC ACID, RED FUMING 705
NITRIC ACID, SILVER(1) SALT 857
NITRIC ACID, SODIUM SALT 883
NITRIC ACID, STRONTIUM SALT 898
NITRIC ACID, THORIUM SALT 944
NITRIC ACID, THORIUM(4) SALT
(8CI,9CI) 944
NITRIC ACID TRIESTER OF GLYCER-
OL 715
NITRIC ACID, WHITE FUMING 705
NITRIC ACID, ZINC SALT 1039

NITRIC OXIDE **706**
NITRILE ACRILICO (ITALIAN) 24
NITRILE ACRYLIQUE (FRENCH) 24
NITRILE ADIPICO (ITALIAN) 27
NITRIL KYSELINY ISOFTALOVE
(CZECH) 784
NITRIL KYSELINY MALONOVE
(CZECH) 606
NITRILOACETONITRILE 316
2-2',2''-NITRILOTRIS-
DODECYLBENZENESULFONATE
(SALT) 968
NITRINE-TDC 715
NITRITO 713
4-NITROANILINE 707
P-NITROANILINE **707**
5-NITRO-O-ANISIDINE **717**
NITROBARITE 132
NITROBENZEEN (DUTCH) 708
NITROBENZEN (POLISH) 708
NITROBENZENE **708**, cix
2-NITRO-1,4-BENZENEDIAMINE 53
NITROBENZOL 708
NITROBENZOL,L 708
4-NITROBIPHENYL **709**
P-NITROBIPHENYL 709
NITROCALCITE 226
NITROCARBOL 716
NITROCELLULOSE xxxv, xxxvii
P-NITROCHLOORBENZEEN (DUTCH) 710
NITROCHLOR 711
P-NITROCHLOROBENZENE **710**
NITROCHLOROBENZENE, PARA-,
SOLID 710
P-NITROCHLOROBENZOL (GERMAN) 710
NITROCHLOROFORM 274
P-NITROCLOROBENZENE (ITALIAN) 710
NITRO COMPOUNDS xciii
2-NITRO-1,4-DIAMINOBENZENE 53
4-NITRODIPHENYL 709
P-NITRODIPHENYL 709
NITROFAN 403
NITROFEN **711**
NITROFENE (FRENCH) 711
4-NITROFENOL (DUTCH) 720
NITROGEN **712**
NITROGEN, COMPRESSED 712
NITROGEN, CRYOGENIC LIQUID 712
NITROGEN DIOXIDE **713**
NITROGEN FLUORIDE 714
NITROGEN GAS 712
NITROGEN LIME 222
NITROGEN MONOXIDE 706
NITROGEN MUSTARD 166
NITROGEN OXIDE 724
NITROGEN PEROXIDE 713
NITROGEN, REFRIGERATED LIQUID 712
NITROGEN TETROXIDE 713
NITROGEN TRIFLUORIDE **714**
NITROGLICERINA (ITALIAN) 715
NITROGLICERYNA (POLISH) 715
NITROGLYCERIN **715**
NITROGLYCERINE 715
NITROGLYCEROL 715
NITROGLYN 715
NITROISOPROPANE 721
NITROL 715
NITROLIME 222
NITROLINGUAL 715
NITRO, LIQUID 708
NITROLOWE 715
2-NITROL-P-PHENYLENEDIAMINE 53
NITROMANNITE 613
NITRO MANNITE 613

Bold numbers denote main chemical entries; roman numerals denote text.

OFTALENT 247
OFTENOL lxii
OHIO 347 424
OIL GREEN 288
OIL OF BITTER ALMOND 142
OIL OF MIRBANE 708
OIL OF MUSTARD, ARTIFICIAL 36
OIL OF MYRBANE 708
OIL OF TURPENTINE 996
OIL OF VITRIOL 905
OIL YELLOW 48
OIL YELLOW 1 48
OIL YELLOW 2R 48
OIL YELLOW 21 48
OIL YELLOW 2681 48
OIL YELLOW A 48
OIL YELLOW AT 48
OIL YELLOW C 48
OIL YELLOW T 48
OK 622 744
OKTANEN (DUTCH) 730
OKTATERR 248
OLAMINE 428
OLEFIANT GAS 448
OLEIC ACID LEAD SALT 573
OLEIC ACID, LEAD(2) SALT (2:1) 573
OLEOAKARITHION 993
OLEOFOS-20 745
OLEOMYCETIN 247
OLEOPARATHION 745
OLEOVOFOTOX 679
OLETETRIN 925
OLEUM **732**
OLEUM SINAPIS VOLATILE 36
OLITREF 976
OLOW (POLISH) 564
OMAHA 564
OMAHA & GRANT 564
OMAL 967
OMCHLOR 357
OMEGA-BROMOTOLUENE 155
OMEGA-CHLOROTOLUENE 156
OMEGAMYCIN 925
OMNITOX 146, 583
OMPA 729
OMPACIDE 729
OMPATOX 729
OMPAX 729
O-IMPC 824
OMS 2 471
OMS-29 234
OMS-33 824
OMS-47 684
OMS-93 638
OMS-771 28
OMS 1328 249
OMS-1394 138
160 DEGREE BENZOL 690
ONION OIL 37
ONT 723
ONYX 854
OOS 913
OPCLOR 247
OPELOR 247
O-PHENYL PHENOL ix, xliii
OPHTHOCHLOR 247
OPHTOCHLOR 247
OPLOSSINGEN (DUTCH) 484
OPTAL 819
OP-THAL-ZIN 1043
ORANGE LEAD 575
ORASONE 815
ORATRAST 137
OREMET 947

ORETON 919
ORETON-F 919
ORGA-414 56
ORGANIC PEROXIDE xciii, xcvii
ORGANIC PHOSPHOROUS COM-
  POUNDS lvii
ORGANOPHOSPHATES cx
ORGANO YELLOW 25 48
ORIMON 760
ORLYCYCLINE 925
ORNAMENTAL WEEDER 246
OROTRAN lxii
ORQUISTERON 919
ORSIN 761
ORTHO 4355 689
ORTHO 5865 232
ORTHO-AMINOTOLUENE 956
ORTHOARSENIC ACID 110
ORTHOBORIC ACID 173
ORTHO C-1 DEFOLIANT & WEED KILL-
  ER 868
ORTHOCIDE 233, lxii
ORTHOCIDE 7.5 233
ORTHOCIDE 50 233
ORTHOCIDE 406 233
ORTHODIBROM 689
ORTHODIBROMO 689
ORTHODICHLOROBENZENE 353
ORTHODICHLOROBENZOL 353
ORTHO EARWIG BAIT 877
ORTHO-KLOR 248
ORTHO L10 DUST 565
ORTHO L40 DUST 565
ORTHO MC 592
ORTHONE lxii
ORTHONITROPHENOL 718
ORTHO PARAQUAT CL 744
ORTHOPHOS 745
ORTHOPHOSPHORIC ACID 775
ORTHOSIL 888
ORTHO TOLUIDINE cv
ORTHO WEEVIL BAIT 877
ORTHOXENOL 768
ORTHO-XYLENE 1016
ORVINYLCARBINOL 31
OS 2046 771
OSMIC ACID 733
OSMIUM(VIII) OXIDE 733
OSMIUM TETROXIDE **733**
OSMOSOL EXTRA 819
OSSALIN 876
OSSIN 876
OTACHRON 247
OTAYLITE 140
OTOPHEN 247
OTTANI (ITALIAN) 730
1,2,4,5,6,7,8,8-OTTOCHLORO-3A,4,7,7A-
  TETRAIDRO-4,7-ENDO-METANO-INDANO
  (ITALIAN) 248
OVADZIAK 146, 583
OVEX lxii
OWADZIAK 146, 583
OXAALZUUR (DUTCH) 734
1-OXA-4-AZACYCLOHEXANE 688
7-OXABICYCLO (4.1.0)HEPTANE, 3-
  OXIRANYL- 1007
OXACYCLOPENTANE 931
OXACYCLOPROPANE 456
3-OXA-1-HEPTANOL 186
OXALIC ACID **734**, xxx
OXALIC ACID, COPPER(2) SALT 312
OXALIC ACID, DIAMMONIUM SALT 74
OXALIC ACID DIHYDRATE 734
OXALIC ACID DINITRILE 316

OXALONITRILE 316
OXALSAEURE (GERMAN) 734
OXALYL CYANIDE 316
OXANE 456
1,2-OXATHIOLANE 2,2-DIOXIDE 818
1,2-OXATHROLANE 2,2-DIOXIDE 818
2-H-1,3,2-OXAZAPHOSPHORINANE 422
2H-1,3,2-OXAZAPHOSPHORINE,2-BIS(2-
  CHLOROETHYL)AINOTETRAHYDRO-2-
  OXIDE 422
2H-1,4-OXAZINE, TETRAHYDRO- 688
OXIDASE, GLUCOSE 495
OXIDATION BASE 22 53
OXIDIZERS xcvii, xcvii
OXIDOETHANE 456
ALPHA,BETA-OXIDOETHANE 456
OXIRAAN (DUTCH) 456
OXIRAN 456
OXIRANE 456
OXIRANE, 2-(CHLOROMETHYL)- 425
OXIRANE, ETHYL- 201
OXIRANEMETHOL 498
OXIRANE, METHYL- 832
OXIRANE, 2,2'-OXYBIS (METHYLENE) BIS-
  383
OXIRANE,((2-PROPENYLOXY)METHYL-) 34
3-OXIRANYL-7-
  OXABICYCLO(4.1.0)HEPTENE 1007
OXIRANYMETHANOL 498
OXIRENE, DIHYDRO- 456
OXITOL 430
OXO 913
OXOLANE 931
OXOMETHANE 484
O-XONAL 768
OXRALOX 370
OXY-5 154
OXY-10 154
OXYBENZENE 759
BUTANE, 1,1'-OXYBIS- 202
1,1'-OXYBIS(BUTANE) 202
1,1'-OXYBIS(2-CHLORO)ETHANE 360
OXYBIS(CHLOROMETHANE) 168
OXYBIS (CHLOROMETHANE) 169
OXYCARBON SULFIDE 242
OXYCHLORURE CHROMIQUE
  (FRENCH) 289
OXYCIL 868
OXYDE D'ALLYLE ET DE GLYCIDYLE
  (FRENCH) 34
OXYDE DE BARYUM (FRENCH) 133
OXYDE DE CALCIUM (FRENCH) 227
OXYDE DE CARBONE (FRENCH) 239
OXYDE DE CHLORETHYLE (FRENCH) 360
OXYDE DE MERCURE (FRENCH) 620
OXYDE DE PROPYLENE (FRENCH) 832
OXYDE D'ETHYLE (FRENCH) 458
OXYDE NITRIQUE (FRENCH) 706
OXYDOL 528
OXYFLUROFEN lxiii
OXYGEN **735**
OXYGEN DIFLUORIDE **736**
OXYGEN FLUORIDE 736
OXYGEN, LIQUID 735
OXYLITE 154
OXYMETHYLENE 484
OXYMURIATE OF POTASH 797
OXYPARATHION 742
OXYPHENIC ACID 243
OXYSULFATOVANADIUM 1003
OXYTOL ACETATE 431
OXY WASH 154
OZIDE 1040
OZLO 1040

Bold numbers denote main chemical entries; roman numerals denote text.

Bold numbers denote main chemical entries; roman numerals denote text.

PHENOLS xxvii, cvi, cviii, cix, cx
PHENOL, 2,4,6-TRICHLORO- 967
PHENOL TRINITRATE 786
PHENOL,2,4,6-TRINITRO-, AMMONIUM SALT (9CI) 80
PHENOL, 2,4,6-TRINITRO-, SILVER(1) SALT. 858
PHENOMERCURIC ACETATE 765
PHENOMERCURY ACETATE 765
PHENOSAN 760
PHENOTHIAZINE **760**
10H-PHENOTHIAZINE 760
PHENOVERM 760
PHENOVIS 760
PHENOX 330
PHENOXUR 760
PHENTHIAZINE 760
PHENYLAMINE 95
2-PHENYLAMINONAPHTHALENE 767
4-PHENYLANILINE 51
N-PHENYLANILINE 410
P-PHENYLANILINE 51
N-PHENYLBENZENAMINE 410
PHENYLBENZENE 164
PHENYL-BETA-NAPHTHYLAMINE 767
N-PHENYL-BETA-NAPHTHYLAMINE 767
PHENYLCARBIMIDE 764
PHENYL CARBAMIDE 764
PHENYL CARBONIMIDE 764
PHENYLCARBOXYLIC ACID 151
PHENYLCHLORIDE 258
PHENYL CHLORIDE 258
PHENYLCHLOROFORM 157
PHENYL CHLOROFORM 157
PHENYL CYANIDE 152
N-PHENYLDIETHYLAMINE 375
O,O'-(THIODI-4,1-PHENYLENE)BIS(O,O-DIMETHYL PHOSPHOROTHIOATE) 917
M-PHENYLENEBIS(METHYLAMINE) 1015
P-PHENYLENEDIAMINE **761**
P-PHENYLENEDIAMINE,2-NITRO- 53
O-PHENYLENEDIOL 243
O,O'-(THIODI-P-PHENYLENE)O,O,O',O'-TETRAMETHYLBIS(PHOSPHOROTHIOATE) 917
PHENYLETHANE 441
PHENYLETHENE 899
PHENYLETHER HEXACHLORO 250
PHENYLETHYLDICHLOROSILANE 465
PHENYLETHYLENE 899
PHENYL FLUORIDE 482
PHENYLFORMIC ACID 151
PHENYL HYDRATE 759
PHENYLHYDRAZIN (GERMAN) 762
PHENYLHYDRAZINE **762**
PHENYL HYDRAZINE 762
PHENYL HYDRIDE 143
PHENYL HYDROXIDE 759
PHENYLHYDROXYLAMINE **763**
N-PHENYLHYDROXYLAMINE 763
PHENYLIC ACID 759
PHENYL ISOCYANATE **764**
PHENYLMERCURIC HYDROXIDE 766
PHENYLMERCURY ACETATE **765**
PHENYLMERCURY HYDROXIDE **766**
PHENYLMETHANE 951
N-PHENYLMETHYLAMINE 647
PHENYLMETHYLDICHLOROSILANE 680
PHENYL-2-NAPHTHYLAMINE 767
N-PHENYL-2-NAPTHTYLAMINE **767**
4-PHENYL-NITROBENZENE 709
P-PHENYL-NITROBENZENE 709
PHENYL PERCHLORYL 506
O-PHENYLPHENOL **768**
2-PHENYLPHENYL 768
PHENYLPHOSPHONOTHIOIC ACID O-(4-BROMO-2,5-BROMO-2,5- DICHLOROPHE-NYL)O-METHYL ESTER 581

2-PHENYLPROPANE 308
PHENYLSILICON TRICHLORIDE 769
(N-PHENYL-P-TOLUENESULFONAMI-DO)ETHYLMERCURY 629
PHENYLTRICHLOROMETHANE 157
PHENYL TRICHLOROSILANE **769**
PHILOSOPHER'S WOOL 1040
PHISODAN 511
PHISOHEX 511
PHLOGOPITE 853
PHORAT (GERMAN) 770
PHORATE **770**
PHORATE-10G 770
PHORSAEURELOESUNGEN (GERMA-NY) 775
PHORTOX 912
PHOSDRIN **771**
PHOSDRIN 24 771
CIS-PHOSDRIN 771
PHOSFENE 771
PHOS-FLUR 876
PHOSGEN (GERMAN) 772
PHOSGENE **772**, cvi
PHOSKIL 745
PHOSMET **773**
PHOSPHACOL 742
PHOSPHAGEL 44
PHOSPHAMID 385
PHOSPHAMIDE 385
PHOSPHATE DE O,O-DIETHYLE ET DEO-2-CHLORO-1-(2,4-DICHLOROPHENYL) VI-NYLE (FRENCH) 249
PHOSPHATE DE O,O-DIMETHLE ET DE O-(1,2-DIBROMO-2,2-DICHLORETHYLE) (FRENCH) 689
PHOSPHATEDE DIMETHYLE ET DE 2-DIMETHYLCARBAMOYL 1-METHYL VI-NYLE (FRENCH) 368
PHOSPHATE DE DIMETHYLE ET DE 2-METHYLCARBAMOYL 1-METHYL VINYLE (FRENCH) 687
PHOSPHATES xxxiii
PHOSPHENE (FRENCH) 771
PHOSPHIDE lxiii
PHOSPHINE **774**, xciii
PHOSPHOMS TRIHYDRIDE, PHOSPHO-ROUS HYDRIDE 774
PHOSPHONOTHIOIC ACID, ETHYL-O-ETHYL S-PHENYL ESTER 483
PHOSPHORAMIDIC ACID, 4-TERT-BUTYL-2-CHLOROPHENYLPHOSPHOR AMI-DATE 307
PHOSPHONOTHIOIC ACID, PHENYL-,O-(4-BROMO-2,5-DICHLOROPHENYL)O-METHYL ESTER 581
PHOSPHORAMIDIC ACID,ISOPROPYL-, 4-(METHYLTHIO)-M-TOLYL ETHYL ESTER 469
PHOSPHORAMIDIC ACID, METHYL-,4-TERT-BUTYL-2-CHLOROPHENYL 307
PHOSPHORAMIDIC ACID, METHYL-,2-CHLORO-4-(1,1-DIMETHYLETHYL)PHENYL METHYL ESTER 307
PHOSPHORAMIDIC ACID, (1-METHYLETHYL)-, ETHYL (3-METHYL-4-(METHYLTHIO)PHENYL)ESTER 469
PHOSPHORATED HYDROGEN 774
PHOSPHORE BLANC (FRENCH) 776
PHOSPHORE(PENTACHLORURE DE) (FRENCH) 778
PHOSPHORE (TRICHLORURE DE) (FRENCH) 782
PHOSPHORIC ACID **775**, cvi
PHOSPHORIC ACID, ALUMINUM SALT (1:1) 44
PHOSPHORIC ACID, DIBUTYL ESTER. 349

PHOSPHORIC ACID, O,O-DIETHYL O-6-METHYL-2-(1-METHYLETHYL)-4-PYRIMIDINYL ESTER 342
PHOSPHORIC ACID, DIETHYL P-NITROPHENYL ESTER 742
PHOSPHORIC ACID, DIMETHYL ESTER, ESTER WITH CIS-3-HYDROXY-N,N-DIMETHYLCROTONAMIDE 368
PHOSPHORIC ACID, DIMETHYL ESTER, ESTER WITH CIS-3-HYDROXY-N-METHYLCROTONAMIDE 687
PHOSPHORIC ACID, DIMETHYL ESTER, ESTER WITH (E)-3-HYDROXY-N,N-DIMETHYLCROTONAMIDE 368
PHOSPHORIC ACID, DIMETHYL ESTER WITH DIMETHYL 3- HYDROXYGLUTA-CONATE 911
PHOSPHORIC ACID, DISODIUM SALT 887
PHOSPHORIC ACID, LEAD SALT 576
PHOSPHORIC ACID, LEAD(2) SALT (2:3) 576
PHOSPHORIC ACID, TETRAETHYL ESTER 918
PHOSPHORIC ACID TRIBUTYL ESTER 960
PHOSPHORIC ACID, TRIS(2,3- DIBROMO-PROPYL)ESTER 992
PHOSPHORIC ACID, TRISODIUM SALT 886
PHOSPHORIC ANHYDRIDE 780
PHOSPHORIC CHLORIDE 777, 778
PHOSPHORIC SULFIDE 779
PHOSPHORIC TRIAMIDE,HEXAMETHYL-515
PHOSPHORIC TRIS(DIMETHYLAMIDE) 515
PHOSPHORODIAMIDIC ACID, N,N-BIS(2-CHLOROETHYL)-N'-(3-HYDROXYPROPYL)-, INTRAMOL. ESTER 422
PHOSPHORODIFLUORIDIC ACID 382
PHOSPHORODIFLUORIDIC ACID (ANHY-DROUS) 382
PHOSPHORODITHIOIC ACID, S-(((P-CHLOROPHENYL)THIO)METHYL) O,O-DIETHYL ESTER 993
PHOSPHORODITHIOIC ACID, O,O-DIETHYL ESTER, S,S-DIESTER WITH P-DIOXANE-2,3-DITHIOL 408
PHOSPHORODITHIOIC ACID, O,O-DIETHYL ESTER, S,S-DIESTER WITH ME-THANEDITHIOL 429
PHOSPHORODITHIOIC ACID, S-((1,3-DIHYDRO-1,3-DIOXO-ISOINDOL-2-YL)METHYL) O,O-DIMETHYL ESTER 773
PHOSPHORODITHIOIC ACID, O,O-DIMETHYL ESTER, S-ESTER WITH N-(MERCAPTOMETHYL)PHTHALIMIDE 773
PHOSPHORODITHIOIC ACID, O,O-DIMETHYL-S-(2-(METHYLAMINO)-2-OXOETHYL) ESTER (9CI) 385
PHOSPHORODITHIOIC ACID-5-5'-1,4-DIOXANE-2,3-DIYL O,O,O',O'-TETRAETHYL ESTER 408
PHOSPHORODITHIOIC ACID, S,S'-1,4-DIOXANE-2,3-DIYL-O,O,O',O'-TETRAETHYL ESTER 408
PHOSPHORODITHIONIC ACID,S-2-(ETHYLTHIO)ETHYL-O,O-DIETHYLESTER 417
PHOSPHORODITHIOIC ACID, O-ETHYLO-(4-(METHYLTHIO)PHENYL)S-PROPYL ESTER 910
PHOSPHOROTHIOIC ACID, O,O-DIETHYL ESTER, O-ESTER WITH 3-CHLORO-7-HYDROXY-4-METHYLCOUMARIN 301
PHOSPHOROTHIOIC ACID,O,O-DIETHYL O-2- (ETHYLTHIO)ETHYL ESTER, MIXED WITH O,O-DIETHYL S-2-(ETHYLTHIO)ETHYL PHOSPHOROTHIO-ATE 334

PHOSPHOROTHIOIC ACID, O,O-DIETHYLO-(P-(METHYLSULFINYL)PHENYL) ESTER 470

PHOSPHOROTHIOIC ACID, O,O,-DIETHYLO-(P-NITROPHENYL)ESTER 745

PHOSPHOROTHIOIC ACID, O,O-DIETHYL O-PYRAZINYL ESTER 1046

PHOSPHOROTHIOIC ACID,O,O-DIMETHYL-,O-(4-METHYLTHIO)M-TOLYL ESTER 471

PHOSPHOROTHIOIC ACID,O,O-DIMETHYL O-(4-NITROPHENYL)ESTER 679

PHOSPHOROTHIOIC ACID, 0-2 (ETHYLTHIO)ETHYL O,O-DIMETHYL ESTER, MIXED WITH S-2- (ETHYLTHIO)ETHYL O,O-DIMETHYL PHOSPHOROTHIOATE 658

PHOSPHOROTHIOIC ACID, O,O'-(THIODI-4,1-PHENYLENE)O,O,O',O'-TETRAMETHYL ESTER 917

PHOSPHOROUS ACID,TRIMETHYL ESTER 986

PHOSPHOROUS CHLORIDE 782

PHOSPHOROUS YELLOW 776

PHOSPHORPENTACHLORID (GERMAN) 778

PHOSPHORTRICHLORID (GERMAN) 782

PHOSPHORUS **776**

PHOSPHORUS, AMORPHOUS, RED 781

PHOSPHORUS CHLORIDE 782

PHOSPHORUS CHLORIDE OXIDE 777

PHOSPHORUS ELEMENTAL, WHITE 776

PHOSPHORUS OXIDE 780

PHOSPHORUS(V) OXIDE 780

PHOSPHORUS OXYCHLORIDE **777**

PHOSPHORUS OXYTRICHLORIDE 777

PHOSPHORUS PENTACHLORIDE **778**

PHOSPHORUS PENTAOXIDE 780

PHOSPHORUS PENTASULFIDE **779**

PHOSPHORUS PENTOXIDE **780**

PHOSPHORUS PERCHLORIDE 778

PHOSPHORUS PERSULFIDE 779

PHOSPHORUS, RED **781**

PHOSPHORUS SULFIDE 779

PHOSPHORUS TRICHLORIDE **782**, xciii

PHOSPHORWASSERSTOFF (GERMAN) 774

PHOSPHORYL CHLORIDE 777

PHOSPHORYLHEXAMETHYLTRIAMIDE 515

PHOSPHOSTIGMINE 745

PHOSPHOTEX 937

PHOSPHOTOX E 429

PHOSPHURES D'ALUMIUM (FRENCH) 45

PHOSTOXIN 45, 774

PHOSVEL 581

PHOTOPHOR 229

PHPH 164

5-PHTHALANACARBOXYLIC ACID, 1,3-DIOXO- 978

PHTHALANDIONE 783

1,3-PHTHALANDIONE 783

PHTHALIC ACID ANHYDRIDE 783

PHTHALIC ACID, BIS(2-METHOXYETHYL)ESTER 986

PHTHALIC ACID DIBUTYL ESTERCELLUFLEX DPB 350

PHTHALIC ACID DIOCTYL ESTER 415

PHTHALIC ACID, DIOCTYL ESTER 406

PHTHALIC ACID METHYL ESTER 398

PHTHALIC ANHYDRIDE **783**, cvi

PHTHALIC ANHYDRIDE, 1,2,3,6-TETRAHYDRO- 932

PHTHALIMIDE,N-(MERCAPTOMETHYL)-, S-ESTER WITH O,O-DIMETHYL PHOSPHORODITHIOATE 773

PHTHALIMIDO O,O-DIMETHYL PHOSPHORODITHIOATE 773

PHTHALIMIDOMETHYL O,O-DIMETHYL PHOSPHORODITHIOATE 773

M-PHTHALODINITRILE **784**

PHTHALOPHOS 773

PHTHALSAEUREANHYDRID (GERMAN) 783

PHTHALSAEUREDIMETHYLESTER (GERMAN) 398

PIC-CHLOR 274

PICFUME 274

PICKEL ALUM 47

PICLORAM **785**

PICOLINIC ACID, 4-AMINO-3,5,6-TRICHLORO- 785

PICRATE SALTS xciii

PICRIC ACID **786**, xciii

PICRIC ACID, AMMONIUM SALT 80

PICRIC ACID, SILVER(I) SALT 858

PICRIDE 274

PICRONITRIC ACID 786

PICRYLNITROMETHYLAMINE 938

PICTAROL 80

PIECIOCHLOREK FOSFORU (POLISH) 778

PIELIK 330

PIKRINEZUUR (DUTCH) 786

PIKRINSAEURE (GERMAN) 786

PIKRYNOWY KWAS (POLISH) 786

PILLARDIN 687

PILLARFURAN 235

PILLARGRAN 824

PILOT HD-90 875

PILOT SF-40 875

PILOT SF-40B 875

PILOT SF-40FG 875

PILOT SF-60 875

PILOT SF-96 875

PILOT SP-60 875

PIMELIC KETONE 320

PINAKON 518

PINE OIL xxxii, cviii, cix

PIPERAZIDINE 787

PIPERAZIN (GERMAN) 787

PIPERAZINE **787**

PIPERAZINE, ANHYDROUS 787

PIPERAZINE, 1,4-BIS (3- AMINOPROPYL)- 165

PIPERAZINE DIHYDROCHLORIDE **788**

PIPERIDINE, 1-ETHYL- 466

PIRIMICARB **789**

PIRIMOR 789

PIROFOS 902

PIRYDYNA (POLISH) 837

PITTCHLOR 225

PITTCIDE 225

PITTSBURGH PX-138 415

PLANOTOX 330

PLANT DITHIO AEROSOL 902

PLANTDRIN 687

PLANTFUME 103 SMOKE GENERATOR 902

PLANTGARD 330, lxii

PLANTIFOG 160M 607

PLASTIC DUST liii

PLATIN (GERMAN) 791

PLATINATE, HEXACHLORO- 275

PLATINATE (2-), HEXACHLORO-, DIAMMONIUM 64

PLATINATE (2-1), HEXACHLORO-, DIAMMONIUM 64

PLATINATE (2-), HEXACHLORO-, DIAMMONIUM, (OC-6-11)-(9CI) 64

PLATINATE(2-),HEXACHLORO-, DIPOTASSIUM 805

PLATINATE(2-),HEXACHLORO-, DISODIUM,TETRAHYDRATE 870

PLATINATE(2-),TETRACHLORO-, DIAMMONIUM 86

PLATINATE(2-),TETRACHLORO- ,DIPOTASSIUM 790

PLATINATE(2-),TETRACHLORO- , DIPOTASSIUM (SP-4-1)- 790

PLATINIC AMMONIUM CHLORIDE 64

PLATINIC CHLORIDE 275, 791

PLATINIC POTASSIUM CHLORIDE 805

PLATINIC SODIUM CHLORIDE 870

PLATINOL 293

PLATINOL AH 415

PLATINOL DOP 415

CIS-PLATINOUS DIAMMINE DICHLORIDE 293

PLATINOUS POTASSIUM CHLORIDE **790**

PLATINUM **791**

CIS-PLATINUM 293

CIS-PLATINUM(II) 293

CIS-PLATINUM(II) DIAMINE DICHLORIDE 293

PLATINUM(II),DIAMMINEDICHLORO-,CIS- 293

PLENUR 587

PLEOPARAPHENE 745

PLIDAN 341

PLOMB FLUORURE (FRENCH) 570

PLUMBANE, TETRAETHYL- 927

PLUMBANE, TETRAMETHYL- 933

PLUMBOPLUMBIC OXIDE 575

PLUMBOUS CHROMATE 567

PLUMBOUS FLUORIDE 570

PLUMBOUS PHOSPHATE 576

PLUMBOUS SULFIDE 580

PMA 765

PMAC 765

PMACETATE 765

PMAL 765

PMAS 765

PMP 773

PNA 707

PNB 709

PNCB 710

PNP 720

POCKET TOUCHWEEDER 914

POINT TWO 876

POLYBOR 866

POLYCHLORCAMPHENE 958

POLYCHLORINATED BIPHENYL **792**

POLYCHLORINATED BIPHENYLS xxv, xxxv, lxxxix, cvi

POLYCHLORINATED CAMPHENE 958

POLYCHLOROBIPHENYL 792

POLYCHLOROCAMPHENE 958

POLYCIZER DBP 350

POLYCYCLINE 925

POLYESTER xxxv

POLYMERS xcii, xciii

POLYOXYMETHYLENE GLYCOLS 484

POLYRAM M 607

POLYRAM ULTRA 943

POLYRAM Z 1045

POLY-SOLV EB 186

POLY-SOLV EE 430

POLY-SOLV EE ACETATE 431

POLY-SOLV EM 454, 649

POLY-SOLVE MPM 830

POLYSTYRENE xcii

POLYVINYL CHLORIDE lxxxix

POMARSOL 943

Bold numbers denote main chemical entries; roman numerals denote text.

2-PROPENE-1-THIOL, 2-CHLORO-, DIETH-
YLDITHIOCARBAMATE   900
2-PROPENE-1-THIOL, 2,3-DICHLORO-
,DIISOPROPYLCARBAMATE   336
2-PROPENIC ACID, 2-METHYL-, BUTYL
ESTER   206
2-PROPENOIC ACID   23
2-PROPENOIC ACID   23
2-PROPENOIC ACID, BUTYL ESTER   192
2-PROPENOIC ACID, ETHYL ESTER   434
PROPENOIC ACID METHYL ESTER   642
2-PROPENOIC ACID, METHYL ESTER   642
2-PROPENOIC ACID, 1-METHYL-, ETHYL
ESTER   464
2-PROPENOIC ACID, 2-METHYL-, METHYL
ESTER   678
2-PROPENOIC ACID, OXIRANYLMETHYL
ESTER (9CI)   499
PROPENOL   31
1-PROPEN-3-OL   832, 31
PROPEN-1-OL-3   31
2-PROPENOL   31
2-PROPEN-1-OL   31
2-PROPEN-1-OL, 2-METHYL-   635
2-PROPEN-1-ONE   21
PROPENYL ALCOHOL   31
2-PROPENYL ALCOHOL   31
2-PROPENYL METHANOATE   30
PROPIONALDEHYDE   **822**
PROPIONE   378
PROPIONIC ACID   **823**
PROPIONIC ACID, 2-CHLORO-, ETHYL
ESTER   446
PROPIONIC ACID GRAIN PRESERVER   823
PROPIONIC ACID, 2-METHYLENE-   633
PROPIONIC ALDEHYDE   822
PROPIONITRILE, BLADEX R   172
PROPOGON   824
PROPOKSURU (POLISH)   824
PROPOXOR   lxii
PROPOXUR   **824**
PROPOXURE   824
PROPROP   366
PROPYLACETATE   825
PROPYL ACETATE   **825**
1-PROPYL ACETATE   825
2-PROPYL ACETATE   553
N-PROPYL ACETATE   825
PROPYLACETONE   648
PROPYL ALCOHOL   819
1-PROPYL ALCOHOL   819
N-PROPYL ALCOHOL   819
SEC-PROPYL ALCOHOL   554
PROPYL ALDEHYDE   822
N-PROPYL ALKOHOL (GERMAN)   819
PROPYL ALLYL DISULFIDE   37
PROPYLAMINE   **826**
2-PROPYLAMINE   555
N-PROPYLAMINE   826
PROPYLCARBINOL   193
PROPYL CARBINOL   193
N-PROPYLCARBINYL CHLORIDE   200
PROPYLENE   821
1-PROPYLENE   821
PROPYLENE ALDEHYDE   306
PROPYLENE CHLORIDE   364
PROPYLENE CHLOROHYDRIN   **827**
PROPYLENECHLOROHYDRIN   827
PROPYLENE DICHLORIDE   364
ALPHA,BETA-PROPYLENE DICHLO-
RIDE   364
PROPYLENE DINITRATE   **828**
PROPYLENE GLYCOL   **829**, xlix
1,2-PROPYLENE GLYCOL   829

PROPYLENE GLYCOL DINITRATE   828
1,2-PROPYLENE GLYCOL DINITRATE   828
PROPYLENE GLYCOL 1,2-DINITRATE   828
PROPYLENE GLYCOL METHYL
ETHER   830
PROPYLENE GLYCOL MONOMETHYL
ETHER   **830**
ALPHA-PROPYLENE GLYCOL MONOMETH-
YL ETHER   830
PROPYLENEIMINE   **831**
PROPYLEN EIMINE   831
PROPYLENE OXIDE   **832**, cvi
1,2-PROPYLENE OXIDE   832
PROPYLENGLYKOL-MONOMETHYLAETHER
(GERMAN)   830
1,2-PROPYLENIMINE   831
1,2-PROPYLEN IMINE   831
PROPYLFORMIC ACID   213
PROPYL HYDRIDE   817
PROPYLIC ALCOHOL   819
PROPYLIC ALDEHYDE   822
PROPYL ISOCYANATE   **833**
1-PROPYLISOCYANATE   833
1-PROPYL ISOCYANATE   833
M-PROPYL ISOCYANATE   833
N-PROPYL ISOCYANATE   833
PROPYL KETONE   413
PROPYLMETHANOL   193
PROPYL NITRATE   834
N-PROPYL NITRATE   **834**
PROPYLOWY ALKOHOL (POLISH)   819
PROPYL OXIRANE   201
PROPYL PARABAN   lxxvii
1-PROPYNE-3-OL   820
3-PROPYNOL   820
2-PROPYN-1-OL   820
2-PROPYNYL ALCOHOL   820
PROPYON   824
PROPYZAMIDE   **835**
PROTEASE 150   120
PROTHROMADIN   1012
PROTOCHLORURE D'IODE (FRENCH)   534
PROTOX TYPE 166   1040
PROTOX TYPE 167   1040
PROTOX TYPE 168   1040
PROTOX TYPE 169   1040
PROTOX TYPE 267   1040
PROTOX TYPE 268   1040
PROWL   lxii
PROXOL   lxii
PROZOIN   823
PRUSSIAN BROWN   538
PRUSSIC ACID   524
PRUSSITE   316
PRYFON   lxii
PS   274
PS 1   43
PS1 (ALUMINA)   43
PSC CO-OP WEEVIL BAIT   877
PSEUDOACETIC ACID   823
PSEUDOCUMENE   983
PSEUDOCUMOL   983
PSEUDOTHIOUREA   559
PSEUDOUREA   997
PSI-CUMENE   983
PST 29084   246
CIS-PT(II)   293
PURALIN   943
PURASAN-SC-10   765
PURATRONIC CHROMIUM TRIOXIDE   290
PURATURF 10   765
PURE QURTZ   854
PURETALC USP   913
PURE ZINC CHROME   1034

PURE ZINC YELLOW   1034
PUROCYCLINA   925
PURPLE SALT   810
PX 104   350
PYRA FOG   836
PYRALENE   792
PYRANOL   792
PYRAPHEN   758
PYRAZINE HEXAHYDRIDE   787
PYRAZINE, HEXAHYDRO-   787
PYRETHRIN I   836
PYRETHRIN II   836
PYRETHRINS PYRETHRUM (ACGIH,
OSHA)   836
PYRETHRUM   **836**, lxvii
PYRETHRUM INSECTICIDE   836
PYRIDIN (GERMAN)   837
2-PYRIDINAMINE   55
ALPHA-PYRIDINAMINE   55
PYRIDINE   **837**
PYRIDINE,2-AMINO   55
PYRIDINE, 2-CHLORO-6-
(TRICHLOROMETHYL)-   704
PYRIDINE, 3-(1-METHYL-2-PYRROLIDINYL)-
703
PYRIDINE, (S)-3-(1-METHYL-2-
PYRROLIDINYL)-AND SALTS   703
2-PYRIDINOL, 3,5,6-TRICHLORO-,O-ESTER
WITH O,O-DIETHYL PHOSPHOROTHIO-
ATE   282
2-PYRIDYLAMINE   55
ALPHA-PYRIDYLAMINE   55
PYRIMICARBE   789
PYRIMIDINE,2-CHLORO-4-
(DIMETHYLAMINO)-6-METHYL-   304
PYRIMOR   789
PYRINEX   282
PYROACETIC ACID   7
PYROACETIC ETHER   7
PYROBENZOL   143
PYROCATECHIN   243
PYROCATECHINE   243
PYROCATECHINIC ACID   243
PYROCATECHOL   243
PYROCATECHUIC ACID   243
PYROLUSITE BROWN   609
PYROMUCIC ALDEHYDE   491
PYROPENTYLENE   327
PYROPHOSPHATE   937
PYROPHOSPHATE DE TETRAETHYLE
(FRENCH)   918
PYROPHOSPHORIC ACID, TETRASODIUM
SALT-   937
PYROPHOSPHORODITHIOIC ACID, TETRA-
ETHYL ESTER   902
PYROPHOSPHORODITHIOIC ACID,O,O,O,O-
TETRAETHYL ESTER   902
PYROSULFUROUS ACID, DISODIUM
SALT   881
PYROSULPHURIC ACID   732
PYROXYLIC SPIRIT   637
PYRROLYLENE   182
PYRROXATE   758
Q-LOID A 30   43
QUADRANGLE DSM   658
QUADRONAL   758
QUARTERNIUM 15   lxxvi
QUARTERNIUM COMPOUNDS   lxxvii
QUARTZ   854
QUARTZ GLASS   852
QUATERNARY AMMONIUM COM-
POUNDS   lxiii
QUATERNIUM-17   64
QUAZO PURO (ITALIAN)   854

Bold numbers denote main chemical entries; roman numerals denote text.

Chemical Name Index

Bold numbers denote main chemical entries; roman numerals denote text.

SODIUM TETRAHYDROALUMINATE(1-) 860
SODIUM THIOCYANATE **890**
SODIUM THIOCYANIDE 890
SODIUM THIOSULFATE **891**
SODIUM THIOSULFATE, ANHYDROUS 891
SODIUM THIOSULPHATE 891
SODIUM TRIPOLYPHOSPHATE xxxi
SODOTHIOL 891
SO- FLO 876
SOFRIL 903
SOHNHOFEN STONE 219
SOILBROM-40 451
SOILBROM-85 451
SOILBROM-90EC 451
SOILBROME-85 451
SOILFUME 451
SOK 234
SOLAESTHIN 663
SOLAR 40 875
SOLAR 90 875
SOLAR WINTER BAN TRIMETHYL GLYCOL 829
SOLBAR 137
SOLFARIN 1012
SOLFURO DI CARBONIO (ITALIAN) 238
SOLMETHINE 663
SOLOZONE 885
SOLPYRON 20
SOL SODOWA KWASU LAURYLOBENZENO-SULFONOWEGO (POLISH) 875
SOLTAIR 743
SOLUBLE VANDYKE BROWN 611
SOLUGLACIT 742
SOLVANOM 398
SOLVARONE 398
SOLVENT 111 652
SOLVENT ETHER 458
SOLVENT NAPHTHA 690
SOLVENTS xvii, xxv, xxxv, xliii, xliv, xlv, xlvi, xlvii, xlvii, xlix, l, lii, lvi, cix
SOLVIREX 417
SOMA 758
SOMALIA YELLOW R 48
SONACIDE 496
SONACON 341
SOPRABEL 565
SOPRANEBE 607
SOPRATHION 429, 745
SOREX GOLDEN FLY BAIT 639
SOUFRAMINE 760
SOUP 715
SOUTHERN BENTONITE 140
SOVOL 792
SPALERITE 1044
SPANNIT 282
SPECIAL TERMITE FLUID 353
SPECTAR 554
SPECTRACIDE 342, lxii
SPECTROKNE BLUE B 338
SPECULAR IRON 538
SPEEDWAY 743
SPERLOX-S 903
SPERLOX-Z 1045
SPERSUL 903
SPERSUL THIOVIT 903
SPIRIT 427
SPIRIT OF HARTSHORN 57
SPIRIT OF SULFUR 905
SPIRITS OF SALT 523
SPIRITS OF TURPENTINE 996
SPIRITS OF WINE 427
SPONTOX 912
SPOR-KIL 765

SPOTRETE 943
SPOTTON 471
SPRAYSET MEKP 667
SPRAY-TROL BRANCH RODEN-TROL 1012
SPRITZ-HORMIN/2,4-D 330
SPRITZ-RAPIDIN 146, 583
SPRUEHPFLANZOL 146, 583
SQ 1489 943
SQUADRON AND QUADRANGLE MANEX 607
SQUILL 842
SR406 233
STABILIZATOR AR 767
STABLE PENTABORANE 746
STAFLEX DBP 350
STAFLEX DOP 415
STAGNO (TETRACLORURO DI) (ITALIAN) 946
STANDARD LEAD ARSENATE 565
STANNANE, TETRAETHYL- 928
STANNIC CHLORIDE 946
STANNIC CHLORIDE, ANHYDROUS 946
STANNIC CHLORIDE, HYDRATED 892
STANNIC CHLORIDE PENTAHYDRATE **892**
STANNIC CHLORIDE, PENTAHYDRATE 892
STANNICHLOR 893
STANNOCHLOR 893
STANNOUS CHLORIDE **893**
STANNOUS CHLORIDE, SOLID 893
STANNOUS FLUORIDE **894**
STANOMYCETIN 247
STARSOL NO. 1 500
STATHION 745
STAUFFER CAPTAN 233
STAUFFER N 2790 483
STAUFFER R-1,303 993
STAUFFER R 1504 773
STAY-FLO 876
STAY-KLEEN 172
STCC 241
STEAM DISTILLED TURPENTINE 996
STEARIC ACID, LEAD SALT 577
STEATITE TALC 913
STECLIN 925
STEINBUHL YELLOW 128
STELLACYL 758
STERAL 511
STERASKIN 511
STESOLID 341
STESOLIN 341
STIBINE **895**
STIBINE, TRIBROMO- 104
STIBINE, TRICHLORO- 105
STIBIUM 99
STICKDIOXYD (GERMAN) 724
STICKMONOXYD (GERMAN) 706
STICKSTOFFDIOXID (GERMAN) 713
STIKSTOFDIOXYDE (DUTCH) 713
STINK DAMP 529
STIROFOS lxii
STIROLO (ITALIAN) 899
STODDARD SOLVENT **896**
STONE DUST liii
STONE RED 538
STOP-GRO 605
STRATHION 745
STREPTOMYCES PEUCETIUS 331
STREUNEX 146, 583
STROBANE lxiii
STROBANE-T 958
STROBANE-T-90 958
STRONTIUM CHROMATE **897**
STRONTIUM CHROMATE (1:1) 897

STRONTIUM CHROMATE (VI) 897
STRONTIUM CHROMATE 12170 897
STRONTIUM CHROMATE A 897
STRONTIUM CHROMATE X-2396 897
STRONTIUM NITRATE **898**
STRONTIUM(II) NITRATE (1:2) 898
STRONTIUM YELLOW 897
STRYCHNINE lxiii
STUDAFLOUR 876
STUNTMAN 605
STYREEN (DUTCH) 899
STYREN (CZECH) 899
STYRENE **899**, xxxvii, xliii, xcii, xciii, xcvii, cvi, cx
STYRENE-BUTADIENE xxxv
STYRENE-BUTADIENE RESINS xxxvii
STYRENE MONOMER 899
STYRENE MONOMER, INHIBITED 899
STYROL (GERMAN) 899
STYROLE 899
STYROLENE 899
STYRON 899
STYROPOL 899
STYROPOR 899
SUBACETATE LEAD 578
SUBCHLORIDE OF MERCURY 624
SUBERANE 317
SUBERYLENE 317
SUBLIMAT (CZECH) 616
SUBLIMED SULFUR 903
SUBTILISIN BPN 120
SUBTILISINS 120
SUCRE EDULCOR 845
SUCRETTE 845
SUDAN YELLOW RRA 48
SULEMA (RUSSIAN) 616
SULFALLATE **900**
SULFAMATE 81
SULFAMIC ACID **901**
SULFAMIC ACID, MONOAMMONIUM SALT 81
SULFAMIDIC ACID 901
SULFAMINSAURE (GERMAN) 81
SULFAN 908
SULFAPOL 875
SULFAPOLU (POLISH) 875
SULFATE DE CUIVRE (FRENCH) 313
SULFATE DE METHYLE (FRENCH) 399
SULFATE DE PLOMB (FRENCH) 579
SULFATE DE ZINC (FRENCH) 1043
SULFATE DIMETHYLIQUE (FRENCH) 399
SULFATE MERCURIQUE (FRENCH) 622
SULFATEP 902
SULFATE WOOD TURPENTINE 996
SULFERROUS 479
SULFIDAL 903
O-SULFOBENZIMIDE 845
O-SULFOBENZOIC ACID IMIDE 845
SULFONIC ACID, MONOCHLORIDE 277
SULFONIMIDE 232
SULFORON 903
SULFOTEP **902**
SULFOTEPP 902
SULFOTHIORIUE 891
SULFOXYL 154
SULFRAMIN 85 875
SULFRAMIN 40 FLAKES 875
SULFRAMIN 90 FLAKES 875
SULFRAMIN 40 GRANULAR 875
SULFRAMIN 40RA 875
SULFRAMIN 1238 SLURRY 875
SULFRAMIN 1250 SLURRY 875
SULFUR **903**, lxii, lxvi
SULFUR DIOXIDE **904**

Bold numbers denote main chemical entries; roman numerals denote text.

Bold numbers denote main chemical entries; roman numerals denote text.

2-TOLIDINE 950
3,3'-TOLIDINE 950
O-TOLIDINE **950**
O,O'-TOLIDINE 950
TOLIT 988
TOLITE 988
TOLL 679
O-TOLUAMIDE, 3,5-DINITRO- 401
TOLUAZOTOLUIDINE 48
TOLUEEN (DUTCH) 951
TOLUEEN-DIISOCYANAAT (DUTCH) 952
TOLUEN (CZECH) 951
TOLUEN-DISOCIANATO (ITALIAN) 952
TOLUENE **951**, xxxv, xliii, xliv, xlvi, lii, lvii, lxx-
    vii, cvi, cviii, cix, cx
O-TOLUENE-AZO-O-TOULUIDENE 48
TOLUENE, ALPHA,ALPHA-DICHLORO- 141
TOLUENE, O-CHLORO- 279
TOLUENE DIISOCYANATE 952
TOLUENE DI-ISOCYANATE 952
2,4-TOLUENEDIISOCYANATE 952
TOLUENE-2,4-DIISOCYANATE **952**, cvi
2,6-TOLUENE DIISOCYANATE 953
TOLUENE-2,6-DIISOCYANATE **953**, cvi
TOLUENE, 2,4-DINITRO- 405
TOLUENE HEXAHYDRIDE 654
P-TOLUENESULFONAMIDE,N-
    (ETHYLMERCURI)-N-PHENYL- 629
P-TOLUENESULFONANILIDE,N-
    (ETHYLMERCURI)- 629
TOLUENE SULFONIC ACID 954
TOLUENE SULFONIC ACID, LIQUID 954
TOLUENESULPHONIC ACID 954
TOLUENE SULPHONIC ACID **954**
TOLUENE SULPHONIC ACID, LIQUID 954
TOLUENE TRICHLORIDE 157
TOLUENE, 2,4,6-TRINITRO,-(WET) 988
TOLUENE, VINYL-(MIXED ISOMERS), INHIB-
    ITED 1011
M-TOLUIDIN (CZECH) 955
O-TOLUIDIN (CZECH) 956
2-TOLUIDINE 956
3-TOLUIDINE 955
M-TOLUIDINE **955**
O-TOLUIDINE **956**, 950
P-TOLUIDINE,ALPHA,ALPHA,ALPHA-
    TRIFLUO RO-2,6-DINITRO-N,N-DIPROPYL-
    976
P-TOLUIDINE, 2-CHLORO- 263
O-TOLUIDINE, 3-CHLORO- 264
O-TOLUIDINE, 6-CHLORO- 280
TREFANOCIDE 265
P-TOLUIDINE, 3-CHLORO- 265
2-TOLUIDINE HYDROCHLORIDE 957
O-TOLUIDINE HYDROCHLORIDE **957**
O-TOLUIDYNA (POLISH) 956
TOLUILENODWUIZOCYJANIAN (POL-
    ISH) 952
TOLUOL 951
TOLUOLO 951
TOLU-SOL 951
TOLUYLENE-2,4-DIISOCYANATE 952
O-TOLYAMINE HYDROCHLORIDE 957
TOLYENE 2,4-DIISOCYANATE 952
TOLYENE-2,6-DIISOCYANATE 953
M-TOLYLAMINE 955
O-TOLYLAMINE 956
TOLYL CHLORIDE 156
O-TOLYLCHLORIDE 279
2,4-TOLYLENE DIISOCYANATE 952
TOLYLENE-2,4-DIISOCYANATE 2,4-
    TOLYLENEDIISOCYANAT E 952
TOMARIN 490
TONOX 664

TOPANOL A-SANTALOL 163
TOPENEL 306
TOPEX 154
TOPICHLOR 20 248
TOPICLOR 248
TOPICLOR 20 248
TOPSHOT 172
TOPZOL 842
TORDON 785
TORDON 10K 785
TORDON 22K 785
TORDON 101 MIXTURE 785
TORMONA 912
TORSITE 768
TOX 47 745
TOXADUST 958
TOXAFEEN (DUTCH) 958
TOXAKIL 958
TOXAPHEN (GERMAN) 958
TOXAPHENE **958**, lxiii
TOXASPRAY 958
TOXICHLOR 248
TOXILIC ACID 603
TOXILIC ANHYDRIDE 604
TOXOL (3) 745
TOXON 63 958
TOXYPHEN 958
TPN 278
TRANIMUL 341
TRANQDYN 341
TRANQUIRIT 341
TRANS-AID 87
TRANSAMINE 330, 912
TRANSMISSION FLUID lvi
TRAPEX 676
TRAPEXIDE 676
TRAVAD 137
TRAVEX 868
TREFICON 976
TREFLAN 976, lxii
TREFLANOCIDE ELANCOLAN 976
TREMOLITE lxxi
TREMOLITE (CAS 14567-73-8) 117
TREMOLITE ASBESTOS (CAS 14567-73-
    8) 117
TREOMICETINA 247
TREUPEL 758
TRI 964
TRI-6 146, 583
2,4,6-TRINITROTOLUOL (GERMAN) 988
TRIACETALDEHYDE (FRENCH) 741
TRIAD 964
TRIADIMEFON lxii, lxvi
TRIAETHYLAMIN (GERMAN) 970
TRIANGLE 313
TRIASOL 964
TRIATOMIC OXYGEN 737
1,4,7-TRIAZAHEPTANE 376
1,2,4-TRIAZIN-5-ONE,4-AMINO-6-TERT-
    BUTYL-3-(METHYLTHIO)- 683
AS-TRIAZIN-5(4H)-ONE,4-AMINO-6-TERT-
    BUTYL-3-(M ETHYLTHIO)- 683
1,2,4-TRIAZIN-5(4H)-ONE,4-AMINO-6-(1,1-
    DIMETHYLETHYL)-3-(ME THYLTHIO)-
    683
S-TRIAZINE, EXAHYDRO-1,3,5-TRINITRO-
    326
S-TRIAZINE-2,4,6(1H,3H,5H)-
    TRIONE,DICHLORO-, POTASSIUM
    DERIV 801
1,3,5-TRIAZINE-2,4,6(1H,3H,5H)-TRIONE, 1,3-
    DICHLORO-, POTASSIUM SALT 801

S-TRIAZINE-2,4,6(1H,3H,5H)-TRIONE, DI-
    CHLORO-, SODIUM SALT 873
1,3,5-TRIAZINE-2,4,6(1H,3H,5H)-TRIONE,
    1,3,5-TRICHLORO- 966
TRIAZOLAMINE 56
TRIBAC 914
TRIBASIC SODIUM PHOSPHATE 886
TRIBEN 914
TRIBROMMETHAAN (DUTCH) 181
TRIBROMMETHAN (GERMAN) 181
TRIBROMOMETAN (ITALIAN) 181
TRIBROMOMETHANE 181
TRIBUTILFOSFATO (ITALIAN) 960
TRIBUTON 330, 912
TRIBUTYL ALUMINUM **959**
TRI-N-BUTYL ALUMINUM 959
TRIBUTYLE (PHOSPHATE DE)
    (FRENCH) 960
TRIBUTYLFOSFAAT (DUTCH) 960
TRIBUTYLPHOSPHAT (GERMAN) 960
TRIBUTYL PHOSPHATE **960**
TRI-N-BUTYL PHOSPHATE 960
TRIBUTYLTIN lxiii
TRICALCIUM DIPHOSPHIDE 229
TRICARNAM 234
TRICHLOORAZIJNZUUR (DUTCH) 961
1,1,1-TRICHLOOR-2,2-BIS(4-
    CHLOORFENYL)-ETHAAN (DUTCH) 332
2,2,2-TRICHLOOR-1,1-BIS(4-
    CHLOORFENYL)-ETHANOL (DUTCH) 367
1,1,1-TRICHLOORETHAAN (DUTCH) 652
TRICHLOORETHEEN (DUTCH) 964
(2,4,5-TRICHLOOR-FENOXY)-AZIJNZUUR
    (DUTCH) 912
TRICHLOORMETHAAN (DUTCH) 262
TRICHLOORMETHYLBENZEEN
    (DUTCH) 157
TRICHLOR 274
TRICHLORACETIC ACID 961
1,1,1-TRICHLORAETHAN (GERMAN) 652
TRICHLORAETHEN (GERMAN) 964
1,1,1-TRICHLOR-2,2-BIS(4-CHLOR-PHENYL)-
    AETHAN (GERMAN) 332
2,2,2-TRICHLOR-1,1-BIS(4-CHLOR-PHENYL)-
    AETHANOL (GERMAN) 367
TRICHLORESSIGSAEURE (GERMAN) 961
1,1,2-TRICHLORETHANE 963
TRICHLORETHENE (FRENCH; ITAL-
    IAN) 964
2,4,6-TRICHLORFENOL (CZECH) 967
TRICHLORFON lxii, lxiv
TRICHLORINATED ISOCYANURIC
    ACID 966
TRICHLORMETHAN (CZECH) 262
TRICHLORMETHYLBENZOL (GER-
    MAN) 157
TRICHLOROACETALDEHYDE 245
TRICHLOROACETIC ACID **961**
TRICHLOROACETIC ACID SOLUTION 961
TRICHLOROALLYLSILANE 38, 211
TRICHLOROALUMINUM 40
3,5,6-TRICHLORO-4-AMINOPICOLINIC
    ACID 785
TRICHLOROARSINE 113
TRICHLOROBENZENE xliv
1,2,4-TRICHLOROBENZENE **962**
TRICHLOROBIS(4-
    CHLOROPHENYL)ETHANE 332
2,2,2-TRICHLORO-1,1-BIS(4-
    CHLOROPHENYL)-ETHANOL
    (FRENCH) 367
1,1,1-TRICHLORO-2,2-BIS(P-
    CHLOROPHENYL)ETHANE 332

Bold numbers denote main chemical entries; roman numerals denote text.

2,4,6-TRIMETHYL-1,3,5-TRIOXAAN (DUTCH) 741
2,4,6-TRIMETHYL-1,3,5-TRIOXANE 741
2,4,6-TRIMETHYL-S-TRIOXANE 741
S-TRIMETHYLTRIOXYMETHYLENE 741
3,5,5- TRIMETIL-2-CICLOESEN-1-ONE (ITALIAN) 549
2,4,6-TRIMETIL-1,3,5-TRIOSSANO (ITALIAN) 741
TRIMETION 385
TRINATRIUMPHOSPHAT (GERMAN) 886
2,4,6-TRINITOPHENOL 786
TRINITRIN 715
TRINITROBENZEEN (DUTCH) 987
TRINITROBENZENE 987
1,3,5-TRINITROBENZENE **987**
SYM-TRINITROBENZENE 987
TRINITROBENZENE, DRY 987
TRINITROBENZOL (GERMAN) 987
TRINITROCYCLOTRIMETHYLENE TRIAMINE 326
2,4,6-TRINITROFENOL (DUTCH) 786
TRINITROGLYCERIN 715
TRINITROGLYCEROL 715
TRINITROPHENYLMETHYLNITRAMINE 938
2,4,6-TRINITROPHENYLMETHYLNITR-AMINE 938
2,4,6-TRINITROPHENYL-N- METHYLNITR-AMINE 938
2,4,6-TRINITROTOLUEEN (DUTCH) 988
TRINITROTOLUENE **988**
2,4,6-TRINITROTOLUENE 988
SYM-TRINITROTOLUENE 988
TRINITROTOLUENE, WET 988
S-TRINITROTOLUOL 988
SYM-TRINITROTOLUOL 988
1,3,5-TRINITRO-1,3,5-TRIAZACYCLOHEXANE 326
1,3,5-TRINITRO-1,3,5-TRIZAZCYCLOHEXANE 326
TRINOXOL 912
TRIOXANE 484
TRIOXON 912
TRIOXONE 912
TRIOXYMETHYLENE 740
TRIPART SYSTEMIC INSECTICIDE 658
TRIPHENYL AMINE **989**
TRI-PLUS 964
TRIPOLI **990**
TRIPOTASSIUM TRICHLORIDE 798
TRIPROPYLALUMINUM **991**
TRIPROPYL ALUMINUM 991
TRIS **992**
TRIS(2,3-DIBROMOPROPYL) PHOSPHATE 992
TRIS(DIMETHYLAMINO)PHOSPHINE OXIDE 515
TRIS(DIMETHYLAMINO)PHOSPHORUS OXIDE 515
TRIS(DIMETHYLCARBAMODTHIOATO-S,S')IRON 472
TRIS(DIMETHYLDITHIOCARBAMA-TO)IRON 472
TRIS(N,N-DIMETHYLDITHIOCARBAMA-TO)IRON(111) 472
TRISODIUM ORTHOPHOSPHATE 886
TRISODIUM PHOSPHATE xxxi
TRISODIUM'-O-PHOSPHATE 886
TRISODIUM PHOSPHATE, TRIBASIC 886
TRISODIUM TRIFLUORIDE 876
TRISTAR 976
TRITHAC 607

TRITHION **993**
TRITHION MITICIDE 993
TRITOFTOROL 1045
TRITOL 988
TRIZLIN 711
TROCLOSENE POTASSIUM 801
TROJCHLOREK FOSFORU (POLISH) 782
TROJCHLOROBENZEN (POLISH) 962
TROJCHLOROETAN(1,1,2) (POLISH) 963
TROJNITROTOLUOL (POLISH) 988
TROLENE 843, lxii
TROMETE 886
TRONA 175, 867, 889
TRONAMANG 608
TRONOX UNITANE 0-110 949
TROTYL 988
TROTYL OIL 988
TROVIDUER 1006
TRUE AMMONIUM SULFIDE 83
TRUFLEX DOP 415
TRYSBEN 914
TRYSBEN 200 914
TS 219 742
TSIKLOMITSIN 925
TSINEB (RUSSIAN) 1045
TSIZP 34 559
TSP 886
TSPP 937
T-STUFF 528
TTD 416, 943
TTE 487
TTS 416
TUADS 943
TUBATOXIN 844
TUBOTHANE 607, lxii
TUEX 943
TUGON FLIEGENKUGEL 824
TULA-BASE FAST GARNET GB 48
TULABASE FAST GARNET GBC 48
TULISAN 943
TULUYLENDIISOCYANAT (GERMAN) 952
TUMBLEAF 868
TUNGSTEN **994**
TUNGSTEN FLUORIDE 995
TUNGSTEN HEXAFLUORIDE **995**
TURCAM 138, lxii
TURFCIDE lxii
TURFLON lxii
TURGEX 511
TURPENTINE **996**, xxxv, l, lii
TURPENTINE STEAM DISTILLED 996
TURPETH MINERAL 632
TURPS 996
TWIN LIGHT RAT AWAY 1012
TWINSPAN 282
TYRANTON 335
U 46 330, 912
UNSYM-TRICHLOROBENZENE 962
U-1149 489
U-3886 863
U-4224 395
U-5043 330
U-5227 108
U-5954 386
U 6020 815
U-6062 247
UC 7744 (UNION CARBIDE) 234
UC21149 (UNION CARBIDE) 28
UCAR BUTYLPHENOL 4-T 209
UCAR SOLVENT 2LM 412
UCET TEXTILE FINISH 11-74 (OBS.) 1007
U-COMPOUND 444
UCON 12 356
UCON 22 261

UCON 113 487
UCON 114 488
UCON FLUOROCARBON 113 487
UCON 12/HALOCARBON 12 356
UCON 22/HALOCARBON 22 261
UCON REFRIGERANT 11 965
U 46DP 330
ULTRACORTEN 815
ULTRACORTENE 815
ULTRAMARINE GREEN 288
ULTRAMARINE YELLOW 128
ULTRAWET 60K 875
ULTRAWET K 875
ULTRAWET KX 875
ULTRAWET SK 875
ULVAIR 687
UMBETHION 301
UMBRIUM 341
UN 2434 344
UNCON 113/HALOCARBON 113 487
UNDEN 824
UNDENE 824
UNIBARYT 137
UNICROP MANEB 607
UNIDRON 418
UNIFUME 451
UNIMYCETIN 247
UNIPON 366
UNISEDIL 341
UNITED CHEMICAL DEFOLIANT NO. 1 868
UNITOX 249
UNIVERM 241
UNOX EPOXIDE 206 1007
URACIL, 5-BROMO-3-SEC-BUTYL-6-METHYL 177
URAGAN 177
URAGON 177
UREA **997**
UREA,1-(1-NAPHTHYL)-2-THIO- 108
UREAPHIL 997
UREOPHIL 997
URETAN ETYLOWY (POLISH) 444
URETHAN 444
URETHANE 444, xxxv, xlii, cvii
URETHANE ALKYDS xxxvii
URETHANE RESINS xxxvii
UREVERT 997
URGENEA MARITIMA 842
UROX 177
UROX B 177
UROX B WATER SOLUBLE CONCENTRATE WEED KILLER 177
UROX D 418
UROX-HX 177
UROX HX GRANULAR WEED KILLER 177
URSOL BROWN RR 53
URSOL D 761
USAF A-4600 606
USAF AM-5 3
USAF AN-7 640
USAF B-15 941
USAF B-30 943
USAF B-33 416
USAF CB-21 940
USAF CB-22 693
USAF CB-27 840
USAF CB-35 942
USAF CY-2 222
USAF DO-41 992
USAF DO-45 1
USAF ED-1995 315
USAF EK-218 838
USAF EK-356 530
USAF EK-394 761

Bold numbers denote main chemical entries; roman numerals denote text.

USAF EK-488 9
USAF EK-497 559
USAF EK-2089 943
USAF EK-1597 428
USAF EK-1719 940
USAF EK-1995 315
USAF EK-P-433 87
USAF EK-P-583 489
USAF EK-P-5976 108
USAF EK-T-434 890
USAF P-5 943
USAF P-7 418
USAF P-220 839
USAF RH-8 8
USAF ST-40 634
USAF XR-22 56
USAF XR-42 418
USEMPAX AP 341
VAC 1004
VAL-DROP 868
VALENTINITE 107
VALEO 341
VALERAL 998
VALERALDEHYDE 998
N-VALERALDEHYDE **998**
VALERIANIC ALDEHYDE 998
VALERIC ACID ALDEHYDE 998
VALERIC ALDEHYDE 998
N-VALERIC ALDEHYDE 998
VALERONE 396
VALINE ALDEHYDE 545
VALITRAN 341
VALIUM 341
VALIUM R 341
VAMPIRINIP 1012
VANADATE (V031-), AMMONIUM 71
VANADIC ACID, AMMONIUM SALT 71
VANADIC ANHYDRIDE 1000
VANADIC OXIDE 1001
VANADIUM **999**, xxxix
VANADIUM CHLORIDE 1002
VANADIUM OXIDE 1000, 1001
VANADIUM(V) OXIDE FUME 1000
VANADIUM, OXYSULFATO (2-)- O- 1003
VANADIUM PENTOXIDE **1000**
VANADIUM PENTOXIDE DUST 1000
VANADIUM PENTOXIDE (FUME) 1000
VANADIUM SESQUIOXIDE **1001**
VANADIUM TETRACHLORIDE **1002**
VANADIUM TRIOXIDE 1001
VANADYL SULFATE **1003**
VANCIDA TM-95 943
VANCIDE 607
VANCIDE 89 233
VANCIDE FE95 472
VANCIDE MANEB 80 607
VANCIDE TM 943
VANDYKE BROWN 611
VANGARD K 233
VANLUBE PCX 163
VANOXIDE 154
VAPONA lxii
VAPOPHOS 745
VAPOTONE 918
VARIOFORM 73
VARIOFORM I 42, 73
VARIOFORM II 997
VARITOX 961
VARNOLINE 896
VAROX 394
VARSOL 690, 896
VASSGRO DSM 658
VASSGRO MANEX 607
VATERITE 219
VATRAN 341
VC 1006
VCL 1006
VCM 1006

VCN 24
V.C.S. 581
VCS-506 581
VDC 1010
VDF 1009
VEGABEN 246
VEGADEX 900
VEGADEX SUPER 900
VEGANINE 758
VEGEDEX 900
VEGETABLE CARBON 25
VEGFRU 770
VEGFRUFOSMITE 429
VEGFRU (INDIAN) 602
VEGFRU MALATOX 602
VEGIBEN 246
VELIUM 341
VELSICOL 104 503
VELSICOL 506 581
VELSICOL 1068 248
VELSICOL HEPTACHLOR 503
VELSICOL VCS 506 581
VENETIAN RED 538
VENTOX 24
VEON 912
VEON 245 912
VERAZINC 1043
VERGEMASTER 330
VERGFRU FORATOX 770
VERMITIN 760
VERMOESTRICID 241
VERSENE 450
VERSENE ACID 450
VERSNELLER NL 63/10 388
VERSTRAN 814
VERTAC 90% 958
VERTAC TOXAPHENE 90 958
VERTON 330
VERTON 2D 330
VERTON 2T 912
VERTON D 330
VERTRON 2D 330
VESTINOL AH 415
VESTROL 964
VETICOL 247
VETIOL 602
VICTOR TSPP 937
VIDDEN D 365
VIDEN 758
VIDON 638 330
VIENNA GREEN 298
VIKANE 909
VIKANE FUMIGANT 909
VILLIAUMITE 46, 876
VINEGAR xxxiii, xxxiv
VINEGAR ACID 5
VINEGAR NAPHTHA 432
VINICIZER 80 415
VINILE (ACETATO DI) (ITALIAN) 1004
VINILE (BROMURO DI) (ITALIAN) 1005
4-VINLYCYCLOHEXENE DIOXIDE 1007
VINYLACETAAT (DUTCH) 1004
VINYLACETAT (GERMAN) 1004
VINYL ACETATE **1004**, xxxvii, xcii, cvii
VINYL ACETATE H.Q. 1004
VINYL A MONOMER 1004
VINYLBENZEN (CZECH) 899
VINYLBENZENE 899
VINYLBENZOL 899
VINYLBROMID (GERMAN) 1005
VINYL BROMIDE **1005**
VINYL CARBINOL 31
VINYL CARBINOL,2-PROPENOL 31
VINYL CHLORIDE **1006**, lxiii, lxxxix, cvii
VINYL CHLORIDE MONOMER 1006
VINYL CHLORIDE MONOMERS xxxiii
VINYL CHLOROIDE 1006
VINYL C MONOMER 1006

VINYL CYANIDE 24
VINYL CYANIDE, PROPENENITRILE 24
VINYL CYCLOHEXENE DIEPOXIDE 1007
4-VINYLCYCLOHEXENE DIEPOXIDE 1007
4-VINYL-1-CYCLOHEXENE DIEPOX-
IDE 1007
4-VINYL-1,2-CYCLOHEXENE DIEPOX-
IDE 1007
VINYL CYCLOHEXENE DIOXIDE **1007**
4-VINYL-1-CYCLOHEXENE DIOXIDE 1007
1-VINYL-3-CYCLOHEXENE DIOXIDE 1007
VINYLE (ACETATE DE) (FRENCH) 1004
VINYLE (BROMURE DE) (FRENCH) 1005
VINYLETHYLENE 182
VINYL FLUORIDE **1008**
VINYLFORMIC ACID 23
VINYLIDENE DICHLORIDE 1010
VINYLIDENE FLUORIDE **1009**
VINYLIDINE CHLORIDE **1010**
VINYLIDENE CHLORIDE (II) 1010
VINYLPHARE 249
VINYLPHATE 249
VINYL RESINS xxxv, xxxvii, cx
VINYLSTYRENE 419
M-VINYLSTYRENE 419
VINYLTOLUENE 1011
VINYL TOLUENE **1011**
VINYL TRICHLORIDE 963
VIOLET 3 1013
VIRORMONE 919
VIROSTERONE 919
VISKO 330
VISKO-RHAP 330
VISKO-RHAP LOW DRIFT HERBICIDES 330
VISKO-RHAP LOW VOLATILE 41 330
VISKO RHAP LOW VOLATILE ESTER 912
VITON 146, 583
VITRAN 964
VITREOUS QUARTZ 852
VITREX 745
VITRIOL BROWN OIL 905
VITROL RED 538
VIVAL 341
VIVOL 341
VOFATOX 679
VOGEL'S IRON RED 538
VOLATILE OIL OF MUSTARD 36
VOLATILE ORGANIC COMPOUNDS xxxv,
lxix, lxx
VOLCAY 140
VOLCLAY 140
VONDALDHYDE 605
VONDCAPTAN 233
VONDRAX 605
VONDURON 418
VORLEX 676
VOROX 56
VORTEX 676
VT 1 947
VUAGT-1-4 943
VULCAFOR TMTD 943
VULKACIT MTIC 943
VULKACIT THIURAM 943
VULKACIT THIURAM/C 943
VULKANOX 4020 761
VULKAUT THIRAM 943
VYAC 1004
W 491 304
WALNUT STAIN 611
WAPNIOWY TLENEK (POLISH) 227
WARF lxii
WARFARIN **1012**, lxii
WARKEELATE ACID 450
WASH OIL 302
WASSERSTOSSPEROXIDE (GERMAN) 528
WATER GLASS 888
WATERSTOFPEROXYDE (DUTCH) 528
WATTLE GUM 500

WAZAKOL YELLOW NL 48
WEED-AG-BAR 330
WEEDAR 330, 912
WEEDAR-64 330
WEEDAR ADS 56
WEEDAZIN 56
WEEDAZOL 56
WEEDAZOL TL 87
WEED-B-GON 330, lxii
WEED DRENCH 31
WEEDEX GRANULAT 56
WEEDEZ WONDER BAR 330
WEED-FEED lxii
WEEDOCLOR 56
WEEDOL 743
WEEDONE 330, 749, 912
WEEDONE LV4 330
WEED-PRO lxii
WEED-RHAP 330, lxii
WEED TOX 330
WEEDTRINE-D 414
WEEDTROL 330
WEEVILTOX 238
WEGLA DWUSIARCZEK (POLISH) 238
WEGLA TLENEK (POLISH) 239
WEISS PHOSPHOR (GERMAN) 776
WEISSSPIESSGLANZ 107
WELDING GAS 15
WESTRON 923
WESTROSOL 964
WFNA 705
WHITE ARSENIC 111, 114, 119
WHITE BO 560
WHITE CAUSTIC 879
WHITE CAUSTIC, SOLUTION 879
WHITE COPPERAS 1043
WHITE FLOWER OF ZINC 1040
WHITE FUMING NITRIC ACID 705
WHITE LEAD 579
WHITE PHOSPHORIC ACID 775
WHITE PHOSPHORUS 776
WHITE SEAL-7 1040
WHITE SPIRITS 896
WHITE TAR 691
WHITE VITRIOL 1043
WIGRAINE 758
WIJS' CHLORIDE 534
WILKNITE 140
WILKONITE 140
WINTERWASH 403
WINYLU CHLORED (POLISH) 1006
WITCIZER 300 350
WITCIZER 312 415
WL 18236 639
WL 19805 172
WN 12 676
WOFATOS 679
WOFATOX 679
WOFOTOX 679
WOJTAB 815
WOLFRAM 994
WOOD ALCOHOL 637
WOOD CARBON 25
WOOD DUST liii
WOOD NAPHTHA 637
WOOD SPIRIT 637
WOOD TURPENTINE 996
WP 776
WURM-THIONAL 760
WURTZITE 1044
WY-3467 341
X-ALL 56
X-ALL LIQUID 56
XARIL 758

XAXA 20
XENENE 164
XENYLAMIN (CZECH) 51
XENYLAMINE 51
XERAC 154
XILENOLI (ITALIAN) 1018
XILIDINE (ITALIAN) 1025
XILOLI (ITALIAN) 1013
XL-50 760
XL ALL INSECTICIDE 703
XYLENE xliii, xliv, xlvi, lii, **1013**, lvii, cvii, cviii, cix, cx
1,2-XYLENE 1016
1,3-XYLENE 1014
1,4-XYLENE 1017
M-XYLENE **1014**
O-XYLENE **1016**
P-XYLENE **1017**, cvi
M-XYLENE-A,A'-DIAMINE 1015
M-XYLENE-ALPHA,ALPHA'-DIAMINE **1015**
XYLENEN (DUTCH) 1013
XYLENOL **1018**
1,2,5-XYLENOL 1021
1,3,5-XYLENOL 1024
2,3-XYLENOL **1019**
2,4-XYLENOL **1020**
2,5-XYLENOL **1021**
2,6-XYLENOL **1022**
3,4-XYLENOL **1023**
3,5-XYLENOL **1024**
O-XYLENOL 1019
M-XYLENOL 1020
P-XYLENOL 1021
3,5-XYLENOL, 4-(DIMETHYLAMINO)-,METHYLCARBAMATE 684
XYLENOLEN (DUTCH) 1018
3,5-XYLENOL,4-(METHYLTHIO)-, METHYL-CARBAMATE 638
XYLIDINE **1025**
2,4-XYLIDINE **1026**
M-XYLIDINE 1026
M-4-XYLIDINE 1026
XYLIDINEN (DUTCH) 1025
XYLOL 1013
M-XYLOL 1014
O-XYLOL 1016
P-XYLOL 1017
XYLOLE (GERMAN) 1013
M-XYLYLENDIAMIN (CZECH) 1015
YALTOX 235
YELLOW CROSS LIQUID 167
YELLOW FERRIC OXIDE 538
YELLOW MERCURIC OXIDE 620
YELLOW MERCURY IODIDE 631
YELLOW OXIDE OF IRON 538
YELLOW OXIDE OF MERCURY 620
YELLOW PHOSPHORUS 776
YELLOW PRECIPITATE 620
YELLOW ULTRAMARINE 221
YOHIMBAN-16-CARBOXYLIC ACID, 11,17-DIMETHOXY-18-(3,4,5-TRIMETHOXYBENXOYL)OXY-, METHYL ESTER 840
YOHIMBAN-16-CARBOXYLIC ACID, 11,17-DIMETHOXY-18-(3,4,5-TRIMETHOXYBENXOYL)OXY-, METHYL ESTER, (3BETA, 16BETA, 17ALPHA, 18BETA, 20ALPHA)- 840
YOSHINOX S 941
YPERITE 167
Z-78 1045
ZACLON DISCOIDS 524
ZACTIRIN COMPOUND 758
ZACTRAN 684

ZAHARINA 845
ZAHLREICHE BEZEICHNUNGEN (GERMAN) 403
ZAPRAWA NASIENNA UNIVERSAL 629
ZAPRAWA NASIENNA UNIVERSAL R 629
ZAPRAWA NASIENNA UNIWERSAL NA 629
ZB 112 1029
ZB 237 1029
ZEBENIDE 1045
ZEBTOX 1045
ZECTANE 684
ZECTRAN 684
ZEIDANE 332
ZELAZA THLENK (POLISH) 538
ZENADRID 815
ZEOLITE **1027**
ZERDANE 332
ZESET T 1004
ZETIFEX ZN 992
ZEXTRAN 684
ZIARNIK 765
ZIDAN 1045
ZIMATE 1045
ZINC **1028**, lvi, lxiii, cvii, cviii, cix
ZINC BLENDE 1044
ZINC BORATE **1029**
ZINC BROMIDE **1030**
ZINC BUTTER 1033
ZINC CARBONATE **1031**
ZINC CHLORATE **1032**
ZINC CHLORIDE **1033**
ZINC CHLORIDE FUME 1033
ZINC (CHLORURE DE) (FRENCH) 1033
ZINC CHROMATE **1034**
ZINC CHROMATE(VI) HYDROXIDE 1034
ZINC CHROMATE C 1034
ZINC CHROMATE O 1034
ZINC CHROMATE T 1034
ZINC CHROMATE Z 1034
ZINC CHROME 1034
ZINC CHROME (ANTI-CORROSION) 1034
ZINC CHROME YELLOW 1034
ZINC CHROMIUM OXIDE 1034
ZINC CYANIDE **1035**
ZINC DIBROMIDE 1030
ZINC DICHLORIDE 1033
ZINC DICYANIDE 1035
ZINC DIETHYL- 380
ZINC DIFORMATE 1038
ZINC DITHIONITE **1036**
ZINC DUST 1028
ZINC ETHIDE 380
ZINC ETHYL 380
ZINC ETHYLENEBIS (DITHIOCARBA-MATE) 1045
ZINC ETHYLENE-1,2-BISDITHIOCARBAMATE 1045
ZINC, (ETHYLENEBIS(DITHIOCARBAMATO))-1045
ZINC FLUORIDE **1037**
ZINC FLUORURE (FRENCH) 1037
ZINC FORMATE **1038**
ZINC HYDROSULFITE 1036
ZINC HYDROXYCHROMATE 1034
ZINCITE 1040
ZINC MONOSULFIDE 1044
ZINC MURIATE SOLUTION 1033
ZINC NAPHTHENATE xli
ZINC NITRATE **1039**
ZINCO (CLORURO DI) (ITALIAN) 1033
ZINCOID 1040
ZINC OXIDE **1040**, xxxix
ZINC OXIDE FUME 1040
ZINC PERMANGANATE **1041**

Bold numbers denote main chemical entries; roman numerals denote text.

Chemical Abstract Service (CAS) Number Index

CAS Number	Substance
82-28-0	1-AMINO-2-METHYL-ANTHRAQUINONE
83-79-4	ROTENONE
84-74-2	DIBUTYL PHTHALATE
85-00-7	DIQUAT
85-43-8	TETRAHYDROPHTHALIC ACID ANHYDRIDE
85-44-9	PHTHALIC ANHYDRIDE
86-88-4	ANTU
87-60-5	3-CHLORO-2-METHYLANILINE
87-63-8	6-CHLORO-2-TOLUIDINE
87-68-3	HEXACHLOROBUTADIENE
87-86-5	PENTACHLOROPHENOL
87-90-1	TRICHLOROISOCYANURIC ACID
88-06-2	2,4,6-TRICHLOROPHENOL
88-72-2	O-NITROTOLUENE
88-75-5	2-NITROPHENOL
88-89-1	PICRIC ACID
89-72-5	O-SEC-BUTYLPHENOL
90-04-0	O-ANISIDINE
90-43-7	O-PHENYLPHENOL
91-08-7	TOLUENE-2,6-DIISOCYANATE
91-20-3	NAPHTHALENE
91-22-5	QUINOLINE
91-59-8	2-NAPHTHYLAMINE
91-66-7	DIETHYL ANILINE
92-52-4	BIPHENYL
92-67-1	4-AMINODIPHENYL
92-84-2	PHENOTHIAZINE
92-87-5	BENZIDINE
92-93-3	4-NITROBIPHENYL
93-76-5	2,4,5-T ACID
94-36-0	BENZOYL PEROXIDE
94-75-7	2,4-D ACID
95-06-7	SULFALLATE
95-13-6	INDENE
95-47-6	O-XYLENE
95-49-8	O-CHLOROTOLUENE
95-50-1	1,2-DICHLOROBENZENE
95-53-4	O-TOLUIDINE
95-57-8	2-CHLOROPHENOL
95-63-6	1,2,4-TRIMETHYL BENZENE
95-65-8	3,4-XYLENOL
95-68-1	2,4-XYLIDINE
95-74-9	3-CHLORO-4-METHYLANILINE
95-79-4	5-CHLORO-2-METHYL ANILINE
95-85-2	2-AMINO-4-CHLOROPHENOL
95-87-4	2,5-XYLENOL
96-10-6	DIETHYLALUMINUM CHLORIDE
96-22-0	DIETHYL KETONE
96-33-3	METHYLACRYLATE
96-37-7	METHYL CYCLOPENTANE
96-45-7	ETHYLENE THIOUREA
96-69-5	4,4'-THIOBIS(6-TERT-BUTYL-M-CRESOL)
97-56-3	2-AMINO-5-AZOTOULENE
97-63-2	ETHYL METHACRYLATE
97-77-8	DISULFIRAM
97-88-1	BUTYL METHACRYLATE
97-93-8	TRIETHYLALUMINIUM
98-00-0	FURFURYL ALCOHOL
98-01-1	FURFURAL
98-07-7	BENZYL TRICHLORIDE
98-09-9	BENZENE SULFONYL CHLORIDE
98-12-4	CYCLOHEXYL TRICHLOROSILANE
98-13-5	PHENYL TRICHLOROSILANE
98-51-1	P-TERT-BUTYLTOLUENE
98-54-4	P-TERT-BUTYLPHENOL
98-82-8	CUMENE
98-87-3	BENZALCHLORIDE
98-88-4	BENZOYL CHLORIDE
98-95-3	NITROBENZENE
99-35-4	1,3,5-TRINITROBENZENE
99-59-2	5-NITRO-O-ANISIDINE
100-00-5	P-NITROCHLOROBENZENE
100-01-6	P-NITROANILINE

CAS Number	Substance
100-02-7	4-NITROPHENOL
100-37-8	DIETHYLAMINOETHANOL
100-39-0	BENZYL BROMIDE
100-41-4	ETHYLBENZENE
100-42-5	STYRENE
100-44-7	BENZYL CHLORIDE
100-47-0	BENZONITRILE
100-52-7	BENZALDEHYDE
100-57-2	PHENYLMERCURY HYDROXIDE
100-61-8	METHYLANILINE
100-63-0	PHENYLHYDRAZINE
100-65-2	PHENYLHYDROXYLAMINE
100-99-2	TRIISOBUTYL ALUMINIUM
101-14-4	4,4'-METHYLENE BIS (2-CHLOROANILINE)
101-61-1	4,4-METHYLENEBIS(N,N-DIMETHYL) ANILINE
101-77-9	4,4-METHYLENEDIANILINE
102-36-3	1,2-DICHLORO-4-PHENYL ISO-CYANATE
102-54-5	FERROCENE
102-67-0	TRIPROPYLALUMINUM
102-81-8	DIBUTYLAMINOETHANOL
103-09-3	ETHYLHEXYL ACETATE
103-23-1	BIS (2-ETHYLHEXYL) ADIPATE
103-69-5	ETHYLANILINE
103-71-9	PHENYL ISOCYANATE
104-94-9	P-ANISIDINE
105-46-4	SEC-BUTYL ACETATE
105-54-4	ETHYL BUTYRATE
105-57-7	ACETAL
105-67-9	2,4-XYLENOL
106-35-4	ETHYL BUTYL KETONE
106-42-3	P-XYLENE
106-46-7	1,4-DICHLOROBENZENE
106-48-9	4-CHLOROPHENOL
106-50-3	P-PHENYLENEDIAMINE
106-51-4	QUINONE
106-87-6	VINYL CYCLOHEXENE DIOXIDE
106-88-7	1,2-BUTYLENE OXIDE
106-89-8	EPICHLOROHYDRIN
106-90-1	GLYCIDYL ACRYLATE
106-92-3	ALLYL GLYCIDYL ETHER
106-93-4	ETHYLENE DIBROMIDE
106-95-6	ALLYL BROMIDE
106-97-8	BUTANE
106-99-0	1,3-BUTADIENE
107-00-6	ETHYL ACETYLENE
107-02-8	ACROLEIN
107-05-1	ALLYL CHLORIDE
107-06-2	1,2-DICHLORETHANE
107-10-8	PROPYLAMINE
107-13-1	ACRYLONITRILE
107-15-3	ETHYLENEDIAMINE
107-18-6	ALLYL ALCOHOL
107-19-7	PROPARGYL ALCOHOL
107-20-0	CHLOROACETALDEHYDE
107-21-1	ETHYLENE GLYCOL
107-27-7	ETHYLMERCURIC CHLORIDE
107-29-9	ACETALDEHYDE OXIME
107-30-2	CHLOROMETHYL METHYL ETHER
107-31-3	METHYL FORMATE
107-37-9	ALLYL TRICHLOROSILANE
107-41-5	HEXYLENE GLYCOL
107-49-3	TEPP
107-66-4	DIBUTYL PHOSPHATE
107-72-2	AMYLTRICHLOROSILANE
107-87-9	2-PENTANONE
107-92-6	BUTYRIC ACID
107-98-2	PROPYLENE GLYCOL MONO-METHYL ETHER
108-05-4	VINYL ACETATE
108-10-1	METHYL ISOBUTYL KETONE
108-11-2	METHYL ISOBUTYL CARBINOL
108-18-9	DIISOPROPYLAMINE
108-20-3	ISOPROPYL ETHER
108-21-4	ISOPROPYL ACETATE

CAS Number	Substance
108-24-7	ACETIC ANHYDRIDE
108-31-6	MALEIC ANHYDRIDE
108-38-3	M-XYLENE
108-43-0	3-CHLOROPHENOL
108-44-1	M-TOLUIDINE
108-46-3	RESORCINOL
108-57-6	DIVINYL BENZENE
108-67-8	1,3,5-TRIMETHYL BENZENE
108-68-9	3,5-XYLENOL
108-83-8	2,6-DIMETHYLHEPTANONE
108-87-2	METHYL CYCLOHEXANE
108-88-3	TOLUENE
108-90-7	CHLOROBENZENE
108-91-8	CYCLOHEXYLAMINE
108-93-0	CYCLOHEXANOL
108-94-1	CYCLOHEXANONE
108-95-2	PHENOL
109-59-1	ISOPROPOXYETHANOL
109-60-4	PROPYL ACETATE
109-65-9	BUTYL BROMIDE
109-66-0	PENTANE
109-69-3	BUTYL CHLORIDE
109-73-9	BUTYLAMINE
109-77-3	MALONONITRILE
109-79-5	BUTYL MERCAPTAN
109-86-4	ETHYLENE GLYCOL METHYL ETHER
109-86-4	METHYL CELLOSOLVE
109-87-5	METHYLAL
109-90-0	ETHYL ISOCYANATE
109-99-9	TETRAHYDROFURAN
110-12-3	METHYL ISOAMYL KETONE
110-16-7	MALEIC ACID
110-17-8	FUMARIC ACID
110-19-0	ISOBUTYL ACETATE
110-22-5	ACETYL PEROXIDE
110-43-0	METHYL AMYL KETONE
110-46-3	AMYL NITRITE
110-49-6	METHYL CELLOSOLVE ACE-TATE
110-54-3	HEXANE
110-62-3	N-VALERALDEHYDE
110-78-1	PROPYL ISOCYANATE
110-80-5	2-ETHOXYETHANOL
110-82-7	CYCLOHEXANE
110-83-8	CYCLOHEXENE
110-85-0	PIPERAZINE
110-86-1	PYRIDINE
110-91-8	MORPHOLINE
111-15-9	2-ETHOXYETHYL ACETATE
111-30-8	GLUTARALDEHYDE
111-40-0	DIETHYLENE TRIAMINE
111-42-2	DIETHANOLAMINE
111-44-4	DICHLOROETHYL ETHER
111-65-9	OCTANE
111-69-3	ADIPONITRILE
111-76-2	2-BUTOXYETHANOL
111-84-2	NONANE
111-87-5	OCTYL ALCOHOL
112-07-2	2-BUTOXYETHYL ACETATE
112-27-6	TRIETHYLENE GLYCOL
112-56-1	LETHANE
112-57-2	TETRAETHYLENEPENTAMINE
114-26-1	PROPOXUR
115-07-1	PROPENE
115-21-9	ETHYL TRICHLOROETHYLSI-LANE
115-25-3	OCTAFLUOROCYCLOBUTANE
115-32-2	DICOFOL
115-90-2	FENSULFOTHION
116-06-3	ALDICARB
116-14-3	TETRAFLUOROETHYLENE
117-52-2	FUMARIN
117-81-7	DI-SEC-OCTYL PHTHALATE
117-84-0	DIOCTYL PHTHALATE
118-52-5	1,3-DICHLORO-5,5-DIMETHYL HYDANTOIN
118-74-1	HEXACHLOROBENZENE

CAS Number Index

CAS Number	Substance
1310-58-3	POTASSIUM HYDROXIDE
1310-66-3	LITHIUM HYDROXIDE MONOHYDRATE
1310-73-2	SODIUM HYDROXIDE
1312-03-4	MERCURY OXIDE SULFATE
1312-73-8	POTASSIUM SULFIDE
1313-13-9	MANGANESE DIOXIDE
1313-27-5	MOLYBDENUM TRIOXIDE
1313-59-3	SODIUM MONOXIDE
1313-60-6	SODIUM PEROXIDE
1314-13-2	ZINC OXIDE
1314-22-3	ZINC PEROXIDE
1314-34-7	VANADIUM SESQUIOXIDE
1314-41-6	LEAD OXIDE, RED
1314-56-3	PHOSPHORUS PENTOXIDE
1314-62-1	VANADIUM PENTOXIDE
1314-80-3	PHOSPHORUS PENTASULFIDE
1314-87-0	LEAD SULFIDE
1314-98-3	ZINC SULFIDE
1317-34-6	MANGANESE(III) OXIDE
1317-65-3	CALCIUM CARBONATE
1317-95-9	TRIPOLI
1318-02-1	ZEOLITE
1319-77-3	CRESOL
1321-64-8	PENTACHLORONAPHTHALENE
1327-33-9	ANTIMONY TRIOXIDE
1327-53-3	ARSENIC OXIDE
1327-53-3	ARSENIC TRIOXIDE
1330-20-7	XYLENE
1332-07-6	ZINC BORATE
1332-21-4	ASBESTOS
1332-58-7	KAOLIN
1333-74-0	HYDROGEN
1333-82-0	CHROMIUM TRIOXIDE
1333-86-4	CARBON BLACK
1335-31-5	MERCURIC OXYCYANIDE
1335-32-6	LEAD SUBACETATE
1335-87-1	HEXACHLORONAPHTHALENE
1335-88-2	TETRACHLORONAPHTHALENE
1336-21-6	AMMONIUM HYDROXIDE
1336-36-3	POLYCHLORINATED BIPHENYL
1338-23-4	METHYL ETHYL KETONE PEROXIDE
1341-49-7	AMMONIUM BIFLUORIDE
1344-28-1	ALUMINUM OXIDE
1395-21-7	BACILLUS SUBTILIS BPN
1461-25-2	TETRABUTYLSTANNINE
1477-55-0	M-XYLENE-ALPHA,ALPHA'-DIAMINE
1563-66-2	CARBOFURAN
1582-09-8	TRIFLURALIN
1600-27-7	MERCURIC ACETATE
1675-54-3	BISPHENOL A DIGLYCIDYL ETHER
1746-01-6	TCDD
1762-95-4	AMMONIUM THIOCYANATE
1789-58-8	ETHYLDICHLOROSILANE
1795-48-8	ISOPROPYL ISOCYANATE
1836-75-5	NITROFEN
1897-45-6	CHLOROTHALONIL
1910-42-5	PARAQUAT DICHLORIDE
1912-24-9	ATRAZINE
1918-02-1	PICLORAM
1929-82-4	NITRAPYRIN
2032-65-7	METHIOCARB
2039-87-4	O-CHLOROSTYRENE
2179-59-1	ALLYL PROPYL DISULFIDE
2234-13-1	OCTACHLORONAPHTHALENE
2238-07-5	DIGLYCIDYL ETHER
2244-21-5	POTASSIUM DICHLOROISOCYANURATE
2303-16-4	DIALLATE
2425-06-1	CAPTAFOL
2426-08-6	BUTYL GLYCIDYL ETHER
2618-77-1	2,5-DIMETHYL-2,5-DI-(BENZOYLPEROXY) HEXANE
2699-79-8	SULFURYL FLUORIDE
2893-78-9	SODIUM DICHLORO-S-TRIAZINETRIONE
2921-88-2	CHLORPYRIFOS
2955-38-6	PRAZEPAM
2971-90-6	CLOPIDOL
3012-65-5	AMMONIUM CITRATE
3054-95-3	DIETHOXYPROPENE
3164-29-2	AMMONIUM TARTRATE
3165-93-3	4-CHLORO-2-TOLUIDINE HYDROCHLORIDE
3173-53-3	CYCLOHEXYL ISOCYANATE
3179-56-4	ACETYL CYCLOHEXANE SULFONYL PEROXIDE
3251-23-8	CUPRIC NITRATE
3333-52-6	TETRAMETHYLSUCCINONITRILE
3383-96-8	TEMEPHOS
3486-35-9	ZINC CARBONATE
3689-24-5	SULFOTEP
3811-04-9	POTASSIUM CHLORATE
4098-71-9	ISOPHORONE DIISOCYANATE
4170-30-3	CROTONALDEHYDE
4435-53-4	BUTOXYL
4484-72-4	DODECYL TRICHLOROSILANE
4685-14-7	PARAQUAT
5124-30-1	METHYLENE BIS(4-CYCLOHEXYLISOCYANATE)
5283-67-0	NONYL TRICHLOROSILANE
5307-14-2	4-AMINO-2-NITROANILINE
5329-14-6	SULFAMIC ACID
6032-29-7	2-PENTANOL
6341-97-5	2,4-DICHLOROPHENOL ACETATE ESTER
6423-43-4	PROPYLENE DINITRATE
6484-52-2	AMMONIUM NITRATE
6834-92-0	SODIUM SILICATE
6923-22-4	MONOCROTOPHOS
7209-38-3	BIS (AMINOPROPYL) PIPERAZINE
7428-48-0	LEAD STEARATE
7429-90-5	ALUMINUM
7439-92-1	LEAD
7439-93-2	LITHIUM
7439-95-4	MAGNESIUM
7439-96-5	MANGANESE
7439-97-6	MERCURY
7439-98-7	MOLYBDENUM
7440-01-9	NEON
7440-02-0	NICKEL
7440-06-4	PLATINUM
7440-09-7	POTASSIUM
7440-22-4	SILVER
7440-23-5	SODIUM
7440-28-0	THALLIUM
7440-31-5	TIN
7440-32-6	TITANIUM
7440-33-7	TUNGSTEN
7440-36-0	ANTIMONY
7440-38-2	ARSENIC
7440-39-3	BARIUM
7440-41-7	BERYLLIUM
7440-43-9	CADMIUM
7440-47-3	CHROMIUM
7440-48-4	COBALT
7440-50-8	COPPER
7440-55-3	GALLIUM
7440-59-7	HELIUM
7440-62-2	VANADIUM
7440-66-6	ZINC
7440-67-7	ZIRCONIUM
7440-70-2	CALCIUM
7440-74-6	INDIUM
7446-09-5	SULFUR DIOXIDE
7446-11-9	SULFUR TRIOXIDE
7446-14-2	LEAD SULFATE
7446-27-7	LEAD PHOSPHATE
7446-34-6	SELENIUM SULFIDE
7446-70-0	ALUMINUM CHLORIDE
7447-40-7	POTASSIUM CHLORIDE
7487-88-9	MAGNESIUM SULFATE
7487-94-7	MERCURIC CHLORIDE
7521-80-4	BUTYL TRICHLOROSILANE
7546-30-7	MERCUROUS CHLORIDE
7550-32-6	TITANIUM CHLORIDE
7553-56-2	IODINE
7558-79-4	SODIUM PHOSPHATE, DIBASIC
7572-29-4	DICHLOROACETYLENE
7580-67-8	LITHIUM HYDRIDE
7601-54-9	SODIUM PHOSPHATE
7601-89-0	SODIUM PERCHLORATE
7601-90-3	PERCHLORIC ACID
7616-94-6	PERCHLOROYL FLUORIDE
7631-86-9	SILICA, AMORPHOUS, HYDRATED
7631-89-2	SODIUM ARSENATE
7631-90-5	SODIUM BISULFITE
7631-99-4	SODIUM NITRATE
7632-51-1	VANADIUM TETRACHLORIDE
7637-07-2	BORON TRIFLUORIDE
7646-69-7	SODIUM HYDRIDE
7646-78-8	TIN(IV) CHLORIDE (1:4)
7646-85-7	ZINC CHLORIDE
7647-01-0	HYDROCHLORIC ACID
7647-18-9	ANTIMONY PENTACHLORIDE
7664-38-2	PHOSPHORIC ACID
7664-39-3	HYDROFLUORIC ACID
7664-41-7	AMMONIA
7664-93-9	SULFURIC ACID
7681-38-1	SODIUM BISULFATE
7681-49-4	SODIUM FLUORIDE
7681-52-9	SODIUM HYPOCHLORITE
7681-57-4	SODIUM METABISULFITE
7697-37-2	NITRIC ACID
7699-45-8	ZINC BROMIDE
7704-34-9	SULFUR
7705-08-0	IRON CHLORIDE
7718-54-9	NICKEL CHLORIDE
7719-12-2	PHOSPHORUS TRICHLORIDE
7720-78-7	FERROUS SULFATE
7722-64-7	POTASSIUM PERMANGANATE
7722-76-1	AMMONIUM PHOSPHATE DIBASIC
7722-84-1	HYDROGEN PEROXIDE
7722-88-5	TETRASODIUM PYROPHOSPHATE
7723-14-0	PHOSPHORUS
7723-14-0	PHOSPHORUS, RED
7726-95-6	BROMINE
7727-21-1	POTASSIUM PERSULFATE
7727-37-9	NITROGEN
7727-43-7	BARIUM SULFATE
7727-54-0	AMMONIUM PERSULFATE
7733-02-0	ZINC SULFATE
7738-94-5	CHROMIC ACID
7757-82-6	SODIUM SULFATE
7758-01-2	POTASSIUM BROMATE
7758-09-0	POTASSIUM NITRITE
7758-19-2	SODIUM CHLORITE
7758-94-3	FERROUS CHLORIDE
7758-97-6	LEAD CHROMATE
7758-98-7	CUPRIC SULFATE
7761-88-8	SILVER NITRATE
7772-98-7	SODIUM THIOSULFATE
7772-99-8	STANNOUS CHLORIDE
7773-06-0	AMMONIUM SULFAMATE
7774-29-0	MERCURIC IODIDE
7775-09-9	SODIUM CHLORATE
7775-11-3	SODIUM CHROMATE
7778-39-4	ARSENIC ACID
7778-50-9	POTASSIUM DICHROMATE
7778-54-3	CALCIUM HYPOCHLORITE
7778-74-7	POTASSIUM PERCHLORATE
7779-86-4	ZINC DITHIONITE
7779-88-6	ZINC NITRATE
7782-41-4	FLUORINE
7782-44-7	OXYGEN

7782-49-2	SELENIUM
7782-50-5	CHLORINE
7782-99-2	SULFUROUS ACID
7783-06-4	HYDROGEN SULFIDE
7783-28-0	AMMONIUM PHOSPHATE
7783-30-4	MERCURY(I) IODIDE
7783-35-9	MERCURIC SULFATE
7783-36-0	MERCUROUS SULFATE
7783-41-7	OXYGEN DIFLUORIDE
7783-46-2	LEAD FLUORIDE
7783-47-3	STANNOUS FLUORIDE
7783-49-5	ZINC FLUORIDE
7783-54-2	NITROGEN TRIFLUORIDE
7783-56-4	ANTIMONY TRIFLUORIDE
7783-60-0	SULFUR TETRAFLUORIDE
7783-66-6	IODINE PENTAFLUORIDE
7783-70-2	ANTIMONY PENTAFLUORIDE
7783-79-1	SELENIUM HEXAFLUORIDE
7783-80-4	TELLURIUM HEXAFLUORIDE
7783-82-6	TUNGSTEN HEXAFLUORIDE
7784-30-7	ALUMINUM PHOSPHATE
7784-34-1	ARSENIC TRICHLORIDE
7784-40-9	LEAD ARSENATE
7784-41-0	POTASSIUM ARSENATE
7784-42-1	ARSINE
7786-30-3	MAGNESIUM CHLORIDE
7786-34-7	PHOSDRIN
7786-81-4	NICKEL SULFATE
7787-32-8	BARIUM FLUORIDE
7787-36-2	BARIUM PERMANGANATE
7787-47-5	BERYLLIUM CHLORIDE
7787-49-7	BERYLLIUM FLUORIDE
7787-71-5	BROMINE TRIFLUORIDE
7788-98-9	AMMONIUM CHROMATE
7789-00-6	POTASSIUM CHROMATE
7789-06-2	STRONTIUM CHROMATE
7789-09-5	AMMONIUM DICHROMATE
7789-23-3	POTASSIUM FLUORIDE
7789-30-2	BROMINE PENTAFLUORIDE
7789-42-6	CADMIUM BROMIDE
7789-47-1	MERCURIC BROMIDE
7789-61-9	ANTIMONY TRIBROMIDE
7789-75-5	CALCIUM FLUORIDE
7790-69-4	LITHIUM NITRATE
7790-91-2	CHLORINE TRIFLUORIDE
7790-94-5	CHLOROSULFURIC ACID
7790-98-9	AMMONIUM PERCHLORATE
7790-99-0	IODINE MONOCHLORIDE
7803-51-2	PHOSPHINE
7803-52-3	STIBINE
7803-55-6	AMMONIUM METAVANADATE
7803-62-5	SILANE
8001-35-2	TOXAPHENE
8001-58-9	CREOSOTE
8002-74-2	PARAFFIN
8003-34-7	PYRETHRUM
8006-61-9	GASOLINE
8006-64-2	TURPENTINE
8008-20-6	KEROSENE
8014-95-7	OLEUM
8022-00-2	METHYL DEMETON
8030-30-6	NAPHTHA
8049-47-6	PANCREATIN
8052-41-3	STODDARD SOLVENT
8052-42-4	ASPHALT
8065-48-3	DEMETON
9000-01-5	GUM ARABIC
9000-92-4	DIASTASE
9001-37-0	GLUCOSE OXIDASE
10022-31-8	BARIUM NITRATE
10024-97-2	NITROUS OXIDE
10025-87-3	PHOSPHORUS OXYCHLORIDE
10025-91-9	ANTIMONY TRICHLORIDE
10025-99-7	PLATINOUS POTASSIUM CHLO-RIDE
10026-06-9	STANNIC CHLORIDE PENTAHY-DRATE

10026-13-8	PHOSPHORUS PENTACHLO-RIDE
10028-15-6	OZONE
10028-22-5	FERRIC SULFATE
10031-13-7	LEAD ARSENITE
10034-81-8	MAGNESIUM PERCHLORATE
10034-85-2	HYDRIODIC ACID
10034-93-2	HYDRAZINE SULFATE
10035-10-6	HYDROGEN BROMIDE
10042-76-9	STRONTIUM NITRATE
10043-01-3	ALUMINUM SULFATE
10043-35-3	BORIC ACID
10045-89-3	FERROUS AMMONIUM SUL-FATE
10045-94-0	MERCURIC NITRATE
10049-04-4	CHLORINE DIOXIDE
10099-74-8	LEAD NITRATE
10101-53-8	CHROMIC SULFATE
10101-63-0	LEAD IODIDE
10102-43-9	NITRIC OXIDE
10102-44-0	NITROGEN DIOXIDE
10124-37-5	CALCIUM NITRATE
10137-69-6	CYCLOHEXENYL TRICHLOROSI-LANE
10137-74-3	CALCIUM CHLORATE
10196-04-0	AMMONIUM SULFITE
10290-12-7	COPPER ARSENITE
10294-33-4	BORON TRIBROMIDE
10294-40-3	BARIUM CHROMATE(VI)
10326-21-3	MAGNESIUM CHLORATE
10361-37-2	BARIUM CHLORIDE
10361-95-2	ZINC CHLORATE
10377-60-3	MAGNESIUM NITRATE
10377-66-9	MANGANESE NITRATE
10415-75-5	MERCUROUS NITRATE
10421-48-4	FERRIC NITRATE
10588-01-9	SODIUM DICHROMATE
11103-86-9	POTASSIUM ZINC CHROMATE
11138-49-1	SODIUM ALUMINUM OXIDE
12001-26-2	SILICA, MICA
12002-03-8	COPPER ACETOARSENITE
12075-68-2	ETHYL ALUMINUM SESQUICHLORIDE
12108-13-3	MANGANESE TRICARBONYL METHYLCYCLOPENTADIENYL
12122-67-7	ZINEB
12124-99-1	AMMONIUM SULFIDE, SOLID
12125-01-8	AMMONIUM FLUORIDE
12125-02-9	AMMONIUM CHLORIDE
12135-76-1	AMMONIUM SULFIDE, SOLU-TION
12136-45-7	POTASSIUM OXIDE
12427-38-2	MANEB
12542-85-7	METHYL ALUMINUM SESQUICHLORIDE
13106-76-8	AMMONIUM MOLYBDATE
13138-45-9	NICKEL(II) NITRATE
13426-91-0	CUPRIETHYLENEDIAMINE
13446-10-1	AMMONIUM PERMANGANATE
13463-39-3	NICKEL CARBONYL
13463-40-6	IRON PENTACARBONYL
13463-67-7	TITANIUM DIOXIDE
13465-95-7	BARIUM PERCHLORATE
13473-90-0	ALUMINUM NITRATE
13477-00-4	BARIUM CHLORATE
13477-10-6	BARIUM HYPOCHLORITE
13530-65-9	ZINC CHROMATE
13548-38-4	CHROMIUM NITRATE
13597-99-4	BERYLLIUM NITRATE
13765-19-0	CALCIUM CHROMATE
13770-96-2	SODIUM ALUMINUM HYDRIDE
13779-41-4	DIFLUOROPHOSPHORIC ACID
13814-96-5	LEAD FLUOBORATE
13820-41-2	AMMONIUM TETRACHLORO-PLATINATE
13823-29-5	THORIUM NITRATE
13838-16-9	ENFLURANE
13967-90-3	BARIUM BROMATE
14452-57-4	MAGNESIUM PEROXIDE

14464-46-1	SILICA, CRYSTALLINE-CRISTOBALITE
14484-64-1	FERBAM
14763-77-0	COPPER CYANIDE
14807-96-6	TALC
14808-60-7	SILICA, QUARTZ
14977-61-8	CHROMIUM OXYCHLORIDE
15096-52-3	ALUMINUM SODIUM FLUORIDE
15663-27-1	CISPLATIN
15699-18-0	NICKEL AMMONIUM SULFATE
15825-70-4	MANNITOL HEXANITRATE
15829-53-5	MERCUROUS OXIDE
16111-62-9	DI(2-ETHYLHEXYL) PEROXYDI-CARBONATE
16219-75-3	ETHYLIDENE NORBORNENE
16752-77-5	METHOMYL
16853-85-3	LITHIUM ALUMINUM HYDRIDE
16871-90-3	POTASSIUM FLUOROSILICATE
16893-85-9	SODIUM FLUOROSILICATE
16919-58-7	AMMONIUM CHLOROPLATI-NATE
16921-30-5	POTASSIUM HEXACHLOROPLA-TINATE(IV)
16941-12-1	CHLOROPLATINIC ACID
16984-48-8	FLUORIDE
17702-41-9	DECABORANE
17804-35-2	BENOMYL
18414-36-3	DIBENZYLDICHLOROSILANE
18810-58-7	BARIUM AZIDE
18972-56-0	MAGNESIUM SILICOFLUORIDE
19287-45-7	DIBORANE
19624-22-7	PENTABORANE
20816-12-0	OSMIUM TETROXIDE
20830-81-3	DAUNOMYCIN
20859-73-8	ALUMINUM PHOSPHIDE
21087-64-9	METRIBUZIN
21351-79-1	CESIUM HYDROXIDE
21609-90-5	LEPTOPHOS
21645-51-2	ALUMINUM HYDROXIDE
21725-46-2	BLADEX
21908-53-2	MERCURIC OXIDE
22224-92-6	FENAMIPHOS
22781-23-3	BENDIOCARB
23103-98-2	PIRIMICARB
23414-72-4	ZINC PERMANGANATE
23950-58-5	PROPYZAMIDE
25013-15-4	VINYL TOLUENE
25154-54-5	DINITROBENZENE
25155-30-0	SODIUM DODECYLBENZENE SULFONATE
25167-67-3	1-BUTENE
25231-46-3	TOLUENE SULPHONIC ACID
25550-58-7	DINITROPHENOL
25551-13-7	TRIMETHYLBENZENE
25639-42-3	METHYL CYCLOHEXANOL
26249-12-7	DIBROMOBENZENE
26571-79-9	CHLOROPHENYL TRICHLORO-SILANE
26628-22-8	SODIUM AZIDE
26952-21-6	ISOOCTYL ALCOHOL
27134-26-5	CHLOROANILINE
27137-85-5	DICHLOROPHENYL TRI-CHLOROSILANE
27156-03-2	DICHLORODIFLUORO-ETHYLENE
27176-87-0	DODECYLBENZENESULFONIC ACID
27323-41-7	TRIETHANOLAMINE DODECYL-BENZENESULFONATE
27598-85-2	AMINOPHENOL
27774-13-6	VANADYL SULFATE
27987-06-0	TRIFLUOROETHANE
28300-74-5	ANTIMONY POTASSIUM TAR-TRATE
28805-86-9	BUTYL PHENOL
30031-64-2	2,3-BUTANEDIONE
30525-89-4	PARAFORMALDEHYDE
34590-94-8	DIPROPYLENE GLYCOL METH-YL ETHER

35400-43-2	SULPROFOS
39156-41-7	2,4-DIAMINOANISOLE SULFATE
39404-03-0	MAGNESIUM SILICIDE
53220-22-7	DIMYRISTYL PEROXYDICAR-
	BONATE

55720-99-5	CHLORINATED DIPHENYL
	OXIDE
58164-88-8	ANTIMONY LACTATE
60616-74-2	MAGNESIUM HYDRIDE
60676-86-0	SILICA, FUSED

61789-51-3	COBALT NAPHTHENATE
64093-79-4	CHROMOSULFURIC ACID
64365-11-3	ACTIVATED CARBON
68476-85-7	LIQUIFIED PETROLEUM GAS
68855-54-9	DIATOMACEOUS EARTH

Resource Index

Hazox, Inc. **2572**
HazTECH News **2278**
HazTech Transfer **2279**
Health Administration; U.S. Occupational Safety and - U.S. Department of Labor **1314**
Health; American Council on Science and **1534**
Health and Consolidated Laboratories; North Dakota Department of - Environmental Health Section **1449**
Health and Environmental Control; South Carolina Department of - Division of Environmental Quality Control **1471**
Health and Environmental Review Division - Office of Toxic Substances - U.S. Environmental Protection Agency (EPA) **1210**
Health and Environmental Sciences; Montana Department of - Environmental Sciences Division **1425**
Health and Environmental Service; Northern Mariana Islands Public - Environmental Quality Division **1318**
Health and Environment; Kansas Department of **1387**
Health and Environment; New Mexico Department of - Environmental Improvement Division **1437**
Health and Hazards Update; Industrial **2284**
Health and Human Services Organizations; National Coalition of Hispanic **1713**
Health and Natural Resources; North Carolina Department of Environment, Division of Environmental Management **1445**
  Division of Solid Waste Management **1446**
Health and Rehabilitative Services; Florida Department of **1362**
Health and Safety Division; Chemical - American Chemical Society **1093**
Health and Safety Letter; Occupational **2310**
Health and Safety News; Canadian Occupational **2193**
Health and Safety Software Version 3.0; Directory of Occupational **1975**
Health and Safety; Wyoming Department of Occupational **1512**
Health and Social Services; Alaska Department of - Division of Public Health **1325**
Health and Social Services; Wisconsin Department of - Division of Health - Section of Radiation Programs **1507**
Health and the Environment; Chemical Deception: The Toxic Threat to **1929**
Health and the Environment; Not in Our Backyards! Community Action for **2061**
Health, and the Environment; Toxic Chemicals, **2106**
Health and Toxicology; Some Publicly Available Sources of Computerized Information on Environmental **2098**
Health and Welfare; Idaho Department of - Division of Environmental Quality **1371**
Health; Arkansas Department of - Division of Radiation Control and Emergency Management **1335**
Health; The Bhopal Syndrome: Pesticides, Environment and **1919**
Health Choices; Environmental **2230**
Health; Colorado Department of **1347**
Health Council; American Industrial **1537**
Health; Department of Environmental - University of Cincinnati **1616**
Health Department; Utah Occupational Safety and **1486**
Health Detective's Handbook: A Guide to the Investigation of Environmental Health Hazards by Nonprofessionals **2034**

Health Division; Oregon Occupational Safety and - Oregon Department of Insurance and Finance **1463**
Health Effects Division - Office of Pesticide Programs - U.S. Environmental Protection Agency (EPA) **1211**
Health Effects Institute—Asbestos Research (HEI-AR) **1662**
Health Effects Laboratory; Air Pollution - University of California—Irvine **1519**
Health Effects of Environmental Pollutants; Abstracts on **2122**
Health Effects of Lawn Care Pesticides Task Force (HELP) **1663**
Health; The Effects of Pesticides on Human **1977**
Health Effects of Toxic Substances; The Toxics Directory: References and Resources on the **2107**
Health; Family Safety and **2247**
Health Foundation; California Public **1566**
Health Fund; Workplace **1891**
Health Guidelines; How to Meet OSHA's Safety and **2282**
Health; Hawaii Department of - Environmental Protection and Health Services Division **1367**
Health Hazardous Substance Fact Sheets; New Jersey Department of **2054**
Health; Hazardous Substances and Public **2269**
Health Hazards Manual for Artists, 3rd ed. **2035**
Health Hazards; The Nontoxic Home: Protecting Yourself and Your Family from Everyday Toxics and **2060**
Health Hotlines: 800 Numbers from DIRLINE **2036**
Health; Illinois Department of Public - Office of Health Regulation **1376**
Health; Indiana State Board of - Division of Industrial Hygiene and Radiological Health **1382**
Health Information Center; National - Office of Disease Prevention and Health Promotion (ODPHP) - U.S. Department of Health and Human Services **1214**
Health Information Clearinghouse; National - Office of Disease Prevention and Health Promotion - U.S. Department of Health and Human Services **1096**
Health; Institute of Environmental and Industrial - University of Michigan **1674**
Health; Institute of Toxicology and Environmental - University of California—Davis **1677**
Health; Iowa Department of Public - Radiological Health Unit **1385**
Health; Job Safety and **2292**
Health; Los Angeles Committee on Occupational Safety and **1690**
Health; Massachusetts Department of Public **1407**
Health; Michigan Department of Public - Bureau of Environmental and Occupational Health **1411**
Health; Minnesota Department of - Environmental Health Division **1413**
Health; Mississippi Department of - Division of Radiological Health **1418**
Health; Missouri Department of - Bureau of Radiological Health **1422**
Health; National Institute for Occupational Safety and - Centers for Disease Control - U.S. Department of Health and Human Services (DHHS) **1216**
Health; Nebraska Department of **1428**
Health Network; Environmental **1633**
Health News; Environmental **2231**
Health News; NCOSH Safety and **2302**
Health; New York Committee on Occupational Safety and **1743**

Health; New York Department of - Environmental Health Center **1442**
Health; Ohio Department of - Division of Technical Environmental Health Services - Radiological Health Program **1451**
Health; Oklahoma Department of - Environmental Health Services **1454**
Health; Oklahoma State Department of - Department of Pollution Control **1457**
Health. Part B: Pesticides, Food Contaminants, and Agricultural Wastes; Journal of Environmental Science and **2159**
Health. Part C: Environmental Carcinogenesis Reviews; Journal of Environmental Science and **2160**
Health Planning Center; Golden Empire **1652**
Health Policy; Public Voice for Food and **1772**
Health Program; Labor Occupational - University of California at Berkeley **1687**
Health Program; Labor Occupational Safety and - University of California at Los Angeles **1688**
Health Project; Alaska **1525**
Health Reporter; Occupational Safety and **2311**
Health Resource Guide; Occupational **2062**
Health; Rhode Island Department of - Division of Occupational Health and Radiation Control **1470**
Health Risks and Reducing Exposure; Hazardous Substances in Our Environment: A Citizen's Guide to Understanding **2030**
Health Risks of Chemicals in Our Environment; Calculated Risks: Understanding the Toxicity and Human **1924**
Health; Rocky Mountain Center for Occupational and Environmental - University of Utah **1778**
Health Science Institute; Environmental and Occupational - University of Medicine and Dentistry of New Jersey **1624**
Health Sciences Center; Environmental - Oregon State University **1634**
Health Sciences Centers; Association of University Environmental **1556**
Health Sciences Research Laboratory; Environmental - Tulane University **1635**
Health Service; Connecticut Department of - Health Systems Regulation Bureau **1349**
Health Services; California Department of **1341**
Health Services; Washington Department of Social and - Environmental Protection Section - Office of Radiation Protection **1501**
Health Strategies; National Center for Environmental **1710**
Health; Texas Department of - Environmental and Consumer Health Protection **1482**
Health; Vermont Department of **1489**
Health; Virginia Department of - Division of Health Hazards Control **1494**
Health; West Virginia Bureau of Public - Environmental Health Office - Industrial Hygiene Division **1502**
Health; Wyoming Division of - Division of Health and Medical Services - Radiological Health Services **1513**
The Healthy Home: An Attic to Basement Guide to Toxin-Free Living **2037**
Hemisphere Publishing Corp. **2402**
Hennepin Regional Poison Center - Hennepin County Medical Center **1147**
Herbicide Coalition; Minnesota **1702**
Herbicide Handbook, 6th ed. **2038**
Herbicide Task Force; Humboldt **1666**
Hispanic Health and Human Services Organizations; National Coalition of **1713**

Glossary and Appendix

Glossary of Terms

absolute 1) Also known as the Kelvin scale, the temperature scale used in engineering and scientific research; at its theoretical zero point, -273.15°C or -459.67°F, molecular motion ceases and substances possess minimal energy. 2) A chemical material free or relatively free from impurities.

absorbent material Commercially packaged clay, kitty litter, or other material used to soak up liquid hazardous materials.

absorption Penetration of a substance across a biologic barrier (such as skin) into either the lymphatic system or bloodstream. *See also:* **adsorption.**

acaricide A chemical substance used to kill ticks and mites.

ACBM Asbestos-containing building material.

accident An unplanned energy transfer causing property damage and/or human injury. *See also:* **incident.**

accumulative effect The effect a chemical substance has on a biologic system when the substance is absorbed at a rate exceeding the body's ability to eliminate it from the system. Excessive accumulation of a substance can lead to toxicity.

acid Any compound containing hydrogen replaceable by metals, and having a pH of 0-6. Strong acids in the pH range of 0-2 are corrosive and will cause chemical burns to the skin, eyes, and mucous membranes. Acids turn litmus red.

acid gas A gas that forms an acid when dissolved in water.

acidosis A pathologic condition resulting from accumulation of acid in or loss of base from the body.

ACL Alternative concentration limits.

acro-osteolysis Progressive destruction of bone tissue at the tip of the fingers and toes.

active ingredient The component that actually performs the primary function of a product. Products generally contain both active and inert ingredients, and both may be harmful. Active ingredients are listed on product labels as percentage by weight or as pounds per gallon of concentrate.

acute Severe symptoms and a rapid change to an organism leading to a crisis in a relatively short period of time, measured in seconds, minutes, hours, or days, following exposure to a health hazard.

acute effect An adverse health effect that usually occurs rapidly, sometimes immediately, as a result of a single, short significant exposure to a health hazard, without implying a degree of severity; may include irritation, corrosiveness, narcosis, and death.

acute exposure A single exposure to a toxic substance that results in death or severe biological harm; are characterized as lasting no longer than one day.

acute toxicity Adverse health effects occurring within a short period of time following exposure, usually a single dose, to a substance. *See also:* **acute effect.**

ADI Acceptable daily intake.

adsorption Attachment of the molecules of a gas or liquid to the surface of another substance, called the adsorbent. *See also:* **absorption.**

aerosol Suspension of fine liquid or solid particles in a gas, with the particle size often being in the 0.01-100 microns range.

AFFF (aqueous film forming foam) An agent designed to extinguish burning liquids.

aflatoxin Cancer-causing molds and liver-killing poisons widely occurring in foodstuffs such as grains, peanuts, and other foods infected with *aspergillus flavus.*

AICS (Australian Inventory of Chemical Substances) A list of chemicals allowed to be used commercially in Australia.

albuminuria The presence of albumin and other serum proteins in the urine; possible symptom of inorganic mercury poisoning.

aliphatic Pertaining to an open-chain hydrocarbon compound; substances such as methane and ethane are typical aliphatic hydrocarbons.

alkali Any acid-destroying compound having a pH of 8-14. Strong alkalis (or bases) in the pH range of 12-14 are considered corrosive and will cause chemical burns to the skin, eyes, and mucous membranes. Alkalis turn litmus blue.

alkaloid An organic nitrogen base of vegetable origin, usually toxic.

allergen A substance that causes the body to produce an antibody and that results in an allergy in hypersensitive people.

allergy A hypersensitive reaction of body tissues to specific substances which, in similar concentrations and circumstances, do not affect other persons. Allergic reactions in the workplace tend to affect the skin (*See:* **dermatitis**) and lungs (*See:* **asthma**).

alopecia Usually temporary partial or total loss of hair on the scalp or other parts of the body.

ALR (Allergenic effects) Systemic reactions such as allergies resulting from a sensitivity to penicillin or bee stings.

ambient An environment's surrounding conditions.

Ames Assay A biochemical screening test capable of detecting mutagenic activity. Also known as the Ames test.

amines Ammonia compounds.

amnesia Total or partial memory loss.

analgesia Insensibility to pain without loss of consciousness.

anemia A deficiency of the blood caused by reduced red blood cell count or a reduction in the amount of hemoglobin per unit volume of blood.

anesthesia Total or partial loss of sensation with or without loss of consciousness.

angina pectoris Pain in chest caused by inadequate supply of blood to the heart.

anhydrous A substance that does not contain water.

anorexia Loss or reduction of appetite for food.

anosmia Total or partial loss of the sense of smell.

anoxia A reduction in the quantity of oxygen supplied by blood to cells or tissues.

antidote A remedy to relieve, prevent, or counteract the effects of a poison; that which counteracts anything noxious.

anuria Absence of urine in the bladder caused by the failure of the kidneys to produce urine; is a possible symptom of chlorate or inorganic mercury poisoning.

apathy Reduced emotions with lack of interest in outside stimuli.

aphicide A chemical substance used to destroy aphids.

aplastic anemia A blood problem in which the bone marrow can no longer produce blood cells.

apnea Temporary cessation of breathing; a possible symptom of poisoning.

aquatic toxicology (AQTX) A branch of toxicology that deals with water pollution and its ecological effects.

aromatic compound A molecular ring structure hydrocarbon compound characterized by the presence of the benzene nucleus.

arrhythmia Disturbed heartbeat.

article A manufactured item which: 1) is formed to a specific shape or design during manufacture; 2) has end use function(s) dependent in whole or in part upon its shape or design during end use; and 3) does not release, or otherwise result in exposure to a hazardous chemical, under normal conditions of use (OSHA).

asbestosis A pulmonary disease resulting from inhaling airborne fibrous dust of the mineral asbestos; also known as asbestos pneumoconiosis.

asphyxia Difficulty in breathing or respiratory arrest; suffocation.

asphyxiant A substance, usually a vapor or gas, that can cause suffocation, unconsciousness, or death by preventing the blood from carrying oxygen. Most simple asphyxiants (which have no inherent toxicity) are harmful to the body only when they become so concentrated that oxygen in the air is reduced (normally about 21%) to dangerous levels (18% or lower).

asthenia Reduced physical and psychological strength.

asthma Respiratory problem caused by contraction of air passages and characterized by attacks of wheezing, shortness of breath, and/or coughing.

ataxia Loss or failure of muscular coordination, voluntary movement, or muscle control; may be a symptom of poisoning.

atm Atmosphere.

atrophy A loss of weight, volume, and activity of an organ, tissue, or cell; shrinkage.

at wt Atomic weight.

autoignition temperature (autoign temp) The minimum temperature at which a substance will ignite spontaneously, or cause self-sustained combustion, in the absence of any heated element, spark, or flame. The closer the auto-ignition temperature is to room temperature, the greater the risk of fire.

BACT Best available control technology.

bactericide A pesticide used to control bacteria.

base A substance that reacts with acids to form salts and water. All bases create solutions having a pH of more than 7, the neutral point, and may be corrosive to skin and other human tissue. The terms **alkali** and **caustic** are closely related in meaning.

BAT Best available technology.

Baumé (Be) An arbitrary scale introduced by the French chemist Antoine Baumé ca. 1800 for use in determining the specific gravity of liquids and in the gradation of hydrometers.

BCM (blood blotting mechanism effects) Any effect that increases or decreases clotting time.

BEI (biological exposure index) The maximum recommended value of a substance in blood, urine, or exhaled air.

bile A yellow-green, bitter fluid secreted by the liver; also called gall.

benign Not harmful.

black lung disease A lung disease caused by inhaling coal dust over a long period of time.

blasting agents A material designed for blasting which has been tested and found to be so insensitive to heat and shock that there is very little probability of accidental initiation to explosion or of transition from deflagration to detonation.

BLD (blood effects) The effect on all blood elements including oxygen carrying or releasing capacity, pH, protein, and electrolytes.

blepharospasm Abnormal contraction of eyelid muscles.

BOD (biological oxygen demand) A test that measures the dissolved oxygen consumed by microbial life while assimilating and oxidizing the organic matter present in organic waste discharges; allows the calculation of the effect of the discharges on the oxygen resources of the receiving water.

boiling point (bp) The temperature at which a product changes from a liquid to a vapor at normal atmospheric pressure.

BPR (blood pressure effects) Any effect that increases or decreases blood pressure.

bradycardia A decrease in the heartbeat rate to less than 60 beats per minute.

b range Boiling range.

breakthrough time The time from initial chemical contact to detection.

breathing zone sample An air sample collected from the area around the nose of a worker to assess exposure to airborne contaminants.

British Anti-Lewisite (BAL) An antidote to treat toxic inhalations of Lewisite, heavy metals, and organic arsenicals; 2,3-dimercaptopropanol; dimercaprol; 1,2-diethiglycerol.

British Thermal Unit (BTU) The quantity of heat required to raise the temperature of one pound of water one degree Fahrenheit, usually at 39-40°F.

bronchitis Inflammatory condition of the bronchia or bronchial tubes, the larger air passages of the lungs.

bronchoconstriction Contraction with narrowing of bronchia.

bronchospasm Spasmodic contraction of the muscles surrounding the bronchia.

brown lung disease A disease of the lungs caused by inhaling cotton dust from textile mills over a period of time.

buddy system A system of organizing employees into work groups so that each employee is designated to be observed by at least one other work group member, to provide rapid assistance to employees in the event of emergency.

bulk density Mass of powdered or granulated solid material per unit of volume.

by-product Any material, other than the principal product, that is generated as a consequence of an industrial process.

byssinosis Affliction of the lung from chronic inhalation of the dust from cotton, flax, and soft hemp; brown lung disease.

canister A personal air cleaning device usually worn by the user that contains sorbents, catalysts, or other materials designed to filter gases, vapors, and liquid and solid particles from air drawn through it. Also known as a cartridge.

CARC Carcinogenic effect.

carcinogen (CAR) Any substance causing the promotion or initiation of malignant or benign neoplasia (cancer) in humans or animals.

carcinoma A malignant tumor; a type of cancer.

cardiovascular Referring to the heart and blood vessel system.

CAS (Chemical Abstract Service) number A unique identifying number assigned to a chemical by this arm of the American Chemical Society. A given chemical has only one CAS number but may have more than one name. Also known as CAS registry number or CAS RN.

catalyst A substance that affects the rate of a chemical reaction but is neither changed nor consumed by the reaction.

cataract A disease of the eye in which the lens becomes gray-white and loses its clearness.

cathartic Substance that aids bowel movement and stimulates evacuation of the intestine.

caustic Any strongly alkaline substance that has a corrosive effect on living tissue.

ceiling value The maximum allowable human exposure limit for an airborne substance which is not to be exceeded even momentarily. Also called ceiling limit.

Centigrade Thermometric scale in which the interval between the freezing point of water and the boiling point is divided into 100 degrees, with 0° representing the freezing point and 100° the boiling point.

CERCLA (Comprehensive Environmental Response, Compensation, and Liability Act) U.S. federal program enacted in 1980 and designed to clean up identified inactive hazardous waste disposal sites. Also significant because it set the first standard for notification of emergencies involving hazardous substances. Commonly referred to as Superfund program.

CERCLA hazard rating A rating system developed as part of CERCLA to rate and evaluate the characteristics of hazardous waste materials based on toxicity, ignitability, reactivity, and persistence.

CFCs (chlorofluorocarbons) Family of inert, non-toxic, and easily liquified chemicals manufactured for use as coolants, cleaning solvents, plastic, aerosol propellants, and foam insulation. Chlorine components of CFCs released into the atmosphere are major contributors to stratospheric ozone depletion.

chelate An organic chemical compound in which the metal ion is bound to atoms of non-metals. Medicinally, chelating agents are used against viruses and fungi and to remove poisonous metals from the blood.

chemical Any element, chemical compound, or mixture of elements and/or compounds.

chemical burn A burn caused by contact with a chemical substance.

chemical family A group of single elements or groups of compounds having a common chemical structure and name. Also known as chemical class.

chemical formula The chemical make-up of a substance using accepted written symbols.

chemically active metals Usually refers to metals that can cause violent reactions with certain other substances and materials; includes sodium, potassium, beryllium, calcium, powdered aluminum, zinc, and magnesium.

chemical name 1) The scientific designation of a chemical as outlined by the rules of nomenclature developed by the Chemical Abstract Service and by the International Union of Pure and Applied Chemistry. 2) A name that clearly identifies the chemical for the purpose of conducting a hazard evaluation (OSHA).

chemical reaction Any chemical change, regardless of rate or whether it occurs naturally or induced by man; includes decomposition, explosion, combustion, condensation, polymerization, and neutralization.

chemical resistance The ability of a material to resist chemical reaction.

chloracne A severe acne-like affliction of the skin resulting from excessive exposure to certain chlorinated or halogenated chemical compounds, such as carbon tetrachloride, chloroform, trichloroethylene, biphenyls, dioxins, napthalenes, and DDT.

chlorhydrate Chemical substance combined with hydrogen chloride in a very specific proportion.

CHRIS (Chemical Hazard Response Information System) Program developed by the U.S. Coast Guard to provide accident prevention and emergency information for responding to spills of hazardous materials, particularly on water transport.

chronic A change to a organism over a long period of time, measured in weeks, months, or years, following repeated exposure to a health hazard.

chronic effect of overexposure An adverse health effect that develops slowly over a long period of time or from prolonged exposure to a health hazard without implying a degree of severity.

chronic toxicity Permanent and irreversible health effects resulting from prolonged exposure to a toxic substance.

cirrhosis Chronic progressive illness affecting the structure and function of the liver.

Clean Air Act Federal guidelines regulating air pollutants and mandating and enforcing toxic emission standards for stationary sources and motor vehicles.

Clean Water Act Federal guidelines regulating the discharge of pollutants into surface water. Also know as the Federal Water Pollution Control Act (FWPCA).

co-carcinogen Any substance, not itself carcinogenic, capable of enhancing the carcinogenic effect of another substance.

COD (chemical oxygen demand) The amount of oxygen required under specified conditions for the oxidation of waterborne organic and inorganic matter.

coden A six-character symbol representing journals and textbooks; assigned by the Chemical Abstract Service.

Code of Federal Regulations A publication of the regulations promulgated under U.S. law.

colic A sharp, crampy, and often painful disorder of the abdomen resulting from blockage, twisting, or muscle spasm.

collapsus Rapid decrease in strength or collapse of an organ.

coma A state of deep unconsciousness from which a victim cannot be wakened by external stimulants.

combustible liquid A material having a flash point at or above 100°F (37.8°C) but below 200°F (93.3°C), excepting any liquid mixture that has one or more components with a flash point above 200°F (93.3°C) that makes up 99% or more of the total volume of the mixture.

common name Any designation or identification such as a code name, code number, trade name, or generic name used to identify a chemical other than by its chemical name (OSHA).

component An ingredient or constituent part.

compound A substance consisting of two or more elements that have united chemically.

compressed gas Any material or mixture having in the container a pressure exceeding 40 pounds per square inch (psi) at 70°F (21.1°C) or a pressure exceeding 104 psi at 130°F (54.4°C) regardless of the pressure at 70°F (21.1 C); or any liquid flammable material having a vapor pressure exceeding 40 psi absolute pressure at 100°F (37.8°C).

compressed gas in solution A non-liquified compressed gas dissolved in a solution.

conc Concentration.

congenital A condition that begins to develop in the uterus and is existing at birth.

congestion Abnormal accumulation of blood in the vessels of tissue, an organ, or another part of the body.

conjunctivitis Irritation and inflammation of the conjunctiva, a part of the inner lining of the eyelids.

container Any bag, barrel, bottle, box, can, cylinder, drum, reaction vessel, storage tank, or the like that contains a hazardous chemical; pipes and piping systems and engines, fuel tanks, or other operating systems in a vehicle are not considered containers (OSHA).

contraindication Any condition that renders some particular treatment of disease improper or undesirable.

convulsions Violent, involuntary spasms or muscle contractions.

corrosive material Any liquid or solid with pH ranges of 2-6 or 12-14 that causes visible destruction or irreversible alteration of living tissue, or a liquid that has a severe corrosion rate on steel.

cryogenic liquid A refrigerated liquefied gas having a boiling point colder than -130°F (-90°C) at one atmosphere, absolute (DOT).

CUM Cumulative effects of increasing amounts of a chemical substance building up in the body.

cutaneous Pertaining to the skin.

CVS Cardiovascular effects.

cyanosis A bluish discoloration of the skin, lips, and mucous membrane resulting from lack of oxygen in the blood hemoglobin or excessive concentration of reduced hemoglobin in the blood.

decomposition Breakdown of a material or substance into parts, elements, or simpler compounds that may be caused by heat, chemical reaction, electrolysis, decay, biodegradation, or other processes.

decontamination The removal of hazardous substances from employees and their equipment to the extent necessary to preclude the occurrence of foreseeable adverse health effects.

degenerescence Abnormal transformation of an organ, tissue, or faculty leading to change in its functioning, not always permanent.

degradation The destructive effect a chemical may have on chemical-protective clothing, reducing its strength and flexibility and permitting a direct route to skin contact.

delayed hazard The potential to cause an adverse effect which may not appear until a long period of time; carcinogenicity, teratogenicity, and certain target organ/system effects are examples of delayed hazards (ANSI).

deliquescent Substance that absorbs moisture from the air to the point of becoming liquid.

delirium A mental state of great excitement or confusion marked by confusion, speech disorders, anxiety, and often hallucinations.

density The mass (weight) per unit volume of a substance.

dermal Pertaining to the skin.

dermatitis Inflammation of the skin from any cause.

dermatosis Generic name for all skin disorders.

desquamation Abnormal elimination of surface layers of skin in small flakes.

diarrhea Abnormally frequent discharge of loose, watery feces from the large intestine (colon); may be a symptom of poisoning.

digital hippocratism Characteristic deformation of the ends of the fingers and often the toes.

dil Dilute.

diuresis Formation and excretion of urine.

diuretic That which increases volume of urine.

DNA (deoxyribonucleic acid) Chromosome of a cell in which the genetic information is encoded.

dose The amount of a chemical substance or drug to which a person has been exposed or absorbed into the body.

DOT ID U.S. Department of Transportation Identification Numbers, used for hazardous materials regulation, safe handling, and emergency response. *See also:* **PIN.**

dysarthria Difficulty in articulating words resulting from damage to a motor nerve.

dysmetria Coordination disorder characterized by loss of ability to judge degree of muscle movements.

dysphagia Difficult or labored swallowing.

dysphonia Change in voice.

dysplasia Malformation; abnormal development.

dyspnea Difficulty or labored breathing and/or shortness of breath; a possible symptom of poisoning.

dysuria Painful or difficulty in urinating.

easily oxidized materials A broad range of materials that can cause violent reactions with other substances and materials; includes combustible materials, organic materials, paper, wood, sulfur, aluminum, acetic acid, alcohols, fuels, oils, plastics, hydrazine, acetic anhydride, sulfuric acid and many other chemicals. *See also:* **oxidizer, strong oxidizer.**

eczema General medical description for swelling of the skin or rash of unknown cause.

edema Swelling caused by infiltration of fluid into the tissues or intercellular spaces.

effusion Collecting of biological liquids in the membrane enfolding the lungs or pleural cavity.

EIS (environmental impact statement) The results of a study to determine the probable effects of a proposed activity on the surrounding environment.

element The simplest form of a pure substance that cannot be broken down into simpler substances by chemical means.

Emergency Planning and Community Right-to-Know Act (EPCRA) Federal program effected in 1986 regulating planning processes concerned with extremely hazardous substances at state and local levels, including requirements for fixed facility owners and operators to inform officials about extremely hazardous substances present at the facilities and mechanisms for making information about extremely hazardous substances available to citizens. Also known as SARA, Title III.

emphysema A condition of the lungs characterized by dilation or destruction of the pulmonary areola causing difficulty in exhaling.

encephalopathy Generic name given to illnesses affecting the brain in general.

environment The water, air, and land and the interrelationship that exists among them and all living things.

epidemiology Science concerned with the study of disease in a general population, especially the incidence (rate of occurrence) and distribution of a particular disease (as by age, sex, or occupation) that may provide information about its cause.

epigastric Relating to the area located between the ribs and the sternum above, the flanks on each side, and the umbilical region below.

epiphora An abnormal flow of tears.

epistaxis Nosebleed.

ergonomics The study of the interactions between workers and their total working environment and their stresses relating to the elements of the environment and their equipment.

erythema Abnormal flushing or redness of the skin due to increase in blood flow.

etiologic agents Non-chemical airborne microorganisms capable of causing disease.

etiology The study of all the factors that contribute to the development of a disease.

euphoria Intense feeling of well-being or elation.

evaporation rate A measure of the time required for a given amount of a substance to evaporate, compared with the time required for an equal amount of butyl acetate or ether to evaporate.

exothermic Heat producing.

expectoration Expulsion from the mouth of secretions from the respiratory tract.

explosive Any chemical compound, mixture, or device that produces a sudden, almost instantaneous release of pressure, gas, and heat when subjected to sudden shock, ignition source, pressure, or high temperature.

explosive limits The boundary-line mixture of vapor or gas with air, which, if ignited, will just propagate the flame. Known as "lower and upper explosive limits," they are usually expressed in terms of percentage by volume of gas or vapor in air. Also known as flammable limits. *See also:* **LEL** and **UEL.**

exposure The harmful effects of a hazardous material concerning people, property, or the environment.

exposure limits Concentrations of substances and conditions under which it is believed that nearly all workers may be repeatedly exposed, day after day, without adverse effects.

Fahrenheit Thermometric scale on which under standard atmospheric pressure the boiling point of water is 212°.

fasciculation Isolated, involuntary, and anarchic contraction of a group of muscle fibers.

fibrillation Rapid and chaotic contractions of many individual muscle fibers of the heart in the area of the ventricles, capable of causing cardiac arrest.

fibrosis Chronic lung affliction or scarring of the lung caused by an unusual increase of fibrous tissue, causing progressive respiratory problems and often occurring following exposure to certain chemical substances.

flammable Catches fire and burns rapidly.

flammable aerosol An aerosol which, under certain conditions, yields a flame projection exceeding 18 inches at full valve opening, or a flashback (a flame extending back to the valve) at any degree of valve opening.

flammable limits Range of gas or vapor concentrations (percent by volume) in air that will burn or explode if an ignition source is present. *See also:* **LEL** and **UEL.**

flash back A phenomenon characterized by vapor ignition and flame traveling back to the source of the vapor.

flash point The minimum temperature at which a substance gives off flammable vapors which, in contact with spark or flame, will easily ignite and burn rapidly. The lower the flash point of a liquid, the higher the risk of fire.

fluorosis Characteristic chronic intoxication caused by fluorine and its derivatives.

fototoxic Toxic effect on the fetus.

forbidden hazardous material A hazardous material that must not be offered or accepted for transportation.

foreseeable emergency Any potential occurrence such as, but not limited to, equipment failure, rupture of containers, or failure of control equipment which could result in an uncontrolled release of a hazardous chemical into the workplace (OSHA).

fume A suspension of very fine solid particle in air or vapors from a volatile liquid.

gas An air-like, formless fluid that uniformly distributes itself throughout a space in air.

gastric lavage Washing out the stomach.

gastroenteritis Simultaneous inflammation of stomach mucosa and intestine.

gastrointestinal (GI) Involving the organs of stomach, intestines, and/or other organs from mouth to anus.

gene Heredity-carrying material that is part of the chromosome.

gestation The development of the fetus in the uterus from conception to birth.

gingivitis Inflammation of the gums.

glycosuria Abnormal presence of glucose in the urine.

granulomatosis Pulmonary lesion characterized by the formation of small nodules.

granulometry Indicates the size of a powder and usually expressed in microns.

halogens Inorganic compounds containing bromine, chlorine, fluorine, and iodine.

hazardous classes A collection of terms established the United Nations Committee of Experts to categorize hazardous materials, as: flammable liquids, explosives, gases, oxidizers, radioactive materials, corrosives, flammable solids, poisonous and infectious substances, and dangerous substances.

hazardous materials Substances or materials which have been determined to be capable of posing an unreasonable risk to health, safety, and property, such as petroleum, natural gas, synthetic gas, acutely toxic chemicals, and other toxic chemicals.

hazard warning Any words, pictures, symbols, or combination thereof appearing on a label or other appropriate form of warning that conveys the hazards of chemicals in containers (OSHA).

hazardous waste Any solid or combination of solid wastes which, because of its physical, chemical, or infectious properties, may pose a health hazard when improperly managed. It must possess at least one of four characteristics: ignitability, corrosiveness, reactivity, or toxicity.

hematemesis Vomiting of blood.

hematoma Localized bleeding into tissue.

hematopoietic system System responsible for formation of blood cells (includes bone marrow and lymphatic organs).

hematuria Presence of blood in the urine.

hemolysis Destruction of red blood cells, releasing hemoglobin.

hemorrhage Loss of a significant amount blood due to external or internal bleeding.

hepatic Concerning the liver.

herbicide A chemical substance used to kill plants.

HHW Household hazardous waste.

highly toxic A chemical falling within any of the following categories: 1) that has a median lethal dose (LD/50) of 50 milligrams or less per kilogram of body weight when administered orally to albino rats weighing between 200 and 300 grams each; 2) that has a LD/50 of 200 milligrams or less per kilogram of body weight when administered by continuous contact for 24 hours (or less if death occurs within 24 hours) with the bare skin of albino rats weighing between 200 and 300 grams each; 3) that has a median lethal concentration (LC/50) in air of 200 parts per million by volume or less of gas or vapor, or 2 milligrams per liter or less of mist, fume, or dust, when administered by continuous inhalation for one hour (or less if death occurs within one hour) to albino rats weighing between 200 and 300 grams each (OSHA).

hydrate Chemical substance combined with water in a specific proportion.

hydrocarbons Chemical compounds that consist entirely of carbon and hydrogen.

hydrolysis Chemical change to a substance in an aqueous environment leading to the formation of new products.

hygroscopic Substances with a tendency to absorb moisture from the air.

hyperkeratosis Increased thickness of the cornified layer of the epidermis, such as a corn.

hyperpigmentation Excessive pigmentation of the skin.

hyperplasia Abnormal growth of normal tissue.

hyperreflexia Excessive reflexes.

hypothermia Lowering of body temperature to below normal.

icterus Yellow coloration of the skin and mucosa.

IDLH (immediately dangerous to life or health) An atmospheric concentration of any toxic, corrosive, or asphyxiant substance that poses an immediate threat to life or would cause irreversible or delayed adverse health effects or interfere with an individual's ability to escape from a dangerous environmental condition within 30 minutes.

ignition temperature The minimum temperature required to initiate or cause self-sustained combustion independent of a heat source.

incident An unplanned event that could have resulted in an accident or which diminishes efficiency or production. *See also:* **accident.**

inhibitor A chemical added to another substance to prevent unwanted chemical change from occurring.

immediate hazard A hazard with immediate effects. *See also:* **acute effect.**

immunosuppression Decrease in the immune response.

impervious A material that cannot be pene-

trated by a chemical substance. Also referred to as impermeable.

incompatibility Indicates whether or not a material can be placed in contact with certain other products or materials; direct contact of incompatible materials can cause dangerous reactions and give off heat and toxic vapors.

inflammation A response of the tissues of the body to injury, infection, or irritation; its chief symptoms are redness, heat, swelling, and pain.

ingestion Taking in by the mouth; swallowing.

inhalation Breathing into the lungs of a (contaminated) substance in the form of a gas, vapor, fume, mist, or dust.

inorganic chemical Chemical substances that do not contain carbon.

irritant A substance or material capable of causing irritation to organs and other body parts.

insecticide A group of chemical substances used to control and kill insects; falls under the broad category of pesticides.

insoluble Products that cannot be dissolved in each other.

interaction Modification of the toxic effects of one substance by another, with effects either amplified or mitigated, depending on the substances involved.

irritability Tendency to anger.

irritating material Any substance which upon contact with fire or air produces irritating fumes.

jaundice Yellowing of the skin or eyes caused by too much bilirubin in the blood; an indication of liver diseases, biliary obstructions, and hemolysis.

keratitis Inflammation of the cornea of the eye.

lacrimation Production or discharge of excess tears from the eyes.

lacrimator A chemical substance that causes the secretion of excess tears from the eyes.

larvacide A chemical substance used to kill insect larvae.

laryngitis Inflammation of the larynx.

LC/50 (Lethal Concentration—50%) The calculated concentration of a material in air that is expected to kill 50% of a group of test animals with a single exposure (usually 1-4 hours); expressed as parts of material per million parts of air, by volume (ppm) for gases and vapors, as milligrams of material per liter of air (mg/1) or milligrams of material per cubic meter of air (mg/m3) for dusts, mists, gases, and vapors.

LC/LO (Lethal Concentration Low) The lowest concentration of a substance in air, other than LC/5O, that has been reported to have caused death in humans or animals. The reported concentrations may be entered for periods of exposure that are less than 24 hours (acute) or greater than 24 hours (subacute and chronic).

LD/50 (Lethal Dose—50%) A single calculated dose of a material expected to kill 50% of a group of test animals; usually expressed as milligrams or grams of material per kilogram of animal body weight (mg/kg or g/kg). The material may be administered by mouth or applied to the skin.

LD/LO (Lethal Dose Low) The lowest dose (other than LD/5O) of a substance introduced by a route other than inhalation over any given period of time in one or more divided portions and reported to have caused death in humans or animals.

LEL 1) Lower explosive limit, or the lowest concentration of gas or vapor (percentage by volume in air) that will burn or explode when a source of ignition is present. Also referred to as lower flammable limit (LFL); 2) Lowest effect level, or the lowest dose used in a toxicology test that produces toxic effects. *See also:* **UEL.**

LEPC Local Emergency Planning Committee.

lethargy Deep and prolonged sleep or extreme indifference.

leucopenia Reduced white blood cell count.

local exhaust A system for capturing and exhausting contaminants from the air at the point where contaminants are produced.

LOC (Level of Concern) The concentration in air of an extremely hazardous substance above which there may be serious immediate health effects to anyone exposed to it for a short period of time.

lymphocytosis Increased lymphocyte count.

MAC Maximum accepted concentration.

manganism Chronic intoxication caused by manganese and its derivatives.

material safety data sheet (MSDS) A worksheet document containing information about hazardous chemicals, including physical characteristics, hazards, safe handling requirements, and actions to be taken in the event of accident, overexposure, fire, or spill.

MCL (maximum contaminant levels) The greatest level of a contaminant permissible in a public water system.

melting point The temperature at which a product changes from the solid to liquid state at normal atmospheric pressure (760 mm Hg).

methemoglobinemia Presence of abnormal concentrations of methemoglobin in the blood, resulting in cyanosis.

micron A unit of length equal to one millionth of a meter; a micrometer.

miscible Products capable of mixing with each other completely and staying mixed under normal conditions.

mist Liquid droplets suspended in air.

mm Hg (millimeters of mercury) A measurement for pressure; at sea level, the earth's atmosphere exerts 760 mm Hg of pressure.

molecular formula A chemical formula that is based on both analysis and molecular weight and gives the total number of atoms of each element in a molecule.

molecular weight Weight (mass) of a molecule based on the sum of the weights of the atoms that make up the molecule.

molluscicide Chemical substances used to control snails and slugs.

mucous membranes Membranes lining body cavities and covered by a viscous substance called mucus.

mydriasis Abnormal dilation of the pupils of the eyes.

myosis Reduction in diameter of pupils of the eye.

mutagen A chemical substance or physical effect capable of inducing transmissible changes in the genetic material of a living cell that results in physical and functional changes in the descendants, which can lead to birth defects, miscarriage, or cancer. Both males and females can be effected.

narcosis A stupor, drowsiness, arrested activity or unconsciousness produced by the influence of narcotics or other chemical substances. Artificially induced sleep.

narcotic Any substance that induces narcosis.

nasopharyngitis Inflammation of the upper part of the throat (pharynx).

nausea The urge to vomit; a feeling of sickness in the stomach.

necrosis Cellular or tissue death.

neoplasm Presence of a new growth of abnormal cells; a tumor.

nephritis Inflammation of the kidney.

nephrotoxic Toxic to the kidneys.

neural Refers to a nerve or the nervous system.

neurasthenia Psychoneurotic affliction characterized by a reduction of physical and psychological strength.

neuritis Inflammation of a nerve.

neuropathy Any affliction of the nervous system.

neurotoxic Toxic to nerve cells and the nervous system; effect may produce emotional or behavioral abnormalities.

neutralize To eliminate potential hazards by inactivating strong acids, caustics, and oxidizers.

nitrogen oxides (NOx) Gasses released primarily from the burning of fossil fuels that are associated with the breakdown of the earth's protective ozone layer.

non-liquified compressed gas A gas, other than gas in solution, which under charged pressure is entirely gaseous at a temperature of 70°F (DOT).

Not in My Backyard (NIMBY) Citizen opinion or sentiment that refers to the desire to place hazardous waste facilities (landfills, incinerators) far from their own community.

nystagmus Involuntary oscillating or rotating movements of the eyeball.

odor threshold The lowest amount of a chemical substance's vapor, in air, that can be smelled; it should be noted that an individual's ability to detect odors varies greatly.

oliguria A reduction in quantity of urine eliminated.

opisthotonos Generalized contraction arching the body backwards.

organic compound A class of chemical compounds containing mainly carbon atoms.

organophosphate Chemical substances that contain a phosphorus atom and are used as insecticides. Also referred to as organophosphorus pesticides.

oxidizer A substance other than an explosive that yields oxygen or other gases readily, thereby causing fire of other, usually organic, materials; includes perchlorates, peroxides, permanganates, chlorates, and nitrates.

oxidizing agent Any chemical substance that can oxidize other substances.

ovicide A chemical substance that kills insect eggs before they hatch.

palpitation A racing, irregular beating or pounding of the heart.

paralysis The loss of function and/or feeling.

paresis Incomplete or mild paralysis.

paresthesia Anomalies in perception of sensation or spontaneous non-painful subjective feelings.

partition coefficient The ratio of a substance's distribution between oil and water when they are in contact. A value of less than 1 indicates better solubility of the substance in oils and greases, and the product is therefore likely to be absorbed by the skin; a value greater than 1 indicates a better solubility in water, therefore likely to be absorbed by the mucosa. This information can be useful in assessing first aid requirements.

PCB (polychlorinated biphenyl) Family of toxic, persistent chemicals used in energy transformers, capacitors, and gas pipelines. Use banned in 1979.

peritonitis Inflammation of the peritoneum or the membrane lining the abdominal cavity and the organs contained within it.

permissible exposure limit (PEL) The limit of allowable exposure to a chemical contaminant expressed as a time-weighted average (TWA) concentration during an 8-hour work day or as a maximum concentration never to be exceeded either instantaneously or in the short term during any maximum period of 15 minutes.

personal protective equipment Safety equipment designed to protect parts or all of the body from workplace hazards, including chemical resistant clothing, gloves, respirators, and eye protection.

pesticide Substances that prevent, destroy, repel, or mitigate any pest; includes insecticides, herbicides, fungicides, acaricides, and rodenticides.

pH Numeric value that indicates the relative acidity or alkalinity of a substance on a scale of 0 to 14. Acid solutions have pH values lower than 7; basic or alkaline solutions have pH values greater than 7.

pharyngitis Inflammation of the throat or pharynx.

phlegm Thick mucous produced in the breathing passages.

photophobia Unpleasant to painful feeling in the eyes, caused by light.

photosensitive Substances that changes in the presence of light.

physical hazard A chemical for which there is scientifically valid evidence that it is a combustible liquid, a compressed gas, explosive, flammable, an organic peroxide, an oxidizer, pyrophoric, unstable (reactive), or water-reactive (OSHA).

PIN (product identification numbers) Four-digit identification numbers used with the Canadian Transportation of Dangerous Goods Regulation.

pneumoconiosis Chronic affliction of the lungs due to the inhaling of certain types of dust.

pneumonia Acute infection of the lung characterized by inflammation.

pneumonitis (chemical) Inflammation of the lung resulting from chemical irritation.

pneumopathy Any pulmonary affliction.

point source contamination Contamination to the environment from a specific source such as a smokestack or sewer pipe.

poise The standard unit for measuring the viscosity of a fluid. The viscosity of water is about 0.01 poise, or 1 centipoise.

poison A A gas or liquid having a toxicity level such that a very small amount of the vapor or gas mixed in air poses a danger to life.

Poison B Substances, other than poison A or irritating materials, that are known to be so toxic to humans as to afford a danger to health during transportation or, in the absence of adequate data on human toxicity, are presumed to be dangerous based on laboratory tests.

polymerization A chemical reaction in which one or more small molecules combine to form larger molecules. A hazardous polymerization is such a reaction that takes place at a rate that releases large amounts of energy.

polyneuritis Inflammation of several nerves.

polyuria Elimination of an abnormally large amount of urine over a given period.

polyvinyl chloride (PVC) An indestructible plastic that releases hydrochloric acid when burned.

pro-carcinogen Product that must be changed by the organism in order to become a carcinogen.

promoter (of carcinogenesis) Substance capable of enhancing the carcinogenic effect of another substance.

PRP (potentially responsible party) An individual, business or organization most likely responsible for a pollution incident.

pulmonary edema A buildup of fluid in the lungs caused by congestive heart failure, lung damage, side effects of drugs, infections, or kidney failure.

psychosis A group of mental illnesses characterized by a change in personality and loss of contact with reality.

pyrophoric A chemical that will ignite spontaneously in air at a temperature of 130°F (54.4°C) or below.

radioactive material Any material or combination of materials that spontaneously emits ions and has a specific activity greater than 0.002 microcuries per gram.

radiomimetic Having an action identical to that of radiation.

radionuclides Radioactive-decay particles emitted from natural and manufactured sources, including cosmic rays, X-rays, radon, and coal-fired utilities.

RCRA (Resource Conservation and Recovery Act) U.S. federal program regulating management and disposal of hazardous materials and wastes currently being generated, treated, stored, disposed, or distributed.

reactive Unstable.

reactivity Chemical reaction with the release of energy. Undesirable effects such as pressure buildup, temperature increase, formation of noxious, toxic or corrosive by–product may occur because of the reactivity of a substance to heating, burning, direct contact with other materials, or other conditions in use or in storage.

reducing agent In a reducing reaction (which occurs simultaneously with an oxidation reaction), the reducing agent is the chemical or substance that either combines with oxygen or loses electrons to the reaction.

regulated material A substance or material that is subject to regulations promulgated by any government agency.

REL (Recommended Exposure Limit) The highest allowable airborne concentration which is not expected to cause injury.

remedial action The actual construction or implementation phase of a Superfund site cleanup that follows remedial design.

residue The material remaining in a packaging, including a tank car, after its contents have been unloaded to the maximum extent practicable and before the packaging is either refilled or cleaned of hazardous material and purged to remove any hazardous vapors (DOT).

respirator A devise worn by a person to filter dust particles or gas out of surrounding air before inhalation air.

rhinitis Inflammation of mucosa of the nasal passages.

right-to-know Applies to laws and regulations enacted under SARA, Title III providing for availability of information on chemical hazards.

risk assessment A detailed examination of any activity or functioning system in which potential adverse effects and their probabilities are calculated and the various risks are quantified or measured.

risk management The process, derived through system safety principles, whereby management decisions are made concerning control and minimization of hazards and acceptance of residual risks.

rodenticide Chemical substances used to kill mice, rats, and other rodents.

route of entry The means or natural route by which hazardous chemicals or other contaminants can penetrate the body; includes the skin, digestive system, and respiratory system. Also known as route of exposure.

SARA (Superfund Amendments and Reauthorization Act) Established in 1986, it is the first national program of emergency planning for dealing with hazardous chemicals.

Sax Rating Hazardous materials rating system devised by the Irving N. Sax.

SCBA Self contained breathing apparatus.

SDWA Safe Drinking Water Act.

sensitization Defense reaction by the organism following exposure to a contaminant, resulting in an allergy.

sensitizer A chemical that causes a substantial proportion of exposed people or animals to develop an allergic reaction in normal tissue after repeated exposure.

SERC (State Emergency Response Commission) Commission appointed by each state governor according to the requirements of SARA, Title III that designates emergency planning districts, appoints local emergency planning committees, and supervises their activities.

shock An abnormal condition resulting from not enough blood flowing through the body causing reduced blood pressure and interference with bodily functions.

short-term exposure limit (STEL) The upper limit of exposure to a substance to which a worker can be exposed for a continuous period of up to 15 minutes without adverse effects to health, performance, and work efficiency.

silicosis Chronic disease of the lungs (fibrosis) provoked by inhaling dust of crystalline silica.

solid waste Any garbage, refuse, or sludge, including solid, liquid, semisolid, or contained gaseous material resulting from industrial, commercial, agricultural, mining operations, and community activities (excluding material in domestic sewage); discharges are subject to regulation as point sources under the Federal Water Pollution Control Act, and nuclear materials or by–products under the Atomic Energy Act of 1954.

solubility in water at saturation A measure of an amount of the minimum quantity of a material that can be dissolved in water at ambient temperature (20°C) expressed in grams per liter. Solubility is normally expressed as negligible (less than 0.1%), slight (0.1-1.0%), moderate (1.0-10.0%), appreciable (more than 10.0%), and complete (soluble in all proportions). Solubility information can be useful in determining spill cleanup methods and extinguishing agents and methods for a material.

solution Any homogeneous liquid mixture of two or more chemical compounds or elements that will not undergo any segregation normal to transportation (DOT).

solvent A chemical liquid capable of dissolving another substance; generally used to describe organic solvents.

small quantity generator (SQG) A facility generating more than 100 kilograms and less than 1,000 kilograms of hazardous wastes per month.

specific chemical identity The chemical name, Chemical Abstract Service (CAS) Registry Number, or any other information that reveals the precise chemical designation of a substance (OSHA).

specific gravity The weight of a material compared to the weight of an equal volume of water is an expression of the density (or heaviness) of a material. Insoluble materials with specific gravity of less than 1.0 will float in (or on) water; those with specific gravity greater than 1.0 will sink in water. Most (but not all) flammable liquids have specific gravity less than 1.0 and, if not soluble, will float on water—an important consideration for fire suppression.

stability Relates to a material's ability to resist change in form or chemical nature.

strong acids Chemicals that can cause violent reactions with certain other substances and materials; includes hydrochloric, sulfuric, and nitric acids.

strong bases Chemicals that can cause violent reactions with certain other substances and materials; includes sodium hydroxide and potassium hydroxide.

strong oxidizers Chemicals that can cause violent reactions with certain other substances and materials; includes chlorine, bromine, fluorine, and many of their compounds.

sublimate To go directly from the solid to the gaseous state without passing through the liquid state.

suspected carcinogen A substance known to cause cancer in test animals but is only suspected of causing cancer in humans. Also referred to as an experimental or potential carcinogen.

syncope Sudden and complete loss of consciousness following cardiac and respiratory arrest.

systemic toxicity Adverse effects caused by a substance that effects the body as a whole rather than local or individual parts or organs.

tachycardia Increased speed of heart beat.

tachypnea Abnormally rapid breathing.

teratogen A substance that has been demonstrated to cause birth defects by causing malformations in the fetus. Teratogenic contaminants can be qualified as being "proven" when an effect has been shown in humans, "possible" when shown in animals or suspected in humans, and "suspected" when suspected in animals.

tetanic Persistence in a muscle contraction.

threshold limit value (TLV) The upper exposure limit of airborne concentrations of substances based on an 8-hour, time-weighted average, representing conditions under which it is believed that nearly all workers may be repeatedly exposed day after day without adverse effects.

threshold limit value, ceiling (TLV-C) The concentration of a substance that should never be exceeded; for hazardous gas substances this level can be extremely important, perhaps critical.

tinnitus Ringing in one or both ears; may be a sign of hearing injury.

toxicology The branch of chemistry dealing with poisons.

toxin Any substance that results in damage to living body tissue, impairment to the central nervous system, severe illness, and, in severe cases, death; poisonous.

toxic pollutant Pollutants which after discharge and upon exposure cause adverse health effects.

TPQ (Threshold Planning Quantity) The quantity of a SARA extremely hazardous substance present at a facility above which the facility's owner/operator must give emergency planning notification to its state emergency response commission (SERC) and local emergency planning committee (LEPC).

tracheobronchitis Simultaneous inflammation of the trachea and bronchia.

trade name The manufacturer's commercial name for a chemical substance or product.

trade secret Confidential formula, pattern, process, device, information or compilation of information that is used in an employer's business which gives the employer an opportunity to obtain an advantage over competitors who do not know the chemical identity or use that chemical (OSHA).

TRI (Toxic Release Inventory) The national inventory of annual toxic chemical releases from manufacturing facilities.

TSCA (Toxic Substances Control Act) Federal law enacted in 1976 authorizing the EPA to gather information on risks related to chemicals.

TWA (time-weighted average) The average concentration of a substance in air over the total time of exposure, usually expressed as an 8-hour day.

UEL (upper explosive limit) As pertaining to a vapor or gas, the highest concentration of the substance in air that will produce a flash of fire when an ignition source (heat, arc, or flame) is present. At higher concentrations, the mixture is too "rich" to burn. *See also:* **LEL.**

UST Underground storage tank.

vapor density Measurement of the number of times vapors of a substance are heavier or lighter than air. If the vapor density is greater than 1, the vapor will tend to collect at floor level; if less than 1, the vapor will rise in air.

vapor pressure Measurement of the pressure exerted by a vapor that is in equilibrium with its solid or liquid form, measured in millimeters of mercury at 20°C and normal atmospheric pressure (760 mm Hg). A vapor pressure above 760 mm Hg indicates a substance in the gaseous state. The higher a product's vapor pressure, the more it tends to evaporate, resulting in a higher concentration of the substance in air and therefore increasing the likelihood of breathing in it.

vascular constriction Constriction with narrowing of blood vessels.

vascular dilation Dilation of the blood vessels.

vertigo Dizziness; giddiness.

VOC (volatile organic compound) Organic materials containing carbon and hydrogen that are used in various industrial applications and are subject to very rapid evaporation.

volatiles A substance, usually a liquid, that easily vaporizes or evaporates to form a gas or vapor.

Common Oxidizing Materials

By Name

ALUMINUM NITRATE	13473-90-0
AMMONIUM BICHROMATE	7789-09-5
AMMONIUM NITRATE	6484-52-2
AMMONIUM PERCHLORATE	7790-98-9
AMMONIUM PERSULFATE	7727-54-0
BARIUM BROMATE	13967-90-3
BARIUM CHLORATE	13477-00-4
BARIUM HYPOCHLORITE	13477-10-6
BARIUM NITRATE	10022-31-8
BARIUM PERCHLORATE TRIHYDRATE	10294-39-0
BARIUM PERMANGANATE	7787-36-2
BARIUM PEROXIDE	1304-29-6
BERYLLIUM NITRATE	13597-99-4
BROMINE PENTAFLUORIDE	7789-30-2
BROMINE TRIFLUORIDE	7787-71-5
CADMIUM NITRATE	10325-94-7
CADMIUM NITRATE TETRAHYDRATE	10022-68-1
CALCIUM CHLORATE	10137-74-3
CALCIUM CHLORITE	14674-72-7
CALCIUM HYPOCHLORITE	7778-54-3
CALCIUM NITRATE	10124-37-5
CALCIUM PERCHLORATE	13477-36-6
CALCIUM PERMANGANATE	10118-76-0
CALCIUM PEROXIDE	1305-79-9
CESIUM NITRATE	7789-18-6
CHLORIC ACID SOLUTION	7790-93-4
CHROMIC ACID	7738-94-5
CHROMIC ANHYDRIDE	11115-74-5
CHROMIUM NITRATE	7789-02-8
CHROMIUM TRIOXIDE, ANHYDROUS	1333-82-0
CHROMIUM(III) NITRATE	13548-38-4
COBALT NITRATE	10141-05-6
COBALT(II) PERCHLORATE, HEXAHYDRATE	13478-33-6
CUPRIC NITRATE	3251-23-8
DICHLOROISOCYANURIC ACID	2782-57-2
FERRIC NITRATE	10421-48-4
GALLIUM(III) NITRATE	13494-90-1
GUANIDINE NITRATE	506-93-4
HYDROGEN PEROXIDE	7722-84-1
INDIUM NITRATE	13770-61-1
IODINE PENTAFLUORIDE	7783-66-6
IRON(II) NITRATE HEXAHYDRATE (1:2:6)	13520-68-8
LANTHANUM NITRATE	10099-59-9
LEAD DIOXIDE	1309-60-0
LEAD NITRATE	10099-74-8
LEAD PERCHLORATE	13637-76-8
LITHIUM HYPOCHLORITE	13840-33-0
LITHIUM NITRATE	7790-69-4
LITHIUM PEROXIDE	12031-80-0
MAGNESIUM BROMATE	7789-36-8
MAGNESIUM CHLORATE	7791-19-7
MAGNESIUM NITRATE	10377-60-3
MAGNESIUM PERCHLORATE	10034-81-8
MAGNESIUM PEROXIDE	14452-57-4
MANGANESE DIOXIDE	1313-13-9
MANGANESE NITRATE	10377-66-9
NICKEL NITRATE	14216-75-2
NICKEL NITRITE	17861-62-0
NICKEL PERCHLORATE	13637-71-3
NICKEL(II) NITRATE	13138-45-9
PALLADIUM DINITRATE	10102-05-3
PERCHLORIC ACID	7601-90-3
POTASSIUM BROMATE	7758-01-2
POTASSIUM CHLORATE	3811-04-9
POTASSIUM DICHLOROISOCYANURATE	2244-21-5
POTASSIUM NITRATE	7757-79-1
POTASSIUM NITRITE	7758-09-0
POTASSIUM PERCHLORATE	7778-74-7
POTASSIUM PERMANGANATE	7722-64-7
POTASSIUM PEROXIDE	17014-71-0
POTASSIUM PERSULFATE	7727-21-1
POTASSIUM SUPEROXIDE	12030-88-5
SILVER NITRATE	7761-88-8
SODIUM BROMATE	7789-38-0
SODIUM CHLORATE	7775-09-9
SODIUM CHLORITE	7758-19-2
SODIUM DICHLORO-S-TRIAZINETRIONE	2893-78-9
SODIUM NITRATE	7631-99-4
SODIUM NITRITE	7632-00-0
SODIUM PERCARBONATE	15630-89-4
SODIUM PERCHLORATE	7601-89-0
SODIUM PERMANGANATE	10101-50-5
SODIUM PEROXIDE	1313-60-6
SODIUM PERSULFATE	7775-27-1
SODIUM SUPEROXIDE	12034-12-7
STRONTIUM CHLORATE	7791-10-8
STRONTIUM NITRATE	10042-76-9
STRONTIUM PERCHLORATE	13450-97-0
STRONTIUM PEROXIDE	1314-18-7
TETRANITROMETHANE	509-14-8
TRICHLOROISOCYANURIC ACID	87-90-1
UREA PEROXIDE	124-43-6
YTTERBIUM NITRATE	13768-67-7
YTTRIUM(III) NITRATE HEXAHYDRATE	13494-98-9
ZINC AMMONIUM NITRITE	63885-01-8
ZINC BROMATE	13517-27-6
ZINC CHLORATE	10361-95-2
ZINC NITRATE	7779-88-6
ZINC PERMANGANATE	23414-72-4
ZINC PEROXIDE	1314-22-3
ZIRCONIUM NITRATE	13746-89-9

By CAS Number

CAS Number	Name
87-90-1	TRICHLOROISOCYANURIC ACID
124-43-6	UREA PEROXIDE
506-93-4	GUANIDINE NITRATE
509-14-8	TETRANITROMETHANE
1304-29-6	BARIUM PEROXIDE
1305-79-9	CALCIUM PEROXIDE
1309-60-0	LEAD DIOXIDE
1313-13-9	MANGANESE DIOXIDE
1313-60-6	SODIUM PEROXIDE
1314-18-7	STRONTIUM PEROXIDE
1314-22-3	ZINC PEROXIDE
1333-82-0	CHROMIUM TRIOXIDE, ANHYDROUS
2244-21-5	POTASSIUM DICHLOROISOCYANURATE
2782-57-2	DICHLOROISOCYANURIC ACID
2893-78-9	SODIUM DICHLORO-S-TRIAZINETRIONE
3251-23-8	CUPRIC NITRATE
3811-04-9	POTASSIUM CHLORATE
6484-52-2	AMMONIUM NITRATE
7601-89-0	SODIUM PERCHLORATE
7601-90-3	PERCHLORIC ACID
7631-99-4	SODIUM NITRATE
7632-00-0	SODIUM NITRITE
7722-64-7	POTASSIUM PERMANGANATE
7722-84-1	HYDROGEN PEROXIDE
7727-21-1	POTASSIUM PERSULFATE
7727-54-0	AMMONIUM PERSULFATE
7738-94-5	CHROMIC ACID
7757-79-1	POTASSIUM NITRATE
7758-01-2	POTASSIUM BROMATE
7758-09-0	POTASSIUM NITRITE
7758-19-2	SODIUM CHLORITE
7761-88-8	SILVER NITRATE
7775-09-9	SODIUM CHLORATE
7775-27-1	SODIUM PERSULFATE
7778-54-3	CALCIUM HYPOCHLORITE
7778-74-7	POTASSIUM PERCHLORATE
7779-88-6	ZINC NITRATE
7783-66-6	IODINE PENTAFLUORIDE
7787-36-2	BARIUM PERMANGANATE
7787-71-5	BROMINE TRIFLUORIDE
7789-02-8	CHROMIUM NITRATE
7789-09-5	AMMONIUM BICHROMATE
7789-18-6	CESIUM NITRATE
7789-30-2	BROMINE PENTAFLUORIDE
7789-36-8	MAGNESIUM BROMATE
7789-38-0	SODIUM BROMATE
7790-69-4	LITHIUM NITRATE
7790-93-4	CHLORIC ACID SOLUTION
7790-98-9	AMMONIUM PERCHLORATE
7791-10-8	STRONTIUM CHLORATE
7791-19-7	MAGNESIUM CHLORATE
10022-31-8	BARIUM NITRATE
10022-68-1	CADMIUM NITRATE TETRAHYDRATE
10034-81-8	MAGNESIUM PERCHLORATE
10042-76-9	STRONTIUM NITRATE
10099-59-9	LANTHANUM NITRATE
10099-74-8	LEAD NITRATE
10101-50-5	SODIUM PERMANGANATE
10102-05-3	PALLADIUM DINITRATE
10118-76-0	CALCIUM PERMANGANATE
10124-37-5	CALCIUM NITRATE
10137-74-3	CALCIUM CHLORATE
10137-74-3	CALCIUM CHLORATE SOLUTION
10141-05-6	COBALT NITRATE
10294-39-0	BARIUM PERCHLORATE TRIHYDRATE
10325-94-7	CADMIUM NITRATE
10361-95-2	ZINC CHLORATE
10377-60-3	MAGNESIUM NITRATE
10377-66-9	MANGANESE NITRATE
10421-48-4	FERRIC NITRATE
11115-74-5	CHROMIC ANHYDRIDE
12030-88-5	POTASSIUM SUPEROXIDE
12031-80-0	LITHIUM PEROXIDE
12034-12-7	SODIUM SUPEROXIDE
13138-45-9	NICKEL(II) NITRATE
13450-97-0	STRONTIUM PERCHLORATE
13473-90-0	ALUMINUM NITRATE
13477-00-4	BARIUM CHLORATE
13477-10-6	BARIUM HYPOCHLORITE
13477-36-6	CALCIUM PERCHLORATE
13478-33-6	COBALT(II) PERCHLORATE, HEXAHYDRATE
13494-90-1	GALLIUM(III) NITRATE
13494-98-9	YTTRIUM(III) NITRATE HEXAHYDRATE
13517-27-6	ZINC BROMATE
13520-68-8	IRON(II) NITRATE HEXAHYDRATE (1:2:6)
13548-38-4	CHROMIUM(III) NITRATE
13597-99-4	BERYLLIUM NITRATE
13637-71-3	NICKEL PERCHLORATE
13637-76-8	LEAD PERCHLORATE
13746-89-9	ZIRCONIUM NITRATE
13768-67-7	YTTERBIUM NITRATE
13770-61-1	INDIUM NITRATE
13840-33-0	LITHIUM HYPOCHLORITE
13967-90-3	BARIUM BROMATE
14216-75-2	NICKEL NITRATE
14452-57-4	MAGNESIUM PEROXIDE
14674-72-7	CALCIUM CHLORITE
15630-89-4	SODIUM PERCARBONATE
17014-71-0	POTASSIUM PEROXIDE
17861-62-0	NICKEL NITRITE
23414-72-4	ZINC PERMANGANATE
63885-01-8	ZINC AMMONIUM NITRITE